Laser Spectroscopy

Advanced Texts in Physics

This program of advanced texts covers a broad spectrum of topics which are of current and emerging interest in physics. Each book provides a comprehensive and yet accessible introduction to a field at the forefront of modern research. As such, these texts are intended for senior undergraduate and graduate students at the MS and PhD level; however, research scientists seeking an introduction to particular areas of physics will also benefit from the titles in this collection.

Springer

Berlin
Heidelberg
New York
Hong Kong
London
Milan
Paris
Tokyo

Physics and Astronomy ONLINE LIBRARY

http://www.springer.de/phys/

Wolfgang Demtröder

Laser Spectroscopy

Basic Concepts and Instrumentation

Third Edition

With 710 Figures, 16 Tables
93 Problems and Hints for Solution

 Springer

Professor Dr. Wolfgang Demtröder
Universität Kaiserslautern
Fachbereich Physik
Erwin-Schrödinger-Strasse
67663 Kaiserslautern, Germany
E-mail: demtroed@physik.uni-kl.de

Library of Congress Cataloging-in-Publication Data: Demtröder, W. Laser spectroscopy: basic concepts and instrumentation/ Wolfgang Demtröder. – 3rd ed. p. cm. ISBN 3540652256 (alk. paper) 1. Laser spectroscopy. I. Title. QC 454.L3 D46 2002 621.36'6–dc21 2002029191

ISSN 1439-2674

ISBN 3-540-65225-6 3rd Edition Springer-Verlag Berlin Heidelberg New York

ISBN 3-540-57171-X 2nd Edition Springer-Verlag Berlin Heidelberg New York

Springer-Verlag Berlin Heidelberg New York
a member of BertelsmannSpringer Science+Business Media GmbH

http://www.springer.de

© Springer-Verlag Berlin Heidelberg 1981, 1996, 2003
Printed in Germany

Typesetting: Data conversion by Fa. Le-TeX, Leipzig
Cover design: *design & production* GmbH, Heidelberg

Printed on acid-free paper   SPIN 10673180   56/3141/ba   5 4 3 2 1 0

Preface to the Third Edition

Laser Spectroscopy continues to develop and expand rapidly. Many new ideas and recent realizations of new techniques based on old ideas have contributed to the progress in this field since the last edition of this textbook appeared. In order to keep up with these developments it was therefore necessary to include at least some of these new techniques in the third edition.

There are, firstly, the improvement of frequency-doubling techniques in external cavities, the realization of more reliable cw-parametric oscillators with large output power, and the development of tunable narrow-band UV sources, which have expanded the possible applications of coherent light sources in molecular spectroscopy. Furthermore, new sensitive detection techniques for the analysis of small molecular concentrations or for the measurement of weak transitions, such as overtone transitions in molecules, could be realized. Examples are Cavity Ringdown Spectroscopy, which allows the measurement of absolute absorption coefficients with great sensitivity or specific modulation techniques that push the minimum detectable absorption coefficient down to $10^{-14}\,\mathrm{cm}^{-1}$!

The most impressive progress has been achieved in the development of tunable femtosecond and subfemtosecond lasers, which can be amplified to achieve sufficiently high output powers for the generation of high harmonics with wavelengths down into the X-ray region and with pulsewidths in the attosecond range. Controlled pulse shaping by liquid crystal arrays allows coherent control of atomic and molecular excitations and in some favorable cases chemical reactions can already be influenced and controlled using these shaped pulses.

In the field of metrology a big step forward was the use of frequency combs from cw mode-locked femtosecond lasers. It is now possible to directly compare the microwave frequency of the cesium clock with optical frequencies, and it turns out that the stability and the absolute accuracy of frequency measurements in the optical range using frequency-stabilized lasers greatly surpasses that of the cesium clock. Such frequency combs also allow the synchronization of two independent femtosecond lasers.

The increasing research on laser cooling of atoms and molecules and many experiments with Bose–Einstein condensates have brought about some remarkable results and have considerably increased our knowledge about the interaction of light with matter on a microscopic scale and the interatomic interactions at very low temperatures. Also the realization of coherent matter waves (atom lasers) and investigations of interference effects between matter waves have proved fundamental aspects of quantum mechanics.

The largest expansion of laser spectroscopy can be seen in its possible and already realized applications to chemical and biological problems and its use in medicine as a diagnostic tool and for therapy. Also, for the solution of technical problems, such as surface inspections, purity checks of samples or the analysis of the chemical composition of samples, laser spectroscopy has offered new techniques.

In spite of these many new developments the representation of established fundamental aspects of laser spectroscopy and the explanation of the basic techniques are not changed in this new edition. The new developments mentioned above and also new references have been added. This, unfortunately, increases the number of pages. Since this textbook addresses beginners in this field as well as researchers who are familiar with special aspects of laser spectroscopy but want to have an overview on the whole field, the author did not want to change the concept of the textbook.

Many readers have contributed to the elimination of errors in the former edition or have made suggestions for improvements. I want to thank all of them. The author would be grateful if he receives such suggestions also for this new edition.

Many thanks go to all colleagues who gave their permission to use figures and results from their research. I thank Dr. H. Becker and T. Wilbourn for critical reading of the manuscript, Dr. H.J. Kölsch and C.-D. Bachem of Springer-Verlag for their valuable assistance during the editing process, and LE-TeX Jelonek, Schmidt and Vöckler for the setting and layout. I appreciate, that Dr. H. Lotsch, who has taken care for the foregoing editions, has supplied his computer files for this new edition. Last, but not least, I would like to thank my wife Harriet who made many efforts in order to give me the necessary time for writing this new edition.

Kaiserslautern,   *Wolfgang Demtröder*
April 2002

Preface to the Second Edition

During the past 14 years since the first edition of this book was published, the field of laser spectroscopy has shown a remarkable expansion. Many new spectroscopic techniques have been developed. The time resolution has reached the femtosecond scale and the frequency stability of lasers is now in the millihertz range.

In particular, the various applications of laser spectroscopy in physics, chemistry, biology, and medicine, and its contributions to the solutions of technical and environmental problems are remarkable. Therefore, a new edition of the book seemed necessary to account for at least part of these novel developments. Although it adheres to the concept of the first edition, several new spectroscopic techniques such as optothermal spectroscopy or velocity-modulation spectroscopy are added.

A whole chapter is devoted to time-resolved spectroscopy including the generation and detection of ultrashort light pulses. The principles of coherent spectroscopy, which have found widespread applications, are covered in a separate chapter. The combination of laser spectroscopy and collision physics, which has given new impetus to the study and control of chemical reactions, has deserved an extra chapter. In addition, more space has been given to optical cooling and trapping of atoms and ions.

I hope that the new edition will find a similar friendly acceptance as the first one. Of course, a texbook never is perfect but can always be improved. I, therefore, appreciate any hint to possible errors or comments concerning corrections and improvements. I will be happy if this book helps to support teaching courses on laser spectroscopy and to transfer some of the delight I have experienced during my research in this fascinating field over the last 30 years.

Many people have helped to complete this new edition. I am grateful to colleagues and friends, who have supplied figures and reprints of their work. I thank the graduate students in my group, who provided many of the examples used to illustrate the different techniques. Mrs. Wollscheid who has drawn many figures, and Mrs. Heider who typed part of the corrections. Particular thanks go to Helmut Lotsch of Springer-Verlag, who worked very hard for this book and who showed much patience with me when I often did not keep the deadlines.

Last but not least, I thank my wife Harriet who had much understanding for the many weekends lost for the family and who helped me to have sufficient time to write this extensive book.

Kaiserslautern,   *Wolfgang Demtröder*
June 1995

Preface to the First Edition

The impact of lasers on spectroscopy can hardly be overestimated. Lasers represent intense light sources with spectral energy densities which may exceed those of incoherent sources by several orders of magnitude. Furthermore, because of their extremely small bandwidth, single-mode lasers allow a spectral resolution which far exceeds that of conventional spectrometers. Many experiments which could not be done before the application of lasers, because of lack of intensity or insufficient resolution, are readily performed with lasers.

Now several thousands of laser lines are known which span the whole spectral range from the vacuum-ultraviolet to the far-infrared region. Of particular interst are the continuously tunable lasers which may in many cases replace wavelength-selecting elements, such as spectrometers or interferometers. In combination with optical frequency-mixing techniques such continuously tunable monochromatic coherent light sources are available at nearly any desired wavelength above 100 nm.

The high intensity and spectral monochromasy of lasers have opened a new class of spectroscopic techniques which allow investigation of the structure of atoms and molecules in much more detail. Stimulated by the variety of new experimental possibilities that lasers give to spectroscopists, very lively research activities have developed in this field, as manifested by an avalanche of publications. A good survey about recent progress in laser spectroscopy is given by the proceedings of various conferences on laser spectroscopy (see "Springer Series in Optical Sciences"), on picosecond phenomena (see "Springer Series in Chemical Physics"), and by several quasi-mongraphs on laser spectroscopy published in "Topics in Applied Physics".

For nonspecialists, however, or for people who are just starting in this field, it is often difficult to find from the many articles scattered over many journals a coherent representation of the basic principles of laser spectroscopy. This textbook intends to close this gap between the advanced research papers and the representation of fundamental principles and experimental techniques. It is addressed to physicists and chemists who want to study laser spectroscopy in more detail. Students who have some knowledge of atomic and molecular physics, electrodynamics, and optics should be able to follow the presentation.

The fundamental principles of lasers are covered only very briefly because many excellent textbooks on lasers already exist.

On the other hand, those characteristics of the laser that are important for its applications in spectroscopy are treated in more detail. Examples are the frequency spectrum of different types of lasers, their linewidths, amplitude and frequency stability, tunability, and tuning ranges. The optical components such as mirrors, prisms, and gratings, and the experimental equipment of spectroscopy, for example, monochromators, interferometers, photon detectors, etc., are discussed extensively because detailed knowledge of modern spectroscopic equipment may be crucial for the successful performance of an experiment.

Each chapter gives several examples to illustrate the subject discussed. Problems at the end of each chapter may serve as a test of the reader's understanding. The literature cited for each chapter is, of course, not complete but should inspire further studies. Many subjects that could be covered only briefly in this book can be found in the references in a more detailed and often more advanced treatment. The literature selection does not represent any priority list but has didactical purposes and is intended to illustrate the subject of each chapter more thoroughly.

The spectroscopic applications of lasers covered in this book are restricted to the spectroscopy of free atoms, molecules, or ions. There exists, of course, a wide range of applications in plasma physics, solid-state physics, or fluid dynamics which are not discussed because they are beyond the scope of this book. It is hoped that this book may be of help to students and researchers. Although it is meant as an introduction to laser spectroscopy, it may also facilitate the understanding of advanced papers on special subjects in laser spectroscopy. Since laser spectroscopy is a very fascinating field of research, I would be happy if this book can transfer to the reader some of my excitement and pleasure experienced in the laboratory while looking for new lines or unexpected results.

I want to thank many people who have helped to complete this book. In particular the students in my research group who by their experimental work have contributed to many of the examples given for illustration and who have spent their time reading the galley proofs. I am grateful to colleages from many laboratories who have supplied me with figures from their publications. Special thanks go to Mrs. Keck and Mrs. Ofiara who typed the manuscript and to Mrs. Wollscheid and Mrs. Ullmer who made the drawings. Last but not least, I would like to thank Dr. U. Hebgen, Dr. H. Lotsch, Mr. K.-H. Winter, and other coworkers of Springer-Verlag who showed much patience with a dilatory author and who tried hard to complete the book in a short time.

Kaiserslautern,   *Wolfgang Demtröder*
March 1981

Contents

1. Introduction

Most of our knowledge about the structure of atoms and molecules is based on spectroscopic investigations. Thus spectroscopy has made an outstanding contribution to the present state of atomic and molecular physics, to chemistry, and to molecular biology. Information on molecular structure and on the interaction of molecules with their surroundings may be derived in various ways from the absorption or emission spectra generated when electromagnetic radiation interacts with matter.

Wavelength measurements of spectral lines allow the determination of energy levels of the atomic or molecular system. The *line intensity* is proportional to the transition probability, which measures how strongly the two levels of a molecular transition are coupled. Since the transition probability depends on the wave functions of both levels, intensity measurements are useful to verify the spatial charge distribution of excited electrons, which can only be roughly calculated from approximate solutions of the Schrödinger equation. The *natural linewidth* of a spectral line may be resolved by special techniques, allowing mean lifetimes of excited molecular states to be determined. Measurements of the *Doppler width* yield the velocity distribution of the emitting or absorbing molecules and with it the temperature of the sample. From *pressure broadening* and *pressure shifts* of spectral lines, information about collision processes and interatomic potentials can be extracted. *Zeemann* and *Stark splittings* by external magnetic or electric fields are important means of measuring magnetic or electric moments and elucidating the coupling of the different angular momenta in atoms or molecules, even with complex electron configurations. The *hyperfine structure* of spectral lines yields information about the interaction between the nuclei and the electron cloud and allows nuclear magnetic dipole moments or electric quadrupole moments to be determined. Time-resolved measurements allow the spectroscopist to follow up dynamical processes in ground-state and excited-state molecules, to investigate collision processes and various energy transfer mechanisms. Laser spectroscopic studies of the interaction of single atoms with a radiation field provide stringent tests of quantum electrodynamics and the realization of high-precision frequency standards allows one to check whether fundamental physical constants show small changes with time.

These examples represent only a small selection of the many possible ways by which spectroscopy provides tools to explore the microworld of atoms and molecules. However, the amount of information that can be extracted from a spectrum depends essentially on the attainable spectral or time resolution and on the detection sensitivity that can be achieved.

The application of new technologies to optical instrumentation (for instance, the production of larger and better ruled gratings in spectrographs, the use of highly reflecting dielectric coatings in interferometers, and the development of optical multichannel analyzers and image intensifiers) has certainly significantly extended the sensitivity limits. Considerable progress was furthermore achieved through the introduction of new spectroscopic techniques, such as Fourier spectroscopy, optical pumping, level-crossing techniques, and various kinds of double-resonance methods and molecular beam spectroscopy.

Although these new techniques have proved to be very fruitful, the really stimulating impetus to the whole field of spectroscopy was given by the introduction of lasers. In many cases these new spectroscopic light sources may increase spectral resolution and sensitivity by several orders of magnitude. Combined with new spectroscopic techniques, lasers are able to surpass basic limitations of classical spectroscopy. Many experiments that could not be performed with incoherent light sources are now feasible or have already been successfully completed recently. This book deals with such new techniques of laser spectroscopy and explains the necessary instrumentation.

The book begins with a discussion of the fundamental definitions and concepts of classical spectroscopy, such as thermal radiation, induced and spontaneous emission, radiation power and intensity, transition probabilities, and the interaction of weak and strong electromagnetic (EM) fields with atoms. Since the coherence properties of lasers are important for several spectroscopic techniques, the basic definitions of coherent radiation fields are outlined and the description of coherently excited atomic levels is briefly discussed.

In order to understand the theoretical limitations of spectral resolution in classical spectroscopy, Chap. 3 treats the different causes of the broadening of spectral lines and the information drawn from measurements of line profiles. Numerical examples at the end of each section illustrate the order of magnitude of the different effects.

The contents of Chap. 4, which covers spectroscopic instrumentation and its application to wavelength and intensity measurements, are essential for the experimental realization of laser spectroscopy. Although spectrographs and monochromators, which played a major rule in classical spectroscopy, may be abandoned for many experiments in laser spectroscopy, there are still numerous applications where these instruments are indispensible. Of major importance for laser spectroscopists are the different kinds of interferometers. They are used not only in laser resonators to realize single-mode operation, but also for line-profile measurements of spectral lines and for very precise wavelength measurements. Since the determination of wavelength is a central problem in spectroscopy, a whole section discusses some modern techniques for precise wavelength measurements and their accuracy.

Lack of intensity is one of the major limitations in many spectroscopic investigations. It is therefore often vital for the experimentalist to choose the proper light detector. Section 4.5 surveys several light detectors and sensitive techniques such as photon counting, which is becoming more commonly used. This chapter concludes the first part of the book, which covers fundamental concepts and basic instrumentation of general spectroscopy. The second part discusses in more detail subjects more specific to laser spectroscopy.

Chapter 5 treats the basic properties of lasers as spectroscopic radiation sources. It starts with a short recapitulation of the fundamentals of lasers, such as threshold conditions, optical resonators, and laser modes. Only those laser characteristics that are important in laser spectroscopy are discussed here. For a more detailed treatment the reader is referred to the extensive laser literature cited in Chap. 5. Those properties and experimental techniqes that make the laser such an attractive spectroscopic light source are discussed more thoroughly. For instance, the important questions of wavelength stabilization and continuous wavelength tuning are treated, and experimental realizations of single-mode tunable lasers and limitations of laser linewidths are presented. The last part of this chapter gives a survey of the various types of tunable lasers that have been developed for different spectral ranges. Advantages and limitations of these lasers are discussed. The available spectral range could be greatly extended by optical frequency doubling and mixing processes. This interesting field of nonlinear optics is briefly presented at the end of Chap. 5 as far as it is relevant to spectroscopy.

The main part of the book presents various applications of lasers in spectroscopy and discusses the different methods that have been developed recently. Chapter 6 starts with Doppler-limited laser absorption spectroscopy with its various high-sensitivity detection techniques such as frequency modulation and intracavity spectroscopy, cavity ring-down techniques, excitation-fluorescence detection, ionization and optogalvanic spectroscopy, optoacoustic and optothermal spectroscopy, or laser-induced fluorescence. A comparison between the different techniques helps to critically judge their merits and limitations.

Really impressive progress toward higher spectral resolution has been achieved by the development of various "Doppler-free" techniques. They rely mainly on nonlinear spectroscopy, which is extensively discussed in Chap. 7. Besides the fundamentals of nonlinear absorption, the techniques of saturation spectroscopy, polarization spectroscopy, and multiphoton absorption are presented, together with various combinations of these methods.

Raman spectroscopy, a very important technique of classical spectroscopy, has been revolutionized by the use of lasers. Not only spontaneous Raman spectroscopy with greatly enhanced sensitivity, but also new techniques such as induced Raman spectroscopy, surface-enhanced Raman spectroscopy, or coherent anti-Stokes Raman spectroscopy (CARS) have contributed greatly to the rapid development of sensitive, high-resolution detection of molecular structure and dynamics, as is outlined in Chap. 8.

The combination of molecular beam methods with laser spectroscopic techniques has brought about a large variety of new methods to study molecules, radicals, loosely bound van der Waals complexes, and clusters. This is discussed extensively in Chap. 9.

Of particular importance for the spectroscopy of highly excited states, such as Rydberg levels of atoms and molecules, and for the assignment of complex molecular spectra are various double-resonance techniques where atoms and molecules are exposed simultaneously to two radiation fields resonant with two transitions sharing a common level. In combination with Doppler-free techniques, these double resonance methods are powerful tools for spectroscopy. Some of these methods, representing modern versions of optical pumping techniques of the prelaser era, are introduced in Chap. 10.

Impressive progress has been achieved in the development of short laser pulses, with pulse durations in the femtosecond range. In 1999 the Nobel Prize in Chemistry was awarded to A. H. Zewail for his work in femtosecond spectroscopy. New techniques allow a time resolution hitherto out of reach. Transient phenomena, such as fast isomerization of excited molecules, relaxation processes by collisional energy transfer, or fast dissociation processes of optically excited molecules can now be investigated. Chapter 11 gives a survey on techniques in the nano-, pico-, and femto-second range to generate and to detect ultrashort light pulses.

Coherent spectroscopy, which is based on the coherent excitation of molecular levels and the detection of coherently scattered light, is treated in Chap. 12, where techniques and applications of this interesting field are illustrated by several examples.

Laser spectroscopy has contributed in an outstanding way to detailed studies of collision processes. Chapter 13 gives some examples of applications of lasers in investigations of elastic, inelastic, and reactive collisions.

The rapid development of laser spectroscopy in recent years is demonstrated in Chap. 14, which compiles some recent fascinating ideas and their methods to further increase spectral resolution and sensitivity. The goal of studying single atoms and their interaction with radiation fields is no longer a dream of theoreticians but can be realized experimentally. Interesting aspects of cooling and trapping of atoms, the achievement of Bose–Einstein condensation for trapped atoms, phase transitions from ordered to chaotic systems, and fundamental limits of detection-sensitivity are briefly outlined. The realization of atom lasers and the fascinating aspects of atom interferometry are also presented.

The last chapter illustrates by some examples the broad field of applications of laser spectroscopy to the solution of scientific, technical, and medical problems.

This book is intended as an *introduction* to the basic methods and instrumentation of laser spectroscopy. The examples in each chapter illustrate the text and may suggest other possible applications. They are mainly concerned with the spectroscopy of free atoms and molecules and are, of course, not complete, but have been selected from the literature or from our own laboratory work for didactic purposes and may not represent the priorities of publication dates. For a far more extensive survey of the latest publications in the broad field of laser spectroscopy, the reader is referred to the proceedings of various conferences on laser spectroscopy [1.1–1.10] and to textbooks or collections of articles on modern aspects of laser spectroscopy [1.11–1.31].

Since scientific achievements in laser physics have been pushed forward by a few pioneers, it is interesting to look back to the historical development and to the people who influenced it. Such a personal view can be found in [1.32, 1.33]. The reference list at the end of the book might be helpful in finding more details of a special experiment or to dig deeper into theoretical and experimental aspects of each chapter. A useful "Encyclopedia of spectroscopy" [1.34] gives a good survey on different aspects of laser spectroscopy.

2. Absorption and Emission of Light

This chapter deals with basic considerations about absorption and emission of electromagnetic waves interacting with matter. Especially emphasized are those aspects that are important for the spectroscopy of gaseous media. The discussion starts with thermal radiation fields and the concept of cavity modes in order to elucidate differences and connections between spontaneous and induced emission and absorption. This leads to the definition of the Einstein coefficients and their mutual relations. The next section explains some definitions used in photometry such as radiation power, intensity, and spectral power density.

It is possible to understand many phenomena in optics and spectroscopy in terms of classical models based on concepts of classical electrodynamics. For example, the absorption and emission of electromagnetic waves in matter can be described using the model of damped oscillators for the atomic electrons. In most cases, it is not too difficult to give a quantum-mechanical formulation of the classical results. The semiclassical approach will be outlined briefly in Sect. 2.7.

Many experiments in laser spectroscopy depend on the coherence properties of the radiation and on the coherent excitation of atomic or molecular levels. Some basic ideas about temporal and spatial coherence of optical fields and the density-matrix formalism for the description of coherence in atoms are therefore discussed at the end of this chapter.

Throughout this text the term "light" is frequently used for electromagnetic radiation in all spectral regions. Likewise, the term "molecule" in general statements includes atoms as well. We shall, however, restrict the discussion and most of the examples to *gaseous* media, which means essentially free atoms or molecules.

For more detailed or more advanced presentations of the subjects summarized in this chapter, the reader is referred to the extensive literature on spectroscopy [2.1–2.11]. Those interested in light scattering from solids are directed to the sequence of Topics volumes edited by Cardona and coworkers [2.12].

2.1 Cavity Modes

Consider a cubic cavity with the sides L at the temperature T. The walls of the cavity absorb and emit electromagnetic radiation. At thermal equilibrium the absorbed power $P_a(\omega)$ has to be equal to the emitted power $P_e(\omega)$ for all frequencies ω. Inside the cavity there is a *stationary radiation field* E, which can be described at the point r by a superposition of plane waves with the amplitudes A_p, the wave vectors k_p, and the angular frequencies ω_p as

$$E = \sum_p A_p \exp[i(\omega_p t - k_p \cdot r)] + \text{compl. conj.} \tag{2.1}$$

The waves are reflected at the walls of the cavity. For each wave vector $k = (k_x, k_y, k_z)$, this leads to eight possible combinations $k = (\pm k_x, \pm k_y, \pm k_z)$ that interfere with each other. A stationary-field configuration only occurs if this superpositions result in *standing waves* (Fig. 2.1a,b). This imposes boundary conditions for the wave vector, namely

$$k = \frac{\pi}{L}(n_1, n_2, n_3), \tag{2.2}$$

where n_1, n_2, n_3 are positive integers.

The magnitudes of the wave vectors allowed by the boundary conditions are

$$|k| = \frac{\pi}{L}\sqrt{n_1^2 + n_2^2 + n_3^3}, \tag{2.3}$$

which can be written in terms of the wavelength $\lambda = 2\pi/|k|$ or the frequency $\omega = c|k|$.

$$\lambda = 2L/\sqrt{n_1^2 + n_2^2 + n_3^2} \quad \text{or} \quad \omega = \frac{\pi c}{L}\sqrt{n_1^2 + n_2^2 + n_3^2}. \tag{2.4}$$

These standing waves are called *cavity modes* (Fig. 2.1b).

Since the amplitude vector A of a transverse wave E is always perpendicular to the wave vector k, it can be composed of two components a_1 and a_2

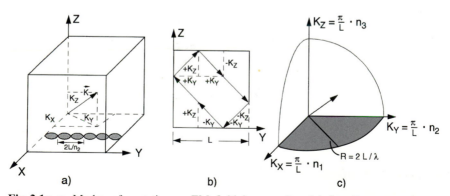

Fig. 2.1a–c. Modes of a stationary EM field in a cavity: (**a**) Standing waves in a cubic cavity; (**b**) superposition of possible k vectors to form standing waves, illustrated in a two-dimensional coordinate system; (**c**) illustration of the calculation of the maximum number of modes in momentum space

with the unit vectors $\hat{e}_1$ and $\hat{e}_2$

$$A = a_1\hat{e}_1 + a_2\hat{e}_2 \qquad (\hat{e}_1 \cdot \hat{e} = \delta_{12} ; \quad \hat{e}_1, \hat{e}_2 \perp k) . \tag{2.5}$$

The complex numbers a_1 and a_2 define the polarization of the standing wave. Equation (2.5) states that any arbitrary polarization can always be expressed by a linear combination of two mutually orthogonal linear polarizations. To each cavity mode defined by the wave vector k_p therefore belong two possible polarization states. This means that *each triple of integers* (n_1, n_2, n_3) *represents two cavity modes. Any arbitrary stationary field configuration can be expressed as a linear combination of cavity modes.*

We shall now investigate how many modes with frequencies $\omega \leq \omega_m$ are possible. Because of the boundary condition (2.2), this number is equal to the number of all integer triples (n_1, n_2, n_3) that fulfil the condition

$$c^2 k^2 = \omega^2 \leq \omega_m^2 .$$

In a system with the coordinates $(\pi/L)(n_1, n_2, n_3)$, see Fig. 2.1c, each triple (n_1, n_2, n_3) represents a point in a three-dimensional lattice with the lattice constant π/L. In this system, (2.4) describes all possible frequencies within a sphere of radius ω/c. If this radius is large compared to π/L, which means that $2L \gg \lambda_m$, the number of lattice points (n_1, n_2, n_3) with $\omega^2 \leq \omega_m^2$ is roughly given by the volume of the octant of the sphere shown in Fig. 2.1c. With the two possible polarization states of each mode, one therefore obtains for the number of allowed modes with frequencies between $\omega = 0$ and $\omega = \omega_m$ in a cubic cavity of volume L^3 with $L \gg \lambda$

$$N(\omega_m) = 2\frac{1}{8}\frac{4\pi}{3}\left(\frac{L\omega_m}{\pi c}\right)^3 = \frac{1}{3}\frac{L^3\omega_m^3}{\pi^2 c^3} . \tag{2.6}$$

and N/L^3 represents the number of modes per unit volume.

It is often interesting to know the number $n(\omega)\,d\omega$ of modes per unit volume within a certain frequency interval $d\omega$, for instance, within the width of a spectral line. The *spectral mode density* $n(\omega)$ can be obtained directly from (2.6) by differentiating $N(\omega)/L^3$ with respect to ω. $N(\omega)$ is assumed to be a continuous function of ω, which is, strictly speaking, only the case for $L \to \infty$. We get

$$n(\omega)\,d\omega = \frac{\omega^2}{\pi^2 c^3}\,d\omega . \tag{2.7a}$$

In spectroscopy the frequency $\nu = \omega/2\pi$ is often used instead of the angular frequency ω. The number of modes per unit volume within the frequency interval $d\nu$ is then

$$n(\nu)\,d\nu = \frac{8\pi\nu^2}{c^3}\,d\nu . \tag{2.7b}$$

Example 2.1

(a) In the visible part of the spectrum ($\lambda = 500\,\text{nm}$; $\nu = 6 \times 10^{14}\,\text{Hz}$), (2.7b) yields for the number of modes per m^3 within the Doppler width of a spectral line ($d\nu = 10^9\,\text{Hz}$)

$$n(\nu)\,d\nu = 3 \times 10^{14}\,\text{m}^{-3} \,.$$

(b) In the microwave region ($\lambda = 1\,\text{cm}$; $\nu = 3 \times 10^{10}\,\text{Hz}$), the number of modes per m^3 within the typical Doppler width $d\nu = 10^5\,\text{Hz}$ is only $n(\nu)\,d\nu = 10^2\,\text{m}^{-3}$.

(c) In the X-ray region ($\lambda = 1\,\text{nm}$; $\nu = 3 \times 10^{17}\,\text{Hz}$), one finds $n(\nu)\,d\nu = 8.4 \times 10^{21}\,\text{m}^{-3}$ within the typical natural linewidth $d\nu = 10^{11}\,\text{Hz}$ of an X-ray transition.

2.2 Thermal Radiation and Planck's Law

In classical thermodynamics each degree of freedom of a system in thermal equilibrium has the mean energy $kT/2$, where k is the Boltzmann constant. Since classical oscillators have kinetic as well as potential energies, their mean energy is kT. If this classical concept is applied to the electromagnetic field discussed in Sect. 2.1, each mode would represent a classical oscillator with the mean energy kT. According to (2.7b), the spectral energy density of the radiation field would therefore be

$$\rho(\nu)\,d\nu = n(\nu)kT\,d\nu = \frac{8\pi\nu^2 k}{c^3}T\,d\nu \,. \tag{2.8}$$

This *Rayleigh–Jeans law* matches the experimental results fairly well at low frequencies (in the infrared region), but is in strong disagreement with experiment at higher frequencies (in the ultraviolet region). The energy density $\rho(\nu)$ actually diverges for $\nu \to \infty$.

In order to explain this discrepancy, M. Planck suggested in 1900 that each mode of the radiation field can only emit or absorb energy in discrete amounts $qh\nu$, which are integer multiples q of a minimum energy quantum $h\nu$. These energy quanta $h\nu$ are called *photons*. Planck's constant h can be determined from experiments. *A mode with q photons therefore has the energy $qh\nu$.*

In thermal equilibrium the partition of the total energy into the different modes is governed by the Maxwell–Boltzmann distribution, so that the probability $p(q)$ that a mode contains the energy $qh\nu$ is

$$p(q) = (1/Z)\,e^{-qh\nu/kT} \,, \tag{2.9}$$

where k is the Boltzmann constant and

$$Z = \sum_q e^{-qh\nu/kT} \tag{2.10}$$

is the partition function summed over all modes containing q photons $h \cdot \nu$. Z acts as a normalization factor which makes $\sum_q p(q) = 1$, as can be seen immediately by inserting (2.10) into (2.9). This means that a mode has to contain with certainty ($p = 1$) some number ($q = 0, 1, 2, \ldots$) of photons.

The mean energy per mode is therefore

$$W = \sum_{q=0}^{\infty} p(q)qh\nu = \frac{1}{Z} \sum_{q=0}^{\infty} qh\nu e^{-qh\nu/kT} . \tag{2.11}$$

The evaluation of the sum yields [2.6]

$$W = \frac{h\nu}{e^{h\nu/kT} - 1} . \tag{2.12}$$

The thermal radiation field has the energy density $\rho(\nu)\,d\nu$ within the frequency interval ν to $\nu + d\nu$, which is equal to the number $n(\nu)\,d\nu$ of modes in the interval $d\nu$ times the mean energy W per mode. Using (2.7b, 2.12) one obtains

$$\boxed{\rho(\nu)\,d\nu = \frac{8\pi\nu^2}{c^3} \frac{h\nu}{e^{h\nu/kT} - 1} d\nu .} \tag{2.13}$$

This is *Planck's* famous *radiation law* (Fig. 2.2), which predicts a spectral energy density of the thermal radiation that is fully consistent with experiments. The expression "thermal radiation" comes from the fact that the spectral energy distribution (2.13) is characteristic of a radiation field that is in thermal equilibrium with its surroundings (in Sect. 2.1 the surroundings are determined by the cavity walls).

The thermal radiation field described by its energy density $\rho(\nu)$ is *isotropic*. This means that through any transparent surface element dA of

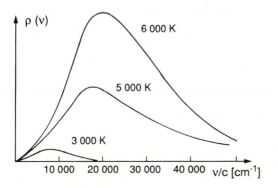

Fig. 2.2. Spectral distribution of the energy density $\rho_\nu(\nu)$ for different temperatures

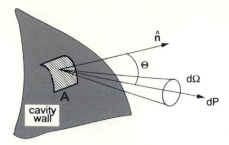

Fig. 2.3. Illustration of (2.14)

a sphere containing a thermal radiation field, the same power flux dP is emitted into the solid angle $d\Omega$ at an angle θ to the surface normal $\hat{n}$ (Fig. 2.3)

$$dP = \frac{c}{4\pi}\rho(\nu)\,dA\,d\Omega\,d\nu\cos\theta\;.\tag{2.14}$$

It is therefore possible to determine $\rho(\nu)$ experimentally by measuring the spectral distribution of the radiation penetrating through a small hole in the walls of the cavity. If the hole is sufficiently small, the energy loss through this hole is negligibly small and does not disturb the thermal equilibrium inside the cavity.

Example 2.2

(a) Examples of real radiation sources with spectral energy distributions close to the Planck distribution (2.13) are the sun, the bright tungsten wire of a light bulb, flash lamps, and high-pressure discharge lamps.

(b) Spectral lamps that emit discrete spectra are examples of *nonthermal* radiation sources. In these gas-discharge lamps, the light-emitting atoms or molecules may be in thermal equilibrium with respect to their translational energy, which means that their velocity distribution is Maxwellian. However, the population of the different excited atomic levels may not necessarily follow a Boltzmann distribution. There is generally no thermal equilibrium between the atoms and the radiation field. The radiation may nevertheless be isotropic.

(c) Lasers are examples of nonthermal and anisotropic radiation sources (Chap. 5). The radiation field is concentrated in a few modes, and most of the radiation energy is emitted into a small solid angle. This means that the laser represents an extreme anisotropic nonthermal radiation source.

2.3 Absorption, Induced, and Spontaneous Emission

Assume that molecules with the energy levels E_1 and E_2 have been brought into the thermal radiation field of Sect. 2.2. If a molecule absorbs a photon of energy $h\nu = E_2 - E_1$, it is excited from the lower energy level E_1 into

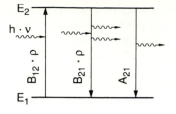

Fig. 2.4. Schematic diagram of the interaction of a two-level system with a radiation field

the higher level E_2 (Fig. 2.4). This process is called *induced absorption*. The probability per second that a molecule will absorb a photon, $d\mathcal{P}_{12}/dt$, is proportional to the number of photons of energy $h\nu$ per unit volume and can be expressed in terms of the spectral energy density $\rho_\nu(\nu)$ of the radiation field as

$$\frac{d}{dt}\mathcal{P}_{12} = B_{12}\rho(\nu) \,. \tag{2.15}$$

The constant factor B_{12} is the *Einstein coefficient of induced absorption*. Each absorbed photon of energy $h\nu$ decreases the number of photons in one mode of the radiation field by one.

The radiation field can also induce molecules in the excited state E_2 to make a transition to the lower state E_1 with simultaneous emission of a photon of energy $h\nu$. This process is called *induced (or stimulated) emission*. The induced photon of energy $h\nu$ is emitted into the same mode that caused the emission. This means that the number of photons in this mode is increased by one. The probability $d\mathcal{P}_{21}/dt$ that one molecule emits one induced photon per second is in analogy to (2.15)

$$\frac{d}{dt}\mathcal{P}_{21} = B_{21}\rho(\nu) \,. \tag{2.16}$$

The constant factor B_{21} is the *Einstein coefficient of induced emission*.

An excited molecule in the state E_2 may also *spontaneously* convert its excitation energy into an emitted photon $h\nu$. This spontaneous radiation can be emitted in the arbitrary direction $\boldsymbol{k}$ and increases the number of photons in the mode with frequency ν and wave vector $\boldsymbol{k}$ by one. In the case of isotropic emission, the probability of gaining a spontaneous photon is equal for all modes with the same frequency ν but different directions $\boldsymbol{k}$.

The probability per second $d\mathcal{P}_{21}^{\text{spont}}/dt$ that a photon $h\nu = E_2 - E_1$ is spontaneously emitted by a molecule, depends on the structure of the molecule and the selected transition $|2\rangle \rightarrow |1\rangle$, *but it is independent of the external radiation field*,

$$\frac{d}{dt}\mathcal{P}_{21}^{\text{spont}} = A_{21} \,. \tag{2.17}$$

A_{21} is the *Einstein coefficient of spontaneous emission* and is often called the *spontaneous transition probability*.

Let us now look for relations between the three Einstein coefficients B_{12}, B_{21}, and A_{21}. The total number N of all molecules per unit volume is distributed among the various energy levels E_i of population density N_i such that $\sum_i N_i = N$. At thermal equilibrium the population distribution $N_i(E_i)$ is given by the Boltzmann distribution

$$N_i = N \frac{g_i}{Z} e^{-E_i/kT} . \tag{2.18}$$

The statistical weight $g_i = 2J_i + 1$ gives the number of degenerate sublevels of the level $|i\rangle$ with total angular momentum J_i and the partition function

$$Z = \sum_i g_i e^{-E_i/kT} ,$$

acts again as a normalization factor which ensures that $\sum_i N_i = N$.

In a stationary field the total absorption rate $N_i B_{12}\rho(v)$, which gives the number of photons absorbed per unit volume per second, has to equal the total emission rate $N_2 B_{21}\rho(v) + N_2 A_{21}$ (otherwise the spectral energy density $\rho(v)$ of the radiation field would change). This gives (Fig. 2.4)

$$[B_{21}\rho(v) + A_{21}]N_2 = B_{12}N_1\rho(v) . \tag{2.19}$$

Using the relation

$$N_2/N_1 = (g_2/g_1)e^{-(E_2-E_1)/kT} = (g_2/g_1)e^{-hv/kT} ,$$

deduced from (2.18), and solving (2.19) for $\rho(v)$ yields

$$\rho(v) = \frac{A_{21}/B_{21}}{\frac{g_1}{g_2}\frac{B_{12}}{B_{21}} e^{hv/kT} - 1} . \tag{2.20}$$

In Sect. 2.2 we derived Planck's law (2.13) for the spectral energy density $\rho(v)$ of the thermal radiation field. Since both (2.13, 2.20) must be valid for an arbitrary temperature T and all frequencies v, comparison of the constant coefficients yields the relations

$$\boxed{B_{12} = \frac{g_2}{g_1} B_{21}} , \tag{2.21}$$

$$\boxed{A_{21} = \frac{8\pi h^3}{c^3} B_{21}} . \tag{2.22}$$

Equation (2.21) states that for levels $|1\rangle$, $|2\rangle$ with the equal statistical weights $g_2 = g_1$, *the probability of induced emission is equal to that of induced absorption.*

From (2.22) the following illustrative result can be extracted: since $n(v) = 8\pi v^2/c^3$ gives the number of modes per unit volume and frequency interval

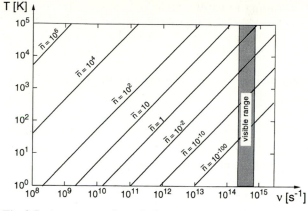

Fig. 2.5. Average number of photons per mode in a thermal radiation field as a function of temperature T and frequency ν

$d\nu = 1$ Hz, (see (2.7b)), (2.22) can be written as

$$\frac{A_{21}}{n(\nu)} = B_{21} h\nu \,, \tag{2.23}$$

which means that the spontaneous emission per mode equals the induced emission that is triggered by one photon. This can be generalized to: *the ratio of the induced- to the spontaneous-emission rate in an arbitrary mode is equal to the number of photons in this mode.*

In Fig. 2.5 the mean number of photons per mode in a thermal radiation field at different absolute temperatures is plotted as a function of frequency ν. The graphs illustrate that in the visible spectrum this number is small compared to unity at temperatures that can be realized in a laboratory. *This implies that in thermal radiation fields, the spontaneous emission per mode exceeds by far the induced emission.* If it is possible, however, to concentrate most of the radiation energy into a few modes, the number of photons in these modes may become exceedingly large and the induced emission in these modes dominates, although the total spontaneous emission into *all* modes may still be larger than the induced rate. Such a selection of a few modes is realized in a laser (Chap. 5).

Comment

Note that the relations (2.21, 2.22) are valid for all kinds of radiation fields. Although they have been derived for stationary fields at thermal equilibrium, the Einstein coefficients are constants that depend only on the molecular properties and not on external fields as far as these fields do not alter the molecular properties. These equations therefore hold for arbitrary $\rho_\nu(\nu)$.

Using the angular frequency $\omega = 2\pi\nu$ instead of ν, the unit frequency interval $d\omega = 1 \,\text{s}^{-1}$ corresponds to $d\nu = 1/2\pi \,\text{s}^{-1}$. The spectral energy density

$\rho_\omega(\omega) = n(\omega)\hbar\omega$ is then, according to (2.7a),

$$\rho_\omega(\omega) = \frac{\omega^2}{\pi^2 c^3} \frac{\hbar\omega}{e^{\hbar\omega/kT} - 1} \ , \tag{2.24}$$

where $\hbar$ is Planck's constant divided by 2π. The ratio of the Einstein coefficients

$$A_{21}/B_{21} = \frac{\hbar\omega^3}{\pi^2 c^3} \ , \tag{2.25}$$

now contains $\hbar$ instead of h, and is smaller by a factor of 2π. However, the ratio $A_{21}/[B_{21}\rho_\omega(\omega)]$, which gives the ratio of the spontaneous to the induced transition probabilities, remains the same.

Example 2.3

(a) In the thermal radiation field of a 100 W light bulb, 10 cm away from the tungsten wire, the number of photons per mode at $\lambda = 500$ nm is about 10^{-8}. If a molecular probe is placed in this field, the induced emission is therefore completely negligible.

(b) In the center spot of a high-current mercury discharge lamp with very high pressure, the number of photons per mode is about 10^{-2} at the center frequency of the strongest emission line at $\lambda = 253.6$ mm. This shows that, even in this very bright light source, the induced emission only plays a minor role.

(c) Inside the cavity of a HeNe laser (output power 1 mW with mirror transmittance $T = 1\%$) that oscillates in a single mode, the number of photons in this mode is about 10^7. In this example the *spontaneous* emission into this mode is completely negligible. Note, however, that the total spontaneous emission power at $\lambda = 632.2$ nm, which is emitted into all directions, is much larger than the induced emission. This spontaneous emission is more or less uniformly distributed among all modes. Assuming a volume of 1 cm^3 for the gas discharge, the number of modes within the Doppler width of the neon transition is about 10^8, which means that the total spontaneous rate is about 10 times the induced rate.

2.4 Basic Photometric Quantities

In spectroscopic applications of light sources, it is very useful to define some characteristic quantities of the emitted and absorbed radiation. This allows a proper comparison of different light sources and detectors and enables one to make an appropriate choice of apparatus for a particular experiment.

2.4.1 Definitions

The *radiant energy W* (measured in joules) refers to the total amount of energy emitted by a light source, transferred through a surface, or collected by a detector. The *radiant power P* (often called *radiant flux Φ*) [W] is the radiant energy per second. The *radiant energy density ρ* [J/m³] is the radiant energy per unit volume of space. Consider a surface element dA of a light source (Fig. 2.6). The radiant power emitted from dA into the solid angle dΩ, around the angle θ against the surface normal $\hat{n}$ is

$$dP = L(\theta)\, dA\, d\Omega \,, \tag{2.26a}$$

where the *radiance L* [W/m² sr⁻¹] is the power emitted per unit surface element d$A = 1\,\text{m}^2$ into the unit solid angle d$\Omega = 1\,\text{sr}$.

The total power emitted by the source is

$$P = \int L(\theta)\, dA\, d\Omega \,. \tag{2.26b}$$

Fig. 2.6. Basic radiant quantities of a light source

The above three quantities refer to the total radiation integrated over the entire spectrum. Their spectral versions $W_\nu(\nu)$, $P_\nu(\nu)$, $\rho_\nu(\nu)$, and $L_\nu(\nu)$ are called the *spectral densities*, and are defined as the amounts of W, P, ρ, and L within the unit frequency interval d$\nu = 1\,\text{s}^{-1}$ around the frequency ν:

$$W = \int_0^\infty W_\nu(\nu)\,d\nu\,; \quad P = \int_0^\infty P_\nu(\nu)\,d\nu\,; \quad \rho = \int_0^\infty \rho_\nu(\nu)\,d\nu\,; \quad L = \int_0^\infty L_\nu(\nu)\,d\nu\,. \tag{2.27}$$

Example 2.4

For a spherical isotropic radiation source of radius R (e.g., a star) with a spectral energy density ρ_ν, the spectral radiance $L_\nu(\nu)$ is independent of θ and can be expressed by

$$L_\nu(\nu) = \rho_\nu(\nu) c/4\pi = \frac{2h\nu^3}{c^2}\frac{1}{e^{h\nu/kT}-1} \rightarrow P_\nu = \frac{8\pi R^2 h\nu^3}{c^2}\frac{1}{e^{h\nu/kT}-1}\,. \tag{2.28}$$

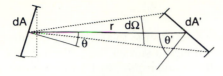

Fig. 2.7. Radiance and irradiance of source and detector

A surface element dA' of a detector at distance r from the source element dA covers a solid angle $d\Omega = dA' \cos\theta'/r^2$ as seen from the source (Fig. 2.7). With $r^2 \gg dA$ and dA', the radiant power received by dA' is

$$dP = L(\theta)\,dA\cos\theta\,d\Omega = L(\theta)\cos\theta\,dA\cos\theta'\,dA'/r^2 ,\qquad(2.29)$$

and $dA\cos\theta$ is the projection of dA, as seen from dA'. For isotropic sources (2.29) is symmetric with regard to θ and θ' or dA and dA'. The positions of detector and source may be interchanged without altering (2.29). Because of this reciprocity, L may be interpreted either as the *radiance of the source* at the angle θ to the surface normal or, equally well, as the *radiance incident onto the detector at the angle θ'*.

For isotropic sources, where L is independent of θ, (2.29) demonstrates that the radiant flux emitted into the unit solid angle is proportional to $\cos\theta$ (*Lambert's law*). An example for such a source is a hole with the area dA in a blackbody radiation cavity (Fig. 2.3).

The radiant flux incident on the unit detector area is called *irradiance I*, while in the spectroscopic literature it is often termed *intensity*. The flux density or intensity I [W/m^2] of a plane wave $\mathbf{E} = \mathbf{E}_0 \cos(\omega t - kz)$ traveling in vacuum in the z-direction is given by

$$I = c\int \rho(\omega)\,d\omega = c\epsilon_0 E^2 = c\epsilon_0 E_0^2 \cos^2(\omega t - kz) .\qquad(2.30a)$$

With the complex notation

$$\mathbf{E} = \mathbf{A}_0 e^{i(\omega t - kz)} + \mathbf{A}_0^* e^{-i(\omega t - kz)} \quad (|\mathbf{A}_0| = \tfrac{1}{2} E_0) ,\qquad(2.30b)$$

the intensity becomes

$$I = c\epsilon_0 E^2 = 4c\epsilon_0 A_0^2 \cos^2(\omega t - kz) .\qquad(2.30c)$$

Most detectors cannot follow the rapid oscillations of light waves with the angular frequencies $\omega \sim 10^{13} - 10^{15}$ Hz in the visible and near-infrared region. With a time constant $T \gg 1/\omega$ they measure, at a fixed position z, the time-averaged intensity

$$\langle I \rangle = \frac{c\epsilon_0 E_0^2}{T}\int_0^T \cos^2(\omega t - kz)\,dt = \frac{1}{2}c\epsilon_0 E_0^2 = 2c\epsilon_0 A_0^2 .\qquad(2.31)$$

2.4.2 Illumination of Extended Areas

In the case of extended detector areas, the total power received by the detector is obtained by integration over all detector elements dA' (Fig. 2.8). The detector receives all the radiation that is emitted from the source element dA within the angles $-u \leq \theta \leq +u$. The same radiation passes an imaginary spherical surface in front of the detector. We choose as elements of this spherical surface circular rings with $dA' = 2\pi r \, dr = 2\pi R^2 \sin\theta \cos\theta \, d\theta$. From (2.29) one obtains for the total flux Φ impinging onto the detector with $\cos\theta' = 1$

$$\Phi = \int_u^0 L \, dA \cos\theta \, 2\pi \sin\theta \, d\theta \,. \tag{2.32}$$

If the source is isotropic, L does not depend on θ and (2.32) yields

$$\Phi = \pi L \sin^2 u \, dA \,. \tag{2.33}$$

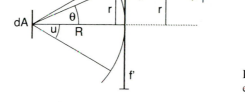

Fig. 2.8. Flux densities of detectors with extended area

Comment

Note that it is impossible to increase the radiance of a source by any sophisticated imaging optics. This means that the image dA^* of a radiation source dA never has a larger radiance than the source itself. It is true that the flux density can be increased by demagnification. The solid angle, however, into which radiation from the image dA^* is emitted is also increased by the same factor. Therefore, the radiance *does not* increase. In fact, because of inevitable reflection, scattering, and absorption losses of the imaging optics, the radiance of the image dA^* is, in practice, always less than that of the source (Fig. 2.9).

A strictly parallel light beam would be emitted into the solid angle $d\Omega = 0$. With a finite radiant power this would imply an infinite radiance L,

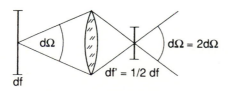

Fig. 2.9. The radiance of a source cannot be increased by optical imaging

which is impossible. This illustrates that such a light beam cannot be realized. The radiation source for a strictly parallel beam anyway has to be a point source in the focal plane of a lens. Such a point source with zero surface cannot emit any power.

For more extensive treatments of photometry see [2.13, 2.14].

Example 2.5

(a) *Radiance of the sun.* An area equal to $1\,\mathrm{m}^2$ of the earth's surface receives at normal incidence without reflection or absorption through the atmosphere an incident radiant flux I_e of about $1.35\,\mathrm{kW/m}^2$ (solar constant). Because of the symmetry of (2.32) we may regard $\mathrm{d}A'$ as emitter and $\mathrm{d}A$ as receiver. The sun is seen from the earth under an angle of $2u = 32$ minutes of arc. This yields $\sin u = 4.7 \times 10^{-3}$. Inserting this number into (2.33), one obtains $L_s = 2 \times 10^7\,\mathrm{W/(m^2\,sr)}$ for the radiance of the sun's surface. The total radiant power Φ of the sun can be obtained from (2.32) or from the relation $\Phi = 4\pi R^2 I_e$, where $R = 1.5 \times 10^{11}$ m is the distance from the earth to the sun. These numbers give $\Phi = 4 \times 10^{26}$ W.

(b) *Radiance of a HeNe laser.* We assume that the output power of 1 mW is emitted from $1\,\mathrm{mm}^2$ of the mirror surface into an angle of 4 minutes of arc, which is equivalent to a solid angle of 1×10^{-6} sr. The maximum radiance in the direction of the laser beam is then $L = 10^{-3}/(10^{-6} \cdot 10^{-6}) = 10^9\,\mathrm{W/(m^2\,sr)}$. This is about 50 times larger than the radiance of the sun. For the spectral density of the radiance the comparison is even more dramatic. Since the emission of a nonstabilized single-mode laser is restricted to a spectral range of about 1 MHz, the laser has a spectral radiance density $L_\nu = 1 \times 10^3\,\mathrm{W \cdot s/(m^2\,sr^{-1})}$, whereas the sun, which emits within a mean spectral range of $\approx 10^{15}$ Hz, only reaches $L_\nu = 2 \times 10^{-8}\,\mathrm{W \cdot s/(m^2\,sr^{-1})}$.

(c) Looking directly into the sun, the retina receives a radiant flux of 1 mW if the diameter of the iris is 1 mm. This is just the same flux the retina receives staring into the laser beam of Example 2.5b. There is, however, a big difference regarding the irradiance of the retina. The image of the sun on the retina is about 100 times as large as the focal area of the laser beam. This means that the power density incident on single retina cells is about 100 times larger in the case of the laser radiation.

2.5 Polarization of Light

The complex amplitude vector A_0 of the plane wave

$$E = A_0 \cdot e^{i(\omega t - kz)} \tag{2.34}$$

can be written in its component representation

$$A_0 = \begin{Bmatrix} A_{0x}\,\mathrm{e}^{\mathrm{i}\phi_x} \\ A_{0y}\,\mathrm{e}^{\mathrm{i}\phi_y} \end{Bmatrix}\ . \tag{2.35}$$

For unpolarized light the phases ϕ_x and ϕ_y are uncorrelated and their difference fluctuates statistically. For linearly polarized light with its electric vector in x-direction $A_{0y} = 0$. When E points into a direction α against the x axis, $\phi_x = \phi_y$ and $\tan\alpha = A_{0y}/A_{0x}$. For circular polarization $A_{0x} = A_{0y}$ and $\phi_x = \phi_y \pm \pi/2$. The different states of polarization can be characterized by their Jones vectors, which are defined as follows:

$$E = \begin{Bmatrix} E_x \\ E_y \end{Bmatrix} = |E| \cdot \begin{Bmatrix} a \\ b \end{Bmatrix}\,\mathrm{e}^{\mathrm{i}(\omega t - kz)} \tag{2.36}$$

where the normalized vector $\{a, b\}$ is the Jones vector. In Table 2.1 the Jones vectors are listed for the different polarization states. For linearly polarized light with $\alpha = 45°$, for example, the amplitude A_0 can be written as

$$A_0 = \sqrt{A_{0x}^2 + A_{0y}^2}\,\frac{1}{\sqrt{2}}\begin{Bmatrix} 1 \\ 1 \end{Bmatrix} = |A_0|\,\frac{1}{\sqrt{2}}\begin{Bmatrix} 1 \\ 1 \end{Bmatrix}\ , \tag{2.37}$$

while for right circular polarization (σ^- light), we obtain

$$A_0 = \frac{1}{\sqrt{2}}\,|A_0|\begin{Bmatrix} 1 \\ -\mathrm{i} \end{Bmatrix} \tag{2.38}$$

because $\exp(-\mathrm{i}\pi/2) = -\mathrm{i}$.

Table 2.1. Jones vectors for light traveling in the z-direction and Jones matrices for polarizers. The x-direction is horizontal, the y-direction vertical

Jones vectors		Jones matrices			
Linear polarizers	**Circular polarizers**	**Linear polarizers**			
		$\longleftrightarrow$ $\quad$ $\updownarrow$ $\quad$ $\nearrow$ $\quad$ $\searrow$			
x-direction $\longleftrightarrow$ $\begin{pmatrix}1\\0\end{pmatrix}$	$\sigma^+:\ \dfrac{1}{\sqrt{2}}\begin{pmatrix}1\\ \mathrm{i}\end{pmatrix}$	$\begin{pmatrix}1&0\\0&0\end{pmatrix}$ $\begin{pmatrix}0&0\\0&1\end{pmatrix}$ $\dfrac{1}{2}\begin{pmatrix}1&1\\1&1\end{pmatrix}$ $\dfrac{1}{2}\begin{pmatrix}1&-1\\-1&1\end{pmatrix}$			
y-direction $\updownarrow$ $\begin{pmatrix}0\\1\end{pmatrix}$	$\sigma^-:\ \dfrac{1}{\sqrt{2}}\begin{pmatrix}1\\ -\mathrm{i}\end{pmatrix}$	**Circular polarizers**			
		$\circlearrowleft$ $\qquad$ $\circlearrowright$ $\qquad$ $\lambda/4$ plate			
$\alpha = 45°$ $\nearrow$ $\dfrac{1}{\sqrt{2}}\begin{pmatrix}1\\1\end{pmatrix}$		$\dfrac{1}{2}\begin{pmatrix}1&\mathrm{i}\\-\mathrm{i}&1\end{pmatrix}$ $\dfrac{1}{2}\begin{pmatrix}1&-\mathrm{i}\\ \mathrm{i}&1\end{pmatrix}$ $\dfrac{1}{2}\begin{pmatrix}-\mathrm{i}&0\\0&\pm\mathrm{i}\end{pmatrix}$			

The Jones representation shows its advantages when we consider the transmission of light through optical elements such as polarizers, $\lambda/4$ plates, or beamsplitters. These elements can be described by 2×2 matrices, which are compiled for some elements in Table 2.1. The polarization state of the transmitted light is then obtained by multiplication of the Jones vector of the incident wave by the Jones matrix of the optical element.

$$E_{\mathrm{t}} = \left\{ \begin{matrix} E_{xt} \\ E_{yt} \end{matrix} \right\} = \begin{pmatrix} a & b \\ c & d \end{pmatrix} \cdot \left\{ \begin{matrix} E_{x0} \\ E_{y0} \end{matrix} \right\} , \tag{2.39}$$

For example, a linearly polarized incident light with $\alpha = 0°$ becomes, after transmission through a circular polarizer,

$$E_{\mathrm{t}} = \frac{1}{2} \begin{pmatrix} 1 & i \\ -i & 1 \end{pmatrix} \cdot \begin{pmatrix} 1 \\ 0 \end{pmatrix} |E| = \frac{1}{2} \left\{ \begin{matrix} 1 \\ -i \end{matrix} \right\} |E| = \frac{1}{2}(E_x \cdot \hat{e}_x - i E_y \hat{e}_y) , \tag{2.40}$$

a right circular polarized σ^- wave. More examples can be found in [2.15–2.17].

2.6 Absorption and Emission Spectra

The spectral distribution of the radiant flux from a source is called its emission *spectrum*. The thermal radiation discussed in Sect. 2.2 has a *continuous* spectral distribution described by its spectral energy density (2.13). *Discrete* emission spectra, where the radiant flux has distinct maxima at certain frequencies ν_{ik}, are generated by transitions of atoms or molecules between two bound states, a higher energy state E_k and a lower state E_i, with the relation

$$h\nu_{ik} = E_k - E_i . \tag{2.41}$$

In a spectrograph (see Sect. 4.1 for a detailed description) the entrance slit is imaged into the focal plane of the camera lens. Because of dispersive elements in the spectrograph, the position of this image depends on the wavelength of the incident radiation. In a discrete spectrum each wavelength λ_{ik} produces a separate line in the imaging plane, provided the spectrograph has a sufficiently high resolving power (Fig. 2.10). Discrete spectra are therefore also called *line spectra*, as opposed to *continuous spectra* where the slit images form a continuous band in the focal plane, even for spectrographs with infinite resolving power.

If radiation with a continuous spectrum passes through a gaseous molecular sample, molecules in the lower state E_i may absorb radiant power at the eigenfrequencies $\nu_{ik} = (E_k - E_i)/h$, which is thus missing in the transmitted power. The difference in the spectral distributions of incident minus transmitted power is the *absorption spectrum* of the sample. The absorbed energy $h\nu_{ik}$ brings a molecule into the higher energy level E_k. If these levels are bound levels, the resulting spectrum is a discrete absorption spectrum. If E_k is above the dissociation limit or above the ionization energy, the absorption spectrum

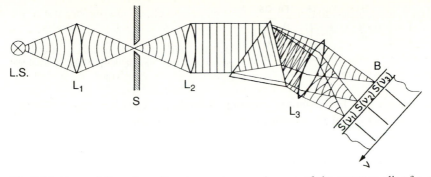

Fig. 2.10. Spectral lines in a discrete spectrum as images of the entrance slit of a spectrograph

becomes continuous. In Fig. 2.11 both cases are schematically illustrated for atoms (a) and molecules (b).

Examples of discrete absorption lines are the Fraunhofer lines in the spectrum of the sun, which appear as dark lines in the bright continuous spectrum (Fig. 2.12). They are produced by atoms in the sun's atmosphere that absorb at their specific eigenfrequencies the continuous blackbody radiation from the sun's photosphere. A measure of the absorption strength is the absorption

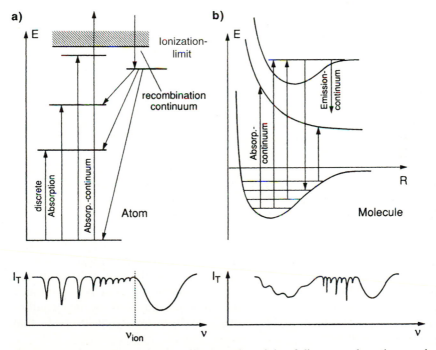

Fig. 2.11a,b. Schematic diagram to illustrate the origin of discrete and continuous absorption and emission spectra for atoms (**a**) and molecules (**b**)

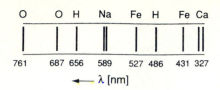

Fig. 2.12. Prominent Fraunhofer absorption lines within the visible and near-UV spectral range

cross section σ_{ik}. Each photon passing through the circular area $\sigma_{ik} = \pi r_{ik}^2$ around the atom is absorbed on the transition $|i\rangle \rightarrow |k\rangle$.

The power

$$\mathrm{d}P_{ik}(\omega)\,\mathrm{d}\omega = P_0\left(N_i - \frac{g_i}{g_k}N_k\right)\sigma_{ik}(\omega)A\Delta z\,\mathrm{d}\omega = P_0\alpha_{ik}(\omega)\Delta V\,\mathrm{d}\omega\,, \quad (2.42)$$

absorbed within the spectral interval $\mathrm{d}\omega$ at the angular frequency ω on the transition $|i\rangle \rightarrow |k\rangle$ within the volume $\Delta V = A\Delta z$ is proportional to the product of incident power P_0, absorption cross section σ_{ik}, the difference $(N_i - N_k)$ of the population densities of absorbing molecules in the upper and lower levels, weighted with their statistical weights g_i, g_k, and the absorption path length Δz. A comparison with (2.15) and (2.21) yields the total power absorbed per cm^3 on the transition $|i\rangle \rightarrow |k\rangle$:

$$P_{ik} = P_0 \int \alpha_{ik}(\omega)\,\mathrm{d}\omega = \frac{\hbar\omega}{c}P_0 B_{ik}\left(N_i - \frac{g_i}{g_k}N_k\right)\,, \quad (2.43)$$

where the integration extends over the absorption profile. This gives the relation

$$B_{ik} = \frac{c}{\hbar\omega}\int \sigma_{ik}(\omega)\,\mathrm{d}\omega\,, \quad (2.44)$$

between the Einstein coefficient B_{ik} and the absorption cross section σ_{ik}.

At thermal equilibrium the population follows a Boltzmann distribution. Inserting (2.18) yields the power absorbed within the volume $\Delta V = A\Delta z$ out of an incident beam with the cross section A

$$P_{ik} = (N/Z)g_i(e^{-E_i/kT} - e^{-E_k/kT})A\Delta z \int P_0\sigma_{ik}\,\mathrm{d}\omega\,,$$

$$= P_0\sigma_{ik}(\omega_0)(N/Z)g_i(e^{-E_i/kT} - e^{-E_k/kT})\Delta V\,, \quad (2.45)$$

for a monochromatic laser with $P_0(\omega) = P_0\delta(\omega - \omega_0)$.

The absorption lines are only measurable if the absorbed power is sufficiently high, which means that the density N or the absorption path length Δz must be large enough. Furthermore, the difference in the two Boltzmann factors in (2.45) should be sufficiently large, which means E_i should be not much larger than kT, but $E_k \gg kT$. Absorption lines in gases at thermal equilibrium are therefore only intense for transitions from low-lying levels E_i that are thermally populated.

It is, however, possible to pump molecules into higher energy states by various excitation mechanisms such as optical pumping or electron excitation.

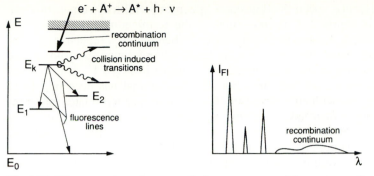

Fig. 2.13. Discrete and continuous emission spectrum and the corresponding level diagram, which also shows radiationless transitions induced by inelastic collisions (*wavy lines*)

This allows the measurement of absorption spectra for transition from these states to even higher molecular levels (Sect. 10.3).

The excited molecules release their energy either by spontaneous or induced emission or by collisional deactivation (Fig. 2.13). The spatial distribution of spontaneous emission depends on the orientation of the excited molecules and on the symmetry properties of the excited state E_k. If the molecules are randomly oriented, the spontaneous emission (often called *fluorescence*) is isotropic.

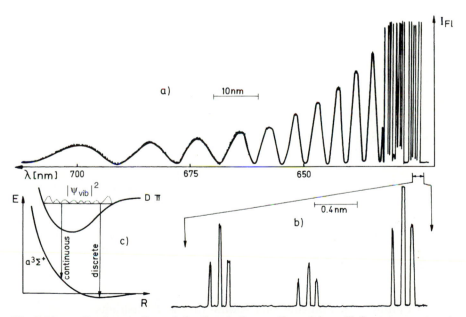

Fig. 2.14a–c. Continuous "bound–free" and discrete "bound–bound" fluorescence transitions of the NaK molecule observed upon laser excitation at $\lambda = 488$ nm: (**a**) part of the spectrum; (**b**) enlargement of three discrete vibrational bands; (**c**) level scheme [2.18]

The fluorescence spectrum (emission spectrum) emitted from a discrete upper level E_k consists of discrete lines if the terminating lower levels E_i are bound states. A continuum is emitted if E_i belongs to a repulsive state of a molecule that dissociates. As an example, the fluorescence spectrum of the $^3\Pi \rightarrow {}^3\Sigma$ transition of the NaK molecule is shown in Fig. 2.14. It is emitted from a selectively excited level in a bound $^3\Pi$ state that has been populated by optical pumping with an argon laser. The fluorescence terminates into a repulsive $^3\Sigma$ state, which has a shallow van der Waals minimum. Transitions terminating to energies E_k above the dissociation energy form the continuous part of the spectrum, whereas transitions to lower bound levels in the van der Waals potential well produce discrete lines. The modulation of the continuum reflects the modulation of the transmission probability due to the maxima and nodes of the vibrational wave function $\psi_{\text{vib}}(R)$ in the upper bound level [2.18].

2.7 Transition Probabilities

The intensities of spectral lines depend not only on the population density of the molecules in the absorbing or emitting level but also on the transition probabilities of the corresponding molecular transitions. If these probabilities are known, the population density can be obtained from measurements of line intensities. This is very important, for example, in astrophysics, where spectral lines represent the main source of information from the extraterrestrial world. Intensity measurements of absorption and emission lines allow the concentration of the elements in stellar atmospheres or in interstellar space to be determined. Comparing the intensities of different lines of the same element (e.g., on the transitions $E_i \rightarrow E_k$ and $E_e \rightarrow E_k$ from different upper levels E_i, E_e to the same lower level E_k) furthermore enables us to derive the temperature of the radiation source from the relative population densities N_i, N_e in the levels E_i and E_e at thermal equilibrium according to (2.18). *All these experiments, however, demand a knowledge of the corresponding transition probabilities.*

There is another aspect that makes measurements of transition probabilities very attractive with regard to a more detailed knowledge of molecular structure. Transition probabilities derived from computed wave functions of upper and lower states are much more sensitive to approximation errors in these functions than are the energies of these states. Experimentally determined transition probabilities are therefore well suited to test the validity of calculated approximate wave functions. A comparison with computed probabilities allows theoretical models of electronic charge distributions in excited molecular states to be improved [2.19, 2.20].

2.7.1 Lifetimes, Spontaneous and Radiationless Transitions

The probability $\mathscr{P}_{ik}$ that an excited molecule in the level E_i makes a transition to a lower level E_k by spontaneous emission of a fluorescence quantum

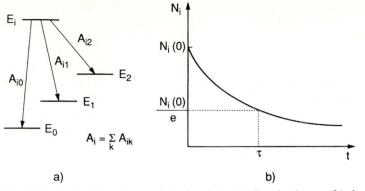

Fig. 2.15a,b. Radiative decay of the level $|i\rangle$: (**a**) Level scheme; (**b**) decay curve $N_i(t)$

$h\nu_{ik} = E_i - E_k$ is, according to (2.17), related to the Einstein coefficient A_{ik} by

$$d\mathscr{P}_{ik}/dt = A_{ik} .$$

When several transition paths from E_i to different lower levels E_k are possible (Fig. 2.15), the total transition probability is given by

$$A_i = \sum_k A_{ik} . \tag{2.46}$$

The decrease dN_i of the population density N_i during the time interval dt due to radiative decay is then

$$dN_i = -A_i N_i \, dt . \tag{2.47}$$

Integration of (2.47) yields

$$N_i(t) = N_{i0} e^{-A_i t} , \tag{2.48}$$

where N_{i0} is the population density at $t = 0$.

After the time $\tau_i = 1/A_i$ the population density N_i has decreased to $1/e$ of its initial value at $t = 0$. The time τ_i represents the *mean spontaneous lifetime* of the level E_i as can be seen immediately from the definition of the mean time

$$\bar{t}_i = \int_0^\infty t\mathscr{P}_i(t)\,dt = \int_0^\infty t A_i e^{-A_i t}\,dt = \frac{1}{A_i} = \tau_i , \tag{2.49}$$

where $\mathscr{P}_i(t)\,dt$ is the probability that one atom in the level E_i makes a spontaneous transition within the time interval between t and $t + dt$.

The *radiant power* emitted from N_i molecules on the transition $E_i \rightarrow E_k$ is

$$P_{ik} = N_i h\nu_{ik} A_{ik} . \tag{2.50}$$

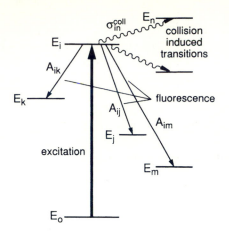

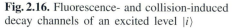

Fig. 2.16. Fluorescence- and collision-induced decay channels of an excited level $|i\rangle$

If several transitions $E_i \to E_k$ from the same upper level E_i to different lower levels E_k are possible, the radiant powers of the corresponding spectral lines are proportional to the product of the Einstein coefficient A_{ik} and the photon energy $h\nu_{ik}$. The relative radiation intensities in a certain direction may also depend on the spatial distribution of the fluorescence, which can be different for the different transitions.

The level E_i of the molecule A can be depopulated not only by spontaneous emission but also by collison-induced *radiationless* transitions (Fig. 2.16). The probability $d\mathcal{P}_{ik}^{coll}/dt$ of such a transition depends on the density N_B of the collision partner B, on the mean relative velocity $\bar{v}$ between A and B, and on the collision cross section σ_{ik}^{coll} for an inelastic collision that induces the transition $E_i \to E_k$ in the molecule A

$$d\mathcal{P}_{ik}^{coll}/dt = \bar{v}N_B\sigma_{ik}^{coll} . \tag{2.51}$$

When the excited molecule $A(E_i)$ is exposed to an intense radiation field, the *induced emission* may become noticeable. It contributes to the depopulation of level E_i in a transition $|i\rangle \to |k\rangle$ with the probability

$$d\mathcal{P}_{ik}^{ind}/dt = \rho(\nu_{ik})B_{ik}[N_i - (g_i/g_k)N_k] . \tag{2.52}$$

The total transition probability that determines the effective lifetime of a level E_i is then the sum of spontaneous, induced, and collisional contributions, and the mean lifetime τ_i^{eff} becomes

$$\frac{1}{\tau_i^{eff}} = \sum_k \left[A_{ik} + \rho(\nu_{ik})B_{ik}(N_i - N_k g_i/g_k) + N_B\sigma_{ik}\bar{v}\right] . \tag{2.53}$$

Measuring the effective lifetime τ_i^{eff} as a function of the exciting radiation intensity and also its dependence on the density N_B of collision partners (Stern–Vollmer plot) allows one to determine the three transition probabilities separately (Sect. 11.3).

2.7.2 Semiclassical Description: Basic Equations

In the semiclassical description, the radiation incident upon an atom is described by a classical electromagnetic (EM) plane wave

$$E = E_0 \cos(\omega t - kz) \,. \tag{2.54a}$$

The atom, on the other hand, is treated quantum-mechanically. In order to simplify the equations, we restrict ourselves to a two-level system with the eigenstates E_a and E_b (Fig. 2.17).

Until now laser spectroscopy was performed in spectral regions where the wavelength λ was large compared to the diameter d of an atom (e.g., in the visible spectrum λ is 500 nm, but d is only about 0.5 nm). For $\lambda \gg d$, the phase of the EM wave does not change much within the volume of an atom because $kz = (2\pi/\lambda)z \ll 1$ for $z \le d$. We can therefore neglect the spatial derivatives of the field amplitude (dipole approximation). In a coordinate system with its origin in the center of the atom, we can assume $kz \simeq 0$ within the atomic volume, and write (2.54a) in the form

$$E = E_0 \cos \omega t = A_0(e^{i\omega t} + e^{-i\omega t}) \qquad \text{with} \quad |A_0| = \tfrac{1}{2}E_0 \,. \tag{2.54b}$$

The Hamiltonian operator

$$\mathcal{H} = \mathcal{H}_0 + \mathcal{V} \,, \tag{2.55}$$

of the atom interacting with the light field can be written as a sum of the unperturbed Hamiltonian $\mathcal{H}_0$ of the free atom plus the perturbation operator $\mathcal{V}$, which reduces in the *dipole approximation* to

$$\mathcal{V} = \boldsymbol{p} \cdot \boldsymbol{E} = \boldsymbol{p} \cdot \boldsymbol{E}_0 \cos \omega t \,, \tag{2.56}$$

where $\mathcal{V}$ is the scalar product of the dipole operator $\boldsymbol{p} = -e \cdot \boldsymbol{r}$ and the electric field $\boldsymbol{E}$.

The general solution $\psi(\boldsymbol{r}, t)$ of the time-dependent Schrödinger equation

$$\mathcal{H}\psi = i\hbar \frac{\partial \psi}{\partial t} \tag{2.57}$$

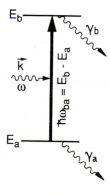

Fig. 2.17. Two-level system with open decay channels into other levels interacting with an EM field

can be expressed as a linear superposition

$$\psi(r, t) = \sum_{n=1}^{\infty} c_n(t) u_n(r) e^{-iE_n t/\hbar} , \tag{2.58}$$

of the eigenfunctions of the unperturbed atom

$$\phi_n(r, t) = u_n(r) e^{-iE_n t/\hbar} . \tag{2.59}$$

The spatial parts $u_n(r)$ of these eigenfunctions are solutions of the time-independent Schrödinger equation

$$\mathcal{H}_0 u_n(r) = E_n u_n(r) , \tag{2.60}$$

and satisfy the orthogonality relations[1]

$$\int u_i^* u_k \, d\tau = \delta_{ik} . \tag{2.61}$$

For our two-level system with the eigenstates $|a\rangle$ and $|b\rangle$ and the energies E_a and E_b, (2.58) reduces to a sum of two terms

$$\psi(r, t) = a(t) u_a e^{-iE_a t/\hbar} + b(t) u_b e^{-iE_b t/\hbar} . \tag{2.62}$$

The coefficients $a(b)$ and $b(t)$ are the time-dependent *probability amplitudes* of the atomic states $|a\rangle$ and $|b\rangle$. This means that the value $|a(t)|^2$ gives the probability of finding the system in level $|a\rangle$ at time t. Obviously, the relation $|a(t)|^2 + |b(t)|^2 = 1$ must hold at all times t, if decay into other levels is neglected.

Substituting (2.62) and (2.55) into (2.57) gives

$$i\hbar \dot{a}(t) u_a e^{-iE_a t/\hbar} + i\hbar \dot{b}(t) u_b e^{-iE_b t/\hbar} = a \mathcal{V} u_a e^{-iE_a t/\hbar} + b \mathcal{V} u_b e^{-iE_b t/\hbar} , \tag{2.63}$$

where the relation $\mathcal{H}_0 u_n = E_n u_n$ has been used to cancel equal terms on both sides. Multiplication with $u_n^* (n = a, b)$ and spatial integration results in the following two equations

$$\dot{a}(t) = -(i/\hbar) \left[a(t) V_{aa} + b(t) V_{ab} e^{i(E_a - E_b)t/\hbar} \right] , \tag{2.64a}$$

$$\dot{b}(t) = -(i/\hbar) \left[b(t) V_{bb} + a(t) V_{ba} e^{-i(E_a - E_b)t/\hbar} \right] , \tag{2.64b}$$

with the spatial integral

$$V_{ab} = \int u_a^* \mathcal{V} u_b \, d\tau = -e E \int u_a^* r u_b \, d\tau . \tag{2.65a}$$

---

[1] Note that in (2.58–2.60) a nondegenerate system has been assumed.

Since r has odd parity, the integrals V_{aa} and V_{bb} vanish when integrating over all coordinates from $-\infty$ to $+\infty$. The quantity

$$\boldsymbol{D}_{ab} = \boldsymbol{D}_{ba} = -e \int u_a^* \boldsymbol{r} u_b \, d\tau \, , \tag{2.65b}$$

is called the atomic *dipole matrix element*. It depends on the stationary wave functions u_a and u_b of the two states $|a\rangle$ and $|b\rangle$ and is determined by the charge distribution in these states.

The expectation value D of the atomic dipole moment for our two-level system, which should be distinguished from the dipole matrix element, is

$$\boldsymbol{D} = -e \int \psi^* \boldsymbol{r} \psi \, d\tau \, . \tag{2.66a}$$

Using (2.62) and the abbreviation $\omega_{ba} = (E_b - E_a)/\hbar = -\omega_{ab}$, this can be expressed by the coefficients $a(t)$ and $b(t)$, and by the matrix element $\boldsymbol{D}_{ab}$ as

$$\boldsymbol{D} = -\boldsymbol{D}_{ab}(a^* b \, e^{-i\omega_{ba}t} + ab^* \, e^{+i\omega_{ba}t}) = D_0 \cos(\omega_{ba}t + \varphi) \, , \tag{2.66b}$$

with

$$D_0 = D_{ab} \left| a^* b \right| \quad \text{and} \quad \tan \varphi = -\frac{\text{Im}\{a^* b\}}{\text{Re}\{a^* b\}} \, .$$

Even without the external field, the expectation value of the atomic dipole moment oscillates with the eigenfrequency ω_{ba} and the amplitude $|a^* \cdot b|$ if the wavefunction of the atomic system can be represented by the superposition (2.65).

Using (2.54b) for the EM field and the abbreviation

$$\Omega_{ab} = D_{ab}E_0/\hbar = 2D_{ab}A_0/\hbar = \Omega_{ba} \tag{2.67}$$

for the Rabi frequency Ω_{ab}, which depends on the field amplitude E_0 and the dipole matrix element D_{ab}, (2.64) reduces to

$$\dot{a}(t) = -(i/2)\Omega_{ab}\big(e^{i(\omega_{ab}-\omega)t} + e^{i(\omega_{ab}+\omega)t}\big)b(t) \, , \tag{2.68a}$$

$$\dot{b}(t) = -(i/2)\Omega_{ab}\big(e^{-i(\omega_{ab}-\omega)t} + e^{-i(\omega_{ab}+\omega)t}\big)a(t) \, . \tag{2.68b}$$

These are the basic equations that must be solved to obtain the probability amplitudes $a(t)$ and $b(t)$.

2.7.3 Weak-Field Approximation

Suppose that at time $t = 0$, the atoms are in the lower state E_a, which implies that $a(0) = 1$ and $b(0) = 0$. We assume the field amplitude A_0 to be sufficiently small so that for times $t < T$ the population of E_b remains small compared with that of E_a, i.e., $|b(t < T)|^2 \ll 1$. Under this *weak-field condition* we can solve (2.68) with an iterative procedure starting with $a = 1$ and $b = 0$. Using thermal radiation sources, the field amplitude A_0 is generally small enough to make the first iteration step already sufficiently accurate.

With these assumptions the first approximation of (2.68) gives

$$\dot{a}(t) = 0 \,, \tag{2.69a}$$

$$\dot{b}(t) = (\mathrm{i}/2)\Omega_{ab}\big(e^{\mathrm{i}(\omega_{ab}-\omega)t} + e^{\mathrm{i}(\omega_{ab}+\omega)t}\big) \,. \tag{2.69b}$$

With the initial conditions $a(0) = 1$ and $b(0) = 0$, integration of (2.69) yields

$$a(t) = a(0) = 1 \,, \tag{2.70a}$$

$$b(t) = \left(\frac{\Omega_{ab}}{2}\right) \left(\frac{e^{\mathrm{i}(\omega_{ba}-\omega)t} - 1}{\omega_{ba} - \omega} + \frac{e^{\mathrm{i}(\omega_{ba}+\omega)t} - 1}{\omega_{ba} + \omega}\right) \,. \tag{2.70b}$$

For $E_b > E_a$ the term $\omega_{ba} = (E_b - E_a)/\hbar$ is positive. In the transition $E_a \to E_b$, the atomic system absorbs energy from the radiation field. Noticeable absorption occurs, however, only if the field frequency ω is close to the eigenfrequency ω_{ba}. In the optical frequency range this implies that $|\omega_{ba} - \omega| \ll \omega_{ba}$. The second term in (2.70b) is then small compared to the first one and may be neglected. This is called the *rotating-wave approximation* for only that term is kept in which the atomic wave functions and the field waves with the phasors $\exp(-\mathrm{i}\omega_{ab}t) = \exp(+\mathrm{i}\omega_{ba}t)$ and $\exp(-\mathrm{i}\omega t)$ rotate together.

In the rotating-wave approximation we obtain from (2.70b) for the probability $|b(t)|^2$ that the system is at time t in the upper level E_b

$$|b(t)|^2 = \left(\frac{\Omega_{ab}}{2}\right)^2 \left(\frac{\sin(\omega_{ba} - \omega)t/2}{(\omega_{ba} - \omega)/2}\right)^2 \,. \tag{2.71}$$

Since we had assumed that the atom was at $t = 0$ in the lower level E_a, (2.71) *gives the transition probability for the atom to go from E_a to E_b during the time t*. Figure 2.18a illustrates this transition probability as a function of the detuning $\Delta\omega = \omega_{ba} - \omega$. Equation (2.71) shows that $|b(t)|^2$ depends on the absolute value of the detuning $\Delta\omega = |\omega_{ba} - \omega|$ of the field frequency ω from the eigenfrequency ω_{ba}. When tuning the frequency ω into resonance with the atomic system ($\omega \to \omega_{ba}$), the second factor in (2.71) approaches the value t^2 because $\lim_{x \to 0}[(\sin^2 xt)/x^2] = t^2$. The transition probability at resonance,

$$|b(t)|^2_{\omega=\omega_{ba}} = \left(\frac{\Omega_{ab}}{2}\right)^2 t^2 \,, \tag{2.72}$$

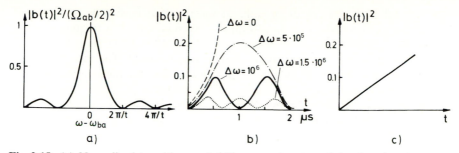

Fig. 2.18. (a) Normalized transition probability as a function of the detuning $(\omega - \omega_{ba})$ in the rotating-wave approximation; **(b)** probability of a transition to the upper level as a function of time for different detuning; **(c)** $|b(t)|^2$ under broadband excitation and weak fields

increases proportionally to t^2. The approximation used in deriving (2.71) has, however, anticipated that $|b(t)|^2 \ll 1$. According to (2.72) and (2.67), this assumption for the resonance case is equivalent to

$$\Omega_{ab}^2 t^2 \ll 1 \quad \text{or} \quad t \ll T = \frac{\hbar}{D_{ab} E_0} \,. \tag{2.73}$$

Our small-signal approximation only holds if the maximum interaction time T of the field (amplitude E_0) with the atom (matrix element D_{ab}) is restricted to $t \ll T$. Because the spectral analysis of a wave with the finite detection time T gives the spectral width $\Delta\omega \simeq 1/T$ (see also Sect. 3.2), we cannot assume monochromaticity, but have to take into account the frequency distribution of the interaction term.

2.7.4 Transition Probabilities with Broad-Band Excitation

In general, thermal radiation sources have the bandwidth $\delta\omega$, which is much larger than the Fourier limit $\Delta\omega = 1/T$. Therefore, the finite interaction time imposes no extra limitation. This may change, however, when lasers are considered (Sects. 2.7.5 and 3.4).

Instead of the field amplitude E_0 (which refers to a unit frequency interval), we introduce the spectral energy density $\rho(\omega)$ within the frequency range of the absorption line by the relation, see (2.30),

$$\int \rho(\omega) \, d\omega = \epsilon_0 E_0^2/2 = 2\epsilon_0 A_0^2 \,.$$

We can now generalize (2.71) to include the interaction of broadband radiation with our two-level system by integrating (2.71) over all frequencies ω of the radiation field. This yields the total transition probability $\mathcal{P}_{ab}(t)$ within the time t, if $D_{ab} \parallel E_0$:

$$\mathcal{P}_{ab}(t) = \int |b(t)|^2 \, d\omega = \frac{(D_{ab})^2}{2\epsilon_0 \hbar^2} \int \rho(\omega) \left(\frac{\sin(\omega_{ba} - \omega)t/2}{(\omega_{ba} - \omega)/2} \right)^2 d\omega \,. \tag{2.74}$$

For thermal light sources or broadband lasers, $\rho(\omega)$ is slowly varying over the absorption line profile. It is essentially constant over the frequency range where the factor $[\sin^2(\omega_{ba} - \omega)t/2]/[(\omega_{ba} - \omega)/2]^2$ is large (Fig. 2.18a). We can therefore replace $\rho(\omega)$ by its resonance value $\rho(\omega_{ba})$. The integration can then be performed, which gives the value $\rho(\omega_{ba})2\pi t$ for the integral because

$$\int_{-\infty}^{\infty} \frac{\sin^2(xt)}{x^2} \, dx = 2\pi t \; .$$

For broadband excitation, the transition probability for the time interval between 0 and t

$$\mathscr{P}_{ab}(t) = \frac{\pi}{\epsilon_0 \hbar^2} D_{ab}^2 \rho(\omega_{ba})t \; , \tag{2.75}$$

is linearly dependent on t (Fig. 2.18c). For broadband excitation the *transition probability per second*

$$\frac{d}{dt} \mathscr{P}_{ab} = \frac{\pi}{\epsilon_0 \hbar^2} D_{ab}^2 \rho(\omega_{ba}) \; , \tag{2.76}$$

becomes independent of time!

To compare this result with the Einstein coefficient B_{ab} derived in Sect. 2.3, we must take into account that the blackbody radiation was isotropic, whereas the EM wave (2.54) used in the derivation of (2.76) propagates into one direction. For randomly oriented atoms with the dipole moment $\boldsymbol{p}$, the averaged component of $\boldsymbol{p}^2$ in the z-direction is $\langle p_z^2 \rangle = p^2 \langle \cos^2 \theta \rangle = p^2/3$.

In the case of isotropic radiation, the interaction term $D_{ab}^2 \rho(\omega_{ba})$ therefore has to be divided by a factor of 3. A comparison of (2.16) with the modified equation (2.76) yields

$$\frac{d}{dt} \mathscr{P}_{ab} = \frac{\pi}{3\epsilon_0 \hbar^2} \rho(\omega_{ba}) D_{ab}^2 = \rho(\omega_{ba}) B_{ab} \; . \tag{2.77}$$

With the definition (2.65) for the dipole matrix element $\boldsymbol{D}_{ik}$, the Einstein coefficient B_{ik} of induced absorption $E_i \rightarrow E_k$ finally becomes

$$B_{ik}^{\omega} = \frac{\pi e^2}{3\epsilon_0 \hbar^2} \left| \int u_i^* \boldsymbol{r} u_k \, d\tau \right|^2 \quad \text{and} \quad B_{ik}^{\nu} = B_{ik}^{\omega}/2\pi \; . \tag{2.78}$$

Equation (2.78) gives the Einstein coefficient for a one-electron system where $\boldsymbol{r} = (x, y, z)$ is the vector from the nucleus to the electron, and $u_n(x, y, z)$ denotes the one-electron wave functions.[2] From (2.78) we learn

---

[2] Note that when using the frequency $\nu = \omega/2\pi$ instead of ω, the spectral energy density $\rho(\nu)$ per unit frequency interval is larger by a factor of 2π than $\rho(\omega)$ because a unit frequency interval $d\nu = 1$ Hz corresponds to $d\omega = 2\pi$ [Hz]. The right-hand side of (2.78) must then be divided by a factor of 2π, since $B_{ik}^{\nu}\rho(\nu) = B_{ik}^{\omega}\rho(\omega)$.

that the Einstein coefficient B_{ik} is proportional to the squared transition dipole moment.

So far we have assumed that the energy levels E_i and E_k are not degenerate, and therefore have the statistical weight factor $g = 1$. In the case of a degenerate level $|k\rangle$, the total transition probability ρB_{ik} of the transition $E_i \to E_k$ is the sum

$$\rho B_{ik} = \rho \sum_n B_{ik_n} ,$$

over all transitions to the sublevels $|k_n\rangle$ of $|k\rangle$. If level $|i\rangle$ is also degenerate, an additional summation over all sublevels $|i_m\rangle$ is necessary, taking into account that the population of each sublevel $|i_m\rangle$ is only the fraction N_i/g_i.

The Einstein coefficient B_{ik} for the transition $E_i \to E_k$ between the two degenerate levels $|i\rangle$ and $|k\rangle$ is therefore

$$B_{ik} = \frac{\pi}{3\epsilon_0 \hbar^2} \frac{1}{g_i} \sum_{m=1}^{g_i} \sum_{n=1}^{g_k} |D_{i_m k_n}|^2 = \frac{\pi}{3\epsilon_0 \hbar^2 g_i} S_{ik} . \qquad (2.79)$$

The double sum is called the *line strength* S_{ik} of the atomic transition $|i\rangle \leftarrow |k\rangle$.

2.7.5 Phenomenological Inclusion of Decay Phenomena

So far we have neglected the fact that the levels $|a\rangle$ and $|b\rangle$ are not only coupled by transitions induced by the external field but may also decay by spontaneous emission or by other relaxation processes such as collision-induced transitions. We can include these decay phenomena in our formulas by adding phenomenological decay terms to (2.68), which can be expressed by the decay constant γ_a and γ_b (Fig. 2.17). A rigorous treatment requires quantum electrodynamics [2.23].

In the rotating-wave approximation, for which the term with the frequency $(\omega_{ba} + \omega)$ is neglected, (2.68) then becomes

$$\dot{a}(t) = -\frac{1}{2}\gamma_a a + \frac{i}{2}\Omega_{ab}\, e^{-i(\omega_{ba}-\omega)t} b(t) , \qquad (2.80a)$$

$$\dot{b}(t) = -\frac{1}{2}\gamma_b b + \frac{i}{2}\Omega_{ab}\, e^{+i(\omega_{ba}-\omega)t} a(t) . \qquad (2.80b)$$

When the field amplitude E_0 is sufficiently small, see (2.73), we can use the weak-signal approximation of Sect. 2.7.3. This means that $|a(t)|^2 = 1$, $|b(t)|^2 \ll 1$, and also $aa^* - bb^* \simeq 1$. With this approximation, one obtains in a similar way as in the derivation of (2.71) the transition probability

$$\mathcal{P}_{ab}(\omega) = |b(t,\omega)|^2 = \int \gamma_{ab}\, e^{-\gamma_{ab}t} |b(t)|^2\, dt = \frac{1}{2} \frac{\Omega_{ab}^2}{(\omega_{ba}-\omega)^2 + (\frac{1}{2}\gamma_{ab})^2} . \qquad (2.80c)$$

This is a Lorentzian line profile (Fig. 2.19) with a full halfwidth $\gamma_{ab} = \gamma_a + \gamma_b$.

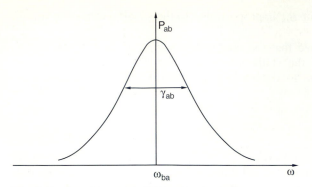

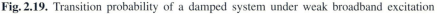

Fig. 2.19. Transition probability of a damped system under weak broadband excitation

After taking the second-time derivative of (2.66b) and using (2.80), the equation of motion for the dipole moment D of the atom under the influence of a radiation field, becomes

$$\ddot{D} + \gamma_{ab}\dot{D} + (\omega_{ba}^2 + \gamma_{ab}^2/4)D$$
$$= (\Omega_{ab})\left[(\omega_{ba} + \omega)\cos\omega t + (\gamma_{ab}/2)\sin\omega t\right]. \qquad (2.81a)$$

The homogeneous equation

$$\ddot{D} + \gamma_{ab}\dot{D} + (\omega_{ba}^2 + \gamma_{ab}^2/4)D = 0, \qquad (2.81b)$$

which describes the atomic dipoles without the driving field ($\Omega_{ab} = 0$), has the solution for weak damping ($\gamma_{ab} \ll \omega_{ba}$)

$$D(t) = D_0 e^{(-\gamma_{ab}/2)t}\cos\omega_{ba}t. \qquad (2.82)$$

The inhomogeneous equation (2.81a) shows that the induced dipole moment of the atom interacting with a monochromatic radiation field behaves like a driven damped harmonic oscillator with $\omega_{ba} = (E_b - E_a)/\hbar$ for the eigenfrequency and $\gamma_{ab} = (\gamma_a + \gamma_b)$ for the damping constant oscillating at the driving field frequency ω.

Using the approximation ($\omega_{ba} + \omega) \simeq 2\omega$ and $\gamma_{ab} \ll \omega_{ba}$, which means weak damping and a close-to-resonance situation, we obtain solutions of the form

$$D = D_1 \cos\omega t + D_2 \sin\omega t, \qquad (2.83)$$

where the factors D_1 and D_2 include the frequency dependence,

$$D_1 = \frac{\Omega_{ab}(\omega_{ba} - \omega)}{(\omega_{ba} - \omega)^2 + (\gamma_{ab}/2)^2}, \qquad (2.84a)$$

$$D_2 = \frac{\frac{1}{2}\Omega_{ab}\gamma_{ab}}{(\omega_{ab} - \omega)^2 + (\gamma_{ab}/2)^2}. \qquad (2.84b)$$

These two equations for D_1 and D_2 describe dispersion and absorption of the EM wave. The former is caused by the phase lag between the radiation field and the induced dipole oscillation, and the latter by the atomic transition from the lower level E_a to the upper level E_b and the resultant conversion of the field energy into the potential energy $(E_b - E_a)$.

The macroscopic polarization P of a sample with N atoms/cm^3 is related to the induced dipole moment D by $P = ND$.

2.7.6 Interaction with Strong Fields

In the previous sections we assumed weak-field conditions where the probability of finding the atom in the initial state was not essentially changed by the interaction with the field. This means that the population in the initial state remains approximately constant during the interaction time. In the case of broadband radiation, this approximation results in a *time-independent transition probability*. Also the inclusion of weak-damping terms with $\gamma_{ab} \ll \omega_{ba}$ did not affect the assumption of a constant population in the initial state.

When intense laser beams are used for the excitation of atomic transitions, the weak-field approximation is no longer valid. In this section, we therefore consider the "strong-field case." The corresponding theory, developed by Rabi, leads to a time-dependent probability of the atom being in either the upper or lower level. The representation outlined below follows that of [2.21].

We consider a monochromatic field of frequency ω and start from the basic equations (2.68) for the probability amplitudes in the rotating wave approximation

$$\dot{a}(t) = \frac{i}{2}\Omega_{ab}\,e^{-i(\omega_{ba}-\omega)t}\,b(t) , \qquad (2.85a)$$

$$\dot{b}(t) = \frac{i}{2}\Omega_{ab}\,e^{+i(\omega_{ba}-\omega)t}\,a(t) . \qquad (2.85b)$$

Inserting the trial solution

$$a(t) = e^{i\mu t} \;\Rightarrow\; \dot{a}(t) = i\mu\,e^{i\mu t} ,$$

into (2.85a) yields

$$b(t) = \frac{2\mu}{\Omega_{ab}}\,e^{i(\omega_{ba}-\omega+\mu)t} \;\Rightarrow\; \dot{b}(t) = \frac{2i\mu(\omega_{ba}-\omega+\mu)}{\Omega_{ab}}\,e^{(\omega_{bu}-\omega+\mu)t} .$$

Substituting this back into (2.85b) gives the relation

$$2\mu(\omega_{ba}-\omega+\mu) = \Omega_{ab}^2/2 .$$

This is a quadratic equation for the unknown quantity μ with the two solutions

$$\mu_{1,2} = -\frac{1}{2}(\omega_{ba}-\omega) \pm \frac{1}{2}\sqrt{(\omega_{ba}-\omega)^2 + \Omega_{ab}^2} . \qquad (2.86)$$

The general solutions for the amplitudes a and b are then

$$a(t) = C_1 e^{i\mu_1 t} + C_2 e^{i\mu_2 t} , \tag{2.87a}$$

$$b(t) = (2/\Omega_{ab}) e^{i(\omega_{ba}-\omega)t} (C_1 \mu_1 e^{i\mu_1 t} + C_2 \mu_2 e^{i\mu_2 t}) . \tag{2.87b}$$

With the initial conditions $a(0) = 1$ and $b(0) = 0$, we find for the coefficients

$$C_1 + C_2 = 1 \quad \text{and} \quad C_1 \mu_1 = -C_2 \mu_2 ,$$

$$\Rightarrow \quad C_1 = -\frac{\mu_2}{\mu_1 - \mu_2} \quad C_2 = +\frac{\mu_1}{\mu_1 - \mu_2} .$$

From (2.86) we obtain $\mu_1 \mu_2 = -\Omega_{ab}^2/4$. With the shorthand

$$\Omega = \mu_1 - \mu_2 = \sqrt{(\omega_{ba} - \omega)^2 + \Omega_{ab}^2} ,$$

we get the probability amplitude

$$b(t) = i(\Omega_{ab}/\Omega) e^{i(\omega_{ba}-\omega)t/2} \sin(\Omega t/2) . \tag{2.88}$$

The probability $|b(t)|^2 = b(t)b^*(t)$ of finding the system in level E_b is then

$$|b(t)|^2 = (\Omega_{ab}/\Omega)^2 \sin^2(\Omega t/2) , \tag{2.89}$$

where

$$\Omega = \sqrt{(\omega_{ba} - \omega)^2 + (\mathbf{D}_{ab} \cdot \mathbf{E}_0/\hbar)^2} \tag{2.90}$$

is called the general Rabi "flopping frequency" for the nonresonant case $\omega \neq \omega_{ba}$. Equation (2.89) reveals that the transition probability is a periodic function of time. Since

$$|a(t)|^2 = 1 - |b(t)|^2 = 1 - (\Omega_{ab}/\Omega)^2 \sin^2(\Omega t/2) , \tag{2.91}$$

the system oscillates with the frequency Ω between the levels E_a and E_b, where the level-flopping frequency Ω depends on the detuning $(\omega_{ba} - \omega)$, on the field amplitude E_0, and the matrix element D_{ab} (Fig. 2.18b).

Note: In the literature the flopping frequency $\Omega_{ab} = D_{ab} E_0/\hbar$ for the resonant case $\omega = \omega_{ba}$ is frequently called the Rabi frequency.

At resonance $\omega_{ba} = \omega$, and (2.89) and (2.91) reduce to

$$|a(t)|^2 = \cos^2(\mathbf{D}_{ab} \cdot \mathbf{E}_0 t/2\hbar) , \tag{2.92a}$$

$$|b(t)|^2 = \sin^2(\mathbf{D}_{ab} \cdot \mathbf{E}_0 t/2\hbar) . \tag{2.92b}$$

After a time

$$T = \pi\hbar/(\mathbf{D}_{ab} \cdot \mathbf{E}_0) = \pi/\Omega_{ab} , \tag{2.93}$$

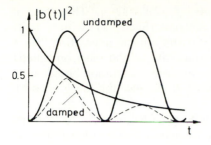

Fig. 2.20. Population probability $|b(t)|^2$ of the levels E_b altering with the Rabi flopping frequency due to the interaction with a strong field. The resonant case is shown without damping and with damping due to decay channels into other levels. The decaying curve represents the factor $\exp[-(\gamma_{ab}/2)t]$

the probability $|b(t)|^2$ of finding the system in level E_b becomes unity. This means that the population probability $|a(0)|^2 = 1$ and $|b(0)|^2 = 0$ of the initial system has been inverted to $|a(T)|^2 = 0$ and $|b(T)|^2 = 1$ (Fig. 2.20).

Radiation with the amplitude A_0, which resonantly interacts with the atomic system for exactly the time interval $T = \pi\hbar/(\mathbf{D}_{ab} \cdot \mathbf{E}_0)$, is called a π-*pulse* because it changes the phases of the probability amplitudes $a(t)$, $b(t)$ by π, see (2.87, 2.88).

We now include the damping terms γ_a and γ_b, and again insert the trial solution

$$a(t) = e^{i\mu t} ,$$

into (2.80a, 2.80b). Similar to the procedure used for the undamped case, this gives a quadratic equation for the parameter μ with the two complex solutions

$$\mu_{1,2} = -\frac{1}{2}\left(\omega_{ba} - \omega - \frac{i}{2}\gamma_{ab}\right) \pm \frac{1}{2}\sqrt{\left(\omega_{ba} - \omega - \frac{i}{2}\gamma\right)^2 + \Omega_{ab}^2} ,$$

where

$$\gamma_{ab} = \gamma_a + \gamma_b \quad \text{and} \quad \gamma = \gamma_a - \gamma_b . \tag{2.94}$$

From the general solution

$$a(t) = C_1 e^{i\mu_2 t} + C_2 e^{i\mu_2 t} ,$$

we obtain from (2.80a) with the initial conditions $|a(0)|^2 = 1$ and $|b(0)|^2 = 0$ the transition probability

$$|b(t)|^2 = \frac{\Omega_{ab}^2 e^{(-\gamma_{ab}/2)t}[\sin(\Omega/2)t]^2}{(\omega_{ba} - \omega)^2 + (\gamma/2)^2 + \Omega_{ab}^2} . \tag{2.95}$$

This is a damped oscillation (Fig. 2.20) with the damping constant $\frac{1}{2}\gamma_{ab} = (\gamma_a + \gamma_b)/2$, the Rabi flopping frequency

$$\Omega = \mu_1 - \mu_2 = \sqrt{\left(\omega_{ba} - \omega + \frac{i}{2}\gamma\right)^2 + \Omega_{ab}^2} , \tag{2.96}$$

and the envelope $\Omega_{ab}^2 e^{-(\gamma_{ab}/2)t}/[(\omega_{ba}-\omega)^2+(\gamma/2)^2+\Omega_{ab}^2]$. The spectral profile of the transition probability is Lorentzian (Sect. 3.1), with a half-width depending on $\gamma = \gamma_a - \gamma_b$ and on the strength of the interaction. Since $\Omega_{ab}^2 = (\boldsymbol{D}_{ab}\cdot\boldsymbol{E}_0/\hbar)^2$ is proportional to the intensity of the electromagnetic wave, the linewidth *increases* with increasing intensity (saturation broadening, Sect. 3.5). Note, that $|a(t)|^2 + |b(t)|^2 < 1$ for $t > 0$, because the levels a and b can decay into other levels.

In some cases the two-level system may be regarded as isolated from its environment. The relaxation processes then occur only between the levels $|a\rangle$ and $|b\rangle$, but do not connect the system with other levels. This implies $|a(t)|^2 + |b(t)|^2 = 1$. Equation (2.80) then must be modified as

$$\dot{a}(t) = -\frac{1}{2}\gamma_a a(t) + \frac{1}{2}\gamma_b b(t) + \frac{i}{2}\Omega_{ab}e^{-i(\omega_{ba}-\omega)t}b(t) , \qquad (2.97a)$$

$$\dot{b}(t) = -\frac{1}{2}\gamma_b b(t) + \frac{1}{2}\gamma_a a(t) + \frac{i}{2}\Omega_{ab}e^{+i(\omega_{ba}-\omega)t}a(t) . \qquad (2.97b)$$

The trial solution $a = \exp(i\mu t)$ yields, for the resonance case $\omega = \omega_{ba}$, the two solutions

$$\mu_1 = \frac{1}{2}\Omega_{ab} + \frac{i}{2}\gamma_{ab} , \quad \mu_2 = -\frac{1}{2}\Omega_{ab} ,$$

and for the transition probability $|b(t)|^2$, one obtains with $|a(0)|^2 = 1$, $|b(0)|^2 = 0$ a damped oscillation that approaches the steady-state value

$$|b(t=\infty)|^2 = \frac{1}{2}\frac{\Omega_{ab}^2+\gamma_a\gamma_b}{\Omega_{ab}^2+(\frac{1}{2}\gamma_{ab})^2} . \qquad (2.98)$$

This is illustrated in Fig. 2.21 for the special case $\gamma_a = \gamma_b$ where $|b(\infty)|^2 = 1/2$, which means that the two levels become equally populated.

For a more detailed treatment see [2.21–2.24].

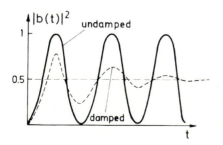

Fig. 2.21. Population of level $|b\rangle$ for a closed two-level system where the relaxation channels are open only for transitions between $|a\rangle$ and $|b\rangle$

2.7.7 Relations Between Transition Probabilities, Absorption Coefficient, and Line Strength

In this section we will summarize important relations between the different quantities discussed so far.

The absorption coefficient $\alpha(\omega)$ for a transition between levels $|i\rangle$ and $|k\rangle$ with population densities N_i and N_k and statistical weights g_i, g_k is related to the absorption cross section $\sigma_{ik}(\omega)$ by

$$\alpha(\omega) = [N_i - (g_i/g_k)N_k]\sigma_{ik}(\omega) . \tag{2.99}$$

The Einstein coefficient for absorption B_{ik} is given by

$$B_{ik} = \frac{c}{\hbar\omega} \int_0^\infty \sigma_{ik}(\omega) \, d\omega . \tag{2.100}$$

And the transition probability per second according to (2.15) is then

$$P_{ik} = B_{ik} \cdot \varrho = \frac{c}{\hbar\omega \cdot \Delta\omega} \int \varrho(\omega) \cdot \sigma_{ik}(\omega) \, d\omega , \tag{2.101}$$

where $\Delta\omega$ is the spectral linewidth of the transition.

The line strength S_{ik} of a transition is defined as the sum

$$S_{ik} = \sum_{m_i,m_k} |D_{m_i,m_k}|^2 = |D_{ik}|^2 , \tag{2.102}$$

over all dipole-allowed transitions between all subcomponents m_i, m_k of levels $|i\rangle$, $|k\rangle$. The oscillator strength f_{ik} gives the ratio of the power absorbed by a molecule on the transition $|i\rangle \rightarrow |k\rangle$ to the power absorbed by a classical oscillator on its eigenfrequency $\omega_{ik} = (E_k - E_i)/h$.

Some of these relations are compiled in Table 2.2.

Table 2.2. Relations between the transition matrix element D_{ik} and the Einstein coefficients A_{ik}, B_{ik}, the oscillator strength f_{ik}, the absorption cross section σ_{ik}, and the line strength S_{ik}. The numerical values are obtained, when λ is given in nm

$$A_{ki} = \frac{1}{g_k} \frac{16\pi^2 \nu^3}{3\varepsilon_0 hc^3} |D_{ik}|^2 \qquad B_{ik}^{(\nu)} = \frac{1}{g_i} \frac{2\pi^2}{3\varepsilon_0 h^2} |D_{ik}|^2 \qquad B_{ik}^{(\omega)} = \frac{1}{g_i} \frac{\pi}{3\varepsilon_0 \hbar^2} |D_{ik}|^2$$

$$= \frac{2.82 \times 10^{73}}{g_k \cdot \lambda^3} |D_{ik}|^2 \, s^{-1} \qquad = 6 \times 10^4 \lambda^3 \frac{g_i}{g_k} A_{ki} \qquad = \frac{g_k}{g_i} B_{ki}$$

$$f_{ik} = \frac{1}{g_i} \frac{8\pi^2 m_e \nu}{e^2 h} |D_{ik}|^2 \qquad S_{ik} = |D_{ik}|^2 \qquad \sigma_{ik} = \frac{1}{\Delta\nu} \frac{2\pi^2 \nu}{3\varepsilon_0 chg_i} \cdot S_{ik}$$

$$= \frac{g_k}{g_i} \cdot 1.5 \times 10^{-14} \lambda^2 A_{ki} \qquad = (2.37 \times 10^{-60} g_i \lambda) f \qquad B_{ik} = \frac{c}{h\nu} \int_0^\infty \sigma_{ik}(\nu) \, d\nu$$

2.8 Coherence Properties of Radiation Fields

The radiation emitted by an extended source S generates a total field amplitude A at the point P that is a superposition of an infinite number of partial waves with the amplitudes A_n and the phases ϕ_n emitted from the different surface elements dS (Fig. 2.22), i.e.,

$$A(P) = \sum_n A_n(P)\,e^{i\phi n(P)} = \sum_n \left[A_n(0)/r_n^2\right] e^{i(\phi n_0 + 2\pi r_n/\lambda)} , \qquad (2.103)$$

where $\phi_{n0}(t) = \omega t + \phi_n(0)$ is the phase of the nth partial wave at the surface element dS of the source. The phases $\phi_n(r_n, t)$ depend on the distances r_n from the source and on the angular frequency ω.

If the phase differences $\Delta\phi_n = \phi_n(P, t_1) - \phi_n(P, t_2)$ at a *given point* P between two different times t_1, t_2 are nearly the same for all partial waves, the radiation field at P is *temporally coherent*. The maximum time interval $\Delta t = t_2 - t_1$ for which $\Delta\phi_n$ for all partial waves differ by less than π is termed the *coherence time* of the radiation source. The path length $\Delta s_c = c\Delta t$ traveled by the wave during the coherence time Δt is the *coherence length*.

If a constant time-independent phase difference $\Delta\phi = \phi(P_1) - \phi(P_2)$ exists for the total amplitudes $A = A_0 e^{i\phi}$ at two different points P_1, P_2, the radiation field is *spatially coherent*. All points P_m, P_n that fulfill the condition that for all times t, $|\phi(P_m, t) - \phi(P_n, t)| < \pi$ have nearly the same optical path difference from the source. They form the *coherence volume*.

The superposition of coherent waves results in interference phenomena that, however, can be observed directly only within the coherence volume. The dimensions of this coherence volume depend on the size of the radiation source, on the spectral width of the radiation, and on the distance between the source and observation point P.

The following examples illustrate these different expressions for the coherence properties of radiation fields.

2.8.1 Temporal Coherence

Consider a point source PS in the focal plane of a lens forming a parallel light beam that is divided by a beam splitter S into two partial beams (Fig. 2.23). They are superimposed in the plane of observation B after reflection from

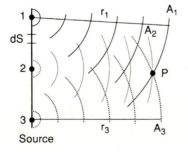

Fig. 2.22. The field amplitudes A_n at a point P in a radiation field as superposition of an infinite number of waves from different surface elements dS_i of an extended source

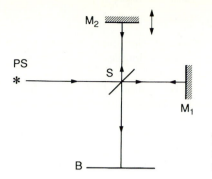

Fig. 2.23. Michelson interferometer for measurement of the temporal coherence of radiation from the source S

the mirrors M_1, M_2. This arrangement is called a *Michelson interferometer* (Sect. 4.2). The two beams with wavelength λ travel different optical path lengths SM_1SB and SM_2SB, and their path difference in the plane B is

$$\Delta s = 2(SM_1 - SM_2) \,.$$

The mirror M_2 is mounted on a carriage and can be moved, resulting in a continuous change of Δs. In the plane B, one obtains maximum intensity when both amplitudes have the same phase, which means $\Delta s = m\lambda$, and minimum intensity if $\Delta s = (2m + 1)\lambda/2$. With increasing Δs, the contrast $(I_{max} - I_{min})/(I_{max} + I_{min})$ decreases and vanishes if Δs becomes larger than the coherence length Δs_c (Sect. 2.8.4). Experiments show that Δs_c is related to the spectral width $\Delta \omega$ of the incident wave by

$$\Delta s_c \simeq c/\Delta \omega = c/(2\pi \Delta \nu) \,. \tag{2.104}$$

This observation may be explained as follows. A wave emitted from a point source with the spectral width $\Delta \omega$ can be regarded as a superposition of many quasi-monochromatic components with frequencies ω_n within the interval $\Delta \omega$. The superposition results in wave trains of finite length $\Delta s_c = c\Delta t = c/\Delta \omega$ because the different components with slightly different frequencies ω_n come out of phase during the time interval Δt and interfere destructively, causing the total amplitude to decrease (Sect. 3.1). If the path difference Δs in the Michelson interferometer becomes larger than Δs_c, the split wave trains no longer overlap in the plane B. The coherence length Δs_c of a light source therefore becomes larger with decreasing spectral width $\Delta \omega$.

Example 2.6

(a) A low-pressure mercury spectral lamp with a spectral filter that only transmits the green line $\lambda = 546$ nm has, because of the Doppler width $\Delta \omega_D = 4 \times 10^9$ Hz, a coherence length of $\Delta s_c \simeq 8$ cm.

(b) A single-mode HeNe laser with a bandwidth of $\Delta \omega = 2\pi \cdot 1$ MHz has a coherence length of about 50 m.

2.8.2 Spatial Coherence

The radiation from an *extended source* LS of size b illuminates two slits S_1 and S_2 in the plane A at a distance d apart (Young's double-slit interference experiment, Fig. 2.24a). The total amplitude and phase at each of the two slits are obtained by superposition of all partial waves emitted from the different surface elements df of the source, taking into account the different paths df–S_1 and df–S_2.

The intensity at the point of observation P in the plane B depends on the path difference S_1P-S_2P and on the phase difference $\Delta\phi = \phi(S_1) - \phi(S_2)$ of the total field amplitudes in S_1 and S_2. If the different surface elements df of the source emit independently with random phases (thermal radiation source), the phases of the total amplitudes in S_1 and S_2 will also fluctuate randomly. However, this would not influence the intensity in P as long as these fluctuations occur in S_1 and S_2 synchronously, because then the phase difference $\Delta\phi$ would remain constant. In this case, the two slits form two coherent sources that generate an interference pattern in the plane B.

For radiation emitted from the central part 0 of the light source, this proves to be true since the paths $0S_1$ and $0S_2$ are equal and all phase fluctuations in 0 arrive simultaneously in S_1 and S_2. For all other points Q of the source, however, path differences $\Delta s_Q = QS_1 - QS_2$ exist, which are largest for the edges R of the source. From Fig. 2.24b one can infer for $b \ll r$ the relation

$$\Delta s_R = R_1 S_2 - R_1 S_1 = R_2 S_1 - R_1 S_1 \simeq b \sin(\theta/2) \, .$$

For $\Delta s_R > \lambda/2$, the phase difference $\Delta\phi$ of the partial amplitudes in S_1 and S_2 exceeds π. With random emission from the different surface elements df of the source, the time-averaged interference pattern in the plane B will be washed out. The condition for coherent illumination of S_1 and S_2 from a light source with the dimension b is therefore

$$\Delta s = b \sin(\theta/2) < \lambda/2 \, .$$

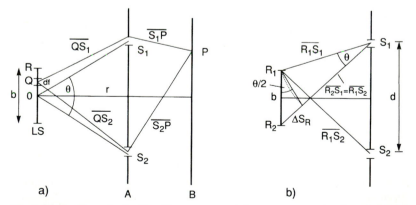

Fig. 2.24a,b. Young's double-slit arrangement for measurements of spatial coherence

With $2\sin(\theta/2) = d/r$, this condition can be written as

$$bd/r < \lambda \,. \tag{2.105a}$$

Extension of this coherence condition to two dimensions yields, for a source area $A_s = b^2$, the following condition for the maximum surface $A_c = d^2$ that can be illuminated coherently:

$$b^2 d^2/r^2 \leq \lambda^2 \,. \tag{2.105b}$$

Since $d\Omega = d^2/r^2$ is the solid angle accepted by the illuminated surface $A_c = d^2$, this can be formulated as

$$A_s \, d\Omega \leq \lambda^2 \,. \tag{2.105c}$$

The source surface $A_s = b^2$ determines the maximum solid angle $d\Omega \leq \lambda^2/A_s$ inside which the radiation field shows spatial coherence. Equation (2.105c) reveals that radiation from a point source (spherical waves) is spatially coherent within the whole solid angle $d\Omega = 4\pi$. The coherence surfaces are spheres with the source in the center. Likewise, a plane wave produced by a point source in the focus of a lens shows spatial coherence over the whole aperture confining the light beam. For given source dimensions, the coherence surface $A_c = d^2$ increases with the square of the distance from the source. Because of the vast distances to stars, the starlight received by telescopes is spatially coherent across the telescope aperture, in spite of the large diameter of the radiation source.

The arguments above may be summarized as follows: the coherence surface S_c, i.e., that maximum area A_c that can be coherently illuminated at a distance r from an extended quasi-monochromatic light source with the area A_s emitting at a wavelength λ is determined by

$$\boxed{S_c = \lambda^2 r^2/A_s}\,. \tag{2.106}$$

2.8.3 Coherence Volume

With the coherence length $\Delta s_c = c/\Delta\omega$ in the propagation direction of the radiation with the spectral width $\Delta\omega$ and the coherence surface $S_c = \lambda^2 r^2/A_s$, the coherence volume $V_c = S_c \Delta s_c$ becomes

$$V_c = \frac{\lambda^2 r^2 c}{\Delta\omega A_s} \,. \tag{2.107}$$

A unit surface element of a source with the spectral radiance L_ω [W/(m^2 sr)] emits $L_\omega/\hbar\omega$ photons per second within the frequency interval $d\omega = 1$ Hz into the unit solid angle 1 sr.

The mean number $\bar{n}$ of photons in the spectral range $\Delta\omega$ within the coherence volume defined by the solid angle $\Delta\Omega = \lambda^2/A_s$ and the coherence length $\Delta s_c = c\Delta t_c$ generated by a source with area A_s is therefore

$$\bar{n} = (L_\omega/\hbar\omega)A_s\Delta\Omega\,\Delta\omega\Delta t_c\,.$$

With $\Delta\Omega = \lambda^2/A_s$ and $\Delta t_c \simeq 1/\Delta\omega$, this gives

$$\bar{n} = (L_\omega/\hbar\omega)\lambda^2\,. \tag{2.108}$$

Example 2.7

For a thermal radiation source, the spectral radiance for linearly polarized light (given by (2.28) divided by a factor 2) is for $\cos\phi = 1$ and $L_\nu\,d\nu = L_\omega\,d\omega$

$$L_\nu = \frac{h\nu^3/c^2}{e^{h\nu/kT} - 1}\,.$$

The mean number of photons within the coherence volume is then with $\lambda = c/\nu$

$$\bar{n} = \frac{1}{e^{h\nu/kT} - 1}\,.$$

This is identical to the mean number of photons per mode of the thermal radiation, as derived in Sect. 2.2. Figure 2.5 and Example 2.3 give values of $\bar{n}$ for different conditions.

The mean number $\bar{n}$ of photons per mode is often called the *degeneracy parameter* of the radiation field. This example shows that the coherence volume is related to the modes of the radiation field. This relation can be also illustrated in the following way:

If we allow the radiation from all modes with the same direction of $\boldsymbol{k}$ to escape through a hole in the cavity wall with the area $A_s = b^2$, the wave emitted from A_s will not be strictly parallel, but will have a diffraction-limited divergence angle $\theta \simeq \lambda/b$ around the direction of $\boldsymbol{k}$. This means that the radiation is emitted into a solid angle $d\Omega = \lambda^2/b^2$. This is the same solid angle (2.105c) that limits the spatial coherence.

The modes with the same direction of $\boldsymbol{k}$ (which we assume to be the z direction) may still differ in $|\boldsymbol{k}|$, i.e., they may have different frequencies ω. The coherence length is determined by the spectral width $\Delta\omega$ of the radiation emitted from A_s. Since $|\boldsymbol{k}| = \omega/c$ the spectral width $\Delta\omega$ corresponds to an interval $\Delta k = \Delta\omega/c$ of the k values. This radiation illuminates a minimum "diffraction surface"

$$A_D = r^2\,d\Omega = r^2\lambda^2/A_s\,.$$

Multiplication with the coherence length $\Delta s_c = c/\Delta\omega$ yields again the coherence volume $V_c = A_D c/\Delta\omega = r^2\lambda^2 c/(\Delta\omega A_s)$ of (2.107). We shall now

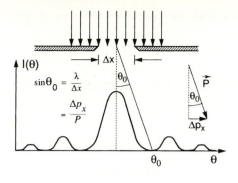

Fig. 2.25. The uncertainty principle applied to the diffraction of light by a slit

demonstrate that the coherence volume is identical with the spatial part of the elementary cell in the general phase space.

As is well known from atomic physics, the diffraction of light can be explained by Heisenberg's uncertainty relation. Photons passing through a slit of width Δx have the uncertainty Δp_x of the x-component p_x of their momentum $\boldsymbol{p}$, given by $\Delta p_x \Delta x \geq \hbar$ (Fig. 2.25).

Generalized to three dimensions, the uncertainty principle postulates that the simultaneous measurements of momentum and location of a photon have the minimum uncertainty

$$\Delta p_x \Delta p_y \Delta p_z \Delta x \Delta y \Delta z \geq \hbar^3 = V_{\text{ph}} \,, \tag{2.109}$$

where $V_{\text{ph}} = \hbar^3$ is the volume of one cell in phase space. *Photons within the same cell of the phase space are indistinguishable and can be therefore regarded as identical.*

Photons that are emitted from the hole $A_s = b^2$ within the diffraction angle $\theta = \lambda/b$ against the surface normal (Fig. 2.26), which may point into the z-direction, have the minimum uncertainty

$$\Delta p_x = \Delta p_y = |\boldsymbol{p}| \lambda/(2\pi b) = (\hbar\omega/c)\lambda/(2\pi b) = (\hbar\omega/c)d/(2\pi r) \,, \tag{2.110}$$

of the momentum components p_x and p_y, where the last equality follows from (2.105b).

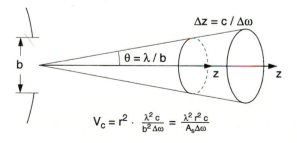

$$V_c = r^2 \cdot \frac{\lambda^2 c}{b^2 \Delta\omega} = \frac{\lambda^2 r^2 c}{A_s \Delta\omega}$$

Fig. 2.26. Coherence volume and phase space cell

The uncertainty Δp_z is mainly caused by the spectral width $\Delta \omega$. Since $p = \hbar \omega / c$, we find

$$\Delta p_z = (\hbar / c) \Delta \omega . \qquad (2.111)$$

Substituting (2.110, 2.111) into (2.109), we obtain for the spatial part of the phase space cell

$$\Delta x \Delta y \Delta z = \frac{\lambda^2 r^2 c}{\Delta \omega A_s} = V_c ,$$

which turns out to be identical with the coherence volume defined by (2.107).

2.8.4 The Coherence Function and the Degree of Coherence

In the previous subsections we have described the coherence properties of radiation fields in a more general way. We now briefly discuss a more quantitative description which allows us to define partial coherence and to measure the degree of coherence.

In the cases of both temporal and spatial coherence, we are concerned with the correlation between optical fields either at the same point P_0 but at different times $[E(P_0, t_1)$ and $E(P_0, t_2)]$, or at the same time t but at two different points $[E(P_1, t)$ and $E(P_2, t)]$. The subsequent description follows the representation in [2.3, 2.25, 2.26].

Suppose we have an extended source that generates a radiation field with a narrow spectral bandwidth $\Delta \omega$, which we shall represent by the complex notation of a plane wave, i.e.,

$$E(r, t) = A_0 \, e^{i(\omega t - k \cdot r)} + \text{c.c.}$$

The field at two points in space S_1 and S_2 (e.g., the two apertures in Young's experiment) is then $E(S_1, t)$ and $E(S_2, t)$. The two apertures serve as secondary sources (Fig. 2.24), and the resultant field at the point of observation P at time t is

$$E(P, t) = k_1 E_1(S_1, t - r_1/c) + k_2 E_2(S_2, t - r_2/c) , \qquad (2.112)$$

where the imaginary numbers k_1 and k_2 depend on the size of the apertures and on the distances $r_1 = S_1 P$ and $r_2 = S_2 P$.

The resulting time averaged irradiance at P measured over a time interval which is long compared to the coherence time is

$$I_p = \epsilon_0 c \langle E(P, t) E^*(P, t) \rangle , \qquad (2.113)$$

where the brackets $\langle \ldots \rangle$ indicate the time average. Using (2.112), this becomes

$$I_p = c \epsilon_0 \big[k_1 k_1^* \langle E_1(t - t_1) E_1^*(t - t_1) \rangle + k_2 k_2^* \langle E_2(t - t_2) E_2^*(t - t_2) \rangle$$
$$+ k_1 k_2^* \langle E_1(t - t_1) E_2^*(t - t_2) \rangle + k_1^* k_2 \langle E_1^*(t - t_1) E_2(t - t_2) \rangle \big] . \qquad (2.114)$$

If the field is stationary, the time-averaged values do not depend on time. We can therefore shift the time origin without changing the irradiances (2.113). Accordingly, the first two time averages in (2.114) can be transformed to $\langle E_1(t)E_1^*(t)\rangle$ and $\langle E_2(t)E_2^*(t)\rangle$. In the last two terms we shift the time origin by an amount t_2 and write them with $\tau = t_2 - t_1$

$$k_1 k_2^* \langle E_1(t+\tau)E_2^*(t)\rangle + k_1^* k_2 \langle E_1^*(t+\tau)E_2(t)\rangle$$
$$= 2\,\mathrm{Re}\left\{k_1 k_2^* \langle E_1(t+\tau)E_2^*(t)\rangle\right\} . \tag{2.115}$$

The term

$$\Gamma_{12}(\tau) = \langle E_1(t+\tau)E_2^*(t)\rangle , \tag{2.116}$$

is called the *mutual coherence function* and describes the cross correlation of the light field amplitudes at S_1 and S_2. When the amplitudes and phases of E_1 and E_2 fluctuate within a time interval $\Delta t < \tau$, the time average $\Gamma_{12}(\tau)$ will be zero if these fluctuations of the two fields at two different points and at two different times are completely uncorrelated. If the field at S_1 at time $t+\tau$ was perfectly correlated with the field at S_2 at time t, the relative phase would be unaltered despite individual fluctuations, and Γ_{12} would become independent of τ.

Inserting (2.116) into (2.114) gives for the irradiance at P (note that k_1 and k_2 are pure imaginary numbers for which $2\,\mathrm{Re}\{k_1 \cdot k_2\} = 2|k_1| \cdot |k_2|$)

$$I_p = \epsilon_0 c \left[|k_1|^2 I_{S1} + |k_2|^2 I_{S2} + 2|k_1||k_2|\,\mathrm{Re}\{\Gamma_{12}(\tau)\}\right] . \tag{2.117}$$

The first term $I_1 = \epsilon_0 c |k_1|^2 I_{S1}$ yields the irradiance at P when only the aperture S_1 is open ($k_2 = 0$); the second term $I_2 = \epsilon_0 c |k_2|^2 I_{S2}$ is that for $k_1 = 0$.

Let us introduce the first-order correlation functions

$$\Gamma_{11}(\tau) = \langle E_1(t+\tau)E_1^*(t)\rangle ,$$
$$\Gamma_{22}(\tau) = \langle E_2(t+\tau)E_2^*(t)\rangle , \tag{2.118}$$

which correlate the field amplitude at the same point but at different times. For $\tau = 0$ the *self-coherence functions*

$$\Gamma_{11}(0) = \langle E_1(t)E_1^*(t)\rangle = I_1/(\epsilon_0 c) ,$$
$$\Gamma_{22}(0) = I_2/(\epsilon_0 c) ,$$

are proportional to the irradiance I at S_1 and S_2, respectively.

With the definition of the normalized form of the mutual coherence function,

$$\gamma_{12}(\tau) = \frac{\Gamma_{12}(\tau)}{\sqrt{\Gamma_{11}(0)\Gamma_{22}(0)}} = \frac{\langle E_1(t+\tau)E_2^*(t)\rangle}{\sqrt{\langle|E_1(t)|^2|E_2(t)|^2\rangle}} , \tag{2.119}$$

(2.117) can be written as

$$\boxed{I_p = I_1 + I_2 + 2\sqrt{I_1 I_2}\,\mathrm{Re}\{\gamma_{12}(\tau)\}} . \tag{2.120}$$

This is the general interference law for partially coherent light; $\gamma_{12}(\tau)$ is called the *complex degree of coherence*. Its meaning will be illustrated by the following: we express the complex quantity $\gamma_{12}(\tau)$ as

$$\gamma_{12}(\tau) = |\gamma_{12}(\tau)|\, e^{i\phi_{12}(\tau)} ,$$

where the phase angle $\phi_{12}(\tau) = \phi_1(\tau) - \phi_2(\tau)$ is related to the phases of the fields E_1 and E_2 in (2.116).

For $|\gamma_{12}(\tau)| = 1$, (2.122) describes the interference of two completely coherent waves out of phase at S_1 or S_2 by the amount $\phi_{12}(\tau)$. For $|\gamma_{12}(\tau)| = 0$, the interference term vanishes. The two waves are said to be completely incoherent. For $0 < |\gamma_{12}(\tau)| < 1$ we have *partial coherence*. $\gamma_{12}(\tau)$ is therefore a measure of the degree of coherence. We illustrate the mutual coherence function $\gamma_{12}(\tau)$ by applying it to the situations outlined in Sects. 2.8.1, 2.8.2.

Example 2.8

In the Michelson interferometer, the incoming nearly parallel light beam is split by S (Fig. 2.23) and recombined in the plane B. If both partial beams have the same amplitude $E = E_0\, e^{i\phi(t)}$, the degree of coherence becomes

$$\gamma_{11}(\tau) = \frac{\langle E(t+\tau)E^*(t)\rangle}{|E(t)|^2} = \left\langle e^{i\phi(t+\tau)}\, e^{-i\phi(t)} \right\rangle .$$

For long averaging times T we obtain with $\Delta\phi = \phi(t+\tau) - \phi(t)$,

$$\gamma_{11}(\tau) = \lim_{T\to\infty} \frac{1}{T} \int\limits_0^T (\cos \Delta\phi + i \sin \Delta\phi)\, dt . \tag{2.121}$$

For a strictly monochromatic wave with infinite coherence length Δs_c, the phase function is $\phi(t) = \omega t - \mathbf{k}\cdot\mathbf{r}$ and $\Delta\phi = +\omega\tau$ with $\tau = \Delta s/c$. This yields

$$\gamma_{11}(\tau) = \cos\omega\tau + i\sin\omega\tau = e^{i\omega\tau} , \qquad |\gamma_{11}(\tau)| = 1 .$$

For a wave with spectral width $\Delta\omega$ so large that $\tau > \Delta s_c/c = 1/\Delta\omega$, the phase differences $\Delta\phi$ vary randomly between 0 and 2π and the integral averages to zero, giving $\gamma_{11}(\tau) = 0$. In Fig. 2.27 the interference pattern $I(\Delta\phi) \propto |E_1(t)\cdot E_2(t+\tau)|^2$ in the observation plane behind a Michelson interferometer is illustrated as a function of the phase difference $\Delta\phi = (2\pi/\lambda)\Delta s$ for equal intensities $I_1 = I_2$ but different values of $|\gamma_{12}(\tau)|$. For completely coherent light ($|\gamma_{12}(\tau)| = 1$) the intensity $I(\tau)$ changes between $4I_1$ and zero, whereas for $|\gamma(t)| = 0$ the interference term vanishes and the total intensity $I = 2I_1$ does not depend on τ.

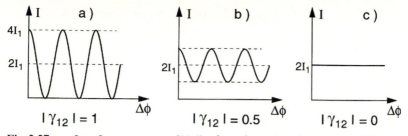

Fig. 2.27a–c. Interference pattern $I(\Delta\phi)$ of two-beam interference for different degrees of coherence

Example 2.9

For the special case of a quasi-monochromatic plane wave $E = E_0 \times \exp(i\omega t - i\boldsymbol{k} \cdot \boldsymbol{r})$, an optical path difference $(\boldsymbol{r}_2 - \boldsymbol{r}_1)$ causes a corresponding phase difference

$$\phi_{12}(\tau) = \boldsymbol{k} \cdot (\boldsymbol{r}_2 - \boldsymbol{r}_1) \,,$$

and (2.120) can be expressed with $\mathrm{Re}\{\gamma_{12}(\tau)\} = |\gamma_{12}(\tau)| \cos\phi_{12}$ by

$$I_p = I_1 + I_2 + 2\sqrt{I_1 I_2}\,|\gamma_{12}(\tau)|\cos\phi_{12}(\tau) \,. \tag{2.122}$$

For $|\gamma_{12}(\tau)| = 1$, the interference term causes a full modulation of the irradiance $I_p(\tau)$. For $\gamma_{12}(\tau) = 0$, the interference vanishes and the total intensity becomes independent of the time delay τ between the two beams.

Example 2.10

Referring to Young's experiment (Fig. 2.24) with a narrow bandwidth but extended source, *spatial* coherence effects will predominate. The fringe pattern in the plane B will depend on $\Gamma(S_1, S_2, \tau) = \Gamma_{12}(\tau)$. In the region about the central fringe $(r_2 - r_1) = 0$, $\tau = 0$, the values of $\Gamma_{12}(0)$ and $\gamma_{12}(0)$ can be determined from the visibility of the interference pattern.

To find the value $\gamma_{12}(\tau)$ for any point P on the screen B in Fig. 2.24, the intensity $I(P)$ is measured when both slits are open, and also when one of the pinholes is blocked. In terms of these observed quantities, the degree of coherence can be determined from (2.120) to be

$$\mathrm{Re}\{\gamma_{12}(P)\} = \frac{I(P) - I_1(P) - I_2(P)}{2\sqrt{I_1(P)I_2(P)}} \,.$$

This yields the desired information about the spatial coherence of the source, which depends on the size of the source and its distance from the pinholes.

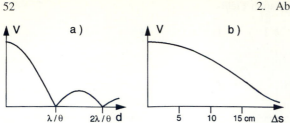

Fig. 2.28. (a) Visibility of the interference pattern behind the two slits of Fig. 2.24 if they are illuminated by a monochromatic extended source. The abscissa gives the slit separation d in units of λ/θ. (b) Visibility of a Doppler-broadened line behind a Michelson interferometer as a function of the path difference Δs

The visibility of the fringes at P is defined as

$$V(P) = \frac{I_{\max} - I_{\min}}{I_{\max} + I_{\min}} = \frac{2\sqrt{I_1(P)}\sqrt{I_2(P)}}{I_1(P) + I_2(P)} |\gamma_{12}(\tau)| \ , \tag{2.123}$$

where the last equality follows from (2.122). If $I_1 = I_2$ (equal size pinholes), we see that

$$V(P) = |\gamma_{12}(\tau)| \ .$$

The visibility is then equal to the degree of coherence. Figure 2.28a depicts the visibility V of the fringe pattern in P as a function of the slit separation d, indicated in Fig. 2.24, when these slits are illuminated by monochromatic light from an extended uniform source with quadratic size $b \times b$ that appears from S_1 under the angle θ. Figure 2.28b illustrates the visibility as a function of path difference Δs in a Michelson interferometer which is illuminated with the Doppler-broadened line $\lambda = 632.8$ nm from a neon discharge lamp.

For more detailed presentations of coherence see the textbooks [2.5, 2.26–2.28].

2.9 Coherence of Atomic Systems

Two levels of an atom are said to be coherently excited if their corresponding wave functions are in phase at the excitation time. With a short laser pulse of duration Δt, which has a Fourier-limited spectral bandwidth $\Delta \omega \simeq 1/\Delta t$, two atomic levels a and b can be excited simultaneously if their energy separation ΔE is smaller than $\hbar \Delta \omega$ (Fig. 2.29). The wave function of the excited atom is then a linear combination of the wave functions ψ_a and ψ_b, and the atom is said to be in a coherent superposition of the two states $|a\rangle$ and $|b\rangle$.

An ensemble of atoms is *coherently* excited if the wave functions of the excited atoms, at a certain time t, have the same phase for all atoms. This phase relation may change with time due to differing frequencies ω in the time-dependent part $\exp(i\omega t)$ of the excited-state wave functions or because of relaxation processes, which may differ for the different atoms. This will result in a "phase diffusion" and a time-dependent decrease of the degree of coherence.

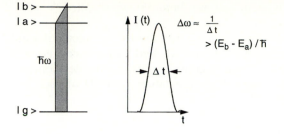

Fig. 2.29. Coherent excitation of two atomic levels $|a\rangle$ and $|b\rangle$ from the same lower level $|g\rangle$ with a broadband laser pulse with $\hbar\Delta\omega \geq (E_b - E_a)$

The realization of such coherent systems requires special experimental preparations that, however, can be achieved with several techniques of coherent laser spectroscopy (Chap. 12). An elegant theoretical way of describing observable quantities of a coherently or incoherently excited system of atoms and molecules is based on the density-matrix formalism.

2.9.1 Density Matrix

Let us assume, for simplicity, that each atom of the ensemble can be represented by a two-level system (Sect. 2.7), described by the wave function

$$\psi(r, t) = \psi_a + \psi_b = a(t)u_a e^{-iE_a t/\hbar} + b(t)u_b e^{-i[(E_b/\hbar)t - \phi]} , \qquad (2.124)$$

where the phase ϕ might be different for each of the atoms. The density matrix $\tilde{\rho}$ is defined by the product of the two state vectors

$$\tilde{\rho} = |\psi\rangle \langle\psi| = \begin{pmatrix} \psi_a \\ \psi_b \end{pmatrix} (\psi_a, \psi_b)$$

$$= \begin{pmatrix} |a(t)|^2 & ab e^{-i[(E_a - E_b)t/\hbar + \phi]} \\ ab e^{+i[(E_a - E_b)t/\hbar + \phi]} & |b(t)|^2 \end{pmatrix} = \begin{pmatrix} \rho_{aa} & \rho_{ab} \\ \rho_{ba} & \rho_{bb} \end{pmatrix} , \qquad (2.125)$$

since the normalized atomic wave functions in vector notation are

$$u_a = \begin{pmatrix} 1 \\ 0 \end{pmatrix} \quad \text{and} \quad u_b = \begin{pmatrix} 0 \\ 1 \end{pmatrix} .$$

The diagonal elements ρ_{aa} and ρ_{bb} represent the probabilities of finding the atoms of the ensemble in the level $|a\rangle$ and $|b\rangle$, respectively.

If the phases ϕ of the atomic wave function (2.124) are randomly distributed for the different atoms of the ensemble, the nondiagonal elements of the density matrix (2.125) average to zero and the incoherently excited system is therefore described by the diagonal matrix

$$\tilde{\rho}_{incoh} = \begin{pmatrix} [a(t)]^2 & 0 \\ 0 & [b(t)]^2 \end{pmatrix} . \qquad (2.126)$$

If definite phase relations exist between the wave functions of the atoms, the system is in a *coherent state*. The nondiagonal elements of (2.125) describe the degree of coherence of the system and are therefore often called "coherences."

Such a coherent state can, for example, be generated by the interaction of the atomic ensemble with a sufficiently strong EM field that induces atomic dipole moments, which add up to a macroscopic oscillating dipole moment if all atomic dipoles oscillate in phase. The expectation value D of such an atomic dipole moment is

$$D = -e \int \psi^* r \psi \, d\tau \, . \tag{2.127}$$

With (2.66b) this becomes

$$D = -D_{ab}(a^* b e^{-i\omega_{ba} t} + a b^* e^{i\omega_{ba} t}) = D_{ab}(\rho_{ab} + \rho_{ba}). \tag{2.128}$$

The nondiagonal elements of the density matrix are therefore proportional to the expectation value of the dipole moment.

2.9.2 Coherent Excitation

We saw in Sect. 2.9.1 that in a coherently excited system of atoms, well-defined phase relations exist between the time-dependent wavefunctions of the atomic levels. In this section we will illustrate such coherent excitations by several examples.

- If identical paramagnetic atoms with magnetic moments μ and total angular momentum J are brought into a homogeneous magnetic field $B_0 = \{0, 0, B_z\}$, the angular momentum vectors J_i of the atoms will precess with the Lamor frequency $\omega_L = \gamma B_0$ around the z-direction, where $\gamma = \mu/|J|$ is the gyromagnetic ratio (Fig. 2.30a). The phases φ_i of this precession will be different for the different atoms and, in general, are randomly distributed. The precession occurs incoherently (Fig. 2.30b). The dipole moments μ of the N atoms add up to a macroscopic "longitudinal"

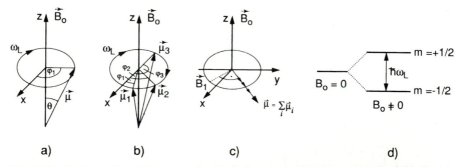

Fig. 2.30a–d. Precession of a magnetic dipole in a homogeneous magnetic field B_0 (**a**); Incoherent precession of the different dipoles (**b**); Synchronization of dipoles by a radio frequency (RF) field (**c**); Coherent superposition of two Zeeman sublevels (**d**) as the quantum-mechanical equivalent to the classical picture (**c**)

magnetization

$$M_z = \sum_{i=1}^{N} \mu \cos \theta_i = N\mu \cos \theta \, ,$$

but the average "transversal" magnetization is zero.

When an additional radio frequency field $B_1 = B_{10} \cos \omega t$ is added with $B_1 \perp B_0$, the dipoles are forced to precess synchroneously with the RF field B_1 in the x–y-plane if $\omega = \omega_L$. This results in a macroscopic magnetic moment $M = N\mu$, which rotates with ω_L in the x–y-plane and has a phase angle $\pi/2$ against B_1 (Fig. 2.30c). The precession of the atoms becomes coherent through their coupling to the RF field. In the quantum-mechanical description, the RF field induces transitions between the Zeeman sublevels (Fig. 2.30d). If the RF field B_1 is sufficiently intense, the atoms are in a coherent superposition of the wave functions of both Zeeman levels.

• Excitation by visible or UV light may also create a coherent superposition of Zeeman sublevels. As an example, we consider the transition $6\,^1S_0 \to 6\,^3P_1$ of the Hg atom at $\lambda = 253.7$ nm (Fig. 2.31). In a magnetic field $B = \{0, 0, B_z\}$, the upper level $6\,^3P_1$ splits into three Zeeman sublevels with magnetic quantum numbers $m_z = 0, \pm 1$. Excitation with linear polarized light ($E \parallel B$) only populates the level $m_J = 0$. The fluorescence emitted from this Zeeman level is also linearly polarized.

However, if the exciting light is polarized perpendicularly to the magnetic field ($E \perp B$), it may be regarded as superposition of σ^+ and σ^- light traveling into the z-direction, which is chosen as the quantization axis.

In this case, the levels with $m = \pm 1$ are populated. As long as the Zeeman splitting is smaller than the homogeneous width of the Zeeman levels (e.g., the natural linewidth $\Delta\omega = 1/\tau$), both components are excited coherently (even with monochromatic light!). The wave function of the excited state is represented by a linear combination $\psi = a\psi_a + b\psi_b$ of the two wavefunctions of the Zeeman sublevels $m = \pm 1$. The fluorescence is nonisotropic,

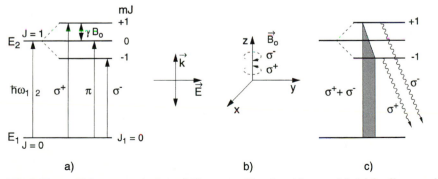

Fig. 2.31a–c. Coherent excitation of Zeeman sublevels with $m = \pm 1$ (**a**) by linear polarized light with $E \perp B$ (**b**). The fluorescence is a superposition of σ^+ and σ^- light (**c**)

but shows an angular distribution that depends on the coefficients a, b (Sect. 12.1).

• A molecule with two closely lying levels $|a\rangle$ and $|b\rangle$ that can both be reached by optical transitions from a common groundstate $|g\rangle$ can be coherently excited by a light pulse with duration ΔT, if $\Delta T < \hbar/(E_a - E_b)$, even if the levels $|a\rangle$ and $|b\rangle$ are different vibrational levels of different electronic states and their separation is larger than their homogeneous width.

The time-dependent fluorescence from these coherently excited states shows, besides the exponential decay $\exp(-t/\tau)$, a beat period $\tau_{QB} = \hbar/(E_a - E_b)$ due to the different frequencies ω_a and ω_b of the two fluorescence components (quantum beats, Sect. 12.2).

2.9.3 Relaxation of Coherently Excited Systems

The time-dependent Schrödinger equation (2.57) is written in the density-matrix formalism as

$$i\hbar\dot{\tilde{\rho}} = [\mathcal{H}, \tilde{\rho}] . \tag{2.129}$$

In order to separate the different contributions of induced absorption or emission and of relaxation processes, we write the Hamiltonian $\mathcal{H}$ as the sum

$$\mathcal{H} = \mathcal{H}_0 + \mathcal{H}_1(t) + \mathcal{H}_R , \tag{2.130}$$

of the "internal" Hamiltonian of the isolated two-level system

$$\mathcal{H}_0 = \begin{pmatrix} E_a & 0 \\ 0 & E_b \end{pmatrix} ,$$

the interaction Hamiltonian of the system with an EM field $E = E_0 \cdot \cos \omega t$

$$\mathcal{H}_1(t) = -\mu E(t) = \begin{pmatrix} 0 & -D_{ab} E_0(t) \\ -D_{ba} E_0(t) & 0 \end{pmatrix} \cos \omega t , \tag{2.131}$$

and a relaxation part

$$\mathcal{H}_R = \hbar \begin{pmatrix} \gamma_a & \gamma_\varphi^a \\ \gamma_\varphi^b & \gamma_b \end{pmatrix} , \tag{2.132}$$

which describes all relaxation processes, such as spontaneous emission or collision-induced transitions. The population relaxation of level $|b\rangle$ with a decay constant γ_b causing an effective lifetime $T_b = 1/\gamma_b$ is, for example, described by

$$i\hbar\rho_{bb}\gamma_b = [\mathcal{H}_R, \tilde{\rho}]_{bb} \Rightarrow T_b = \frac{1}{\gamma_b} = \frac{i\hbar\rho_{bb}}{[\mathcal{H}_R, \tilde{\rho}]_{bb}} . \tag{2.133}$$

The decay of the off-diagonal elements ρ_{ab}, ρ_{ba} describes the decay of the coherence, i.e., of the phase relations between the atomic dipoles.

The dephasing rate is represented by the phase-relaxation constants γ_φ^a, γ_φ^b and the decay of the nondiagonal elements is governed by

$$\frac{i\hbar\rho_{ab}}{T_2} = -[\mathcal{H}_R, \rho]_{ab} , \tag{2.134}$$

where the "transverse" relaxation time T_2 (dephasing time) is defined by

$$\frac{1}{T_2} = \frac{1}{2}\left(\frac{1}{T_a} + \frac{1}{T_b}\right) + \gamma_\phi . \tag{2.135}$$

In general, the phase relaxation is faster than the population relaxation defined by the relaxation time T_1, which means that the nondiagonal elements decay faster than the diagonal elements (Chap. 12).

For more information on coherent excitation of atomic and molecular systems see [2.29–2.31] and Chap. 12.

Problems

2.1 The angular divergence of the output from a 1-W argon laser is assumed to be 4×10^{-3} rad. Calculate the radiance L and the radiant intensity I^* of the laser beam and the irradiance I (intensity) at a surface 1 m away from the output mirror, when the laser beam diameter at the mirror is 2 mm. What is the spectral power density $\rho(\nu)$ if the laser bandwidth is 1 MHz?

2.2 Unpolarized light of intensity I_0 is transmitted through a dichroic polarizer with thickness 1 mm. Calculate the transmitted intensity when the absorption coefficients for the two polarizations are $\alpha_\| = 100\,\text{cm}^{-1}$ and $\alpha_\perp = 5\,\text{cm}^{-1}$.

2.3 Assume the isotropic emission of a pulsed flashlamp with spectral bandwidth $\Delta\lambda = 100\,\text{nm}$ around $\lambda = 400\,\text{nm}$ amounts to 100-W peak power out of a volume of $1\,\text{cm}^3$. Calculate the spectral power density $\rho(\nu)$ and the spectral intensity $I(\nu)$ through a spherical surface 2 cm away from the center of the emitting sphere. How many photons per mode are contained in the radiation field?

2.4 The beam of a monochromatic laser passes through an absorbing atomic vapor with path length $L = 5\,\text{cm}$. If the laser frequency is tuned to the center of an absorbing transition $|i\rangle \rightarrow |k\rangle$ with absorption cross section $\sigma_0 = 10^{-14}\,\text{cm}^2$, the attenuation of the transmitted intensity is 10%. Calculate the atomic density N_i in the absorbing level $|i\rangle$.

2.5 An excited molecular level E_i is connected with three lower levels E_n by radiative transitions with spontaneous probabilities $A_{i1} = 3 \times 10^7\,\text{s}^{-1}$, $A_{i2} = 1 \times 10^7\,\text{s}^{-1}$, and $A_{i3} = 5 \times 10^7\,\text{s}^{-1}$. Calculate the spontaneous lifetime τ_i and the relative populations N_n/N_i under cw excitation of E_i, when $\tau_1 = 5 \times 10^{-7}\,\text{s}$, $\tau_2 = 6 \times 10^{-9}\,\text{s}$, and $\tau_3 = 10^{-8}\,\text{s}$. Determine the Einstein coefficient B_{0i} for the excitation $|0\rangle \rightarrow |i\rangle$ from a ground state $|0\rangle$ with $\tau_0 = \infty$

at the wavelength $\lambda = 400$ nm. How large is the absorption cross section σ_{0i} when the absorption line profile has a bandwidth of $\Delta\omega = 9 \times 10^7$ Hz?

2.6 Which pumping rates (molecules/second) are necessary on the transition $|n\rangle \rightarrow |i\rangle$ of Exercise 2.5 to achieve equal population densities $N_i = N_n$ for $n = 1, 2, 3$?

2.7 Expansion of a laser beam is accomplished by two lenses with different focal lengths (Fig. 2.32). Why does an aperture in the focal plane improve the quality of the wave fronts in the expanded beam by eliminating perturbations due to diffraction effects by dust and other imperfections on the lens surfaces?

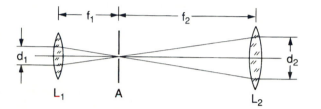

Fig. 2.32. Beam-expanding telescope with an aperture in the focal plane

2.8 Calculate the maximum slit separation in Young's interference experiments that still gives distinct interference fringes, if the two slits are illuminated

(a) by incoherent light of $\lambda = 500$ nm from a hole with 1-mm diameter, 1 m away from the slits;
(b) by a star with 10^6-km diameter, at a distance of 4 light-years;
(c) by the expanded beam from an HeNe laser with diffraction-limited divergence, emitted from a spot size of 1 mm at the output mirror.

2.9 A sodium atom is placed in a cavity with walls at the temperature T, producing a thermal radiation field with spectral energy density $\rho(v)$. At what temperature T are the spontaneous and induced transition probabilities equal

(a) for the transition $3P \rightarrow 3S$ ($\lambda = 589$ nm) with $\tau(3P) = 16$ ns;
(b) for the hyperfine transition $3S$ ($F = 3 \rightarrow F = 2$) with $\tau(3F) \simeq 1$ s and $v = 1772$ MHz?

2.10 An optically excited sodium atom Na($3P$) with a spontaneous lifetime $\tau(3P) = 16$ ns is placed in a cell filled with 10 mbar nitrogen gas at a temperature of $T = 400$ K. Calculate the effective lifetime $\tau_{\text{eff}}(3P)$ if the quenching cross section for Na($3P$)–N_2 collisions is $\sigma_q = 4 \times 10^{-15}$ cm^2.

3. Widths and Profiles of Spectral Lines

Spectral lines in discrete absorption or emission spectra are never strictly monochromatic. Even with the very high resolution of interferometers, one observes a spectral distribution $I(\nu)$ of the absorbed or emitted intensity around the central frequency $\nu_0 = (E_i - E_k)/h$ corresponding to a molecular transition with the energy difference $\Delta E = E_i - E_k$ between upper and lower levels. The function $I(\nu)$ in the vicinity of ν_0 is called the *line profile* (Fig. 3.1). The frequency interval $\delta\nu = |\nu_2 - \nu_1|$ between the two frequencies ν_1 and ν_2 for which $I(\nu_1) = I(\nu_2) = I(\nu_0)/2$ is the *full-width at half-maximum* of the line (FWHM), often shortened to the *linewidth* or *halfwidth* of the spectral line.

The halfwidth is sometimes written in terms of the angular frequency $\omega = 2\pi\nu$ with $\delta\omega = 2\pi\delta\nu$, or in terms of the wavelength λ (in units of nm or Å) with $\delta\lambda = |\lambda_1 - \lambda_2|$. From $\lambda = c/\nu$, it follows that

$$\delta\lambda = -(c/\nu^2)\delta\nu \ . \tag{3.1}$$

The *relative* halfwidths, however, are the same in all three schemes:

$$\left|\frac{\delta\nu}{\nu}\right| = \left|\frac{\delta\omega}{\omega}\right| = \left|\frac{\delta\lambda}{\lambda}\right| \ . \tag{3.2}$$

The spectral region within the halfwidth is called the *kernel of the line*, the regions outside ($\nu < \nu_1$ and $\nu > \nu_2$) are the *line wings*.

In the following sections we discuss various origins of the finite linewidth. Several examples illustrate the order of magnitude of different line-broadening effects in different spectral regions and their importance for high-resolution

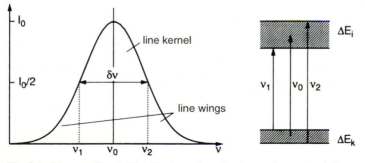

Fig. 3.1. Line profile, halfwidth, kernel, and wings of a spectral line

spectroscopy [3.1–3.4]. Following the usual convention we shall often use the angular frequency $\omega = 2\pi\nu$ to avoid factors of 2π in the equations.

3.1 Natural Linewidth

An excited atom can emit its excitation energy as spontaneous radiation (Sect. 2.7). In order to investigate the spectral distribution of this spontaneous emission on a transition $E_i \rightarrow E_k$, we shall describe the excited *atomic electron* by the classical model of a damped harmonic oscillator with frequency ω, mass m, and restoring force constant k. The radiative energy loss results in a damping of the oscillation described by the damping constant γ. We shall see, however, that for real atoms the damping is extremely small, which means that $\gamma \ll \omega$.

The amplitude $x(t)$ of the oscillation can be obtained by solving the differential equation of motion

$$\ddot{x} + \gamma\dot{x} + \omega_0^2 x = 0 , \tag{3.3}$$

where $\omega_0^2 = k/m$.

The real solution of (3.3) with the initial values $x(0) = x_0$ and $\dot{x}(0) = 0$ is

$$x(t) = x_0 e^{-(\gamma/2)t} [\cos \omega t + (\gamma/2\omega) \sin \omega t] . \tag{3.4}$$

The frequency $\omega = (\omega_0^2 - \gamma^2/4)^{1/2}$ of the damped oscillation is slightly lower than the frequency ω_0 of the undamped case. However, for small damping ($\gamma \ll \omega_0$) we can set $\omega \simeq \omega_0$ and also may neglect the second term in (3.4). With this approximation, which is still very accurate for real atoms, we obtain the solution of (3.3) as

$$x(t) = x_0 e^{-(\gamma/2)t} \cos \omega_0 t . \tag{3.5}$$

The frequency $\omega_0 = 2\pi\nu_0$ of the oscillator corresponds to the central frequency $\omega_{ik} = (E_i - E_k)/\hbar$ of an atomic transition $E_i \rightarrow E_k$.

3.1.1 Lorentzian Line Profile of the Emitted Radiation

Because the amplitude $x(t)$ of the oscillation decreases gradually, the frequency of the emitted radiation is no longer monochromatic as it would be for an oscillation with constant amplitude. Instead, it shows a frequency distribution related to the function $x(t)$ in (3.5) by a Fourier transformation (Fig. 3.2).

The damped oscillation $x(t)$ can be described as a superposition of monochromatic oscillations $\exp(i\omega t)$ with slightly different frequencies ω and amplitudes $A(\omega)$

$$x(t) = \frac{1}{2\sqrt{2\pi}} \int_0^\infty A(\omega) e^{i\omega t} d\omega . \tag{3.6}$$

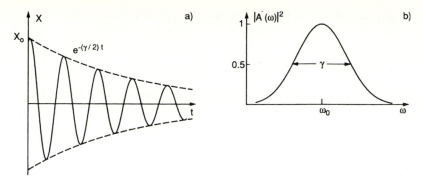

Fig. 3.2. Damped oscillation: (**a**) the frequency distribution $A(\omega)$ of the amplitudes obtained by the Fourier transform of $x(t)$ yields the intensity profile $I(\omega - \omega_0) \propto |A(\omega)|^2$ (**b**)

The amplitudes $A(\omega)$ are calculated from (3.6) as the Fourier transform

$$A(\omega) = \frac{1}{\sqrt{2\pi}} \int_{-\infty}^{+\infty} x(t)\,e^{-i\omega t}\,dt = \frac{1}{\sqrt{2\pi}} \int_{0}^{\infty} x_0\,e^{-(\gamma/2)t} \cos(\omega_0 t)\,e^{-i\omega t}\,dt \,.$$

(3.7)

The lower integration limit is taken to be zero because $x(t) = 0$ for $t < 0$. Equation (3.7) can readily be integrated to give the complex amplitudes

$$A(\omega) = \frac{x_0}{\sqrt{8\pi}} \left(\frac{1}{i(\omega - \omega_0) + \gamma/2} + \frac{1}{i(\omega + \omega_0) + \gamma/2} \right) \,.$$

(3.8)

The real intensity $I(\omega) \propto A(\omega)A^*(\omega)$ contains terms with $(\omega - \omega_0)$ and $(\omega + \omega_0)$ in the denominator. In the vicinity of the central frequency ω_0 of an atomic transition where $(\omega - \omega_0)^2 \ll \omega_0^2$, the terms with $(\omega + \omega_0)$ can be neglected and the intensity profile of the spectral line becomes

$$I(\omega - \omega_0) = \frac{C}{(\omega - \omega_0)^2 + (\gamma/2)^2} \,.$$

(3.9)

The constant C can be defined in two different ways:
(a) For comparison of different line profiles it is useful to define a normalized intensity profile $L(\omega - \omega_0) = I(\omega - \omega_0)/I_0$ with $I_0 = \int I(\omega)\,d\omega$ such that

$$\int_{0}^{\infty} L(\omega - \omega_0)\,d\omega = \int_{-\infty}^{+\infty} L(\omega - \omega_0)\,d(\omega - \omega_0) = 1 \,.$$

With this normalization, the integration of (3.9) yields $C = I_0 \gamma / 2\pi$.

$$L(\omega - \omega_0) = \frac{1}{2\pi} \frac{\gamma}{(\omega - \omega_0)^2 + (\gamma/2)^2} , \qquad (3.10)$$

is called the *normalized Lorentzian profile*. Its full halfwidth at half-maximum (FWHM) is

$$\delta \omega_n = \gamma \quad \text{or} \quad \delta \nu_n = \gamma / 2\pi . \qquad (3.11)$$

Any intensity distribution with a Lorentzian profile is then

$$I(\omega - \omega_0) = I_0 \frac{\gamma/2\pi}{(\omega - \omega_0)^2 + (\gamma/2)^2} = I_0 L(\omega - \omega_0) , \qquad (3.10a)$$

with a peak intensity $I(\omega_0) = 2I_0/(\pi\gamma)$.

(b) Often in the literature the normalization of (3.9) is chosen in such a way that $I(\omega_0) = I_0$; furthermore, the full halfwidth is denoted by 2Γ. In this notation the Lorentzian profile of a transition $|k\rangle \leftarrow |i\rangle$ is

$$I(\omega) = I_0 L^*(\omega - \omega_{ik}) \quad \text{with} \quad I_0 = I(\omega_0) ,$$

and

$$L^*(\omega - \omega_{ik}) = \frac{\Gamma^2}{(\omega_{ik} - \omega)^2 + \Gamma^2} \quad \text{with} \quad \Gamma = \gamma/2 . \qquad (3.10b)$$

With $x = (\omega_{ik} - \omega)/\Gamma$ this can be abbreviated as

$$L^*(\omega - \omega_{ik}) = \frac{1}{1 + x^2} \quad \text{with} \quad L^*(\omega_{ik}) = 1 . \qquad (3.10c)$$

In this notation the area under the line profile becomes

$$\int_0^\infty I(\omega) \, d\omega = \Gamma \int_{-\infty}^{+\infty} I(x) \, dx = \pi I_0 \Gamma . \qquad (3.10d)$$

3.1.2 Relation Between Linewidth and Lifetime

The radiant power of the damped oscillator can be obtained from (3.3) if both sides of the equation are multiplied by $m\dot{x}$, which yields after rearranging

$$m\ddot{x}\dot{x} + m\omega_0^2 x\dot{x} = -\gamma m\dot{x}^2 . \qquad (3.12)$$

The left-hand side of (3.12) is the time derivative of the total energy W (sum of kinetic energy $\frac{1}{2}m\dot{x}^2$ and potential energy $Dx^2/2 = m\omega_0^2 x^2/2$), and can therefore be written as

$$\frac{d}{dt}\left(\frac{m}{2}\dot{x}^2 + \frac{m}{2}\omega_0^2 x^2\right) = \frac{dW}{dt} = -\gamma m\dot{x}^2 . \qquad (3.13)$$

Inserting $x(t)$ from (3.5) and neglecting terms with γ^2 yields

$$\frac{dW}{dt} = -\gamma m x_0^2 \omega_0^2 e^{-\gamma t} \sin^2 \omega_0 t . \tag{3.14}$$

Because the time average $\overline{\sin^2 \omega t} = 1/2$, the time-averaged radiant power $\overline{P} = \overline{dW/dt}$ is

$$\overline{\frac{dW}{dt}} = -\frac{\gamma}{2} m x_0^2 \omega_0^2 e^{-\gamma t} . \tag{3.15}$$

Equation (3.15) shows that $\overline{P}$ and with it the intensity $I(t)$ of the spectral line decreases to $1/e$ of its initial value $I(t=0)$ after the decay time $\tau = 1/\gamma$.

In Sect. 2.8 we saw that the mean lifetime τ_i of a molecular level E_i, which decays exponentially by spontaneous emission, is related to the Einstein coefficient A_i by $\tau_i = 1/A_i$. Replacing the classical damping constant γ by the spontaneous transition probability A_i, we can use the classical formulas (3.9–3.11) as a correct description of the frequency distribution of spontaneous emission and its linewidth. The natural halfwidth of a spectral line spontaneously emitted from the level E_i is, according to (3.11),

$$\delta \nu_n = A_i/2\pi = (2\pi \tau_i)^{-1} \quad \text{or} \quad \delta \omega_n = A_i = 1/\tau_i . \tag{3.16}$$

The radiant power emitted from N_i excited atoms on a transition $E_i \to E_k$ is given by

$$dW_{ik}/dt = N_i A_{ik} \hbar \omega_{ik} . \tag{3.17}$$

If the emission of a source with volume ΔV is isotropic, the radiation power received by a detector of area A at a distance r through the solid angle $d\Omega = A/r^2$ is

$$P_{ik} = \left(\frac{dW_{ik}}{dt} \right) \frac{d\Omega}{4\pi} = N_i A_{ik} \hbar \omega_{ik} \Delta V \frac{A}{4\pi r^2} . \tag{3.18}$$

This means that the density N_i of emitters can be inferred from the measured power, if A_{ik} is known (Sect. 11.3).

Note: Equation (3.16) can also be derived from the uncertainty principle (Fig. 3.3). With the mean lifetime τ_i of the excited level E_i, its energy E_i can be determined only with an uncertainty $\Delta E_i \simeq \hbar/\tau_i$ [3.5]. The frequency $\omega_{ik} = (E_i - E_k)/\hbar$ of a transition terminating in the stable ground state E_k has therefore the uncertainty

$$\delta \omega = \Delta E_i/\hbar = 1/\tau_i . \tag{3.19}$$

If the lower level E_k is not the ground state but also an excited state with the lifetime τ_k, the uncertainties ΔE_i and ΔE_k of the two levels both contribute to the linewidth. This yields for the total uncertainty

$$\Delta E = \Delta E_i + \Delta E_k \to \delta \omega_n = (1/\tau_i + 1/\tau_k) . \tag{3.20}$$

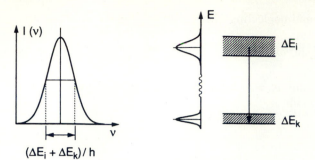

(ΔE$_i$ + ΔE$_k$)/ h

Fig. 3.3. Illustration of the uncertainty principle, which relates the natural linewidth to the energy uncertainties of the upper and lower levels

3.1.3 Natural Linewidth of Absorbing Transitions

In a similar way, the spectral profile of an *absorption line* can be derived for atoms *at rest*: the intensity I of a plane wave passing in the z-direction through an absorbing sample decreases along the distance dz by

$$dI = -\alpha I \, dz \; . \tag{3.21}$$

The absorption coefficient α_{ik} [cm^{-1}] for a transition $|i\rangle \rightarrow |k\rangle$ depends on the population densities N_i, N_k of the lower and upper levels, and on the optical absorption cross section σ_{ik} [cm^2] of each absorbing atom, see (2.42):

$$\alpha_{ik}(\omega) = \sigma_{ik}(\omega)[N_i - (g_i/g_k)N_k] \; , \tag{3.22}$$

which reduces to $\alpha_{ik} = \sigma N_i$ for $N_k \ll N_i$ (Fig. 3.4). For sufficiently small intensities I, the induced absorption rate is small compared to the refilling rate of level $|i\rangle$ and the population density N_i does not depend on the intensity I (linear absorption). Integration of (3.21) then yields Beer's law

$$I = I_0 e^{-\alpha(\omega)z} \; . \tag{3.23}$$

The absorption profile $\alpha(\omega)$ can be obtained from our classical model of a damped oscillator with charge q under the influence of a driving force qE

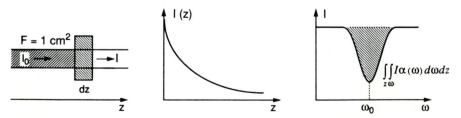

Fig. 3.4. Absorption of a parallel light beam passing through an optically thin absorbing layer

caused by the incident wave with amplitude $E = E_0 e^{i\omega t}$. The corresponding differential equation

$$m\ddot{x} + b\dot{x} + kx = qE_0 e^{i\omega t} , \tag{3.24}$$

has the solution

$$x = \frac{qE_0 e^{i\omega t}}{m(\omega_0^2 - \omega^2 + i\gamma\omega)} , \tag{3.25}$$

with the abbreviations $\gamma = b/m$, and $\omega_0^2 = k/m$. The forced oscillation of the charge q generates an induced dipole moment

$$p = qx = \frac{q^2 E_0 e^{i\omega t}}{m(\omega_0^2 - \omega^2 + i\gamma\omega)} . \tag{3.26}$$

In a sample with N oscillators per unit volume, the macroscopic polarization P, which is the sum of all dipole moments per unit volume, is therefore

$$P = Nqx . \tag{3.27}$$

On the other hand, the polarization can be derived in classical electrodynamics from Maxwell's equations using the dielectric constant ϵ_0 or the susceptibility χ, i.e.,

$$\boldsymbol{P} = \epsilon_0(\epsilon - 1)\boldsymbol{E} = \epsilon_0\chi\boldsymbol{E} . \tag{3.28}$$

The relative dielectric constant ϵ is related to the refractive index n by

$$n = \epsilon^{1/2} . \tag{3.29}$$

This can be easily verified from the relations

$$v = (\epsilon\epsilon_0\mu\mu_0)^{-1/2} = c/n \text{ and } c = (\epsilon_0\mu_0)^{-1/2} \quad \Rightarrow \quad n = \sqrt{\epsilon\mu} ,$$

for the velocity of light, which follows from Maxwell's equations in media with the dielectric constant $\epsilon_0\epsilon$ and the magnetic permeability $\mu_0\mu$. Except for ferromagnetic materials, the relative permeability is $\mu \simeq 1 \rightarrow n = \epsilon^{1/2}$.

Combining (3.25–3.29), the refractive index n can be written as

$$n^2 = 1 + \frac{Nq^2}{\epsilon_0 m(\omega_0^2 - \omega^2 + i\gamma\omega)} . \tag{3.30}$$

In gaseous media at sufficiently low pressures, the index of refraction is close to unity (for example, in air at atmospheric pressure, $n = 1.00028$ for $\lambda = 500$ nm). In this case, the approximation

$$n^2 - 1 = (n+1)(n-1) \simeq 2(n-1) ,$$

is sufficiently accurate for most purposes. Therefore (3.30) can be reduced to

$$n = 1 + \frac{Nq^2}{2\epsilon_0 m(\omega_0^2 - \omega^2 + i\gamma\omega)} . \tag{3.31}$$

In order to make clear the physical implication of this complex index of refraction, we separate the real and the imaginary parts and write

$$n = n' - i\kappa . \tag{3.32}$$

An EM wave $E = E_0 \exp[i(\omega t - kz)]$ passing in the z-direction through a medium with the refractive index n has the same frequency $\omega_n = \omega_0$ as in vacuum, but a different wave vector $k_n = k_0 n$. Inserting (3.32) with $|k| = 2\pi/\lambda$ yields

$$E = E_0 e^{-k_0\kappa z} e^{i(\omega t - k_0 n' z)} = E_0 e^{-2\pi\kappa z/\lambda_0} e^{ik_0(ct - n'z)} . \tag{3.33}$$

Equation (3.33) shows that the imaginary part $\kappa(\omega)$ of the complex refractive index n describes the *absorption* of the EM wave. At a penetration depth of $\Delta z = \lambda_0/(2\pi\kappa)$, the amplitude $E_0 \exp(-k_0\kappa z)$ has decreased to $1/e$ of its value at $z = 0$. The real part $n'(\omega)$ represents the *dispersion* of the wave, i.e., the dependence of the phase velocity $v(\omega) = c/n'(\omega)$ on the frequency. The intensity $I \propto EE^*$ then decreases as

$$I = I_0 e^{-2\kappa k_0 z} . \tag{3.34}$$

Comparison with (3.23) yields the relation

$$\alpha = 2\kappa k_0 = 4\pi\kappa/\lambda_0 . \tag{3.35}$$

The absorption coefficient α is proportional to the imaginary part κ of the complex refractive index $n = n' - i\kappa$.

The frequency dependence of α and n' can be obtained by inserting (3.32, 3.35) into (3.31). Separating the real and imaginary parts, we get

$$\boxed{\alpha = \frac{Nq^2\omega_0}{c\epsilon_0 m} \frac{\gamma\omega}{(\omega_0^2 - \omega^2)^2 + \gamma^2\omega^2} ,} \tag{3.36a}$$

$$\boxed{n' = 1 + \frac{Nq^2}{2\epsilon_0 m} \frac{\omega_0^2 - \omega^2}{(\omega_0^2 - \omega^2)^2 + \gamma^2\omega^2} .} \tag{3.37a}$$

The equations (3.36a) and (3.37) are the *Kramers–Kronig dispersion relations*. They relate absorption and dispersion through the complex refractive index $n = n' - i\kappa = n' - i\alpha/(2k_0)$.

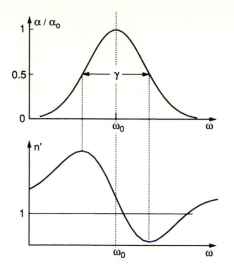

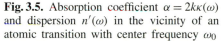

Fig. 3.5. Absorption coefficient $\alpha = 2k\kappa(\omega)$ and dispersion $n'(\omega)$ in the vicinity of an atomic transition with center frequency ω_0

In the neighborhood of a molecular transition frequency ω_0 where $|\omega_0 - \omega| \ll \omega_0$, the dispersion relations reduce with $q = e$ to

$$\alpha(\omega) = \frac{Ne^2}{4\epsilon_0 mc} \frac{\gamma}{(\omega_0 - \omega)^2 + (\gamma/2)^2} \,, \tag{3.36b}$$

$$n' = 1 + \frac{Ne^2}{4\epsilon_0 m\omega_0} \frac{\omega_0 - \omega}{(\omega_0 - \omega)^2 + (\gamma/2)^2} \,. \tag{3.37b}$$

The absorption profile $\alpha(\omega)$ is Lorentzian with a FWHM of $\Delta\omega_n = \gamma$, which equals the natural linewidth. The difference $n' - n_0 = n' - 1$ between the refractive indices in a gas and in vacuum yields a dispersion profile.

Figure 3.5 shows the frequency dependence of $\alpha(\omega)$ and $n'(\omega)$ in the vicinity of the eigenfrequency ω_0 of an atomic transition.

Note: The relations derived in this section are only valid for oscillators at rest in the observer's coordinate system. The thermal motion of real atoms in a gas introduces an additional broadening of the line profile, the *Doppler broadening*, which will be discussed in Sect. 3.2. The profiles (3.36, 3.37) can therefore be observed only with Doppler-free techniques (Chaps. 7 and 9).

Example 3.1

(a) The natural linewidth of the sodium D_1 line at $\lambda = 589.1$ nm, which corresponds to a transition between the $3P_{3/2}$ level ($\tau = 16$ ns) and the $3S_{1/2}$ ground state, is

$$\delta\nu_n = \frac{10^9}{16 \times 2\pi} = 10^7 \, s^{-1} = 10 \, \text{MHz} \,.$$

Note that with a central frequency $v_0 = 5 \times 10^{14}\,\mathrm{Hz}$ and a lifetime of 16 ns, the damping of the corresponding classical oscillator is extremely small. Only after 8×10^6 periods of oscillation has the amplitude decreased to $1/e$ of its initial value.

(b) The natural linewidth of a molecular transition between two vibrational levels of the electronic ground state with a wavelength in the infrared region is very small because of the long spontaneous lifetimes of vibrational levels. For a typical lifetime of $\tau = 10^{-3}\,\mathrm{s}$, the natural linewidth becomes $\delta v_n = 160\,\mathrm{Hz}$.

(c) Even in the visible or ultraviolet range, atomic or molecular electronic transitions with very small transition probabilities exist. In a dipole approximation these are "forbidden" transitions. One example is the $2s \leftrightarrow 1s$ transition for the hydrogen atom. The upper level $2s$ cannot decay by electric dipole transition, but a two-photon transition to the $1s$ ground state is possible. The natural lifetime is $\tau = 0.12\,\mathrm{s}$ and the natural linewidth of such a two-photon line is therefore $\delta v_n = 1.3\,\mathrm{Hz}$.

3.2 Doppler Width

Generally, the Lorentzian line profile with the natural linewidth δv_n, as discussed in Sect. 3.1, cannot be observed without special techniques, because it is completely concealed by other broadening effects. One of the major contributions to the spectral linewidth in gases at low pressures is the Doppler width, which is due to the thermal motion of the absorbing or emitting molecules.

Consider an excited molecule with a velocity $\boldsymbol{v} = \{v_x, v_y, v_z\}$ relative to the rest frame of the observer. The central frequency of a molecular emission line that is ω_0 in the coordinate system of the molecule is Doppler shifted to

$$\omega_e = \omega_0 + \boldsymbol{k} \cdot \boldsymbol{v},\tag{3.38}$$

for an observer looking toward the emitting molecule (that is, against the direction of the wave vector $\boldsymbol{k}$ of the emitted radiation; Fig. 3.6a). For the ob-

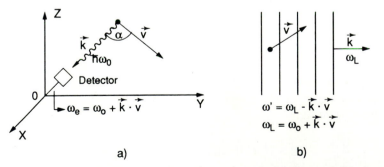

Fig. 3.6. (a) Doppler shift of a monochromatic emission line and (b) absorption line

server, the apparent emission frequency ω_e is increased if the molecule moves toward the observer ($\boldsymbol{k} \cdot \boldsymbol{v} > 0$), and decreased if the molecule moves away ($\boldsymbol{k} \cdot \boldsymbol{v} < 0$).

Similarly, one can see that the absorption frequency ω_0 of a molecule moving with the velocity $\boldsymbol{v}$ across a plane EM wave $\boldsymbol{E} = \boldsymbol{E}_0 \exp(i\omega t - \boldsymbol{k} \cdot \boldsymbol{r})$ is shifted. The wave frequency ω in the rest frame appears in the frame of the moving molecule as

$$\omega' = \omega - \boldsymbol{k} \cdot \boldsymbol{v} .$$

The molecule can only absorb if ω' coincides with its eigenfrequency ω_0. The absorption frequency $\omega = \omega_a$ is then

$$\omega_a = \omega_0 + \boldsymbol{k} \cdot \boldsymbol{v} . \tag{3.39a}$$

As in the emission case, the absorption frequency ω_a is increased for $\boldsymbol{k} \cdot \boldsymbol{v} > 0$ (Fig. 3.6b). This happens, for example, if the molecule moves parallel to the wave propagation. It is decreased if $\boldsymbol{k} \cdot \boldsymbol{v} < 0$, e.g., when the molecule moves against the light propagation. If we choose the $+z$-direction to coincide with the light propagation, with $\boldsymbol{k} = \{0, 0, k_z\}$ and $|k| = 2\pi/\lambda$, (3.39a) becomes

$$\omega_a = \omega_0(1 + v_z/c) . \tag{3.39b}$$

Note: Equations (3.38) and (3.39) describe the *linear* Doppler shift. For higher accuracies, the quadratic Doppler effect must also be considered (Sect. 14.1).

At thermal equilibrium, the molecules of a gas follow a Maxwellian velocity distribution. At the temperature T, the number of molecules $n_i(v_z) \, dv_z$ in the level E_i per unit volume with a velocity component between v_z and $v_z + dv_z$ is

$$n_i(v_z) \, dv_z = \frac{N_i}{v_p \sqrt{\pi}} e^{-(v_z/v_p)^2} \, dv_z , \tag{3.40}$$

where $N_i = \int n_i(v_z) \, dv_z$ is the density of all molecules in level E_i, $v_p = (2kT/m)^{1/2}$ is the most probable velocity, m is the mass of a molecule, and k is Boltzmann's constant. Inserting the relation (3.39b) between the velocity component and the frequency shift with $dv_z = (c/\omega_0) \, d\omega$ into (3.40) gives the number of molecules with absorption frequencies shifted from ω_0 into the interval from ω to $\omega + d\omega$

$$n_i(\omega) \, d\omega = N_i \frac{c}{\omega_0 v_p \sqrt{\pi}} \exp\left[-\left(\frac{c(\omega - \omega_0)}{\omega_0 v_p}\right)^2\right] d\omega . \tag{3.41}$$

Since the emitted or absorbed radiant power $P(\omega) \, d\omega$ is proportional to the density $n_i(\omega) \, d\omega$ of molecules emitting or absorbing in the interval $d\omega$, the intensity profile of a Doppler-broadened spectral line becomes

$$I(\omega) = I_0 \exp\left[-\left(\frac{c(\omega - \omega_0)}{\omega_0 v_p}\right)^2\right] . \tag{3.42}$$

This is a Gaussian profile with a full halfwidth

$$\delta\omega_D = 2\sqrt{\ln 2}\,\omega_0 v_p/c = \left(\frac{\omega_0}{c}\right)\sqrt{8kT\ln 2/m}\,, \tag{3.43a}$$

which is called the *Doppler width*. Inserting (3.43) into (3.42) with $1/(4\ln 2) = 0.36$ yields

$$I(\omega) = I_0 \exp\left(-\frac{(\omega-\omega_0)^2}{0.36\delta\omega_D^2}\right). \tag{3.44}$$

Note that $\delta\omega_D$ increases linearly with the frequency ω_0 and is proportional to $(T/m)^{1/2}$. The largest Doppler width is thus expected for hydrogen ($M=1$) at high temperatures and a large frequency ω for the Lyman α line.

Equation (3.43) can be written more conveniently in terms of the Avogadro number N_A (the number of molecules per mole), the mass of a mole, $M = N_A m$, and the gas constant $R = N_A k$. Inserting these relations into (3.43) for the Doppler width gives

$$\delta\omega_D = (2\omega_0/c)\sqrt{2RT\ln 2/M}\,. \tag{3.43b}$$

or, in frequency units, using the values for c and R,

$$\delta v_D = 7.16 \times 10^{-7} v_0\sqrt{T/M}\quad[\text{Hz}]\,. \tag{3.43c}$$

Example 3.2

(a) Vacuum ultraviolet: for the Lyman α line ($2p \rightarrow 1s$ transition in the H atom) in a discharge with temperature $T = 1000\,\text{K}$, $M = 1$, $\lambda = 121.6\,\text{nm}$, $v_0 = 2.47 \times 10^{15}\,\text{s}^{-1}$ $\rightarrow$ $\delta v_D = 5.6 \times 10^{10}\,\text{Hz}$, $\delta\lambda_D = 2.8 \times 10^{-3}\,\text{nm}$.

(b) Visible spectral region: for the sodium D line ($3p \rightarrow 3s$ transition of the Na atom) in a sodium-vapor cell at $T = 500\,\text{K}$, $\lambda = 589.1\,\text{nm}$, $v_0 = 5.1 \times 10^{14}\,\text{s}^{-1}$ $\rightarrow$ $\delta v_D = 1.7 \times 10^9\,\text{Hz}$, $\delta\lambda_D = 1 \times 10^{-3}\,\text{nm}$.

(c) Infrared region: for a vibrational transition $(J_i, v_i) \leftrightarrow (J_k, v_k)$ between two rovibronic levels with the quantum numbers J, v of the CO_2 molecule in a CO_2 cell at room temperature ($T = 300\,\text{K}$), $\lambda = 10\,\mu\text{m}$, $v = 3 \times 10^{13}\,\text{s}^{-1}$, $M = 44$ $\rightarrow$ $\delta v_D = 5.6 \times 10^7\,\text{Hz}$, $\delta\lambda_D = 1.9 \times 10^{-2}\,\text{nm}$.

These examples illustrate that in the visible and UV regions, the Doppler width exceeds the natural linewidth by about two orders of magnitude. Note, however, that the intensity I approaches zero for large arguments $(v - v_0)$ much faster for a Gaussian line profile than for a Lorentzian profile (Fig. 3.7). It is therefore possible to obtain information about the Lorentzian profile from the extreme line wings, even if the Doppler width is much larger than the natural linewidth (see below).

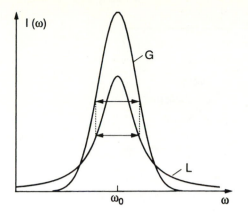

Fig. 3.7. Comparison between Lorentzian (L) and Gaussian (G) line profiles of equal halfwidths

More detailed consideration shows that a Doppler-broadened spectral line cannot be strictly represented by a pure Gaussian profile as has been assumed in the foregoing discussion, since not all molecules with a definite velocity component v_z emit or absorb radiation at the same frequency $\omega' = \omega_0(1+v_z/c)$. Because of the finite lifetimes of the molecular energy levels, the frequency response of these molecules is represented by a Lorentzian profile, see (3.10)

$$L(\omega-\omega') = \frac{\gamma/2\pi}{(\omega-\omega')^2+(\gamma/2)^2} \ ,$$

with a central frequency ω' (Fig. 3.8). Let $n(\omega')\,d\omega' = n(v_z)\,dv_z$ be the number of molecules per unit volume with velocity components within the interval v_z to $v_z + dv_z$. The spectral intensity distribution $I(\omega)$ of the total absorption or

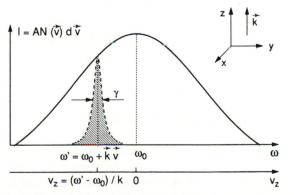

Fig. 3.8. Lorentzian profile centered at $\omega' = \omega_0 + \mathbf{k}\cdot\mathbf{v} = \omega_0(1+v_z/c)$ for molecules with a definite velocity component v_z

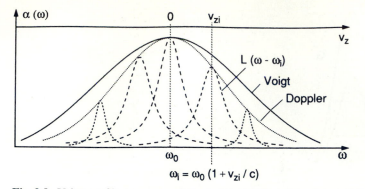

Fig. 3.9. Voigt profile as a convolution of Lorentzian line shapes $L(\omega_0 - \omega_i)$ with $\omega_i = \omega_0(1 + v_{zi}/c)$

emission of all molecules at the transition $E_i \rightarrow E_k$ is then

$$I(\omega) = I_0 \int n(\omega')L(\omega - \omega')\, d\omega' \ . \tag{3.45}$$

Inserting (3.10) for $L(\omega - \omega')\, d\omega'$ and (3.41) for $n(\omega')$, we obtain

$$I(\omega) = C \int_0^\infty \frac{\exp\{-[(c/v_p)(\omega_0 - \omega')/\omega_0]^2\}}{(\omega - \omega')^2 + (\gamma/2)^2} \, d\omega' \tag{3.46}$$

with

$$C = \frac{\gamma N_i c}{2v_p \pi^{3/2}\omega_0} \ .$$

This intensity profile, which is a convolution of Lorentzian and Gaussian profiles (Fig. 3.9), is called a *Voigt profile*. Voigt profiles play an important role in the spectroscopy of stellar atmospheres, where accurate measurements of line wings allow the contributions of Doppler broadening and natural linewidth or collisional line broadening to be separated (see [3.6] and Sect. 3.3). From such measurements the temperature and pressure of the emitting or absorbing layers in the stellar atmospheres may be deduced [3.7].

3.3 Collisional Broadening of Spectral Lines

When an atom A with the energy levels E_i and E_k approaches another atom or molecule B, the energy levels of A are shifted because of the interaction between A and B. This shift depends on the electron configurations of A and B and on the distance $R(A, B)$ between both collision partners, which we define as the distance between the centers of mass of A and B.

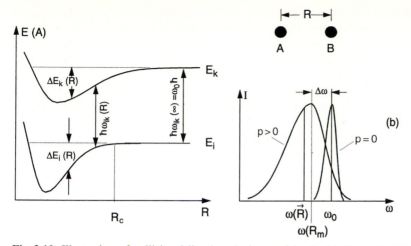

Fig. 3.10. Illustration of collisional line broadening explained with the potential curves of the collision pair AB

The energy shifts ΔE are, in general, different for the levels E_i and E_k and may be positive as well as negative. The energy shift ΔE is positive if the interaction between A and B is repulsive, and negative if it is attractive. When plotting the energy $E(R)$ for the different energy levels as a function of the interatomic distance R the potential curves of Fig. 3.10 are obtained.

This mutual interaction of both partners at distances $R \leq R_c$ is called a *collision* and $2R_c$ is the *collision diameter*. If no internal energy of the collision partners is transferred during the collision by nonradiative transitions, the collision is termed *elastic*. Without additional stabilizing mechanisms (recombination), the partners will separate again after the collision time $\tau_c \simeq R_c/v$, which depends on the relative velocity v.

Example 3.3
At thermal velocities of $v = 5 \times 10^2$ m/s and a typical collision radius of $R_c = 1$ nm, we obtain the collision time $\tau_c = 2 \times 10^{-12}$ s. During this time the electronic charge distribution generally follows the perturbation "adiabatically", which justifies the potential curve model of Fig. 3.10.

3.3.1 Phenomenological Description

If atom A undergoes a *radiative* transition between levels E_i and E_k during the collision time, the frequency

$$\omega_{ik} = |E_i(R) - E_k(R)| / \hbar \tag{3.47}$$

of absorbed or emitted radiation depends on the distance $R(t)$ at the time of the transition. We assume that the radiative transition takes place in a time interval that is short compared to the collision time, so that the distance R does not change during the transition. In Fig. 3.10 this assumption leads to vertical radiative transitions.

In a gas mixture of atoms A and B, the mutual distance $R(A, B)$ shows random fluctuations with a distribution around a mean value R that depends on pressure and temperature. According to (3.47), the fluorescence yields a corresponding frequency distribution around a most probable value $\omega_{ik}(R_m)$, which may be shifted against the frequency ω_0 of the unperturbed atom A. The shift $\Delta\omega = \omega_0 - \omega_{ik}$ depends on how differently the two energy levels E_i and E_k are shifted at a distance $R_m(A, B)$ where the emission probability has a maximum. The intensity profile $I(\omega)$ of the collision-broadened and shifted emission line can be obtained from

$$I(\omega) \propto \int A_{ik}(R) P_{col}(R) [E_i(R) - E_k(R)] \, dR \, , \tag{3.48}$$

where $A_{ik}(R)$ is the spontaneous transition probability, which depends on R because the electronic wave functions of the collision pair (AB) depend on R, and $P_{col}(R)$ is the probability per unit time that the distance between A and B lies in the range from R to $R + dR$.

From (3.48) it can be seen that the intensity profile of the collision-broadened line reflects the difference of the potential curves

$$E_i(R) - E_k(R) = V[A(E_i), B] - V[A(E_k), B] \, .$$

Let $V(R)$ be the interaction potential between the ground-state atom A and its collision partner B. The probability that B has a distance between R and $R + dR$ is proportional to $4\pi R^2 \, dR$ and (in thermal equilibrium) to the Boltzmann factor $\exp[-V(R)/kT]$. The number $N(R)$ of collision partners B with distance R from A is therefore

$$N(R) \, dR = N_0 4\pi R^2 \, e^{-V(R)/kT} \, dR \, , \tag{3.49}$$

where N_0 is the average density of atoms B. Because the intensity of an absorption line is proportional to the density of absorbing atoms while they are forming collision pairs, the intensity profile of the absorption line can be written as

$$I(\omega) \, d\omega = C^* \left\{ R^2 \exp\left(-\frac{V_i(R)}{kT} \right) \frac{d}{dR} [V_i(R) - V_k(R)] \right\} dR \, , \tag{3.50}$$

where $\hbar\omega(R) = [V_i(R) - V_k(R)] \to \hbar \, d\omega/dR = d[V_i(R) - V_k(R)]/dR$ has been used. Measuring the line profile as a function of temperature yields

$$\frac{dI(\omega, T)}{dT} = \frac{V_i(R)}{kT^2} I(\omega, T) \, ,$$

and therefore the ground-state potential $V_i(R)$ separately.

Frequently, different *spherical model potentials* $V(R)$ are substituted in (3.50), such as the Lennard–Jones potential

$$V(R) = a/R^{12} - b/R^6 , \qquad \qquad (3.51)$$

The coefficients a, b are adjusted for optimum agreement between theory and experiment [3.8–3.16].

The line shift caused by elastic collisions corresponds to an energy shift $\Delta E = \hbar \Delta \omega$ between the excitation energy $\hbar \omega_0$ of the free atom A^* and the photon energy $\hbar \omega$. It is supplied from the kinetic energy of the collision partners. This means that in case of positive shifts ($\Delta \omega > 0$), the kinetic energy is smaller after the collision than before.

Besides elastic collisions, inelastic collisions may also occur in which the excitation energy E_i of atom A is either partly or completely transferred into internal energy of the collision partner B, or into translational energy of both partners. Such inelastic collisions are often called *quenching collisions* because they decrease the number of excited atoms in level E_i and therefore quench the fluorescence intensity. The total transition probabiltiy A_i for the depopulation of level E_i is a sum of radiative and collision-induced probabilities (Fig. 2.16)

$$A_i = A_i^{\text{rad}} + A_i^{\text{coll}} \quad \text{with} \quad A_i^{\text{coll}} = N_B \sigma_i v . \qquad (3.52)$$

Inserting the relations

$$v = \sqrt{\frac{8kT}{\pi \mu}} , \qquad \mu = \frac{M_A \cdot M_B}{M_A + M_B} , \qquad p_B = N_B kT ,$$

between the mean relative velocity v, the responsible pressure p_B, and the gas temperature T into (3.52) gives the total transition probability

$$A_i = \frac{1}{\tau_{\text{sp}}} + a p_B \quad \text{with} \quad a = 2\sigma_{ik}\sqrt{\frac{2}{\pi \mu kT}} . \qquad (3.53)$$

It is evident from (3.16) that this pressure-dependent transition probability causes a corresponding pressure-dependent linewidth $\delta \omega$, which can be described by a sum of two damping terms

$$\delta \omega = \delta \omega_n + \delta \omega_{\text{col}} = \gamma_n + \gamma_{\text{col}} = \gamma_n + a p_B . \qquad (3.54)$$

The collision-induced additional line broadening $a p_B$ is therefore often called *pressure broadening*.

From the derivation in Sect. 3.1, one obtains a Lorentzian profile (3.9) with a halfwidth $\gamma = \gamma_n + \gamma_{\text{col}}$ for the line broadened by inelastic collisions:

$$I(\omega) = \frac{C}{(\omega - \omega_0)^2 + [(\gamma_n + \gamma_{\text{col}})/2]^2} . \qquad (3.55)$$

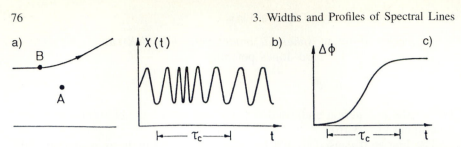

Fig. 3.11a–c. Phase perturbation of an oscillator by collisions: (**a**) classical path approximation of colliding particles; (**b**) frequency change of the oscillator $A(t)$ during the collision; (**c**) resulting phase shift

The *elastic collisions* do not change the amplitude, but the *phase* of the damped oscillator is changed due to the frequency shift $\Delta\omega(R)$ during the collisions. They are often termed *phase-perturbing collisions* (Fig. 3.11).

When taking into account line shifts $\Delta\omega$ caused by elastic collisions, the line profile for cases where it still can be described by a Lorentzian becomes

$$I(\omega) = \frac{C^*}{(\omega - \omega_0 - \Delta\omega)^2 + (\gamma/2)^2} \,, \tag{3.56}$$

where the line shift $\Delta\omega = N_B \cdot \bar{v} \cdot \sigma_s$ and the line broadening $\gamma = \gamma_n + N_B \cdot \bar{v}$ $\cdot \sigma_b$ are determined by the number density N_B of collision parameters B and by the collision cross sections σ_s for line shifts and σ_b for broadening (Fig. 3.12). The constant $C^* = (I_0/2\pi)(\gamma + N_B\bar{v}\sigma_b)$ becomes $I_0\gamma/2\pi$ for $N_B = 0$, when (3.56) becomes identical to (3.10).

Note: The real collision-induced line profile depends on the interaction potential between A and B. In most cases it is no longer Lorentzian, but has an asymmetric profile because the transition probability depends on the internuclear distance and because the energy difference $\Delta E(R) = E_i(R) - E_k(R)$ is generally not a uniformly rising or falling function but may have extrema.

Figure 3.13 depicts as examples pressure broadening and shifts in [MHz/torr] of the lithium resonance line perturbed by different noble gas atoms. Table 3.1 compiles pressure-broadening and line shift data for different alkali resonance lines.

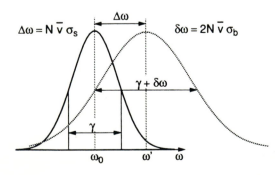

Fig. 3.12. Shift and broadening of a Lorentzian line profile by collisions

Table 3.1. Broadening and shift of atomic alkali resonance lines by noble gases and N_2. (All numbers are given in units of 10^{-20} cm^{-1} cm$^{-3} \approx$ 10 MHz/torr at $T = 300$ K)

Transition	λ (nm)	Self-broadening	Helium width	Shift	Neon width	Shift	Argon width	Shift	Krypton width	Shift	Xenon width	Shift	Nitrogen width	Shift
Li $2S-2P$	670.8		1.0	−0.08	0.6	−0.2	1.2	−0.7	1.46	−0.8	1.66	−1.0		
Na $3S_{1/2}-3P_{1/2}$	598.6		0.9	0.00	0.6	−0.3	1.4	−0.75	1.4	−0.6	1.5	−0.6		
$-P_{3/2}$	598.0		1.1	0.06	0.6	−0.4	1.2	−0.7	1.3	−0.7	1.5	−0.6		
K $4S_{1/2}-4P_{1/2}$	769.9	3×10^3	0.8	0.24	0.4	−0.2	1.3	−1.2	1.2	−0.9	1.5	−1.0	1.3	−1.0
$-4P_{3/2}$	766.5		1.1	0.13	0.6	−0.3	1.1	−0.8	1.2	−0.6	1.5	−1.0	1.3	−0.7
$-5P_{1/2}$	404.7	2×10^3	1.9	0.7	0.8	0	3.6	−2.0	3.3	−2.0	3.3			
Rb $5S_{1/2}-5P_{1/2}$	794.7	1×10^4	1.0		0.5	−0.04	1.0	−0.8	1.0	−0.8	1.2	−0.8		
$-6P_{1/2}$	421.6	8×10^3	5.0				3.8							
$-10P_{1/2}$	315.5	4×10^3	5.0	7.0			12.0	−9.5			30	−6		
Cs $6S_{1/2}-6P_{1/2}$	894.3		1.0	0.67	0.5	−0.29	1.0	−0.9	1.0	−0.27	1.1	−0.8	1.5	−0.7
$-7P_{1/2}$	459.0		4.2	1.5	4.1		4.3			−1.5	3.1	−1.7		

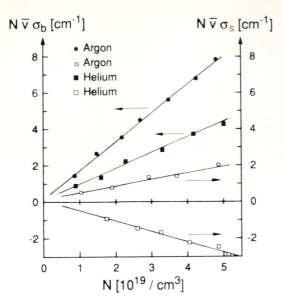

Fig. 3.13. Pressure broadening (*left scale*) and shifts (*right scale*) of the lithium resonance line by different noble gases

3.3.2 Relations Between Interaction Potential, Line Broadening, and Shifts

In order to gain more insight into the physical meaning of the cross sections σ_s and σ_b, we have to discover the relation between the phase shift $\eta(R)$ and the potential $V(R)$. Assume potentials of the form

$$V_i(R) = C_i/R^n , \qquad V_k(R) = C_k/R^n , \tag{3.57}$$

between the atom in level E_i or E_k and the perturbing atom B. The frequency shift $\Delta\omega$ for the transition $E_i \rightarrow E_k$ is then

$$\hbar\Delta\omega(R) = \frac{C_i - C_k}{R^n} . \tag{3.58}$$

The corresponding phase shift for a collision with impact parameter R_0, where we neglect the scattering of B and assume that the path of B is not deflected

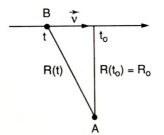

Fig. 3.14. Linear path approximation of a collision between A and B

but follows a straight line (Fig. 3.14), is

$$\Delta\phi(R_0) = \int\limits_{-\infty}^{+\infty} \Delta\omega\,dt = \frac{1}{\hbar}\int \frac{(C_i - C_k)\,dt}{[R_0^2 + \bar{v}^2(t-t_0)^2]^{n/2}} = \frac{\alpha_n(C_i - C_k)}{v R_0^{n-1}} . \quad (3.59)$$

Equation (3.59) provides the relation between the phase shift $\Delta\phi(R_0)$ and the difference (3.58) of the interaction potentials, where α_n is a numerical constant depending on the exponent n in (3.58).

The phase shifts may be positive ($C_i > C_k$) or negative depending on the relative orientation of spin and angular momenta. This is illustrated by Fig. 3.15, which shows the phase shifts of the Na atom, oscillating on the 3s–3p transition for Na–H collisions at large impact parameters [3.12].

It turns out that the main contribution to σ_b comes from collisions with *small* impact parameters, whereas σ_s still has large values for *large* impact parameters. This means that elastic *collisions at large distances do not cause noticeable broadening of the line, but can still very effectively shift the line center* [3.17]. Figure 3.16 exhibits broadening and shift of the Cs resonance line by argon atoms.

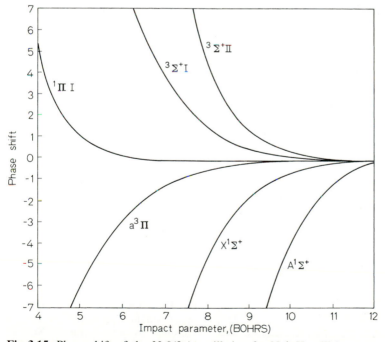

Fig. 3.15. Phase shift of the Na*(3p) oscillation for Na*–H collisions versus impact parameter. The various adiabatic molecular states for Na*H are indicated [3.12]

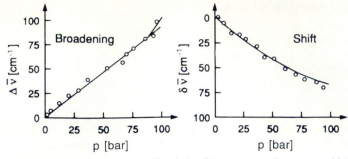

Fig. 3.16. Broadening and shift of the Cs resonance line at $\lambda = 894.3$ nm by argon

Nonmonotonic interaction potentials $V(R)$, such as the Lennard–Jones potential (3.51), cause satellites in the wings of the broadened profiles (Fig. 3.17) From the satellite structure the interaction potential may be deduced [3.18].

Because of the long-range Coulomb interactions between charged particles (electrons and ions) described by the potential (3.57) with $n = 1$, pressure broadening and shift is particularly large in plasmas and gas discharges [3.19, 3.20]. This is of interest for gas discharge lasers, such as

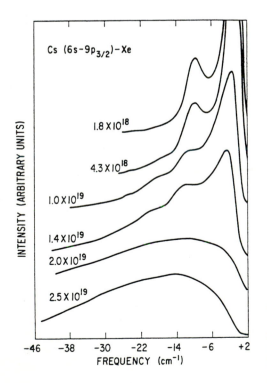

Fig. 3.17. Satellites in the pressure-broadened line profile of the cesium transition $6s \to 9p_{3/2}$ for Cs–Xe collisions at different xenon densities [atoms/cm^3] [3.14]

the HeNe laser or the argon-ion laser [3.21, 3.22]. The interaction between charged particles can be described by the linear and quadratic Stark effects. It can be shown that the linear Stark effect causes only line broadening, while the quadratic effect also leads to line shifts. From measurements of line profiles in plasmas, very detailed plasma characteristics, such as electron or ion densities and temperatures, can be determined. Plasma spectroscopy has therefore become an extensive field of research [3.23], of interest not only for astrophysics, but also for fusion research in high-temperature plasmas [3.24]. Lasers play an important role in accurate measurements of line profiles in plasmas [3.25–3.28].

The classical models used to explain collisional broadening and line shifts can be improved by using quantum mechanical calculations. These are, however, beyond the scope of this book, and the reader is referred to the literature [3.1, 3.14, 3.21–3.33].

Example 3.4

(a) The pressure broadening of the sodium D line $\lambda = 589$ nm by argon is 2.3×10^{-5} nm/mbar, equivalent to 0.228 MHz/Pa. The shift is about -1 MHz/torr. The self-broadening of 150 MHz/torr due to collisions between Na atoms is much larger. However, at pressures of several torr, the pressure broadening is still smaller than the Doppler width.

(b) The pressure broadening of molecular vibration–rotation transitions with wavelengths $\lambda \simeq 5\,\mu$m is a few MHz/torr. At atmospheric pressure, the collisional broadening therefore exceeds the Doppler width. For example, the rotational lines of the ν_2 band of H_2O in air at normal pressure (760 torr) have a Doppler width of 150 MHz, but a pressure-broadened linewidth of 930 MHz.

(c) The collisional broadening of the red neon line at $\lambda = 633$ nm in the low-pressure discharge of a HeNe laser is about $\delta\nu = 150$ MHz/torr; the pressure shift $\Delta\nu = 20$ MHz/torr. In high-current discharges, such as the argon laser discharge, the degree of ionization is much higher than in the HeNe laser and the Coulomb interaction between ions and electrons plays a major role. The pressure broadening is therefore much larger: $\delta\nu = 1500$ MHz/torr. Because of the high temperature in the plasma, the Doppler width $\delta\nu_D \simeq 5000$ MHz is even larger [3.22].

3.3.3 Collisional Narrowing of Lines

In the infrared and microwave ranges, collisions may sometimes cause a narrowing of the linewidth instead of a broadening (*Dicke narrowing*) [3.34]. This can be explained as follows: if the lifetime of the upper molecular level (e.g., an excited vibrational level in the electronic ground state) is long compared to the mean time between successive collisions, the velocity of the oscillator is often altered by elastic collisions and the mean velocity compo-

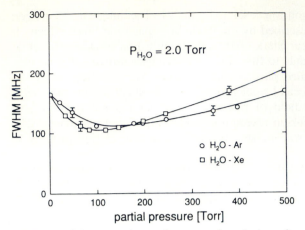

Fig. 3.18. Dicke narrowing and pressure broadening of a rotational transition in H_2O at $1871\,cm^{-1}$ ($\lambda = 5.3\,\mu m$) as a function of Ar and Xe pressure [3.35]

nent is smaller than without these collisions, resulting in a smaller Doppler shift. When the Doppler width is larger than the pressure-broadened width, this effect causes a narrowing of the lines, if the mean-free path is smaller than the wavelength of the molecular transition [3.35]. Figure 3.18 illustrates this Dicke narrowing for a rotational transition of the H_2O molecule at $\lambda = 5.34\,\mu m$. The linewidth decreases with increasing pressure up to pressures of about $100-150$ torr, depending on the collision partner, which determines the mean-free path Λ. For higher pressures, the pressure broadening overcompensates the Dicke narrowing, and the linewidth increases again.

There is a second effect that causes a collisional narrowing of spectral lines. In the case of very long lifetimes of levels connected by an EM transition, the linewidth is determined by the diffusion time of the atoms out of the laser beam (Sect. 3.4). Inserting a noble gas into the sample cell decreases the diffusion rate and therefore increases the interaction time of the sample atoms with the laser field, which results in a decrease of the linewidth with pressure [3.36] until the pressure broadening overcompensates the narrowing effect.

3.4 Transit-Time Broadening

In many experiments in laser spectroscopy, the interaction time of molecules with the radiation field is small compared with the spontaneous lifetimes of excited levels. Particularly for transitions between rotational–vibrational levels of molecules with spontaneous lifetimes in the millisecond range, the transit time $T = d/|v|$ of molecules with a mean thermal velocity v passing through a laser beam of diameter d may be smaller than the spontaneous lifetime by several orders of magnitude.

Example 3.5

(a) Molecules in a molecular beam with thermal velocities $|v| = 5 \times 10^4$ cm/s passing through a laser beam of 0.1-cm diameter have the mean transit time $T = 2\,\mu s$.
(b) For a beam of fast ions with velocities $\bar{v} = 3 \times 10^8$ cm/s, the time required to traverse a laser beam with $d = 0.1$ cm is already below 10^{-9} s, which is shorter than the spontaneous lifetimes of most atomic levels.

In such cases, the linewidth of a Doppler-free molecular transition is no longer limited by the spontaneous transition probabilities (Sect. 3.1), but by the time of flight through the laser beam, which determines the interaction time of the molecule with the radiation field. This can be seen as follows: consider an undamped oscillator $x = x_0 \cos \omega_0 t$ that oscillates with constant amplitude during the time interval T and then suddenly stops oscillating. Its frequency spectrum is obtained from the Fourier transform

$$A(\omega) = \frac{1}{\sqrt{2\pi}} \int_0^T x_0 \cos(\omega_0 t)\, e^{-i\omega t}\, dt \ . \tag{3.60}$$

The spectral intensity profile $I(\omega) = A^*A$ is, for $(\omega - \omega_0) \ll \omega_0$,

$$I(\omega) = C\frac{\sin^2[(\omega - \omega_0)T/2]}{(\omega - \omega_0)^2} \ , \tag{3.61}$$

according to the discussion in Sect. 3.1. This is a function with a full halfwidth $\delta\omega_T = 5.6/T$ of its central maximum (Fig. 3.19a) and a full width $\delta\omega_b = 4\pi/T \simeq 12.6/T$ between the zero points on both sides of the central maximum.

This example can be applied to an atom that traverses a laser beam with a rectangular intensity profile (Fig. 3.19a). The oscillator amplitude $x(t)$ is proportional to the field amplitude $E = E_0(r) \cos \omega t$. If the interaction time $T = d/v$ is small compared to the damping time $T = 1/\gamma$, the oscillation amplitude can be regarded as constant during the time T. The full halfwidth of the absorption line is then $\delta\omega = 5.6v/d \rightarrow \delta v \simeq v/d$.

In reality, the field distribution across a laser beam that oscillates in the fundamental mode is given by (Sect. 5.3)

$$E = E_0 e^{-r^2/w^2} \cos \omega t \ ,$$

in which $2w$ gives the diameter of the Gaussian beam profile across the points where $E = E_0/e$. Substituting the forced oscillator amplitude $x = \alpha E$ into (3.60), one obtains instead of (3.61) a Gaussian line profile (Fig. 3.19b)

$$I(\omega) = I_0 \exp\left(-(\omega - \omega_0)^2 \frac{w^2}{2v^2}\right) \ , \tag{3.62}$$

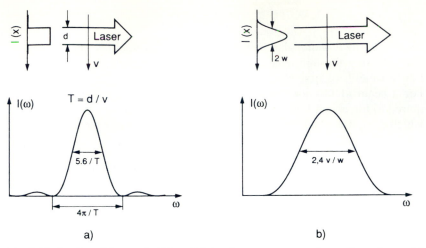

a)
b)

Fig. 3.19a,b. Transition probability $\mathcal{P}(\omega)$ of an atom traversing a laser beam (**a**) with a rectangular intensity profile $I(x)$; and (**b**) with a Gaussian intensity profile for the case $\gamma < 1/T = v/d$. The intensity profile $I(\omega)$ of an absorption line is proportional to $\mathcal{P}(\omega)$

with a transit-time limited halfwidth (FWHM)

$$\delta\omega_{tt} = 2(v/w)\sqrt{2\ln(2)} \simeq 2.4v/w \rightarrow \delta v \simeq 0.4v/w \,. \tag{3.63}$$

There are two possible ways of reducing the transit-time broadening: one may either enlarge the laser beam diameter $2w$, or one may decrease the molecular velocity v. Both methods have been verified experimentally and will be discussed in Sects. 7.3 and 14.2. The most efficient way is to directly reduce the atomic velocity by optical cooling (Chap. 14).

Example 3.6

(a) A beam of NO_2 molecules with $\bar{v} = 600\,m/s$ passes through a focused laser beam with $w = 0.1\,mm$. Their transit time broadening $\delta v \simeq 1.2\,MHz$ is large compared to their natural linewidth $\delta v_n \simeq 10\,kHz$ of optical transitions.

(b) For frequency standards the rotational–vibrational transition of CH_4 at $\lambda = 3.39\,\mu m$ is used (Sect. 7.3). In order to reduce the transit-time broadening for CH_4 molecules with $\bar{v} = 7 \times 10^4\,cm/s$ below their natural linewidth of $\delta v = 10\,kHz$, the laser-beam diameter must be enlarged to $2w \geq 6\,cm$.

So far, we have assumed that the wave fronts of the laser radiation field are planes and that the molecules move parallel to these planes. However, the phase surfaces of a focused Gaussian beam are curved except at the focus. As Fig. 3.20 illustrates, the spatial phase shift $\Delta\phi = x2\pi/\lambda$ experienced by an atom moving along the r-direction perpendicular to the laser beam z-axis

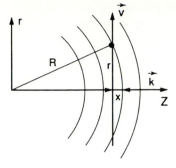

Fig. 3.20. Line broadening caused by the curvature of wave fronts

with $r^2 = R^2 - (R-x)^2 \rightarrow x \simeq r^2/2R$ for $x \ll R$ is

$$\Delta\phi = kr^2/2R = \omega r^2/(2cR) , \tag{3.64}$$

where $k = \omega/c$ is the magnitude of the wave vector, and R is the radius of curvature of the wave front. This phase shift depends on the location of an atom and is therefore different for the different atoms, and causes additional line broadening (Sect. 3.3.1). The calculation [3.37] yields for the transit-time broadened halfwidth, including the wave-front curvature,

$$\delta\omega = \frac{2v}{w}\sqrt{2\ln 2}\left[1+\left(\frac{\pi w^2}{R\lambda}\right)^2\right]$$

$$= \delta\omega_{tt}\left[1+\left(\frac{\pi w^2}{R\lambda}\right)^2\right] \approx \delta\omega_{tt}(1+\Delta\phi^2)^{1/2} . \tag{3.65}$$

In order to minimize this additional broadening, the radius of curvature has to be made as large as possible. If $\Delta\phi \ll \pi$ for a distance $r = w$, the broadening by the wave-front curvature is small compared to the transit-time broadening. This imposes the condition $R \gg w^2/\lambda$ on the radius of curvature.

Example 3.7

With $w = 1\,\text{cm}$, $\lambda = 1\,\mu\text{m} \rightarrow \omega = 2 \times 10^{15}\,\text{Hz}$. This gives, according to (3.64), a maximum phase shift $\Delta\phi = 2 \times 10^{15} / (6 \times 10^{10} R\ [\text{cm}])$. In order to keep $\Delta\phi \ll 2\pi$, the radius of curvature should be $R \gg 5 \times 10^3\,\text{cm}$. For $R = 5 \times 10^3\,\text{cm} \rightarrow \Delta\phi = 2\pi$ and the phase-front curvature causes an additional broadening by a factor of about 6.5.

3.5 Homogeneous and Inhomogeneous Line Broadening

If the probability $\mathscr{P}_{ik}(\omega)$ of absorption or emission of radiation with frequency ω causing a transition $E_i \rightarrow E_k$ is equal for all the molecules of a sample that are in the same level E_i, we call the spectral line profile of this transition *homogeneously broadened*. Natural line broadening is an example that yields a homogeneous line profile. In this case, the probability for emission of light with frequency ω on a transition $E_i \to E_k$ with the normalized Lorentzian profile $L(\omega - \omega_0)$ and central frequency ω_0 is given by

$$\mathcal{P}_{ik}(\omega) = A_{ik} L(\omega - \omega_0) \ .$$

It is equal for all atoms in level E_i.

The standard example of inhomogeneous line broadening is Doppler broadening. In this case, the probability of absorption or emission of monochromatic radiation $E(\omega)$ is not equal for all molecules, but depends on their velocity $\bar{v}$ (Sect. 3.2). We divide the molecules in level E_i into subgroups such that all molecules with a velocity component within the interval v_z to $v_z + \Delta v_z$ belong to one subgroup. If we choose Δv_z to be $\delta \omega_n / k$ where $\delta \omega_n$ is the natural linewidth, we may consider the frequency interval $\delta \omega_n$ to be homogeneously broadened inside the much larger inhomogeneous Doppler width. That is to say, all molecules in the subgroup can absorb or emit radiation with wave vector $\boldsymbol{k}$ and frequency $\omega = \omega_0 + v_z |\boldsymbol{k}|$ (Fig. 3.8), because in the coordinate system of the moving molecules, this frequency is within the natural width $\delta \omega_n$ around ω_0 (Sect. 3.2).

In Sect. 3.3 we saw that the spectral line profile is altered by two kinds of collisions. Inelastic collisions cause additional damping, resulting in pure broadening of the Lorentzian line profile. This broadening by inelastic collisions brings about a homogeneous Lorentzian line profile. The elastic collisions could be described as phase-perturbing collisions. The Fourier transform of the oscillation trains with random phase jumps again yields a Lorentzian line profile, as derived in Sect. 3.3. Summarizing, we can state that elastic and inelastic collisions that only perturb the phase or amplitude of an oscillating atom without changing its velocity cause homogeneous line broadening.

So far, we have neglected the fact that collisions also change the velocity of both collision partners. If the velocity component v_z of a molecule is altered by an amount u_z during the collision, the molecule is transferred from one subgroup $(v_z \pm \Delta v_z)$ within the Doppler profile to another subgroup $(v_z + u_z \pm \Delta v_z)$. This causes a shift of its absorption or emission frequency from ω to $\omega + k u_z$ (Fig. 3.21). This shift should not be confused with the line shift caused by phase-perturbing elastic collisions that also occurs when the velocity of the oscillator does not noticeably change.

At thermal equilibrium, the changes u_z of v_z by velocity-changing collisions are randomly distributed. Therefore, the whole Doppler profile will, in general, not be affected and the effect of these collisions is canceled out in Doppler-limited spectroscopy. In Doppler-free laser spectroscopy, however, the velocity-changing collisions may play a non-negligible role. They cause effects that depend on the ratio of the mean time $T = \Lambda / \bar{v}$ between collisions to the interaction time τ_c with the radiation field. For $T > \tau_c$, the redistribution of molecules by velocity-changing collisions causes only a small change of the population densities $n_i(v_z) \, dv_z$ within the different subgroups, without noticeably changing the homogeneous width of this subgroup. If $T \ll \tau_c$, the

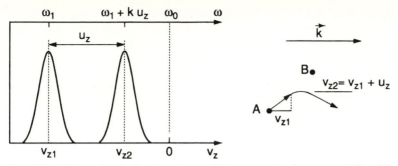

Fig. 3.21. Effect of velocity-changing collisions on the frequency shift of homogeneous subgroups within a Doppler-broadened line profile

different subgroups are uniformly mixed. This results in a broadening of the homogeneous linewidth associated with each subgroup. The effective interaction time of the molecules with a monochromatic laser field is shortened because the velocity-changing collisions move a molecule out of resonance with the field. The resultant change of the line shape can be monitored using saturation spectroscopy (Sect. 7.3).

Under certain conditions, if the mean free path Λ of the molecules is smaller than the wavelength of the radiation field, velocity-changing collisions may also result in a narrowing of a Doppler-broadening line profile (Dicke narrowing, Sect. 3.3.3).

3.6 Saturation and Power Broadening

At sufficiently large laser intensities, the optical pumping rate on an absorbing transition becomes larger than the relaxation rates. This results in a noticeable decrease of the population in the absorbing levels. This saturation of the population densities also causes additional line broadening. The spectral line profiles of such partially saturated transitions are different for homogeneously and for inhomogeneously broadened lines [3.38]. Here we treat the homogeneous case, while the saturation of inhomogeneous line profiles is discussed in Chap. 7.

3.6.1 Saturation of Level Population by Optical Pumping

The effect of optical pumping on the saturation of population densities is illustrated by a two-level system with population densities N_1 and N_2. The two levels are coupled to each other by absorption or emission and by relaxation processes, but have no transitions to other levels (Fig. 3.22). Such a "true" two-level system is realized by many atomic resonance transitions without hyperfine structure.

With the probability $P = B_{12}\rho(\omega)$ for a transition $|1\rangle \rightarrow |2\rangle$ by absorption of photons $\hbar\omega$ and the relaxation probability R_i for level $|i\rangle$, the rate equation

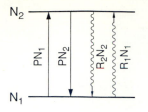

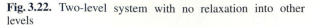

Fig. 3.22. Two-level system with no relaxation into other levels

for the level population is

$$\frac{dN_1}{dt} = -\frac{dN_2}{dt} = -PN_1 - R_1 N_1 + PN_2 + R_2 N_2 , \qquad (3.66)$$

where we have assumed nondegenerate levels with statistical weight factors $g_1 = g_2 = 1$. Under stationary conditions $(dN_i/dt = 0)$ we obtain with $N_1 + N_2 = N$ from (3.66):

$$(P + R_1)N_1 = (P + R_2)(N - N_1) \Rightarrow N_1 = N\frac{P + R_2}{2P + R_1 + R_2} . \qquad (3.67)$$

When the pump rate P becomes much larger than the relaxation rates R_i, the population N_1 approaches $N/2$, i.e., $N_1 = N_2$. This means that the absorption coefficient $\alpha = \sigma(N_1 - N_2)$ goes to zero (Fig. 3.23). The medium becomes completely transparent.

Without a radiation field $(P = 0)$, the population densities at thermal equilibrium according to (3.67) are

$$N_{10} = \frac{R_2}{R_1 + R_2}N ; \quad N_{20} = \frac{R_2}{R_1 + R_2}N . \qquad (3.68)$$

With the abbreviations

$$\Delta N = N_1 - N_2 \quad \text{and} \quad \Delta N_0 = N_{10} - N_{20}$$

we obtain from (3.67) and (3.68)

$$\Delta N = \frac{\Delta N_0}{1 + 2P/(R_1 + R_2)} = \frac{\Delta N_0}{1 + S} . \qquad (3.69)$$

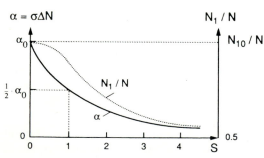

Fig. 3.23. Saturation of population density N_1 and absorption coefficient $\alpha = \sigma(N_1 - N_2)$ as functions of the saturation parameter S (see text)

The *saturation parameter*

$$S = 2P/(R_1 + R_2) = P/\overline{R} = B_{12}\rho(\omega)/\overline{R} \tag{3.70}$$

represents the ratio of pumping rate P to the average relaxation rate $\overline{R} = (R_1 + R_2)/2$. If the spontaneous emission of the upper level $|2\rangle$ is the only relaxation mechanism, we have $R_1 = 0$ and $R_2 = A_{21}$. Since the pump rate due to a monochromatic wave with intensity $I(\omega)$ is $P = \sigma_{12}(\omega)I(\omega)/\hbar\omega$, we obtain for the saturation parameter

$$S = \frac{2\sigma_{12}I(\omega)}{\hbar\omega A_{12}} . \tag{3.71}$$

The saturated absorption coefficient $\alpha(\omega) = \sigma_{12}\Delta N$ is, according to (3.69),

$$\boxed{\alpha = \frac{\alpha_0}{1 + S} ,} \tag{3.72}$$

where α_0 is the unsaturated absorption coefficient without pumping.

3.6.2 Saturation Broadening of Homogeneous Line Profiles

According to (2.15) and (3.69), the power absorbed per unit volume on the transition $|1\rangle \rightarrow |2\rangle$ by atoms with the population densities N_1, N_2 in a radiation field with a broad spectral profile and spectral energy density ρ is

$$\frac{dW_{12}}{dt} = \hbar\omega B_{12}\rho(\omega)\Delta N = \hbar\omega B_{12}\rho(\omega)\frac{\Delta N_0}{1 + S} . \tag{3.73}$$

With $S = B_{12}\rho(\omega)/\overline{R}$, see (3.70), this can be written as

$$\frac{dW_{12}}{dt} = \hbar\omega\overline{R}\frac{\Delta N_0}{1 + S^{-1}} . \tag{3.74}$$

Since the absorption profile $\alpha(\omega)$ of a homogeneously broadened line is Lorentzian, see (3.36b), the induced absorption probability of a monochromatic wave with frequency ω follows a Lorentzian line profile $B_{12}\rho(\omega) \cdot L(\omega - \omega_0)$. We can therefore introduce a frequency-dependent spectral saturation parameter S_ω for the transition $E_1 \rightarrow E_2$,

$$S_\omega = \frac{B_{12}\rho(\omega)}{\langle R\rangle}L(\omega - \omega_0) . \tag{3.75}$$

We can assume that the mean relaxation rate $\langle R\rangle$ is independent of ω within the frequency range of the line profile. With the definition (3.36b) of the Lorentzian profile $L(\omega - \omega_0)$, we obtain for the spectral saturation parameter S_ω

$$S_\omega = S_0\frac{(\gamma/2)^2}{(\omega - \omega_0)^2 + (\gamma/2)^2} \quad \text{with} \quad S_0 = S_\omega(\omega_0) . \tag{3.76}$$

Substituting (3.76) into (3.74) yields the frequency dependence of the absorbed radiation power per unit frequency interval $d\omega = 1\,s^{-1}$

$$\frac{d}{dt}W_{12}(\omega) = \frac{\hbar\omega R\Delta N_0 S_0(\gamma/2)^2}{(\omega-\omega_0)^2+(\gamma/2)^2(1+S_0)} = \frac{C}{(\omega-\omega_0)^2+(\gamma_s/2)^2} . \qquad (3.77)$$

This a Lorentzian profile with the increased halfwidth

$$\gamma_s = \gamma\sqrt{1+S_0} . \qquad (3.78)$$

The halfwidth $\gamma_s = \delta\omega_s$ of the saturation-broadened line increases with the saturation parameter S_0 at the line center ω_0. If the induced transition rate at ω_0 equals the total relaxation rate R, the saturation parameter $S_0 = [B_{12}\rho(\omega_0)]/R$ becomes $S_0 = 1$, which increases the linewidth by a factor $\sqrt{2}$, compared to the unsaturated linewidth $\delta\omega_0$ for weak radiation fields ($\rho \to 0$).

Since the power dW_{12}/dt absorbed per unit volume equals the intensity decrease per centimeter, $dI = -\alpha_s I$, of an incident wave with intensity I, we can derive the absorption coefficient α from (3.77). With $I = c\rho$ and $S_\omega = [B_{12}\rho(\omega)]/R$ we obtain

$$\alpha_s(\omega) = \alpha_0(\omega_0)\frac{(\gamma/2)^2}{(\omega-\omega_0)^2+(\gamma_s/2)^2} = \frac{\alpha_0(\omega)}{1+S_\omega} , \qquad (3.79)$$

where the unsaturated absorption profile is

$$\alpha_0(\omega) = \frac{2\hbar\omega B_{12}\Delta N_0}{\pi c\gamma}\frac{(\gamma/2)^2}{(\omega-\omega_0)^2+(\gamma/2)^2} . \qquad (3.80)$$

This shows that the saturation decreases the absorption coefficient $\alpha(\omega)$ by the factor $(1+S_\omega)$. At the line center, this factor has its maximum value $(1+S_0)$, while it decreases for increasing $(\omega-\omega_0)$ to 1, see (3.76). The saturation is therefore strongest at the line center, and approaches zero for $(\omega-\omega_0) \to \infty$ (Fig. 3.24). This is the reason why the line broadens. For a more detailed discussion of saturation broadening, see Chap. 7 and [3.37–3.39].

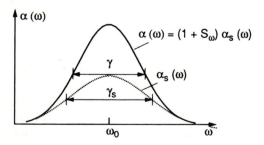

α (ω)

$\alpha(\omega) = (1 + S_\omega)\,\alpha_s(\omega)$

γ

$\alpha_s(\omega)$

γ_s

ω_0 ω

Fig. 3.24. Saturation broadening of a homogeneous line profile

3.6.3 Power Broadening

The broadening of homogeneous line profiles by intense laser fields can also be regarded from another viewpoint compared to Sect. 3.6.2. When a two-level system is exposed to a radiation field $E = E_0 \cos \omega t$, the population probability of the upper level $|b\rangle$ is, according to (2.89),

$$|b(\omega, t)|^2 = \frac{D_{ab}^2 E_0^2}{\hbar^2 (\omega_{ab} - \omega)^2 + D_{ab}^2 E_0^2}$$
$$\times \sin^2\left[\frac{1}{2}\sqrt{(\omega_{ab} - \omega)^2 + (D_{ab}E_0/\hbar)^2}\, t\right] , \tag{3.81}$$

an oscillatory function of time, which oscillates at exact resonance $\omega = \omega_{ab}$ with the Rabi flopping frequency $\Omega_R = D_{ab}E_0/\hbar$.

If the upper level $|b\rangle$ can decay by spontaneous processes with a relaxation constant γ, its mean population probability is

$$\mathcal{P}_b(\omega) = \overline{|b(\omega, t)|^2} = \int_0^\infty \gamma e^{-\gamma t}\, |b(\omega_1, t)|^2 \, dt . \tag{3.82}$$

Inserting (3.81) and integrating yields

$$\mathcal{P}_b(\omega) = \frac{1}{2} \frac{D_{ab}^2 E_0^2/\hbar^2}{(\omega_{ab} - \omega)^2 + \gamma^2(1 + S)} , \tag{3.83}$$

with $S = D_{ab}^2 E_0^2/(\hbar^2 \gamma^2)$. Since $\mathcal{P}_2(\omega)$ is proportional to the absorption line profile, we obtain as in (3.77) a power-broadened Lorentzian line profile with the linewidth

$$\gamma_S = \gamma \sqrt{1 + S} .$$

Since the induced absorption rate within the spectral interval γ is, according to (2.41) and (2.77)

$$B_{12}\rho\gamma = B_{12}I\gamma/c \simeq D_{12}^2 E_0^2/\hbar^2 , \tag{3.84}$$

the quantity S in (3.83) turns out to be identical with the saturation parameter S in (3.70).

If both levels $|a\rangle$ and $|b\rangle$ decay with the relaxation constants γ_a and γ_b, respectively, the line profile of the homogeneously broadened transition $|a\rangle \to |b\rangle$ is again described by (3.83), where now (Sect. 7.1 and [3.39])

$$\gamma = \frac{1}{2}(\gamma_a + \gamma_b) \quad \text{and} \quad S = D_{ab}^2 E_0^2/(\hbar^2 \gamma_a \gamma_b) . \tag{3.85}$$

If a strong pump wave is tuned to the center $\omega_0 = \omega_{ab}$ of the transition and the absorption profile is probed by a tunable weak probe wave, the absorption profile looks different: due to the population modulation with the

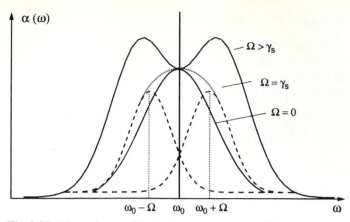

Fig. 3.25. Absorption profile of a homogeneous transition pumped by a strong pump wave kept at ω_0 and probed by a weak tunable probe wave for different values of the ratio Ω/γ_s of the Rabi frequency Ω to the linewidth γ_s

Rabi flopping frequency Ω, sidebands are generated at $\omega_0 \pm \Omega$ that have the homogeneous linewidth γ_s. The superposition of these sidebands (Fig. 3.25) gives a line profile that depends on the ratio Ω/γ_s of the Rabi flopping frequency Ω and the saturated linewidth γ_s. For a sufficiently strong pump wave ($\Omega > \gamma_s$), the separation of the sidebands becomes larger than their width and a dip appears at the center ω_0.

3.7 Spectral Line Profiles in Liquids and Solids

Many different types of lasers use liquids or solids as amplifying media. Since the spectral characteristics of such lasers play a significant role in applications of laser spectroscopy, we briefly outline the spectral linewidths of optical transitions in liquids and solids. Because of the large densities compared with the gaseous state, the mean relative distances $R(A, B_j)$ between an atom or molecule A and its surrounding partners B_j are very small (typically a few tenths of a millimeter), and the interaction between A and the adjacent partners B_j is accordingly large.

In general, the atoms or molecules used for laser action are diluted to small concentrations in liquids or solids. Examples are the dye laser, where dye molecules are dissolved in organic solutions at concentrations of 10^{-4} to 10^{-3} moles/liter, or the ruby laser, where the concentration of the active Cr^{3+} ions in Al_3O_3 is on the order of 10^{-3}. The optically pumped laser molecules A^* interact with their surrounding host molecules B. The resulting broadening of the excited levels of A^* depends on the total electric field produced at the location of A by all adjacent molecules B_j, and on the dipole moment or the polarizability of A^*. The linewidth $\Delta\omega_{ik}$ of a transition $A^*(E_i) \rightarrow A^*(E_k)$ is determined by the difference in the level shifts $(\Delta E_i - \Delta E_k)$.

In liquids, the distances $R_j(A^*, B_j)$ show random fluctuations analogous to the situation in a high-pressure gas. The linewidth $\Delta \omega_{ik}$ is therefore determined by the probability distribution $P(R_j)$ of the mutal distances $R_j(A^*, B_j)$ and the correlation between the phase perturbations at A^* caused by elastic collisions during the lifetime of the levels E_i, E_k (see the analogous discussion in Sect. 3.3).

Inelastic collisions of A^* with molecules B of the liquid host may cause radiationless transitions from the level E_i populated by optical pumping to lower levels E_n. These radiationless transitions shorten the lifetime of E_i and cause collisional line broadening. In liquids the mean time between successive inelastic collisions is of the order of 10^{-11} to 10^{-13} s. Therefore the spectral line $E_i \rightarrow E_k$ is greatly broadened with a homogeneously broadened profile. When the line broadening becomes larger than the separation of the different spectral lines, a broad continuum arises. In the case of molecular spectra with their many closely spaced rotational–vibrational lines within an electronic transition, such a continuum inevitably appears since the broadening at liquid densities is always much larger than the line separation.

Examples of such continuous absorption and emission line profiles are the optical dye spectra in organic solvents, such as the spectrum of Rhodamine 6G shown in Fig. 3.26, together with a schematic level diagram [3.40]. The optically pumped level E_i is collisionally deactivated by radiationless transitions to the lowest vibrational level E_m of the excited electronic state. The fluorescence starts therefore from E_m instead of E_i and ends on various vibrational levels of the electronic ground state (Fig. 3.26a). The emission spectrum is therefore shifted to larger wavelengths compared with the absorption spectrum (Fig. 3.26b).

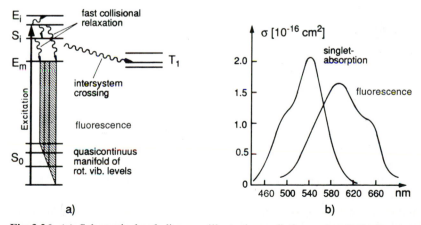

Fig. 3.26. (a) Schematic level diagram illustrating radiative and radiationless transitions. **(b)** Absorption and emission cross section of rhodamine 6G dissolved in ethanol

In crystalline solids the electric field $E(R)$ at the location R of the excited molecule A^* has a symmetry depending on that of the host lattice. Because the lattice atoms perform vibrations with amplitudes depending on the temperatur T, the electric field will vary in time and the time average $\langle E(T, t, R) \rangle$ will depend on temperature and crystal structure [3.41–3.43]. Since the oscillation period is short compared with the mean lifetime of $A^*(E_i)$, these vibrations cause homogeneous line broadening for the emission or absorption of the atom A. If all atoms are placed at completely equivalent lattice points of an ideal lattice, the total emission or absorption of all atoms on a transition $E_i \rightarrow E_k$ would be homogeneously broadened.

However, in reality it often happens that the different atoms A are placed at nonequivalent lattice points with nonequal electric fields. This is particularly true in amorphous solids or in supercooled liquids such as glass, which have no regular lattice structure. For such cases, the line centers ω_{0j} of the homogeneously broadened lines for the different atoms A_j are placed at different frequencies. *The total emission or absorption forms an inhomogeneously broadened line profile,* which is composed of homogeneous subgroups. This is completely analogous to the gaseous case of Doppler broadening, although the resultant linewidth in solids may be larger by several orders of magnitude. An example of such inhomogeneous line broadening is the emission of excited neodymium ions in glass, which is used in the Nd-glass laser. At sufficiently low temperatures, the vibrational amplitudes decrease and the linewidth becomes narrower. For $T < 4\,\text{K}$ it is possible to obtain, even in solids under favorable conditions, linewidths below $10\,\text{MHz}$ for optical transitions [3.44, 3.45].

Problems

3.1 Determine the natural linewidth, the Doppler width, pressure broadening and shifts for the neon transition $3s_2 \rightarrow 2p_4$ at $\lambda = 632.8\,\text{nm}$ in a HeNe discharge at $p_{\text{He}} = 2\,\text{mbar}$, $p_{\text{Ne}} = 0.2\,\text{mbar}$ at a gas temperature of $400\,\text{K}$. The relevant data are: $\tau(3s_2) = 58\,\text{ns}$, $\tau(2p_4) = 18\,\text{ns}$, $\sigma_B(\text{Ne} - \text{He}) \,\hat{=}\, 6 \times 10^{-14}\,\text{cm}^2$, $\sigma_S(\text{Ne} - \text{He}) \simeq 1 \times 10^{-14}\,\text{cm}^2$, $\sigma_B(\text{Ne} - \text{Ne}) = 1 \times 10^{-13}\,\text{cm}^2$, $\sigma_S(\text{Ne} - \text{Ne}) = 1 \times 10^{-14}\,\text{cm}^2$.

3.2 What is the dominant broadening mechanism for absorption lines in the following examples:

(a) The output from a CO_2 laser with $50\,\text{W}$ at $\lambda = 10\,\mu\text{m}$ is focused into an absorbing probe of SF_6 molecules. The beam waist has a 0.5-mm diameter, $T = 300\,\text{K}$, $p = 100\,\text{mbar}$, $\sigma_b = 5 \times 10^{-14}\,\text{cm}^2$, absorption cross section $\sigma_a = 10^{-12}\,\text{cm}^2$.

(b) Radiation from a star passes through an absorbing atomic hydrogen cloud with $N = 100\,\text{cm}^{-3}$, $T = 10\,\text{K}$, and a path length of $3 \times 10^9\,\text{km}$. The Einstein coefficient for the $\lambda = 21\,\text{cm}$ line is $A_{ik} = 4 \times 10^{-15}\,\text{s}^{-1}$, that for the Lyman-$\alpha$ line at $\lambda = 121.6$ nm is $A_{ik} = 1 \times 10^9$ s^{-1}. What is the attenuation of the light for the two transistions? Note that the cloud is optically thick for the Lyman-α radiation.

(c) The expanded beam from a HeNe laser at $\lambda = 3.39\,\mu$m with 10 mW power is sent through a methane cell ($T = 300$ K, $p = 0.1$ mbar, beam diameter: 1 cm). The absorbing CH$_4$ transition is from the vibrational ground state ($\tau \simeq \infty$) to an excited vibrational level with $\tau \simeq 20$ ms. Give the ratios of Doppler width to transit-time width to natural width to pressure-broadened linewidth.

(d) Calculate the minimum beam diameter that is necessary to bring about the transit-time broadening in Exercise 3.2c below the natural linewidth. Is saturation broadening important, if the absorption coefficient is $\sigma = 10^{-14}$ cm^2?

3.3 The sodium D-line at $\lambda = 589$ nm has a natural linewidth of 20 MHz.

(a) How far away from the line center do the wings of the Lorentzian line profile exceed the Doppler profile at $T = 500$ K?

(b) Calculate the intensity $I(\omega - \omega_0)$ of the Lorentzian at this frequency ω_c relative to the line center ω_0.

(c) Compare the intensities of both profiles normalized to 1 at $\omega = \omega_0$ at a distance $10(\omega_0 - \omega_c)$ from the line center.

(d) At what laser intensity is the power broadening equal to half of the Doppler width at $T = 500$ K, when the laser frequency is tuned to the line center ω_0?

3.4 Estimate the collision-broadened width of the Na D$_2$ line at $\lambda = 589\,\mu$m due to

(a) Na$-$Ar collisions at $p(\mathrm{Ar}) = 1$ mbar (Fig. 3.13);

(b) Na$-$Na collisions at $p(\mathrm{Na}) = 1$ mbar. This "resonance broadening" is caused by the interaction potential $V(r) \propto r^{-3}$ and can be calculated as $\gamma_{res} = N e^2 f_{ik}/(4\pi\epsilon_0 m \omega_{ik})$, where the oscillator strength f_{ik} is 0.65.

3.5 An excited atom with spontaneous lifetime τ suffers quenching collisions. Show that the line profile stays Lorentzian and doubles its linewidth if the mean time between two collisions is $\bar{t}_c = \tau$. Calculate the pressure of N$_2$ molecules at $T = 400$ K for which $\bar{t}_c = \tau$ for collisions Na$^* + N_2$ with the quenching cross section $\sigma_a = 4 \times 10^{-15}$ cm^2.

3.6 A pulsed laser with a spectral bandwidth of 200 MHz and a pulse duration of $\Delta T = 10$ ns excites K atoms at low potassium pressures in a cell with 1 bar neon as a buffer gas. Estimate the different contributions to the total linewidth. At what laser powers does the power broadening of the K resonance line at 404.7 nm exceed the pressure broadening or the Doppler width? Use the data of Table 3.1.

4. Spectroscopic Instrumentation

This chapter is devoted to a discussion of instruments and techniques that are of fundamental importance for the measurements of wavelengths and line profiles, or for the sensitive detection of radiation. The optimum selection of proper equipment or the application of a new technique is often decisive for the success of an experimental investigation. Since the development of spectroscopic instrumentation has shown great progress in recent years, it is most important for any spectroscopist to be informed about the state-of-the-art regarding sensitivity, spectral resolving power, and signal-to-noise ratios attainable with modern equipment.

At first we discuss the basic properties of *spectrographs* and *monochromators*. Although for many experiments in laser spectroscopy these instruments can be replaced by monochromatic tunable lasers (Chaps. 5–6), they are still indispensible for the solution of quite a number of problems in spectroscopy.

Probably the most important instruments in laser spectroscopy are *interferometers*, which are applicable in various modifications to numerous problems. We therefore treat these devices in somewhat more detail. Recently, new techniques of measuring laser wavelengths with high accuracy have been developed; they are mainly based on interferometric devices. Because of their relevance in laser spectroscopy they will be discussed in a separate section.

Great progress has also been achieved in the field of low-level signal detection. Apart from new photomultipliers with an extended spectral sensivity range and large quantum efficiencies, new detection instruments have been developed such as image intensifiers, infrared detectors, or optical multichannel analyzers, which could move from classified military research into the open market. For many spectroscopic applications they prove to be extremely useful.

4.1 Spectrographs and Monochromators

Spectrographs, the first instruments for measuring wavelengths, still hold their position in spectroscopic laboratories, particularly when equipped with modern accessories such as computerized microdensitometers or optical multichannel analyzers. Spectrographs are optical instruments that form images $S_2(\lambda)$ of the entrance slit S_1; the images are laterally separated for different wavelengths λ of the incident radiation (Fig. 2.10). This lateral dispersion is due to either spectral dispersion in prisms or diffraction on plane or concave reflection gratings.

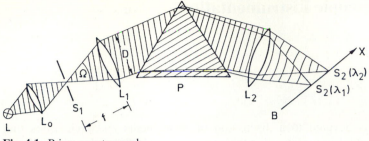

Fig. 4.1. Prism spectrograph

Figure 4.1 depicts the schematic arrangement of optical components in a *prism spectrograph*. The light source L illuminates the entrance slit S_1, which is placed in the focal plane of the collimator lens L_1. Behind L_1 the parallel light beam passes through the prism P, where it is diffracted by an angle $\theta(\lambda)$ depending on the wavelength λ. The camera lens L_2 forms an image $S_2(\lambda)$ of the entrance slit S_1. The position $x(\lambda)$ of this image in the focal plane of L_2 is a function of the wavelength λ. The *linear dispersion* $dx/d\lambda$ of the spectrograph depends on the spectral dispersion $dn/d\lambda$ of the prism material and on the focal length of L_2.

When a reflecting diffraction grating is used to separate the spectral lines $S_2(\lambda)$, the two lenses L_1 and L_2 are commonly replaced by two spherical mirrors M_1 and M_2, which image the entrance slit onto the plane of observation (Fig. 4.2). Both systems can use either photographic or photoelectric recording. According to the kind of detection, we distinguish between *spectrographs* and *monochromators*.

In spectrographs a charge-coupled device (CCD) diode array is placed in the focal plane of L_2 or M_2. The whole spectral range $\Delta\lambda = \lambda_1(x_1) - \lambda_2(x_2)$ covered by the lateral extension $\Delta x = x_1 - x_2$ of the diode array can be simultaneously recorded. The cooled CCD array can accumulate the incident radiant power over long periods (up to 20 h). CCD detection can be employed for both pulsed and cw light sources. The spectral range is limited by the

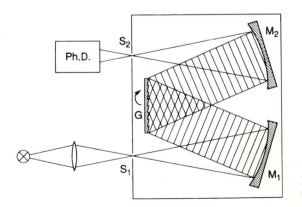

Fig. 4.2.
Grating monochromator

spectral sensitivity of available CCD materials and covers the region between about $200-1000$ nm.

Monochromators, on the other hand, use photoelectric recording of a selected small spectral interval. An exit slit S_2, selecting an interval Δx_2 in the focal plane B, lets only the limited range $\Delta\lambda$ through to the photoelectric detector. Different spectral ranges can be detected by shifting S_2 in the x-direction. A more convenient solution (which is also easier to construct) turns the prism or grating by a gear-box drive, which allows the different spectral regions to be tuned across the fixed exit slit S_2. Modern devices uses a direct drive of the grating axis by step motors and measure the turning angle by electronic angle decoders. This avoids backlash of the driving gear. Unlike the spectrograph, different spectral regions are not detected simultaneously but successively. The signal received by the detector is proportional to the product of the area $h\Delta x_2$ of the exit slit with height h with the spectral intensity $\int I(\lambda)\,d\lambda$, where the integration extends over the spectral range dispersed within the width Δx_2 of S_2.

Whereas the spectrograph allows the simultaneous measurement of a large region with moderate time resolution, photoelectric detection allows high time resolution but permits, for a given spectral resolution, only a small wavelength interval $\Delta\lambda$ to be measured at a time. With integration times below some minutes, photoelectric recording shows a higher sensitivity, while for very long detection times of several hours, photoplates may still be more convenient, although cooled CCD arrays currently allow integration times up to several hours.

In spectroscopic literature the name *spectrometer* is often used for both types of instruments. We now discuss the basic properties of spectrometers, relevant for laser spectroscopy. For a more detailed treatment see for instance [4.1–4.10].

4.1.1 Basic Properties

The selection of the optimum type of spectrometer for a particular experiment is guided by some basic characteristics of spectrometers and their relevance to the particular application. The basic properties that are important for all dispersive optical instruments may be listed as follows:

a) Speed of a Spectrometer

When the spectral intensity I_λ^* within the solid angle $d\Omega = 1$ sr is incident on the entrance slit of area A, a spectrometer with an acceptance angle Ω transmits the radiant flux within the spectral interval $d\lambda$

$$\phi_\lambda\,d\lambda = I_\lambda^*(A/A_s)T(\lambda)\Omega\,d\lambda \,, \tag{4.1}$$

where $A_s \geq A$ is the area of the source image at the entrance slit (Fig. 4.3), and $T(\lambda)$ the transmission of the spectrometer.

The product $U = A\Omega$ is often named *étendue*. For the prism spectrograph the maximum solid angle of acceptance, $\Omega = F/f_1^2$, is limited by the effec-

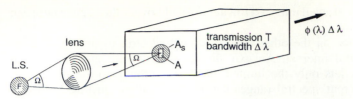

Fig. 4.3. Light-gathering power of a spectrometer

tive area $F = hD$ of the prism, which represents the limiting aperture with height h and width D for the light beam (Fig. 4.1). For the grating spectrometer the sizes of the grating and mirrors limit the acceptance solid angle Ω.

Example 4.1

For a prism with height $h = 6\,cm$, $D = 6\,cm$, $f_1 = 30\,cm \rightarrow D/f = 1:5$ and $\Omega = 0.04\,sr$. With an entrance slit of $5 \times 0.1\,mm^2$, the étendue is $U = 5 \times 10^{-3} \times 4 \times 10^{-2} = 2 \times 10^{-4}\,cm^2\,sr$.

In order to utilize the optimum speed, it is advantageous to image the light source onto the entrance slit in such a way that the acceptance angle Ω is fully used (Fig. 4.4). Although more radiant power from an extended source can pass the entrance slit by using a converging lens to reduce the source image on the entrance slit, the divergence is increased. The radiation outside the acceptance angle Ω cannot be detected, but may increase the background by scattering from lens holders and spectrometer walls.

Often the wavelength of lasers is measured with a spectrometer. In this case, it is not recommended to direct the laser beam directly onto the entrance slit, because the prism or grating would be not uniformly illuminated. This decreases the spectral resolution. Furthermore, the symmetry of the optical path with respect to the spectrometer axis is not guaranteed with such

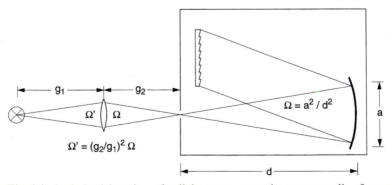

Fig. 4.4. Optimized imaging of a light source onto the entrance slit of a spectrometer is achieved when the solid angle Ω' of the incoming light matches the acceptance angle $\Omega = (a/d)^2$ of the spectrometer

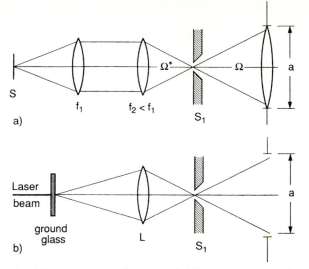

Fig. 4.5. (a) Imaging of an extended light source onto the entrance slit of a spectrometer with $\Omega^* = \Omega$. **(b)** Correct imaging optics for laser wavelength measurements with a spectrometer. The laser light, scattered by the ground glass, forms the source that is imaged onto the entrance slit

an arrangement, resulting in systematic errors of wavelengths measurements. It is better to illuminate a ground-glass plate with the laser and to use the incoherently scattered laser light as a secondary source, which is imaged in the usual way (Fig. 4.5).

b) Spectral Transmission

For prism spectrometers, the spectral transmission depends on the material of the prism and the lenses. Using fused quartz, the accessible spectral range spans from about 180 to 3000 nm. Below 180 nm (vacuum-ultraviolet region), the whole spectrograph must be evacuated, and lithium fluoride or calcium fluoride must be used for the prism and the lenses, although most VUV spectrometers are equipped with reflection gratings and mirrors.

In the infrared region, several materials (for example, CaF_2, NaCl, and KBr crystals) are transparent up to 30 μm (Fig. 4.6). However, because of the high reflectivity of metallic coated mirrors and gratings in the infrared region, grating spectrometers with mirrors are preferred over prism spectrographs.

Many vibrational–rotational transitions of molecules such as H_2O or CO_2 fall within the range 3−10 μm, causing selective absorption of the transmitted radiation. Infrared spectrometers therefore have to be either evacuated or filled with dry nitrogen. Because dispersion and absorption are closely related, prism materials with low absorption losses also show low dispersion, resulting in a limited resolving power (see below).

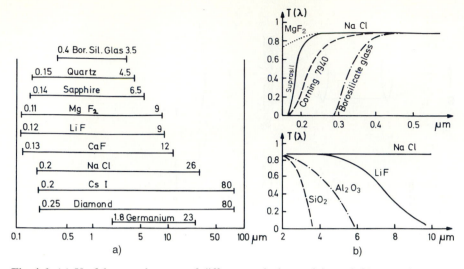

Fig. 4.6. (a) Useful spectral ranges of different optical materials; and (b) transmittance of different materials with 1-cm thicknesses [4.5b]

Since the ruling or holographic production of high-quality gratings has reached a high technological standard, most spectrometers used today are equipped with diffraction gratings rather than prisms. The spectral transmission of grating spectrometers reaches from the VUV region into the far infrared. The design and the coatings of the optical components as well as the geometry of the optical arrangement are optimized according to the specified wavelength region.

c) Spectral Resolving Power

The spectral resolving power of any dispersing instrument is defined by the expression

$$R = |\lambda/\Delta\lambda| = |v/\Delta v| \, , \tag{4.2}$$

where $\Delta\lambda = \lambda_1 - \lambda_2$ stands for the minimum separation of the central wavelengths λ_1 and λ_2 of two closely spaced lines that are considered to be just resolved. It is possible to recognize that an intensity distribution is composed of two lines with the intensity profiles $I_1(\lambda - \lambda_1)$ and $I_2(\lambda - \lambda_2)$ if the total intensity $I(\lambda) = I_1(\lambda - \lambda_1) + I_2(\lambda - \lambda_2)$ shows a pronounced dip between two maxima (Fig. 4.7). The intensity distribution $I(\lambda)$ depends, of course, on the ratio I_1/I_2 and on the profiles of both components. Therefore, the minimum resolvable interval $\Delta\lambda$ will differ for different profiles.

Lord Rayleigh introduced a criterion of resolution for diffraction-limited line profiles, where two lines are considered to be just resolved if the central diffraction maximum of the profile $I_1(\lambda - \lambda_1)$ coincides with the first minimum of $I_2(\lambda - \lambda_2)$ [4.3].

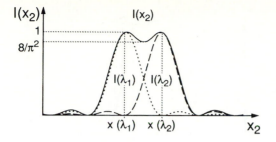

Fig. 4.7. Rayleigh's criterion for the resolution of two nearly overlapping lines

Let us consider the attainable spectral resolving power of a spectrometer. When passing the dispersing element (prism or grating), a parallel beam composed of two monochromatic waves with wavelengths λ and $\lambda + \Delta\lambda$ is split into two partial beams with the angular deviations θ and $\theta + \Delta\theta$ from their initial direction (Fig. 4.8). The angular separation is

$$\Delta\theta = (\mathrm{d}\theta/\mathrm{d}\lambda)\Delta\lambda ,\qquad(4.3)$$

where $\mathrm{d}\theta/\mathrm{d}\lambda$ is called the *angular dispersion* [rad/nm]. Since the camera lens with focal length f_2 images the entrance slit S_1 into the plane B (Fig. 4.1), the distance Δx_2 between the two images $S_2(\lambda)$ and $S_2(\lambda + \Delta\lambda)$ is, according to Fig. 4.8,

$$\Delta x_2 = f_2\Delta\theta = f_2\frac{\mathrm{d}\theta}{\mathrm{d}\lambda}\Delta\lambda = \frac{\mathrm{d}x}{\mathrm{d}\lambda}\Delta\lambda .\qquad(4.4)$$

The factor $\mathrm{d}x/\mathrm{d}\lambda$ is called the *linear dispersion* of the instrument. It is generally measured in mm/nm. In order to resolve two lines at λ and $\lambda + \Delta\lambda$, the separation Δx_2 in (4.4) has to be at least the sum $\delta x_2(\lambda) + \delta x_2(\lambda + \Delta\lambda)$ of the widths of the two slit images. Since the width δx_2 is related to the width δx_1 of the entrance slit according to geometrical optics by

$$\delta x_2 = (f_2/f_1)\delta x_1 ,\qquad(4.5)$$

the resolving power $\lambda/\Delta\lambda$ can be increased by decreasing δx_1. Unfortunately, there is a theoretical limitation set by diffraction. Because of the fundamental importance of this resolution limit, we discuss this point in more detail.

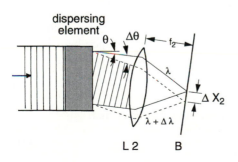

Fig. 4.8. Angular dispersion of a parallel beam

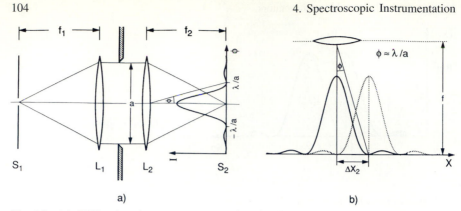

a)   b)

Fig. 4.9. (**a**) Diffraction in a spectrometer by the limiting aperture with diameter a. (**b**) Limitation of spectral resolution by diffraction

When a parallel light beam passes a limiting aperture with diameter a, a Fraunhofer diffraction pattern is produced in the plane of the focusing lens L_2 (Fig. 4.9). The intensity distribution $I(\phi)$ as a function of the angle ϕ with the optical axis of the system is given by the well-known formula [4.3]

$$I(\phi) = I_0 \left(\frac{\sin(a\pi \sin\phi/\lambda)}{(a\pi \sin\phi)/\lambda} \right)^2 \simeq I_0 \left(\frac{\sin(a\pi\phi/\lambda)}{a\pi\phi/\lambda} \right)^2 . \tag{4.6}$$

The first two diffraction minima at $\phi = \pm\lambda/a \ll \pi$ are symmetrical to the central maximum (zeroth diffraction order) at $\phi = 0$. The central maximum contains about 90% of the total intensity.

Even an infinitesimally small entrance slit therefore produces a slit image of width

$$\delta x_s^{\text{diffr}} = f_2(\lambda/a) , \tag{4.7}$$

defined as the distance between the central diffraction maximum and the first minimum, which is approximately equal to the FWHM of the central maximum.

According to the Rayleigh criterion, two equally intense spectral lines with wavelengths λ and $\lambda + \Delta\lambda$ are just resolved if the central diffraction maximum of $S_2(\lambda)$ coincides with the first minimum of $S_2(\lambda + \Delta\lambda)$ (see above). This means that their maxima are just separated by $\delta x_S^{\text{diffr}} = F_2(\lambda/a)$. From (4.6) one can compute that, in this case, both lines partly overlap with a dip of $(8/\pi^2)I_{\text{max}} \approx 0.8 I_{\text{max}}$ between the two maxima. The distance between the centers of the two slit images is then obtained from (4.7) (see Fig. 4.9b) as

$$\Delta x_2 = f_2(\lambda/a) . \tag{4.8a}$$

With (4.4) the fundamental limit on the resolving power is then

$$|\lambda/\Delta\lambda| \leq a(\mathrm{d}\theta/\mathrm{d}\lambda) , \tag{4.9}$$

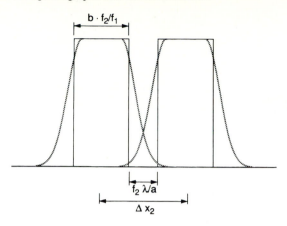

Fig. 4.10. Intensity profiles of two monochromatic lines measured in the focal plane of L_2 with an entrance slit width $b \gg f_1 \cdot \lambda/a$ and a magnification factor f_2/f_1 of the spectrograph. *Solid line*: without diffraction; *dashed line*: with diffraction. The minimum resolvable distance between the line centers is $\Delta x_2 = f_2(b/f_1 + \lambda/a)$

which clearly depends only on the size a of the limiting aperture and on the angular dispersion of the instrument.

For a finite entrance slit with width b the separation Δx_2 between the central peaks of the two images $I(\lambda - \lambda_1)$ and $I(\lambda - \lambda_2)$ must be

$$\Delta x_2 \geq f_2 \frac{\lambda}{a} + b \frac{f_2}{f_1} , \tag{4.8b}$$

in order to meet the Rayleigh criterion (Fig. 4.10). With $\Delta x_2 = f_2(\mathrm{d}\theta/\mathrm{d}\lambda)\Delta\lambda$, the smallest resolvable wavelength interval $\Delta\lambda$ is then

$$\Delta\lambda \geq \left(\frac{\lambda}{a} + \frac{b}{f_1} \right) \left(\frac{\mathrm{d}\theta}{\mathrm{d}\lambda} \right)^{-1} . \tag{4.10}$$

Note: The spectral resolution is limited, *not* by the diffraction due to the entrance slit, but by the diffraction caused by the much larger aperture a, determined by the size of the prism or grating.

Although it does not influence the spectral resolution, the much larger diffraction by the entrance slit imposes a limitation on the transmitted intensity at small slit widths. This can be seen as follows: when illuminated with parallel light, the entrance slit with width b produces a Fraunhofer diffraction pattern analogous to (4.6) with a replaced by b. The central diffraction maximum extends between the angles $\delta\phi = \pm\lambda/b$ (Fig. 4.11) and can completely pass the limiting aperture a only if $2\delta\phi$ is smaller than the acceptance angle a/f_1 of the spectrometer. This imposes a lower limit to the useful width $b_{\min}$ of the entrance slit,

$$b_{\min} \geq 2\lambda f_1/a . \tag{4.11}$$

In all practical cases, the incident light is divergent, which demands that the sum of the divergence angle and the diffraction angle has to be smaller than a/f and the minimum slit width b correspondingly larger.

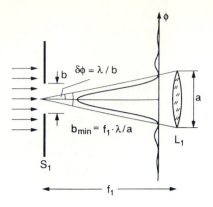

Fig. 4.11. Diffraction by the entrance slit

Figure 4.12a illustrates the intensity distribution $I(x)$ in the plane B for different slit widths b. Figure 4.12b shows the dependence of the width $\Delta x_2(b)$ of the slit image S_2 on the slit width b, taking into account the diffraction caused by the aperture a. This demonstrates that the resolution cannot be increased much by decreasing b below b_{min}. The peak intensity $I(b)_{x=0}$ is plotted in Fig. 4.12c as a function of the slit width. According to (4.1), the transmitted radiation flux $\phi(\lambda)$ depends on the product $U = A\Omega$ of the entrance slit area A and the acceptance angle $\Omega = (a/f_1)^2$. The flux in B would therefore depend linearly on the slit width b if diffraction were not present. This means that for monochromatic radiation the peak intensity $[\text{W}/\text{m}^2]$ in the plane B should then be constant (*curve 1m*), while for a spectral continuum it should decrease linearly with decreasing slit width (*curve 1c*). Because of the diffraction by S_1, the intensity decreases with the slit width b both for monochromatic radiation (*2m*) and for a spectral continuum (*2c*). Note the steep decrease for $b < b_{min}$.

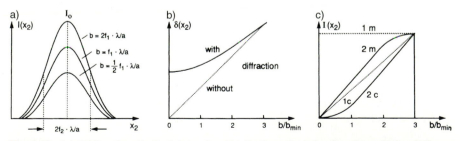

Fig. 4.12. (a) Diffraction limited intensity distribution $I(x_2)$ in the plane B for different widths b of the entrance slit. (b) The width $\delta x_2(b)$ of the entrance slit image $S_2(x_2)$ with and without diffraction by the aperture a. (c) Intensity $I(x_2)$ in the observation plane as a function of entrance slit width b for a spectral continuum c and for a monochromatic spectral line (*m*) with diffraction (*solid curves*) and without diffraction (*dashed curves*)

Substituting $b = b_{\min} = 2f\lambda/a$ into (4.10) yields the practical limit for $\Delta\lambda$ imposed by diffraction by S_1 and by the limiting aperture with width a

$$\Delta\lambda = 3f(\lambda/a)\,d\lambda/dx\ . \tag{4.12}$$

Instead of the theoretical limit (4.9) given by the diffraction through the aperture a, a smaller practically attainable resolving power is obtained from (4.12), which takes into account a finite minimum entrance slit width $b_{\min}$ imposed by intensity considerations and which yields:

$$\boxed{R = \lambda/\Delta\lambda = (a/3)\,d\theta/d\lambda\ . \tag{4.13}}$$

Example 4.2

For $a = 10\,\text{cm}$, $\lambda = 5 \times 10^{-5}\,\text{cm}$, $f = 100\,\text{cm}$, $d\lambda/dx = 1\,\text{nm/mm}$, with $b = 10\,\mu\text{m}$, $\rightarrow \Delta\lambda = 0.015\,\text{nm}$; with $b = 5\,\mu\text{m}$, $\rightarrow \Delta\lambda = 0.01\,\text{nm}$. However, from Fig. 4.12 one can see that the transmitted intensity with $b = 5\,\mu\text{m}$ is only 25% of that with $b = 10\,\mu\text{m}$.

Note: For photographic detection of line spectra, it is actually better to really use the lower limit $b_{\min}$ for the width of the entrance slit, because the density of the developed photographic layer depends only on the spectral irradiance [W/m^2] rather than on the radiation power [W]. Increasing the slit width beyond the diffraction limit $b_{\min}$, in fact, does not significantly increase the density contrast on the plate, but does decrease the spectral resolution.

Using photoelectric recording, the detected signal depends on the radiation power $\phi_\lambda\,d\lambda$ transmitted through the spectrometer and therefore increases with increasing slit width. In the case of completely resolved line spectra, this increase is proportional to the slit width b since $\phi_\lambda \propto b$. For continuous spectra it is even proportional to b^2 because the transmitted spectral interval $d\lambda$ also

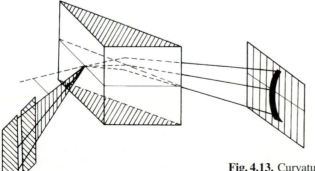

Fig. 4.13. Curvature of the image of a straight entrance slit caused by astigmatic imaging errors

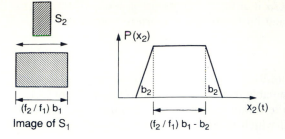

Fig. 4.14. Signal profile $P(t) \propto P(x_2(t))$ at the exit slit of a monochromator with $b \gg b_{min}$ and $b_2 < (f_2/f_1)b_1$ for monochromatic incident light with uniform turning of the grating

increases proportional to b and therefore $\phi_\lambda \, d\lambda \propto b^2$. Using diode arrays as detectors, the image $\Delta x_2 = (f_2/f_1)b$ should have the same width as one diode in order to obtain the optimum signal at maximum resolution.

The obvious idea of increasing the product of ΩA without loss of spectral resolution by keeping the width b constant but increasing the height h of the entrance slit is of limited value because imaging defects of the spectrometer cause a curvature of the slit image, which again decreases the resolution. Rays from the rim of the entrance slit pass the prism at a small inclination to the principal axis. This causes a larger angle of incidence α_2, which exceeds that of miniumum deviation. These rays are therefore refracted by a larger angle θ, and the image of a straight slit becomes curved toward shorter wavelengths (Fig. 4.13). Since the deviation in the plane B is equal to $f_2\theta$, the radius of curvature is of the same order of magnitude as the focal length of the camera lens and increases with increasing wavelength because of the decreasing spectral dispersion. In grating spectrometers, curved images of straight slits are caused by astigmatism of the spherical mirrors. The astigmatism can be partly compensated by using curved entrance slits [4.9]. Another solution is based on astigmatism-corrected imaging by using an asymmetric optical setup where the first mirror M_1 in Fig. 4.2 is placed at a distance $d_1 < f_1$ from the entrance slit and the exit slit at a distance $d_2 > f_2$ from M_2. In this arrangement [4.11] the grating is illuminated with slightly divergent light.

When the spectrometer is used as a monochromator with an entrance slit width b_1 and an exit slit width b_2, the power $P(t)$ recorded as a function of time while the grating is uniformly turned has a trapezoidal shape for $b_1 \gg b_{min}$ (Fig. 4.14a) with a baseline $(f_2/f_1)b_1 + b_2$. Optimum resolution at maximum transmitted power is achieved for $b_2 = (f_2/f_1)b_1$. The line profile $P(t) = P(x_2)$ then becomes a triangle.

d) Free Spectral Range

The free spectral range of a spectrometer is the wavelength interval $\delta\lambda$ of the incident radiation for which a one-valued relation exists between λ and the position $x(\lambda)$ of the entrance slit image. Two spectral lines with wavelengths λ_1 and $\lambda_2 = \lambda_1 \pm \delta\lambda$ cannot be distinguished without further information. This means that the wavelength λ measured by the instrument must be known beforehand with an uncertainty $\Delta\lambda < \delta\lambda$. While for prism spectrometers the free spectral range covers the whole region of normal dispersion of the prism material, for grating spectrometers $\delta\lambda$ is determined by the diffraction order m and decreases with increasing m (Sect. 4.1.3).

Interferometers, which are generally used in very high orders ($m = 10^4 - 10^8$), have a high spectral resolution but a small free spectral range $\delta\lambda$. For unambiguous wavelength determination they need a preselector, which allows one to measure the wavelength within $\delta\lambda$ of the high-resolution instrument (Sect. 4.2.3).

4.1.2 Prism Spectrometer

When passing through a prism, a light ray is refracted by an angle θ that depends on the prism angle ϵ, the angle of incidence α_1, and the refractive index n of the prism material (Fig. 4.15). The minimum deviation θ is obtained when the ray passes the prism parallel to the base g (symmetrical arrangement with $\alpha_1 = \alpha_2 = \alpha$). In this case, one can derive [4.5]

$$\frac{\sin(\theta + \epsilon)}{2} = n\sin(\epsilon/2) \,. \tag{4.14}$$

From (4.14) the derivation $d\theta/dn = (dn/d\theta)^{-1}$ is

$$\frac{d\theta}{dn} = \frac{2\sin(\epsilon/2)}{\cos[(\theta + \epsilon)/2]} = \frac{2\sin(\epsilon/2)}{\sqrt{1 - n^2\sin^2(\epsilon/2)}} \,. \tag{4.15}$$

The *angular dispersion* $d\theta/d\lambda = (d\theta/dn)(dn/d\lambda)$ is therefore

$$\frac{d\theta}{d\lambda} = \frac{2\sin(\epsilon/2)}{\sqrt{1 - n^2\sin^2(\epsilon/2)}}\frac{dn}{d\lambda} \,. \tag{4.16}$$

This shows that the angular dispersion increases with the prism angle ϵ, *but does not depend on the size of the prism.*

For the deviation of laser beams with small beam diameters, small prisms can therefore be used without losing angular dispersion. In a prism spectrometer, however, the size of the prism determines the limiting aperture a and therefore the diffraction; it has to be large in order to achieve a large spectral resolving power (see previous section). For a given angular dispersion, an

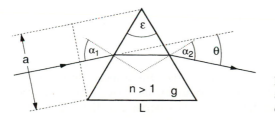

Fig. 4.15. Refraction of light by a prism at minimum deviation where $\alpha_1 = \alpha_2 = \alpha$ and $\theta = 2\alpha - \epsilon$

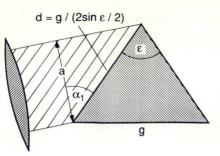

$d = g / (2\sin \varepsilon / 2)$

Fig. 4.16. Limiting aperture in a prism spectrometer

equilateral prism with $\epsilon = 60°$ uses the smallest quantity of possibly expensive prism material. Because $\sin 30° = 1/2$, (4.16) then reduces to

$$\frac{d\theta}{d\lambda} = \frac{dn/d\lambda}{\sqrt{1 - (n/2)^2}} \ . \tag{4.17}$$

The diffraction limit for the resolving power $\lambda/\Delta\lambda$ according to (4.9) is

$$\lambda/\Delta\lambda \leq a(d\theta/d\lambda) \ . $$

The diameter a of the limiting aperture in a prism spectrometer is (Fig. 4.16)

$$a = d\cos\alpha_1 = \frac{g\cos\alpha}{2\sin(\epsilon/2)} \ . \tag{4.18}$$

Substituting $d\theta/d\lambda$ from (4.16) gives

$$\lambda/\Delta\lambda = \frac{g\cos\alpha_1}{\sqrt{1 - n^2\sin^2(\epsilon/2)}} \frac{dn}{d\lambda} \ . \tag{4.19}$$

At minimum deviation, (4.14) gives $n\sin(\epsilon/2) = \sin(\theta + \epsilon)/2 = \sin\alpha_1$ and therefore (4.19) reduces to

$$\lambda/\Delta\lambda = g(dn/d\lambda) \ . \tag{4.20a}$$

According to (4.20a), the theoretical maximum resolving power depends solely on the base length g and on the spectral dispersion of the prism material. Because of the finite slit width $b \geq b_{min}$, the resolution reached in practice is somewhat lower. The corresponding resolving power can be derived from (4.11) to be at most

$$\boxed{R = \frac{\lambda}{\Delta\lambda} \leq \frac{1}{3} g \left(\frac{dn}{d\lambda} \right) \ .} \tag{4.20b}$$

The spectral dispersion $dn/d\lambda$ is a function of prism material and wavelength λ. Figure 4.17 shows dispersion curves $n(\lambda)$ for some materials commonly used for prisms. Since the refractive index increases rapidly in the vicinity of absorption lines, glass has a larger disperison in the visible and

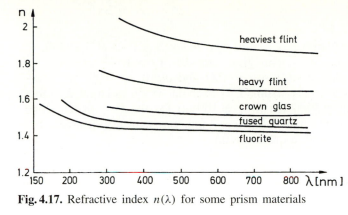

Fig. 4.17. Refractive index $n(\lambda)$ for some prism materials

near-ultraviolet regions than quartz, which, on the other hand, can be used advantageously in the UV down to 180 nm. In the vacuum-ultraviolet range CaF, MgF, or LiF prisms are sufficiently transparent. Table 4.1 gives a summary of the optical characteristics and useful spectral ranges of some prism materials.

    If achromatic lenses (which are expensive in the infrared and ultraviolet region) are not employed, the focal length of the two lenses decreases with the wavelength. This can be partly compensated by inclining the plane B against the principal axis in order to bring it at least approximately into the focal plane of L_2 for a large wavelength range (Fig. 4.1).

In Summary: The advantage of a prism spectrometer is the unambiguous assignment of wavelengths, since the position $S_2(\lambda)$ is a monotonic function of λ. Its drawback is the moderate spectral resolution. It is mostly used for survey scans of extended spectral regions.

Table 4.1. Refractive index and dispersion of some materials used in prism spectrometers

Material	Useful spectral range [µm]	Refractive index n	Dispersion $-dn/d\lambda[\text{nm}^{-1}]$		
Glass (BK7)	0.35−3.5	1.516	4.6×10^{-5}	at	589 nm
		1.53	1.1×10^{-4}	at	400 nm
Heavy flint	0.4−2	1.755	1.4×10^{-4}	at	589 nm
		1.81	4.4×10^{-4}	at	400 nm
Fused quartz	0.15−4.5	1.458	3.4×10^{-5}	at	589 nm
		1.470	1.1×10^{-4}	at	400 nm
NaCl	0.2−26	1.79	6.3×10^{-3}	at	200 nm
		1.38	1.7×10^{-5}	at	20 µm
LiF	0.12−9	1.44	6.6×10^{-4}	at	200 nm
		1.09	8.6×10^{-5}	at	10 µm

Example 4.3

(a) Suprasil (fused quartz) has a refractive index $n = 1.47$ at $\lambda = 400\,\text{nm}$ and $\mathrm{d}n/\mathrm{d}\lambda = 1100\,\text{cm}^{-1}$. This gives $\mathrm{d}\theta/\mathrm{d}\lambda = 1.6 \times 10^{-4}\,\text{rad/nm}$.

(b) For heavy flint glass at 400 nm $n = 1.81$ and $\mathrm{d}n/\mathrm{d}\lambda = 4400\,\text{cm}^{-1}$, giving $\mathrm{d}\theta/\mathrm{d}\lambda = 1.0 \times 10^{-3}\,\text{rad/nm}$. This is about six times larger than that for quartz. With a focal length $f = 100\,\text{cm}$ for the camera lens, one achieves a linear dispersion $\mathrm{d}x/\mathrm{d}\lambda = 1\,\text{mm/nm}$ with a flint prism, but only 0.15 mm/nm with a quartz prism.

4.1.3 Grating Spectrometer

In a grating spectrometer (Fig. 4.2) the collimating lens L_1 is replaced by a spherical mirror M_1 with the entrance slit S_1 in the focal plane of M_1. The collimated parallel light is reflected by M_1 onto a reflection grating consisting of many straight grooves (about 10^5) parallel to the entrance slit. The grooves have been ruled onto an optically smooth glass substrate or have been produced by holographic techniques [4.12–4.17]. The whole grating surface is coated with a highly reflecting layer (metal or dielectric film). The light reflected from the grating is focused by the spherical mirror M_2 onto the exit slit S_2 or onto a photographic plate in the focal plane of M_2.

a) Basic considerations

The many grooves, which are illuminated coherently, can be regarded as small radiation sources, each of them diffracting the light incident onto this small

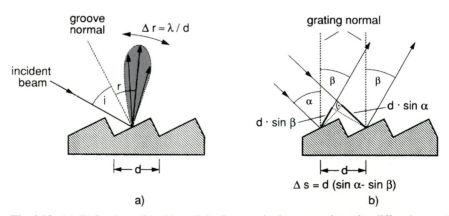

Fig. 4.18. (**a**) Reflection of incident light from a single groove into the diffraction angle λ/d around the specular reflection angle $r = i$. (**b**) Illustration of the grating equation (4.21)

groove with a width $d \approx \lambda$ into a large range $\Delta r \approx \lambda/d$ of angles r around the direction of geometrical reflection (Fig. 4.18a). The total reflected light consists of a coherent superposition of these many partial contributions. Only in those directions where all partial waves emitted from the different grooves are in phase will constructive interference result in a large total intensity, while in all other directions the different contributions cancel by destructive interference.

Figure 4.18b depicts a parallel light beam incident onto two adjacent grooves. At an angle of incidence α to the grating normal (which is normal to the grating surface, but not necessarily to the grooves) one obtains constructive interference for those directions β of the reflected light for which the path difference $\Delta s = \Delta s_1 - \Delta s_2$ is an integer multiple m of the wavelength λ. This yields the grating equation

$$d(\sin \alpha \pm \sin \beta) = m\lambda , \qquad (4.21)$$

the plus sign has to be taken if β and α are on the same side of the grating normal; otherwise the minus sign, which is the case shown in Fig. 4.18b.

The reflectivity $R(\beta, \theta)$ of a ruled grating depends on the diffraction angle β and on the blaze angle θ of the grating, which is the angle between the groove normal and the grating normal (Fig. 4.19). If the diffraction angle β coincides with the angle r of specular reflection from the groove surfaces, $R(\beta, \theta)$ reaches its optimum value R_0, which depends on the reflectivity of the groove coating. From Fig. 4.19 one infers for the case where α and β are on opposite sides of the grating normal, $i = \alpha - \theta$ and $r = \theta + \beta$, which yields, for specular reflection $i = r$, the condition for the optimum blaze angle θ

$$\theta = (\alpha - \beta)/2 . \qquad (4.22)$$

Because of the diffraction of each partial wave into a large angular range, the reflectivity $R(\beta)$ will not have a sharp maximum at $\beta = \alpha - 2\theta$, but will rather show a broad distribution around this optimum angle. The angle of incidence α is determined by the particular construction of the spectrometer, while the angle β for which constructive interference occurs depends on the

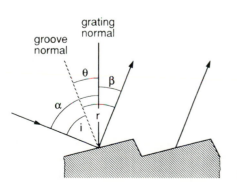

Fig. 4.19. Illustration of the blaze angle θ

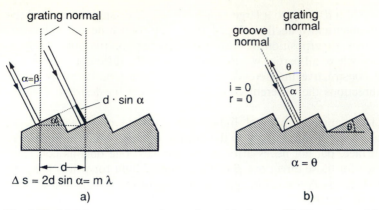

Fig. 4.20. (a) Littrow mount of a grating with $\beta = \alpha$. (b) Illustration of blaze angle for a Littrow grating

wavelength λ. Therefore the blaze angle θ has to be specified for the desired spectral range and the spectrometer type.

In laser-spectroscopic applications the case $\alpha = \beta$ often occurs, which means that the light is reflected back into the direction of the incident light. For such an arrangement, called a *Littrow-grating mount* (shown in Fig. 4.20), the grating equation (4.21) for constructive interference reduces to

$$2d \sin \alpha = m\lambda \ . \tag{4.21a}$$

Maximum reflectivity of the Littrow grating is achieved for $i = r = 0 \rightarrow \theta = \alpha$ (Fig. 4.20b). The Littrow grating acts as a wavelength-selective reflector because light is only reflected if the incident wavelength satisfies the condition (4.21a).

b) Intensity Distribution of Reflected Light

We now examine the intensity distribution $I(\beta)$ of the reflected light when a monochromatic plane wave is incident onto an arbitrary grating.

The path difference between partial waves reflected by adjacent grooves is $\Delta s = d(\sin \alpha \pm \sin \beta)$ and the corresponding phase difference is

$$\phi = \frac{2\pi}{\lambda} \Delta s = \frac{2\pi}{\lambda} d(\sin \alpha \pm \sin \beta) \ . \tag{4.23}$$

The superposition of the amplitudes reflected from all N grooves in the direction β gives the total reflected amplitude

$$A_R = \sqrt{R} \sum_{m=0}^{N-1} A_g \, e^{im\phi} = \sqrt{R} A_g \frac{1 - e^{iN\phi}}{1 - e^{-i\phi}} \ , \tag{4.24}$$

where $R(\beta)$ is the reflectivity of the grating, which depends on the reflection angle β, and A_g is the amplitude of the partial wave incident onto each

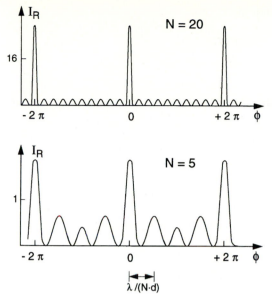

Fig. 4.21. Intensity distribution $I(\beta)$ for two different numbers N of illuminated grooves. Note the different scales of the ordinates!

groove. Because the intensity of the reflected wave is related to its amplitude by $I_R = \epsilon_0 c A_R A_R^*$, see (2.30c), we find from (4.24)

$$I_R = RI_0 \frac{\sin^2(N\phi/2)}{\sin^2(\phi/2)} \quad \text{with} \quad I_0 = c\epsilon_0 A_g A_g^* . \tag{4.25}$$

This intensity distribution is plotted in Fig. 4.21 for two different values of the total groove number N. The principal maxima occur for $\phi = 2m\pi$, which is, according to (4.23), equivalent to the grating equation (4.21) and means that at a fixed angle α the path difference between partial beams from adjacent grooves is for certain angles β_m an integer multiple of the wavelength, where the integer m is called the *order of the interference*. The function (4.25) has $(N-1)$ minima with $I_R = 0$ between two successive principal maxima. These minima occur at values of ϕ for which $N\phi/2 = \ell\pi$, $\ell = 1, 2, \ldots, N-1$, and mean that for each groove of the grating, another one can be found that emits light into the direction β with a phase shift π, such that all pairs of partial waves just cancel.

The line profile $I(\beta)$ of the principal maximum of order m around the diffraction angle β_m can be derived from (4.25) by substituting $\beta = \beta_m + \epsilon$. Because for large N, $I(\beta)$ is very sharply centered around β_m, we can assume $\epsilon \ll \beta_m$. With the relation

$$\sin(\beta_m + \epsilon) = \sin\beta_m \cos\epsilon + \cos\beta_m \sin\epsilon \sim \sin\beta_m + \epsilon\cos\beta_m ,$$

and because $(2\pi d/\lambda)(\sin\alpha + \sin\beta_m) = 2m\pi$, we obtain from (4.23)

$$\phi(\beta) = 2m\pi + 2\pi(d/\lambda)\epsilon\cos\beta_m = 2m\pi + \delta_1 \quad \text{with} \quad \delta_1 \ll 1 . \tag{4.26}$$

Furthermore, (4.25) can be written as

$$I_R = RI_0 \frac{\sin^2(N\delta_1/2)}{\sin^2(\delta_1/2)} \simeq RI_0 N^2 \frac{\sin^2(N\delta_1/2)}{(N\delta_1/2)^2} \, , \tag{4.27}$$

with $\delta_1 = 2\pi(d/\lambda)\epsilon \cos \beta_m$.

The first two minima on both sides of the central maximum at β_m are at

$$N\delta_1 = \pm 2\pi \; \Rightarrow \; \epsilon_{1,2} = \frac{\pm \lambda}{Nd \cos \beta_m} \, . \tag{4.28}$$

The central maximum of mth order therefore has a line profile (4.27) with a base halfwidth $\Delta\beta = \lambda/(Nd \cos \beta_m)$. This corresponds to a diffraction pattern produced by an aperture with width $b = Nd \cos \beta_m$, which is just the size of the whole grating projected onto a plane, normal to the direction normal of β_m (Fig. 4.18).

Example 4.4

For $N \cdot d = 10 \, \text{cm}$, $\lambda = 5 \times 10^{-5} \, \text{cm}$, $\cos \beta_m = \frac{1}{2}\sqrt{2} \Rightarrow \varepsilon_{1/2} = 7 \times 10^{-6} \, \text{rad}$.

Note: According to (4.28) the angular halfwidth $\Delta\beta = 2\epsilon$ of the interference maxima decreases as $1/N$, while according to (4.27) the peak intensity increases with the number N of illuminated grooves proportional to $N^2 I_0$, where I_0 is the power incident onto a single groove. The area under the main maxima is therefore proportional to NI_0, which is due to the increasing concentration of light into the directions β_m. Of course, the incident power per groove decreases as $1/N$. The total reflected power is therefore independent of N.

The intensity of the $N-2$ small side maxima, which are caused by incomplete destructive interference, decreases proportional to $1/N$ with increasing groove number N. Figure 4.21 illustrates this point for $N = 5$ and $N = 20$. For gratings used in practical spectroscopy with groove numbers of about 10^5, the reflected intensity $I_R(\lambda)$ at a given wavelength λ has very sharply defined maxima only in those directions β_m, as defined by (4.21). The small side maxima are completely negligible at such large values of N, provided the distance d between the grooves is exactly constant over the whole grating area.

c) Spectral Resolving Power

Differentiating the grating equation (4.21) with respect to λ, we obtain at a given angle α the angular dispersion

$$\frac{d\beta}{d\lambda} = \frac{m}{d \cos \beta} \, . \tag{4.29a}$$

Substituting $m/d = (\sin\alpha \pm \sin\beta)/\lambda$ from (4.21), we find

$$\frac{d\beta}{d\lambda} = \frac{\sin\alpha \pm \sin\beta}{\lambda\cos\beta}. \tag{4.29b}$$

This illustrates that the angular dispersion is determined solely by the angles α and β and *not by the number of grooves!* For the Littrow mount with $\alpha = \beta$, we obtain

$$\frac{d\beta}{d\lambda} = \frac{2\tan\alpha}{\lambda}. \tag{4.29c}$$

The resolving power can be immediately derived from (4.29a) and the base halfwidth $\Delta\beta = \epsilon = \lambda/(Nd\cos\beta)$ of the principal diffraction maximum (4.28), if we apply the Rayleigh criterion (see above) that two lines λ and $\lambda + \Delta\lambda$ are just resolved when the maximum of $I(\lambda)$ falls into the adjacent minimum for $I(\lambda + \Delta\lambda)$. This is equivalent to the condition

$$\frac{d\beta}{d\lambda}\Delta\lambda = \frac{\lambda}{Nd\cos\beta},$$

or

$$\frac{\lambda}{\Delta\lambda} = \frac{Nd(\sin\alpha \pm \sin\beta)}{\lambda}, \tag{4.30}$$

which reduces with (4.21) to

$$\boxed{R = \frac{\lambda}{\Delta\lambda} = mN.} \tag{4.31}$$

The theoretical spectral resolving power is the product of the diffraction order m with the total number N of illuminated grooves. If the finite slit width b_1 and the diffraction at limiting aperatures are taken into account, the practically achievable resolving power according to (4.13) is about $2-3$ times lower.

Often it is advantageous to use the spectrometer in second order ($m = 2$), which increases the spectral resolution by a factor of 2 without losing much intensity, if the blaze angle θ is correctly choosen to satisfy (4.21) and (4.22) with $m = 2$.

Example 4.5

A grating with a ruled area of $10 \times 10\,\text{cm}^2$ and 10^3 grooves/mm allows in second order ($m = 2$) a theoretical spectral resolution of $R = 2 \times 10^5$. This means that at $\lambda = 500\,\text{nm}$ two lines that are separated by $\Delta\lambda = 2.5 \times 10^{-3}\,\text{nm}$ should be resolvable. Because of diffraction, the practical limit is about $5 \times 10^{-3}\,\text{nm}$. The dispersion for $\alpha = \beta = 30°$ and a focal length $f = 1\,\text{m}$ is $dx/d\lambda = f\,d\beta/d\lambda = 2\,\text{mm/nm}$. With a slit width

$b_1 = b_2 = 50 \,\mu$m a spectral resolution of $\Delta\lambda = 0.025$ nm can be achieved. In order to decrease the slit image width to 5×10^{-3} nm, the entrance slit width b has to be narrowed to $10 \,\mu$m. Lines around $\lambda = 1 \,\mu$m in the spectrum would appear in 1st order at the same angles β. They have to be suppressed by filters.

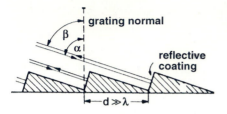

Fig. 4.22. Echelle grating

A special design is the so-called *echelle grating*, which has very widely spaced grooves forming right-angled steps (Fig. 4.22). The light is incident normal to the small side of the grooves. The path difference between two reflected partial beams incident on two adjacent grooves with an angle of incidence $\alpha = 90° - \theta$ is $\Delta s = 2d \cos\theta$. The grating equation (4.21) gives for the angle β of the mth diffraction order

$$d(\cos\theta + \sin\beta) \approx 2d \cos\theta = m\lambda \,, \tag{4.32}$$

where β is close to $\alpha = 90° - \theta$.

With $d \gg \lambda$ the grating is used in a very high order ($m \simeq 10-100$) and the resolving power is very high according to (4.31). Because of the larger distance d between the grooves, the relative ruling accuracy is higher and large gratings (up to 30 cm) can be ruled. The disadvantage of the echelle is the small free spectral range $\delta\lambda = \lambda/m$ between successive diffraction orders.

Example 4.6

$N = 3 \times 10^4$, $d = 10 \,\mu$m, $\theta = 30°$, $\lambda = 500$ nm, $m = 34$. The spectral resolving power is $R = 10^6$, but the free spectral range is only $\delta\lambda = 15$ nm. This means that the wavelengths λ and $\lambda + \delta\lambda$ overlap in the same direction β.

Minute deviations of the distance d between adjacent grooves, caused by inaccuracies during the ruling process, may result in constructive interference from parts of the grating for "wrong" wavelengths. Such unwanted maxima, which occur for a given angle of incidence α into "wrong" directions β, are called *grating ghosts*. Although the intensity of these ghosts is generally very small, intense incident radiation at a wavelength λ_i may cause ghosts with intensities comparable to those of other weak lines in the spectrum. This problem is particularly serious in laser spectroscopy when the intense light at the laser wavelength, which is scattered by cell walls or windows, reaches the entrance slit of the monochromator.

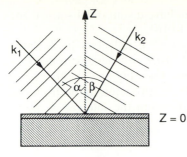

Fig. 4.23. Photographic production of a holographic grating

In order to illustrate the problematic nature of achieving the ruling accuracy that is required to avoid these ghosts, let us assume that the carriage of the ruling engine expands by only $1\,\mu m$ during the ruling of a $10 \times 10\,cm^2$ grating, e.g., due to temperature drifts. The groove distance d in the second half of the grating differs therefore from that of the first half by $5 \times 10^{-6}d$. With $N = 10^5$ grooves, the waves from the second half are then completely out of phase with those from the first half. The condition (4.21) is then fulfilled for different wavelengths in both parts of the grating, giving rise to unwanted wavelengths at the wrong positions β. Such ghosts are particularly troublesome in laser Raman spectroscopy (Chap. 8) or low-level fluorescence spectroscopy, where very weak lines have to be detected in the presence of extremely strong excitation lines. The ghosts from these excitation lines may overlap with the fluorescence or Raman lines and complicate the assignment of the spectrum.

d) Holographic Gratings

Although modern ruling techniques with interferometric length control have greatly improved the quality of ruled gratings [4.12–4.14] the most satisfactory way of producing completely ghost-free gratings is with holography. The production of holographic gratings proceeds as follows: a photosensitive layer on the grating's blank surface is illuminated by two coherent plane waves with the wave vectors k_1 and k_2 ($|k_1| = |k_2|$), which form the angles α and β against the surface normal (Fig. 4.23). The intensity distribution of the superposition in the plane $z = 0$ of the photolayer consists of parallel dark and bright fringes imprinting an ideal grating into the layer, which becomes visible after developing the photoemulsion. The grating constant

$$d = \frac{\lambda/2}{\sin\alpha + \sin\beta}$$

depends on the wavelength $\lambda = 2\pi/|k|$ and on the angles α and β. Such holographic gratings are essentially free of ghosts. Their reflectivity R, however, is lower than that of ruled gratings and is furthermore strongly dependent on the polarization of the incident wave. This is due to the fact that holographically produced grooves are no longer planar, but have a sinusoidal surface and the "blaze angle" θ varies across each groove [4.16].

For Littrow gratings used as wavelength-selective reflectors, it is desirable to have a high reflectivity in a selected order m and low reflections for all other orders. This can be achieved by selecting the width of the grooves and the blaze angle correctly. Because of diffraction by each groove with a width d, light can only reach angles β within the interval $\beta_0 \pm \lambda/d$ (Fig. 4.18a).

Example 4.7

With a blaze angle $\theta = \alpha = \beta = 30°$ and a step height $h = \lambda$, the grating can be used in second order, while the third order appears at $\beta = \beta_0 + 37°$. With $d = \lambda/\tan\theta = 2\lambda$, the central diffraction lobe extends only to $\beta_0 \pm 30°$, the intensity in the third order is very small.

Summarizing the considerations above, we find that the grating acts as a wavelength-selective mirror, reflecting light of a given wavelength only into definite directions β_m, called the mth diffraction orders, which are defined by (4.21). The intensity profile of a diffraction order corresponds to the diffraction profile of a slit with width $b = Nd \cos\beta_m$ representing the size of the whole grating projection as seen in the direction β_m. *The spectral resolution* $\lambda/\Delta\lambda = mN = Nd(\sin\alpha + \sin\beta)/\lambda$ *is therefore limited by the effective size of the grating measured in units of the wavelength.*

For a more detailed discussion of special designs of grating monochromators, such as the concave gratings used in VUV spectroscopy, the reader is referred to the literature on this subject [4.12–4.17]. An excellent account of the production and design of ruled gratings can be found in [4.12].

4.2 Interferometers

For the investigation of the various line profiles discussed in Chap. 3, interferometers are preferentially used because, with respect to the spectral resolving power, they are superior even to large spectrometers. In laser spectroscopy the different types of interferometers not only serve to measure emission – or absorption – line profiles, but they are also essential devices for narrowing the spectral width of lasers, monitoring the laser linewidth, and controlling and stabilizing the wavelength of single-mode lasers (Chap. 5).

In this section we discuss some basic properties of interferometers with the aid of some illustrating examples. The characteristics of the different types of interferometers that are essential for spectroscopic applications are discussed in more detail. Since laser technology is inconceivable without dielectric coatings for mirrors, interferometers, and filters, an extra section deals with such dielectric multilayers. The extensive literature on interferometers [4.19–4.22] informs about special designs and applications.

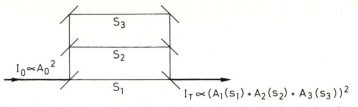

Fig. 4.24. Schematic illustration of the basic principle for all interferometers

4.2.1 Basic Concepts

The basic principle of all interferometers may be summarized as follows (Fig. 4.24). The indicent lightwave with intensity I_0 is divided into two or more partial beams with amplitudes A_k, which pass different optical path lengths $s_k = nx_k$ (where n is the refractive index) before they are again super-imposed at the exit of the interferometer. Since all partial beams come from the same source, they are coherent as long as the maximum path difference does not exceed the coherence length (Sect. 2.8). The total amplitude of the transmitted wave, which is the superposition of all partial waves, depends on the amplitudes A_k and on the phases $\phi_k = \phi_0 + 2\pi s_k/\lambda$ of the partial waves. *It is therefore sensitively dependent on the wavelength* λ.

The maximum transmitted intensity is obtained when all partial waves interfere constructively. This gives the condition for the optical path difference $\Delta s_{ik} = s_i - s_k$, namely

$$\Delta s_{ik} = m\lambda \, , \quad m = 1, 2, 3, \dots \, . \tag{4.33}$$

The condition (4.33) for maximum transmission of the interferometer applies not only to a single wavelength λ but to all λ_m for which

$$\lambda_m = \Delta s/m \, , \quad m = 1, 2, 3, \dots \, .$$

The wavelength interval

$$\delta\lambda = \lambda_m - \lambda_{m+1} = \frac{\Delta s}{m} - \frac{\Delta s}{m+1} = \frac{\Delta s}{m^2 + m} = \frac{2\lambda}{2m+1} \tag{4.34a}$$

where $\lambda = \frac{1}{2}(\lambda_m + \lambda_{m+1})$, is called the *free spectral range* of the interferometer. It is more conveniently expressed in terms of frequency. With $\nu = c/\lambda$, (4.33) yields $\Delta s = mc/\nu_m$ and the free spectral frequency range

$$\delta\nu = \nu_{m+1} - \nu_m = c/\Delta s \, , \tag{4.34b}$$

becomes independent of the order m.

It is important to realize that from one interferometric measurement alone one can only determine λ modulo $m \cdot \delta\lambda$ because all wavelengths $\lambda = \lambda_0 + m\delta\lambda$ are equivalent with respect to the transmission of the interferometer. One therefore has at first to measure λ within one free spectral range using other techniques before the absolute wavelength can be obtained with an interferometer.

Examples of devices in which only *two* partial beams interfere are the Michelson interferometer and the Mach–Zehnder interferometer. *Multiple-beam* interference is used, for instance, in the grating spectrometer, the Fabry–Perot interferometer, and in multilayer dielectric coatings of highly reflecting mirrors.

Some interferometers utilize the optical birefringence of specific crystals to produce two partial waves with mutually orthogonal polarization. The phase difference between the two waves is generated by the different refractive index for the two polarizations. An example of such a "polarization interferometer" is the *Lyot filter* [4.23] used in dye lasers to narrow the spectral linewidth (Sect. 4.2.9).

4.2.2 Michelson Interferometer

The basic principle of the Michelson interferometer (MI) is illustrated in Fig. 4.25. The incident plane wave

$$E = A_0 \, e^{i(\omega t - kx)}$$

is split by the beam splitter S (with reflectivity R and transmittance T) into two waves

$$E_1 = A_1 \exp\left[i(\omega t - kx + \phi_1)\right] \quad \text{and} \quad E_2 = A_2 \exp\left[i(\omega t - ky + \phi_2)\right] \ .$$

If the beam splitter has negligible absorption ($R + T = 1$), the amplitudes A_1 and A_2 are determined by $A_1 = \sqrt{R} A_0$ with $A_0^2 = A_1^2 + A_2^2$.

After being reflected at the plane mirrors M_1 and M_2, the two waves are superimposed in the plane of observation B. In order to compensate for the dispersion that beam 1 suffers by passing twice through the glass plate of beam splitter S, often an appropriate compensation plate P is placed in one side arm of the interferometer. The amplitudes of the two waves in the plane B are $\sqrt{TR} A_0$, because each wave has been transmitted and reflected once at the beam splitter surface S. The phase difference ϕ between the two waves is

$$\phi = \frac{2\pi}{\lambda} 2(SM_1 - SM_2) + \Delta\phi \ , \tag{4.35}$$

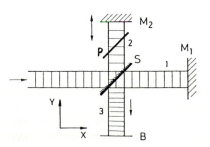

Fig. 4.25. Two-beam interference in a Michelson interferometer

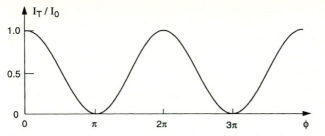

Fig. 4.26. Intensity transmitted through the Michelson interferometer as a function of the phase difference ϕ between the two interfering beams for $R = T = 0.5$

where $\Delta\phi$ accounts for additional phase shifts that may be caused by reflection. The total complex field amplitude in the plane B is then

$$E = \sqrt{RT}\, A_0\, e^{i(\omega t + \phi_0)}(1 + e^{i\phi}) \,. \tag{4.36}$$

The detector in B cannot follow the rapid oscillations with frequency ω but measures the time-averaged intensity $\bar{I}$, which is, according to (2.30c),

$$\bar{I} = \frac{1}{2}c\epsilon_0 A_0^2 RT(1 + e^{i\phi})(1 + e^{-i\phi}) = c\epsilon_0 A_0^2 RT(1 + \cos\phi)$$

$$= \frac{1}{2}I_0(1 + \cos\phi) \quad \text{for} \quad R = T = \frac{1}{2} \quad \text{and} \quad I_0 = \frac{1}{2}c\epsilon_0 A_0^2 \,. \tag{4.37}$$

If mirror M_2 (which is mounted on a carriage) moves along a distance Δy, the optical path difference changes by $\Delta s = 2n\Delta y$ (n is the refractive index between S and M_2) and the phase difference ϕ changes by $2\pi\Delta s/\lambda$. Figure 4.26 shows the intensity $I_T(\phi)$ in the plane B as a function of ϕ for a monochromatic incident plane wave. For the maxima at $\phi = 2m\pi$ ($m = 0, 1, 2, \dots$), the transmitted intensity I_T becomes equal to the incident intensity I_0, which means that the transmission of the interferometer is $T_I = 1$ for $\phi = 2m\pi$. In the minima for $\phi = (2m + 1)\pi$ the transmitted intensity I_T is zero! The incident plane wave is being reflected back into the source.

This illustrates that the MI can be regarded either as a wavelength-dependent filter for the transmitted light, or as a wavelength-selective reflector. In the latter function it is often used for mode selection in lasers (Fox–Smith selector, Sect. 5.4.3).

For divergent incident light the path difference between the two waves depends on the inclination angle (Fig. 4.27). In the plane B an interference pattern of circular fringes, concentric to the symmetry axis of the system, is produced. Moving the mirror M_2 causes the ring diameter to change. The intensity behind a small aperture still follows approximately the function $I(\phi)$ in Fig. 4.26. With parallel incident light but slightly tilted mirrors M_1 or M_2, the interference pattern consists of parallel fringes, which move into a direction perpendicular to the fringes when Δs is changed.

The MI can be used for absolute wavelength measurements by counting the number N of maxima in B when the mirror M_2 is moved along a known

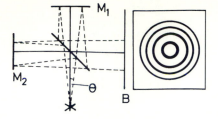

Fig. 4.27. Circular fringe pattern produced by the MI with divergent incident light

distance Δy. The wavelength λ is then obtained from

$$\lambda = 2n\Delta y/N \ .$$

This technique has been applied to very precise determinations of laser wavelengths (Sect. 4.4).

The MI may be described in another equivalent way, which is quite instructive. Assume that the mirror M_2 in Fig. 4.25 moves with a constant velocity $v = \Delta y/\Delta t$. A wave with frequency ω and wave vector $\boldsymbol{k}$ incident perpendicularly on the moving mirror suffers a Doppler shift

$$\Delta\omega = \omega - \omega' = 2\boldsymbol{k}\cdot\boldsymbol{v} = (4\pi/\lambda)v \ , \tag{4.38}$$

on reflection.

Inserting the path difference $\Delta s = \Delta s_0 + 2vt$ and the corresponding phase difference $\phi = (2\pi/\lambda)\Delta s$ into (4.37) gives, with (4.38) and $\Delta s_0 = 0$,

$$\bar{I} = \frac{1}{2}\bar{I}_0(1 + \cos \Delta\omega t) \quad \text{with} \quad \Delta\omega = 2\omega v/c \ . \tag{4.39}$$

We recognize (4.39) as the time-averaged beat signal, obtained from the superposition of two waves with frequencies ω and $\omega' = \omega - \Delta\omega$, giving the averaged intensity of

$$\bar{I} = I_0\overline{(1 + \cos \Delta\omega t)\cos^2[(\omega' + \omega)t/2]x} = \frac{1}{2}\bar{I}_0(1 + \cos \Delta\omega t) \ .$$

Note that the frequency $\omega = (c/v)\Delta\omega/2$ of the incoming wave can be measured from the beat frequency $\Delta\omega$, provided the velocity v of the moving mirror is known. The MI with uniformly moving mirror M_2 can be therefore regarded as a device that transforms the high frequency ω ($10^{14}-10^{15}\,\text{s}^{-1}$) into an easily accessible audio range $(v/c)\omega$.

Example 4.8

$v = 3\,\text{cm/s} \rightarrow (v/c) = 10^{-10}$. The frequency $\omega = 3 \times 10^{15}\,\text{Hz}$ ($\lambda = 0.6\,\mu\text{m}$) is transformed to $\Delta\omega = 6 \times 10^5\,\text{Hz} \simeq \Delta\nu \sim 100\,\text{kHz}$.

The maximum path difference Δs that still gives interference fringes in the plane B is limited by the coherence length of the incident radiation (Sect. 2.8). Using spectral lamps, the coherence length is limited by the

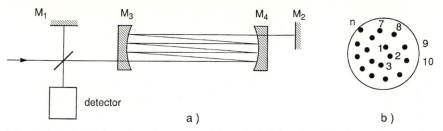

Fig. 4.28a,b. Michelson interferometer with optical delay line allowing a large path difference between the two interfering beams: (**a**) schematic arrangement; (**b**) spot positions of the reflected beams on mirror M_3

Doppler width of the spectral lines and is typically a few centimeters. With stabilized single-mode lasers, however, coherence lengths of several kilometers can be achieved. In this case, the maximum path difference in the MI is, in general, not restricted by the source but by technical limits imposed by laboratory facilities.

The attainable path difference Δs can be considerably increased by an *optical delay line*, placed in one arm of the interferometer (Fig. 4.28). It consists of a pair of mirrors, M_3, M_4, which reflect the light back and forth many times. In order to keep diffraction losses small, spherical mirrors, which compensate by collimation the divergence of the beam caused by diffraction, are preferable. With a stable mounting of the whole interferometer, optical path differences up to 350 m could be realized [4.24], allowing a spectral resolution of $\nu/\Delta\nu \simeq 10^{11}$. This was demonstrated by measuring the linewidth of a HeNe laser oscillating at $\nu = 5 \times 10^{14}$ Hz as a function of discharge current. The accuracy obtained was better than 5 kHz.

For gravitational-wave detection [4.25], a MI with side arms of about 1-km length has been built where the optical path difference can be increased to $\Delta s > 100$ km by using highly reflective spherical mirrors and an ultrastable solid-state laser with a coherence length of $\Delta s_c \gg \Delta s$ (see Sect. 14.8) [4.26].

When the incoming radiation is composed of several components with frequencies ω_k, the total amplitude in the plane B of the detector is the sum of all interference amplitudes (4.36),

$$E = \sum_k A_k \, e^{i(\omega_k t + \phi_{0k})}(1 + e^{i\phi_k}) \, . \tag{4.40}$$

A detector with a large time constant compared with the maximum period $1/(\omega_i - \omega_k)$ does not follow the rapid oscillations of the amplitude at frequencies ω_k or at the difference frequencies $(\omega_i - \omega_k)$, but gives a signal proportional to the sum of the intensities I_k in (4.37). We therefore obtain for the time-dependent total intensity

$$\bar{I}(t) = \sum_k \frac{1}{2}\bar{I}_{k0}(1 + \cos\phi_k) = \sum_k \frac{1}{2}\bar{I}_{k0}(1 + \cos\Delta\omega_k t) \, , \tag{4.41a}$$

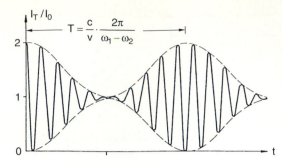

Fig. 4.29. Interference signal behind the MI with uniformly moving mirror M_2 when the incident wave consists of two components with frequencies ω_1 and ω_2 and equal amplitudes

where the audio frequencies $\Delta\omega_k = 2\omega_k v/c$ are determined by the frequencies ω_k of the components and by the velocity v of the moving mirror. Measurements of these frequencies $\Delta\omega_k$ allows one to reconstruct the spectral components of the incoming wave with frequencies ω_k (Fourier transform spectroscopy [4.27, 4.28]).

For example, when the incoming wave consists of two components with frequencies ω_1 and ω_2, the interference pattern varies with time according to

$$\bar{I}(t) = \frac{1}{2}\bar{I}_{10}[1 + \cos 2\omega_1(v/c)t] + \frac{1}{2}I_{20}[1 + \cos 2\omega_2(v/c)t]$$
$$= \bar{I}_0\{1 + \cos[(\omega_1 - \omega_2)vt/c]\cos[(\omega_1 + \omega_2)vt/c]\}, \tag{4.41b}$$

where we have assumed $I_{10} = I_{20} = I_0$. This is a beat signal, where the amplitude of the interference signal at $(\omega_1 + \omega_2)(v/c)$ is modulated at the difference frequency $(\omega_1 - \omega_2)v/c$ (Fig. 4.29).

The spectral resolution can roughly be estimated as follows: if Δy is the path difference traveled by the moving mirror, the number of interference maxima that are counted by the detector is $N_1 = 2\Delta y/\lambda_1$ for an incident wave with the wavelength λ_1, and $N_2 = 2\Delta y/\lambda_2$ for $\lambda_2 < \lambda_1$. The two wavelengths can be clearly distinguished when $N_2 \geq N_1 + 1$. This yields with $\lambda_1 = \lambda_2 + \Delta\lambda$ and $\Delta\lambda \ll \lambda$ for the spectral resolving power

$$\frac{\lambda}{\Delta\lambda} = \frac{2\Delta y}{\lambda} = N \quad \text{with} \quad \lambda = (\lambda_1 + \lambda_2)/2 \quad \text{and} \quad N = \frac{1}{2}(N_1 + N_2).$$
$$\tag{4.42a}$$

The equivalent consideration in the frequency domain follows. In order to determine the two frequencies ω_1 and ω_2, one has to measure at least over one modulation period

$$T = \frac{c}{v}\frac{2\pi}{\omega_1 - \omega_2} = \frac{c}{v}\frac{1}{v_1 - v_2}.$$

The frequency difference that can be resolved is then

$$\Delta v = \frac{c}{vT} = \frac{c}{\Delta s} = \frac{c}{N\lambda} \quad \Rightarrow \quad \frac{\Delta v}{c/\lambda} = \frac{1}{N} \quad \text{or} \quad \frac{v}{\Delta v} = N. \tag{4.42b}$$

The spectral resolving power $\lambda/\Delta\lambda$ of the Michelson interferometer equals the maximum path difference $\Delta s/\lambda$ measured in units of the wavelength λ.

Example 4.9

(a) $\Delta y = 5\,\text{cm}, \lambda = 10\,\mu\text{m} \to N = 10^4,$
(b) $\Delta y = 100\,\text{cm}, \lambda = 0.5\,\mu\text{m} \to N = 4 \times 10^6$
   where the latter example can be realized only with lasers that have a sufficiently large coherence length (Sect. 4.4).
(c) $\lambda_1 = 10\,\mu\text{m}, \lambda_2 = 9.8\,\mu\text{m} \to (\nu_2 - \nu_1) = 6 \times 10^{11}\,\text{Hz};$ with $v = 1\,\text{cm/s} \to T = 50\,\text{ms}.$

4.2.3 Mach–Zehnder Interferometer

Analogous to the Michelson interferometer, the Mach–Zehnder interferometer is based on the two-beam interference by amplitude splitting of the incoming wave. The two waves travel along different paths with a path difference $\Delta s = 2a\cos\alpha$ (Fig. 4.30b). Inserting a transparent object into one arm of the interferometer alters the optical path difference between the two beams. This results in a change of the interference pattern, which allows a very accurate determination of the refractive index of the sample and its local variation. The Mach–Zehnder interferometer may be regarded therefore as a sensitive refractometer.

If the beam splitters B_1, B_2 and the mirrors M_1, M_2 are all strictly parallel, the path difference between the two split beams does not depend on the angle of incidence α because the path difference between the beams 1 and 3 is exactly compensated by the same path length of beam 4 between M_2 and B_2 (Fig. 4.30a). This means that the interfering waves in the symmetric interferometer (without sample) experience the same path difference on the solid path as on the dashed path in Fig. 4.30a. Without the sample, the total path difference is therefore zero; it is $\Delta s = (n - 1)L$ *with* the sample having the refractive index n in one arm of the interferometer.

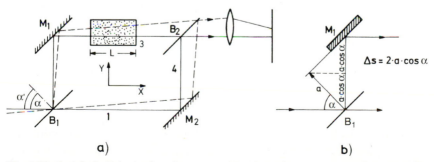

Fig. 4.30a,b. Mach–Zehnder interferometer: (**a**) schematic arrangement (**b**) path difference between the two parallel beams

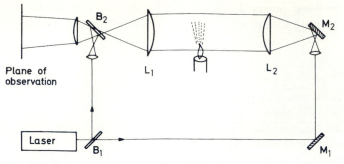

Fig. 4.31. Laser interferometer for sensitive measurements of local variations of the index of refraction in extended samples, for example, in air above a candle flame

Expanding the beam on path 3 gives an extended interference-fringe pattern, which reflects the local variation of the refractive index. Using a laser as a light source with a large coherence length, the path lengths in the two interferometer arms can be made different without losing the contrast of the interference pattern (Fig. 4.31). With a beam expander (lenses L_1 and L_2), the laser beam can be expanded up to $10-20$ cm and large objects can be tested. The interference pattern can either be photographed or may be viewed directly with the naked eye or with a television camera [4.29]. Such a laser interferometer has the advantage that the laser beam diameter can be kept small everywhere in the interferometer, except between the two expanding lenses. Since the illuminated part of the mirror surfaces should not deviate from an ideal plane by more than $\lambda/10$ in order to obtain good interferograms, smaller beam diameters are advantageous.

The Mach–Zehnder interferometer has found a wide range of applications. Density variations in laminar or turbulent gas flows can be seen with this technique and the optical quality of mirror substrates or interferometer plates can be tested with high sensitivity [4.29, 4.30].

In order to get quantitative information of the local variation of the optical path through the sample, it is useful to generate a fringe pattern for calibration purposes by slightly tilting the plates B_1, M_1 and B_2, M_2 in Fig. 4.31, which makes the interferometer slightly asymmetric. Assume that B_1 and M_1 are tilted clockwise around the z-direction by a small angle β and the pair B_2, M_2 is tilted counterclockwise by the same angle β. The optical path between B_1 and M_1 is then $\Delta_1 = 2a\cos(\alpha+\beta)$, whereas $B_2 M_2 = \Delta_2 = 2a\cos(\alpha-\beta)$. After being recombined, the two beams therefore have the path difference

$$\Delta = \Delta_2 - \Delta_1 = 2a[\cos(\alpha-\beta) - \cos(\alpha+\beta)] = 4a\sin\alpha\sin\beta , \qquad (4.43)$$

which depends on the angle of incidence α. In the plane of observation, an interference pattern of parallel fringes with path differences $\Delta = m \cdot \lambda$ is observed with an angular separation $\Delta\epsilon$ between the fringes m and $m+1$ given by $\Delta\epsilon = \alpha_m - \alpha_{m+1} = \lambda/(4a\sin\beta\cos\alpha)$.

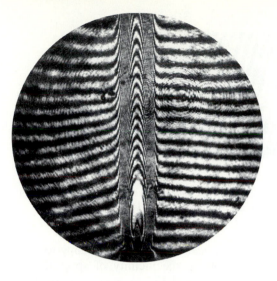

Fig. 4.32. Interferogram of the density profile in the convection zone above a candle flame [4.29]

A sample in path 3 introduces an additional path difference

$$\Delta s(\beta) = (n-1)L/\cos\beta$$

depending on the local refractive index n and the path length L through the sample. The resulting phase difference shifts the interference pattern by an angle $\gamma = (n-1)(L/\lambda)\Delta\varepsilon$. Using a lens with a focal length f, which images the interference pattern onto the plane O, gives the spatial distance $\Delta y = f\Delta\varepsilon$ between neighboring fringes. The additional path difference caused by the sample shifts the interference pattern by $N = (n-1)(L/\lambda)$ fringes.

Figure 4.32 shows for illustration the interferogram of the convection zone of hot air above a candle flame, placed below one arm of the laser interferometer in Fig. 4.31. It can be seen that the optical path through this zone changes by many wavelengths.

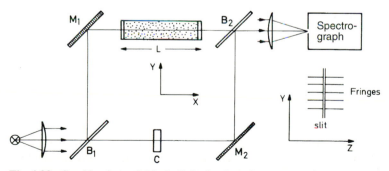
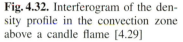

Fig. 4.33. Combination of Mach–Zehnder interferometer and spectrograph used for the hook method

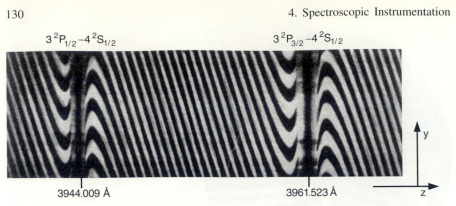

Fig. 4.34. Position of fringes as a function of wavelength around the absorption line doublet of aluminum atoms, as observed behind the spectrograph [4.31]

The Mach–Zehnder interferometer has been used in spectroscopy to measure the refractive index of atomic vapors in the vicinity of spectral lines (Sect. 3.1). The experimental arrangement (Fig. 4.33) consists of a combination of a spectrograph and an interferometer, where the plates B_1, M_1 and B_2, M_2 are tilted in such a direction that without the sample the parallel interference fringes with the separation $\Delta y(\lambda) = f \Delta \varepsilon$ are perpendicular to the entrance slit. The spectrograph disperses the fringes with different wavelengths λ_i in the z-direction. Because of the wavelength-dependent refractive index $n(\lambda)$ of the atomic vapor (Sect. 3.1.3), the fringe shift follows a dispersion curve in the vicinity of the spectral line (Fig. 4.34). The dispersed fringes look like hooks around an absorption line, which gave this technique the name *hook method*. To compensate for background shifts caused by the windows of the absorption cell, a compensating plate is inserted into the second arm. For more details of the Hook method, see [4.30, 4.31].

4.2.4 Multiple-Beam Interference

In a grating spectrometer, the interfering partial waves emitted from the different grooves of the grating all have the same amplitude. In contrast, in multiple-beam interferometers these partial waves are produced by multiple reflection at plane or curved surfaces and their amplitude decreases with increasing number of reflections. The resultant total intensity therefore differs from (4.25).

a) Transmitted and Reflected Intensity

Assume that a plane wave $E = A_0 \exp[i(\omega t - kx)]$ is incident at the angle α on a plane transparent plate with two parallel, partially reflecting surfaces (Fig. 4.35). At each surface the amplitude A_i is split into a reflected component $A_R = A_i \sqrt{R}$ and a refracted component $A_T = A_i \sqrt{1 - R}$, neglecting absorption. The reflectivity $R = I_R / I_i$ depends on the angle of incidence α and on the polarization of the incident wave. Provided the refractive index n

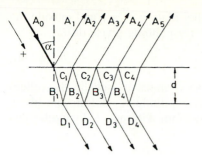

Fig. 4.35. Multiple-beam interference at two plane-parallel partially reflecting surfaces

is known, R can be calculated from Fresnel's formulas [4.3]. From Fig. 4.35, the following relations are obtained for the amplitudes A_i of waves reflected at the upper surface, B_i of refracted waves, C_i of waves reflected at the lower surface, and D_i of transmitted waves

$$|A_1| = \sqrt{R}\,|A_0|\,, \qquad\qquad\qquad |B_1| = \sqrt{1-R}\,|A_0|\,,$$
$$|C_1| = \sqrt{R(1-R)}\,|A_0|\,, \qquad\qquad |D_1| = (1-R)\,|A_0|\,,$$
$$|A_2| = \sqrt{1-R}\,|C_1| = (1-R)\sqrt{R}\,|A_0|\,, \qquad\qquad (4.44)$$
$$|C_2| = R\sqrt{R(1-R)}\,|A_0|\,, \qquad\qquad |D_2| = R(1-R)\,|A_0|\,,$$
$$|A_3| = \sqrt{1-R}\,|C_2| = R^{3/2}(1-R)\,|A_0|\,,\dots\,.$$

This scheme can be generalized to the equations

$$|A_{i+1}| = R|A_i|\,, \quad i \geq 2\,, \qquad\qquad\qquad (4.45a)$$
$$|D_{i+1}| = R|D_i|\,, \quad i \geq 1\,. \qquad\qquad\qquad (4.45b)$$

Two successively reflected partial waves E_i and E_{i+1} have the optical path difference (Fig. 4.36)

$$\Delta s = (2nd/\cos\beta) - 2d\tan\beta\sin\alpha\,.$$

Because $\sin\alpha = n\sin\beta$, this can be reduced to

$$\Delta s = 2nd\cos\beta = 2dn\sqrt{1-\sin^2\beta}\,, \qquad\qquad (4.46a)$$

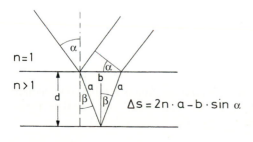

Fig. 4.36. Optical path difference between two beams being reflected from the two surfaces of a plane-parallel plate

if the refractive index within the plane-parallel plate is $n > 1$ and outside the plate $n = 1$. This path difference causes a corresponding phase difference

$$\phi = 2\pi \Delta s/\lambda + \Delta\phi \, , \tag{4.46b}$$

where $\Delta\phi$ takes into account possible phase changes caused by the reflections. For instance, the incident wave with amplitude A_1 suffers the phase jump $\Delta\phi = \pi$ while being reflected at the medium with $n > 1$. Including this phase jump, we can write

$$A_1 = \sqrt{R}A_0 \exp(i\pi) = -\sqrt{R}A_0 \, .$$

The total amplitude A of the reflected wave is obtained by summation over all partial amplitudes A_i, taking into account the different phase shifts,

$$A = \sum_{m=1}^{p} A_m e^{i(m-1)\phi} = -\sqrt{R}A_0 + \sqrt{R}A_0(1-R)e^{i\phi} + \sum_{m=3}^{p} A_m e^{i(m-1)\phi}$$

$$= -\sqrt{R}A_0 \left[1 - (1-R)e^{i\phi} \sum_{m=0}^{p-2} R^m e^{im\phi} \right] \, . \tag{4.47}$$

For vertical incidence ($\alpha = 0$), or for an infinitely extended plate, we have an infinite number of reflections. The geometrical series in (4.47) has the limit $(1 - Re^{i\phi})^{-1}$ for $p \to \infty$. We obtain for the total amplitude

$$A = -\sqrt{R}A_0 \frac{1 - e^{i\phi}}{1 - Re^{i\phi}} \, . \tag{4.48}$$

The intensity $I = 2c\epsilon_0 AA^*$ of the reflected wave is then

$$\boxed{I_R = I_0 R \frac{4\sin^2(\phi/2)}{(1-R)^2 + 4R\sin^2(\phi/2)} \, .} \tag{4.49a}$$

In an analogous way, we find for the total transmitted amplitude

$$D = \sum_{m=1}^{\infty} D_m e^{i(m-1)\phi} = (1-R)A_0 \sum_{0}^{\infty} R^m e^{im\phi} \, ,$$

which gives the total transmitted intensity

$$\boxed{I_T = I_0 \frac{(1-R)^2}{(1-R)^2 + 4R\sin^2(\phi/2)} \, .} \tag{4.50a}$$

Equations (4.49, 4.50) are called the *Airy formulas*. Since we have neglected absorption, we should have $I_R + I_T = I_0$, as can easily be verified from (4.49, 4.50).

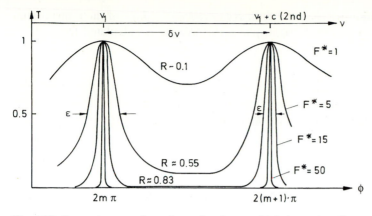

Fig. 4.37. Transmittance of an absorption-free multiple-beam interferometer as a function of the phase difference ϕ for different values of the finesse F^*

The abbreviation $F = 4R/(1-R)^2$ is often used, which allows the Airy equations to be written in the form

$$I_R = I_0 \frac{F \sin^2(\phi/2)}{1 + F \sin^2(\phi/2)} \, , \tag{4.49b}$$

$$I_T = I_0 \frac{1}{1 + F \sin^2(\phi/2)} \, . \tag{4.50b}$$

Figure 4.37 illustrates (4.50) for different values of the reflectivity R. The maximum transmittance is $T = 1$ for $\phi = 2m\pi$. At these maxima $I_T = I_0$, therefore the reflected intensity I_R is zero. The minimum transmittance is

$$T^{\min} = \frac{1}{1+F} = \left(\frac{1-R}{1+R}\right)^2 \, .$$

b) Free Spectral Range and Finesse

The frequency range $\delta\nu$ between two maxima is the *free spectral range of the interferometer*. With $\phi = 2\pi\Delta s/\lambda$ and $\lambda = c/\nu$, we obtain from (4.46a)

$$\delta\nu = \frac{c}{\Delta s} = \frac{c}{2d\sqrt{n^2 - \sin^2\alpha}} \, . \tag{4.51a}$$

For vertical incidence ($\alpha = 0$), the free spectral range becomes

$$|\delta\nu|_{\alpha=0} = \frac{c}{2nd} \, . \tag{4.51b}$$

The full halfwidth $\epsilon = |\phi_1 - \phi_2|$ with $I(\phi_1) = I(\phi_2) = I_0/2$ of the transmission maxima in Fig. 4.37 expressed in phase differences is calculated from (4.50) as

$$\epsilon = 4 \arcsin \left(\frac{1-R}{2\sqrt{R}} \right) , \tag{4.52a}$$

which reduces for $R \approx 1 \Rightarrow (1-R) \ll R$ to

$$\epsilon = \frac{2(1-R)}{\sqrt{R}} = \frac{4}{\sqrt{F}} . \tag{4.52b}$$

In frequency units, the halfwidth becomes with (4.51b)

$$\Delta\nu = \frac{\epsilon}{2\pi} \delta\nu \simeq \frac{2\delta\nu}{\pi\sqrt{F}} = \frac{c}{2nd} \frac{1-R}{\pi\sqrt{R}} . \tag{4.52c}$$

The ratio $\delta\nu/\Delta\nu$ of free spectral range $\delta\nu$ to the halfwidth $\Delta\nu$ of the transmission maxima is called the *finesse* F^* of the interferometer. From (4.51b) and (4.52c) we obtain for the finesse

$$\boxed{F^* = \frac{\pi\sqrt{R}}{1-R} = \frac{\pi}{2}\sqrt{F} .} \tag{4.53a}$$

The full halfwidth of the transmission peaks is then

$$\Delta\nu = \frac{\delta\nu}{F^*} . \tag{4.53b}$$

The finesse is a measure for the effective number of interfering partial waves in the interferometer. This means that the maximum path difference between interfering waves is $\Delta S_{\max} = F^* 2nd$.

Since we have assumed an ideal plane-parallel plate with a perfect surface quality, the finesse (4.53a) is determined only by the reflectivity R of the surfaces. In practice, however, deviations of the surfaces from an ideal plane and slight inclinations of the two surfaces cause imperfect superposition of the interfering waves. This results in a decrease and a broadening of the transmission maxima, which decreases the total finesse. If, for instance, a reflecting surface deviates by the amount λ/q from an ideal plane, the finesse cannot be larger than q. One can define the total finesse F^* of an interferometer by

$$\frac{1}{F^{*2}} = \sum_i \frac{1}{F_i^{*2}} , \tag{4.53c}$$

where the different terms F_i^* give the contributions to the decrease of the finesse caused by the different imperfections of the interferometer.

Example 4.10

A plane, nearly parallel plate has a diameter $D = 5\,\text{cm}$, a thickness $d = 1\,\text{cm}$, and a wedge angle of $0.2''$. The two reflecting surfaces have the reflectivity $R = 95\%$. The surfaces are flat to within $\lambda/50$, which means that no point of the surface deviates from an ideal plane by more than $\lambda/50$. The different contributions to the finesse are:

- Reflectivity finesse: $F_R^* = \pi\sqrt{R}/(1 - R) \simeq 60$;
- Surface finesse: $F_S \simeq 50$;
- Wedge finesse: with a wedge angle of $0.2''$ the optical path between the two reflecting surfaces changes by about $0.1\lambda(\lambda = 0.5\,\mu\text{m})$ across the diameter of the plate. For a monochromatic incident wave this causes imperfect interference and broadens the maxima corresponding to a finesse of about 20.

The total finesse is then $F^{*2} = 1/(1/60^2 + 1/50^2 + 1/20^2) \rightarrow F^* \simeq 17.7$.

This illustrates that high-quality optical surfaces are necessary to obtain a high total finesse [4.32]. It makes no sense to increase the reflectivity without a corresponding increase of the surface finesse. In our example the imperfect parallelism was the main cause for the low finesse. Decreasing the wedge angle to $0.1''$ increases the wedge finesse to 40 and the total finesse to 27.7.

A much larger finesse can be achieved using spherical mirrors, because the demand for parallelism is dropped. With sufficiently accurate alignment and high reflectivities, values of $F^* > 50\,000$ are possible (Sect. 4.2.8).

c) Spectral Resolution

The spectral resolution, $\nu/\Delta\nu$ or $\lambda/\Delta\lambda$, of an interferometer is determined by the free spectral range $\delta\nu$ and by the finesse F^*. Two incident waves with frequencies ν_1 and $\nu_2 = \nu_1 + \Delta\nu$ can still be resolved if their frequency separation $\Delta\nu$ is larger than $\delta\nu/F^*$, which means that their peak separation should be larger than their full halfwidth.

Quantitatively this can be seen as follows: assume the incident radiation consists of two components with the intensity profiles $I_1(\nu - \nu_1)$ and $I_2(\nu - \nu_2)$ and equal peak intensities $I_1(\nu_1) = I_2(\nu_2) = I_0$. For a peak separation $\nu_2 - \nu_1 = \delta\nu/F^* = 2\delta\nu/\pi\sqrt{F}$, the total transmitted intensity $I(\nu) = I_1(\nu) + I_2(\nu)$ is obtained from (4.50a) as

$$I(\nu) = I_0 \left(\frac{1}{1 + F\sin^2(\pi\nu/\delta\nu)} + \frac{1}{1 + F\sin^2[\pi(\nu + \delta\nu/F^*)/\delta\nu]} \right), \quad (4.54)$$

where the phase shift $\phi = 2\pi\Delta s\nu/c = 2\pi\nu/\delta\nu$ in (4.50b) has been expressed by the free spectral range $\delta\nu$. The function $I(\nu)$ is plotted in Fig. 4.38 around the frequency $\nu = (\nu_1 + \nu_2)/2$. For $\nu = \nu_1 = mc/2nd$, the first term in (4.54) becomes 1 and the second term can be derived with $\sin[\pi(\nu_1 + \delta\nu/F^*)/\delta\nu] = \sin\pi/F^* \simeq \pi/F^*$ and $F(\pi/F^*)^2 = 4$ to become 0.2. Inserting this into

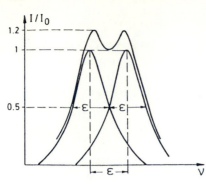

Fig. 4.38. Transmitted intensity $I_T(\nu)$ for two closely spaced spectral lines at the limit of spectral resolution where the linespacing equals the halfwidth of the lines

(4.54) yields $I(\nu = \nu_1) = 1.2I_0$, $I(\nu = (\nu_1 + \nu_2)/2) \simeq I_0$, and $I(\nu = \nu_2) = 1.2I_0$. This just corresponds to the Rayleigh criterion for the resolution of two spectral lines. The spectral resolving power of the interferometer is therefore

$$\nu/\Delta\nu = (\nu/\delta\nu)F^* \to \Delta\nu = \delta\nu/F^* \,. \tag{4.55}$$

This can be also expressed by the optical path differences Δs between two successive partial waves

$$\frac{\nu}{\Delta\nu} = \frac{\lambda}{\Delta\lambda} = F^* \frac{\Delta s}{\lambda} \,. \tag{4.56}$$

The resolving power of an interferometer is the product of finesse F^* and optical path difference $\Delta s/\lambda$ in units of the wavelength λ.

A comparison with the resolving power $\nu/\Delta\nu = mN = N\Delta s/\lambda$ of a grating spectrometer with N grooves shows that the finesse F^* can indeed be regarded as the effective number of interfering partial waves and $F^*\Delta s$ can be regarded as the maximum path difference between these waves.

Example 4.11

$d = 1$ cm, $n = 1.5$, $R = 0.98$, $\lambda = 500$ nm. An interferometer with negligible wedge and high-quality surfaces, where the finesse is mainly determined by the reflectivity, achieves with $F^* = \pi\sqrt{R}/(1 - R) = 155$ a resolving power of $\lambda/\Delta\lambda = 10^7$. This means that the instrument's linewidth is about $\Delta\lambda \sim 5 \times 10^{-5}$ nm or, in frequency units, $\Delta\nu = 60$ MHz.

d) Influence of Absorption Losses

Taking into account the absorption $A = (1 - R - T)$ of each reflective surface, (4.50) must be modified to

$$I_T = I_0 \frac{T^2}{(A+T)^2} \frac{1}{[1 + F \sin^2(\delta/2)]} \,, \tag{4.57a}$$

where $T^2 = T_1 T_2$ is the product of the transmittance of the two reflecting surfaces. The absorption causes three effects:

(a) The maximum transmittance is decreased by the factor

$$\frac{I_T}{I_0} = \frac{T^2}{(A+T)^2} = \frac{T^2}{(1-R)^2} < 1 \,. \tag{4.57b}$$

Note that even a small absorption of each reflecting surface results in a drastic reduction of the total transmittance. For $A = 0.05$, $R = 0.9 \rightarrow T = 0.05$ and $T^2/(1-R)^2 = 0.25$.

(b) For a given transmission factor T, the reflectivity $R = 1 - A - T$ decreases with increasing absorption. The quantity

$$F = \frac{4R}{(1-R)^2} = \frac{4(1-T-A)}{(T+A)^2} \tag{4.57c}$$

decreases with increasing A. For the example above we obtain $F = 360$. This makes the transmission peaks broader because of the decreasing number of interfering partial waves. The *contrast*

$$\frac{I_T^{max}}{I_T^{min}} = 1 + F \tag{4.57d}$$

of the transmitted intensity also decreases.

(c) The absorption causes a phase shift $\Delta\phi$ at each reflection, which depends on the wavelength λ, the polarization, and the angle of incidence α [4.3]. This effect causes a wavelength-dependent *shift* of the maxima.

4.2.5 Plane Fabry–Perot Interferometer

A practical realization of the multiple beam-interference discussed in this section may use either a solid plane-parallel glass or fused quartz plate with two coated reflecting surfaces (Fabry–Perot etalon, Fig. 4.39a) or two separate plates, where one surface of each plate is coated with a reflection layer. The two reflecting surfaces are opposed and are aligned to be as parallel as achievable (Fabry–Perot interferometer (FPI), Fig. 4.39b). The outer surfaces are coated with antireflection layers in order to avoid reflections from these surfaces that might overlap the interference pattern. Furthermore, they have a slight angle against the inner surfaces (wedge).

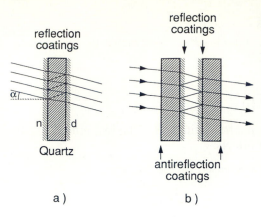

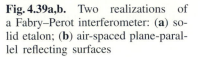

Fig. 4.39a,b. Two realizations of a Fabry–Perot interferometer: (a) solid etalon; (b) air-spaced plane-parallel reflecting surfaces

Both devices can be used for parallel as well as for divergent incident light. We now discuss them in more detail, first considering their illumination with *parallel* light.

a) The Plane FPI as a Transmission Filter

In laser spectroscopy, etalons are mainly used as wavelength-selective transmission filters within the laser resonator to narrow the laser bandwidth (Sect. 5.4). The wavelength λ_m or frequency ν_m for the transmission maximum of mth order, where the optical path between successive beams is $\Delta s = m\lambda$, can be deduced from (4.46a) and Fig. 4.36 to be

$$\lambda_m = \frac{2d}{m}\sqrt{n^2 - \sin^2\alpha} = \frac{2nd}{m}\cos\beta , \tag{4.58a}$$

$$\nu_m = \frac{mc}{2nd\cos\beta} . \tag{4.58b}$$

For all wavelengths $\lambda = \lambda_m$ ($m = 0, 1, 2, \ldots$) in the incident light, the phase difference between the transmitted partial waves becomes $\delta = 2m\pi$ and the transmitted intensity is, according to (4.57),

$$I_T = \frac{T^2}{(1 - R)^2} I_0 = \frac{T^2}{(A + T)^2} I_0 , \tag{4.59}$$

where $A = 1 - T - R$ is the absorption of the etalon (substrate absorption plus absorption of one reflecting surface). The reflected waves interfere destructively for $\lambda = \lambda_m$ and the reflected intensity becomes zero.

Note, however, that this is only true for $A \ll 1$ and infinitely extended plane waves, where the different reflected partial waves completely overlap. If the incident wave is a laser beam with the finite diameter D, the different reflected partial beams do *not* completely overlap because they are laterally shifted by $b = 2d\tan\beta\cos\alpha$ (Fig. 4.40). For a rectangular intensity profile of the laser beam, the fraction b/D of the reflected partial amplitudes does not

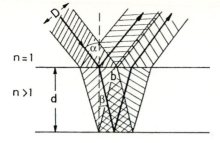

Fig. 4.40. Incomplete interference of two reflected beams with finite diameter D, causing a decrease of the maximum transmitted intensity

overlap and cannot interfere destructively. This means that, even for maximum transmission, the reflected intensity is not zero but a background reflection remains, which is missing in the transmitted light. For small angles α, one obtains for the intensity loss per transit due to reflection [4.33] for a rectangular beam profile

$$\frac{I_R}{I_0} = \frac{4R}{(1-R)^2}\left(\frac{2\alpha d}{nD}\right)^2 . \tag{4.60a}$$

For a Gaussian beam profile the calculation is more difficult, and the solution can only be obtained numerically. The result for a Gaussian beam with the radius w (Sect. 5.3) is [4.34]

$$\frac{I_R}{I_0} \simeq \frac{8R}{(1-R)^2}\left(\frac{2d\alpha}{nw}\right)^2 . \tag{4.60b}$$

A parallel light beam with the diameter D passing a plane-parallel plate with the angle of incidence α therefore suffers reflection losses in addition to the eventual absorption losses. The reflection losses increase with α^2 and are proportional to the ratio $(d/D)^2$ of the etalon thickness d and the beam diameter D (walk-off losses).

Example 4.12

$d = 1\,\text{cm}$, $D = 0.2\,\text{cm}$, $n = 1.5$, $R = 0.3$, $\alpha = 1° \hat{=} 0.017\,\text{rad} \to I_R/I_0 = 0.05$, which means 5% walk-off losses.

The transmission peak λ_m of the etalon can be shifted by tilting the etalon. According to (4.58) the wavelength λ_m *decreases* with increasing angle of incidence α. The walk-off losses limit the tuning range of tilted etalons within a laser resonator. With increasing angle α, the losses may become intolerably large.

b) Illumination with Divergent Light

Illuminating the FPI with divergent monochromatic light (e.g., from an extended source or from a laser beam behind a diverging lens), a continuous

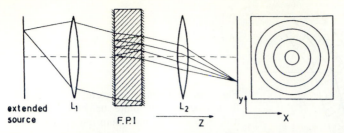

Fig. 4.41. The interference ring system of the transmitted intensity may be regarded as wavelength-selective imaging of corresponding ring areas of an extended light source

range of incident angles α is offered to the FPI, which transmits those directions α_m that obey (4.58a). We then observe an interference pattern of bright rings in the transmitted light (Fig. 4.41). Since the reflected intensity $I_R = I_0 - I_T$ is complementary to the transmitted one, a corresponding system of dark rings appears in the reflected light at the same angles of incidence α_m.

When β is the angle of inclination to the interferometer axis inside the FPI, the transmitted intensity is maximum, according to (4.58), for

$$m\lambda = 2nd \cos \beta , \tag{4.61}$$

where n is the refractive index between the reflecting planes. Let us number the rings by the integer p, beginning with $p = 0$ for the central ring. With $m = m_0 - p$, we can rewrite (4.61) for small angles β_p as

$$(m_0 - p)\lambda = 2nd \cos \beta_p \sim 2nd(1 - \beta_p^2/2) = 2nd \left[1 - \frac{1}{2} \left(\frac{n_0 \alpha_p}{n} \right)^2 \right] , \tag{4.62}$$

where n_0 is the refractive index of air, and Snell's law $\sin \alpha \simeq \alpha = n\beta$ has been used (Fig. 4.42).

When the interference pattern is imaged by a lens with the focal length f into the plane of the photoplate, we obtain for the ring diameters $D_p = 2f\alpha_p$

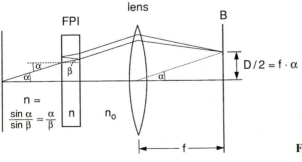

Fig. 4.42. Illustration of (4.63)

the relations

$$(m_0 - p)\lambda = 2nd[1 - (n_0/n)^2 D_p^2/(8f^2)] \,, \tag{4.63a}$$

$$(m_0 - p - 1)\lambda = 2nd[1 - (n_0/n)^2 D_{p+1}^2/(8f^2)] \,. \tag{4.63b}$$

Subtracting the second equation from the first one yields

$$D_{p+1}^2 - D_p^2 = \frac{4nf^2}{n_0^2 d}\lambda \,. \tag{4.64}$$

For the smallest ring with $p = 0$, (4.62) becomes

$$m_0\lambda = 2nd(1 - \beta_0^2/2) \,, \tag{4.65}$$

which can be written as

$$(m_0 + \epsilon)\lambda = 2nd \,. \tag{4.66}$$

The "excess" $\epsilon < 1$, also called *fractional interference order*, can be obtained from a comparison of (4.65) and (4.66) as

$$\epsilon = nd\beta_0^2/\lambda \,. \tag{4.67}$$

Inserting ϵ into (4.63a) yields the relation

$$D_p^2 = \frac{8n^2 f^2}{n_0^2(m_0 + \epsilon)}(p + \epsilon) \,. \tag{4.68}$$

A linear fit of the squares D_p^2 of the measured ring diameters versus the ring number p yields the excess ϵ and therefore from (4.66) the wavelength λ, provided the refractive index n and the value of d are known from a previous calibration of the interferometer. However, the wavelength is determined by (4.66) only modulo a free spectral range $\delta\lambda = \lambda^2/(2nd)$. This means that all wavelengths λ_m differing by m free spectral ranges produce the same ring systems. For an absolute determination of λ, the integer order m_0 must be known.

The experimental scheme for the absolute determination of λ utilizes a combination of FPI and spectrograph in a so-called *crossed arrangement* (Fig. 4.43), where the ring system of the FPI is imaged onto the entrance slit of a spectrograph. The spectrograph disperses the slit images $S(\lambda)$ with a medium dispersion in the x-direction (Sect. 4.1), the FPI provides high dispersion in the y-direction. The resolution of the spectrograph must only be sufficiently high to separate the images of two wavelengths differing by one free spectral range of the FPI. Figure 4.44 shows, for illustration, a section of the Na_2 fluorescence spectrum excited by an argon laser line. The ordinate corresponds to the FPI dispersion and the abscissa to the spectrograph dispersion [4.35].

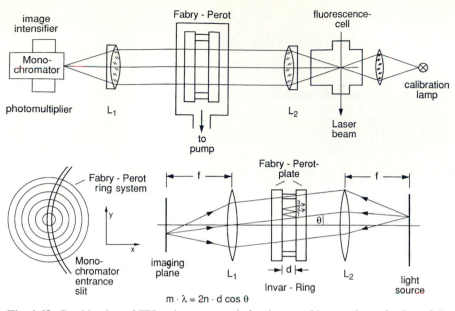

$$m \cdot \lambda = 2n \cdot d \cos \theta$$

Fig. 4.43. Combination of FPI and spectrograph for the unambiguous determination of the integral order m_0

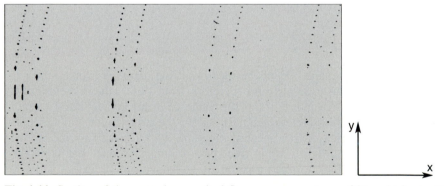

Fig. 4.44. Section of the argon laser-excited fluorescence spectrum of Na_2 obtained with the arrangement of crossed FPI and spectrograph shown in Fig. 4.43 [4.35]

The angular dispersion $d\beta/d\lambda$ of the FPI can be deduced from (4.62)

$$\frac{d\beta}{d\lambda} = \left(\frac{d\lambda}{d\beta}\right)^{-1} = m/(2nd \sin \beta) = \frac{1}{\lambda_m \sin \beta} \quad \text{with} \quad \lambda_m = 2nd/m \,.$$

$$(4.69)$$

Equation (4.69) shows that the angular dispersion becomes infinite for $\beta \to 0$. The linear dispersion of the ring system on the photoplate is

$$\frac{\mathrm{d}D}{\mathrm{d}\lambda} = f\frac{\mathrm{d}\beta}{\mathrm{d}\lambda} = \frac{f}{\lambda_m \sin \beta}. \tag{4.70}$$

Example 4.13

$f = 50\,\mathrm{cm}$, $\lambda = 0.5\,\mu\mathrm{m}$. At a distance of $1\,\mathrm{mm}$ from the ring center is $\beta = 0.1/50$ and we obtain a linear dispersion of $\mathrm{d}D/\mathrm{d}\lambda = 500\,\mathrm{mm/nm}$. This is at least one order of magnitude larger than the dispersion of a large spectrograph.

c) The Air-Spaced FPI

Different from the solid etalon, which is a plane-parallel plate coated on both sides with reflecting layers, the plane FPI consists of two wedged plates, each having one high-reflection and one antireflection coating (Fig. 4.39b). The finesse of the FPI critically depends, apart from the reflectivity R and the optical surface quality, on the parallel alignment of the two reflecting surfaces. The advantage of the air-spaced FPI, that any desired free spectral range can be realized by choosing the corresponding plate separation d, must be paid for by the inconvenience of careful alignment. Instead of changing the angle of incidence α, wavelength tuning can be also achieved for $\alpha = 0$ by variation of the optical path difference $\Delta s = 2nd$, either by changing d with piezoelectric tuning of the plate separation, or by altering the refractive index by a pressure change in the container enclosing the FPI.

The tunable FPI is used for high-resolution spectroscopy of line profiles. The transmitted intensity $I_T(p)$ as a function of the optical path difference nd is given by the convolution

$$I_T(\nu) = I_0(\nu)T(nd, \lambda),$$

where $T(nd, \lambda) = T(\phi)$ can be obtained from (4.50).

With photoelectric recording (Fig. 4.45), the large dispersion at the ring center can be utilized. The light source LS is imaged onto a small pinhole P1, which serves as a point source in the focal plane of L1. The parallel light beam passes the FPI, and the transmitted intensity is imaged by L2 onto another pinhole P2 in front of the detector. All light rays within the cone $\cos \beta_0 \leq m_0\lambda/(nd)$, where β is the angle against the interferometer axis, contribute according to (4.62) to the central fringe. If the optical path length nd is tuned, the different transmission orders with $m = m_0, m_0 + 1, m_0 + 2, \ldots$ are successively transmitted for a wavelength λ according to $m\lambda = 2nd$. Light sources that come close to being a point source, can be realized when a focused laser beam crosses a sample cell and the laser-induced fluorescence emitted from a small section of the beam length is imaged through the FPI

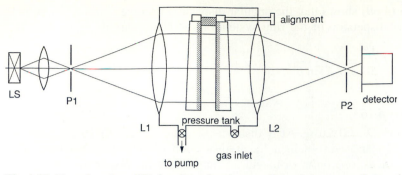

Fig. 4.45. Use of a plane FPI for photoelectric recording of the spectrally resolved transmitted intensity $I_T(n \cdot d, \lambda)$ emitted from a point source

onto the entrance slit of a monochromator, which is tuned to the desired wavelength interval $\Delta\lambda$ around λ_m (Fig. 4.43). If the spectral interval $\Delta\lambda$ resolved by the monochromator is smaller then the free spectral range $\delta\lambda$ of the FPI, an unambigious determination of λ is possible. For illustration, Fig. 4.46 shows a Doppler-broadened fluorescence line of Na_2 molecules excited by a single-mode argon laser at $\lambda = 488$ nm, together with the narrow line profile of the scattered laser light. The pressure change $\Delta p \triangleq 2d\Delta n_L = a$ corresponds to one free spectral range of the FPI, i.e., $2d\Delta n_L = \lambda$.

For Doppler-free resolution of fluorescence lines (Chap. 9), the laser-induced fluorescence of molecules in a collimated molecular beam can be imaged through a FPI onto the entrance slit of the monochromator (Fig. 4.47). In this case, the crossing point of laser and molecular beams, indeed, represents nearly a point source.

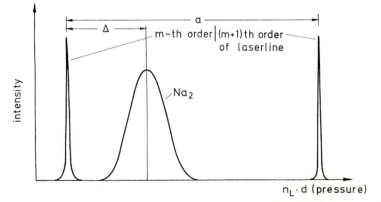

Fig. 4.46. Photoelectric recording of a Doppler-broadened laser-excited fluorescence line of Na_2 molecules in a vapor cell and the Doppler-free scattered laser line. The pressure scan $\Delta p = a$ corresponds to one free spectral range of the FPI

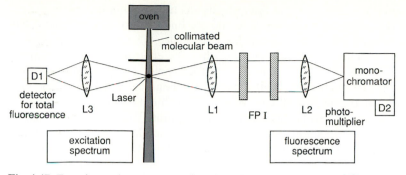

Fig. 4.47. Experimental arrangement for photoelectric recording of high-resolution fluorescence lines excited by a single-mode laser in a collimated molecular beam and observed through FPI plus monochromator

4.2.6 Confocal Fabry–Perot Interferometer

A confocal interferometer, sometimes called incorrectly a *spherical FPI*, consists of two spherical mirrors M_1, M_2 with equal curvatures (radius r) that are opposed at a distance $d = r$ (Fig. 4.48a) [4.36–4.40]. These interferometers

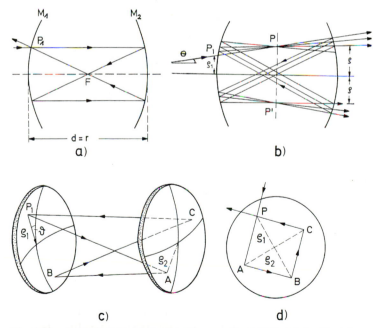

Fig. 4.48a–d. Trajectories of rays in a confocal FPI: (**a**) incident beam parallel to the FPI axis; (**b**) inclined incident beam; (**c**) perspective view for illustrating the skew angle; (**d**) projection of the skewed rays onto the mirror surfaces

have gained great importance in laser physics as high-resolution spectrum analyzers for detecting the mode structure and the linewidth of lasers [4.38–4.40],
and, in the nearly confocal form, as laser resonators (Sect. 5.2).

Neglecting spherical aberration, all light rays entering the interferometer
parallel to its axis would pass through the focal point F and would reach the
entrance point P1 again after having passed the confocal FPI four times. Figure 4.48 illustrates the general case of a ray which enters the confocal FPI at
a small inclination θ and passes the successive points P1, A, B, C, P1, shown
in Fig. 4.48d in a projection. Angle θ is the skew angle of the entering ray.

Because of spherical aberration, rays with different distances ρ_1 from the
axis will not all go through F but will intersect the axis at different positions F' depending on ρ_1 and θ. Also, each ray will not exactly reach the
entrance point P_1 after four passages through the confocal FPI since it is
slightly shifted at successive passages. However, it can be shown [4.36, 4.39]
that for sufficiently small angles θ, all rays intersect at a distance $\rho(\rho_1, \theta)$
from the axis in the vicinity of the two points P and P' located in the central
plane of the confocal FPI (Fig. 4.48b).

The optical path difference Δs between two successive rays passing
through P can be calculated from geometrical optics. For $\rho_1 \ll r$ and $\theta \ll 1$,
one obtains for the near confocal case $d \approx r$ [4.39]

$$\Delta s = 4d + \rho_1^2 \rho_2^2 \cos 2\theta / r^3 + \text{higher-order terms} . \tag{4.71}$$

An incident light beam with diameter $D = 2\rho_1$ therefore produces, in the
central plane of a confocal FPI, an interference pattern of concentric rings.
Analogous of the treatment in Sect. 4.2.5, the intensity $I(\rho, \lambda)$ is obtained by
adding all amplitudes with their correct phases $\delta = \delta_0 + (2\pi/\lambda)\Delta s$. According
to (4.50) we get

$$I(\rho, \lambda) = \frac{I_0 T^2}{(1 - R)^2 + 4R \sin^2[(\pi/\lambda)\Delta s]} , \tag{4.72}$$

where $T = 1 - R - A$ is the transmission of each of the two mirrors. The intensity has maxima for $\delta = 2m\pi$, which is equivalent to

$$4d + \rho^4/r^3 = m\lambda , \tag{4.73}$$

when we neglect the higher-order terms in (4.71) and set $\theta = 0$ and $\rho^2 = \rho_1\rho_2$.

The free spectral range $\delta\nu$, i.e., the frequency separation between successive interference maxima, is for the near-confocal FPI with $\rho \ll d$

$$\delta\nu = \frac{c}{4d + \rho^4/r^3} , \tag{4.74}$$

which is *different* from the expression $\delta\nu = c/2d$ for the plane FPI.

The radius ρ_m of the mth-order interference ring is obtained from (4.73),

$$\rho_m = [(m\lambda - 4d)r^3]^{1/4} , \tag{4.75}$$

which reveals that ρ_m depends critically on the separation d of the spherical mirrors. Changing d by a small amount ϵ from $d = r$ to $d = r + \epsilon$ changes the path difference to

$$\Delta s = 4(r + \epsilon) + \rho^4/(r + \epsilon)^3 \sim 4(r + \epsilon) + \rho^4/r^3 . \tag{4.76}$$

For a given wavelength λ, the value of ϵ can be chosen such that $4(r + \epsilon) = m_0\lambda$. In this case, the radius of the central ring becomes zero. We can number the outer rings by the integer p and obtain with $m = m_0 + p$ for the radius of the pth ring the expression

$$\rho_p = (p\lambda r^3)^{1/4} . \tag{4.77}$$

The radial dispersion deduced from (4.75),

$$\frac{d\rho}{d\lambda} = \frac{mr^3/4}{[(m\lambda - 4d)r^3]^{3/4}} , \tag{4.78}$$

becomes infinite for $m\lambda = 4d$, which occurs according to (4.75) at the center with $\rho = 0$.

This large dispersion can be used for high-resolution spectroscopy of narrow line profiles with a scanning confocal FPI and photoelectric recording (Fig. 4.49).

If the central plane of the near-confocal FPI is imaged by a lens onto a circular aperture with sufficiently small radius $b < (\lambda r^3)^{1/4}$ only the central interference order is transmitted to the detector while all other orders are stopped. Because of the large radial dispersion for small ρ one obtains a high spectral resolving power. With this arrangement not only spectral line profiles but also the instrumental bandwidth can be measured, when an incident monochromatic wave (from a stabilized single-mode laser) is used. The mirror separation $d = r + \epsilon$ is varied by the small amount ϵ and the power

$$P(\lambda, b, \epsilon) = 2\pi \int_{\rho=0}^{b} \rho I(\rho, \lambda, \epsilon)\, d\rho , \tag{4.79}$$

transmitted through the aperture is measured as a function of ϵ at fixed values of λ and b.

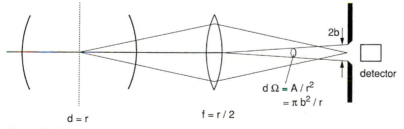

Fig. 4.49. Photoelectric recording of the spectral light power transmitted of a scanning confocal FPI

The integrand $I(\rho, \lambda, \epsilon)$ can be obtained from (4.72), where the phase difference $\delta(\epsilon) = 2\pi \Delta s / \lambda$ is deduced from (4.76).

The optimum choice for the radius of the aperture is based on a compromise between spectral resolution and transmitted intensity. When the interferometer has the finesse F^*, the spectral halfwidth of the transmission peak is $\delta \nu / F^*$, see (4.53b), and the maximum spectral resolving power becomes $F^* \Delta s / \lambda$ (4.56). For the radius $b = (r^3 \lambda / F^*)^{1/4}$ of the aperture, which is just $(F^*)^{1/4}$ times the radius ρ_1 of a fringe with $p = 1$ in (4.77), the spectral resolving power is reduced to about 70% of its maximum value. This can be verified by inserting this value of b into (4.79) and calculating the halfwidth of the transmission peak $P(\lambda_1, F^*, \epsilon)$.

The total finesse of the confocal FPI is, in general, higher than that of a plane FPI for the following reasons:

- The alignment of spherical mirrors is far less critical than that of plane mirrors, because tilting of the spherical mirrors does not change (to a first approximation) the optical path length $4r$ through the confocal FPI, which remains approximately the same for all incident rays (Fig. 4.50). For the plane FPI, however, the path length increases for rays below the interferometer axis, but decreases for rays above the axis.
- Spherical mirrors can be polished to a higher precision than plane mirrors. This means that the deviations from an ideal sphere are less for spherical mirrors than those from an ideal plane for plane mirrors. Furthermore, such deviations do not wash out the interference structure but cause only a distortion of the ring system because a change of r allows the same path difference Δs for another value of ρ according to (4.71).

The total finesse of a confocal FPI is therefore mainly determined by the reflectivity R of the mirrors. For $R = 0.99$, a finesse $F^* = \pi \sqrt{R/(1-R)} \approx 300$ can be achieved, which is much higher than that obtainable with a plane FPI, where other factors decrease F^*. With the mirror separation $r = d = 3$ cm, the free spectral range is $\delta = 2.5$ GHz and the spectral resolution is $\Delta \nu = 7.5$ MHz at the finesse $F^* = 300$. This is sufficient to measure the natural linewidth of many optical transitions. With modern high-reflection coatings, values of $R = 0.9995$ can be obtained and confocal FPI with a finesse $F^* \geq 10^4$ have been realized [4.41].

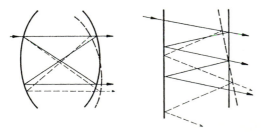

Fig. 4.50. Illustration of the larger sensitivity against misalignment for the plane FPI compared with the spherical FPI

From Fig. 4.49 we see that the solid angle accepted by the detector behind the aperture with radius b is $\Omega = \pi b^2 / r^2$. The light power transmitted to the detector is proportional to the product of the solid angle Ω and area A in the central plane, which is imaged by the lens onto the aperture (often called the *étendue U*). With the aperture radius $b = (r^3 \lambda / F^*)^{1/4}$ (see above) the étendue becomes

$$U = A\Omega = \pi^2 b^4 / r^2 = \pi^2 r \lambda / F^* . \tag{4.80}$$

For a given finesse F^*, the étendue of the confocal FPI increases with the mirror separation $d = r$. The spectral resolving power

$$\frac{\nu}{\Delta\nu} = 4F^* \frac{r}{\lambda} , \tag{4.81}$$

of the confocal FPI is proportional to the product of finesse F^* and the ratio of mirror separation $r = d$ to the wavelength λ. With a given étendue $U = \pi^2 r \lambda / F^*$, the spectral resolving power is

$$\frac{\nu}{\Delta\nu} = \left(\frac{2F^*}{\pi\lambda}\right)^2 U , \quad \text{(confocal FPI)} . \tag{4.82}$$

Let us compare this with the case of a plane FPI with the plate diameter D and the separation d, which is illuminated with nearly parallel light (Fig. 4.45). According to (4.62), the path difference between a ray parallel to the axis and a ray with an inclination β is, for small β, given by $\Delta s = n d \beta^2$. To achieve a finesse F^* with photoelectric recording, this variation of the path length should not exceed λ / F^*, which restricts the solid angle $\Omega = \beta^2$ acceptable by the detector to $\Omega \leq \lambda / (d \cdot F^*)$. The étendue is therefore

$$U = A\Omega = \pi \frac{D^2}{4} \frac{\lambda}{d \cdot F^*} . \tag{4.83}$$

Inserting the value of d given by this equation into the spectral resolving power $\nu / \Delta\nu = 2dF^* / \lambda$, we obtain

$$\frac{\nu}{\Delta\nu} = \frac{\pi D^2}{2U} , \quad \text{(plane FPI)} . \tag{4.84}$$

While the spectral resolving power is proportional to U for the confocal FPI, it is *inversely proportional to U for the plane FPI*. This is because the étendue increases with the mirror separation d for the confocal FPI but decreases proportional to $1/d$ for the plane FPI. For a mirror radius $r > D^2/4d$, the étendue of the confocal FPI is larger than that of a plane FPI with equal spectral resolution.

Example 4.14

A confocal FPI with $r = d = 5$ cm has for $\lambda = 500$ nm the étendue $U = (2.47 \times 10^{-3} / F^*)$ cm^2/sr. This is the same étendue as that of a plane FPI with $d = 5$ cm and $D = 10$ cm. However, the diameter of the spherical mirrors can be much smaller (less than 5 mm). With a finesse $F^* = 100$, the étendue is $U = 2.5 \times 10^{-5}$ [cm^2 sr] and the spectral resolving power is $\nu/\Delta\nu = 4 \times 10^7$. With this étendue the resolving power of the plane FPI is 6×10^6, provided the whole plane mirror surface has a surface quality to allow a surface finesse of $F^* \geq 100$. In practice, this is difficult to achieve for a flat plane with $D = 10$ cm diameter, while for the small spherical mirrors even $F^* > 10^4$ is feasible.

This example shows that for a given light-gathering power, the confocal FPI can have a much higher spectral resolving power than the plane FPI.

More detailed information on the history, theory, practice, and application of plane and spherical Fabry–Perot interferometers may be found in [4.42–4.44].

4.2.7 Multilayer Dielectric Coatings

The constructive interference found for the reflection of light from plane-parallel interfaces between two regions with different refractive indices can be utilized to produce highly reflecting, essentially absorption-free mirrors. The improved technology of such dielectric mirrors has greatly supported the development of visible and ultraviolet laser systems.

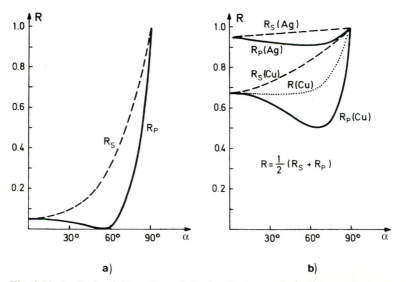

a)   b)

Fig. 4.51a,b. Reflectivities R_p and R_s for the two polarization components parallel and perpendicular to the plane of incidence as a function of the angle of incidence α: (**a**) air–glass boundary ($n = 1, 5$); (**b**) air–metal boundary for Cu($n' = 0.76$, $\kappa = 3.32$) and Ag($n' = 0.055$, $\kappa = 3.32$)

The reflectivity R of a plane interface between two regions with complex refractive indices $n_1 = n_1' - i\kappa_1$ and $n_2 = n_2' - i\kappa_2$ can be calculated from Fresnel's formulas [4.15]. It depends on the angle of incidence α and on the direction of polarization. Reflectivity $R(\alpha)$ is illustrated in Fig. 4.51 for three different materials for incident light polarized parallel (R_p) and perpendicular (R_s) to the plane of incidence.

For vertical incidence $(\alpha = 0)$, one obtains from Fresnel's formulas for both polarizations

$$R|_{\alpha=0} = \left(\frac{n_1 - n_2}{n_1 + n_2}\right)^2 . \tag{4.85}$$

Since this case represents the most common situation for laser mirrors, we shall restrict the following discussion to vertical incidence.

To achieve maximum reflectivities, the numerator $(n_1 - n_2)^2$ should be maximized and the denominator minimized. Since n_1 is always larger than one, this implies that n_2 should be as large as possible. Unfortunately, the dispersion relations (3.36), (3.37) imply that a large value of n also causes large absorption. For instance, highly polished metal surfaces have a maximum reflectivity of $R = 0.95$ in the visible spectral range. The residual 5% of the incident intensity is absorbed and therefore lost.

The situation can be improved by selecting reflecting materials with low absorption (which then necessarily also have low reflectivity), but using many layers with alternating high and low refractive index n. Choosing the proper optical thickness nd of each layer allows constructive interference between the different reflected amplitudes to be achieved. Reflectivities of up to $R = 0.9995$ have been reached [4.45–4.48].

Figure 4.52 illustrates such constructive interference for the example of a two-layer coating. The layers with refractive indices n_1, n_2 and thicknesses d_1, d_2 are evaporated onto an optically smooth substrate with the refractive index n_3. The phase differences between all reflected components have to be $\delta_m = 2m\pi$ $(m = 1, 2, 3, \dots)$ for constructive interference. Taking into account the phase shift $\delta = \pi$ at reflection from an interface with a larger refractive

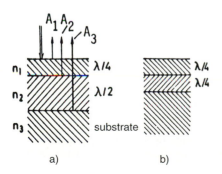

Fig. 4.52a,b. Maximum reflection of light with wavelength λ by a two-layer dielectric coating: (**a**) $n_1 > n_2 > n_3$; (**b**) $n_1 > n_2 < n_3$

index, we obtain the conditions

$$n_1 d_1 = \lambda/4 \quad \text{and} \quad n_2 d_2 = \lambda/2 \quad \text{for} \quad n_1 > n_2 > n_3 \,, \tag{4.86a}$$

and

$$n_1 d_1 = n_2 d_2 = \lambda/4 \quad \text{for} \quad n_1 > n_2, n_3 > n_2 \,. \tag{4.86b}$$

The reflected amplitudes can be calculated from Fresnel's formulas. The total reflected intensity is obtained by summation over all reflected amplitudes taking into account the correct phase. The refractive indices are now selected such that $\sum A_i$ becomes a maximum. The calculation is still feasible for our example of a two-layer coating and yields for the three reflected amplitudes (double reflections are neglected)

$$A_1 = \sqrt{R_1} A_0; \quad A_2 = \sqrt{R_2}(1 - \sqrt{R_1}) A_0 \,,$$
$$A_3 = \sqrt{R_3}(1 - \sqrt{R_2})(1 - \sqrt{R_1}) A_0 \,,$$

where the reflectivities R_i are given by (4.85).

Example 4.15

$|n_1| = 1.6$, $|n_2| = 1.2$, $|n_3| = 1.45$; $A_1 = 0.231 A_0$, $A_2 = 0.143 A_0$, $A_3 = 0.094 A_0$. $A_R = \sum A_i = 0.468 A_0 \to I_R = 0.22 I_0 \to R = 0.22$, provided the path differences have been choosen correctly.

This example illustrates that for materials with low absorption, many layers are necessary to achieve a high reflectivity. Figure 4.53a depicts schematically the composition of a dielectric multilayer mirror. The calculation and optimization of multilayer coatings with up to 20 layers becomes

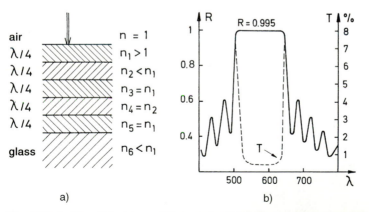

a)   b)

Fig. 4.53a,b. The dielectric multilayer mirror: **(a)** Composition of multilayers; **(b)** Reflectivity of a high-reflectance multilayer mirror with 17 layers as a function of the incident wavelength λ

very tedious and time consuming, and is therefore performed using computer programs [4.46, 4.48]. Figure 4.53b illustrates the reflectivity $R(\lambda)$ of a high-reflectance mirror with 17 layers.

By proper selection of different layers with slightly different optical path lengths, one can achieve a high reflectivity over an extended spectral range. Currently, "broad-band" reflectors are available with reflectivity of $R \geq 0.99$ within the spectral range $(\lambda_0 \pm 0.2\lambda_0)$, while the absorption losses are less than 0.2% [4.45, 4.47]. At such low absorption losses, the scattering of light from imperfect mirror surfaces may become the major loss contribution. When total losses of less than 0.5% are demanded, the mirror substrate must be of high optical quality (better than $\lambda/20$), the dielectric layers have to be evaporated very uniformly, and the mirror surface must be clean and free of dust or dirty films [4.48]. The best mirrors are produced by ion implantation techniques.

Instead of maximizing the reflectivity of a dielectric multilayer coating through *constructive* interference, it is, of course, also possible to minimize it by destructive interference. Such *antireflection coatings* are commonly used to minimize unwanted reflections from the many surfaces of multiple-lens camera objectives, which would otherwise produce an annoying background illumination of the photomaterial. In laser spectroscopy such coatings are important for minimizing reflection losses of optical components inside the laser resonator and for avoiding reflections from the back surface of output mirrors, which would introduce undesirable couplings, thereby causing frequency instabilities of single-mode lasers.

Using a single layer (Fig. 4.54a), the reflectivity reaches a minimum only for a selected wavelength λ (Fig. 4.55). We obtain $I_R = 0$ for $\delta = (2m + 1)\pi$, if the two amplitudes A_1 and A_2 reflected by the interfaces (n_1, n_2) and (n_2, n_3) are equal. For vertical incidence this gives the condition

$$R_1 = \left(\frac{n_1 - n_2}{n_1 + n_2}\right)^2 = R_2 = \left(\frac{n_2 - n_3}{n_2 + n_3}\right)^2 , \tag{4.87}$$

which can be reduced to

$$n_2 = \sqrt{n_1 n_3} . \tag{4.88}$$

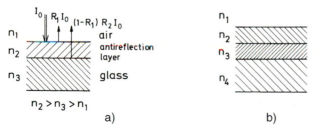

Fig. 4.54. Antireflection coating: (**a**) single layer; (**b**) multilayer coating

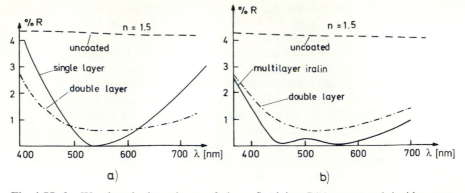

Fig. 4.55a,b. Wavelength dependence of the reflectivity $R(\lambda)$ at normal incidence on a substrate with $n = 1.5$ at $\lambda = 550\,\text{nm}$: (**a**) for uncoated substrate, single and double, antireflection coatings; (**b**) comparison of double and multilayer coatings

For a single layer on a glass substrate the values are $n_1 = 1$ and $n_3 = 1.5$. According to (4.88), n_2 should be $n_2 = \sqrt{1.5} = 1.23$. Durable coatings with such low refractive indices are not available. One often uses MgF_2 with $n_2 = 1.38$, giving a reduction of reflection from 4% to 1.2% (Fig. 4.55).

With multilayer antireflection coatings the reflectivity can be decreased below 0.2% for an extended spectral range [4.49]. For instance, with three $\lambda/4$ layers (MgF_2, SiO, and CeF_3) the reflection drops to below 1% for the whole range between 420 nm and 840 nm [4.45, 4.50, 4.51].

4.2.8 Interference Filters

Interference filters are used for selective transmission in a narrow spectral range. Incident radiation of wavelengths outside this transmission range is either reflected or absorbed. One distinguishes between line filters and bandpass filters.

A line filter is essentially a Fabry–Perot etalon with a very small optical path nd between the two reflecting surfaces. The technical realization uses two highly reflecting coatings (either silver coatings or dielectric multilayer coatings) that are separated by a nonabsorbing layer with a low refractive index (Fig. 4.56). For instance, for $nd = 0.5\,\mu\text{m}$ the transmission maxima for vertical incidence are obtained from (4.58a) at $\lambda_1 = 1\,\mu\text{m}$, $\lambda_2 = 0.5\,\mu\text{m}$, $\lambda_3 = 0.33\,\mu\text{m}$, etc. In the visible range this filter has therefore only one transmission peak at $\lambda = 500\,\text{nm}$, with a halfwidth that depends on the finesse $F^* = \pi\sqrt{R}/(1 - R)$ (Fig. 4.37).

The interference filter is characterized by the following quantities:

- The wavelength λ_m at peak transmission;
- The maximum transmission;
- The contrast factor, which gives the ratio of maximum to minimum transmission;
- The bandwidth at half-transmitted peak intensity.

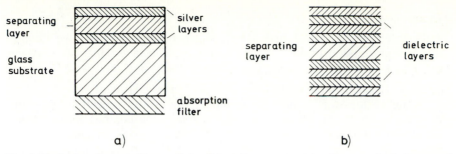

Fig. 4.56a,b. Interference filters of the Fabry–Perot type: (a) with two single layers of silver; (b) with dielectric multilayer coatings

The maximum transmission according to (4.57) is $T^*_{max} = T^2/(1-R)^2$. Using thin silver or aluminum coatings with $R = 0.8$, $T = 0.1$, and $A = 0.1$, the transmission of the filter is only $T^* = 0.25$ and the finesse $F^* = 15$. For our example this means a halfwidth of $660\,\text{cm}^{-1}$ at a free spectral range of $10^4\,\text{cm}^{-1}$. At $\lambda = 500\,\text{nm}$ this corresponds to a halfwidth of about $16\,\text{nm}$. For many applications in laser spectroscopy, the low peak transmission of interference filters with absorbing metal coatings is not tolerable. One has to use absorption-free dielectric multilayer coatings (Fig. 4.56b) with high reflectivity, which allows a large finesse and therefore a smaller bandwidth and a larger peak transmission (Fig. 4.57).

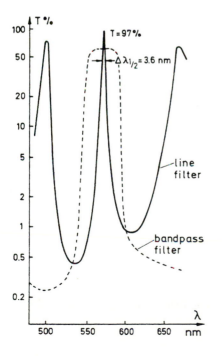

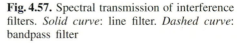

Fig. 4.57. Spectral transmission of interference filters. *Solid curve*: line filter. *Dashed curve*: bandpass filter

Example 4.16

With $R = 0.95$, $A = 0.01$ and $T = 0.04$, according to (4.57) we obtain a peak transmission of 64%, which increases with $A = 0.005$, $T = 0.045$ to 81%. The contrast becomes $\gamma = I_T^{max}/I_T^{min} = (1 + F) = 1 + 4F^{*2}/\pi^2 = 1520$. The free spectral range is $\delta\nu = 3 \times 10^{12}$ Hz $\hat{=} 4\,\mu$m.

A higher finesse F^* caused larger reflectivities of the reflecting films not only decreases the bandwidth but also increases the contrast factor. With $R = 0.98 \rightarrow F = 4R/(1 - R)^2 = 9.8 \times 10^3$, which means that the intensity at the transmission minimum is only about 10^{-4} of the peak transmission.

The bandwidth can be further decreased by using two interference filters in series. However, it is preferable to construct a double filter that consists of three highly-reflecting surfaces, separated by two nonabsorbing layers of the same optical thickness. If the thickness of these two layers is made slightly different, a bandpass filter that has a flat transmission curve but steep slopes to both sides results. Commercial interference filters are currently available with a peak transmission of at least 90% and a bandwidth of less than 2 nm [4.46, 4.52]. Special narrow-band filters even reach 0.3 nm, however, with reduced peak transmission.

The wavelength λ_m of the transmission peak can be shifted to lower values by tilting the interference filter, which increases the angle of incidence α, see (4.58a). The tuning range is, however, restricted, because the reflectivity of the multilayer coatings also depends on the angle α and is, in general, optimized for $\alpha = 0$. The transmission bandwidth increases for divergent incident light with the divergence angle.

In the ultraviolet region, where the absorption of most materials used for interference filters becomes large, the selective *reflectance* of interference filters can be utilized to achieve narrow-band filters with low losses (Fig. 4.58). For more detailed treatment, see [4.45–4.52].

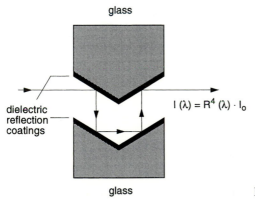

Fig. 4.58. Reflection interference filter

4.2.9 Birefringent Interferometer

The basic principle of the birefringent interferometer or *Lyot filter* [4.23, 4.53] is founded on the interference of polarized light that has passed through a birefringent crystal. Assume that a linearly polarized plane wave

$$\boldsymbol{E} = \boldsymbol{A} \cdot \cos(\omega t - kx) \,,$$

with

$$\boldsymbol{A} = \{0, A_y, A_z\} \,, \quad A_y = |A| \sin\alpha \,, \quad A_z = |A| \cos\alpha \,,$$

is incident on the birefringent crystal (Fig. 4.59). The electric vector $\boldsymbol{E}$ makes an angle α with the optical axis, which points into the z-direction. Within the crystal, the wave is split into an ordinary beam with the wave number $k_o = n_o k$ and the phase velocity $v_o = c/n_o$, and an extraordinary beam with $k_e = n_e k$ and $v_e = c/n_e$. The partial waves have mutually orthogonal polarization in directions parallel to the z- and y-axis, respectively. Let the crystal with length L be placed between $x = 0$ and $x = L$. Because of the different refractive indices n_o and n_e for the ordinary and the extraordinary beams, the two partial waves at $x = L$

$$E_y(L) = A_y \cos(\omega t - k_e L) \quad \text{and} \quad E_z(L) = A_z \cos(\omega t - k_0 L) \,,$$

show a phase difference of

$$\phi = k(n_0 - n_e)L = (2\pi/\lambda)\Delta n L \quad \text{with} \quad \Delta n = n_0 - n_e \,. \tag{4.89}$$

The superposition of these two waves results, in general, in elliptically polarized light, where the principal axis of the ellipse is turned by an angle $\beta = \phi/2$ against the direction of $\boldsymbol{A}_0$.

For phase differences $\phi = 2m\pi$, linearly polarized light with $\boldsymbol{E}(L) \parallel \boldsymbol{E}(0)$ is obtained. However, for $\phi = (2m + 1)\pi$ and $\alpha = 45°$, the transmitted wave is also linearly polarized, but now $\boldsymbol{E}(L) \perp \boldsymbol{E}(0)$.

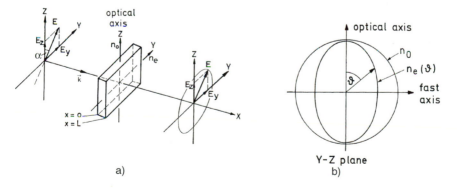

Fig. 4.59a,b. Lyot filter: (**a**) schematic arrangement; (**b**) index ellipsoid of the birefringent crystal

The elementary Lyot filter consists of a birefringent crystal placed between two linear polarizers (Fig. 4.59a). Assume that the two polarizers are both parallel to the electric vector $E(0)$ of the incoming wave. The second polarizer parallel to $E(0)$ transmits only the projection

$$E = E_y \sin \alpha + E_z \cos \alpha$$
$$= A[\sin^2 \alpha \cos(\omega t - k_e L) + \cos^2 \alpha \cos(\omega t - k_0 L)] \, ,$$

of the amplitudes, which yields the transmitted time averaged intensity

$$\bar{I}_T = \frac{1}{2} c \epsilon_0 \bar{E}^2 = \bar{I}_0 (\sin^4 \alpha + \cos^4 \alpha + 2 \sin^2 \alpha \cos^2 \alpha \cos \phi) \, . \tag{4.90}$$

Using the relations $\cos \phi = 1 - 2 \sin^2 \frac{1}{2} \phi$, and $2 \sin \alpha \cos \alpha = \sin 2\alpha$, this reduces to

$$\bar{I}_T = I_0 [1 - \sin^2 \tfrac{1}{2} \phi \sin^2 (2\alpha)] \, , \tag{4.91}$$

which gives for $\alpha = 45°$

$$I_T = I_0 [1 - \sin^2 \tfrac{1}{2} \phi] = I_0 \cos^2 \tfrac{1}{2} \phi \, . \tag{4.91a}$$

The transmission of the Lyot filter is therefore a function of the phase retardation, i.e.,

$$T(\lambda) = \frac{I_T}{I_0} = T_0 \cos^2 \left(\frac{\pi \Delta n L}{\lambda} \right) \, . \tag{4.92}$$

Taking into account absorption and reflection losses, the maximum transmission $I_T / I_0 = T_0 < 1$ becomes less than 100%. Within a small wavelength interval, the difference $\Delta n = n_0 - n_e$ can be regarded as constant. Therefore (4.92) gives the wavelength-dependent transmission function, $\cos^2 \phi$, typical of a two-beam interferometer (Fig. 4.26). For extended spectral ranges the different dispersion of $n_o(\lambda)$ and $n_e(\lambda)$ has to be considered, which causes a wavelength dependence, $\Delta n(\lambda)$.

The free spectral range $\delta \nu$ is obtained from (4.92) as

$$\frac{\Delta n \cdot L}{\lambda_1} - \frac{\Delta n \cdot L}{\lambda_2} = 1 \, .$$

With $\nu = c/\lambda$, this becomes

$$\delta \nu = \frac{c}{(n_0 - n_e) L} \, . \tag{4.93}$$

Note: According to (4.91) the maximum modulation of the transmittance with $T_{max} = T_0$ and $T_{min} = 0$ is only achieved for $\alpha = 45°$!

Example 4.17

For a crystal of potassium dihydrogen phosphate (KDP), $n_e = 1.51$, $n_0 = 1.47 \rightarrow \Delta n = 0.04$ at $\lambda = 600\,\text{nm}$. A crystal with $L = 2\,\text{cm}$ then has a free spectral range $\delta\nu = 3.75 \times 10^{11}\,\text{Hz} \;\hat{=}\; \delta\bar{\nu} = 12.5\,\text{cm}^{-1} \rightarrow \Delta\lambda = 0.45\,\text{nm}$ at $\lambda = 600\,\text{nm}$.

If N elementary Lyot filters with different lengths L_m are placed in series, the total transmission T is the product of the different transmissions T_m, i.e.,

$$T(\lambda) = \prod_{m=1}^{N} T_{0m} \cos^2\left(\frac{\pi \Delta n L_m}{\lambda}\right).
\tag{4.94}$$

Figure 4.60 illustrates a possible experimental arrangement and the corresponding transmission for a Lyot filter composed of three components with the lengths $L_1 = L$, $L_2 = 2L$, and $L_3 = 4L$. The free spectral range $\delta\nu$ of this filter equals that of the shortest component; the halfwidth $\Delta\nu$ of the transmission peaks is, however, mainly determined by the longest component. When we define, analogous to the Fabry–Perot interferometer, the finesse F^* of the Lyot filter as the ratio of the free spectral range $\delta\nu$ to the halfwidth $\Delta\nu$, we

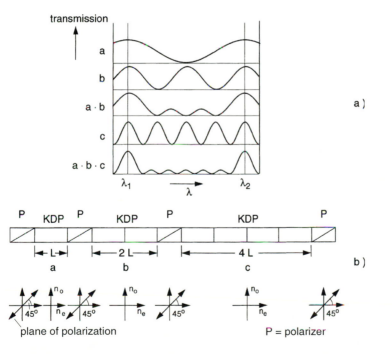

Fig. 4.60. (a) Transmitted intensity $I_T(\lambda)$ of a Lyot filter composed of three birefringent crystals with lengths L, $2L$, and $4L$ between polarizers. (b) Arrangement of the crystals and the state of polarization of the transmitted wave

obtain, for a composite Lyot filter with N elements of lengths $L_m = 2^{m-1}L_1$, a finesse that is approximately $F^* = 2^N$.

The wavelength of the transmission peak can be tuned by changing the difference $\Delta n = n_o - n_e$. This can be realized in two different ways:

- By changing the angle θ between the optical axis and the wave vector $\mathbf{k}$, which alters the index n_e. This can be illustrated with the index ellipsoid (Fig. 4.59b), which gives both refractive indices for a given wavelength as a function of θ. The difference $\Delta n = n_o - n_e$ therefore depends on θ. The two axes of the ellipsoid with minimum n_e ($\theta = 90°$ for a negative bire-fringent crystal) and maximum n_o ($\theta = 0°$) are often called the *fast* and the *slow* axes. Turning the crystal around the y-axis in Fig. 4.59a, which is perpendicular to the x–z-plane of Fig. 4.59b, results in a continuous change of Δn and a corresponding tuning of the peak transmission wavelength λ (Sect. 5.7.4).
- By using the different dependence of the refractive indices n_o and n_e on an applied electric field [4.55]. This "induced birefringence" depends on the orientation of the crystal axis in the electric field. A common arrangement employs a potassium dihydrogen phosphate (KDP) crystal with an orien-tation where the electric field is parallel to the optical axis (z-axis) and the wave vector $\mathbf{k}$ of the incident wave is perpendicular to the z-direction (transverse electro-optic effect, Fig. 4.61). Two opposite sides of the rect-angular crystal with the side length d are coated with gold electrodes and the electric field $E = U/d$ is controlled by the applied voltage.

In the external electric field the uniaxial crystal becomes biaxial. In ad-dition to the natural birefringence of the uniaxial crystal, a field-induced birefringence is generated, which is approximately proportional to the field strength E [4.56]. The changes of n_o or n_e by the electric field depend on the symmetry of the crystal, the direction of the applied field, and on the magnitude of the electro-optic coefficients. For the KDP crystal only one electro-optic coefficient $d_{36} = -10.7 \times 10^{-12}$ [m/V] (see Sect. 5.8.1) is effec-tive if the field is applied parallel to the optical axis.

The difference $\Delta n = n_o - n_e$ then becomes

$$\Delta n(E_z) = \Delta n(E = 0) + \frac{1}{2}n_1^3 d_{36} E_z \ . \tag{4.95}$$

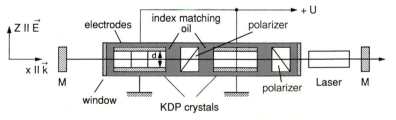

Fig. 4.61. Electro-optic tuning of a Lyot filter [4.54]

Maximum transmittance is obtained for

$$\Delta n L = m\lambda \quad (m = 0, 1, 2 \ldots),$$

which gives the wavelength λ at the maximum transmittance

$$\lambda = (\Delta n(E=0) + 0.5\,n_1{}^3 d_{36} E_z)L/m,\tag{4.96}$$

as a function of the applied field.

While this electro-optic tuning of the Lyot filter allows rapid switching of the peak transmission, for many applications, where a high tuning speed is not demanded, mechanical tuning is more convenient and easier to realize.

4.2.10 Tunable Interferometers

For many applications in laser spectroscopy it is advantageous to have a high-resolution interferometer that is able to scan, in a given time interval Δt, through a limited spectral range $\Delta\nu$. The scanning speed $\Delta\nu/\Delta t$ depends on the method used for tuning, while the spectral range $\Delta\nu$ is limited by the free spectral range $\delta\nu$ of the instrument. All techniques for tuning the wavelength $\lambda_m = 2nd/m$ at the transmission peak of an interferometer are based on a continuous change of the optical path difference between successive interfering beams. This can be achieved in different ways:

(a) Change the refractive index n by altering the pressure between the reflecting plates of a FPI (pressure-scanned FPI);
(b) Change the distance d with piezoelectric or magnetostrictive elements;
(c) Tilt the solid etalons with a given thickness d against the direction of the incoming plane wave;
(d) Change the optical path difference $\Delta s = \Delta n L$ in birefringent crystals by electro-optic tuning or by turning the optical axis of the crystal (Lyot filter).

While method (a) is often used for high-resolution fluorescence spectroscopy with slow scan rates or for tuning pulsed dye lasers, method (b) is realized in a scanning confocal FPI (used as an optical spectrum analyzer) for monitoring the mode structure of lasers.

With a commercial spectrum analyzer, the transmitted wavelength λ can be repetitively scanned over more than one free spectral range with a saw-tooth voltage applied to the piezoelectric distance holder [4.38, 4.57]. Scanning rates up to several kilohertz are possible. Although the finesse of such devices may exceed 10^3, the hysteresis of piezoelectric crystals limits the accuracy of absolute wavelength calibration. Here a pressure-tuned FPI may be advantageous. The pressure change has to be sufficiently slow to avoid turbulence and temperature drifts. With a digitally pressure-scanned FPI, where the pressure of the gas in the interferometer chamber is changed by small, discrete steps, repetitive scans are reproduced within about 10^{-3} of the free spectral range [4.58].

For fast wavelength tuning of dye lasers, Lyot filters with electro-optic tuning are employed within the laser resonator. A tuning range of a few nanometers can be repetitively scanned with rates up to 10^5 per second [4.59].

4.3 Comparison Between Spectrometers and Interferometers

When comparing the advantages and disadvantages of different dispersing devices for spectroscopic analysis, the characteristic properties of the instruments discussed in the foregoing sections, such as *spectral resolving power, étendue, spectral transmission*, and *free spectral* range, are important for the optimum choice. Of equal significance is the question of how accurately the wavelengths of spectral lines can be measured. To answer this question, further specifications are necessary, such as the backlash of monochromator drives, imaging errors in spectrographs, and hysteresis in piezo-tuned interferometers. In this section we shall treat these points in a comparison for different devices in order to give the reader an impression of the capabilities and limitations of these instruments.

4.3.1 Spectral Resolving Power

The spectral resolving power discussed for the different instruments in the previous sections can be expressed in a more general way, which applies to all devices with spectral dispersion based on interference effects. Let Δs_m be the maximum path difference between interfering waves in the instrument, e.g., between the rays from the first and the last groove of a grating (Fig. 4.62a) or between the direct beam and a beam reflected m times in a Fabry–Perot interferometer (Fig. 4.62b). Two wavelengths λ_1 and $\lambda_2 = \lambda_1 + \Delta\lambda$ can still be

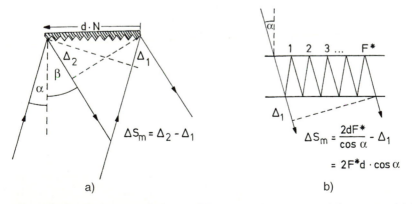

Fig. 4.62a,b. Maximum optical path difference and spectral resolving power: (**a**) in a grating spectrometer; (**b**) in a Fabry–Perot interferometer

resolved if the number of wavelengths over this maximum path difference

$$\Delta s_m = 2m\lambda_2 = (2m+1)\lambda_1 , \quad m = \text{integer} ,$$

differs for the two wavelengths by at least one unit. In this case, an interference maximum for λ_1 coincides with the first minimum for λ_2. From the above equation we obtain the theoretical upper limit for the resolving power

$$\boxed{\frac{\lambda}{\Delta\lambda} = \frac{\Delta s_m}{\lambda}} , \tag{4.97}$$

which is equal to the maximum path difference measured in units of the wavelength λ.

With the maximum time difference $\Delta T_m = \Delta s_m/c$ for traversing the two paths with the path difference Δs_m, we obtain with $\nu = c/\lambda$ from (4.97) for the minimum resolvable interval $\Delta\nu = -(c/\lambda^2)\Delta\lambda$,

$$\Delta\nu = 1/\Delta T_m \;\Rightarrow\; \Delta\nu \cdot \Delta T_m = 1 . \tag{4.98}$$

The product of the minimum resolvable frequency interval $\Delta\nu$ and the maximum difference in transit times through the spectral apparatus is equal to 1.

Example 4.18

(a) **Grating Spectrometer:** The maximum path difference is, according to (4.30) and Fig. 4.62,

$$\Delta s_m = Nd(\sin\alpha - \sin\beta) = mN\lambda .$$

The upper limit for the resolving power is therefore, according to (4.97),

$$R = \lambda/\Delta\lambda = mN \quad (m : \text{diffraction order,}$$
$$N : \text{number of illuminated grooves}) .$$

For $m = 2$ and $N = 10^5$ this gives $R = 2 \times 10^5$, or $\Delta\lambda = 5 \times 10^{-6}\lambda$. Because of diffraction, which depends on the size of the grating (Sect. 4.1.3), the realizable resolving power is 2–3 times lower. This means that at $\lambda = 500\,\text{nm}$, two lines with $\Delta\lambda \geq 10^{-2}\,\text{nm}$ can still be resolved.

(b) **Michelson Interferometer:** The path difference Δs between the two interfering beams is changed from $\Delta s = 0$ to $\Delta s = \Delta s_m$. The numbers of interference maxima are counted for the two components λ_1 and λ_2 (Sect. 4.2.3). A distinction between λ_1 and λ_2 is possible if the number $m_1 = \Delta s/\lambda_1$ differs by at least 1 from $m_2 = \Delta s/\lambda_2$; this immediately gives (4.97). With a modern design, maximum path differences Δs up to several meters have been realized for wavelength measurements of stabilized lasers (Sect. 4.5.3). For $\lambda = 500\,\text{nm}$ and $\Delta s = 1\,\text{m}$, we obtain

$\lambda/\Delta\lambda = 2 \times 10^6$, which is one order of magnitude better than for the grating spectrometer.

(c) **Fabry–Perot Interferometer:** The path difference is determined by the optical path difference $2nd$ between successive partial beams times the effective number of reflections, which can be expressed by the reflectivity finesse $F^* = \pi\sqrt{R}/(1-R)$. With ideal reflecting planes and perfect alignment, the maximum path difference would be $\Delta s_m = 2ndF^*$ and the spectral resolving power, according to (4.97), would be

$$\lambda/\Delta\lambda = F^*2nd/\lambda .$$

Because of imperfections of the alignment and deviations from ideal planes, the effective finesse is lower than the reflectivity finesse. With a value of $F^*_{\text{eff}} = 50$, which can be achieved, we obtain for $nd = 1\,\text{cm}$

$$\lambda/\Delta\lambda = 2 \times 10^6 ,$$

which is comparable with the Michelson interferometer having $\Delta s_m = 100\,\text{cm}$. However, with a confocal FPI, a finesse of $F^*_{\text{eff}} = 1000$ can be achieved. With $r = d = 4\,\text{cm}$ we then obtain

$$\lambda/\Delta\lambda = F^*4d/\lambda \approx 5 \times 10^8 ,$$

which means that for $\lambda = 500\,\text{nm}$, two lines with $\Delta\lambda = 1 \times 10^{-6}\,\text{nm}$ ($\Delta\nu = 1\,\text{MHz}$ at $\nu = 5 \times 10^{14}\,\text{s}^{-1}$) are still resolvable, provided that their linewidth is sufficiently small. With high-reflection mirror coatings a finesse of $F^*_{\text{eff}} = 10^5$ has been realized. With $r = d = 1\,\text{m}$ this yields $\lambda/\Delta\lambda = 8 \times 10^{11}$ [4.44].

4.3.2 Light-Gathering Power

The *light-gathering power*, or *étendue*, has been defined in Sect. 4.1.1 as the product $U = A\Omega$ of entrance area A and solid angle of acceptance Ω of the spectral apparatus. For most spectroscopic applications it is desirable to have an étendue U as large as possible to gain intensity. An equally important goal is to reach a maximum resolving power R. However, the two quantitites U and R are not independent of each other but are related, as can be seen from the following examples.

Example 4.19

(a) **Spectrometer:** The area of the entrance slit with width b and height h is $A = b \cdot h$. The acceptance angle $\Omega = (a/f)^2$ is determined by the fo-

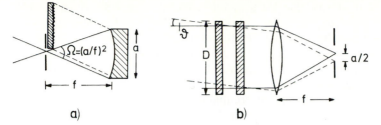

Fig. 4.63a,b. Acceptance angle of a spectrometer (**a**); and a Fabry–Perot interferometer (**b**)

cal length f of the collimating lens or mirror and the diameter a of the limiting aperture in the spectrometer (Fig. 4.63a). We can write the étendue,

$$U = bha^2/f^2 ,$$

as the product of the area $A = bh$ and the solid angle $\Omega = (a/f)^2$. Using typical figures for a medium-sized spectrometer ($b = 10\,\mu\text{m}$, $h = 0.5\,\text{cm}$, $a = 10\,\text{cm}$, $f = 100\,\text{cm}$) we obtain $\Omega = 0.01$, $A = 5 \times 10^{-4}\,\text{cm}^2$ $\rightarrow U = 5 \times 10^{-6}\,\text{cm}^2$ sr. With the resolving power $R = mN$, the product

$$RU = mNA\Omega \approx mN\frac{bha^2}{f^2} , \tag{4.99a}$$

increases with the diffraction order m, the size a of the grating, the number of illuminated grooves N, and the slit area bh (as long as imaging errors can be neglected). For $m = 1$, $N = 10^5$, and the above figures for h, b, a, and f, we obtain $RU = 0.5\,\text{cm}^2$ sr.

(b) **Interferometer:** For the Michelson and Fabry–Perot interferometers, the allowable acceptance angle for photoelectric recording is limited by the aperture in front of the detector, which selects the central circular fringe. From Figs. 4.49 and 4.63b we see that the fringe images at the center and at the edge of the limiting aperture with diameter a are produced by incoming beams that are inclined by an angle ϑ against each other. With $a/2 = f\vartheta$, the solid angle accepted by the FPI is $\Omega = a^2/(4f^2)$. For a plate diameter D the étendue is then $U = \pi(D^2/4)\Omega$. According to (4.84) the spectral resolving power $R = \nu/\Delta\nu$ of a plane FPI is correlated with the étendue U by $R = \pi D^2(2U)^{-1}$. The product

$$RU = \pi D^2/2 , \tag{4.99b}$$

is, for a plane FPI, therefore solely determined by the plate diameter. For $D = 5\,\text{cm}$, RU is about $40\,\text{cm}^2$ sr, and therefore two orders of magnitude larger than for a grating spectrometer.

In Sect. 4.2.10 we saw that for a given resolving power the spherical FPI has a larger étendue for mirror separations $r > D^2/4d$. For Example 4.19 with $D = 5$ cm, $d = 1$ cm, the confocal FPI therefore gives the largest product RU of all interferometers for $r > 6$ cm. Because of the higher total finesse, however, the confocal FPI may be superior to all other instruments even for smaller mirror separations.

In summary, we can say that at comparable resolving power interferometers have a larger light-gathering power than spectrometers.

4.4 Accurate Wavelength Measurements

One of the major tasks for spectroscopists is the measurement of wavelengths of spectral lines. This allows the determination of molecular energy levels and of molecular structure. The attainable accuracy of wavelength measurements depends not only on the spectral resolution of the measuring device but also on the achievable signal-to-noise ratio and on the reproducibility of measured absolute wavelength values.

With the ultrahigh resolution, which can, in principle, be achieved with single-mode tunable lasers (Chaps. 6–10), the accuracy of absolute wavelength measurements attainable with conventional techniques may not be satisfactory. New methods have been developed that are mainly based on interferometric measurements of laser wavelengths. For applications in molecular spectroscopy, the laser can be stabilized on the center of a molecular transition. Measuring the wavelength of such a stabilized laser yields simultaneously the wavelength of the molecular transition with a comparable accuracy. We shall briefly discuss some of these devices, often called *wavemeters*, that measure the unknown laser wavelength by comparison with a reference wavelength λ_R of a stabilized reference laser. Most proposals use for reference a HeNe laser, stabilized on a hyperfine component of a molecular iodine line, which has been measured by direct comparison with the primary wavelength standard to an accuracy of better than 10^{-10} [4.60].

Another method measures the absolute frequency ν_L of a stabilized laser and deduces the wavelength λ_L from the relation $\lambda_L = c/\nu_L$ using the *best* average of experimental values for the speed of light [4.61–4.63], which has been chosen to *define* the meter and thus the wavelength λ by the definition: 1 m *is the distance traveled by light in vacuum during the time* $\Delta t = 1/299\,792\,458\,\text{s}^{-1}$. *This defines the speed of light* as

$$c = 299\,792\,458 \text{ m/s} . \tag{4.100}$$

Such a scheme reduces the determination of lengths to the measurements of times or frequencies, which can be measured much more accurately than lengths [4.64]. Recently, the direct comparison of optical frequencies with the Cs standard in the microwave region has become possible with broad frequency combs generated by visible femtosecond lasers. This method will be discussed in Sect. 14.7.

4.4.1 Precision and Accuracy of Wavelength Measurements

Resolving power and light-gathering power are not the only criteria by which a wavelength-dispersing instrument should be judged. A very important question is the attainable *precision* and *accuracy* of absolute wavelength measurements.

To measure a physical quantity means to *compare* it with a reference standard. This comparison involves statistical and systematic errors. Measuring the same quantity n times will yield values X_i that scatter around the mean value

$$\overline{X} = \frac{1}{n} \sum_{i=1}^{n} X_i \,.$$

The attainable **precision** for such a set of measurements is determined by statistical errors and is mainly limited by the signal-to-noise ratio for a single measurement and by the number n of measurements (i.e., by the total measuring time). The precision can be characterized by the *standard deviation* [4.65–4.66],

$$\sigma = \left(\sum_{i=1}^{n} \frac{(\overline{X} - X_i)^2}{n} \right)^{1/2} . \tag{4.101}$$

The adopted mean value $\overline{X}$, averaged over many measured values X_i, is claimed to have a certain *accuracy*, which is a measure of the reliability of this value, expressed by its probable deviation $\Delta \overline{X}$ from the unknown "true" value X. A stated accuracy of $\overline{X}/\Delta\overline{X}$ means a certain confidence that the true value X is within $\overline{X} \pm \Delta\overline{X}$. Since the accuracy is determined not only by statistical errors but, particularly, by systematic errors of the apparatus and measuring procedure, it is always lower than the precision. It is also influenced by the precision with which the reference standard can be measured and by the accuracy of its comparison with the value $\overline{X}$. Although the attainable accuracy depends on the experimental efforts and expenditures, the skill, imagination, and critical judgement of the experimentalist always have a major influence on the ultimate achieved and stated accuracy.

We shall characterize precision and accuracy by the relative uncertainties of the measured quantity X, expressed by the ratios

$$\frac{\sigma}{X} \quad \text{or} \quad \frac{\Delta\overline{X}}{X} \,,$$

respectively. A series of measurements with a standard deviation $\sigma = 10^{-8}\overline{X}$ has a relative uncertainty of 10^{-8} or a precision of 10^8. Often one says that the precision is 10^{-8}, although this statement has the disadvantage that a high precision is expressed by a small number.

Let us now briefly examine the attainable precision and accuracy of wavelength measurements with the different instruments discussed above.

Although both quantities are correlated with the resolving power and the attainable signal-to-noise ratio, they are furthermore influenced by many other instrumental conditions, such as backlash of the monochromator drive, or asymmetric line profiles caused by imaging errors, or shrinking of the photographic film during the developing process. Without such additional error sources, the precision could be much higher than the resolving power, because the center of a symmetric line profile can be measured to a small fraction ϵ of the halfwidth. The value of ϵ depends on the attainable signal-to-noise ratio, which is determined, apart from other factors, by the étendue of the spectrometer. We see that for the precision of wavelength measurements, the product of resolving power R and étendue U, RU, discussed in the previous section, plays an important role.

For scanning monochromators with photoelectric recording, the main limitation for the attainable accuracy is the backlash of the grating-drive and nonuniformities of the gears, which limits the reliability of linear extrapolation between two calibration lines. Carefully designed monochromators have errors due to the drive that are less than $0.1\,\mathrm{cm}^{-1}$, allowing a relative uncertainty of 10^{-5} or an accuracy of about 10^5 in the visible range.

In absorption spectroscopy with a tunable laser, the accuracy of line positions is also limited by the nonuniform scan speed $\mathrm{d}\lambda/\mathrm{d}t$ of the laser (Sect. 5.6). One has to record reference wavelength marks simultaneously with the spectrum in order to correct for the nonuniformities of $\mathrm{d}\lambda/\mathrm{d}t$.

A serious source of error with scanning spectrometers or scanning lasers is the distortion of the line profile and the shift of the line center caused by the time constant of the recording device. If the time constant τ is comparable with the time $\Delta t = \Delta\lambda/v_{sc}$ needed to scan through the halfwidth $\Delta\lambda$ of the line profile (which depends on the spectral resolution), the line becomes broadened, the maximum decreases, and the center wavelength is shifted. The line shift $\delta\lambda$ depends on the scanning speed v_{sc} [nm/min] and is approximately $\delta\lambda = v_{sc}\tau = (\mathrm{d}\lambda/\mathrm{d}t)\tau$ [4.9].

Example 4.20

With a scanning speed $v_{sc} = 10\,\mathrm{nm/min}$ and a time constant of the recorder $\tau = 1\,\mathrm{s}$ the line shift is already $\delta\lambda = 0.15\,\mathrm{nm/min}$!

Because of the additional line broadening, the resolving power is reduced. If this reduction is to be less than 10%, the scanning speed must be below $v_{sc} < 0.24\Delta\lambda/\tau$. With $\Delta\lambda = 0.02\,\mathrm{nm}$, $\tau = 1\,\mathrm{s} \to v_{sc} < 0.3\,\mathrm{nm/min}$.

Photographic recording avoids these problems and therefore allows a more accurate wavelength determination at the expense of an inconvenient developing process of the photoplate and the subsequent measuring procedure to determine the line positions. A typical figure for the standard deviation for a 3-m spectrograph is $0.01\,\mathrm{cm}^{-1}$. Imaging errors causing curved lines, asymmetric line profiles due to misalignment, and backlash of the microdensitometer used for measuring the line positions on the photoplate are the main sources of errors.

Modern devices use photodiodes or CCD arrays (Sect. 4.5.2) instead of photoplates. With a diode width of 25 μm, the peak of a symmetric line profile extending over 3−5 diodes can be determined by a least-squares fit to a model profile within 1−5 μm, depending on the S/N ratio. When the array is placed behind a spectrometer with a dispersion of 1 mm/nm, the center of the line can be determined within 10^{-3} nm. Since the signals are read electronically, there are no moving parts in the device and any mechanical error source (backlash) is eliminated.

The highest accuray (i.e., the lowest uncertainty) can be achieved with modern *wavemeters*, which we shall discuss in Sect. 4.4.2.

4.4.2 Today's Wavemeters

The different types of wavemeters for very accurate measurements of laser wavelengths are based on modifications of the Michelson interferometer [4.67], the Fizeau interferometer [4.68], or on a combination of several Fabry–Perot interferometers with different free spectral ranges [4.69–4.71]. The wavelength is measured either by monitoring the spatial distribution of the interference pattern with photodiode arrays, or by using traveling devices with electronic counting of the interference fringes.

a) The Michelson Wavemeter

Figure 4.64 illustrates the principle of a traveling-wave Michelson-type interferometer as used in our laboratory. Such a wavemeter was first demonstrated in a slightly different version by Hall and Lee [4.67] and by Kowalski et al. [4.72]. The beams B_R of a reference laser and B_x of a laser with unknown wavelength λ_x traverse the interferometer on identical paths, but in opposite directions. Both incoming beams are split into two partial beams by

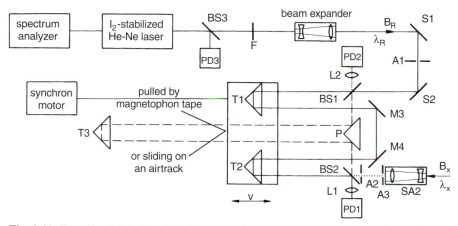

Fig. 4.64. Traveling Michelson interferometer for accurate measurements of wavelengths of single-mode cw lasers

the beam splitters BS1 and BS2, respectively. One of the partial beams travels the constant path BS1–P–T3–P–BS2 for the reference beam, and in the opposite direction for the beam B_X. The second partial beam travels the variable path BS1–T1–M3–M4–T2–BS2 for B_R, and in the opposite direction for B_X. The moving corner-cube reflectors T1 and T2 are mounted on a carriage, which either travels with wheels on rods or slides on an airtrack.

The *corner-cube reflectors* guarantee that the incoming light beam is always reflected exactly parallel to its indicent direction, irrespective of slight misalignments or movements of the traveling reflector. The two partial beams (BS1–T1–M3–M4–T2–BS2 and BS1–P–T3–P–BS2) for the reference laser interfere at the detector PD1, and the two beams BS2–T2–M4–M3–T1–BS1 and BS2–P–T3–P–BS1 from the unknown laser interfere at the detector PD2. When the carriage is moving at a speed $v = dx/dt$ the phase difference $\delta(t)$ between the two interfering beams changes as

$$\delta(t) = 2\pi \frac{\Delta s}{\lambda} = 2\pi \cdot 4 \frac{dx}{dt} \frac{t}{\lambda} = 8\pi \frac{vt}{\lambda} \,, \qquad (4.102)$$

where the optical path difference Δs has been doubled by introducing two corner-cube reflectors. The rates of interference maxima, which occur for $\delta = m2\pi$, are counted by PD2 for the unknown wavelength λ_X and by PD1 for the reference wavelength λ_R. The unknown wavelength λ_X can be obtained from the ratio of both counting rates if proper corrections are made for the dispersion $n(\lambda_R) - n(\lambda_X)$ of air.

The signal lines to both counters are simultaneously opened at the time t_0 when the detector PD2 just delivers a trigger signal. Both counters are simultaneously stopped at the time t_1 when PD2 has reached the preset number N_0. From

$$\Delta t = t_1 - t_0 = N_0 \lambda_X / 4v = (N_R + \epsilon) \lambda_R / 4v \,,$$

we obtain for the vacuum wavelength λ_X^0

$$\lambda_X^0 = \frac{N_R + \epsilon}{N_0} \lambda_R^0 \frac{n(\lambda_X, P, T)}{n(\lambda_R, P, T)} \,. \qquad (4.103a)$$

The unknown fractional number $\epsilon < 2$ takes into account that the trigger signals from PD1, which define the start and stop times t_0 and t_1 (Fig. 4.64), may not exactly coincide with the pulse rise times in channel 2. The two worst cases are shown in Fig. 4.65. For case a, the trigger pulse at t_0 just misses the rise of the signal pulse, but the trigger at t_1 just coincides with the rise of a signal pulse. This means that the signal channel counts one pulse less than it should. In case b, the start pulse at t_0 coincides with the rise time of a signal pulse, but the stop pulse just misses a signal pulse. In this case, the signal channel counts one pulse more than it should.

For a maximum optical path difference $\Delta s = 4\,\text{m}$, the number of counts for $\lambda = 500\,\text{nm}$ is 8×10^6, which allows a precision of about 10^7, if the counting error is not larger than 1. Provided the signal-to-noise ratio is sufficiently

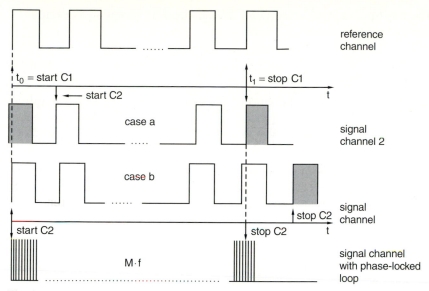

Fig. 4.65. Signal sequences in the two detection channels of the traveling Michelson wavemeter. The grey signal pulses are not counted

high, the attainable precision can, however, be enhanced by interpolations between two successive counts using a *phase-locked loop* [4.73–4.74]. This is an electronic device that multiplies the frequency of the incoming signal by a factor M while always being locked to the phase of the incoming signal. Assume that the counting rate $f_R = 4v/\lambda_R$ in the reference channel is multiplied by M. Then the unknown wavelength λ_X is determined by

$$\lambda_X^0 = \frac{MN_R + \epsilon}{MN_0} \lambda_R^0 \frac{n_X}{n_R} = \frac{N_R + \varepsilon/M}{N_0} \lambda_R^0 \frac{n_X}{n_R} . \qquad (4.103b)$$

For $M = 100$ the limitation of the accuracy by the counting error due to the unknown fractional number ϵ is reduced by a factor of 100.

Instead of the phase-locked loop a coincidence circuit may be employed. Here the signal paths to both counters are opened and closed at selected times t_0 and t_1, when both trigger signals from PD2 and PD1 coincide within a small time interval, say 10^{-8} s. Both techniques reduce the counting uncertainty to a value below 2×10^{-9}.

In general, the attainable accuracy, however, is lower because it is influenced by several sources of systematic errors. One is a misalignment of the interferometer, which causes both beams to travel slightly different path lengths. Another point that has to be considered is the curvature of the wavefronts in the diffraction-limited Gaussian beams (Sect. 5.3). This curvature can be reduced by expanding the beams through telescopes (Fig. 4.64). The uncertainty of the reference wavelength λ_R and the accuracy of measuring the refractive index $n(\lambda)$ of air are further error sources.

The maximum relative uncertainty of the absolute vacuum wavelength λ_X can be written as a sum of five terms:

$$\left|\frac{\Delta\lambda_X}{\lambda_X}\right| \le \left|\frac{\Delta\lambda_R}{\lambda_R}\right| + \left|\frac{\epsilon}{MN_R}\right| + \left|\frac{\Delta r}{r}\right| + \left|\frac{\delta s}{\Delta s}\right| + \left|\frac{\delta\phi}{2\pi N_0}\right| , \tag{4.104}$$

where $r = n(\lambda_X)/n(\lambda_R)$ is the ratio of the refractive indices, δs is the difference of the travel paths for reference and signal beams, and $\delta\phi$ is the phase front variation in the detector plane. Let us briefly estimate the magnitude of the different terms in (4.104):

- The wavelength λ_R of the I_2-stabilized HeNe laser is known within an uncertainty $|\Delta\lambda_R/\lambda_R| < 10^{-10}$ [4.64]. Its frequency stability is better than 100 kHz, i.e., $|\Delta\nu/\nu| < 2 \times 10^{-10}$. This means that the first term in (4.104) contributes at most 3×10^{-10} to the uncertainty of λ_X.
- With $\epsilon = 1.5$, $M = 100$, and $N_R = 8 \times 10^6$, the second term is about 2×10^{-9}.
- The index of refraction, $n(\lambda, p, T)$, depends on the wavelength λ, on the total air pressure, on the partial pressures of H_2O and CO_2, and on the temperature. If the total pressure is measured within 0.5 mbar, the temperature T within 0.1 K, and the relative humidity within 5%, the refractive index can be calculated from formulas given by Edlen [4.75] and Owens [4.76].

  With the stated accuracies, the third term in (4.104) becomes

$$|\Delta r/r| \approx 1 \times 10^{-3} |n_0(\lambda_X) - n_0(\lambda_R)| , \tag{4.105}$$

where n_0 is the refractive index for dry air under standard conditions ($T_0 = 15°C$, $p_0 = 1013$ hPa). The contribution of the third term depends on the wavelength difference $\Delta\lambda = \lambda_R - \lambda_X$. For $\Delta\lambda = 1$ nm one obtains $|\Delta r/r| < 10^{-11}$, while for $\Delta\lambda = 200$ nm this term becomes, with $|\Delta r/r| \approx 5 \times 10^{-9}$, a serious limitation of the accuracy of $|\Delta\lambda_X/\lambda_X|$.

- The magnitude of the fourth term $|\delta s/\Delta s|$ depends on the effort put into the alignment of the two laser beams within the interferometer. If the two beams are tilted against each other by a small angle α, the two path lengths for λ_X and λ_R differ by

$$\delta s = \Delta s(\lambda_R) - \Delta s(\lambda_X) = \Delta s_R(1 - \cos\alpha) \approx (\alpha^2/2)\Delta s_R .$$

With $\alpha = 10^{-4}$ rad, the systematic relative error becomes

$$|\delta s/\Delta s| \approx 5 \times 10^{-9} .$$

It is therefore necessary to align both beams very carefully.

- With a surface quality of $\lambda/10$ for all mirrors and beam splitters, the distortions of the wavefront are already visible in the interference pattern. However, plane waves are focused onto the detector area and the phase of the detector signal is due to an average over the cross section of the enlarged beam ($\approx 1\,\mathrm{cm}^2$). This averaging minimizes the effect of wave-front distortion on the accuracy of λ_X. If the modulation of the interference intensity (4.37) exceeds 90%, this term may be neglected.

With careful alignment, good optical quality of all optical surfaces and accurate recording of p, T, and P_{H_2O}, the total uncertainty of λ_X can be pushed below 10^{-8}. This gives an absolute uncertainty $\Delta\nu_x \approx 3\,\mathrm{MHz}$ of the optical frequency $\nu_x = 5 \times 10^{14}\,\mathrm{s}^{-1}$ for a wavelength separation between λ_R and λ_x of $\Delta\lambda \approx 120\,\mathrm{nm}$. This has been proved by a comparison of independently measured wavelengths $\lambda_x = 514.5\,\mathrm{nm}$ (I_2-stabilized argon laser) and $\lambda_R = 632.9\,\mathrm{nm}$ (I_2-stabilized HeNe laser) [4.77].

When cw dye laser wavelengths are measured, another source of error arises. Due to air bubbles in the dye jet or dust particles within the resonator beam waist, the dye laser emission may be interrupted for a few microseconds. If this happens while counting the wavelength a few counts are missing. This can be avoided by using an additional phase-locked loop with a multiplication factor $M_x = 1$ in the counting channel of PD_x. If the time constant of the phase-locked loop is larger than $10\,\mu\mathrm{s}$, it continues to oscillate at the counting frequency during the few microseconds of dye laser beam interruptions.

There are several different designs of Michelson wavemeters that are commercially available and are described in [4.78–4.80].

b) Sigmameter

While the traveling Michelson is restricted to cw lasers, a motionless Michelson interferometer was designed by Jacquinot, et al. [4.81], which includes no moving parts and can be used for cw as well as for pulsed lasers. Figure 4.66 illustrates its operation. The basic element is a Michelson interferometer with a *fixed* path difference δ. The laser beam enters the interferometer polarized at 45° with respect to the plane of Fig. 4.66. When inserting a prism into one arm of the interferometer, where the beam is totally reflected at the prism base, a phase difference $\Delta\varphi$ is introduced between the two components polarized parallel and perpendicular to the totally reflecting surface. The value of $\Delta\varphi$ depends, according to Fresnel's formulas [4.15], on the incidence angle α and can be made $\pi/2$ for $\alpha = 55°19'$ and $n = 1.52$. The interference signal at the exit of the interferometer is recorded separately for the two polarizations and one obtains, because of the phase shifts $\pi/2$, $I_{||} = I_0(1 + \cos 2\pi\delta/\lambda)$ and $I_\perp = I_0(1 + \sin 2\pi\delta/\lambda)$. From these signals it is possible to deduce the wave number $\sigma = 1/\lambda$ modulo $1/\delta$, since all wave numbers $\sigma_m = \sigma_0 + m/\delta$ ($m = 1, 2, 3, \ldots$) give the same interference signals. Using several interferometers of the same type with a common mirror M1 but different positions of M2, which have path differences in geometric ratios, such as 50 cm, 5 cm, 0.5 cm, and 0.05 cm, the wave number σ can be deduced unambiguously with an accuracy determined by the interferometer with the highest path difference. The actual path differences δ_i are calibrated with a reference line and are servo-locked to this line. The precision obtained with this instrument is

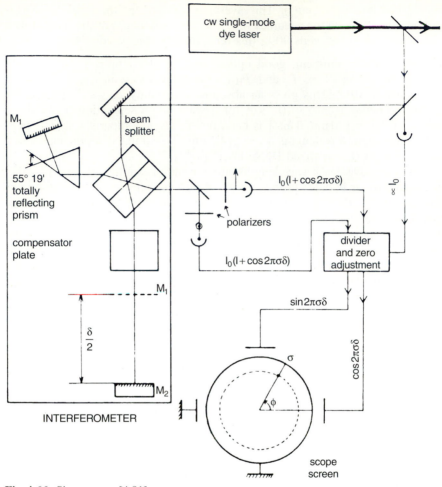

Fig. 4.66. Sigmameter [4.81]

about 5 MHz, which is comparable with that of the traveling Michelson interferometer. The measuring time, however, is much less since the different δ_i can be determined simultaneously. This instrument is more difficult to build but easier to handle. Since it measures wave numbers $\sigma = 1/\lambda$, the inventors called it a *sigmameter*.

c) Computer-Controlled Fabry–Perot Wavemeter

Another approach to accurate wavelength measurements of pulsed and cw lasers, which can be also applied to incoherent sources, relies on a combination of a small grating monochromator and three Fabry–Perot etalons [4.69–4.71]. The incoming laser beam is sent simultaneously through the monochromator and three temperature-stabilized Fabry–Perot interferom-

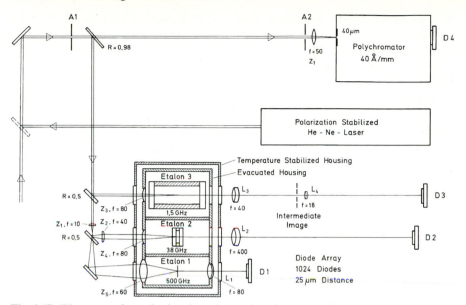

Fig. 4.67. Wavemeter for pulsed and cw lasers, based on a combination of a small poly-chromator and three FPI with widely differing free spectral ranges [4.77]

eters with different free spectral ranges δv_i (Fig. 4.67). In order to match the laser beam profile to the sensitive area of the linear diode arrays (25 mm $\times$ 50 μm), focusing with cylindrical lenses Z_i is utilized. The divergence of the beams in the plane of Fig. 4.67 is optimized by the spherical lenses L_i in such a way that the diode arrays detect 4−6 FPI fringes (Fig. 4.68). The linear arrays have to be properly aligned so that they coincide with a diameter through the center of the ring system. According to (4.68), the wavelength λ can be determined from the ring diameters D_p and the excess ϵ, provided the integer order m_0 is known, which means that λ must already be known at least within one-half of a free spectral range (Sect. 4.3).

The device is calibrated with different lines from a cw dye laser that are simultaneously measured with the traveling Michelson wavemeter (see above). This calibration allows:

Correctly aligned: a)

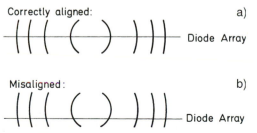

Diode Array

Misaligned: b)

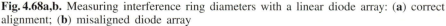

Diode Array

Fig. 4.68a,b. Measuring interference ring diameters with a linear diode array: (**a**) correct alignment; (**b**) misaligned diode array

- The unambiguous correlation between wavelength λ and the position of the illuminated diode of array 1 behind the monochromator with an accuracy of ± 0.1 nm, which is sufficient to determine λ within 0.5 of the free spectral range of etalon 1;
- The accurate determination of nd for all three FPI.

If the free spectral range $\delta \nu_1$ of the thin FPI is at least twice as large as the uncertainty $\Delta \nu$ of the monochromator measurement, the integer order m_0 of FPI1 can be unambiguously determined. The measurement of the ring diameters improves the accuracy by a factor of about 20. This is sufficient to determine the larger integer order m_0 of FPI2; from its ring diameters, λ can be measured with an accuracy 20 times higher than that from FPI1. The final wavelength determination uses the ring diameters of the large FPI3. Its accuracy reaches about 1% of the free spectral range of FPI3.

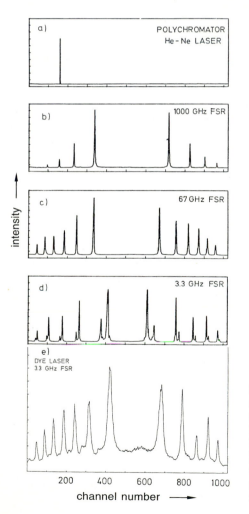

Fig. 4.69a–e. Output signals at the polychromator and the three diode arrays of the FPI wavemeter, which had been illuminated by a cw HeNe laser oscillating on two axial modes (a–d). The lowest figure shows the ring intensity pattern of an excimer-pumped single-mode dye laser measured behind a FPI with 3.3 GHz free spectral range [4.70]

The whole measuring cycle is controlled by a computer. For pulsed lasers, one pulse (with an energy of $\geq 5\,\mu J$) is sufficient to initiate the device, while for cw lasers, a few microwatts input power are sufficient. The arrays are read out by the computer and the signals can be displayed on a screen. Such signals for the arrays D1–D4 are shown in Fig. 4.69 for a HeNe laser oscillating on two longitudinal modes and for a pulsed dye laser.

Since the optical distances $n_i d_i$ of the FPI depend critically on temperature and pressure, all FPI must be kept in a temperature-stabilized pressure-tight box. Furthermore, a stabilized HeNe laser can be used to control long-term drift of the FPI [4.77].

> **Example 4.21**
>
> With a free spectral range of $\delta\nu = 1\,\text{GHz}$, the uncertainty of calibration and of the determination of an unknown wavelength are both about 10 MHz. This gives an absolute uncertainty of less than 20 MHz. For the optical frequency $\nu = 6 \times 10^{14}\,\text{Hz}$, the relative accuracy is then $\Delta\nu/\nu \leq 3 \times 10^{-8}$.

d) Fizeau Wavemeter

The Fizeau wavemeter constructed by Snyder [4.82] can be used for pulsed and cw lasers. While its optical design is simpler than that of the sigmameter and the FPI wavemeter, its accuracy is slightly lower. Its basic principle is shown in Fig. 4.70b. The incident laser beam is focused by an achromatic

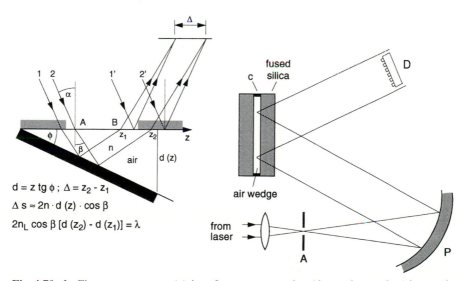

$d = z\,\text{tg}\,\phi\,;\; \Delta = z_2 - z_1$

$\Delta s \approx 2n \cdot d\,(z) \cdot \cos\beta$

$2n_L \cos\beta\,[d\,(z_2) - d\,(z_1)] = \lambda$

Fig. 4.70a,b. Fizeau wavemeter: (**a**) interference at a wedge (the wedge angle ϕ is greatly exaggerated); (**b**) schematic design; A, aperture as spatial filter; P, parabolic mirror; C, distance holder of cerodur; D, diode array

microscope lens system onto a small pinhole, which represents a nearly point-like light source. The divergent light is transformed by a parabolic mirror into an enlarged parallel beam, which hits the Fizeau interferometer (FI) under an incident angle α (Fig. 4.70a). The FI consists of two fused quartz plates with a slightly wedged air gap ($\phi \approx 1/20°$). For small wedge angles ϕ, the optical path difference Δs between the constructively interfering beams 1 and 1' is approximately equal to that of a plane-parallel plate according to (4.46a), namely

$$\Delta s_1 = 2nd(z_1) \cos \beta = m\lambda .$$

The path difference between the beams 2 and 2', which belong to the next interference order, is $\Delta s_2 = (m+1)\lambda$. The interference of the reflected light produces a pattern of parallel fringes with the separation

$$\Delta = z_2 - z_1 = \frac{d(z_2) - d(z_1)}{\tan \phi} = \frac{\lambda}{2n \tan \phi \cos \beta} , \qquad (4.106)$$

which depends on the wavelength λ, the wedge angle ϕ, the angle of incidence α, and the refractive index n of air.

Changing the wavelength λ causes a shift Δz of the fringe pattern and a slight change of the fringe separation Δ. For a change of λ by one free spectral range

$$\delta\lambda = \frac{\lambda^2}{2nd \cos \beta} . \qquad (4.107)$$

and Δz is equal to the fringe separation Δ. Therefore the two fringe patterns for λ and $\lambda + \delta\lambda$ look identical, apart from the slight change of Δ. It is therefore essential to know λ at least within $\pm\delta\lambda/2$. This is possible from a measurement of Δ. With a diode array of 1024 diodes, the fringe separation Δ can be obtained from a least-squares fit to the measured intensity distribution $I(z)$ with a relative accuracy of 10^{-4}, which yields an absolute value of λ within $\pm 10^{-4}\lambda$ [4.83].

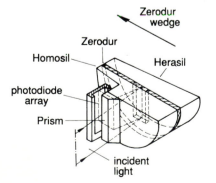

Fig. 4.71. Compact design of a Fizeau waveme-ter [4.84]

With a value $d = 1$ mm of the air gap, the order of interference m is about 3000 at $\lambda = 500$ nm. An accuracy of 10^{-4} is therefore sufficient for the unambiguous determination of m. Since the *position* of the interference fringes can be measured within 0.3% of the fringe separation, the wavelength λ can be obtained within 0.3% of a free spectral range, which gives the accuracy $\lambda/\Delta\lambda \approx 10^{7}$. The preliminary value of λ, deduced from the fringe separation Δ, and the final value, determined from the *fringe position*, are both obtained from the same FI after having calibrated the system with lines of known wavelengths.

The advantage of the Fizeau wavemeter is its compact design and its low price. A very elegant construction by Gardner [4.84, 4.85] is sketched in Fig. 4.71. The wedge air gap is fixed by a Zerodur spacer between the two interferometer plates and forms a pressure tight volume. Variations of air pressure in the surroundings therefore do not cause changes of n within the air gap. The reflected light is sent to the diode array by a totally reflecting prism. The data are processed by a small computer.

4.5 Detection of Light

For many applications in spectroscopy the sensitive detection of light and the accurate measurement of its intensity are of crucial importance for the successful performance of an experiment. The selection of the proper detector for optimum *sensitivity* and *accuracy* for the detection of radiation must take into account the following characteristic properties, which may differ for the various detector types:

- The spectral relative response $R(\lambda)$ of the detector, which determines the wavelength range in which the detector can be used. The knowledge of $R(\lambda)$ is essential for the comparison of the true relative intensities $I(\lambda_1)$ and $I(\lambda_2)$ at different wavelengths.
- The absolute sensitivity $S(\lambda) = V_s/P$, which is defined as the ratio of output signal V_s to incident radiation power P. If the output is a voltage, as in photovoltaic devices or in thermocouples, the sensitivity is expressed in units of volts per watt. In the case of photocurrent devices, such as photomultipliers, $S(\lambda)$ is given in amperes per watt. With the detector area A the sensitivity S can be expressed in terms of the irradiance I:

$$S(\lambda) = V_s/(AI) . \tag{4.108}$$

- The achievable signal-to-noise ratio V_s/V_n, which is, in principle, limited by the noise of the incident radiation. It may, in practice, be further reduced by inherent noise of the detector. The detector noise is often expressed by the *noise equivalent input power* (NEP), which means an incident radiation power that generates the same output signal as the detector noise itself, thus yielding the signal-to-noise ratio $S/N = 1$. In infrared physics a figure of merit for the infrared detector is the detectivity

$$D^* = \frac{\sqrt{A\Delta f}}{P} \frac{V_s}{V_n} = \frac{\sqrt{A\Delta f}}{\text{NEP}} \; . \tag{4.109}$$

The detectivity D^* [cm s$^{-1/2}$ W^{-1}] gives the obtainable signal-to-noise ratio V_s/V_n of a detector with the sensitive area $A = 1\,\text{cm}^2$ and the detector bandwidth $\Delta f = 1\,\text{Hz}$, at an incident radiation power of $P = 1\,\text{W}$. Because the noise equivalent input power is NEP $= P \cdot V_n/V_s$, the detectivity of a detector with the area $1\,\text{cm}^2$ and a bandwidth of $1\,\text{Hz}$ is $D^* = 1/\text{NEP}$.

- The maximum intensity range in which the detector response is linear. It means that the output signal V_s is proportional to the incident radiation power P. This point is particularly important for applications where a wide range of intensities is covered. Examples are output-power measurements of pulsed lasers, Raman spectroscopy, and spectroscopic investigations of line broadening, when the intensities in the line wings may be many orders of magnitude smaller than at the center.
- The time or frequency response of the detector, characterized by its time constant τ. Many detectors show a frequency response that can be described by the model of a capacitor, which is charged through a resistor R_1 and discharged through R_2 (Fig. 4.72a). When a very short light pulse falls onto the detector, its output pulse is smeared out. If the output is a current $i(t)$ that is proportional to the incident radiation power $P(t)$ (as, for example, in photomultipliers), the output capacitance C is charged by this current and shows a voltage rise and fall, determined by

$$\frac{dV}{dt} = \frac{1}{C}\left[i(t) - \frac{V}{R_2}\right] . \tag{4.110}$$

If the current pulse $i(t)$ lasts for the time T, the voltage $V(t)$ at the capacitor increases up to $t = T$ and for $R_2 C \gg T$ reaches the peak voltage

$$V_{\text{max}} = \frac{1}{C}\int_0^T i(t)\,dt \; ,$$

which is determined by C and not by R_2! After the time T the voltage decays exponentially with the time constant $\tau = CR_2$. Therefore, the value of R_2 limits the repetition frequency f of pulses to $f < (R_2 C)^{-1}$.

The time constant τ of the detector causes the output signal to rise slower than the incident input pulse. It can be determined by modulating the continuous input radiation at the frequency f. The output signal of such a device is characterized by (see Exercise 4.12)

$$V_s(f) = \frac{V_s(0)}{\sqrt{1 + (2\pi f\tau)^2}} \; , \tag{4.111}$$

where $\tau = CR_1 R_2/(R_1 + R_2)$. At the modulation frequency $f = 1/(2\pi\tau)$, the output signal has decreased to $1/\sqrt{2}$ of its dc value. The knowledge of the detector time constant τ is essential for all applications where fast

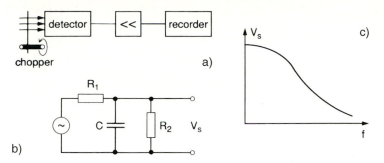

Fig. 4.72a–c. Typical detector: (**a**) schematic setup; (**b**) equivalent electrical circuit; (**c**) frequency response $V_s(f)$

transient phenomena are to be monitored, such as atomic lifetimes or the time dependence of fast laser pulses (Chap. 11).

• The price of a detector is another factor that cannot be ignored, since unfortunately it often restricts the optimum choice.

In this section we briefly discuss some detectors that are commonly used in laser spectroscopy. The different types can be divided into two categories, *thermal detectors* and *direct photodetectors*. In thermal detectors, the energy absorbed from the incident radiation raises the temperature and causes changes in the temperature-dependent properties of the detector, which can be monitored. Direct photodetectors are based either on the emission of photoelectrons from photocathodes, or on changes of the conductivity of semiconductors due to incident radiation, or on photovoltaic devices where a voltage is generated by the internal photoeffect. Whereas thermal detectors have a *wavelength-independent* sensitivity, photodetectors show a spectral response that depends on the work function of the emitting surface or on the band gap in semiconductors.

During recent years the development of image intensifiers, image converters, and vidicon detectors has made impressive progress. At first pushed by military demands, these devices are now coming into use for light detection at low levels, e.g., in Raman spectroscopy, or for monitoring the faint fluorescence of spurious molecular constituents. Because of their increasing importance we give a short survey of the principles of these devices and their application in laser spectroscopy. In time-resolved spectroscopy, subnanosecond detection can now be performed with fast phototubes in connection with transient digitizers, which resolve time intervals of less than 100 ps. Since such time-resolved experiments in laser spectroscopy with streak cameras and correlation techniques are discussed in Chap. 11, we confine ourselves here to discussing only some of these modern devices from the point of view of spectroscopic instrumentation. A more extensive treatment of the characteristics and the performance of various detectors can be found in special monographs on detectors [4.86–4.95]. For reviews on photodetection techniques relevant in laser physics, see also [4.96–4.98].

4.5.1 Thermal Detectors

Because of their wavelength-independent sensitivity, thermal detectors are useful for calibration purposes, e.g., for an absolute measurement of the radiation power of cw lasers, or of the output energy of pulsed lasers. In the rugged form of medium-sensitivity calibrated calorimeters, they are convenient devices for any laser laboratory. With more sophisticated and delicate design, they have been developed as sensitive detectors for the whole spectral range, particularly for the infrared region, where other sensitive detectors are less abundant than in the visible range.

For a simple estimate of the sensitivity and its dependence on the detector parameters, such as the heat capacitance and thermal losses, we shall consider the following model [4.99]. Assume that the fraction β of the incident radiation power P is absorbed by a thermal detector with heat capacity H, which is connected to a heat sink at constant temperature T_s (Fig. 4.73a). When G is the thermal conductivity of the link between the detector and the heat sink, the temperature T of the detector under illumination can be obtained from

$$\beta P = H \frac{dT}{dt} + G(T - T_s) . \tag{4.112}$$

If the time-independent radiation power P_0 is switched on at $t = 0$, the time-dependent solution of (4.112) is

$$T = T_s + \frac{\beta P_0}{G}(1 - e^{-(G/H)t}) . \tag{4.113}$$

The temperature T rises from the initial value T_s at $t = 0$ to the temperature $T = T_s + \Delta T$ for $t = \infty$. The temperature rise

$$\Delta T = \frac{\beta P_0}{G} \tag{4.114}$$

is inversely proportional to the thermal losses G and does not depend on the heat capacity H, while the time constant of the rise $\tau = H/G$ depends on the ratio of both quantities. Small values of G make a thermal detector sensitive, but slow!

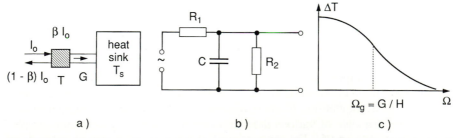

Fig. 4.73a–c. Model of a thermal detector: (**a**) schematic diagram; (**b**) equivalent electrical circuit; (**c**) frequency response $\Delta T(\Omega)$

In general, P will be time dependent. When we assume the periodic function

$$P = P_0(1 + a \cos \Omega t) , \quad |a| \leq 1 , \tag{4.115}$$

we obtain, inserting (4.115) into (4.112), a detector temperature of

$$T(\Omega) = T_s + \Delta T (1 + \cos(\Omega t + \varphi)) , \tag{4.116}$$

which depends on the modulation frequency Ω, and which shows a phase lag ϕ determined by

$$\tan \phi = \Omega H / G = \Omega \tau , \tag{4.117a}$$

and a modulation amplitude

$$\Delta T = \frac{a \beta P_0 G}{\sqrt{G^2 + \Omega^2 H^2}} . \tag{4.117b}$$

At the frequency $\Omega_g = G/H$, the amplitude ΔT decreases by a factor of $\sqrt{2}$ compared to its DC value.

Note: The problem is equivalent to the analogous case of charging a capacitor ($C \leftrightarrow H$) through a resistor R_1 that discharges through R_2 ($R_2 \leftrightarrow 1/G$) (the charging current i corresponds to the radiation power P). The ratio $\tau = H/G$ ($H/G \leftrightarrow R_2C$) determines the time constant of the device (Fig. 4.73b).

We learn from (4.116) that the sensitivity $S = \Delta T/P_0$ becomes large if G and H are made as small as possible. For modulation frequencies $\Omega > G/H$, the amplitude ΔT will decrease approximately inversely to Ω. Since the time constant $\tau = H/G$ limits the frequency response of the detector, *a fast and sensitive detector should have a minimum heat capacity H.*

For the calibration of the output power from cw lasers, the demand for high sensitivity is not as relevant since, in general, sufficiently large radiation power is available. Figure 4.74 depicts a simple home-made calorimeter and its circuit diagram. The radiation falls through a hole into a metal cone with a black inner surface. Because of the many reflections, the light has only a small chance of leaving the cone, ensuring that all light is absorbed. The absorbed power heats a thermocouple or a temperature-dependent resistor (thermistor) embedded in the cone. For calibration purposes, the cone can be heated by an electric wire. If the detector represents one part of a bridge (Fig. 4.74c) that is balanced for the electric input $W = UI$, but without incident radiation, the heating power has to be reduced by $\Delta W = P$ to maintain the balance with the incident radiation power P. A system with higher accuracy uses the difference in output signals of two identical cones, where only one is irradiated (Fig. 4.74b).

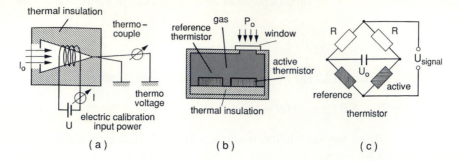

Fig. 4.74a–c. Calorimeter for measuring the output power of cw lasers or the output energy of pulsed lasers: (**a**) experimental design; (**b**) calorimeter with active irradiated thermistor and nonirradiated reference thermistor; (**c**) balanced bridge circuit

For the measurement of output energies from pulsed lasers, the calorimeter should integrate the absorbed power at least over the pulse duration. From (4.112) we obtain

$$\int_0^{t_0} \beta P \, dt = H \Delta T + \int_0^{t_0} G(T - T_s) \, dt \; . \tag{4.118}$$

When the detector is thermally isolated, the heat conductivity G is small, therefore the second term may be completely neglected for sufficiently short pulse durations t_0. The temperature rise

$$\Delta T = \frac{1}{H} \int_0^{t_0} \beta P \, dt \; , \tag{4.119}$$

is then directly proportional to the input energy. For calibration, a charged capacitor C is discharged through the heating coil (Fig. 4.74a). If the discharge time is matched to the laser pulse time, the heat conduction is the same for both cases and does not enter into calibration. If the temperature rise caused by the discharge of the capacitor equals that caused by the laser pulse, the pulse energy is $\frac{1}{2}CU^2$.

For more sensitive detection of low incident powers, *bolometers* and *Golay cells* are used. A bolometer consists of N thermocouples in series, where one junction touches the backside of a thin electrically insulating foil that is exposed to the incident radiation (Fig. 4.75a). The other junction is in contact with a heat sink. The output voltage is

$$U = N \frac{dU}{dT} \Delta T \; ,$$

where dU/dT is the sensitivity of a single thermocouple.

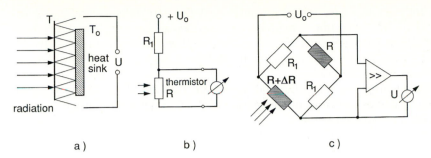

Fig. 4.75a–c. Schematic circuit diagram of a bolometer: (**a**) thermopile; (**b**) thermistor; and (**c**) bridge circuit with difference amplifier

Another version utilizes a thermistor that consists of a material with a large temperature coefficient $\alpha = (dR/dT)/R$ of the resistance R. If a constant current i is fed through R (Fig. 4.75b), the incident power P that causes a temperature increase ΔT produces the voltage output signal

$$\Delta U = i\Delta R = iR\alpha\Delta T = \frac{V_0 R}{R+R_1}\alpha\Delta T \;, \tag{4.120}$$

where ΔT is determined from (4.116) as $\Delta T = \beta P(G^2 + \Omega^2 H^2)^{-1/2}$. The response $\Delta U/P$ of the detector is therefore proportional to i, R, and α, and decreases with increasing H and G. At a constant supply voltage V_0, the current change Δi caused by the irradiation is, for $\Delta R \ll R + R_1$,

$$\Delta i = V_0 \left(\frac{1}{R_1 + R} - \frac{1}{R_1 + R + \Delta R} \right) \Delta T \approx V_0 \frac{\alpha R \Delta T}{(R_1 + R)^2} \;, \tag{4.121}$$

and can be generally neglected.

Since the input impedance of the following amplifier has to be larger than R, this puts an upper limit on R. Because any fluctuation of i causes a noise signal, the current i through the bolometer has to be extremely constant. This and the fact that the temperature rise due to Joule's heating should be small limits the maximum current through the bolometer.

Equations (4.120 and 4.116) demonstrate again that small values of G and H are desirable. Even with perfect thermal isolation, heat radiation is still present and limits the lower value of G. At the temperature difference ΔT between a bolometer and its surroundings, the Stefan–Boltzmann law gives for the net radiation flux ΔP to the surroundings from the detector with the emitting area A^* and the emissivity $\epsilon \le 1$

$$\Delta P = 4A\epsilon\sigma T^3 \Delta T \;, \tag{4.122}$$

where $\sigma = 5.77 \times 10^{-8}\,\mathrm{W/m^2\,K^{-4}}$ is the Stefan–Boltzmann constant. The minimum thermal conductivity is therefore

$$G_\mathrm{m} = 4A\sigma\epsilon T^3 \;, \tag{4.123}$$

even for the ideal case where no other heat links to the surroundings exist. This limits the detection sensitivity to a minimum input radiation of about 10^{-10} W for detectors operating at room temperatures and with a bandwidth of 1 Hz. It is therefore advantageous to cool the bolometer, which furthermore decreases the heat capacity.

This cooling has the additional advantage that the slope of the function dR/dT becomes larger at low temperatures T. Two different materials can be utilized, as discussed below.

In semiconductors the electrical conductivity is proportional to the electron density n_e in the conduction band. With the band gap ΔE_G this density is, according to the Boltzmann relation

$$\frac{n_e(T)}{n_e(T+\Delta T)} = \exp\left(-\frac{\Delta E_G \Delta T}{2kT^2}\right), \tag{4.124}$$

and is very sensitively dependent on temperature.

The quantity dR/dT becomes exceedingly large at the critical temperature T_c of superconducting materials. If the bolometer is always kept at this temperature T_c by a temperature control, the incident radiation power P can be very sensitively measured by the magnitude of the feedback control signal used to compensate for the absorbed radiation power [4.100–4.102].

Example 4.22
With $\int P\,dt = 10^{-12}$ Ws, $\beta = 1$, $H = 10^{-11}$ Ws/K we obtain from (4.119): $\Delta T = 0.1$ K. With $\alpha = 10^{-4}$/K and $R = 10\,\Omega$, $R_1 = 10\,\Omega$, $V_0 = 1$ V, the current change is $\Delta i = 2.5 \times 10^{-6}$ A and the voltage change is $\Delta V = R\Delta i = 2.5 \times 10^{-5}$ V, which is readily detected.

Another method of thermal detection of radiation used in a Golay cell is the absorption of radiation in a closed capsule. According to the ideal gas law, the temperature rise ΔT causes the pressure rise $\Delta p = N(R/V)\Delta T$ (where N is the number of moles and R the gas constant), which expands a flexible

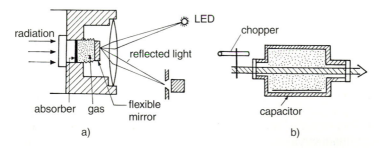

Fig. 4.76a,b. Golay cell: (**a**) using deflection of light by a flexible mirror; (**b**) monitoring the capacitance change ΔC of a capacitor C with a flexible membrane (spectraphone)

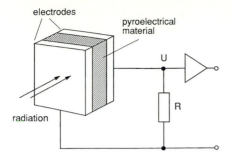

electrodes

pyroelectrical material

U

R

radiation

Fig. 4.77. Pyroelectric detector

membrane on which a mirror is mounted (Fig. 4.76a). The movement of the mirror is monitored by observing the deflection of a light beam from a light-emitting diode [4.103].

In modern devices the flexible membrane is part of a capacitor with the other plate fixed. The pressure rise causes a corresponding change of the capacitance, which can be converted to an AC voltage (Fig. 4.76b). This sensitive detector, which is essentially a *capacitor microphone*, is now widely used in photoacoustic spectroscopy (Sect. 6.3) to detect the absorption spectrum of molecular gases by the pressure rise proportional to the absorption coefficient.

A recently developed thermal detector for infrared radiation is based on the pyroelectric effect [4.104–4.106]. Pyroelectric materials are good electrical insulators that possess an internal macroscopic electric-dipole moment, depending on the temperature. The crystal neutralizes the electric field of this dielectric polarization by a corresponding surface-charge distribution. A change of the internal polarization caused by a temperature rise will produce a measurable change in surface charge, which can be monitored by a pair of electrodes applied to the sample (Fig. 4.77). Because of the capacitive transfer of the change of the electric dipole moments, pyroelectric detectors monitor only *changes* of input power. Any incident cw radiation therefore has to be chopped.

While the sensitivity of good pyroelectric detectors is comparable to that of Golay cells or high-sensitivity bolometers, they are more robust and therefore less delicate to handle. They also have a much better time resolution down into the nanosecond range [4.105].

4.5.2 Photodiodes

Photodiodes are doped semiconductors that can be used as photovoltaic or photoconductive devices. When the *p–n* junction of the diode is irradiated, the photovoltage V_{ph} is generated at the open output of the diode (Fig. 4.78a); within a restricted range it is proportional to the absorbed radiation power. Diodes used as photoconductive elements change their internal resistance upon irradiation and can therefore be used as photoresistors in combination with an external voltage source (Fig. 4.78b).

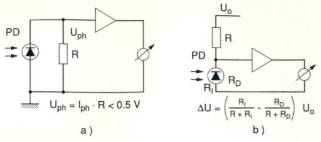

Fig. 4.78a,b. Use of a photodiode: (**a**) as a photovoltaic device; and (**b**) as a photoconductive resistor

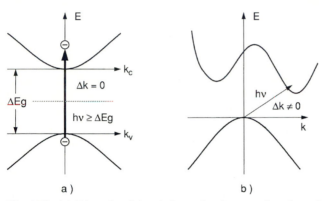

Fig. 4.79. (**a**) Direct band–band absorption in an undoped semiconductor; and (**b**) indirect transitions, illustrated in a $E(k)$ band diagram

For their use as radiation detectors the spectral dependence of their absorption coefficient is of fundamental importance. In an undoped semiconductor the absorption of one photon $h\nu$ causes an excitation of an electron from the valence band into the conduction band (Fig. 4.79a). With the energy gap $\Delta E_g = E_c - E_v$ between the valence and conduction band, only photons with $h\nu \geq \Delta E_g$ are absorbed. The intrinsic absorption coefficient

$$
\alpha_{\text{intr}}(\nu) = \begin{cases} \alpha_0 (h\nu - \Delta E_g)^{1/2}, & \text{for } h\nu > \Delta E_g , \\ 0 , & \text{for } h\nu < \Delta E_g , \end{cases} \tag{4.125}
$$

is shown in Fig. 4.80 for different undoped materials. The quantity α_0 depends on the material and is generally larger for semiconductors with direct transitions ($\Delta k = 0$) than for indirect transitions with $\Delta k \neq 0$. The steep rise of $\alpha(\nu)$ for $h\nu > E_g$ has only been observed for direct transitions, while it is much flatter for indirect transitions.

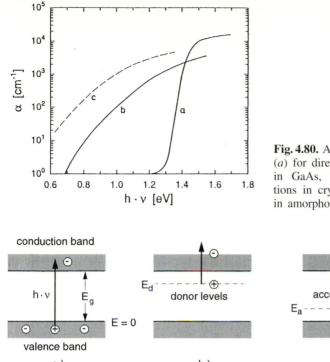

Fig. 4.80. Absorption coefficient $\alpha(\nu)$ (*a*) for direct band–band transitions in GaAs, (*b*) for indirect transitions in crystalline silicon, and (*c*) in amorphous silicon

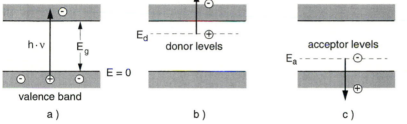

Fig. 4.81a–c. Photoabsorption in undoped semiconductors (**a**) and by donors (**b**) and acceptors (**c**) in *n*- or *p*-doped semiconductors

In doped semiconductors photon-induced electron transitions can occur between the donor levels and the conduction band, or between the valence band and the acceptor levels (Fig. 4.81). Since the energy gaps $\Delta E_d = E_c - E_d$ or $\Delta E_a = E_v - E_a$ are much smaller than the gap $E_c - E_v$, doped semiconductors absorb even at smaller photon energies $h\nu$ and can therefore be employed for the detection of longer wavelengths in the mid-infrared. In order to minimize thermal excitation of electrons, these detectors must be operated at low temperatures. For $\lambda \leq 10\,\mu m$ generally liquid-nitrogen cooling is sufficient, while for $\lambda > 10\,\mu m$ liquid-helium temperatures around $4-10\,K$ are required.

Figure 4.82 plots the detectivity of commonly used photodetector materials with their spectral dependence, while Fig. 4.83 illustrates their useful spectral ranges and their dependence on the energy gap ΔE_g.

a) Photoconductive Diodes

When a photodiode is illuminated, its electrical resistance decreases from a "dark value" R_D to a value R_I under illumination. In the circuit shown in

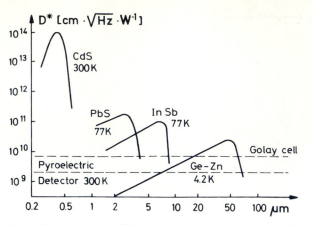

Fig. 4.82. Detectivity $D^*(\lambda)$ of some photodetectors [4.96]

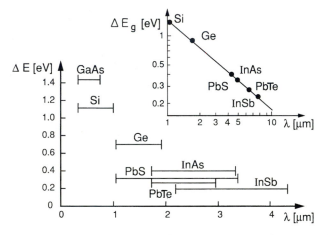

Fig. 4.83. Energy gaps and useful spectral ranges of some semiconducting materials

Fig. 4.78b, the change of the output voltage is given by

$$\Delta U = \left(\frac{R_D}{R_D + R} - \frac{R_I}{R_I + R} \right) U_0 = \frac{R(R_D - R_I)}{(R + R_D)(R + R_I)} U_0 , \qquad (4.126)$$

which becomes, at a given illumination, maximum for

$$R \approx \sqrt{R_D R_I} .$$

The time constant of the photoconductive diode is determined by $\tau \geq RC$, where C is the capacitance of the diode plus the input capacitance of the circuit. Its lower limit is set by the diffusion time of the electrons on their way from the p–n junction where they are generated to the electrodes. Detectors from PbS, for example, have typical time constants of $0.1-1$ ms, while InSb detectors are much faster ($\tau \simeq 10^{-7}-10^{-6}\,\mathrm{s}$). Although photoconductive detectors are generally more sensitive, photovoltaic detectors are better suited for the detection of fast signals.

b) Photovoltaic Detector

While photoconductors are passive elements that need an external power supply, photovoltaic diodes are active elements that generate their own photovoltage upon illumination, although they are often used with an external bias voltage. The principle of the photogenerated voltage is shown in Fig. 4.84.

In the nonilluminated diode, the diffusion of electrons from the n-region into the p-region (and the opposite diffusion of the holes) generates a space charge, which results in the diffusion voltage V_D and a corresponding electric field across the p–n junction (Fig. 4.84b). Note that this diffusion voltage cannot be detected across the electrodes of the diode, because it is just compensated by the different contact potentials between the two ends of the diode and the connecting leads.

When the detector is illuminated, electron–hole pairs are created by photon absorption within the p–n junction. The electrons are driven by the diffusion voltage into the n-region, the holes into the p-region. This leads to a *decrease* ΔV_D of the diffusion voltage, which appears as the photovoltage $V_{\mathrm{ph}} = \Delta V_D$ across the open electrodes of the photodiode. If these electrodes are connected through an Ampére-meter, the photoinduced current

$$i_{\mathrm{ph}} = -\eta e \phi A \,, \tag{4.127}$$

is measured, which equals the product of quantum efficiency η, the illuminated active area A of the photoiode, and the incident photon flux density $\phi = I/h\nu$.

The illuminated p–n photodetector can therefore be used either as a current generator or a voltage source, depending on the external resistor between the electrodes.

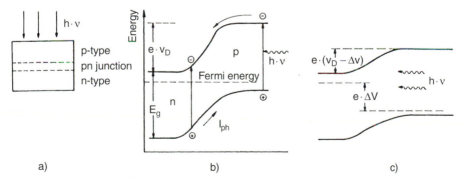

a)   b)   c)

Fig. 4.84a–c. Photovoltaic diode: (**a**) schematic structure and (**b**) generation of an electron–hole pair by photon absorption within the p–n junction. (**c**) Reduction of the diffusion voltage V_D for an open circuit

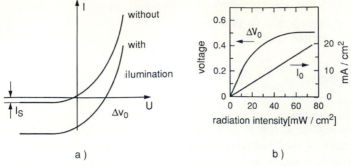

Fig. 4.85. (**a**) Current–voltage characteristics of a dark and an illuminated diode; (**b**) photovoltage at the open ends and photocurrent in a shortened diode as a function of incident radiation power

Note: The photon-induced voltage $U_{ph} < \Delta E_g/e$ is always limited by the energy gap ΔE_g. The voltage U_{ph} across the open ends of the photodiode is reached even at relatively small photon fluxes, while the photocurrent is linear over a large range (Fig. 4.85b). When using photovoltaic detectors for measuring radiation power, the load resistor R_L must be sufficiently low to keep the output voltage $U_{ph} = i_{ph}R_L < U_s = \Delta E_g/e$ always below its saturation value U_s. Otherwise, the output signal is no longer proportional to the input power.

If an external voltage U is applied to the diode, the diode current without illumination

$$i_D(U) = CT^2 e^{-eV_D/kT}(e^{eU/kT} - 1) , \tag{4.128}$$

shows the typical diode characteristics (Fig. 4.85a). During illumination this current is superimposed by the opposite photocurrent

$$i_{ill}(U) = i_D(U) - i_{ph} . \tag{4.129}$$

With open ends of the diode we obtain $i = 0$, and therefore from (4.128, 4.129) the photovoltage

$$U_{ph}(i = 0) = \frac{kT}{e}\left[\ln\left(\frac{i_{ph}}{i_s}\right) + 1\right] , \tag{4.130}$$

where $i_s = CT^2 \exp(-eV_D/kT)$ is the saturation dark current for large negative values of U.

Fast photodiodes are always operated at a reverse bias voltage $U < 0$, where the saturated reverse current i_s of the dark diode is small (Fig. 4.85a). From (4.128) we obtain, with $[\exp(eU/kT) \ll 1]$ for the total diode current,

$$i = -i_s - i_{ph} = -CT^2 e^{-eV_D/kT} - i_{ph} , \tag{4.131}$$

which becomes independent of the external voltage U.

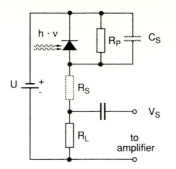

Fig. 4.86. Equivalent circuit of a photodiode with internal capacity C_S, series internal resistor R_S, parallel internal resistor R_P, and external load resistor R_L

Materials used for photovoltaic detectors are, e.g., silicon, cadmium sulfide (Cds), and gallium arsenide (GaAs). Silicon detectors deliver photovoltages up to $550\,\text{mV}$ and photocurrents up to $40\,\text{mA/cm}^2$ [4.86]. The efficiency $\eta = P_{\text{el}}/P_{\text{ph}}$ of energy conversion reaches $10-14\%$. Gallium arsenide (GaAs) detectors show larger photovoltages up to $1\,\text{V}$, but slightly lower photocurrents of about $20\,\text{mA/cm}^2$.

c) Fast Photodiodes

The photocurrent generates a signal voltage $V_s = U_{\text{ph}} = R_L i_{\text{ph}}$ across the load resistor R_L that is proportional to the absorbed radiation power over a large intensity range of several decades, as long as $V_s < \Delta E_g/e$ (Fig. 4.85b). From the circuit diagram in Fig. 4.86 with the capacitance C_s of the semiconductor and its series and parallel resistances R_s and R_p, one obtains for the upper frequency limit [4.107]

$$f_{\max} = \frac{1}{2\pi C_s(R_s + R_L)(1 + R_s/R_p)} \,, \tag{4.132}$$

which reduces, for diodes with large R_p and small R_s, to

$$f_{\max} = \frac{1}{2\pi C_s R_L} \,. \tag{4.133}$$

With small values of the resistor R_L, a high-frequency response can be achieved, which is limited only by the drift time of the carriers through the boundary layer of the p–n junction. This drift time can be reduced by an external bias voltage. Using diodes with large bias voltages and a 50-Ω load resistor matched to the connecting cable, rise times in the subnanosecond range can be obtained.

Example 4.23

$C_s = 10^{-11}\,\text{F}, \quad R_L = 50\,\Omega \ \Rightarrow\ f_{\max} = 300\,\text{MHz}, \quad \tau = \dfrac{1}{2\pi f_{\max}} \simeq 0.6\,\text{ns}.$

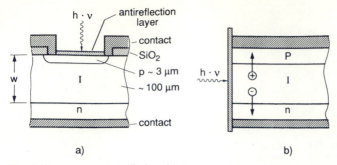

Fig. 4.87a,b. PIN photodiode with head-on (**a**) and side-on (**b**) illumination

For photon energies $h\nu$ close to the band gap, the absorption coefficient decreases, see (4.125). The penetration depth of the radiation, and with it the volume from which carriers have to be collected, becomes large. This increases the collection time and makes the diode slow.

Definite collection volumes can be achieved in *PIN diodes*, where an undoped zone I of an intrinsic semiconductor separates the *p*- and *n*-regions (Fig. 4.87). Since no space charges exist in the intrinsic zone, the bias voltage applied to the diode causes a constant electric field, which accelerates the carriers. The intrinsic region may be made quite wide, which results in a low capacitance of the *p–n* junction and provides the basis for a very fast and sensitive detector. The limit for the response time is, however, also set by the transit time $\tau = w/v_{th}$ of the carriers in the intrinsic region, which is determined by the width w and the thermal velocity v_{th} of the carriers. Silicon PIN diodes with a 700-µm wide zone I have response times of about 10 ns and a sensitivity maximum at $\lambda = 1.06\,\mu m$, while diodes with a 10-µm wide zone I reach 100 ps with a sensitivity maximum at a shorter wavelength–around $\lambda = 0.6\,\mu m$ [4.108]. Fast response combined with high sensitivity can be achieved when the incident radiation is focused from the side into the zone I (Fig. 4.87b). The only experimental disadvantage is the critical alignment necessary to hit the small active area.

Very fast response times can be reached by using the photoeffect at the metal–semiconductor boundary known as the Schottky barrier [4.109]. Because of the different work functions ϕ_m and ϕ_s of the metal and the semiconductor, electrons can tunnel from the material with low ϕ to that with high ϕ (Fig. 4.88). This causes a space-charge layer and a potential barrier

$$V_B = \phi_B/e\,, \quad \text{with} \quad \phi_B = \phi_m - \chi\,, \tag{4.134}$$

between metal and semiconductor. The electron affinity is given by $\chi = \phi_s - (E_c - E_F)$. If the metal absorbs photons with $h\nu > \phi_B$, the metal electrons gain sufficient energy to overcome the barrier and "fall" into the semiconductor, which thus acquires a negative photovoltage. The *majority* carriers are responsible for the photocurrent, which ensures fast response times.

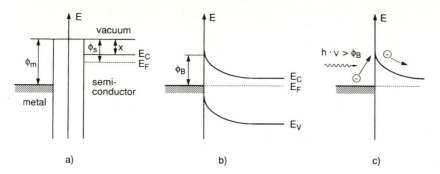

Fig. 4.88. (a) Work functions ϕ_m of metal and ϕ_s of semiconductor and electron affinity χ. E_c is the energy at the bottom of the conduction band and E_F is the Fermi energy. **(b)** Schottky barrier at the contact layer between metal and n-type semiconductor. **(c)** Generation of a photocurrent

For measurements of optical frequencies, ultrafast metal–insulator–metal (MIM) diodes have been developed [4.110], which can be operated up to 66 THz ($\lambda = 3.39\,\mu m$). In these diodes, a 25-μm diameter tungsten wire, electrochemically etched to a point less than 200 nm in radius, serves as the point contact element, while the optically polished end of a nickel plate with a thin oxide layer forms the base element of the diode (Fig. 4.89).

These MIM diodes can be used as mixing elements at optical frequencies. When illuminating the contact point with a focused CO_2 laser, a response time of 10^{-14} s or better has been demonstrated by the measurement of the 88-THz emission from the third harmonic of the CO_2 laser. If the beams of two lasers with the frequencies f_1 and f_2 are focused onto the junction between the nickel oxide layer and the sharp tip of a tungsten wire, the MIM diode acts as a rectifier and the wire as an antenna, and a signal with the difference frequency $f_1 - f_2$ is generated. Difference frequencies up into the terahertz range can be monitored [4.111] (see Sect. 5.8.7). The basic processes in these MIM diodes represent very interesting phenomena of solid-state physics. They could be clarified only recently [4.111].

Difference frequencies up to 900 GHz between two visible dye lasers have been measured with Schottky diodes by mixing this frequency with har-

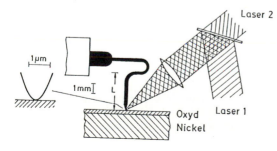

Fig. 4.89. Arrangement of a metal–insulator–metal (MIM) diode used for optical frequency mixing of laser frequencies

monics of 90-GHz microwave radiation which was also focused onto the diode [4.112]. Meanwhile, Schottky-barrier mixer diodes have been developed that cover the frequency range $1-10$ THz [4.112].

d) Avalanche-Diodes

Internal amplification of the photocurrent can be achieved with avalanche diodes, which are reverse-biased semiconductor diodes, where the free carriers acquire sufficient energy in the accelerating field to produce additional carriers on collisions with the lattice (Fig. 4.90). The multiplication factor M, defined as the average number of electron–hole pairs after avalanche multiplication initiated by a single photoproduced electron–hole pair, increases with the reverse-bias voltage. Values of M up to 10^6 have been reported in silicon, which allows sensitivities comparable with those of a photomultiplier. The advantage of these avalanche diodes is their fast response time, which decreases with increasing bias voltage. In this device the product of gain times bandwidth may exceed 10^{12} Hz if the breakdown voltage is sufficiently high [4.87].

In order to avoid electron avalanches induced by holes accelerated into the opposite direction, which would result in additional background noise, the amplification factor for holes must be kept considerably smaller than for electrons. This is achieved by a specially tailored layer structure, which yields a sawtooth-like graded band-gap dependence $\Delta E_g(x)$ in the field x-direction (Fig. 4.90c,d). In an external field this structure results in an amplification factor M that is $50-100$ times larger for electrons than for holes [4.113].

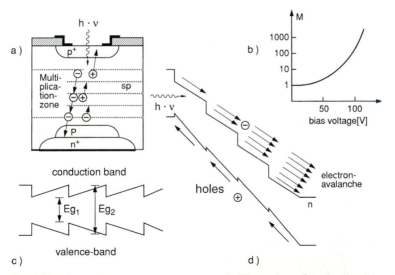

Fig. 4.90a–d. Avalanche diode: (**a**) schematic illustration of avalanche formation (n^+, p^+ are heavily doped layers); (**b**) amplification factor $M(V)$ as a function of the bias voltage V for a Si-avalanche diode; (**c**) spatial variation of band edges and bandgap without external field; and (**d**) within an external electric field

Such modern avalanche diodes may be regarded as the solid-state analog to photomultipliers (Sect. 4.5.4). Their advantages are a high quantum efficiency (up to 40%) and a low supply voltage (10−20 V). Their disadvantage for fluorescence detection is the small active area compared to the much larger cathode area of photomultipliers [4.114–4.115].

4.5.3 Photodiode Arrays

Many small photodiodes can be integrated on a single chip, forming a *photodiode array*. If all diodes are arranged in a line we have a one-dimensional diode array, consisting of up to 2048 diodes. With a diode width $b = 15 \, \mu m$ and a spacing of $d = 10 \, \mu m$ between two diodes, the length L of an array of 1024 diodes becomes 25 mm with a height of about 40 μm [4.116].

The basic principle and the electronic readout diagram is shown in Fig. 4.91. An external bias voltage U_0 is applied to p–n diodes with the sensitive area A and the internal capacitance C_s. Under illumination with an intensity I the photocurrent $i_{ph} = \eta A I$, which is superimposed on the dark current i_D, discharges the diode capacitance C_s during the illumination time ΔT by

$$\Delta Q = \int_t^{t+\Delta T} (i_D + \eta A I) \, dt = C_s \Delta U . \tag{4.135}$$

Every photodiode is connected by a multiplexing MOS switch to a voltage line and is recharged to its original bias voltage U_0. The recharging pulse $\Delta U = \Delta Q / C_s$ is sent to a video line connected with all diodes. These pulses are, according to (4.135), a measure for the incident radiation energy $\int A I \, dt$, if the dark current i_D is subtracted and the quantum efficiency η is known.

The maximum integration time ΔT is limited by the dark current i_D, which therefore also limits the attainable signal-to-noise ratio. At room temperature typical integration times are in the millisecond range. Cooling of the diode array by Peltier cooling down to −40°C drastically reduces the dark current and allows integration times of 1−100 s. The minimum detectable

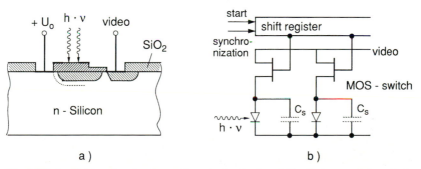

Fig. 4.91a,b. Schematic structure of a single diode within the array (**a**) and electronic circuit diagram of a one-dimensional diode array (**b**)

incident radiation power is determined by the minimum voltage pulse ΔU that can be safely distinguished from noise pulses. The detection sensitivity therefore increases with decreasing temperature because of the possible increasing integration time. At room temperature typical sensitivity limits are about 500 photons per second and diode.

If such a linear diode array with N diodes and a length $L = N(b+d)$ is placed in the observation plane of a spectrograph (Fig. 4.1), the spectral interval

$$\delta\lambda = \frac{d\lambda}{dx}L \ ,$$

which can be detected simultaneously, depends on the linear dispersion $dx/d\lambda$ of the spectrograph. The smallest resolvable spectral interval

$$\Delta\lambda = \frac{d\lambda}{dx}b \ ,$$

is limited by the width b of the diode. Such a system of spectrograph plus diode array is called an *optical multichannel analyzer* (OMA) or an *optical spectrum analyzer* (OSA) [4.116, 4.117].

Example 4.24
$b+d = 25\,\mu m, \qquad L = 25\,mm, \qquad d\lambda/dx = 5\,nm/mm$
$\Rightarrow \ \delta\lambda = 125\,nm, \qquad \Delta\lambda = 0.125\,nm.$

The diodes can be also arranged in a two-dimensional array, which allows the detection of two-dimensional intensity distributions. This is, for instance, important for the observation of spatial distributions of light-emitting atoms in gas discharges or flames (Sect. 15.4) or of the ring pattern behind a Fabry–Perot interferometer.

Another version of optical multichannel analyzers is based on the *vidicon principle*. The diode array is irradiated on one side by the image to be detected (Fig. 4.92) and the absorbed photons produce electron–hole pairs. The diodes are reverse biased and the holes that diffuse into the p-zone partially discharge the p–n junction. If the other side of the array is scanned with a focused electron beam (20-μm diameter), the electrons recharge the anodes until they have nearly cathode potential. This recharging, which just compensates for the discharge by the incident photons, can be determined as a current pulse in the external circuit and is measured as the video signal.

Photodiode arrays are now increasingly replaced by charge-coupled device (CCD) arrays, which consist of an array of small MOS junctions on a doped silicon substrate (Fig. 4.93) [4.118–4.121]. The incident photons generate electrons and holes in the n- or p-type silicon. The electrons or holes are collected and change the charge of the MOS capacitances. These changes of the charge can be shifted to the next MOS capacitance by applying a sequence of suitable voltage steps according to Fig. 4.93b. The charges are thus shifted from one diode to the next until they reach the last diode of a row, where they cause the voltage change ΔU, which is sent to a video line.

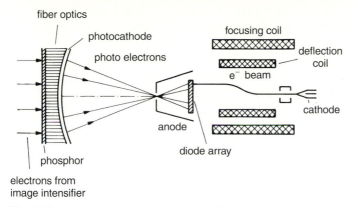

Fig. 4.92. Schematic arrangement of a vidicon

The quantum efficiency η of CCD arrays depends on the material used for the substrate, it reaches peak values over 90%. The efficiency $\eta(\lambda)$ is generally larger than 20% over the whole spectral range from 350–900 nm (Fig. 4.93c) and exceeds the efficiencies of most photocathodes (Sect. 4.5.4). The spectral range of special CCDs ranges from 0.1–1000 nm. They can therefore be used in the VUV and X-ray regions, too. The highest sensitivity up to 90% efficiency is achieved with backward-illuminated devices (Fig. 4.94). Table 4.2 compiles some relevant data of commercial CCD devices.

The dark current of cooled CCD arrays may be below 10^{-2} electrons per second and diode. The readout dark pulses are smaller than those of

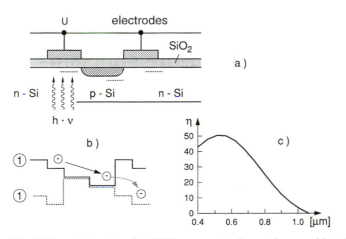

Fig. 4.93a–c. Principle of a CCD array: (**a**) alternately, a positive (*solid line*) and a negative (*dashed line*) voltage are applied to the electrodes. (**b**) This causes the charged carriers generated by photons to be shifted to the next diode. This shift occurs with the pulse frequency of the applied voltage. (**c**) Spectral sensitivity of CCD diodes

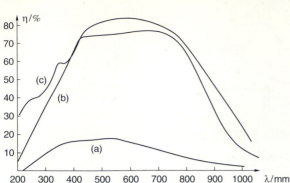

Fig. 4.94. Spectral dependence of the quantum efficiency $\eta(\lambda)$ of front-illuminated (a) and backward-illuminated CCD arrays with visible-AR coatings (b) and UV–AR coatings (c)

Table 4.2. Characteristic data of CCD arrays

Active area [mm^2]	24.6×24.6
Pixel size [μm]	7.5×15 up to 24×24
Number of pixels	1024×1024 up to 2048×2048
Dynamic range [bits]	16
Readout noise at 50 kHz [electron charges]	$4-6$
Dark charge [electrons/(h pixel)]	< 1
Hold time at $-120\,°C$ [h]	> 10
Quantum efficiency peak	
Front illuminated	45%
Backward illuminated	$> 80\%$

photodiode arrays. Therefore, the sensitivity is high and may exceed that of good photomultipliers. A particular advantage is their large dynamic range, which covers about five orders of magnitude. More information about CCD detectors, which are of increasing importance in spectroscopy, can be found in [4.121, 4.122].

4.5.4 Photoemissive Detectors

Photoemissive detectors, such as the photocell or the photomuliplier, are based on the external photoeffect. The photocathode of such a detector is covered with one or several layers of materials with a low work function ϕ (e.g., alkali metal compounds or semiconductor compounds). Under illumination with monochromatic light of wavelength $\lambda = c/v$, the emitted photoelectrons leave the photocathode with a kinetic energy given by the Einstein relation

$$E_{\text{kin}} = h v - \phi \,. \tag{4.136}$$

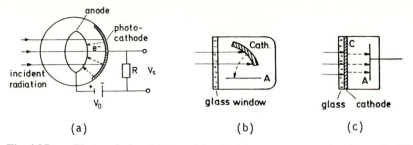

Fig. 4.95a–c. Photoemissive detector: (**a**) principle arrangement of a photocell; (**b**) opaque photocathode; and (**c**) semitransparent photocathode

They are further accelerated by the voltage V_0 between the anode and cathode and are collected at the anode. The resultant photocurrent is measured either directly or by the voltage drop across a resistor (Fig. 4.95a).

a) Photocathodes

The quantum efficiency $\eta = n_e/n_{ph}$ is defined as the ratio of the rate of photoelectrons n_e to the rate of incident photons n_{ph}. It depends on the cathode material, on the form and thickness of the photoemissive layer, and on the wavelength λ of the incident radiation. The quantum efficiency $\eta = n_a n_b n_c$ can be represented by the product of three factors. The first factor n_a gives the probability that an incident photon is actually absorbed. For materials with a large absorption coefficient, such as pure metals, the reflectivity R is high (e.g., for metallic surfaces $R \geq 0.8$–0.9 in the visible region), and the factor n_a cannot be larger than $(1 - R)$. For semitransparent photocathodes of thickness d, on the other hand, the absorption must be large enough to ensure that $\alpha d > 1$. The second factor n_b gives the probability that the absorbed photon really produces a photoelectron instead of heating the cathode material. Finally, the third factor n_c stands for the probability that this photoelectron reaches the surface and is emitted instead of being backscattered into the interior of the cathode.

Two types of photoelectron emitters are manufactured: opaque layers, where light is incident on the same side of the photocathode from which the photoelectrons are emitted (Fig. 4.95b); and semitransparent layers (Fig. 4.95c), where light enters at the opposite side to the photoelectron emission and is absorbed throughout the thickness d of the layer. Because of the two factors n_a and n_c, the quantum efficiency of semitransparent cathodes and its spectral change are critically dependent on the thickness d, and reach that of the reflection-mode cathode only if the value of d is optimized.

Figure 4.96 shows the spectral sensitivity $S(\lambda)$ of some typical photocathodes, scaled in milliamperes of photocurrent per watt incident radiation. For comparison, the quantum efficiency curves for $\eta = 0.001$, 0.01 and 0.1 are

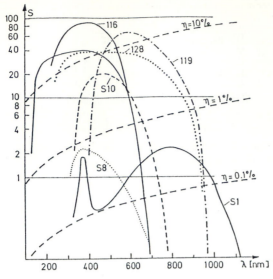

Fig. 4.96. Spectral sensitivity curves of some commercial cathode types. The *solid lines* give $S(\lambda)$ [mA/W], whereas the *dashed curves* given quantum efficiencies $\eta = n_e/n_{ph}$

also drawn (dashed curves). Both quantities are related by

$$S = \frac{i}{P_{in}} = \frac{n_e\,e}{n_{ph}h\nu} \;\Rightarrow\; S = \frac{\eta e\lambda}{hc} \;.\tag{4.137}$$

For most emitters the threshold wavelength for photoemission is below $0.85\,\mu m$, corresponding to a work function $\phi \geq 1.4\,eV$. An example for such a material with $\phi \sim 1.4\,eV$ is a surface layer of NaKSb [4.123]. Only some complex cathodes consisting of two or more separate layers have an extended sensitivity up to about $\lambda \leq 1.2\,\mu m$. The spectral response of the most commonly fabricated photocathodes is designated by a standard nomenclature, using the symbols S1 to S20. Some newly developed types are labeled by special numbers, which differ for the different manufacturers [4.124].

Recently, a new type of photocathode has been developed that is based on photoconductive semiconductors whose surfaces have been treated to obtain a state of negative electron affinity (NEA). In this state an electron at the bottom of the conduction band inside the semiconductor has a higher energy than the zero energy of a free electron in vacuum [4.125]. When an electron is excited by absorption of a photon into such an energy level within the bulk, it may travel to the surface and leave the photocathode. These NEA cathodes have the advantage of a high sensitivity, which is fairly constant over an extended spectral range and even reaches into the infrared up to about $1.2\,\mu m$. Since these cathodes represent cold-electron emission devices, the dark current is very low. Until now, their main disadvantage has been the complicated fabrication procedure and the resulting high price.

Different devices of photoemissive detectors are of major importance in modern spectroscopy. These are the *the photomultiplier, the image intensifier, and the streak camera.*

b) Photomultipliers

Photomultipliers are a good choice for the detection of low light levels. They overcome some of the noise limitations by internal amplification of the photocurrent using secondary-electron emission from internal dynodes to mulitply the number of photoelectrons (Fig. 4.97). The photoelectrons emitted from the cathode are accelerated by a voltage of a few hundred volts and are focused onto the metal surface (e.g., Cu–Be) of the first "dynode" where each impinging electron releases, on the average, q secondary electrons. These electrons are further accelerated to a second dynode where each secondary electron again produces about q tertiary electrons, and so on. The amplification factor q depends on the accelerating voltage U, on the incidence angle α, and on the dynode material. Typical figures for $U = 200\,\text{V}$ are $q = 3\text{–}5$. A photomulitplier with ten dynodes therefore has a total current amplification of $G = q^{10} \sim 10^5 \text{–} 10^7$. Each photoelectron in a photomultiplier with N dynodes produces a charge avalanche at the anode of $Q = q^N e$ and a corresponding voltage pulse of

$$V = \frac{Q}{C} = \frac{q^N \text{e}}{C} = \frac{G\,\text{e}}{C}\,, \tag{4.138}$$

where C is the capacitance of the anode (including connections).

Example 4.25
$G = 2 \times 10^6, \quad C = 30\,\text{pf} \;\Rightarrow\; V = 10\,\text{mV}.$

For experiments demanding high time resolution, the rise time of this anode pulse should be as small as possible. Let us consider which effects may contribute to the anode pulse rise time, caused by the spread of transit times

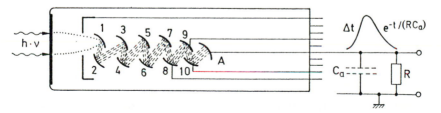

Fig. 4.97. Photomultiplier with time-dependent output voltage pulse induced by an electron avalanche that was triggered by a delta-function light pulse

for the different electrons [4.126, 4.127]. Assume that a single photoelectron is emitted from the photocathode, and is accelerated to the first dynode. The initial velocities of the secondary electrons vary because these electrons are released at different depths of the dynode material and their initial energies, when leaving the dynode surface, are between 0 and 5 eV. The transit time between two parallel electrodes with distance d and potential difference V is

$$t = d\sqrt{\frac{2m}{eV}} \,, \tag{4.139}$$

for electrons with mass m starting with zero initial energy. Electrons with the initial energy E_{kin} reach the next electrode earlier by the time difference

$$\Delta t_1 = \frac{d}{eV}\sqrt{2m E_{kin}} \,. \tag{4.140}$$

Example 4.26
$E_{kin} = 0.5\,\text{eV}, \quad d = 1\,\text{cm}, \quad V = 250\,\text{V} \;\Rightarrow\; \Delta t_1 = 0.1\,\text{ns}.$

The electrons travel slightly different path lengths through the tube, which causes an additional time spread of

$$\Delta t_2 = \Delta d \sqrt{\frac{2m}{eV}} \,, \tag{4.141}$$

which is of the same magnitude as Δt_1. The rise time of an anode pulse started by a single photoelectron therefore decreases with increasing voltage proportional to $V^{-1/2}$. It depends on the geometry and form of the dynode structures.

When a short intense light pulse produces many photoelectrons simultaneously, the time spread is further increased by two phenomena:

- The initial velocities of the emitted photoelectrons differ, e.g., for a cesium antimonide S5 cathode between 0 and 2 eV. This spread depends on the wavelength of the incoming light [4.128a].
- The time of flight between the cathode and the first dynode strongly depends on the locations of the spot on the cathode where the photoelectron is emitted. The resulting time spread may be larger than that from the other effects, but may be reduced by a focusing electrode between the cathode and the first dynode with careful optimization of its potential. Typical anode rise times of photomultipliers range from 0.5−20 ns. For specially designed tubes with optimized side-on geometry, where the curved opaque cathode is illuminated from the side of the tube, rise times of 0.4 ns have been achieved [4.128b]. Shorter rise times can be reached with channel plates and channeltrons [4.128c].

Example 4.27

Photomultiplier type 1P28: $N = 9$, $q = 5.1$ at $V = 1250\,\text{V}$ $\Rightarrow$ $G = 2.5 \times 10^6$; anode capacitance and input capacitance of the amplifier $C_a = 15\,\text{pF}$. A single photoelectron produces an anode pulse of 27 mV with a rise time of 2 ns. With a resistor $R = 10^5\,\Omega$ at the PM exit, the trailing edge of the output pulse is $C_a = 1.5 \times 10^{-6}\,\text{s}$.

For low-level light detection, the question of noise mechanisms in photomultipliers is of fundamental importance [4.129]. There are three main sources of noise:

- Photomultiplier dark current;
- Noise of the incoming radiation;
- Shot noise and Johnson noise caused by fluctuations of the amplification and by noise of the load resistor.

We shall discuss these contributions separately:

- When a photomultiplier is operated in complete darkness, electrons are still emitted from the cathode. This dark current is mainly due to thermionic emission and is only partly caused by cosmic rays or by radioactive decay of spurious radioactive isotopes in the multiplier material. According to Richardson's law, the thermionic emission current

$$i = C_1 T^2 e^{-C_2 \phi / T} , \tag{4.142}$$

strongly depends on the cathode temperature T and on its work function ϕ. If the spectral sensitvity extends into the infrared, the work function ϕ must be small, which increases the dark current. In order to decrease the dark current, the temperature T of the cathode must be reduced. For instance, cooling a cesium–antimony cathode from 20°C to 0°C reduces the dark current by a factor of about ten. The optimum operation temperature depends on the cathode type (because of ϕ). For S1 cathodes, e.g., those with a high infrared sensitivity and therefore a low work function ϕ, it is advantageous to cool the cathode down to liquid nitrogen temperatures. For other types with maximum sensitivity in the green, cooling below -40°C gives no significant improvement because the thermionic part of the dark current has already dropped below other contributions, e.g., caused by high-energy β-particles from disintegration of ^{40}K nuclei in the window material. Excessive cooling can even cause undesirable effects, such as a reduction of the signal photocurrent or voltage drops across the cathode, because the resistance of the cathode film increases with decreasing temperature [4.130].
  For many spectroscopic applications only a small fraction of the cathode area is illuminated, e.g., for photomultipliers behind the exit slit of a monochromator. In such cases, the dark current can be futher reduced either by using photomulitpliers with a small effective cathode area or by placing small magnets around an extended cathode. The magnetic field de-

focuses electrons from the outer parts of the cathode area. These electrons cannot reach the first dynode and do not contribute to the dark current.

- The shot noise

$$\langle i_n \rangle_s = \sqrt{2e \cdot i \cdot \Delta f}$$ (4.143a)

of the photocurrent [4.129] is amplified in a photomultiplier by the gain factor G. The root-mean-square (rms) noise voltage across the anode load resistor R is therefore

$$\langle V \rangle_s = GR\sqrt{2ei_c\Delta f}, \quad i_c : \text{cathode current},$$

$$= R\sqrt{2eGi_a\Delta f}, \quad i_a : \text{anode current}.$$ (4.143b)

The gain factor G is not constant, but shows fluctuations due to random variations of the secondary-emission coefficient q, which is a small integer. This contributes to the total noise and multiplies the rms shot noise voltage by a factor $a > 1$, which depends on the mean value of q [4.131]. The Johnson noise of the load resistor R at the temperature T gives an rms-noise current

$$\langle i_n \rangle_J = \sqrt{4kT\Delta f/R}$$ (4.144a)

- From (4.143) we obtain with (4.144a) for the sum of shot noise and Johnson noise across the anode load resistor R at room temperature, where $4kT/e \approx 0.1\,\text{V}$

$$\langle V \rangle_{J+s} = \sqrt{eR\Delta f(2RGa^2 i_a + 0.1)} \quad [\text{Volt}].$$ (4.144)

For $GRi_a a^2 \gg 0.05\,\text{V}$, the Johnson noise can be neglected. With the gain factor $G = 10^6$ and the load resistor of $R = 10^5\,\Omega$, this implies that the anode current i_a should be larger than $5 \times 10^{-13}\,\text{A}$. Since the anode dark current is already much larger than this limit, we see that *the Johnson noise does not contribute to the total noise of photomultipliers*.

A significant improvement of the signal-to-noise ratio in detection of low levels of radiation can be achieved with single-photon counting techniques, which enable spectroscopic investigations to be performed at incident radiation fluxes down to $10^{-17}\,\text{W}$. These techniques are discussed in Sect. 4.5.5. More details about photomultipliers and optimum conditions of performance can be found in excellent introductions issued by EMI or RCA [4.131, 4.132]. An extensive review of photoemissive detectors has been given by Zwicker [4.123]; see also the monographs [4.133, 4.134].

c) Microchannel plates

Photomultipliers are now often replaced by microchannel plates. They consist of a photocathode layer on a thin semiconductive glass plate (0.5−1.5 mm) that is perforated by millions of small holes with diameters in the range

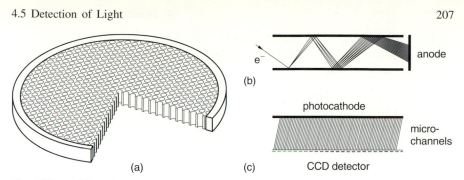

Fig. 4.98a–c. Microchannel plate (MCP): (**a**) schematic construction; (**b**) electron avalanche in one channel; (**c**) schematic arrangement of MCP detector with spatial resolution

$10-25\,\mu$m (Fig. 4.98). The total area of the holes covers about 60% of the glass plate area. The inner surface of the holes (channels) has a high secondary emission coefficient for electrons that enter the channels from the photocathode and are accelerated by a voltage applied between the two sides of the glass plate. The amplification factor is about 10^3 at an electric field of $500\,\mathrm{V/mm}$. Placing two microchannel plates in series allows an amplification of 10^6, which is comparable to that of photomultipliers.

The advantage of the microchannel plates is the short rise time ($< 1\,$ns) of the electron avalanches generated by a single photon, the small size, and the possibility of spatial resolution [4.135].

d) Photoelectric Image Intensifiers

Image intensifiers consist of a photocathode, an electro-optical imaging device, and a fluorescence screen, where an intensified image of the irradiation pattern at the photocathode is reproduced by the accelerated photoelectrons. Either magnetic or electric fields can be used for imaging the cathode pattern onto the fluorescent screen. Instead of the intensified image being viewed on a phosphor screen, the electron image can be used in a camera tube to generate picture signals, which can be reproduced on the television screen and can be stored either photographically or on a recording tape [4.136, 4.137].

For applications in spectroscopy, the following characteristic properties of image intensifiers are important:

- The intensity magnification factor M, which gives the ratio of output intensity to input intensity;
- The dark current of the system, which limits the minimum detectable input power;
- The spatial resolution of the device, which is generally given as the maximum number of parallel lines per millimeter of a pattern at the cathode which can still be resolved in the intensified output pattern;
- The time resolution of the system, which is essential for recording of fast transient input signals.

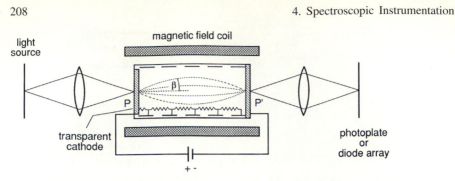

Fig. 4.99. Single-stage image intensifier with magnetic focusing

Figure 4.99 illustrates a simple, single-stage image intensifier with a magnetic field parallel to the accelerating electric field. All photoelectrons starting from the point P at the cathode follow helical paths around the magnetic field lines and are focused into P' at the phosphor screen after a few revolutions. The location of P' is, to a first approximation, independent of the direction β of the initial photoelectron velocities. To get a rough idea about the possible magnification factor M, let us assume a quantum efficiency of 20% for the photocathode and an accelerating potential of 10 kV. With an efficiency of 20% for the conversion of electron energy to light energy in the phosphor screen, each electron produces about 1000 photons with $h\nu = 2\,\text{eV}$. The amplification factor M giving the number of output photons per incoming photon is then $M = 200$. However, light from the phosphor is emitted into all directions and only a small fraction of it can be collected by an optical system. This reduces the total gain factor.

The collection efficiency can be enhanced when a thin mica window is used to support the phosphor screen and photographic contact prints of the image are made. Another way is the use of fiber-optic windows.

Larger gain factors can be achieved with cascade intensifier tubes (Fig. 4.100), where two or more stages of simple image intensifiers are coupled in series [4.138]. The critical components of this design are the phosphor–photocathode sandwich screens, which influence the sensitivity and the spatial resolution. Since light emitted from a spot around P on the phosphor should release photelectrons from the opposite spot around P' of the photocathode, the distance between P and P' should be as small as possible in order to preserve the spatial resolution. Therefore, a thin layer of phosphor (a few microns) of very fine grain-size is deposited by electrophoresis on a mica sheet with a few microns thickness. An aluminum foil reflects the light from the phosphor back onto the photocathode (Fig. 4.100b) and prevents optical feedback to the preceding cathode.

The spatial resolution depends on the imaging quality, which is influenced by the thickness of the phosphor-screen–photocathode sandwiches, by the homogeneity of the magnetic field, and by the lateral velocity spread of the photoelectrons. Red-sensitive photocathodes generally have a lower spatial resolution since the initial velocities of the photoelectrons are larger. The res-

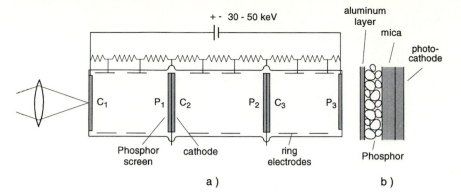

Fig. 4.100a,b. Cascade image intensifier: (**a**) schematic diagram with cathodes C_i, fluorescence screens P_i, and ring electrodes providing the acceleration voltage; (**b**) detail of phosphor–cathode sandwich structure

Table 4.3. Characteristic data of image intensifiers

Type	Useful diameter [mm]	Resolution [linepairs/mm]	Gain
RCA 4550	18	32	3×10^4
RCA C33085DP	38	40	6×10^5
EMI 9794	48	50	2×10^5
Multichannel plate	40	35	1×10^7

olution is highest at the center of the screen and decreases toward the edges. Table 4.3 compiles some typical data of commercial three-stage image intensifiers [4.139]. In Fig. 4.101 a modern version of an image intensifier is shown. It consists of a photocathode, two short proximity-focused image intensifiers, and a fiber-optic coupler, which guides the intensified light generated at the exit of the second stage onto a CCD array.

Image intensifiers can be advantageously employed behind a spectrograph for the sensitive detection of extended spectral ranges [4.140]. Let us assume a linear dispersion of 1 mm/nm of a medium-sized spectrograph. An image intensifier with a useful cathode size of 30 mm and a spatial resolution of 30 lines/mm allows simultaneous detection of a spectral range of 30 nm with a spectral resolution of 3×10^{-2} nm. This sensitivity exceeds that of a photographic plate by many orders of magnitude. With cooled photocathodes, the thermal noise can be reduced to a level comparable with that of a photomultiplier, therefore incident radiation powers of a few photons can be detected. A combination of image intensifiers and vidicons or special diode arrays has been developed (optical multichannel analyzers, OMA) that has proved to be very useful for fast and sensitive measurements of extended spectral ranges, in particular for low-level incident radiation (Sect. 4.5.3).

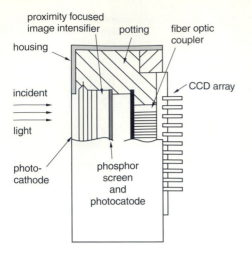

proximity focused
image intensifier   potting   fiber optic
                                coupler
housing

incident   CCD array

light

photo-   phosphor
cathode   screen
                    and
                    photocatode

Fig. 4.101. Modern version of a compact image intensifier

Such intensified OMA systems are commercially available. Their advantages may be summarized as follows [4.141, 4.142]:

- The vidicon targets store optical signals and allow integration over an extended period, whereas photomultipliers respond only while the radiation falls on the cathode.
- All channels of the vidicon acquire optical signals simultaneously. Mounted behind a spectrometer, the OMA can measure an extended spectral range simultaneously, while the photomultiplier accepts only the radiation passing through the exit slit, which defines the resolution. With a spatial resolution of 30 lines per mm and a linear dispersion of 0.5 nm/mm of the spectrometer, the spectral resolution is 1.7×10^{-2} nm. A vidicon target with a length of 16 mm can detect a spectral range of 8 nm simultaneously.
- The signal readout is performed electronically in digital form. This allows computers to be used for signal processing and data analyzing. The dark current of the OMA, for instance, can be automatically substracted, or the program can correct for background radiation superimposed on the signal radiation.
- Photomultipliers have an extended photocathode where the dark current from all points of the cathode area is summed up and adds to the signal. In the image intensifier in front of the vidicon, only a small spot of the photocathode is imaged onto a single diode. Thus the whole dark current from the cathode is distributed over the spectral range covered by the OMA. The image intensifier can be gated and allows detection of signals with high time resolution [4.143]. If the time dependence of a spectral distribution is to be measured, the gate pulse can be applied with variable delay and the whole system acts like a boxcar integrator with additional spectral display. The two-dimensional diode arrays also allow the time de-

pendence of single pulses and their spectral distribution to be displayed, if the light entering the entrance slit of the spectrometer is swept (e.g., by a rotating mirror) parallel to the slit. The OMA or OSA systems therefore combine the advantages of high sensitivity, simultaneous detection of extended spectral ranges, and the capability of time resolution. These merits have led to their increased popularity in spectroscopy [4.141, 4.142].

4.5.5 Detection Techniques and Electronic Equipment

In addition to the radiation detectors, the detection technique and the optimum choice of electronic equipment are also essential factors for the success and the accuracy of spectroscopic measurements. This subsection is devoted to some modern detection techniques and electronic devices.

a) Photon Counting

At very low incident radiation powers it is advantageous to use the photomultiplier for counting single photoelectrons emitted at a rate n per second rather than to measure the photocurrent $i = n e G \Delta t / \Delta t$ averaged over a period Δt [4.144]. The electron avalanches with the charge $Q = G e$ generated by a single photoelectron produce voltage pulses $U = eG/C$ at the anode with the capacitance C. With $C = 1.5 \times 10^{-11}$ F, $G = 10^6 \rightarrow U = 10$ mV. These pulses with rise times of about 1 ns trigger a fast discriminator, which delivers a TTL-norm pulse of 5 V to a counter or to a digital–analog converter (DAC) driving a rate meter with variable time constant (Fig. 4.102) [4.145].

Compared with the conventional analog measurement of the anode current, the photon-counting technique has the following advantages:

- Fluctuations of the photomultiplier gain G, which contribute to the noise in analog measurements, see (4.144), are not significant here, since each photoelectron induces the same normalized pulse from the discriminator as long as the anode pulse exceeds the discriminator threshold.
- Dark curent generated by thermal electrons from the various dynodes can be suppressed by setting the discriminator threshold correctly. This discrimination is particularly effective in photomultipliers with a large conversion efficiency q at the first dynode, covered with a GaAsP layer.

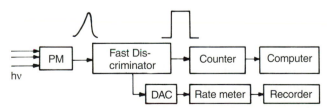

Fig. 4.102. Schematic block-diagram of photon-counting electronics

- Leakage currents between the leads in the photomulitplier socket contribute to the noise in current measurements, but are not counted by the discriminator if it is correctly biased.
- High-energy β-particles from the disintegration of radioactive isotopes in the window material and cosmic ray particles cause a small, but nonnegligible, rate of electron bursts from the cathode with a charge $n \cdot e$ of each burst ($n \gg 1$). The resulting large anode pulses cause additional noise of the anode current. They can, however, be completely suppressed by a window discriminator used in photon counting.
- The digital form of the signal facilitates its further processing. The discriminator pulses can be directly fed into a computer that analyzes the data and may control the experiment [4.146].

The upper limit of the counting rate depends on the time resolution of the discriminator, which may be below 10 ns. This allows counting of randomly distributed pulse rates up to about 10 MHz without essential counting errors.

The lower limit is set by the dark pulse rate [4.144]. With selected low-noise photomultipliers and cooled cathodes, the dark pulse rate may be below 1 per second. Assuming a quantum efficiency of $\eta = 0.2$, it should therefore be possible to achieve, within a measuring time of 1 s, a signal-to-noise ratio of unity even at a photon flux of 5 photons/s. At these low photon fluxes, the probability $p(N)$ of N photoelectrons being detected within the time interval Δt follows a Poisson distribution

$$p(N) = \frac{\overline{N}^N \, e^{-\overline{N}}}{N!} \, , \tag{4.145}$$

where $\overline{N}$ is the average number of photoelectrons detected within Δt [4.147]. If the probability that at least one photoelectron will be detected within Δt is 99, then $1 - p(0) = 0.99$ and

$$p(0) = e^{-\bar{N}} = 0.01 \, , \tag{4.146}$$

which yields $\overline{N} \geq 4.6$. This means that we can expect a pulse with 99% during the observation time certainty only if at least 20 photons fall onto the photocathode with a quantum efficiency of $\eta = 0.2$. For longer detection times, however, the detectable photoelectron rate may be even lower than the dark current rate if, for instance, lock-in detection is used. It is not the dark pulse rate N_D itself that limits the signal-to-noise ratio, but rather its fluctuations, which are proportional to $N_D^{1/2}$.

b) Measurements of Fast Transient Events

Many spectroscopic investigations require the observation of fast transient events. Examples are lifetime measurements of excited atomic or molecular states, investigations of collisional relaxation, and studies of fast laser pulses (Chap. 11). Another example is the transient response of molecules when the incident light frequency is switched into resonance with molecular eigenfrequencies (Chap. 12). Several techniques are used to observe and to analyze such events and recently developed instruments help to optimize the measuring procedure. The combination of a CCD detector and a gated microchannel plate, which acts as an image intensifier with nanosecond resolution, allows the time-resolved sensitive detection of fast events. In addition, there are several devices that are particularly suited for the electronic handling of short pulses. We briefly present three examples of such equipment: the *boxcar integrator* with signal averaging, the *transient recorder*, and the *fast transient digitizer* with subnanosecond resolution.

The *boxcar integrator* measures the amplitudes of signals with a constant repetition rate integrated over a specific sampling interval Δt. It records these signals repetitively over a selected number of pulses and computes the average value of those measurements. With a synchronized trigger signal it can be assured that one looks each time at the identical time interval of each sampled waveform. A delay circuit permits the sampled time interval Δt (called *aperture*) to be shifted to any portion of the waveform under investigation. Figure 4.103 illustrates a possible way to perform this sampling and averaging. The aperture delay is controlled by a ramp generator, which is synchronized to the signal repetition rate and which provides a sawtooth voltage at the signal repetition frequency. A slow aperture-scan ramp shifts the gating time interval Δt, where the signal is sampled between two successive signals by an amount τ, which depends on the slope of the ramp. This slope has to be sufficiently slow in order to permit a sufficient number of samples to be taken in each segment of the waveform. The output signal is then averaged over several scans of the time ramp by a signal averager [4.148]. This increases the

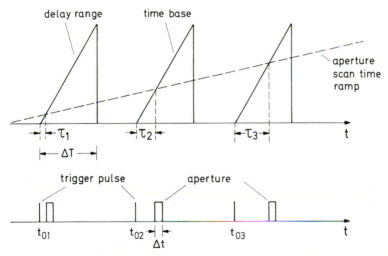

Fig. 4.103. Principle of boxcar operation with synchronization of the repetitive signals. The time base determines the opening times of the gate with width Δt. The slow scan-time ramp shifts the delay times τ_i continuously over the signal–pulse time profile

signal-to-noise ratio and smooths the dc output, which follows the shape of the waveform under study.

The slow ramp is generally not a linearly increasing ramp as shown in Fig. 4.103, but rather a step function where the time duration of each step determines the number of samples taken at a given delay time τ. If the slow ramp is replaced by a constant selectable voltage, the system works as a gated integrator.

The integration of the input signal $U_s(t)$ over the sampling time interval Δt can be performed by charging a capacitance C through a resistor R (Fig. 4.104), which gives a current $i(t) = U_s(t)/R$. The output is then

$$U(\tau) = \frac{1}{C} \int\limits_{\tau}^{\tau+\Delta t} i(t)\, dt = \frac{1}{RC} \int\limits_{\tau}^{\tau+\Delta t} U_s(t)\, dt \; . \tag{4.147}$$

For repetitive scans, the voltages $U(\tau)$ can be summed (Fig. 4.104). Because of inevitable leakage currents, however, unwanted discharge of the capacitance occurs if the signal under study has a low duty factor and the time between successive samplings becomes large. This difficulty may be overcome by a digital output, consisting of a two-channel analog-to-digital-to-analog converter. After a sampling switch opens, the acquired charge is digitized and loaded into a digital storage register. The digital register is then read by a digital-to-analog converter producing a dc voltage equal to the voltage $U(\tau) = Q(\tau)/C$ on the capacitor. This dc voltage is fed back to the integrator to maintain its output potential until the next sample is taken.

The boxcar integrator needs repetitive waveforms because it samples each time only a small time interval Δt of the input pulse and composes the whole period of the repetitive waveform by adding many sampling points with different delays. For many spectroscopic applications, however, only single-shot

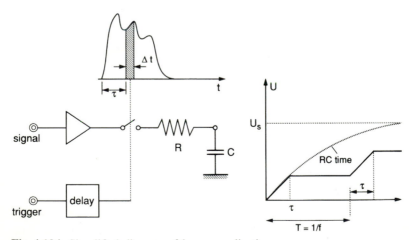

Fig. 4.104. Simplified diagram of boxcar realization

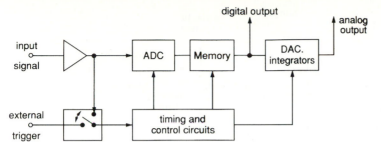

Fig. 4.105. Block diagram of a transient recorder

signals are available. Examples are shock-tube experiments or spectroscopic studies in laser-induced fusion. In such cases, the boxcar integrator is not useful and a *transient recorder* is a better choice. This instrument uses digital techniques to record a preselected section of an analog signal as it varies with time. The wave shape during the selected period of time is recorded and held in the instrument's memory until the operator instructs the instrument to make a new recording. The operation of a transient recorder is illustrated in Fig. 4.105 [4.149, 4.150]. A trigger, derived from the input signal or provided externally, initiates the sweep. The amplified input signal is converted at equidistant time intervals to its digital equivalent by an analog-to-digital converter and stored in a semiconductor memory in different channels. With 100 channels, for instance, a single-shot signal is recorded by 100 equidistant sampling intervals. The time resolution depends on the sweep time and is limited by the frequency response of the transient recorder. Sample intervals between 10 ns up to 20 s can be selected. This allows sweep times of 20 µs to 5 h for 2000 sampling points.

Acquisition and analysis beyond 500 MHz has become possible by combining the features of a transient recorder with the fast response time of an electron beam that writes and stores information on a diode matrix target in a scan converter tube. Figure 4.106 illustrates the basic principle of the *transient digitizer* [4.151]. The diode array of about 640 000 diodes is scanned by the reading electron beam, which charges all reverse-biased *p–n* junctions until the diodes reach a saturation voltage. The writing electron beam impinges on the other side of the 10 µm thick target and creates electron–hole pairs, which diffuse to the anode and partially discharge it. When the reading beam hits a discharged diode, it becomes recharged, subsequently a current signal is generated at the target lead, which can be digitally processed.

The instrument can be used in a nonstoring mode where the operation is similar to that of a conventional television camera with a video signal, which can be monitored on a TV monitor. In the digital mode the target is scanned by the reading beam in discrete steps. The addresses of points on the target are transferred and stored in memory only when a trace has been written at those points on the target. This transient digitizer allows one to monitor fast transient signals with a time resolution of 100 ps and to process the data in

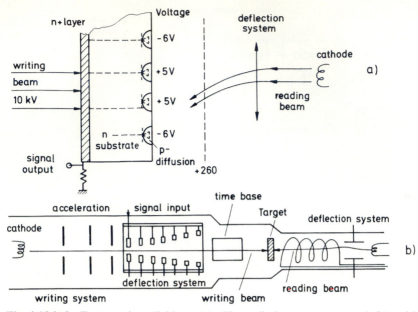

Fig. 4.106a,b. Fast transient digitizer: **(a)** silicon diode-array target; and **(b)** writing and reading gun [4.151]

digital form in a computer. It is, for instance, possible to obtain the frequency distribution of the studied signal from its time distribution by a Fourier transformation performed by the computer.

c) Optical Oscilloscope

The optical oscilloscope represents a combination of a streak camera and a sampling oscilloscope. Its principle of operation is illustrated by Fig. 4.107 [4.152]: The incident light $I(t)$ is focused onto the photocathode of the streak camera. The electrons released from the cathode pass between two deflecting electrodes toward the sampling slit. Only those electrons that traverse the deflecting electric field at a given selectable time can pass through the slit. They impinge on a phosphor screen and produce light that is detected by a photomultiplier (PM). The PM output is amplified and fed into a sampling oscilloscope, where it is stored and processed. The sampling operation can be repeated many times with different delay times t between the trigger and the sampling, similar to the principle of a boxcar operation (Fig. 4.103). Each sampling interval yields the signal

$$S(t, \Delta t) = \int\limits_{t}^{t+\Delta t} I(t)\, dt \ . \tag{4.148}$$

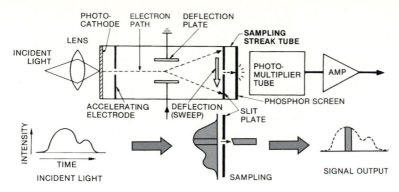

Fig. 4.107. Optical oscilloscope [4.152]

The summation over all sampled time intervals Δt gives the total signal

$$S(t) = \sum_{n=1}^{N} \int_{t=(n-1)\Delta t}^{t=n\cdot\Delta t} I(t)\,dt \; , \tag{4.149}$$

which reflects the input time profile $I(t)$ of the incident light.

The spectral response of the system depends on that of the first photocathode and reaches from 350 to 850 μm for the visible version and from 400 to 1550 μm for the extended infrared version. The time resolution is better than 10 ps and the sampling rate can be selected up to 2 MHz. The limitation is given by the time jitter, which was stated to be less than 20 ps.

4.6 Conclusions

The aim of this chapter was to provide a general background in spectroscopic instrumentation, to summarize some basic ideas of spectroscopy, and to present some important relations between spectroscopic quantities. This background should be helpful in understanding the following chapters that deal with the main subject of this textbook: the applications of lasers to the solution of spectroscopic problems. Although until now we have only dealt with general spectroscopy, the examples given were selected with special emphasis on laser spectroscopy. This is especially true in Chap. 4, which is, of course, not a complete account of spectroscopic equipment, but is intended to give a survey on modern instrumentation used in laser spectroscopy.

There are several excellent and more detailed presentations of special instruments and spectroscopic techniques, such as spectrometers, interferometry, and Fourier spectroscopy. Besides the references given in the various sections, several series on optics [4.2], optical engineering [4.1], advanced optical techniques [4.154], and the monographs [4.4, 4.6, 4.153–4.158] may help to give more extensive information about special problems. Useful practical hints can be found in the handbook [4.159].

Problems

4.1 Calculate the spectral resolution of a grating spectrometer with an entrance slit width of $10\,\mu$m, focal lengths $f_1 = f_2 = 2$ m of the mirrors M_1 and M_2, and a grating with 1800 grooves/mm. What is the useful minimum slit width if the size of grating is 100×100 mm^2?

4.2 The spectrometer in Problem 4.1 shall be used in first order for a wavelength range around 500 nm. What is the optimum blaze angle, if the geometry of the spectrometer allows an angle of incidence α about $20°$?

4.3 Calculate the number of grooves/mm for a Littrow grating for a $25°$ incidence at $\lambda = 488$ nm (i.e., the first diffraction order is being reflected back into the incident beam at an angle $\alpha = 25°$ to the grating normal).

4.4 A prism can be used for expansion of a laser beam if the incident beam is nearly parallel to the prism surface. Calculate the angle of incidence α for which a HeNe laser beam ($\lambda = 632.8$ nm) transmitted through a $60°$ equilateral flint glass prism is expanded tenfold.

4.5 Assume that a signal-to-noise ratio of 50 has been achieved in measuring the fringe pattern of a Michelson interferometer with one continuously moving mirror. Estimate the minimum path length ΔL that the mirror has to travel in order to reach an accuracy of 10^{-4} nm in the measurement of a laser wavelength at $\lambda = 600$ nm.

4.6 The dielectric coatings of a Fabry–Perot interferometer have the following specifications: $R = 0.98$, $A = 0.3\%$. The flatness of the surfaces is $\lambda/100$ at $\lambda = 500$ nm. Estimate the finesse, the maximum transmission, and the spectral resolution of the FPI for a plate separation of 5 mm.

4.7 A fluorescence spectrum shall be measured with a spectral resolution of 10^{-2} nm. The experimentor decides to use a crossed arrangement of grating spectrometer (linear dispersion: 5×10^{-2} nm/mm) and an FPI with coatings of $R = 0.98$ and $A = 0.3\%$. Estimate the optimum combination of spectrometer slit width and FPI plate separation.

4.8 An interference filter shall be designed with peak transmission at $\lambda = 550$ nm and a bandwidth of 5 nm. Estimate the reflectivity R of the dielectric coatings and the thickness of the etalon, if no further transmission maximum is allowed between 350 and 750 nm.

4.9 A confocal FPI shall be used as optical spectrum analyzer, with a free spectral range of 3 GHz. Calculate the mirror separation d and the finesse that is necessary to resolve spectral features in the laser output within 10 MHz. What is the minimum reflectivity R of the mirrors, if the surface finesse is 500?

4.10 Calculate the transmission of a Lyot filter with two plates ($d_1 = 1\,$mm, $d_2 = 4\,$mm) with $n = 1.40$ in the fast axis and $n = 1.45$ in the slow axis (a) as a function of λ for $\alpha = 45°$ in (4.92); and (b) as a function of α for a fixed wavelength λ. What is the contrast of the transmitted intensity $I(\alpha)$ for arbitrary values of λ?

4.11 Derive (4.111) for the equivalent electrical circuit of Fig. 4.73b.

4.12 A thermal detector has a heat capacity $H = 10^{-8}\,$J/K and a thermal conductivity to a heat sink of $G = 10^{-9}\,$W/K. What is the temperature increase ΔT for $10^7\,$W incident cw radiation if the efficiency $\beta = 0.8$? If the radiation is switched on at a time $t = 0$, how long does it take before the detector reaches a temperature increase $\Delta T(t) = 0.9\Delta T_\infty$? What is the time constant of the detector and at which modulation frequency ω of the incident radiation has the response decreased to 0.5 of its dc value?

4.13 A bolometer is operated at the temperature $T = 8\,$K between superconducting and normal conducting states, where dR/dT is $10^3\,\Omega$/K. The heat capacity is $H = 10^{-8}\,$J/K and the dc electrical current 1 mA. What is the change Δi of the heating current in order to keep the temperature constant when the bolometer is irradiated with $10^{-10}\,$W?

4.14 The anode of a photomultiplier tube is connected by a resistor of $R = 1\,$kΩ to ground. The stray capacitance is 10 pf, the current amplification 10^6, and the anode rise time 1.5 ns. What is the peak amplitude and the halfwidth of the anode output pulse produced by a single photoelectron? What is the dc output current produced by $10^{-12}\,$W cw radiation at $\lambda = 500\,$nm, if the quantum efficiency of the cathode is $\eta = 0.2$? Estimate the necessary voltage amplification of a preamplifier (a) to produce 1 V pulses for single-photon counting; and (b) to read 1 V on a dc meter of the cw radiation?

4.15 A manufacturer of a two-stage optical image intensifier states that incident intensities of $10^{-17}\,$W at $\lambda = 500\,$nm can still be "seen" on the phosphor screen of the output state. Estimate the minimum intensity amplification, if the quantum efficiency of the cathodes and the conversion efficiency of the phosphor screens are both 0.2 and the collection efficiency of light emitted by the phosphor screens is 0.1. The human eye needs at least 20 photons/s to observe a signal.

4.16 Estimate the maximum output voltage of an open photovoltaic detector at room temperature under $10\,\mu$W irradiation when the photocurrent of the shortened output is $50\,\mu$A and the dark current is $50\,$nA.

5. Lasers as Spectroscopic Light Sources

In this chapter we summarize basic laser concepts with regard to their applications in spectroscopy. A sound knowledge of laser physics with regard to passive and active optical cavities and their mode spectra, the realization of single-mode lasers, or techniques for frequency stabilization will help the reader to gain a deeper understanding of many subjects in laser spectroscopy and to achieve optimum performance of an experimental setup. Of particular interest for spectroscopists are the various types of tunable lasers, which are discussed in Sect. 5.7. Even in spectral ranges where no tunable lasers exist, optical frequency-doubling and mixing techniques may provide tunable coherent radiation sources, as outlined in Sect. 5.8.

5.1 Fundamentals of Lasers

This section gives a short introduction to the basic physics of lasers in a more intuitive than mathematical way. A more detailed treatment of laser physics and an extensive discussion of various types of lasers can be found in textbooks (see, for instance, [5.1–5.10]). For more advanced presentations based on the quantum-mechanical description of lasers, the reader is referred to [5.11–5.15].

5.1.1 Basic Elements of a Laser

A laser consists of essentially three components (Fig. 5.1a):

- The active medium, which amplifies an incident electromagnetic (EM) wave;
- The energy pump, which selectively pumps energy into the active medium to populate selected levels and to achieve population inversion;
- The optical resonator composed, for example, of two opposite mirrors, which stores part of the induced emission that is concentrated within a few resonator modes.

The energy pump (e.g., flashlamps, gas discharges, or even other lasers) generates a population distribution $N(E)$ in the laser medium, which strongly deviates from the Boltzmann distribution (2.18) that exists for thermal equilibrium. At sufficiently large pump powers the population density $N(E_k)$ of the specific level E_k may exceed that of the lower level E_i (Fig. 5.1b).

For such a population inversion, the induced emission rate $N_k B_{ki} \rho(v)$ for the transition $E_k \rightarrow E_i$ exeeds the absorption rate $N_i B_{ik} \rho(v)$. An EM wave

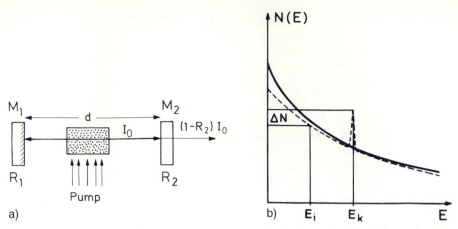

Fig. 5.1. (a) Schematic setup of a laser and (b) population inversion (*dashed curve*), compared with a Boltzmann distribution at thermal equilibrium (*solid curve*)

passing through this active medium is amplified instead of being attenuated according to (3.22).

The function of the optical resonator is the selective feedback of radiation emitted from the excited molecules of the active medium. Above a certain pump threshold this feedback converts the laser *amplifier* into a laser *oscillator*. When the resonator is able to store the EM energy of induced emission within a few resonator modes, the spectral energy density $\rho(\nu)$ may become very large. This enhances the induced emission into these modes since, according to (2.22), the induced emission rate already exceeds the spontaneous rate for $\rho(\nu) > h\nu$. In Sect. 5.1.3 we shall see that this concentration of induced emission into a small number of modes can be achieved with open resonators, which act as spatially selective and frequency-selective optical filters.

5.1.2 Threshold Condition

When a monochromatic EM wave with the frequency ν travels in the z-direction through a medium of molecules with energy levels E_i and E_k and $(E_k - E_i)/h = \nu$, the intensity $I(\nu, z)$ is, according to (3.23), given by

$$I(\nu, z) = I(\nu, 0)\,e^{-\alpha(\nu)z} , \tag{5.1}$$

where the frequency-dependent absorption coefficient

$$\alpha(\nu) = [N_i - (g_i/g_k)N_k]\sigma(\nu) , \tag{5.2}$$

is determined by the absorption cross section $\sigma(\nu)$ for the transition $(E_i \rightarrow E_k)$ and by the population densities N_i, N_k in the energy levels E_i, E_k with the statistical weights g_i, g_k, see (2.44). We infer from (5.2) that for

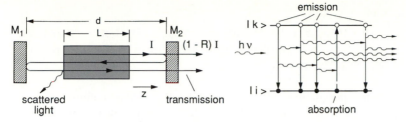

Fig. 5.2. Gain and losses of an EM wave traveling back and forth along the resonator axis

$N_k > (g_k/g_i)N_i$, the absorption coefficient $\alpha(v)$ becomes negative and the incident wave is amplified instead of attenuated.

If the active medium is placed between two mirrors (Fig. 5.2), the wave is reflected back and forth, and traverses the amplifying medium many times, which increases the total amplification. With the length L of the active medium the total gain factor per single round-trip without losses is

$$G(v) = \frac{I(v, 2L)}{I(v, 0)} = e^{-2\alpha(v)L} \; . \tag{5.3}$$

A mirror with reflectivity R reflects only the fraction R of the incident intensity. The wave therefore suffers at each reflection a fractional reflection loss of $(1 - R)$. Furthermore, absorption in the windows of the cell containing the active medium, diffraction by apertures, and scattering due to dust particles in the beam path or due to imperfect surfaces introduce additional losses. When we summarize all these losses by a loss coefficient γ, which gives the fractional energy loss $\Delta W/W$ per round-trip time T, the intensity I decreases without an active medium per round-trip (if we assume the loss to be equally distributed along the resonator length d) as

$$I = I_0 e^{-\gamma} \; . \tag{5.4}$$

Including the amplification by the active medium with length L, we obtain for the intensity after a single round-trip through the resonator with length d, which may be larger than L:

$$I(v, 2d) = I(v, 0) \exp[-2\alpha(v)L - \gamma] \; . \tag{5.5}$$

The wave is amplified if the gain overcomes the losses per round-trip. This implies that

$$-2L\alpha(v) > \gamma \; . \tag{5.6}$$

With the absorption cross section $\sigma(v)$ from (5.2), this can be written as the *threshold condition* for the population difference, i.e.,

$$\Delta N = N_k(g_i/g_k) - N_i > \Delta N_{\text{thr}} = \frac{\gamma}{2\sigma(v)L} \; . \tag{5.7}$$

Example 5.1

$L = 10$ cm, $\gamma = 10\%$, $\sigma = 10^{-12}$ cm^2 $\rightarrow \Delta N = 5 \times 10^9$/cm^3. At a neon pressure of 0.2 mbar, ΔN corresponds to about 10^{-6} of the total density of neon atoms in a HeNe laser.

If the inverted population difference ΔN of the active medium is larger than ΔN_{thr}, a wave that is reflected back and forth between the mirrors will be amplified in spite of losses, therefore its intensity will increase.

The wave is initiated by spontaneous emission from the excited atoms in the active medium. Those spontaneously emitted photons that travel into the right direction (namely, parallel to the resonator axis) have the longest path through the active medium and therefore the greater chance of creating new photons by induced emission. Above the threshold they induce a photon avalanche, which grows until the depletion of the population inversion by stimulated emission just compensates the repopulation by the pump. Under steady-state conditions the inversion decreases to the threshold value ΔN_{thr}, the saturated net gain is zero, and the laser intensity limits itself to a finite value I_L. This laser intensity is determined by the pump power, the losses γ, and the gain coefficient $\alpha(v)$ (Sects. 5.7, 5.9).

The frequency dependence of the gain coefficient $\alpha(v)$ is related to the line profile $g(v - v_0)$ of the amplifying transition. Without saturation effects (i.e., for small intensities), $\alpha(v)$ directly reflects this line shape, for homogeneous as well as for inhomogeneous profiles. According to (2.83) we obtain with the Einstein coefficienct B_{ik}

$$\alpha(v) = \Delta N(hv/c)B_{ik}g(v - v_0) = \Delta N\sigma_{ik}(v) , \tag{5.8}$$

which shows that the amplification is largest at the line center. For high intensities, saturation of the inversion occurs, which is different for homogeneous and for inhomogeneous line profiles (Sects. 7.1, 7.2).

The loss factor γ also depends on the frequency v because the resonator losses are strongly dependent on v. The frequency spectrum of the laser therefore depends on a number of parameters, which we discuss in more detail in Sect. 5.2.

5.1.3 Rate Equations

The photon number inside the laser cavity and the population densities of atomic or molecular levels under stationary conditions of a laser can readily be obtained from simple rate equations. Note, however, that this approach does *not* take into account coherence effects (Chap. 12).

With the pump rate P (which equals the number of atoms that are pumped per second and per cm^3 into the upper laser level $|2\rangle$), the relaxation rates $R_i N_i$ (which equal the number of atoms that are removed per second and cm^3 from the level $|i\rangle$ by collision or spontaneous emission), and the spontaneous emission probability A_{21} per second, we obtain from (2.21) for equal statisti-

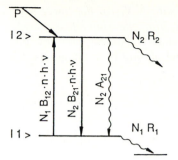

Fig. 5.3. Level diagram for pumping process P, relaxation rates $N_i R_i$, spontaneous and induced transitions in a four-level system

cal weights $g_1 = g_2$ the rate equations for the population densities N_i and the photon densities n (Fig. 5.3):

$$\frac{dN_1}{dt} = (N_2 - N_1)B_{21}nh\nu + N_2 A_{21} - N_1 R_1 , \tag{5.9a}$$

$$\frac{dN_2}{dt} = P - (N_2 - N_1)B_{21}nh\nu - N_2 A_{21} - N_2 R_2 , \tag{5.9b}$$

$$\frac{dn}{dt} = -\beta n + (N_2 - N_1)B_{21}nh\nu . \tag{5.9c}$$

The loss coefficient β determines the loss rate of the photon density $n(t)$ stored inside the optical resonator. Without an active medium ($N_1 = N_2 = 0$), we obtain from (5.9c)

$$n(t) = n(0)\,e^{-\beta t} . \tag{5.10}$$

A comparison with the definition (5.4) of the loss coefficient γ per round-trip yields for a resonator with length d and round-trip time $T = 2d/c$

$$\gamma = \beta T = \beta(2d/c) . \tag{5.11}$$

Under stationary conditions we have $dN_1/dt = dN_2/dt = dn/dt = 0$. Adding (5.9a and 5.9b) then yields

$$P = N_1 R_1 + N_2 R_2 , \tag{5.12}$$

which means that the pump rate P just compensates the loss rates $N_1 R_1 + N_2 R_2$ of the atoms in the two laser levels caused by relaxation processes into other levels. Further insight can be gained by adding (5.9b and 5.9c), which gives for stationary conditions

$$P = \beta n + N_2(A_{21} + R_2) . \tag{5.13}$$

In a continuous-wave (cw) laser the pump rate equals the sum of photon loss rate βn plus the total relaxation rate $N_2(A_{21} + R_2)$ of the upper laser level. A comparison of (5.12 and 5.13) shows that for a cw laser the relation holds

$$N_1 R_1 = \beta n + N_2 A_{21} . \tag{5.14}$$

Under stationary laser operation the relaxation rate $N_1 R_1$ of the lower laser level must always be larger than its feeding rate from the upper laser level!

The stationary inversion ΔN_{stat} can be obtained from the rate equation when multiplying (5.9a) by R_2, (5.9b) by R_1, and adding both equations. We find

$$\Delta N_{\text{stat}} = \frac{(R_1 - A_{21})P}{B_{12}nh\nu(R_1 + R_2) + A_{21}R_1 + R_1 R_2} . \tag{5.15}$$

This shows that a stationary inversion $\Delta N_{\text{stat}} > 0$ can only be maintained for $R_1 > A_{21}$. The relaxation probability R_1 of the lower laser level $|1\rangle$ must be larger than its refilling probability A_{21} by spontaneous transitions from the upper laser level $|2\rangle$. In fact, during the laser operation the induced emission mainly contributes to the population N_1 and therefore the more difficult condition $R_1 > A_{21} + B_{21}\rho$ must be satisfied. Continuous-wave lasers can therefore be realized on the transitions $|2\rangle \rightarrow |1\rangle$ only if the effective lifetime $\tau_{\text{eff}} = 1/R_1$ of level $|1\rangle$ is smaller than $(A_2 + B_{21}\rho)^{-1}$.

When starting a laser, the photon density n increases until the inversion density ΔN has decreased to the threshold density ΔN_{thr}. This can immediately be concluded from (5.9c), which gives for $dn/dt = 0$ and $d = L$

$$\Delta N = \frac{\beta}{B_{21}h\nu} = \frac{\gamma}{2LB_{21}h\nu/c} = \frac{\gamma}{2L\sigma} = \Delta N_{\text{thr}} , \tag{5.16}$$

where the relation (5.8) with

$$\int \alpha(\nu)\,d\nu = \Delta N(h\nu/c)B_{12} = \Delta N\sigma_{12} ,$$

has been used.

Example 5.2

With $N_2 = 10^{10}/\text{cm}^3$ and $(A_{21} + R_2) = 2 \times 10^7\,\text{s}^{-1}$, the total incoherent loss rate is $2 \times 10^{17}/\text{cm}^3 \cdot \text{s}$. In a HeNe laser discharge tube with $L = 10\,\text{cm}$ and 1 mm diameter, the active volume is about $0.075\,\text{cm}^3$. The total loss rate of the last two terms in (5.9c) then becomes $1.5 \times 10^{16}\,\text{s}^{-1}$.

For a laser output power of 3 mW at $\lambda = 633\,\text{nm}$, the rate of emitted photons is $\beta n = 10^{16}\,\text{s}^{-1}$. In this example the total pump rate has to be $P = (1.5 + 1) \times 10^{16}\text{s}^{-1} = 2.5 \times 10^{16}\,\text{s}^{-1}$, where the fluorescence represents a larger loss than the mirror transmission.

5.2 Laser Resonators

In Sect. 2.1 it was shown that in a closed cavity a radiation field exists with a spectral energy density $\rho(\nu)$ that is determined by the temperature T of the cavity walls and by the eigenfrequencies of the cavity modes. In the optical region, where the wavelength λ is small compared with the dimension L of the cavity, we obtained the *Planck distribution* (2.13) at thermal equilibrium for $\rho(v)$. The number of modes per unit volume,

$$n(v)\,\mathrm{d}v = 8\pi(v^2/c^3)\,\mathrm{d}v \;,$$

within the spectral interval $\mathrm{d}v$ of a molecular transition turns out to be very large (Example 2.1a). When a radiation source is placed inside the cavity, its radiation energy will be distributed among all modes; the system will, after a short time, again reach thermal equilibrium at a correspondingly higher temperature. Because of the large number of modes in such a closed cavity, the mean number of photons per mode (which gives the ratio of induced to spontaneous emission rate in a mode) is very small in the optical region (Fig. 2.5). *Closed cavities with $L \gg \lambda$ are therefore not suitable as laser resonators.*

In order to achieve a concentration of the radiation energy into a small number of modes, the resonator should exhibit a strong feedback for these modes but large losses for all other modes. This would allow an intense radiation field to be built up in the modes with low losses but would prevent the system from reaching the oscillation threshold in the modes with high losses.

Assume that the kth resonator mode with the loss factor β_k contains the radiation energy W_k. The energy loss per second in this mode is then

$$\frac{\mathrm{d}W_k}{\mathrm{d}t} = -\beta_k W_k \;. \tag{5.17}$$

Under stationary conditions the energy in this mode will build up to a stationary value where the losses equal the energy input. If the energy input is switched off at $t = 0$, the energy W_k will decrease exponentially since integration of (5.17) yields

$$W_k(t) = W_k(0)\,\mathrm{e}^{-\beta_k t} \;. \tag{5.18}$$

When we define the quality factor Q_k of the kth cavity mode as 2π times the ratio of energy stored in the mode to the energy loss per oscillation period $T = 1/v$

$$Q_k = -\frac{2\pi v W_k}{\mathrm{d}W_k/\mathrm{d}t} \;, \tag{5.19}$$

we can relate the loss factor β_k and the qualtiy factor Q_k by

$$Q_k = -2\pi v/\beta_k \;. \tag{5.20}$$

After the time $\tau = 1/\beta_k$, the energy stored in the mode has decreased to $1/\mathrm{e}$ of its value at $t = 0$. This time can be regarded as the mean lifetime of a photon in this mode. If the cavity has large loss factors for most modes but a small β_k for a selected mode, the number of photons in this mode will be larger than in the other modes, even if at $t = 0$ the radiation energy in all modes was the same. If the unsaturated gain coefficient $\alpha(\nu)L$ of the active medium is larger than the loss factor $\gamma_k = \beta_k(2d/c)$ per round-trip but smaller than the losses of all other modes, the laser will oscillate only in this selected mode.

5.2.1 Open Optical Resonators

A resonator that concentrates the radiation energy of the active medium into a few modes can be realized with *open* cavities, which consist of two plane or curved mirrors aligned in such a way that light traveling along the resonator axis may be reflected back and forth between the mirrors. Such a ray traverses the active medium many times, resulting in a larger total gain. Other rays may leave the resonator after a few reflections before the intensity has reached a noticeable level (Fig. 5.4).

Besides these *walk-off losses*, *reflection losses* also cause a decrease of the energy stored in the resonator modes. With the reflectivities R_1 and R_2 of the resonator mirrors M_1 and M_2, the intensity of a wave in the passive resonator has decreased after a single round-trip to

$$I = R_1 R_2 I_0 = I_0 e^{-\gamma_R} , \tag{5.21}$$

with $\gamma_R = -\ln(R_1 R_2)$. Since the round-trip time is $T = 2d/c$, the decay constant β in (5.18) due to reflection losses is $\beta_R = \gamma_R c/2d$, therefore the mean lifetime of a photon in the resonator becomes without any additional losses

$$\tau = \frac{1}{\beta_R} = \frac{2d}{\gamma_R c} = -\frac{2d}{c \ln(R_1 R_2)} . \tag{5.22}$$

These open resonators are, in principle, the same as the Fabry–Perot interferometers discussed in Chap. 4; we shall see that several relations derived in Sect. 4.2 apply here. However, there is an essential difference with regard to the geometrical dimensions. While in a common FPI the distance between both mirrors is small compared with their diameter, the relation is generally reversed for laser resonators. The mirror diameter $2a$ is small compared with the mirror separation d. This implies that diffraction losses of the wave, which

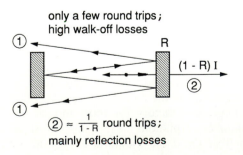

only a few round trips;
high walk-off losses

$(1 - R)\,I$

② $\approx \frac{1}{1-R}$ round trips;
mainly reflection losses

Fig. 5.4. Walk-off losses of inclined rays and reflection losses in an open resonator

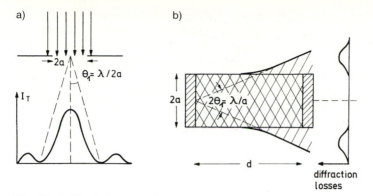

Fig. 5.5a,b. Equivalence of diffraction at an aperture (**a**) and at a mirror of equal size (**b**). The diffraction pattern of the transmitted light in (a) equals that of the reflected light in (b). The case $\theta_1 d = a \to N = 0.5$ is shown

is reflected back and forth between the mirrrors, play a major role in laser resonators, while they can be completely neglected in the conventional FPI.

In order to estimate the magnitude of diffraction losses let us make use of a simple example. A plane wave incident onto a mirror with diameter $2a$ exhibits, after being reflected, a spatial intensity distribution that is determined by diffraction and that is completely equivalent to the intensity distribution of a plane wave passing through an aperture with diameter $2a$ (Fig. 5.5). The central diffraction maximum at $\theta = 0$ lies between the two first minima at $\theta_1 = \pm\lambda/2a$ (for circular apertures a factor 1.2 has to be included, see, e.g., [5.16]). About 16% of the total intensity transmitted through the aperture is diffracted into higher orders with $|\theta| > \lambda/2a$. Because of diffraction the outer part of the reflected wave misses the second mirror M_2 and is therefore lost. This example demonstrates that the diffraction losses depend on the values of a, d, λ, and on the amplitude distribution $A(x, y)$ of the incident wave across the mirror surface. The influence of diffraction losses can be characterized by the Fresnel number

$$N = \frac{a^2}{\lambda d},$$
(5.23)

where N gives the number of Fresnel zones [5.16, 5.17] across a resonator mirror, as seen from the center A of the opposite mirror. For the mirror separation d these zones have radii $s_m = \sqrt{m\lambda d}$ and the distances $\rho = \frac{1}{2}(m + q)\lambda$ ($m = 0, 1, 2, ... \ll q$) from A (Fig. 5.6).

If a photon makes n transits through the resonator, the maximum diffraction angle 2θ should be smaller than $a/(nd)$. With $2\theta = \lambda/a$ we obtain the condition

$$N > n ,$$
(5.24)

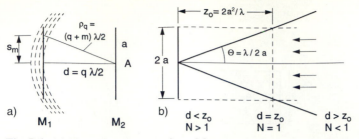

Fig. 5.6. (a) Fresnel zones on mirror M_1, as seen from the center A of the other mirror M_2; (b) the three regions of d/a with the Fresnel number $N > 1$, $N = 1$, and $N < 1$

which states *that the Fresnel number of a plane mirror resonator with negligible diffraction losses should be larger than the number of transits through the resonator.*

Example 5.3

(a) A plane Fabry–Perot interferometer with $d = 1\,\text{cm}$, $a = 3\,\text{cm}$, $\lambda = 500\,\text{nm}$ has a Fresnel number $N = 1.8 \times 10^5$.
(b) The resonator of a gas laser with plane mirrors at a distance $d = 50\,\text{cm}$, $a = 0.1\,\text{cm}$, $\lambda = 500\,\text{nm}$ has a Fresnel number $N = 4$.

The fractional energy loss per transit due to diffraction of a plane wave reflected back and forth between the two plane mirrors is approximately given by

$$\gamma_D \sim \frac{1}{N} \,. \tag{5.25}$$

For our first example the diffraction losses of the plane FPI are about 5×10^{-6} and therefore completely negligible, whereas for the second example they reach 25% and may already exceed the gain for many laser transitions. This means that a plane wave would not reach threshold in such a resonator. However, these high diffraction losses cause nonnegligible distortions of a plane wave and the amplitude $A(x, y)$ is no longer constant across the mirror surface (Sect. 5.2.2), but decreases towards the mirror edges. This decreases the diffraction losses, which become, for example, $\gamma_{\text{Diffr}} \leq 0.01$ for $N \geq 20$.

It can be shown [5.18] *that all resonators with plane mirrors that have the same Fresnel number also have the same diffraction losses, independent of the special choice of a, d, or λ.*

Resonators with curved mirrors may exhibit much lower diffraction losses than the plane mirror resonator because they can refocus the divergent diffracted waves of Fig. 5.5 (Sect. 5.2.5).

5.2.2 Spatial Field Distributions in Open Resonators

In Sect. 2.1 we have seen that any stationary field configuration in a closed cavity (called a *mode*) can be composed of plane waves. Because of diffraction, plane waves cannot give stationary fields in open resonators, since the diffraction losses depend on the coordinates (x, y) and increase from the axis of the resonator towards its edges. This implies that the distribution $A(x, y)$, which is independent of x and y for a plane wave, will be altered with each round-trip for a wave traveling back and forth between the mirrors of an open resonator until it approaches a stationary distribution. Such a stationary field configuration, called a *mode of the open resonator*, is reached when $A(x, y)$ no longer changes its form, although, of course, the losses result in a decrease of the total amplitude if they are not compensated by the gain of the active medium.

The mode configurations of open resonators can be obtained by an iterative procedure using the Kirchhoff–Fresnel diffraction theory [5.17]. Concerning the diffraction losses, the resonator with two plane square mirrors can be replaced by the equivalent arrangement of apertures with size $(2a)^2$ and a distance d between successive apertures (Fig. 5.7). When an incident plane wave is traveling into the z-direction, its amplitude distribution is successively altered by diffraction, from a constant amplitude to the final stationary distribution $A_n(x, y)$. The spatial distribution $A_n(x, y)$ in the plane of the nth aperture is determined by the distribution $A_{n-1}(x, y)$ across the previous aperture.

From Kirchhoff's diffraction theory we obtain (Fig. 5.8)

$$A_n(x, y) = -\frac{i}{\lambda} \int\!\!\int A_{n-1}(x', y')\frac{1}{\rho} e^{-ik\rho} \cos \vartheta \, dx' \, dy' \,. \tag{5.26}$$

A stationary field distribution is reached if

$$A_n(x, y) = CA_{n-1}(x, y) \quad \text{with} \quad C = \sqrt{1 - \gamma_D}\, e^{i\phi} \,, \tag{5.27}$$

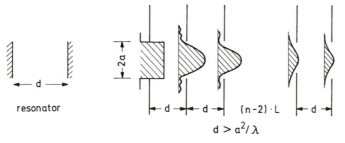

$$d > a^2/\lambda$$

equivalent system of equidistant apertures

Fig. 5.7. The diffraction of an incident plane wave at successive apertures separated by d is equivalent to the diffraction by successive reflections in a plane-mirror resonator with mirror separation d

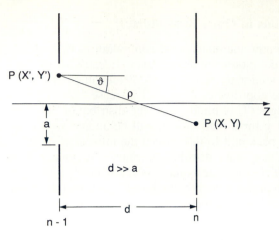

Fig. 5.8. Illustration of (5.26), showing the relations $\rho^2 = d^2 + (x - x')^2 + (y - y')^2$ and $\cos \vartheta = d/\rho$

The amplitude attenuation factor $|C|$ does not depend on x and y. The quantity γ_D represents the diffraction losses and ϕ the corresponding phase shift caused by diffraction.

Inserting (5.27) into (5.26) gives the following integral equation for the stationary field configuration

$$A(x, y) = -\frac{i}{\lambda}(1 - \gamma_D)^{-1/2} e^{-i\phi} \iint A(x', y') \frac{1}{\rho} e^{-ik\rho} \cos \vartheta \, dx' \, dy' . \quad (5.28)$$

Because the arrangement of successive apertures is equivalent to the plane-mirror resonator, the solutions of this integral equation also represent the stationary modes of the open resonator. The diffraction-dependent phase shifts ϕ for the modes are determined by the condition of resonance, which requires that the mirror separation d equals an integer multiple of $\lambda/2$.

The general integral equation (5.28) cannot be solved analytically, therefore one has to look for approximate methods. For two identical plane mirrors of quadratic shape $(2a)^2$, (5.28) can be solved numerically by splitting it into two one-dimensional equations, one for each coordinate x and y, if the Fresnel number $N = a^2/(d\lambda)$ is small compared with $(d/a)^2$, which means if $a \ll (d^3\lambda)^{1/4}$. Such numerical iterations for the "infinite strip" resonator have been performed by Fox and Li [5.19]. They showed that stationary field configurations do exist and computed the field distributions of these modes, their phase shifts, and their diffraction losses.

5.2.3 Confocal Resonators

The analysis has been extended by Boyd, Gordon, and Kogelnik to resonators with confocally-spaced spherical mirrors [5.20, 5.21] and later by others to general laser resonators [5.22–5.30]. For the symmetric confocal case (the two foci of the two mirrors with equal radii $R_1 = R_2 = R$ coincide, i.e., the mirror separation d is equal to the radius of curvature R), the integral equation (5.28) can be solved with the acceptable approximation $a \ll d$, which implies $\rho \approx d$ in the denominator and $\cos \vartheta \approx 1$. In the phase term $\exp(-ik\rho)$, the distance ρ cannot be replaced by d, since the phase is sensitive even to small changes in the exponent. One can, however, for $x', x, y', y \ll d$ expand ρ into a power series

$$\rho \approx d \left[1 + \frac{1}{2} \left(\frac{x'-x}{d} \right)^2 + \frac{1}{2} \left(\frac{y'-y}{d} \right)^2 \right] . \tag{5.29}$$

Inserting (5.29) into (5.28) allows the two-dimensional equation to be separated into two one-dimensional homogeneous Fredholm equations that can be solved analytically [5.20, 5.24].

The stationary amplitude distribution is obtained from the solutions. For the confocal resonator it can be represented by the product of Hermitian polynomials, a Gaussian function, and a phase factor:

$$A_{mn}(x, y, z) = C^* H_m(x^*) H_n(y^*) \exp(-r^2/w^2) \exp[-i\phi(z, r, R)] . \tag{5.30}$$

Here, C^* is a normalization factor. The function H_m is the Hermitian polynomial of mth order. The last factor gives the phase $\phi(z_0, r)$ in the plane $z = z_0$ at a distance $r = (x^2 + y^2)^{1/2}$ from the resonator axis. The arguments x^* and y^* depend on the mirror separation d and are related to the coordinates x, y by $x^* = \sqrt{2}x/w$ and $y^* = \sqrt{2}y/w$, where

$$w^2(z) = \frac{\lambda d}{2\pi} \left[1 + (2z/d)^2 \right] , \tag{5.31}$$

is a measure of the radial intensity distribution.

From the definition of the Hermitian polynomials [5.31], one can see that the indices m and n give the number of nodes for the amplitude $A(x, y)$ in the x- (or the y-) direction. Figures 5.9, 5.10 illustrate some of these "transverse electromagnetic standing waves," which are called TEM$_{m,n}$ modes. The diffraction effects do not essentially influence the transverse character of the waves. While Fig. 5.9a shows the one-dimensional amplitude distribution $A(x)$ for some modes, Fig. 5.9b depicts the two-dimensional field amplitude $A(x, y)$ in Cartesian coordinates and $A(r, \vartheta)$ in polar coordinates. Modes with $m = n = 0$ are called *fundamental modes* or *axial modes* (often zero-order transverse modes as well), while configurations with $m > 0$ or $n > 0$ are transverse modes of higher order. The intensity distribution of the fundamental mode $I_{00} \propto A_{00} A_{00}^*$ can be derived from (5.30). With $H_0(x^*) = H_0(y^*) = 1$ we obtain

$$I_{00}(x, y, z) = I_0 e^{-2r^2/w^2} . \tag{5.32}$$

The fundamental modes have a Gaussian profile. For $r = w(z)$ the intensity decreases to $1/e^2$ of its maximum value $I_0 = C^{*2}$ on the axis ($r = 0$). The

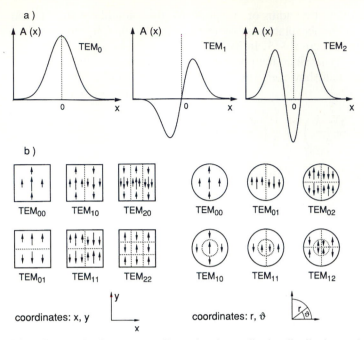

Fig. 5.9. (a) Stationary one-dimensional amplitude distributions $A_m(x)$ in a confocal resonator; (b) two-dimensional presentation of linearly polarized resonator modes $\mathrm{TEM}_{m,n}(x, y)$ for square and $\mathrm{TEM}_{m,n}(r, \vartheta)$ for circular apertures

value $r = w(z)$ is called the *beam radius* or *mode radius*. The smallest beam radius w_0 within the confocal resonator is the *beam waist*, which is located at the center $z = 0$. From (5.31) we obtain with $d = R$

$$w_0 = (\lambda R/2\pi)^{1/2} . \tag{5.33}$$

At the mirrors $(z = d/2)$ the beam radius $w_s = w(d/2) = \sqrt{2}w_0$ is increased by a factor $\sqrt{2}$.

Example 5.4

(a) For a HeNe laser with $\lambda = 633\,\mathrm{nm}$, $R = d = 30\,\mathrm{cm}$, (5.33) gives $w_0 = 0.17\,\mathrm{mm}$ for the beam waist.
(b) For a CO_2 laser with $\lambda = 10\,\mu\mathrm{m}$, $R = d = 2\,\mathrm{m}$ is $w_0 = 1.8\,\mathrm{mm}$.

Note that w_0 and w do not depend on the mirror size. Increasing the mirror width $2a$ reduces, however, the diffraction losses as long as no other limiting aperture exists inside the resonator.

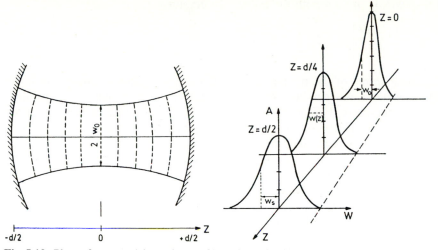

Fig. 5.10. Phase fronts and intensity profiles of the fundamental TEM$_{00}$ mode at several locations z in a confocal resonator with the mirrors at $z = \pm d/2$

For the phase $\phi(r, z)$ in the plane $z = z_0$, one obtains with the abbreviation $\xi_0 = 2z_0/R$ [5.20]

$$\phi(r, z) = \frac{2\pi}{\lambda} \left[\frac{R}{2}(1 + \xi_0) + \frac{x^2 + y^2}{R} \frac{\xi_0}{1 + \xi_0^2} \right]$$
$$- (1 + m + n) \left[\frac{\pi}{2} - \arctan\left(\frac{1 - \xi_0}{1 + \xi_0} \right) \right]. \tag{5.34}$$

Inside the resonator $0 < |\xi_0| < 1$, outside $|\xi_0| > 1$.

The equations (5.30) and (5.34) show that the field distributions $A_{mn}(x, y)$ and the form of the phase fronts depend on the location z_0 within the resonator.

From (5.34) we can deduce the phase fronts inside the confocal resonator, i.e., all points (x, y, z) for which $\phi(r, z)$ is constant. For the fundamental mode with $m = n = 0$ and for points close to the resonator axis, i.e., for $r \ll R$, the variation of the arctan term with $z - z_0$ along the phase front can be neglected. We obtain as a condition for the curved phase front intersecting the resonator axis at $z = z_0$ that the first bracket in (5.34) must be constant, i.e., independent of x and y, which means: $[\ldots]_{x,y\neq0} - [\ldots]_{x,y=0} = 0$, or

$$\frac{R}{2}(1 + \xi) + \frac{x^2 + y^2}{R} \frac{\xi}{1 + \xi^2} = \frac{R}{2}(1 + \xi_0), \tag{5.35}$$

with the shorthand $\xi = 2z/R$. This yields the equation

$$z_0 - z = \frac{x^2 + y^2}{R} \frac{\xi}{1 + \xi^2}, \tag{5.36}$$

of a spherical surface with the radius of curvature

$$R' \approx \left| \frac{1+\xi_0^2}{2\xi_0} \right| R = \left(\frac{R}{4z_0} + \frac{z_0}{R} \right) R \,. \tag{5.37}$$

The phase fronts of the fundamental modes inside a confocal resonator close to the resonator axis can be described as spherical surfaces with a z-dependent radius of curvature. For $z_0 = R/2 \to \xi_0 = 1 \Rightarrow R' = R$. This means that at the mirror surfaces of the confocal resonator close to the resonator axis the wavefronts are identical with the mirror surfaces. Due to diffraction this is not quite true at the mirror edges, (i.e., at larger distances r from the axis), where the approximation (5.35) is not correct.

At the center of the resonator $z = 0 \to \xi_0 = 0 \to R' = \infty$. The radius R' becomes infinite. At the beam waist the constant phase surface becomes a plane $z = 0$. This is illustrated by Fig. 5.10, which depicts the phase fronts and intensity profiles of the fundamental mode at different locations inside a confocal resonator.

5.2.4 General Spherical Resonators

It can be shown [5.1, 5.24] that in nonfocal resonators with large Fresnel numbers N the field distribution of the fundamental mode can also be described by the Gaussian profile (5.32). The confocal resonator with $d = R$ can be replaced by other mirror configurations without changing the field configurations if the radius R_i of each mirror at the position z_0 equals the radius R' of the wavefront in (5.37) at this position. This means that any two surfaces of constant phase can be replaced by reflectors, which have the same radius of curvature as the wave front – in the approximation outlined above.

For symmetrical resonators with $R_1 = R_2 = R^*$ and the mirror separation d^*, we find from (5.37) with $z_0 = d^*/2 \to \xi_0 = d^*/R$ for the possible mirror separations

$$d^* = R^* \pm \sqrt{R^{*2} - R^2} = R^* \left[1 \pm \sqrt{1 - (R/R^{*2})} \right] \,. \tag{5.38}$$

These resonators are equivalent, with respect to the field distribution, to the confocal resonator with the mirror radii R and mirror separation $d = R$.

The beam radius w_s of the TEM$_{00}$ mode at the mirror surfaces (called the *spot size*) is, according to (5.31) and (5.38), with $z_0 = d/2$ for a symmetric resonator, mirror separation d, and the mirror radii R

$$w_s = \left(\frac{\lambda d}{\pi} \right)^{1/2} \left[\frac{2d}{R} - \left(\frac{d}{R} \right)^2 \right]^{-1/4} \,. \tag{5.39}$$

The second factor is a function of d and becomes maximum for $d = R$. *This reveals that of all symmetrical resonators with a given mirror separation d the confocal resonator with $d = R$ has the smallest spot sizes at the mirrors.*

5.2.5 Diffraction Losses of Open Resonators

The diffraction losses of a resonator depend on its Fresnel number $N = a^2/d\lambda$
(Sect. 5.2.1) and also on the field distribution $A(x, y, z = \pm d/2)$ at the mirror. The fundamental mode, where the field energy is concentrated near the
resonator axis, has the lowest diffraction losses, while the higher transverse
modes, where the field amplitude has larger values toward the mirror edges,
exhibit large diffraction losses. Using (5.31) with $z = d/2$ the Fresnel number
$N = a^2/(d\lambda)$ can be expressed as

$$N = \frac{1}{\pi}\frac{\pi a^2}{\pi w_s^2} = \frac{1}{\pi}\frac{\text{effective resonator-mirror surface area}}{\text{confocal TEM}_{00}\text{ mode area on the mirror}} , \qquad (5.40)$$

which illustrates that the diffraction losses decrease with increasing N. Figure 5.11 presents the diffraction losses of a confocal resonator as a function
of the Fresnel number N for the fundamental mode and some higher-order
transverse modes. For comparison, the much higher diffraction losses of
a plane-mirror resonator are also shown in order to illustrate the advantages of
curved mirrors, which refocus the waves otherwise diverging by diffraction.
From Fig. 5.11 it is obvious that higher-order transverse modes can be suppressed by choosing a resonator with a suitable Fresnel number, which may
be realized, for instance, by a limiting aperture with the diameter $D < 2a$ inside the laser resonator. If the losses exceed the gain for these modes they do
not reach threshold, and the laser oscillates only in the fundamental mode.

The confocal resonator with the smallest spot sizes at a given mirror
separation d according to (5.39) also has the lowest diffraction losses per
round-trip, which can be approximated for circular mirrors and Fresnel numbers $N > 1$ by [5.1]

$$\gamma_D \sim 16\pi^2 N\,e^{-4\pi N} . \qquad (5.41)$$

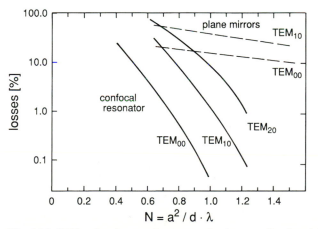

Fig. 5.11. Diffraction losses of some modes in a confocal and in a plane-mirror resonator,
plotted as a function of the Fresnel number N

5.2.6 Stable and Unstable Resonators

In a stable resonator the field amplitude $A(x, y)$ reproduces itself after each round-trip apart from a constant factor C, which represents the total diffraction losses but does not depend on x or y, see (5.27).

The question is now how the field distribution $A(x, y)$ and the diffraction losses change with varying mirror radii R_1, R_2 and mirror separation d for a general resonator. We will investigate this problem for the fundamental TEM_{00} mode, described by the Gaussian beam intensity profile. For a stationary field distribution, where the Gaussian beam profile reproduces itself after each round-trip, one obtains for a resonator consisting of two spherical mirrors with the radii R_1, R_2, separated by the distance d, the spot sizes πw_1^2 and πw_2^2 on the mirror surfaces [5.1, 5.24]

$$\pi w_1^2 = \lambda d \left[\frac{g_2}{g_1(1 - g_1 g_2)} \right]^{1/2} \, , \quad \pi w_2^2 = \lambda d \left[\frac{g_1}{g_2(1 - g_1 g_2)} \right]^{1/2} \, , \quad (5.42)$$

with the parameters ($i = 1, 2$)

$$g_i = 1 - d/R_i \, . \tag{5.43}$$

This reveals that for $g_1 = 0$ or $g_2 = 0$ and for $g_1 g_2 = 1$, the spot sizes become infinite at one or at both mirror surfaces, which implies that the Gaussian beam diverges: the resonator becomes unstable. An exception is the confocal resonator with $g_1 = g_2 = 0$, which is "metastable", because it is only stable if both parameters g_i are exactly zero. For $g_1 g_2 > 1$ or $g_1 g_2 < 0$, the right-hand sides of (5.42) become imaginary, which means that the resonator is unstable. The condition for a stable resonator is therefore

$$\boxed{0 < g_1 g_2 < 1.} \tag{5.44}$$

With the general stability parameter $G = 2g_1 g_2 - 1$ we can distinguish stable resonators: $0 < |G| < 1$, unstable resonators: $|G| > 1$, metastable res-

Table 5.1. Some commonly used optical resonators with their stability parameters $g_i = 1 - d/R_i$, $R_1 = 2d$, $R_2 = \infty$ and the resonator parameters $G = 2g_1 g_2 - 1$

Type of resonator	Mirror radii	Stability parameter	
Confocal	$R_1 + R_2 = 2d$	$g_1 + g_2 = 2g_1 g_2$	$|G| > 1$
Concentric	$R_1 + R_2 = d$	$g_1 g_2 = 1$	$G = 1$
Symmetric	$R_1 = R_2$	$g_1 = g_2 = g$	$|G| < 1$
Symmetric confocal	$R_1 = R_2 = d$	$g_1 = g_2 = 0$	$G = -1$
Symmetric concentric	$R_1 = R_2 = 1/2d$	$g_1 = g_2 = -1$	$G = 1$
Semiconfocal	$R_1 = 2d$, $R_2 = \infty$	$g_1 = 1, g_2 = 1/2$	$G = 0$
Plane	$R_1 = R_2 = \infty$	$g_1 = g_2 = +1$	$G = 1$

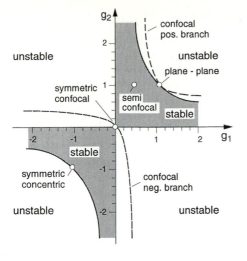

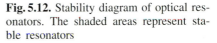

Fig. 5.12. Stability diagram of optical resonators. The shaded areas represent stable resonators

onators: $|G| = 1$. In Table 5.1 some resonators are compiled with their corresponding parameters g_i. Figure 5.12 displays the stability diagram in the g_1–g_2-plane. According to (5.44) the plane resonator ($R_1 = R_2 = \infty \Rightarrow g_1 = g_2 = 1$) is not stable, because the spot size of a Gaussian beam would increase after each round-trip. As was shown above, there are, however, other non-Gaussian field distributions, which form stable eigenmodes of a plane resonator, although their diffraction losses are much higher than those of resonators within the stability region. The symmetric confocal resonator with $g_1 = g_2 = 0$ might be called "metastable," since it is located between unstable regions in the stability diagram and even a slight deviation of g_1, g_2 into the direction $g_1 g_2 < 0$ makes the resonator unstable. For illustration, some commonly used resonators are depicted in Fig. 5.13.

For some laser media, in particular those with large gain, unstable resonators with $g_1 g_2 < 0$ may be more advantageous than stable ones for the following reason: in stable resonators the beam waist $w_0(z)$ of the fundamental mode is given by the mirror radii R_1, R_2 and the mirror separation d, see (5.33), and is generally small (Example 5.4). If the cross section of the active volume is larger than πw^2, only a fraction of all inverted atoms can contribute to the laser emission into the TEM_{00} mode, while in unstable resonators the beam fills the whole active medium. This allows extraction of the maximum output power. One has, however, to pay for this advantage by a large beam divergence.

Let us consider the simple example of a symmetric unstable resonator depicted in Fig. 5.14 formed by two mirrors with radii R_i separated by the distance d. Assume that a spherical wave with its center at F_1 is emerging from mirror M_1. The spherical wave geometrically reflected by M_2 has its center in F_2. If this wave after ideal reflection at M_1 is again a spherical wave with its center at F_1, the field configuration is stationary and the mirrors image the local point F_1 into F_2, and vice versa.

a) plane resonator

b) confocal resonator

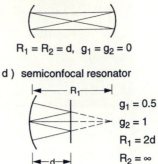

$R_1 = R_2 = \infty$, $g_1 = g_2 = 1$ $R_1 = R_2 = d$, $g_1 = g_2 = 0$

c) concentric resonator

d) semiconfocal resonator

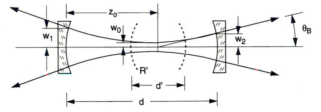

$R_1 + R_2 = d$, $g_1 \cdot g_2 = 1$

$g_1 = 0.5$
$g_2 = 1$
$R_1 = 2d$
$R_2 = \infty$

e) general spherical resonator with TEM$_{00}$- mode

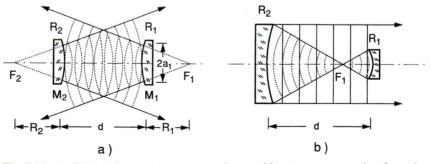

Fig. 5.13a–e. Some examples of commonly used open resonators

For the magnification of the beam diameter on the way from mirror M_1 to M_2 or from M_2 to M_1, we obtain from Fig. 5.14 the relations

$$M_{12} = \frac{d + R_1}{R_1} , \qquad M_{21} = \frac{d + R_2}{R_2} . \tag{5.45}$$

Fig. 5.14. (a) Spherical waves in a symmetric unstable resonator emerging from the virtual focal points F_1 and F_2; **(b)** asymmetric unstable resonator with a real focal point between the two mirrors

We define the magnification factor $M = M_{12}M_{21}$ per round-trip as the ratio of the beam diameter after one round-trip to the initial one:

$$M = M_{12}M_{21} = \frac{d+R_1}{R_1}\frac{d+R_2}{R_2}. \qquad (5.46)$$

For $R_i > 0$ $(i = 1, 2)$ the virtual focal points are outside the resonator and the magnification factor becomes $M > 1$ (Fig. 5.14a).

In the resonator of Fig. 5.14a the waves are coupled out of both sides of the resonator. The resultant high resonator losses are generally not tolerable and for practical applications the resonator configurations of Fig. 5.14b and Fig. 5.15 consisting of one large and one small mirror are better suited. Two types of nonsymmetric spherical unstable resonators are possible with $g_1 g_2 > 1 \Rightarrow G > 1$ (Fig. 5.15a) with the virtual beam waist outside the resonator and with $g_1 g_2 < 0 \Rightarrow G < -1$ (Fig. 5.15b) where the focus lies inside the resonator.

For these spherical resonators the magnification factor M can be expressed by the resonator parameter G [5.25]:

$$M_{\pm} = |G| \pm \sqrt{G^2 - 1}, \qquad (5.47a)$$

where the $+$ sign holds for $g_1 g_2 > 1$ and the $-$ sign for $g_1 g_2 < 0$.

If the intensity profile $I(x_1, y_1, z_0)$ in the plane $z = z_0$ of the outcoupling mirror does not change much over the mirror size, the fraction P_2/P_0 of the power P_0 incident on M_2 that is reflected back to M_1 equals the ratio of the areas

$$\frac{P_2}{P_0} = \frac{\pi w_2^2}{\pi w_1^2} = \frac{1}{M^2}. \qquad (5.47b)$$

The loss factor per round-trip is therefore

$$V = \frac{P_0 - P_2}{P_0} = 1 - \frac{1}{M^2} = \frac{M^2 - 1}{M^2}. \qquad (5.48)$$

For the two unstable resonators of Fig. 5.15 the near-field pattern of the outcoupled wave is an annular ring (Fig. 5.16). The spatial far-field intensity distribution can be obtained as a numerical solution of the corresponding

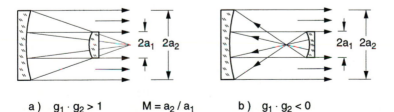

a) $g_1 \cdot g_2 > 1$ $M = a_2 / a_1$ b) $g_1 \cdot g_2 < 0$

Fig. 5.15a,b. Two examples of unstable confocal resonators: **(a)** $g_1 \cdot g_2 > 1$; **(b)** $g_1 \cdot g_2 < 0$, with a definition of the magnification factor

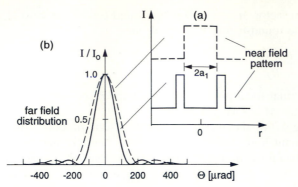

Fig. 5.16a,b. Diffraction pattern of the output intensity of a laser with an unstable resonator: (**a**) near field just at the output coupler and (**b**) far-field distribution for a resonator with $a = 0.66$ cm, $g_1 = 1.21$, $g_2 = 0.85$. The patterns obtained with a circular output mirror (*solid curve*) are compared with those of a circular aperture (*dashed curves*)

Kirchhoff–Fresnel integro-differential equation analog to (5.26). For illustration, the near-field and far-field patterns of an unstable resonator of the type shown in Fig. 5.15a is compared with the diffraction pattern of a circular aperture.

Note that the angular divergence of the central diffraction order in the far field is smaller for the annular-ring near-field distribution than that of a circular aperture with the same size as the small mirror of the unstable resonator. However, the higher diffraction orders are more intense, which means that the angular intensity distribution has broader wings.

In unstable resonators the laser beam is divergent and only a fraction of the divergent beam area may be reflected by the mirrors. The losses are therefore high and the effective number of round-trips is small. Unstable resonators are therefore suited only for lasers with a sufficiently large gain per round-trip [5.26–5.29].

In recent years, specially designed optics with slabs of cylindrical lenses have been used to make the divergent output beam more parallel, which allows one to focus the beam into a smaller spot size [5.30].

5.2.7 Ring Resonators

A ring resonator consists of at least three reflecting surfaces, which may be provided by mirrors or prisms. Two possible arrangements are illustrated in Fig. 5.17. Instead of the *standing* waves in a Fabry–Perot-type resonator, the ring resonator allows *traveling* waves, which may run clockwise or counterclockwise through the resonator. With an "optical diode" inside the ring resonator unidirectional traveling waves can be enforced. Such an "optical diode" is a device that has low losses for light passing into one direction but sufficiently high losses to prevent laser oscillation for light traveling into the opposite direction. It consists of a Faraday rotator, which turns the plane

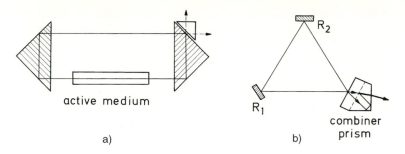

Fig. 5.17a,b. Two examples of possible ring resonators, using total reflection: (**a**) with corner-cube prism reflectors and frustrated total reflection for output coupling; (**b**) three-mirror arrangement with beam-combining prism

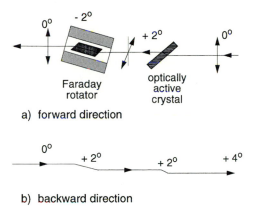

Fig. 5.18a,b. Optical diode consisting of a Faraday rotator, a birefringent crystal, and Brewster windows. Tilting of the polarization vector for the forward (**a**) and backward (**b**) directions

of polarization by the angle $\pm\alpha$ (Fig. 5.18), a birefringent crystal, which also turns the plane of polarization by α, and elements with a polarization-dependent transmission, such as Brewster windows [5.32]. For the wanted direction the turning angles $-\alpha+\alpha=0$ just cancel, and for the other direction they add to 2α, causing reflection losses at the Brewster windows. If these are larger than the gain this direction cannot reach the threshold.

The unidirectional ring laser has the advantage that spatial hole burning, which impedes single-mode oscillation of lasers (Sect. 5.3.3), can be avoided. In the case of homogeneous gain profiles, the ring laser can utilize the total population inversion within the active mode volume contrary to a standing-wave laser, where the inversion at the nodes of the standing wave cannot be utilized. One therefore expects larger output powers in single-mode operation than from standing-wave cavities at comparable pump powers.

5.2.8 Frequency Spectrum of Passive Resonators

The stationary field configurations of open resonators, discussed in the previous sections, have an eigenfrequency spectrum that can be directly derived from the condition that the phase fronts at the reflectors have to be identical with the mirror surfaces. Because these stationary fields represent standing waves in the resonators, the mirror separation d must be an integer multiple of $\lambda/2$ and the phase factor in (5.30) becomes unity at the mirror surfaces. This implies that the phase ϕ has to be an integer multiple of π. Inserting the condition $\phi = q\pi$ into (5.34) gives the eigenfrequencies ν_r of the confocal resonator with $R = d$, $\xi = 1$, $x = y = 0$

$$\nu_r = \frac{c}{2d}\left[q + \frac{1}{2}(m+n+1)\right].\tag{5.49}$$

The fundamental axial modes TEM_{00q} $(m = n = 0)$ have the frequencies $\nu = (q+\frac{1}{2})c/2d$ and the frequency separation of adjacent axial modes is

$$\boxed{\delta\nu = \frac{c}{2d}.}\tag{5.50}$$

Equation (5.49) reveals that the frequency spectrum of the confocal resonator is degenerate because the transverse modes with $q = q_1$ and $m+n = 2p$ have the same frequency as the axial mode with $m = n = 0$ and $q = q_1 + p$. Between two axial modes there is always another transverse mode with $m + n + 1 = $ odd. The free spectral range of a *confocal resonator* is therefore

$$\delta\nu_{\text{confocal}} = \frac{c}{4d}.\tag{5.51}$$

If the mirror separation d deviates slightly from the radius of the mirror curvature R, the degeneracy is removed. We obtain from (5.34) with $\phi = q\pi$ and $\xi_0 = d/R \neq 1$ for a symmetric nonconfocal resonator with two equal mirror radii $R_1 = R_2 = R$

$$\nu_r = \frac{c}{2d}\left\{q + \frac{1}{2}(m+n+1)\left[1 + \frac{4}{\pi}\arctan\left(\frac{d-R}{d+R}\right)\right]\right\}.\tag{5.52}$$

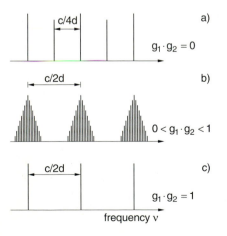

Fig. 5.19a–c. Degenerate mode frequency spectrum of a confocal resonator $(d = R)$ (a), degeneracy lifting in a near-confocal resonator $(d = 1.1R)$ (b), and the spectrum of fundamental modes in a plane-mirror resonator (c)

Now the higher-order transverse modes are no longer degenerate with axial modes. The frequency separation depends on the ratio $(d-R)/(d+R)$. Figure 5.19 illustrates the frequency spectrum of the plane-mirror resonator, the confocal resonator $(R=d)$, and of a nonconfocal resonator where d is slightly larger than R. Due to higher diffraction losses the amplitudes of the higher transverse modes decrease.

As has been shown in [5.21] the frequency spectrum of a general resonator with unequal mirror curvatures R_1 and R_2 can be represented by

$$\nu_r = \frac{c}{2d}\left[q + \frac{1}{\pi}(m+n+1)\arccos^{-1}\sqrt{g_1 g_2}\right], \tag{5.53}$$

where $g_i = 1 - d/R_i$ $(i = 1, 2)$ are the resonator parameters. The eigenfrequencies of the axial modes $(m = n = 0)$ are no longer at $(c/2d)(q+\frac{1}{2})$, but are slightly shifted. The free spectral range, however, is again $\delta\nu = c/2d$.

Example 5.5

(a) Consider a nonconfocal symmetric resonator: $R_1 = R_2 = 75\,\mathrm{cm}$, $d = 100\,\mathrm{cm}$. The free spectral range $\delta\nu$, which is the frequency separation of the adjacent axial modes q and $q+1$, is $\delta\nu = (c/2d) = 150\,\mathrm{MHz}$. The frequency separation $\Delta\nu$ between the $(q,0,0)$ mode and the $(q,1,0)$ mode is $\Delta\nu = 87\,\mathrm{MHz}$ from (5.52).

(b) Consider a confocal resonator: $R = d = 100\,\mathrm{cm}$. The frequency spectrum consists of equidistant frequencies with $\delta\nu = 75\,\mathrm{MHz}$. If, however, the higher-order transverse modes are suppressed, only axial modes oscillate with a frequency separation $\delta\nu = 150\,\mathrm{MHz}$.

Now we briefly discuss the spectral width $\Delta\nu$ of the resonator resonances. The problem will be approached in two different ways.

Since the laser resonator is a Fabry–Perot interferometer, the spectral distribution of the transmitted intensity follows the Airy formula (4.57). According to (4.53b), the halfwidth $\Delta\nu_r$ of the resonances, expressed in terms of the free spectral range $\delta\nu$, is $\Delta\nu_r = \delta\nu/F^*$. If diffraction losses can be neglected, the finesse F^* is mainly determined by the reflectivity R of the mirrors, therefore the halfwidth of the resonance becomes

$$\Delta\nu = \frac{\delta\nu}{F^*} = \frac{c}{2d}\frac{1-R}{\pi\sqrt{R}}. \tag{5.54}$$

Example 5.6

With the reflectivity $R = 0.98 \Rightarrow F^* = 150$. A resonator with $d = 1\,\mathrm{m}$ has the free spectral range $\delta\nu = 150\,\mathrm{MHz}$. The halfwidth of the resonator modes then becomes $\Delta\nu_r = 1\,\mathrm{MHz}$ if the mirrors are perfectly aligned and have nonabsorptive ideal surfaces.

Generally speaking, other losses such as diffraction, absorption, and scattering losses decrease the total finesse. Realistic values are $F^* = 50-100$, giving for Example 5.6 a resonance halfwidth of the passive resonator of about 2 MHz.

The second approach for the estimate of the resonance width starts from the quality factor Q of the resonator. With total losses β per second, the energy W stored in a mode of a passive resonator decays exponentially according to (5.18). The Fourier transform of (5.18) yields the frequency spectrum of this mode, which gives a Lorentzian (Sect. 3.1) with the halfwidth $\Delta \nu_r = \beta/2\pi$. With the mean lifetime $T = 1/\beta$ of a photon in the resonator mode, the frequency width can be written as

$$\Delta \nu_r = \frac{1}{2\pi T} \,. \tag{5.55}$$

If reflection losses give the main contribution to the loss factor, the photon lifetime is, with $R = \sqrt{R_1 R_2}$, see (5.22), $T = -d/(c \ln R)$. The width $\Delta \nu$ of the resonator mode becomes

$$\Delta \nu_r = \frac{c |\ln R|}{2\pi d} = \frac{\delta\nu(|\ln R|)}{\pi} \,, \tag{5.56}$$

which yields with $|\ln R| \approx 1 - R$ the same result as (5.54), apart from the factor $\sqrt{R} \approx 1$. The slight difference of the two results stems from the fact that in the second estimation we distributed the reflection losses uniformly over the resonator length.

5.3 Spectral Characteristics of Laser Emission

The frequency spectrum of a laser is determined by the spectral range of the active laser medium, i.e., its gain profile, and by the resonator modes falling within this spectral gain profile (Fig. 5.20). All resonator modes for which the gain exceeds the losses can participate in the laser oscillation. The active medium has two effects on the frequency distribution of the laser emission:

- Because of its index of refraction $n(\nu)$, it shifts the eigenfrequencies of the passive resonator (mode-pulling).
- Due to spectral gain saturation competition effects between different oscillating laser modes occur; they may influence the amplitudes and frequencies of the laser modes.

In this section we shall briefly discuss spectral characteristics of multimode laser emission and the effects that influence it.

5.3.1 Active Resonators and Laser Modes

Introducing the amplifying medium into the resonator changes the refractive index between the mirrors and with it the eigenfrequencies of the resonator.

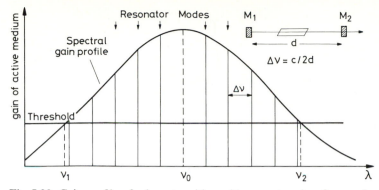

Fig. 5.20. Gain profile of a laser transition with resonator eigenfrequencies of axial modes

We obtain the frequencies of the *active resonator* by replacing the mirror separation d in (5.52) by

$$d^* = (d - L) + n(\nu)L = d + (n - 1)L \, , \tag{5.57}$$

where $n(\nu)$ is the refractive index in the active medium with length L. The refractive index $n(\nu)$ depends on the frequency ν of the oscillating modes within the gain profile of a laser transition where anomalous dispersion is found. Let us at first consider how laser oscillation builds up in an active resonator.

If the pump power is increased continuously, the threshold is reached first at those frequencies that have a maximum net gain. According to (5.5) the net gain factor per round-trip

$$G(\nu, 2d) = \exp[-2\alpha(\nu)L - \gamma(\nu)] \, , \tag{5.58}$$

is determined by the amplification factor $\exp[-2\alpha(\nu)L]$, which has the frequency dependence of the gain profile (5.8) and also by the loss factor $\exp(-2\beta d/c) = \exp[-\gamma(\nu)]$ per round-trip. While absorption or diffraction losses do not strongly depend on the frequency within the gain profiles of a laser transition, the transmission losses exhibit a strong frequency dependence, which is closely connected to the eigenfrequency spectrum of the resonator. This can be illustrated as follows:

Assume that a wave with the spectral intensity distribution $I_0(\nu)$ traverses an interferometer with two mirrors, each having the reflectivity R and transmission factor T (Fig. 5.21). For the passive interferometer we obtain a frequency spectrum of the transmitted intensity according to (4.50). With an amplifying medium inside the resonator, the incident wave experiences the amplification factor (5.58) per round-trip and we obtain, analogous to (4.61) by summation over all interfering amplitudes, the total transmitted intensity

$$I_T = I_0 \frac{T^2 G(\nu)}{[1 - G(\nu)]^2 + 4G(\nu)\sin^2(\phi/2)} \, . \tag{5.59}$$

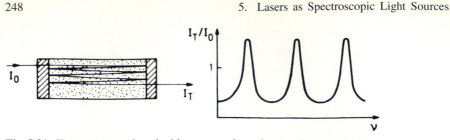

Fig. 5.21. Transmission of an incident wave through an active resonator

The total amplification I_T/I_0 has maxima for $\phi = 2q\pi$, which corresponds to the condition (5.53) for the eigenfrequencies of the resonator with the modification (5.57). For $G(\nu) \to 1$, the total amplification I_T/I_0 becomes infinite for $\phi = 2q\pi$. This means that even an infinitesimally small input signal results in a finite output signal. Such an input is always provided, for instance, by the spontaneous emission of the excited atoms in the active medium. *For $G(\nu) = 1$ the laser amplifier converts to a laser oscillator.* This condition is equivalent to the threshold condition (5.7). Because of gain saturation (Sect. 5.3), the amplification remains finite and the total output power is determined by the pump power rather than by the gain.

According to (5.8) the gain factor $G_0(\nu) = \exp[-2\alpha(\nu)L]$ depends on the line profile $g(\nu - \nu_0)$ of the molecular transition $E_i \to E_k$. The threshold condition can be illustrated graphically by subtracting the frequency-dependent losses from the gain profile. Laser oscillation is possible at all frequencies ν_L where this subtraction gives a positive net gain (Fig. 5.22).

Example 5.7

(a) In gas lasers, the gain profile is the Doppler-broadened profile of a molecular transition (Sect. 3.2) and therefore shows a Gaussian distribution with the Doppler width $\delta\omega_D$ (see Sect. 3.2),

$$\alpha(\omega) = \alpha(\omega_0) \exp\left(-\frac{\omega - \omega_0}{0.6\delta\omega_D}\right)^2 .$$

With $\alpha(\omega_0) = -0.01\,\text{cm}^{-1}$, $L = 10\,\text{cm}$, $\delta\omega_0 = 1.3 \times 10^9\,\text{Hz}\cdot 2\pi$, and $\gamma = 0.03$, the gain profile extends over a frequency range of $\delta\omega = 2\pi \cdot 3\,\text{GHz}$ where $-2\alpha(\omega)L > 0.03$. In a resonator with $d = 50\,\text{cm}$, the mode spacing is 300 MHz and ten axial modes can oscillate.

(b) Solid-state or liquid lasers generally exhibit broader gain profiles because of additional broadening mechanisms (Sect. 3.7). A dye laser has, for example, a gain profile with a width of about 10^{13} Hz. Therefore, in a resonator with $d = 50\,\text{cm}$ about 3×10^4 resonator modes fall within the gain profile.

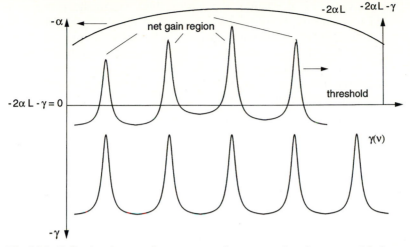

Fig. 5.22. Reflection losses of a resonator (*lower curve*), gain curve $\alpha(\nu)$ (*upper curve*), and net gain $\Delta\alpha(\nu) = -2L\alpha(\nu) - \gamma(\nu)$ as difference between gain ($\alpha < 0$) and losses (*middle curve*). Only frequencies with $\Delta\alpha(\nu) > 0$ reach the oscillation threshold

The preceding example illustrates that the passive resonance halfwidth of typical resonators for gas lasers is very small compared with the linewidth of a laser transition, which is generally determined by the Doppler width. The active medium inside a resonator compensates the losses of the passive resonator resonances resulting in an exceedingly high quality factor Q. The linewidth of an oscillating laser mode should therefore be much smaller than the passive resonance width.

From (5.59) we obtain for the halfwidth $\Delta\nu$ of the resonances for an active resonator with a free spectral range $\delta\nu$ the expression

$$\Delta\nu_a = \delta\nu \frac{1 - G(\nu)}{2\pi\sqrt{G(\nu)}} = \delta\nu / F_a^* \;. \tag{5.60}$$

The finesse F_a^* of the active resonator approaches infinity for $G(\nu) \to 1$. Although the laser linewidth $\Delta\nu_L$ may become much smaller than the halfwidth of the passive resonator, it does not approach zero. This will be discussed in Sect. 5.6.

For frequencies between the resonator resonances, the losses are high and the threshold will not be reached. In the case of a Lorentzian resonance profile, for instance, the loss factor has increased to about ten times $\beta(\nu_0)$ at frequencies that are $3\Delta\nu_r$ away from the resonance center ν_0.

5.3.2 Gain Saturation

When the pump power of a laser is increased beyond its threshold value, laser oscillation will start at first at a frequency where the net gain, that is, the difference between total gain minus total losses, has a maximum. During the buildup time of the laser oscillation, the gain is larger than the losses and the stimulated wave inside the resonator is amplified during each round-trip until the radiation energy is sufficiently large to deplete the population inversion ΔN down to the threshold value ΔN_{thr}. Under stationary conditions the increase of ΔN due to pumping is just compensated by its decrease due to stimulated emission. The gain factor of the active medium saturates from the value $G_0(I=0)$ at small intensities to the threshold value

$$G_{thr} = e^{-2L\alpha_{sat}(\nu)} = e^{+\gamma} , \tag{5.61}$$

where the gain just equals the total losses per round-trip. This gain saturation is different for homogeneous and for inhomogeneous line profiles of laser transitions (Sect. 3.6).

In the case of a homogeneous profile $g(\nu - \nu_0)$, all molecules in the upper level can contribute to stimulated emission at the laser frequency ν_a with the probability $B_{ik}\rho g(\nu_a - \nu_0)$, see (5.8). Although the laser may oscillate only with a single frequency ν, the whole homogeneous gain profile $\alpha(\nu) = \Delta N\sigma(\nu)$ saturates until the inverted population difference ΔN has decreased to the threshold value ΔN_{thr} (Fig. 5.23a). The saturated amplification coefficient $\alpha_{sat}(\nu)$ at the intracavity laser intensity I is, according to Sect. 3.6,

$$\alpha_s^{hom}(\nu) = \frac{\alpha_0(\nu)}{1+S} = \frac{\alpha_0(\nu)}{1+I/I_s} , \tag{5.62}$$

where $I = I_s$ is the intensity for which the saturation parameter $S = 1$, which means that the induced transition rate equals the relaxation rate. For homoge-

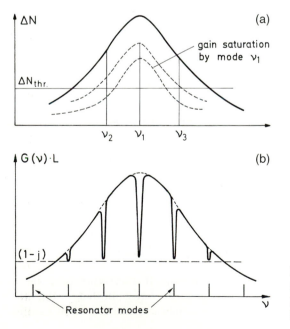

Fig. 5.23a,b. Saturation of gain profiles: (a) for a homogeneous profile; (b) for an inhomogeneous profile

neous gain profiles, the saturation caused by one laser mode also diminishes the gain for adjacent modes (mode competition).

In the case of inhomogeneous laser transitions, the whole line profile can be divided into homogeneously broadened subsections with the spectral width $\Delta \nu^{\text{hom}}$ (for example, the natural linewidth or the pressure- or power-broadened linewidth). Only those molecules in the upper laser level that belong to the subgroup in the spectral interval $\nu_L \pm \frac{1}{2}\Delta\nu^{\text{hom}}$, centered at the laser frequency ν_L, can contribute to the amplification of the laser wave. A monochromatic wave therefore causes selective saturation of this subgroup and burns a hole into the inhomogeneous distribution $\Delta N(\nu)$ (Fig. 5.23b). At the bottom of the hole, the inversion $\Delta N(\nu_L)$ has decreased to the threshold value ΔN_{thr}, but several homogeneous widths $\Delta\nu^{\text{hom}}$ away from ν_L, ΔN remains unsaturated. According to (3.78), the homogeneous width $\Delta\nu^{\text{hom}}$ of this hole increases with increasing saturating intensity as

$$\Delta\nu_s = \Delta\nu_0\sqrt{1+S} = \Delta\nu_0\sqrt{1+I/I_s}\,. \tag{5.63}$$

This implies that with increasing saturation *more* molecules from a larger spectral interval $\Delta\nu_s$ can contribute to the amplification. The gain factor decreases by the factor $1/(1+S)$ because of a decrease of ΔN caused by saturation. It increases by the factor $(1+S)^{1/2}$ because of the increased homogeneous width. The combination of both phenomena gives (Sect. 7.2)

$$\alpha_s^{\text{inh}}(\nu) = \alpha_0(\nu)\frac{\sqrt{1+S}}{1+S} = \frac{\alpha_0(\nu)}{\sqrt{1+I/I_s}}\,. \tag{5.64}$$

5.3.3 Spatial Hole Burning

A resonator mode represents a standing wave in the laser resonator with a z-dependent field amplitude $E(z)$, as illustrated in Fig. 5.24a. Since the saturation of the inversion ΔN, discussed in the previous section, depends on the intensity $I \propto |E|^2$, the inversion saturated by a single laser mode exhibits a spatial modulation $\Delta N(z)$, as sketched in Fig. 5.24c. Even for a completely homogeneous gain profile, there are always spatial regions of unsaturated inversion at the nodes of the standing wave $E_1(z)$. This may give sufficient gain for another laser mode $E_2(z)$ that is spatially shifted by $\lambda/4$ against $E_1(z)$, or even for a third mode with a shift of $\lambda/3$ of its amplitude maximum (Fig. 5.24b).

If the mirror separation d changes by only one wavelength (e.g., caused by acoustical vibrations of the mirrors), the maxima and nodes of the standing waves are shifted and the gain competition, governed by spatial hole burning, is altered. Therefore, every fluctuation of the laser wavelength caused by changes of the refractive index or the cavity length d results in a corresponding fluctuation of the coupling strength between the modes and changes the gain relations and the intensities of the simultaneously oscillating modes.

If the length L of the active medium is small compared to the resonator length (e.g., in cw dye lasers), it is possible to minimize the spatial hole-

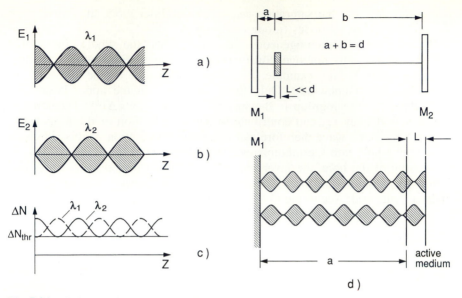

Fig. 5.24a–d. Spatial intensity distribution for two standing waves with slightly different wavelengths λ_1 and λ_2 (**a**), (**b**), and their corresponding saturation of the inversion $\Delta N(z)$ (**c**). Explanation of spatial hole-burning modes in the active medium (**d**) with a small length L, close to a resonator mirror M1 ($a \ll b$)

burning phenomenon by placing the active medium close to one cavity mirror (Fig. 5.24d). Consider two standing waves with the wavelengths λ_1 and λ_2. Their maxima in the active medium, placed at a distance a from the end mirror, are shifted by λ/p ($p = 2, 3, ...$). Since all standing waves must have nodes at the mirror surface, we obtain for two waves with the minimum possible wavelength difference $\Delta\lambda = \lambda_1 - \lambda_2$ the relation

$$m\lambda_1 = a = (m + 1/p)\lambda_2 , \tag{5.65}$$

or for their frequencies

$$\nu_1 = m\frac{c}{a}, \quad \nu_2 = \frac{c}{a}(m + 1/p) \;\Rightarrow\; \delta\nu_{sp} = \frac{c}{ap} . \tag{5.66}$$

In terms of the spacing $\delta\nu = c/2d$ of the longitudinal resonator modes, the spacing of the spatial hole-burning modes is

$$\delta\nu_{sp} = \frac{2d}{ap}\delta\nu . \tag{5.67}$$

Even when the net gain is sufficiently large to allow oscillation of, e.g., up to three spatially separated standing waves ($p = 1, 2, 3$), only one mode can oscillate if the spectral width of the homogeneous gain profile is smaller than $(2/3)(d/a)\delta\nu$ [5.33].

Example 5.8

$d = 100$ cm, $L = 0.1$ cm, $a = 5$ cm, $p = 3$, $\delta\nu = 150$ MHz, $\delta\nu_{sp} = 2000$ MHz. Single-mode operation could be achieved if the spectral gain profile is smaller than 2000 MHz.

In gas lasers the effect of spatial hole burning is partly averaged out by diffusion of the excited molecules from nodes to maxima of a standing wave. It is, however, important in solid-state and in liquid lasers such as the ruby laser or the dye laser. Spatial hole burning can be completely avoided in unidirectional ring lasers (Sect. 5.2.7) where no standing waves exist. Waves propagating in one direction can saturate the entire spatially distributed inversion.

5.3.4 Multimode Lasers and Gain Competition

The different gain saturation of homogeneous and inhomogeneous transitions strongly affects the frequency spectrum of multimode lasers, as can be understood from the following arguments:

Let us first consider a laser transition with a purely *homogeneous* line profile. The resonator mode that is next to the center of the gain profile starts oscillating when the pump power exceeds the threshold. Since this mode experiences the largest net gain, its intensity grows faster than that of the other laser modes. This causes partial saturation of the whole gain profile (Fig. 5.23a), mainly by this strongest mode. This saturation, however, decreases the gain for the other weaker modes and their amplification will be slowed down, which further increases the differences in amplification and favors the strongest mode even more. This mode competition of different laser modes within a homogeneous gain profile will finally lead to a complete suppression of all but the strongest mode. Provided that no other mechanism disturbs the predominance of the strongest mode, this saturation coupling results in single-frequency oscillation of the laser, even if the homogeneous gain profile is broad enough to allow, in principle, simultaneous oscillation of several resonator modes [5.34].

In fact, such single-mode operation without further frequency-selecting elements in the laser resonator can be observed only in a few exceptional cases because there are several phenomena, such as spatial hole burning, frequency jitter, or time-dependent gain fluctuations, that interfere with the pure case of mode competition discussed above. These effects, which will be discussed below, prevent the unperturbed growth of one definite mode, introduce time-dependent coupling phenomena between the different modes, and cause in many cases a frequency spectrum of the laser which consists of a random superposition of many modes that fluctuate in time.

In the case of a purely *inhomogeneous* gain profile, the different laser modes do not share the same molecules for their amplification, and no mode competition occurs. Therefore all laser modes within that part of the gain profile, which is above the threshold, can oscillate simultaneously. The laser

a)   b)

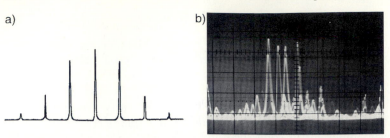

Fig. 5.25. (a) Stable multimode operation of a HeNe laser (exposure time: 1 s); (b) two short-time exposures of the multimode spectrum of an argon laser superimposed on the same film to demonstrate the randomly fluctuating mode distribution

output consists of all axial and transverse modes for which the total losses are less than the gain (Fig. 5.25a).

Real lasers do not represent these pure cases, but exhibit a gain profile that is a convolution of inhomogeneous and homogeneous broadening. It is the ratio of mode spacing δv to the homogeneous width Δv^{hom} that governs the strength of mode competition and that is crucial for the resulting single- or multi-mode operation. There is another reason why many lasers oscillate on many modes: if the gain exceeds the losses for higher transverse modes, mode competition between the modes TEM_{m_1,n_1} and TEM_{m_2,n_2} with $(m_1, n_1) \neq (m_2, n_2)$ is restricted because of their different spatial amplitude distributions. They gain their amplification from different regions of the active medium. This applies to laser types such as solid-state lasers (ruby or Nd:YAG lasers), flash-lamp-pumped dye lasers, or excimer lasers. In a nonconfocal resonator the frequencies of the transverse modes fill the gap between the TEM_{00} frequencies $v_a = (q + \frac{1}{2})c/(2nd)$ (Fig. 5.19). These transverse modes lead to a larger divergence of the laser beam, which is no longer a Gaussian-shaped beam.

The suppression of higher-order $\text{TEM}_{m,n}$ modes can be achieved by a proper choice of the resonator geometry, which has to be adapted to the cross section and the length L of the active medium (Sect. 5.4.2).

If only the axial modes TEM_{00} participate in the laser oscillation, the laser beam transmitted through the output mirrors has a Gaussian intensity profile (5.32), (5.42). It may still consist of many frequencies $v_a = qc/(2nd)$ within the spectral gain profile. The spectral bandwidth of a multimode laser oscillating on an atomic or molecular transition is comparable to that of an incoherent source emitting on this transition!

We illustrate this discussion by some examples:

Example 5.9
HeNe Laser at $\lambda = 632.8$ nm: The Doppler width of the Ne transition is about 1500 MHz, and the width of the gain profile above the threshold, which depends on the pump power, may be 1200 MHz. With a resonator length of $d = 100$ cm, the spacing of the longitudinal modes is $\delta v = c/2d = 150$ MHz.

If the higher transverse modes are suppressed by an aperture inside the resonator, seven to eight longitudinal modes reach the threshold. The homogeneous width $\Delta\nu^{\text{hom}}$ is determined by several factors: the natural linewidth $\Delta\nu_{\text{n}} = 20\,\text{MHz}$; a pressure broadening of about the same magnitude; and a power broadening, which depends on the laser intensity in the different modes. With $I/I_{\text{s}} = 10$, for example, we obtain with $\Delta\nu_0 = 30\,\text{MHz}$ a power-broadened linewidth of about $100\,\text{MHz}$, which is still smaller than the longitudinal modes spacing. The modes will therefore not compete strongly, and simultaneous oscillation of all longitudinal modes above threshold is possible. This is illustrated by Fig. 5.25a, which exhibits the spectrum of a HeNe laser with $d = 1\,\text{m}$, monitored with a spectrum analyzer and integrated over a time interval of $1\,\text{s}$.

Example 5.10

Argon Laser: Because of the high temperature in the high-current discharge (about $10^3\,\text{A/cm}^2$), the Doppler width of the Ar^+ transitions is very large (about 8 to $10\,\text{GHz}$). The homogeneous width $\Delta\nu^{\text{hom}}$ is also much larger than for the HeNe laser for two reasons: the long-range Coulomb interaction causes a large pressure broadening from electron–ion collisions and the high laser intensity ($10-100\,\text{W}$) in a mode results in appreciable power broadening. Both effects generate a homogeneous linewidth that is large compared to the mode spacing $\delta\nu = 125\,\text{MHz}$ for a commonly used resonator length of $d = 120\,\text{cm}$. The resulting mode competition in combination with the perturbations mentioned above cause the observed randomly fluctuating mode spectrum of the multimode argon laser. Figure 5.25b illustrates this by the superposition of two short-time exposures of the oscilloscope display of a spectrum analyzer taken at two different times.

Example 5.11

Dye Laser: The broad spectral gain profile of dye molecules in a liquid are predominantly homogeneously broadened (Sect. 3.7). About 10^5 modes of a laser resonator with $L = 75\,\text{cm}$ fall within a typical spectral width of $20\,\text{nm}$ ($\hat{=} 2 \times 10^{13}\,\text{Hz}$ at $\lambda = 600\,\text{nm}$). Without spectral hole burning and fluctuations of the optical length nd of the resonator, the laser would oscillate in a single mode at the center of the gain profile, despite the large number of possible modes. However, fluctuations of the refractive index n in the dye liquid cause corresponding perturbations of the frequencies and the coupling of the laser modes, which results in a time-dependent multimode spectrum; the emission jumps in a random way between different mode frequencies. In the case of pulsed lasers, the time-averaged spectrum of the dye laser emission fills more or less uniformly a broader spectral interval (about $1\,\text{nm}$) around the maximum of the gain profile. The spatial hole burning may result in oscillation of several groups of lines centered around the spatial hole-burning modes. In this case, the time-averaged frequency distribution generally does *not* result in a uniformly smoothed intensity profile $I(\lambda)$. In order to achieve tunable single-mode operation, extra wavelength-selective elements have to be inserted into the laser resonator (Sect. 5.4).

For spectroscopic applications of multimode lasers one has to keep in mind that the spectral interval $\Delta\nu$ within the bandwidth of the laser is, in general, not uniformly filled. This means that, contrary to an incoherent source, the intensity $I(\nu)$ is not a smooth function within the laser bandwidth but exhibits holes. This is particularly true for multimode dye lasers with Fabry–Perot-type resonators where standing waves are present and hole burning occurs (Sect. 5.3.4).

The spectral intensity distribution of the laser output is the superposition

$$I_{\mathrm{L}}(\omega, t) = \left| \sum_k A_k(t) \cos[\omega_k t + \phi_k(t)] \right|^2 , \tag{5.68}$$

of the oscillating modes, where the phases $\phi_k(t)$ and the amplitudes $A_k(t)$ may randomly fluctuate in time because of mode competition and mode-pulling effects.

The time average of the spectral distribution of the output intensity

$$\langle I(\omega) \rangle = \frac{1}{T} \int_0^T \left| \sum_k A_k(t) \cos[\omega_k t + \phi_k(t)] \right|^2 \mathrm{d}t , \tag{5.69}$$

reflects the gain profile of the laser transition. The necessary averaging time T depends on the buildup time of the laser modes. It is determined by the unsaturated gain and the strength of the mode competition. In the case of gas lasers, the average spectral width $\langle \Delta\nu \rangle$ corresponds to the Doppler width of the laser transition. The coherence length of such a multimode laser is comparable to that of a conventional spectral lamp where a single line has been filtered out.

If such a multimode laser is used for spectroscopy and is scanned, for instance, with a grating or prism inside the laser resonator (Sect. 5.5), through the spectral range of interest, this nonuniform spectral structure $I_{\mathrm{L}}(0)$ may cause artificial structures in the measured spectrum. In order to avoid this problem and to obtain a smooth intensity profile $I_{\mathrm{L}}(\nu)$, the length d of the laser resonator can be wobbled at the frequency $f > 1/\tau$, which should be larger than the inverse scanning time τ over a line in the investigated spectrum. This wobbling modulates all oscillating frequencies of the laser and results in a smoother time average, particularly, if $\tau > T$.

5.3.5 Mode Pulling

We now briefly discuss the frequency shift (called *mode pulling*) of the passive resonator frequencies by the presence of an active medium [5.35]. The phase shift for a standing wave with frequency ν_p and round-trip time T_p through a resonator with mirror separation d without an active medium is

$$\phi_p = 2\pi\nu_p T_p = 2\pi\nu_p 2d/c = m\pi \, , \qquad (5.70)$$

where the integer m determines the oscillating resonator mode. On insertion of an active medium with refractive index $n(\nu)$, the frequency ν_p changes to ν_a in such a way that the phase shift per round-trip remains

$$\phi_a = 2\pi\nu_a T_a = 2\pi\nu_a n(\nu_a)2d/c = m\pi \, . \qquad (5.71)$$

This gives the condition

$$\frac{\partial\phi}{\partial\nu}(\nu_a - \nu_p) + [\phi_a(\nu_a) - \phi_p(\nu_a)] = 0 \, . \qquad (5.72)$$

The index of refraction $n(\nu)$ is related to the absorption coefficient $\alpha(\nu)$ of a homogeneous absorption profile by the dispersion relation

$$n(\nu) = 1 + \frac{\nu_0 - \nu}{\Delta\nu_m}\frac{c}{2\pi\nu}\alpha(\nu) \, , \qquad (5.73)$$

where $\Delta\nu_m$ is the linewidth of the amplifying transition in the active medium. In case of inversion ($\Delta N < 0$), $\alpha(\nu)$ becomes negative and $n(\nu) < 1$ for $\nu < \nu_0$, while $n(\nu) > 1$ for $\nu > \nu_0$ (Fig. 5.26). Under stationary conditions, the total gain per pass $\alpha(\nu)L$ saturates to the threshold value, which equals the total losses γ. These losses determine the resonance width $\Delta\nu_r = \beta/2\pi = c\gamma/(4\pi d)$ of the cavity, see (5.54). We obtain from (5.70, 5.73) the final result for the frequency ν_a of a laser mode for laser transitions with homogeneous line broadening

$$\nu_a = \frac{\nu_r\Delta\nu_m + \nu_0\Delta\nu_r}{\Delta\nu_m + \Delta\nu_r} \, . \qquad (5.74)$$

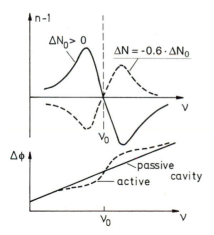

Fig. 5.26. Dispersion curves for absorbing transitions ($\Delta N < 0$) and amplifying transitions ($\Delta N > 0$) and phase shifts $\Delta\phi$ per round-trip in the passive and active cavity

The resonance width $\Delta \nu_r$ of gas laser resonators is of the order of 1 MHz, while the homogeneous width of the amplifying medium is about 100 MHz. Therefore, when $\Delta \nu_r \ll \Delta \nu_m$, (5.74) reduces to

$$\nu_a = \nu_r + \frac{\Delta \nu_r}{\Delta \nu_m}(\nu_0 - \nu_r) . \tag{5.75}$$

This demonstrates that the mode-pulling effect increases proportionally to the difference of cavity resonance frequency ν_r and central frequency ν_0 of the amplifying medium. At the slopes of the gain profile, the laser frequency is pulled towards the center.

5.4 Experimental Realization of Single-Mode Lasers

In the previous sections we have seen that without specific manipulation a laser generally oscillates in many modes, for which the gain exceeds the total losses. In order to select a single wanted mode, one has to suppress all others by increasing their losses to such an amount that they do not reach the oscillation threshold. The suppression of higher-order transverse TEM_{mn} modes demands actions other than the selection of a single longitudinal mode out of many other TEM_{00} modes.

Many types of lasers, in particular, gaseous lasers, may reach oscillation threshold for several atomic or molecular transitions. The laser can then simultaneously oscillate on these transitions [5.36]. In order to reach single-mode operation, one has to first select a single transition.

5.4.1 Line Selection

In order to achieve single-line oscillation in laser media that exhibit gain for several transitions, wavelength-selecting elements inside or outside the laser resonator can be used. If the different lines are widely separated in the spectrum, the selective reflectivity of the dielectric mirrors may already be sufficient to select a single transition. In the case of broadband reflectors or closely spaced lines, prisms, gratings, or Lyot filters are commonly utilized for wavelength selection. Figure 5.27 illustrates line selection by a prism in an argon laser. The different lines are refracted by the prism, and only the line that is vertically incident upon the end mirror is reflected back into itself and can reach the oscillation threshold, while all other lines are reflected out of the resonator. Turning the end reflector M_2 allows the desired line to be selected. To avoid reflection losses at the prism surfaces, a Brewster prism with $\tan \phi = 1/n$ is used, with the angle of incidence for both prism surfaces being Brewster's angle. The prism and the end mirror can be combined by coating the end face of a Brewster prism reflector (Fig. 5.27b). Such a device is called a *Littrow prism*.

Because most prism materials such as glass or quartz absorb in the infrared region, it is more convenient to use a Littrow grating (Sect. 4.1) as wavelength selector in this wavelength range. Figure 5.28 illustrates the line selection in

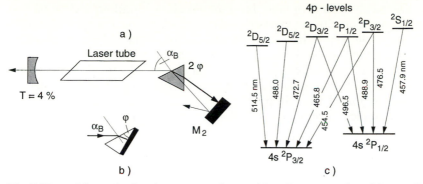

Fig. 5.27a–c. Line selection in an argon laser with a Brewster prism (**a**) or a Littrow prism reflector (**b**). Term diagram of laser transition in Ar$^+$ (**c**)

a CO_2 laser, which can oscillate on many rotational lines of a vibrational transition. Often the laser beam is expanded by a proper mirror configuration in order to cover a larger number of grating grooves, thus increasing the spectral resolution (Sect. 4.1). This has the further advantage that the power density is lower and damage of the grating is less likely.

If some of the simultaneously oscillating laser transitions share a common upper or lower level, such as the lines 1, 2, and 3 in Fig. 5.27c and Fig. 5.29a, gain competition diminishes the output of each line. In this case, it is advantageous to use *intracavity* line selection in order to suppress all but one of the competing transitions. Sometimes, however, the laser may oscillate on cascade transitions (Fig. 5.29b). In such a case, the laser transition $1 \rightarrow 2$ increases the population of level 2 and therefore *enhances* the gain for the transition $2 \rightarrow 3$

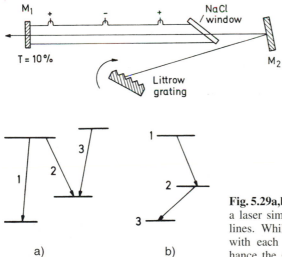

Fig. 5.28. Selection of CO_2 laser lines corresponding to different rotational transitions by a Littrow grating

Fig. 5.29a,b. Schematic level diagram for a laser simultaneously oscillating on several lines. While in (a) the transitions compete with each other for gain, those in (b) enhance the gain for the other line

[5.37]. Obviously, it is then more favorable to allow multiline oscillation and to select a single line by an external prism or grating. Using a special mounting design, it can be arranged so that no deflection of the output beam occurs when the multiline output is tuned from one line to the other [5.38].

For lasers with a broad continuous spectral gain profile, the preselecting elements inside the laser resonator restrict laser oscillation to a spectral interval, which is a fraction of the gain profile.

Some examples illustrate the situation (see also Sect. 5.7):

Example 5.12

HeNe Laser: The HeNe laser is probably the most thoroughly investigated gas laser [5.39]. From the level scheme (Fig. 5.30), which uses the Paschen notation [5.40], we see that two transitions around $\lambda = 3.39\,\mu$m and the visible transitions at $\lambda = 0.6328\,\mu$m share a common *upper* level. Suppression of the $3.39\,\mu$m lines therefore enhances the output power at $0.6328\,\mu$m. The $1.15\,\mu$m and the $0.6328\,\mu$m lines, on the other hand, share a common *lower* level and also compete for gain, since both laser transitions increase the lower-level population and therefore decrease the inversion. If the 3.3903-μm transition is suppressed, e.g., by placing an absorbing CH_4 cell inside the resonator, the population of the upper $3s_2$ level increases, and a new line at $\lambda = 3.3913\,\mu$m reaches the threshold.

This laser transition populates the $3p_4$ level and produces gain for another line at $\lambda = 2.3951\,\mu$m. This last line only oscillates together with the 3.3913-μm one, which acts as pumping source. This is an example of cascade transitions in laser media [5.37], as depicted in Fig. 5.29b.

The homogeneous width of the laser transitions is mainly determined by pressure and power broadening. At total pressures of above 5 mb and an

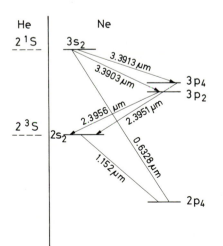

Fig. 5.30. Level diagram of the HeNe laser system in Paschen notation showing the most intense laser transitions

intracavity power of 200 mW, the homogeneous linewidth for the transition $\lambda = 632.8$ nm is about 200 MHz, which is still small compared with the Doppler width $\Delta\nu_D = 1500$ MHz. In single-mode operation, one can obtain about 20% of the multimode power [5.41]. This roughly corresponds to the ratio $\Delta\nu_h/\Delta\nu_D$ of homogeneous to inhomogeneous linewidth above the threshold. The mode spacing $\delta\nu = \frac{1}{2}c/d$ equals the homogeneous linewidth for $d = d^* = \frac{1}{2}c/\Delta\nu_h$. For $d < d^*$, stable multimode oscillation is possible; for $d > d^*$, mode competition occurs.

Example 5.13

Argon Laser: The discharge of a cw argon laser exhibits gain for more than 15 different transitions. Figure 5.27c shows part of the energy level diagram, illustrating the coupling of different laser transitions. Since the lines at 514.5 nm, 488.0 nm, and 465.8 nm share the same lower level, suppression of the competing lines enhances the inversion and the output power of the selected line. The mutual interaction of the various laser transitions has therefore been studied extensively [5.42, 5.43] in order to optimize the ouput power. Line selection is generally achieved with an internal Brewster prism (Fig. 5.27 and Fig. 5.39b). The homogeneous width $\Delta\nu_h$ is mainly caused by collision broadening due to electron–ion collisions and saturation broadening. Additional broadening and shifts of the ion lines result from ion drifts in the field of the discharge. At intracavity intensities of 350 W/cm^2, which correspond to about 1 W output power, appreciable saturation broadening increases the homogeneous width, which may exceed 1000 MHz. This explains why the output at single-mode operation may reach 30% of the multimode output on a single line [5.44].

Example 5.14

CO$_2$ Laser: A section of the level diagram is illustrated in Fig. 5.31. The vibrational levels (v_1, v_2^l, v_3) are characterized by the number of quanta in the three normal vibrational modes. The upper index of the degenerate vibration v_2 gives the quantum number of the corresponding vibrational angular momentum l [5.45]. Laser oscillation is achieved on many rotational lines within two vibrational transitions $(v_1, v_2^l, v_3) = 00^01 \rightarrow 10^00$ and $00^01 \rightarrow 02^00$ [5.46–5.48]. Without line selection, generally only the band around 961 cm^{-1} (10.6 µm) appears because these transitions exhibit larger gain. The laser oscillation depletes the population of the 00^01 vibrational level and suppresses laser oscillation on the second transition, because of to gain competition. With internal line selection (Fig. 5.28), many more lines can successively be optimized by turning the wavelength-selecting grating. The output power of each line is then higher than that of the same line in multiline operation. Because of the small Doppler width (66 MHz), the free spectral range $\delta\nu = \frac{1}{2}c/d^*$ is already larger than the width of the gain profile for $d^* < 200$ cm. For such resonators, the mirror separation d has to be ad-

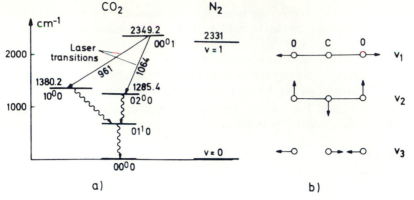

Fig. 5.31a,b. Level diagram and laser transitions in the CO_2 molecule (**a**) and normal vibrations (ν_1, ν_2, ν_3) (**b**)

justed to tune the resonator eigenfrequency $\nu_R = \frac{1}{2}qc/d^*$ (where q is an integer) to the center of the gain profile. If the resonator parameters are properly chosen to suppress higher transverse modes, the CO_2 laser then oscillates on a single longitudinal mode.

5.4.2 Suppression of Transverse Modes

Let us first consider the selection of *transverse* modes. In Sect. 5.2.3 it was shown that the higher transverse TEM_{mnq} modes have radial field distributions that are less and less concentrated along the resonator axis with increasing transverse order n or m. This means that their diffraction losses are much higher than those of the fundamental modes TEM_{00q} (Fig. 5.11). The field distribution of the modes and therefore their diffraction losses depend on the resonator parameters such as the radii of curvature of the mirrors R_i, the mirror separation d, and, of course, the Fresnel number N (Sect. 5.2.1). Only those resonators that fulfill the stability condition [5.1, 5.24]

$$0 < g_1 g_2 < 1 \quad \text{or} \quad g_1 g_2 = 0 \quad \text{with} \quad g_i = (1 - d/R_i)$$

have finite spot sizes of the TEM_{00} field distributions inside the resonator (Sect. 5.2.6). The choice of proper resonator parameters therefore establishes the beam waist w of the fundamental TEM_{00q} mode and the radial extension of the higher-order TEM_{mn} modes. This, in turn, determines the diffraction losses of the modes.

In Fig. 5.32, the ratio γ_{10}/γ_{00} of the diffraction losses for the TEM_{10} and the TEM_{00} modes in a symmetric resonator with $g_1 = g_2 = g$ is plotted for different values of g as a function of the Fresnel number N. From this diagram one can obtain, for any given resonator, the diameter $2a$ of an aperture

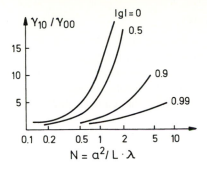

Fig. 5.32. Ratio γ_{10}/γ_{00} of diffraction losses for the TEM_{10} and TEM_{00} modes in symmetric resonators as a function of the Fresnel number N for different resonator parameters $g = 1 - d/R$

that suppresses the TEM_{10} mode but still has sufficiently small losses for the fundamental TEM_{00} mode with beam radius w. In gas lasers, the diameter $2a$ of the discharge tube generally forms the limiting aperture. One has to choose the resonator parameters in such a way that $a \simeq 3w/2$ because this assures that the fundamental mode nearly fills the whole active medium, but still suffers less than 1% diffraction losses (Sect. 5.2.6).

Because the frequency separation of the transverse modes is small and the TEM_{10q} mode frequency is separated from the TEM_{00q} frequency by less than the homogeneous width of the gain profile, the fundamental mode can partly saturate the inversion at the distance r_m from the axis, where the TEM_{10q} mode has its field maximum. The resulting transverse mode competition (Fig. 5.33) reduces the gain for the higher transverse modes and may suppress their oscillation even if the unsaturated gain exceeds the losses. The restriction for the maximum-allowed aperture diameter is therefore less stringent. The resonator geometry of many commercial lasers has already been designed in such a way that "single-transverse-mode" operation is obtained. The laser can, however, still oscillate on several longitudinal modes, and for true single-mode operation, the next step is to suppress all but one of the longitudinal modes.

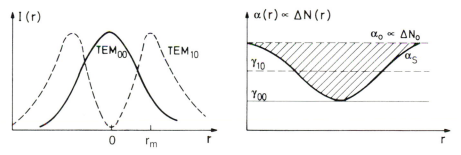

Fig. 5.33. Transverse gain competition between the TEM_{00} and TEM_{10} modes

5.4.3 Selection of Single Longitudinal Modes

From the discussion in Sect. 5.3 it should have become clear that simultaneous oscillation on several longitudinal modes is possible when the inhomogeneous width $\Delta \nu_g$ of the gain profile exceeds the mode spacing $\frac{1}{2}c/d$ (Fig. 5.20). A simple way to achieve single-mode operation is therefore the reduction of the resonator length $2d$ such that the width $\Delta \nu_g$ of the gain profile above threshold becomes smaller than the free spectral range $\delta \nu = \frac{1}{2}c/d$ [5.49].

If the resonator frequency can be tuned to the center of the gain profile, single-mode operation can be achieved even with the double length $2d$, because then the two neighboring modes just fail to reach the threshold (Fig. 5.34). However, this solution for the achievement of single-mode operation has several drawbacks. Since the length L of the active medium cannot be larger than d ($L \leq d$), the threshold can only be reached for transitions with a high gain. The output power, which is proportional to the active mode volume, is also small in most cases. For single-mode lasers with higher output powers, other methods are therefore preferable. We distinguish between *external* and *internal* mode selection.

When the output of a multimode laser passes through an external spectral filter, such as an interferometer or a spectrometer, a single mode can be selected. For perfect selection, however, high suppression of the unwanted modes and high transmission of the wanted mode by the filter are required. This technique of external selection has the further disadvantage that only part of the total laser output power can be used. *Internal* mode selection with spectral filters inside the laser resonator completely suppresses the unwanted modes even when the losses exceed the gain. Furthermore, the output power of a single-mode laser is generally higher than the power in this mode at multimode oscillation because the total inversion $V \cdot \Delta N$ in the active volume V is no longer shared by many modes, as is the case for multimode operation with gain competition.

In single-mode operation with internal mode selection, we can expect output powers that reach the fraction $\Delta \nu_{\mathrm{hom}}/\Delta \nu_g$ of the multimode power, where

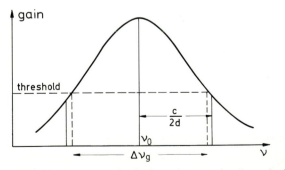

Fig. 5.34. Single longitudinal mode operation by reducing the cavity length d to a value where the mode spacing exceeds half of the gain profile width above threshold

$\Delta\nu_{hom}$ is the homogeneous width within the inhomogeneous gain profile. This width $\Delta\nu_{hom}$ becomes even larger for single-mode operation because of power broadening by the more intense mode. In an argon-ion laser, for example, one can obtain up to 30% of the multimode power in a single mode with internal mode selection.

This is the reason why virtually all single-mode lasers use *internal* mode selection. We now discuss some experimental possibilities that allow stable single-mode operation of lasers with internal mode selection. As pointed out in the previous section, all methods for achieving single-mode operation are based on mode suppression by increasing the losses beyond the gain for all but the wanted mode. A possible realization of this idea is illustrated in Fig. 5.35, which shows longitudinal mode selection by a tilted plane-parallel etalon (thickness t and refractive index n) inside the laser resonator [5.50]. In Sect. 4.2.5, it was shown that such an etalon has transmission maxima at those wavelengths λ_m for which

$$m\lambda_m = 2nt\cos\theta , \tag{5.76}$$

for all other wavelengths the reflection losses should dominate the gain.

If the free spectral range of the etalon,

$$\delta\lambda = 2nt\cos\theta\left(\frac{1}{m} - \frac{1}{m+1}\right) = \frac{\lambda_m}{m+1} , \tag{5.77}$$

is larger than the spectral width $|\lambda_1 - \lambda_2|$ of the gain profile above the threshold, only a single mode can oscillate (Fig. 5.36). Since the wavelength λ is also determined by the resonator length d, the tilting angle θ has to be adjusted so that

$$2nt\cos\theta/m = 2d/q \quad \text{(where } q \text{ is an integer)}$$

$$\Rightarrow \cos\theta = \frac{m}{q}\cdot\frac{d}{n\cdot t} , \tag{5.78}$$

which means that the transmission peak of the etalon has to coincide with an eigenresonance of the laser resonator.

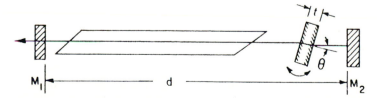

Fig. 5.35. Single-mode operation by inserting a tilted etalon inside the laser resonator

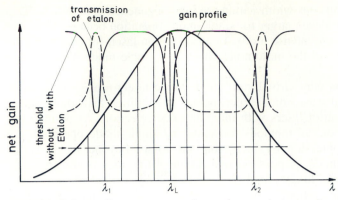

Fig. 5.36. Gain profile, resonator modes, and transmission peaks of the intracavity etalon (*dashed curve*). Also shown are the threshold curves with and without etalon

Example 5.15

In the argon-ion laser the width of the gain profile is about 8 GHz. With a free spectral range of $\Delta \nu = c/(2nt) = 10$ GHz of the intracavity etalon, single-mode operation can be achieved. This implies with $n = 1.5$ a thickness $t = 1$ cm.

The finesse F^* of the etalon has to be sufficiently high to ensure for the modes adjacent to the selected mode losses that overcome their gain (Fig. 5.36). Fortunately, in many cases their gain is already reduced by the oscillating mode due to gain competition. This allows the less stringent demand that the losses of the etalon must only exceed the saturated gain at a distance $\Delta \nu \geq \Delta \nu_{\mathrm{hom}}$ away from the transmission peak.

Often a Michelson interferometer coupled by a beam splitter BS to the laser resonator is used for mode selection (Fig. 5.37). The free spectral range

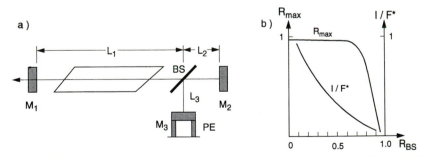

Fig. 5.37a,b. Mode selection with a Fox–Smith selector: (a) experimental setup; (b) maximum reflectivity and inverted finesse $1/F^*$ of the Michelson-type reflector as a function of the reflectivity R_{BS} of the beam splitter for $R_2 = R_3 = 0.99$ and $A_{\mathrm{BS}} = 0.5\%$

$\delta v = \frac{1}{2}c/(L_2 + L_3)$ of this *Fox–Smith cavity* [5.51] again has to be broader than the width of the gain profile. With a piezoelement PE, the mirror M_3 can be translated by a few microns to achieve resonance between the two coupled resonators. For the resonance condition

$$(L_1 + L_2)/q = (L_2 + L_3)/m \quad \text{(where } m \text{ and } q \text{ are integers)}, \tag{5.79}$$

the partial wave $M_1 \rightarrow BS$, reflected by BS, and the partial wave $M_3 \rightarrow BS$, transmitted through BS, interfere destructively. This means that for the resonance condition (5.79) the reflection losses by BS have a minimum (in the ideal case they are zero). For all other wavelengths, however, these losses are larger, and if they exceed the gain, single-mode oscillation is achieved [5.52].

In a more detailed discussion the absorption losses A_{BS}^2 of the beam splitter BS cannot be neglected, since they cause the maximum reflectance R of the Fox–Smith cavity to be less than 1. Similar to the derivation of (4.76), the reflectance of the Fox–Smith selector, which acts as a wavelength-selecting laser reflector, can be calculated to be [5.53]

$$R = \frac{T_{BS}^2 R_2 (1 - A_{BS})^2}{1 - R_{BS}\sqrt{R_2 R_3} + 4 R_{BS}\sqrt{R_2 R_3}\sin^2 \phi/2}. \tag{5.80}$$

Figure 5.37b exhibits the reflectance R_{max} for $\phi = 2m\pi$ and the additional losses of the laser resonator introduced by the Fox–Smith cavity as a function of the beam splitter reflectance R_{BS}. The finesse F^* of the selecting device is also plotted for $R_2 = R_3 = 0.99$ and $A_{BS} = 0.5\%$. The spectral width Δv of the reflectivity maxima is determined by

$$\Delta v = \delta v / F^* = c/[2F^*(L_2 + L_3)]. \tag{5.81}$$

There are several other resonator-coupling schemes that can be utilized for mode selection. Figure 5.38 compares some of them, together with their frequency-selective losses [5.54].

In case of multiline lasers (e.g., argon or krypton lasers), line selection and mode selection can be simultaneously achieved by a combination of prism and Michelson interferometers. Figure 5.39 illustrates two possible realizations. The first replaces mirror M_2 in Fig. 5.37 by a Littrow prism reflector (Fig. 5.39a). In Fig. 5.39b, the front surface of the prism acts as beam splitter, and the two coated back surfaces replace the mirrors M_2 and M_3. The incident wave is split into the partial beams 4 and 2. After being reflected by M_2, beam 2 is again split into 3 and 1. Destructive interference between beams 4 and 3, after reflection from M_3, occurs if the optical path difference $\Delta s = 2n(S_2 + S_3) = m\lambda$. If both beams have equal amplitudes, no light is emitted in the direction of beam 4. This means that all the light is reflected back into the incident direction and the device acts as a wavelength-selective reflector, analogous to the Fox–Smith cavity [5.55]. Since the wavelength λ depends on the optical path length $n(L_2 + L_3)$, the prism has to be temperature stabilized to achieve wavelength-stable, single-mode operation. The whole prism is therefore embedded in a temperature-stabilized oven.

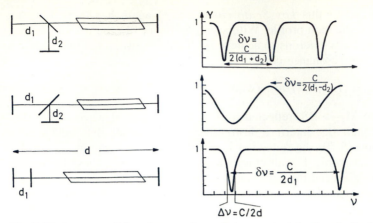

Fig. 5.38. Some possible schemes of coupled resonators for longitudinal mode selection, with their frequency-dependent losses. For comparison the eigenresonances of the long laser cavity with a mode spacing $\Delta v = c/2d$ are indicated

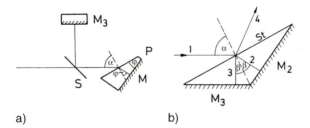

a)   b)

Fig. 5.39. (a) Simultaneous line selection and mode selection by a combination of prism selector and Michelson-type interferometer; **(b)** compact arrangement

For lasers with a broad gain profile, one wavelength-selecting element alone may not be sufficient to achieve single-mode operation, therefore one has to use a proper combination of different dispersing elements. With preselectors, such as prisms, gratings, or Lyot filters, the spectral range of the effective gain profile is narrowed down to a width that is comparable to that of the Doppler width of fixed-frequency gas lasers. Figure 5.40 represents a possible scheme that has been realized in practice. Two prisms are used as preselector to narrow the spectral width of a cw dye laser [5.56]; two etalons with different thicknesses t_1 and t_2 are used to achieve stable single-mode operation. Figure 5.40b illustrates the mode selection, depicting schematically the gain profile narrowed by the prisms and the spectral transmission curves of the two etalons. In the case of the dye laser with its homogeneous gain profile, not every resonator mode can oscillate, but only those that draw gain from the spatial hole-burning effect (Sect. 5.3.3). The "suppressed modes" at the bottom of Fig. 5.40 represent these spatial hole-burning modes that would

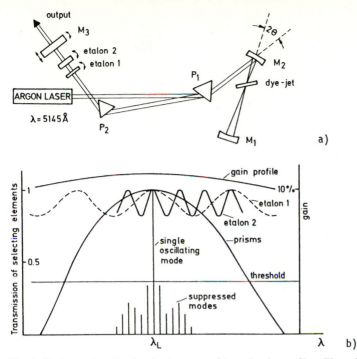

Fig. 5.40a,b. Mode selection in the case of broad gain profiles. The prisms narrow the net gain profile and two etalons enforce single-mode operation: (**a**) experimental realization for a jet stream cw dye laser; (**b**) schematic diagram of gain profile and transmission curves of the two etalons

simultaneously oscillate without the etalons. The transmission maxima of the two etalons have, of course, to be at the same wavelength λ_L. This can be achieved by choosing the correct tilting angles θ_1 and θ_2 such that

$$nt_1 \cos \theta_1 = m_1 \lambda_L , \quad \text{and} \quad nt_2 \cos \theta_2 = m_2 \lambda_L . \tag{5.82}$$

Example 5.16

The two prisms narrow the spectral width of the gain profile above threshold to about 100 GHz. If the free spectral range of the thin etalon 1 is 100 GHz ($\hat{=} \Delta\lambda \sim 1$ nm at $\lambda = 600$ nm) and that of the thick etalon 2 is 10 GHz, single-mode operation of the cw dye laser can be achieved. This demands $t_1 = 0.1$ cm and $t_2 = 1$ cm for $n = 1.5$.

Commercial cw dye laser systems (Sect. 5.5) generally use a different realization of single-mode operation (Fig. 5.41). The prisms are replaced by a birefringent filter, which is based on the combination of three Lyot filters (Sect. 4.2.9), and the thick etalon is substituted by a Fabry–Perot interferome-

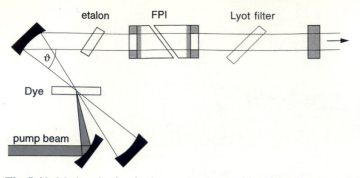

Fig. 5.41. Mode selection in the cw dye laser with a folded cavity using a birefringent filter, a tilted etalon, and a prism FPI (Coherent model 599). The folding angle ϑ is chosen for optimum compensation of astigmatism introduced by the dye jet

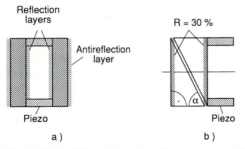

Fig. 5.42a,b. Fabry–Perot interferometer tuned by a piezocylinder: (**a**) two plane-parallel plates with inner reflecting surfaces; (**b**) two Brewster prisms with the outer coated surfaces forming the FPI reflecting planes

ter with the thickness t controllable by piezocylinders (Fig. 5.42). This is done because the walk-off losses of an etalon increase according to (4.60) with the square of the tilting angle α and the etalon thickness t. They may become intolerably high if a large, uninterrupted tuning range shall be achieved by tilting of the etalon. Therefore the long intracavity FPI (Fig. 5.41) is kept at a fixed, small tilting angle while its transmission peak is tuned by changing the separation t between the reflecting surfaces.

In order to minimize the air gap between the reflecting surfaces of the FPI, the prism construction of Fig. 5.42b is often used, in which the small air gap is traversed by the laser beam at Brewster's angle to avoid reflection losses [5.57]. This design minimizes the influence of air pressure variations on the transmission peak wavelength λ_L.

Figure 5.43 depicts the experimental arrangement for narrow-band operation of an excimer laser-pumped dye laser; the beam is expanded to fill the whole grating. Because of the higher spectral resolution of the grating (compared with a prism) and the wider mode spacing from the short cavity, a single etalon inside or outside the laser resonator may be sufficient to select a single mode [5.58].

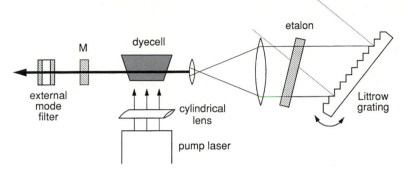

Fig. 5.43. Short Hänsch-type dye laser cavity with Littrow grating and mode selection either with an internal etalon or an external FPI as "mode filter" [5.58]

There are many more experimental possibilities for achieving single-mode operation. For details, the reader is referred to the extensive literature on this subject, which can be found, for instance, in the excellent reviews on mode selection and single-mode lasers by Smith [5.54] or Goldsborough [5.59] and in [5.60, 5.61].

5.4.4 Intensity Stabilization

The intensity $I(t)$ of a cw laser is not completely constant, but shows periodic and random fluctuations and also, in general, long-term drifts. The reasons for these fluctuations are manifold and may, for example, be due to an insufficiently filtered power supply, which results in a ripple on the discharge current of the gas laser and a corresponding intensity modulation. Other noise sources are instabilities of the gas discharge, dust particles diffusing through the laser beam inside the resonator, and vibrations of the resonator mirrors. In multimode lasers, internal effects, such as mode competition, also contribute to noise. In cw dye lasers, density fluctuations in the dye jet stream and air bubbles are the main cause of intensity fluctuations.

Long-term drifts of the laser intensity may be caused by slow temperature or pressure changes in the gas discharge, by thermal detuning of the resonator, or by increasing degradation of the optical quality of mirrors, windows, and other optical components in the resonator. All these effects give rise to a noise level that is well above the theoretical lower limit set by the photon noise. Since these intensity fluctuations lower the signal-to-noise ratio, they may become very troublesome in many spectroscopic applications, therefore one should consider steps that reduce these fluctuations by stabilizing the laser intensity.

Of the various possible methods, we shall discuss two that are often used for intensity stabilization. They are schematically depicted in Fig. 5.44. In the first method, a small fraction of the output power is split by the beam splitter

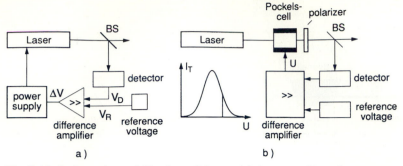

Fig. 5.44a,b. Intensity stabilization of lasers (**a**) by controlling the power supply, and (**b**) by controlling the transmission of a Pockels cell

BS to a detector (Fig. 5.44a). The detector output V_D is compared with a reference voltage V_R and the difference $\Delta V = V_D - V_R$ is amplified and fed to the power supply of the laser, where it controls the discharge current. The servo loop is effective in a range where the laser intensity increases with increasing current.

The upper frequency limit of this stabilization loop is determined by the capacitances and inductances in the power supply and by the time lag between the current increase and the resulting increase of the laser intensity. The lower limit for this time delay is given by the time required by the gas discharge to reach a new equilibrium after the current has been changed. It is therefore not possible with this method to stabilize the system against fluctuations of the gas discharge. For most applications, however, this stabilization technique is sufficient; it provides an intensity stability where the fluctuations are less than 05%.

To compensate fast intensity fluctuations, another technique, illustrated in Fig. 5.44b, is more suitable. The output from the laser is sent through a Pockels cell, which consists of an optically anisotropic crystal placed between two linear polarizers. An external voltage applied to the electrodes of the crystal causes optical birefringence, which rotates the polarization plane of the transmitted light and therefore changes the transmittance through the second polarizer. If part of the transmitted light is detected, the amplified detector signal can be used to control the voltage U at the Pockels cell. Any change of the transmitted intensity can be compensated by an opposite transmission change of the Pockels cell. This stabilization control works up to frequencies in the megahertz range if the feedback-control electronics are sufficiently fast. Its disadvantage is an intensity loss of 20% to 50% because one has to bias the Pockels cell to work on the slope of the transmission curve (Fig. 5.44b).

Figure 5.45 sketches how the electronic system of a feedback control can be designed to optimize the response over the whole frequency spectrum of the input signals. In principle, three operational amplifiers with different frequency responses are put in parallel. The first is a common *proportional* amplifier, with an upper frequency determined by the resonance frequency of

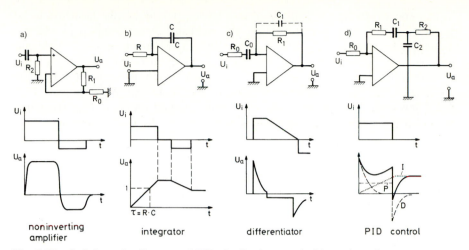

Fig. 5.45a–d. Schematic diagram of PID feedback control: (**a**) noninverting proportional amplifier; (**b**) integrator; (**c**) differentiating amplifier; (**d**) complete PID circuit that combines the functions (a–c)

the mirror piezosystem. The second is an integral amplifier with the output

$$U_{\text{out}} = \frac{1}{RC} \int_0^T U_{\text{in}}(t)\,dt \ .$$

This amplifier is necessary to bring the signal, which is proportional to the wavelength deviation ($U_{\text{in}} \approx \Delta\lambda$), really back to zero. This cannot be performed with a proportional amplifier. The third amplifier is a differentiating device that takes care of fast peaks in the perturbations. All three functions can be combined in a system called *PID control* [5.62, 5.63], which is widely used for intensity stabilization and wavelength stabilization of lasers.

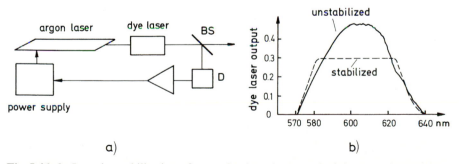

Fig. 5.46a,b. Intensity stabilization of a cw dye laser by control of the argon laser power: (**a**) experimental arrangement; (**b**) stabilized and unstabilized dye laser output $P(\lambda)$ when the dye laser is tuned across its spectral gain profile

For spectroscopic applications of dye lasers, where the dye laser has to be tuned through a large spectral range, the intensity change caused by the decreasing gain at both ends of the gain profile may be inconvenient. An elegant way to avoid this change of $I_L(\lambda)$ with λ is to stabilize the dye laser output by controlling the argon laser power (Fig. 5.46). Since the servo control must not be too fast, the stabilization scheme of Fig. 5.46a can be employed. Figure 5.46b demonstrates how effectively this method works if one compares the stabilized with the unstabilized intensity profile $I(\lambda)$ of the dye laser.

5.4.5 Wavelength Stabilization

For many applications in high-resolution laser spectroscopy, it is essential that the laser wavelength stays as stable as possible at a preselected value λ_0. This means that the fluctuations $\Delta\lambda$ around λ_0 should be smaller than the molecular linewidths that are to be resolved. For such experiments only *single-mode* lasers can, in general, be used, because in most multimode lasers the momentary wavelengths fluctuate and only the time-averaged envelope of the spectral output profile is defined, as has been discussed in the previous sections. This stability of the wavelength is important both for fixed-wavelength lasers, where the laser wavelength has to be kept a time-independent value λ_0, as well as for tunable lasers, where the fluctuations $\Delta\lambda = |\lambda_L - \lambda_R(t)|$ around a controlled tunable wavelength $\lambda_R(t)$ have to be smaller than the resolvable spectral interval.

In this section we discuss some methods of wavelength stabilization with their advantages and drawbacks. Since the laser frequency $\nu = c/\lambda$ is directly related to the wavelength, one often speaks about *frequency* stabilization, although for most methods in the visible spectral region, it is not the frequency but the wavelength that is directly measured and compared with a reference standard. There are, however, new stabilization methods that rely directly on absolute frequency measurements (Sect. 14.7).

In Sect. 5.3 we saw that the wavelength λ or the frequency ν of a longitudinal mode in the active resonator is determined by the mirror separation d and the refractive indices n_2 of the active medium with length L and n_1 outside the amplifying region. The resonance condition is

$$q\lambda = 2n_1(d-L) + 2n_2L . \tag{5.83}$$

For simplicity, we shall assume that the active medium fills the whole region between the mirrors. Thus (5.83) reduces, with $L = d$ and $n_2 = n_1 = n$, to

$$q\lambda = 2nd , \quad \text{or} \quad \nu = qc/(2nd) . \tag{5.84}$$

Any fluctuation of n or d causes a corresponding change of λ and ν. We obtain from (5.84)

$$\frac{\Delta\lambda}{\lambda} = \frac{\Delta d}{d} + \frac{\Delta n}{n} , \quad \text{or} \quad -\frac{\Delta\nu}{\nu} = \frac{\Delta d}{d} + \frac{\Delta n}{n} . \tag{5.85}$$

Example 5.17

To illustrate the demands of frequency stabilization, let us assume that we want to keep the frequency $\nu = 6 \times 10^{14}$ Hz of an argon laser constant within 1 MHz. This means a relative stability of $\Delta\nu/\nu = 1.6 \times 10^{-9}$ and implies that the mirror separation of $d = 1$ m has to be kept constant within 1.6 nm!

From this example it is evident that the requirements for such stabilization are by no means trivial. Before we discuss possible experimental solutions, let us consider the causes of fluctuations or drifts in the resonator length d or the refractive index n. If we could reduce or even eliminate these causes, we would already be well on the way to achieving a stable laser frequency. We shall distinguish between *long-term drifts* of d and n, which are mainly caused by temperature drifts or slow pressure changes, and *short-term fluctuations* caused, for example, by acoustic vibrations of mirrors, by acoustic pressure waves that modulate the refractive index, or by fluctuations of the discharge in gas lasers or of the jet flow in dye lasers.

To illustrate the influence of long-term drifts, let us make the following estimate. If α is the thermal expansion coefficient of the material (e.g., quartz or invar rods), which defines the mirror separation d, the relative change $\Delta d/d$ for a possible temperature change ΔT is, under the assumption of linear thermal expansion,

$$\Delta d/d = \alpha \Delta T .\tag{5.86}$$

Table 5.2 compiles the thermal expansion coefficients for some commonly used materials.

Example 5.18

For invar, with $\alpha = 1 \times 10^{-6}$ K^{-1}, we obtain from (5.86) for $\Delta T = 0.1$ K a relative distance change of $\Delta d/d = 10^{-7}$, which gives for Example 5.17 a frequency drift of 60 MHz.

Table 5.2. Linear thermal expansion coefficient of some relevant materials at room temperature $T = 20°C$

Material	$\alpha[10^{-6}$ K$^{-1}]$	Material	$\alpha[10^{-6}$ K$^{-1}]$
Aluminum	23	BeO	6
Brass	19	Invar	1.2
Steel	11–15	Soda-lime glass	5–8
Titanium	8.6	Pyrex glass	3
Tungsten	4.5	Fused quartz	0.4–0.5
Al$_2$O$_3$	5	Cerodur	< 0.1

If the laser wave inside the cavity travels a path length $d - L$ through air at atmospheric pressure, any change Δp of the air pressure results in the change

$$\Delta s = (d - L)(n - 1)\Delta p/p \,, \quad \text{with} \quad \Delta p/p = \Delta n/(n - 1) \,, \tag{5.87}$$

of the optical path length between the resonator mirrors.

Example 5.19

With $n = 1.00027$ and $d - L = 0.2d$, which is typical for gas lasers, we obtain from (5.85) and (5.87) for pressure changes of $\Delta p = 3$ mbar (which can readily occur during one hour, particularly in air-conditioned rooms)

$$\Delta\lambda/\lambda = -\Delta\nu/\nu \approx (d - L)\Delta n/(nd) \geq 1.5 \times 10^{-7} \,.$$

For our example above, this means a frequency change of $\Delta\nu \geq 90$ MHz. In cw dye lasers, the length L of the active medium is negligible compared with the resonator length d, therefore we can take $d - L \simeq d$. This implies for the same pressure change a frequency drift that is five times larger than estimated above.

To keep these long-term drifts as small as possible, one has to choose distance holders for the resonator mirrors with a minimum thermal expansion coefficient α. A good choice is, for example, the recently developed cerodur–quartz composition with a temperature-dependent $\alpha(T)$ that can be made zero at room temperature [5.64]. Often massive granite blocks are used as support for the optical components; these have a large heat capacity with a time constant of several hours to smoothen temperature fluctuations. To minimize pressure changes, the whole resonator must be enclosed by a pressure-tight container, or the ratio $(d - L)/d$ must be chosen as small as possible. However, we shall see that such long-term drifts can be mostly compensated by electronic servo control if the laser wavelength is locked to a constant reference wavelength standard.

A more serious problem arises from the short-term fluctuations, since these may have a broad frequency spectrum, depending on their causes, and the frequency response of the electronic stabilization control must be adapted to this spectrum. The main contribution comes from acoustical vibrations of the resonator mirrors. The whole setup of a wavelength-stabilized laser should therefore be vibrationally isolated as much as possible. There are commercial optical tables with pneumatic damping, in their more sophisticated form even electronically controlled, which guarantee a stable setup for frequency-stabilized lasers. A homemade setup is considerably cheaper: Fig. 5.47 illustrates a possible table mount for the laser system as employed in our laboratory. The optical components are mounted on a heavy granite plate, which rests in a flat container filled with sand to damp the eigenresonances of the granite block. Styrofoam blocks and acoustic damping elements prevent room vibrations from being transferred to the system. The optical system is protected against direct sound waves through the air, air turbulence, and dust

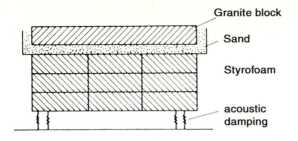

Granite block

Sand

Styrofoam

acoustic
damping

Fig. 5.47. Experimental realization of an acoustically isolated table for a wavelength-stabilized laser system

by a dust-free solid cover resting on the granite plate. A filtered laminar air flow from a flow box above the laser table avoids dust and air turbulence and increases the passive stability of the laser system considerably.

The high-frequency part of the noise spectrum is mainly caused by fast fluctuations of the refractive index in the discharge region of gas lasers or in the liquid jet of cw dye lasers. These perturbations can only be reduced partly by choosing optimum discharge conditions in gas lasers. In jet-stream dye lasers, density fluctuations in the free jet, caused by small air bubbles or by pressure fluctuations of the jet pump and by surface waves along the jet surfaces, are the main causes of fast laser frequency fluctuations. Careful fabrication of the jet nozzle and filtering of the dye solution are essential to minimize these fluctuations.

All the perturbations discussed above cause fluctuations of the optical path length inside the resonator that are typically in the nanometer range. In order to keep the laser wavelength stable, these fluctuations must be compensated by corresponding changes of the resonator length d. For such controlled and fast length changes in the nanometer range, piezoceramic elements are mainly used [5.65, 5.66]. They consist of a piezoelectric material whose length in an external electric field changes proportionally to the field strength. Either cylindrical plates are used, where the end faces are covered by silver coatings that provide the electrodes or a hollow cylinder is used, where the coatings cover the inner and outer wall surfaces (Fig. 5.48a). Typical parameters of such piezoelements are a few nanometers of length change per volt. With stacks of many thin piezodisks, one reaches length changes of 100 nm/V. When a resonator mirror is mounted on such a piezoelement (Fig. 5.48b,c), the resonator length can be controlled within a few microns by the voltage applied to the electrodes of the piezoelement.

The frequency response of this length control is limited by the inertial mass of the moving system consisting of the mirror and the piezoelement, and by the eigenresonances of this system. Using small mirror sizes and carefully selected piezos, one may reach the 100 kHz range [5.67]. For the compensation of faster fluctuations, an optical anisotropic crystal, such as potassium-dihydrogen-phosphate (KDP), can be utilized inside the laser resonator. The optical axis of this crystal must be oriented in such a way that a voltage applied to the crystal electrodes changes its refractive index along the resonator axis without turning the plane of polarization. This allows the

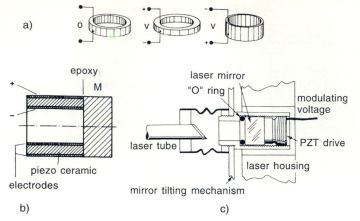

a)

epoxy
M
laser mirror
"O" ring
modulating voltage
+
−
laser tube
PZT drive
piezo ceramic
electrodes
laser housing
mirror tilting mechanism

b)   c)

Fig. 5.48. (**a**) Piezocylinders and their (exaggerated) change of length with applied voltage; (**b**) laser mirror epoxide on a piezocylinder; (**c**) mirror plus piezomount on a single-mode tunable argon laser

optical path length nd, and therefore the laser wavelength, to be controlled with a frequency response up into the megahertz range.

The wavelength stabilization system consists essentially of three elements (Fig. 5.49):

(a) The wavelength reference standard with which the laser wavelength is compared. One may, for example, use the wavelength λ_R at the maximum or at the slope of the transmission peak of a Fabry–Perot interferometer that is maintained in a controlled environment (temperature and pressure stabilization). Alternately, the wavelength of an atomic or molecular transition may serve as reference. Sometimes another stabilized laser is used as a standard and the laser wavelength is locked to this standard wavelength.

(b) The controlled system, which is in this case the resonator length nd defining the laser wavelength λ_L.

(c) The electronic control system with the servo loop, which measures the deviation $\Delta\lambda = \lambda_L - \lambda_R$ of the laser wavelength λ_L from the reference value λ_R and which tries to bring $\Delta\lambda$ to zero as quickly as possible (Fig. 5.45).

A schematic diagram of a commonly used stabilization system is shown in Fig. 5.50. A few percent of the laser output are sent from the two beam splitters BS$_1$ and BS$_2$ into two interferometers. The first FPI1 is a scanning

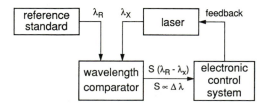

Fig. 5.49. Schematic of laser wavelength stabilization

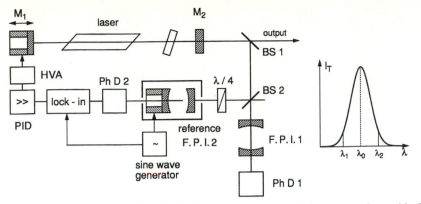

Fig. 5.50. Laser wavelength stabilization onto the transmission peak of a stable Fabry–Perot interferometer as reference

confocal FPI and serves as spectrum analyzer for monitoring the mode spectrum of the laser. The second interferometer FPI2 is the wavelength reference and is therefore placed in a pressure-tight and temperature-controlled box to keep the optical path nd between the interferometer mirrors and with it the wavelength $\lambda_R = 2nd/m$ of the transmission peak as stable as possible (Sect. 4.2). One of the mirrors is mounted on a piezoelement. If a small ac voltage with the frequency f is fed to the piezo, the transmission peak of FPI2 is periodically shifted around the center wavelength λ_0, which we take as the required reference wavelength λ_R. If the laser wavelength λ_L is within the transmission range λ_1 to λ_2 in Fig. 5.50, the photodiode PD2 behind FPI2 delivers a dc signal that is modulated at the frequency f. The modulation amplitude depends on the slope of the transmission curve $dI_T/d\lambda$ of FPI2 and the phase is determined by the sign of $\lambda_L - \lambda_0$. Whenever the laser wavelength λ_L deviates from the reference wavelength λ_R, the photodiode delivers an ac amplitude that increases as the difference $\lambda_L - \lambda_R$ increases, as long as λ_L stays within the transmission range between λ_1 and λ_2. This signal is fed to a lock-in amplifier, where it is rectified, passes a PID control (Fig. 5.45), and a high-voltage amplifier (HVA). The output of the HVA is connected with the piezoelement of the laser mirror, which moves the resonator mirror M1 until the laser wavelength λ_L is brought back to the reference value λ_R.

Instead of using the maximum λ_0 of the transmission peak of $I_T(\lambda)$ as reference wavelength, one may also choose the wavelength λ_t at the turning point of $I_T(\lambda)$ where the slope $dI_T(\lambda)/d\lambda$ has its maximum (Fig. 5.51). This has the advantage that a modulation of the FPI transmission curve is not necessary and the lock-in amplifier can be dispensed with. The cw laser intensity $I_T(\lambda)$ transmitted through FPI2 is compared with a reference intensity I_R split by BS$_2$ from the same partial beam. The output signals S$_1$ and S$_2$ from the two photodiodes D1 and D2 are fed into a difference amplifier, which is adjusted so that its output voltage becomes zero for $\lambda_L = \lambda_t$. If the laser wavelength λ_L deviates from $\lambda_R = \lambda_t$, S$_1$ becomes smaller or larger, depending on the sign of $\lambda_L - \lambda_R$; the output of the difference amplifier is, for small dif-

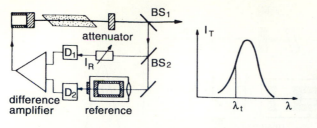

Fig. 5.51. Wavelength stabilization onto the slope of the transmission $T(\lambda)$ of a stable reference FPI

ferences $\lambda - \lambda_R$, proportional to the deviation. The output signal again passes a PID control and a high-voltage amplifier, and is fed into the piezoelement of the resonator mirror. The advantages of this difference method are the larger bandwidth of the difference amplifier (compared with a lock-in amplifier), and the simpler and less expensive composition of the whole electronic control system. Furthermore, the laser frequency does not need to be modulated which represents a big advantage for many spectroscopic applications [5.68]. Its drawback lies in the fact that different dc voltage drifts in the two branches of the difference amplifier result in a dc output, which shifts the zero adjustment and, with it, the reference wavelength λ_R. Such dc drifts are much more critical in dc amplifiers than in the ac-coupled devices used in the first method.

The stability of the laser wavelength can, of course, never exceed that of the reference wavelength. Generally it is worse because the control system is not ideal. Deviations $\Delta\lambda(t) = \lambda_L(t) - \lambda_R$ cannot be compensated immediately because the system has a finite frequency response and the inherent time constants always cause a phase lag between deviation and response.

Most methods for wavelength stabilization use a stable FPI as reference standard [5.69]. This has the advantage that the reference wavelength λ_0 or λ_t can be tuned by tuning the reference FPI. This means that the laser can be stabilized onto any desired wavelength within its gain profile. Because the signals from the photodiodes D1 and D2 in Fig. 5.51 have a sufficiently large amplitude, the signal-to-noise ratio is good, therefore the method is suitable for correcting short-term fluctuations of the laser wavelength.

For long-term stabilization, however, stabilization onto an external FPI has its drawbacks. In spite of temperature stabilization of the reference FPI, small drifts of the transmission peak cannot be eliminated completely. With a thermal expansion coefficient $\alpha = 10^{-6}$ of the distance holder for the FPI mirrors, even a temperature drift of 0.01°C causes, according to (5.86), a relative frequency drift of 10^{-8}, which gives 6 MHz for a laser frequency of $\nu_L = 6 \times 10^{14}$ Hz. For this reason, an atomic or molecular laser transition is more suitable as a *long-term frequency standard*. A good reference wavelength should be reproducible and essentially independent of external perturbations, such as electric or magnetic fields and temperature or pressure changes. Therefore, transitions in atoms or molecules without permanent dipole moments, such as CH_4 or noble gas atoms, are best suited to serve as reference wavelength standards (Chap. 14).

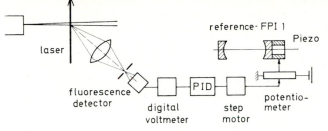

Fig. 5.52. Long-term stabilization of the laser wavelength locked to a reference FPI that in turn is locked by a digital servo loop to a molecular transition

The accuracy with which the laser wavelength can be stabilized onto the center of such a transition depends on the linewidth of the transition and on the attainable signal-to-noise ratio of the stabilization signal. Doppler-free line profiles are therefore preferable. They can be obtained by some of the methods discussed in Chaps. 7 and 9. In the case of small line intensities, however, the signal-to-noise ratio may be not good enough to achieve satisfactory stabilization. It is therefore advantageous to continue to lock the laser to the reference FPI, but to lock the FPI itself to the molecular line. In this double servo control system, the short-term fluctuations of λ_L are compensated by the fast servo loop with the FPI as reference, while the slow drifts of the FPI are stabilized by being locked to the molecular line.

Figure 5.52 illustrates a possible arrangement. The laser beam is crossed perpendicularly with a collimated molecular beam. The Doppler width of the absorption line is reduced by a factor depending on the collimation ratio (Sect. 9.1). The intensity $I_F(\lambda_L)$ of the laser-excited fluorescence serves as a monitor for the deviation $\lambda_L - \lambda_c$ from the line center λ_c. The output signal of the fluorescence detector after amplification can be fed directly to the piezoelement of the laser resonator or to the reference FPI.

To decide whether λ_t drifts to lower or to higher wavelengths, one must either modulate the laser frequency or use a digital servo control, which shifts the laser frequency in small steps. A comparator compares whether the intensity has increased or decreased by the last step and activates accordingly a switch determining the direction of the next step. Since the drift of the reference FPI is slow, the second servo control can also be slow, and the fluorescence intensity can be integrated. This allows the laser to be stabilized for a whole day, even onto faint molecular lines where the detected fluorescence intensity is less than 100 photons per second [5.70].

Recently, cryogenic optical sapphire resonators with a very high finesse operating at $T = 4$ K have proven to provide very stable reference standards [5.71]. They reach a relative frequency stability of 3×10^{-15} at an integration time of 20 s.

Since the accuracy of wavelength stabilization increases with decreasing molecular linewidth, spectroscopists have looked for particularly narrow lines that could be used for extremely well-stabilized lasers. It is very common to stabilize onto a hyperfine component of a visible transition in the I_2 molecule

using Doppler-free saturated absorption inside [5.72] or outside [5.73] the
laser resonator (Sect. 7.3). The stabilization record was held for a long time by
a HeNe laser at $\lambda = 3.39\,\mu m$ that was stabilized onto a Doppler-free infrared
transition in CH_4 [5.74, 5.75].

Using the dispersion profiles of Doppler-free molecular lines in polarization spectroscopy (Sect. 7.4), it is possible to stabilize a laser to the line center
without frequency modulation. An interesting alternative for stabilizing a dye
laser on atomic or molecular transitions is based on Doppler-free two-photon
transitions (Sect. 7.5) [5.77]. This method has the additional advantage that
the lifetime of the upper state can be very long, and the natural linewidth
may become extremely small. The narrow $1s - 2s$ two-photon transition in the
hydrogen atom with a natural linewidth of 1.3 Hz provides the best known
optical frequency reference to date [5.76].

Often the narrow Lamb dip at the center of the gain profile of a gas laser
transition is utilized (Sect. 7.2) to stabilize the laser frequency [5.78, 5.79].
However, due to collisional line shifts the frequency ν_0 of the line center
slightly depends on the pressure in the laser tube and may therefore change
in time when the pressure is changing (for instance, by He diffusion out of
a HeNe laser tube).

A simple technique for wavelength stabilization uses the orthogonal polarization of two adjacent axial modes in a HeNe laser [5.80]. The two-mode
output is split by a polarization beam splitter BS1 in the two orthogonally polarized modes, which are monitored by the photodetectors PD1 and PD2. The
difference amplification delivers a signal that is used to heat the laser tube,
which expands until the two modes have equal intensities (Fig. 5.53). They are
then kept at the frequencies $\nu_\pm = \nu_0 \pm \Delta\nu/2 = \nu_0 \pm c/(4nd)$. Only one of the
modes is transmitted to the experiment.

So far we have only considered the stability of the laser resonator itself. In
the previous section we saw that wavelength-selecting elements inside the resonator are necessary for single-mode operation to be achieved, and that their
stability and the influence of their thermal drifts on the laser wavelength must
also be considered. We illustrate this with the example of single-mode selec-

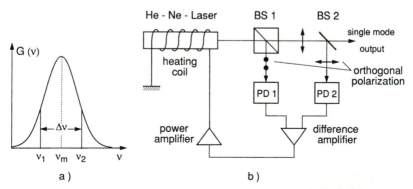

Fig. 5.53a,b. Schematic diagram of a polarization-stabilized HeNe laser: (**a**) symmetric
cavity modes ν_1 and ν_2 within the gain profile; (**b**) experimental setup

tion by a tilted intracavity etalon. If the transmission peak of the etalon is shifted by more than one-half of the cavity mode spacing, the total gain becomes more favorable for the next cavity mode, and the laser wavelength will jump to the next mode. This implies that the optical pathlength of the etalon nt must be kept stable so that the peak transmission drifts by less than $c/4d$, which is about 50 MHz for an argon laser. One can use either an air-spaced etalon with distance holders with very small thermal expansion or a solid etalon in a temperature-stabilized oven. The air-spaced etalon is simpler but has the drawback that changes of the air pressure influence the transmission peak wavelength.

The actual stability obtained for a single-mode laser depends on the laser system, on the quality of the electronic servo loop, and on the design of the resonator and mirror mounts. With moderate efforts, a frequency stability of about 1 MHz can be achieved, while extreme precautions and sophisticated equipment allow a stability of better than 1 Hz to be achieved for some laser types [5.81].

A statement about the stability of the laser frequency depends on the averaging time and on the kind of perturbations. For short time periods the frequency stability is mainly determined by random fluctuations. The best way to describe short-term frequency fluctuations is the statistical root Allan variance. For longer time periods ($\Delta t \gg 1$ s), the frequency stability is limited by predictable and measurable fluctuations, such as thermal drifts and aging of materials. The stability against short-term fluctuations, of course, becomes better if the averaging time is increased, while long-term drifts increase with the sampling time. Figure 5.54 illustrates the stability of a single-mode argon laser, stabilized with the arrangement of Fig. 5.50. With more expenditure, a stability of better than 3 kHz has been achieved for this laser [5.82], with novel techniques even better than 1 Hz (Sect. 14.7).

The residual frequency fluctuations of a stabilized laser can be represented in an *Allan plot*. The Allan variance [5.81, 5.83, 5.84]

$$\sigma = \frac{1}{\nu}\left(\sum_{i=1}^{N}\frac{(\Delta\nu_i - \Delta\nu_{i-1})^2}{2(N-1)}\right)^{1/2} \tag{5.88}$$

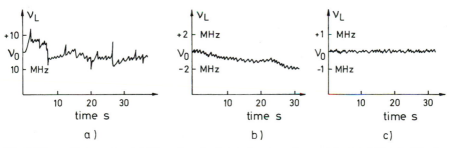

Fig. 5.54a–c. Frequency stability of a single-mode argon laser: (**a**) unstabilized; (**b**) stabilized with the arrangement of Fig. 5.50; (**c**) additional long-term stabilization onto a molecular transition. Note the different ordinate scales!

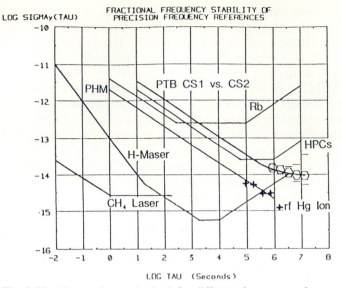

Fig. 5.55. Allan variance obtained for different frequency-reference devices [5.83]

is comparable to the relative standard deviation. It is determined by measuring at N times $t_i = t_0 + i \Delta t$ the frequency difference $\Delta \nu_i$ between two lasers stabilized onto the same reference frequency ν_R averaged over equal time intervals Δt. Figure 5.55 illustrates the Allan variance of the beat of two HeNe lasers locked to the same laser cavity [5.85]. The relative frequency fluctuations go down to $\Delta \nu / \nu_0 < 10^{-16}$, which implies an absolute stability of about 0.05 Hz.

Such extremely stable lasers are of great importance in metrology since they can provide high-quality wavelength or frequency standards with an accuracy approaching or even surpassing that of present-day standards [5.86]. For most applications in high-resolution laser spectroscopy, a frequency stability of 100 kHz to 1 MHz is sufficiently good because most spectral linewidths exceed that value by several orders of magnitude.

For a more complete survey of wavelength stabilization, the reader is referred to the reviews by Baird and Hanes [5.87], Ikegami [5.88], Hall et al. [5.89], Bergquist et al. [5.90] and Ohtsu [5.91].

5.5 Controlled Wavelength Tuning of Single-Mode Lasers

Although fixed-wavelength lasers have proved their importance for many spectroscopic applications (Sect. 6.7 and Chaps. 8, 10, and 13), it was the development of continuously tunable lasers that really revolutionized the whole field of spectroscopy. This is demonstrated by the avalanche of publications on tunable lasers and their applications (e.g., [5.92]). We shall therefore treat in this section some basic techniques for controlled tuning of single-mode lasers, while Sect. 5.7 gives a survey on tunable coherent sources developed in various spectral regions.

5.5.1 Continuous Tuning Techniques

Since the laser wavelength λ_L of a single-mode laser is determined by the optical path length nd between the resonator mirrors,

$$q\lambda = 2nd ,$$

either the mirror separation d or the refractive index n can be continuously varied to obtain a corresponding tuning of λ_L. This can be achieved, for example, by a linear voltage ramp $U = U_0 + at$ applied to the piezoelement on which the resonator mirror is mounted, or by a continuous pressure variation in a tank containing the resonator or parts of it. However, as has been discussed in Sect. 5.4.3, most lasers need additional wavelength-selecting elements inside the laser resonator to ensure single-mode operation. When the resonator length is varied, the frequency ν of the oscillating mode is tuned away from the transmission maximum of these elements (Fig. 5.36). During this tuning the neighboring resonator mode (which is not yet oscillating) approaches this transmission maximum and its losses may now become smaller than those of the oscillating mode. As soon as this mode reaches the threshold, it will start to oscillate and will suppress the former mode because of mode competition (Sect. 5.3). This means that the single-mode laser will jump back from the selected resonator mode to that which is next to the transmission peak of the wavelength-selecting element. Therefore the continuous tuning range is restricted to about half of the free spectral range $\delta\nu = \frac{1}{2}c/t$ of the selecting interferometer with thickness t, if no additional measures are taken. Similar but smaller mode hops occur when the wavelength-selecting elements are continuously tuned but the resonator length is kept constant.

Such discontinuous tuning of the laser wavelength will be sufficient if the mode hops $\delta\nu = \frac{1}{2}c/d$ are small compared with the spectral linewidths under investigation. As illustrated by Fig. 5.56a, which shows part of the neon spectrum excited in a HeNe gas discharge with a discontinuously tuned single-mode dye laser, the mode hops are barely seen and the spectral resolution is limited by the Doppler width of the neon lines. In sub-Doppler spectroscopy, however, the mode jumps appear as steps in the line profiles, as is depicted in Fig. 5.56b, where a single-mode argon laser is tuned with mode hops through some absorption lines of Na_2 molecules in a slightly collimated molecular beam where the Doppler width is reduced to about 200 MHz.

In order to enlarge the tuning range, the transmission maxima of the wavelength selectors have to be tuned synchronously with the tuning of the resonator length. When a tilted etalon with the thickness t and refractive index n is employed, the transmission maximum λ_m that, according to (5.76), is given by

$$m\lambda_m = 2nt \cos\theta ,$$

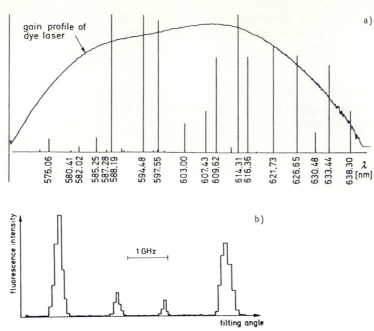

Fig. 5.56a,b. Discontinuous tuning of lasers: (**a**) part of the neon spectrum excited by a single-mode dye laser in a gas discharge with Doppler-limited resolution, which conceals the cavity mode hops of the laser; (**b**) excitation of Na_2 lines in a weakly collimated beam by a single-mode argon laser. In both cases the intracavity etalon was continuously tilted but the cavity length was kept constant

can be continuously tuned by changing the tilting angle θ. In all practical cases, θ is very small, therefore we can use the approximation $\cos\theta \approx 1 - \frac{1}{2}\theta^2$. The wavelength shift $\Delta\lambda = \lambda_0 - \lambda$ is

$$\Delta\lambda = \frac{2nt}{m}(1 - \cos\theta) \approx \frac{1}{2}\lambda_0\theta^2 , \quad \lambda_0 = \lambda(\theta = 0) . \tag{5.89}$$

Equation (5.89) reveals that the wavelength shift $\Delta\lambda$ is proportional to θ^2 but is *independent of the thickness* t. Two etalons with different thicknesses t_1 and t_2 can be mounted on the same tilting device, which may simply be a lever that is tilted by a micrometer screw driven by a small motor gearbox. The motor simultaneously drives a potentiometer, which provides a voltage proprotional to the tilting angle θ. This voltage is electronically squared, amplified, and fed into the piezoelement of the resonator mirror. With properly adjusted amplification, one can achieve an exact sychronization of the resonator wavelength shift $\Delta\lambda_L = \lambda_L \Delta d/d$ with the shift $\Delta\lambda_1$ of the etalon transmission maximum. This can be readily realized with computer control.

Unfortunately, the reflection losses of an etalon increase with increasing tilting angle θ (Sect. 4.2 and [5.50, 5.93]). This is due to the finite beam radius w of the laser beam, which prevents a complete overlap of the partial beams reflected from the front and back surfaces of the etalon. These "walk-off losses" increases with the square of the tilting angle θ, see (4.60) and Fig. 4.40.

Example 5.20

With $w = 1\,\text{mm}$, $t = 1\,\text{cm}$, $n = 1.5$, $R = 0.4$, we obtain for $\theta = 0.01$ ($\approx 0.6°$) transmission losses of 13%. The frequency shift is, see (5.89): $\Delta v = \frac{1}{2} v_0 \theta^2 \approx 30\,\text{GHz}$. For a dye laser with the gain factor $G < 1.13$ the tuning range would therefore be smaller than 30 GHz.

For wider tuning ranges interferometers with a variable air gap can be used at a fixed tilting angle θ (Fig. 5.42a). The thickness t of the interferometer and with it the transmitted wavelength $\lambda_m = 2nt \cos\theta / m$ can be tuned with a piezocylinder. This keeps the walk-off losses small. However, the extra two surfaces have to be antireflection-coated in order to minimize the reflection losses.

An elegant solution is shown in Fig. 5.42b, where the interferometer is formed by two prisms with coated backsides and inner Brewster surfaces. The air gap between these surfaces is very small in order to minimize shifts of the transmission peaks due to changes of air pressure.

The continuous change of the resonator length d is limited to about $5-10\,\mu\text{m}$ if small piezocylinders are used ($5-10\,\text{nm/V}$). A further drawback of piezoelectric tuning is the hysteresis of the expansion of the piezocylinder when tuning back and forth. Larger tuning ranges can be obtained by tilting a plane-parallel glass plate around the Brewster angle inside the laser resonator (Fig. 5.57). The additional optical path length through the plate with refractive index n at an incidence angle α is

$$s = (n\overline{AB} - \overline{AC}) = \frac{d}{\cos\beta}[n - \cos(\alpha - \beta)] = d\left[\sqrt{n^2 - \sin^2\alpha} - \cos\alpha\right].$$

$$(5.90)$$

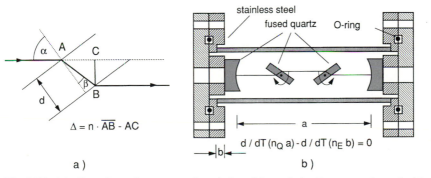

Fig. 5.57. (**a**) Changing of resonator length by tilting of the Brewster plates inside the resonator; (**b**) temperature-compensated reference cavity with tiltable Brewster plates for wavelength tuning

If the plate is tilted by the angle $\Delta\alpha$, the optical path length changes by

$$\delta s = \frac{\mathrm{d}s}{\mathrm{d}\alpha}\Delta\alpha = d\sin\alpha\left(1 - \frac{\cos\alpha}{\sqrt{n^2 - \sin^2\alpha}}\right)\Delta\alpha\ . \tag{5.91}$$

Example 5.21

A tilting of the plate with $d = 3$ mm, $n = 1.5$ from $\alpha = 51°$ to $\alpha = 53°$ around the Brewster angle $\alpha_B = 52°$ yields with $\Delta\alpha = 3 \times 10^{-2}$ rad a change $\delta s = 35\ \mu$m of the optical pathlength.

The reflection losses per surface from the deviation from Brewster's angle are less than 0.01% and are therefore completely negligible.

If the free spectral range of the resonator is $\delta\nu$, the frequency-tuning range is

$$\Delta\nu = 2(\delta s/\lambda)\delta\nu \approx 116\,\delta\nu \quad \text{at} \quad \lambda = 600\ \text{nm}\ . \tag{5.92}$$

With a piezocylinder with $\mathrm{d}s/\mathrm{d}V = 3$ nm/V only a change of $\Delta\nu = 5\delta\nu$ can be realized at $V = 500$ V.

The Brewster plate can be tilted in a controllable way by a galvo-drive [5.94], where the tilting angle is determined by the strength of the magnetic field. In order to avoid a translational shift of the laser beam when tilting the plate, two plates with $\alpha = \pm\alpha_\beta$ can be used (Fig. 5.57b), which are tilted into opposite directions. This gives twice the shift of (5.91).

For many applications in high-resolution spectroscopy where the wavelength $\lambda(t)$ should be a linear function of the time t, it is desirable that the fluctuations of the laser wavelength λ_L around the programmed tunable value $\lambda(t)$ are kept as small as possible. This can be achieved by stabilizing λ_L to the reference wavelength λ_R of a stable external FPI (Sect. 5.4), while this reference wavelength λ_R is synchronously tuned with the wavelength-selecting elements of the laser resonator. The synchronization utilizes an electronic feedback system. A possible realization is shown in Fig. 5.58. A digital voltage ramp provided by a computer through a digital–analog converter (DAC) activates the galvo-drive and results in a controlled tilting of the Brewster plates in a temperature-stabilized FPI. The laser wavelength is locked via a PID feedback control (Sect. 5.4.4) to the slope of the transmission peak of the reference FPI (Fig. 5.50). The output of the PID control is split into two parts: the low-frequency part of the feedback is applied to the galvo-plate in the laser resonator, while the high-frequency part is given to a piezoelement, which translates one of the resonator mirrors.

5.5.2 Wavelength Calibration

An essential goal of laser spectroscopy is the accurate determination of energy levels in atoms or molecules and their splittings due to external fields or internal couplings. This goal demands the precise knowledge of wavelengths and distances between spectral lines while the laser is scanned through the spectrum. There are several techniques for the solution of this problem: part of the laser beam is sent through a long FPI with mirror separation d, which is pressure-tight (or evacuated) and temperature stabilized. The equidistant transmission peaks with distances $\delta v = \frac{1}{2}c/(nd)$ serve as frequency markers and are monitored simultaneously with the spectral lines.

Most tunable lasers show an optical frequency $v(V)$ that deviates to a varying degree from the linear relation $v = \alpha V + b$ between laser frequency v and input voltage V to the scan electronics. For a visible dye laser the deviations may reach 100 MHz over a 20-GHz scan. These deviations can be monitored and corrected for by comparing the measured frequency markers with the linear expression $v = v_0 + mc/(2nd)$ $(m = 0, 1, 2, ...)$.

For *absolute* wavelength measurements of spectral lines the laser is stabilized onto the center of the line and its wavelength λ is measured with one of the wavemeters described in Sect. 4.4. For Doppler-free lines (Chaps. 7–11), one may reach absolute wavelength determinations with an uncertainty of smaller than $10^{-3}\,\text{cm}^{-1}$ ($\hat{=} 20\,\text{pm}$ at $\lambda = 500\,\mu\text{m}$).

Often calibration spectra that are taken simultaneously with the unknown spectra are used. Examples are the I_2 spectrum, which has been published in the iodine atlas by Gerstenkorn and Luc [5.95] in the range of 14 800 to 20 000 cm^{-1}. For wavelengths below 500 nm, thorium lines [5.96] measured in a hollow cathode by optogalvanic spectroscopy (Sect. 6.5) or uranium lines [5.97] can be utilized.

If no wavemeter is available, two FPIs with slightly different mirror separations d_1 and d_2 can be used for wavelength determination (Fig. 5.58b). Assume $d_1/d_2 = p/q$ equals the ratio of two rather large integers p and q with no common divisor and both interferometers have a transmission peak at λ_1:

$$\left. \begin{array}{l} m_1\lambda_1 = 2d_1 \\ m_2\lambda_1 = 2d_2 \end{array} \right\} \quad \text{with} \quad \frac{m_1}{m_2} = p/q . \tag{5.93a}$$

Let us assume that λ_1 is known from calibration with a spectral line. When the laser wavelength is tuned, the next coincidence appears at $\lambda_2 = \lambda_1 + \Delta\lambda$ where

$$(m_1 - p)\lambda_2 = 2d_1 \quad \text{and} \quad (m_2 - q)\lambda_2 = 2d_2 . \tag{5.93b}$$

From (5.93a, 5.93b) we obtain

$$\frac{\Delta\lambda}{\lambda_1} = \frac{p}{m_1 - p} = \frac{q}{m_2 - q} \Rightarrow \lambda_2 = R_1 \frac{m_1}{m_1 - p} = R_1 \frac{m_2}{m_2 - q} ,$$

where p and q are known integers that can be counted by the number of transmission maxima when λ is tuned from λ_1 to λ_2.

Between these two wavelengths λ_1 and λ_2 the maximum of a spectral line with the unknown wavelength λ_x may appear in a linear wavelength scan at the distance δ_x from the position of λ_1. Then we obtain from Fig. 5.59

$$\lambda_x = \lambda_1 + \frac{\delta_x}{\delta}\Delta\lambda = \lambda_1 \left(1 + \frac{\delta_x}{\delta}\frac{p}{m_1 - p}\right) . \tag{5.93c}$$

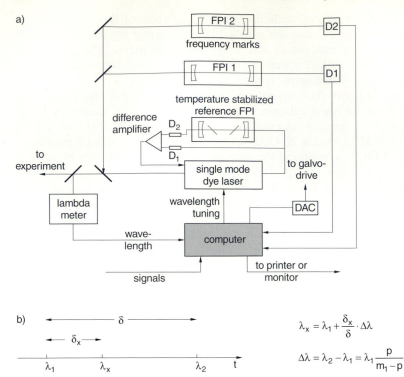

a)

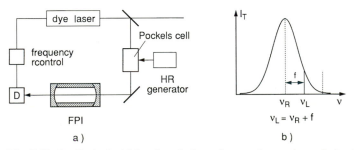

b)

$$\lambda_x = \lambda_1 + \frac{\delta_x}{\delta} \cdot \Delta\lambda$$

$$\Delta\lambda = \lambda_2 - \lambda_1 = \lambda_1 \frac{p}{m_1 - p}$$

Fig. 5.58. (a) Schematic diagram of computer-controlled laser spectrometer with frequency marks provided by two FPI with slightly different free spectral ranges and a lambdameter for absolute wavelength measurement. **(b)** Scheme for wavelength determination according to (5.93c)

a)

b)

$$I_T$$

$$\nu_R \quad \nu_L \qquad \nu$$

$$\nu_L = \nu_R + f$$

Fig. 5.59a,b. Optical sideband technique for precise tuning of the laser wavelength λ: **(a)** experimental setup; **(b)** stabilization of the sideband ν_R onto the transmission peak of the FPI

With the inputs for λ_1, p, q, d_1, and d_2 a computer can readily calculate λ_x from the measured value δ_x.

For a very precise measurement of small spectral invervals between lines a sideband technique is very useful. In this technique part of the laser beam is sent through a Pockels cell (Fig. 5.59), which modulates the transmitted intensity and generates sidebands at the frequencies $\nu_R = \nu_L \pm f$. When $\nu_R^+ = \nu_L + f$ is stabilized onto an external FPI, the laser frequency $\nu_L = \nu_R^+ - f$ can be continuously tuned by varying the modulation frequency f. This method does not need a tunable interferometer and its accuracy is only limited by the accuracy of measuring the modulation frequency f [5.98a].

This controllable shift of a laser frequency ν_L against a reference frequency ν_R can be also realized by electronic elements in the stabilization feedback circuit. This omits the Pockels cell of the previous method. A tunable laser is "frequency-offset locked" to a stable reference laser in such a way that the difference frequency $f = \nu_L - \nu_R$ can be controlled electronically. This technique has been described by Hall [5.98b] and is used in many laboratories.

5.6 Linewidths of Single-Mode Lasers

In the previous sections we have seen that the frequency fluctuations of single-mode lasers caused by fluctuations of the product nd of the refractive index n and the resonator length d can be greatly reduced by appropriate stabilization techniques. The output beam of such a single-mode laser can be regarded for most applications as a *monochromatic wave* with a radial Gaussian amplitude profile, see (5.32).

For some tasks in ultrahigh-resolution spectroscopy, the residual finite linewidth $\Delta\nu_L$, which may be small but nonzero, still plays an important role and must therefore be known. Furthermore, the question *why* there is an ultimate lower limit for the linewidth of a laser is of fundamental interest, since this leads to basic problems of the nature of electromagnetic waves. Any fluctuation of amplitude, phase, or frequency of our "monochromatic" wave results in a finite linewidth, as can be seen from a Fourier analysis of such a wave (see the analogous discussion in Sects. 3.1, 3.2). Besides the "technical noise" caused by fluctuations of the product nd, there are essentially three noise sources of a *fundamental* nature, which cannot be eliminated, even by an ideal stabilization system. These noise sources are, to a different degree, responsible for the residual linewidth of a single-mode laser.

The first contribution to the noise results from the spontaneous emission of excited atoms in the upper laser level E_i. The total power P_{sp} of the fluorescence spontaneously emitted on the transition $E_i \rightarrow E_k$ is, according to Sect. 2.3, proportional to the population density N_i, the active mode volume V_m, and the transition probability A_{ik}, i.e.,

$$P_{sp} = N_i V_m A_{ik} . \tag{5.94}$$

This fluorescence is emitted into all modes of the EM field within the spectral width of the fluorescence line. According to Example 2.1 in Sect. 2.1, there are about 3×10^8 modes/cm^3 within the Doppler-broadened linewidth $\Delta\nu_D = 10^9$ Hz at $\lambda = 500$ nm. The mean number of fluorescence photons per mode is therefore small.

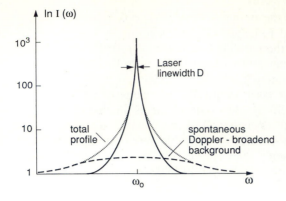

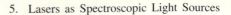

Fig. 5.60. Linewidth of a single-mode laser just above threshold with Doppler-broadened background due to spontaneous emission. Note the logarithmic scale!

Example 5.22

In a HeNe laser the stationary population density of the upper laser level is $N_i \simeq 10^{10}$ cm^{-3}. With $A_{ik} = 10^8$ s^{-1}, the number of fluorescence photons is 10^{18} s^{-1} cm^{-3}, which are emitted into 3×10^8 modes. Into each mode a photon flux $\phi = 3 \times 10^9$ photons/s is emitted, which corresponds to a mean photon density of $\langle n_{ph} \rangle = \phi/c \le 10^{-1}$ in one mode. This has to be compared with 10^7 photons per mode due to induced emission inside the resonator at a laser output power of 1 mW through a mirror with $R = 0.99$.

When the laser reaches threshold, the number of photons in the laser mode increases rapidly by stimulated emission and the narrow laser line grows from the weak but Doppler-broadened background radiation (Fig. 5.60). Far above the threshold, the laser intensity is larger than this background by many orders of magnitude and we may therefore neglect the contribution of spontaneous emission to the laser linewidth.

The second contribution to the noise resulting in line broadening is due to amplitude fluctuations caused by the statistical distribution of the number of photons in the oscillating mode. At the laser output power P, the average number of photons that are transmitted per second through the output mirror is $\bar{n} = P/h\nu$. With $P = 1$ mW and $h\nu = 2$ eV ($\hat{=} \lambda = 600$ nm), we obtain $n = 8 \times 10^{15}$. If the laser operates far above threshold, the probability $p(n)$ that n photons are emitted per second is given by the Poisson distribution [5.14, 5.15]

$$p(n) = \frac{e^{-\bar{n}}(\bar{n}^n)}{n!} . \tag{5.95}$$

The average number $\bar{n}$ is mainly determined by the pump power P_p (Sect. 5.1.3). If at a given value of P_p the number of photons increases because of an amplitude fluctuation of the induced emission, saturation of the amplifying transition in the active medium reduces the gain and decreases the field amplitude. Thus saturation provides a self-stabilizing mechanism for amplitude fluctuations and keeps the laser field amplitude at a value $E_s \sim (\bar{n})^{1/2}$.

The main contribution to the residual laser linewidth comes from *phase fluctuations*. Each photon that is spontaneously emitted into the laser mode can be amplified by induced emission; this amplified contribution is super-imposed on the oscillating wave. This does not essentially change the total amplitude of the wave because these additional photons decrease the gain for the other photons (by gain saturation) such that the average photon number $\bar{n}$ remains constant. However, the phases of these spontaneously initiated pho-ton avalanches show a random distribution, as does the phase of the total wave. There is no such stabilizing mechanism for the total phase as there is for the amplitude. In a polar diagram, the total field amplitude $E = A e^{i\varphi}$ can be described by a vector with the amplitude A, which is restricted to a nar-row range δA, and a phase angle φ that can vary from 0 to 2π (Fig. 5.61). In the course of time, a *phase diffusion* φ occurs that can be described in a thermodynamic model by the diffusion coefficient D [5.99, 5.100].

For the spectral distribution of the laser emission in the ideal case in which all technical fluctuations of nd are totally eliminated, this model yields from a Fourier transform of the statistically varying phase the Lorentzian line pro-file

$$|E(\nu)|^2 = E_0^2 \frac{(D/2)^2}{(\nu - \nu_0)^2 + (D/2)^2} , \quad \text{with} \quad E_0 = E(\nu_0) , \tag{5.96}$$

with the center frequency ν_0, which may be compared with the Lorentzian line profile of a classical oscillator broadened by phase-perturbing collisions.

The full halfwidth $\Delta\nu = D$ of this intensity profile $I(\nu) \propto |E(\nu)|^2$ de-creases with increasing output power because the contributions of the sponta-neously initiated photon avalanches to the total amplitude and phase become less and less significant with increasing total amplitude.

Furthermore, the halfwidth $\Delta\nu_c$ of the resonator resonance must influence the laser linewidth, because it determines the spectral interval where the gain exceeds the losses. The smaller the value of $\Delta\nu_c$, the smaller is the fraction of spontaneously emitted photons (which are emitted within the full Doppler width) with frequencies within the interval $\Delta\nu_c$ that find enough gain to build up a photon avalanche. When all these factors are taken into account, one

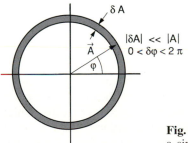

Fig. 5.61. Polar diagram of the amplitude vector A of a single-mode laser, for illustration of phase diffusion

obtains for the theoretical lower limit $\Delta \nu_L = D$ for the laser linewidth the relation [5.101]

$$\Delta \nu_L = \frac{\pi h \nu_L (\Delta \nu_c)^2 (N_{sp} + N_{th} + 1)}{2 P_L} , \qquad (5.97)$$

where N_{sp} is the number of photons spontaneously emitted per second into the oscillating laser mode, N_{th} is the number of photons in this mode due to the thermal radiation field, and P_L is the laser output power. At room temperature in the visible region, $N_{th} \ll 1$ (Fig. 2.5). With $N_{sp} = 1$ (at least one spontaneous photon starts the induced photon avalanche), we obtain from (5.97) the famous Schwalow–Townes relation [5.101]

$$\Delta \nu_L = \frac{\pi h \nu_L \Delta \nu_c^2}{P_L} . \qquad (5.98)$$

Example 5.23

(a) For a HeNe laser with $\nu_L = 5 \times 10^{14}$ Hz, $\Delta \nu_c = 1$ MHz, $P = 1$ mW, we obtain $\Delta \nu_L = 5 \times 10^{-4}$ Hz.
(b) For an argon laser with $\nu_L = 6 \times 10^{14}$ Hz, $\Delta \nu_c = 3$ MHz, $P = 1$ W, the theoretical lower limit of the linewidth is $\Delta \nu_L = 3 \times 10^{-5}$ Hz.

However, even for lasers with a very sophisticated stabilization system, the residual uncompensated fluctuations of nd cause frequency fluctuations that are large compared with this theoretical lower limit. With moderate efforts, laser linewidths of $\Delta \nu_L = 10^4 - 10^6$ Hz have been realized for gas and dye lasers. With very great effort, laser linewidths of a few Hertz or even below 1 Hz [5.81, 5.102] can be achieved. However, several proposals have been made how the theoretical lower limit may be approached more closely [5.103, 5.104].

This linewidth should not be confused with the attainable frequency stability, which means the stability of the center frequency of the line profile. For dye lasers, stabilities of better than 1 Hz have been achieved, which means a relative stability $\Delta \nu / \nu \leq 10^{-15}$ [5.81]. For gas lasers, such as the stabilized HeNe laser or specially designed solid-state lasers, even values of $\Delta \nu / \nu \leq 10^{-16}$ are possible [5.105, 5.106].

5.7 Tunable Lasers

In this section we discuss experimental realizations of some tunable lasers, which are of particular relevance for spectroscopic applications. A variety of tuning methods have been developed for different spectral regions, which will be illustrated by several examples. While semiconductor lasers, color-center lasers, and vibronic solid-state lasers are the most widely used tunable *infrared* lasers to date, the dye laser in its various modifications and the titanium:sapphire laser are still by far the most important tunable lasers in the *visible* region. Great progress has recently been made in the development of new types of ultraviolet lasers as well as in the generation of coherent UV radiation by frequency-doubling or frequency-mixing techniques (Sect. 5.8). In particular, great experimental progress in optical parametric oscillators has been made; they are discussed in Sect. 5.8.8 in more detail. Meanwhile, the whole spectral range from the far infrared to the vacuum ultraviolet can be covered by a variety of tunable coherent sources. Of great importance for basic research on highly ionized atoms and for a variety of applications is the development of X-ray lasers, which is briefly discussed in Sect. 5.7.7.

This section can give only a brief survey of those tunable devices that have proved to be of particular importance for spectroscopic applications. For a more detailed discussion of the different techniques, the reader is referred to the literature cited in the corresponding subsections. A review of tunable lasers that covers the development up to 1974 has been given in [5.108], while more recent compilations can be found in [5.92, 5.109]. For a survey on infrared spectroscopy with tunable lasers see [5.110–5.112].

5.7.1 Basic Concepts

Tunable coherent light sources can be realized in different ways. One possibility, which has already been discussed in Sect. 5.5, relies on lasers with a *broad gain profile*. Wavelength-selecting elements inside the laser resonator restrict laser oscillation to a narrow spectral interval, and the laser wavelength may be continuously tuned across the gain profile by varying the transmission maxima of these elements. Dye lasers, color-center lasers, and excimer lasers are examples of this type of tunable device.

Another possibility of wavelength tuning is based on the shift of energy levels in the active medium by external perturbations, which cause a corresponding spectral shift of the gain profile and therefore of the laser wavelength. This level shift may be effected by an external magnetic field (spin-flip Raman laser and Zeeman-tuned gas laser) or by temperature or pressure changes (semiconductor laser).

A third possibility for generating coherent radiation with tunable wavelength uses the principle of optical frequency mixing, which is discussed in Sect. 5.8.

The experimental realization of these tunable coherent light sources is, of course, determined by the spectral range for which they are to be used. For the particular spectroscopic problem, one has to decide which of the possibilities summarized above represents the optimum choice. The experimental expenditure depends substantially on the desired tuning range, on the achievable output power, and, last but not least, on the realized spectral bandwidth $\Delta \nu$. Coherent light sources with bandwidths $\Delta \nu \simeq 30\,\mathrm{MHz}$ to $30\,\mathrm{GHz}$ $(0.001-1\,\mathrm{cm}^{-1})$, which can be continuously tuned over a larger range, are already commercially available. In the visible region, single-mode

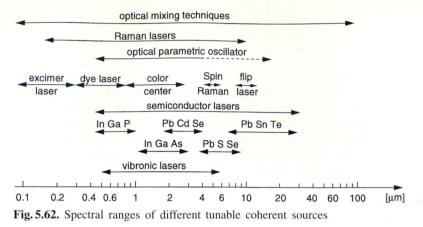

Fig. 5.62. Spectral ranges of different tunable coherent sources

dye lasers are offered with a bandwidth down to about 1 MHz. These lasers are continuously tunable over a restricted tuning range of about 30 GHz ($1\,\mathrm{cm}^{-1}$). Computer control of the tuning elements allows a successive continuation of such ranges. In principle, "continuous" scanning of a single-mode laser over the whole gain profile of the laser medium, using automatic resetting of all tuning elements at definite points of a scan, is now possible. Examples are single-mode semiconductor lasers, dye lasers, or vibronic solid-state lasers.

We briefly discuss the most important tunable coherent sources, arranged according to their spectral region. Figure 5.62 illustrates the spectral ranges covered by the different devices.

5.7.2 Semiconductor-Diode Lasers

Many of the most widely used tunable coherent infrared sources use various semiconductor materials, either directly as the active laser medium (semiconductor lasers) or as the nonlinear mixing device (frequency-difference generation).

The basic principle of semiconductor lasers [5.113–5.117] may be summarized as follows. When an electric current is sent in the forward direction through a *p–n* semiconductor diode, the electrons and holes can recombine within the *p–n* junction and may emit the recombination energy in the form of EM radiation (Fig. 5.63). The linewidth of this spontaneous emission amounts to several cm^{-1}, and the wavelength is determined by the energy difference between the energy levels of electrons and holes, which is essentially determined by the band gap. The spectral range of spontaneous emission can therefore be varied within wide limits (about $0.4-40\,\mu\mathrm{m}$) by the proper selection of the semiconductor material and its composition in binary compounds (Fig. 5.64).

Above a certain threshold current, determined by the particular semiconductor diode, the radiation field in the junction becomes sufficiently intense

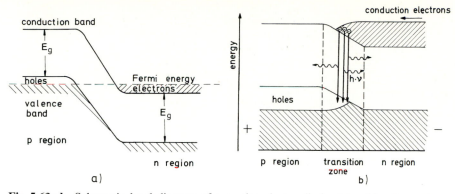

Fig. 5.63a,b. Schematic level diagram of a semiconductor diode: (**a**) unbiased *p–n* junction and (**b**) inversion in the zone around the *p–n* junction and recombination radiation when a forward voltage is applied

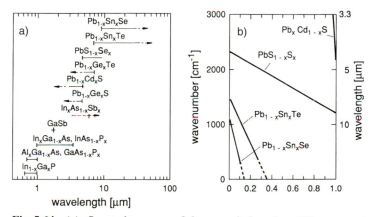

Fig. 5.64. (**a**) Spectral ranges of laser emission for different semiconductor materials [5.115]; (**b**) dependence of the emission wave number on the composition x of $Pb_{1-x}Sn_xTe$, Se, or S–lead-salt lasers (courtesy of Spectra-Physics)

to make the induced-emission rate exceed the spontaneous or radiationless recombination processes. The radiation can be amplified by multiple reflections from the plane end faces of the semiconducting medium and may become strong enough that induced emission occurs in the *p–n* junction before other relaxation processes deactivate the population inversion.

The wavelengths of the laser radiation are determined by the spectral gain profile and by the eigenresonances of the laser resonator (Sect. 5.3). If the polished end faces of the semiconducting medium are used as resonator mirrors (Fig. 5.65a), the free spectral range

$$\delta \nu = \frac{c}{2nd\left(1+(\nu/n)\,dn/d\nu\right)}, \quad \text{or} \quad \delta\lambda = \frac{\lambda^2}{2nd\left(1-(\lambda/n)\,dn/d\lambda\right)}, \quad (5.99)$$

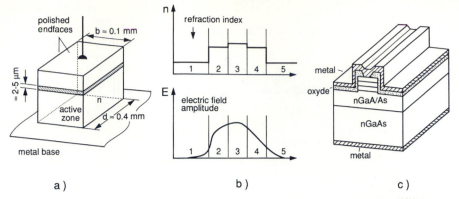

Fig. 5.65a–c. Schematic diagram of a diode laser: (**a**) geometrical structure; (**b**) index profile of a heterostructure diode laser around the p–n junction; (**c**) concentration of the injection current in order to reach high current densities in the inversion zone

is very large, because of the short resonator length d. Note that $\delta\nu$ depends not only on d but also on the dispersion $dn/d\nu$ of the active medium.

Example 5.24

With $d = 0.5$ mm, $n = 2.5$ and $(\nu/n)\,dn/d\nu = 1.5$, the free spectral range is $\delta\nu = 48$ GHz $\hat{=} 1.6$ cm^{-1}, or $\delta\lambda = 0.16$ nm at $\lambda = 1$ μm.

This illustrates that only a few axial resonator modes fit within the gain profile, which has a spectral width of several cm^{-1} (Fig. 5.65b).

For wavelength tuning, all those parameters that determine the energy gap between the upper and lower laser levels may be varied. A temperature change produced by an external cooling system or by a current change is most frequently utilized to generate a wavelength shift (Fig. 5.66a). Sometimes an external magnetic field or a mechanical pressure applied to the semiconductor is also employed for wavelength tuning. In general, however, no truly continuous tuning over the whole gain profile is possible. After a continuous tuning over about one wavenumber, mode hops occur because the resonator length is not altered synchronously with the maximum of the gain profile (Fig. 5.66b). In the case of temperature tuning this can be seen as follows:

The temperature difference ΔT changes the energy difference $E_g = E_1 - E_2$ between upper and lower levels in the conduction and valence band, the index of refraction by $\Delta n = (dn/dT)\Delta T$, and the length L of the cavity by $\Delta L = (dL/dT)\Delta T$.

The frequency $\nu_c = mc/(2nL)$ (m: integer) of a cavity mode is then shifted by

$$\Delta\nu_c = \frac{\partial\nu_c}{\partial n}\frac{dn}{dT}\Delta T + \frac{\partial\nu_c}{\partial L}\frac{dL}{dT}\Delta T = -\nu\left(\frac{1}{n}\frac{dn}{dT} + \frac{1}{L}\frac{dL}{dT}\right)\Delta T\,, \qquad (5.100)$$

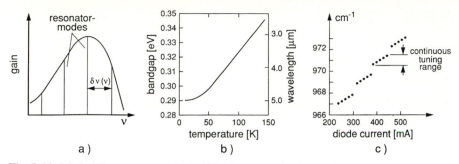

Fig. 5.66. (a) Axial resonator modes within the spectral gain profile; (b) temperature tuning of the gain maximum; and (c) mode hops of a quasi-continuously tunable cw PbS$_n$Te diode laser in a helium cryostat. The points correspond to the transmission maxima of an external Ge etalon with a free spectral range of 1.955 GHz [5.112]

while the maximum of the gain profile is shifted by

$$\Delta \nu_g = \frac{1}{h} \frac{\partial E_g}{\partial T} \Delta T \,. \tag{5.101}$$

Although the first term in (5.100) is much larger than the second, the total shift $\Delta \nu_c / \Delta T$ amounts to only about $10-20\%$ of the shift $\Delta \nu_g / \Delta T$.

As soon as the maximum of the gain profile reaches the next resonator mode, the gain for this mode becomes larger than that of the oscillating one and the laser frequency jumps to this mode.

For a realization of continuous tuning over a wider range, it is therefore necessary to use external resonator mirrors with the distance d that can be independently controlled. Because of technical reasons this implies, however, a much larger distance d than the small length L of the diode and therefore a much smaller free spectral range. To achieve single-mode oscillation, additional wavelength-selecting elements, such as optical reflection gratings or etalons, have to be inserted into the resonator. Furthermore, one end face of the semiconducting medium must be antireflection coated because the large reflection coefficient of the uncoated surfaces (with $n = 3.5$ the reflectivity becomes 0.3) causes large reflection losses. Such single-mode semiconductor lasers have been built [5.118–5.120].

An example is presented in Fig. 5.67. The etalon E enforces single-mode operation. The resonator length is varied by tilting a Brewster plate and the maximum of the gain profile is synchronously shifted through a change of the diode current. The laser wavelength is stabilized onto an external Fabry–Perot interferometer and can be controllably tuned by tilting a galvo-plate in this external cavity. Tuning ranges up to 100 GHz without mode hops have been achieved for a GaAlAs laser around 850 nm [5.120].

Another realization of tunable single-mode diode lasers uses a Littrow grating, which couples part of the laser output back into the gain medium (Fig. 5.68) [5.121]. When the grating with a groove spacing d_g is tilted by an

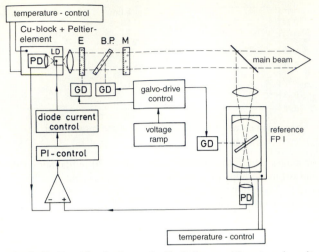

Fig. 5.67. Tunable single-mode diode laser with external cavity. The etalon allows single-mode operation and the Brewster plate tunes the optical length of the cavity synchronized with etalon tilt and gain profile shift [5.120]

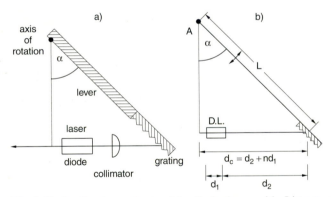

Fig. 5.68a,b. Continuously tunable diode laser with Littrow grating: (**a**) experimental setup, and (**b**) geometric condition for the location of the tilting axis for the grating. The rotation around point R_1 compensates only in first order, around R_2 in second order [5.122]

angle $\Delta\alpha$, the wavelength shift is according to (4.21a)

$$\Delta\lambda = (2d_g)\cos\alpha \cdot \Delta\alpha \ . \tag{5.102}$$

Tilting of the grating is realized by mounting the grating on a lever of length L. If the tilting axis A in Fig. 5.68 is chosen correctly, the change $\Delta d_c = L\cdot\cos\alpha\cdot\Delta\alpha$ of the cavity length d_c results in the same wavelength change $\Delta\lambda = (\Delta d_c/d_c)\lambda$ of the cavity modes, as given by (5.102). This gives the condition $d_c/L = \sin\alpha$, which shows that the tilting axis should be lo-

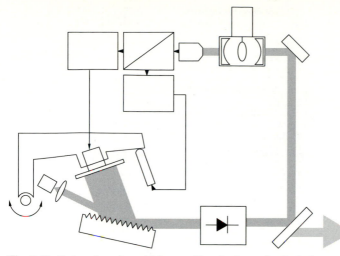

Fig. 5.69. External-cavity widely tunable single-mode diode laser with Littman resonator

cated at the crossing of the plane through the grating surface and the plane indicated by the dashed line that intersects the resonator axis at a distance $d_c = d_1 + n \cdot d_2$ from the grating, where n is the refractive index of the diode (Fig. 5.68b).

An improved version with a fixed Littman grating configuration and a tiltable end mirror (Fig. 5.69) allows a wider tuning range up to 500 GHz, which is only limited by the maximum expansion of the piezo used for tilting the mirror lever [5.122a]. A novel compact external-cavity diode laser with a transmission grating (Fig. 5.70 in Littrow configuration allows an extremely compact mechanical design with a good passive frequency stability [5.122b].

Tilting of the etalon or grating tunes the laser wavelength across the spectral gain profile $G(\lambda)$, where the maximum $G(\lambda_m)$ is determined by the temperature. A change ΔT of the temperature shifts this maximum λ_m. Temperature changes are used for coarse tuning, whereas the mechanical tilting allows fine-tuning of the single-mode laser.

A complete commercial diode laser spectrometer for convenient use in infrared spectroscopy is depicted in Fig. 5.71.

Meanwhile tunable diode lasers in the visible region down to below 0.4 μm are available [5.123].

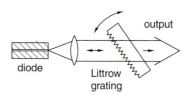

output

diode

Littrow grating

Fig. 5.70. External-cavity diode laser with transmission grating

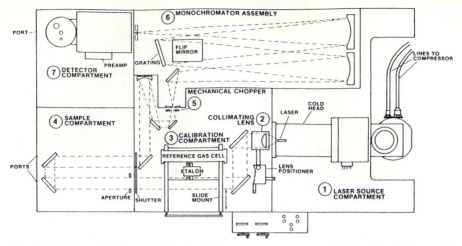

Fig. 5.71. Schematic diagram of a diode laser spectrometer tunable from 3 to 25 µm with different diodes. (courtesy of Spectra-Physics)

Besides their applications as tunable light sources, diode lasers are more and more used as pump lasers for tunable solid-state lasers and optical parametric amplifiers. Monolithic diode laser arrays can now deliver up to 100 W cw pump powers [5.124].

5.7.3 Tunable Solid-State Lasers

The absorption and emission spectra of crystalline or amorphous solids can be varied within wide spectral ranges by doping them with atomic or molecular ions [5.125–5.127]. The strong interaction of these ions with the host lattice causes broadenings and shifts of the ionic energy levels. The absorption spectrum shown in Fig. 5.72b for the example of alexandrite depends on the polarization direction of the pump light. Optical pumping of excited states generally leads to many overlapping fluorescence bands terminating on many higher "vibronic levels" in the electronic ground state, which rapidly relax by ion–phonon interaction back into the original ground state (Fig. 5.72a). These lasers are therefore often called *vibronic lasers*. If the fluorescence bands overlap sufficiently, the laser wavelength can be continously tuned over the corresponding spectral gain profile (Fig. 5.72c).

Vibronic solid-state laser materials are, e.g., alexandrite ($BeAl_2O_4$ with Cr^{3+} ions) titanium–sapphire ($Al_2O_3:Ti^+$) fluoride crystals doped with transition metal ions (e.g., $MgF_2:Co^{++}$ or $CsCaF_3:V^{2+}$) [5.109, 5.126–5.129].

The tuning range of vibronic solid-state lasers can be widely varied by a proper choice of the implanted ions and by selecting different hosts. This is illustrated in Fig. 5.73a, which shows the spectral ranges of laser-excited fluorescence of the same Cr^{3+} ion in different host materials [5.128] and different metal ions in a MgF_2 crystal.

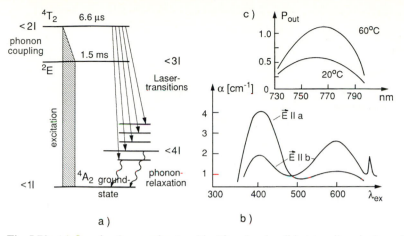

Fig. 5.72. (a) Level scheme of a tunable "four-level solid-state vibronic laser "; (b) absorption spectrum for two different polarization directions of the pump laser; (c) output power $P_{out}(\lambda)$ for the example of the alexandrite laser

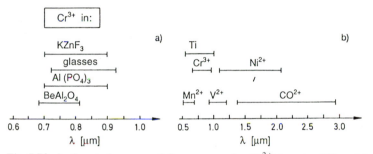

Fig. 5.73a,b. Spectral ranges of fluorescence for Cr^{3+} ions in different host materials (a) and different metal ions in MgF_2 (b)

A particularly efficient cw vibronic laser is the emerald laser $(Be_3Al_2Si_6O_{18}:Cr^{3+})$. When pumped by a 3.6-W krypton laser at $\lambda_p = 641$ nm, it reaches an output power of up to 1.6 W and can be tuned between 720 and 842 nm [5.131]. The slope efficiency dP_{out}/dP_{in} reaches 64%! The erbium:YAG laser, tunable around $\lambda = 2.8\,\mu m$, has found a wide application range in medical physics.

A very important vibronic laser is the titanium:sapphire (Ti:sapphire) laser, which has a large tuning range between 650 nm and 1000 nm when pumped by an argon laser. The effective tuning range is limited by the reflectivity curve of the resonator mirrors, and for an optimum output power over the whole spectral range three different sets of mirrors are used. For spectral ranges with $\lambda > 700$ nm, the Ti:sapphire laser is superior to the dye laser (Sect. 5.7.4) because it has higher output power, better frequency stability and

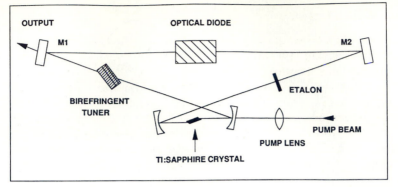

Fig. 5.74. Experimental setup of a Ti:sapphire laser. (courtesy of Schwartz Electro-Optics)

a smaller linewidth. The experimental setup of a titanium-sapphire laser is depicted in Fig. 5.74.

The different vibronic solid-state lasers cover the red and near-infrared spectral range from 0.65 to 2.5 μm (Fig. 5.75). Most of them can run at room temperature in a pulsed mode, some of them also run in cw operation.

The future importance of these lasers is derived from the fact that many of them may be pumped by diode laser arrays. This has already been demonstrated for Nd:YAG and alexandrite lasers, where very high total energy conversion efficiencies were achieved. For the diode laser-pumped Nd:YAG laser, values of $\eta = 0.3$ for the ratio of laser output power to electrical input power have been reported (30% plug-in efficiency) [5.132].

Intracavity frequency doubling of these lasers (Sect. 5.8) covers the visible and near-ultraviolet range [5.133]. Although dye lasers are still the most important tunable lasers in the visible range, these compact and handy solid-state devices present attractive alternatives and have started to replace dye lasers for many applications.

For more details about tunable solid-state lasers and their pumping by high-power diode lasers, the reader is referred to [5.109, 5.134–5.137].

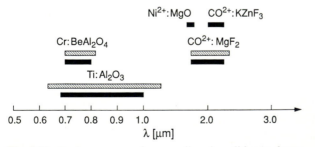

Fig. 5.75. Tuning ranges of some vibronic solid-state lasers

5.7.4 Color-Center Lasers

Color centers in alkali halide crystals are based on a halide ion vacancy in the crystal lattice of rock-salt structure (Fig. 5.76). If a single electron is trapped at such a vacancy, its energy levels result in new absorption lines in the visible spectrum, broadened to bands by the interaction with phonons. Since these visible absorption bands, which are caused by the trapped electrons and which are absent in the spectrum of the ideal crystal lattice, make the crystal appear colored, these imperfections in the lattice are called *F-centers* (from the German word "Farbe" for color) [5.138]. These F-centers have very small oscillator strengths for electronic transitions, therefore they are not suited as active laser materials.

If *one* of the six positive metal ions that immediately surround the vacancy is foreign (e.g., a Na^+ ion in a KCl crystal, Fig. 5.76b), the F-center is specified as an F_A-center [5.139], while F_B-centers are surrounded by *two* foreign ions (Fig. 5.76c). A pair of two adjacent F-centers along the (110) axis of the crystal is called an F_2-center (Fig. 5.76d). If one electron is taken away from an F_2-center, an F_2^+-center is created (Fig. 5.76e).

The F_A- and F_B-centers can be further classified into two categories according to their relaxation behavior following optical excitation. While centers of type I retain the single vacancy and behave in this respect like ordinary F-centers, the type-II centers relax to a double-well configuration (Fig. 5.77) with energy levels completely different from the unrelaxed counterpart. The oscillator strength for an electric-dipole transition between upper level $|k\rangle$ and lower level $|i\rangle$ in the relaxed double-well configuration is quite large. The relaxation times T_{R1} and T_{R2} for the transitions to the upper level $|k\rangle$ and from the lower level $|i\rangle$ back to the initial configuration are below 10^{-12} s. The lower level $|i\rangle$ is therefore nearly empty, which also allows sufficient inversion for cw laser operation. All these facts make the F_A- and F_B-type-II color centers – or, in shorthand, F_A(II) and F_B(II) – very suitable for tunable laser action [5.140–5.142].

The quantum efficiency η of F_A(II)-center luminescence decreases with increasing temperature. For a KCl:Li crystal, for example, η amounts to 40%

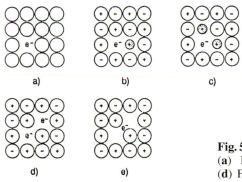

Fig. 5.76a–e. Color centers in alkali halides: **(a)** F-center; **(b)** F_A-center; **(c)** F_B-center; **(d)** F_2-center; and **(e)** F_2^+-center

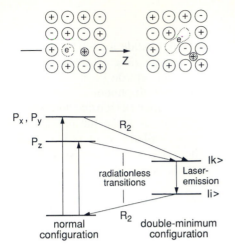

Fig. 5.77. Structural change and level diagram of optical pumping, relaxation, and lasing of a $F_A(II)$-center

at liquid nitrogen temperatures (77 K) and approaches zero at room temperature (300 K). This implies that color-center lasers should be operated at low temperatures, generally at 77 K.

Two possible experimental arrangements of color-center lasers are shown schematically in Fig. 5.78. The folded astigmatically compensated three-mirror cavity design is identical to that of cw dye lasers of the Kogelnik type [5.143] (Sect. 5.7.5). A collinear pump geometry allows optimum overlap between the pump beam and the waist of the fundamental resonator mode in the crystal. The mode-matching parameter (i.e., the ratio of pump-beam waist to resonator-mode waist) can be chosen by appropriate mirror curvatures. The optical density of the active medium, which depends on the preparation of the F_A centers [5.140], has to be carefully adjusted to achieve optimum absorption of the pump wavelength. The crystal is mounted on a cold finger cooled with liquid nitrogen in order to achieve a high quantum efficiency η.

Coarse wavelength tuning can be accomplished by turning mirror M_3 of the resonator with an intracavity dispersing sapphire Brewster prism. Because of the homogeneous broadening of the gain profile, single-mode operation would be expected without any further selecting element (Sect. 5.3). This is, in fact, observed except that neighboring spatial hole-burning modes appear, which are separated from the main mode by

$$\Delta\nu = \frac{c}{4a} ,$$

where a is the distance between the end mirror M_1 and the crystal (Sect. 5.3). With one Fabry–Perot etalon of 5-mm thickness and a reflectivity of 60−80%, stable single-mode operation without other spatial hole-burning modes can be achieved [5.144]. With a careful design of the low-loss optical components inside the cavity (made, e.g., of sapphire or of CaF_2), single-mode powers

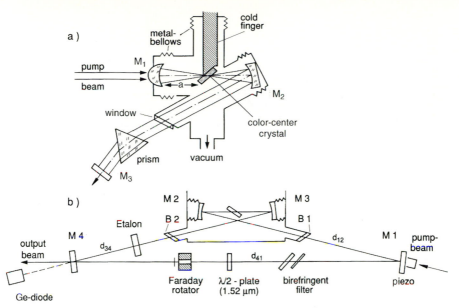

Fig. 5.78a,b. Two possible resonator designs for cw color-center lasers: (**a**) folded linear resonator with astigmatic compensation; and (**b**) ring resonator with optical diode for enforcing only one direction of the traveling laser wave and tuning elements (birefringent filter and etalon) [5.145]

up to 75% of the multimode output can be reached, since the gain profile is homogeneous.

Spatial hole burning can be avoided in the ring resonator (Fig. 5.78b). This facilitates stable single-mode operation and yields higher output powers. For example, a NaCl:OH color-center laser with a ring resonator yields 1.6 W output power at $\lambda = 1.55\,\mu\text{m}$ when pumped by 6 W of a cw YAG laser at $\lambda = 1.065\,\mu\text{m}$ [5.145].

When an $F_A(\text{II})$- or F_2^+-color-center laser is pumped by a linearly polarized cw YAG laser, the output power degrades within a few minutes to a few percent of its initial value. The reason for this is as follows: many of the laser-active color centers possess a symmetry axis, for example, the (110) direction. Two-photon absorption of pump photons brings the system into an excited state of another configuration. Fluorescence releases the excited centers back into a ground state that, however, differs in its orientation from the absorbing state and therefore does not absorb the linearly polarized pump wave. This optical pumping process with changing orientation leads to a gradual bleaching of the original ground-state population, which could absorb the pump light. This orientation bleaching can be avoided when the crystal is irradiated during laser operation by the light of a mercury lamp or an argon laser, which "repumps" the centers with "wrong" orientation back into the initial ground state [5.141].

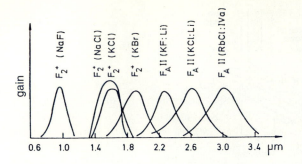

Fig. 5.79. Spectral ranges of emission bands for different color-center crystals

With different color-center crystals the total spectral range covered by existing color-center lasers extends from $0.65-3.4\,\mu m$. The luminescence bands of some color-center alkali halide crystals are exhibited in Fig. 5.79. Typical characteristics of some commonly used color-center lasers are compiled in Table 5.3 and are compared with some vibronic solid-state lasers. Recently room-temperature color-center lasers have been realized which are pumped by diode lasers [5.142].

The linewidth $\Delta\nu$ of a single-mode color-center laser is mainly determined by fluctuations of the optical path length in the cavity (Sect. 5.4). Besides the contribution $\Delta\nu_m$ caused by mechanical instabilities of the resonator, temperature fluctuations in the crystal, caused by pump power variations or by temperature variations of the cooling system, further increase the linewidth by adding contributions $\Delta\nu_p$ and $\Delta\nu_t$. Since all three contributions are indepen-

Table 5.3. Characteristic data of some tunable solid-state lasers

Laser	Composition	Tuning range [nm]	Operation temperature [K]	Pump
Ti:sapphire	$Al_2O_3:Ti^{3+}$	660−986	300	Ar laser
Alexandrite	$BeAl_2O_4:Cr^{3+}$	710−820	300−600	Flashlamp
		720−842	300	Kr laser
Emerald	$Be_3Al_2(SiO_3)_6:Cr^{3+}$	660−842	300	Kr^+ laser
Olivine	$Mg_2SiO_4:Cr^{4+}$	1160−1350	300	YAG laser
Flouride	$SrAlF_5:Cr^{3+}$	825−1010	300	Kr laser
laser	$KZnF_3:Cr^{3+}$	1650−2070	77	cw Nd:YAG laser
Magnesium fluoride	$Ni:MgF_2$	1600−1740	77	YAG laser
F_2^+ F-center	$NaCl/F_2^+$	1400−1750	77	cw Nd:YAG laser
Holmium laser	Ho:YLF	2000−2100	300	Flashlamp
Erbium laser	Er:YAG	2900−2950	300	Flashlamp

dent, we obtain for the total frequency fluctuations

$$\Delta \nu = \sqrt{\Delta \nu_m^2 + \Delta \nu_p^2 + \Delta \nu_t^2} \, . \tag{5.103}$$

The linewidth of the unstabilized single-mode laser has been measured to be smaller than 260 kHz, which was the resolution limit of the measuring system [5.144]. An estimated value for the overall linewidth $\Delta \nu$ is 25 kHz [5.146]. This extremely small linewidth is ideally suited to perform high-resolution Doppler-free spectroscopy (Chaps. 7–10).

More examples of color-center lasers in different spectral ranges are given in [5.147–5.149]. Good surveys on color-center lasers can be found in [5.141, 5.149] and, in particular, in [5.92], Chap. 6. All these lasers, which provide tunable sources with narrow bandwidths, have serious competition from cw optical parametric oscillators (see Sect. 5.8.8), which are now available within the tuning range $0.4-4 \, \mu$m.

5.7.5 Dye Lasers

Although tunable solid-state lasers and optical parametric oscillators are more and more competitive, dye lasers in their various modifications in the visible and UV range are still the most widely used types of tunable lasers. Dye lasers were invented independently by P. Sorokin and F.P. Schäfer in 1966 [5.150]. Their active media are organic dye molecules solved in liquids. They display strong broadband fluorescence spectra under excitation by visible or UV light. With different dyes, the overall spectral range where cw or pulsed laser operation has been achieved extends from 300 nm to $1.2 \, \mu$m (Fig. 5.80). Combined with frequency-doubling or mixing techniques (Sect. 5.8), the range of tunable devices where dye lasers are involved ranges from the VUV at 100 nm to the infrared at about $4 \, \mu$m. In this section we briefly summarize the basic physical background and the most important experimental realizations of dye lasers used in high-resolution spectroscopy. For a more extensive treatment the reader is referred to the laser literature [5.1, 5.8, 5.151, 5.152].

When dye molecules in a liquid solvent are irradiated with visible or ultraviolet light, higher vibrational levels of the first excited singlet state S_1 are populated by optical pumping from thermally populated rovibronic levels in the S_0 ground state (Fig. 5.81). Induced by collisions with solvent molecules, the excited dye molecules undergo very fast radiationless transitions into the lowest vibrational level v_0 of S_1 with relaxation times of 10^{-11} to 10^{-12} s. This level is depopulated either by spontaneous emission into the different rovibronic levels of S_0, or by radiationless transitions into a lower triplet state T_1 (intersystem crossing). Since the levels populated by optical pumping are generally above v_0 and since many fluorescence transitions terminate at higher rovibronic levels of S_0, the fluorescence spectrum of a dye molecule is redshifted against its absorption spectrum. This is shown in Fig. 5.81b for rhodamine 6G, the most widely used laser dye.

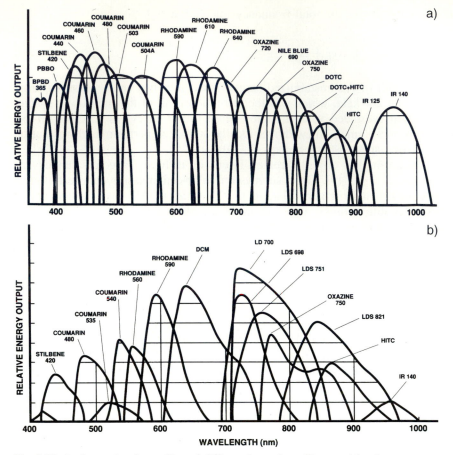

Fig. 5.80a,b. Spectral gain profiles of different laser dyes, illustrated by the output power of pulsed lasers (**a**) and cw dye lasers (**b**) (Lambda Physik and Spectra-Physics information sheets)

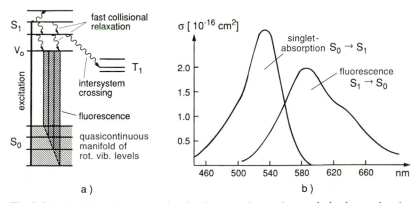

Fig. 5.81. (**a**) Schematic energy level scheme and pumping cycle in dye molecules; (**b**) absorption and fluorescence spectrum of rhodamine 6G dissolved in ethanol

Because of the strong interaction of dye molecules with the solvent, the closely spaced rovibronic levels are collision broadened to such an extent that the different fluorescence lines completely overlap. The absorption and fluorescence spectra therefore consist of a broad continuum, which is homogeneously broadened (Sect. 3.3).

At sufficiently high pump intensity, population inversion may be achieved between the level v_0 in S_1 and higher rovibronic levels v_k in S_0, which have a negligible population at room temperature, due to the small Boltzmann factor $\exp[-E(v_k)/kT]$. As soon as the gain on the transition $v_0(S_1) \rightarrow v_k(S_0)$ exceeds the total losses, laser oscillation starts. The lower level $v_k(S_0)$, which now becomes populated by stimulated emission, is depleted very rapidly by collisions with the solvent molecules. The whole pumping cycle can therefore be described by a four-level system.

According to Sect. 5.2, the spectral gain profile $G(v)$ is determined by the population difference $N(v_0) - N(v_k)$, the absorption cross section $\sigma_{0k}(v)$ at the frequency $v = E(v_0) - E(v_k)/h$, and the length L of the active medium. The net gain coefficient at the frequency v is therefore

$$-2\alpha(v)L = +2L[N(v_0) - N(v_k)] \int \sigma_{0k}(v - v')\,\mathrm{d}v' - \gamma(v) \,,$$

where $\gamma(v)$ is the total losses per round-trip, which may depend on the frequency v.

The spectral profile of $\sigma(v)$ is essentially determined by the Franck–Condon factors for the different transitions ($v_0 \rightarrow v_k$). The total losses are determined by resonator losses (mirror transmission and absorption in optical components) and by absorption losses in the active dye medium. The latter are mainly caused by two effects:

(a) The intersystem crossing transitions $S_1 \rightarrow T_1$ not only diminish the population $N(v_0)$ and therefore the attainable inversion, but they also lead to an increased population $N(T_1)$ of the triplet state. The triplet absorption spectrum due to the transitions $T_1 \rightarrow T_m$ into higher triplet states T_m partly overlaps with the singlet fluorescence spectrum. This results in additional absorption losses $N(T_1)\alpha_T(v)L$ for the dye laser radiation. Because of the long lifetimes of molecules in this lowest triplet state, which can only relax into the S_0 ground state by slow phosphorescence or by collisional deactivation, the population density $N(T_1)$ may become undesirably large. One therefore has to take care that these triplet molecules are removed from the active zone as quickly as possible. This may be accomplished by mixing *triplet-quenching additives* to the dye solution. These are molecules that quench the triplet population effectively by spin-exchange collisions enhancing the intersystem crossing rate $T_1 \rightarrow S_0$. Examples are O_2 or cyclo-octotetraene (COT). Another solution of the triplet problem is *mechanical quenching*, used in cw dye lasers. This means that the triplet molecules are transported very rapidly through the active zone. The transit time should be much smaller than the triplet lifetime. This is achieved, e.g., by fast-flowing free jets, where the molecules pass the active zone in the focus of the pump laser in about 10^{-6} s.

(b) For many dye molecules the absorption spectra $S_1 \rightarrow S_m$, corresponding to transitions from the optically pumped singlet state S_1 to still higher states S_m, partly overlap with the gain profile of the laser transition $S_1 \rightarrow S_0$. These inevitable losses often restrict the spectral range where the net gain is larger than the losses [5.151].

The essential characteristic of dye lasers is their broad homogeneous gain profile. Under ideal experimental conditions, homogeneous broadening allows *all* excited dye molecules to contribute to the gain at a single frequency. This implies that under single-mode operation the output power should not be much lower than the multimode power (Sect. 5.3), provided that the selecting intracavity elements do not introduce large additional losses.

The experimental realizations of dye lasers employ either flashlamps, pulsed lasers, or cw lasers as pumping sources. Recently, several experiments on pumping of dye molecules in the gas phase by high-energy electrons have been reported [5.153–5.155].

We now present the most important types of dye lasers in practical use for high-resolution spectroscopy.

a) Flashlamp-Pumped Dye Lasers

Flashlamp-pumped dye lasers [5.156, 5.157] have the advantage that they do not need expensive pump lasers. Figure 5.82 displays two commonly used pumping arrangements. The linear flashlamp, which is filled with xenon, is placed along one of the focal lines of a cylindric reflector with elliptical cross section. The liquid dye solution flowing through a glass tube in the second focal line is pumped by the focused light of the flashlamp. The useful maximum pumping time is again limited by the triplet conversion rate. By using additives as triplet quenchers, the triplet absorption is greatly reduced and long pulse emission has been obtained. Low-inductance pulsed power supplies have been designed to achieve short flashlamp pulses below 1 μs. A pulse-forming network of several capacitors is superior to the single energy storage capacitor because it matches the circuit impedance to that of the lamps, therefore a constant flashlight intensity over a period of 60–70 μs can be achieved [5.158]. With two linear flashlamps in a double-elliptical reflector, a reliable rhodamine 6G dye laser with 60-μs pulse duration, and a repetition rate up to 100 Hz, an *average* power of 4 W has been demonstrated. With the pumping geometry of Fig. 5.82b, which takes advantage of four linear flashlamps, a very high collection efficiency for the pump light is achieved. The light rays parallel to the plane of the figure are collected into an angle of about 85° by the rear reflector, the aplanatic lens directly in front of the flashlamp, the condenser lens, and the cylindrical mirrors. An average laser output power of 100 W is possible with this design [5.159].

Similar to the laser-pumped dye lasers, reduction of the linewidth and wavelength tuning can be accomplished by prisms, gratings, interference filters [5.160], Lyot filters [5.161], and interferometers [5.162, 5.163].

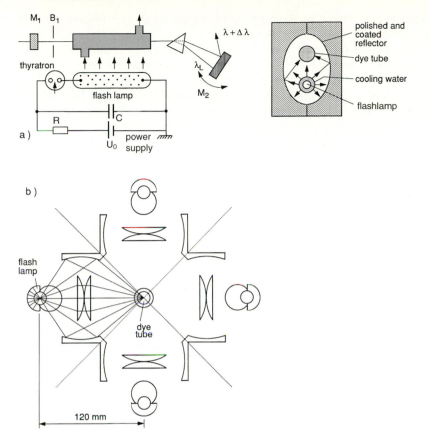

Fig. 5.82a,b. Two possible pumping designs for flashlamp-pumped dye lasers: (**a**) elliptical reflector geometry for pumping of a flowing dye solution by one linear xenon flashlamp; (**b**) arrangement of four flashlamps for higher pump powers [5.159]

One drawback of flashlamp-pumped dye lasers is the bad optical quality of the dye solution during the pumping process. Local variations of the refractive index due to schlieren in the flowing liquid, and temperature gradients due to the nonuniform absorption of the pump light deteriorate the optical homogeneity. The frequency jitter of narrow-band flashlamp-pumped dye lasers is therefore generally larger than the linewidth obtained in a single shot and they are mainly used in multimode operation. However, with three FPI inside the laser cavity, single-mode operation of a flashlamp-pumped dye laser has been reported [5.164]. The linewidth achieved was 4 MHz, stable to within 12 MHz. A better and more reliable solution for achieving single-mode operation is injection seeding. If a few milliwatts of narrow-band radiation from a single-mode cw dye laser is injected into the resonator of the flashlamp-pumped dye laser, the threshold is reached earlier for the injected wavelength than for the others. Due to the homogeneous gain profile, most of the induced emission power will then be concentrated at the injected wavelength [5.165].

A convenient tuning method of flashlamp-pumped dye lasers is based on intracavity electro-optically tunable Lyot filters (Sect. 4.2), which have the advantage that the laser wavelength can be tuned in a short time over a large spectral range [5.166, 5.167]. This is of particular importance for the spectroscopy of fast transient species, such as radicals formed in intermediate stages of chemical reactions. A single-element electro-optical birefringent filter can be used to tune a flashlamp-pumped dye laser across the entire dye emission band. With an electro-optically tunable Lyot filter (Sect. 4.2.9) in combination with a grating a spectral bandwidth of below 10^{-3} nm was achieved even without injection seeding [5.161].

b) Pulsed Laser-Pumped Dye Lasers

The first dye laser, developed independently by Schäfer [5.168] and Sorokin [5.169] in 1966, was pumped by a ruby laser. In the early days of dye laser development, giant-pulse ruby lasers, frequency-doubled Nd:glass lasers, and nitrogen lasers were the main pumping sources. All these lasers have sufficiently short pulse durations T_p, which are shorter than the intersystem crossing time constant $T_{IC}(S_1 \rightarrow T_1)$.

The short wavelength $\lambda = 337$ nm of the nitrogen laser permits pumping of dyes with fluorescence spectra from the near UV up to the near infrared. The high pump power available from this laser source allows sufficient inversion, even in dyes with lower quantum efficiency [5.170–5.174]. At present the most important dye laser pumps are the excimer laser [5.175, 5.176], the frequency-doubled or -tripled output of high-power Nd:YAG or Nd:glass lasers [5.177, 5.178], or copper-vapor lasers [5.179].

Various pumping geometries and resonator designs have been proposed or demonstrated [5.151]. In transverse pumping (Fig. 5.83), the pump laser beam is focused by a cylindrical lens into the dye cell. Since the absorption coefficient for the pump radiation is large, the pump beam is strongly attenuated

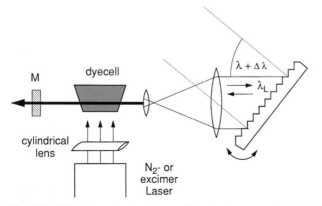

Fig. 5.83. Hansch-type dye laser with transverse pumping and beam expander [5.171]. The wavelength is tuned by turning the Littrow grating. Light with a different wavelength $\lambda_D + \Delta\lambda$ is diffracted out of the resonator

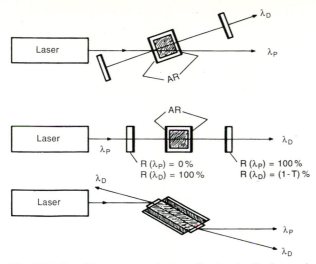

Fig. 5.84. Possible resonator designs for longitudinal pumping of dye lasers [5.151]

and the maximum inversion in the dye cell is reached in a thin layer directly behind the entrance window along the focal line of the cylindrical lens. This geometrical restriction to a small gain zone gives rise to large diffraction losses and beam divergence. This divergent beam is converted by a telescope of two lenses into a parallel beam with enlarged diameter and is then reflected by a Littrow grating, which acts as wavelength selector (Hänsch-type arrangement) [5.171].

In longitudinal pumping schemes (Fig. 5.84), the pump beam enters the dye laser resonator through one of the mirrors, which are transparent for the pump wavelength. This arrangement avoids the drawback of nonuniform pumping, present in the transverse pumping scheme. However, it needs a good beam quality of the pump laser and is therefore not suitable for excimer lasers as pump sources, but is used more and more frequently for pumping with frequency-doubled Nd:YAG lasers [5.177].

If wavelength selection is performed with a grating, it is preferable to expand the dye laser beam for two reasons.

(a) The resolving power of a grating is proportional to the product Nm of the number of illuminated grooves N times the diffraction order m (Sect. 4.1). The more grooves that are hit by the laser beam, the better is the spectral resolution and the smaller is the resulting laser linewidth.

(b) The power density without beam expansion might be high enough to damage the grating surface.

The enlargement of the beam can be accomplished either with a beam-expanding telescope (Hänsch-type laser [5.171, 5.172], Fig. 5.83) or by using grazing incidence under an angle of $\alpha \simeq 90°$ against the grating normal (Littman-type laser, Fig. 5.85). The latter arrangement [5.180] allows very

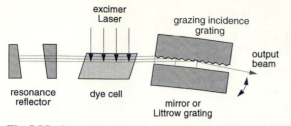

Fig. 5.85. Short dye laser cavity with grazing incidence grating. Wavelength tuning is accomplished by turning the end mirror, which may also be replaced by a Littrow grating

short resonator lengths (below 10 cm). This has the advantage that even for short pump pulses, the induced dye laser photons can make several transits through the resonator during the pumping time. A further, very important advantage is the large spacing $\delta \nu = \frac{1}{2} c/d$ of the resonator modes, which allows single-mode operation with only one etalon or even without any etalon but with a fixed grating position and a turnable mirror M_2 (Fig. 5.86) [5.181, 5.182]. At the wavelength λ the first diffraction order is reflected from the grazing incidence grating ($\alpha \approx 88-89°$) into the direction β determined by the grating equation (4.21)

$$\lambda = d(\sin \alpha + \sin \beta) \simeq d(1 + \sin \beta) .$$

For $d = 4 \times 10^{-5}$ cm (2500 lines/mm) and $\lambda = 400$ nm $\rightarrow \beta = 0°$, which means that the first diffraction order is reflected normal to the grating surface onto mirror M_2. With the arrangement in Fig. 5.86, a single-shot linewidth of less than 300 MHz and a time-averaged linewidth of 750 MHz have been achieved. Wavelength tuning is accomplished by tilting the mirror M_2.

For reliable single-mode operation of the Littman laser longitudinal pumping is better than transverse pumping, because the dye cell is shorter and inhomogenities of the refractive index caused by the pump process are less severe [5.183].

The reflectivity of the grating is very low at grazing incidence and the round-trip losses are therefore high. Using Brewster prisms for preexpansion of the laser beam (Fig. 5.87), the angle of incidence α at the grazing incidence

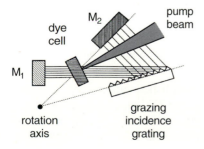

Fig. 5.86. Littman laser with grazing incidence grating and Littrow grating using longitudinal pumping

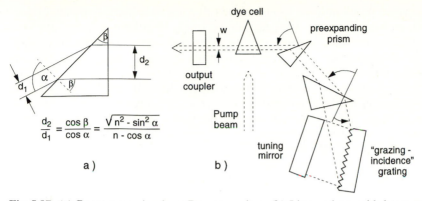

$$\frac{d_2}{d_1} = \frac{\cos\beta}{\cos\alpha} = \frac{\sqrt{n^2 - \sin^2\alpha}}{n - \cos\alpha}$$

Fig. 5.87. (a) Beam expansion by a Brewster prism; **(b)** Littman laser with beam-expanding prisms and grazing incidence grating

grating can be decreased from 89° to 85−80° achieving the same total expansion factor. This reduces the reflection losses considerably [5.184, 5.185].

Example 5.25

Assume a reflectivity of $R(\alpha = 89°) = 0.05$ into the wanted first order at $\beta = 0°$. The attenuation factor per round-trip is then $(0.05)^2 \simeq 2.5 \times 10^{-3}$! The gain factor per round-trip must be larger than 4×10^2 in order to reach threshold. With preexpanding prisms and an angle $\alpha = 85°$, the reflectivity of the grating increases to $R(\alpha = 85°) = 0.25$, which yields the attenuation factor 0.06. Threshold is now reached if the gain factor exceeds 16.

In order to increase the laser power the output beam of the dye laser oscillator is sent through one or more amplifying dye cells, which are pumped by the same pump laser (Fig. 5.88).

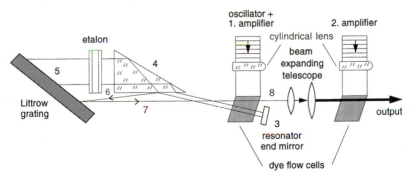

Fig. 5.88. Excimer laser-pumped dye laser with oscillator and two amplifier stages. This design suppresses effectively the ASE (Lambda Physik FL 3002)

A serious problem in all laser-pumped dye lasers is the spontaneous background, emitted from the pumped volume of the oscillator and the amplifier cells. This spontaneous emission is amplified when passing through the gain medium. It represents a perturbing, spectrally broad background of the narrow laser emission. This amplified spontaneous emission (ASE) can partly be suppressed by prisms and apertures between the different amplifying cells. An elegant solution is illustrated in Fig. 5.88. The end face of a prism expander serves as beam splitter. Part of the laser beam is refracted, expanded, and spectrally narrowed by the Littrow grating and an etalon [5.175] before it is sent back into the oscillator traversing the path 3–4–5–4–3. The spectral bandwidth of the oscillator is thus narrowed and only a small fraction of the ASE is coupled back into the oscillator. The partial beam 6 reflected at the prism end face is sent to the same grating before it passes through another part of the first dye cell, where it is further amplified (path: 3–6–7–8). Again only a small fraction of the ASE can reach the narrow gain region along the focal lines of the cylindrical lenses used for pumping the amplifiers. The newly developed "super pure" design shown in Fig. 5.88 further decreases the ASE by a factor of 10 compared to the former device [5.186].

For high-resolution spectroscopy the bandwidth of the dye laser should be as small as possible. With two etalons having different free spectral ranges, single-mode operation of the Hänsch-type laser (Fig. 5.83) can be achieved. For continuous tuning both etalons and the optical length of the laser resonator must be tuned synchronously. This can be realized with computer control (Sect. 5.4.5)

A simple mechanical solution for wavelength tuning of the dye laser in Fig. 5.85 without mode hops has been realized by Littman [5.182] for a short

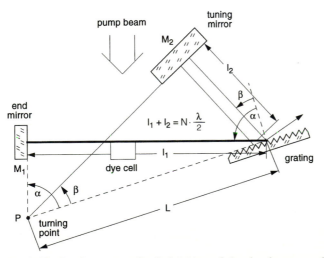

Fig. 5.89. Continuous mechanical tuning of the dye laser wavelength without mode hops by tilting mirror M_2 around an axis through the intersection of two planes through the grating surface and the surface of mirror M_2

laser cavity (Fig. 5.89). If the turning axis of mirror M_2 coincides with the intersection of the two planes through mirror M_2 and the grating surface, the two conditions for the resonance wavelength (cavity length $l_1 + l_2 = N \cdot \lambda/2$ and the diffracted light must always have vertical incidence on mirror M_2) can be simultaneously fulfilled. In this case we obtain from Fig. 5.89 the relations:

$$N\lambda = 2L(\sin\alpha + \sin\beta) , \quad \text{and}$$
$$\lambda = d(\sin\alpha + \sin\beta) \;\Rightarrow\; L = Nd/2 . \tag{5.104}$$

With such a system single-mode operation without etalons has been achieved. The wave number could be tuned over a range of $100\,\mathrm{cm}^{-1}$ without mode hops.

The spectral bandwidth of a single-mode pulsed laser with pulse duration ΔT is, in principle, limited by the Fourier limit, that is,

$$\Delta \nu = a/\Delta T , \tag{5.105}$$

where the constant $a \simeq 1$ depends on the time profile $I(t)$ of the laser pulse. This limit is, however, generally not reached because the center frequency ν_0 of the laser pulse shows a jitter from pulse to pulse, due to fluctuations and thermal instabilities. This is demonstrated by Fig. 5.90 where the spectral profile of a Littman-type single-mode pulsed laser was measured with a Fabry–Perot wavemeter for a single shot and compared with the average

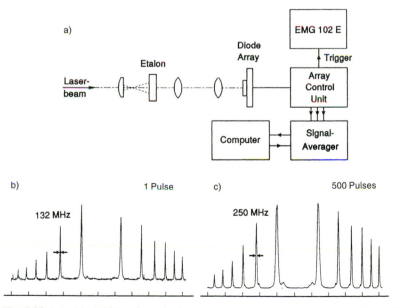

Fig. 5.90a–c. Linewidth of a single-mode pulsed laser measured with a Fabry–Perot wavemeter: (**a**) experimental setup; (**b**) single shot; and (**c**) signal averaged over 500 pulses

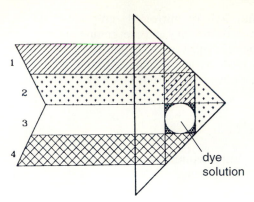

Fig. 5.91. Transversely pumped prismatic amplifier cell (Berthune cell) for more uniform isotropic pumping. The laser beam should have a diameter about four times larger than the bore for the dye. The partial beam 1 traverses the bore from above, beam 2 from behind, beam 4 from below, and beam 3 from the front

over 500 shots. A very stable resonator design and, in particular, temperature stabilization of the dye liquid, which is heated by absorption of the pump laser, decreases both the jitter and the drift of the laser wavelength.

A more reliable technique for achieving really Fourier-limited pulses is based on the amplification of a cw single-mode laser in several pulsed amplifier cells. The expenditure for this setup is, however, much larger because one needs a cw dye laser with a pump laser and a pulsed pump laser for the amplifier cells. Since the Fourier limit $\Delta \nu = 1/\Delta T$ decreases with increasing pulse width ΔT, copper-vapor lasers with $\Delta T = 50$ ns are optimum for achieving spectrally narrow and frequency-stable pulses. A further advantage of copper-vapor lasers is their high repetition frequency up to $f = 20$ kHz.

In order to maintain the good beam quality of the cw dye laser during its amplification by transversely pumped amplifier cells, the spatial distribution of the inversion density in these cells should be as uniform as possible. Special designs (Fig. 5.91) of prismatic cells, where the pump beam traverses the dye several times after being reflected from the prism end faces, considerably improves the quality of the amplified laser beam profile.

Example 5.26

When the output of a stable cw dye laser ($\Delta \nu \simeq 1$ MHz) is amplified in three amplifier cells, pumped by a copper-vapor laser with a Gaussian time profile $I(t)$ with the halfwidth Δt, Fourier-limited pulses with $\Delta \nu \simeq 40$ MHz and peak powers of 500 kW can be generated. These pulses are wavelength tunable with the wavelength of the cw dye laser.

c) Continuous-Wave Dye Lasers

For sub-Doppler spectroscopy, single-mode cw dye lasers represent the most important laser types besides cw tunable solid-state lasers. Great efforts have therefore been undertaken in many laboratories to increase the output power, tuning range, and frequency stability. Various resonator configurations, pump geometries, and designs of the dye flow system have been successfully tried

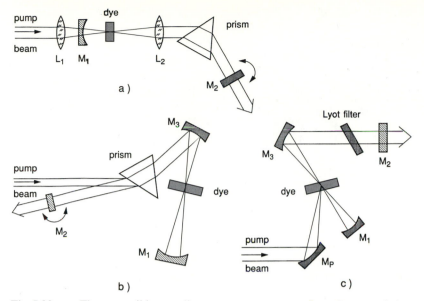

Fig. 5.92a–c. Three possible standing-wave resonator configurations used for cw dye lasers: (**a**) collinear pumping geometry; (**b**) folded astigmatically compensated resonator of the Kogelnik type [5.143] with a Brewster prism for separation of pump beam and dye-laser beam; and (**c**) the pump beam is focused by an extra pump mirror into the dye jet and is tilted against the resonator axis

to realize optimum dye-laser performance. In this section we can only present some examples of the numerous arrangements used in high-resolution spectroscopy.

Figure 5.92 illustrates three possible resonator configurations. The pump beam from an argon or krypton laser enters the resonator either collinearly through the semitransparent mirror M1 and is focused by L1 into the dye (Fig. 5.92a), or the pump beam and dye laser beam are separated by a prism (Fig. 5.92b). In both arrangements the dye laser wavelength can be tuned by tilting the flat end mirror M2. In another commonly used arrangement (Fig. 5.92c), the pump beam is focused by the spherical mirror M_p and crosses the dye medium under a small angle against the resonator axis.

In all these configurations the active zone consists of the focal spot of the pump laser within the dye solution streaming in a laminar free jet of about 0.5–1-mm thickness, which is formed through a carefully designed polished nozzle. At flow velocities of 10 m/s the time of flight for the dye molecules through the focus of the pump laser (about 10 µm) is about 10^{-6} s. During this short period the intersystem crossing rate cannot build up a large triplet concentration, and the triplet losses are therefore small.

For free-running dye jets the viscosity of the liquid solvent must be sufficiently large to ensure the laminar flow necessary for high optical quality of the gain zone. Most jet-stream dye lasers use ethylene glycol or propylene glycol as solvents. Since these alcohols decrease the quantum efficiency

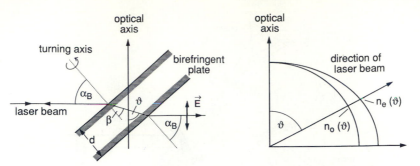

Fig. 5.93. Birefringent plane-parallel plate as wavelength selector inside the laser resonator. For wavelength tuning the plate is turned around an axis parallel to the surface normal. This changes the angle ϑ against the optical axis and thus the difference $n_{\mathrm{e}}(\vartheta) - n_{\mathrm{o}}(\vartheta)$

of several dyes and also do not have optimum thermal properties, the use of water-based dye solutions with viscosity-raising additives can improve the power efficiency and frequency stability of jet-stream cw dye lasers [5.187]. Output powers of more than 30 W have been reported for cw dye lasers [5.188].

In order to achieve a symmetric beam waist profile of the dye laser mode in the active medium, the astigmatism produced by the spherical folding mirror M_3 in the folded cavity design has to be compensated by the plane-parallel liquid slab of the dye jet, which is tilted under the Brewster angle against the resonator axis [5.143]. The folding angle for optimum compensation depends on the optical thickness of the jet and on the curvature of the folding mirror.

The threshold pump power depends on the size of the pump focus and on the resonator losses, and varies between 1 mW and several watts. The size of the pump focus should be adapted to the beam waist in the dye laser resonator (mode matching). If it is too small, less dye molecules are pumped and the maximum output power is smaller. If it is too large, the inversion for transverse modes exceeds threshold and the dye laser oscillates on several transverse modes. Under optimum conditions, pump efficiencies (dye laser output/pump power input) up to $\eta = 35\%$ have been achieved, yielding dye output powers of 2 W for only 8 W pump power.

Coarse wavelength tuning can be accomplished with a birefringent filter (Lyot filter, see Sect. 4.2.9) that consists of three birefringent plates with thicknesses d, $q_1 d$, $q_2 d$ (where q_1, q_2 are integers), placed under the Brewster angle inside the dye laser resonator (Fig. 5.93). Contrary to the Lyot filter discussed in Sect. 4.2.9, no polarizers are necessary here because the many Brewster faces inside the resonator already define the direction of the polarization vector, which lies in the plane of Fig. 5.93.

When the beam passes through the birefringent plate with thickness d under the angle β against the plate normal, a phase difference $\Delta\varphi = (2\pi/\lambda) \cdot (n_{\mathrm{e}} - n_{\mathrm{o}})\Delta s$ with $\Delta s = d/\cos\beta$ develops between the ordinary and the extraordinary waves. Only those wavelengths λ_{m} can reach oscillation threshold

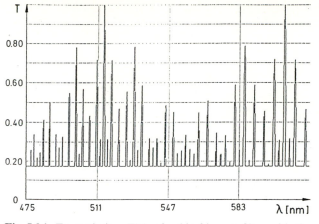

Fig. 5.94. Transmission $T(\lambda)$ of a birefringent filter with three Brewster plates of KDP, with plate thickness $d_1 = 0.34$ mm, $d_2 = 4d_1$, $d_3 = 16d_1$ [5.189]

for which this phase difference is $2m\pi$ ($m = 1, 2, 3, ...$). In this case, the plane of polarization of the incident wave has been turned by $m\pi$ and the transmitted wave is again linearly polarized in the same direction as the incident wave. For all other wavelengths the transmitted wave is elliptically polarized and suffers reflection losses at the Brewster end faces. The transmission curve $T(\lambda)$ of a three-stage birefringent filter is depicted in Fig. 5.94. The laser will oscillate on the transmission maximum that is closest to the gain maximum of the dye medium [5.189, 5.190]. Turning the Lyot filter around the axis in Fig. 5.93 will shift all these maxima.

For single-mode operation additional wavelength-selecting elements have to be inserted into the resonator (Sect. 5.4.3). In most designs two FPI etalons with different free spectral ranges are employed [5.191, 5.192]. Continuous tuning of the single-mode laser demands synchronous control of the cavity length and the transmission maxima of the selecting elements (Sect. 5.5). Figure 5.95 shows a commercial version of a single-mode cw dye laser. The optical path length of the cavity can be conveniently tuned by turning a tilted plane-parallel glass plate inside the resonator (galvo-plate). If the tilting range is restricted to a small interval around the Brewster angle, the reflection losses remain negligible (see Sect. 5.5.1).

The scanning etalon can be realized by the piezo-tuned prism FPI etalon in Fig. 5.42 with a free spectral range of about 10 GHz. It can be locked to the oscillating cavity eigenfrequency by a servo loop: if the transmission maximum ν_T of the FPI is slightly modulated by an ac voltage fed to the piezoelement, the laser intensity will show this modulation with a phase depending on the difference $\nu_c - \nu_T$ between the cavity resonance ν_c and the transmission peak ν_T. This phase-sensitive error signal can be used to keep the difference $\nu_c - \nu_T$ always zero. If only the prism FPI is tuned synchronously with the cavity length, tuning ranges of about 30 GHz ($\hat{=} 1$ cm^{-1}) can be covered without mode hops. For larger tuning ranges the second thin etalon and

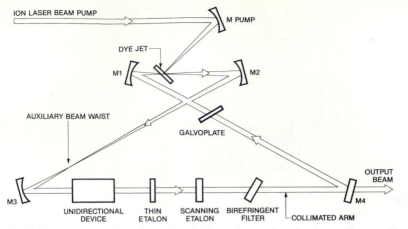

Fig. 5.95. Commercial version of a single-mode cw ring dye laser (Spectra-Physics)

the Lyot filter must also be tuned synchronously. This demands a more sophisticated servo system, which can, however, be provided by computer control.

A disadvantage of cw dye lasers with standing-wave cavities is spatial hole burning (Sect. 5.3.3), which impedes single-mode operation and prevents all of the molecules within the pump region from contributing to laser emission. This effect can be avoided in ring resonators, where the laser wave propagates in only one direction (Sect. 5.2.7). Ring lasers therefore show, in principle, higher output powers and more stable single-mode operation [5.193]. However, their design and their alignment are more critical than for standing-wave resonators.

In order to avoid laser waves propagating in both directions through the ring resonator, losses must be higher for one direction than for the other. This can be achieved with an optical diode [5.32]. This diode essentially consists of a birefringent crystal and a Faraday rotator (Fig. 5.18), which turns the bifringent rotation back to the input polarization for the wave incident in one direction but increases the rotation for the other direction.

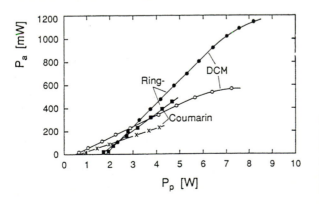

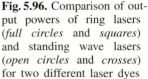

Fig. 5.96. Comparison of output powers of ring lasers (*full circles* and *squares*) and standing wave lasers (*open circles* and *crosses*) for two different laser dyes

Table 5.4. Characteristic parameters of some dye lasers pumped by different sources

Pump	Tuning range [nm]	Pulse width [ns]	Peak power [W]	Pulse energy [mJ]	Repetition rate [s^{-1}]	Average output [W]
Excimer laser	370−985	10−200	$\leq 10^7$	≤ 300	20−200	0.1−10
N$_2$ laser	370−1020	1−10	$< 10^5$	< 1	$< 10^3$	0.01−0.1
Flashlamp	300−800	300−10^4	10^2−4	< 5000	1−100	0.1−200
Ar$^+$ laser	350−900	CW	CW	−	CW	0.1−5
Kr$^+$ laser	400−1100	CW	CW	−		0.1−3
Nd:YAG laser $\lambda/2$: 530 nm $\lambda/3$: 355 nm	400−920	10−20	10^5−10^7	10−100	10−30	0.1−1
Copper-vapor laser	530−890	30−50	$\simeq 10^4$−10^5	≈ 1	$\simeq 10^4$	≤ 10

The specific characteristics of a cw ring dye laser regarding output power and linewidth have been studied in [5.193]. A theoretical treatment of mode selection in Fabry–Perot-type and in ring resonators can be found in [5.194]. Because of the many optical elements in the ring resonator, the losses are generally slightly higher than in standing-wave resonators. This causes a higher threshold. Since more molecules contribute to the gain, the slope efficiency $\eta_{\mathrm{al}} = \mathrm{d}P_{\mathrm{out}}/\mathrm{d}P_{\mathrm{in}}$ is, however, higher. At higher input powers well above threshold, the output power of ring lasers is therefore higher (Fig. 5.96).

The characteristic data of different dye laser types are compiled in Table 5.4 for "typical" operation conditions in order to give a survey on typical orders of magnitude for these figures. The tuning ranges depend not only on the dyes but also on the pump lasers. They are slightly different for pulsed lasers pumped by excimer lasers from that of cw lasers pumped by argon or krypton lasers. Meanwhile, frequency-doubled Nd:YAG lasers are used more and more frequently as pump sources for dye lasers. Many data on dye laser wavelengths, tuning ranges and possible pump lasers can be found in [5.6].

5.7.6 Excimer Lasers

Excimers (that is, excited dimers) are molecules that are bound in excited states but are unstable in their electronic ground states. Examples are diatomic molecules composed of closed-shell atoms with $^1S_0$ ground states, such as the rare gases, which form stable *excited dimers* He$_2^*$, Ar$_2^*$, etc., but have a mainly repulsive potential in the ground state with a very shallow van der Waals minimum (Fig. 5.97). The well depth ϵ of this minimum is small compared to the thermal energy kT at room temperature, which prevents the stable formation of ground-state molecules. Mixed excimers such as KF or XeNa can be formed from combinations of closed-shell/open-shell atoms (for example, combination of atomic states $^1S + {}^2S$, $^1S + {}^2P$, $^1S + {}^3P$, etc.), which lead to repulsive ground-state potentials [5.195, 5.196].

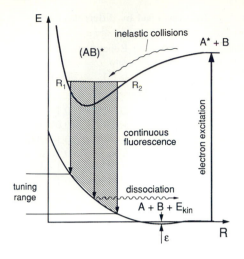

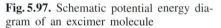

Fig. 5.97. Schematic potential energy diagram of an excimer molecule

These excimers are ideal candidates for forming the active medium of tunable lasers since inversion between the upper bound state and the dissociating lower state is automatically maintained. This inversion is maintained because the lower state dissociates very rapidly ($\simeq 10^{-12}-10^{-13}$ s) and the frequently occurring bottleneck caused by a small depletion rate of the lower laser level is prevented. The output power of excimer lasers mainly depends on the excitation rate of the *upper* state.

The tunability range depends on the slope of the repulsive potential and on the internuclear distances R_1 and R_2 of the classical turning points in the excited vibrational levels. The spectral gain profile is determined by the Franck–Condon factors for bound–free transitions. The corresponding intensity distribution $I(\omega)$ of the fluorescence from the upper vibrational levels shows a modulatory structure (see Fig. 2.14) reflecting the R dependence $|\psi_{\mathrm{vib}}(R)|^2$ of the vibrational wave function in these levels [5.197].

The gain of the active medium at the frequency $\omega = (E_k - E_i)/\hbar$ is, according to (5.2), given by

$$\alpha(\omega) = [N_i - (g_i/g_k)N_k]\sigma(\omega) , \tag{5.106}$$

where the absorption cross section $\sigma(\omega)$ is related to the spontaneous transition probability $A_{ki} = 1/\tau_k$ [5.195] by

$$\int_{\omega_1}^{\omega_2} \sigma(\omega)\,\mathrm{d}\omega = (\lambda/2)^2 A_{ki} = \frac{(\lambda/2)^2}{\tau_k} . \tag{5.107}$$

Because of the broad spectral range $\Delta\omega = \omega_1 - \omega_2$, the cross section $\sigma(\omega)$ may be very small in spite of the large overall transition probability indicated by the short upper-state lifetime τ_k. Consequently, a high population density N_k is necessary to achieve sufficient gain. Since the pumping rate R_p has to compete with the spontaneous transition rate, which is proportional to the third power of the transition frequency ω, the pumping power $R_p \hbar \omega$ at laser threshold scales at least as the fourth power of the lasing frequency. *Short-wavelength lasers therefore require high pumping powers* [5.198, 5.199].

Pumping sources are provided by high-voltage, high-current electron beam sources, such as the FEBETRON [5.200] or by fast transverse discharges [5.201]. The primary step is the excitation of atoms by electron impact. Since the excitation of the upper excimer states needs collisions between these excited atoms and ground-state atoms (remember that there are no ground-state excimer molecules), high atom densities are required to form a sufficient number N^* of excimers in the upper state. These high pressures impede a uniform discharge along the whole active zone in the channel. Preionization by fast electrons or by ultraviolet radiation is required to achieve a large and uniform density of excimers, and specially formed electrodes are used [5.202]. Fast switches, such as magnetically confined thyratrons have been developed, and the inductances of the discharge circuits must be matched to the discharge time [5.203].

Up to now the rare-gas halide excimers, such as KrF, ArF, or XeCl, form the active medium of the most advanced UV excimer lasers. Similar to the nitrogen laser, these rare-gas halide lasers can be pumped by fast transverse discharges, and lasers of this type are the most common commercial excimer lasers (Table 5.5).

Inversion is reached by a sufficiently fast and large population increase of the upper laser level. This is achieved through a chain of different collision processes that are still not been completely understood for all excimer lasers. As an example of the complexity of these processes, some possible paths to inversion in XeCl excimer lasers, which use a mixture of Xe, HCl, and He or Ne as gas filling, are given by

$$
\begin{aligned}
\mathrm{Xe} + \mathrm{e}^- &\begin{cases} \rightarrow \mathrm{Xe}^* + \mathrm{e}^- \ , \\ \rightarrow \mathrm{Xe}^+ + 2\mathrm{e}^- \ , \end{cases} \\
\mathrm{Xe}^* + \mathrm{Cl}_2 &\rightarrow \mathrm{XeCl}^* + \mathrm{Cl} \ , \\
\mathrm{Xe}^* + \mathrm{HCl} &\rightarrow \mathrm{XeCl}^* + \mathrm{H} \ , \\
\mathrm{Xe}^+ + \mathrm{Cl}^- + \mathrm{M} &\rightarrow \mathrm{XeCl}^* + \mathrm{M} \ .
\end{aligned} \tag{5.108}
$$

Table 5.5. Characteristic data of some excimer lasers. (Pulse width: $10-200\,\mu s$; repetition frequency: $1-200\,s^{-1}$, depending on the model; output beam divergence: $2 \times 4\,\mathrm{mrad}$; jitter of the pulse energy: $3-10$; time jitter: $1-10\,\mu s$, depending on the model)

Laser medium	F_2	ArF	KrCl	KrF	XeCl	XeF
Wavelength [nm]	157	193	222	248	308	357
Pulse energy [mJ]	15	≤ 500	≤ 60	≤ 1000	≤ 500	200

All these formation processes of XeCl* occur very rapidly on a time scale of $10^{-8}-10^{-9}$ s and have to compete with quenching processes such as

$$XeCl^* + He \rightarrow Xe + Cl + He ,$$

which diminish the inversion.

The pulse width of most excimer lasers lies within $5-20$ ns. Recently, long-pulse XeCl lasers have been developed, which have pulse widths of $T > 300$ ns [5.204]. They allow amplification of single-mode cw dye lasers with Fourier-limited bandwidths of $\Delta\nu < 2$ MHz at peak powers of $P > 10$ kW.

More details on experimental designs and on the physics of excimer lasers can be found in [5.196, 5.204–5.206].

5.7.7 Free-Electron Lasers

In recent years a completely novel concept of a tunable laser has been developed that does not use atoms or molecules as an active medium, but rather "free" electrons in a specially designed magnetic field. The first free-electron laser (FEL) was realized by Madey and coworkers [5.207]. A schematic diagram of the FEL is shown in Fig. 5.98. The high-energy relativistic electrons from an accelerator pass along a static, spatially periodic magnetic field $\boldsymbol{B}$, which can be realized, for example, by a doubly-wound helical superconducting magnet (wiggler) providing a circularly polarized $\boldsymbol{B}$ field.

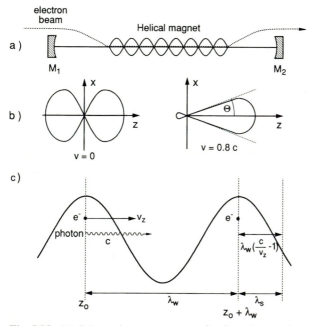

Fig. 5.98. (a) Schematic arrangement of a free-electron laser; **(b)** radiation of a dipole at rest $(v = 0)$ and a moving dipole with $v \simeq c$; **(c)** phase-matching condition

The basic physics of the FEL and the process in which FEL radiation originates can be understood in a classical model, following the representation in [5.208]. Because of the Lorentz force, the electrons passing through the wiggler undergo periodic oscillations, resulting in the emission of radiation. For an electron oscillating in the x-direction around a point at rest, the angular distribution of such a dipole radiation is $I(\theta) = I_0 \cdot \sin^2 \theta$ (Fig. 5.98b). In contrast, for the relativistic electron with the velocity $v \simeq c$, it is sharply peaked in the forward direction (Fig. 5.98b) within a cone of solid angle $\theta \simeq (1 - v^2/c^2)^{1/2}$. For electrons of energy $E = 100 \, \text{MeV}$, for instance, θ is about 2 mrad. This relativistic dipole radiation is the analog to the spontaneous emission in conventional lasers and can be used to initiate induced emission in the FEL.

The wavelength λ of the emitted light is determined by the wiggler period Λ_w and the following *phase-matching* condition: assume the oscillating electron at the position z_0 in the wiggler emits radiation of all wavelengths. However, the light moves faster than the electron (velocity v_z) in the z-direction. After one wiggler period at $z_1 = z_0 + \Lambda_\text{w}$, there will be a time lag

$$\Delta t = \Lambda_\text{w} \left(\frac{1}{v_z} - \frac{1}{c} \right) ,$$

between the electron and the light emitted at z_0. The light emitted by the electron in z_1 will therefore not be in phase with the light emitted in z_0 unless the time difference $\Delta t = m \cdot T = m \cdot \lambda/c$ is an integer multiple of the light period T. Phase matching can be therefore only be achieved for certain wavelengths

$$\lambda_\text{m} = \frac{\Delta L}{m} = \frac{\Lambda_\text{w}}{m} \left(\frac{c}{v_z} - 1 \right) \quad (m = 1, 2, 3, ...) . \tag{5.109}$$

Only for these wavelengths λ_m are the contributions emitted by the electron at different locations in phase and therefore interfere constructively. The lowest harmonic λ_1 ($m = 1$) of the emitted light has therefore the wavelength $\lambda_1 = \Lambda_\text{w}(c/v_z - 1)$ and can be tuned with the velocity v_z of the electron.

Example 5.27

With $\Lambda_\text{w} = 3 \, \text{cm}$, $E_\text{el} = 10 \, \text{MeV} \rightarrow v_z \simeq 0.999c$, we obtain $\lambda = 40 \, \mu\text{m}$ for $m = 1$ and $\lambda = 13 \, \mu\text{m}$ for $m = 3$, which lies in the mid-infrared. For $E_\text{el} = 100 \, \text{MeV} \Rightarrow v_z = (1 - 1.25) \times 10^{-5} c$ and the phase-matching wavelength has decreased to $\lambda_1 = 1.25 \times 10^{-5} \Lambda_\text{w} = 375 \, \text{nm}$, which is in the UV range.

When the field amplitude of the radiation emitted by a single electron is E_j, the total intensity radiated by N independent electrons is

$$I_\text{tot} = \left| \sum_{j=1}^{N} E_j \, e^{i\varphi_j} \right|^2 , \tag{5.110a}$$

where the phases φ_j of the different contributions may be randomly distributed.

If somehow all electrons emit with the same phase, the total intensity for the case of equal amplitudes $E_j = E_0$ becomes

$$I_{\text{tot}}^{\text{coherent}} = \left| \sum E_j \, e^{i\varphi_j} \right|^2 \propto |NE_0|^2 \propto N^2 I_{\text{el}} , \qquad (5.110b)$$

when $I_{\text{el}} \propto E_0^2$ is the intensity emitted by a single electron. This coherent emission with equal phases therefore yields N times the intensity of the incoherent emission with random phases. It is realized in the FEL.

In order to understand how this can be achieved, we first consider a laser beam with the correct wavelength λ_m that passes along the axis of the wiggler. Electrons that move at the critical velocity $v_c = c\Lambda_w/(\Lambda_w + m\lambda_m)$ are in phase with the laser wave and can be induced to emit a photon that amplifies the laser wave (stimulated Compton scattering). The electron loses the emitted radiation energy and becomes slower. All electrons that are a little bit faster than v_c can lose energy by adding radiation to the laser wave without coming out of phase as long as they are not slower than v_c. On the other hand, electrons that are slower than v_c can *absorb* photons, which makes them faster until they reach the velocity v_c.

This means that the faster electrons contribute to the amplification of the incident laser wave, whereas the slower electrons attenuate it. This stimulated emission of the faster electrons and the absorption of photons by the slower electrons leads to a velocity bunching of the electrons toward the critical velocity v_c and enhances the coherent superposition of their contributions to the radiation field. The energy pumped by the electrons into the radiation field comes from their kinetic energy and has to be replaced by acceleration in RF cavities, if the same electrons in storage rings are to be used for multiple traversions through the wiggler.

This free-electron radiation amplifier can be converted into a laser by providing reflecting mirrors for optical feedback. Such FELs are now in operation at several places in the world. Their advantages are their tunability over a large spectral range from millimeter waves into the VUV region by changing the electron energy. Their potential high output power represents a further plus for FELs. Their definitive disadvantage is the large experimental expenditure that demands, besides a delicate wiggler structure, a high-energy accelerator or a storage ring.

At present FELs with output powers of several kilowatts in the infrared and several watts in the visible have been realized. The Stanford FEL reaches, for example, 130 kW at 3.4 µm, whereas from a cooperation between TRW and Stanford University, peak powers of 1.2 MW at $\lambda = 500$ mm were reported. There are plans to build FELs that cover all wavelengths in the UV down to 10 nm. The spectral brilliance of these sources will be three to four orders of magnitude higher than the advanced third-generation synchrotron radiation sources. More details can be found in the literature [5.208–5.211].

5.8 Nonlinear Optical Mixing Techniques

Besides the various types of tunable lasers discussed in the foregoing sections, sources of tunable coherent radiation have been developed that are based on the nonlinear interaction of intense radiation with atoms or molecules in crystals or in liquid and gaseous phases. Second-harmonic generation, sum- or difference-frequency generation, parametric processes, or stimulated Raman scattering are examples of such nonlinear optical mixing techniques. These techniques cover the whole spectral range from the vacuum ultraviolet (VUV) up to the far infrared (FIR) with sufficiently intense tunable coherent sources. After a brief summary of the basic physics of these devices, we exemplify their applications by presenting some experimentally realized systems [5.212–5.220].

5.8.1 Physical Background

The dielectric polarization P of a medium with nonlinear susceptibility χ, subject to an electric field E, can be written as an expansion in powers of the applied field

$$P = \epsilon_0(\tilde{\chi}^{(1)} E + \tilde{\chi}^{(2)} E^2 + \tilde{\chi}^{(3)} E^3 + ...) , \tag{5.111}$$

where $\tilde{\chi}^{(k)}$ is the kth-order susceptibility tensor of rank $k+1$.

Example 5.28

Consider, for example, the EM wave

$$E = E_1 \cos(\omega_1 t - k_1 z) + E_2 \cos(\omega_2 t - k_2 z) , \tag{5.112}$$

composed of two components incident on the nonlinear medium. The induced polarization at a fixed position (say, $z = 0$) in the crystal is generated by the combined action of both components. The linear time describes the Rayleigh scattering. The quadratic term $\chi^{(2)} E^2$ in (5.111) gives the contributions

$$
\begin{aligned}
P^{(2)} &= \epsilon_0 \tilde{\chi}^{(2)} E^2(z=0) \\
&= \epsilon_0 \tilde{\chi}^{(2)} \left(E_1^2 \cos^2 \omega_1 t + E_2^2 \cos^2 \omega_2 t + 2 E_1 E_2 \cos \omega_1 t \cdot \cos \omega_2 t \right) \\
&= \epsilon_0 \tilde{\chi}^{(2)} \left\{ \frac{1}{2}(E_1^2 + E_2^2) + \frac{1}{2} E_1^2 \cos 2\omega_1 t \right. \\
&\quad \left. + \frac{1}{2} E_2^2 \cos 2\omega_2 t + E_1 \cdot E_2 [\cos(\omega_1 + \omega_2)t + \cos(\omega_1 - \omega_2)t] \right\} ,
\end{aligned}
\tag{5.113}
$$

which represents dc polarization, ac components at the second harmonics $2\omega_1$, $2\omega_2$, and components at the sum or difference frequencies $\omega_1 \pm \omega_2$.

Taking into account that the field amplitudes $\boldsymbol{E}_1$, $\boldsymbol{E}_2$ are vectors and that the second-order susceptibility $\tilde{\chi}^{(2)}$ is a tensor of rank 3 with components χ_{ijk} depending on the symmetry properties of the nonlinear crystal [5.217], we can write the second-order term in the explicit form

$$P_i^{(2)} = \epsilon_0 \sum_{j,k=1}^{3} \chi_{ijk}^{(2)} E_j E_k \quad (1 \triangleq x, 2 \triangleq y, 3 \triangleq z), \qquad (5.114)$$

where P_i $(i = x, y, z)$ gives the ith component of the dielectric polarization $\boldsymbol{P} = \{P_x, P_y, P_z\}$.

Note: The direction of the polarization vector $\boldsymbol{P}$ may be different from those of $\boldsymbol{E}_1$ and $\boldsymbol{E}_2$. The components χ_{ijk} are generally complex and the phase of the polarization differs from that of the driving fields.

Equation (5.114) demonstrates that the components of the induced polarization $\boldsymbol{P}$ are determined by the tensor components χ_{ijk} and the components of the incident fields. Since the sequence $E_j E_k$ produces the same polarization as $E_k E_j$, we obtain

$$\chi_{ijk} = \chi_{ikj} \; .$$

This reduces the 27 components of the susceptibility tensor $\tilde{\chi}^{(2)}$ to 18 independent components.

In isotropic media the reflection of all vectors at the origin should not change the nonlinear susceptibility. This yields $\chi_{ijk} = -\chi_{ijk}$, which could be only fulfilled by $\chi_{ijk} \equiv 0$. In all media with an inversion center the second-order susceptibility tensor vanishes! This means, for instance, that optical frequency doubling in gases is not possible.

In order to reduce the number of indices in the formulas, the components χ_{ijk} are often written in the reduced *Voigt notation*. For the first index the convention $x = 1$, $y = 2$, $z = 3$ is used, whereas the second and third indices are combined as follows: $xx = 1$, $yy = 2$, $zz = 3$, $yz = zy = 4$, $xz = zx = 5$, $xy = yx = 6$. The coefficients in this Voigt notation are named d_{im}. Equation (5.114) can then be written as:

$$\begin{pmatrix} P_1^{(2)} \\ P_2^{(2)} \\ P_3^{(2)} \end{pmatrix} = \varepsilon_0 \begin{pmatrix} d_{11} & d_{12} & d_{13} & d_{14} & d_{15} & d_{16} \\ d_{21} & d_{22} & d_{23} & d_{24} & d_{25} & d_{26} \\ d_{31} & d_{32} & d_{33} & d_{34} & d_{35} & d_{36} \end{pmatrix} \begin{pmatrix} E_1^2 \\ E_2^2 \\ E_3^2 \\ 2E_2 E_3 \\ 2E_1 E_3 \\ 2E_1 E_2 \end{pmatrix} . \qquad (5.115)$$

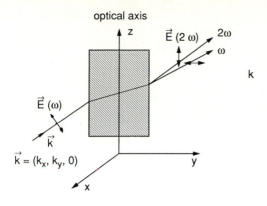

Fig. 5.99. Coordinate system for the description of nonlinear optics in a uniaxial birefringent crystal. An incident wave with wavevector k_λ and $\boldsymbol{k} = (k_x, k_y, 0)$ electric field vector $\boldsymbol{E} = \{E_x, E_y, 0\}$ generates in a KDP crystal the polarization $\boldsymbol{P} = \{0, 0, P_z(2\omega)\}$

Example 5.29

In potassium dihydrogen phosphate (KDP) the only nonvanishing components of the susceptibility tensor are

$$\chi^{(2)}_{xyz} = d_{14} = \chi^{(2)}_{yxz} = d_{25} \quad \text{and} \quad \chi^{(2)}_{zxy} = d_{36} \,.$$

The components of the induced polarization are therefore with $d_{25} = d_{14}$

$$P_x = 2\epsilon_0 d_{14} E_y E_z \,, \quad P_y = 2\epsilon_0 d_{14} E_x E_z \,, \quad P_z = 2\epsilon_0 d_{36} E_x E_y \,.$$

Suppose there is only one incident wave traveling in a direction $\boldsymbol{k}$ with the polarization vector $\boldsymbol{E}$ normal to the optical axis of a uniaxial birefringent crystal, which we choose to be the z-axis (Fig. 5.99). In this case, $E_z = 0$ and the only nonvanishing component of $P(2\omega)$,

$$P_z(2\omega) = 2\epsilon_0 d_{36} E_x(\omega) E_y(\omega) \,,$$

is perpendicular to the polarization plane of the incident wave.

5.8.2 Phase Matching

The nonlinear polarization induced in an atom or molecule acts as a source of new waves at frequencies $\omega = \omega_1 \pm \omega_2$, which propagate through the nonlinear medium with the phase velocity $v_{ph} = \omega/k = c/n(\omega)$. However, the microscopic contributions generated by atoms at different positions (x, y, z) in the nonlinear medium can only add up to a macroscopic wave with appreciable intensity if the vectors of the phase velocities of incident inducing waves and the polarization waves are properly matched. This means that the phases of the contributions $\boldsymbol{P}_i(\omega_1 \pm \omega_2, \boldsymbol{r}_i)$ to the polarization wave generated by all atoms at different locations $\boldsymbol{r}_i$ within the pump beam must be equal at a given point within the pump beam. In this case, the amplitudes $E_i(\omega_1 \pm \omega_2)$ add up in phase in the direction of the pump beam and the intensity increases with the

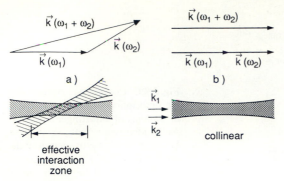

Fig. 5.100a,b. Phase-matching condition as momentum conservation for (**a**) noncollinear and (**b**) collinear propagation of the three waves

length of the interaction zone. This *phase-matching condition* can be written as

$$k(\omega_1 \pm \omega_2) = k(\omega_1) \pm k(\omega_2) \,, \tag{5.116}$$

which may be interpreted as *momentum conservation for the three photons participating in the mixing process.*

The phase-matching condition (5.116) is illustrated by Fig. 5.100. If the angles between the three wave vectors are too large, the overlap region between focused beams becomes too small and the efficiency of the sum- or difference-frequency generation decreases. Maximum overlap is achieved for collinear propagation of all three waves. In this case, $k_1 \| k_2 \| k_3$ and we obtain with $c/n = \omega/k$ and $\omega_3 = \omega_1 \pm \omega_2$ the condition

$$n_3\omega_3 = n_1\omega_1 \pm n_2\omega_2 \quad \Rightarrow \quad n_3 = n_1 = n_2 \,, \tag{5.117}$$

for the refractive indices n_1, n_2, and n_3.

This condition can be fulfilled in unaxial birefringent crystals that have two different refractive indices n_o and n_e for the ordinary and the extraordinary waves. The ordinary wave is polarized in the x–y-plane perpendicular to the optical axis, while the extraordinary wave has its E-vector in a plane defined by the optical axis and the incident beam. While the ordinary index n_o does not depend on the propagation direction, the extraordinary index n_e depends on the directions of both E and k. The refractive indices n_o, n_e and their dependence on the propagation direction in uniaxial birefringent crystals can be illustrated by the index ellipsoid defined by the three principal axes of the dielectric tensor. If these axes are aligned with the x-, y-, z-axes, we obtain with $n = \sqrt{\varepsilon}$ the *index ellipsoid.*

$$\frac{1}{\epsilon_0}\left(\frac{x^2}{n_1^2} + \frac{y^2}{n_2^2} + \frac{z^2}{n_3^2}\right) = \hat{R} \quad \text{with} \quad |\hat{R}| = 1 \,. \tag{5.118}$$

For uniaxial crystals $n_1 = n_2$ and the index ellipsoid becomes symmetric with respect to the optical axis, which we choose to be the z-axis (Fig. 5.101a). If

optical axis   optical axis

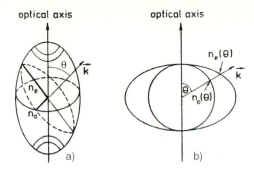

$n_e(\theta)$

$n_o(\theta)$

a)   b)

Fig. 5.101. (a) Index ellipsoid and refractive indices n_o and n_e for two directions of the electric vector of the wave in a plane perpendicular to the wave propagation k. **(b)** Dependence of n_o and n_e on the angle θ between the wave vector k and the optical axis of a uniaxial positive birefringent crystal

we specify a propagation direction k, we can illustrate the refractive indices n_o and n_e experienced by the EM wave $E = E_0 \times \cos(\omega t - k \cdot r)$ in the following way: consider a plane through the center of the index ellipsoid with its normal in the direction of k. The intersection of this plane with the ellipsoid forms an ellipse. The principal axes of this ellipse give the ordinary and extraordinary indices of refraction n_o and n_e, respectively. These principal axes are plotted in Fig. 5.101b as a function of the angle θ between the optical axis and the wave vector k. If the angle θ between k and the optical axis (which is assumed to coincide with the z-axis) is varied, n_o remains constant, while the extraordinary index $n_e(\theta)$ changes according to

$$\frac{1}{n_e^2(\theta)} = \frac{\cos^2 \theta}{n_o^2} + \frac{\sin^2 \theta}{n_e^2(\theta = \pi/2)} \ . \tag{5.119}$$

The uniaxial crystal is called *positively* birefringent if $n_e \geq n_o$ and *negatively* birefringent if $n_e \leq n_o$. It is possible to find nonlinear birefringent crystals where the phase-matching condition (5.117) for collinear phase matching can be fulfilled if one of the three waves at ω_1, ω_2, and $\omega_1 \pm \omega_2$ propagates as an extraordinary wave and the others as ordinary waves through the crystal in a direction θ specified by (5.119) [5.218].

One distinguishes between type-I and type-II phase-matching depending on which of the three waves with ω_1, ω_2, $\omega_3 = \omega_1 \pm \omega_2$ propagates as an ordinary or as an extraordinary wave. Type I corresponds to (1 → e, 2 → e, 3 → o) in positive uniaxial crystals and to (1 → o, 2 → o, 3 → e) in negative uniaxial crystals, whereas type II is characterized by (1 → o, 2 → e, 3 → o) for positive and (1 → e, 2 → o, 3 → e) for negative uniaxial crystals [5.220]. Let us now illustrate these general considerations with some specific examples.

5.8.3 Second-Harmonic Generation

For the case $\omega_1 = \omega_2 = \omega$, the phase-matching condition (5.116) for second-harmonic generation (SHG) becomes

$$k(2\omega) = 2k(\omega) \quad \Rightarrow \quad v_{\mathrm{ph}}(2\omega) = v_{\mathrm{ph}}(\omega) \ , \tag{5.120}$$

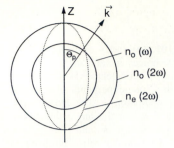

Fig. 5.102. Index matching for SHG in a uniaxial negatively birefringent crystal

which implies that *the phase velocities of the incident and SH wave must be equal.* This can be achieved in a negative birefringent uniaxial crystal (Fig. 5.102) in a certain direction θ_p against the optical axis if in this direction the extraordinary refractive index $n_e(2\omega)$ for the SH wave equals the ordinary index $n_o(\omega)$ for the fundamental wave. When the incident wave propagates as an ordinary wave in this direction θ_p through the crystal, the local contributions of $P(2\omega, r)$ can all add up in phase and a macroscopic SH wave at the frequency 2ω will develop as an extraordinary wave. The polarization direction of this SH wave is orthogonal to that of the fundamental wave. In uniaxial positive birefringent crystals, the phase-matching condition can be fulfilled for type-I phase matching when the fundamental wave at ω travels as an extraordinary wave through the crystal and the second harmonic at 2ω travels as an ordinary wave.

In favorable cases phase-matching is achieved for $\theta = 90°$. This has the advantage that both the fundamental and the SH beams travel collinearly through the crystal, whereas for $\theta \neq 90°$ the power flow direction of the extraordinary wave differs from the propagation direction k_e. This results in a decrease of the overlap region between both beams.

Let us estimate how a possible slight phase mismatch $\Delta n = n(\omega) - n(2\omega)$ affects the intensity of the SH wave. The nonlinear polarization $P(2\omega)$ generated at the position r by the driving field $E_0 \cos[\omega t - k(\omega) \cdot r]$ can be deduced from (5.113) as

$$P(2\omega) = \frac{1}{2} \epsilon_0 \chi_{\text{eff}}^{(2)} E_0^2(\omega)[1 + \cos(2\omega t)] . \tag{5.121}$$

This nonlinear polarization generates a wave

$$P(2\omega, r) = E_0(2\omega) \cdot \cos(2\omega t - k(2\omega) \cdot r) ,$$

with amplitude $E(2\omega)$, which travels with the phase velocity $v(2\omega) = 2\omega/k(2\omega)$ through the crystal. The effective nonlinear coefficient $\chi_{\text{eff}}^{(2)}$ depends on the nonlinear crystal and on the propagation direction.

Assume that the pump wave propagates in the z-direction. Over the path length z a phase difference

$$\Delta\varphi = \Delta k \cdot z = [2k(\omega) - k(2\omega)] \cdot z ,$$

between the fundamental wave at ω and the second-harmonic wave at 2ω has developed. If the field amplitude $E(2\omega)$ always remains small compared to $E(\omega)$ (low conversion efficiency), we may neglect the decrease of $E(\omega)$ with increasing z. Therefore we obtain the total amplitude of the SH wave summed over the path length $z = 0$ to $z = L$ through the nonlinear crystal by integration over the microscopic contribution $dE(2\omega, z)$ generated by $P(2\omega, z)$. From (5.121), one obtains with $\Delta k = |2k(\omega) - k(2\omega)|$ and $dE(2\omega)/dz = [2\omega/\epsilon_0 nc]P(2\omega)$ [5.220]

$$E(2\omega, L) = \int\limits_{z=0}^{L} \chi_{\text{eff}}^{(2)}(\omega/nc)E_0^2(\omega)\cos(\Delta kz)\,dz$$

$$= \chi_{\text{eff}}^{(2)}(\omega/nc)E_0^2(\omega)\frac{\sin \Delta kL}{\Delta k} \; . \tag{5.122a}$$

The intensity $I = (nc\epsilon_0/2n)\,|E(2\omega)|^2$ of the SH wave is then

$$I(2\omega, L) = I^2(\omega)\frac{2\omega^2|\chi_{\text{eff}}^{(2)}|^2 L^2}{n^3 c^3 \epsilon_0}\frac{\sin^2(\Delta kL)}{(\Delta kL)^2} \; . \tag{5.122b}$$

If the length L exceeds the coherence length

$$L_{\text{coh}} = \frac{\pi}{2\Delta k} = \frac{\lambda}{4(n_{2\omega} - n_\omega)} \; , \tag{5.123}$$

the fundamental wave (λ) and the SH wave ($\lambda/2$) have a phase difference $\Delta\varphi > \pi/2$, and destructive interference begins, which diminishes the amplitude of the SH wave. The difference $n_{2\omega} - n_\omega$ should therefore be sufficiently small to provide a coherence length larger than the crystal length L.

According to the definition at the end of Sect. 5.8.1, type-I phase matching is achieved in uniaxial negatively birefringent crystals when $n_{\text{e}}(2\omega, \theta) = n_{\text{o}}(\omega)$. The polarizations of the fundamental wave and the SH wave are then orthogonal. From (5.119) and the condition $n_{\text{e}}(2\omega, \theta) = n_{\text{o}}(\omega)$, we obtain the phase-matching angle θ as

$$\sin^2 \theta = \frac{v_{\text{o}}^2(\omega) - v_{\text{o}}^2(2\omega)}{v_{\text{e}}^2(2\omega, \pi/2) - v_{\text{o}}^2(2\omega)} \; . \tag{5.124a}$$

For type-II phase matching the polarization of the fundamental wave does not fall into the plane defined by the optical axis and the k-vector. It therefore has one component in the plane, which travels with $v = c/n_{\text{o}}$, and another component with $v = c/n_{\text{e}}$ perpendicular to the plane. The phase-matching condition now becomes

$$n_{\text{e}}(2\omega, \theta) = \frac{1}{2}[n_{\text{e}}(\omega, \theta) + n_{\text{o}}(\omega)] \; . \tag{5.124b}$$

The choice of the nonlinear medium depends on the wavelength of the pump laser and on its tuning range (Table 5.6). For SHG of lasers around

Table 5.6. Characteristic data of nonlinear crystals used for frequency doubling or sum-frequency generation

Material	Transparency range [nm]	Spectral range of phase matching of type I or II	Damage threshold [GW/cm^2]	Relative doubling efficiency	Reference
ADP	220–2000	500–1100	0.5	1.2	[5.5]
KDP	200–2500	517–1500 (I)	8.4	1.0	[5.239]
		732–1500 (II)		8.4	
Urea	210–1400	473–1400 (I)	1.5	6.1	[5.249]
BBO	197–3500	410–3500 (I)	9.9	26.0	[5.224–5.229]
		750–1500 (II)			
LiJO$_3$	300–5500	570–5500 (I)	0.06	50.0	[5.250, 5.233]
KTP	350–4500	1000–2500 (II)	1.0	215.0	[5.248]
LiNbO$_3$	400–5000	800–5000 (II)	0.05	105.0	[5.239]
LiB$_3$O$_5$	160–2600	550–2600	18.9	3	[5.240]
CdGeAs$_2$	1–20 μm	2–15 μm	0.04	9	[5.253]
Te	3.8–32 μm		0.045	270	[5.239]

Table 5.7. Abbreviations for some commonly used nonlinear crystals

ADP	= Ammonium dihydrogen phosphate	$NH_4H_2PO_4$
KDP	= Potassium dihydrogen phosphate	KH_2PO_4
KD*P	= Potassium dideuterium phosphate	KD_2PO_4
KTP	= Potassium titanyl phosphate	$KTiOPO_4$
KNbO$_3$	= Potassium niobate	
LBO	= Lithium triborate	LiB_3O_5
LiIO$_3$	= Lithium iodate	
LiNbO$_3$	= Lithium niobate	
BBO	= Beta-barium borate	β-BaB_2O_4

$\lambda = 1\,\mu m$, 90° phase matching can be achieved with LiNbO$_3$ crystals, while for SHG of dye lasers around $\lambda = 0.5$–$0.6\,\mu m$, KDP crystals or ADA can be used. Figure 5.103 illustrates the dispersion curves $n_o(\lambda)$ and $n_e(\lambda)$ of ordinary and extraordinary waves in KDP and LiNbO$_3$, which show that 90° phase matching can be achieved in LiNbO$_3$ for $\lambda_p = 1.06\,\mu m$ and in KDP for $\lambda_p \simeq 515\,nm$ [5.218].

Since the intensity $I(2\omega)$ of the SH wave is proportional to the square of the pump intensity $I(\omega)$, most of the work on SHG has been performed with pulsed lasers, which offer high peak powers.

Focusing of the pump wave into the nonlinear medium increases the power density and therefore enhances the SHG efficiency. However, the resulting divergence of the focused beam decreases the coherence length because the wave vectors $\boldsymbol{k}_p$ are spread out over an interval $\Delta\boldsymbol{k}_p$, which depends on the divergence angle. The partial compensation of both effects leads to an optimum focal length of the focusing lens, which depends on the angular dispersion

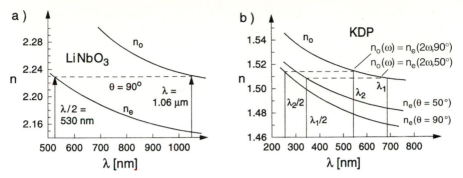

Fig. 5.103a,b. Refractive indices $n_o(\lambda)$ and $n_e(\lambda)$: (**a**) for $\theta = 90°$ in LiNbO$_3$ [5.220] and (**b**) for $\theta = 50°$ and 90° KDP [5.217]. Collinear phase matching can be achieved in LiNbO$_3$ for $\theta = 90°$ and $\lambda = 1.06\,\mu$m (Nd$^+$ laser) and in KDP for $\theta = 50°$ at $\lambda = 694\,$nm (ruby laser) or for $\theta = 90°$ at $\lambda = 515\,$nm (argon laser)

$dn_e/d\theta$ of the refractive index n_e and on the spectral bandwidth $\Delta\omega_p$ of the pump radiation [5.223].

If the wavelength λ_p of the pump laser is tuned, phase matching can be maintained either by turning the crystal orientation θ against the pump beam propagation $\boldsymbol{k}_p$ (angle tuning) or by temperature control (temperature tuning), which relies on the temperature dependence $\Delta n(T, \lambda) = n_o(T, \lambda) - n_e(T, \lambda/2)$. The tuning range $2\omega \pm \Delta_2\omega$ of the SH wave depends on that of the pump wave ($\omega \pm \Delta_1\omega$) and on the range where phase matching can be maintained. Generally, $\Delta_2\omega < 2\Delta_1\omega$ because of the limited phase-matching range. With frequency-doubled pulsed dye lasers and different dyes the whole tuning range between $\lambda = 195-500\,$nm can be completely covered. The strong optical absorption of most nonlinear crystals below 220 nm causes a low damage threshold, and the shortest wavelength achieved by SHG is, at present, $\lambda = 200\,$nm [5.221–5.226].

Example 5.30

The refractive indices $n_o(\lambda)$ and $n_e(\lambda)$ of ADP (ammonium dihydrogen phosphate) for $\theta = 90°$ are plotted in Fig. 5.104, together with the phase–matching curve: $\Delta(T, \lambda) = n_o(T, \lambda) - n_e(T, \lambda/2) = 0$. This plot shows that at $T = -11°$C, the phase-matching condition $\Delta(T, \lambda) = 0$ is fulfilled for $\lambda = 514.5\,$mm, and thus 90° phase matching for SHG of the powerful green argon laser line at $\lambda = 514.5\,$nm is possible.

Limitations of the SH output power generated by pulsed lasers are mainly set by the damage threshold of available nonlinear crystals. Very promising new crystals are the negative uniaxial BBO (beta-barium borate) β-BaB$_2$O$_4$ [5.225–5.228] and lithium borate LiBO, which have high damage thresholds and which allow SHG from 205 nm to above 3000 nm.

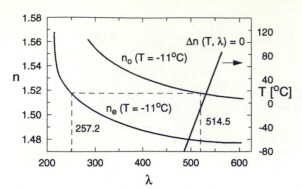

Fig. 5.104. Wavelength dependence for n_o and n_e in ADP at $\theta = 90°$ and temperature dependence of the phase-matching condition $\Delta n(T, \lambda) = n_o(T, \lambda) - n_e(T, \lambda/2) = 0$

Example 5.31

The five nonvanishing nonlinear coefficients of BBO are d_{11}, d_{22}, d_{31}, d_{13}, d_{14}, where the largest coefficient d_{11} is about 6 times larger than d_{36} of KDP. The transmission range of BBO is $195-3500$ nm. It has a low temperature dependence of its birefringence and a high optical homogeneity. Its damage threshold is about $10\,\mathrm{GW/cm^2}$.

Type-I phase matching is possible in the range $410-3500$ nm, type-II phase-matching in the range $750-1500$ nm.

The effective nonlinear coefficient for type-I phase-matching is

$$d_{\mathrm{eff}} = d_{31} \sin\theta + (d_{11} \cos 3\phi - d_{22} \sin 3\phi) \cos\phi ,$$

where θ and ϕ are the polar angles between the k-vector of the incident wave and the $z(= c)$-axis and the $x(= a)$-axis of the crystal, respectively. For $\phi = 0$ d_{eff} becomes maximum.

With cw dye lasers in the visible (output power ≤ 1 W), generally UV powers of only a few milliwatts are achieved by frequency doubling. The doubling efficiency $\eta = I(2\omega)/I(\omega)$ can be greatly enhanced when the doubling crystal is placed inside the laser cavity where the power of the fundamental wave is much higher [5.231–5.235]. The auxiliary beam waist in a ring laser resonator is the best location for placing the crystal (Fig. 5.95). With an intracavity $LiIO_3$ crystal, for example, UV powers in the range $20-50$ mW have been achieved at $\lambda/2 = 300$ nm [5.233].

If the dye laser must be used for visible as well as for UV spectroscopy, a daily change of the configuration is troublesome, therefore it is advantageous to apply an extra external ring resonator for frequency doubling [5.236–5.238]. This resonator must, of course, always be kept in resonance with the dye laser wavelength λ_L and therefore must be stabilized by a feedback control to the wavelength λ_L when the dye laser is tuned.

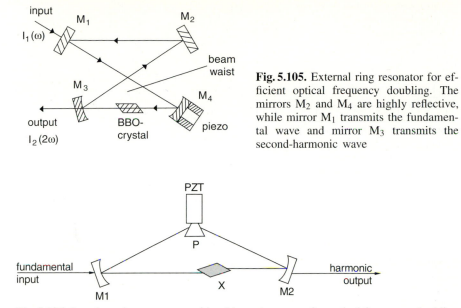

Fig. 5.105. External ring resonator for efficient optical frequency doubling. The mirrors M_2 and M_4 are highly reflective, while mirror M_1 transmits the fundamental wave and mirror M_3 transmits the second-harmonic wave

Fig. 5.106. Low-loss ring resonator with wide tuning range for optical frequency doubling with astigmatic compensation [5.221]

One example is illustrated in Fig. 5.105. In order to avoid feedback into the laser, ring resonators are used and the crystal is placed under the Brewster angle in the beam waist of the resonator. Since the enhancement factor for $I(\omega)$ depends on the resonator lasers, the mirrors should be highly reflective for the fundamental wave, but the output mirror should have a high transmission for the second-harmonic wave. An elegant solution is shown in Fig. 5.106, where only two mirrors and a Brewster prism form the ring resonator. The resonator length can be conveniently tuned by shifting the prism with a piezo-translating device in the z-direction.

Many more examples of external and intracavity frequency doubling with different nonlinear crystals [5.239] can be found in the literature [5.241–5.243]. Table 5.6 compiles some optical properties of commonly utilized nonlinear crystals.

5.8.4 Quasi Phase Matching

Recently, optical frequency doubling devices have been developed that consist of many thin slices of crystal with periodically varying directions of their optical axes. This can be achieved by producing many thin electrodes with lithographic techniques on the two side faces of the crystal and then placing the crystal at higher temperatures in a spatially periodic electric field. This results in a corresponding anisotropy of the charge distribution (induced

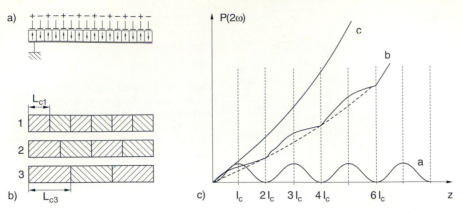

Fig. 5.107a–c. Quasi phase matching: (**a**) periodic poling of crystal orientation; (**b**) array of crystals with different period lengths for choosing the optimum doubling efficiency for a given wavelength; (**c**) second-harmonic output power as a function of total length $L = n \cdot L_c$ for one crystal with slight phase mismatch (curve a), for a periodically poled crystal (curve b), and for a single crystal with ideal phase matching

electric dipole moments), which determines the optical axis of the crystal (Fig. 5.107a). If there is a phase mismatch

$$\Delta k = \frac{2\pi}{\lambda}[n(2\omega - n(\omega))] ,$$

the phases of fundamental and second-harmonic waves differ by π after the coherence length

$$L_c = \frac{\pi}{k(2\omega) - 2k(\omega)} = \frac{\lambda}{4[n(2\omega) - n(\omega)]} .$$

A nonlinear crystal with a length $L \gg L_c$ shows the output power $P(2\omega)$ of the second-harmonic wave as a function of the propagation length z depicted by curve a in Fig. 5.107c. After one coherence length the power decreases again because of destructive interference between the second-harmonic and the out-of-phase fundamental wave.

If, however, the crystal has length $L = L_c$ followed by a second crystal with $L = 2L_c$ but opposite orientation of its optical axis, then the phase mismatch is reversed and the phase difference decreases from π to $-\pi$. Now the next layer follows with the orientation of the first one and the phase difference again increases from $-\pi$ to $+\pi$, and so on. This yields the output power of the second harmonic as shown in Fig. 5.107c, curve b.

For comparison, the curve c of a perfectly phase-matched long crystal is shown in Fig. 5.107c. This demonstrates that the quasi-phase-matching device gives a lower output power than the perfectly matched crystal, but a much larger power than for a single crystal in the case of slight phase mismatches.

For frequency doubling of tunable lasers, it is difficult to maintain perfect phase matching for all wavelengths; therefore phase mismatches cannot be avoided. Furthermore, for angle tuning of the crystal, noncollinear propagation of the fundamental and the second-harmonic wave occurs. This limits the effective interaction length and therefore the doubling efficiency. With correctly designed quasi-phase-matched devices, collinear noncritical phase matching can be realized, which allows long interaction lengths. Furthermore, fundamental and second-harmonic waves can have the same polarization; therefore one can use the largest nonlinear coefficient for the doubling efficiency by choosing the correct electro-optic poling of the slices. The greatest advantage is the large tuning range, where either temperature tuning can be utilized or an array of periodic slices with different slice thicknesses $L = L_c$ adapted to the wavelength-dependent phase mismatch is used (Fig. 5.107b). In the latter case the different devices, all on the same chip, can be shifted into the laser beam by a translational stage.

For these reasons many modern nonlinear frequency-doubling or mixing devices, in particular, optical parametric oscillators, use quasi phase matching [5.244, 5.245].

5.8.5 Sum-Frequency and Higher-Harmonic Generation

In the case of laser-pumped dye lasers, it is often more advantageous to generate tunable UV radiation by optical mixing of the pump laser and the tunable dye laser outputs rather than by frequency doubling of the dye laser. Since the intensity $I(\omega_1 + \omega_2)$ is proportional to the product $I(\omega_1)I(\omega_2)$, the larger intensity $I(\omega_1)$ of the pump laser allows enhanced UV intensity $I(\omega_1 + \omega_2)$. Furthermore, it is often possible to choose the frequencies ω_1 and ω_2 in such a way that 90° phase matching can be achieved. The range $(\omega_1 + \omega_2)$ that can be covered by sum-frequency generation is generally wider than that accessible to SHG. Radiation at wavelengths too short to be produced by frequency doubling can be generated by the mixing of two different frequencies ω_1 and ω_2. This is illustrated by Fig. 5.108, which depicts possible wavelength combinations λ_1 and λ_2 that allow 90° phase-matched sum-frequency mixing in KDP and ADP at room temperature or along the b-axis of biaxial KB5 crystals [5.246].

Some examples are given to demonstrate experimental realizations of the sum-frequency mixing technique [5.247–5.255].

Example 5.32

(a) The output of a cw rhodamine 6G dye laser pumped with 15 W on all lines of an argon laser is mixed with a selected line of the same argon laser (Fig. 5.109). The superimposed beams are focused into the temperature-stabilized KDP crystal. Tuning is accomplished by simultaneously tuning the dye laser wavelength and the orientation of the KDP crystal. The entire wavelength range from 257 to 320 nm can be covered by using different argon lines with a single Rhodamine 6G dye laser without changing dyes [5.247].

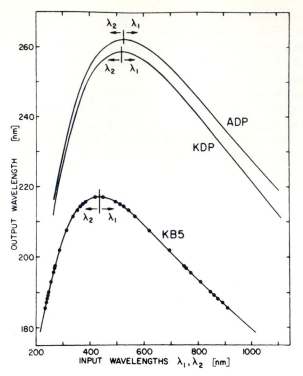

Fig. 5.108. Possible combinations of wavelength pairs (λ_1, λ_2) that allow 90° phase-matched sum-frequency generation in ADP, KDP, and KB5 [5.247, 5.252]

(b) The generation of intense tunable radiation in the range 240−250 nm has been demonstrated by mixing in a temperature-tuned 90° phase-matched ADP crystal the second harmonic of a ruby laser with the output of an infrared dye laser pumped by the ruby laser's fundamental output [5.246].

(c) UV radiation tunable between 208 and 259 nm has been generated efficiently by mixing the fundamental output of a Nd:YAG laser and the output of a frequency-doubled dye laser. Wavelengths down to 202 nm can be obtained with a refrigerated ADP crystal because ADP is particularly sensitive to temperature tuning [5.253].

(d) In lithium borate (LBO) noncritical phase-matched sum-frequency generation at $\theta = 90°$ can be achieved over a wide wavelength range. Starting with $\lambda_1 < 220$ nm and $\lambda_2 \geq 1064$ nm, sum-frequency radiation down to wavelengths of $\lambda_3 = (1/\lambda_1 + 1/\lambda_2)^{-1} = 160$ nm can be generated. The lower limit is set by the transmission cutoff of LBO [5.254].

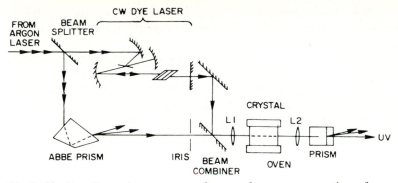

Fig. 5.109. Experimental arrangement for sum-frequency generation of cw radiation in a KDP crystal [5.247]

A novel device for efficiently generating intense radiation at wavelengths around 202 nm is shown in Fig. 5.110. A laser diode-pumped Nd:YVO$_4$ laser is frequency doubled and delivers intense radiation at $\lambda = 532$ nm, which is again frequency doubled to $\lambda = 266$ nm in a BBO crystal inside a ring resonator. The output from this resonator is superimposed in a third enhancement cavity with the output from a diode laser at $\lambda = 850$ nm to generate radiation at $\lambda = 202$ nm by sum-frequency mixing. This 202-nm radiation is polarized perpendicularly to that at the two other waves and can be therefore efficiently coupled out of the cavity by a Brewster plate [5.255].

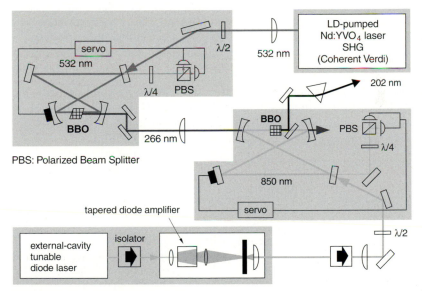

Fig. 5.110. Sum-frequency generation in an enhancement cavity down to $\lambda = 202$ nm [5.255]

The lower-wavelength limit for nonlinear processes in crystals (SHG or sum-frequency mixing) is generally given by the absorption (transmission cutoff) of the crystals.

For shorter wavelengths sum-frequency mixing or higher-harmonic generation in homogeneous mixtures of rare gases and metal vapors can be achieved. Because in centro-symmetric media the second-order susceptibility must vanish, SHG is not posssible, but all third-order processes can be utilized for the generation of tunable ultraviolet radiation. Phase matching is achieved by a proper density ratio of raregas atoms to metal atoms. Several examples illustrate the method.

Example 5.33

(a) Third-harmonic generation of Nd:YAG laser lines around $\lambda = 1.05\,\mu m$ can be achieved in mixtures of xenon and rubidium vapor in a heat pipe. Figure 5.111 is a schematic diagram for the refractive indices $n(\lambda)$ for Xe and rubidium vapor. Choosing the proper density ratio $N(Xe)/N(Rb)$, phase matching is obtained for $n(\omega) = n(3\omega)$, where the refractive index $n = n(Xe) + n(Rb)$ is determined by the rubidium and Xe densities. Figure 5.111 illustrates that this method utilizes the compensation of the normal dispersion in Xe by the anomalous dispersion for rubidium [5.256].

(b) A second example is the generation of tunable VUV ratiation between 110 and 130 nm by phase-matched sum-frequency generation in a xenon–krypton mixture [5.257]. This range covers the Lyman-α line of hydrogen and is therefore particularly important for many experiments in plasma diagnostics and in fundamental physics. A frequency-doubled dye laser at $\omega_{UV} = 2\omega_1$ and a second tunable dye laser at ω_2 are focused into a cell that contains a proper mixture of Kr/Xe. The sum frequency $\omega_3 = 2\omega_{UV} + \omega_2$ can be tuned by synchronous tuning of ω_2 and the variation of the Kr/Xe mixture.

Because of the lower densities of gases compared with solid crystals, the efficiency $I(3\omega)/I(\omega)$ is much smaller than in crystals. However, there is no short-wavelength limit as in crystals, and the spectral range accessible by optical mixing can be extended far into the VUV range [5.258].

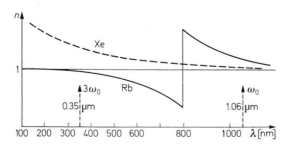

Fig. 5.111. Schematic diagram of the refractive indices $n(\lambda)$ for rubidium vapor and xenon, illustrating phase matching for third-harmonic generation

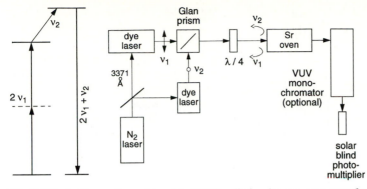

Fig. 5.112a,b. Generation of tunable VUV radiation by resonant sum-frequency mixing in metal vapors: (**a**) level scheme; (**b**) experimental arrangement

The efficiency may be greatly increased by resonance enhancement if, for example, a resonant two-photon transition $2\hbar\omega_1 = E_1 \rightarrow E_k$ can be utilized as a first step of the sum-frequency generation $\omega = 2\omega_1 + \omega_2$. This is demonstrated by an early experiment shown in Fig. 5.112. The orthogonally polarized outputs from two N_2 laser-pumped dye lasers are spatially overlapped in a Glan–Thompson prism. The collinear beams of frequencies ω_1 and ω_2 are then focused into a heat pipe containing the atomic metal vapor. One laser is fixed at half the frequency of an appropriate two-photon transition and the other is tuned. For a tuning range of the dye laser between 700 and 400 nm achievable with different dyes, tunable VUV radiation at the frequencies $\omega = 2\omega_1 + \omega_2$ is generated, which can be tuned over a large range. Third-harmonic generation can be eliminated in this experiment by using circularly polarized ω_1 and ω_2 radiation, since the angular momentum will not be conserved for frequency tripling in an isotropic medium under these conditions. The sum frequency $\omega = 2\omega_1 + \omega_2$ corresponds to an energy level beyond the ionization limit [5.259–5.263].

Windows cannot be used for wavelengths below 120 nm because all materials absorb the radiation, therefore apertures and differential pumping is needed. An elegant solution is the VUV generation in pulsed laser jets (Fig. 5.113), where the density of wanted molecules within the focus of the incident lasers can be made large without having too much absorption for the generated VUV radiation because the molecular density is restricted to the small path length across the molecular jet close to the nozzle [5.264–5.266]. The output of a tunable dye laser is frequency doubled in a BBO crystal. Its UV radiation is then focused into the gas jet where frequency tripling occurs. The VUV radiation is now collimated by a parabolic mirror and imaged into a second molecular beam within the same vacuum chamber, where the experiment is performed.

More information on the generation of VUV radiation by nonlinear mixing techniques can be found in [5.254–5.269].

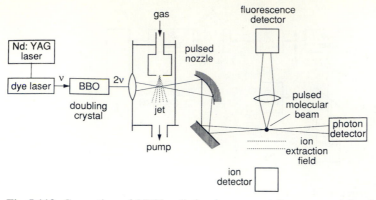

Fig. 5.113. Generation of VUV radiation by resonant frequency mixing in a jet [5.264]

5.8.6 X-Ray Lasers

For many problems in atomic, molecular, and solid-state physics intense sources of tunable X-rays are required. Examples are inner-shell excitation of atoms and molecules or spectroscopy of multiply charged ions. Until now, these demands could only partly be met by X-ray tubes or by synchrotron radiation. The development of lasers in the spectral range below 100 nm is therefore of great interest.

According to (2.22), the spontaneous transition probability A_i scales with the third power ν^3 of the emitted frequency. The losses of upper-state population N_i by fluorescence are therefore proportional to $A_i h\nu \propto \nu^4$! This means

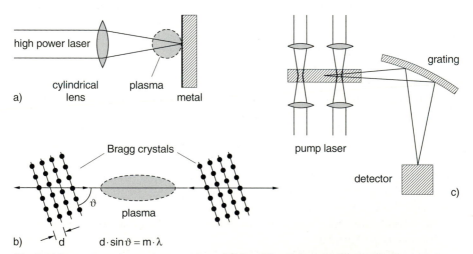

Fig. 5.114a–c. Experimental setup for realizing X-ray lasers: (**a**) production of high-temperature plasma; (**b**) X-ray resonator using Bragg reflection by crystals; (**c**) measurement of single-pass gain and line narrowing

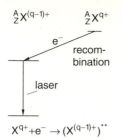

$$X^{q+} + e^- \rightarrow (X^{(q-1)+})^{**}$$

Fig. 5.115. Level scheme for inversion by ion–electron recombination

that high pumping powers are required to achieve inversion. Therefore only pulsed operation has a chance to be realized where ultrashort laser pulses with high peak powers are used as pumping sources. Possible candidates that can serve as active media for X-ray lasers are highly excited multiply charged ions. They can be produced in a laser-induced high-temperature plasma (Fig. 5.114). If the pump laser beam is focused by a cylindrical lens onto the target, a high-temperature plasma is produced along the focal line. The q-fold ionized species with nuclear charge $z \cdot e$ in the plasma plume recombine with electrons to form Rydberg states of ions with electron charge $Q_{el} = -(z - q + 1)$. In favorable cases these high Rydberg levels are more strongly populated than lower states of this ion and inversion is achieved (Fig. 5.115). The conditions for achieving inversion and thus amplification of X-ray radiation can only be maintained for very short times (on the order of picoseconds).

Such soft X-ray lasers have also been realized [5.265–5.272]. The shortest wavelength reported to date is 6 nm [5.271]. Resonators for X-ray lasers can be composed of Bragg reflectors, which consist of suitable crystals that can be tilted to fulfill the Bragg condition $2d \cdot \sin \vartheta = m \cdot \lambda$ for constructive interference between the partial waves reflected by the crystal planes with distance d (Fig. 5.114b). More detailed information on this interesting subject can be found in [5.270–5.276].

5.8.7 Difference-Frequency Spectrometer

While generation of *sum frequencies* yields tunable ultraviolet radiation by mixing the output from two lasers in the visible range, the phase-matched generation of *difference* frequencies allows one to construct tunable coherent *infrared* sources. One early example is the difference-frequency spectrometer of Pine [5.277], which has proved to be very useful for high-resolution infrared spectroscopy.

Two collinear cw beams from a stable single-mode argon laser and a tunable single-mode dye laser are mixed in a $LiNbO_3$ crystal (Fig. 5.116). For 90° phase matching of collinear beams, the phase-matching condition

$$\boldsymbol{k}(\omega_1 - \omega_2) = \boldsymbol{k}(\omega_1) - \boldsymbol{k}(\omega_2) \,,$$

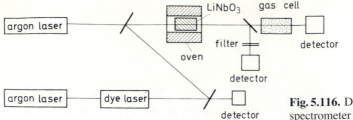

Fig. 5.116. Difference-frequency spectrometer [5.278]

can be written as $|k(\omega_1 - \omega_2)| = |k(\omega_1)| - |k(\omega_2)|$, which gives for the refractive index $n = c(k/\omega)$ the relation

$$n(\omega_1 - \omega_2) = \frac{\omega_1 n(\omega_1) - \omega_2 n(\omega_2)}{\omega_1 - \omega_2} . \tag{5.125}$$

The whole spectral range from 2.2 to 4.2 µm can be continuously covered by tuning the dye laser and the phase-matching temperature of the $LiNbO_3$ crystal ($-0.12\,°C/cm^{-1}$). The infrared power is, according to (5.114), (5.122b), proportional to the product of the incident laser powers and to the square of the coherence length. For typical operating powers of 100 mW (argon laser) and 10 mW (dye laser), a few microwatts of infrared radiation is obtained. This is 10^4 to 10^5 times higher than the noise equivalent input power of standard IR detectors.

The spectral linewidth of the infrared radiation is determined by that of the two pump lasers. With frequency stabilization of the pump lasers, a linewidth of a few megahertz has been reached for the difference-frequency spectrometer. In combination with a multiplexing scheme devised for calibration, monitoring, drift compensation, and absolute stabilization of the difference spectrometer, a continuous scan of 7.5 cm^{-1} has been achieved with a reproducibility of better than 10 MHz [5.278].

A very large tuning range has been achieved with a cw laser spectrometer based on difference-frequency generation in $AgGaS_2$ crystals. By mixing a single-mode tunable dye laser with a Ti:sapphire laser, infrared powers up to 250 µW have been generated in the spectral range 4−9 µm (Fig. 5.117) [5.279]. Even more promising is the difference-frequency generation of two tunable diode lasers, which allows the construction of a very compact and much cheaper difference-frequency spectrometer [5.281, 5.282].

Of particular interest are tunable sources in the far infrared region where no microwave generators are available and incoherent sources are very weak. With selected crystals such as proustite (Ag_3AsS_3), $LiNbO_3$, or GaAs, phase matching for difference-frequency generation can be achieved for the middle infrared using CO_2 lasers and spin-flip Raman lasers. The search for new nonlinear materials will certainly enhance the spectroscopic capabilities in the whole infrared region [5.283].

A very useful frequency-mixing device is the MIM diode (Sect. 4.5.2), which allows the realization of continuously tunable FIR radiation covering the difference-frequency range from the microwave region (GHz) to the

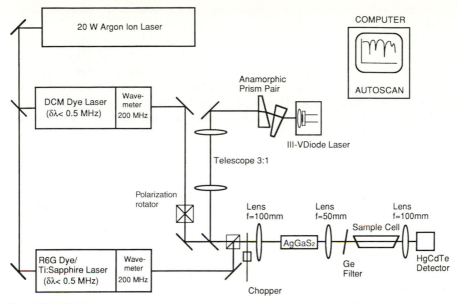

Fig. 5.117. Difference-frequency spectrometer based on mixing a cw Ti:sapphire ring laser with a single-frequency III–V diode laser in the nonlinear crystal AgGaS$_2$ [5.280]

submillimeter range (THz) [5.284–5.286]. It consists of a specially shaped tungsten wire with a very sharp tip that is pressed against a nickel surface covered with a thin layer of nickel oxide (Fig. 4.89). If the beams of two lasers with freqencies ν_1 and ν_2 are focused onto the contact point (Fig. 5.118), frequency mixing due to the nonlinear response of the diode occurs. The tungsten wire acts as an antenna that radiates waves at the difference frequency $(\nu_1 - \nu_2)$ into a narrow solid angle corresponding to the antenna lobe. These

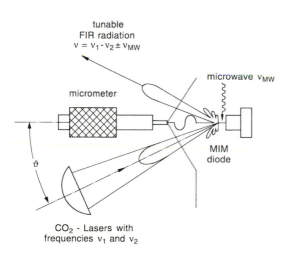

Fig. 5.118. Generation of tunable FIR radiation by frequency mixing of two CO$_2$ laser beams with a microwave in a MIM diode

waves are collimated by a parabolic mirror with a focus at the position of the diode.

Using CO_2 lasers with different isotope mixtures, laser oscillation on several hundred lines within the spectral range between 9 and 10 µm can be achieved. This laser oscillation can be fine-tuned over the pressure-broadened gain profiles. Therefore their difference frequencies cover the whole FIR region with only small gaps. These gaps can be closed when the radiation of a tunable microwave generator is additionally focused onto the MIM mixing diode. The waves at frequencies

$$\nu = \nu_1 - \nu_2 \pm \nu_{MW}$$

represent continuous tunable collimated coherent radiation, which can be used for absorption spectroscopy in the far infrared [5.286, 5.287].

5.8.8 Optical Parametric Oscillator

The optical parametric oscillator (OPO) [5.288–5.295] is based on the parametric interaction of a strong pump wave $E_p \cos(\omega_p t - k_p \cdot r)$ with molecules in a crystal that have a sufficiently large nonlinear susceptibility. This interaction can be described as an inelastic scattering of a pump photon $\hbar\omega_p$ by a molecule where the pump photon is absorbed and two new photons $\hbar\omega_s$ and $\hbar\omega_i$ are generated. Because of energy conservation, the frequencies ω_i and ω_s are related to the pump frequency ω_p by

$$\omega_p = \omega_i + \omega_s \ . \tag{5.126}$$

Analogous to the sum-frequency generation, the parametrically generated photons ω_i and ω_s can add up to a macroscopic wave if the phase-matching condition

$$k_p = k_i + k_s \tag{5.127}$$

is fulfilled, which may be regarded as the conservation of momentum for the three photons involved in the parametric process. Simply stated, parametric generation splits a pump photon into two photons that satisfy conservation of energy and momentum at every point in the nonlinear crystal. For a given wave vector k_p of the pump wave, the phase-matching condition (5.127) selects, out of the infinite number of possible combinations $\omega_1 + \omega_2$ allowed by (5.126), a single pair (ω_i, k_i) and (ω_s, k_s) that is determined by the orientation of the nonlinear crystal with respect to k_p. The two resulting macroscopic waves $E_s \cos(\omega_s t - k_s \cdot r)$ and $E_i \cos(\omega_i t - k_i \cdot r)$ are called the *signal* wave and *idler* wave. The most efficient generation is achieved for collinear phase matching where $k_p || k_i || k_s$. For this case, the relation (5.117) between the refractive indices gives

$$n_p \omega_p = n_s \omega_s + n_i \omega_i \ . \tag{5.128}$$

If the pump is an extraordinary wave, collinear phase matching can be achieved for some angle θ against the optical axis, if $n_p(\theta)$, defined by (5.119), lies between $n_o(\omega_p)$ and $n_e(\omega_p)$.

The gain of the signal and idler waves depends on the pump intensity and on the effective nonlinear suceptibility. Analogous to the sum- or difference-frequency generation, one can define a parametric gain coefficient per unit pathlength $\Gamma = I_s/I_p$ or I_i/I_p

$$\Gamma = \frac{\omega_i \omega_s |d|^2 |E_p|^2}{n_i n_s c^2} = \frac{2\omega_i \omega_s |d|^2 I_p}{n_i n_s n_p \epsilon_0 c^3} , \qquad (5.129)$$

which is proportional to the pump intensity I_p and the square of the effective nonlinear susceptibility $|d| = \chi_{\text{eff}}^{(2)}$. For $\omega_i = \omega_s$, (5.129) becomes identical with the gain coefficient for SHG in (5.122b).

If the nonlinear crystal that is pumped by the incident wave E_p is placed inside a resonator, oscillation on the idler or signal frequencies can start when the gain exceeds the total losses. The optical cavity may be resonant for both the idler and signal waves (doubly-resonant oscillator) or for only one of the waves (singly-resonant oscillator) [5.292]. Often, the cavity is also resonant for the pump wave in order to increase I_p and thus the gain coefficient Γ.

Figure 5.119 shows schematically the experimental arrangement of a collinear optical parametric oscillator. Due to the much higher gain, pulsed operation is generally preferred where the pump is a Q-switched laser source. The threshold of a doubly-resonant oscillator occurs when the gain equals the product of the signal and idler losses. If the resonator mirrors have high reflectivities for both the signal and idler waves, the losses are small, and even cw parametric oscillators can reach threshold [5.296]. For singly-resonant cav-

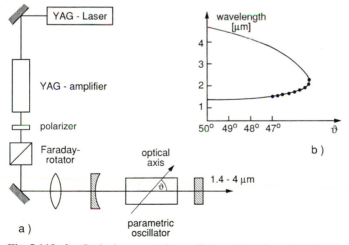

Fig. 5.119a,b. Optical parametric oscillator: (**a**) schematic diagram of experimental arrangement; (**b**) pairs of wavelengths (λ_1, λ_2) for idler and signal wave for collinear phase matching in LiNbO$_3$ as a function of angle θ [5.290]

ities, however, the losses for the nonresonant waves are high and the threshold increases.

Example 5.34

For a 5-cm long 90° phase-matched LiNbO$_3$ crystal pumped at $\lambda_p = 0.532\,\mu$m, threshold is at 38-mW pump power for the doubly-resonant cavity with 2% losses at ω_i and ω_s. For the singly-resonant cavity, threshold increases by a factor of 100 to 3.8 W [5.293].

Tuning of the OPO can be accomplished either by crystal rotation or by controlling the crystal temperature. The tuning range of a LiNbO$_3$ OPO, pumped by various frequency-doubled wavelengths of a Q-switched Nd:YAG laser, extends from 0.55 to about 4 μm. Turning the crystal orientation by only 4° covers a tuning range between 1.4 and 4.4 μm (Fig. 5.119b). Figure 5.120 shows temperature tuning curves for idler and signal waves generated in

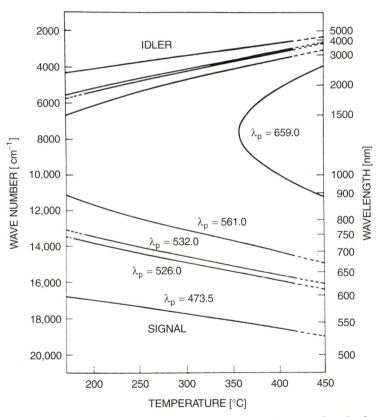

Fig. 5.120. Temperature tuning curves of signal and idler wavelengths for a LiNbO$_3$ optical parametric oscillator pumped by different pump wavelengths [5.292]

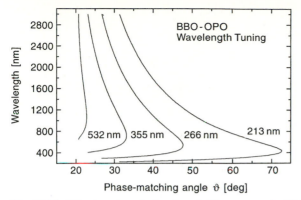

Fig. 5.121. Wavelengths of signal and idler waves in BBO as a function of the phase-matching angle ϑ for different pump wavelengths λ_p [5.295]

LiNbO$_3$ by different pump wavelengths. Angle tuning has the advantage of faster tuning rates than in the case of temperature tuning.

Previously, one of the drawbacks of the OPO was the relatively low damage threshold of available nonlinear crystals. The growth of advanced materials with high damage thresholds, large nonlinear coefficients, and broad transparency spectral ranges has greatly aided the development of widely tunable and stable OPOs [5.294]. Examples are BBO (β-barium borate) and lithium borate (LBO) [5.295]. For illustration of the wide tuning range, Fig. 5.121 displays wavelength tuning of the BBO OPO for different pump wavelengths.

The bandwidth of the OPO depends on the parameters of the resonator, on the linewidth of the pump laser, on the pump power, and, because of the different slopes of the tuning curves in Figs. 5.120, 5.121, also on the wavelength. Typical bandwidths are $0.1-5\,\mathrm{cm}^{-1}$. Detailed spectral properties depend on the longitudinal mode structure of the pump and on the resonator mode spacing $\Delta\nu = (c/2L)$ for the idler and signal standing waves. For the singly-resonant oscillator the cavity has to be adjusted to only one frequency, while the nonresonant frequency can be adjusted so that $\omega_p = \omega_i + \omega_s$ is satisfied. There are several ways to narrow the bandwidths of the OPO. With a tilted etalon inside the resonator of a singly-resonant cavity, single-mode operation can be achieved. Frequency stability of a few MHz has been demonstrated [5.297]. Another possibility is injection seeding. Stable single-mode operation was, for example, obtained by injecting the beam of a single-mode Nd:YAG pump-laser into the OPO cavity [5.298]. Using a single mode cw dye laser as the injection seeding source, tunable pulsed OPO-radiation with linewidths below 500 MHz have been achieved. A seed power of 0.3 mW(!) was sufficient for stable single-mode OPO operation. The pump threshold can be lowered with a doubly-resonant resonator. However, the simple cavity of Fig. 5.119 cannot be kept in resonance for two different wavelengths, if these wavelengths are tuned. Here the three-mirror cavity of Fig. 5.122 solves this problem. Since the polarizations of the pump wave and the idler wave are gen-

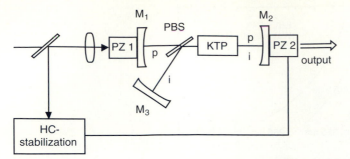

Fig. 5.122. Three-mirror resonator for tunable cw OPO, resonant for pump and idler with polarization beam splitter and separately controlled cavity lengths $\overline{M_1 M_2}$ and $\overline{M_3 M_2}$ [5.300]

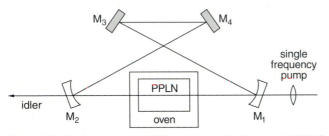

Fig. 5.123. High-power cw OPO with periodically poled LiNbO$_3$ crystal with temperature control in a ring cavity [5.302]

erally orthogonal, a polarization beam splitter PBS splits both waves, which now experience resonant enhancement in the resonator $M_1 M_2$ or $M_3 M_2$. When the pump wavelength λ_p (a dye laser is used as pump source) is tuned, both cavities can be controlled by piezos to keep in resonance [5.300]. Frequency stabilities of below the 1-kHz level can be achieved [5.301]. The tuning range for collinear phase matching can be greatly extended by quasi phase matching in periodically poled LiNbO$_3$ (PPLN) (Fig. 5.123). Meanwhile, cw OPOs are commercially available [5.302].

A good survey on different aspects of OPOs can be found in [5.299].

5.8.9 Tunable Raman Lasers

The tunable "Raman laser" may be regarded as a parametric oscillator based on stimulated Raman scattering. Since stimulated Raman scattering is discussed in more detail in Sect. 8.3, we here summarize only very briefly the basic concept of these devices.

The ordinary Raman effect can be described as an inelastic scattering of pump photons $\hbar\omega_p$ by molecules in the energy level E_i. The energy loss $\hbar(\omega_p - \omega_s)$ of the scattered *Stokes photons* $\hbar\omega_s$ is converted into excitation

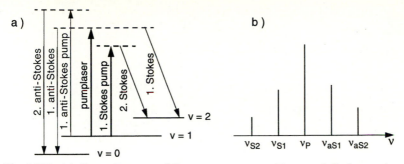

Fig. 5.124. (a) Term diagram of Raman processes with several Stokes and anti-Stokes lines at frequencies $\nu = \nu_p \pm m\nu_v$; (b) spectral distribution of Raman lines and their overtones

energy (vibrational, rotational, or electronic energy) of the molecules

$$\hbar\omega_p + M(E_i) \rightarrow M^*(E_f) + \hbar\omega_s , \qquad (5.130)$$

where $E_f - E_i = \hbar(\omega_p - \omega_s)$. For the vibrational Raman effect this process can be interpreted as parametric splitting of the pump photon $\hbar\omega_p$ into a Stokes photon $\hbar\omega_s$ and an optical phonon $\hbar\omega_v$ representing the molecular vibrations (Fig. 5.124a). The contributions $\hbar\omega_s$ from all molecules in the interaction region can add up to macroscopic waves when the phase-matching condition

$$\boldsymbol{k}_p = \boldsymbol{k}_s + \boldsymbol{k}_v ,$$

is fulfilled for the pump wave, the Stokes wave, and the phonon wave. In this case, a strong Stokes wave $E_s \cos(\omega_s t - \boldsymbol{k}_s \cdot \boldsymbol{r})$ develops with a gain that depends on the pump intensity and on the Raman scattering cross section. If the active medium is placed in a resonator, oscillation arises on the Stokes component as soon as the gain exceeds the total losses. Such a device is called a *Raman oscillator* or *Raman laser*, although, strictly speaking, it is not a laser but a parametric oscillator.

Those molecules that are initially in *excited* vibrational levels can give rise to superelastic scattering of *anti-Stokes radiation*, which has gained energy $(\hbar\omega_s - \hbar\omega_p) = (E_i - E_f)$ from the deactivation of vibrational energy.

The Stokes and the anti-Stokes radiation have a constant frequency shift against the pump radiation, which depends on the vibrational eigenfrequencies ω_v of the molecules in the active medium.

$$\omega_s = \omega_p - \omega_v , \qquad \omega_{as} = \omega_p + \omega_n \dots .$$

If the Stokes or anti-Stokes wave becomes sufficiently strong, it can again produce another Stokes or anti-Stokes wave at $\omega_s^{(2)} = \omega_s^{(1)} - \omega_v = \omega_p - 2\omega_v$ and $\omega_{as}^{(2)} = \omega_p + 2\omega_v$. Therefore, several Stokes and anti-Stokes waves are generated at frequencies $\omega_s^{(n)} = \omega_p - n\omega_v : \omega_{as}^{(n)} = \omega_p + n\omega_v$ $(n = 1, 2, 3, \dots)$

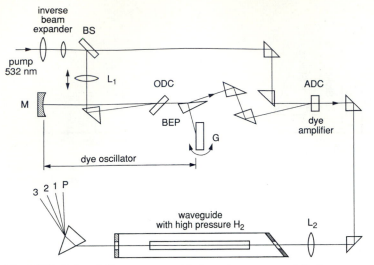

Fig. 5.125. Infrared Raman waveguide laser in compressed hydrogen gas H_2, pumped by a tunable dye laser. The frequency-doubled output beam of a Nd:YAG laser is split by BS in order to pump a dye laser oscillator and amplifier. The dye laser oscillator is composed of mirror M, grating G, and beam-expanding prism BEP. The different Stokes lines are separated by the prism P (ODC: oscillator dye cell) [5.304]

(Fig. 5.124b). Tunable lasers as pumping sources therefore allow one to transfer the tunability range ($\omega_p \pm \Delta\omega$) into the other spectral regions ($\omega_p \pm \Delta\omega \pm n\omega_v$).

Stimulated Raman scattering (SRS) of dye laser radiation in hydrogen gas can cover the whole spectrum between 185 and 880 nm without any gaps, using three different laser dyes and frequency doubling the dye laser radiation [5.303]. A broadly tunable IR waveguide Raman laser pumped by a dye laser can cover the infrared region from 0.7 to 7 μm without gaps, using SRS up to the third Stokes order ($\omega_s = \omega_p - 3\omega_v$) in compressed hydrogen gas. Energy conversion efficiencies of several percent are possible and output powers in excess of 80 kW for the third Stokes component ($\omega_p - 3\omega_v$) have been achieved [5.304]. Figure 5.125 depicts part of the experimental arrangement of an IR waveguide Raman laser.

For infrared spectroscopy, Raman lasers pumped by the numerous intense lines of CO_2, CO, HF, or DF lasers may be advantageous. Besides the vibrational Raman scattering, the rotational Raman effect can be utilized, although the gain is much lower than for vibrational Raman scattering, due to the smaller scattering cross section. For instance, H_2 and D_2 Raman lasers excited with a CO_2 laser can produce many Raman lines in the spectral range from 900 to 400 cm^{-1}, while liquid N_2 and O_2 Raman lasers pumped with an HF laser cover a quasi-continuous tuning range between 1000 and 2000 cm^{-1}. With high-pressure gas lasers as pumping sources, the small gaps between the many rotational–vibrational lines can be closed by pressure broadening (Sect. 3.3) and a true continuous tuning range of IR Raman lasers in the far infrared region becomes possible. Recently, a cw tunable Raman oscillator has been realized that utilizes as active medium a 650-m long single-mode silica fiber pumped by a 5-W cw Nd:YAG laser. The first Stokes radiation is tunable from 1.08 to 1.13 µm, the second Stokes from 1.15 to 1.175 µm [5.305]. With stimulated Raman scattering up to the seventh anti-Stokes order, efficient tunable radiation down to 193 nm was achieved when an excimer-laser pumped dye laser tunable around 440 nm was used [5.306].

A more detailed presentation of IR Raman lasers may be found in the review by Grasiuk et al. [5.307] and in [5.92, 5.308–5.310].

5.9 Gaussian Beams

In Sect. 5.2 we saw that the radial intensity distribution of a laser oscillating in the fundamental mode has a Gaussian profile. The laser beam emitted through the output mirror therefore also exhibits this Gaussian intensity profile. Although such a nearly parallel laser beam is in many respects similar to a plane wave, it shows several features that are different but that are important when the Gaussian beam is imaged by optical elements, such as lenses or mirrors. Often the problem arises of how to match the laser output to the fundamental mode of a passive resonator, such as a confocal spectrum analyzer or external enhancement cavities (Sect. 4.3). We therefore briefly discuss some properties of Gaussian beams; our presentation follows that of the recommendable review by Kogelnik and Li [5.24].

A laser beam traveling into the z-direction can be represented by the field amplitude

$$E = A(x, y, z)\,\mathrm{e}^{-\mathrm{i}(\omega t - kz)}\,. \tag{5.131}$$

While $A(x, y, z)$ is constant for a plane wave, it is a slowly varying complex function for a Gaussian beam. Since every wave obeys the general wave equation

$$\Delta E + k^2 E = 0\,, \tag{5.132}$$

we can obtain the amplitude $A(x, y, z)$ of our particular laser wave by inserting (5.131) into (5.132). We assume the trial solution

$$A = \mathrm{e}^{-\mathrm{i}[\varphi(z) + (k/2q)r^2]}\,, \tag{5.133}$$

where $r^2 = x^2 + y^2$, and $\varphi(z)$ represents a complex phase shift. In order to understand the physical meaning of the complex parameter $q(z)$, we express it in terms of two real parameters $w(z)$ and $R(z)$

$$\frac{1}{q} = \frac{1}{R} - \mathrm{i}\frac{\lambda}{\pi w^2}\,. \tag{5.134}$$

With (5.134) we obtain from (5.133) the amplitude $A(x, y, z)$ in terms of R, w, and φ

$$A = \exp\left(-\frac{r^2}{w^2}\right)\exp\left[-\mathrm{i}\frac{kr^2}{2R(z)} - \mathrm{i}\varphi(z)\right] . \tag{5.135}$$

This illustrates that $R(z)$ represents the radius of curvature of the wavefronts intersecting the axis at z, and $w(z)$ gives the distance $r = (x^2 + y^2)^{1/2}$ from the axis where the amplitude has decreased to $1/\mathrm{e}$ and thus the intensity has decreased to $1/\mathrm{e}^2$ of its value on the axis (Sect. 5.2.3 and Fig. 5.10). Inserting (5.135) into (5.132) and comparing terms of equal power in r yields the relations

$$\frac{dq}{dz} = 1 , \quad \text{and} \quad \frac{d\varphi}{dz} = -\mathrm{i}/q , \tag{5.136}$$

which can be integrated and gives, with $R(z = 0) = \infty$

$$q = q_0 + z = \mathrm{i}\frac{\pi w_0^2}{\lambda} + z , \tag{5.137}$$

where $w_0 = w(z = 0)$ (Fig. 5.126). When we measure z from the beam waist at $z = 0$, we obtain

$$w^2(z) = w_0^2\left[1 + \left(\frac{\lambda z}{\pi w_0^2}\right)^2\right] , \tag{5.138}$$

$$R(z) = z\left[1 + \left(\frac{\pi w_0^2}{\lambda z}\right)^2\right] . \tag{5.139}$$

Integration of the phase relation (5.136)

$$\frac{d\varphi}{dz} = -\mathrm{i}/q = -\frac{\mathrm{i}}{z + \mathrm{i}\pi w_0^2/\lambda} ,$$

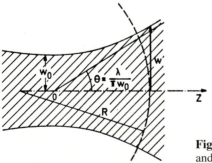

Fig. 5.126. Gaussian beam with beam waist w_0 and phase-front curvature $R(z)$ [5.24]

yields the z-dependent phase factor

$$i\varphi(z) = \ln\sqrt{1+(\lambda z/\pi w_0^2)} - i\arctan(\lambda z/\pi w_0^2) .\tag{5.140}$$

Having found the relations between φ, R, and w, we can finally express the Gaussian beam (5.131) by the real beam parameters R and w. From (5.140) and (5.135), we get

$$E = C_1 \frac{w_0}{w} e^{(-r^2/w^2)} e^{[ik(z-r^2/2R)-i\phi]} e^{-i\omega t} .\tag{5.141}$$

The first exponential factor gives the radial Gaussian distribution, the second the phase, which depends on z and r. We have used the abbreviation

$$\phi = \arctan(\lambda z/\pi w_0^2) .$$

The factor C_1 is a normalization factor. When we compare (5.141) with the field distribution (5.30) of the fundamental mode in a laser resonator, we see that both formulas are identical for $m = n = 0$.

The radial intensity distribution (Fig. 5.127) is

$$I(r, z) = \frac{c\epsilon_0}{2}|E|^2 = C_2 \frac{w_0^2}{w^2}\exp\left(-\frac{2r^2}{w^2}\right) .\tag{5.142}$$

The normalization factor C_2 allows

$$\int\limits_{r=0}^{\infty} 2\pi r I(r)\,\mathrm{d}r = P_0 \tag{5.143}$$

to be normalized, which yields $C_2 = (2/\pi w_0^2)P_0$, where P_0 is the total power in the beam. This yields

$$I(r, z) = \frac{2P_0}{\pi w^2}\exp\left(-\frac{2r^2}{w(z)^2}\right) .\tag{5.144}$$

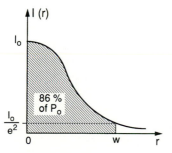

Fig. 5.127. Radial intensity profile of a Gaussian beam

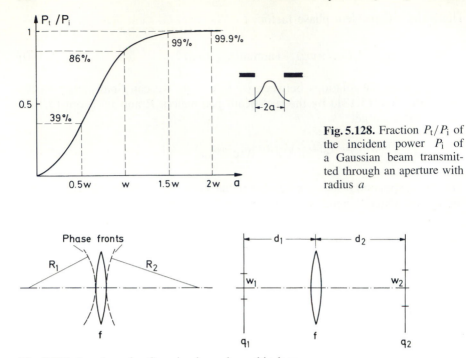

Fig. 5.128. Fraction P_t/P_i of the incident power P_i of a Gaussian beam transmitted through an aperture with radius a

Fig. 5.129. Imaging of a Gaussian beam by a thin lens

When the Gaussian beam is sent through an aperture with diameter $2a$, the fraction

$$\frac{P_t}{P_i} = \frac{2}{\pi w^2} \int_{r=0}^{a} 2r\pi\, \mathrm{e}^{-2r^2/w^2}\, \mathrm{d}r = 1 - \mathrm{e}^{-2a^2/w^2} \;, \tag{5.145}$$

of the incident power is transmitted through the aperture. Figure 5.128 illustrates this fraction as a function of a/w. For $a = (3/2)w$ 99% of the incident power is transmitted, and for $a = 2w$ more than 99.9% of the incident power is transmitted. In this case diffraction losses are therefore negligible.

A Gaussian beam can be imaged by lenses or mirrors, and the imaging equations are similar to those of spherical waves. When a Gaussian beam passes through a focusing thin lens with focal length f, the spot size w_s is the same on both sides of the lens (Fig. 5.129). The radius of curvature R of the phase fronts changes from R_1 to R_2 in the same way as for a spherical wave, so that

$$\frac{1}{R_2} = \frac{1}{R_1} - \frac{1}{f} \;. \tag{5.146}$$

The beam parameter q therefore satisfies the imaging equation

$$\frac{1}{q_2} = \frac{1}{q_1} - \frac{1}{f} . \tag{5.147}$$

If q_1 and q_2 are measured at the distances d_1 and d_2 from the lens, we obtain from (5.147) and (5.137) the relation

$$q_2 = \frac{(1 - d_2/f)q_1 + (d_1 + d_2 - d_1 d_2/f)}{(1 - d_1/f) - q_1/f} , \tag{5.148}$$

which allows the spot size w and radius of curvature R at any distance d_2 behind the lens to be calculated.

If, for instance, the laser beam is focused into the interaction region with absorbing molecules, the beam waist of the laser resonator has to be transformed into a beam waist located in this region. The beam parameters in the waists are purely imaginary, that is,

$$q_1 = i\pi w_1^2/\lambda , \quad q_2 = i\pi w_2^2/\lambda . \tag{5.149}$$

The beam diameters in the waists are $2w_1$ and $2w_2$, and the radius of curvature is infinite. Inserting (5.149) into (5.148) and equating the imaginary and the real parts yields the two equations

$$\frac{d_1 - f}{d_2 - f} = \frac{w_1^2}{w_2^2} , \tag{5.150}$$

$$(d_1 - f)(d_2 - f) = f^2 - f_0^2 , \quad \text{with} \quad f_0 = \pi w_1 w_2/\lambda . \tag{5.151}$$

Since $d_1 > f$ and $d_2 > f$, this shows that any lens with $f > f_0$ can be used. For a given f, the position of the lens is determined by solving the two equations for d_1 and d_2,

$$d_1 = f \pm \frac{w_1}{w_2}\sqrt{f^2 - f_0^2} , \tag{5.152}$$

$$d_2 = f \pm \frac{w_2}{w_1}\sqrt{f^2 - f_0^2} . \tag{5.153}$$

From (5.150) we obtain the beam waist radius w_2 in the collimated region

$$w_2 = w_1 \left(\frac{d_2 - f}{d_1 - f}\right)^{1/2} . \tag{5.154}$$

When the Gaussian beam is mode matched to another resonator, the beam parameter q_2 at the mirrors of this resonator must match the curvature R and the spot size w (5.39). From (5.148), the correct values of f, d_1, and d_2 can be calculated.

We define the collimated or *waist region* as the range $|z| \leq z_R$ around the beam waist at $z = 0$, where at $z = \pm z_R$ the spot size $w(z)$ has increased by

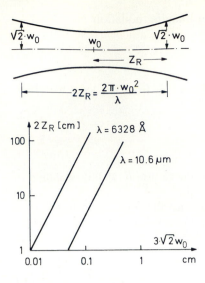

Fig. 5.130. Beam waist region and Rayleigh length z_R of a Gaussian beam

Fig. 5.131. Full Rayleigh lengths $2z_R$ as a function of the beam waist w_0 for two different wavelengths $\lambda_1 = 632.8\,\text{nm}$ (HeNe laser) and $\lambda_2 = 10.6\,\mu\text{m}$ (CO_2 laser)

a factor of $\sqrt{2}$ compared with the value w_0 at the waist. Using (5.138) we obtain

$$w(z) = w_0 \left[1 + \left(\frac{\lambda z_R}{\pi w_0^2} \right)^2 \right]^{1/2} = \sqrt{2} w_0 \,, \tag{5.155}$$

which yields for the *waist length* or *Rayleigh length*

$$\boxed{z_R = \pi w_0^2 / \lambda \,.} \tag{5.156}$$

The waist region extends about one Rayleigh distance on either side of the waist (Fig. 5.130). The length of the Rayleigh distance depends on the spot size and therefore on the focal length of the focusing lens. Figure 5.131 depicts the dependence on w_0 of the full Rayleigh length $2z_R$ for two different wavelengths.

At large distances $z \gg z_R$ from the waist, the Gaussian beam wavefront is essentially a spherical wave emitted from a point source at the waist. This region is called the *far field*. The divergence angle θ (far-field half angle) of the beam can be obtained from (5.138) and Fig. 5.126 with $z \gg z_R$ as

$$\theta = \frac{w(z)}{z} = \frac{\lambda}{\pi w_0} \,. \tag{5.157}$$

Note, however, that in the near-field region the center of curvature *does not* coincide with the center of the beam waist (Fig. 5.126). When a Gaussian beam is focused by a lens or a mirror with focal length f, the spot size in

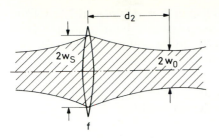

Fig. 5.132. Focusing of a Gaussian beam by a lens

the beam waist is for $f \gg w_s$

$$w_0 = \frac{f\lambda}{\pi w_s} , \qquad (5.158)$$

where w_s is the spot size at the lens (Fig. 5.132).

To avoid diffraction losses the diameter of the lens should be $d \geq 3w_s$.

Example 5.35

A lens with $f = 5$ cm is imaging a Gaussian beam with a spot size of $w_s = 0.2$ cm at the lens. For $\lambda = 623$ nm the focal spot has the waist radius $w_0 = 5\,\mu$m.

In order to achieve a smaller waist radius, one has to increase w_s or decrease f (Fig. 5.132).

Problems

5.1 Calculate the necessary threshold inversion of a gas laser transition at $\lambda = 500$ nm with the transition probability $A_{ik} = 5 \times 10^7\,\text{s}^{-1}$ and a homogeneous linewidth $\Delta\nu_{\text{hom}} = 20$ MHz. The active length is $L = 20$ cm and the resonator losses per round-trip are 5%.

5.2 A laser medium has a Doppler-broadened gain profile of halfwidth 2 GHz. The homogeneous width is 50 MHz, and the transition probability $A_{ik} = 1 \times 10^8\,\text{s}^{-1}$. Assume that one of the resonator modes ($L = 40$ cm) coincides with the center frequency ν_0 of the gain profile. What is the threshold inversion for the central mode, and at which inversion does oscillation start on the two adjacent longitudinal modes if the resonator losses are 10%?

5.3 The frequency of a passive resonator mode ($L = 15$ cm) lies $0.5\Delta\nu_D$ away from the center of the Gaussian gain profile of a gas laser at $\lambda = 632.8$ nm. Estimate the mode pulling if the cavity resonance width is 2 MHz and $\Delta\nu_D = 1$ GHz.

5.4 Assume a laser transition with a homogeneous width of 100 MHz, while the inhomogeneous width of the gain profile is 1 GHz. The resonator length is $d = 200$ cm and the active medium with length $L \ll d$ is placed 20 cm from one end mirror. Estimate the spacing of the spatial hole-burning modes. How many modes can oscillate simultaneously if the unsaturated gain at the line center exceeds the losses by a factor of 10%?

5.5 Estimate the optimum transmission of the laser output mirror if the unsaturated gain is 2 and the internal resonator losses are 10.

5.6 The output beam from an HeNe laser with a confocal resonator ($R = L = 30$ cm) is focused by a lens of $f = 30$ cm, 50 cm away from the output mirror. Calculate the location of the focus, the Rayleigh length, and the beam waist in the focal plane.

5.7 A nearly parallel Gaussian beam is expanded by a telescope with two lenses of focal lengths $f_1 = 1$ cm and $f_2 = 10$ cm. The spot size at the entrance lens is $w = 1$ mm. An aperture in the common focal plane of the two lenses acts as a spatial filter to improve the quality of the wavefront in the expanded beam (why?). What is the diameter of this aperture, if 95% of the intensity is transmitted?

5.8 A HeNe laser with an unsaturated gain of $G_0(\nu_0) = 1.3$ at the center of the Gaussian gain profile has a resonator length of $d = 50$ cm and total losses of 4%. Single-mode operation at ν_0 is achieved with a coated tilted etalon inside the resonator. Design the optimum combination of etalon thickness and finesse.

5.9 An argon laser with resonator length $d = 100$ cm and two mirrors with radius $r_1 = \infty$ and $r_2 = 400$ cm has an intracavity circular aperture close to the spherical mirror to prevent oscillation on transversal modes. Estimate the maximum diameter of the aperture that introduces losses $\gamma_{\mathrm{diffr}} < 1\%$ for the TEM_{00} mode, but prevents oscillation of higher transverse modes, which without the aperture have a net gain of 10%.

5.10 A single-mode HeNe laser with resonator length $L = 15$ cm is tuned by moving a resonator mirror mounted on a piezo. Estimate the maximum tuning range before a mode hop will occur, assuming an unsaturated gain of 10% at the line center and resonator losses of 3%. What voltage has to be applied to the piezo (expansion 1 nm/V) for this tuning range?

5.11 Estimate the frequency drift of a laser oscillating at $\lambda = 500$ nm because of thermal expansion of the resonator at a temperature drift of 1°C/h, when the resonator mirrors are mounted on distance-holder rods (a) made of invar and (b) made of fused quartz.

5.12 Mode selection in an argon laser is often accomplished with an intracavity etalon. What is the frequency drift of the transmission maximum

(a) for a solid fused quartz etalon with thickness $d = 1$ cm due to a temperature change of 2°C?

(b) For an air-space etalon due to an air pressure change of 4 mb?

(c) Estimate the average time between two mode hopes (cavity length $L = 100$ cm) for a temperature drift of 1°C/h or a pressure drift of 2 mbar/h.

5.13 Assume that the output power of a laser shows random fluctuations of about 5%. Intensity stabilization is accomplished by a Pockels cell with a half-wave voltage of 600 V. Estimate the ac output voltage of the amplifier driving the Pockels cell that is necessary to stabilize the transmitted intensity if the Pockels cell is operated around the maximum slope of the transmission curve.

5.14 A single-mode laser is frequency stabilized onto the slope of the transmission maximum of an external reference Fabry–Perot interferometer made of invar with a free spectral range of 8 GHz. Estimate the frequency stability of the laser

(a) against temperature drifts, if the FPI is temperature stabilized within 0.01°C,

(b) against acoustic vibrations of the mirror distance d in the FPI with amplitudes of 1 nm.

(c) Assume that the *intensity* fluctuations are compensated to 1% by a difference amplifier. Which *frequency* fluctuations are still caused by the residual intensity fluctuations, if a FPI with a free spectral range of 10 GHz and a finesse of 50 is used for frequency stabilization at the slope of the FPI transmission peak?

6. Doppler-Limited Absorption and Fluorescence Spectroscopy with Lasers

In the previous chapter we presented the different realizations of tunable lasers; we now discuss their applications in absorption and fluorescence spectroscopy. First we discuss those methods where the spectral resolution is limited by the Doppler width of the molecular absorption lines. This limit can in fact be reached if the laser linewidth is small compared with the Doppler width. In several examples, such as optical pumping or laser-induced fluorescence spectroscopy, multimode lasers may be employed, although in most cases single-mode lasers are superior. In general, however, these lasers may not necessarily be frequency stabilized as long as the frequency jitter is small compared with the absorption linewidth. We compare several detection techniques of molecular absorption with regard to their sensitivity and their feasibility in the different spectral regions. Some examples illustrate the methods to give the reader a feeling of what has been achieved. After the discussion of Doppler-limited spectroscopy, Chaps. 7–10 give an extensive treatment of various techniques which allow sub-Doppler spectroscopy.

6.1 Advantages of Lasers in Spectroscopy

In order to illustrate the advantages of absorption spectroscopy with tunable lasers, we first compare it with conventional absorption spectroscopy, which uses incoherent radiation sources. Figure 6.1 presents schematic diagrams for both methods.

In classical absorption spectroscopy, radiation sources with a *broad emission continuum* are preferred (e.g., high-pressure Hg arcs, Xe flash lamps, etc.). The radiation is collimated by the lens L_1 and passes through the absorption cell. Behind a dispersing instrument for wavelength selection (spectrometer or interferometer) the intensity $I_T(\lambda)$ of the transmitted light is measured as a function of the wavelength λ (Fig. 6.1a). By comparison with the reference beam $I_R(\lambda)$, which can be realized, for instance, by shifting the absorption cell alternatively out of the light beam, the absorption spectrum

$$I_A(\lambda) = a[I_0(\lambda) - I_T(\lambda)] = a[bI_R(\lambda) - I_T(\lambda)],$$

can be obtained, where the constants a and b take into account wavelength-independent losses of I_R and I_T (e.g., reflections of the cell walls).

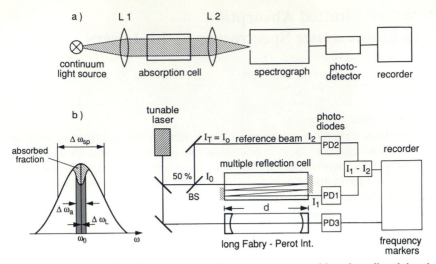

Fig. 6.1a,b. Comparison between absorption spectroscopy with a broadband incoherent source (**a**) and with a tunable single-mode laser (**b**)

The *spectral resolution* is generally limited by the resolving power of the dispersing spectrometer. Only with large and expensive instruments (e.g., Fourier spectrometers) may the Doppler limit be reached [6.1].

The *detection sensitivity* of the experimental arrangement is defined by the minimum absorbed power that can still be detected. In most cases it is limited by the detector noise and by intensity fluctuations of the radiation source. Generally, the limit of the detectable absorption is reached at relative absorptions $\Delta I/I \geq 10^{-4}-10^{-5}$. This limit can be pushed further down only in favorable cases by using special sources and lock-in detection or signal-averaging techniques.

Contrary to radiation sources with broad emission continua used in conventional spectroscopy, tunable lasers offer radiation sources in the spectral range from the UV to the IR with extremely narrow bandwidths and with spectral power densities that may exceed those of incoherent light sources by many orders of magnitude (Sects. 5.7, 5.8).

In several regards laser absorption spectroscopy corresponds to microwave spectroscopy, where klystrons or carcinotrons instead of lasers represent tunable coherent radiation sources. Laser spectroscopy transfers many of the techniques and advantages of microwave spectroscopy to the infrared, visible, and ultraviolet spectral ranges.

The advantages of absorption spectroscopy with tunable lasers may be summarized as follows:

- No monochromator is needed, since the absorption coefficient $\alpha(\omega)$ and its frequency dependence can be directly measured from the difference $\Delta I(\omega) = a[I_R(\omega) - I_T(\omega)]$ between the intensities of the reference beam with $I_R = I_2$ and transmitted beam with $I_T = I_1$ (Fig. 6.1b). The spectral resolution is higher than in conventional spectroscopy. With tunable single-mode lasers it is only limited by the linewidths of the absorbing molecular transitions. Using Doppler-free techniques (Chaps. 7–10), even sub-Doppler resolution can be achieved.

- Because of the high spectral power density of many lasers, the detector noise is generally negligible. Intensity fluctuations of the laser, which limit the detection sensitivity, may essentially be suppressed by intensity stabilization (Sect. 5.4). This furthermore increases the signal-to-noise ratio and therefore enhances the sensitivity.

- The detection sensitivity increases with increasing spectral resolution $\omega/\Delta\omega$ as long as $\Delta\omega$ is still larger than the linewidth $\delta\omega$ of the absorption line. This can be seen as follows:

The relative intensity attenuation per absorption path length $x = 1\,\mathrm{cm}$ on the transition with center frequency ω_0 is, for small absorption $\alpha x \ll 1$,

$$\Delta I/I = \int_{\omega_0 - \frac{1}{2}\Delta\omega}^{\omega_0 + \frac{1}{2}\Delta\omega} \alpha(\omega) I(\omega)\,d\omega \bigg/ \int_{\omega_0 - \frac{1}{2}\delta\omega}^{\omega_0 + \frac{1}{2}\delta\omega} I(\omega)\,d\omega \ . \tag{6.1}$$

If $I(\omega)$ does not change much within the interval $\Delta\omega$, we can write

$$\int_{\omega_0 - \frac{1}{2}\Delta\omega}^{\omega_0 + \frac{1}{2}\Delta\omega} I(\omega)\,d\omega = \bar{I}\Delta\omega \quad \text{and} \quad \int \alpha(\omega) I(\omega)\,d\omega = \bar{I} \int \alpha(\omega)\,d\omega \ .$$

This yields

$$\Delta I/I = \frac{1}{\Delta\omega} \int_{\omega_0 - \frac{1}{2}\delta\omega}^{\omega_0 + \frac{1}{2}\delta\omega} \alpha(\omega)\,d\omega \simeq \bar{\alpha}\frac{\delta\omega}{\Delta\omega} \quad \text{for} \quad \Delta\omega > \delta\omega \ . \tag{6.2}$$

The measured relative intensity attenuation per centimeter absorption pathlength is therefore the product of absorption coefficient α and the ratio of absorption linewidth $\delta\omega$ and spectral resolution bandwidth $\Delta\omega$, as long as $\delta\omega < \Delta\omega$!

For $\Delta\omega < \delta\omega$, the line profile $\alpha(\omega)$ of the absorption line can be measured, although the ratio $\Delta I/I$ does not exceed $\alpha(\omega) \cdot L$.

Example 6.1

The spectral resolution of a 1-m spectrograph is about 0.01 nm, which corresponds at $\lambda = 500\,\mathrm{nm}$ to $\Delta\omega = 2\pi \cdot 12\,\mathrm{GHz}$. The Doppler width of gas molecules with $M = 30$ at $T = 300\,\mathrm{K}$ is, according to (3.43) $\delta\omega \simeq 2\pi \cdot 1\,\mathrm{GHz}$. With a single-mode laser the value of $\delta\omega$ becomes smaller than $\Delta\omega$ and the observable signal $\Delta I/I$ becomes 12 times larger than in conventional spectroscopy with the same absorption cell.

- Because of the good collimation of a laser beam, long absorption paths can be realized by multiple reflection back and forth through the multiple-path absorption cell. Disturbing reflections from cell walls or windows, which may influence the measurements, can essentially be avoided (for example, by using Brewster end windows). Such long absorption paths enable measurements of transitions even with small absorption coefficients. Furthermore, pressure broadening can be reduced by using low gas pressure. This is especially important in the infrared region, where the Doppler width is small and pressure broadening may become the limiting factor for the spectral resolution (Sect. 3.3).

Example 6.2

With an intensity-stabilized light source and lock-in detection the minimum relative absorption that may be safely detected is about $\Delta I/I \geq 10^{-6}$, which yields the minimum measurable absorption coefficient α_{min} for an absorption pathlength L,

$$\alpha_{min} = \frac{10^{-6}}{L}\frac{\Delta\omega}{\delta\omega} \quad [\text{cm}^{-1}].$$

With conventional spectroscopy for a pathlength $L = 10\,\text{cm}$ and $\delta\omega = 10\Delta\omega$ we obtain $\alpha_{min} = 10^{-6}\,\text{cm}^{-1}$. With single-mode lasers one may reach $\Delta\omega < \delta\omega$ and a long absorption path with $L = 10\,\text{m}$, which yields a minimum detectable absorption coefficient of $\alpha_{min} = 10^{-9}\,\text{cm}^{-1}$, i.e., an improvement of a factor of 1000 !

- If a small fraction of the laser output is sent through a long Fabry–Perot interferometer with a separation d of the mirrors (Fig. 6.1b), the photodetector PD3 receives intensity peaks each time the laser frequency ν_L is tuned to a transmission maximum at $\nu = \frac{1}{2}mc/d$ (Sects. 4.2–4.4). These peaks serve as accurate wavelength markers, which allow one to calibrate the separation of adjacent absorption lines (Fig. 6.1). With $d = 1\,\text{m}$ the frequency separation $\Delta\nu_p$ between successive transmission peaks is $\Delta\nu_p = c/2d = 150\,\text{MHz}$, corresponding to a wavelength separation of $10^{-4}\,\text{nm}$ at $\lambda = 550\,\text{nm}$. With a semiconfocal FPI the free spectral range is $c/8d$, which gives $\Delta\nu_p = 75\,\text{MHz}$ for $d = 0.5\,\text{m}$.
- The laser frequency may be stabilized onto the center of an absorption line. With the methods discussed in Sect. 4.4 it is possible to measure the wavelength λ_L of the laser with an absolute accuracy of 10^{-8} or better. This allows determination of the molecular absorption lines with the same accuracy.
- It is possible to tune the laser wavelength very rapidly over a spectral region where molecular absorption lines have to be detected. With electro-optical components, for instance, pulsed dye lasers can be tuned over several wave numbers within a microsecond. This opens new perspectives for spectroscopic investigations of short-lived intermediate

radicals in chemical reactions. The capabilities of classical flash photolysis may be considerably extended using such rapidly tunable laser sources.

- An important advantage of absorption spectroscopy with tunable single-mode lasers stems from their capabilites to measure line profiles of absorbing molecular transitions with high accuracy. In case of pressure broadening, the determination of line profiles allows one to derive information about the interaction potential of the collision partners (Sects. 3.3 and 13.1). In plasma physics this technique is widely used to determine electron and ion densities and temperatures.

- In fluorescence spectroscopy and optical pumping experiments, the high intensity of lasers allows an appreciable population in selectively excited states to be achieved that may be comparable to that of the absorbing ground states. The small laser linewidth favors the selectivity of optical excitation and results in favorable cases in the exclusive population of single molecular levels. These advantageous conditions allow one to perform absorption and fluorescence spectroscopy of *excited* states and to transform spectroscopic methods, such as microwave or RF spectroscopy, which has until now been restricted to electronic ground states, also to excited states.

This brief overview of some of the advantages of lasers in spectroscopy will be outlined in more detail in the following chapters and several examples will illustrate their relevance.

6.2 High-Sensitivity Methods of Absorption Spectroscopy

The general method for measuring absorption spectra is based on the determination of the absorption coefficient $\alpha(\omega)$ from the spectral intensity

$$I_T(\omega) = I_0 \exp[-\alpha(\omega)x] \,, \tag{6.3}$$

which is transmitted through an absorbing path length x. For small absorption $\alpha x \ll 1$, we can use the approximation $\exp(-\alpha x) \simeq 1 - \alpha x$, and (6.3) can be reduced to

$$I_T(\omega) \simeq I_0[1 - \alpha(\omega)x] \,. \tag{6.4}$$

With the reference intensity $I_R = I_0$, as produced, for example, by a 50% beam splitter with the reflectivity $R = 0.5$ (Fig. 6.1b), one can measure the absorption coefficient

$$\alpha(\omega) = \frac{I_R - I_T(\omega)}{I_R x} \,, \tag{6.5}$$

from the difference $\Delta I = I_R - I_T(\omega)$.

The absorption coefficient $\alpha_{ik}(\omega)$ of the transition $|i\rangle \to |k\rangle$ with an absorption cross section σ_{ik} is determined by the density N_i of absorbing molecules (see Sect. 5.1.2)

$$\alpha_{ik}(\omega) = [N_i - (g_i/g_k)N_k]\sigma_{ik}(\omega) = \Delta N \sigma_{ik}(\omega) . \qquad (6.6)$$

If the population N_k is small compared to N_i, we obtain from (6.6) for the minimum detectable density N_i over the absorption path length $x = L$:

$$N_i \geq \frac{\Delta I}{I_0 L \sigma_{ik}} . \qquad (6.7)$$

The minimum still detectable concentration N_i of absorbing molecules is determined by the absorption path length L, the absorption cross section σ_{ik}, and the minimum detectable relative intensity change $\Delta I/I_0$ caused by absorption.

In order to reach a high detection sensitivity for absorbing molecules, $L\sigma_{ik}$ should be large and the minimum detectable value of $\Delta I/I_0$ as small as possible.

In the case of very small values of αx, this method in which the attenuation of the transmitted light is measured cannot be very accurate, since it must determine a small difference $I_0 - I_T$ of two large quantities I_0 and I_T. Small fluctuations of I_0 or of the splitting ratio of the beam splitter BS in Fig. 6.1b can severely influence the measurement. Therefore, other techniques have been developed leading to an increase in sensitivity and accuracy of absorption measurements by several orders of magnitude compared to direct absorption measurements. These sensitive detection methods represent remarkable progress, since their sensitivity limits have been pushed from relative absorptions $\Delta\alpha/\alpha \simeq 10^{-5}$ down to about $\Delta\alpha/\alpha \geq 10^{-17}$. We now discuss these different methods in more detail.

6.2.1 Frequency Modulation

The first scheme to be treated is based on a frequency modulation of the monochromatic incident wave. It was not designed specifically for laser spectroscopy, but was taken from microwave spectroscopy where it is a standard method. The laser frequency ω_L is modulated at the modulation frequency Ω, which tunes ω_L periodically from ω_L to $\omega_L + \Delta\omega_L$. When the laser is tuned through the absorption spectrum, the difference $I_T(\omega_L) - I_T(\omega_L + \Delta\omega_L)$ is detected with a lock-in amplifier (phase-sensitive detector) tuned to the modulation frequency Ω (Fig. 6.2). If the modulation sweep $\Delta\omega_L$ is sufficiently small, the first term of the Taylor expansion

$$I_T(\omega_L + \Delta\omega_L) - I_T(\omega_L) = \frac{dI_T}{d\omega}\Delta\omega_L + \frac{1}{2!}\frac{d^2 I_T}{d\omega^2}\Delta\omega_L^2 + \ldots, \qquad (6.8)$$

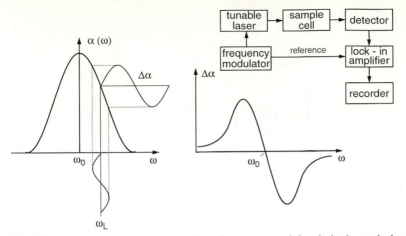

Fig. 6.2. Absorption spectroscopy with a frequency-modulated single-mode laser

is dominant. This term is proportional to the first derivative of the absorption spectrum, as can be seen from (6.5): when I_R is independent of ω we obtain for the absorption length L

$$\frac{d\alpha(\omega)}{d\omega} = -\frac{1}{I_R L}\frac{dI_T}{d\omega}. \tag{6.9}$$

If the laser frequency

$$\omega_L(t) = \omega_0 + a\sin\Omega t,$$

is sinusoidally modulated at a modulation frequency Ω, the Taylor expansion yields

$$I_T(\omega_L) = I_T(\omega_0) + \sum_n \frac{a^n}{n!}\sin^n\Omega t\left(\frac{d^n I_T}{d\omega^n}\right)_{\omega_0}. \tag{6.10}$$

For $\alpha L \ll 1$, we obtain from (6.4)

$$\left(\frac{d^n I_T}{d\omega^n}\right)_{\omega_0} = -I_0 x\left(\frac{d^n\alpha(\omega)}{d\omega^n}\right)_{\omega_0}.$$

The terms $\sin^n\Omega t$ can be converted into linear functions of $\sin(n\Omega t)$ and $\cos(n\Omega t)$ using known trigonometric formulas.

Inserting these relations into (6.10), one finds after rearrangement of the terms the expression

$$
\begin{aligned}
\frac{I_T(\omega_L) - I_T(\omega_0)}{I_0} = -aL \Bigg\{ &\left[\frac{a}{4} \left(\frac{d^2\alpha}{d\omega^2} \right)_{\omega_0} + \frac{a^3}{64} \left(\frac{d^4\alpha}{d\omega^4} \right)_{\omega_0} + \cdots \right] \\
&+ \left[\left(\frac{d\alpha}{d\omega} \right)_{\omega_0} + \frac{a^2}{8} \left(\frac{d^3\alpha}{d\omega^3} \right)_{\omega_0} + \cdots \right] \sin(\Omega t) \\
&+ \left[-\frac{a}{4} \left(\frac{d^2\alpha}{d\omega^2} \right)_{\omega_0} + \frac{a^3}{48} \left(\frac{d^4\alpha}{d\omega^4} \right)_{\omega_0} + \cdots \right] \cos(2\Omega t) \\
&+ \left[-\frac{a^2}{24} \left(\frac{d^3\alpha}{d\omega^3} \right)_{\omega_0} + \frac{a^4}{384} \left(\frac{d^5\alpha}{d\omega^5} \right)_{\omega_0} + \cdots \right] \sin(3\Omega t) \\
&+ \cdots \Bigg\} .
\end{aligned}
$$

For a sufficiently small modulation amplitude $(a/\omega_0 \ll 1)$, the first terms in each bracket are dominant. Therefore we obtain for the signal $S(n\Omega)$ behind a lock-in amplifier tuned to the frequency $n\Omega$ (Fig. 6.3):

$$
S(n\Omega) = \left(\frac{I_T(\omega_L) - I_T(\omega_0)}{I_0} \right)_{n\Omega} = aL \begin{cases} b_n \sin(n\Omega t) , & \text{for} \quad n = 2m+1 , \\ c_n \cos(n\Omega t) , & \text{for} \quad n = 2m . \end{cases}
$$

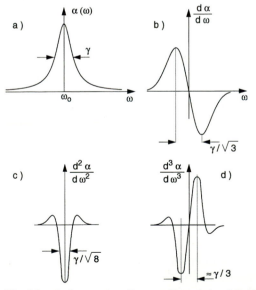

Fig. 6.3a–d. Lorentzian line profile $\alpha(\omega)$ of halfwidth γ (FWHM) (**a**), with first (**b**), second (**c**), and third (**d**) derivatives

In particular, the signals for the first three derivatives of the absorption coefficient $\alpha(\omega)$, shown in Fig. 6.3, are

$$S(\Omega) = -aL\frac{d\alpha}{d\omega}\sin(\Omega t) \, ,$$

$$S(2\Omega) = +\frac{a^2 L}{4}\frac{d^2\alpha}{d\omega^2}\cos(2\Omega t) \, , \qquad (6.11)$$

$$S(3\Omega) = +\frac{a^3 L}{24}\frac{d^3\alpha}{d\omega^3}\sin(3\Omega t) \, .$$

The advantage of this "derivative spectroscopy" [6.2] with a frequency-modulated laser is the possibility for phase-sensitive detection, which restricts the frequency response of the detection system to a narrow frequency interval centered at the modulation frequency Ω. Frequency-independent background absorption from cell windows and background noise from fluctuations of the laser intensity or of the density of absorbing molecules are essentially reduced. Regarding the signal-to-noise ratio and achievable sensitivity, the *frequency* modulation technique is superior to an *intensity* modulation of the incident radiation. The frequency of a single-mode laser can readily be modulated when an ac voltage is applied to the piezo onto which a resonator mirror is mounted (Sect. 5.5).

The technical noise, which represents the major limitation, decreases with increasing frequency. It is therefore advantageous to choose the modulation frequency as high as possible. For diode lasers this can be achieved by modulation of the diode current. For other lasers, often electro-optical modulators outside the laser resonator are used as phase modulators, resulting in a frequency modulation of the transmitted laser beam [6.3].

The phase modulation has an additional advantage: the first two sidebands at frequencies $\omega + \Omega$ and $\omega - \Omega$ have equal amplitudes but opposite phases (Fig. 6.4). A lock-in detector tuned to the modulation frequency Ω therefore receives the superposition of two beat signals between the carrier and the two sidebands, which cancel to zero if no absorption is present. Any fluctuation of the laser intensity appears equally on both signals and is therefore also cancelled. If the laser wavelength is tuned over an absorption line, one sideband is absorbed, if $\omega + \Omega$ or $\omega - \Omega$ coincides with the absorption frequency ω_0. This perturbs the balance and gives raise to a signal with a profile that is similar to the profile of the second derivative in Fig. 6.3c.

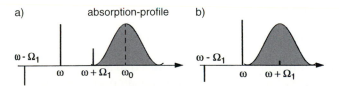

Fig. 6.4a,b. Principle of phase-modulated absorption spectroscopy

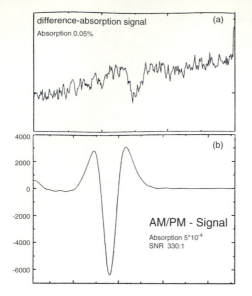

Fig. 6.5. (**a**) Water overtone absorption line measured with unmodulated (**b**) and with phase-modulated single-mode diode laser

The sensitivity of this technique is demonstrated by Fig. 6.5, which shows an overtone absorption line of the water molecule H_2O, recorded with an unmodulated laser and with this modulation technique. The signal-to-noise ratio of the phase modulation is about 2 orders of magnitude higher if the modulation frequency is chosen to be equal to the width of the absorption line.

If the modulation frequency Ω is chosen sufficiently high ($\Omega > 1000\,\mathrm{MHz}$), the technical noise may drop below the quantum-noise limit set by the statistical fluctuations of detected photons. In this case, the detection limit is mainly due to the quantum limit [6.4]. Since lock-in detectors cannot handle such high frequencies, the signal input has to be downconverted in a mixer, where the difference frequency between a local oscillator and the signal is generated.

A new method of high-frequency modulation spectroscopy using low-frequency detection is two-tone frequency-modulation spectroscopy [6.5]. In this method the laser output is phase-modulated in an electro-optic $LiTaO_3$ crystal that is driven by a high-frequency voltage in the GHz range, and that is amplitude-modulated at frequencies in the MHz range. The detector output is fed into a frequency mixer (downconverter) and the final signal is received with a lock-in amplifier in the kilohertz range [6.6, 6.7].

A comparison of different modulation techniques can be found in [6.8–6.10].

6.2.2 Intracavity Absorption

When the absorbing sample is placed inside the laser cavity (Fig. 6.6), the detection sensitivity can be enhanced considerably, in favorable cases by several orders of magnitude. Four different effects can be utilized to achieve this "amplified" sensitivity. The first two are based on single-mode operation, the last two on multimode oscillation of the laser.

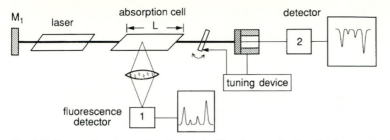

Fig. 6.6. Intracavity absorption detected either by monitoring the laser output $P(\omega_L)$ with detector 2 or the laser-induced fluorescence $I_{Fl}(\omega_L)$ with detector 1

(1) Assume that the reflectivities of the two resonator mirrors are $R_1 = 1$ and $R_2 = 1 - T_2$ (mirror absorption is neglected). At the laser output power P_{out}, the power inside the cavity is $P_{int} = qP_{out}$ with $q = 1/T_2$. For $\alpha L \ll 1$, the power $\Delta P(\omega)$ absorbed at the frequency ω in the absorption cell (length L) is

$$\Delta P(\omega) = \alpha(\omega) L P_{int} = q\alpha(\omega) L P_{out} . \tag{6.12}$$

If the absorbed power can be measured directly, for example, through the resulting pressure increase in the absorption cell (Sect. 6.3.3) or through the laser-induced fluorescence (Sect. 6.3.1), the signal will be q times larger than for the case of single-pass absorption outside the cavity.

Example 6.3

With a resonator transmission $T_2 = 0.02$ (a figure that can be readily realized in practice) the enhancement factor becomes $q = 50$, as long as saturation effects can be neglected and provided the absorption is sufficiently small that it does not noticeably change the laser intensity.

This q-fold amplification of the sensitivity can be also understood from the simple fact that a laser photon travels on the average q times back and forth between the resonator mirrors before it leaves the resonator. It therefore has a q-fold chance to be absorbed in the sample.

This sensitivity enhancement in detecting small absorptions has no direct correlation with the gain medium and can be also realized in external passive resonators. If the laser output is mode matched (Sect. 5.2.3) by lenses or mirrors into the fundamental mode of the passive cavity containing the absorbing sample (Fig. 6.7), the radiation power inside this cavity is q times larger. The enhancement factor q may become larger if the internal losses of the cavity can be kept low.

Intracavity absorption cells are particularly advantageous if the absorption is monitored via the laser-induced fluorescence. Since the radiation field inside the active resonator or inside the mode-matched passive cavity is concentrated

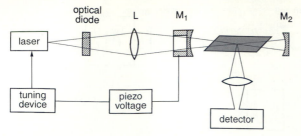

Fig. 6.7. Spectroscopy inside an external resonator, which is synchronously tuned with the laser frequency ω_L in order to be always in resonance

within the region of the Gaussian beam (Sect. 5.9), the laser-excited fluorescence can be effectively imaged onto the entrance slit of a spectrometer with a larger efficiency than in the commonly used multipass cells. If minute concentrations of an absorbing component have to be selectively detected in the presence of other constituents with overlapping absorption lines but different fluorescence spectra, the use of a spectrometer for dispersing the fluorescence can solve the problem.

External passive resonators may become advantageous when the absorption cell cannot be placed directly inside the active laser resonator. However, there also exist some drawbacks: the cavity length has to be changed synchronously with the tunable-laser wavelength in order to keep the external cavity always in resonance. Furthermore, one has to take care to prevent optical feedback from the passive to the active cavity, which would cause a coupling of both cavities with resulting instabilities. This feedback can be avoided by an optical diode (Sect. 5.2.7).

(2) Another way of detecting intracavity absorption with a very high sensitivity relies on the dependence of the single-mode laser output power on absorption losses inside the laser resonator (detector 2 in Fig. 6.6). At constant pump power just above the threshold, minor changes of the intracavity losses may result in drastic changes of the laser output. In Sect. 5.3 we saw that under steady-state conditions, the laser ouput power essentially depends on the pump power and reaches the value P_s, where the gain factor $G = \exp[-2L_1\alpha_s - \gamma]$ becomes $G = 1$. This implies that the saturated gain $g_s = 2L_1\alpha_s$ of the active medium with the length L_1 equals the total losses γ per cavity round trip (Sect. 5.1).

The saturated gain $g_s = 2L_1\alpha_s$ depends on the intracavity intensity I. According to (3.61) we obtain

$$g_s = \frac{g_0}{1 + I/I_s} = \frac{g_0}{1 + P/P_s} , \tag{6.13}$$

where I_s is the saturation intensity, which decreases g_0 to $g_s = g_0/2$ (Sect. 3.6). At a constant pump power the laser power P stabilizes itself at the value where $g_s = \gamma$. This gives with (6.13):

$$P = P_s \frac{g_0 - \gamma}{\gamma} \, . \tag{6.14}$$

If small additional losses $\Delta\gamma$ are introduced by the absorbing sample inside the cavity, the laser power drops to a value

$$P_\alpha = P - \Delta P = P_s \cdot \frac{g_0 - \gamma - \Delta\gamma}{\gamma + \Delta\gamma} \, . \tag{6.15}$$

From (6.13–6.15) we obtain for the relative change $\Delta P/P$ of the laser output power by the absorbing sample

$$\Delta P/P = \frac{g_0}{g_0 - \gamma} \frac{\Delta\gamma}{\gamma + \Delta\gamma} \simeq \frac{g_0}{\gamma} \frac{\Delta\gamma}{g_0 - \gamma} \, , \quad \text{for} \quad \Delta\gamma \ll \gamma \, , \tag{6.16}$$

where g_0 is the unsaturated gain.

Compared with the single-pass absorption of a sample with the absorption coefficient α and the absorption pathlength L_2 outside the laser resonator where $\Delta P/P = -\alpha L_2 = -\Delta\gamma$, the intracavity absorption represents a sensitivity enhancement by the factor

$$Q = \frac{g_0}{\gamma(g_0 - \gamma)} \, . \tag{6.17}$$

At pump powers far above threshold, the unsaturated gain g_0 is large compared to the losses γ, and (6.17) reduces to

$$Q \simeq 1/\gamma \, , \quad \text{for} \quad g_0 \gg \gamma \, .$$

If the resonator losses are mainly due to the transmission T_2 of the output mirror, the enhancement factor Q then becomes $Q = 1/\gamma = 1/T_2 = q$, which is equal to the enhancement of the previous detection method 1.

Just above threshold, however, g_0 is only slightly larger than γ and the denominator in (6.16) becomes very small, which means that the enhancement

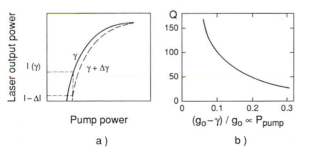

a)   b)

Fig. 6.8. (a) Laser output power $P_L(P_{pump})$ for two slightly different losses γ and $\gamma + \Delta\gamma$. (b) Enhancement factor Q as a function of pump power P_{pump} above threshold

factor Q may reach very large values (Fig. 6.8). At first, it might appear that the sensitivity could be made arbitrarily large for $2\alpha_0 L \to \gamma$. However, there are experimental as well as fundamental limitations that restrict the maximum achievable value of Q. The increasing instability of the laser output, for instance, limits the detection sensitivity when the threshold is approached. Just above threshold, the spontaneous radiation that is emitted into the solid angle accepted by the detector cannot be neglected. It represents a constant background intensity, nearly independent of γ, which puts a principal upper limit to the relative change $\Delta I / I$ and thus to the sensitivity.

Note: If the gain medium is inhomogeneously broadened, the saturated gain g_s becomes

$$g_s = \frac{g_0}{\sqrt{1 + I/I_s}} \,,$$

(Sect. 3.6 and 7.2). The derivation analogue to (6.13–6.16) yields

$$\frac{\Delta P}{P} = \frac{g_0^2 (2\gamma \Delta\gamma + \Delta\gamma^2)}{(g_0^2 - \gamma^2)(\gamma + \Delta\gamma)^2} \simeq \frac{g_0^2}{\gamma^2} \frac{\Delta\gamma}{g_0 - \gamma} \,, \tag{6.18a}$$

instead of (6.16).

(3) In the preceding discussion of the sensitivity enhancement by intracavity absorption, we have implicitly assumed that the laser oscillates in a single mode. Even larger enhancement factors Q can be achieved, however, with lasers oscillating simultaneously in several competing modes. Pulsed or cw dye lasers without additional mode selection are examples of lasers with mode competition. As discussed in Sect. 5.3, the active dye medium exhibits a broad homogeneous spectral gain profile, which allows the same dye molecules to simultaneously contribute to the gain of all modes with frequencies within the homogeneous linewidth (see the discussion in Sects. 5.3, 5.8). This means that the different oscillating laser modes may share the same molecules to achieve their gains. This leads to mode competition and bring about the following mode-coupling phenomena:

Assume that the laser oscillates simultaneously on N modes, which may have equal gain and equal losses, and therefore equal intensities. When the laser wavelength is tuned across the absorption spectrum of an absorbing sample inside the laser resonator, one of the oscillating modes may be tuned into resonance with an absorption line (frequency ω_k) of the sample molecules. This mode now suffers additional losses $\Delta\gamma = \alpha(\omega_k)L$, which cause the decrease ΔI of its intensity. Because of this decreased intensity the inversion of the active medium is less depleted by this mode, and the gain at ω_k *increases*. Since the other $(N-1)$ modes can participate in the gain at ω_k, their intensity will increase. This, however, again depletes the gain at ω_k and *decreases* the intensity of the mode oscillating at ω_k. With sufficiently strong coupling between the modes, this mutual interaction may finally result in a total suppression of the absorbing mode.

Frequency fluctuations of the modes caused by external perturbations (Sect. 5.4) limit the mode coupling. Furthermore, in dye lasers with standing-wave resonators, the spatial hole-burning effect (Sect. 5.8) weakens the coupling between modes. Because of their slightly different wavelengths, the maxima and nodes of the field distributions in the different modes are located at different positions in the active medium. This has the effect that the volumes of the active dye from which the different modes extract their gain overlap only partly. With sufficiently high pump power, the absorbing mode has an adequate gain volume on its own and is not completely suppressed but suffers a large intensity loss.

A more detailed calculation [6.11, 6.12] yields for the relative power change of the absorbing mode

$$\frac{\Delta P}{P} = \frac{g_0 \Delta \gamma}{\gamma(g_0 - \gamma)}(1 + KN) , \tag{6.18b}$$

where $K(0 \leq K \leq 1)$ is a measure of the coupling strength.

Without mode coupling ($K = 0$), (6.18) gives the same result as (6.16) for a single-mode laser. In the case of strong coupling ($K = 1$) and a large number of modes ($N \gg 1$), the relative intensity change of the absorbing mode increases proportionally to the number of simultaneously oscillating modes.

If several modes are simultaneously absorbed, the factor N in (6.18) stands for the ratio of all coupled modes to those absorbed. If all modes have equal frequency spacing, their number N gives the ratio of the spectral width of the homogeneous gain profile to the width of the absorption profile.

In order to detect the intensity change of one mode in the presence of many others, the laser output has to be dispersed by a monochromator or an interferometer. The absorbing molecules may have many absorption lines within the broadband gain profile of a multimode dye laser. Those laser modes that overlap with absorption lines are attenuated or are even completely quenched. This results in "spectral holes" in the output spectrum of the laser and allows the sensitive simultaneous recording of the whole absorption spectrum within the laser bandwidth, if the laser output is photographically recorded behind a spectrograph or if an optical multichannel analyzer (Sect. 4.5) is used.

(4) The discussion in item 3 has assumed that the mode coupling and mode frequencies were time independent. In real laser systems this is not true. Fluctuations of mode frequencies due to density fluctuations of the dye liquid or caused by external perturbations prevent stationary conditions in a multimode laser. We can define a mean "mode lifetime" t_m, which represents the average time a specific mode exists in a multimode laser. If the measuring time for intracavity absorption exceeds the mode lifetime t_m, no quantitatively reliable information on the magnitude of the absorption coefficient $\alpha(\omega)$ of the intracavity sample can be obtained.

It is therefore better to pump the "intracavity laser" with a step-function pump laser, which starts pumping at $t = 0$ and then remains constant. The intracavity absorption is then measured at times t with $0 < t < t_m$, which are

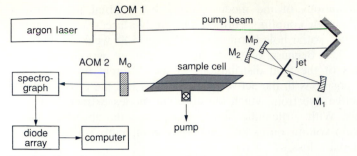

Fig. 6.9. Experimental arrangement for intracavity laser spectroscopy using a step-function pump intensity and a definite delay for the detection [6.13]

shorter than the mean mode lifetime t_m. The experimental arrangement is depicted in Fig. 6.9 [6.13]. The cw broadband dye laser is pumped by an argon laser. The pump beam can be switched on by the acousto-optic modulator AOM1 at $t = 0$. The dye laser output passes a second modulator AOM2, which transmits the dye laser beam for the time interval Δt at the selectable time t to the entrance slit of a high-resolution spectrograph. A photodiode array with a length D allows simultaneous recording of the spectral interval $\Delta\lambda = (\mathrm{d}x/\mathrm{d}\lambda)D$. The repetition rate f of the pump cycles must be smaller than the inverse delay time t between AOM1 and AOM2, i.e., $f < 1/t$.

A detailed consideration [6.11–6.17] shows that the time evolution of the laser intensity in a specific mode $q(\omega)$ with frequency ω after the start of the pump pulse depends on the gain profile of the laser medium, the absorption $\alpha(\omega)$ of the intracavity sample, and the mean mode lifetime t_m. If the broad gain profile with the spectral width $\Delta\omega_g$ and the center frequency ω_0 can be approximated by the parabolic function

$$g(\omega) = g_0 \left[1 - \left(\frac{\omega - \omega_0}{\Delta\omega_g} \right)^2 \right] \, ,$$

the time evolution of the output power $P_q = P(\omega_a)$ in the qth mode for a constant pumping rate and times $t < t_m$ is, after saturation of the gain medium

$$P_q(t) = P_q(0) \sqrt{\frac{t}{\pi t_m}} \exp\left[-\left(\frac{\omega - \omega_0}{\Delta\omega_g} \right)^2 t/t_m \right] e^{-\alpha(\omega_q)ct} \, . \tag{6.19}$$

The first exponential factor describes the spectral narrowing of the gain profile with increasing time t from laser mode competition, and the second factor can be recognized as the Beer–Lambert absorption law for the transmitted laser power in the qth mode with the effective absorption length $L_{eff} = ct$. In practice, effective absorption lengths up to 70 000 km have been realized [6.15]. The spectral width of the laser output becomes narrower with increasing time, but the absorption dips become more pronounced (Fig. 6.10).

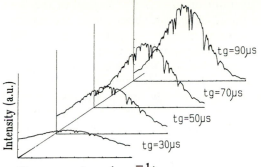

Fig. 6.10. Time evolution of the spectral profile of the laser output measured with time-resolved intracavity absorption spectroscopy [6.13]

Example 6.4

With typical delay times $t = 10^{-4}$ s, the effective absorption pathlength becomes $L_{\text{eff}} = 3 \times 10^8 \times 10^{-4}$ m $= 30$ km! If dips of 1% can still be detected, this gives a sensitivity limit of $\alpha \geq 3 \times 10^{-9}$ cm^{-1} for the absorption coefficient $\alpha(\omega)$. For $L_{\text{eff}} = 7 \times 10^7$ m, even a limit $\alpha_{\min} = 10^{-12}$ cm^{-1} is possible.

In ring lasers (Sect. 5.6) the spatial hole-burning effect does not occur if the laser is oscillating in unidirectional traveling-wave modes. If no optical diode is inserted into the ring resonator, the unsaturated gain is generally equal for clockwise and counterclockwise running waves. In such a bistable operational mode, slight changes of the net gain, which might be different for both waves because of their opposite Doppler shifts, may switch the laser from clockwise to counterclockwise operation, and vice versa. Such a bistable multimode ring laser with strong gain competition between the modes therefore represents an extremely sensitive detector for small absorptions inside the resonator [6.18].

The enhanced sensitivity of intracavity absorption may be utilized either to detect minute concentrations of absorbing components or to measure very weak forbidden transitions in atoms or molecules at sufficently low pressures to study the unperturbed absorption line profiles. With intracavity absorption cells of less than 1 m, absorbing transitions have been measured that would demand a path length of several kilometers with conventional single-pass absorption at a comparable pressure [6.15, 6.19].

Some examples illustrate the various applications of the intracavity absorption technique.

- With an iodine cell inside the resonator of a cw multimode dye laser, an enhancement factor of $Q = 10^5$ could be achieved, allowing the detection of I_2 molecules at concentrations down to $n \leq 10^8$ /cm^3 [6.20].

This corresponds to a sensitivity limit of $\alpha L \leq 10^{-7}$. Instead of the laser output power, the laser-induced fluorescence from a second iodine cell outside the laser resonator was monitored as a function of wavelength. This experimental arrangement (Fig. 6.11) allows demonstration of the isotope-specific absorption. When the laser beam passes through two external iodine cells filled with the isotopes $^{127}I_2$ and $^{129}I_2$, tiny traces of $^{127}I_2$ inside the laser cavity are sufficient to completely quench the laser-induced fluorescence from the external $^{127}I_2$ cell, while the $^{129}I_2$ fluorescence is not affected [6.21]. This demonstrates that those modes of the broadband dye laser that are absorbed by the internal $^{127}I_2$ are completely suppressed.

- Detection of absorbing transitions with very small oscillator strength (Sect. 2.7.2) has been demonstrated by Bray et al. [6.22], who measured the extremely weak ($v' = 2$, $v'' = 0$) infrared overtone absorption band of the red-atmospheric system of molecular oxygen and also the ($v' = 6 \leftarrow v'' = 0$) overtone band of HCl, with a cw rhodamine B dye laser (0.3-nm bandwidth) using a 97-cm-long intracavity absorption cell. Sensitivity tests indicated that even transitions with oscillator strengths down to $f \leq 10^{-12}$ could be readily detected. An example is the $P(11)$ line in the $(2 \leftarrow 0)$ band of the $b\,^1\Sigma_g^+ \leftarrow x\,^3\Sigma_g^-$ system of O_2 with the oscillator strength $f = 8.4 \times 10^{-13}$ [6.23]!

- The overtone spectrum $\Delta v = 6$ of SiH_4 was measured with the effective absorption pathlength $L_{eff} = 5.25$ km [6.13]. The spectrum shows well-resolved rotational structure, and local Fermi resonances were observed.

Although most experiments have so far been performed with dye lasers, the color-center lasers or the newly developed vibronic solid-state lasers with broad spectral-gain profiles (Sect. 5.7.3) are equally well suited for intracavity spectroscopy in the near infrared. An example is the spectroscopy of rovibronic transitions between higher electronic states of the H_3 molecule with a color-center laser [6.24].

Instead of absorption, weak emission lines can also be detected with the intracavity techniques [6.25]. If this light is injected into specific modes of the multimode laser, the intensity of these modes will increase for observation times $t < t_m$ before they can share their intensity by mode-coupling with other modes.

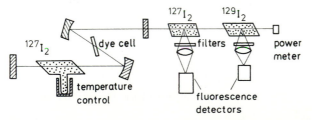

Fig. 6.11. Isotope-selective intracavity absorption spectroscopy. The frequencies ω_k absorbed by the $^{127}I_2$ isotope inside the laser cavity are missing in the laser output, which therefore does not excite any fluorescence in the same isotope outside the laser resonator [6.20]

A detailed discussion of intracavity absorption, its dynamics, and its limitations can be found in the articles of Baev et al. [6.15, 6.26], in the thesis of Atmanspacher [6.23], and in several review articles [6.13, 6.27–6.30].

6.2.3 Cavity Ring-Down Spectroscopy (CRDS)

During the last ten years a new, very sensitive detection technique for measuring small absorptions, cavity ring-down spectroscopy (CRDS), has been developed and gradually improved. It is based on measurements of the decay times of optical resonators filled with the absorbing species [6.31]. We can understand its general principle as follows:

Assume a short laser pulse with input power P_0 is sent through an optical resonator with two highly reflecting mirrors (reflectivities $R_1 = R_2 = R$, and transmission $T = 1 - R - A \ll 1$, where A includes all losses of the cavity from absorption, scattering, and diffraction, except those losses introduced by the absorbing sample). The pulse will be reflected back and forth between the mirrors (Fig. 6.12), while for each round-trip a small fraction will be transmitted through the end mirror and reach the detector. The transmitted power of the first output pulse is

$$P_1 = T^2 e^{-\alpha L} \cdot P_0 , \tag{6.20}$$

where α is the absorption coefficient of the gas sample inside the resonator with length L. For each round-trip the pulse power decreases by an additional factor $R^2 \cdot \exp(-2\alpha L)$. After n round-trips the power of the transmitted pulse has decreased to

$$P_n = \left[R \cdot e^{-\alpha L} \right]^{2n} P_1 = \left[(1 - T - A) e^{-\alpha L} \right]^{2n} P_1 , \tag{6.21}$$

which can be written as

$$P_n = P_1 \cdot e^{-2n(\ln R - \alpha \cdot L)} \approx P_1 \cdot e^{-[2n(T + A + \alpha \cdot L)]} . \tag{6.22}$$

The time delay between successive transmitted pulses equals the cavity round-trip time $T_R = 2L/c$. The nth pulse therefore is detected at the time $t = 2nL/c$. If the time constant of the detector is large compared to T_R, the detector averages over subsequent pulses and the detected signals give the exponential function

$$P(t) = P_1 \cdot e^{-t/\tau_1} , \tag{6.23}$$

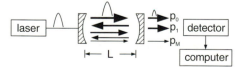

Fig. 6.12. Principle of cavity ring-down spectroscopy with pulsed lasers

with the decay time

$$\tau_1 = \frac{L/c}{T + A + \alpha \cdot L} . \tag{6.24}$$

Without an absorbing sample inside the resonator ($\alpha = 0$), the decay time of the resonator will be lengthened to

$$\tau_2 = \frac{L/c}{T + A} . \tag{6.25}$$

From the difference $\tau_2 - \tau_1$

$$\Delta\tau = \tau_2 - \tau_1 = \frac{\alpha \cdot L/c}{(T+A)(T+A+\alpha L)} = \frac{\alpha \cdot L^2/c}{(1-R)(1-R+\alpha L)} , \tag{6.26}$$

the product

$$\alpha \cdot L = (1 - R) \cdot \Delta\tau/\tau_1 \tag{6.27}$$

of absorption coefficient α and cavity length L can be determined as a function of the laser wavelength λ. The minimum detectable absorption $\alpha \cdot L$ is limited by the reflection R of the cavity mirrors and the accuracy of the decay time measurements, which in turn depends on the achievable signal-to-noise ratio. Similar to intracavity absorption, this technique takes advantage of the increased effective absorption length $L_{\text{eff}} = L/(1-R)$, because the laser pulse traverses the absorbing sample $1/(1-R)$ times.

Example 6.5
$R = 99.99\%$, $\Delta\tau/\tau = 10^{-4} \Rightarrow (\alpha \cdot L)_{\min} = 10^{-8}$. This implies for a cavity length $L = 1$ m the possibility to measure absorption coefficients as small as $\alpha = 10^{-10}$ cm^{-1}.

The experimental setup is shown in Fig. 6.13. The laser pulses are coupled into the resonator by carefully designed mode-matching optics, which ensure that only the TEM$_{00}$ modes of the cavity are excited. Diffraction losses are minimized by spherical mirrors, which also form the end windows of the absorption cell. If the absorbing species are in a molecular beam inside the cavity, the mirrors form the windows of the vacuum chamber. For a sufficiently short input pulse ($T_p < T_R$), the output consists of a sequence of pulses with a time separation of T_R and with exponentially decreasing intensities, which are detected with a boxcar integrator. For longer pulses ($T_p > T_R$), these pulses overlap in time and one observes a quasi-continuous exponential decay of the transmitted intensity. Instead of input pulses, the resonator can also be illuminated with cw radiation, which is suddenly switched off at $t = 0$.

If several resonator modes within the bandwidth of the laser pulse are excited, beat signals are superimposed onto the exponential decay curve. These

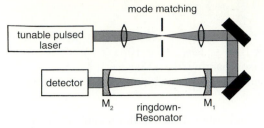

Fig. 6.13. Experimental setup with mode-matching optics

beats are due to interference between the different modes with differing frequencies. They depend on the relative phases between the excited resonator modes. Since these phase differences vary from pulse to pulse when the cavity is excited by a train of input pulses, averaging over many excitation pulses smears out the interference pattern, and again a pure exponential decay curve is obtained.

When the laser wavelength λ is tuned over the absorption range of the molecules inside the resonator, the measured differences $\Delta\tau(\lambda)$ of the decay times yield the absorption spectrum [6.32]. For illustration, the rotationally resolved vibrational overtone band $(2, 0, 5) \leftarrow (0, 0, 0)$ of the HCN molecule, obtained with cavity ring-down spectroscopy, is shown in Fig. 6.14.

The possible high sensitivity of cavity ring-down spectroscopy can be achieved only if several conditions are met:

(a) the bandwidth $\delta\omega_R$ of all excited cavity modes should be smaller than the width $\delta\omega_a$ of the absorption lines. This implies that the laser pulsewidth $\delta\omega_L < \delta\omega_a$.

(b) The relaxation time of the cavity must be longer than that of the excited molecules, which means $\delta\omega_R = 1/T_R < \delta\omega_a = 1/T_{exc}$.

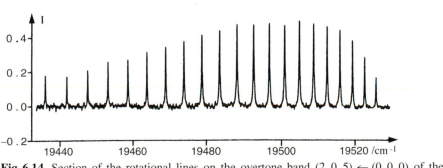

Fig. 6.14. Section of the rotational lines on the overtone band $(2, 0, 5) \leftarrow (0, 0, 0)$ of the HCN molecule, measured with CRDS [6.32]

Example 6.6

The resonator length is $L = 0.5\,\mathrm{m}$, the mirror reflectivity is $R = 0.995$; $\Rightarrow T_R = 3.3 \times 10^{-7}\,\mathrm{s}$, $\tau_1 = 3.3 \times 10^{-5}\,\mathrm{s}$, $\delta\omega_R = 3 \times 10^4\,\mathrm{s}^{-1}$. With a laser pulse duration of $10^{-8}\,\mathrm{s}$, the laser bandwidth is $\delta\omega_L = 10^8\,\mathrm{s}^{-1} \to \delta\nu = 1.5 \times 10^7\,\mathrm{s}^{-1}$. This is smaller than the Doppler width of absorption lines.

The major source of noise in single-shot decay is the technical noise introduced by the detection electronics and by fluctuations of the cavity length. Here an optical heterodyne detection technique can greatly enhance the signal-to-noise ratio. The experimental arrangement [6.33] is illustrated by Fig. 6.15. The output of a cw tunable single mode diode laser is split into two parts. One part is directly sent into the cavity ring-down ring resonator and one of the TEM$_{00}$ modes of the cavity is locked to the laser frequency. This part serves as the local oscillator. The other part is frequency shifted in an AOM (acousto-optic modulator) by one free spectral range of the cavity. It is therefore resonant with the adjacent TEM$_{00}$ mode. It is chopped by the AOM at 40 kHz. Both parts are superimposed, pass through the ring-down cavity, and the total transmitted intensity

$$I_T \alpha \left| E_s(t) + E_{LO} \cdot \mathrm{e}^{\mathrm{i}(2\pi\delta\nu t + \phi)} \right|^2 = |E_s(t)|^2 + |E_{LO}|^2$$
$$+ 2E_s \cdot E_{LO} \cdot \cos(2\pi\delta\nu t + \phi)\,, \qquad (6.28)$$

is measured by the detector. Since the signal enters the cavity as pulses of 12.5-μs duration at a repetition rate of 40 kHz, the transmitted signal intensity decays exponentially after the end of each pulse, while the local oscillator has constant intensity.

The interference term in (6.28) contains the product of the large amplitude E_{LO} (its transmission through the ring-down cavity is $T \approx 1$) and the small amplitude $E_s(t)$ and is, therefore, much larger than $(E_s(t))^2$. Measuring the decay time of the interference amplitude at the frequency $\delta\nu$ (= free spectral range of the cavity), therefore, gives a larger signal (which decays with 2τ) and thus a larger signal-to-noise ratio.

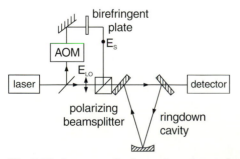

Fig. 6.15. Apparatus for heterodyne-detected cavity ring-down spectroscopy

Because a single-mode cw laser is used, the spectral resolution is generally higher than in case of pulsed lasers. It is only limited by the linewidth of the absorption lines.

Instead of measuring the differences of decay times τ_1 and τ_2 with and without absorbing sample inside the cavity, the time-integrated transmitted intensity can be also monitored as a function of the wavelength of the incident laser pulse [6.37]. This method is similar to the absorption in external cavities, discussed in Sect. 6.2. More information on CRDS can be found in [6.34–6.36] and in several reviews [6.34, 6.38, 6.39].

6.3 Direct Determination of Absorbed Photons

In the methods discussed in the preceeding sections, the attenuation of the transmitted light (or of the laser power in intracavity spectroscopy) is monitored to determine the absorption coefficient $\alpha(\omega)$ or the number density of absorbing species. For small absorptions this means the measurement of a small difference of two large quantities, which limits, of course, the signal-to-noise ratio.

Several different techniques have been developed where the absorbed radiation power, i.e., the number of absorbed photons, can be directly monitored. These techniques belong to the most sensitive detection methods in spectroscopy, and it is worthwhile to know about them.

6.3.1 Fluorescence Excitation Spectroscopy

In the visible and ultraviolet regions a very high sensitivity can be achieved, if the absorption of laser photons is monitored through the laser-induced fluorescence (Fig. 6.16). When the laser wavelength λ_L is tuned to an absorbing molecular transition $E_i \rightarrow E_k$, the number of photons absorbed per second along the path length Δx is

$$n_a = N_i n_L \sigma_{ik} \Delta x , \tag{6.29}$$

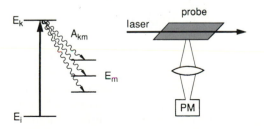

Fig. 6.16. Level scheme and experimental arrangement for fluorescence excitation spectroscopy

where n_L is the number of incident laser photons per second, σ_{ik} the absorption cross section per molecule, and N_i the density of molecules in the absorbing state $|i\rangle$.

The number of fluorescence photons emitted per second from the excited level E_k is

$$n_{Fl} = N_k A_k = n_a \eta_k \, , \tag{6.30}$$

where $A_k = \sum_m A_{km}$ stands for the total spontaneous transition probability (Sect. 2.8) to all levels with $E_m < E_k$. The quantum efficiency of the excited state $\eta_k = A_k/(A_k + R_k)$ gives the ratio of the spontaneous transition rate to the total deactivation rate, which may also include the radiationless transition rate R_k (e.g., collision-induced transitions). For $\eta_k = 1$, the number n_{Fl} of fluorescence photons emitted per second equals the number n_a of photons absorbed per second under stationary conditions.

Unfortunately, only the fraction δ of the fluorescence photons emitted into all directions can be collected on the photomultiplier cathode, where again only the fraction $\eta_{ph} = n_{pe}/n_{ph}$ of these photons produces on the average n_{pe} photoelectrons. The quantity η_{ph} is called the *quantum efficiency of the photocathode* (Sect. 4.5.2). The number n_{pe} of photoelectrons counted per second is then

$$n_{pe} = n_a \eta_k \eta_{ph} \delta = (N_i \sigma_{ik} n_L \Delta x) \eta_k \eta_{ph} \delta \, . \tag{6.31}$$

Example 6.7

Modern photomultipliers reach quantum efficiencies of $\eta_{ph} = 0.2$. With carefully designed optics it is possible to achieve a collection factor $\delta = 0.1$, which implies that the collecting optics cover a solid angle $d\Omega = 0.4\,\pi$. Using photon-counting techniques and cooled multipliers (dark pulse rate ≤ 10 counts/s), counting rates of $n_{pe} = 100$ counts/s are sufficient to obtain a signal-to-noise ratio $S/R \sim 8$ at integration times of 1 s.

Inserting this figure for n_{PE} into (6.31) illustrates that with $\eta_k = 1$ absorption rates of $n_a = 5 \times 10^3$ /s can be measured quantitatively. Assuming a laser power of 1 W at the wavelength $\lambda = 500$ nm, which corresponds to a photon flux of $n_L = 3 \times 10^{18}$ /s, this implies *that it is possible to detect a relative absorption of* $\Delta I/I \leq 10^{-14}$. When placing the absorbing probe inside the cavity where the laser power is q times larger ($q \sim 10$ to 100, Sect. 6.2.2), this impressive sensitivity may be even further enhanced.

Since the attainable signal is proportional to the fluorescence collection efficiency δ, it is important to design collection optics with optimum values of δ. Two possible designs are shown in Fig. 6.17 that are particularly useful if the excitation volume is small (e.g., the crossing volume of a laser beam with a collimated molecular beam). One of these collection optics uses a parabolic mirror, which collects the light from a solid angle of nearly 2π. A lens images the light source onto the photomultiplier cathode. The design of Fig. 6.17b

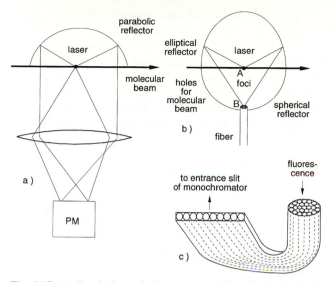

Fig. 6.17. (a) Parabolic optical system and **(b)** elliptical/spherical mirror system with high collection efficiency of fluorescence light. **(c)** Imaging of the fluorescence onto the entrance slit of a monochromator with a properly adjusted shape of the fiber bundle

uses an elliptical mirror where the light source is placed in one focal point A and the polished end of a multifiber bundle into the other B. A half-sphere reflector with its center in A reflects light, which is emitted into the lower half space back into the source A, which is then further reflected by the elliptical mirror and focused into B.

The exit end of the fiber bundle can be arranged to have a rectangular shape in order to match it to the entrance slit of a spectrograph (Fig. 6.17c).

When the laser wavelength λ_L is tuned across the spectral range of the absorption lines, the total fluorescence intensity $I_{Fl}(\lambda_L) \propto n_L \sigma_{ik} N_i$ monitored as a function of laser wavelength λ_L represents an image of the absorption spectrum called the *excitation spectrum*. According to (6.31) the photoelectron rate n_{PE} is directly proportional to the absorption coefficient $N_i \sigma_{ik}$, where the proportionality factor depends on the quantum efficiency η_{ph} of the photomultiplier cathode and on the collection efficiency δ of the fluorescence photons.

Although the excitation spectrum directly reflects the absorption spectrum with respect to the line positions, the relative *intensities* of different lines $I(\lambda)$ are identical in both spectra only if the following conditions are guaranteed:

- The quantum efficiency η_k must be the same for all excited states E_k. Under collision-free conditions, i.e., at sufficiently low pressures, the excited molecules radiate before they can collide, and we obtain $\eta_k = 1$ for all levels E_k.
- The quantum efficiency η_{ph} of the detector should be constant over the whole spectral range of the emitted fluorescence. Otherwise, the spectral distribution of the fluorescence, which may be different for different

excited levels E_k, will influence the signal rate. Some modern photomultipliers can meet this requirement.

• The geometrical collection efficiency δ of the detection system should be identical for the fluorescence from different excited levels. This demand excludes, for example, excited levels with very long lifetimes, where the excited molecules may diffuse out of the observation region before they emit the fluorescence photon. Furthermore, the fluorescence may not be isotropic, depending on the symmetry of the excited state. In this case δ will vary for different upper levels.

However, even if these requirements are not strictly fulfilled, excitation spectroscopy is still very useful to measure absorption lines with extremely high sensitivity, although their relative intensities may not be recorded accurately.

The technique of excitation spectroscopy has been widely used to measure very small absorptions. One example is the determination of absorption lines in molecular beams where both the pathlength Δx and the density N_i of absorbing molecules are small.

The LIF method is illustrated by Fig. 6.18, which shows a section of the excitation spectrum of Ag_2, taken in a collimated molecular beam under conditions comparable to that of Example 6.8.

Example 6.8

With $\Delta x = 0.1$ cm, $\delta = 0.5$, $\eta = 1$, a density $N_i = 10^7$ /cm^3 of absorbing molecules yields for an absorption cross section $\sigma_{ik} = 10^{-17}$ cm^2 and an incident flux of $n_L = 10^{16}$ photons/s ($= 3$ mW at $\lambda = 500$ nm) about 5×10^4 fluorescence photons imaged onto the photomultiplier cathode, which emits about 1×10^4 photoelectrons per second.

The extremely high sensitivity of this technique has been demonstrated impressively by Fairbanks et al. [6.40], who performed absolute density measurements of sodium vapor in the density range $N = 10^2 - 10^{11}$ cm^{-3} using laser-excited fluorescence as a monitor. The lower detection limit of $N = 10^2$ cm^{-3} was imposed by background stray light scattered out of the incident laser beam by windows and cell walls.

Because of the high sensitivity of excitation spectroscopy, it can be successfully used to monitor minute concentrations of radicals and short-lived intermediate products in chemical reactions [6.42]. Besides measurements of small concentrations, detailed information on the internal state distribution $N_i(v_i'', J_i'')$ of reaction products can be extracted, since the fluorescence signal is, according to (6.31), proportional to the number N_i of absorbing molecules in the level $|i\rangle$ (Sect. 6.8.4).

If in atoms a transition $|i\rangle \rightarrow |k\rangle$ can be selected, which represents a true two level system (i.e., the fluorescence from $|k\rangle$ terminates only in $|i\rangle$), the atom may be excited many times while it flies through the laser beam. At a spontaneous lifetime τ and a travel time T, a maximum of $n = T/(2\tau)$ excitation-fluorescence cycles can be achieved (photon burst).

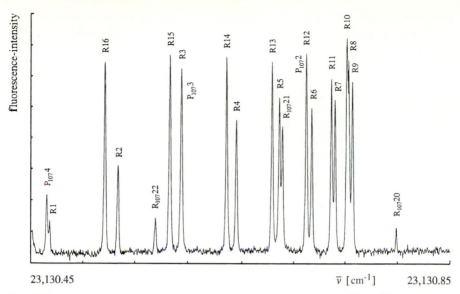

Fig. 6.18. Section of the fluorescence excitation spectrum of the ^{107}Ag^{109}Ag isotope, showing the band head of the $v' = 1 \leftarrow v'' = 0$ band in the $A\,^1\Sigma_u \leftarrow X\,^1\Sigma_g$ system, superimposed by some lines of the ^{107}Ag^{107}Ag isotope [6.41]

With $T = 10^{-5}$ s and $\tau = 10^{-8}$ s we can expect $n \cong 500$ fluorescence photons per atom! This allows single atom detection.

If molecules are diluted in solutions or in solids, single molecules can be excited if the focused laser beam diameter is smaller than the average distance between the molecules. Although the fluorescence from the upper state terminates in many rovibrational levels of the electronic ground state, these levels are rapidly quenched by collisions with the solvent molecules, which brings them back into the initial level, from where they can again be excited by the laser. This allows the detection of many photons per second from a single molecule and gives sufficient sensitivity to follow the diffusion of molecules into and out of the laser spot. This technique of "single-molecule detection" has developed into a very useful and sensitive method in chemistry and biology [6.43–6.45].

Excitation spectroscopy has its highest sensitivity in the visible, ultraviolet, and near-infrared regions. With increasing wavelength λ the sensitivity decreases for the following reasons: (6.31) shows that the detected photoelectron rate n_{PE} decreases with η_k, η_{ph}, and δ. All these numbers generally decrease with increasing wavelength. The quantum efficiency η_{ph} and the attainable signal-to-noise ratio are much lower for infrared than for visible photodetectors (Sect. 4.5). By absorption of infrared photons, vibrational–rotational levels of the electronic ground state are excited with radiative lifetimes that are generally several orders of magnitude larger than those of excited *electronic* states. At sufficiently low pressures the molecules diffuse out of the observation region before they radiate. This diminishes the collection efficiency δ. At higher pressures the quantum efficiency η_k of the excited level E_k is decreased because collisional deactivation competes with radiative transitions. Under these conditions photoacoustic detection may become preferable.

6.3.2 Photoacoustic Spectroscopy

Photoacoustic spectroscopy is a sensitive technique for measuring small absorptions that is mainly applied when minute concentrations of molecular species have to be detected in the presence of other components at higher pressure. An example is the detection of spurious pollutant gases in the atmosphere. Its basic principle may be summarized as follows:

The laser beam is sent through the absorber cell (Fig. 6.19). If the laser is tuned to the absorbing molecular transition $E_i \rightarrow E_k$, part of the molecules in the lower level E_i will be excited into the upper level E_k. By collisions with other atoms or molecules in the cell, these excited molecules may transfer their excitation energy $(E_k - E_i)$ completely or partly into translational, rotational, or vibrational energy of the collision partners. At thermal equilibrium, this energy is randomly distributed onto all degrees of freedom, causing an increase of thermal energy and with it a rise in temperature and pressure at a constant density in the cell.

When the laser beam is chopped at frequencies $\Omega < 1/T$, where T is the mean relaxation time of the excited molecules, periodic pressure variations appear in the absorption cell, which can be detected with a sensitive microphone placed inside the cell. The output signal S [Volt] of the microphone is proportional to the pressure change Δp induced by the absorbed radiation power ΔW. If saturation can be neglected, the absorbed energy per cycle

$$\Delta W = N_i \sigma_{ik} \Delta x (1 - \eta_K) P_L \Delta t , \tag{6.32}$$

is proportional to the density N_i [cm^{-3}] of the absorbing molecules in level $|i\rangle$, the absorption cross section σ_{ik} [cm^2], the absorption pathlength Δx, the cycle period Δt, and the incident laser power P_L. The signal decreases with increasing quantum efficiency η_k (which gives the ratio of emitted fluorescence energy to absorbed laser energy) unless the fluorescence is absorbed inside the cell and contributes to the temperature rise.

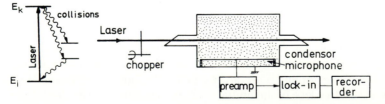

Fig. 6.19a,b. Photoacoustic spectroscopy: (a) level scheme; (b) schematic experimental arrangement

Since the absorbed energy ΔW is transferred into kinetic or internal energy of all N molecules per cm^3 in the photoacoustic cell with the volume V, the temperature rise ΔT is obtained from

$$\Delta W = \tfrac{1}{2} fVNk\Delta T , \tag{6.33}$$

where f is the number of degrees of freedom that are accessible for each of the N molecules at the temperature T [K]. If the chopping frequency of the laser is sufficiently high, the heat transfer to the walls of the cell during the pressure rise time can be neglected. From the equation of state $pV = NVkT$, we finally obtain

$$\Delta p = Nk\Delta T = \frac{2\Delta W}{fV} . \tag{6.34}$$

The output signal S from the microphone is then

$$S = \Delta p S_m = \frac{2N_i\sigma_{ik}}{fV}\Delta x(1-\eta_k)P_L\Delta t S_m , \tag{6.35}$$

where the sensitivity S_m [Volt/Pascal] of the microphone not only depends on the characteristics of the microphone but also on the geometry of the photoacoustic cell.

With infrared lasers the molecules are generally excited into higher vibrational levels of the electronic ground state. Assuming cross sections of $10^{-18}-10^{-19}$ cm^2 for the collisional deactivation of the vibrationally excited molecules, the equipartition of energy takes only about 10^{-5} s at pressures around 1 mbar. Since the spontaneous lifetimes of these excited vibrational levels are typically around $10^{-2}-10^{-5}$ s, it follows that at pressures above 1 mbar, the excitation energy absorbed from the laser beam will be almost completely transferred into thermal energy, which implies that $\eta_k \sim 0$.

The idea of the spectraphone is very old and was demonstrated by Bell and Tyndal [6.46] in 1881. However, the impressive detection sensitivity obtained today could only be achieved with the development of lasers, sensitive capacitance microphones, low-noise amplifiers, and lock-in techniques. Concentrations down to the parts per billion range (ppb, or 10^{-9}) at total pressures of 1 mbar up to several atmospheres are readily detectable with a modern spectraphone (Fig. 6.20).

Modern condensor microphones with a low-noise FET preamplifier and phase-sensitive detection achieve signals of larger than 1 V/mbar ($\widehat{=}$ 10 mV/Pa) with a background noise of 3×10^{-8} V at integration times of 1 s. This sensitivity allows detection of pressure amplitudes below 10^{-7} mbar and is, in general, not limited by the electronic noise but by another disturbing effect: laser light reflected from the cell windows or scattered by aerosols in the cell may partly be absorbed by the walls and contributes to a temperature increase. The resulting pressure rise is, of course, modulated at the chopping frequency and is therefore detected as background signal. There are several ways to reduce this phenomenon. Antireflection coatings of the cell windows or, in case

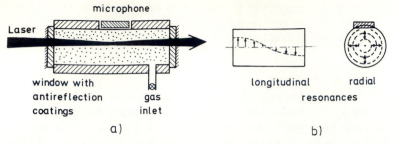

Fig. 6.20. (a) Spectrophone with capacitance microphone; **(b)** longitudinal and radial acoustic resonance modes

of linearly polarized laser light, the use of Brewster windows minimize the reflections. An elegant solution chooses the chopping frequency to coincide with an acoustic resonance of the cell. This results in a resonant amplification of the pressure amplitude, which may be as large as 1000-fold. This experimental trick has the additional advantage that those acoustic resonances can be selected that couple most efficiently to the beam profile but are less effectively excited by heat conduction from the walls. The background signal caused by wall absorption can thus be reduced and the true signal is enhanced. Figure 6.20b shows longitudinal and radial acoustic resonances in a cylindrical cell.

Example 6.9

With $N_i = 2.5 \times 10^{11}$ cm^{-3} ($\hat{=} 10^{-8}$ bar), $\sigma_{ik} = 10^{-16}$ cm^2, $\Delta x = 10$ cm, $V = 50$ cm^3, $\eta_k = 0$, $f = 6$, we obtain the pressure change $\Delta p = 1.5$ Pa ($\hat{=} 0.015$ mbar) for the incident laser power $P_L = 100$ mW. With a microphone sensitivity of $S_m = 10^{-2}$ V/Pa the output signal becomes $S = 15$ mV.

The sensitivity can be further enhanced by frequency modulation of the laser (Sect. 6.2.1) and by intracavity absorption techniques. With the spectraphone inside the laser cavity, the photoacoustic signal due to nonsaturating transitions is increased by a factor q as a result of a q-fold increase of the laser intensity inside the resonator (Sect. 6.2.2). The optoacoustic cell can be placed inside a multipath optical cell (Fig. 6.21) where an absorption pathlength of about 50 m can be readily realized [6.56].

According to (6.35) the optoacoustic signal decreases with increasing quantum efficiency because the fluorescence carries energy away without heating the gas, as long as the fluorescence light is not absorbed within the cell. Since the quantum efficiency is determined by the ratio of spontaneous to collision-induced deactivation of the excited level, it decreases with increasing spontaneous lifetime and gas pressure. Therefore, the optoacoustic method is particularly favorable to monitor vibrational spectra of molecules in the infrared region (because of the long lifetimes of excited vibrational levels) and to detect small concentrations of molecules in the presence of other gases

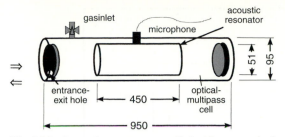

Fig. 6.21. Acoustic resonance cell inside an optical multipass cell. Dimensions are in millimeters

at higher pressures (because of the large collisional deactivation rate). It is even possible to use this technique for measuring *rotational* spectra in the microwave region as well as electronic molecular spectra in the visible or ultraviolet range, where electronic states with short spontaneous lifetimes are excited. However, the sensitivity in these spectral regions is not quite as high and there are other methods that are superior.

Some examples illustrate this very useful spectroscopic technique. For a more detailed discussion of optoacoustic spectroscopy, its experimental tricks, and its various applications [6.50], the reader is referred to recently published monographs [6.47–6.49,6.62] and to conference proceedings [6.51–6.53].

Example 6.10

(a) The sensitivity of the spectraphone has been demonstrated by Kreutzer et al. [6.54]. At a total air pressure of 660 mbar in the absorption cell these researchers could detect concentrations of ethylene down to 0.2 ppb, of NH_3 down to 0.4 ppb, and of NO pollutants down to 10 ppb. The feasibility of determining certain important isotope abundances or ratios by simple and rapid infrared spectroscopy with the spectraphone and also the ready control of small leaks of polluting or poison gases have been demonstrated [6.55].

(b) The optoacoustic method has been applied with great success to high-resolution spectroscopy of rotational–vibrational bands of numerous molecules [6.56]. Figure 6.22 illustrates as an example a section of the visible overtone absorption spectrum of the C_2H_2 molecule where, in spite of the small absorption coefficient, a good signal-to-noise ratio could be achieved.

(c) A general technique for the optoacoustic spectroscopy of *excited* molecular vibrational states has been demonstrated by Patel [6.57]. This technique involves the use of vibrational energy transfer between two dissimilar molecules A and B. When A is excited to its first vibrational level by absorption of a laser photon $h\nu_1$, it can transfer its excitation

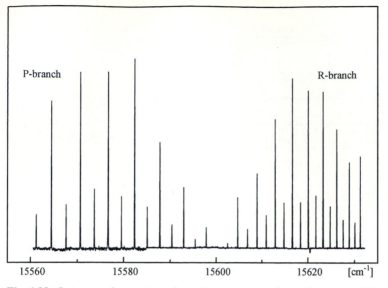

Fig. 6.22. Optoacoustic overtone absorption spectrum of acetylene around $\bar{\nu} = 15\,600\,\mathrm{cm}^{-1}$ corresponding to the excitation of a local mode by five quantum vibrations [6.56]

energy by a near-resonant collision to molecule B. Because of the large cross section for such collisions, a high density of vibrationally excited molecules B can be achieved also for those molecules that cannot be pumped directly by existing powerful laser lines. The excited molecule B can absorb a photon $h\nu_2$ from a second, weak tunable laser, which allows spectroscopy of all accessible transitions ($v = 1 \rightarrow v = 2$). The technique has been proved for the NO molecule, where the frequency of the four transitions in the $^2\Pi_{1/2}$ and $^2\Pi_{3/2}$ subbands of ^{15}NO and the Λ doubling for the $v = 1 \rightarrow 2$ transition have been accurately measured. The following scheme illustrates the method:

$$^{14}\mathrm{NO} + h\nu_1(\mathrm{CO_2\ laser}) \longrightarrow {}^{14}\mathrm{NO}^*(v = 1)\,,$$

$$^{14}\mathrm{NO}^*(v = 1) + {}^{15}\mathrm{NO}(v = 0) \longrightarrow {}^{14}\mathrm{NO}(v = 0) + {}^{15}\mathrm{NO}^*(v = 1)$$

$$+ \Delta E(35\,\mathrm{cm}^{-1})\,,$$

$$^{15}\mathrm{NO}^*(v = 1) + h\nu_2(\mathrm{spin\text{-}flip\ laser}) \longrightarrow {}^{15}\mathrm{NO}^*(v = 2)\,.$$

The last process is detected by optoacoustic spectroscopy.

(d) The application of photoacoustic detection to the visible region has been reported by Stella et al. [6.58]. They placed the spectraphone inside the cavity of a cw dye laser and scanned the laser across the absorption bands of the CH_4 and NH_3 molecules. The high-quality spectra with resolving power of over 2×10^5 proved to be adequate to resolve single rotational features of the very weak vibrational overtone transitions in these molecules. The experimental results are very useful for the investigation of the planetary atmospheres, where such weak overtone transitions are induced by the sun light.

(e) An interesting application of photoacoustic detection is in the measurement of dissociation energies of molecules [6.59]. When the laser wavelength is tuned across the dissociation limit, the photoacoustic signal drops drastically because beyond this limit the absorbed laser energy is used for dissociation. (This means that it is converted into potential energy and cannot be transferred into kinetic energy as in the case of excited-state deactivation. Only the kinetic energy causes a pressure increase.)

(f) With a special design of the spectraphone employing a coated quartz membrane for the condensor microphone, even corrosive gases can be measured [6.60]. This extends the applications of optoacoustic spectroscopy to aggressive pollutant gases such as NO_2 or SO_2, which are important constituents of air pollution.

Photoacoustic spectroscopy can be also applied to liquids and solids [6.61]. An interesting application is the determination of species adsorbed on surfaces. Optoacoustic spectroscopy allows a time-resolved analysis of adsorption and desorption processes of atoms or molecules at liquid or solid surfaces, and their dependence on the surface characteristics and on the temperature [6.63].

6.3.3 Optothermal Spectroscopy

For the spectroscopy of vibrational–rotational transitions in molecules the laser-excited fluorescence is generally not the most sensitive tool, as was discussed at the end of Sect. 6.3.1. Optoacoustic spectroscopy, on the other hand, is based on collisional energy transfer and is therefore not applicable to molecular beams, where collisions are rare or even completely absent. For the infrared spectroscopy of molecules in a molecular beam therefore a new detection technique has been developed, which relies on the collision-free conditions in a beam and on the long radiative lifetimes of vibrational–rotational levels in the electronic ground state [6.64–6.66].

Optothermal spectroscopy uses a cooled bolometer (Sect. 4.5) to detect the excitation of molecules in a beam (Fig. 6.23). When the molecules hit the bolometer they transfer their kinetic and their internal thermal energy, thereby increasing the bolometer temperature T_0 by an amount ΔT. If the molecules are excited by a tunable laser (for example, a color-center laser or a diode laser) their vibrational–rotational energy increases by $\Delta E = h\nu \gg E_{\mathrm{kin}}$. If the lifetime τ of the excited levels is larger than the flight time $t = d/v$ from the excitation region to the bolometer, they transfer this extra energy to the bolometer. If N excited molecules hit the bolometer per second, the additional

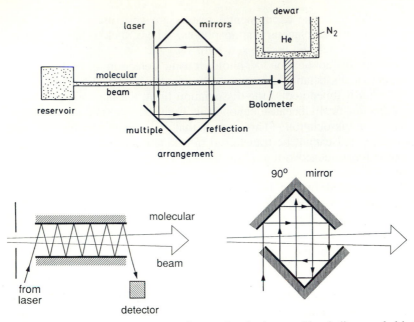

Fig. 6.23. Optothermal spectroscopy in a molecular beam with a helium-cooled bolometer as detector and two optical systems, which increase the absorption path length

rate of heat transfer is

$$\frac{\mathrm{d}Q}{\mathrm{d}t} = N\Delta E = Nh\nu . \tag{6.36}$$

With the heat capacity C of the bolometer and a heat conduction $G(T - T_0)$ the temperature T is determined by

$$Nh\nu = C\frac{\mathrm{d}T}{\mathrm{d}t} + G(T - T_0) . \tag{6.37}$$

Under stationary conditions $(\mathrm{d}T/\mathrm{d}t = 0)$, we obtain from (6.37) the temperature rise

$$\Delta T = T - T_0 = \frac{Nh\nu}{G} . \tag{6.38}$$

In general, the exciting laser beam is chopped in order to increase the signal-to-noise ratio by lock-in detection. The time constant $\tau = C/G$ of the bolometer (Sect. 4.5) should be smaller than the chopping period. Therefore the bolometer must be constructed in such a way that both C and G are as small as possible.

The bolometer consists of a small $(0.25 \times 0.25 \times 0.25\,\mathrm{mm}^3)$ doped silicon semiconductor crystal (Fig. 6.24), which has a low heat capacity at $T = 1.5\,\mathrm{K}$.

If a small electric current i is sent through the crystal with resistance R, a voltage $U = R \cdot i$ is generated across the bolometer. The resistance decreases

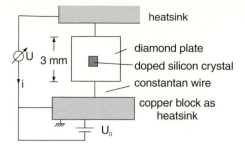

Fig. 6.24. Central part of the bolometer. The diamond plate increases the sensitivity area without contributing much to the heat capacity

with increasing temperature. The temperature change ΔT can be measured by the resulting resistance change

$$\Delta R = \frac{\mathrm{d}R}{\mathrm{d}T}\Delta T \;,$$

which is a function of the temperature dependence $\mathrm{d}R/\mathrm{d}T$ of the bolometer material. Large values of $\mathrm{d}R/\mathrm{d}T$ are achieved with doped semiconductor materials at low temperatures (a few Kelvin!). Even larger values can be realized with materials around their critical temperature T_c for the transition from the superconducting to the normal conducting state. In this case, however, one always has to keep the temperature at T_c. This can be achieved by a temperature feedback control, where the feedback signal is a measure for the rate $\mathrm{d}Q/\mathrm{d}t$ of energy transfer to the bolometer by the excited molecules.

With such a detector at $T = 1.5$ K, energy transfer rates $\mathrm{d}Q/\mathrm{d}t \geq 10^{-14}$ W are still detectable [6.65]. This means that an absorbed laser power of $\Delta P \geq 10^{-14}$ W is measurable. In order to maximize the absorbed power, the absorption path length can be increased by an optical device consisting of two $90°$ reflectors, which reflect the laser beam many times back and forth through the molecular beam (Fig. 6.23). An even higher sensitivity is achieved with an optical enhancement cavity with spherical mirrors, where the optical beam waist has to be matched to the molecular beam diameter (Fig. 6.25).

The sensitivity of the optothermal technique is illustrated by the comparison of the same sections of the overtone spectrum of C_2H_4 molecules,

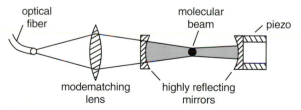

Fig. 6.25. Excitation of molecules in a collimated beam inside an optical enhancement cavity

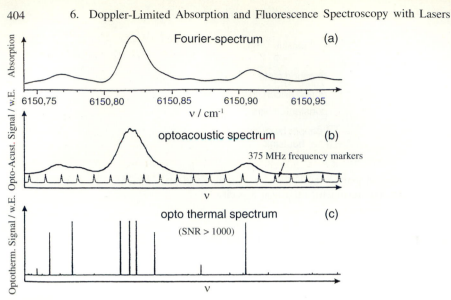

Fig. 6.26a–c. Section of the overtone band $(v_5 + v_9)$ of C_2H_4, comparison of (a) Fourier spectrum; (b) optoacoustic spectrum; (c) Doppler-free optothermal spectrum

measured with Fourier, optoacoustic, and optothermal spectroscopy, respectively (Fig. 6.26). Note the increase of spectral resolution and signal-to-noise ratio in the optothermal spectrum compared to the two other Doppler-limited techniques [6.67]. More examples can be found in [6.68].

The term photothermal spectroscopy is also used in the literature for the generation of thermal-wave phenomena in samples that are illuminated with a time-dependent light intensity [6.69]. In many aspects this is equivalent to photoacoustic spectroscopy. An interesting modification of this technique for

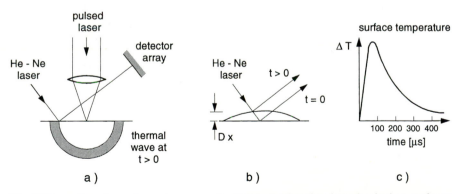

Fig. 6.27a–c. Optothermal spectroscopy of solids and of molecules adsorbed at surfaces: (a) propagation of thermal wave induced by a pulsed laser; (b) deformation of a surface detected by the deflection of a HeNe laser beam; (c) time profile of surface temperature change following pulsed illumination [6.70]

the study of molecules adsorbed at surfaces is illustrated in Fig. 6.27. A small spot of a surface is irradiated by a pulsed laser beam. The absorbed power results in a temperature rise of the illuminated spot. This temperature travels as a thermal shock wave through the solid. It leads to a slight time-dependent deformation of the surface from thermal expansion. This deformation is probed by the time-dependent deflection of a low-power HeNe laser. In the case of adsorbed molecules, the absorbed power varies when the laser wavelength is tuned over the absorption spectrum of the molecules. Time-resolved measurements of the probe-beam deflection allow one to determine the kind and amount of adsorbed molecules and to follow their desorption, caused by the irradiating light, with time [6.70].

6.4 Ionization Spectroscopy

6.4.1 Basic Techniques

Ionization spectroscopy monitors the absorption of photons on the molecular transition $E_i \rightarrow E_k$ by detecting the ions or electrons produced by ionization of the excited state E_k. The necessary ionization of the excited molecule may be performed by photons, by collisions, or by an external electric field.

a) Photoionization

The excited molecules are ionized by absorption of a second photon, i.e.,

$$M^*(E_k) + h\nu_2 \longrightarrow M^+ + e^- + E_{kin} .\qquad(6.39a)$$

The ionizing photon may come either from the same laser that has excited the level E_k or from a separate light source, which can be another laser or even an incoherent source (Fig. 6.28a).

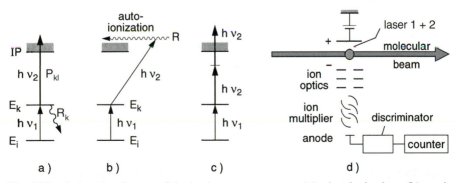

Fig. 6.28a–d. Level schemes of ionization spectroscopy: (**a**) photoionization; (**b**) excitation of autoionizing Rydberg levels; (**c**) two-photon ionization of excited molecules; (**d**) experimental arrangement for photoionization spectroscopy in a molecular beam

A very efficient photoionization process is the excitation of high-lying Rydberg levels above the ionization limit (Fig. 6.28b), which decay by autoionization into lower levels of the ion M^+

$$M^*(E_k) + h\nu_2 \longrightarrow M^{**} \to M^+ + e^- + E_{kin} . \tag{6.39b}$$

The absorption cross section for this process is generally much larger than that of the bound–free transition described by (6.39a) (Sect. 10.4.2).

The excited molecule may also be ionized by a nonresonant two-photon process (Fig. 6.28c)

$$M^*(E_k) + 2h\nu_2 \longrightarrow M^+ + e^- + E_{kin} . \tag{6.39c}$$

b) Collision-Induced Ionization

Ionizing collisions between excited atoms or molecules and electrons represent the main ionization process in gas

$$M^*(E_k) + e^- \longrightarrow M^+ + 2e^- . \tag{6.40a}$$

If the excited level E_k is not too far from the ionization limit, the molecule may also be ionized by thermal collisions with other atoms or molecules. If E_k lies above the ionization limit of the collision partners A, Penning ionization [6.71] becomes an efficient process and proceeds as

$$M^*(E_k) + A \longrightarrow M + A^+ + e^- . \tag{6.40b}$$

c) Field Ionization

If the excited level E_k lies closely below the ionization limit, the molecule $M^*(E_k)$ can be ionized by an external electric dc field (Fig. 6.29a). This

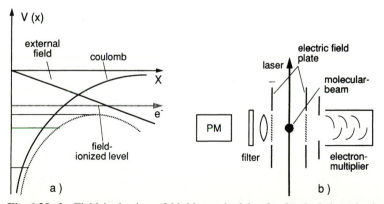

Fig. 6.29a,b. Field ionization of highly excited levels closely below the ionization limit: (a) Potential diagram and (b) experimental arrangement for field ionization in a molecular beam. The photomultiplier monitors the LIF from the intermediate level populated in a two-step excitation process

method is particularly efficient if the excited level is a long-lived, highly excited Rydberg state with principal quantum number n, quantum defect δ, and ionization energy Ry/n^{*2}, which can be expressed by the Rydberg constant Ry and the effective quantum number $n^* = n - \delta$. The required minimum electric field can readily be estimated from Bohr's atomic model, which gives a good approximation for atomic levels with large principal quantum number n. Without an external field the ionization energy for the outer electron at the mean radius r from the nucleus is determined by the Coulomb field of the nucleus shielded by the inner electron core.

$$IP = \int\limits_{r}^{\infty} \frac{Z_{\text{eff}} e^2}{4\pi\epsilon_0 r^2}\, dr = \frac{Z_{\text{eff}} e^2}{4\pi\epsilon_0 r} = \frac{Ry}{(n-\delta)^2} = \frac{Ry}{n^{*2}}$$

where eZ_{eff} is the effective nuclear charge, i.e., the nuclear charge eZ partly screened by the electron cloud. If an external field $E_{\text{ext}} = -E_0 x$ is applied, the effective ionization potential is lowered to the value (see Problem 6.9)

$$IP_{\text{eff}} = IP - \sqrt{\frac{Z_{\text{eff}} e^3 E_0}{\pi\epsilon_0}}\ . \tag{6.41}$$

If the energy E of the excited level is above IP_{eff}, it will be field-ionized.

Techniques where the laser excites the atoms but the ionizing step is performed by field ionization have found increasing applications in the detection of Rydberg atoms in molecular beams (Sect. 14.6), in analytical chemistry for trace elements, or for small concentrations of pollutants [6.72].

Example 6.11

For levels 10 meV below the ionization limit, (6.41) gives $E_0 \geq 1.7 \times 10^4$ V/m for the ionizing external field. However, because of the quantum-mechanical tunnel effect the fields required for complete ionization are even lower.

6.4.2 Sensitivity of Ionization Spectroscopy

The following estimation illustrates the possible sensitivity of ionization spectroscopy (Fig. 6.28a). Let N_k be the density of excited molecules in level E_k, P_{kI} the probability per second that a molecule in level E_k is ionized, and n_a the number of photons absorbed per second on the transition $E_i \rightarrow E_k$. If R_k is the total relaxation rate of level E_k, besides the ionization rate (spontaneous transitions plus collision-induced deactivation) the signal rate in counts per second for the absorption path length Δx and for n_L incident laser photons per second under steady state conditions is:

$$S_I = N_k P_{kI} \delta \cdot \eta = n_a \frac{P_{kI}}{P_{kI} + R_k} \delta \cdot \eta = N_i n_L \sigma_{ik} \Delta x \frac{P_{kI}}{P_{kI} + R_k} \delta \cdot \eta\ . \tag{6.42}$$

With a proper design the *collection* efficiency δ for the ionized electrons or ions can reach $\delta = 1$. If the electrons or ions are accelerated to several keV and detected by electron multipliers or channeltrons, a *detection* efficiency of $\eta = 1$ can also be achieved. If the ionization probability P_{kI} can be made large compared to the relaxation rate R_k of the level $|k\rangle$, the signal S_I then becomes with $\delta = \eta = 1$:

$$S_I \sim n_a .$$

This means that every laser photon absorbed in the transition $E_i \to E_k$ gives rise to a detected ion or electron. It implies that *single absorbed photons* can be detected with an overall efficiency close to unity (or 100%). In experimental practice there are, of course, additional losses and sources of noise, which limit the detection efficiency to a somewhat lower level. However, for all absorbing transitions $E_i \to E_k$ where the upper level E_k can readily be ionized, ionization spectroscopy is the most sensitive detection technique and is superior to all other methods discussed so far [6.73, 6.74].

6.4.3 Pulsed Versus CW Lasers for Photoionization

In case of photoionization of the excited level $|k\rangle$, the ionization probability per second

$$P_{kI} = \sigma_{kI} n_{L_2} \quad [\text{s}^{-1}]$$

equals the product of the ionization cross section σ_{kI} [cm^2] and the photon flux density n_{L_2} [cm$^{-2} \cdot$ s^{-1}] of the ionizing laser. We can then write (6.42) as

$$S_I = N_i \left[\frac{\sigma_{ik} n_{L_1} \delta \cdot \eta}{1 + R_k/(\sigma_{kI} n_{L_2})} \right] \Delta x . \tag{6.43}$$

The maximum ion rate S_I^{max}, which is achieved if

$$\sigma_{kI} n_{L_2} \gg R_k , \quad \text{and} \quad \delta = \eta = 1 ,$$

becomes equal to the rate n_a of photons absorbed in the transition $|i\rangle \to |k\rangle$:

$$S_I^{\text{max}} = N_i \sigma_{ik} n_{L_1} \Delta x = n_a . \tag{6.44}$$

The following estimation illustrates under which conditions this maximum ion rate can be realized:
Typical cross sections for photoionization are $\sigma_{kI} \sim 10^{-17}$ cm^2. If radiative decay is the only deactivation mechanism of the excited level $|k\rangle$, we have $R_k = A_k \approx 10^8$ s^{-1}. In order to achieve $n_{L_2} \sigma_{kI} > A_k$, we need a photon flux $n_{L_2} > 10^{25}$ cm^{-2}s^{-1} of the ionizing laser. With pulsed lasers this condition can be met readily.

Example 6.12

Excimer laser: $100 \, \text{mJ/pulse}$, $\Delta T = 10 \, \text{ns}$, the cross section of the laser beam may be $1 \, \text{cm}^2 \rightarrow n_{L_2} = 2 \times 10^{25} \, \text{cm}^{-2}\text{s}^{-1}$. With the numbers above we can reach an ionization probability of $P_{ik} = 2 \times 10^8 \, \text{s}^{-1}$ for all molecules within the laser beam. This gives an ion rate S_I that is 2/3 of the maximum rate $S_I = n_a$.

The advantage of pulsed lasers is the large photon flux during the pulse time ΔT, which allows the ionization of the excited molecules *before they decay* by relaxation into lower levels where they are lost for further ionization. Their disadvantages are their large spectral bandwidth, which is generally larger than the Fourier-limited bandwidth $\Delta \nu \geq 1/\Delta T$, and their low duty cycle. At typical repetition rates of $f_L = 10$ to $100 \, \text{s}^{-1}$ and a pulse duration of $\Delta T = 10^{-8} \, \text{s}$, the duty cycle is only $10^{-7} - 10^{-6}$!

If the diffusion time t_D of molecules out of the excitation-ionization region is smaller than $1/f_L$, at most the fraction $f_L t_D$ of molecules can be ionized, even if the ionization probability during the laser pulse ΔT approaches 100.

Example 6.13

Assume that the two laser beams L1 and L2 for excitation and ionization have the diameter $D = 1 \, \text{cm}$ and traverse a collimated molecular beam with 1-cm^2 cross section perpendicularly. During the pulse time $\Delta T = 10^{-8} \, \text{s}$ the distance traveled by the molecules at the mean velocity $\bar{v} = 500 \, \text{m/s}$ is $d = \Delta T \bar{v} \sim 5 \times 10^{-4} \, \text{cm}$. This means that all molecules in the excitation volume of $1 \, \text{cm}^3$ can be ionized during the time ΔT. During the "dark" time $T = 1/f_L$, however, the molecules travel the distance $d = \bar{v} T \sim 500 \, \text{cm}$ at $f_L = 10^2 \, \text{s}^{-1}$. Therefore, only the fraction $1/500 = 2 \times 10^{-3}$ of all molecules in the absorbing level $|i\rangle$ are ionized in a continuous molecular beam.

There are two solutions:

(a) If pulsed molecular beams can be employed having the pulse time $\Delta T_B \leq D/\bar{v}$ and the repetition rate $f_B = f_L$, an optimum detection probability can be achieved.

(b) In continuous molecular beams the two laser beams may propagate antiparallel to the molecular beam axis. If lasers with a high-repetition frequency f_L are used (for example, copper-laser-pumped dye lasers with $f_L \leq 10^4 \, \text{s}^{-1}$), then the distance traveled by the molecules during the dark time is only $d = \bar{v}/f \geq 5 \, \text{cm}$. They can therefore still be detected by the next pulse [6.76, 6.77].

With cw lasers the duty cycle is 100% and the spectral resolution is not limited by the laser bandwidth. However, their intensity is smaller and the laser beam must be focused in order to meet the requirement $P_{kI} > R_k$.

Example 6.14

If a cw argon laser with $10\,\mathrm{W}$ at $\lambda = 488\,\mathrm{nm}(\widehat{=}\ 2.5 \times 10^{19}$ photon/s) is used for the ionizing step, it has to be focused down to an area of $2.5 \times 10^{-6}\,\mathrm{cm}^2$, i.e., a diameter of $17\,\mu\mathrm{m}$, in order to reach a photon flux density of $n_{L_2} = 10^{25}\,\mathrm{cm}^{-2}\mathrm{s}^{-1}$.

Here the following problem arises: the molecules excited by laser L1 into level $|k\rangle$ travel during their spontaneous lifetime of $\tau = 10\,\mathrm{ns}$ at around thermal velocities of about $\overline{v} = 5 \times 10^4\,\mathrm{m/s}$ only a distance of $d = 5\,\mu\mathrm{m}$ before they decay into lower levels. The second laser L2 must therefore be focused in a similar way as L1 and its focus must overlap that of L1 within a few microns.

A technical solution of this problem is depicted in Fig. 6.30. The dye laser L1 is guided through a single-mode optical fiber. The divergent light emitted out of the fiber end is made parallel by a spherical lens, superimposed with the beam of the argon laser L2 by a dichroic mirror M. Both beams are then focused by a cylindrical lens into the molecular beam, forming a rectangular "sheet of light" with a thickness of about $5-10\,\mu\mathrm{m}$ and a height of about $1\,\mathrm{mm}$, adapted to the dimensions of the molecular beam [6.78]. All molecules in the beam have to pass through the two laser beams. Since the transition probability for the first step $|i\rangle \rightarrow |k\rangle$ is generally larger by some orders of magnitude than that of the ionizing transition, the first transition can be readily saturated (Sect. 7.1). It is therefore advantageous to adjust the relative positions of the two beams in such a way that the maximum intensity of L2 coincides spatially with the slope of the Gaussian intensity profile of L1 (see insert of Fig. 6.30).

Often it is possible to tune the ionizing laser L2 to transitions from $|k\rangle$ into autoionizing Rydberg levels (Sect. 10.4). For such transitions the probability may be from two to three orders of magnitude larger than for bound–free tran-

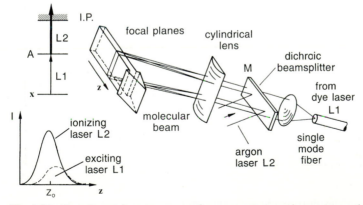

Fig. 6.30. Experimental arrangement for resonant two-photon two-color ionization with two cw lasers. The insert shows the optimum overlap of the Gaussian intensity profiles in the focal plane

sitions into the ionization continuum. In these cases the requirement (6.44) can be met even at much lower intensities of L2.

Resonant two-step ionization with two laser photons from pulsed or cw lasers represents the most versatile and sensitive detection technique. If laser L1 excites all atoms or molecules that fly through the laser beam, single atoms or molecules can be detected [6.73, 6.79, 6.80] if the condition (6.44) can be fulfilled.

6.4.4 Resonant Two-Photon Ionization Combined with Mass Spectrometry

In combination with mass spectrometry mass- and wavelength-selective spectroscopy becomes possible, even if the spectral lines of the different species overlap. This is particularly important for *molecular* isotopes with dense spectra, which overlap for the different isotopes (Fig. 6.31). Such isotope-selective spectra give detailed information on isotope shifts of vibrational and rotational levels and facilitate the correct assignment of the spectral lines considerably. Furthermore, they yield the relative isotopic abundances.

Time-of-flight mass spectrometers with pulsed lasers are convenient since they allow the simultaneous but separate recording of the spectra of different isotopes [6.81, 6.82]. For ionization by cw lasers, quadrupole mass spectrom-

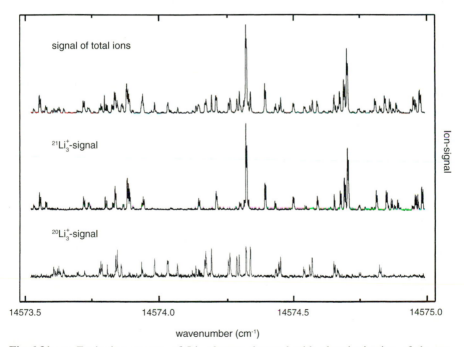

Fig. 6.31a–c. Excitation spectra of Li$_3$-clusters detected with photoionization of the excited states: (**a**) no mass selection; (**b**) spectrum of the ^{21}Li$_3$ isotopomer; (**c**) spectrum of ^{20}Li$_3$, recorded with doubled sensitivity [6.75]

eters are generally used. Their disadvantage is the lower transmittance and the fact that different masses cannot be recorded simultaneously but only sequentially. At sufficiently low ion rates, delayed coincidence techniques in combination with time-of-flight spectrometers can be utilized if both the photoion and the corresponding photoelectron are detected. The detected electron provides the zero point of the time scale and the ions with different masses are separated by their differences $\Delta t_a = t_{ion} - t_{el}$ in arrival times at the ion detector.

For measurements of cluster-size distributions in cold molecular beams (Sect. 9.3), or for monitoring the mass distribution of laser-desorbed molecules from surfaces, these combined techniques of laser ionization and mass spectrometry are very useful [6.83, 6.84]. For the detection of rare isotopes in the presence of other much more abundant isotopes, the double discrimination of isotope-selective excitation by the first laser L1 and the subsequent mass separation by the mass spectrometer is essential to completely separate the isotopes, even if the far wings of their absorption lines overlap [6.85]. The combination of resonant multiphoton ionization (REMPI) with mass spectrometry for the investigation of molecular dynamics and fragmentation is discussed in Chap. 10.

6.4.5 Thermionic Diode

The collisional ionization of high-lying Rydberg levels is utilized in the thermionic diode [6.86], which consists in its simplest form of a metallic cylindrical cell filled with a gas or vapor, a heated wire as cathode, and the walls as anode (Fig. 6.32). If a small voltage of a few volts is applied, the diode operates in the space-charge limited region and the diode current is restricted by the electron space-charge around the cathode. The laser beam passes through this space-charge region close to the cathode and excites molecules into Rydberg states, which are ionized by electron impact. Because of its much larger mass, the ion stays on the average a much longer time Δt_{ion} within the space-charge region than an electron. During this period it compensates one negative charge and therefore allows $n = \Delta t_{ion} / \Delta t_{el}$ extra electrons

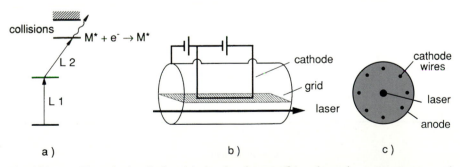

Fig. 6.32a–c. Thermionic diode: (**a**) level scheme; (**b**) schematic arrangement; and (**c**) field-free excitation scheme, where the laser beam passes through the field-free central region in a symmetric arrangement of cathode wires

to leave the space-charge region. If N ions are formed per second, the diode current therefore increases by

$$\Delta i = eN\Delta t_{ion}/\Delta t_{el} = eMN . \tag{6.45}$$

The current magnification factor M can reach values of up to $M = 10^5$.

Sensitive and accurate measurements of atomic and molecular Rydberg levels have been performed [6.87–6.89] with thermionic diodes. With a special arrangement of the electrodes, a nearly field-free excitation zone can be realized that allows the measurement of Rydberg states up to the principal quantum numbers $n = 300$ [6.89] without noticeable Stark shifts.

A more detailed representation of ionization spectroscopy and its various applications to sensitive detection of atoms and molecules can be found in [6.72–6.74, 6.79–6.81].

6.5 Optogalvanic Spectroscopy

Optogalvanic spectroscopy is an excellent and simple technique to perform laser spectroscopy in gas discharges. Assume that the laser beam passes through part of the discharge volume. When the laser frequency is tuned to the transition $E_i \rightarrow E_k$ between two levels of atoms or ions in the discharge, the population densities $n_i(E_i)$ and $n_k(E_k)$ are changed by optical pumping. Because of the different ionization probabilities from the two levels, this population change will result in a change ΔI of the discharge current that is detected as a voltage change $\Delta U = R\Delta I$ across the resistor R at a constant supply voltage U_0 (Fig. 6.33). When the laser intensity is chopped, an ac voltage is obtained, which can be directly fed into a lock-in amplifier.

Even with moderate laser powers (a few milliwatts) large signals (μV to mV) can be achieved in gas discharges of several milliamperes. Since the absorbed laser photons are detectd by the optically induced current change, this very sensitive technique is called *optogalvanic spectroscopy* [6.90–6.92].

Both positive and negative signals are observed, depending on the levels E_i, E_k involved in the laser-induced transition $E_i \rightarrow E_k$. If $IP(E_i)$ is the total

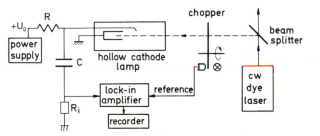

Fig. 6.33. Experimental arrangement of optogalvanic spectroscopy in a hollow cathode lamp

ionization probability of an atom in level E_i, the voltage change ΔU produced by the laser-induced change $\Delta n_i = n_{i0} - n_{iL}$ is given by

$$\Delta U = R\Delta I = a[\Delta n_i IP(E_i) - \Delta n_k IP(E_k)] \,. \tag{6.46}$$

There are several competing processes that may contribute to the ionization of atoms in level E_i, such as direct ionization by electron impact $A(E_i) + e^- \rightarrow A^+ + 2e$, collisional ionization by metastable atoms $A(E_i) + B^* \rightarrow A^+ + B + e^-$, or, in particular for highly excited levels, the direct photoionization by laser photons $A(E_i) + h\nu \rightarrow A^+ + e^-$. The competition of these and other processes determines whether the population changes Δn_i and Δn_k cause an increase or a decrease of the discharge current. Figure 6.34a shows the optogalvanic spectrum of a Ne discharge (5 mA) recorded in a fast scan with 0.1-s time constant. The good signal-to-noise ratio demonstrates the sensitivity of the method.

In hollow cathodes the cathode material can be sputtered by ion bombardment in the discharge. The metal vapor, consisting of atoms and ions, can be investigated by optogalvanic spectroscopy. Figure 6.34b illustrates a section of the optogalvanic spectrum of aluminum, copper, and iron atoms, and ions Al^+, Fe^+, measured simultanously in two hollow cathodes irradiated with a tunable pulsed dye laser [6.93].

Molecular spectra can also be measured by optogalvanic spectroscopy [6.94]. In particular, transitions from highly excited molecular states that are not accessible to optical excitation but are populated by electron impact in the gas discharge can be investigated. Furthermore, molecular ions and radicals can be studied. Some molecules, called *excimers* (Sect. 5.7.6) are only stable in their excited states. They are therefore appropriate for this technique because they do not exist in the ground state and cannot be studied in neutral-gas cells. Examples are He_2^* or H_2^* [6.95, 6.96].

Optogalvanic spectroscopy is a suitable technique for studies of excitation and ionization processes in flames, gas discharges, and plasmas [6.97]. Of par-

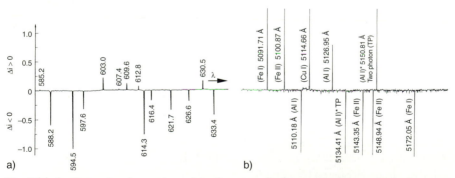

Fig. 6.34a,b. Optogalvanic spectrum (**a**) of a neon discharge (1 mA, $p = 1$ mbar), generated with a broadband cw dye laser [6.92]; (**b**) of Al, Cu, and Fe vapor sputtered in a hollow cathode that was illuminated with a pulsed dye laser [6.93]

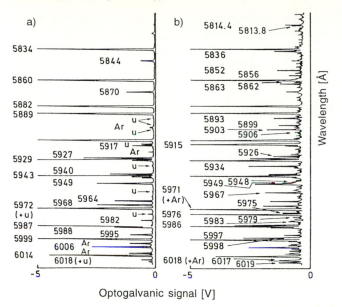

Optogalvanic signal [V]

Fig. 6.35a,b. Optogalvanic spectrum of a uranium hollow-cathode lamp filled with argon buffer gas. In the upper spectrum (**a**) taken at 7 mA discharge current, most of the lines are argon transitions, while in the lower spectrum (**b**) at 20 mA many more uranium lines appear, because of sputtering of uranium from the hollow cathode walls [6.98]

ticular interest is the investigation of radicals and unstable reaction products that are formed by electron-impact fragmentation in gas discharges. These species play an important role in the extremely rarefied plasma in molecular clouds in the interstellar medium.

Besides its applications to studies of collision processes and ionization probabilities in gas discharges and combustion processes, this technique has the very useful aspect of simple wavelength calibration in laser spectroscopy [6.98]. A small fraction of the output from a tunable laser is split into a hollow-cathode spectral lamp and the optogalvanic spectrum of the discharge is recorded simultaneously with the unknown spectrum under investigation. The numerous lines of thorium or uranium are nearly equally distributed throughout the visible and ultraviolet spectral regions (Fig. 6.35). They are recommended as secondary wavelength standards since they have been measured interferometrically to a high precision [6.99, 6.100]. They can therefore serve as convenient absolute wavelength markers, accurate to about $0.001\,\mathrm{cm}^{-1}$.

If the discharge cell has windows of optical quality, it can be placed inside the laser resonator to take advantage of the q-fold laser intensity (Sect. 6.2.2). With such an intracavity arrangement, Doppler-free saturation spectroscopy can also be performed with the optogalvanic technique (Sect. 7.2 and [6.101]). An increased sensitivity can be achieved by optogalvanic spectroscopy in thermionic diodes under space-charge-limited conditions (Sect. 6.4.5). Here the internal space-charge amplification is utilized to generate signals in the millivolt to volt range without further external amplification [6.87, 6.102].

For more details on optogalvanic spectroscopy see the reviews [6.72, 6.90–6.92] and the book [6.103], which also give extensive reference lists.

6.6 Velocity-Modulation Spectroscopy

The analysis of absorption spectra in molecular gas discharges is by no means simple, because the gas discharge produces a large variety of neutral and ionized fragments from the neutral parent molecules. The spectra of these different species may overlap, and often an unambiguous identification is not possible if the spectra are not known. An elegant technique, developed by Saykally et al. [6.104, 6.105], is very helpful in distinguishing between spectra of ionized and neutral species.

The external voltage applied to the gas discharge accelerates the positive ions toward the cathode and the negative ions toward the anode. They therefore acquire a drift velocity v_D, which causes a corresponding Doppler shift $\Delta\omega = \omega - \omega_0 = \boldsymbol{k} \cdot \boldsymbol{v}_D$ of their absorption frequency ω_0. If an ac voltage of frequency f is applied instead of a dc voltage, the drift velocity is periodically changed and the absorption frequency $\omega = \omega_0 + \boldsymbol{k} \cdot \boldsymbol{v}_D$ is modulated at the frequency f around the unshifted frequency ω_0. When the absorption spectrum is recorded with a lock-in amplifier, the spectra of the ions can be immediately distinguished from those of the neutral species. This velocity-modulation technique has the same effect as the frequency modulation discussed in Sect. 6.2.1. When the laser is scanned through the spectrum, the first derivatives of the ion lines are seen where the phase of the signals is opposite for positive and negative ions, respectively (Fig. 6.36). The two species can therefore be distinguished by the sign of the lock-in output signal.

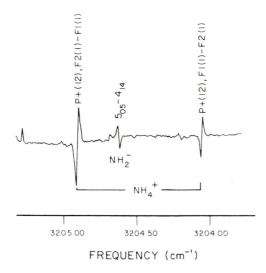

Fig. 6.36. Opposite signal phases of the derivative line profiles of negative and positive ions in velocity-modulation spectroscopy [6.104]

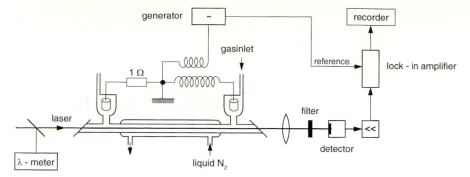

Fig. 6.37. Experimental arrangement for velocity-modulation spectroscopy [6.104]

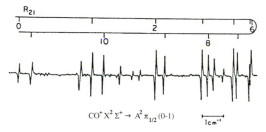

$$CO^+ X^2\Sigma^+ \to A^2\pi_{1/2}(0\text{-}1) \qquad \vert 1\,cm^{-1}$$

Fig. 6.38. Rotational lines of CO^+ around the band head of the R_{21} branch in the $A^2\Pi_{1/2}(v'=1) \leftarrow X^2\Sigma^+(v''=0)$ band measured with velocity modulation [6.102]

A typical experimental arrangement [6.105] is depicted in Fig. 6.37. With specially designed electron-switching circuits, polarity modulation frequencies up to 50 kHz can be realized for gas discharges of 300 V and 3 A [6.106]. The attainable signal-to-noise ratio is illustrated by Fig. 6.38, which shows the band head of a vibrational band of the $A^2\Pi_{1/2} \leftarrow X^2\Sigma_g^+$ transition of the CO^+-ion [6.105].

This technique was first applied to the infrared region where many vibrational–rotational transitions of ions were measured with color-center lasers or diode lasers [6.105, 6.109]. Meanwhile, electronic transitions have also been studied with dye lasers [6.110].

A modification of this velocity-modulation technique in fast ion beams is discussed in Sect. 9.5.

6.7 Laser Magnetic Resonance and Stark Spectroscopy

In all methods discussed in the Sects. 6.1–6.6, the laser frequency ω_L was tuned across the constant frequencies $\omega_i k$ of molecular absorption lines. For molecules with permanent magnetic or electric dipole moments, it is often preferable to tune the absorption lines by means of external magnetic or electric fields across a fixed-frequency laser line. This is particularly advantageous if intense lines of fixed-frequency lasers exist in the spectral region of interest but no tunable source with sufficient intensity is available. Such spectral regions of interest are, for example, the $3-5\,\mu m$ and the $10\,\mu m$ ranges, where numerous intense lines of HF, DF, CO, N_2O, and CO_2 lasers can be utilized. Since many of the vibrational bands fall into this spectral range, it is often called the *fingerprint* region of molecules.

Another spectral range of interest is the far infrared, where the rotational lines of polar molecules are found. Here a large number of lines from H_2O or D_2O lasers ($125\,\mu m$) and from HCN lasers ($330\,\mu m$) provide intense sources. The successful development of numerous optically pumped molecular lasers [6.111] has considerably increased the number of FIR lines.

6.7.1 Laser Magnetic Resonance

The molecular level E_0 with the total angular momentum J splits in an external magnetic field $\boldsymbol{B}$ into $(2J+1)$ Zeeman components. The sublevel with the magnetic quantum number M shifts from the energy E_0 at zero field to

$$E = E_0 - g\mu_0 \boldsymbol{B} M \,, \tag{6.47}$$

where μ_0 is the Bohr magneton, and g is the Landé factor, which depends on the coupling scheme of the different angular momenta (electronic angular momentum, electron spin, molecular rotation, and nuclear spin). The frequency ω of a transition $(v'', J'', M'') \rightarrow (v', J', M')$ is therefore tuned by the magnetic field from its unperturbed position ω_0 to

$$\omega = \omega_0 - \mu_0(g'M' - g''M'')\boldsymbol{B}/\hbar \,, \tag{6.48}$$

and we obtain on the transition $(v'', J'', M'') \rightarrow (v', J', M')$ three groups of lines with $\Delta M = M'' - M' = 0, \pm 1$, which degenerate to three single lines if $g'' = g'$ (normal Zeeman effect). The tuning range depends on the magnitude of $g'' - g'$ and is larger for molecules with a large permanent dipole moment. These are, in particular, radicals with an unpaired electron, which have a large spin moment. In favorable cases a tuning range of up to $2\,cm^{-1}$ can be reached in magnetic fields of $2\,T\ (\widehat{=}\,20\,kG)$.

Figure 6.39a explains schematically the appearance of resonances between the fixed frequency ω_L and the different Zeeman components when the magnetic field $\boldsymbol{B}$ is tuned. The experimental arrangement is illustrated in Fig. 6.39b. The sample is placed inside the laser cavity and the laser output is monitored as a function of the magnetic field. The cell is part of a flow system in which radicals are generated either directly in a microwave discharge or by adding reactants to the discharge close to the laser cavity. A polyethylene membrane beam splitter separates the laser medium from the sample. The beam splitter polarizes the radiation and transitions with either $\Delta M = 0$ or ± 1, which can be selected by rotation of the tube about the laser axis. For illustration, Fig. 6.39c shows the laser magnetic resonance (LMR) spectrum of the CH radical with some OH lines overlapping. Concentrations of 2×10^8 molecules/cm^3 could be still detected with reasonable signal-to-noise ratio for the detector time constant of $1\,s$ [6.112, 6.113].

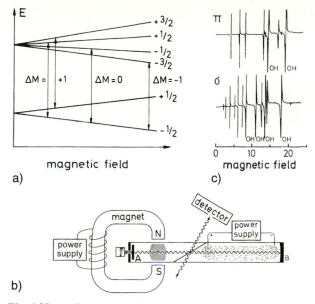

Fig. 6.39a–c. Laser magnetic resonance spectroscopy: (**a**) level diagram; (**b**) experimental set-up with an intracavity sample; and (**c**) LMR spectrum of CH radicals, superimposed by some OH-lines, measured in a low-pressure oxygen–acetylene flame with an H_2O laser [6.113]

The sensitivity of this intracavity technique (Sect. 6.2.2) can even be enhanced by modulating the magnetic field, which yields the first derivative of the spectrum (Sect. 6.2.1). When a tunable laser is used it can be tuned to the center ν_0 of a molecular line at zero field $\boldsymbol{B} = 0$. If the magnetic field is now modulated around zero, the phase of the zero-field LMR resonances for $\Delta M = +1$ transitions is opposite to that for $\Delta M = -1$ transitions. The advantages of this zero-field LMR spectroscopy have been proved for the NO molecule by Urban et al. [6.114] using a spin-flip Raman laser.

Because of its high sensitivity LMR spectroscopy is an excellent method to detect radicals in very low concentrations and to measure their spectra with high precision. If a sufficient number of resonances with laser lines can be found, the rotational constant, the fine structure parameters, and the magnetic moments can be determined very accurately. The identification of the spectra and the assignment of the lines are often possible even if the molecular constant was not known before [6.115]. All radicals observed in interstellar space by radio astronomy have been found and measured in the laboratory with LMR spectroscopy.

Often, a combination of LMR spectroscopy with a fixed-frequency laser and absorption spectroscopy at zero magnetic field with a tunable laser is helpful for the identification of spectra.

Instead of inside the laser cavity, the sample can also be placed outside between two crossed polarizers (Fig. 6.40). In a *longitudinal* magnetic field the plane of polarization of the transmitted light is turned due to the Faraday

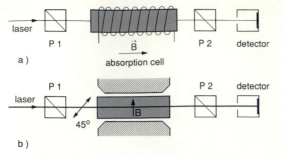

Fig. 6.40a,b. Schematic arrangement for LMR spectroscopy using the Faraday effect in a longitudinal magnetic field (**a**), or the Voigt effect in a transverse magnetic field (**b**) [6.116]

effect if its frequency ω coincides with one of the allowed Zeeman transitions. The detector receives a signal only for these resonance cases, while the nonresonant background is blocked by the crossed polarizers [6.116]. This technique is similar to polarization spectroscopy (Sect. 7.4). Modulation of the magnetic field and lock-in detection further enhances the sensitivity. In a transverse magnetic field, the plane of polarization of the incident beam is chosen 45° inclined against B. Due to the Voigt effect, the plane of polarization is turned if ω_L coincides with a Zeeman transition [6.116].

6.7.2 Stark Spectroscopy

Analogously to the LMR technique, Stark spectroscopy utilizes the Stark shift of molecular levels in *electric* fields to tune molecular absorption lines into resonance with lines of fixed-frequency lasers. A number of small molecules with permanent electric dipole moments and sufficiently large Stark shifts have been investigated, in particular, those molecules that have rotational spectra outside spectral regions accessible to conventional microwave spectroscopy [6.117].

To achieve large electric fields, the separation of the Stark electrodes is made as small as possible (typically about 1 mm). This generally excludes an intracavity arrangement because the diffraction by this narrow aperture would introduce intolerably large losses. The Stark cell is therefore placed outside the resonator, and for enhanced sensitivity the electric field is modulated while the dc field is tuned. This modulation technique is also common in microwave spectroscopy. The accuracy of 10^{-4} for the Stark field measurements allows a precise determination of the absolute value for the electric dipole moment.

Figure 6.41 illustrates the obtainable sensitivity by a $\Delta M = 0$ Stark spectrum of the ammonia isotope $^{14}NH_2D$ composed of measurements with several laser lines [6.117]. Since the absolute frequency of many laser lines was measured accurately within $20-40\,\text{kHz}$ (Sect. 14.7), the absolute frequency of the Stark components at resonance with the laser line can be measured with the same accuracy. The total accuracy in the determination

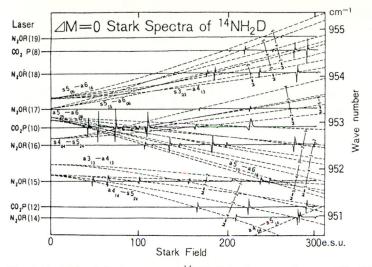

Fig. 6.41. $\Delta M = 0$ Stark spectra of $^{14}NH_2D$ in the spectral range $950-955\ cm^{-1}$ obtained with different fixed-frequency laser lines [6.117]

of the molecular parameters is therefore mainly limited by the accuracy of 10^{-4} for the electric field measurements. To date numerous molecules have been measured with laser Stark spectroscopy [6.117–6.120]. The number of molecules accessible to this technique can be vastly enlarged if tunable lasers in the relevant spectral regions are used, which can be stabilized close to a molecular line with sufficient accuracy and long-term stability. Stark spectroscopy with constant electric fields and tunable lasers has been performed in molecular beams at sub-Doppler resolution to measure the electric dipole moments of polar molecules in excited vibrational states [6.119].

An efficient way to generate coherent, tunable radiation in the far infrared is the difference frequency generation by mixing the output of an CO_2 laser kept on a selected line with the output of a tunable CO_2 waveguide laser in a MIM diode (Sect. 5.8). With this technique Stark spectra of $^{13}CH_3OH$ were measured over a broad spectral range [6.120].

Reviews about more recent investigations in LMR and Stark-spectroscopy, including the visible and UV range, can be found in [6.121–6.123].

6.8 Laser-Induced Fluorescence

Laser-induced fluorescence (LIF) has a large range of applications in spectroscopy. First, LIF serves as a sensitive monitor for the absorption of laser photons in fluorescence excitation spectroscopy (Sect. 6.3.1).

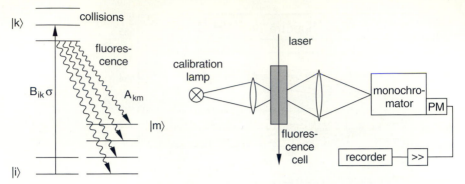

Fig. 6.42a,b. Laser-induced fluorescence: (**a**) level scheme and (**b**) experimental arrangement for measuring LIF spectra

Second, it is well suited to gain information on molecular states if the fluorescence spectrum excited by a laser on a selected absorption transition is dispersed by a monochromator. The fluorescence spectrum emitted from a selectively populated rovibronic level (v'_k, J'_k) consists of all allowed transitions to lower levels (v''_m, J''_m) (Fig. 6.42). The wavenumber differences of the fluorescence lines immediately yield the term differences of these terminating levels (v''_m, J''_m).

A third aspect of LIF is the spectroscopic study of collision processes. If the excited molecule is transferred by inelastic collisions from the level (v'_k, J'_k) into other rovibronic levels, the fluorescence spectrum emitted from these collisionally populated levels gives quantitative information on the collision cross sections (Sect. 13.4).

Another aspect of LIF concerns its application to the determination of the internal-state distribution in molecular reaction products of chemical reactions (Sects. 6.8.4 and 13.6). Under certain conditions the intensity I_{Fl} of LIF excited on the transition $|i\rangle \rightarrow |k\rangle$ is a direct measure of the population density N_i in the absorbing level $|i\rangle$.

Let us now consider some basic facts of LIF in molecules. Extensive information on LIF can be found in [6.124–6.127].

6.8.1 Molecular Spectroscopy by Laser-Induced Fluorescence

Assume a rovibronic level (v'_k, J'_k) in an excited electronic state of a diatomic molecule has been selectively populated by optical pumping. With a mean lifetime $\tau_k = 1/\sum_m A_{km}$, the excited molecules undergo spontaneous transitions to lower levels $E_m(v''_m, J''_m)$ (Fig. 6.42). At a population density $N_k(v'_k, J'_k)$ the radiation power of a fluorescence line with frequency $v_{km} = (E_k - E_m)/h$ is given by (Sect. 2.7.1)

$$P_{km} \propto N_k A_{km} v_{km} . \tag{6.49}$$

The spontaneous transition probability A_{km} is proportional to the square of the matrix element (Sect. 2.7.2)

$$A_{km} \propto \left| \int \psi_k^* r \psi_m \, d\tau_n \, d\tau_{el} \right|^2 , \tag{6.50}$$

where r is the vector of the excited electron and the integration extends over all nuclear and electronic coordinates. Within the Born–Oppenheimer approximation [6.128, 6.129] the total wave function can be separated into a product

$$\psi = \psi_{el} \psi_{vib} \psi_{rot} , \tag{6.51}$$

of electronic, vibrational, and rotational factors. If the electronic transition moment does not critically depend on the internuclear separation R, the total transition probability is then proportional to the product of three factors

$$A_{km} \propto |M_{el}|^2 |M_{vib}|^2 |M_{rot}|^2 , \tag{6.52a}$$

where the first factor

$$M_{el} = \int \psi_{el}^* r \psi_{el}'' \, d\tau_{el} , \tag{6.52b}$$

represents the electronic matrix element that depends on the coupling of the two electronic states. The second integral

$$M_{vib} = \int \psi_{vib}' \psi_{vib}'' \, d\tau_{vib} , \quad \text{with} \quad d\tau_{vib} = R^2 \, dR , \tag{6.52c}$$

is the Franck–Condon factor, which depends on the overlap of the vibrational wave functions $\psi_{vib}(R)$ in the upper and lower state. The third integral

$$M_{rot} = \int \psi_{rot}' \psi_{rot}'' g_i \, d\tau_{rot} , \quad \text{with} \quad d\tau_{rot} = d\vartheta \, d\varphi , \tag{6.52d}$$

is called the Hönl–London factor. It depends on the orientation of the molecular axis relative to the electric vector of the observed fluorescence wave. This is expressed by the factor g_i $(i = x, y, z)$, where $g_x = \sin\vartheta \cos\varphi$, $g_y = \sin\vartheta \sin\varphi$, and $g_z = \cos\vartheta$, where ϑ and φ are the polar and azimuthal angles [6.128].

Only those transitions for which all three factors are nonzero appear as lines in the fluorescence spectrum. The Hönl–London factor is always zero unless

$$\Delta J = J_k' - J_m'' = 0, \pm 1 . \tag{6.53}$$

If a *single* upper level (v_k', J_k') has been selectively excited, each vibrational band $v_k' \rightarrow v_m''$ consists of at most *three* lines: a *P line* $(\Delta J = -1)$, a *Q line*

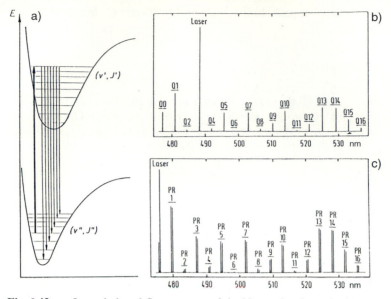

Fig. 6.43a–c. Laser-induced fluorescence of the Na_2 molecule excited by argon laser lines: (**a**) energy level diagram; (**b**) fluorescence lines with $\Delta J = 0$ (Q-lines) emitted from the upper level ($v' = 3$, $J' = 43$) of the $B^1\Pi_u$ state, excited at $\lambda = 488$ nm; (**c**) P and R doublets, emitted from the upper level ($v' = 6$, $J' = 27$)

($\Delta J = 0$), and an R *line* ($\Delta J = +1$). For diatomic *homonuclear* molecules additional symmetry selection rules may further reduce the number of possible transitions. A selectively excited level (v'_k, J'_k) in a Π state, for example, emits on a $\Pi \to \Sigma$ transition either only Q lines or only P and R lines, depending on the symmetry of the rotational levels, while on a $\Sigma_u \to \Sigma_g$ transition only P and R lines are allowed [6.129].

The fluorescence spectrum emitted from selectively excited molecular levels of a diatomic molecule is therefore very simple compared with a spectrum obtained under broadband excitation. Figure 6.43 illustrates this by two fluorescence spectra of the Na_2 molecule, excited by two different argon laser lines. While the $\lambda = 488$ nm line excites a positive Λ component in the ($v' = 3$, $J' = 43$) level, which emits only Q lines, the $\lambda = 476.5$ nm line populates the negative Λ component in the ($v' = 6$, $J' = 27$) level of the $^1\Pi_u$ state, resulting in P and R lines.

6.8.2 Experimental Aspects of LIF

In atomic physics the selective excitation of single atomic levels was achieved with atomic resonance lines from hollow cathode lamps even before the invention of lasers. However, in molecular spectroscopy only fortuitous coincidences between atomic resonance lines and molecular transitions could be

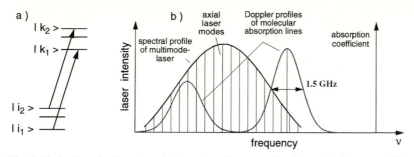

Fig. 6.44a,b. Doppler-broadened absorption lines overlapping with the spectral line profile of a laser: (**a**) level scheme and (**b**) line profiles

used in some cases. Molecular lamps generally emit on many lines and are therefore not useful for the selective excitation of molecular levels.

Tunable narrow-band lasers can be tuned to every wanted molecular transition $|i\rangle \rightarrow |k\rangle$ within the tuning range of the laser. However, selective excitation of a single upper level can only be achieved if neighboring absorption lines do not overlap within their Doppler width (Fig. 6.44). In the case of atoms this can generally be achieved, whereas for molecules with complex absorption spectra, many absorption lines often overlap. In this latter case the laser simultaneously excites several upper levels, which are not necessarily energetically close to each other (Fig. 6.44a). In many cases, however, the fluorescence spectra of these levels can readily be separated by a spectrometer of medium size [6.130].

In order to achieve selective excitation of single levels even in complex molecular spectra, one may use collimated cold molecular beams where the Doppler width is greatly decreased and the number of absorbing levels is drastically reduced due to the low internal temperature of the molecules (Sect. 9.2).

A very elegant technique combines selective laser excitation with high-resolution Fourier transform spectroscopy of the LIF spectrum. This combination takes advantage of the simultaneous recording of all fluorescence lines and has therefore, at the same total recording time, a higher signal-to-noise ratio. It has also been applied to the spectroscopy of visible and infrared fluorescence spectra of a large variety of molecules [6.131–6.133].

When the beam of a single-mode laser passes in the z-direction through a molecular absorption cell, only molecules with velocity components $v_z = 0 \pm \gamma$ are excited if the laser is tuned to the center frequency of an absorption line with the homogeneous width γ. The fluorescence collected within a narrow cone around the z-axis then shows sub-Doppler linewidths, which may be resolved with Fourier transform spectroscopy (Fig. 6.45) [6.133].

The advantages of LIF spectroscopy for the determination of molecular parameters may be summarized as follows:

(a) The relatively simple structure of the fluorescence spectra allows ready assignment. The fluorescence lines can be resolved with medium-sized

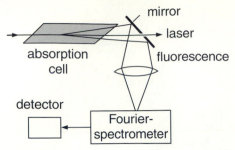

Fig. 6.45. Measurement of LIF spectrum with reduced Doppler width, excited by a single-mode laser

spectrometers. The demands of experimental equipment are much less stringent than those necessary for complete resolution and analysis of absorption spectra of the same molecule. This advantage still remains if a few upper levels are simultaneously populated under Doppler-limited excitation [6.130].

(b) The large intensities of many laser lines allow achievement of large population densities N_k in the excited level. This yields, according to (6.49), correspondingly high intensities of the fluorescence lines and enables detection even of transitions with small Franck–Condon factors (FCF). A fluorescence progression $v'_k \to v''_m$ may therefore be measured with sufficiently good signal-to-noise ratio up to very high vibrational quantum numbers v''_m. The potential curve of a diatomic molecule can be determined very accurately from the measured term energies $E(v''_m, J''_m)$ using the Rydberg–Klein–Rees (RKR) method, which is based on a modified WKB procedure [6.134–6.137]. Since the term values $E(v''_m, J''_m)$ can be immediately determined from the wave numbers of the fluorescence lines, the RKR potential can be constructed up to the highest measured level v''_{max}. In some cases fluorescence progressions are found up to levels v'' just below the dissociation limit [6.138, 6.139]. This allows the spectroscopic determination of the dissociation energy by an extrapolation of the decreasing vibrational spacings $\Delta E_{vib} = E(v''_{m+1}) - E(v''_m)$ to $\Delta E_{vib} = 0$ (Birge–Sponer plot) [6.140–6.143].

(c) The relative intensities of the fluorescence lines $(v'_k, J'_k \to v''_m, J''_m)$ are proportional to the Franck–Condon factors. The comparison between calculated FCF obtained with the RKR potential from the Schrödinger equation and the measured relative intensities allows a very sensitive test for the accuracy of the potential. In combination with lifetime measurements, these intensity measurements yield absolute values of the electronic transition moment $M_{el}(R)$ and its dependence on the internuclear distance R [6.144].

(d) In several cases discrete molecular levels have been excited that emit continuous fluorescence spectra terminating on repulsive potentials of dissociating states [6.145]. The overlap integral between the vibrational eigenfunctions ψ'_{vib} of the upper discrete level and the continuous function $\psi_{cont}(R)$ of the dissociating lower state often shows an intensity modulation of the continuous fluorescence spectrum that reflects the square $|\psi'_{vib}(R)|^2$ of the upper-state wave function (Fig. 2.14). If the upper potential is known, the repulsive part of the lower potential can be accurately determined [6.146, 6.147]. This is of particular relevance for excimer spectroscopy (Sect. 5.7) [6.148].

(e) For transitions between high vibrational levels of two bound states, the main contribution to the transition probability comes from the internuclear distances R close to the classical turning points R_{min} and R_{max} of the vibrating oscillator. There is, however, a nonvanishing contribution from the internuclear distance R between R_{min} and R_{max}, where the vibrating molecule has kinetic energy $E_{kin} = E(v, J) - V(R)$. During a radiative transition this kinetic energy has to be preserved. If the total energy $E'' = E(v', J') - h\nu = V''(R) + E_{kin} = U(R)$ in the lower state is below the dissociation limit of the potential $V''(R)$, the fluorescence terminates in bound levels and the fluorescence spectrum consists of discrete lines. If it is above this limit, the fluorescence transitions end in the dissociation continuum (Fig. 6.46). The intensity distribution of these Condon internal-diffraction bands [6.149, 6.150] is very sensitive to the difference potential $V''(R) - V'(R)$, and therefore allows an accurate determination of one of the potential curves if the other is known [6.151].

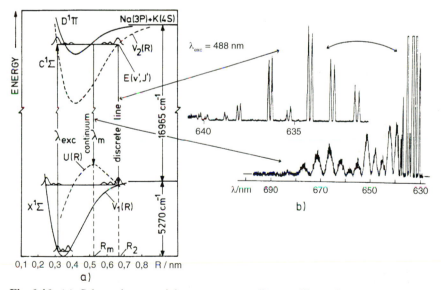

Fig. 6.46. (a) Schematic potential energy curve diagram illustrating bound–bound transitions between the discrete levels and bound–free transitions from an upper discrete level into the lower continuum above the dissociation energy of the electronic ground state. (b) Two sections of the NaK fluorescence spectrum showing both kinds of transitions [6.149]

6.8.3 LIF of Polyatomic Molecules

The technique of LIF is, of course, not restricted to diatomic molecules, but has also been applied to the investigation of triatomic molecules such as NO_2, SO_2, BO_2, NH_2, and many other polyatomic molecules. In combination with excitation spectroscopy, it allows the assignment of transitions and the identification of complex spectra. Examples of such measurements have been cited in [6.124–6.127].

The LIF of polyatomic molecules opens possibilities to recognize perturbations both in excited and in ground electronic states. If the upper state is perturbed its wave function is a linear combination of the BO wave functions of the mutually perturbing levels. The admixture of the perturber wave function opens new channels for fluorescence into lower levels with different symmetries, which are forbidden for unperturbed transitions. The LIF spectrum of NO_2 depicted in Fig. 6.47 is an example where the "forbidden" vibrational bands terminating on the vibrational levels (v_1, v_2, v_3) with an odd number v_3 of vibrational quanta in the asymmetric stretching vibrational mode in the electronic ground state are marked by an asterisk [6.152].

Due to nonlinear coupling between high-lying vibrational levels of polyatomic molecules, the level structure may become quite complex. The classical motion of the nuclei in such a highly vibrationally excited molecule can often no longer be described by a superposition of normal vibrations [6.153]. In such cases a statistical model of the vibrating molecule may be more adequate when the distribution of the energy spacings of neighboring vibrational levels is investigated. For several molecules a transition regime has been found from *classical oscillations* to *chaotic behavior* when the vibrational

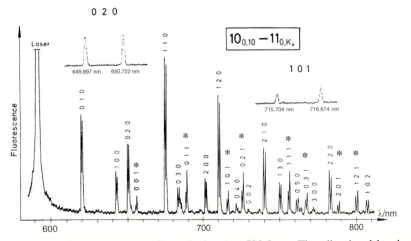

Fig. 6.47. LIF spectrum of NO_2 excited at $\lambda = 590.8$ nm. The vibrational bands terminating on ground-state vibrational levels (v_1, v_2, v_3) marked with an asterisk are forbidden by symmetry selection rules but are made allowable by an admixture of a perturbing level with other symmetry to the excited-state wave function [6.152]

energy is increasing. The chaotic region corresponds to a Wigner distribution of the energy separations of neighboring vibrational levels, while the classical nonchaotic regime leads to a Poisson distribution of the vibrational energy spacings. With LIF spectroscopy such high-lying levels in the electronic ground state can be reached. The analysis of such LIF spectra therefore yields information on the dynamics of molecules in vibronically coupled levels [6.154–6.156].

6.8.4 Determination of Population Distributions by LIF

An interesting application of LIF is the measurement of relative population densities $N(v_i'', J_i'')$ and their distribution over the different vibrational–rotational levels (v_i'', J_i'') under situations that differ from thermal equilibrium. Examples are chemical reactions of the type $AB + C \rightarrow AC^* + B$, where a reaction product AC^* with internal energy is formed in a reactive collision between the partners AB and C. The measurement of the internal state distribution $N_{AC}(v'', J'')$ can often provide useful information on the reaction paths and on the potential surfaces of the collision complex $(ABC)^*$. The initial internal state distribution $N_{AC}(v, J)$ is often far away from a Boltzmann distribution. There even exist reactions that produce population inversions allowing the realization of chemical lasers [6.157]. Investigations of these population distributions may finally allow the optimization and better control of these reactions.

The population density $N_K(v_K, J_K)$ in the excited state $|k\rangle$ can be determined from the measurement of the fluorescence rate

$$n_{Fl} = N_K A_K V_R , \tag{6.54}$$

which represents the number of fluorescence photons n_{Fl} emitted per second from the reaction volume V_R.

In order to obtain the population densities $N_i(v_i, J_i)$ in the electronic ground state, the laser is tuned to the absorbing transitions $|i\rangle \rightarrow |k\rangle$ starting from levels (v_i'', J_i'') of the reaction products under investigation, and the total fluorescence rate (6.54) is measured for different upper levels $|k\rangle$. Under stationary conditions these rates are obtained from the rate equation

$$\frac{dN_k}{dt} = 0 = N_i B_{ik} \rho - N_k (B_{ki} \rho + A_K + R_K) \tag{6.55a}$$

which gives, according to (6.54) with $B_{ik} = B_{ki}$:

$$n_{Fl} = N_k A_k V_R = N_i A_k V_R \frac{B_{ik} \rho}{B_{ik} \rho + A_k + R_k} , \tag{6.55b}$$

where R_K is the total nonradiative deactivation rate of the level $|k\rangle$ (Fig. 6.48). If the collisional deactivation of level $|k\rangle$ is negligible compared with its radiative depopulation by fluorescence, (6.55) reduces to

$$n_{Fl} = N_i \cdot A_k \cdot V_R \cdot \frac{B_{ik} \rho}{A_k + B_{ik} \rho} = \frac{N_i V_R B_{ik} \rho}{1 + B_{ik} \rho / A_k} . \tag{6.56}$$

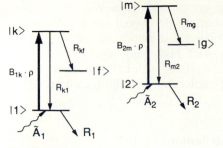

Fig. 6.48. Level scheme for measurements of population distributions in rotational–vibrational levels of molecular reaction products in their electronic ground state

We distinguish two limiting cases:

(a) The laser intensity is sufficiently low to ensure $B_{ik}\rho \ll A_K$. Then the ratio $n_{Fl}(k)/n_{Fl}(m)$ of fluorescence rates observed under excitation on the two transitions $|1\rangle \to |k\rangle$ and $|2\rangle \to |m\rangle$, respectively, with the same laser intensity $I_L = c\rho$ becomes (Fig. 6.48)

$$\frac{n_{Fl}(k)}{n_{Fl}(m)} = \frac{N_1}{N_2}\frac{B_{1k}}{B_{2m}} = \frac{N_1\sigma_{1k}}{N_2\sigma_{2m}} = \frac{\alpha_{1k}}{\alpha_{2m}} , \tag{6.57}$$

where σ is the optical absorption cross section and $\alpha = N\sigma$ the absorption coefficient. Therefore the ratio N_1/N_2 of the lower-level population can be obtained from the measured ratio of fluorescence rates or of absorption coefficients if the absorption cross sections are known.

(b) If the laser intensity is sufficiently high to saturate the absorbing transition we have the case $B_{ik}\rho \gg A_K$. Then (6.56) yields

$$\frac{n_{Fl}(k)}{n_{Fl}(m)} = \frac{N_1 A_k}{N_2 A_m} . \tag{6.58}$$

Under stationary conditions the population densities N_i are determined by the rate equations

$$dN_i/dt = 0 = \widetilde{A}_i - N_i(R_i + B_{ik}\rho) + \sum_m N_m R_{mi} , \tag{6.59}$$

where $\widetilde{A}_i$ is the feeding rate of the level $|i\rangle$ by the reaction, or by diffusion of molecules in the level $|i\rangle$ into the detection volume. The rate $N_i(R_i + B_{ik}\rho)$ is the total deactivation rate of the level $|i\rangle$, whereas $\sum N_m R_{mi}$ is the sum of all transition rates from other levels $|m\rangle$ into the level $|i\rangle$. If $\widetilde{A}_i \gg \sum N_m R_{mi}$ and the main depletion rate of N_i is due to laser absorption ($B_{ik}\rho \gg R_i$), the stationary level population becomes $N_i = \widetilde{A}_i/B_{ik}\rho$ and the ratio (6.58) of LIF rates becomes

$$\frac{n_{Fl}(k)}{n_{Fl}(m)} = \frac{\widetilde{A}_1 B_{2m} A_k}{\widetilde{A}_2 B_{1k} A_m} . \tag{6.60}$$

With time-resolved nonlinear LIF the branching ratio of different fluorescence transitions from a selectively excited upper level can be determined [6.158].

First measurements of internal-state distributions of reaction products using LIF have been performed by Zare et al. [6.159–6.161]. One example is the formation of BaCl in the reaction of barium with halogens

$$Ba + HCl \rightarrow BaCl^{*}(X^{2}\Sigma^{+}v'', J'') + H \,. \tag{6.61}$$

Figure 6.49a exhibits the vibrational population distribution of BaCl for two different collision energies of the reactants Ba and HCl. Figure 6.49b illustrates that the total *rotational* energy of BaCl barely depends on the collision energy in the center of mass system of Ba + HCl, while the *vibrational* energy increases with increasing collision energy.

The interesting question of how the internal-state distribution of the products is determined by the internal energy of the reacting molecules can be answered experimentally with a second laser that pumps the reacting molecule into excited levels (v'', J''). The internal-state distribution of the product is measured with and without the pump laser. An example that has been studied [6.160] is the reaction

$$Ba + HF(v'' = 1) \rightarrow BaF^{*}(v = 0-12) + H \,, \tag{6.62}$$

where a chemical HF laser has been used to excite the first vibrational level of HF, and the internal state distrubtion of BaF* is measured with a tunable dye laser.

An example where LIF is used for the optimization of the production of thin amorphous layers of Si:H from the deposition of gaseous silane SiH_4 in a gas discharge is the spectroscopy of SiH radicals that are formed in the discharge [6.162].

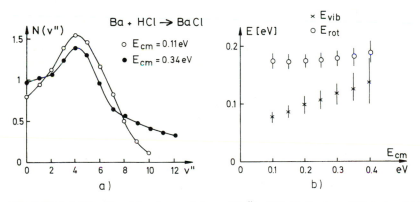

Fig. 6.49. (a) Vibrational level populations $N(v'')$ of BaCl for two different collision energies of the reactants Ba + HCl. **(b)** Mean vibrational and rotational energy of reactively formed BaCl as a function of the relative collision energy [6.161]

Another example is the determination of vibrational–rotational population distributions in supersonic molecular beams [6.163], where molecules are cooled down to rotational temperatures of a few Kelvin (Sect. 9.2).

The method is not restricted to neutral molecules but can also be applied to ionic species. This is of importance when LIF is used for diagnostics of combustion processes [6.164, 6.165] or plasmas [6.166].

6.9 Comparison Between the Different Methods

The different sensitive techniques of Doppler-limited laser spectroscopy discussed in the previous sections supplement each other in an ideal way. In the *visible* and *ultraviolet* range, where *electronic* states of atoms or molecules are excited by absorption of laser photons, *excitation spectroscopy* is generally the most suitable technique, particularly at low molecular densities. Because of the short spontaneous lifetimes of most excited electronic states E_k, the quantum efficiency η_k reaches 100% in many cases. For the detection of the laser-excited fluorescence, sensitive photomultipliers or intensified CCD cameras are available that allow, together with photon-counting electronics (Sect. 4.5), the detection of single fluorescence photons with an overall efficiency of $10^{-3}-10^{-1}$ including the collection efficiency $\delta \approx 0.01-0.3$ (Sect. 6.3.1).

Excitation of very high-lying states close below the ionization limit, e.g., by ultraviolet lasers or by two-photon absorption, enables the detection of absorbed laser photons by monitoring the ions. Because of the high collection efficiency of these ions, *ionization spectroscopy* represents the most sensitive detection method, superior to all other techniques in all cases where it can be applied.

In the *infrared* region, excitation spectroscopy is less sensitive because of the lower sensitivity of infrared photodetectors and because of the longer lifetimes of excited vibrational levels. These long lifetimes bring about either at low pressures a diffusion of the excited molecules out of the observation region, or at high pressures a collision-induced radiationless deactivation of the excited states. Here *photoacoustic spectroscopy* becomes superior, since this technique utilizes this collision-induced transfer of excitation energy into thermal energy. A specific application of this technique is the quantitative determination of small concentrations of molecular components in gases at higher pressures. Examples are measurements of air pollution or of poisonous constituents in auto engine exhausts, where sensitivities in the ppb range have been successfully demonstrated. For pure gases at lower pressures, where collisions are less significant, absorption spectroscopy with wavelength-modulation may be even superior to optoacoustic spectroscopy, Fig. 6.50.

For infrared spectroscopy in molecular beams optothermal spectroscopy is a very good choice (Sect. 6.3.3).

For the spectroscopy of atoms or ions in gas discharges, *optogalvanic spectroscopy* (Sect. 6.5) is a very convenient and experimentally simple alternative to fluorescence detection. In favorable cases it may even reach the

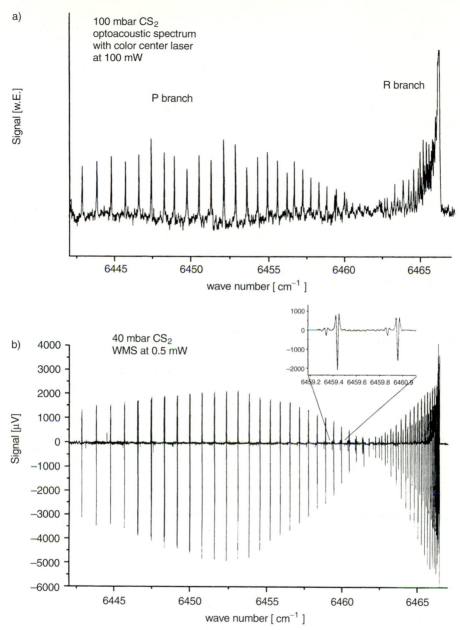

Fig. 6.50a,b. Section of CS_2-overtone spectrum. Comparison of optoacoustic detection (**a**) and absorption spectroscopy with wavelength modulation (**b**) [G.H. Wenz, Thesis, K.L.]

sensitivity of excitation spectroscopy. For the distinction between spectra of ions and neutral species *velocity-modulation spectroscopy* (Sect. 6.6) offers an elegant solution.

Regarding detection sensitivity, LMR and Stark spectroscopy can compete with the other methods. However, their applications are restricted to molecules with sufficiently large permanent dipole moments to achieve the necessary tuning range. They are therefore mainly applied to the spectroscopy of free radicals with an unpaired electron. The magnetic moment of these radicals is predominantly determined by the electron spin and is therefore several orders of magnitude larger than that of stable molecules in $^1\Sigma$ ground states. The advantage of LMR or Stark spectroscopy is the direct determination of Zeeman or Stark splittings from which the Landé factors and therefore the coupling schemes of the different angular momenta can be deduced. A further merit is the higher accuracy of the *absolute* frequency of molecular absorption lines because the frequencies of fixed-frequency laser lines can be absolutely measured with a higher accuracy than is possible with tunable lasers.

All these methods represent modifications of *absorption* spectroscopy, whereas LIF spectroscopy is based on *emission* of fluorescence from selectively populated upper levels. The absorption spectra depend on both the upper and lower levels. The absorption transitions start from thermally populated lower levels. If their properties are known (e.g., from microwave spectroscopy) the absorption spectra yield information on the *upper* levels. On the other hand, the LIF spectra start from one or a few upper levels and terminate on many rotational–vibrational levels of a lower electronic state. They give information about this lower state.

All these techniques may be combined with intracavity absorption when the sample molecules are placed inside the laser resonator to enhance the sensitivity. Cavity ring-down spectroscopy yields absorption spectra with a detection sensitivity that is comparable to the most advanced modulation techniques in multipass absorption spectroscopy.

A serious competitor of laser absorption spectroscopy is Fourier spectroscopy [6.1, 6.167], which offers the advantage of simultaneous recording of a large spectral interval and with it a short measuring time. In contrast, in laser absorption spectroscopy the whole spectrum must be measured sequentially, which takes much more time. However, laser techniques offer the two advantages, namely, higher spectral resolution and higher sensitivity. This is demonstrated by examples in two different spectral regimes. Figure 6.51 exhibits part of the submillimeter, purely rotational spectrum of $^{16}O_3$ taken with a high-resolution ($0.003\,\text{cm}^{-1}$) Fourier spectrometer (Fig. 6.51a) and a small section of Fig. 6.51a in an expanded scale, recorded with the tunable far-infrared laser spectrometer shown in Fig. 5.118. Although the resolution is still limited by the Doppler width of the absorption lines, this imposes no real limitation because at $\nu = 1.5 \times 10^{12}\,\text{Hz}$ the Doppler width of ozone is only $\Delta\nu_D \approx 2\,\text{MHz}$, compared with a resolution of $90\,\text{MHz}$ of the Fourier spectrometer. Another example illustrates the overtone spectrum of acetylene C_2H_2 around $\lambda = 1.5\,\mu\text{m}$, recorded with a Fourier

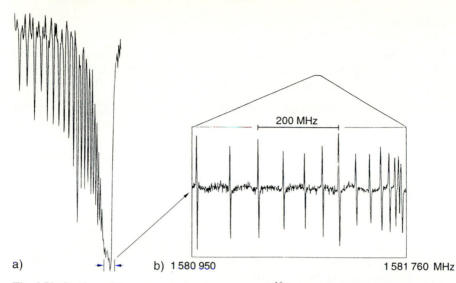

200 MHz

a)   b) 1 580 950   1 581 760 MHz

Fig. 6.51. Section of the pure rotational spectrum of $^{16}O_3$ recorded with a high-resolution Fourier spectrometer (*left spectrum*) and a tunable far-infrared laser spectrometer (*right spectrum*), demonstrating the superior resolution of the laser spectrometer [6.167]

spectrometer (Fig. 6.52a) and with a color-center laser and photo-acoustic spectroscopy (Fig. 6.52b). Weak lines, which are barely seen in the Fourier spectrum, still have a large signal-to-noise ratio in the optoacoustic spectrum as shown by the insert in Fig. 6.52b [6.56].

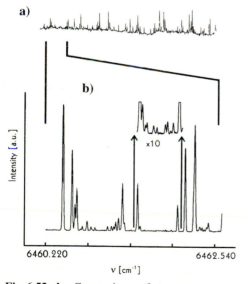

a)

b)

Intensity [a.u.]

×10

6460.220   6462.540

ν [cm⁻¹]

Fig. 6.52a,b. Comparison of overtone spectra of C_2H_2 around $\lambda = 1.5\,\mu m$, measured (**a**) with a Fourier spectrometer and (**b**) with a color-center laser and intracavity opto-acoustic spectroscopy [6.56]

Problems

6.1 A monochromatic laser beam is sent through a sample of diatomic molecules. The laser wavelength is tuned to a vibration–rotation transition $(v'', J'') \rightarrow (v', J')$ with an absorption cross section of $\sigma_{ik} = 10^{-18}\,\text{cm}^2$.

(a) Estimate the fraction n_i/n of molecules in the level $(v''_i = 0, J''_i = 20)$ at $T = 300\,\text{K}$ (vibrational constant $\omega_e = 200\,\text{cm}^{-1}$, rotational constant $B_e = 1.5\,\text{cm}^{-1}$).
(b) Calculate the absorption coefficient for a total gas pressure of 10 mbar.
(c) What is the transmitted laser power P_t behind an absorption path length of 1 m for an incident power $P_0 = 100\,\text{mW}$?

6.2 A focused laser beam ($\varnothing = 0.4\,\text{mm}$) with 1-mW input power at $\lambda = 623\,\text{nm}$ crosses a molecular beam ($\varnothing = 1\,\text{mm}$) perpendicularly. The absorbing molecules with a partial flux density $N_i = n_i \cdot \bar{v} = 10^{12}/(\text{s}\,\text{cm}^2)$ have the absorption cross section $\sigma = 10^{-16}\,\text{cm}^2$.

How many photoelectrons are counted with a photomultiplier if the fluorescence emitted from the crossing volume $V_c = 10^{-4}\,\text{cm}^{-3}$ of both the laser and molecular beams is imaged by a lens with $L = 1\,\text{cm}$ and $D = 4\,\text{cm}$ onto the photocathode with quantum efficiency $\eta_{ph} = 0.2$?

6.3 A monochromatic laser beam with $P = 1\,\text{mW}$ is sent through a 1-m long sample cell filled with absorbing molecules. The absorbing transition has the Doppler width $\Delta\omega_D = 2\pi \cdot 10^9\,\text{s}^{-1}$ and a peak absorption $\alpha(\omega_0) = 10^{-8}\,\text{cm}^{-1}$. The laser frequency $\omega_L = \omega_0 + \Delta\omega \cdot \cos 2\pi ft$ is modulated ($\Delta\omega = 2\pi \cdot 10\,\text{MHz}$). Calculate the maximum ac amplitude of the detector output signal for a detector with a sensitivity of $1\,\text{V/mW}$. How large is the dc background signal?

6.4 How many ions per second are formed by resonant two-step photoionization if the transition $|i\rangle \rightarrow |k\rangle$ of the first step is saturated, the lifetime τ_k of the level $|k\rangle$ is $\tau_k = 10^{-8}\,\text{s}$, the ionization probability by the second laser is $10^{-7}\,\text{s}^{-1}$, and the diffusion rate of molecules in the absorbing level $|i\rangle$ into the excitation volume is $dN_i/dt = 10^5/\text{s}$?

6.5 A thermal detector (bolometer area $3 \times 3\,\text{mm}^2$) is hit by a molecular beam with a density of $n = 10^{-8}$ molecules per cm^3, having a mean velocity $\langle v \rangle = 4 \times 10^4\,\text{cm/s}$. All impinging molecules are assumed to stick at the surface of the detector.

(a) Calculate the energy transferred per second to the bolometer.
(b) What is the temperature rise ΔT, if the heat losses of the bolometer are $G = 10^{-10}\,\text{W/K}$?
(c) The molecular beam is crossed by an infrared laser ($\lambda = 1.5\,\mu\text{m}$) with $P_0 = 10\,\text{mW}$. The absorption coefficient is $\alpha = 10^{-10}\,\text{cm}^{-1}$ and the ab-

sorption path length $L = 10\,\text{cm}$ (by the multiple-reflection techniques). Calculate the additional temperature rise of the bolometer.

6.6 The frequency ν_0 of a molecular transition to an upper level $|k\rangle$ is $10^8\,\text{Hz}$ away from the frequency ν_L of a fixed laser line. Assume that the molecules have no magnetic moment in the ground state, but $\mu = 0.5\mu_B$ ($\mu_B = 9.27 \times 10^{-24}\,\text{J/T}$) in the upper state. What magnetic field $\boldsymbol{B}$ is necessary to tune the absorption line into resonance with the laser line? How many Zeeman components are observed, if the lower level has the rotational quantum number $J = 1$ and the upper $J = 2$, and the laser is (a) linearly and (b) circularly polarized?

6.7 For velocity-modulation spectroscopy an ac voltage of $2\,\text{kV}$ peak-to-peak and $f = 1\,\text{kHz}$ is applied along a discharge tube with 1-m length in the z-direction.

(a) How large is the mean electric field strength?
(b) Estimate the value of Δv_t for the ion velocity $v = v_0 + \Delta v_z \sin 2\pi f t$ for ions with the mass $m = 40\,\text{amu}$ ($1\,\text{amu} = 1.66 \times 10^{-27}\,\text{kg}$) if their mean free path length is $\Lambda = 10^{-3}\,\text{m}$ and the density of neutral species in the discharge tube is $n = 10^{17}/\text{cm}^3$.
(c) How large is the maximum modulation $\Delta \nu$ of the absorption frequency $\nu(v_t) = \nu_0 + \Delta \nu(v_t)$ for $v_0 = 10^{14}\,\text{s}^{-1}$?
(d) How large is the ac signal for an absorbing transition with $\alpha(\nu_0) = 10^{-6}\,\text{cm}^{-1}$ if the incident laser power is $10\,\text{mW}$ and the detector sensitivity $1\,\text{V/mW}$?

6.8 An absorbing sample in a cell with $L = 10\,\text{cm}$ is placed within the laser resonator with a total resonator loss of 1% per round trip and a transmission $T = 0.5\%$ of the output coupler.

(a) Calculate the relative decrease of the laser output power of $10\,\text{mW}$ when the laser is tuned from a frequency ν where $\alpha = 0$ to a frequency ν_0 where the sample has an absorption coefficient $\alpha = 5 \times 10^{-4}\,\text{cm}^{-1}$, while the gain of the active laser medium stays constant.
(b) How many fluorescence photons are emitted if the fluorescence yield of the sample is 0.5 and the laser wavelength is $500\,\text{nm}$?
(c) Could you design the optimum collection optics that image a maximum ratio of the fluorescence light onto the photomultiplier cathode of 40-mm $\varnothing$? How many photon counts are observed if the quantum yield of the cathode is $\eta = 0.15$?
(d) Compare the detection sensitivity of (c) (dark current of the PM is $10^{-9}\,\text{A}$) with that of method (a) if the laser output is monitored by a photodiode with a sensitivity of $10\,\text{V/W}$ and a noise level of $10^{-9}\,\text{V}$.

6.9 Derive (6.41).

7. Nonlinear Spectroscopy

One of the essential advantages that single-mode lasers can offer for high-resolution spectroscopy is the possibility of overcoming the limitation set by Doppler broadening. Several techniques have been developed that are based on selective saturation of atomic or molecular transitions by sufficiently intense lasers.

The population density of molecules in the absorbing level is decreased by optical pumping. This results in a nonlinear dependence of the absorbed radiation power on the incident power. Such techniques are therefore summarized as *nonlinear spectroscopy*, which also includes methods that are based on the simultaneous absorption of two or more photons during an atomic or molecular transition. In the following sections the basic physics and the experimental realization of some important methods of nonlinear spectroscopy are discussed. At first we shall, however, treat the saturation of population densities by intense incident radiation.

7.1 Linear and Nonlinear Absorption

Assume that a monochromatic plane lightwave

$$E = E_0 \cos(\omega t - kz) ,$$

with the mean intensity

$$I = \frac{1}{2} c \epsilon_0 E_0^2 \quad [\text{W/m}^2] ,$$

passes through a sample of molecules, which absorb on the transition $E_i \rightarrow E_k$ ($E_k - E_i = \hbar \omega$). The power dP absorbed in the volume $dV = A\,dz$ is then

$$dP = A I \sigma_{ik} \Delta N \, dz \quad [\text{W}] , \tag{7.1a}$$

where $\Delta N = [N_i - (g_i/g_k)N_k]$ is the difference of the population densities, and $\sigma_{ik}(\nu)$ is the absorption cross section per molecule at the light frequency $\nu = \omega/2\pi$, see (5.2).

If the incident plane wave with the spectral energy density $\rho_\nu(\nu) = I_\nu(\nu)/c$ [W s^2/m^3] has the spectral width $\Delta \nu_L$, which is large compared to the halfwidth $\delta \nu_L$ of the absorption profile, the total intensity becomes

$$I = \int I_\nu(\nu)\,d\nu \approx I_\nu(\nu_0) \cdot \Delta \nu_L .$$

The absorbed power is then

$$dP = \Delta N \cdot dV \cdot \int I_\nu(\nu) \cdot \sigma_{ik}(\nu) \, d\nu ,$$

$$\approx \Delta N \cdot dV \cdot I \cdot \sigma_{ik}(\nu_0) \cdot \frac{\delta\nu}{\Delta\nu_L} . \tag{7.1b}$$

This absorbed power corresponds to $n_{ph} = \delta P/h\nu$ absorbed photons. From (2.15) we can deduce

$$n_{ph} = B_{ik} \cdot \rho(\nu)\Delta N \cdot dV . \tag{7.1c}$$

The comparison of (7.1b) and (7.1c) yields the relation

$$B_{ik} = \frac{c}{h\nu} \int\limits_{\nu=0}^{\infty} \sigma_{ik}(\nu) \, d\nu . \tag{7.2}$$

$B_{ik}\rho_\nu$ gives the net probability for absorbing a photon per molecule and per second within the unit volume $dV = 1\,m^3$, see (2.15, 2.78).

The absorption of the incident wave causes population changes of the levels involved in the absorbing transition. This can be described by the rate equations for the population densities N_1, N_2 of the nondegenerate levels $|1\rangle$ and $|2\rangle$ with $g_1 = g_2 = 1$ (Fig. 7.1):

$$\frac{dN_1}{dt} = B_{12}\rho_\nu(N_2 - N_1) - R_1 N_1 + C_1 , \tag{7.3a}$$

$$\frac{dN_2}{dt} = B_{12}\rho_\nu(N_1 - N_2) - R_2 N_2 + C_2 , \tag{7.3b}$$

where $R_i N_i$ represents the total relaxation rate (including spontaneous emission) that depopulates the level $|i\rangle$ and

$$C_i = \sum_k R_{ki} N_k + D_i , \tag{7.3c}$$

takes care of all relaxation paths from the other levels $|k\rangle$ that contribute to the repopulation of the level $|i\rangle$ and also of the diffusion rate D_i of molecules

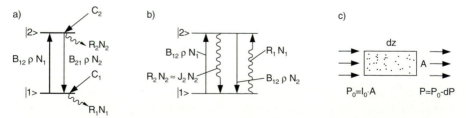

Fig. 7.1a–c. Level diagram of an open two-level system with open relaxation channels into other levels and population paths from outside the system (**a**), a closed system (**b**), and schematic illustration of absorption (**c**)

in level $|i\rangle$ into the excitation volume dV. We call the system described by (7.3a, 7.3b) an *open two-level system* because optical pumping occurs only between the two levels $|1\rangle$ and $|2\rangle$, which, however, may decay into other levels. That is, channels are open for transitions out of the system and for the population of the system from outside.

If the quantities C_i are not noticeably changed by the radiation field, we obtain from (7.3) under stationary conditions $(dN/dt = 0)$ the unsaturated population difference for $\rho = 0$

$$\Delta N^0 = \Delta N(\rho = 0) = N_2^0 - N_1^0 = \frac{C_2 R_1 - C_1 R_2}{R_1 R_2} , \tag{7.4}$$

and for the saturated population difference $(\rho \neq 0)$:

$$\Delta N = \frac{\Delta N^0}{1 + B_{12}\rho_v(1/R_1 + 1/R_2)} = \frac{\Delta N^0}{1 + S} , \tag{7.5}$$

where the *saturation parameter*

$$S = \frac{B_{12}\rho_v}{R^*} = \frac{B_{12}I_v}{R^* \cdot C} , \quad \text{with} \quad R^* = \frac{R_1 R_2}{R_1 + R_2} , \tag{7.6}$$

gives the ratio of the induced transition probability $B_{12}\rho$ to the "mean" relaxation probability R^*.

The power decrease of the incident light wave from absorption along the length dz of the path is, according to (7.1 and 7.5)

$$dP = -A \cdot I \cdot \sigma_{12} \frac{\Delta N^0}{1 + S} \cdot \frac{\delta v_a}{\Delta v_L} \, dz . \tag{7.7}$$

The intensity $I = I_s$ at which the saturation parameter S becomes $S = 1$ is called the *saturation intensity*. From (7.5) we derive that for $S = 1$ the population-density difference ΔN decreases to one half of its unsaturated value ΔN^0. The saturation power P_s is $P_s = I_s A$, where A is the cross section of the laser beam at the absorbing molecular sample.

In case of incoherent light sources, such as spectral lamps, the intensity I_v is so small that $S \ll 1$. We can then approximate (7.7) by

$$dP = -P\sigma_{12}\Delta N^0 \frac{\delta v_a}{\Delta v_L} \, dz , \tag{7.8}$$

where the unsaturated population difference as given by (7.4) is independent of the intensity I, and the absorbed power is proportional to the incident power (linear absorption), i.e., the relative absorbed power dP/P is constant. Integration of (7.8) gives

$$P = P_0 \exp(-\sigma_{12}\Delta N^0 z) = P_0 e^{-\alpha z} . \tag{7.9}$$

This is the *Lambert–Beer law* of linear absorption.

With lasers incident intensities I_ν are achievable, where $S \ll 1$ may be no longer valid and (7.7) instead of (7.8) has to be used. Because of the decreasing population difference the absorbed power increases less than linearly with increasing incident intensity. The decreasing relative absorption dP/P with increasing intensity I can readily be demonstrated when the absorbed power as a function of the incident intensity is measured via the laser-induced fluorescence. If the depopulation of the absorbing level $|1\rangle$ by absorption of laser photons becomes a noticeable fraction of the repopulation rate (Fig. 7.2), the population N_1 decreases and the laser-induced fluorescence intensity increases less than linearly with the incident laser intensity I_L.

Using the relations (2.30, 2.78 and 2.90) with the homogeneous spectral width $\delta\omega_a = 2\pi\delta\nu_a = \gamma = \gamma_1 + \gamma_2 = R_1 + R_2$ (see Sect. 3.1.2) and the Rabi frequency Ω_{12} we obtain from (7.6)

$$S = \frac{\Omega_{12}^2}{R^*\gamma} = \frac{\Omega_{12}^2}{\gamma_1\gamma_2}, \quad \text{for} \quad R_1 = \gamma_1, R_2 = \gamma_2. \tag{7.10}$$

This demonstrates that the saturation parameter can also be expressed as the square of the ratio $\Omega/\sqrt{\gamma_1\gamma_2}$ of the Rabi frequency Ω_{12} at resonance ($\omega = \omega_{12}$) and the geometric mean of the relaxation rates of $|1\rangle$ and $|2\rangle$. In other words, when the atoms are exposed to light with intensity $I = I_s$, their Rabi frequency is $\Omega_{12} = \sqrt{\gamma_1 \cdot \gamma_2}$. The saturation parameter defined by (7.6) for the open two-level system is more general than that defined in (3.69) for a closed two-level system. The difference lies in the definition of the mean relaxation probability, which is $R = (R_1 + R_2)/2$ in the closed system but $R^* = R_1 R_2/(R_1 + R_2)$ in the open system. We can close our open system defined by the rate equations (7.3) by setting $C_1 = R_2 N_2$, $C_2 = R_1 N_1$, and $N_1 + N_2 = N = \text{const.}$ (see Fig. 7.1b). The rate equations then become identical to (3.66) and R^* converts to R.

Regarding the saturation of the absorbing-level population N_1, there exists an important difference between a closed and an open two-level system. For the closed two-level system the stationary population density N_1 of the absorbing level $|1\rangle$ is, according to (7.3) and (3.67) with $C_1 = R_2 N_2$,

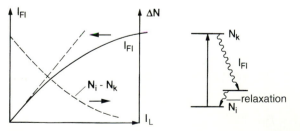

Fig. 7.2. Laser-induced fluorescence power and population difference ΔN as a function of laser intensity I_L

$C_2 = R_1 N_1, \; N_1 + N_2 = N = \text{const.}$:

$$N_1 = \frac{B_{12} I/c + R_2}{2 B_{12} I_\nu/c + R_1 + R_2} N \,, \quad \text{with} \quad N = N_1 + N_2 \,. \tag{7.11a}$$

N_1 can never drop below $N/2$ because

$$N_1 \geq \lim_{I \to \infty} N_1 = N/2 \to N_1 \geq N_2 \,.$$

For our open system, however, we obtain from (7.3)

$$N_1 = \frac{(C_1 + C_2) B_{12} I_\nu/c + R_2 C_1}{(R_1 + R_2) B_{12} I_\nu/c + R_1 R_2} \,. \tag{7.11b}$$

For large intensities I_ν ($S \gg 1$), the population density $N_1(I)$ approaches the limit

$$N_1(S \to \infty) = \frac{C_1 + C_2}{R_1 + R_2} \,. \tag{7.11c}$$

If the repopulation rates C_1, C_2 are small compared to the depopulation rates R_1, R_2 the saturated population density N_1 may become quite small.

This case applies, for instance, to the saturation of molecular transitions in a molecular beam, where collisions are generally negligible. The excited level $|2\rangle$ decays by spontaneous emission at the rate $N_2 A_2$ into many other rotational–vibrational levels $|m\rangle \neq |1\rangle$, and only a small fraction $N_2 A_{21}$ comes back to the level $|1\rangle$. The only repopulation mechanisms of level $|1\rangle$ are the diffusion of molecules into the excitation volume and the radiative decay rate $N_2 A_{21}$. For $E_2 \gg kT$, the upper level $|2\rangle$ can be populated only by optical pumping. If $|1\rangle$ is the ground state, its "lifetime" is given by the transit time $T = d/v$ through the excitation region of length d. We therefore have to replace in (7.3): $C_1 = D_1 + N_2 A_{21}$, $D_2 = 0$, $R_1 = 1/T$, $R_2 = A_2 + 1/T$, and find

$$N_1 = \frac{D_1 (B_{12} \rho + A_2 + 1/T)}{B_{12} \rho (A_2 - A_{21} + 2/T) + 1/T^2} \,. \tag{7.12a}$$

Without the laser excitation ($\rho = 0$, $A_2 = A_{21} = 0$), we obtain $N_1^0 = D_1 T$. This is the stationary population due to the diffusion of molecules into the excitation region. For large laser intensities, (7.12a) yields

$$\lim_{I \to \infty} N_1 = \frac{D_1}{A_2 - A_{21} + 2/T} \,. \tag{7.12b}$$

Example 7.1

For $d = 1\,\text{mm}$, $v = 5 \times 10^4\,\text{cm/s} \rightarrow T = 2 \times 10^{-6}\,\text{s}$. With $D_1 = 10^{14}\,\text{s}^{-1}$ $\cdot\text{cm}^{-3}$ we have the stationary population density $N_1^0 = 2 \times 10^8\,\text{cm}^{-3}$. With typical figures of $A_2 = 10^8\,\text{s}^{-1}$, $A_{21} = 10^7\,\text{s}^{-1}$ we obtain for the completely saturated population density $N_1 \approx 10^6\,\text{cm}^{-3}$. The saturated population N_1 has therefore decreased to 0.5% of its unsaturated value N_1^0.

We shall now briefly illustrate for two different situations at which intensities the saturation becomes noticeable:

(a) The bandwidth Δv_L of the cw laser is larger than the spectral width δv_a of the absorbing transition. In this case, we get the same results both for homogeneous and inhomogeneous line broadening. The saturation intensity $I_s = c\rho_s(v)\Delta v_L$ is then according to (7.6)

$$I_s \approx \frac{R^* c}{B_{12}} \Delta v_L \quad [\text{W/m}^2]\,. \tag{7.13a}$$

Example 7.2

Saturation of a molecular transition in a molecular beam by a broadband cw laser with $\Delta v_L = 3 \times 10^9\,\text{s}^{-1}$ $(\hat{=} 0.1\,\text{cm}^{-1})$:

With $R_1 = 1/T$, $R_2 = A_2 + 1/T$, we obtain with $B_{12} = (c^3/8\pi h v^3)A_{21}$ the saturation intensity

$$I_s(\Delta v_L) = \frac{(A_2 + 1/T)8\pi h v^3 \cdot \Delta v_L}{(TA_2 + 2)A_{21}c^2} \quad [\text{W/m}^2]\,. \tag{7.13b}$$

With the values of our previous example $A_2 = 10^8\,\text{s}^{-1}$, $T = 2 \times 10^{-6}\,\text{s}$, $v = 5 \times 10^{14}\,\text{s}^{-1}$, $A_{21} = 10^7\,\text{s}^{-1}$, and $\Delta v_L = 3 \times 10^9\,\text{s}^{-1}$, which gives

$$I_s \approx 3 \times 10^3\,\text{W/m}^2\,.$$

If the laser beam is focused to a cross section of $A = 1\,\text{mm}^2$, a laser power of 3 mW is sufficient for our example to reach the value $S = 1$ of the saturation parameter.

If the laser bandwidth Δv_L matches the homogeneous width $\gamma/2\pi$ of the absorption line, we find with $\gamma = A_2 + 2/T$ from (7.13b) the saturation intensity

$$I_s = \frac{4h v^3}{TA_{21}c^2}(A_2 + 1/T) \approx 100\,\text{W/m}^2\,. \tag{7.13c}$$

(b) The second case deals with a cw single-mode laser with frequency $\nu = \nu_0$ tuned to the center frequency ν_0 of a homogeneously broadened atomic resonance transition. If spontaneous emission into the ground state is the only relaxation process of the upper level, the relaxation rate is $R^* = A_{21}/2$ for $S = 1$, the saturation-broadened linewidth is $\delta\nu_a = \sqrt{2}A_{21}/2\pi$, and the saturation intensity is, according to (7.6) and (2.22)

$$I_s = c\rho_s \delta\nu_a = \frac{cR^* A_{21}}{\sqrt{2}\pi B_{12}} = \frac{2\sqrt{2}h\nu A_{21}}{\lambda^2} . \tag{7.14}$$

The same result could have been obtained from (3.71) $I_s = h\nu A_{21}/2\sigma_{12}$ with

$$\int \sigma_{12}\,d\nu \sim \sigma(\nu_0)\Delta\nu_a = (h\nu/c)B_{12} = (c^2/8\pi\nu^2)A_{21} , \tag{7.15a}$$

which yields without saturation broadening

$$\sigma(\nu_0) \sim c^2/4\nu^2 = (\lambda/2)^2 , \quad \text{and} \quad I_s = \frac{2h\nu A_{21}}{\lambda^2} . \tag{7.15b}$$

For $A_{21} = 10^8\,\text{s}^{-1}$ the relaxation due to diffusion out of the excitation volume can be neglected. At sufficiently low pressures the collision-induced transition probability is small compared to A_{21}.

Example 7.3
$\lambda = 500\,\text{nm} \rightarrow \nu = 6 \times 10^{14}\,\text{s}^{-1}$, $A_{21} = 10^8\,\text{s}^{-1} \rightarrow I_s \approx 380\,\text{W/m}^2$. Focusing the beam to a focal area of $1\,\text{mm}^2$ means a saturation power of only $265\,\mu\text{W}$! With the value $A_{21} = 10^7\,\text{s}^{-1}$ of Example 7.2, the saturation intensity drops to $38\,\text{W/m}^2$.

If collision broadening is essential, the linewidth increases and the saturation intensity increases roughly proportionally to the homogeneous linewidth. If pulsed lasers are used, the saturation peak powers are much higher because the system does generally not reach the stationary conditions. For the optical pumping time the laser pulse duration T_L often is the limiting time interval. Only for long laser pulses (e.g., copper-vapor laser-pumped dye lasers) the transit time of the molecules through the laser beam may be shorter than T_L.

7.2 Saturation of Inhomogeneous Line Profiles

In Sect. 3.6 we saw that the saturation of *homogeneously broadened* transitions with Lorentzian line profiles results again in a Lorentzian profile with the halfwidth

$$\Delta\omega_s = \Delta\omega_0\sqrt{1 + S_0} , \quad S_0 = S(\omega_0) , \tag{7.16}$$

which is increased by the factor $(1+S_0)^{1/2}$ compared to the unsaturated halfwidth $\Delta\omega_0$. The saturation broadening is due to the fact that the absorption coefficient

$$\alpha(\omega) = \frac{\alpha_0(\omega)}{1+S(\omega)}$$

decreases by the factor $[1+S(\omega)]^{-1}$, whereas the saturation parameter $S(\omega)$ itself has a Lorentzian line profile and the saturation is stronger at the line center than in the line wings (Fig. 3.24).

We will now discuss saturation of inhomogeneous line profiles. As an example we treat Doppler-broadened transitions, which represent the most important case in saturation spectroscopy.

7.2.1 Hole Burning

When a monochromatic light wave

$$E = E_0 \cos(\omega t - kz), \quad \text{with} \quad k = k_z,$$

passes through a gaseous sample of molecules with a Maxwell–Boltzmann velocity distribution, only those molecules that move with such velocities $\mathbf{v}$ that the Doppler-shifted laser frequency in the frame of the moving molecule $\omega' = \omega - \mathbf{k} \cdot \mathbf{v}$ with $\mathbf{k} \cdot \mathbf{v} = kv_z$ falls within the homogeneous linewidth γ around the center absorption frequency ω_0 of a molecule at rest, i.e., $\omega' = \omega_0 \pm \gamma$, can significantly contribute to the absorption. The absorption cross section for a molecule with the velocity component v_z on a transition $|1\rangle \rightarrow |2\rangle$ is

$$\sigma_{12}(\omega, v_z) = \sigma_0 \frac{(\gamma/2)^2}{(\omega - \omega_0 - kv_z)^2 + (\gamma/2)^2}, \tag{7.17}$$

where $\sigma_0 = \sigma(\omega = \omega_0 + kv_z)$ is the maximum absorption cross section at the line center of the molecular transition.

Due to saturation, the population density $N_1(v_z)\,dv_z$ decreases within the velocity interval $dv_z = \gamma/k$, while the population density $N_2(v_z)\,dv_z$ of the upper level $|2\rangle$ increases correspondingly (Fig. 7.3a). From (7.5) and (3.76) we obtain

$$N_1(\omega, v_z) = N_1^0(v_z) - \frac{\Delta N^0}{\gamma_1 \tau}\left[\frac{S_0(\gamma/2)^2}{(\omega - \omega_0 - kv_z)^2 + (\gamma_s/2)^2}\right], \tag{7.18a}$$

$$N_2(\omega, v_z) = N_2^0(v_z) + \frac{\Delta N^0}{\gamma_2 \tau}\left[\frac{S_0(\gamma/2)^2}{(\omega - \omega_0 - kv_z)^2 + (\gamma_s/2)^2}\right], \tag{7.18b}$$

where $\gamma = \gamma_1 + \gamma_2$ denotes the homogeneous width of the transition. The quantity

$$\tau = \frac{1}{\gamma_1} + \frac{1}{\gamma_2} = \frac{\gamma}{\gamma_1 \cdot \gamma_2}, \tag{7.18c}$$

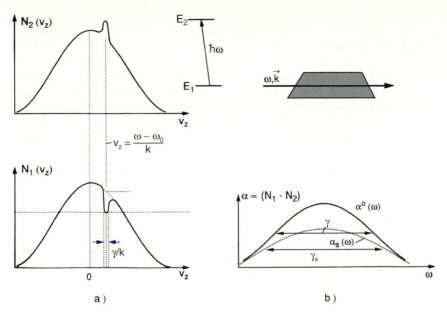

Fig. 7.3a,b. Velocity-selective saturation of a Doppler-broadened transition: (**a**) the Bennet hole in the lower- and the Bennet peak in the upper-state population distribution $N_i(v_z)$. (**b**) Saturated absorption profile when the saturating laser is tuned across the Doppler-profile of a molecular transition (*dashed curve*)

is called the *longitudinal relaxation time*, while

$$T = \frac{1}{\gamma_1 + \gamma_2} = \frac{1}{\gamma} \tag{7.18d}$$

is the *transverse relaxation time*.

According to (7.6) the saturation intensity can also be written as

$$I_s = \frac{c}{\tau \cdot B_{12}} \cdot \tag{7.18e}$$

Note that for $\gamma_1 \neq \gamma_2$ the depth of the hole in $N_1(v_z)$ and the heights of the peak in $N_2(v_z)$ are different. Subtracting (7.18b) from (7.18a) yields for the saturated population difference

$$\Delta N(\omega_s, v_z) = \Delta N^0(v_z)\left[1 - \frac{S_0(\gamma/2)^2}{(\omega - \omega_0 - kv_z)^2 + (\gamma_s/2)^2}\right] \cdot \tag{7.18f}$$

The velocity-selective minimum in the velocity distribution $\Delta N(v_z)$ at $v_z = (\omega - \omega_0)/k$, which is often called a *Bennet hole* [7.1], has the homogeneous width (according to Sect. 3.6)

$$\gamma_s = \gamma\sqrt{1 + S_0} \,,$$

and a depth at the hole center at $\omega = \omega_0 + kv_z$ of

$$\Delta N^0(v_z) - \Delta N(v_z) = \Delta N^0(v_z)\frac{S_0}{1 + S_0} \ . \tag{7.19}$$

For $S_0 = 1$ the hole depth amounts to 50% of the unsaturated population difference. Molecules with velocity components in the interval v_z to $v_z + dv_z$ give the contribution

$$\frac{d\alpha(\omega, v_z)}{dv_z}\, dv_z = \Delta N(v_z)\sigma(\omega, v_z)\, dv_z \ , \tag{7.20}$$

to the absorption coefficient $\alpha(\omega, v_z)$. The total absorption coefficient caused by all molecules in the absorbing level is then

$$\alpha(\omega) = \int \Delta N(v_z)\sigma_{12}(\omega, v_z)\, dv_z \ . \tag{7.21}$$

Inserting $\Delta N(v_z)$ from (7.18), $\sigma(\omega, v_z)$ from (7.17) and $\Delta N^0(v_z)$ from (3.40) yields

$$\alpha(\omega) = \frac{\Delta N^0 \sigma_0}{v_p\sqrt{\pi}} \int \frac{e^{-(v_z/v_p)^2}\, dv_z}{(\omega - \omega_0 - kv_z)^2 + (\gamma_s/2)^2} \ , \tag{7.22}$$

with the most probable velocity $v_p = (2k_B T/m)^{1/2}$ and the total unsaturated population difference $\Delta N^0 = \int \Delta N^0(v_z)\, dv_z$. Despite the saturation, one obtains again a Voigt profile for $\alpha(\omega)$, similarly to (3.46). The only difference is the saturation-broadened homogeneous linewidth γ_s in (7.22) instead of γ in (3.46).

Since for $S_0 < 1$ the Doppler width is generally large compared to the homogeneous width γ_s, the nominator in (7.22) at a given frequency ω does not vary much within the interval $\Delta v_z = \gamma_s/k$, where the integrand contributes significantly to $\alpha(\omega)$. We therefore can take the factor $\exp[-(v_z/v_p)^2]$ outside the integral. The residual integrals can be solved analytically and one obtains with $v_z = (\omega - \omega_0)/k$ and (3.44) the saturated absorption coefficient

$$\alpha_s(\omega) = \frac{\alpha^0(\omega_0)}{\sqrt{1 + S_0}} \exp\left\{-\left[\frac{\omega - \omega_0}{0.6\delta\omega_D}\right]^2\right\} \ , \tag{7.23}$$

with the unsaturated absorption coefficient

$$\alpha^0(\omega_0) = \Delta N_0 \frac{\sigma_0 \gamma c\sqrt{\pi}}{v_p \omega_0} \ , \quad \text{and the Doppler width} \quad \delta\omega_D = \frac{\omega_0}{c}\sqrt{\frac{8kT \ln 2}{m}} \ ,$$

Equation (7.23) illustrates a remarkable result: **Although at each frequency ω the monochromatic laser burns a Bennet hole into the velocity distribution $N_1(v_z)$, this hole cannot be detected just by tuning the laser through the absorption profile.** The absorption coefficient

$$\alpha(\omega) = \frac{\alpha^0(\omega)}{\sqrt{1 + S_0}} \ , \tag{7.24}$$

of the inhomogeneous profile still shows a Voigt profile *without any hole* but is reduced by the constant factor $(1 + S_0)^{-1/2}$, which is independent of ω (Fig. 7.3b).

Note: The difference to the homogeneous absorption profile where $\alpha(\omega)$ is reduced by the *frequency-dependent* factor $(1 + S(\omega))^{-1}$, see (3.79) and Fig. 3.24.

The Bennet hole can, however, be detected if *two lasers* are used:

- The saturating pump laser with the wave vector $\mathbf{k}_1$, which is kept at the frequency ω_1 and which burns a hole, according to (7.18), into the velocity class $v_z \pm \Delta v_z / 2$ with $v_z = (\omega_0 - \omega_1)/k_1$ and $\Delta v_z = \gamma / k_1$ (Fig. 7.4).
- A weak probe laser with the wave vector $\mathbf{k}_2$ and a frequency ω tunable across the Voigt profile. This probe laser is sufficiently weak to cause no extra saturation. The absorption coefficient for the tunable probe laser is then

$$\alpha_s(\omega_1, \omega_2) = \frac{\sigma_0 \Delta N^0}{v_p \sqrt{\pi}} \int \frac{e^{-(v_z/v_p)^2}}{(\omega_0 - \omega - k_2 v_z)^2 + (\gamma/2)^2}$$
$$\times \left[1 - \frac{S_0(\gamma/2)^2}{(\omega_0 - \omega - k_1 v_z)^2 + (\gamma_s/2)^2} \right] dv_z . \tag{7.25}$$

The integration over the velocity distribution yields analoguously to (7.23)

$$\alpha_s(\omega_1, \omega_2) = \alpha^0(\omega) \left[1 - \frac{S_0}{\sqrt{1 + S_0}} \frac{(\gamma/2)^2}{(\omega - \omega')^2 + (\Gamma_s/2)^2} \right] . \tag{7.26}$$

This is an unsaturated Doppler profile $\alpha^0(\omega)$ with a saturation dip at the probe frequency

$$\omega = \omega' = \omega_0 \pm (\omega_1 - \omega_0) k_1 / k_2 ,$$

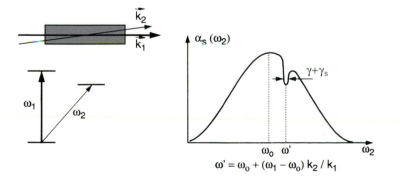

Fig. 7.4. Dip in the Doppler-broadened absorption profile $\alpha(\omega)$ burned by a strong pump laser at $\omega_p = \omega_1$ and detected by a weak probe laser at $\omega' = \omega_0^{\pm}(\omega_1 - \omega_0) k_2 / k_1$

where the $+$ sign holds for collinear and the $-$ sign for anticollinear propagation of the pump and probe waves.

The halfwidth $\Gamma_s = \gamma + \gamma_s = \gamma[1 + (1 + S_0)^{1/2}]$ of the absorption dip at $\omega = \omega'$ equals the sum of the saturated dip (due to the strong pump) and the unsaturated homogeneous absorption width γ of the weak probe wave. The depth of the dip at $\omega = \omega'$ is

$$\Delta\alpha(\omega') = \alpha^0(\omega') - \alpha_s(\omega') = \alpha^0(\omega')\frac{S_0}{\sqrt{1 + S_0}(1 + \sqrt{1 + S_0})} ,$$

$$\approx \frac{S_0}{2}\alpha^0(\omega') , \quad \text{for} \quad S_0 \ll 1 . \tag{7.26a}$$

Note: In the derivation of (7.26) we have only regarded population changes due to saturation effects. We have neglected coherence phenomena that may, for instance, result from interference between the two waves. These effects, which differ for copropagating waves from those of counterpropagating waves, have been treated in detail in [7.2–7.4].

For sufficiently small laser intensities ($S \ll 1$) they do not strongly affect the results derived above, but add finer details to the spectral structures obtained.

7.2.2 Lamb Dip

Pump and probe waves may also be generated by a single laser when the incident beam is reflected back into the absorption cell (Fig. 7.5).

The saturated population difference in such a case of equal intensities $I_1 = I_2 = I$ of the two counterpropagating waves with wavevectors $k_1 = -k_2$ is then

$$\Delta N(v_z) = \Delta N^0(v_z)$$

$$\times \left[1 - \frac{S_0(\gamma/2)^2}{(\omega_0 - \omega - kv_z)^2 + (\gamma_s/2)^2} - \frac{S_0(\gamma/2)^2}{(\omega_0 - \omega + kv_z)^2 + (\gamma_s/2)^2}\right] , \tag{7.27}$$

where $S_0 = S_0(I)$ is the saturation parameter due to one of the running waves. Because of the opposite Doppler shifts the two waves with frequency ω burn two Bennet holes at the velocity components $v_z = \pm(\omega_0 - \omega)/k$ into the population distribution $\Delta N(v_z)$ (Fig. 7.5b).

The saturated absorption coefficient then becomes

$$\alpha_s(\omega) = \int \Delta N(v_z)[\sigma(\omega_0 - \omega - kv_z) + \sigma(\omega_0 - \omega + kv_z)]dv_z . \tag{7.28}$$

Inserting (7.27 and 7.17) into (7.28) yields in the weak-field approximation ($S_0 \ll 1$) after some elaborate calculations [7.2] the saturated-absorption co-

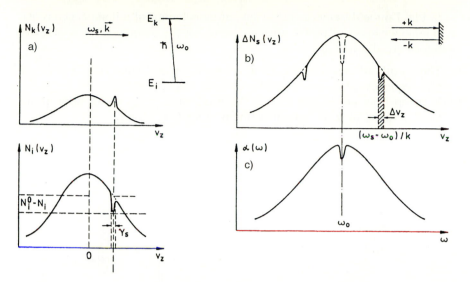

Fig. 7.5a–c. Saturation of an inhomogeneous line profile: (**a**) Bennet hole and dip produced by a monochromatic running wave with $\omega \neq \omega_0$; (**b**) Bennet holes caused by the two counterpropagating waves for $\omega \neq \omega_0$ and for $\omega = \omega_0$ (*dashed curve*); (**c**) Lamb dip in the absorption profile $\alpha_s(\omega)$

efficient for a sample in a standing-wave field

$$\alpha_s(\omega) = \alpha^0(\omega)\left[1 - \frac{S_0}{2}\left(1 + \frac{(\gamma_s/2)^2}{(\omega - \omega_0)^2 + (\gamma_s/2)^2}\right)\right], \tag{7.29a}$$

with

$$\gamma_s = \gamma\sqrt{1 + S_0}, \quad \text{and} \quad S_0 = S_0(I, \omega_0).$$

This represents the Doppler-broadened absorption profile $\alpha^0(\omega)$ with a dip at the line center $\omega = \omega_0$, which is called a *Lamb dip* after W.E. Lamb, who first explained it theoretically [7.5]. The depth of the Lamb dip is $S_0 = B_{ik}I/(c\gamma_s)$, $I = I_1 = I_2$ being the intensity of one of the counterpropagating waves that form the standing wave. For $\omega_0 - \omega \gg \gamma_s$ the saturated absorption coefficient becomes $\alpha_s = \alpha_0(1 - S_0/2)$, which corresponds to the saturation by one of the two waves.

The Lamb dip can be understood in a simple, conspicuous way: for $\omega \neq \omega_0$ the incident wave is absorbed by molecules with the velocity components $v_z = +(\omega - \omega_0 \mp \gamma_s/2)/k$, the reflected wave by other molecules with $v_z = -(\omega - \omega_0) \pm \gamma_s/2$. For $\omega = \omega_0$ both waves are absorbed by the same molecules with $v_z = (0 \pm \gamma_s/s)/k$, which essentially move perpendicularly to the laser beams. The intensity per molecule absorbed is now twice as large, and the saturation accordingly higher.

In Fig. 7.6 the differences in the saturation behavior of a homogeneous and an inhomogeneous line profile are illustrated. For the inhomogeneous case two situations are illustrated:

(a) The absorbing sample is placed in a standing-wave field ($I_1 = I_2 = I$) and the frequency ω is tuned over the line profiles (Fig. 7.6b).
(b) A pump laser ($I = I_1$) is kept at the line center at ω_0 and a weak probe laser is tuned over the saturated line profiles (Fig. 7.6c).

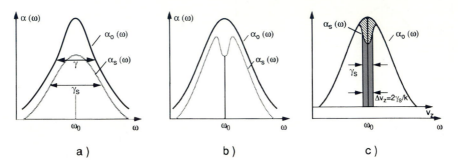

a)   b)   c)

Fig. 7.6a–c. Comparison of the saturation of a homogeneous absorption line profile (**a**) and an inhomogeneous profile (**b**) in a standing-wave field. (**c**) The saturating wave is kept at $\omega = \omega_0$ and a weak probe wave is tuned across the line profile

In the first case (Fig. 7.6b), the saturation of the inhomogeneous line profile is $S_0 = B_{ik} I/(c\gamma_s)$ at the center of the Lamb dip and $S_0/2$ for $(\omega - \omega_0) \gg \gamma_s$. In the second case (Fig. 7.6b), the depth of the Bennet hole is $S_0/2 = B_{ik} I_1/(c\gamma_s)$.

For strong laser fields the approximation $S_0 \ll 1$ no longer holds. Instead of (7.29) one obtains, neglecting coherent effects [7.2]:

$$\alpha_s(\omega) = \alpha^0(\omega) \frac{\gamma/2}{B\left[1 - \left(\dfrac{2(\omega - \omega_0)}{A+B}\right)^2\right]^{1/2}}, \qquad (7.29b)$$

with

$$A = [(\omega - \omega_0)^2 + (\gamma/2)^2]^{1/2} \quad \text{and} \quad B = [(\omega - \omega_0)^2 + (\gamma/2)^2(1+2S)]^{1/2}.$$

This yields $\alpha_s(\omega_0) = \alpha^0(\omega_0)/\sqrt{1+2S}$ at the line center $\omega = \omega_0$, and $\alpha_s(\omega) = \alpha^0(\omega)/\sqrt{1+S}$ for $(\omega - \omega_0) \gg \gamma$. The maximum depth of the Lamb dip is achieved when

$$\frac{\alpha(\omega - \omega_0 \gg \gamma_s) - \alpha(\omega_0)}{\alpha(\omega_0)} = \frac{1}{\sqrt{1+S_0}} - \frac{1}{\sqrt{1+2S_0}},$$

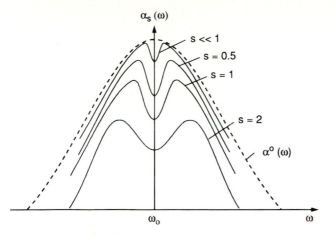

Fig. 7.7. Lamb dips for several values of the saturation parameter S_0

becomes maximum, which occurs for $S_0 \approx 1.4$. In Fig. 7.7 the saturated absorption profile is depicted for some values of S_0.

Note: The width of the Lamb dip in (7.29a) is $\delta\omega_{LD} = \gamma_s$. This corresponds, however, to the velocity interval $\Delta v_z = 2\gamma_s/k$ because the opposite Doppler shifts $\Delta\omega = (\omega_0 - \omega) = \pm kv_z$ of the two Bennet holes add when the laser frequency ω is tuned.

If the intensity of the reflected wave in Fig. 7.5 is very small ($I_2 \ll I_1$), we obtain instead of (7.29) a formula similar to (7.26). However, we must replace Γ_s by $\Gamma_s^* = (\gamma + \gamma_s)/2$ since the pump and probe waves are simultaneously tuned. For $S_0 \ll 1$ the result is

$$\alpha_s(\omega) = \alpha^0(\omega) \left[1 - \frac{S_0}{2} \frac{(\gamma_s/2)^2}{(\omega - \omega_0)^2 + (\Gamma_s^*/2)^2}\right] . \tag{7.30}$$

7.3 Saturation Spectroscopy

Saturation spectroscopy is based on the velocity-selective saturation of Doppler-broadened molecular transitions, treated in Sect. 7.2. Here the spectral resolution is no longer limited by the Doppler width but only by the much narrower width of the Lamb dip. The gain in spectral resolution is illustrated by the example of two transitions from a common lower level $|c\rangle$ to two closely spaced levels $|a\rangle$ and $|b\rangle$ (Fig. 7.8). Even when the Doppler profiles of the two transitions completely overlap, their narrow Lamb dips can clearly be separated, as long as $\Delta\omega = \omega_{ca} - \omega_{cb} > 2\gamma_s$.

Saturation spectroscopy is therefore often called *Lamb-dip spectroscopy*.

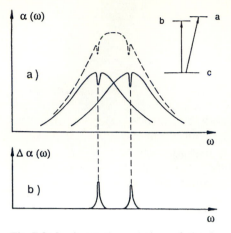

Fig. 7.8a,b. Spectral resolution of the Lamb dips of two transitions with overlapping Doppler profiles

7.3.1 Experimental Schemes

A possible experimental realization of saturation spectroscopy is illustrated in Fig. 7.9. The output beam from a tunable laser is split by the beam splitter BS into a strong pump beam with the intensity I_1 and a weak probe beam with intensity $I_2 \ll I_1$, which pass through the absorbing sample in opposite directions. When the transmitted probe-beam intensity $I_{t2}(\omega)$ is measured as a function of the laser frequency ω, the detection signal $DS(\omega) \propto I_2 - I_{t2}$ shows the Doppler-broadened absorption profiles with Lamb dips at their centers.

The Doppler-broadened background can be eliminated if the pump beam is chopped and the transmitted probe intensity is monitored through a lock-in amplifier that is tuned to the chopping frequency. According to (7.30), we ob-

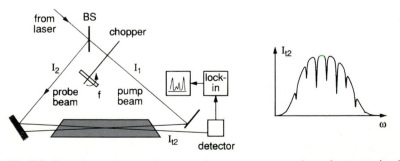

Fig. 7.9. Experimental setup for saturation spectroscopy where the transmitted probe intensity $I_2(\omega)$ is monitored

tain for a sufficiently weak probe intensity the Doppler-free absorption profile

$$\alpha_0 - \alpha_s = \frac{\alpha_0 S_0}{2} \frac{(\gamma_s/2)^2}{(\omega - \omega_0)^2 + (\Gamma_s^*/2)^2} \cdot \tag{7.31}$$

To enhance the sensitivity, the probe beam can again be split into two parts. One beam passes the region of the sample that is saturated by the pump beam, and the other passes the sample cell at an unsaturated region (Fig. 7.10). The difference between the two probe-beam outputs monitored with the detectors D1 and D2 yields the saturation signal if this difference has been set to zero with the pump beam off. The pump-beam intensity measured by D3 can be used for normalization of the saturation signal. Figure 7.11 shows as an example the saturation spectrum of a mixture of different cesium isotopes

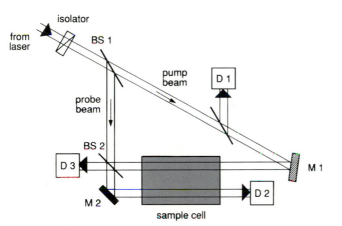

Fig. 7.10. Sensitive version of saturation spectroscopy. The probe is split by BS2 into two parallel beams, which pass through a pumped and an unpumped region, respectively. Pump and probe beams are strictly antiparallel. A Faraday isolator prevents feedback of the probe beam into the laser

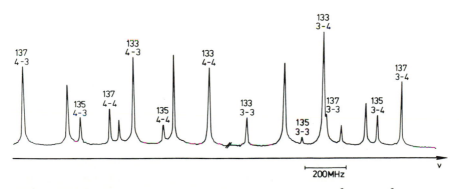

Fig. 7.11. Saturation spectrum of all hyperfine components of the $6^2S_{1/2} \rightarrow 7^2P$ transition at $\lambda = 459.3$ nm in a mixture of isotopes ^{133}Cs, ^{135}Cs, and ^{137}Cs [7.6]

contained in a glass cell heated to about 100°C [7.6]. The hyperfine structure and the isotope shifts of the different isotopes can be derived from these
measurements with high accuracy. The small crossing angle α [rad] of the
two beams in Fig. 7.9 results in a residual Doppler width $\delta\omega_1 = \Delta\omega_D\alpha$. If the
pump and probe beams are strictly anticollinear ($\alpha = 0$) the probe beam is
coupled back into the laser, resulting in laser instabilities. This can be prevented by an optical isolator, such as a Faraday isolator [7.7], which turns the
plane of polarization by 90° and suppresses the reflected beam by a polarizer
(Fig. 7.10).

Instead of measuring the *attenuation* of the probe beam, the absorption
can also be monitored by the laser-induced *fluorescence*, which is proportional
to the absorbed laser power. This technique is, in particular, advantageous
when the density of the absorbing molecules is low and the absorption accordingly small. The change in the attenuation of the probe beam is then difficult
to detect and the small Lamb dips may be nearly buried under the noise
of the Doppler-broadened background. Sorem and Schawlow [7.8, 7.9] have
demonstrated a very sensitive *intermodulated fluorescence technique*, where
the pump beam and the probe beam are chopped at two different frequencies f_1 and f_2 (Fig. 7.12a). Assume the intensities of the two beams to be
$I_1 = I_{10}(1 + \cos \Omega_1 t)$ and $I_2 = I_{20}(1 + \cos \Omega_2 t)$ with $\Omega_i = 2\pi f_i$. The intensity of the laser-induced fluorescence is then

$$I_{FL} = C\Delta N_s(I_1 + I_2) , \tag{7.32}$$

where ΔN_s is the difference between the saturated population densities of the
absorbing and upper levels, and the constant C includes the transition probability and the collection efficiency of the fluorescence detector. According to
(7.27), we obtain at the center of an absorption line

$$\Delta N_s = \Delta N^0[1 - a(I_1 + I_2)] .$$

Inserting this into (7.32) gives

$$I_{FL} = C[\Delta N^0(I_1 + I_2) - a\Delta N^0(I_1 + I_2)^2] . \tag{7.33}$$

The quadratic expression $(I_1 + I_2)^2$ contains the frequency-dependent term

$$I_{10}I_{20} \cos \Omega_1 t \cdot \cos \Omega_2 t = \frac{1}{2}I_{10}I_{20}[\cos(\Omega_1 + \Omega_2)t + \cos(\Omega_1 - \Omega_2)t] ,$$

which reveals that the fluorescence intensity contains linear terms, modulated at the chopping frequencies f_1 and f_2, respectively, and quadratic
terms with the modulation frequencies $(f_1 + f_2)$ and $(f_1 - f_2)$, respectively.
While the linear terms represent the normal laser-induced fluorescence with
a Doppler-broadened excitation line profile, the quadratic ones describe the
saturation effect because they depend on the decrease of the population density ΔN ($v_z = 0$) from the simultaneous interaction of the molecules with
both fields. When the fluorescence is monitored through a lock-in amplifier
tuned to the sum frequency $f_1 + f_2$, the linear background is suppressed and

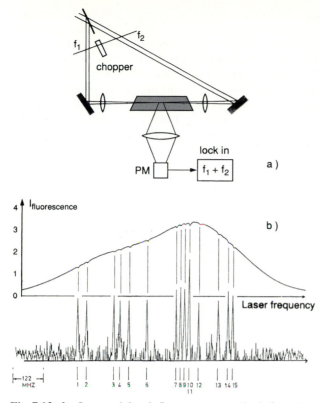

Fig. 7.12a,b. Intermodulated fluorescence method for saturation spectroscopy at small densities of the sample molecules: (**a**) experimental arrangement; (**b**) hyperfine spectrum of the $(v'' = 1, J'' = 98) \rightarrow (v' = 58, J' = 99)$ line in the $X^1\Sigma_g \rightarrow {}^3\Pi_{ou}$ system of I_2 at $\lambda = 514.5$ nm, monitored at the chopping frequency f_1 of the pump beam (*upper spectrum* with the Lamb dips) and at $(f_1 + f_2)$ (*lower spectrum*) [7.10]

only the saturation signals are detected. This is demonstrated by Fig. 7.12b, which shows the 15 hyperfine components of the rotational line $(v'' - 1, J'' = 98)$ $(v' = 58, J' = 99)$ in the $X^1\Sigma_g^+ \rightarrow B^3\Pi_{u0}$ transition of the iodine molecule I_2 [7.9, 7.10]. The two laser beams were chopped by a rotating disc with two rows of a different number of holes, which interrupted the beams at $f_1 = 600 \, \text{s}^{-1}$ and $f_2 = 900 \, \text{s}^{-1}$. The upper spectrum was monitored at the chopping frequency f_1 of the pump beam. The Doppler-broadened background caused by the linear terms in (7.33) and the Lamb dips both show a modulation at the frequency f_1 and are therefore recorded simultaneously. The center frequencies of the hyperfine structure (hfs) components, however, can be obtained more accurately from the intermodulated fluorescence spectrum (lower spectrum), which was monitored at the sum frequency $(f_1 + f_2)$ where the linear background is suppressed. This technique has found wide applications for sub-Doppler modulation spectroscopy of molecules and radicals at low pressures [7.11–7.13].

7.3.2 Cross-Over Signals

If two molecular transitions with a common lower or upper level overlap within their Doppler width, extra resonances, called *cross-over* signals, occur in the Lamb-dip spectrum. Their generation is explained in Fig. 7.13a.

Assume that for the center frequencies ω_1 and ω_2 of the two transitions $|\omega_1 - \omega_2| < \Delta\omega_D$ holds. At the laser frequency $\omega = (\omega_1 + \omega_2)/2$, the incident wave saturates the velocity class $(v_z \pm dv_z) = (\omega_2 - \omega_1)/2k \pm \gamma k$ on the transition 1 with center frequency ω_1, while the reflected wave saturates the same velocity class on the transition 2 with the center frequency ω_2. One therefore observes, besides the saturation signals at ω_1 and ω_2 (where the velocity class $v_z = 0$ is saturated), an additional saturation signal (cross-over) at $\omega = (\omega_1 + \omega_2)/2$ because one of the two waves causes a decrease $-\Delta N_1$ of the population density N_1 in the common lower level, which is probed

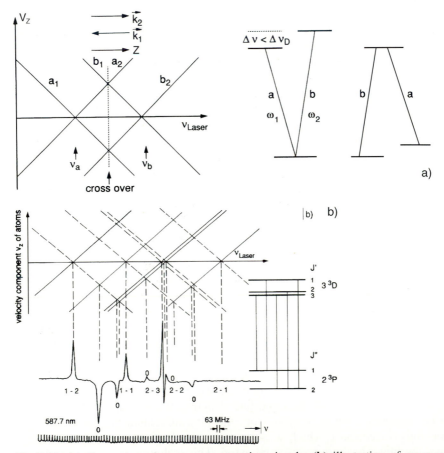

Fig. 7.13. (**a**) Generation of cross-over saturation signals; (**b**) illustration of cross-overs in the helium transition $3^3D \leftarrow 2^3P$. The cross-over signals are marked by 0 above the lines [7.14]

by the second wave on another transition. In case of a common upper level, both waves contribute at $\omega = (\omega_1 + \omega_2)/2$ to the increase ΔN_2 of the population N_2; one wave on transition a, the other on b. The sign of the cross-over signals is negative for a common lower level and positive for a common upper level. Their frequency position $\omega_c = (\omega_1 + \omega_2)/2$ is just at the center between the two transitions 1 and 2.

Although these cross-over signals increase the number of observed Lamb dips and may therefore increase the complexity of the spectrum, they have the great advantage that they allow one to assign pairs of transitions with a common level. This may also facilitate the assignment of the whole spectrum; see, for example, Fig. 7.13b and [7.4, 7.13–7.15].

7.3.3 Intracavity Saturation Spectroscopy

When the absorbing sample is placed inside the resonator of a tunable laser, the Lamb dip in the absorption coefficient $\alpha(\omega)$ causes a corresponding peak of the laser output power $P(\omega)$ (Fig. 7.14).

The power $P(\omega)$ depends on the spectral gain profile $G(\omega)$ and on the absorption profile $\alpha(\omega)$ of the intracavity sample, which is generally Doppler broadened. The Lamb peaks therefore sit on a broad background (Fig. 7.14a). With the center frequency ω_1 of the gain profile and an absorption Lamb dip

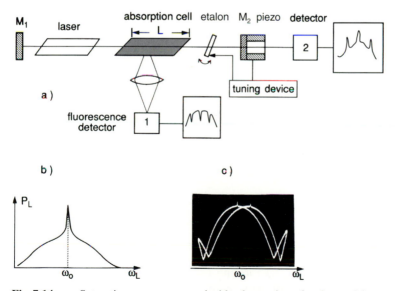

Fig. 7.14a–c. Saturation spectroscopy inside the cavity of a laser: (**a**) experimental arrangement; (**b**) output power $P(\omega)$; (**c**) experimental detection of the Lamb peak in the output power of a HeNe laser tunable around $\lambda = 3.39\,\mu m$, caused by the Lamb dip of a CH_4 transition in a methane cell inside the laser cavity [7.20]

at ω_0 we obtain, according to (5.58) and (7.29)

$$P_L(\omega) \propto \left\{ G(\omega - \omega_1) - \alpha^0(\omega) \left[1 - \frac{S_0}{2} \left(1 + \frac{(\gamma_s/2)^2}{(\omega - \omega_0)^2 + (\gamma_s/2)^2} \right) \right] \right\} .$$
(7.34)

In a small interval around ω_0 we may approximate the gain profile $G(\omega - \omega_1)$ and the unsaturated absorption profile $\alpha^0(\omega)$ by a quadratic function of ω, yielding for (7.34) the approximation

$$P_L(\omega) = A\omega^2 + B\omega + C + \frac{D}{(\omega - \omega_0)^2 + (\gamma_s/2)^2} ,$$
(7.35)

where the constants A, B, C, and D depend on ω_0, ω_1, γ, and S_0. The derivatives

$$P_L^{(n)}(\omega) = \frac{d^n P_L(\omega)}{d\omega^n} , \quad (n = 1, 2, 3, \dots),$$

of the laser output power with respect to ω are

$$P_L^{(1)}(\omega) = 2A\omega + B - \frac{2D(\omega - \omega_0)}{[(\omega - \omega_0)^2 + (\gamma_s/2)^2]^2} ,$$
$$P_L^{(2)}(\omega) = 2A + \frac{6D(\omega - \omega_0)^2 - 2D(\gamma_s/2)^2}{[(\omega - \omega_0)^2 + (\gamma_s/2)^2]^3} ,$$
(7.36)
$$P_L^{(3)}(\omega) = \frac{24D(\omega - \omega_0)[(\omega - \omega_0)^2 - (\gamma_s/2)^2]}{[(\omega - \omega_0)^2 + (\gamma_s/2)^2]^4} .$$

These derivatives are exhibited in Fig. 7.15, which illustrates that the broad background disappears for the higher derivatives. If the absorptive medium is the same as the gain medium, the Lamb peak appears at the center of the gain profile (Fig. 7.14b,c).

If the laser frequency ω is modulated at the frequency Ω, the laser output

$$P_L(\omega) = P_L(\omega_0 + a \sin \Omega t) ,$$

can be expanded into a Taylor series around ω_0. The derivation in Sect. 6.2.1 shows that the output $P_L(\omega, 3\Omega)$ measured with a lock-in device at the frequency 3Ω is proportional to the third derivative, see (6.11).

The experimental performance is depicted in Fig. 7.16. The modulation frequency $\Omega = 2\pi f$ is tripled by forming rectangular pulses, where the third harmonic is filtered and fed into the reference input of a lock-in amplifier that is tuned to 3Ω. Figure 7.17 illustrates this technique by the third-derivative spectrum of the same hfs components of I_2 as obtained with the intermodulated fluorescence technique in Fig. 7.12.

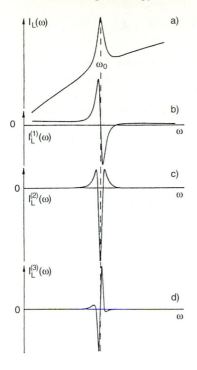

Fig. 7.15a–d. Lamb peak at the slope of the Doppler-broadened gain profile and the first three derivatives, illustrating the suppression of the Doppler background

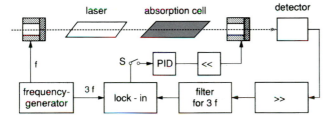

Fig. 7.16. Schematic arrangement for third-derivative intracavity saturation spectroscopy

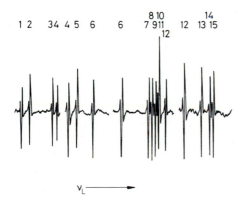

Fig. 7.17. The third-derivative intracavity absorption spectrum of I_2 around $\lambda = 514$ nm, showing the same hfs components as Fig. 7.12 [7.15]

7.3.4 Lamb-Dip Frequency Stabilization of Lasers

The steep zero crossing of the third derivative of narrow Lamb dips gives
a good reference for accurate stabilization of the laser frequency onto an
atomic or molecular transition. Either the Lamb dip in the gain profile of the
laser transition or Lamb dips of absorption lines of an intracavity sample can
be used.

In the infrared spectral range, Lamb dips of a vibration–rotation transition of CH_4 at $\lambda = 3.39\,\mu m$ or of CO_2 around $10\,\mu m$ are commonly used
for frequency stabilization of the HeNe laser at $3.39\,\mu m$ or the CO_2 laser. In
the visible range various hyperfine components of rotational lines within the
$^1\Sigma_g \rightarrow {}^3\Pi_{ou}$ system of the I_2 molecule are mainly chosen. The experimental
setup is the same as that shown in Fig. 7.16. The laser is tuned to the wanted
hfs component and then the feedback switch S is closed in order to lock the
laser onto the zero crossing of this component [7.16].

Using a double servo loop for fast stabilization of the laser frequency
onto the transmission peak of a Fabry–Perot Interferometer (FPI) and a slow
loop to stabilize the FPI onto the first derivative of a forbidden narrow calcium transition, Barger et al. constructed an ultrastable cw dye laser with
a short-term linewidth of approximately $800\,Hz$ and a long-term drift of
less than $2\,kHz/h$ [7.17]. Stabilities of better than $1\,Hz$ have also been realized [7.18, 7.19].

This extremely high stability can be transferred to tunable lasers by a special *frequency-offset locking* technique (Sect. 6.9) [7.20]. Its basic principle is
illustrated in Fig. 7.18. A reference laser is frequency stabilized onto the Lamb
dip of a molecular transition. The output from a second, more powerful laser
at the frequency ω is mixed in detector D1 with the output from the reference
laser at the frequency ω_0. An electronic device compares the difference frequency $\omega_0 - \omega$ with the frequency ω' of a stable but tunable RF oscillator, and
controls the piezo P_2 such that $\omega_0 - \omega = \omega'$ at all times. The frequency ω of

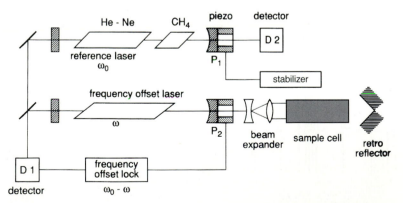

Fig. 7.18. Schematic diagram of a frequency-offset locked laser spectrometer

the powerful laser is therefore always locked to the *offset frequency* $\omega_0 - \omega'$, which can be controlled by tuning the RF frequency ω'.

The output beam of the powerful laser is expanded before it is sent through the sample cell in order to minimize transit-time broadening (Sect. 3.4). A retroreflector provides the counterpropagating probe wave for Lamb-dip spectroscopy. The real experimental setup is somewhat more complicated. A third laser is used to eliminate the troublesome region near the zero-offset frequency. Furthermore, optical decoupling elements have to be inserted to avoid optical feedback between the three lasers. A detailed description of the whole system can be found in [7.21].

An outstanding example of the amount of information on interactions in a large molecule that can be extracted from a high-resolution spectrum is represented by the work of Bordé et al. on saturation spectroscopy of SF_6 [7.22]. Many details of the various interactions, such as spin-rotation coupling, Coriolis coupling, and hyperfine structure, which are completely masked at lower resolution, can be unravelled when sufficiently high resolution is achieved. For illustration Fig. 7.19 depicts a section of the saturation spectrum of SF_6 taken by this group.

Meanwhile, many complex molecular spectra have been resolved by saturation spectroscopy. One example are the ultranarrow overtone transitions of the acetylene isotopomer $^{13}C_2H_2$ [7.23].

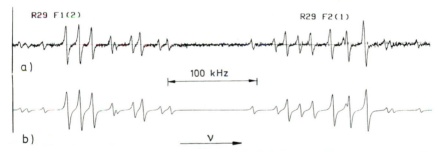

Fig. 7.19a,b. Hyperfine and super-hyperfine structures of a rotational–vibrational transition in SF_6, showing the molecular transitions and cross-over signals: (**a**) experimental spectrum; (**b**) calculated spectrum [7.22]

7.4 Polarization Spectroscopy

While saturation spectroscopy monitors the decrease of *absorption* of a probe beam caused by a pump wave that has selectively depleted the absorbing level, the signals in polarization spectroscopy come mainly from the change of the *polarization state* of the probe wave induced by a *polarized* pump wave. Because of optical pumping, the pump wave causes a change of refractive index n and absorption coefficient α.

This very sensitive Doppler-free spectroscopic technique has many advantages over conventional saturation spectroscopy and will certainly gain increasing attention [7.24, 7.25]. We therefore discuss the basic principle and some of its experimental modifications in more detail.

7.4.1 Basic Principle

The basic idea of polarization spectroscopy can be understood in a simple way (Fig. 7.20). The output from a monochromatic tunable laser is split into a weak probe beam with the intensity I_1 and a stronger pump beam with the intensity I_2. The probe beam passes through a linear polarizer P_1, the sample cell, and a second linear polarizer P_2, which is crossed with P_1. Without the pump laser, the sample is isotropic and the detector D behind P_2 receives only a very small signal caused by the residual transmission of the crossed polarizer, which might be as small as $10^{-8} I_1$.

After having passed through a $\lambda/4$-plate, which produces a circular polarization, the pump beam travels in the opposite direction through the sample cell. When the laser frequency ω is tuned to a molecular transition $(J'', M'') \to (J', M')$, molecules in the lower level (J'', M'') can absorb the pump wave. The quantum number M that describes the projection of J onto the direction of light propagation follows the selection rule $\Delta M = +1$ for the transitions $M'' \to M'$ induced by σ^+ circularly polarized

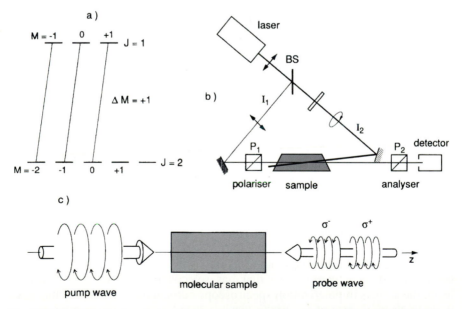

Fig. 7.20a–c. Polarization spectroscopy: (**a**) level scheme for a P transition $J = 2 \to J = 1$; (**b**) experimental setup; (**c**) linearly polarized probe wave as superposition of σ^+ (angular momentum $+5$ in z-direction) and σ^- components

light ($M'' \rightarrow M' = M'' + 1$). The degenerate M'' sublevels of the rotational level J'' become partially or completely depleted because of saturation. The degree of depletion depends on the pump intensity I_2, the absorption cross section $\sigma(J'', M'' \rightarrow J', M')$, and on possible relaxation processes that may repopulate the level (J'', M''). The cross section σ depends on J'', M'', J', and M'. From Fig. 7.20a it can be seen that in the case of P or R transitions $\Delta J = -1$ or $+1$, not all of the M sublevels are pumped. For example, from levels with $M'' = +J''$ no P transitions with $\Delta M = +1$ are possible, while for R transitions the levels $M' = -J'$ are not populated. *This implies that the pumping process produces an unequal saturation and with it a nonuniform population of the M sublevels*, which is equivalent to an anisotropic distribution for the orientations of the angular momentum vector J.

Such an anisotropic sample becomes birefringent for the incident linearly polarized probe beam. Its plane of polarization is slightly rotated after having passed the anisotropic sample. This effect is quite analogous to the Faraday effect, where the nonisotropic orientation of J is caused by an external magnetic field. For polarization spectroscopy no mangetic field is needed. Contrary to the Faraday effect where all molecuels are oriented, here only those molecules that interact with the monochromatic pump wave show this nonisotropic orientation. As has already been discussed in Sect. 7.2, this is the subgroup of molecules with the velocity components

$$v_z \pm \Delta v_z = (\omega_0 - \omega)/k \pm \gamma/k \; ,$$

where Δv_z is determined by the homogeneous linewidth $\delta\omega = \gamma$.

For $\omega \neq \omega_0$ the probe wave that passes in the opposite direction through the sample interacts with a *different group* of molecules in the velocity interval $v_z \pm \Delta v_z = -(\omega_0 - \omega \pm \delta\omega)/k$, and will therefore not be influenced by the pump wave. If, however, the laser frequency ω coincides with the center frequency ω_0 of the molecular transition within its homogeneous linewidth $\delta\omega$ (i.e., $\omega = \omega_0 \pm \delta\omega \rightarrow v_z = 0 \pm \Delta v_z$), both waves can be absorbed by the same molecules and the probe wave experiences a birefringence from the nonisotropic M distribution of the molecules in the absorbing lower rotational level J'' or in the upper level J'.

Only in this case will the plane of polarization of the probe wave be slightly rotated by $\Delta\theta$, and the detector D will receive a Doppler-free signal every time the laser frequency ω is tuned across the center of a molecular absorption line.

7.4.2 Line Profiles of Polarization Signals

Let us now discuss the generation of this signal in a more quantitative way. The linearly polarized probe wave

$$\boldsymbol{E} = \boldsymbol{E}_0 \, e^{i(\omega t - kz)} \; , \quad \boldsymbol{E}_0 = \{E_0, 0, 0\} \; ,$$

can always be composed of a σ^+ and a σ^- circularly polarized component (Fig. 7.20c)

$$E^+ = E_0^+ \, e^{i(\omega t - k^+ z)}, \qquad E_0^+ = \frac{1}{2} E_0(\hat{x} + i\hat{y}), \tag{7.37a}$$

$$E^- = E_0^- \, e^{i(\omega t - k^- z)}, \qquad E_0^- = \frac{1}{2} E_0(\hat{x} - i\hat{y}), \tag{7.37b}$$

where $\hat{x}$ and $\hat{y}$ are unit vectors in the x- and y-direction, respectively. While passing through the sample the two components experience different absorption coefficients α^+ and α^- and different refractive indices n^+ and n^- from the nonisotropic saturation caused by the σ^+-polarized pump wave. After a path length L through the pumped region of the sample, the two components are

$$E^+ = E_0^+ \, e^{i[\omega t - k^+ L + i(\alpha^+/2)L]},$$
$$E^- = E_0^- \, e^{i[\omega t - k^- L + i(\alpha^-/2)L]}. \tag{7.38}$$

Because of the differences $\Delta n = n^+ - n^-$ and $\Delta \alpha = \alpha^+ - \alpha^-$ caused by the nonisotropic saturation, a phase difference

$$\Delta \phi = (k^+ - k^-)L = (\omega L/c)(n^+ - n^-),$$

has developed between the two components, as has a small amplitude difference

$$\Delta E = \frac{E_0}{2} \left[e^{-(\alpha^+/2)L} - e^{-(\alpha^-/2)L} \right].$$

The windows of the absorption cell with thickness d also show a small absorption and a pressure-induced birefringence from the atmospheric pressure on one side and vacuum on the other side. Their index of refraction n_w and their absorption α_w can be expressed by the complex quantity

$$n_w^{*\pm} = b_r^\pm + i b_i^\pm, \quad \text{with} \quad n_w = b_r, \quad \text{and} \quad \alpha_w = 2k \cdot b_i \cdot 2d.$$

Behind the exit window the superposition of the σ^+ and σ^- components of the linearly polarized probe wave traveling into the z-direction yields the elliptically polarized wave

$$E(z = L) = E^+ + E^- \tag{7.39}$$
$$= \frac{1}{2} E_0 \, e^{i\omega t} \, e^{-i[\omega(nL + b_r)/c - i\alpha L/2 - i\alpha_w/2]} \left[(\hat{x} + i\hat{y}) \, e^{-i\Delta} + (\hat{x} - i\hat{y}) \, e^{+i\Delta} \right],$$

where $\alpha_w = (2\omega/c)b_i$ is the window absorption and

$$n = \frac{1}{2}(n^+ + n^-); \quad \alpha = \frac{1}{2}(\alpha^+ + \alpha^-); \quad b = \frac{1}{2}(b^+ + b^-)$$

are the average quantities. The phase factor

$$\Delta = \omega(L\Delta n + \Delta b_{\mathrm{r}})/2c - \mathrm{i}(L\Delta\alpha/4 + \Delta\alpha_{\mathrm{w}}/2) ,$$

depends on the differences

$$\Delta n = n^+ - n^- , \quad \Delta\alpha = \alpha^+ - \alpha^- , \quad \Delta b = b^+ - b^- .$$

If the transmission axis of the analyzer P_2 is tilted by the small angle $\theta \ll 1$ against the y-axis (Fig. 7.21) the transmitted amplitude becomes

$$E_{\mathrm{t}} = E_x \sin\theta + E_y \cos\theta .$$

For most practical cases, the differences $\Delta\alpha$ and Δn caused by the pump wave are very small. Also, the birefringence of the cell windows can be minimized (for example, by compensating the air pressure by a mechanical pressure onto the edges of the windows). With

$$L\Delta\alpha \ll 1 , \quad L\Delta k \ll 1 , \quad \text{and} \quad \Delta b \ll 1 ,$$

we can expand $\exp(\mathrm{i}\Delta)$ in (7.39). We then obtain for small angles $\theta \ll 1$ ($\cos\theta \approx 1$, $\sin\theta \approx \theta$) the transmitted amplitude

$$E_{\mathrm{t}} = E_0 \, \mathrm{e}^{\mathrm{i}\omega t} \exp[\mathrm{i}\omega(nL + b_{\mathrm{r}})/c - \frac{\mathrm{i}}{2}(\alpha \cdot L + \alpha_{\mathrm{w}})](\theta + \Delta) . \tag{7.40}$$

The detector signal $S(\omega)$ is proportional to the transmitted intensity

$$S(\omega) \propto I_{\mathrm{T}}(\omega) = c\epsilon_0 E_{\mathrm{t}} E_{\mathrm{t}}^* .$$

Even for $\theta = 0$ the crossed polarizers have a small residual transmission $I_{\mathrm{t}} = \xi I_0$ ($\xi \approx 10^{-6}$ to 10^{-8}). Taking this into account, we obtain with the incident probe intensity I_0 and the abbreviation $\theta' = \theta + \omega/(2c) \cdot \Delta b_{\mathrm{r}}$, the

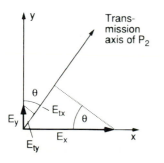

Fig. 7.21. Transmission of the elliptically polarized probe wave through the analyzer, uncrossed by the angle θ

transmitted intensity

$$I_t = I_0 e^{-\alpha L - \alpha_w} \left(\xi + |\theta + \Delta|^2 \right) ,$$

$$= I_0 e^{-\alpha L - \alpha_w} \left[\xi + \theta'^2 + \left(\frac{1}{2} \Delta \alpha_w \right)^2 + \frac{1}{4} \Delta \alpha_w L \Delta \alpha + \frac{\omega}{c} \theta' L \Delta n \right.$$

$$\left. + \left(\frac{\omega}{2c} L \Delta n \right)^2 + \left(\frac{L \Delta \alpha}{4} \right)^2 \right] . \tag{7.41}$$

The change of absorption $\Delta \alpha$ is caused by those molecules within the velocity interval $\Delta v_z = 0 \pm \gamma_s / k$ that simultaneously interact with the pump and probe waves. The line profile of $\Delta \alpha(\omega)$ is therefore, analogous to the situation in saturation spectroscopy (Sect. 7.2) with a Lorentzian profile

$$\Delta \alpha(\omega) = \frac{\Delta \alpha_0}{1 + x^2} , \quad \text{with} \quad x = \frac{\omega_0 - \omega}{\gamma_s / 2} , \quad \text{and} \quad \alpha_0 = \alpha(\omega_0) , \tag{7.42}$$

having the halfwidth γ_s, which corresponds to the homogeneous width of the molecular transition saturated by the pump wave.

The absorption coefficient $\alpha(\omega)$ and the refractive index $n(\omega)$ are related by the Kramers–Kronig dispersion relation, see (3.36b, 3.37b) in Sect. 3.1.

We therefore obtain for $\Delta n(\omega)$ the dispersion profile

$$\Delta n(\omega) = \frac{c}{\omega_0} \frac{\Delta \alpha_0 x}{1 + x^2} . \tag{7.43}$$

Inserting (7.42 and 7.43) into (7.41) gives for the circularly polarized pump beam the line profile of the detector signal

$$I_t(\omega) = I_0 e^{-\alpha L - \alpha_w} \left\{ \xi + \theta'^2 + \frac{1}{4} \Delta \alpha_w^2 + \theta' \Delta \alpha_0 L \frac{x}{1 + x^2} \right.$$

$$\left. + \left[\frac{1}{4} \Delta \alpha_0 \Delta \alpha_w L + \left(\frac{\Delta \alpha_0 L}{4} \right)^2 \right] \frac{1}{1 + x^2} + 3 \left(\frac{\Delta \alpha_0 2x}{4(1 + x^2)} \right)^2 \right\} . \tag{7.44}$$

The signal contains a constant background term $\xi + \theta'^2 + \Delta \alpha_w^2 / 4$ that is independent of the frequency ω. The quantity $\xi = I_T / I_0$ ($\theta = 0$, $\Delta \alpha_w = \Delta b_r = 0$, $\Delta_\alpha = \Delta n_0 = 0$) gives the residual transmission of the completely crossed analyzer P_2. With normal Glan–Thomson polarizers $\xi < 10^{-6}$ can be realized, with selected samples even $\xi < 10^{-8}$ can be obtained. The third term is due to the absorption part of the window birefringence. All these three terms are approximately independent of the frequency ω.

The next three terms contribute to the line profile of the polarization signal. The first frequency-dependent term in (7.44) gives the additional transmitted intensity when the angle $\theta' = \theta + (\omega/2c)\Delta b_r$ is not zero. It has a dispersion-type profile. For $\theta' = 0$ the dispersion term vanishes. The two Lorentzian terms depend on the product of $\Delta\alpha_w$ and $\Delta\alpha_0 \cdot L$ and on $(\Delta\alpha_0 L)^2$. By squeezing the windows their dichroism (that is, $\Delta\alpha_w$) can be increased and the amplitude of the first Lorentzian term can be enlarged. Of course, the background term $\Delta\alpha_w^2$ also increases and one has to find the optimum signal-to-noise ratio. Since in most cases $\Delta\alpha_0 L \ll 1$, the last term in (7.44), which is proportional to $\Delta\alpha_0^2 L^2$, is generally negligible. Nearly pure dispersion signals can be obtained if $\Delta\alpha_w$ is minimized and θ' is increased until the optimum signal-to-noise ratio is achieved. Therefore by controlling the birefringence of the window either dispersion-shaped or Lorentzian line profilescan be obtained for the signals.

The sensitivity of polarization spectroscopy compared with saturation spectroscopy is illustrated by Fig. 7.22a, which shows the same hfs transitions of I_2 as in Fig. 7.12 taken under comparable experimental conditions. A section of the same spectrum is depicted in Fig. 7.22b with $\theta' \neq 0$, optimized for dispersion-line profiles.

If the pump wave is *linearly polarized* with the electric field vector 45° against the x-axis, one obtains in an analogous derivation for the polarization signal instead of (7.44) the expression

$$S^{LP}(\omega) = I_0 e^{-\alpha L - \alpha_w} \left\{ \xi + \frac{1}{4}\theta^2 \Delta\alpha_w^2 + \left(\frac{\omega}{2c}\Delta b_r\right)^2 + \frac{\Delta b_r}{4}\frac{\omega}{c}\Delta\alpha_0 L \frac{x}{1+x^2} \right.$$
$$\left. + \left[-\frac{1}{4}\theta\Delta\alpha_w\Delta\alpha_0 L + \left(\frac{\Delta\alpha_0 L}{4}\right)^2 \right]\frac{1}{1+x^2} \right\} , \qquad (7.45)$$

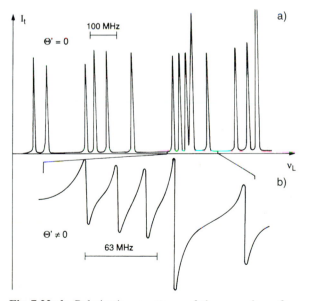

Fig. 7.22a,b. Polarization spectrum of the same hyperfine components of I_2 as shown in Fig. 7.12 with circularly polarized pump: (**a**) with $\theta' = 0$; (**b**) with $\theta' \neq 0$

where $\Delta\alpha = \alpha_\parallel - \alpha_\perp$ and $\Delta b = b_\parallel - b_\perp$ are now defined by the difference of the components parallel or perpendicular to the E vector of the pump wave. Dispersion and Lorentzian terms as well as $\Delta\alpha_w$ and Δb_r are interchanged compared with (7.44). For illustration, in Fig. 7.23 two identical sections of the Cs_2 polarization spectrum around $\lambda = 627.8$ nm are shown, pumped with a linearly or circularly polarized pump beam, respectively. As shown below, the magnitude of the polarization signals for linearly polarized pump is large for transitions with $\Delta J = 0$, whereas for a circularly polarized pump they are maximum for $\Delta J = \pm 1$. While the Q lines are prominent in the upper spectrum, they appear in the lower spectrum only as small dispersion-type signals.

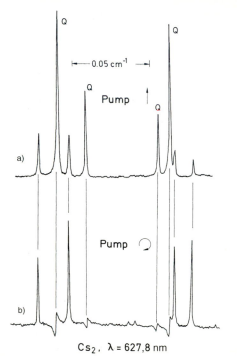

Fig. 7.23a,b. Two identical sections of the Cs_2 polarization spectrum recorded with linear (*upper spectrum*) and circular pump polarization (*lower spectrum*)

7.4.3 Magnitude of Polarization Signals

In order to understand the different magnitudes and line profiles, we have to investigate the magnitudes of the differences $\Delta\alpha_0 = \alpha^+ - \alpha^-$ for a circularly polarized pump wave and $\Delta\alpha_0 = \alpha_\parallel - \alpha_\perp$ for a linearly polarized pump wave, and their relation to the absorption cross section of the molecular transitions.

When a linearly polarized weak probe wave propagating through a ensemble of molecules is tuned to a molecular transition $|J, M\rangle \rightarrow |J_1, M \pm 1\rangle$, the difference $\Delta\alpha = \alpha^+ - \alpha^-$ of the absorption coefficients of its left- and right-hand circularly polarized components is

$$\Delta\alpha(M) = N_M \left(\sigma^+_{JJ_1M} - \sigma^-_{JJ_1M}\right) , \qquad (7.46)$$

where $N_M = N_J/(2J+1)$ represents the population density of one of the $(2J+1)$ degenerate sublevels $|J, M\rangle$ of a rotational level $|J\rangle$ in the lower state, and $\sigma^{\pm}_{JJ_1M}$ the absorption cross section for transitions starting from the level $|J, M\rangle$ and ending on $|J_1, M\pm 1\rangle$.

Because of the saturation by the pump beam, the population N_M decreases from its unsaturated value N^0_M to

$$N^S_M = \frac{N^0_M}{1+S_0} ,$$

where the saturation parameters $S_0 = S(\omega_0)$ at the line center

$$S_0 = \frac{2S}{\pi\gamma_s} = \frac{8\sigma_{JJ_1M}}{\gamma_s R^*} \frac{I_2}{\hbar\omega} , \qquad (7.47)$$

depends on the absorption cross section for the pump transition, the saturated homogeneous linewidth γ_s, the mean relaxation rate R^*, which refills the saturated level population (for example, by collisions) and the number of pump photons $I_2/\hbar\omega$ per cm^2 and second.

The absorption cross section

$$\sigma_{JJ_1M} = \sigma_{JJ_1} C(J, J_1, M, M_1) \qquad (7.48)$$

can be separated into a product of two factors: the first factor is independent of the molecular orientation, but depends solely on the internal transition probability of the molecular transitions, which differs for P, Q, and R lines [7.26, 7.27]. The second factor in (7.48) is the Clebsch–Gordan coefficient, which depends on the rotational quantum numbers J, J_1 of the lower and upper levels, and on the orientation of the molecule with respect to the quantization axis. In the case of a σ^+ pump wave, this is the propagation direction k_p; in the case of a π pump wave, it is the direction of the electric vector E, see Fig. 7.24.

The total change $\Delta\alpha$ $(J \to J_1)$ on a rotational transition $J \to J_1$ is due to the saturation of all allowed transitions $(M_J \to M_{J1}$ with $\Delta M = \pm 1$ for a circular pump polarization or $\Delta M = 0$ for a linearly polarized pump) between the $(2J+1)$ degenerate sublevels M in the lower level J and the $(2J_1+1)$ sublevels in the upper level J_1.

$$\Delta\alpha(J, J_1) = \sum_M N_M \left(\sigma^+_{JJ_1M} - \sigma^-_{JJ_1M}\right) . \qquad (7.49)$$

From (7.4.3–7.49) we finally obtain for the quantity $\Delta\alpha_0$ in (7.44, 7.45)

$$\Delta\alpha_0 = \alpha_0 S_0 \Delta C^*_{JJ_1} , \qquad (7.50)$$

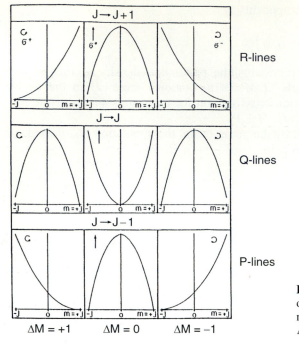

Fig. 7.24. Dependence of σ_{JJ_1M} on the orientational quantum number M for transitions with $\Delta M = 0, \pm 1$

Table 7.1. Values of $2/3\Delta C^*_{J_1 J_2}$ for linear pump polarization (**a**) and circular pump polarization (**b**). For the two-level system $J_1 = J_2$ and $r = (\gamma_J - \gamma_{J_2})/(\gamma_J + \gamma_{J_2})$. For the three-level system J is the rotational quantum number of the common level and $r = -1$

	$J_2 = J+1$	$J_2 = J$	$J_2 = J-1$
(a)	Linear pump polarization		
$J_1 = J+1$	$\dfrac{2J^2 + J(4+5r) + 5 + 5r}{5(J+1)(2J+3)}$	$\dfrac{-(2J-1)}{5(J+1)}$	$\dfrac{1}{5}$
$J_1 = 1$	$\dfrac{-(2J-1)}{5(J+1)}$	$\dfrac{(2J-3)(2J-1)}{5J(J+1)}$	$-\dfrac{2J+3}{5J}$
$J_1 = J-1$	$\dfrac{1}{5}$	$-2J + \dfrac{3}{5J}$	$\dfrac{2J^2 - 5rJ + 3}{5J(2J-1)}$
(b)	Circular pump polarization		
$J_1 = J+1$	$\dfrac{2J^2 + J(4+r) + r + 1}{(2J+3)(J+1)}$	$\dfrac{-1}{J+1}$	-1
$J_1 = 1$	$\dfrac{-1}{J+1}$	$\dfrac{1}{J(J+1)}$	$\dfrac{1}{J}$
$J_1 = J-1$	-1	$\dfrac{1}{J}$	$\dfrac{2J^2 - rJ - 1}{J(2J-1)}$

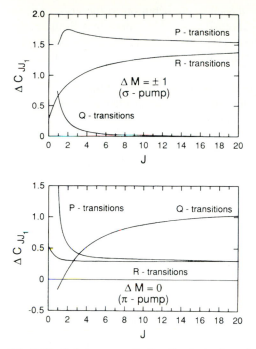

Fig. 7.25. Relative magnitude of polarization signals expressed by the factor $\Delta C^*_{JJ_1}$ as a function of rotational quantum number J for circular and linear pump polarization

where $\alpha_0 = N_J \sigma_{JJ_1}$ is the unsaturated absorption coefficient of the probe wave at the line center, and $\Delta C^*_{JJ_1}$ is a numerical factor that is proportional to the sum $\sum \Delta \sigma_{JJ_1 M} C(JJ_1 MM_1)$ with $\Delta \sigma = \sigma^+ - \sigma^-$. The numerical values of $\Delta C^*_{JJ_1}$ are compiled in Table 7.1 for P, Q, and R transitions. Their dependence on the rotational quantum number J is illustrated by Fig. 7.25.

7.4.4 Sensitivity of Polarization Spectroscopy

In the following we briefly discuss the sensitivity and the signal-to-noise ratio achievable with polarization spectroscopy. The amplitude of the dispersion signal in (7.44) is approximately the difference $\Delta I_T = I_T(x = +1) - I_T(x = -1)$ between the maximum and the minimum of the dispersion curve. From (7.44) we obtain

$$\Delta S_{\max} = I_0 e^{-\alpha L - \alpha_w} \theta' \Delta \alpha_0 L ,$$
(7.51a)

while for the Lorentzian profiles $(I_T(x = 0) - I_T(x = \infty), \theta' = 0)$ we derive from (7.44)

$$\Delta S_{\max} = \frac{1}{4} \Delta \alpha_0 \cdot L \left(\Delta \alpha_w + \frac{1}{4} \Delta \alpha_0 L \right) .$$
(7.51b)

Under general laboratory conditions the main contribution to the noise comes from fluctuations of the probe-laser intensity, while the principal limit set by shot noise (Chap. 4) is seldom reached. The noise level is therefore essentially proportional to the transmitted intensity, which is given by the background term in (7.44).

Because the crossed polarizers greatly reduce the background level, we can expect a better signal-to-noise ratio than in saturation spectroscopy, where the full intensity of the probe beam is detected.

In the absence of window birefringence (that is, $\Delta b_r = \Delta b_i = 0$, $\theta' = \theta$) the signal-to-noise (S/N) ratio, which is, besides a constant factor a, equal to the signal-to-background ratio, becomes with (7.50) and (7.51a) for the dispersion signals

$$\frac{S}{N} = a \frac{\theta \alpha_0 L S_0}{\xi + \theta^2} \Delta C^*_{JJ_1} , \tag{7.52}$$

where a is a measure of the intensity stability of the probe laser, i.e., I_1/a is the mean noise of the incident probe wave, and S_0 is the saturation parameter at $\omega = \omega_0$.

This ratio is a function of the uncrossing angle θ of the two polarizers (Fig. 7.21) and has a maximum for $d(S/N)/d\theta = 0$, which yields $\theta^2 = \xi$ and

$$\left(\frac{S}{N}\right)_{max} = a \frac{\alpha_0 L S_0 \Delta C^*_{JJ_1}}{4\sqrt{\xi}} . \tag{7.53}$$

In this case of ideal windows ($b_r = b_i = 0$) the quality of the two polarizers, given by the residual transmission ξ of the completely crossed polarizations ($\theta = 0$), limits the achievable S/N ratio. In saturation spectroscopy the S/N ratio is, according to (7.31), given by

$$\frac{S}{N} = \frac{1}{2} a \alpha_0 L S_0 . \tag{7.54}$$

Polarization spectroscopy therefore increases the S/N ratio by the factor $\Delta C^*_{JJ_1}/2\sqrt{\xi}$ for optimized dispersion-type signals.

Example 7.4

For $\Delta C^*_{JJ_1} = 0.5$, $\xi = 10^{-6}$, the S/N ratio in polarization spectroscopy with ideal cell windows becomes 500 times better than saturation spectroscopy.

For windows with birefringence the situation is more complex: for $b_r = 0$ we obtain for the dispersion signals the optimum uncrossing angle

$$\theta' = \sqrt{\xi + \left(\frac{1}{2}\Delta\alpha_w\right)^2} . \tag{7.55}$$

This yields the maximum obtainable signal-to-noise ratio for dispersion signals

$$\left(\frac{S}{N}\right)^{disp}_{max} = a \cdot \alpha_0 \cdot LS_0 \cdot \Delta C^*_{JJ_1} \frac{1}{2 \cdot \sqrt{\xi + (1/2\Delta\alpha_w)}} , \tag{7.56}$$

which converts to (7.52) for $\alpha_w = 0$.

Example 7.5

For $\xi = 10^{-6}$ and $\Delta\alpha_w = 10^{-3}$, the S/N ratio is smaller than in Example 7.4 by a factor of 0.89. For $\xi = 10^{-8}$ one reaches $S/N = 5000$ with $\Delta\alpha_w = 0$, but only $S/N = 980$ for $\Delta\alpha_w = 10^{-3}$.

For the Lorentzian line profiles with $\theta' = 0$ in (7.44) we obtain

$$\left(\frac{S}{N}\right)_{max} = \frac{a}{4} \frac{\Delta\alpha_0 L(\Delta\alpha_w + \Delta\alpha_0 L/4)}{\xi + (\Delta\alpha_w/2)^2} . \tag{7.57}$$

Here we have to optimize $\Delta\alpha_w$ to achieve the maximum S/N ratio. The differentiation of (7.57) with respect to $\Delta\alpha_w$ gives for $\xi \ll \Delta_\alpha L$ the optimum window birefringence

$$\Delta\alpha_w \approx \frac{4\xi}{\Delta\alpha_0 \cdot L} , \tag{7.58}$$

which yields for the maximum S/N ratio for Lorentzian signals

$$\left(\frac{S}{N}\right)_{max} \approx a \cdot \alpha_0 L S_0 \Delta C^*_{JJ_1} \cdot \frac{\Delta\alpha_0 L}{4\xi} \left(1 + \frac{12\xi}{(\Delta\alpha_0 L)^2} - \frac{64\xi^2}{(\Delta\alpha_0 L)^3}\right) . \tag{7.59}$$

Example 7.6

With $a = 10^2$, $\xi = 10^{-6}$, $\Delta C^*_{JJ_1} = 0.5$, $\alpha_0 L = 10^{-2}$, and $S_0 = 0.1$, we obtain a signal-to-noise ratio of $5\alpha_0 \cdot L = 5 \times 10^{-2}$ without lock-in detection for the saturation spectroscopy signal measured with the arrangement in Fig. 7.9, but $S/N = 1.25 \times 10^3 \alpha_0 \cdot L$ for the dispersion signal in polarization spectroscopy and for the optimized Lorentzian profiles. In practice, these figures are somewhat lower because $\Delta\alpha_w \neq 0$. One also has to take into account that the length L of the interaction zone is smaller in the arrangement of Fig. 7.20 (because of the finite angle between pump and probe beams) than for saturation spectroscopy with completely overlapping pump and probe beams (Fig. 9.10). For $\alpha_0 L = 10^{-2}$ and $\xi = 10^{-6}$, one still achieves $S/N \approx 12.5$ for dispersion and for Lorentzian signals. For $\xi = 10^{-8}$ and $\Delta\alpha_w = 10^{-3}$, the improvement factor over saturation spectroscopy is 2×10^3 for dispersion signals and 2×10^4 for Lorentzian signals.

7.4.5 Advantages of Polarization Spectroscopy

Let us briefly summarize the advantages of polarization spectroscopy, discussed in the previous sections:

- Along with the other sub-Doppler techniques, it has the advantage of high spectral resolution, which is mainly limited by the residual Doppler width due to the finite angle between the pump beam and the probe beam. This limitation corresponds to that imposed to linear spectroscopy in collimated molecular beams by the divergence angle of the molecular beam. The transit-time broadening can be reduced if the pump and probe beams are less tightly focused.
- The sensitivity is 2−3 orders of magnitude larger than that of saturation spectroscopy. It is surpassed only by that of the intermodulated fluorescence technique at very low sample pressures (Sect. 7.3.1).
- The possibility of distinguishing between P, R, and Q lines is a particular advantage for the assignment of complex molecular spectra.
- The dispersion profile of the polarization signals allows a stabilization of the laser frequency to the line center without any frequency modulation. The large achievable signal-to-noise ratio assures an excellent frequency stability.

Meanwhile polarization spectroscopy has been applied to the measurement of many high-resolution atomic and molecular spectra. Examples can be found in [7.3, 7.28–7.34].

7.5 Multiphoton Spectroscopy

In this section we consider the simultaneous absorption of two or more photons by a molecule that undergoes a transition $E_i \to E_f$ with $(E_f - E_i) = \hbar \sum_i \omega_i$. The photons may either come from a single laser beam passing through the absorbing sample or they may be provided by two or more beams emitted from one or several lasers.

The first detailed theoretical treatment of two-photon processes was given in 1929 by Göppert-Mayer [7.35], but the experimental realization had to wait for the development of sufficiently intense light sources, now provided by lasers [7.36, 7.37].

7.5.1 Two-Photon Absorption

Two-photon absorption can be formally described by a two-step process from the initial level $|i\rangle$ via a "virtual level" $|v\rangle$ to the final level $|f\rangle$ (Fig. 7.26b). This fictitious virtual level is represented by a linear combination of the wave functions of all real molecular levels $|k_n\rangle$ that combine with $|i\rangle$ by allowed one-photon transitions. The excitation of $|v\rangle$ is equivalent to the off-resonance excitation of all these real levels $|k_n\rangle$. The probability amplitude for a transition $|i\rangle \to |v\rangle$ is then represented by the sum of the amplitudes of all allowed

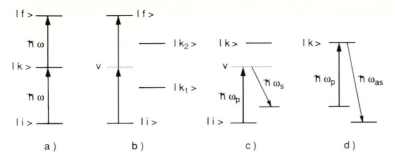

Fig. 7.26a–d. Level schemes of different two-photon transitions: (**a**) resonant two-photon absorption with a real intermediate level $|k\rangle$; (**b**) nonresonant two-photon absorption; (**c**) Raman transition; (**d**) resonant anti-Stokes Raman scattering

transitions $|i\rangle \rightarrow |k\rangle$ with off-resonance detuning $(\omega - \omega_{ik})$. The same arguments hold for the second step $|v\rangle \rightarrow |f\rangle$.

For a molecule moving with a velocity $\boldsymbol{v}$, the probability A_{if} for a two-photon transition between the ground state E_i and an excited state E_f induced by the photons $\hbar\omega_1$ and $\hbar\omega_2$ from two light waves with the wave vectors $\boldsymbol{k}_1$, $\boldsymbol{k}_2$, the polarization unit vectors $\hat{\boldsymbol{e}}_1$, $\hat{\boldsymbol{e}}_2$, and the intensities I_1, I_2 can be written as

$$A_{if} \propto \frac{\gamma_{if} I_1 I_2}{[\omega_{if} - \omega_1 - \omega_2 - \boldsymbol{v} \cdot (\boldsymbol{k}_1 + \boldsymbol{k}_2)]^2 + (\gamma_{if}/2)^2}$$
$$\times \left| \sum_k \frac{\boldsymbol{R}_{ik} \cdot \hat{\boldsymbol{e}}_1 \cdot \boldsymbol{R}_{kf} \cdot \hat{\boldsymbol{e}}_2}{\omega_{ki} - \omega_1 - \boldsymbol{v} \cdot \boldsymbol{k}_1} + \frac{\boldsymbol{R}_{ik} \cdot \hat{\boldsymbol{e}}_2 \cdot \boldsymbol{R}_{kf} \cdot \hat{\boldsymbol{e}}_1}{\omega_{ki} - \omega_2 - \boldsymbol{v} \cdot \boldsymbol{k}_2} \right|^2 . \tag{7.60}$$

The first factor gives the spectral line profile of the two-photon transition of a single molecule. It corresponds exactly to that of a single-photon transition of a moving molecule at the center frequency $\omega_{if} = \omega_1 + \omega_2 + \boldsymbol{v} \cdot (\boldsymbol{k}_1 + \boldsymbol{k}_2)$ with a homogeneous linewidth γ_{if} (Sects. 3.1, 3.6). Integration over all molecular velocities v gives a *Voigt profile* with a halfwidth that depends on the relative orientations of $\boldsymbol{k}_1$ and $\boldsymbol{k}_2$. If both light waves are parallel, the Doppler width, which is proportional to $|\boldsymbol{k}_1 + \boldsymbol{k}_2|$, becomes maximum and is, in general, large compared to the homogeneous width γ_{if}. For $\boldsymbol{k}_1 = -\boldsymbol{k}_2$ the Doppler broadening vanishes and we obtain a pure Lorentzian line profile with a homogeneous linewidth γ_{if} provided that the laser linewidth is small compared to γ_{if}. This *Doppler-free two-photon spectroscopy* is discussed in Sect. 7.5.2.

Because the transition probability (7.60) is proportional to the product of the intensities $I_1 I_2$ (which has to be replaced by I^2 in the case of a single laser beam), *pulsed* lasers, which deliver sufficiently large peak powers, are generally used. The spectral linewidth of these lasers is often comparable to or even larger than the Doppler width. For nonresonant transitions $|\omega_{ki} - \omega_i| \gg$

$\boldsymbol{v} \cdot \boldsymbol{k}_i$, and the denominators $(\omega_{ki} - \omega - \boldsymbol{k} \cdot \boldsymbol{v})$ in the sum in (7.60) can then be approximated by $(\omega_{ki} - \omega_i)$.

The second factor in (7.60) describes the transition probability for the two-photon transition. It can be derived quantum mechanically by second-order perturbation theory (see, for example, [7.38, 7.39]). This factor contains a sum of products of matrix elements $R_{ik} R_{kf}$ for the transitions between the initial level i and intermediate molecular levels k or between these levels k and the final state f, see (2.110). The summation extends over all molecular levels k that are accessible by allowed one-photon transitions from the initial state $|i\rangle$. The denominator shows, however, that only those levels k that are not too far off resonance with one of the Doppler-shifted laser frequencies $\omega_n' = \omega_n - \boldsymbol{v} \cdot \boldsymbol{k}_n$ $(n = 1, 2)$ will mainly contribute.

Often the frequencies ω_1 and ω_2 can be selected in such a way that the virtual level is close to a real molecular eigenstate, which greatly enhances the transition probability. It is therefore generally advantageous to excite the final level E_f by two different photons with $\omega_1 + \omega_2 = (E_f - E_i)/\hbar$ rather than by two photons out of the same laser with $2\omega = (E_f - E_i)/\hbar$.

The second factor in (7.60) describes quite generally the transition probability for all possible two-photon transitions such as Raman scattering or two-photon absorption and emission. Figure 7.26 illustrates schematically three different two-photon processes. The *important point is that the same selection rules are valid for all these two-photon processes.* Equation (7.60) reveals that both matrix elements R_{ik} and R_{kf} must be nonzero to give a non-vanishing transition probability A_{if}. This means that two-photon transitions can only be observed between two states $|i\rangle$ and $|f\rangle$ that are both connected to intermediate levels $|k\rangle$ by allowed single-photon optical transitions. Because the selection rule for single-photon transitions demands that the levels $|i\rangle$ and $|k\rangle$ or $|k\rangle$ and $|f\rangle$ have opposite parity, *the two levels $|i\rangle$ and $|f\rangle$ connected by a two-photon transition* must have the same parity. In atomic two-photon spectroscopy $s \rightarrow s$ or $s \rightarrow d$ transitions are allowed, and in diatomic homonuclear molecules $\Sigma_g \rightarrow \Sigma_g$ transitions are allowed.

It is therefore possible to reach molecular states that cannot be populated by single-photon transitions from the ground state. In this regard two-photon absorption spectroscopy is complementary to one-photon absorption spectroscopy, and its results are of particular interest because they yield information about states, which often had not been found before [7.40]. Although excited molecular states are often perturbed by nearby states of opposite parity, it is generally difficult to deduce the structure of these perturbing states from the degree of perturbations, while two-photon spectroscopy allows direct access to such states.

Since the matrix elements $\boldsymbol{R}_{ik} \cdot \hat{\boldsymbol{e}}_1$ and $\boldsymbol{R}_{kf} \cdot \hat{\boldsymbol{e}}_2$ depend on the polarization characteristics of the incident radiation, it is possible to select the accessible upper states by a proper choice of the polarization. While for single-photon transitions the total transition probability (summed over all M sublevels) is *independent* of the polarization of the incident radiation, there is a distinct polarization effect in multiphoton transitions, which can be understood by applying known selection rules to the two matrix elements in (7.60). For example, two parallel laser beams, which both have right-hand circular polarization, induce two-photon transitions in atoms with $\Delta L = 2$. This allows, for instance, $s \rightarrow d$ transitions, but not $s \rightarrow s$ transitions. When a circularly polarized wave is reflected back on itself, the right-hand circular polarization changes into a left-hand one, and if a two-photon transition is induced by one photon from each wave, only $\Delta L = 0$ transitions are selected. Figure 7.27 illustrates the different atomic transitions that are possible by multiphoton absorption of linearly polarized light and of right or left circularly polarized light.

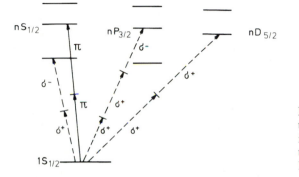

Fig. 7.27. Different two-photon transitions in an atom, depending on the polarization characteristics $\hat{e}_1$ and $\hat{e}_2$ of the two laser fields

Different upper states can therefore be selected by a proper choice of the polarization. In many cases it is possible to gain information about the symmetry properties of the upper states from the known symmetry of the ground state and the polarization of the two light waves. Since the selection rules of two-photon absorption and Raman transitions are identical, one can utilize the group-theoretical techniques originally developed for Raman scattering to analyze the symmetry properties of the excited states [7.41, 7.42].

7.5.2 Doppler-Free Multiphoton Spectroscopy

In the methods discussed in Sects. 7.3 and 7.4, the Doppler width had been reduced or even completely eliminated by proper selection of a *velocity subgroup* of molecules with the velocity components $v_z = 0 \pm \Delta v_z$, due to selective saturation. The technique of Doppler-free multiphoton spectroscopy does not need such a velocity selection because *all* molecules in the absorbing state, regardless of their velocities, can contribute to the Doppler-free transition.

While the general concepts and the transition probability of multiphoton transitions are discussed in Sect. 7.5.1, we concentrate in this subsection on *Doppler-free* multiphoton spectroscopy [7.43–7.47].

Assume a molecule moves with a velocity $\boldsymbol{v}$ in the laboratory frame. In the reference frame of the moving molecule the frequency ω of an EM wave with the wave vector $\boldsymbol{k}$ is Doppler shifted to (Sect. 3.2)

$$\omega' = \omega - \boldsymbol{k} \cdot \boldsymbol{v} \, . \tag{7.61}$$

The resonance condition for the simultaneous absorption of two photons is

$$(E_f - E_i)/\hbar = (\omega_1' + \omega_2') = \omega_1 + \omega_2 - \boldsymbol{v} \cdot (\boldsymbol{k}_1 + \boldsymbol{k}_2) \, . \tag{7.62}$$

If the two photons are absorbed out of two light waves with equal frequencies $\omega_1 = \omega_2 = \omega$, which travel in opposite directions, we obtain $\boldsymbol{k}_1 = -\boldsymbol{k}_2$ and (7.62) shows that the Doppler shift of the two-photon transition becomes zero. This means that *all* molecules, *independent of their velocities*, absorb at the same sum frequency $\omega_1 + \omega_2 = 2\omega$.

Although the probability of a two-photon transition is generally much lower than that of a single-photon transition, the fact that *all* molecules in the absorbing state can contribute to the signal may outweigh the lower transition probability, and the signal amplitude may even become, in favorable cases, larger than that of the saturation signals.

The considerations above can be generalized to many photons. When the moving molecule is simultaneously interacting with several plane waves with wave vectors $\boldsymbol{k}_i$ and one photon is absorbed from each wave, the total Doppler shift $\boldsymbol{v} \cdot \sum_i \boldsymbol{k}_i$ becomes zero for $\sum \boldsymbol{k}_i = 0$.

Figure 7.28 shows a possible experimental arrangement for the observation of Doppler-free two-photon absorption. The two oppositely traveling waves are formed by reflection of the output beam from a single-mode tunable dye laser. The Faraday rotator prevents feedback into the laser. The two-photon absorption is monitored by the fluorescence emitted from the final state E_f into other states E_m. From (7.60) it follows that the probability of two-photon absorption is proportional to the square of the power density. Therefore, the two beams are focused into the sample cell by the lens L and the spherical mirror M.

For illustration the examples in Fig. 7.29 show the Doppler-free two-photon spectra of the $3S \to 5S$ transition in the sodium atom with resolved hyperfine structure [7.43].

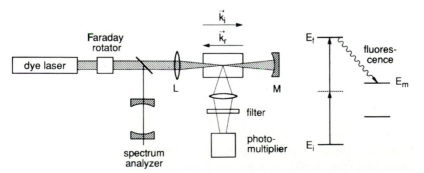

Fig. 7.28. Experimental arrangement for Doppler-free two-photon spectroscopy

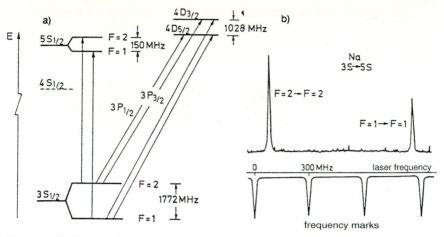

Fig. 7.29a,b. Doppler-free two-photon spectrum of the 3S → 5S and 3S → 4D transitions in the Na atom: (**a**) level scheme; (**b**) 3S → 5S transition with resolved hyperfine structure [7.43]

The line profile of two-photon transitions can be deduced from the following consideration: assume the reflected beam in Fig. 7.28 to have the same intensity as the incident beam. In this case the two terms in the second factor of (7.60) become identical, while the first factor, which describes the line profile, differs for the case when both photons are absorbed out of the same beam from the case when they come from different beams. The probability for the latter case is twice as large as for the former case. This can be seen as follows:

Let (a, a) be the probability for the case that both photons are provided by the incident beam and (b, b) by the reflected beam. The total probability for a two-photon absorption with both photons of the same beam (which gives a Doppler-broadened contribution to the signal) is then $(a, a)^2 + (b, b)^2$.

The probability amplitude of the Doppler-free absorption is the sum $(a, b) + (b, a)$ of two nondistinguishable events. The total probability for this case is then $|(a, b) + (b, a)|^2$. For equal intensities of the two beams this is twice the probability $(a, a)^2 + (b, b)^2$.

We therefore obtain from (7.60) for the two-photon absorption probability

$$W_{if} \propto \left| \sum_m \frac{(R_{im}\hat{e}_1)(R_{mf}\hat{e}_2)}{\omega - \omega_{im} - k_1 \cdot v} + \sum_m \frac{(R_{im}\hat{e}_2)(R_{mf}\hat{e}_1)}{\omega - \omega_{im} - k_2 \cdot v} \right|^2 I^2$$

$$\times \left[\frac{4\gamma_{if}}{(\omega_{if} - 2\omega)^2 + (\gamma_{if}/2)^2} + \frac{\gamma_{if}}{(\omega_{if} - 2\omega - 2k \cdot v)^2 + (\gamma_{if}/2)^2} \right.$$

$$\left. + \frac{\gamma_{if}}{(\omega_{if} - 2\omega + 2k \cdot v)^2 + (\gamma_{if}/2)^2} \right]. \tag{7.63}$$

Integration over the velocity distribution yields the absorption profile

$$\alpha(\omega) \propto \Delta N^0 I^2 \left| \sum_m \frac{(\boldsymbol{R}_{im}\boldsymbol{\hat{e}})(\boldsymbol{R}_{mf}\boldsymbol{\hat{e}})}{\omega - \omega_{im}} \right|^2$$

$$\times \left\{ \exp\left[-\left(\frac{\omega_{if} - 2\omega}{2kv_p} \right)^2 \right] + \frac{kv_p}{\sqrt{\pi}} \frac{\gamma_{if}/2}{(\omega_{if} - 2\omega)^2 + (\gamma_{if}/2)^2} \right\},$$

$$(7.64)$$

where $v_p = (2kT/m)^{1/2}$ is the most probable velocity, and $\Delta N_0 = N_i^0 - N_f$ the nonsaturated population difference. The absorption profile (7.64) represents a superposition of a Doppler-broadened background and a narrow Lorentzian profile with the linewidth $\gamma_{if} = \gamma_i + \gamma_f$ (Fig. 7.30).

The area under the Doppler profile with halfwidth $\Delta\omega_D$ is half of that under the narrow Lorentzian profile with halfwidth γ_{if}. However, its peak intensity amounts only to the fraction

$$\epsilon = \frac{\gamma_{if}\sqrt{\pi}}{2kv_p} \simeq \frac{\gamma_{if}}{2\Delta\omega_D},$$

of the Lorentzian peak heights.

Example 7.7

With $\gamma_{if} = 20\,\text{MHz}$, $\Delta\omega_D = 2\,\text{GHz}$, the Doppler-free signal in Fig. 7.30 is about 200 times higher than the maximum of the Doppler-broadened background.

By choosing the proper polarization of the two laser waves, the background can often be completely suppressed. For example, if the incident laser

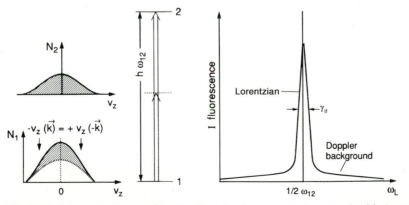

Fig. 7.30. Schematic line profile of a Doppler-free two-photon signal with (greatly exaggerated) Doppler-broadened background

beam has σ^+ polarization, a $\lambda/4$-plate between mirror M and the sample in Fig. 7.28 converts the reflected beam to σ^- polarization. Two-photon absorption on a $S \to S$ transition with $\Delta M = 0$ is then possible only if one photon comes from the incident wave and the other from the reflected wave because two photons out of the same beam would induce transitions with $\Delta M = \pm 2$.

For the resonance case $2\omega = \omega_{if}$, the second term in (7.63) becomes $2kv_p/(\gamma_{if}\sqrt{\pi}) \gg 1$. We can neglect the contribution of the Doppler term and obtain the maximum two-photon absorption:

$$\alpha(\omega = \tfrac{1}{2}\omega_{if}) \propto I^2 \frac{\Delta N^0 k v_p}{\sqrt{\pi}\gamma_{if}} \left| \sum_m \frac{(\boldsymbol{R}_{im} \cdot \hat{\boldsymbol{e}}) \cdot (\boldsymbol{R}_{mf} \cdot \hat{\boldsymbol{e}})}{\omega - \omega_{im}} \right|^2 . \tag{7.65}$$

Note: Although the product of matrix elements in (7.65) is generally much smaller than the corresponding one-photon matrix elements, the magnitude of the Doppler-free two-photon signal (7.62) may exceed that of Doppler-free saturation signals. This is due to the fact that *all* molecules in level $|i\rangle$ contribute to two-photon absorption, whereas the signal in Doppler-free saturation spectroscopy is provided only by the subgroup of molecules out of a narrow velocity interval $\Delta v_z = 0 \pm \gamma/k$. For $\gamma = 0.01\Delta\omega_D$, this subgroup represents only about 1% of all molecules.

For *molecular* transitions the matrix elements R_{im} and R_{mf} are composed of three factors: the electronic transition dipole element, the Franck–Condon factor, and the Hönl–London factor (Sect. 6.7). Within the two-photon dipole approximation $\alpha(\omega)$ becomes zero if one of these factors is zero. The calculation of linestrengths for two- or three-photon transitions in diatomic molecules can be found in [7.41, 7.42].

7.5.3 Influence of Focusing on the Magnitude of Two-Photon Signals

Since the two-photon absorption probability is proportional to the square of the incident laser intensity, the signal can generally be increased by focusing the laser beam into the sample cell. However, the signal is also proportional to the number of absorbing molecules within the interaction volume, which decreases with decreasing focal volume. With pulsed lasers the two-photon transition may already be saturated at a certain incident intensity. In this case stronger focusing will decrease the signal. The optimization of the focusing conditions can be based on the following estimation:

Assume the laser beams propagate into the $\pm z$-directions. The two-photon absorption is monitored via the fluorescence from the upper level $|f\rangle$ and can be collected from a sample volume $\Delta V = \pi r^2(z)\Delta z$ (Fig. 7.31), where $r(z)$ is the beam radius (Sect. 5.3) and Δz the maximum length of the interaction volume seen by the collecting optics. The two-photon signal is then for $N_f \ll N_i$,

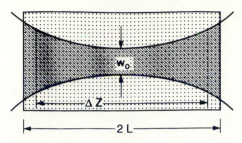

Fig. 7.31. Beam waist w_0 and Rayleigh length L for optimum focusing in two-photon spectroscopy

that is, $\Delta N \approx N_i$, according to (7.65)

$$S(\tfrac{1}{2}\omega_{if}) \propto N_i \left| \sum_m \frac{(R_{im}\hat{e}) \cdot (R_{mf}\hat{e})}{\omega - \omega_{im}} \right|^2 \int_z \int_r I^2(r, z) 2\pi r \, dr \, dz \,, \qquad (7.66)$$

where N_i is the density of the molecules in the absorbing level $|i\rangle$ and

$$I(r, z) = \frac{2P_0}{\pi w^2} e^{2r^2/w^2} \qquad (7.67)$$

gives the radial intensity distribution of the laser beam with power P_0 in the TEM$_{00}$ mode with a Gaussian intensity profile. The beam waist

$$w(z) = w_0 \sqrt{1 + \frac{\lambda z}{\pi \gamma w_0^2}} \,, \qquad (7.68)$$

gives the radius of the Gaussian beam in the vicinity of the focus at $z = 0$ (Sect. 5.4). The Rayleigh length $L = \pi w_0^2/\lambda$ gives that distance from the focus at $z = 0$ where the beam cross section $\pi w^2(L) = 2\pi w_0^2$ has increased by a factor of two compared to the area πw_0^2 at $z = 0$.

Inserting (7.67, 7.68) into (7.66) shows that the signal S_{if} becomes maximum if

$$\int_{-\Delta z/2}^{+\Delta z/2} \frac{dz}{1 + \left(\lambda z/\pi w_0^2\right)^2} = 2L \arctan[\Delta z/(2L)] \,, \qquad (7.69)$$

has a maximum value. Equation (7.69) shows that the signal cannot be increased any longer by stronger focusing when the Rayleigh length L becomes smaller than the interval Δz from which the fluorescence can be collected.

The two-photon signal can be greatly enhanced if the sample is placed either inside the laser resonator or in an external resonator, which has to be tuned synchroneously with the laser wavelength (Sect. 6.4).

7.5.4 Examples of Doppler-Free Two-Photon Spectroscopy

The first experiments on Doppler-free two-photon spectroscopy were performed on the alkali atoms [7.43–7.47] because their two-photon transitions
can be induced by cw dye lasers or diode lasers in convenient spectral ranges.
Furthermore, the first excited P state is not too far away from the virtual level
in Fig. 7.26b. This enlarges the two-photon transition probabilities for such
"near-resonant" transitions. Meanwhile, there are numerous further applications of this sub-Doppler technique in atomic and molecular physics. We shall
illustrate them by a few examples only.

The isotope shift between different stable lead isotopes shown in Fig. 7.32
has been determined by Doppler-free two-photon absorption of a cw dye laser
beam at $\lambda = 450.4$ nm on the transition $6p^2\,^3P_0 \to 7p\,^3P_0$ [7.48] using the
experimental arrangement of Fig. 7.28.

Doppler-free two-photon transitions to atomic Rydberg levels [7.49] allow
the accurate determination of quantum defects and of level shifts in external fields. Hyperfine structures in Rydberg states of two-electron atoms, such
as calcium and singlet–triplet mixing of the valence state 4s and ^{1}D and
^{3}D Rydberg levels have been thoroughly studied by Doppler-free two-photon
spectroscopy [7.50].

The application of two-photon spectroscopy to molecules has brought
a wealth of new insight to excited molecular states. One example is the
two-photon excitation of CO in the fourth positive system $A\,^1\Pi \leftarrow X\,^1\Sigma_g$
and of N_2 in the Lyman–Birge–Hopfield system with a narrow-band pulsed
frequency-doubled dye laser. Doppler-free spectra of states with excitation energies between $8-12$ eV can be measured with this technique [7.51].

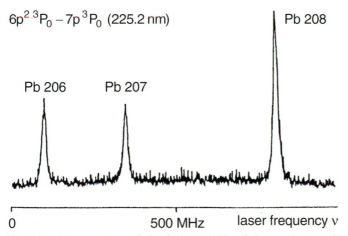

Fig. 7.32. Measurement of the isotope shift of the stable lead isotopes measured with
Doppler-free two-photon spectroscopy at $\lambda_{exc} = 450$ nm and monitored via fluorescence
[7.48]

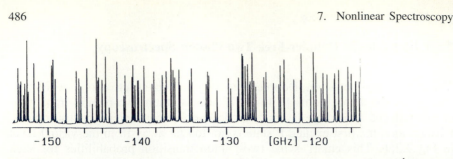

Fig. 7.33. Section of the Doppler-free two-photon excitation spectrum of the 14_0^1 Q_Q band of C_6H_6 [7.53]

For larger molecules, rotationally resolved absorption spectra could, for the first time, be measured, as has been demonstrated for the UV spectra of benzene C_6H_6. Spectral features, which had been regarded as true continua in former times, could now be completely resolved (Fig. 7.33) and turned out to be dense but discrete rotational-line spectra [7.52, 7.53]. The lifetimes of these upper levels could be determined from the natural linewidths of these transitions [7.54]. It was proven that these lifetimes strongly decrease with increasing vibrational–rotational energy in the excited electronic state because of the increasing rate of radiationless transitions [7.55].

Molecular two-photon spectroscopy can also be applied in the infrared region to induce transitions between rotational–vibrational levels within the electronic ground state. One example is the Doppler-free spectroscopy of rotational lines in the ν_2 vibrational bands of NH_3 [7.56]. This allows the study of the collisional properties of the ν_2 vibrational manifold from pressure broadening and shifts (Sect. 3.3) and Stark shifts.

Near-resonant two-photon spectroscopy of NH_3 has been reported by Winnewisser et al. [7.57], where a diode laser was tuned to a frequency ν_1 and a two-photon excitation $\nu_1 + \nu_2$ of the $2\nu_2$ (1,1) level of NH_3 is observed with the second photon $h\nu_2$ coming from a fixed-frequency CO_2 laser.

The possibilities of Doppler-free two-photon spectroscopy for metrology and fundamental physics has been impressively demonstrated by precision measurements of the $1S - 2S$ transition in atomic hydrogen [7.58–7.61]. Precise measurements of this one-photon forbidden transition with a very narrow natural linewidth of 1.3 Hz yield accurate values of fundamental constants and can provide stringent tests of quantum electrodynamic theory (Sect. 14.7). A comparison of the $1S - 2S$ transition frequency with the $2S - 3P$ frequency allows the precise determination of the Lamb shift of the $1S$ ground state [7.59], whereas the $2S$ Lamb shift was already measured long ago by the famous Lamb–Rutherford experiments where the RF transition between $2S_{1/2}$ and $2P_{1/2}$ were observed. Because of the isotope shift the $1S - 2S$ transitions of ^1H and $^2H = {}_1^2D$ differ by 670.99433464 GHz (Fig. 7.34). From this measurement a value $r_D^2 - r_P^2 = 3.8212 \, \text{fm}^2$ for the difference of the mean square charge radii of proton and deuteron is derived. The Rydberg constant has been determined within a relative uncertainty of 10^{-10} [7.60–7.62].

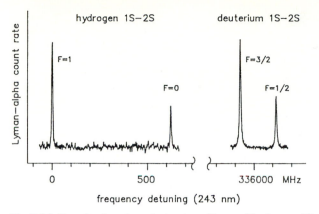

Fig. 7.34. Doppler-free two-photon transitions of hydrogen $1S - 2S$ and deuterium $1S - 2S$ [7.58]

7.5.5 Multiphoton Spectroscopy

If the incident intensity is sufficiently large, a molecule may absorb several photons simultaneously. The probability for absorption of a photon $\hbar\omega_k$ on the transition $|i\rangle \rightarrow |f\rangle$ with $E_f - E_i = \sum \hbar\omega_k$ can be obtained from a generalization of (7.60). The first factor in (7.60) now contains the product $\prod_k I_k$ of the intensities I_k of the different beams. In the case of n-photon absorption of a single laser beam this product becomes I^n. The second factor in the generalized formula (7.60) includes the sum over products of n one-photon matrix elements.

In the case of *Doppler-free* multiphoton absorption, momentum conservation

$$\sum_k \boldsymbol{p}_k = \hbar \sum_k \boldsymbol{k}_k = 0 , \qquad (7.70)$$

has to be fulfilled in addition to energy conservation $\sum \hbar\omega_k = E_f - E_i$. Each of the absorbed photons transfers the momentum $\hbar\boldsymbol{k}_k$ to the molecule. If (7.70) holds, the total transfer of momentum is zero, which implies that the velocity of the absorbing molecule has not changed. This means that the photon energy $\hbar \sum \omega_k$ is completely converted into excitation energy of the molecule without changing its kinetic energy. This is independent of the initial velocity of the molecule, that is, the transition is Doppler-free.

A possible experimental arrangement for Doppler-free three-photon absorption spectroscopy is depicted in Fig. 7.35. The three laser beams generated by beam splitting of a single dye laser beam cross each other under $120°$ in the absorbing sample.

The three-photon absorption can be used for the excitation of high-lying molecular levels with the same parity as accessible to one-photon transitions. However, for a one-photon absorption, lasers with a wavelength $\lambda/3$

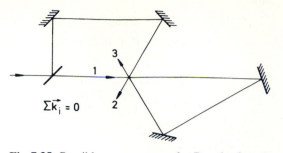

Fig. 7.35. Possible arrangements for Doppler-free three-photon spectroscopy

have to be available in order to reach the same excitation energy. An example of Doppler-limited collinear three-photon spectroscopy is the excitation of high-lying levels of xenon and CO with a narrow-band pulsed dye laser at $\lambda = 440$ nm (Fig. 7.36). For one-photon transitions light sources at $\lambda = 146.7$ nm in the VUV would have been necessary.

The absorption probability increases if a two-photon resonance can be found. One example is illustrated in Fig. 7.37a, where the $4D$ level of the Na atom is excited by two-photons of a dye laser at $\lambda = 578.7$ nm and further excitation by a third photon reaching high-lying Rydberg levels nP or nF with the electronic orbital momentum $\ell = 1$ or $\ell = 3$. For Doppler-free excitation the wave-vector diagram of Fig. 7.35 has been used [7.64].

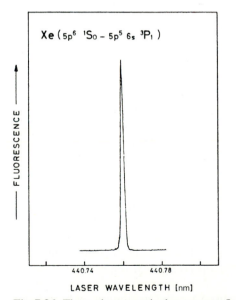

Fig. 7.36. Three-photon excited resonance fluorescence in Xe at $\lambda = 147$ nm, excited with a pulsed dye laser at $\lambda = 441$ nm with 80-kW peak power. The Xe pressure was 8 mtorr [7.63]

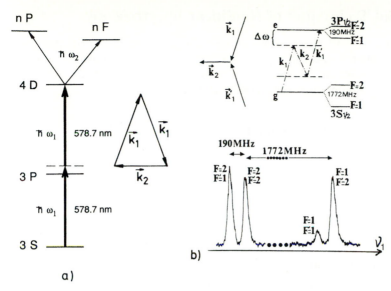

Fig. 7.37a,b. Level schemes for Doppler-free three-photon spectroscopy of the Na atom: (**a**) stepwise excitation of Rydberg states; (**b**) Raman-type process shown for the example of the 3S3P excitation of the Na atom. Laser 1 is tuned while L2 is kept 30 GHz below the Na D_1 line [7.64]

Three-photon excitation may also be used for a Raman-type process, depicted in Fig. 7.37b, which proceeds via two virtual levels. This Doppler-free technique was demonstrated for the $3\,^2S_{1/2} \rightarrow 3\,^2P_{1/2}$ transition of the Na atom, where the photons with the momentum $\hbar\,\boldsymbol{k}_1$ and $\hbar\boldsymbol{k}_1'$ are absorbed while the photon with the momentum $\hbar k_2$ is emitted. The hyperfine structure of the upper and lower states could readily be resolved [7.64].

Multiphoton absorption of visible photons may result in ionization of atoms or molecules. At a given laser intensity the ion rate $N_{\mathrm{ion}}(\lambda_{\mathrm{L}})$ recorded as a function of the laser intensity shows narrow maxima if one-, two-, or three-photon resonances occur. If, for instance, the ionization potential (IP) is smaller than $3\hbar\omega$, resonances in the ionization yield are observed, either when the laser frequency ω is in resonance with a two-photon transition between the levels $|i\rangle$ and $|f\rangle$, i.e. ($E_f - E_i = 2\hbar\omega$), or when autoionizing Rydberg states can be reached by three-photon transitions [7.65].

Multiphoton absorption has also been observed on transitions within the electronic ground states of molecules, induced by infrared photons of a CO_2 laser [7.66, 7.67]. At sufficient intensities this multiphoton excitation of high vibrational–rotational states may lead to the dissociation of the molecule [7.68].

If the first step of the multiphoton excitation can be chosen isotope selectively so that a wanted isotopomer has a larger absorption probability than the other isotopomers of the molecular species, selective dissociation may be achieved [7.68], which can be used for isotope separation by chemical reactions with the isotope-selective dissociation products (Sect. 15.2).

7.6 Special Techniques of Nonlinear Spectroscopy

In this section we will briefly discuss some variations of saturation, polarization, or multiphoton spectroscopy that either increase the sensitivity or are adapted to the solution of special spectroscopic problems. They are often based on combinations of several nonlinear techniques.

7.6.1 Saturated Interference Spectroscopy

The higher sensitivity of polarization spectroscopy compared with conventional saturation spectroscopy results from the detection of *phase differences* rather than amplitude differences. This advantage is also used in a method that monitors the interference between two probe beams where one of the beams suffers saturation-induced phase shifts. This saturated interference spectroscopy was independently developed in different laboratories [7.69, 7.70]. The basic principle can easily be understood from Fig. 7.38. We follow here the presentation in [7.69].

The probe beam is split by the plane-parallel plate Pl1 into two beams. One beam passes through that region of the absorbing sample that is saturated by the pump beam; the other passes through an unsaturated region of the same sample cell. The two beams are recombined by a second plane-parallel plate Pl2. The two carefully aligned parallel plates form a Jamin interferometer [7.71], which can be adjusted by a piezoelement in such a way that without the saturating pump beam the two probe waves with intensities I_1 and I_2 interfere destructively.

If the saturation by the pump wave introduces a phase shift φ, the resulting intensity at the detector becomes

$$I = I_1 + I_2 - 2\sqrt{I_1 I_2}\cos\varphi . \tag{7.71}$$

The intensities I_1 and I_2 of the two interfering probe waves can be made equal by placing a polarizer P1 into one of the beams and a second polarizer P2 in

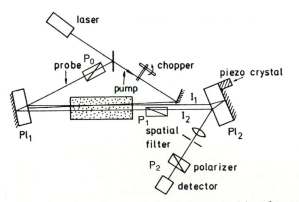

Fig. 7.38. Schematic arrangement of saturated interference spectroscopy

front of the detector. Due to a slight difference δ in the absorptions of the two beams by the sample molecules, their intensities at the detector are related by

$$I_1 = I_2(1+\delta) \quad \text{with} \quad \delta \ll 1 .$$

For small phase shifts φ $(\varphi \ll 1 \rightarrow \cos\varphi \approx 1 - \frac{1}{2}\varphi^2)$, we can approximate (7.71) by

$$I \approx \left(\frac{1}{4}\delta^2 + \varphi^2\right) I_2 , \tag{7.72}$$

when we neglect the higher-order terms $\delta\varphi^2$ and $\delta^2\varphi^2$. The amplitude difference δ and the phase shift φ are both caused by selective saturation of the sample through the monochromatic pump wave, which travels in the opposite direction. Analogous to the situation in polarization spectroscopy, we therefore obtain Lorentzian and dispersion profiles for the frequency dependence of both quantities

$$\delta(\omega) = \frac{\delta_0}{1+x^2} , \quad \varphi(\omega) = \frac{1}{2}\delta_0\frac{x}{1+x^2} , \tag{7.73}$$

where $\delta_0 = \delta(\omega_0)$, $x = 2(\omega-\omega_0)/\gamma$, and γ is the homogeneous linewidth (FWHM).

Inserting (7.73) into (7.72) yields, for the total intensity I at the minimum of the interference patterns, the Lorentzian profile,

$$I = \frac{1}{4}\frac{I_2\delta_0^2}{1+x^2} . \tag{7.74}$$

According to (7.73), the phase differences $\varphi(\omega)$ depends on the laser frequency ω. However, it can always be adjusted to zero while the laser frequency is scanned. This can be accomplished by a sine-wave voltage at the piezoelement, which causes a modulation

$$\varphi(\omega) = \varphi_0(\omega) + a\sin(2\pi f_1 t) .$$

When the detector signal is fed to a lock-in amplifier that is tuned to the modulation frequency f_1, the lock-in output can drive a servo loop to bring the phase difference φ_0 back to zero. For $\varphi(\omega) \equiv 0$, we obtain from (7.72, 7.73)

$$I(\omega) = \frac{1}{4}\delta(\omega)^2 I_2 = \frac{1}{4}\frac{\delta_0^2 I_2}{(1+x^2)^2} . \tag{7.75}$$

The halfwidth of this signal is reduced from γ to $(\sqrt{2}-1)^{1/2}\gamma \approx 0.64\gamma$.

Contrary to the situation in polarization spectroscopy, where for slightly uncrossed polarizers the line shape of the polarization signal is a superposition of Lorentzian and dispersion profiles, with saturated interference spectroscopy pure Lorentzian profiles are obtained because the phase shift is compensated by the feedback control. Measuring the first derivative of the profiles, pure dispersion-type signals appear. To achieve this, the output of the lock-in amplifier that controls the phase is fed into another lock-in that is tuned to a frequency f_2 ($f_2 \ll f_1$) at which the saturating pump beam is chopped.

The method has been applied so far to the spectroscopy of Na$_2$ [7.69] and I$_2$ [7.70]. Figure 7.39a shows saturated absorption signals in I$_2$ obtained with conventional saturation spectroscopy using a dye laser at $\lambda = 600$ nm with 10-mW pump power and 1-mW probe power. Figure 7.39b displays the first derivative of the spectrum in Fig. 7.39a and Fig. 7.39c the first derivative of the saturated interference signal.

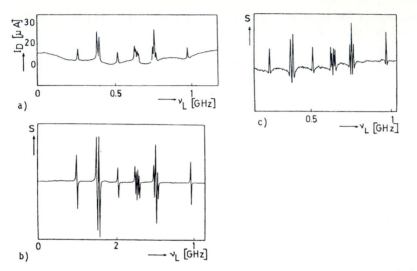

Fig. 7.39a–c. Saturated interference spectra of I$_2$ at $\lambda = 600$ nm: (**a**) saturated absorption signal of the hfs components; (**b**) first-derivative spectrum of (a); (**c**) first derivative of saturated dispersion signal [7.70]

The sensitivity of the saturated interference technique is comparable to that of polarization spectroscopy. While the latter can be applied only to transitions from levels with a rotational quantum number $J \geq 1$, the former works also for $J = 0$. An experimental drawback may be the critical alignment of the Jamin interferometer and its stability during the measurements.

7.6.2 Doppler-Free Laser-Induced Dichroism and Birefringence

A slight modification of the experimental arrangement used for polarization spectroscopy allows the simultaneous observation of saturated absorption *and* dispersion [7.72]. While in the setup of Fig. 7.20 the probe beam had linear polarization, here a circularly polarized probe and a linearly polarized pump beam are used (Fig. 7.40). The probe beam can be composed of two components with linear polarization parallel and perpendicular to the pump beam polarization. Due to anisotropic saturation by the pump, the absorption coefficients $\alpha_\parallel$ and $\alpha_\perp$ and the refractive indices $n_\parallel$ and $n_\perp$ experienced by the

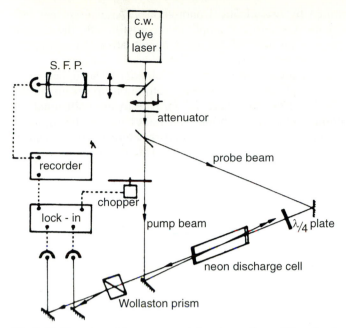

Fig. 7.40. Experimental arrangement for observation of Doppler-free laser-induced dichroism and birefringence [7.72]

probe beam are different for the parallel and the perpendicular polarizations. This causes a change of the probe beam polarization, which is monitored behind a linear analyzer rotated through the angle β from the reference direction π. Analogous to the derivation in Sect. 7.4.2, one can show that the transmitted intensity of a circularly polarized probe wave with incident intensity I is for $\alpha L \ll 1$ and $\Delta n(L/\lambda) \ll 1$

$$I_t(\beta) = \frac{I}{2}\left(1 - \frac{\alpha_\| + \alpha_\perp}{2}L - \frac{L}{2}\Delta\alpha\cos 2\beta - \frac{\omega L}{c}\Delta n \sin 2\beta\right),\qquad(7.76)$$

with $\Delta\alpha = \alpha_\| - \alpha_\perp$ and $\Delta n = n_\| - n_\perp$.

The difference of the two transmitted intensities

$$\Delta_1 = I_t(\beta = 0°) - I_t(\beta = 90°) = IL\Delta\alpha/2,\qquad(7.77)$$

gives the pure dichroism signal (anisotropic saturated absorption), while the difference

$$\Delta_2 = I_t(45°) - I_t(-45°) = I(\omega L/c)\Delta n,\qquad(7.78)$$

yields the pure birefringence signal (saturated dispersion). A birefringent Wollaston prism behind the interaction region allows the spatial separation of the two probe beam components with mutually orthogonal polarizations. The

two beams are monitored by two identical photodiodes. After a correct balance of the output signals, a differential amplifier records directly the desired
differences Δ_1 and Δ_2 if the axes of the birefringent prism have suitable
orientations.

Figure 7.41 illustrates the advantages of this technique. The upper spectrum represents a Lamb peak in the intracavity saturation spectrum of the
neon line $(1s \rightarrow 2p)$ at $\lambda = 588.2$ nm (Sect. 7.3.3). Due to the collisional redistribution of the atomic velocities, a broad and rather intense background
appears in addition to the narrow peak. This broad structure is not present in
the dichroism and birefringent curves (Fig. 7.41b,c). This improves the signal-
to-noise ratio and the spectral resolution.

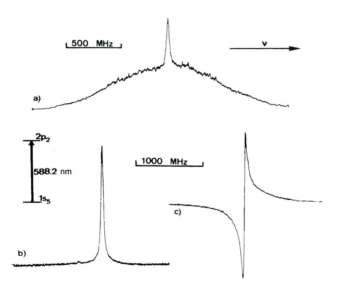

Fig. 7.41a–c. Comparison of different techniques for measuring the neon transition $1s_2 -$
$2p_2$ at $\lambda = 588.2$ nm: (**a**) intracavity saturation spectroscopy (Lamb peak of the laser
output $I_L(\omega)$ with Doppler-broadened background); (**b**) laser-induced dichroism; and
(**c**) laser-induced birefringence [7.72]

7.6.3 Heterodyne Polarization Spectroscopy

In most detection schemes of saturation or polarization spectroscopy the intensity fluctuations of the probe laser represent the major contribution to
the noise. Generally, the noise power spectrum $P_{Noise}(f)$ shows a frequen-
cy-dependence, where the spectral power density decreases with increasing
frequency (e.g., $1/f$-noise). It is therefore advantageous for a high S/N ratio
to detect the signal S behind a lock-in amplifier at high frequencies f.

This is the basic idea of heterodyne polarization spectroscopy [7.73, 7.74],
where the pump wave at the frequency ω_p passes through an acousto-optical
modulator, driven at the modulation frequency f, which generates sidebands
at $\omega = \omega_p \pm 2\pi f$ (Fig. 7.42). The sideband at $\omega_t = \omega_p + 2\pi f$ is sent as a pump

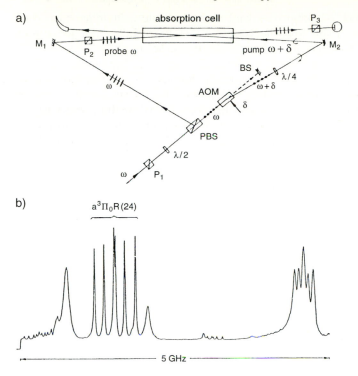

Fig. 7.42. (a) Experimental arrangement of heterodyne polarization spectroscopy; **(b)** section of the Na$_2$ polarization spectrum showing the hyperfine splitting of the R(24) rotational line of the spin-forbidden transition $X^1\Sigma_g \to {}^3\Pi_u$ [7.74]

beam through the sample cell, while the probe beam at frequency ω_p is split off the laser beam in front of the modulator. Otherwise the setup is similar to that of Fig. 7.20.

The signal intensity transmitted by the analyzer P$_3$ is described by (7.44). However, the quantity $x = 2(\omega - \omega_0 - 2\pi f)/\gamma$ now differs from (7.42) by the frequency shift f, and the amplitude of the polarization signal is modulated at the difference frequency f between pump and probe beam. The signal can therefore be detected through a lock-in device tuned to this frequency, which is chosen in the MHz range. No further chopping of the pump beam is necessary.

7.6.4 Combination of Different Nonlinear Techniques

A combination of Doppler-free two-photon spectroscopy and polarization spectroscopy was utilized by Grützmacher et al. [7.75] for the measurement of the Lyman-λ line profile in a hydrogen plasma at low pressure.

A famous example of such a combination is the first precise measurement of the Lamb shift in the 1S ground state of the hydrogen atom by

Hänsch et al. [7.59]. Although new and more accurate techniques have been developed by Hänsch and coworkers (Sect. 14.7), the "old" technique is quite instructive and shall therefore be briefly discussed here.

The experimental arrangement is shown in Fig. 7.43. The output of a tunable dye laser at $\lambda = 486$ nm is frequency-doubled in a nonlinear crystal. While the fundamental wave at 486 nm is used for Doppler-free saturation spectroscopy [7.59] or polarization spectroscopy [7.76] of the Balmer transition $2S_{1/2} - 4P_{1/2}$, the second harmonics of the laser at $\lambda = 243$ nm induce the Doppler-free two-photon transition $1S_{1/2} \to 2S_{1/2}$. In the simple Bohr model [7.77], both transitions should be induced at the same frequency since in this model $\nu(1S - 2S) = 4\nu(2S - 4P)$. The measured frequency difference $\Delta\nu = \nu(1S - 2S) - 4\nu(2S - 4P)$ yields the Lamb shift $\delta\nu_L(1S) = \Delta\nu - \delta\nu_L(2S)$ $- \Delta\nu_{fs}(4S_{1/2} - 4P_{1/2}) - \delta\nu_L(4S)$. The Lamb shift $\delta\nu_L(2S)$ is known and $\Delta\nu_{fs}$ $(4S_{1/2} - 4P_{1/2})$ can be calculated within the Dirac theory. The frequency markers of the FPI allow the accurate determination of the hfs splitting of the 1S state and the isotope shift $\Delta\nu_{Is}(^1H - {}^2H)$ between the $1S - 2S$ transitions of hydrogen ^1H and deuterium ^2H (Fig. 7.34).

The more recent version of this precision measurement of the $1S - 2S$ transition is shown in Fig. 7.44. The hydrogen atoms are formed in a microwave discharge and effuse through a cold nozzle into the vacuum, forming a collimated beam. The laser beam is sent through the nozzle and is reflected back anticollinear to the atomic beam axis. The metastable 2S atoms fly through an electric field where the 2S state is mixed with the 2P state. The 2P atoms

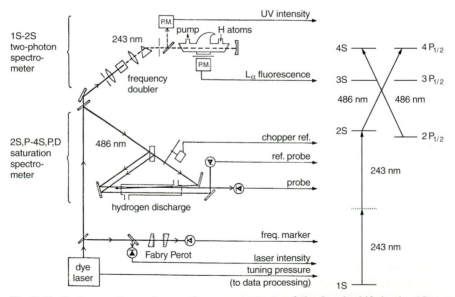

Fig. 7.43. Early experimental setup for measurements of the Lamb shift in the 1S state and of the fine structure in the $^2P_{1/2}$ state of the H atom by combination of Doppler-free two-photon and saturation spectroscopy [7.59]

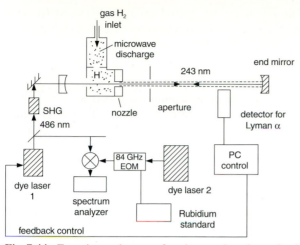

Fig. 7.44. Experimental setup for the precise determination of H(1S → 2S) frequency [7.78]

emit Lyman-α fluorescence, which is detected by a solar blind photomultiplier. The achievable $1S - 2S$ linewidth is about 1 kHz, limited by transit-time broadening. The uncertainty of determining the line center is below 30 Hz!

7.7 Conclusion

The few examples shown above illustrate that nonlinear spectroscopy represents an important branch of laser spectroscopy of atoms and molecules. Its advantages are the Doppler-free spectral resolution if narrow-band lasers are used and the possibilities to reach high-lying states by multiphoton absorption with pulsed or cw lasers. Because of its relevance for molecular physics, numerous books and reviews cover this field. The references [7.2, 7.4, 7.79–7.87] represent only a small selection.

In combination with double-resonance techniques, nonlinear spectroscopy has contributed particularly to the assignment of complex spectra and has therefore considerably increased our knowledge about molecular structure and dynamics. This subject is covered in Chap. 10.

Problems

7.1

(a) A collimated sodium beam is crossed by a single-mode cw dye laser, tuned to the D_1 transition $3\,^2P_{1/2} \leftarrow 3\,^2S_{1/2}$ of Na. Calculate the saturation intensity I_s if the flux of sodium atoms is $N = n \cdot \bar{v} = 10^{15}$ atoms [cm^{-2} s^{-1}]. The lifetime τ_K of the upper level is $\tau_K = 16\,\mu$s.

(b) How large is I_s in a sodium cell at $P_{Na} = 10^{-6}$ mbar with $P_{An} = 10$ mbar additional argon pressure? The pressure broadening is 25 MHz/mbar for Na $-$ Ar collisions.

7.2 A pulsed dye laser with the pulse length $\Delta T = 10^{-8}$ s and with a peak power of $P = 1$ kW illuminates a molecular sample in a cell at $p = 1$ mbar and $T = 300$ K. A rectangular intensity profile is assumed with a laser-beam cross section of 1 cm^2. Which fraction of all molecules N_i in the absorbing lower level $|i\rangle$ is excited when the laser is tuned to a weak absorbing transition $|i\rangle \rightarrow |k\rangle$ with the absorption cross section $\sigma_{ik} = 10^{-18}$ cm^2? The laser bandwidth is assumed to be 3 times the Doppler width.

7.3 In an experiment on polarization spectroscopy the circularly polarized pump laser causes a change $\Delta\alpha = \alpha^+ - \alpha^- = 10^{-2}\alpha_0$ of the absorption coefficient. By which angle is the plane of polarization of the linearly polarized probe laser beam at $\lambda = 600\,\mu$m tuned after passing through the pumped region with length L, if the absorption without pump laser $\alpha_0 L = 5 \times 10^{-2}$?

7.4 Estimate the fluorescence rate (number of fluorescence photons/s) on the Na transition $5s \rightarrow 3p$, obtained in the Doppler-free free-photon experiment of Fig. 7.28, when a single-mode dye laser is timed to $\nu/2$ of the transition $3s \rightarrow 5s$ in a cell with a Na density of $n = 10^{12}$ cm^{-3}. The laser power is $P = 100$ mW, the beam is focused to the beam waist $w_0 = 10^{-2}$ cm and a length $L = 1$ cm around the focus is imaged with a collection efficiency of 5% onto the fluorescence detector.

7.5 The saturation spectrum of the Na D_1 transition $3\,^2S_{1/2} \rightarrow 3\,^2P_{1/2}$ shows the resolved hyperfine components. Estimate the relative magnitude of the cross-over signal between the two transitions $3\,^2S_{1/2}(F = 1) \rightarrow 3\,P_{1/2}$ ($F = 1$ and $F = 2$) sharing the same laser level, if the laser intensity is 2 times the saturation intensity I_s.

7.6 Estimate from the known matrix elements of an atomic-hydrogen transition the fraction of H atoms in the $1\,^2S_{1/2}$ ground state that can be excited by a Doppler-free two-photon transition into the $2\,^2S_{1/2}$ state in a collimated H atomic beam with $\bar{v} = 10^3$ m/s, when a laser with $I = 10^3$ W/cm^2 and a rectangular beam cross section of 1×1 mm^2 crosses the atomic beam perpendicularly.

8. Laser Raman Spectroscopy

For many years Raman spectroscopy has been a powerful tool for the investigation of molecular vibrations and rotations. In the pre-laser era, however, its main drawback was a lack of sufficiently intense radiation sources. The introduction of lasers, therefore, has indeed revolutionized this classical field of spectroscopy. Lasers have not only greatly enhanced the sensitivity of spontaneous Raman spectroscopy but they have furthermore initiated new spectroscopic techniques, based on the stimulated Raman effect, such as coherent anti-Stokes Raman scattering (CARS) or hyper-Raman spectroscopy. The research activities in laser Raman spectroscopy have recently shown an impressive expansion and a vast literature on this field is available. In this chapter we summarize only briefly the basic background of the Raman effect and present some experimental techniques that have been developed for Raman spectroscopy of gaseous media. For more thorough studies of this interesting field the textbooks and reviews given in [8.1–8.12] and the conference proceedings [8.13, 8.14] are recommended. More information on Raman spectroscopy of liquids and solids can be found in [8.11, 8.15–8.18].

8.1 Basic Considerations

Raman scattering may be regarded as an inelastic collision of an incident photon $\hbar\omega_i$ with a molecule in the initial energy level E_i (Fig. 8.1a). Following the collision, a photon $\hbar\omega_s$ with lower energy is detected and the molecule is found in a higher-energy level E_f

$$\hbar\omega_i + M(E_i) \rightarrow M^*(E_f) + \hbar\omega_s , \quad \text{with} \quad \hbar(\omega_i - \omega_s) = E_f - E_i > 0 . \quad (8.1a)$$

The energy difference $\Delta E = E_f - E_i$ may appear as vibrational, rotational, or electronic energy of the molecule.

If the photon $\hbar\omega_i$ is scattered by a vibrationally excited molecule, it may gain energy and the scattered photon has a higher frequency ω_{as} (Fig. 8.1c), where

$$\hbar\omega_{as} = \hbar\omega_i + E_i - E_f , \quad \text{with} \quad E_i > E_f . \quad (8.1b)$$

This "superelastic" photon scattering is called *anti-Stokes radiation*.

In the energy level scheme (Fig. 8.1b), the intermediate state $E_v = E_i + \hbar\omega_i$ of the system "during" the scattering process is often formally described as

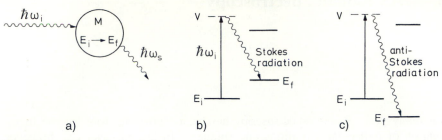

Fig. 8.1a–c. Schematic level diagram of Raman scattering

a *virtual* level, which, however, is not necessarily a "real" stationary eigen-state of the molecule. If the virtual level coincides with one of the molecular eigenstates, one speaks of the *resonance Raman effect*.

A classical description of the *vibrational Raman effect* (which was the main process studied before the introduction of lasers) has been developed by Placek [8.8]. It starts from the relation

$$p = \mu_0 + \tilde{\alpha} E , \tag{8.2}$$

between the electric field amplitude $E = E_0 \cos \omega t$ of the incident wave and the dipole moment p of a molecule. The first term μ_0 represents a possible *permanent* dipole moment while $\tilde{\alpha} E$ is the *induced* dipole moment. The polar-izability is generally expressed by the tensor (α_{ij}) of rank two, which depends on the molecular symmetry. Dipole moment and polarizability are functions of the coordinates of the nuclei and electrons. However, as long as the frequency of the incident radiation is far off resonance with electronic or vibrational transitions, the nuclear displacements induced by the polarization of the elec-tron cloud are sufficiently small. Since the electronic charge distribution is determined by the nuclear positions and adjusts "instantaneously" to changes in these positions, we can expand the dipole moment and polarizability into Taylor series in the normal coordinates q_n of the nuclear displacements

$$\mu = \mu(0) + \sum_{n=1}^{Q} \left(\frac{\partial \mu}{\partial q_n} \right)_0 q_n + \dots ,$$

$$\alpha_{ij}(q) = \alpha_{ij}(0) + \sum_{n=1}^{Q} \left(\frac{\partial \alpha_{ij}}{\partial q_n} \right)_0 q_n + \dots , \tag{8.3}$$

where $Q = 3N - 6$ (or $3N - 5$ for linear molecules) gives the number of nor-mal vibrational modes for N nuclei, and $\mu(0) = \mu_0$ and $\alpha_{ij}(0)$ are the dipole moment and the polarizability at the equilibrium configuration $q_n = 0$. For small vibrational amplitudes the normal coordinates $q_n(t)$ of the vibrating molecule can be approximated by

$$q_n(t) = q_{n0}\cos(\omega_n t) ,\qquad(8.4)$$

where q_{n0} gives the amplitude, and ω_n the vibrational frequency of the nth normal vibration. Inserting (8.4 and 8.3) into (8.2) yields the total dipole moment

$$\boldsymbol{p} = \boldsymbol{\mu}_0 + \sum_{n=1}^{Q}\left(\frac{\partial \boldsymbol{\mu}}{\partial q_n}\right)_0 q_{n0}\cos(\omega_n t) + \alpha_{ij}(0)E_0\cos(\omega t)$$

$$+ \frac{1}{2}E_0\sum_{n=1}^{Q}\left(\frac{\partial \alpha_{ij}}{\partial q_n}\right)_0 q_{n0}[\cos(\omega+\omega_n)t + \cos(\omega-\omega_n)t] .\qquad(8.5)$$

The second term describes the infrared spectrum, the third term the Rayleigh scattering, and the last term represents the Raman scattering. In Fig. 8.2 the dependence of $\partial\boldsymbol{\mu}/\partial q$ and $\partial\alpha/\partial q$ is shown for the three normal vibrations of the CO_2 molecule. This illustrates that $\partial\boldsymbol{\mu}/\partial q \neq 0$ for the bending vibration ν_2 and for the asymmetric stretch ν_3. These two vibrational modes are called "infrared active." The polarizability change is $\partial\alpha/\partial q \neq 0$ only for the symmetric stretch ν_1, which is therefore called "Raman active."

Since an oscillating dipole moment is a source of new waves generated at each molecule, (8.5) shows that an elastically scattered wave at the frequency ω is produced (Rayleigh scattering) as are inelastically scattered components with the frequencies $\omega - \omega_n$ (*Stokes waves*) and superelastically scattered waves with the frequencies $\omega + \omega_n$ (*anti-Stokes components*). The microscopic contributions from each molecule add up to macroscopic waves

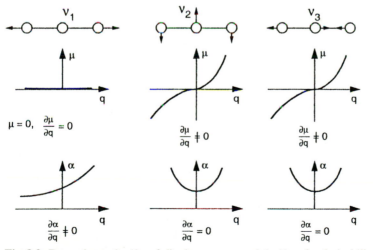

Fig. 8.2. Dependence $\partial\boldsymbol{\mu}/\partial q$ of dipole moment and $\partial\alpha/\partial q$ of polarizability on the normal vibrations of the CO_2 molecule

with intensities that depend on the population $N(E_i)$ of the molecules in the initial level E_i, on the intensity of the incident radiation, and on the expression $(\partial\alpha_{ij}/\partial q_n)q_n$, which describes the dependence of the polarizability components on the nuclear displacements.

Although the classical theory correctly describes the frequencies $\omega \pm \omega_n$ of the Raman lines, it fails to give the correct intensities and a quantum mechanical treatment is demanded. The expectation value of the component α_{ij} of the polarizability tensor is given by

$$\langle\alpha_{ij}\rangle_{ab} = \int u_b^*(q)\alpha_{ij}u_a(q)\,dq \;, \tag{8.6}$$

where the functions $u(q)$ represent the molecular eigenfunctions in the initial level a and the final level b. The integration extends over all nuclear coordinates. This shows that a computation of the intensities of Raman lines is based on the knowledge of the molecular wave functions of the initial and final states. In the case of vibrational–rotational Raman scattering these are the rotational–vibrational eigenfunctions of the electronic ground state.

For small displacements q_n, the molecular potential can be approximated by a harmonic potential, where the coupling between the different normal vibrational modes can be neglected. The functions $u(q)$ are then separable into a product

$$u(q) = \prod_{n=1}^{Q} w_n(q_n, v_n) \;, \tag{8.7}$$

of vibrational eigenfunction of the nth normal mode with v_n vibrational quanta. Using the orthogonality relation

$$\int w_n w_m \, dq = \delta_{nm} \;, \tag{8.8}$$

of the functions $w_n(q_n)$, one obtains from (8.6 and 8.3)

$$\langle\alpha_{ij}\rangle_{ab} = (\alpha_{ij})_0 + \sum_{n=1}^{Q} \left(\frac{\partial\alpha_{ij}}{\partial q_n}\right)_0 \int w_n(q_n, v_a)q_n w_n(q_n, v_b)\,dq_n \;. \tag{8.9}$$

The first term is a constant and is responsible for the Rayleigh scattering. For nondegenerate vibrations the integrals in the second term vanish unless $v_a = v_b \pm 1$. In these cases it has the value $[\frac{1}{2}(v_a+1)]^{1/2}$ [8.18]. The basic intensity parameter of vibrational Raman spectroscopy is the derivative $(\partial\alpha_{ij}/\partial q)$, which can be determined from the Raman spectra.

The intensity of a Raman line at the Stokes or anti-Stokes frequency $\omega_s = \omega \pm \omega_n$ is determined by the population density $N_i(E_i)$ in the initial level $E_i(v, J)$, by the intensity I_L of the incident pump laser, and by the

Raman scattering cross section $\sigma_R(i \to f)$ for the Raman transition $E_i \to E_f$:

$$I_s = N_i(E_i)\sigma_R(i \to f)I_L .\tag{8.10}$$

At thermal equilibrium the population density $N_i(E_i)$ follows the Boltzmann distribution

$$N_i(E_i, v, J) = \frac{N}{Z} g_i\, e^{-E_i/kT} , \quad \text{with} \quad N = \sum N_i .\tag{8.11a}$$

The statistical weight factors g_i depend on the vibrational state $v = (n_1 v_1, n_2 v_2, \dots)$, the rotational state with the rotational quantum number J, the projection K onto the symmetry axis in the case of a symmetric top, and furthermore on the nuclear spins I of the N nuclei. The partition function

$$Z = \sum_i g_i\, e^{-E_i/kT} ,\tag{8.11b}$$

is a normalization factor, which makes $\sum N_i(v, J) = N$, as can be verified by inserting (8.11b) into (8.11a).

In case of Stokes radiation the initial state of the molecules may be the vibrational ground state, while for the emission of anti-Stokes lines the molecules must have initial excitation energy. Because of the lower population density in these excited levels, the intensity of the anti-Stokes lines is lower by a factor $\exp(-\hbar\omega_v/kT)$.

> **Example 8.1**
> $\hbar\omega_v = 1000\,\text{cm}^{-1}$, $T = 300\,\text{K} \to kT \sim 250\,\text{cm}^{-1} \to \exp(-E_i/kT) \approx e^{-4} \approx 0.018$. With comparable cross sections σ_R the intensity of anti-Stokes lines is therefore lower by two orders of magnitude compared with those of Stokes lines.

The scattering cross section depends on the matrix element (8.9) of the polarizability tensor and furthermore contains the ω^4 frequency dependence derived from the classical theory of light scattering. One obtains [8.19] analogously to the two-photon cross section (Sect. 7.5)

$$\sigma_R(i \to f) = \frac{8\pi\omega_s^4}{9\hbar c^4} \left| \sum_j \frac{\langle\alpha_{ij}\rangle\hat{e}_L\langle\alpha_{jf}\rangle\hat{e}_s}{\omega_{ij} - \omega_L - i\gamma_j} + \frac{\langle\alpha_{ji}\rangle\hat{e}_L\langle\alpha_{jf}\rangle\hat{e}_s}{\omega_{jf} - \omega_L - i\gamma_j} \right|^2 ,\tag{8.12}$$

where $\hat{e}_L$ and $\hat{e}_s$ are unit vectors representing the polarization of the incident laser beam and the scattered light. The sum extends over all molecular levels j with homogeneous width γ_j accessible by single-photon transitions from the initial state i. We see from (8.12) that the initial and final states are connected by *two-photon* transitions, which implies that both states have the same parity. For example, the vibrational transitions in homonuclear diatomic molecules, which are forbidden for single-photon infrared transitions, are accessible to Raman transitions.

The matrix elements $\langle \alpha_{ij} \rangle$ depend on the symmetry characteristics of the molecular states. While the theoretical evaluation of the magnitude of $\langle \alpha_{ij} \rangle$ demands a knowledge of the corresponding wave functions, the question whether $\langle \alpha_{ij} \rangle$ is zero or not depends on the symmetry properties of the molecular wave functions for the states $|i\rangle$ and $|f\rangle$ and can therefore be answered by group theory without explicitly calculating the matrix elements (8.9).

According to (8.12), the Raman scattering cross section increases considerably if the laser frequency ω_L matches a transition frequency ω_{ij} of the molecule (resonance Raman effect) [8.20, 8.21]. With tunable dye lasers and optical frequency doubling this resonance condition can often be realized. The enhanced sensitivity of resonant Raman scattering can be utilized for measurements of micro-samples or of very small concentrations of molecules in solutions, where the absorption of the pump wave is small in spite of resonance with a molecular transition.

If the frequency difference $\omega_L - \omega_s$ corresponds to an electronic transition of the molecule, we speak of electronic Raman scattering [8.22, 8.23], which gives complementary information to electronic-absorption spectroscopy. This is because the initial and final states must have the same parity, and therefore a direct dipole-allowed electronic transition $|i\rangle \to |f\rangle$ is not possible.

In paramagnetic molecules Raman transitions between different fine-structure components (spin-flip Raman transitions) can occur [8.9]. If the molecules are placed in a longitudinal magnetic field parallel to the laser beam, the Raman light is circularly polarized and is measured as σ^+ light for $\Delta M = +1$ and as σ^- for $\Delta M = -1$ transitions.

8.2 Experimental Techniques of Linear Laser Raman Spectroscopy

The scattering cross sections in spontaneous Raman spectroscopy are very small, typically on the order of 10^{-30} cm^2. The experimental problems of detecting weak signals in the presence of intense background radiation are by no means trivial. The achievable signal-to-noise ratio depends both on the pump intensity and on the sensitivity of the detector. Recent years have brought remarkable progress on the source as well as on the detector side [8.24]. The incident light intensity can be greatly enhanced by using multiple reflection cells, intracavity techniques (Sect. 6.2.2), or a combination of both. Figure 8.3 depicts as an example of such advanced equipment a Raman spectrometer with a multiple-reflection Raman cell inside the resonator of an argon laser. The laser can be tuned by the prism LP to the different laser lines [8.25]. A sophisticated system of mirrors CM collects the scattered light, which is further imaged by the lens L_1 onto the entrance slit of the spectrometer. A Dove prism [8.26] turns the image of the line source by 90° to make it parallel to the entrance slit. Figure 8.4, which shows the pure rotational Raman

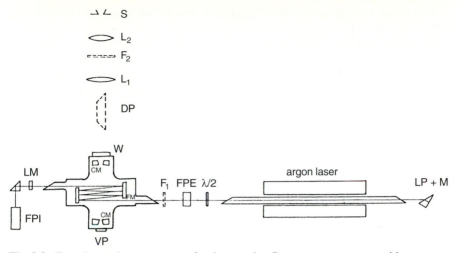

Fig. 8.3. Experimental arrangement for intracavity Raman spectroscopy with an argon laser: CM, multiple reflection four-mirror system for efficient collection of scattered light; LM, laser-resonator mirror; DP, Dove prism, which turns the image of the horizontal interaction plane by 90° in order to match it to the vertical entrance slit S of the spectrograph; FPE, Fabry–Perot etalon to enforce single-mode operation of the argon laser; LP, Littrow prism for line selection [8.25]

spectrum of C_2N_2, illustrates the sensivity that can be obtained with this setup [8.25].

In earlier days of Raman spectroscopy the photographic plate was the only detector used to record the Raman spectra. The introduction of sensitive photomultipliers and, in particular, the development of image intensifiers and optical multichannel analyzers with cooled photocathodes (Sect. 4.5) have greatly enhanced the detection sensitivity. Image intensifiers and instrumentation such as optical multichannel analyzers (OMAs) or CCD arrays (Sect. 4.5.3) allow

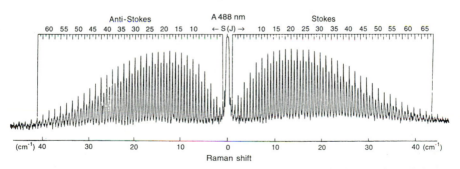

Fig. 8.4. Rotational Raman spectrum of C_2N_2 excited with the 488-nm line of the argon laser in the experimental setup of Fig. 8.3 and recorded on a photographic plate with 10-min exposure time [8.25]

simultaneous recording of extended spectral ranges with sensitivities comparable to those of photomultipliers [8.26].

The third experimental component that has contributed to the further improvement of the quality of Raman spectra is the introduction of digital computers to control the experimental procedure, to calibrate the Raman spectra, and to analyze the data. This has greatly reduced the time spent for preparing the data for the interpretation of the results [8.27].

Because of the increased sensitivity of an intracavity arrangement, even weak vibrational overtone bands can be recorded with rotational resolution. For illustration, Fig. 8.5 shows the rotationally resolved Q-branch of the D_2 molecule for the transitions $(v' = 2 \leftarrow v'' = 0)$ [8.28]. The photon counting rate for the overtone transitions was about 5000 times smaller than those for the fundamental $(v' = 1 \leftarrow v'' = 0)$ band. This overtone Raman spectroscopy can also be applied to large molecules, as has been demonstrated for the overtone spectrum of the torsional vibration of CH_3CD_3 and C_2H_6, where the torsional splittings could be measured up to the 5th torsional level [8.29].

Just as in absorption spectroscopy, the sensitivity may be enhanced by difference laser Raman spectroscopy, where the pump laser passes alternately through a cell containing the sample molecules dissolved in some liquid and through a cell containing only the liquid. The basic advantages of this difference technique are the cancellation of unwanted Raman bands of the solvent in the spectrum of the solution and the accurate determination of small frequency shifts due to interactions with the solvent molecules.

In the case of strongly absorbing Raman samples, the heat production at the focus of the incident laser may become so large that the molecules under investigation may thermally decompose. A solution to this problem is

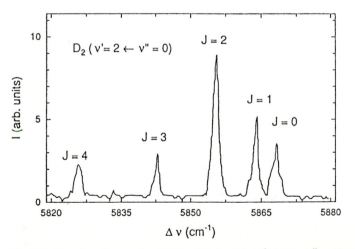

Fig. 8.5. Rotationally resolved Q-branch in the $(v' = 2 \leftarrow v'' = 0)$ overtone spectrum of the D_2 molecule, measured with the sample inside the resonator of a 250-W argon laser at $\lambda = 488$ nm [8.28]

the rotating sample technique [8.30], where the sample is rotated with an angular velocity Ω. If the interaction point with the laser beam is R centimeters away from the axis, the time T spent by the molecules within the focal region with diameter d [cm] is $T = d/(R\Omega)$. This technique which allows much higher input powers and therefore better signal-to-noise ratios, can be combined with the difference technique by mounting a cylindrical cell onto the rotation axis. One half of the cell is filled with the liquid Raman sample in solution, the other half is only filled with the solvent (Fig. 8.6).

A larger increase of sensitivity in linear Raman spectroscopy of liquids has been achieved with optical-fiber Raman spectroscopy. This technique uses a capillary optical fiber with the refractive index n_f, filled with a liquid with refractive index $n_e > n_f$. If the incident laser beam is focused into the fiber, the laser light as well as the Raman light is trapped in the core due to internal reflection and therefore travels inside the capillary. With sufficiently long capillaries $(1-30\,\text{m})$ and low losses, very high spontaneous Raman intensities can be achieved, which may exceed those of conventional techniques by factors of 10^3 [8.31]. Figure 8.7 shows schematically the experimental ar-

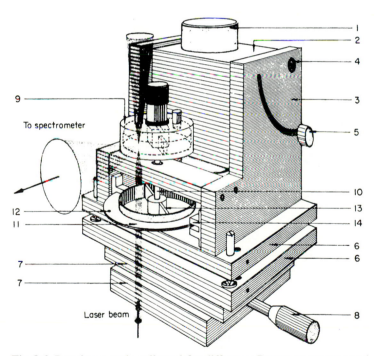

Fig. 8.6. Rotating sample cell used for difference Raman spectroscopy: 1, motor; 2, motor block, 3, side parts; 4, motor axis; 5, set screw; 6, kinematic mount; 7, x–y precision ball glider; 8, adjustment screw; 9, divided liquid cell for difference Raman spectroscopy; 10, axis for trigger wheel; 11, trigger wheel; 12, trigger hole; 13, bar; 14, optoelectronic array consisting of a photodiode and transistor [8.30]

rangement where the fiber is wound on a drum. Because of the increased sensitivity, this fiber technique also allows one to record second- and third-order Raman bands, which facilitates complete assignments of vibrational spectra [8.32].

The sensitivity of Raman spectroscopy in the gas phase can be greatly enhanced by combination with one of the detection techniques discussed in Chap. 6. For example, the vibrationally excited molecules produced by Raman–Stokes scattering can be selectively detected by resonant two-photon ionization with two visible lasers or by UV ionization with a laser frequency ω_{UV}, which can ionize molecules in level E_f but not in E_i (Fig. 8.8).

The information obtained from linear Raman spectroscopy is derived from the following experimental data:

- The linewidth of the scattered radiation, which represents for gaseous samples a convolution of the Doppler width, collisional broadening, spectral profile of the exciting laser, and natural linewidth, depending on the lifetimes of molecular levels involved in the Raman transition.

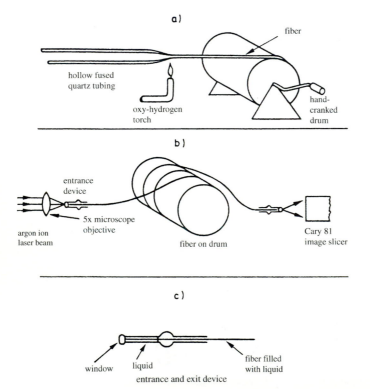

Fig. 8.7a–c. Raman spectroscopy of liquid samples in a thin capillary fiber: (**a**) production of the fiber; (**b**) incoupling of an argon laser beam with a microscope objective into the fiber and imaging of the outcoupled radiation into a spectrometer; (**c**) fiber with liquid [8.30]

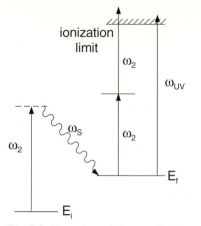

Fig. 8.8. Detection of Raman–Stokes scattering by photoionization of the excited level E_f either by one UV photon ($IP(E_f) < \hbar\omega_{UV} < IP(E_i)$) or by resonant two-photon ionization

- The degree of polarization ρ of the scattered light, defined as

$$\rho = \frac{I_\| - I_\perp}{I_\| + I_\perp} \, , \tag{8.13}$$

where $I_\|$ and $I_\perp$ are the intensities of the scattered light with a polarization parallel and perpendicular, respectively, to that of a linearly polarized excitation laser. A more detailed calculation shows that for statistically oriented molecules the degree of polarization

$$\rho = \frac{3\beta^2}{45\overline{\alpha}^2 + 4\beta^2} \, , \tag{8.14}$$

depends on the mean value $\overline{\alpha} = (\alpha_{xx} + \alpha_{yy} + \alpha_{zz})/3$ of the diagonal components of the polarizibility tensor $\tilde{\alpha}$ and on the anisotropy

$$\beta^2 = \frac{1}{2}\Big[(\alpha_{xx} - \alpha_{yy})^2 + (\alpha_{yy} - \alpha_{zz})^2 + (\alpha_{zz} - \alpha_{xx})^2$$
$$+ 6(\alpha_{xy}^2 + \alpha_{xz}^2 + \alpha_{zx}^2) \Big] \, . \tag{8.15}$$

Measurements of ρ and β therefore allow the determination of the polarizability tensor [8.33].
It turns out that

$$\overline{\alpha_{xx}^2} = \overline{\alpha_{yy}^2} = \overline{\alpha_{zz}^2} = \frac{1}{45}\left(45\overline{\alpha}^2 + 4\beta^2\right) ,$$
$$\overline{\alpha_{xy}^2} = \overline{\alpha_{xz}^2} = \overline{\alpha_{yz}^2} = \frac{1}{15}\beta^2 \, . \tag{8.16}$$

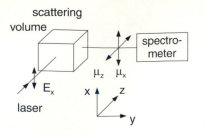

Fig. 8.9. Possible scattering geometry for measurements of the components α_{xx} and α_{zx} of the polarizability tensor

With the experimental arrangement of Fig. 8.9, where the exciting laser is polarized in the x-direction and the Raman light is observed in the y-direction without polarizer ($\mu_x + \mu_z$), the measured intensity becomes

$$I_{x(x+z)} = \frac{\omega^4 \cdot I_0}{16\pi^2 \varepsilon_0^2 c^4} \left(\alpha_{xx}^2 + \alpha_{zx}^2 \right) . \tag{8.17}$$

- The intensity of the Raman lines is proportional to the product of the Raman scattering cross section σ_R, which depends according to (8.12) on the matrix elements $\langle \alpha_{ij} \rangle$ of the polarizability tensor and the density N_i of molecules in the initial state. If the cross sections σ_R have been determined elsewhere, the intensity of the Raman lines can be used for measurements of the population densities $N(v, J)$. Assuming a Boltzmann distribution (8.11a), the temperature T of the sample can be derived from measured values of $N(v, J)$. This is frequently used for the determination of unknown temperature profiles in flames [8.34] or of unknown density profiles in liquid or gaseous flows [8.35] at a known temperature (Sect. 8.5).

One example is intracavity Raman spectroscopy of molecules in a supersonic jet, demonstrated by van Helvoort et al. [8.36]. If the intracavity beam waist of an argon-ion laser is shifted to different locations of the molecular jet (Fig. 8.10), the vibrational and rotational temperatures of the molecules (Sect. 9.2) and their local variations can be derived from the Raman spectra. More details of recent techniques in linear laser Raman spectroscopy can be found in [8.11, 8.37].

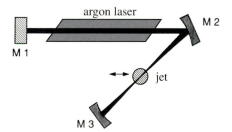

Fig. 8.10. Intracavity Raman spectroscopy of molecules in a cold jet with spatial resolution

8.3 Nonlinear Raman Spectroscopy

When the intensity of the incident light wave becomes sufficiently large, the induced oscillation of the electron cloud surpasses the linear range assumed in Sect. 8.1. This implies that the induced dipole moments p of the molecules are no longer proportional to the electric field E and we have to generalize (8.2). The function $p(E)$ can be expanded into a power series of E^n ($n = 0, 1, 2, \ldots$), which is generally written as

$$p(E) = \mu + \tilde{\alpha} E + \tilde{\beta} E \cdot E + \tilde{\gamma} E \cdot E \cdot E , \tag{8.18a}$$

where $\tilde{\alpha}$ is the polarizability tensor, $\tilde{\beta}$ is named *hyper-polarizability*, and $\tilde{\gamma}$ is called the *second hyper-polarizability*. The quantities α, β, and γ are tensors of rank two, three, and four, respectively.

In component notation ($i = x, y, z$) (8.18a) can be written as

$$p_i(E) = \mu_i + \sum_k \alpha_{ik} E_k + \sum_k \sum_j \beta_{ikj} E_k E_j$$

$$+ \sum_k \sum_j \sum_l \gamma_{ikjl} E_k E_j E_l . \tag{8.18b}$$

This gives for the polarization $P = Np$ of a medium with oriented dipoles

$$P_i(E) = \epsilon_0 \left(\chi_i + \sum_k \chi_{ik} E_k + \sum_{k,j} \chi_{ikj} E_k E_j + \ldots \right) , \tag{8.18c}$$

which corresponds to (5.114) discussed in the section on nonlinear optics and frequency conversion, if we define the susceptibilities $\chi_i = N\mu_i/\epsilon_0$, $\chi_{ik} = N\alpha_{ik}/\epsilon_0$, and so on.

For sufficiently small electric field amplitudes E the nonlinear terms in (8.18a) can be neglected, and we then obtain (8.2) for the linear Raman spectroscopy.

8.3.1 Stimulated Raman Scattering

If the incident laser intensity I_L becomes very large, an appreciable fraction of the molecules in the initial state E_i is excited into the final state E_f and the intensity of the Raman-scattered light is correspondingly large. Under these conditions we have to consider the simultaneous interaction of the molecules with *two* EM waves: the laser wave at the frequency ω_L and the Stokes wave at the frequency $\omega_S = \omega_L - \omega_V$ or the anti-Stokes wave at $\omega_a = \omega_L + \omega_V$. Both waves are coupled by the molecules vibrating with the frequencies ω_V. This parametric interaction leads to an energy exchange between the pump wave and the Stokes or anti-Stokes waves. This phenomenon of *stimulated* Raman scattering, which was first observed by Woodbury et al. [8.38] and then explained by Woodbury and Eckhardt [8.39], can be described in classical terms [8.40, 8.41].

The Raman medium is taken as consisting of N harmonic oscillators per unit volume, which are independent of each other. Because of the combined action of the incident laser wave and the Stokes wave, the oscillators experience a driving force F that depends on the total field amplitude E

$$E(z, t) = E_L e^{i(\omega_L t - k_L z)} + E_S e^{i(\omega_S t - k_S z)} , \tag{8.19}$$

where we have assumed plane waves traveling in the z-direction. The potential energy W_{pot} of a molecule with induced dipole moment $p = \alpha E$ in an EM field with amplitude E is, according to (8.2, 8.3) with $\mu = 0$

$$W_{pot} = -p \cdot E = -\alpha(q) E^2 . \tag{8.20}$$

The force $F = -\text{grad } W_{pot}$ acting on the molecule gives

$$F(z, t) = +\frac{\partial}{\partial q}\{[\alpha(q)]E^2\} = \left(\frac{\partial \alpha}{\partial q}\right)_0 E^2(z, t) . \tag{8.21}$$

The equation of motion for the molecular oscillator with oscillation amplitude q, mass m, and vibrational eigenfrequency ω_v is then

$$\frac{\partial^2 q}{\partial t^2} - \gamma \frac{\partial q}{\partial t} + \omega_v^2 q = \left(\frac{\partial \alpha}{\partial q}\right)_0 E^2/m , \tag{8.22}$$

where γ is the damping constant that is responsible for the linewidth $\Delta\omega = \gamma$ of spontaneous Raman scattering. Inserting the complex ansatz

$$q = \frac{1}{2}\left(q_v e^{i\omega t} + q_v^* e^{-i\omega t}\right) , \tag{8.23}$$

into (8.22) we get with the field amplitude (8.19)

$$(\omega_v^2 - \omega^2 + i\gamma\omega)q_v e^{i\omega t} = \frac{1}{2m}\left(\frac{\partial \alpha}{\partial q}\right)_0 E_L E_S e^{i[(\omega_L - \omega_S)t - (k_L - k_S)z]} . \tag{8.24}$$

Comparison of the time-dependent terms on both sides of (8.24) shows that $\omega = \omega_L - \omega_S$. The molecular vibrations are therefore driven at the difference frequency $\omega_L - \omega_S$. Solving (8.24) for q_v yields

$$q_v = \frac{(\partial \alpha/\partial q)_0 E_L E_S}{2m[\omega_v^2 - (\omega_L - \omega_S)^2 + i(\omega_L - \omega_S)\gamma]} e^{-i(k_L - k_S)z} . \tag{8.25}$$

The oscillation-induced molecular dipoles $p(\omega, z, t)$ result in a macroscopic polarization $P = Np$. According to (8.5), the polarization $P_S = P(\omega_S)$ at the Stokes frequency ω_S, which is responsible for Raman scattering, is given by

$$P_S = \frac{1}{2}N\left(\frac{\partial \alpha}{\partial q}\right)_0 qE . \tag{8.26}$$

Inserting q from (8.23, 8.25) and E from (8.19) yields the *nonlinear* polarization

$$P_S^{NL}(\omega_S) = N \frac{(\partial\alpha/\partial q)_0^2 E_L^2 E_S}{4m[\omega_v^2 - (\omega_L - \omega_S)^2 + i\gamma(\omega_L - \omega_S)]} e^{-i(\omega_S t - k_S z)} . \qquad (8.27)$$

This shows that a *polarization wave* travels through the medium with an amplitude proportional to the product $E_L^2 \cdot E_S$. It has the same wave vector k_S as the Stokes wave and can therefore amplify this wave. The amplification can be derived from the wave equation

$$\Delta E = \mu_0 \sigma \frac{\partial}{\partial t} E + \mu_0 \epsilon \frac{\partial^2}{\partial t^2} E + \mu_0 \frac{\partial^2}{\partial t^2} \left(P_S^{NL}\right) , \qquad (8.28)$$

for waves in a medium with the conductivity σ, where P_S^{NL} acts as the driving term.

For the one-dimensional problem ($\partial/\partial y = \partial/\partial x = 0$) with the approximation $d^2 E/dz^2 \ll k\, dE/dz$ and with (8.26), the equation for the Stokes wave becomes

$$\frac{dE_S}{dz} = -\frac{\sigma}{2}\sqrt{\mu_0/\epsilon}\, E_S + N\frac{k_S}{2\epsilon}\left(\frac{\partial\alpha}{\partial q}\right)_0 q_v E_L . \qquad (8.29)$$

Substituting q_v from (8.25) gives the final result for the case $\omega_v = \omega_L - \omega_S$

$$\frac{dE_S}{dz} = \left[-\frac{\sigma}{2}\sqrt{\mu_0/\epsilon} + N\frac{(\partial\alpha/\partial q)_0^2 E_L^2}{4m\epsilon i\gamma(\omega_L - \omega_S)}\right] E_S = (-f + g)E_S . \qquad (8.30)$$

Integration of (8.30) yields

$$E_S = E_S(0)\, e^{(g-f)z} . \qquad (8.31)$$

The Stokes wave is amplified if g exceeds f. The amplification factor g depends on the square of the laser amplitude E_L and on the term $(\partial\alpha/\partial q)^2$. Stimulated Raman scattering is therefore observed only if the incident laser intensity exceeds a threshold value that is determined by the nonlinear term $(\partial\alpha_{ij}/\partial q)_0$ in the polarization tensor of the Raman-active normal vibration and by the loss factor $f = \frac{1}{2}\sigma(\mu_0/\epsilon)^{1/2}$.

While the intensity of anti-Stokes radiation is very small in spontaneous Raman scattering due to the low thermal population density in excited molecular levels (Sect. 8.1), this is not necessarily true in stimulated Raman scattering. Because of the strong incident pump wave, a large fraction of all interacting molecules is excited into higher vibrational levels, and strong anti-Stokes radiation at frequencies $\omega_L + \omega_v$ has been found.

According to (8.26), the driving term in the wave equation (8.28) for an anti-Stokes wave at $\omega_a = \omega_L + \omega_v$ is given by

$$P_{\omega_a}^{NL} = \frac{1}{2}N\left(\frac{\partial\alpha}{\partial q}\right)_0 q_v E_L\, e^{i[(\omega_L + \omega_v)t - k_L z]} , \qquad (8.32)$$

For small amplitudes $E_a \ll E_L$ of the anti-Stokes waves, we can assume that the molecular vibrations are independent of E_a and can replace q_v by its solution (8.25). This yields an equation for the amplification of E_a that is analogous to (8.29) for E_S

$$\frac{dE_a}{dz} = -\frac{f}{2} E_a e^{i(\omega_a t - k_a z)}$$

$$+ N_v \left[\frac{\omega_a \sqrt{\mu_0/\epsilon}}{8m_v} \left(\frac{\partial \alpha}{\partial q}\right)_0^2 \right] E_L^2 E_S^* e^{i(2k_L - k_S - k_a)z} , \qquad (8.33)$$

where N_v is the density of vibrationally excited molecules. This shows that, analogously to sum- or difference-frequency generation (Sect. 5.8), a macroscopic wave can build up only if the phase-matching condition

$$k_a = 2k_L - k_S , \qquad (8.34)$$

can be satisfied. In a medium with normal dispersion this condition cannot be met for collinear waves. From a three-dimensional analysis, however, one obtains the vector equation

$$2\boldsymbol{k}_L = \boldsymbol{k}_S + \boldsymbol{k}_a , \qquad (8.35)$$

which reveals that the anti-Stokes radiation is emitted into a cone whose axis is parallel to the beam-propagation direction (Fig. 8.11). This is exactly what has been observed [8.38, 8.41].

Let us briefly summarized the differences between the linear (spontaneous) and the nonlinear (induced) Raman effect:

- While the intensity of spontaneous Raman lines is proportional to the incident pump intensity, but lower by several orders of magnitude compared with the pump intensity, the stimulated Stokes or anti-Stokes radiation depend in a nonlinear way on I_p but have intensities comparable to that of the pump wave.

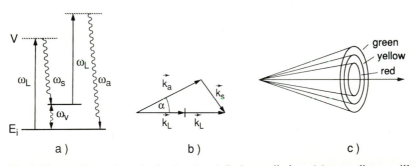

Fig. 8.11a–c. Generation of stimulated anti-Stokes radiation: (**a**) term diagram illustrating energy conservation; (**b**) vector diagram of momentum conservation for the collinear and noncollinear case; (**c**) radiation cone for different values of $\boldsymbol{k}_S$ showing red, yellow, and green rings of anti-Stokes radiation excited by a ruby laser at 694 nm

- The stimulated Raman effect is observed only above a threshold pump intensity, which depends on the gain of the Raman medium and the length of the pump region.
- Most Raman-active substances show only one or two Stokes lines at the frequencies $\omega_S = \omega_L - \omega_V$ in stimulated emission. At higher pump intensities, however, lines at the frequencies $\omega = \omega_L - n\omega_V$ $(n = 1, 2, 3)$, which do not correspond to overtones of vibrational frequencies, have been observed besides these Stokes lines. Because of the anharmonicity of molecular vibrations, the spontaneous Raman lines from vibrational overtones are shifted against ω_L by $\Delta\omega = n\omega_v - (n^2 + n)x_k$, where x_k represent the anharmonicity constants. In Fig. 8.12 is illustrated that these higher-order Stokes lines are generated by consecutive Raman processes, induced by the pump wave, the Stokes wave, etc.
- The linewidths of spontaneous and stimulated Raman lines depend on the linewidth of the pump laser. For narrow linewidths, however, the width of the stimulated Raman lines becomes smaller than that of the spontaneous lines, which are Doppler-broadened by the thermal motion of the scattering molecules. A Stokes photon $\hbar\omega_s$, which is scattered into an angle ϕ against the incident laser beam by a molecule moving with the velocity v, has a Doppler-shifted frequency

$$\omega_s = \omega_L - \omega_V - (k_L - k_S) \cdot v \qquad (8.36)$$
$$= \omega_L - \omega_V - [1 - (k_S/k_L)\cos\phi]k_L \cdot v \, .$$

In the case of spontaneous Raman scattering we have $0 \leq \phi \leq 2\pi$, and the spontaneous Raman lines show a Doppler width that is $(k_S/k_L) = (\omega_S/\omega_L)$ times that of fluorescence lines at ω_L. For induced Raman scattering $k_S \parallel k_L \rightarrow \cos\phi = 1$, and the bracket in (8.36) has the value $(1 - k_S/k_L) \ll 1$, if $\omega_V \ll \omega_L$.

- The main merit of the stimulated Raman effect for molecular spectroscopy may be seen in the much higher intensities of stimulated Raman lines. During the same measuring time one therefore achieves a much better

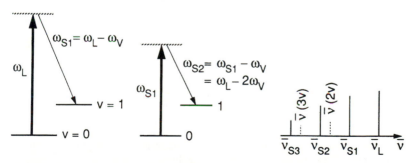

Fig. 8.12. Level diagram for the generation of higher-order Stokes sidebands, which differ from the vibrational overtone frequencies

signal-to-noise ratio than in linear Raman spectroscopy. The experimental realization of stimulated Raman spectroscopy is based on two different techniques:

(a) Stimulated Raman gain spectroscopy (SRGS), where a strong pump laser at ω_1 is used to produce sufficient gain for the Stokes radiation at ω_2 according to (8.31). This gain can be measured with a weak tunable probe laser tuned to the Stokes wavelengths [8.42].
(b) Inverse Raman spectroscopy (IRS), where the attenuation of the weak probe laser at ω_1 is measured when the strong pump laser at ω_2 is tuned through Stokes or anti-Stokes transitions [8.43].

Several high-resolution stimulated Raman spectrometers have been built [8.42–8.44] that are used for measurements of linewidths and Raman line positions in order to gain information on molecular structure and dynamics. A good compromise for obtaining a high resolution and a large signal-to-noise ratio is to use a pulsed pump laser and a single-mode cw probe laser (quasi-cw spectrometer [8.44]). The narrow-band pulsed laser can be realized by pulsed amplification of a single-mode cw laser (Sect. 5.7). For illustration Fig. 8.13 shows a typical quasi-cw stimulated Raman spectrometer, where the wavelengths of the tunable dye laser are measured with a traveling Michelson wavemeter (Sect. 4.4)

There is another important application of stimulated Raman scattering in the field of *Raman lasers*. With a tunable pump laser at the frequency ω_L, intense coherent radiation sources at frequencies $\omega_L \pm n\omega_v$ ($n = 1, 2, 3, \ldots$) can be realized that cover the UV and infrared spectral range if visible pump lasers are used (Sect. 5.8).

More details on the stimulated Raman effect and reference to experiments in this field can be found in [8.11, 8.45–8.51].

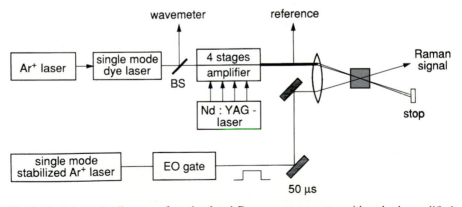

Fig. 8.13. Schematic diagram of a stimulated Raman spectrometer with pulsed, amplified cw pump laser and a single-mode cw probe laser [8.46]

8.3.2 Coherent Anti-Stokes Raman Spectroscopy

In Sect. 8.3.1 we discussed the observation that a sufficiently strong incident pump wave at the frequency ω_L can generate an intense Stokes wave at
$\omega_S = \omega_L - \omega_V$. Under the combined action of both waves, the nonlinear polarization P_{NL} of the medium is generated that contains contributions at the
frequencies $\omega_V = \omega_L - \omega_S$, $\omega_S = \omega_L - \omega_V$, and $\omega_a = \omega_L + \omega_V$. These contributions act as driving forces for the generation of new waves. Provided that the
phase-matching conditions $2k_L = k_S + k_a$ can be satisfied, a strong anti-Stokes
wave $E_a \cos(\omega_a t - k_a \cdot r)$ is observed in the direction of k_a.

Despite the enormous intensities of stimulated Stokes and anti-Stokes
waves, stimulated Raman spectroscopy has been of little use in molecular
spectroscopy. The high threshold, which, according to (8.30), depends on the
molecular density N, the incident intensity $I \propto E_L^2$, and the square of the
small polarizability term $(\partial \alpha_{ij}/\partial q)$ in (8.27), limits stimulated emission to
only the strongest Raman lines in materials of high densities N.

The recently developed technique of coherent anti-Stokes Raman spectroscopy (CARS), however, combines the advantages of signal strength obtained in stimulated Raman spectroscopy with the general applicability of
spontaneous Raman spectroscopy [8.45–8.56]. In this technique *two* lasers
are needed. The frequencies ω_1 and ω_2 of the two incident laser waves are
chosen such that their difference $\omega_1 - \omega_2 = \omega_V$ coincides with a Raman-active vibration of the molecules under investigation. These two incident waves
correspond to the pump wave ($\omega_1 = \omega_L$) and the Stokes wave ($\omega_2 = \omega_S$) in
stimulated Raman scattering. The advantage is that the Stokes wave at ω_2 is
already present and does not need to be generated in the medium. These two
waves are considered in (8.29).

Because of the nonlinear interaction discussed in Sect. 8.3.1, new Stokes
and anti-Stokes waves are generated (Fig. 8.14). The waves ω_1 and ω_2 generate a large population density of vibrationally excited molecules by stimulated
Raman scattering. These excited molecules act as the nonlinear medium for

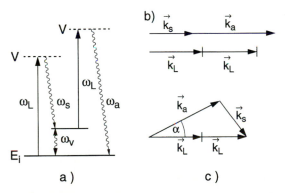

Fig. 8.14. (a) Level diagram of CARS; (b) vector diagrams for phase matching in gases
with negligible dispersion and (c) in liquids or solids with noticeable dispersion

the generation of anti-Stokes radiation at $\omega_a = 2\omega_1 - \omega_2$ by the incident wave with frequency ω_1. In a similar way, a Stokes wave with frequency $\omega_s = 2\omega_2 - \omega_1$ is generated by the incident waves at ω_1 and ω_2. Since four waves are involved in the generation of the anti-Stokes wave, CARS is called a *four-wave parametric mixing process*.

It can be derived that from (8.33) the power S of the CARS signal (which is proportional to the square of the amplitude E_a)

$$S \propto N^2 I_1^2 I_2 \,, \tag{8.37}$$

increases with the square N^2 of the molecular density N and is proportional to the product $I_1(\omega_1)^2 I_2(\omega_2)$ of the pump laser intensities. It is therefore necessary to use either high densities N or large intensities I. If the two pump beams are focused into the sample, most of the CARS signal is generated within the local volume where the intensities are maximum. CARS spectroscopy therefore offers high spatial resolution.

If the incident waves are at *optical* frequencies, the difference frequency $\omega_R = \omega_1 - \omega_2$ is small compared with ω_1 in the case of rotational–vibrational frequencies ω_R. In gaseous Raman samples the dispersion is generally negligible over small ranges $\Delta\omega = \omega_1 - \omega_2$ and satisfactory phase matching is obtained for *collinear* beams. The Stokes wave at $\omega_S = 2\omega_2 - \omega_1$ and the anti-Stokes wave at $\omega_a = 2\omega_1 - \omega_2$ are then generated in the same direction as the incoming beams (Fig. 8.14b). In liquids, dispersion effects are more severe and the phase-matching condition can be satisfied over a sufficiently long coherence length only, if the two incoming beams are crossed at the phase-matching angle (Fig. 8.14c).

In the collinear arrangement the anti-Stokes wave at $\omega_a = 2\omega_1 - \omega_2$ ($\omega_a > \omega_1$!) is detected through filters that reject both incident laser beams as well as the fluorescence that may be generated in the sample. Figure 8.15 illustrates an early experimental setup used for rotational–vibrational spectroscopy of gases by CARS [8.57]. The two incident laser beams are provided

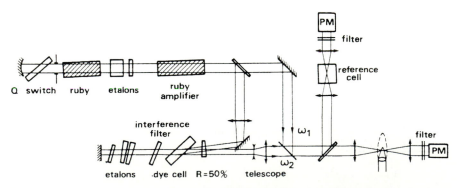

Fig. 8.15. Experimental setup of an early CARS experiment in gases using a ruby laser and a dye laser pumped by the ruby laser [8.57]

by a Q-switched ruby laser and a tunable dye laser that pumped by this ruby laser. Because the gain of the anti-Stokes wave depends quadratically on the molecular density N (see (8.37)), megawatt-range power levels of the incident beams are required for gaseous samples, while kilowatt powers are sufficient for liquid samples [8.58].

The most common pump system for pulsed CARS experiments are two dye lasers pumped by the same pump laser (N_2 laser, excimer laser, or frequency-doubled Nd:YAG laser). This system is very flexible because both frequencies ω_1 and ω_2 can be varied over large spectral ranges. Since both the frequency and intensity fluctuations of the dye lasers result in strong intensity fluctuations of the CARS signal, the stability of the dye lasers needs particular attention. With compact and stable systems the signal fluctuations can be reduced below 10% [8.59].

In addition, many CARS experiments have been performed with cw dye lasers with liquid samples as well as with gaseous ones. An experimental setup for cw CARS of liquid nitrogen is shown in Fig. 8.16, where the two incident collinear pump waves are provided by the 514.5-nm argon laser line (ω_1) and a cw dye laser (ω_2) pumped by the same argon laser [8.58].

The advantage of cw CARS is its higher spectral resolution because the bandwidth $\Delta\nu$ of single-mode cw lasers is several orders of magnitude below that of pulsed lasers. In order to obtain sufficiently high intensities, intracavity excitation has been used. A possible experimental realization (Fig. 8.17) places the sample cell inside the ring resonator of an argon ion laser, where the cw dye laser is coupled into the resonator by means of a prism [8.43]. The CARS signal generated in the sample cell at the beam waist of the resonator is transmitted through the dichroic mirror M2 and is spectrally purified by a filter or a prism and a monochromator before it reaches a photomultiplier.

High-resolution CARS can be also performed with injection-seeded pulsed dye lasers [8.43, 8.60]. If the radiation of a single-mode cw dye laser with frequency ω is injected into the cavity of a pulsed dye laser that has been mode matched to the Gaussian beam of the cw laser (Sect. 5.8), the amplification of the gain medium is enhanced considerably at the frequency ω and

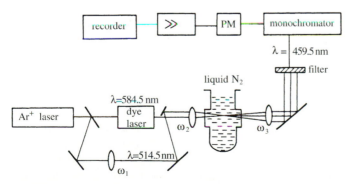

Fig. 8.16. Experimental setup for cw CARS in liquids [8.58]

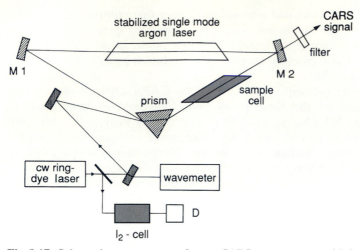

Fig. 8.17. Schematic arrangement of a cw CARS spectrometer with intracavity excitation of the sample [8.43]

the pulsed laser oscillates on a single cavity mode at the frequency ω. Only some milliwatts of the cw laser are needed for injection, while the output of the single-mode pulsed laser reaches several kilowatts, which may be further amplified (Sect. 5.5).

8.3.3 Resonant CARS and BOX CARS

If the frequencies ω_1 and ω_2 of the two incident laser waves are chosen to match a molecular transition, one or even two of the virtual levels in Fig. 8.14 coincide with a real molecular level. In this case of *resonant CARS*, the sensitivity may be increased by several orders of magnitude. Because of the larger absorption of the incident waves, the absorption path length must be sufficiently short or the density of absorbing molecules must be correspondingly small [8.61, 8.62].

A certain disadvantage of collinear CARS in gases is the spatial overlap of the two parallel incident beams with the signal beam. This overlap must be separated by spectral filters. This disadvantage can be overcome by the BOX CARS technique [8.63], where the pump beam of laser L1 (k_1, ω_1) is split into two parallel beams that are focused by a lens into the sample (Fig. 8.18b), where the directions of the three incoming beams match the vector diagram of Fig. 8.18a. The CARS signal beam can now be separated by geometrical means (beam stop and apertures). For comparison the vector diagrams of the phase-matching condition (8.35) are shown in Fig. 8.19 for the general case ($k_1 \neq k_2$), the collinear CARS arrangement, and for BOX CARS, where the vector diagram has the form of a box. From Fig. 8.18a with the relation

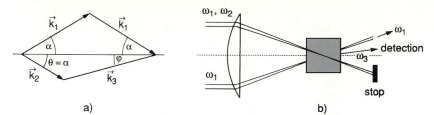

a)   b)

Fig. 8.18a,b. Wave vector diagram for BOX CARS (**a**) and experimental realization (**b**)

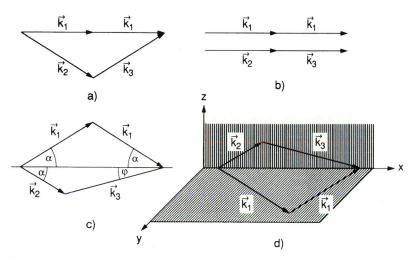

Fig. 8.19a–d. Comparison of phase-matching diagrams for CARS: (**a**) for the general case $k_1 \nparallel k_2$; (**b**) the collinear case $k_1 \parallel k_2$; (**c**) for BOX CARS; and (**d**) for folded BOX CARS

$|k| = n \cdot \omega/c$, the phase-matching conditions

$$n_2\omega_2 \sin\theta = n_3\omega_3 \sin\varphi \,,$$
$$n_2\omega_2 \cos\theta + n_3\omega_3 \cos\varphi = 2n_1\omega_1 \cos\alpha \,, \tag{8.38}$$

can be derived, which yield for $\theta = \alpha$ the relation

$$\sin\varphi = \frac{n_2\omega_2}{n_3\omega_3} \sin\alpha \,, \tag{8.39}$$

between the angle φ of the CARS signal beam $k_s = k_3$ against the z-direction and the angle α of the incident beams.

The CARS signal is generated within the common overlap volume of the three incident beams. This BOX-CAR technique considerably increases the spatial resolution, which may reach values below 1 mm.

In another beam configuration (folded BOX CARS), the pump beam with ω_2 is split into two parallel beams, which are directed by the focusing lens in such a way that the wave vectors k_2 and $k_3 = k_{as}$ are contained in a plane orthogonal to that of the two k_1 vectors (Fig. 8.19d). This has the advantage that neither of the two pump beams overlaps the signal beam at the detector [8.64]. If the Raman shifts are small, spectral filtering becomes difficult and the advantage of this folded BOX CAR technique is obvious.

The advantages of CARS may be summarized as follows:

- The signal levels in CARS may exceed those obtained in spontaneous Raman spectroscopy by a factor of $10^4 - 10^5$.
- The higher frequency $\omega_3 > \omega_1$, ω_2 of the anti-Stokes waves allows one to use filters that reject the incident light as well as fluorescence light.
- The small beam divergence allows one to place the detector far away from the sample, which yields excellent spatial discrimination against fluorescence or thermal luminous background such as occurs in flames, discharges, or chemiluminescent samples.
- The main contribution to the anti-Stokes generation comes from a small volume around the focus of the two incident beams. Therefore very small sample quantities (microliters for liquid samples or millibar pressures for gaseous samples) are required. Furthermore, a high spatial resolution is possible, which allows one to probe the spatial distribution of molecules in definite rotational–vibrational levels. The measurements of local temperature variations in flames from the intensity of anti-Stokes lines in CARS is an example where this advantage is utilized.
- A high spectral resolution can be achieved without using a monochromator. The Doppler width $\Delta\omega_D$, which represents a principal limitation in $90°$ spontaneous Raman scattering, is reduced to $[(\omega_2 - \omega_1)/\omega_1]\Delta\omega_D$ for the collinear arrangement used in CARS. While a resolution of 0.3 to 0.03 cm^{-1} is readily obtained in CARS with pulsed lasers, even linewidths down to 0.001 cm^{-1} have been achieved with single-mode lasers.

The main disadvantages of CARS are the expensive equipment and strong fluctuations of the signals caused by instabilities of intensities and alignments of the incident laser beams. The sensitivity of detecting low relative concentrations of specified sample molecules is mainly limited by interference with the nonresonant background from the other molecules in the sample. This limitation can be overcome, however, by resonant CARS.

8.3.4 Hyper-Raman Effect

The higher-order terms βEE, γEEE in the expansion (8.18a) of $p(E)$ represent the hyper-Raman effect. Analogous to (8.3), we can expand β in a Taylor series in the normal coordinates $q_n = q_{n0} \cos(\omega_n t)$

$$\beta = \beta_0 + \sum_{n=1}^{2Q} \left(\frac{\partial \beta}{\partial q_n}\right)_0 q_n + \dots . \tag{8.40}$$

Assume that two laser waves $E_1 = E_{01} \cos(\omega_1 t - k_1 z)$ and $E_2 = E_{02} \cos(\omega_2 t - k_2 t)$ are incident on the Raman sample. From the third term in (8.18a) we then obtain with (8.40) contributions to $p(E)$ due to β_0

$$\beta_0 E_{01}^2 \cos(2\omega_1 t) , \quad \text{and} \quad \beta_0 E_{02}^2 \cos(2\omega_2 t) , \tag{8.41}$$

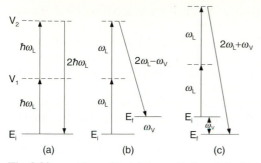

Fig. 8.20a–c. Hyper-Rayleigh scattering (**a**), Stokes hyper-Raman (**b**), and anti-Stokes hyper-Raman scattering (**c**)

which give rise to *hyper-Rayleigh scattering* (Fig. 8.20a). The term $(\partial\beta/\partial q_n)$ $q_{n0}\cos(\omega_n t)$ of (8.40) inserted into (8.18a) yields contributions

$$p^{HR} \propto \left(\frac{\partial\beta}{\partial q}\right)_0 q_{n0}[\cos(2\omega_1 \pm \omega_n)t + \cos(2\omega_2 \pm \omega_n)t] \,, \tag{8.42}$$

which are responsible for *hyper-Raman scattering* (Fig. 8.20b,c) [8.65].

Since the coefficients $(\partial\beta/\partial q)_0$ are very small, one needs large incident intensities to observe hyper-Raman scattering. Similar to second-harmonic generation (Sect. 5.8), hyper-Rayleigh scattering is forbidden for molecules with a center of inversion. The hyper-Raman effect obeys selection rules that differ from those of the linear Raman effect. It is therefore very attractive to molecular spectroscopists since molecular vibrations can be observed in the hyper-Raman spectrum that are forbidden for infrared as well as for linear Raman transitions. For example, spherical molecules such as CH_4 have no pure rotational Raman spectrum but a hyper-Raman spectrum, which was found by Maker [8.66]. A general theory for rotational and rotational–vibrational hyper-Raman scattering has been worked out by Altmann and Strey [8.67].

The intensity of hyper-Raman lines can be considerably enhanced when the molecules under investigation are adsorbed at a surface [8.68], because the surface lowers the symmetry and increases the induced dipole moments.

Similar to the induced Raman effect, the hyper-Raman effect can also be used to generate coherent radiation in spectral ranges where no intense lasers exist. One example is the generation of tunable radiation around $16\,\mu m$ by the stimulated hyper-Raman effect in strontium vapor [8.69].

8.3.5 Summary of Nonlinear Raman Spectroscopy

In the previous subsections we briefly introduced some nonlinear techniques of Raman spectroscopy. Besides stimulated Raman spectroscopy, Raman gain spectroscopy, inverse Raman spectroscopy, and CARS, several other special techniques such as the Raman-induced Kerr effect [8.70] or coherent Raman ellipsometry [8.71] also offer attractive alternatives to conventional Raman spectroscopy.

All these nonlinear techniques represent coherent third-order processes analogous to saturation spectroscopy, polarization spectroscopy, or two-photon absorption (Chap. 7), because the magnitude of the nonlinear signal is proportional to the third power of the involved field amplitudes (8.18).

The advantages of these nonlinear Raman techniques are the greatly increased signal-to-noise ratio and thus the enhanced sensitivity, the higher spectral and spatial resolution, and in the case of the hyper-Raman spectroscopy, the possibility of measuring higher-order contributions of molecules in the gaseous, liquid, or solid state to the susceptibility.

Additionally, there are several good books and reviews on nonlinear Raman spectroscopy. For more thorough information the reader is therefore referred to [8.11, 8.37, 8.43, 8.47, 8.48, 8.54, 8.56, 8.72].

8.4 Special Techniques

In this section we briefly discuss some special techniques of linear and nonlinear Raman spectroscopy that have special advantages for different applications. These are the resonance Raman effect, surface-enhanced Raman signals, Raman microscopy, and time-resolved Raman spectroscopy.

8.4.1 Resonance Raman Effect

The Raman scattering cross section can be increased by several orders of magnitude if the excitation wavelength matches an electronic transition of the molecule, that is, when it coincides with or comes close to a line in an electronic absorption band. In this case, the denominator in (8.12) becomes very small for the neighboring lines in this band, and several terms in the sum (8.12) give large contributions to the signal.

The Raman lines appear predominantly at those downward transitions that have the largest Franck–Condon factors, that is, the largest overlap of the vibrational wavefunctions in the upper and lower state (Fig. 8.21). Unlike the nonresonant case, in spontaneous resonance Raman scattering a larger number of Raman lines may appear that are shifted against the excitation line by several vibrational quanta. They correspond to Raman transitions terminating on higher vibrational levels of the ground electronic state. This is quite similar to laser-excited fluorescence and opens the possibility to determine the anharmonicity constants of the molecular potential curve or potential surface of the lower electronic state.

In stimulated Raman scattering the strongest Raman transition will have the largest gain and reaches threshold before the other transitions can develop. Just above threshold we therefore expect only a single Raman line in the stimulated Raman spectum, while at higher pump powers more lines will appear.

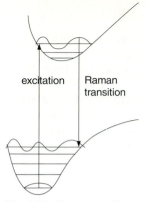

Fig. 8.21. Resonance Raman effect

Resonance Raman scattering is particularly advantageous for samples with small densities, for example, for gases at low pressures, where the absorption of the incident radiation is not severe and where nonresonant Raman spectroscopy might not be sufficiently sensitive.

If the excited state lies above the dissociation limit of the upper electronic state, the scattered Raman light shows a continuous spectrum. The intensity profile of this spectrum yields information on the repulsive part of the potential in the upper state.

8.4.2 Surface-Enhanced Raman Scattering

The intensity of Raman scattered light may be enhanced by several orders of magnitude if the molecules are adsorbed on a surface [8.73]. There are several mechanisms that contribute to this enhancement. Since the amplitude of the scattered radiation is proportional to the induced dipole moment

$$\boldsymbol{P}_{\mathrm{ind}} = \alpha \cdot \boldsymbol{E} \ ,$$

the increase of the polarizability α by the interaction of the molecule with the surface is one of the causes for the enhancement. In the case of metal surfaces, the electric field $\boldsymbol{E}$ may also be much larger than that of the incident radiation, which also leads to an increase of the induced dipole moment. Both effects depend on the orientation of the molecule relative to the surface normal, on its distance from the surface, and on the morphology, in particular the roughness of the surface. Small metal clusters on the surface increase the intensity of the molecular Raman lines. The frequency of the exciting light also has a large influence on the enhancement factor. In the case of metal surfaces it becomes maximum if it is close to the plasma frequency of the metal.

Because of these dependencies, surface-enhanced Raman spectroscopy has been successfully applied for surface analysis and also for tracing small concentrations of adsorbed molecules [8.73].

8.4.3 Raman Microscopy

For the nondestructive investigation of very small samples, for example, parts of living cells or inclusions in crystals, the combination of microscopy and Raman spectroscopy turns out to be a very useful technique. The laser beam is focused into the sample and the Raman spectrum that is emitted from the small focal spot is monitored through the microscope with subsequent spectrometer. This also applies to measurements of phase transitions in molecular crystals under high pressures. These pressures, which reach up to several gigapascals, can be realized with moderate efforts in a small volume between two diamonds with area A that are pressed together by a force F producing a pressure $p = F/A$. For example, with $F = 10^3$ Pa and $A = 10^{-6}$ cm^2, one obtains a pressure of 10^9 Pa. The phase transitions lead to a frequency shift of molecular vibrations, which are detected by the corresponding shift of the Raman lines [8.74].

One example of the application of this technique is the investigation of different phases of "fluid inclusions" in a quartz crystal found in the Swiss Alps. Their Raman spectra permitted determination of CO_2 as a gaseous inclusion and water as a liquid inclusion, and they showed that the mineral thought to be $CaSO_4$ was in fact $CaCO_3$ [8.75].

A typical experimental arrangement for Raman microscopy is shown in Fig. 8.22. The output beam of an argon laser or a dye laser is focused by a microscope objective into the microsample. The backscattered Raman light is imaged onto the entrance slit of a double or triple monochromator, which effectively supresses scattered laser light. A CCD camera at the exit of the monochromator records the wanted spectral range of the Raman radiation [8.73, 8.76, 8.77].

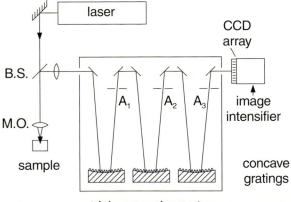

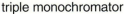

Fig. 8.22. Raman microscopy with suppression of scattered laser light by three apertures A_i

8.4.4 Time-Resolved Raman Spectroscopy

Both linear and nonlinear Raman spectroscopy can be combined with time-resolved detection techniques when pumping with short laser pulses [8.78]. Since Raman spectroscopy allows the determination of molecular parameters from measurements of frequencies and populations of vibrational and rotational energy levels, time-resolved techniques give information on energy transfer between vibrational levels or on structural changes of short-lived intermediate species in chemical reactions. One example is the vibrational excitation of molecules in liquids and the collisional energy transfer from the excited vibrational modes into other levels or into translational energy of the collision partners. These processes proceed on picosecond to femtosecond time scales [8.77, 8.79].

Time-resolved Raman spectroscopy has proved to be a very useful tool to elucidate fast processes in biological molecules, for instance, to follow the fast structural changes during the visual process where, after photoexcitation of rhodopsin molecules, a sequence of energy transfer processes involving isomerization and proton transfer takes place. This subject is treated in more detail in Chap. 11 in comparison with other time-resolved techniques.

8.5 Applications of Laser Raman Spectroscopy

The primary object of Raman spectroscopy is the determination of molecular energy levels and transition probabilities connected with molecular transitions that are not accessible to infrared spectroscopy. Linear laser Raman spectroscopy, CARS, and hyper-Raman scattering have very successfully collected many spectroscopic data that could not have been obtained with other techniques. Besides these basic applications to molecular spectroscopy there are, however, a number of scientific and technical applications of Raman spectroscopy to other fields, which have become feasible with the new methods discussed in the previous sections. We can give only a few examples.

Since the intensity of spontaneous Raman lines is proportional to the density $N(v_i, J_i)$ of molecules in the initial state (v_i, J_i), Raman spectroscopy can provide information on the population distribution $N(v_i, J_i)$, its local variation, and on concentrations of molecular constituents in samples. This allows one, for instance, to probe the temperature in flames or hot gases from the rotational Raman spectra [8.80–8.83] and to detect deviations from thermal equilibrium.

With CARS the spatial resolution is greatly increased, in particular if BOX CARS is used. The focal volume from which the signal radiation is generated can be made smaller than $0.1 \, \text{mm}^3$ [8.53]. The local density profiles of reaction products formed in flames or discharges can therefore be accurately probed without disturbing the sample conditions. The intensity of the stimu-

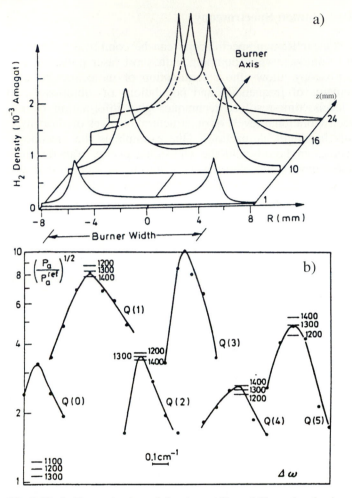

Fig. 8.23a,b. Determination of density profiles of H_2 molecules in a flame. R is the distance from the burner axis, z the distance along this axis. The profiles in (**a**) have been obtained from the spatial variations of the Q line intensities shown in (**b**). The relative intensities of $Q(J)$ furthermore allow the determinations of the temperature profiles [8.57]

lated anti-Stokes radiation is proportional to N^2 (8.31). Figure 8.23 shows for illustration the H_2 distribution in a horizontal Bunsen flame, measured from the CARS spectrum of the Q branch in H_2. The H_2 molecules are formed by the pyrolysis of hydrocarbon molecules [8.57]. Another example is the measurement of CARS spectra of water vapor in flames, which allowed one to probe the temperature in the postflame region of a premixed CH_4 air flame [8.81].

With a detection sensitivity of 10 to 100 ppm, CARS is not as good as some other techniques in monitoring pollutant gases at low concentrations

(Sect. 6.2), but its advantage is the capability to examine a large number of species quickly by tuning the dye lasers. The good background rejection allows the use of this technique under conditions of bright background radiation where other methods may fail [8.82]. Examples are temperature and concentration measurements of molecular nitrogen, oxygen, and methane in a high-temperature furnace up to 2000 K [8.83], where the thermal radiation is much stronger than laser-induced fluorescence. Therefore CARS is the best choice.

A further example of the scientific application of CARS is the investigation of cluster formation in supersonic beams (Sect. 9.3), where the decrease in the rotational and vibrational temperatures during the adiabatic expansion (Sect. 9.2) and the degree of cluster formation in dependence on the distance from the nozzle can be determined [8.84].

CARS has been successfully used for the spectroscopy of chemical reactions (Sect. 13.4). The BOX CARS technique with pulsed lasers offers spectral, spatial, and time-resolved investigations of collision processes and reactions, not only in laboratory experiments but also in the tougher surroundings of factories, in the reaction zone of car engines, and in atmospheric research (Sect. 15.2 and [8.85, 8.86]).

The detection sensitivity of CARS ranges from 0.1−100 ppm ($\hat{=}$ 10^{-7} − 10^{-4} relative concentrations) depending on the Raman cross sections. Although other spectroscopic techniques such as laser-induced fluorescence or resonant two-photon ionization (Sect. 6.2) may reach higher sensitivities, there are enough examples where CARS is the best or even the only choice, for instance, when the molecules under investigation are not infrared active or have no electronic transitions within the spectral range of available lasers.

Problems

8.1 What is the minimum detectable concentration N_i of molecules with a Raman scattering cross section $\sigma = 10^{-30}$ cm^2, if the incident cw laser has 10 W output power at $\lambda = 500$ nm that is focused into a scattering volume of 5 mm $\times$ 1 mm^2, which can be imaged with 10% collection efficiency onto a photomultiplier with the quantum efficiency $\eta = 25\%$? The multiplier dark current is 10 photoelectrons per second and a signal-to-noise ratio of 3:1 should be achieved.

8.2 The linear acetylene molecule C_2H_2 has seven normal vibrations. Which of these are Raman active and which are infrared active? Are there also vibrations that are both infrared as well as Raman active?

8.3 A small molecular sample of 10^{21} molecules in a volume of 5 mm $\times$ 1 mm^2 is illuminated by 10 W of argon laser radiation. The Raman cross section is $\sigma = 10^{-29}$ cm^2 and the Stokes radiation is shifted by 1000 cm^{-1}. Calculate the heat energy dW_H/dt generated per second in the sample, if the molecules do not absorb the laser radiation or the the Stokes radiation. How much is dW_H/dt increased if the laser wavelength is close to resonance of an absorbing transition, causing an absorption coefficient $\alpha = 10^{-1}\,\mathrm{cm}^{-1}$?

8.4 Estimate the intensity of Raman radiation emerging out of the endface of an optical fiber of 100-m length and $0.1 - \mathrm{mm}\ \varnothing$, filled with a Raman-active medium with a molecular density $N_i = 10^{21}\,\mathrm{cm}^{-3}$ and a Raman scattering cross section of $\phi_R = 10^{-30}\,\mathrm{cm}^2$, if the laser radiation (1 W) and the Raman light are both kept inside the fiber by total internal reflection.

8.5 The two parallel incident laser beams with a Gaussian intensity profile are focused by a lens with $f = 5\,\mathrm{cm}$ in a BOX CARS arrangement into the sample. Estimate the spatial resolution, defined by the halfwidth $S_A(z)$ of the CARS signal, when the beam diameter of each beam at the lens is 3 mm and their separation is 20 mm.

9. Laser Spectroscopy in Molecular Beams

For many years molecular beams were mainly employed for scattering experiments. The combination of new spectroscopic methods with molecular beam techniques has brought about a wealth of new information on the structure of atoms and molecules, on details of collision processes, and on fundamentals of quantum optics and the interaction of light with matter.

There are several aspects of laser spectroscopy performed with molecular beams that have contributed to the success of these combined techniques. First, the spectral resolution of absorption and fluorescence spectra can be increased by using collimated molecular beams with reduced transverse velocity components (Sect. 9.1). Second, the internal cooling of molecules during the adiabatic expansion of supersonic beams compresses their population distribution into the lowest vibrational–rotational levels. This greatly reduces the number of absorbing levels and results in a drastic simplification of the absorption spectrum (Sect. 9.2).

The low translational temperature achieved in supersonic beams allows the generation and observation of loosely bound van der Waals complexes and clusters (Sect. 9.3). The collision-free conditions in molecular beams after their expansion into a vacuum chamber facilitates saturation of absorbing levels, since no collisions refill a level depleted by optical pumping. This makes Doppler-free saturation spectroscopy feasible even at low cw laser intensities (Sect. 9.4).

New techniques of high-resolution laser spectroscopy in beams of positive or negative ions have been developed. These techniques are discussed in Sects. 9.5 and 9.6.

Several examples illustrate the advantages of molecular beams for spectroscopic investigation. The wide, new field of laser spectroscopy of collision processes in crossed molecular beams is discussed in Chap. 13.

9.1 Reduction of Doppler Width

Let us assume molecules effusing into a vacuum tank from a small hole A in an oven that is filled with a gas or vapor at pressure p (Fig. 9.1). The molecular density behind A and the background pressure in the vacuum tank are sufficiently low to assure a large mean free path of the effusing molecules, such that collisions can be neglected. The number $N(\theta)$ of molecules that travel into the cone $\theta \pm d\theta$ around the direction θ against the symmetry axis (which we choose to be the z-axis) is proportional to $\cos\theta$. A slit B with width b, at a distance d from the point source A, selects a small angular in-

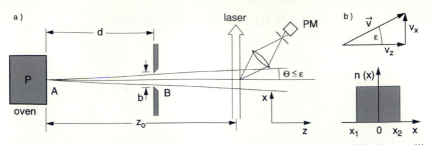

Fig. 9.1a,b. Laser excitation spectroscopy with reduced Doppler width in a collimated molecular beam: **(a)** schematic experimental arrangement; **(b)** collimation ratio and density profile $n(x)$ in a collimated beam effusing from a point source A

terval $-\epsilon \leq \theta \leq +\epsilon$ around $\theta = 0$ (Fig. 9.1). The molecules passing through the slit B, which is parallel to the y-axis, form a molecular beam in the z-direction, collimated with respect to the x-direction. The collimation ratio is defined by (Fig. 9.1b)

$$\frac{v_x}{v_z} = \tan \epsilon = \frac{b}{2d} . \tag{9.1}$$

If the source diameter is small compared with the slit width b and if $b \ll d$ (which means $\epsilon \ll 1$), the flux density behind the slit B is approximately constant across the beam diameter, since $\cos \theta \simeq 1$ for $\theta \ll 1$. For this case, the density profile of the molecular beam is illustrated in Fig. 9.1b.

The density $n(v)\, dv$ of molecules with velocities $v = |\boldsymbol{v}|$ inside the interval v to $v + dv$ in a molecular beam at thermal equilibrium, which effuses with the most probable velocity $v_p = (2kT/m)^{1/2}$ into the z-direction, can be described at the distance $r = (z^2 + x^2)^{1/2}$ from the source A as

$$n(v, r, \theta)\, dv = C \frac{\cos \theta}{r^2} nv^2\, e^{-(v/v_p)^2}\, dv , \tag{9.2}$$

where the normalization factor $C = (4/\sqrt{\pi})v_p^{-3}$ assures that the total density n of the molecules is $n = \int n(v)\, dv$.

Note: The mean flux density is $N = n\bar{v} = \int vn(v)\, dv$.

If the collimated molecular beam is crossed perpendicularly with a monochromatic laser beam with frequency ω propagating into the x-direction, the absorption probability for each molecule depends on its velocity component v_x. In Sect. 3.2 it was shown that the center frequency of a molecular transition, which is ω_0 in the rest frame of the moving molecule, is Doppler shifted to a frequency ω_0' according to

$$\omega_0' = \omega_0 - \boldsymbol{k} \cdot \boldsymbol{v} = \omega_0 - kv_x , \quad k = |\boldsymbol{k}| . \tag{9.3}$$

Only those molecules with velocity components v_x in the interval $dv_x = \delta\omega_n/k$ around $v_x = (\omega - \omega_0)/k$ essentially contribute to the absorption of the monochromatic laser wave, because these molecules are shifted into resonance with the laser frequency ω within the natural linewidth $\delta\omega_n$ of the absorbing transition.

When the laser beam in the x–z-plane ($y = 0$) travels along the x-direction through the molecular beam, its power decreases as

$$P(\omega) = P_0 \exp\left[-\int_{x_1}^{x_2} \alpha(\omega, x)\,dx \right] . \tag{9.4}$$

The absorption within the distance $\Delta x = x_2 - x_1$ inside the molecular beam is generally extremely small. Typical figures for $\Delta P(\omega) = P_0 - P(x_2, \omega)$ range from 10^{-4} to 10^{-15} of the incident power. We can therefore use the approximation $e^{-x} \simeq 1 - x$ and obtain with the absorption coefficient

$$\alpha(\omega, x) = \int n(v_x, x)\sigma(\omega, v_x)\,dv_x , \tag{9.5}$$

the spectral profile of the absorbed power

$$\Delta P(\omega) = P_0 \int_{-\infty}^{+\infty}\left[\int_{x_1}^{x_2} n(v_x, x)\sigma(\omega, v_x)\,dx \right] dv_x . \tag{9.6}$$

With $v_x = (x/r)v \rightarrow dv_x = (x/r)\,dv$ and $\cos\theta = z/r$, we derive from (9.2) for the molecular density

$$n(v_x, x)\,dv_x = Cn\frac{z}{x^3}v_x^2 \exp\left[-(rv_x/xv_p)^2 \right] dv_x . \tag{9.7}$$

The absorption cross section $\sigma(\omega, v_x)$ describes the absorption of a monochromatic wave of frequency ω by a molecule with a velocity component v_x. Its spectral profile is represented by a Lorentzian (Sect. 3.6), namely

$$\sigma(\omega, v_x) = \sigma_0 \frac{(\gamma/2)^2}{(\omega - \omega_0 - kv_x)^2 + (\gamma/2)^2} = \sigma_0 L(\omega - \omega_0, \gamma) . \tag{9.8}$$

Inserting (9.7) and (9.8) into (9.6) yields the absorption profile

$$\Delta P(\omega) = a_1 \int_{-\infty}^{+\infty}\left[\int_{x_1}^{x_2} \Delta\omega_0^2 \frac{\exp\left[-c^2\Delta\omega_0^2 \left(1 + z^2/x^2\right)/\omega_0^2 v_p^2 \right]}{(\omega - \omega_0 - kv_x)^2 + (\gamma/2)^2} dx \right] d\Delta\omega_0 ,$$

with $a_1 = P_0 n\sigma_0\gamma c^3 z/(\sqrt{\pi}v_p^3\omega_0^3)$, and $\Delta\omega_0 = \omega_0' - \omega_0 = v_x\omega_0/c$, where $\omega_0' = \omega_0 + kv_x$ is the Doppler-shifted eigenfrequency ω_0. The integration over $\Delta\omega_0$ extends from $-\infty$ to $+\infty$ since the velocities v are spread from 0 to ∞.

The integration over x is analytically possible and yields with $x_1 = -r \sin \epsilon$, $x_2 = +r \sin \epsilon$

$$\Delta P(\omega) = a_2 \int_{-\infty}^{+\infty} \frac{\exp\left[-\left(\frac{c(\omega-\omega_0')}{\omega_0' v_p \sin \epsilon}\right)^2\right]}{(\omega-\omega_0')^2 + (\gamma/2)^2} d\omega_0', \quad \text{with} \quad a_2 = a_1 \left(\frac{c\gamma}{2z\omega_0}\right)^2.$$

$$(9.9)$$

This represents a *Voigt profile*, that is, a convolution product of a Lorentzian function with halfwidth γ and a Doppler function. A comparison with (3.33) shows, however, that the Doppler width is reduced by the factor $\sin \epsilon = v_x/v = b/2d$, which equals the collimation ratio of the beam. The collimation of the molecular beam therefore reduces the Doppler width $\Delta \omega_0$ of the absorption lines to the width

$$\boxed{\Delta \omega_D^* = \Delta \omega_D \sin \epsilon,} \quad \text{with} \quad \Delta \omega_D = 2\omega_0 (v_p/c)\sqrt{\ln 2}, \qquad (9.10)$$

where $\Delta \omega_D$ is the corresponding Doppler width in a gas at thermal equilibrium.

Example 9.1

Typical figures of $b = 1\,\text{mm}$ and $d = 5\,\text{cm}$ yield a collimation ratio $b/(2d) = 1/100$. This brings the Doppler width $\Delta \nu_0 = \Delta \omega_0/2\pi \approx 1500\,\text{MHz}$ down to $\Delta \nu_D^* = \Delta \omega_D^*/2\pi \approx 15\,\text{MHz}$, which is of the same order of magnitude as the natural linewidth γ of many molecular transitions.

Note: For larger diameters of the oven hole A, the density profile $n(x)$ of the molecular beam is no longer rectangular but decreases gradually beyond the limiting angles $\theta = \pm\epsilon$. For $\Delta \omega_D^* > \gamma$, the absorption profile is then altered compared to that in (9.9), while for $\Delta \omega_D^* \ll \gamma$ the difference is negligible because the Lorentzian profile is dominant in the latter case [9.1].

The technique of reducing the Doppler width by the collimation of molecular beams was employed before the invention of lasers to produce light sources with narrow emission lines [9.2]. Atoms in a collimated beam were excited by electron impact. The fluorescence lines emitted by the excited atoms showed a reduced Doppler width if observed in a direction perpendicular to the atomic beam. However, the intensity of these atomic beam light sources was very weak and only the application of intense monochromatic, tunable lasers has allowed one to take full advantage of this method of *Doppler-free spectroscopy*.

A typical laser spectrometer for sub-Doppler excitation spectroscopy in a collimated molecular beam is shown in Fig. 9.2. The laser wavelength λ_L is controlled by a computer, which also records the laser-induced fluorescence $I_{Fl}(\lambda_L)$. Spectral regions in the UV can be covered by frequency-doubling the

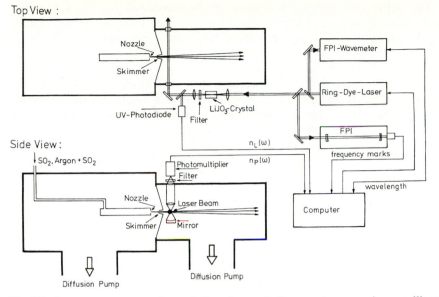

Fig. 9.2. Laser spectrometer for sub-Doppler excitation spectroscopy in a collimated molecular beam

visible laser frequency in a nonlinear optical crystal, such as $LiIO_3$. For effective collection of the fluorescence, the optical system of Fig. 6.17 can be utilized. The transmission peaks of a long FPI give frequency marks separated by the free spectral range $\delta \nu = c/2d$ of the FPI (Sect. 4.3). The absolute laser wavelength is measured by a wavemeter described in Sect. 4.4.

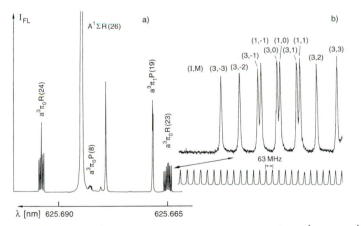

Fig. 9.3. (a) Hyperfine structure of rotational lines in the $A\,^1\Sigma_u \leftarrow X\,^1\Sigma_g^+$ system of Na_2, caused by spin–orbit coupling between the $A\,^1\Sigma_u$ and $a\,^3\Pi_u$ state [9.3]; and (b) enlarged scan of the HF multiplet of the R(23) line

The achievable spectral resolution is demonstrated by Fig. 9.3, which shows a small section of the Na_2 spectrum of the $A\,^1\Sigma_u \leftarrow X\,^1\Sigma_g$ system. Because of spin–orbit coupling with a $^3\Pi_u$ state, some rotational levels of the $A\,^1\Sigma_u$ state are mixed with levels of the $^3\Pi_u$ state and therefore show hyperfine splittings [9.3].

Particularly for polyatomic molecules with their complex visible absorption spectra, the reduction of the Doppler width is essential for the resolution of single lines [9.5]. This is illustrated by a section from the excitation spectrum of the SO_2 molecule, excited with a single-mode frequency-doubled dye laser tunable around $\lambda = 304$ nm (Fig. 9.4b). For comparison the same section of the spectrum as obtained with Doppler-limited laser spectroscopy in an SO_2 cell is shown in Fig. 9.4a [9.4].

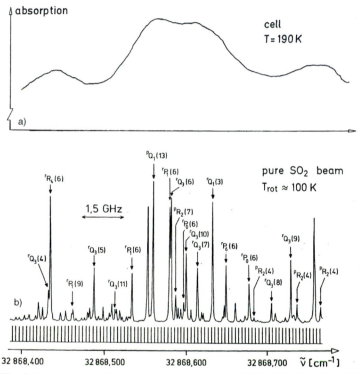

Fig. 9.4a,b. Section of the excitation spectrum of SO_2: (**a**) taken in a SO_2 cell at 0.1 mbar with Doppler-limited resolution; (**b**) taken in a collimated SO_2 beam [9.4]

The possibilities of molecular beam spectroscopy can be enhanced by allowing for spectrally resolved fluorescence detection or for resonant two-photon ionization in combination with a mass spectrometer. Such a molecular beam apparatus is shown in Fig. 9.5. The photomultiplier PM1 monitors the total fluorescence $I_{Fl}(\lambda_L)$ as a function of the laser wavelength λ_L (excitation spectrum, Sect. 6.3). Photomultiplier PM2 records the dispersed fluorescence spectrum excited at a fixed laser wavelength, where the laser is stabilized onto

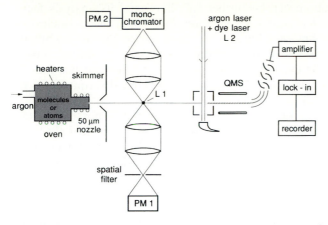

Fig. 9.5. Experimental setup for sub-Doppler spectroscopy in a collimated molecular beam. Photomultiplier PM1 monitors the total undispersed fluorescence, while PM2 behind a monochromator measures the dispersed fluorescence spectrum. The mass-specific absorption can be monitored by resonant two-color two-photon ionization in the ion source of a mass spectrometer

a selected molecular absorption line. In a second crossing point of the molecular beam with two laser beams within the ion chamber of a quadrupole mass spectrometer, the molecules are selectively excited by laser L1 and the excited molecules are ionized by L2. This allows the selection of spectra of specific molecules (for example, isotopomers) when several species are present in the molecular beam.

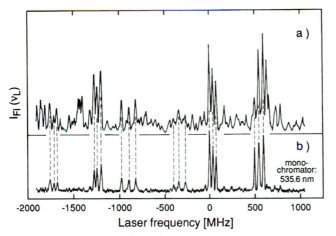

Fig. 9.6a,b. Section of the spectrum of NO_2 excited at $\lambda_{ex} = 488$ nm in a collimated NO_2 beam with a collimation ratio of $\sin \epsilon = 1/80$: (**a**) Total fluorescence monitored; and (**b**) filtered excitation spectrum, where instead of the total fluorescence only the fluorescence band at $\lambda = 535.6$ nm for the lower vibrational level (0,10) was monitored by PM2 behind a monochromator [9.6]

The excitation spectrum can be further simplified and its analysis facilitated by recording a *filtered excitation spectrum* (Fig. 9.6b). The monochromator is set to a selected vibrational band of the fluorescence spectrum while the
laser is tuned through the absorption spectrum. Then only transitions to those
upper levels appear in the excitation spectrum that emit fluorescence into the
selected band. These are levels with a certain symmetry, determined by the
selected fluorescence band.

The selective excitation of single upper levels, which is possible in molecular beams with sufficiently good collimation, results even in polyatomic
molecules in astonishingly simple fluorescence spectra. This is, for example,
demonstrated in Fig. 6.47 for the NO_2 molecule, which is excited at a fixed
wavelength $\lambda_L = 592$ nm. The fluorescence spectrum consists of readily as-
signed vibrational bands that are composed of three rotational lines (strong
P and R lines, and weak Q lines).

Instead of measuring the fluorescence intensity $I_{Fl}(\lambda_L)$, the excitation
spectrum can also be monitored via resonant two-photon ionization (RTPI).
This is illustrated by Fig. 9.7, which shows a RTPI spectrum of a band head
of the Cs_2 molecule, excited by a tunable cw dye laser and ionized by a cw
argon laser [9.7] using the arrangement of Fig. 9.5.

Besides these three examples, a large number of atoms and molecules have
been studied in molecular beams with high spectral resolution. For atoms
mainly hyperfine structure splittings, isotope shifts, and Zeeman splittings
have been investigated by this technique, because these splittings are generally
so small that they may be completely masked in Doppler-limited spectroscopy
[9.6, 9.7]. An impressive illustration of the sensitivity of this technique is the

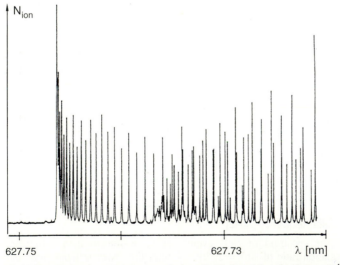

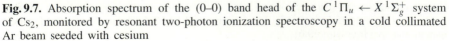

Fig. 9.7. Absorption spectrum of the (0–0) band head of the $C\,^1\Pi_u \leftarrow X\,^1\Sigma_g^+$ system
of Cs_2, monitored by resonant two-photon ionization spectroscopy in a cold collimated
Ar beam seeded with cesium

measurement of nuclear-charge radii and nuclear moments of stable and radioactive unstable isotopes through the resolution of optical hfs splittings and
isotope shifts performed by several groups [9.8, 9.9]. Even spurious concentrations of short-lived radioactive isotopes could be measured in combination
with an on-line mass separator.

For molecules the line densities are much higher, and often the rotational structure can only be resolved by sub-Doppler spectroscopy. Limiting
the collimation angle of the molecular beam below 2×10^{-3} rad, the residual
Doppler width can be reduced to values below 500 kHz. Such high-resolution
spectra with linewidths of less than 150 KHz could be, for instance, achieved
in a molecular iodine beam since the residual Doppler width of the heavy I_2
molecules, which is proportional to $m^{-1/2}$, is already below this value for a
collimation ratio $\epsilon \leq 4 \times 10^{-4}$ [9.10]. At such small linewidths the transit-time
broadening from the finite interaction time of the molecules with a focused
laser beam is no longer negligible, since the spontaneous lifetime already exceeds the transit time.

More examples of sub-Doppler spectroscopy in atomic or molecular beams
can be found in reviews on this field by Jacquinot [9.11], and Lange et al.
[9.12], in the two-volume edition on molecular beams by Scoles [9.13], as
well as in [9.14–9.17].

9.2 Adiabatic Cooling in Supersonic Beams

For effusive beams discussed in the previous section, the pressure in the
reservoir is so low that the mean free path Λ of the molecules is large compared with the diameter a of the hole A. This implies that collisions during
the expansion can be neglected. Now we will treat the case where $\Lambda \ll a$.
This means that the molecules suffer many collisions during their passage
through A and in the spatial region behind A. In this case the expanding gas
may be described by a hydrodynamic-flow model [9.18]: the expansion occurs
so rapidly that essentially no heat exchange occurs between the gas and the
walls, the expansion is adiabatic, and the enthalpy per mole of the expanding
gas is conserved.

The total energy E of a mole with mass M is the sum of the internal
energy $U = U_{\text{trans}} + U_{\text{rot}} + U_{\text{vib}}$ of a gas volume at rest in the reservoir, its potential energy pV, and the kinetic-flow energy $\frac{1}{2}Mu^2$ of the gas expanding
with the mean flow $u(z)$ in the z-direction into the vacuum. Energy conservation demands for the total energy before and after the expansion

$$U_0 + p_0 V_0 + \frac{1}{2}Mu_0^2 = U + pV + \frac{1}{2}Mu^2 . \tag{9.11}$$

If the mass flow dM/dt through A is small compared to the total mass of
the gas in the reservoir, we can assume thermal equilibrium inside the reservoir, which implies $u_0 = 0$. Since the gas expands into the vacuum chamber,
the pressure after the expansion is small ($p \ll p_0$). Therefore we may approx-

imate (9.11) by setting $p = 0$, which yields

$$U_0 + p_0 V_0 = U + \frac{1}{2} M u^2 . \tag{9.12}$$

This equation illustrates that a "cold beam" with small internal energy U is obtained, if most of the initial energy $U_0 + p_0 V_0$ is converted into kinetic-flow energy $\frac{1}{2} M u^2$. The flow velocity u may exceed the local velocity of sound $c(p, T)$. In this case a supersonic beam is produced. In the limiting case of total conversion we would expect $U = 0$, which means $T = 0$. We will later discuss several reasons why this ideal case cannot be reached in reality.

The decrease of the internal energy means also a decrease of the relative velocities of the molecules. In a microscopic model this may be understood as follows (Fig. 9.8): during the expansion the faster molecules collide with slower ones flying ahead and transfer kinetic energy.

The energy transfer rate decreases with decreasing relative velocity and decreasing density and is therefore important only during the first stage of the expansion. Head-on collisions with impact parameter zero narrow the velocity distribution $n(v_\parallel)$ of velocity components $v_\parallel = v_z$ parallel to the flow velocity u in the z-direction. This results in a modified Maxwellian distribution

$$n(v_z) = C_1 \exp\left(-\frac{m(v_z - u)^2}{2kT_\parallel}\right) , \tag{9.13}$$

around the flow velocity u. This distribution may be characterized by the *translational temperature* $T_\parallel$, which is a measure of the width of the distribution (9.13).

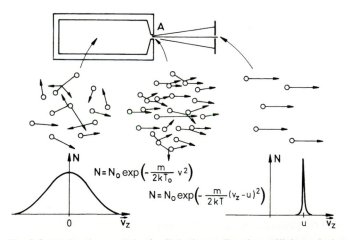

Fig. 9.8. Molecular model of adiabatic cooling by collisions during the expansion from a reservoir with a Maxwellian velocity distribution into the directed molecular flow with a narrow distribution around the flow velocity u [9.20]

For collisions with nonzero impact parameter both collision partners are deflected. If the deflection angle is larger than the collimation angle ϵ, these molecules can no longer pass the collimating aperture B in Fig. 9.1. The aperture causes for effusive as well as for supersonic beams a reduction of the transverse velocity components. Along the beam axis z the width of the distribution $n(v_x)$ measured within a fixed spatial interval Δx, which can be set by the optics for LIF detection, decreases proportionally to $\Delta x/z$. This is often named *geometrical cooling* because the reduction of the width of $n(v_z)$ is not caused by collision but by a pure geometrical effect. The transverse velocity distribution

$$n(v_x) = C_2 \exp\left(-\frac{mv_x^2}{2kT_\perp}\right) = C_2 \exp\left(-\frac{mv^2 \sin^2 \epsilon}{2kT_\perp}\right) , \qquad (9.14)$$

is often characterized by a *transverse temperature* $T_\perp$, which is determined by the velocity distribution $n(v)$, the collimation ratio $\epsilon = v_x/v_z = b/2d$, and the distance z from the nozzle.

The reduction of the velocity distribution can be measured with different spectroscopic techniques. The first method is based on measurements of the Doppler profiles of absorption lines (Fig. 9.9). The beam of a single-mode dye laser is split into one beam that crosses the molecular beam perpendicularly, and another that is directed anticollinearly to the molecular beam. The maximum ω_m of the Doppler-shifted absorption profile yields the most probable velocity $v_p = (\omega_0 - \omega_m)/k$, while the absorption profiles of the two arrangements give the distribution $n(v_\parallel)$ and $n(v_\perp)$ [9.19, 9.20].

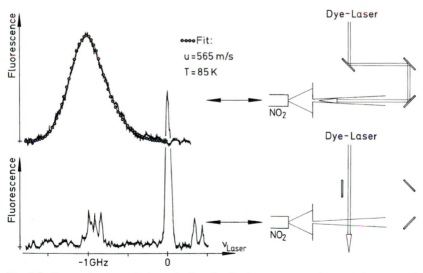

Fig. 9.9. Determination of the velocity distributions $n(v_\parallel)$ and $n(v_\perp)$ by measuring the Doppler profile of absorption lines in a thermal, effusive NO_2 beam [9.20]

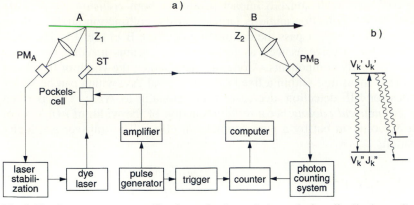

Fig. 9.10. Quantum-state-specific determination of the velocity distribution $n(i, v_{\parallel})$ of molecules in an absorbing level $|i\rangle = (v_i'', J_i'')$ by time-of-flight measurements

The second method is based on a time-of-flight measurement. The laser beam is again split, but now both partial beams cross the molecular beam perpendicularly at different positions z_1 and z_2 (Fig. 9.10). When the laser is tuned to a molecular transition $|i\rangle \to |k\rangle$ the lower level $|i\rangle = (v_i'', J_i'')$ is partly depleted due to optical pumping. The second laser beam therefore experiences a smaller absorption and produces a smaller fluorescence signal. In the case of molecules even small intensities are sufficient to saturate a transition and to completely deplete the lower level (Sect. 7.1).

If the first laser beam is interrupted at the time t_0 for a time interval Δt that is short compared to the transit time $T = (z_2 - z_1)/v$ (this can be realized by a Pockels cell or a fast mechanical chopper), a pulse of molecules in level $|i\rangle$ can pass the pump region without being depleted. Because of their velocity distribution, the different molecules reach z_2 at different times $t = t_0 + T$. The time-resolved detection of the fluorescence intensity $I_{\mathrm{Fl}}(t)$ induced by the second, noninterrupted laser beam yields the distribution

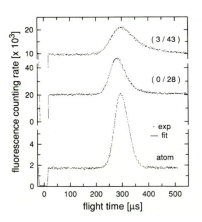

Fig. 9.11. Time-of-flight distribution of Na atoms and Na$_2$ molecules in two different vibrational–rotational levels ($v'' = 3$, $J'' = 43$) and ($v'' = 0$, $J'' = 28$) [9.21]

$n(T) = n(\Delta z/v)$, which can be converted by a Fourier transformation into the velocity distribution $n(v)$. Figure 9.11 shows as an example the velocity distribution of Na atoms and Na_2 molecules in a sodium beam in the intermediate range between effusive and supersonic conditions. If the molecules Na_2 had been formed in the reservoir before the expansion, one would expect the relation $v_p(Na) = \sqrt{2}v_p(Na_2)$. The result of Fig. 9.11 proves that the Na_2 molecules have a larger most probable velocity v_p. This implies that most of the dimers are formed during the adiabatic expansion [9.21].

The laser spectroscopic techniques provide much more detailed information about the state-dependent velocity distribution than measurements with mechanical velocity selectors. Note that in Fig. 9.11 not only $v_p(Na_2) > v_p(Na)$ but the velocity distribution of the Na_2 molecules differs for different vibration–rotation levels (v, J). This is due to the fact that molecules are being formed by stabilizing collisions during the adiabatic expansion. Molecules in lower states have suffered more collisions with atoms of the "cold bath." Their distribution $n(v)$ becomes narrower and their most probable velocity v_p more closely approaches the flow velocity u.

The main advantage of cold molecular beams for molecular spectroscopy is the decrease of rotational energy U_{rot} and vibrational energy U_{vib}, which results in a compression of the population distribution $n(v, J)$ into the lowest vibrational and rotational levels. This energy transfer proceeds via collisions during the adiabatic expansion (Fig. 9.12). Since the cross sections for collisional energy transfer $U_{rot} \rightarrow U_{trans}$ are generally smaller than for elastic collisions ($U_{trans} \rightarrow U_{trans}$), the rotational energy of the molecules before the expansion cannot be completely transferred into flow energy during the short time interval of the expansion, where collisions are important. This implies that after the expansion $U_{rot} > U_{trans}$, which means that the translational energy of the relative velocities (internal kinetic energy) has cooled faster than the rotational energy. The rotational degrees of freedom are not completely in thermal equilibrium with the translation. However, for molecules with sufficiently small rotational constants, the cross sections $\sigma_{rot-rot}$ for collisional redistribution within the manifold of rotational levels may be larger than $\sigma_{rot-trans}$. In these cases it is often possible to describe the rotational population $n(J)$ approximately by a Boltzmann distribution

$$n(J) = C_2(2J+1) \exp\left(-\frac{E_{rot}}{kT_{rot}}\right) , \qquad (9.15)$$

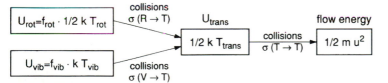

Fig. 9.12. Energy transfer diagram for adiabatic cooling in supersonic molecular beams

where $C_2 = n_v/Z$ is a constant, depending on the partition function Z and the total population density n_v in a vibrational level. This defines a *rotational temperature* T_{rot}, which is higher than the translational temperature $T_\parallel$ defined by (9.13).

The rotational temperature can be determined experimentally by measuring relative intensities of the absorption lines

$$I_{abs} = C_1 n_i(v_i, J_i) B_{ik} \rho_L , \tag{9.16a}$$

for different rotational lines in the same vibrational band, starting from different rotational levels $|J_i\rangle$ in the lower state. If the laser intensity $I_L = \rho_L c$ is sufficiently low, saturation can be neglected and the line intensities (monitored, for instance, by laser-induced fluorescence, LIF) are proportional to the unsaturated population densities $n_1(v_i J_i)$.

For unperturbed transitions the relative transition probabilities of different rotational lines in the same vibrational bands are given by the corresponding Hönl–London factors [9.22] and can thus be readily calculated. In case of perturbed spectra this may be no longer true. Here the following alternatives can be used:

- If the laser intensity is sufficiently large the molecular transitions are saturated. In the case of complete saturation ($S \gg 1$, Sect. 7.1) every molecule in the absorbing level $|i\rangle$ passing through the laser beam will absorb a photon. The intensity of the absorption line is then *independent* of the transition probability B_{ik} but only depends on the molecular flux $N(v_i, J_i) = u(v_i, J_i) \times n(v_i, J_i)$. We then obtain instead of (9.16a)

$$I_{abs} = C_2 u(v_i, J_i) n(v_i, J_i) . \tag{9.16b}$$

- Another possibility for the determination of T_{rot} in the case of perturbed spectra is based on the following procedure: at first the relative intensities $I_{th}(v, J)$ are measured in an effusive thermal beam with a sufficiently low pressure p_0 in the reservoir. Here the population distribution $n(v, J)$ can still be described by a Boltzmann distribution

$$n_{th}(v_i, J_i) = g_i \exp(-E_i/kT_0) ,$$

with the reservoir temperature T_0 since cooling is negligible. Then the changes in the relative intensities are observed while the pressure p_0 is increased and adiabatic cooling starts. This yields the dependence $T_{rot}(p_0, T_0)$ of rotational temperatures in the supersonic beam on the reservoir parameters p_0 and T_0.

If the intensities in the supersonic beam are named $I_s(v, J)$, we obtain

$$\frac{I_s(v_i, J_i)/I_s(v_k, J_k)}{I_{th}(v_i, J_i)/I_{th}(v_k, J_k)} = \frac{n_s(v_i, J_i)/n_s(v_k, J_k)}{n_{th}(v_i, J_i)/n_{th}(v_k, J_k)} . \tag{9.17}$$

Since the relative Boltzmann distribution in the thermal beam is

$$\frac{n_{th}(v_i, J_i)}{n_{th}(v_k, J_k)} = \frac{g_i}{g_k} e^{-(E_i - E_k)/kT_0} , \tag{9.18}$$

we can determine the rotational and vibrational temperatures T_{rot} and T_{vib} with (9.15)–(9.18) from the equation

$$n(v_i J_i) = g_i \, e^{-E_{vib}/kT_{vib}} \, e^{-E_{rot}/kT_{rot}} , \tag{9.19}$$

where $g_i = g_{vib} g_{rot}$ is the statistical weight factor.

The cross sections $\sigma_{vib-trans}$ or $\sigma_{vib-rot}$ are generally much smaller than $\sigma_{rot-trans}$. This implies that the cooling of vibrational energy is less effective than that of E_{rot}. Although the population distribution $n(v)$ deviates more or less from a Boltzmann distribution, it is often described by the *vibrational temperature* T_{vib}. From the discussion above we can deduce the relation

$$T_{trans} < T_{rot} < T_{vib} . \tag{9.20}$$

The lowest translational temperatures $T_{\parallel} < 1\,\mathrm{K}$ can be reached with supersonic beams of noble gas atoms. The reason for this fact is the following:

If two atoms A recombine during the expansion to form a dimer A_2, the binding energy is transferred to a third collision partner. This results in heating of the cold beam and prevents the translational temperature from reaching its lowest possible value. Since the binding energy of noble gas atoms is very small, this heating effect is generally negligible in beams of noble gas atoms.

In order to reach low values of T_{rot} for molecules, it is advantageous to use noble gas atomic beams that are "seeded" with a few percent of the wanted molecules. The cold bath of the atoms acts as a heat sink for the transfer of rotational energy of the molecules to translational energy of the atoms. This effect is demonstrated by Fig. 9.13, which shows the rotational temperature T_{rot} as a function of the pressure p_0 in the reservoir for a pure NO_2 beam and an argon beam seeded with 5% NO_2 molecules.

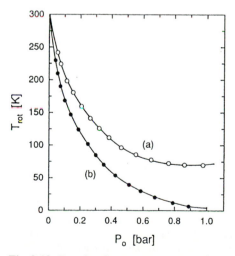

Fig. 9.13. Rotational temperature T_{rot} of NO_2 molecules in a pure NO_2 beam (*a*) and in an argon beam seeded with 5% NO_2 (*b*) as a function of the pressure p_0 in the reservoir

Example 9.2

In a molecular beam of 3% NO_2 diluted in argon one measures at a total pressure of $p_0 = 1$ bar in the reservoir and a nozzle diameter $a = 100\,\mu m$

$$T_{trans} \approx 1\,K, \quad T_{rot} \approx 5-10\,K, \quad T_{vib} \approx 50-100\,K \; .$$

Because of the small cross sections $\sigma_{vib-trans}$, vibrational cooling in seeded noble gas beams is not as effective. Here a beam of cold inert molecules such

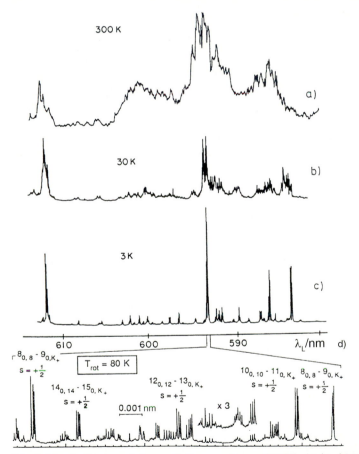

Fig. 9.14a–d. Section of the excitation spectrum of NO_2 obtained under different experimental conditions: (**a**) in a vapor cell at $T = 300\,K$, $p(NO_2) = 0.05$ mbar; (**b**) in a pure NO_2 beam at $T_{rot} = 30\,K$; (**c**) in a supersonic argon beam seeded with 5% NO_2 at $T_{rot} = 3\,K$, where (**a–c**) were excited with a dye laser with 0.05-nm bandwidth [9.24]; (**d**) 0.01-nm section of (b) recorded with a single-mode dye laser (1-MHz bandwidth) [9.25]

as nitrogen N_2 or SF_6 seeded with the molecules M under investigation may be better for cooling T_{vib} by vibrational–vibrational energy transfer [9.23].

The reduction of T_{rot} and T_{vib} results in a drastic simplification of the molecular absorption spectrum because only the lowest, still populated levels contribute to the absorption. Transitions from low rotational levels become stronger, those from higher rotational levels are nearly completely eliminated. Even complex spectra, where several bands may overlap at room temperature, reduce at sufficiently low rotational temperatures in cold beams to a few rotational lines for each band, which are grouped around the band head. This greatly facilitates their assignments and allows one to determine the band origins more reliably. For illustration, Fig. 9.14 depicts the same section of the visible NO_2 spectrum recorded under different experimental conditions. While in the congested spectrum at room temperature no spectral lines can be recognized, the spectrum at $T_{rot} = 3\,K$ in a cold He jet seeded with NO_2 [9.24] clearly demonstrates the cooling effect and shows the well-separated manifold of vibronic bands. The spectrum in Fig. 9.14d shows one band recorded with sub-Doppler resolution at $T_{rot} = 80\,K$, where rotational levels up to $J = 12$ are populated.

In addition, a large number of molecules has been investigated in cold molecular beams. Even large biomolecules become accessible to laser spectroscopic techniques [9.24–9.29].

Rotational temperatures $T_{rot} < 1\,K$ have been achieved with pulsed supersonic beams. A valve between the reservoir and nozzle opens for times $\Delta t \approx 0.1-1\,ms$ with repetition frequencies f adapted to that of the pulsed lasers. Pressures p_0 up to 100 bar are used, which demand only modest pumping speeds because of the small duty cycle $\Delta t \cdot f \ll 1$.

9.3 Formation and Spectroscopy of Clusters and Van der Waals Molecules in Cold Molecular Beams

Because of their small relative velocities Δv (Fig. 9.8), atoms A or molecules M with mass m may recombine to bound systems A_n or M_n ($n = 2, 3, 4, \ldots$) if the small translational energy $\frac{1}{2}m\Delta v^2$ of their relative motion can be transferred to a third collision partner (which may be another atom or molecule or the wall of the nozzle). This results in the formation of loosely bound atomic or molecular complexes (e.g., NaHe, I_2He_4) or clusters that are bound systems of n equal atoms or molecules (e.g., Na_n, Ar_n, $(H_2O)_n$, where $n = 2, 3, \ldots$).

In a thermodynamic model condensation takes place if the vapor pressure of the condensating substance falls below the total local pressure. The vapor pressure

$$p_s = A\,e^{-B/T}\,,$$

in the expanding beam decreases exponentially with decreasing temperature T, while the total pressure p_t decreases because of the decreasing density

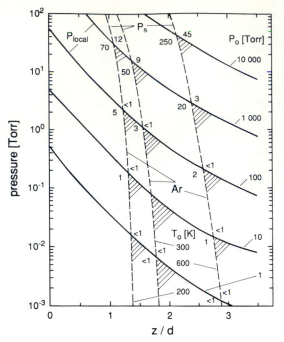

Fig. 9.15. Vapor pressure p_s of argon and local total pressure p_{loc} as a function of normalized distance $z^* = z/d$ from the nozzle in units of the nozzle diameter d for different stagnation pressures p_0 in the reservoir. Condensation can take place in the hatched areas. The numbers of three-body collisions at the points where $p_s = p_{loc}$ are also given [9.30]

in the expanding beam and the falling temperature (Fig. 9.15). If sufficient three-body collisions occur in regions where $p_s \leq p_t$ recombination can take place.

Clusters represent a transition regime between molecules and small liquid drops or small solid particles, and they have therefore found increasing interest [9.31–9.34]. Laser spectroscopy has contributed in an outstanding way to the investigation of cluster structures and dynamics. A typical experimental arrangement for the study of small metal clusters is shown in Fig. 9.5. The metal vapor is produced in an oven and mixed with argon. The mixture expands through a small nozzle ($\sim$50-μm diameter) and cools to rotational temperatures of a few Kelvin. Resonant two-photon ionization with a tunable dye laser and an argon-ion laser converts the clusters $A_n(i)$ in a specified level $|i\rangle$ into ions, which are monitored behind a quadrupole mass spectrometer in order to select the wanted cluster species A_n [9.35]. If the wavelength of the ionizing laser is chosen properly, fragmentation of the ionized clusters A_n^+ can be avoided or at least minimized, and the measured mass distribution $N(A_n^+)$ represents the distribution of the neutral clusters $N(A_n)$, which can be studied as a function of the oven parameters (p_0, T_0), seed-gas concentration, and the nozzle diameter [9.36]. This gives information about the nucleation process of clusters during the expansion of a supersonic beam.

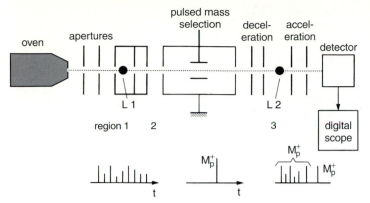

Fig. 9.16. Time-of-flight molecular beam apparatus for the measurement of cluster fragmentation

The unimolecular dissociation of cluster ions after photoexcitation can be used to determine their binding energy as a function of cluster size [9.37] with the apparatus shown schematically in Fig. 9.16. The clusters, formed during the adiabatic expansion through the nozzle, are photoionized by a pulsed UV laser L1 in region 1. After acceleration the ions fly through a field-free region 2 and pass deflection plates at a time t that depends on their mass. When the wanted cluster mass arrives at the deflection region the deflecting voltage is switched off, allowing this species to fly straight on. In region 3 the related cluster ions are photoionized by a second tunable laser and their fragments are detected by a time-of-flight mass spectrometer.

Clusters are often *floppy systems*, which do not have a rigid molecular geometry. Their spectroscopy therefore offers the possibility to study new behavior of quantum systems between regular, "well-behaved" molecules with normal vibrations and classically chaotic systems with irregular movements of the nuclei [9.38].

The most intensively studied clusters are alkali-metal clusters [9.39–9.42], where the stability and the ionization energies have been measured as have the electronic spectra and their transition from localized molecular orbitals to delocalized band structure of solids [9.43, 9.44].

Molecular clusters are also formed during the adiabatic expansion of free jets. Examples are the production of benzene clusters $(C_6H_6)_n$ and their analysis by two-photon ionization in a reflectron [9.45], or the determination of the structure and ionization potential of benzene–argon complexes by two-color resonance-enhanced two-photon ionization techniques [9.46].

The structure of molecular complexes in their electronic ground state can be obtained from direct IR laser absorption spectroscopy in pulsed supersonic-slit jet expansions [9.47]. This allows one to follow the formation rate of clusters and complexes during the adiabatic expansion [9.48]. Selective photodissociation of van der Waals clusters by infrared lasers may be used for isotope separation [9.49].

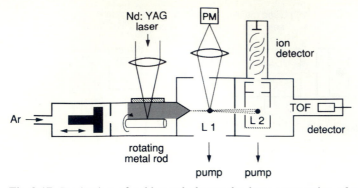

Fig. 9.17. Production of cold metal clusters by laser evaporation of metal vapors

An interesting technique for the production of metal clusters has been developed by Smalley et al. [9.50]. A slowly rotating metal rod is irradiated by pulses from a Nd:YAG laser (Fig. 9.17). The metal vapor produced by evaporation of material in the focal spot on the rod surface is mixed with a noble gas, which is let into the evaporation chamber through a pulsed nozzle synchronized with the laser pulses. The resulting mixture of noble gas and metal vapor expands through a narrow nozzle. In the resulting supersonic pulsed jet, metal clusters are formed, which can be analyzed by their fluorescence induced by laser L1 or by two-photon ionization with L2. The mass distribution can be measured with a time-of-flight spectrometer.

With this technique metal clusters of materials with high melting points could be produced, which are more difficult to realize by evaporation in hot furnaces. A famous example of carbon clusters formed by this technique for the first time were the fullerenes C_{60}, C_{70}, etc. [9.51].

An elegant technique for studying van der Waals complexes at low temperatures was developed by Toennies and coworkers [9.52]. A beam of large He clusters ($10^4 - 10^5$ He atoms) passes through a region with a suffient vapor pressure of molecules. The He droplets pick up a molecule, which then diffuses into the central part of the droplet, where it is cooled down to the low temperature of a few Kelvin. Since the interaction with the four He atoms is very small, the spectrum of this trapped molecule does not differ much from that of a free cold molecule. However, in this case $T_{rot} = T_{vib}$ [9.53–9.55].

A similar technique, where alkali molecules and clusters in high spin states are formed, was invented by Scoles and coworkers [9.56]. Here the He-clusters pass through a region with high alkali vapor pressure. The alkali atoms condense at the surface of the He clusters. Alkali atoms meet and recombine by migration along the surface. Since the singlet states of alkali dimers have a large binding energy, which is transferred to the He cluster, this leads to the evaporation of many He atoms and may completely destroy the cluster. However, the formation of triplet states with a much smaller binding energy will evaporate only a few He atoms. The alkali dimers and multimers in high spin states rapidly adjust this temperature to that of the He cluster.

They can be studied by laser spectroscopy, thus giving access to states that are difficult to produce in normal gas phase spectroscopy [9.57].

Van der Waals diatomics formed through the weak interaction between a noble gas atom and an alkali atom have been studied by noble gas beams seeded with alkali vapor [9.58].

9.4 Nonlinear Spectroscopy in Molecular Beams

The residual Doppler width from the finite collimation ratio ϵ of the molecular beam can be completely eliminated when nonlinear Doppler-free techniques are applied. Since collisions can generally be neglected at the crossing point of the molecular and laser beam, the lower molecular level $|i\rangle$ depleted by absorption of laser photons can be only refilled by diffusion of new, un-pumped molecules into the interaction zone and by the small fraction of the fluorescence terminating on the initial level $|i\rangle$. The saturation intensity I_s is therefore lower in molecular beams than in gas cells (Example 7.2).

A possible arrangement for saturation spectroscopy in a molecular beam is depicted in Fig. 9.18. The laser beam crosses the molecular beam perpendicularly and is reflected by the mirror M1. The incident and the reflected beam can only be absorbed by the same molecules within the transverse velocity group $v_x = 0 \pm \gamma k$ if the laser frequency $\omega_L = \omega_0 \pm \gamma$ matches the molecular absorption frequency ω_0 within the homogeneous linewidth γ. When tuning the laser frequency ω_L one observes narrow Lamb dips (Fig. 9.19) with a saturation-broadened width γ_s at the center of broader profiles with a reduced Doppler-width $\epsilon \Delta \omega_D$, from the collimation ratio $\epsilon \ll 1$ of the molecular beam (Sect. 9.1).

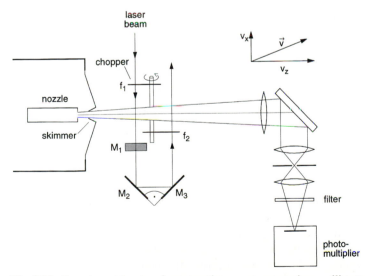

Fig. 9.18. Experimental setup for saturation spectroscopy in a collimated molecular beam

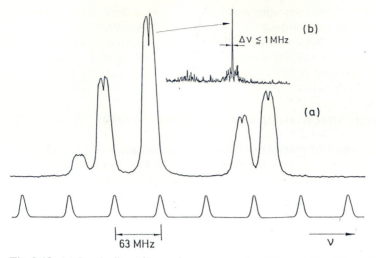

Fig. 9.19. (a) Lamb dips of hyperfine components of the rotational transition $J' = 1 \leftarrow J''$ $= 0$ in a collimated NO_2 beam. The residual Doppler width is 15 MHz. **(b)** The insert shows one component with suppression of the Doppler background by chopping the laser beams and using lock-in techniques [9.59]

It is essential that the two laser beams cross the molecular beam exactly perpendicularly; otherwise, the opposite Doppler shifts $\pm\delta\omega_D$ observed for the two beams result in a broadening of the Lamb dips for $2\delta\omega_D \leq \gamma$, while no Lamb dips can be observed for $2\delta\omega_D \gg \gamma$!

Example 9.3

In a supersonic beam with $u = 10^3$ m/s, $\epsilon = 10^{-2}$, the residual Doppler width for a visible transition with $\nu = 6 \times 10^{14}$ is $\epsilon\Delta\omega_D \sim 2\pi \cdot 20$ MHz. For a crossing angle of 89° the Doppler shift between the two beams is $2\delta\omega_D = 2\pi \cdot 60$ MHz. In this case the Doppler shift $2\delta\omega_D$ is larger than the residual Doppler width, and even in the linear spectrum with reduced Doppler width the lines would be doubled. No Lamb dips can be observed with opposite laser beams.

If the width γ_s of the Lamb dips is very narrow, the demand for exactly perpendicular crossing becomes very stringent. In this case, an arrangement where the mirror M1 is removed and replaced by the retroreflectors M2, M3 is experimentally more convenient. The two laser beams intersect the molecular beam at two closely spaced locations z_1 and z_2. This arrangement eliminates the reflection of the laser beam back into the laser.

The Doppler-broadened background with the residual Doppler width from the divergence of the molecular beam can completely be eliminated by chopping the two laser beams at two different frequencies f_1, f_2, and monitoring the signal at the sum frequency $f_1 + f_2$ (intermodulated fluorescence,

Sect. 7.3.1). This is demonstrated by the insert in Fig. 9.19. The linewidth of the Lamb dips in Fig. 9.19 is below 1 MHz and is mainly limited by frequency fluctuations of the cw single-mode dye laser [9.59].

Additionally, several experiments on saturation spectroscopy of molecules and radicals in molecular beams have been reported [9.60–9.61] where finer details of congested molecular spectra, such as hyperfine structure or Λ-doubling can be resolved. Another alternative is Doppler-free two-photon spectroscopy in molecular beams, where high-lying molecular levels with the same purity as the absorbing ground state levels are accessible [9.62].

The improvements in the sensitivity of CARS (Sect. 8.3) have made this nonlinear technique an attractive method for the investigation of molecular beams. Its spectral and spatial resolution allow the determination of the internal-state distributions of molecules in effusive or in supersonic beams, and their dependence on the location with respect to the nozzle (Sect. 8.5). An analysis of rotationally-resolved CARS spectra and their variation with increasing distance z from the nozzle allows the determination of rotational and vibrational temperatures $T_{rot}(z)$, $T_{vib}(z)$, from which the cooling rates can be obtained [9.63]. With cw CARS realized with focused cw laser beams the main contribution to the signal comes from the small focal volume, and a spatial resolution below $1\,mm^3$ can be achieved [9.64].

Another example of the application of CARS is the investigation of cluster formation in a supersonic beam. The formation rate of clusters can be inferred from the increasing intensity $I(z)$ of characteristic cluster bands in the CARS spectra.

The advantage of CARS compared to infrared absorption ion spectroscopy is the higher sensitivity and the fact that nonpolar molecules, such as N_2, can also be studied [9.65]. With pulsed CARS short-lived transient species produced by photo dissociation in molecular beams can also be investigated [9.66, 9.67].

9.5 Laser Spectroscopy in Fast Ion Beams

In the examples considered so far, the laser beam was crossed *perpendicularly* with the molecular beam, and the reduction of the Doppler width was achieved through the limitation of the maximum transverse velocity components v_x by geometrical apertures. One therefore often calls this reduction of the transverse velocity components geometrical cooling. Kaufmann [9.68] and Wing et al. [9.69] have independently proposed another arrangement, where the laser beam travels *collinearly* with a fast ion or atom beam and the narrowing of the *longitudinal* velocity distribution is achieved by an acceleration voltage (*acceleration cooling*). This fast-ion-beam laser spectroscopy (FIBLAS) can be understood as follows:

Assume that two ions start from the ion source (Fig. 9.20) with different thermal velocities $v_1(0)$ and $v_2(0)$. After being accelerated by the voltage U

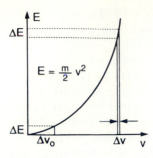

Fig. 9.20. Acceleration cooling

their kinetic energies are

$$E_1 = \frac{m}{2}v_1^2 = \frac{1}{2}mv_1^2(0) + eU \ ,$$

$$E_2 = \frac{m}{2}v_2^2 = \frac{1}{2}mv_2^2(0) + eU \ .$$

Subtracting the first equation from the second yields

$$v_2^2 - v_1^2 = v_2^2(0) - v_1^2(0) \ \Rightarrow \ \Delta v = v_1 - v_2 = \frac{v_0}{v}\Delta v_0 \ ,$$

with

$$v = \frac{1}{2}(v_1 + v_2) \ , \quad \text{and} \quad v_0 = \frac{1}{2}[v_1(0) + v_2(0)] \ .$$

Since $E_{th} = (m/2)v_0^2$ and $v = (2eU/m)^{1/2}$, we obtain for the final velocity spread

$$\Delta v = \Delta v_0 \sqrt{E_{th}/eU} \ . \tag{9.21}$$

For $E_{th} \ll eU \Rightarrow \Delta v \ll \Delta v_0$.

Example 9.4

$\Delta E_{th} = 0.1\,\mathrm{eV}$, $eU = 10\,\mathrm{keV} \to \Delta v = 3 \times 10^{-3}\Delta v_0$. This means that the Doppler width of the ions in the ion source has been decreased by acceleration cooling by a factor of 300! If the laser crosses the ion beam perpendicularly, the transverse velocity components for ions with $v = 3 \times 10^5\,\mathrm{m/s}$ at a collimation ratio $\epsilon = 10^{-2}$ are $v_x = v_y \leq 3 \times 10^3\,\mathrm{m/s}$. This would result in a residual Doppler width of $\Delta v \sim 3\,\mathrm{GHz}$, which illustrates that for fast beams the longitudinal arrangement is superior to the transverse one.

This reduction of the velocity spread results from the fact that *energies* rather than velocities are added (Fig. 9.20). If the energy $eU \gg E_{th}$, the velocity change is mainly determined by U, but is hardly affected by the fluctuations of the initial thermal velocity. This implies, however, that the acceleration voltage has to be extremely well stabilized to take advantage of this acceleration cooling.

A voltage change ΔU results, according to (9.21) and $\nu = \nu_0(1 + \nu/c)$, in the frequency change

$$\Delta \nu = \frac{\nu_0}{c}\Delta \nu = \nu_0 \sqrt{\frac{eU}{2mc^2} \frac{\Delta U}{U}}.\qquad(9.22)$$

Example 9.5

At the acceleration voltage $U = 10\,\mathrm{kV}$, which is stable within $\pm 1\,\mathrm{V}$, an absorption line of neon ions ($m = 21\,\mathrm{AMU}$) at $\nu_0 = 5 \times 10^{14}$ suffers Doppler broadening from the voltage instability according to (9.22) of $\Delta \nu \approx 25\,\mathrm{MHz}$.

A definite advantage of this coaxial arrangement of laser beam and ion beam in Fig. 9.21 is the longer interaction zone between the two beams because the laser-induced fluorescence can be collected by a lens from the path length Δz of several centimeters, compared to a few millimeters in the perpendicular arrangement. This increases the sensitivity because of the longer absorption path. Furthermore, transit-time broadening for an interaction length L of $10\,\mathrm{cm}$ becomes $\delta \nu_{\mathrm{tr}} \approx 0.4\,\nu/L \approx 2\,\mathrm{MHz}$, while for the perpendicular intersection of a laser beam with diameter $2w = 1\,\mathrm{mm}$, the transit time broadening $\delta \nu_{\mathrm{tr}} = 400\,\mathrm{MHz}$ gives a nonnegligible contribution.

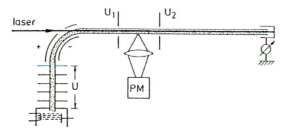

Fig. 9.21. Collinear laser spectroscopy in ion beams

A further advantage of collinear laser spectroscopy is the possibility of "electric Doppler tuning". The absorption spectrum of the ions can be scanned across a *fixed* laser frequency ν_0 simply by tuning the acceleration voltage U. This allows one to use high-intensity, fixed-frequency lasers, such as the argon-ion laser. Because of their high gain the interaction zone may even be placed inside the laser cavity. Instead of tuning the acceleration voltage U (which influences the beam collimation), the velocity of the ions in the interaction zone with the laser beam is tuned by retarding or accelerating potentials U_1 and U_2 (Fig. 9.22).

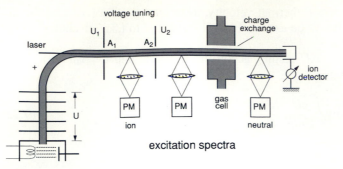

Fig. 9.22. Arrangement for "electric Doppler tuning" applied to the spectroscopy of fast ionized or neutral species

Example 9.6

A voltage shift of $\Delta U = 100\,\text{V}$ at $U = 10\,\text{kV}$ causes for H_2^+ ions a relative frequency shift of $\Delta v/v \approx 1.5 \times 10^{-5}$. At an absorption frequency of $v = 6 \times 10^{14}\,\text{s}^{-1}$ this gives an absolute shift of $\Delta v \approx 10\,\text{GHz}$.

If the ion beam passes through a differentially pumped alkali-vapor cell the ions can suffer charge-exchange collisions, where an electron is transferred from an alkali atom to the ion. Because such charge-exchange collisions show very large cross sections they occur mainly at large impact parameters, and the transfer of energy and momentum is small. This means that the fast beam of neutralized atoms has nearly the same narrow velocity distribution as the ions before the collisions. The charge exchange produces neutral atoms or molecules in highly excited states. This offers the possibility to investigate electronically excited atoms or neutral molecules and to study their structure and dynamics.

With this technique it is, for instance, possible to investigate excimers (molecular dimers that are stable in excited states but are unstable in their electronic ground state, see Sect. 5.7) in more detail. Examples are high-resolution studies of fine structure and barrier tunneling in excited triplet states of He_2 [9.70, 9.71].

9.6 Applications of FIBLAS

Some special techniques and possible applications of fast-ion-beam laser spectroscopy (FIBLAS) are illustrated by four different groups of experiments.

9.6.1 Spectroscopy of Radioactive Elements

The first group comprises high-resolution laser spectroscopy of short-lived radioactive isotopes with lifetimes in the millisecond range. The ions are

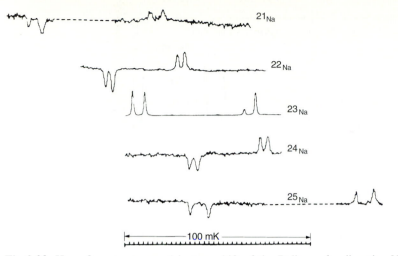

Fig. 9.23. Hyperfine structure and isotope shift of the D lines of radioactive Na isotopes [9.74]

produced by nuclear reactions induced by bombardment of a thin foil with neutrons, protons, γ-quanta, or other particles inside the ion source of a mass spectrometer. They are evaporated and enter after mass selection the interaction zone of the collinear laser [9.72].

Precision measurements of hyperfine structure and isotope shifts yield information on nuclear spins, quadrupole moments, and nuclear deformations. The results of these experiments allow tests of nuclear models of the spatial distribution of protons and neutrons in highly deformed nuclei [9.73]. In Fig. 9.23 the hyperfine spectra of different Na isotopes are depicted, which had been produced by spallation of aluminum nuclei by proton bombardment according to the reaction ^{27}Al(p,3p,xn) $^{25-x}$Na [9.74]. Such precision measurements have been performed in several laboratories for different families of isotopes [9.72–9.75].

9.6.2 Photofragmentation Spectroscopy of Molecular Ions

Besides excitation spectroscopy of bound–bound transitions, photofragmentation spectroscopy has gained increasing interest. Here predissociating upper levels of parent molecular ions M^+, which decay into neutral and ionized fragments, are excited. The ionized fragments can be detected with a mass spectrometer, while the neutral fragments need to be ionized by laser photons or by electron impact.

For illustration, Fig. 9.24 shows the number of O^+ ions formed in the photodissociation reaction

$$O_2^+ + h\nu \rightarrow O_2^{+*} \rightarrow O^+ + O ,$$

as a function of the absorption wavelength λ [9.76].

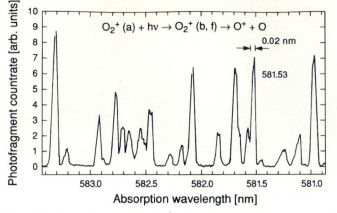

Fig. 9.24. Dependance of the O^+ photofragment signal on the absorption wavelength of O_2^+, obtained by Doppler tuning at a fixed laser wavelength [9.76]

With a properly selected polarization of the laser, the photofragments are ejected into a direction perpendicular to the ion beam direction. Their transverse energy distribution can be measured with a position-sensitive detector, because their impact position x, y at the ion detector centered around the position $x = y = 0$ of the parent ion beam is given by $x = (v_x/v_z)z$, where z is the distance between the excitation zone and the detector [9.77, 9.78].

A special version of ion spectroscopy is the Coulomb-explosion technique (Fig. 9.25). A collimated beam of molecular ions with several MeV kinetic energy pass through a thin foil where all valence electrons are stripped. By Coulomb explosion the fragments are ejected and are detected by a position-sensitive detector. The geometry and structure of the original parent ion M^+ in its electronic ground state can be inferred from the measured pattern. Excitation of the ions M^+ by a laser just before they enter the foil allows the determination of the molecular structure in excited states [9.79].

If the kinetic energy distribution of fragments produced by direct photodissociation of molecular ions is measured, the form of the repulsive potential may be deduced. This can be realized with the apparatus shown in Fig. 9.26. In the long interaction zone with the collinear laser beam, part of the parent

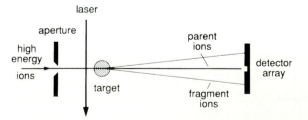

Fig. 9.25. Schematic illustration of Coulomb-explosion technique [9.79]

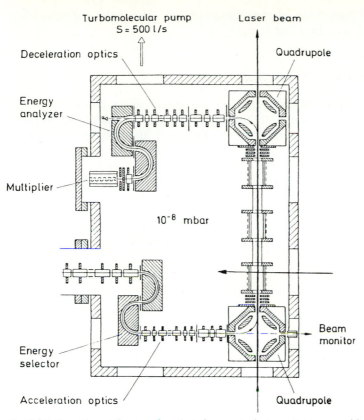

Fig. 9.26. Experimental setup for photofragmentation spectroscopy with energy and mass selection of the parent ions and the fragment ions [9.80]

ions are photodissociated. Both the parent ions and the fragment ions are deflected by a quadrupole field. The fragments are separated and their energies are measured by two 180° energy analyzers. The parent ions M^+ are mass and energy selected before they enter the interaction zone [9.80, 9.81].

Of particular interest is the multiphoton dissociation by infrared lasers, which can be studied in more detail by such an arrangement. One example is the dissociation of SO_2^+ ions induced by multiphoton absorption of CO_2 laser photons. The relative probability for the two channels

$$SO_2^+ \begin{cases} \rightarrow SO^+ + O\,, \\ \rightarrow S^+ + O_2\,, \end{cases}$$

depends on the wavelength and on the intensity of the CO_2 laser [9.82].

The combination of fast ion beam photofragmentation with field-dissociation spectroscopy opens interesting new possibilities for studying long-range ion–atom interactions. This has been demonstrated by Bjerre and Keiding [9.83], who measured the $O^+ - O$ potential in the internuclear distance range $1-2$ nm from electric field-induced dissociation of selectively laser-excited O_2^+ ions in a fast beam.

9.6.3 Laser Photodetachment Spectroscopy

Another group of experiments based on the coaxial arrangement deals with photodetachment spectroscopy of the negative molecular ions [9.84]. Negative molecular ions play an important role in the upper atmosphere and in many chemical reactions. Although hundreds of bound molecular negative ions are known, very few have been measured with rotational resolution.

Since the extra electron generally has a low binding energy, most negative ions can be ionized (photodetachment) by visible or infrared lasers. The remaining ions are separated from the neutral molecules formed in the photodetachment process by a deflecting electric field. An example of sub-Doppler photodetachment spectroscopy of C_2^- can be found in [9.85].

9.6.4 Saturation Spectroscopy in Fast Beams

For the elimination of the residual Doppler width, the FIBLAS technique allows an elegant realization of saturation spectroscopy with a single fixed-frequency laser (Fig. 9.27). The ions are accelerated by the voltage U, which is tuned to a value where the laser radiation is absorbed in the first part of the interaction zone on a transition $|i\rangle \leftarrow |k\rangle$ at the fixed laser frequency

$$\nu_L = \nu_0 \sqrt{1 + (2eU)/(mc^2)} \,. \tag{9.23}$$

In the second part of the interaction zone an additional voltage ΔU is applied, which changes the velocity of the ions. If the laser-induced fluorescence is monitored by PM2 as a function of ΔU, a Lamb dip will be observed at $\Delta U = 0$ because the absorbing level $|i\rangle$ has already been partly depleted in the first zone.

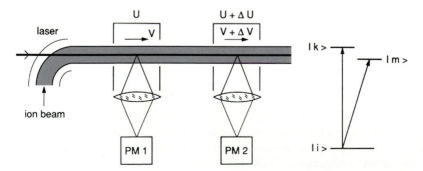

Fig. 9.27. Saturation spectroscopy in fast ion beams

If several transitions are possible from level $|i\rangle$ with frequencies ω within the Doppler tuning range

$$\Delta v(\Delta U) = v_0 \left(1 \pm \sqrt{1 + \Delta U/U} \right) , \tag{9.24}$$

a Lamb-dip spectrum of all transitions starting from the level $|i\rangle$ is obtained when U is kept constant and ΔU is tuned [9.86].

Fast ion and neutral beams are particularly useful for very accurate measurements of lifetimes of highly excited ionic and neutral molecular levels (Sect. 11.3).

9.7 Spectroscopy in Cold Ion Beams

Although the velocity spread of ions in fast beams is reduced by acceleration cooling, their internal energy (E_{vib}, E_{rot}, E_{e}), which they acquired in the ion source, is generally not decreased unless the ions can undergo radiative transitions to lower levels on their way from the ion source to the laser interaction zone. Therefore, other techniques have been developed to produce "cold ions" with low internal energies. Three of them are shown in Fig. 9.28.

In the first method (Fig. 9.28a), a low current discharge is maintained through a glass nozzle between a thin tungsten wire acting as a cathode and an anode ring on the vacuum side. The molecules are partly ionized or dissociate in the discharge during their adiabatic expansion into the vacuum [9.87]. If the expanding beam is crossed by a laser beam just behind the nozzle,

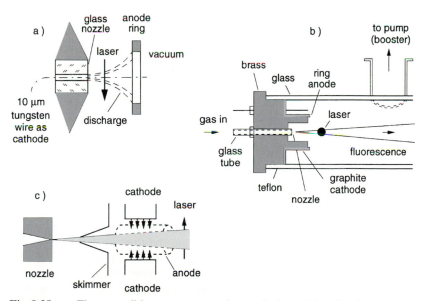

Fig. 9.28a–c. Three possible arrangements for producing cold molecular ions

excitation spectroscopy of cold molecular ions or short-lived radicals can be performed [9.88]. Erman et al. [9.89] developed a simple hollow-cathode supersonic beam arrangement, which allows sub-Doppler spectroscopy of cold ions (Fig. 9.28b).

Instead of gas discharges, electrons emitted from hot cathodes can be used for ionization (Fig. 9.28c). With several cathodes arranged cylindrically around a cylindrical grid acting as an anode, a large electron current can be focused into the cold molecular beam. Because of the low electron mass, electron impact ionization at electron energies closely above ionization threshold does not much increase the rotational energy of the ionized molecules, and rotationally cold molecular ions can be formed from cold neutral molecules. Rotational temperatures of about 20 K have been reached, for instance, when supersonically cooled neutral triacetylene molecules were ionized by 200-eV electrons in a seeded free jet of helium [9.90]. The vibrational energy depends on the Franck–Condon factors of the ionizing transitions. If the electron beam is modulated, lock-in detection allows separation of the spectra of neutral and ionized species [9.91]. When pulsed lasers are used, the electron gun can also be pulsed in order to reach high peak currents.

Cold ions can also be formed by two-photon ionization directly behind the nozzle of a supersonic neutral molecular beam [9.92]. These cold ions can then be further investigated by one of the laser spectroscopic techniques discussed above. The combination of pulsed lasers and pulsed nozzles with time-of-flight spectrometers gives sufficiently large signals to study not only molecular excitation but also the different fragmentation processes [9.93–9.95].

9.8 Combination of Molecular Beam Laser Spectroscopy and Mass Spectrometry

The combination of pulsed lasers, pulsed molecular beams, and time-of-flight mass spectrometry represents a powerful technique for studying the selective excitation, ionization, and fragmentation of wanted molecules out of a large variety of different molecules or species in a molecular beam [9.93–9.99]. The technique, developed by Boesl et al. [9.93] is illustrated by Fig. 9.29: rotationally and vibrationally cold neutral parent molecules M in a supersonic molecular beam pass through the ion source of a time-of-flight mass spectrometer. A pulsed laser L1 forms molecular ions M^+ by resonant enhanced multiphoton ionization. By selecting special intermediate states of M, the molecular ion M^+ can often be preferentially prepared in a selected vibrational level.

In the second step, after a time delay Δt that is long compared with typical lifetimes of excited states of M^+ but shorter than the time of flight out of the excitation region, a pulsed tunable dye laser L2 excites the molecular ions M^+ from their electronic ground state into selected electronic states of interest.

The detection of spectroscopic excitation is performed by photofragmentation of $(M^+)^*$ with a pulsed laser L3. For discrimination of the resulting

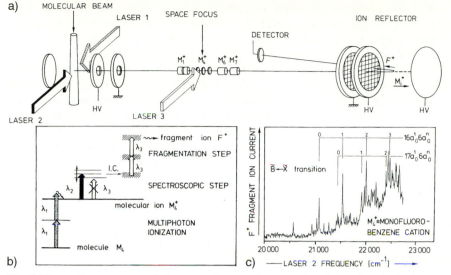

Fig. 9.29a–c. Spectroscopy of fragmentation products produced by photoionization and excitation of large molecules: (**a**) experimental setup; (**b**) level scheme; and (**c**) spectrum of the fragment ion [9.93]

secondary fragment ions F^+ from unwanted molecular ions and other fragment ions produced by laser L1, the laser L3 crosses the time-of-flight spectrometer (reflectron [9.97]) at the space focus in the field-free drift region, where ions of the same mass (in the example of Fig. 9.29b this is M_4^+) are compressed to ion bunches, while ion clouds of different masses are already separated by several microseconds due to their different flight times. This allows mass-selective excitation of M_4^+ at the space focus by choosing the correct delay time of laser pulse 3.

If the wavelength λ_3 of laser L3 is chosen properly, the ions M_4^+ cannot be excited by L3 if they are in their electronic ground state. Therefore, the excitation by L2 can be monitored by L3.

Behind the space focus, the whole assembly of molecular primary and secondary ions passes a field-free drift region and enters the ion mirror of the reflectron. This mirror has two functions: by setting the potential of the mirror end plate lower than the potential in the ion source where all primary ions are formed, these primary ions hit the end plate and are eliminated. All secondary fragment ions F^+ formed at the space focus have considerably less kintic energy (with the neutral fragments carrying away the residual kinetic energy). They are reflected in the ion mirror and reach the ion detector. By choosing the right reflecting field strength, secondary fragment ions F^+ (within the mass range of interest) are focused in time and appear in a narrow time window (for example, $\Delta t = 10\,\mathrm{ns}$) at the detector. This allows an excellent discrimination against most sources of noise, which may produce signals at different times or in very wide time ranges.

The spectroscopic technique described above is applicable to most ionic states, but is particularly useful for nonfluorescing or nonpredissociating molecular ion states, such as those of many radical cations. Because of the considerably lower energies of their first excited electronic states in comparison to their neutral parent molecules, internal conversion is enhanced, thus suppressing fluorescence. Typical examples are the cations of all mono- and of many di- and trihalogenated benzenes as well as of the benzene cation itself. For illustration, the UV/VIS spectrum of the monofluorobenzene cation shown in Fig. 9.29c was measured by the method described above and revealed for the first time vibrational resolution for this molecular cation.

The whole experimental arrangement is very flexible due to an uncomplicated mechanical setup; it allows some more ion optical variations as well as laser excitation schemes as presented above and thus several additional possibilities to perform spectroscopy and analysis of ionized and of neutral molecules [9.93].

Further information on molecular multiphoton ionization and ion fragmentation spectroscopy can be found in [9.99].

Problems

9.1 A collimated effusive molecular beam with a rectangular density profile behind the collimating aperture has a thermal velocity distribution at $T = 500$ K. Calculate the intensity profile $\alpha(\omega)$ of an absorption line, centered at ω_0 for molecules at rest, if the beam of a weak tunable monochromatic laser crosses the molecular beam under 45° against the molecular beam axis

(a) for negligible divergence of the molecular beam, and
(b) for a collimation angle of $\epsilon = 5°$.

9.2 A monochromatic laser beam in the x-direction crosses a supersonic divergent atomic beam perpendicularly to the beam axis at a distance $d = 10$ cm away from the nozzle. Calculate the halfwidth Δx of the spatial fluorescence distribution $I(x)$, if the laser frequency is tuned to the center of an atomic transition with a homogeneous linewidth of $\Delta \nu_h = 10$ MHz (saturation effects shall be negligible).

9.3 A laser beam from a tunable monochromatic laser is directed along the z-axis against a collimated thermal molecular beam with a Maxwell–Boltzmann velocity distribution. Calculate the spectral profiles of the absorption $\alpha(\omega)$

(a) for a weak laser (no saturation);
(b) for a strong laser (complete saturation $s \gg 1$), where the saturated homogeneous linewidth is still small compared to the Doppler width.

9.4 The beam of a monochromatic cw laser is split into two beams that intersect an atomic beam perpendicularly at the positions z_1 and $z_2 = z_1 + d$. The laser frequency is tuned to the center of the absorption transition $|k\rangle \leftarrow |i\rangle$ where the laser depletes the level $|i\rangle$. Calculate the time profile $I(z_2, t)$ of the LIF signal measured at z_2, if the first beam at z_1 is interrupted for a time interval $\Delta t = 10^{-7}$ s, which is short compared to the mean transit time $\bar{t} = d/\bar{v}$ with $d = 0.4$ m

(a) for a supersonic beam with the approximate velocity distribution $N(v) = a(u - 10|u - v|)$, $u = \bar{v} = 10^3$ m/s, $0.9u \leq v \leq 1.1u$;
(b) for a thermal beam with a translational temperature of $T = 600$ K and $\bar{v} = 10^3$ m/s.

9.5 Calculate the population distribution $N(v'')$ and $N(J'')$ of vibrational and rotational levels of Na_2 molecules in a supersonic beam with $T_{vib} = 100$ K and $T_{rot} = 10$ K. Which fraction of all molecules is in the levels $(v'' = 0, J'' = 20)$ and $(v'' = 1, J'' = 20)$? At which value of J'' is the maximum of $N(J'')$? The rotational constant is $B_e = 0.15$ cm^{-1}, and the vibrational constant is $\omega_e = 150$ cm^{-1}.

10. Optical Pumping and Double-Resonance Techniques

Optical pumping means selective population or depletion of atomic or molecular levels by aborption of radiation, resulting in a population change ΔN in these levels, which causes a noticeable deviation from the thermal equilibrium population. With intense atomic resonance lines emitted from hollow-cathode lamps or from microwave discharge lamps, optical pumping had successfully been used for a long time in *atomic* spectroscopy, even before the invention of the laser [10.1, 10.2]. However, the introduction of lasers as very powerful pumping sources with narrow linewidths has substantially increased the application range of optical pumping. In particular, lasers have facilitated the transfer of this well-developed technique to *molecular* spectroscopy. While early experiments on optical pumping of molecules [10.3, 10.4] were restricted to accidental coincidences between molecular absorption lines and atomic resonance lines from incoherent sources, the possibility of tuning a laser to the desired molecular transition provides a much more selective and effective pumping process. It allows, because of the larger intensity, a much larger change $\Delta N_i = N_{i0} - N_i$ of the population density in the selected level $|i\rangle$ from its unsaturated value N_{i0} at thermal equilibrium to a nonequilibrium value N_i.

This change ΔN_i of the population density can be probed by a second EM wave, which may be a radio frequency (RF) field, a microwave, or another laser beam. If this "probe wave" is tuned into resonance with a molecular transition sharing one of the two levels $|i\rangle$ or $|k\rangle$ with the pump transition, the pump laser and the probe wave are simultaneously in resonance with the coupled atomic or molecular transitions (Fig. 10.1). This situation is therefore called *optical–RF, optical–microwave* or *optical–optical double resonance*.

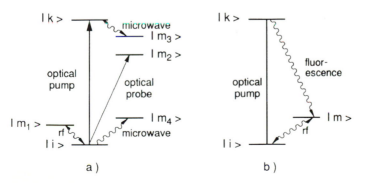

Fig. 10.1. Schematic level schemes of optical pumping and double-resonance transitions

Even this double-resonance spectroscopy has already been applied to the study of atomic transitions before lasers were available. In these pre-laser experiments incoherent atomic resonance lamps served as pump sources and a radio frequency field provided probe transitions between Zeeman levels of optically excited atomic states [10.5]. However, with tunable lasers as pump sources, these techniques are no longer restricted to some special favorable cases, and the achievable signal-to-noise ratio of the double-resonance signals may be increased by several orders of magnitude [10.6].

In this chapter we shall treat the most important laser double-resonance techniques by illustration with several examples. While the pump transition is always induced by a pulsed or cw laser, the probe field may be provided by any coherent source in the spectral range between the RF region and the $-65\,\mathrm{mmUV}$.

10.1 Optical Pumping

The effect of optical pumping on a molecular system depends on the characteristics of the pump laser, such as intensity, spectral bandwidth, and polarization, and on the linewidth and the transition probability of the absorbing transition. If the bandwidth $\Delta\omega_L$ of the pump laser is larger than the linewidth $\Delta\omega$ of the molecular transition, all molecules in the absorbing level $|i\rangle$ can be pumped. In case of dominant Doppler broadening this means that molecules within the total velocity range can simultaneously be pumped into a higher level $|k\rangle$. If the laser bandwidth is small compared to the inhomogeneous width of a molecular transition, only a subgroup of molecules with a matching absorption frequency $\omega = \omega_0 - \mathbf{k}\cdot\mathbf{v} = \omega_L$ is pumped (Sect. 7.2).

There are several different aspects of optical pumping that are related to a number of spectroscopic techniques based on optical pumping. The *first aspect* concerns the *increase or decrease of the population* in selected levels. At sufficiently high laser intensities the molecular transition can be saturated. This means that a maximum change $\Delta N = N_{is} - N_{i0}$ of the population densities can be achieved, where ΔN is negative for the lower level and positive for the upper level of the transition (Sect. 7.1). In case of *molecular* transitions, where only a small fraction of all excited molecules returns back into the initial level $|i\rangle$ by fluorescence, this level may be depleted rather completely.

Since the fluorescent transitions must obey certain selection rules, it is often possible to populate a selected level $|m\rangle$ by fluorescence from the laser-pumped upper level (Fig. 10.1b). Even with a weak pumping intensity large population densities in the level $|m\rangle$ may be achieved. In the pre-laser era, the term "optical pumping" was used for this special case because this scheme was the only way to achieve an appreciable population change with incoherent pumping sources.

Excited molecular levels $|k\rangle$ with $E_k \gg kT$ are barely populated at thermal equilibrium. With lasers as pumping sources large population densities N_k can be achieved, which may become comparable to those of the absorbing ground states. This opens several possibilities for new experimental techniques:

- The selectively excited molecular levels emit fluorescence spectra that are much simpler than those emitted in gas discharges, where many upper levels are populated. These laser-induced fluorescence spectra are therefore more readily assignable and allow the determination of molecular constants for all levels in lower states into which fluorescence transitions terminate (Sect. 6.8).
- A sufficiently large population of the upper state furthermore allows the measurement of absorption spectra for transitions from this state to still higher-lying levels (excited-state spectroscopy, stepwise excitation) (Sect. 10.4). Since all absorbing transitions start from this selectively populated level, the absorption spectrum is again much simpler than in gas discharges.

The selectivity of optical pumping depends on the laser bandwidth and on the line density of the absorption spectrum. If several absorption lines overlap within their Doppler width with the spectral profile of the laser, more than one transition is simultaneously pumped, which means that more than one upper level is populated (Fig. 10.2). In such cases of dense absorption spectra, optical pumping with narrow-band lasers in collimated cold molecular beams can be utilized to achieve the wanted selectivity for populating a single upper level (Sects. 9.3, 10.5).

The situation is different when narrow-band lasers are used as pumping sources:

If the beam of a *single-mode* laser with the frequency ω is sent in the z-direction through an absorption cell, only molecules within the velocity group $v_z = (\omega - \omega_0 \pm \gamma)/k$ can absorb the laser photons $\hbar\omega$ (Sect. 7.2) on the transition $|i\rangle \to |k\rangle$ with $E_k - E_i = \hbar\omega_0$. Therefore only molecules within this velocity group are excited. This implies that the absorption of a tunable narrow-band probe laser by these excited molecules yields a Doppler-free double-resonance signal.

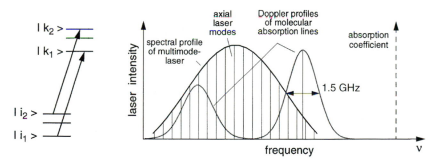

Fig. 10.2. Overlap of several Doppler-broadened absorption lines with the laser line profile leads to simultaneous optical pumping of several levels

A further important aspect of optical pumping with a *polarized* laser is the selective population or depletion of degenerate M sublevels $|J, M\rangle$ of a level with angular momentum J. These sublevels differ by the projection $M\hbar$ of J onto the quantization axis. Atoms or molecules with a nonuniform population density $N(J, M)$ of these sublevels are oriented because their angular momentum J has a preferred spatial distribution while under thermal equilibrium conditions J points into all directions with equal probability, that is, the orientational distribution is uniform. The highest degree of orientation is reached if only one of the $(2J+1)$ possible M sublevels is selectively populated.

By choosing the appropriate polarization of the pump laser, it is possible to achieve equal population densities within the pairs of sublevels $|\pm M\rangle$ with the same value of $|M|$ while levels with different $|M|$ may have different populations. This situation is called *alignment*.

Note that orientation or alignment can be produced in both the upper state of a pump transition due to a M-selective population as well as in the lower state because of the corresponding M-selective depletion (Fig. 10.3).

Example 10.1

Let us illustrate the above consideration by some specific examples: if the pump beam propagating into the z-direction has σ^+ polarization (left-circularly polarized light inducing transitions $\Delta M = +1$), we choose the direction of the k vector (i.e., the z-axis) as the quantization axis. The photon spin $\sigma = +\hbar k/k$ points into the propagation direction and absorption of these photons induces transitions with $\Delta M = +1$. For a transition $J'' = 0 \rightarrow J' = 1$ optical pumping with σ^+ light can only populate the upper sublevel with $M = +1$. This generates orientation of the atoms in the upper state since their angular momentum precesses around the $+z$-direction with the projection $+\hbar$ (Fig. 10.3a). Optical pumping with σ^+ light on a P-transition $J'' = 2 \rightarrow J' = 1$ causes M-selective depletion of the lower state and therefore orientation of molecules in the lower state (Fig. 10.3c).

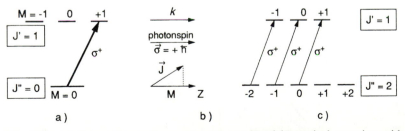

Fig. 10.3. (a) Orientation in the upper state produced by optical pumping with σ^+ light on an R transition $J'' = 0 \rightarrow J' = 1$. (b) Classical model of orientation where J precesses around the z-direction with a projection $M\hbar$, (c) orientation in the lower state caused by partial depletion for the example of a P pump transition $J'' = 2 \rightarrow J' = 1$

Example 10.2

Optical pumping with linearly polarized light (π-polarization) propagating into the z-direction can be regarded as a superposition of σ^+ and σ^- light (Sect. 7.4). This means that $\Delta M = \pm 1$ transitions can simultaneously be pumped with equal probability. The upper state becomes aligned because both sublevels $M = \pm 1$ are equally populated (Fig. 10.4).

Note: If the direction of the E vector of the linearly polarized light is selected as the quantization axis (which we chose as the x-axis), the two sublevels with $M_z = \pm 1$ now transform into the sublevel $M_x = 0$ (Fig. 10.4). The linearly polarized pump induces transitions with $\Delta M_x = 0$ and again produces alignment since only the component $M_x = 0$ is populated. Of course, the selection of the quantization axis cannot change the physical situation but only its description.

For a quantitative treatment of optical pumping we consider a pump transition between the levels $|J_1 M_1\rangle \rightarrow |J_2 M_2\rangle$. Without an external magnetic field all $(2J+1)$ sublevels $|M\rangle$ are degenerate and their population densities at thermal equilibrium are, without the pump laser

$$N^0(J, M) = \frac{N^0(J)}{2J+1}, \quad \text{with} \quad N^0(J) = \sum_{M=-J}^{+J} N^0(M, J). \tag{10.1}$$

The decrease of $N_1^0(J, M)$ by optical pumping $P_{12} = P(|J_1 M_1\rangle \rightarrow |J_2 M_2\rangle)$ can be described by the rate equation

$$\frac{\mathrm{d}}{\mathrm{d}t} N_1(J_1, M_1) = \sum_{M_2} P_{12}(N_2 - N_1) + \sum_{k}(R_{k1} N_k - R_{1k} N_1). \tag{10.2}$$

Included are the optical pumping (induced absorption and emission) and all relaxation processes that refill level $|1\rangle$ from other levels $|k\rangle$ or that

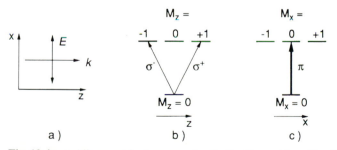

Fig. 10.4a–c. Alignment in the upper level of a R transition $J'' = 0 \rightarrow J' = 1$ by a linearly polarized pump wave: (**a**) direction of E and k; (**b**) level scheme with with z-quantization axis; and (**c**) with x-quantization axis

deplete $|1\rangle$. The optical pumping probability

$$P_{12} \propto |\langle J_1 M_1 | \boldsymbol{D} \cdot \boldsymbol{E} | J_2 M_2 \rangle|^2, \tag{10.3}$$

is proportional to the square of the transition matrix element (Sect. 2.7) and depends on the scalar product $\boldsymbol{D} \cdot \boldsymbol{E}$, of transition dipole moment $\boldsymbol{D}$, and electric field vector $\boldsymbol{E}$, that is, on the polarization of the pump laser. Molecules that are oriented with their transition dipole moment parallel to the electric field vector have the highest optical pumping probability.

When the laser intensity is sufficiently high, the population difference ΔN^0 decreases according to (7.5) to its saturated value

$$\Delta N^s = \frac{\Delta N^0}{1 + S}.$$

Since the transition probability and therefore also the saturation parameter S have a maximum value for molecules with $\boldsymbol{D} \parallel \boldsymbol{E}$, the population density N will decrease with increasing laser intensity for molecules with $\boldsymbol{D} \parallel \boldsymbol{E}$ more than for those with $\boldsymbol{D} \perp \boldsymbol{E}$. This means that the degree of orientation decreases with increasing saturation (Fig. 10.5).

The pump rate

$$P_{12}(N_2 - N_1) = \sigma_{12} N_{\mathrm{ph}}(N_2 - N_1), \tag{10.4}$$

is proportional to the optical absorption cross section σ_{12} and the photon flux rate N_{ph} [number of photons/cm^2 s]. The absorption cross section can be written as a product

$$\sigma(J_1 M_1, J_2 M_2) = \sigma_{J_1 J_2} \cdot C(J_1 M_1, J_2 M_2), \tag{10.5}$$

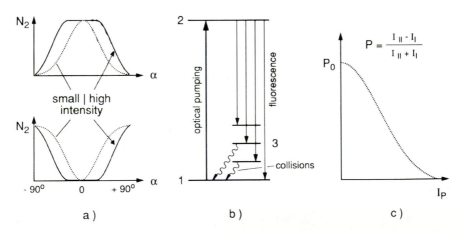

Fig. 10.5. (a) Decrease of molecular orientation with increasing pump intensity from saturation in the lower and upper state; **(b)** schematic level scheme monitored through the polarization P_0 of laser-induced fluorescence**(c)**

of two factors. The cross section $\sigma_{J_1 J_2}$ is independent of the molecular orientation and is essentially equal to the product of the electronic transition
probability times the Franck–Condon factor times the Hönl–London factor.
The second factor is the Clebsch–Gordan coefficient $C(J_1 M_1 J_2 M_2)$, which
depends on the rotational quantum numbers *and on the molecular orientation* in both levels of the pump transition [10.7]. In the case of molecules
the orientation is partly canceled by rotation. The maximum achievable degree of orientation obtained with optical pumping depends on the orientation
of the transition-dipole vector with respect to the molecular rotation axis and
is different for P, Q, or R transitions [10.8]. For molecules the maximum
orientation is therefore generally smaller than in case of atoms.

The coupling of the molecular angular momentum J with a possible nuclear spin I leads to a precession of J around the total angular momentum
$F = J + I$, which further reduces the molecular orientation [10.9]. A careful
analysis of experiments on optical pumping of molecules gives detailed information on the various coupling mechanisms between the different angular
momenta in selected molecular levels [10.10].

Another aspect of optical pumping is related to the coherent excitation of
two or more molecular levels. This means that the optical excitation produces
definite phase relations between the wave functions of these levels. This leads
to interference effects, which influence the spatial distribution and the time
dependence of the laser-induced fluorescence. This subject of coherent spectroscopy is covered in Chap. 12.

A thorough theoretical treatment of optical pumping can be found in
the review of Happer [10.11]. Specific aspects of optical pumping by lasers
with particular attention to problems arising from the spectral intensity distribution of the pump laser and from saturation effects were treated in
[10.12, 10.13]. Applications of optical pumping methods to the investigation
of small molecules were discussed in [10.4].

10.2 Optical–RF Double-Resonance Technique

The combination of laser-spectroscopic techniques with molecular beams and
RF spectroscopy has considerably enlarged the application range of optical–
RF double-resonance schemes. This optical–RF double-resonance method has
now become a very powerful technique for high-precision measurements of
electric or magnetic dipole moments, of Landé factors, and of fine or hyperfine splitting in atoms and molecules. It is therefore used in many laboratories.

10.2.1 Basic Considerations

Two different levels $|i\rangle$ and $|k\rangle$ that are connected by an optical transition may
be split into closely spaced sublevels $|i_n\rangle$ and $|k_m\rangle$. A narrow-band laser that
is tuned to the transition $|i_n\rangle \to |k_m\rangle$ between specific sublevels selectively de-
pletes the level $|i_n\rangle$ and increases the population of the level $|k_m\rangle$. Examples

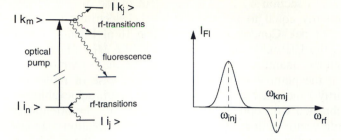

Fig. 10.6. Optical–radio frequency double resonance for lower and upper level of the optical transition

of this situation are hyperfine components of two rotational–vibrational levels in two different electronic states of a molecule or Zeeman sublevels of atomic electronic states (Fig. 10.6).

If the optically pumped sample is placed inside an RF field with the frequency ω_{rf} tuned into resonance with the transition $|i_j\rangle \rightarrow |i_n\rangle$ between two sublevels of the lower state, the level population $N(i_n)$ that was depleted by optical pumping will increase again. This leads to an increased absorption of the optical pump beam, which may be monitored by the corresponding increase of the laser-induced fluorescence intensity. Measuring $I_{Fl}(\omega_{rf})$ while ω_{rf} is tuned yields a double-resonance signal at $\omega_{rf} = \omega_{inj} = [E(i_n) - E(i_j)]/\hbar$ (Fig. 10.6).

Each absorbed RF photon leads to an extra absorbed optical photon of the pump beam. This optical–RF double resonance therefore yields an internal energy amplification factor $V = \omega_{opt}/\omega_{rf}$ for the detection of an RF transition. With $\omega_{opt} = 3 \times 10^{15}$ Hz and $\omega_{rf} = 10^7$ Hz, we obtain $V = 3 \times 10^8$! Since optical photons can be detected with a much higher efficiency than RF quanta, this inherent energy amplification results in a corresponding increase in the detection sensitivity.

RF transitions between sublevels of the upper state result in a change of the polarization and the spatial distribution of the laser-induced fluorescence. They can therefore be monitored through polarizers in front of the photomultiplier. Since the RF transitions deplete the optically pumped upper level, the RF double-resonance signal at $\omega_2 = \omega_{kmj} = [E(k_j) - E(k_m)]/\hbar$ has an opposite sign to that of the lower state at $\omega_1 = \omega_{inj}$ (Fig. 10.6).

In case of Zeeman sublevels or hfs levels, the allowed RF transitions are magnetic dipole transitions. Optimum conditions are then achieved if the sample is placed at the maximum of the magnetic field amplitude of the RF. This can be realized, for instance, inside a coil that is fed by an RF current. For *electric* dipole transitions (for example, between Stark components in an external dc electric field) the electric amplitude of the RF field should be maximum in the optical pumping region.

A typical experimental arrangement for measuring RF transitions between Zeeman levels in the upper state of the optical transition is shown in Fig. 10.7.

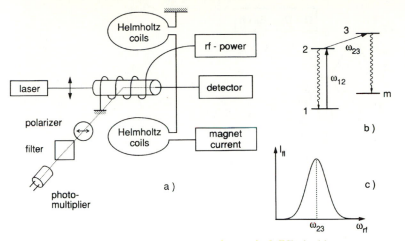

Fig. 10.7a–c. Experimental arrangement for optical–RF double-resonance spectroscopy (**a**), level scheme (**b**) and double-resonance signal monitored through the LIF (**c**)

A coil around the sample cell provides the RF field, while the dc magnetic field is produced by a pair of Helmholtz coils. The fluorescence induced by a polarized dye laser beam is monitored by a photomultiplier through a polarizer as a function of the radio frequency ω_{rf} [10.10].

Instead of tuning ω_{rf} one may also vary the dc magnetic or electric field at a fixed value of ω_{rf} thus tuning the Zeeman or Stark splittings into resonance with the fixed radio frequency (Sect. 6.7). This has the experimental advantage that the RF coil can be better impedance-matched to the RF generator.

The fundamental advantage of the optical–RF double-resonance technique is its high spectral resolution, which is not limited by the optical Doppler width. Although the optical excitation may occur on a Doppler-broadened transition, the optical–RF double-resonance signal $I_{Fl}(\omega_{rf})$ is measured at a low radio frequency ω_{rf}, and the Doppler width, which is according to (3.43) porportional to the frequency, is reduced by the factor $(\omega_{rf}/\omega_{opt})$. This makes the residual Doppler width of the double-resonance signal completely negligible compared with other broadening effects such as collisional or saturation broadening. In the absence of these additional line-broadening effects, the halfwidth of the double-resonance signal for the RF transition $|2\rangle \rightarrow |3\rangle$

$$\Delta\omega_{23} = (\Delta E_2 + \Delta E_3)/\hbar , \tag{10.6}$$

is essentially determined by the energy level widths ΔE_i of the corresponding levels $|2\rangle$ and $|3\rangle$, which are related to their spontaneous lifetime τ_i by $\Delta E_i = \hbar/\tau_i$. For transitions between sublevels in the ground state, the radiative lifetimes may be extremely long and the linewidth is only limited by the transit time of the molecules through the RF field.

With increasing RF intensity, however, saturation broadening is observed (Sect. 3.6) and the double-resonance signal may even exhibit a minimum at

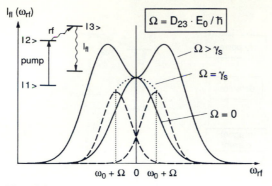

Fig. 10.8. Saturation broadening and Rabi splitting of double-resonance signals with increasing RF power P

the center frequency ω_{23} (Fig. 10.8). This can readily be understood from the semiclassical model of Sect. 2.7: for large RF field amplitudes E_{rf} the Rabi flopping frequency (2.90)

$$\Omega = \sqrt{(\omega_{23} - \omega_{rf})^2 + D_{23}^2 E_{rf}^2/\hbar} \, ,$$

becomes comparable to the natural linewidth $\delta\omega_n$ of the RF transition. The resulting modulation of the time-dependent population densities $N_2(t)$ and $N_3(t)$ causes a splitting of the line profile into two components with $\omega = \omega_{23} \pm \Omega$, which can be observed for $\Omega > \delta\omega_n$ (Fig. 10.8). If the halfwidth Δ_{23} of the double-resonance signal is plotted against the RF power P_{rf}, the extrapolation toward $P_{rf} = 0$ yields the linewidth $\gamma = 1/\tau$ of those levels of which the natural linewidth represents the dominant contribution to broadening; it allows one to determine the natural lifetime τ.

The achievable accuracy is mainly determined by the signal-to-noise ratio of the double-resonance signal, which limits the accuracy of the exact determination of the center frequency ω_{23}. However, the *absolute* accuracy in the determination of the RF frequency ω_{23} is generally several orders of magnitude higher than in conventional optical spectroscopy, where the level splitting $\Delta E = \hbar\omega_{23} = h(\nu_{31} - \nu_{21})$ is indirectly deduced from a small difference $\Delta\lambda$ between two directly measured wavelengths $\lambda_{31} = c/\nu_{31}$ and $\lambda_{21} = c/\nu_{21}$.

10.2.2 Laser–RF Double-Resonance Spectroscopy in Molecular Beams

The *Rabi technique* of radio frequency or microwave spectroscopy in atomic or molecular beams [10.14–10.17] has made outstanding contributions to the accurate determination of ground state parameters, such as the hfs splittings in atoms and molecules, small Coriolis splitting in rotating and vibrating molecules, or the narrow rotational structures of weakly bound van der Waals complexes [10.18]. Its basic principle is illustrated in Fig. 10.9. A collimated beam of molecules with a permanent dipole moment is deflected in a static

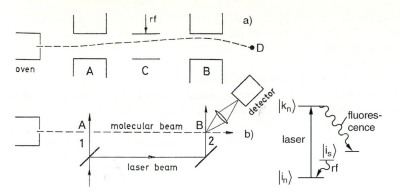

Fig. 10.9a,b. Comparison between the conventional Rabi method (**a**) and its laser version (**b**)

inhomogeneous magnetic field A but is deflected back onto the detector D in a second static field B with opposite field gradient. An RF field C is applied between A and B; which induces transitions between the molecular levels $|i_n\rangle$ and $|i_j\rangle$. Since the magnetic moment generally differs for the two levels, the deflection in B will change after such an RF transition and the detector output $S(\omega_{rf})$ will show a resonance signal for the resonance radio frequency $\omega_{rf} = [E(i_n) - E(i_j)]/\hbar$. For the measurement of Zeeman or Stark components an additional dc magnetic or electric field is applied in the RF region C. A famous example of such a device is the cesium clock [10.19].

The technique is restricted to atoms or molecules with a sufficiently large difference of the dipole moments in the two levels since the change of deflection in the inhomogeneous fields must be detectable. Furthermore, the detection of neutral particles with a universal ionization detector is generally not very sensitive except for those molecules (for example, alkali atoms or dimers) that can be monitored with a Langmuir–Taylor detector.

The laser version of the Rabi method (Fig. 10.9b) overcomes both limitations. The two magnets A and B are replaced by the two partial beams 1 and 2 of a laser that cross the molecular beam perpendicularly at the positions A and B. If the laser frequency ω_L is tuned to the molecular transition $|i_n\rangle \rightarrow |k_n\rangle$, the lower level $|i_n\rangle$ is partly depleted due to optical pumping in the first crossing point A. Therefore the absorption of the second beam in the crossing point B is decreased, which can be monitored through the laser-induced fluorescence. The RF transition $|i_j\rangle \rightarrow |i_n\rangle$ induced in C increases the population density $N(i_n)$ and with it the fluorescence signal in B.

This laser version has the following advantages:

- It is not restricted to molecules with a permanent dipole moment but may be applied to all molecules that can be excited by existing lasers.
- Even at moderate laser powers the optical transition may become saturated (Sect. 7.1) and the lower level $|i_n\rangle$ can be appreciably depleted. This considerably increases the population difference $\Delta N = N(i_j) - N(i_n)$ and thus

the absorption of the RF field on the transition $|i_j\rangle \rightarrow |i_n\rangle$, which is proportional to ΔN. In the conventional Rabi technique the population $N(E)$ follows a Boltzmann distribution, and for $\Delta E \ll kT$ the difference ΔN becomes very small.

- The detection of the RF transitions through the change in the LIF intensity is much more sensitive than the universal ionization detector. Using resonant two-photon ionization (Sect. 6.3), the sensitivity may further be enhanced.
- In addition, the signal-to-noise ratio is higher because only molecules in the level $|i_n\rangle$ contribute to the RF–double-resonance signal, while for the conventional Rabi technique the difference in deflection is generally so small that molecules in other levels also reach the detector and the signal represents a small difference of two large background currents.

These advantages allow the extension of the laser Rabi technique to a large variety of different problems [10.20], which shall be illustrated by three examples:

Example 10.3. Measurements of the hfs in the $^1\Sigma_g^+$ state of Na$_2$

The small magnetic hfs in the $^1\Sigma$ state of a homonuclear diatomic molecule is caused by the interaction of the nuclear spins I with the weak magnetic field produced by the rotation of the molecule. For the Na$_2$ molecular ground state the hyperfine splittings are smaller than the natural linewidth of the optical transitions. They could nevertheless be measured by the laser version of the Rabi technique [10.21]. A polarized argon laser beam at $\lambda = 476.5$ nm crosses the sodium beam and excites the Na$_2$ molecules on the transition $X\,^1\Sigma_g^+(v'' = 0, J'' = 28) \rightarrow B\,^1\Pi_u(v' = 3, J' = 27)$. The hfs splittings are much smaller than the laser linewidth. Therefore all hfs components are pumped by the laser. However, the optical transition probability depends on the hfs components of the lower and upper states. Therefore the depletion of the lower levels $|i_n\rangle$ differs for the different hfs components. The RF transitions change the population distribution and affect the monitored LIF signal. The hfs constant for the level $(v'' = 0, J'' = 28)$ was determined to be 0.17 ± 0.03 kHz, while the quadrupole coupling constant was found to be eQq $= 463.7 \pm 0.9$ kHz.

With a tunable single-mode cw dye laser any wanted transition can be selected and the hfs constants can be obtained for arbitrary rotational levels [10.22]. The hyperfine splittings are very small compared to the natural linewidth of the optical transition $X\,^1\Sigma_g \rightarrow B\,^1\Pi_u$, which is about 20 MHz because of the short lifetime $\tau = 7$ ns of the upper levels. At larger laser intensities optical pumping of the overlapping hfs components generates a coherent superposition of the lower-state hfs components. This results in a drastic change of line profiles and center frequencies of the RF double-resonance signals with increasing laser intensity. One therefore has to measure at different laser intensities I_L and to extrapolate to $I_L \rightarrow 0$.

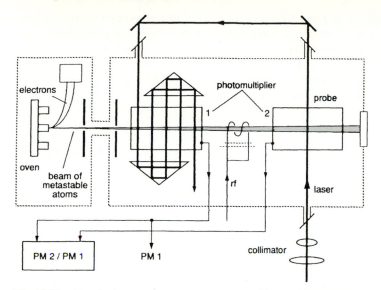

Fig. 10.10. Atomic beam resonance apparatus with combined electron-impact and laser pumping for the sensitive detection of optical–RF double resonance in highly excited states [10.23]

Example 10.4. Hyperfine structure of highly excited atomic levels

The combination of electron-impact excitation with the laser version of the Rabi technique allows measurements of hfs splittings in higher excited states. Penselin and coworkers [10.23, 10.24] have developed the atomic beam RF resonance apparatus shown in Fig. 10.10. Metals with a high condensation temperature are evaporated from a source heated by an electron gun and are further excited by electron impact into metastable states. Behind a collimating aperture optical pumping with a single-mode dye laser in a multiple-path arrangement results in a selective depletion of single hfs sublevels of the metastable states, which are refilled by RF transitions. In order to enhance the sensitivity, the laser beam is reflected by a prism arrangement back and forth and crosses the atomic beam several times. With difference and ratio recording the influence of fluctuations of the laser intensity or of the atomic-beam intensity could be minimized. The high sensitivity allowed a good signal-to-noise ratio even in cases where the population density of the metastable levels was less than 1% of the ground-state population.

10.3 Optical–Microwave Double Resonance

Microwave spectroscopy has contributed in an outstanding way to the precise determination of molecular parameters, such as bond lengths and bond angles, nuclear equilibrium configurations of polyatomic molecules, fine and hyperfine splittings, or Coriolis interactions in rotating molecules. Its application was, however, restricted to transitions between thermally populated levels, which generally means transitions within electronic ground states [10.26].

At room temperatures ($T = 300\,\text{K} \rightarrow kT \sim 250\,\text{cm}^{-1}$) the ratio $h\nu/kT$ is very small for typical microwave frequencies $\nu \sim 10^{10}\,\text{Hz}$ ($\hat{=}\,0.3\,\text{cm}^{-1}$). For the thermal population distribution

$$N_k/N_i = (g_k/g_i)\,e^{-h\nu_{ik}/kT}\ ,$$

the power of a microwave absorbed along the path length Δx through the sample is

$$\Delta P = -P_0\sigma_{ik}[N_i - (g_i/g_k)N_k]\Delta x \approx -P_0 N_i \sigma_{ik}\Delta x h\nu_{ik}/kT\ . \tag{10.7}$$

Since $h\nu/kT \ll 1$, the absorbed power is small, because induced absorption and emission nearly balance one another. Furthermore, the absorption coefficient σ_{ik}, which scales with ν^3, is many orders of magnitude smaller than in the optical region.

Optical–microwave double resonance (OMDR) can considerably improve the situation and extends the advantages of microwave spectroscopy to excited vibrational or electronic states, because selected levels in these states can be populated by optical pumping. Generally dye lasers or tunable diode lasers are used for optical pumping. However, even fixed frequency lasers can often be used. Many lines of intense infrared lasers (for example, CO_2, N_2O, CO, HF, and DF lasers) coincide with rotational–vibrational transitions of polyatomic molecules. Even for lines that are only close to molecular transitions the molecular lines may be tuned into resonance by external magnetic or electric fields (Sect. 6.6). The advantages of this OMDR may be summarized as follows:

- Optical pumping of a level $|i\rangle$ increases the population difference $\Delta N = N_i - N_k$ for a microwave transition by several orders of magnitude compared to the value $N_i h\nu/kT$ of (10.7).
- The selective population or depletion of a single level simplifies the microwave spectrum considerably. The difference of the microwave spectra with and without optical pumping yields directly those microwave transitions that start from one of the two levels connected by the pump transition.
- The detection of the microwave transition does not rely on the small absorption of the microwave but can use the higher sensitivity of optical or infrared detectors.

Some examples illustrate these advantages.

Example 10.5. Laser version of the cesium clock

The cesium clock provides our present time or frequency standard that is based on the hfs transition $7\,^2S_{1/2}$ ($F = 3 \rightarrow F = 4$) in the electronic ground state of Cs. The accuracy of the frequency standard depends on the achievable signal-to-noise ratio and on the symmetry of the line profile of the microwave transition. With optical pumping and probing, the signal-to-noise ratio can be increased by more than one order of magnitude. A further advantage is the absence of the two inhomogeneous magnetic fields, which may cause stray fields in the RF zone [10.25].

Example 10.6

Figure 10.11 illustrates the enhancement of microwave signals on transitions between the inversion doublets of NH_3 in rotational levels of the vibrational ground state and of the excited state $v'_2 = 1$ from to infrared pumping with an N_2O-laser line. This pumping causes selective depletion of the upper inversion component of the ($J'' = 8$, $K'' = 7$) level and an increased population of the lower inversion component into the upper vibrational level. The selective change ΔN of the populations is partly transferred by collisions to neighboring levels (Sect. 13.2), resulting in secondary collision-induced double-resonance signals [10.27].

Example 10.7

Figure 10.12 gives an example of microwave spectroscopy in an excited vibrational state, where the ($N_{Ka,Kc} = 2_{1,2}$) rotational level in the excited vibrational state $v_2 = 1$ of DCCCHO has been selectively populated by infrared pumping with a HeXe laser [10.28]. The solid arrows represent the direct microwave transitions from the pumped level, while the wavy arrows correspond to "triple-resonance" transitions starting from levels that have been populated either by the first microwave quantum or by collision-induced transitions from the laser-pumped level.

Further examples and experimental details on infrared–microwave double-resonance spectroscopy can be found in [10.29–10.32].

The electronically excited states of most molecules are far less thoroughly investigated than their ground states. On the other hand, their level structure is generally more complex because of interactions between electron and nuclear motions, which are more pronounced in excited states (breakdown of the Born–Oppenheimer approximation, perturbations). It is therefore most desirable to apply spectroscopic techniques that are sensitive and selective and that facilitate assignment. This is just what the optical–microwave double-resonance technique can provide.

Example 10.8

One example of this technique is the excited-state spectroscopy of the molecule BaO [10.33]. In the volume that is crossed by an atomic barium beam and a molecular O_2 beam, BaO molecules are formed in various vibronic levels (v'', J'') of their electronic $X\,^1\Sigma_g$ ground state by the reaction $Ba + O_2 \rightarrow BaO + O$ (Fig. 10.13). With a dye laser tuned to the transitions

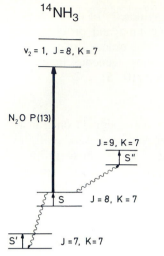

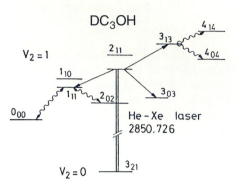

Fig. 10.11. Infrared–microwave double resonance in NH$_3$. The microwave transitions between the inversion doublets can start from the laser-pumped level (signal S) or from levels where the depletion has been transferred by collisions (secondary DR signals S′, S″) [10.27]

Fig. 10.12. Infrared–microwave double resonance in the vibrationally excited $v_2 = 1$ state of DCCCOH. The *solid arrows* indicate the microwave transitions, the *wavy arrows* secondary MW transitions [10.28]

$A\,^1\Sigma(v', J') \leftarrow X\,^1\Sigma(v'', J'')$ different levels (v', J') in the excited A state are selectively populated. Their quantum numbers (v', J') can be determined from the LIF spectrum (Sect. 6.7).

When the crossing volume is irradiated with a microwave produced by a tunable clystron, transitions $J'' \to J'' \pm 1$ or $J' \to J' \pm 1$ between adjacent rotational levels in the X or the A state can be induced by tuning the microwave to the corresponding frequencies.

Since optical pumping by the laser has *decreased* the population $N(v'', J'')$ below its thermal equilibrium value, microwave transitions $J'' \to J'' \pm 1$ *increase* $N(v'', J'')$, which result in a corresponding increase of the total fluorescence intensity. The transitions $J' \to J' \pm 1$, on the other hand, *decrease* the population $N(v', J')$ in the optically excited levels. Therefore they decrease the fluorescence intensity emitted from (v', J'), but generate new lines in the fluorescence spectrum originating from the levels $(J' \pm 1)$. They can be separated by a monochromator or by interference filters.

The use of OMDR spectroscopy allows very accurate measurements of rotational spacings in both the ground state and in electronically exited states [10.34]. The sensitivity is sufficiently high to measure even very small concentrations of molecules or radicals that are formed as intermediate products

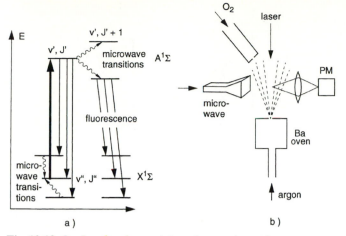

Fig. 10.13a,b. Level scheme (**a**) and experimental arrangement (**b**) for detection of optical-microwave DR in BaO molecules [10.33]

in chemical reactions. This has been proven, for example, for the case of NH_2 radicals in a discharge flow system [10.35]

Example 10.9

An impressive example of the OMDR version of the Rabi technique is the measurement of the hyperfine splittings in the electronic ground state of CaCl [10.36], where linewidths as small as 15 kHz could be achieved for the OMDR signals. They were only limited by the transit time of the molecules flying through the microwave zone. With an additional dc electric field in the microwave region, the resulting Stark splittings of the OMDR signals were observed and the eletric dipole moments could be accuracely determined [10.37, 10.38].

10.4 Optical–Optical Double Resonance

The optical–optical double-resonance (OODR) technique is based on the simultaneous interaction of a molecule with two optical waves that are tuned to two molecular transitions sharing a common level. This may be the lower or the upper level of the pump transition. The three possible level schemes of OODR are depicted in Fig. 10.14.

The "V-type" double-resonance method of Fig. 10.14a (named after its V-shaped appearance in the level diagram) may be regarded as the inversion of laser-induced fluorescence (Fig. 10.15). With the LIF technique one *excited* level $|2\rangle$ is selectively populated. The resulting fluorescence spectrum corresponds to all allowed optical transitions from this level into the lower levels $|m\rangle$. The analysis of the LIF spectrum mainly yields information on the

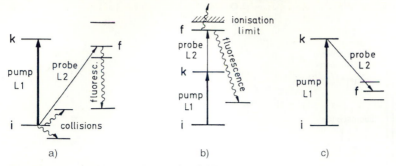

Fig. 10.14a–c. Different schemes for OODR: **(a)** V-type OODR; **(b)** stepwise excitation; **(c)** Λ-type OODR

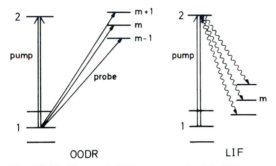

Fig. 10.15. V-type OODR compared with LIF

lower state. With V-type OODR, on the other hand, a single *lower* level $|i\rangle$ is selectively *depleted*. The OODR transitions, monitored through the difference in the absorption of the probe wave with and without the pump, start from the lower level $|1\rangle$ and end in all upper levels $|m\rangle$, which can be reached by probe transitions from $|1\rangle$. They give mainly information about these *upper* levels.

The second OODR scheme (Fig. 10.14b) represents stepwise excitation of high-lying levels via a common intermediate level $|2\rangle$, which is the upper level of the pump transition but the lower level of the probe transition.

The last scheme in Fig. 10.14c, named Λ-type OODR, represents a stimulated resonant Raman process where the molecules are coherently transferred from level $|1\rangle$ to levels $|m\rangle$ by absorption of the pump laser and stimulated emission by the probe laser.

We will now discuss these three schemes in more detail.

10.4.1 Simplification of Complex Absorption Spectra

The V-type OODR scheme can be utilized for the simplification and analysis of infrared, visible, or UV molecular spectra. This is particularly helpful if the spectra are perturbed and might not show any regular pattern. Assume the pump laser L1, with intensity I_1, which has been tuned to the transition $|1\rangle \rightarrow |2\rangle$ is chopped at the frequency f_1. The population densities N_1, N_2 then show a corresponding modulation

$$N_1(t) = N_1^0 \{1 - aI_1[1 + \cos(2\pi f_1 t)]\} \,,$$

$$N_2(t) = N_2^0 \{1 + bI_1[1 + \cos(2\pi f_1 t)]\} \,, \tag{10.8}$$

which has an opposite phase for the two levels. The modulation amplitudes a and b depend on the transition probability of the pump transition and on possible relaxation processes such as spontaneous emission, collisional relaxation, or diffusion of molecules into or out of the pump region (Sect. 7.1).

The LIF intensity $I_{Fl}(\lambda_2)$ induced by the tunable probe laser L2 will be modulated at the frequency f_1 if the wavelength λ_2 coincides with an absorbing transition $|1\rangle \rightarrow |m\rangle$ from the optically pumped lower level $|1\rangle$ to an upper level $|m\rangle$ or with a downward transition $|2\rangle \rightarrow |m\rangle$ from the upper pump level $|2\rangle$ to a lower level $|m\rangle$. If the probe laser-induced fluorescence $I_{Fl}(\lambda_2)$ is monitored through a lock-in amplifier at the frequency f_1, one obtains negative OODR signals for all transitions $|1\rangle \rightarrow |m\rangle$ and positive signals for the transitions $|2\rangle \rightarrow |m\rangle$ (Fig. 10.16). From the phase of the lock-in signal it is therefore, in principle, possible to decide which of the two possible types of probe transitions is detected. Since this double-resonance technique selectively detects transitions that start from or terminate at levels labeled by the pump lever, it is often called *labeling spectroscopy*.

In reality, the situation is generally more complex. If the OODR experiments are performed on molecules in a cell that have a thermal velocity distribution and that may suffer collisions, double-resonance signals are also observed for probe transitions starting from other levels than $|1\rangle$ or $|2\rangle$. This is due to the following facts:

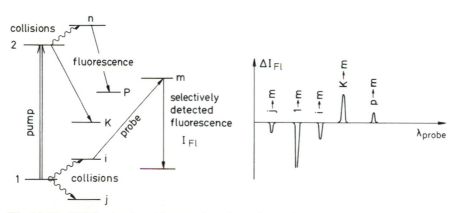

Fig. 10.16. OODR signals with opposite phase for probe transitions starting from the lower or the upper pump level, respectively

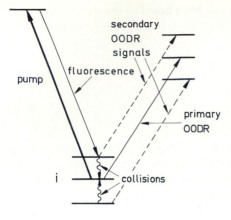

Fig. 10.17. Secondary OODR signals induced by collisions or by fluorescence from the optically pumped levels

(a) Even with a narrow-band pump laser several absorbing transitions may simultaneously be pumped if the absorption profiles overlap within their Doppler width (Fig. 10.2).

(b) Collisions may transfer the modulation of the population density N_1 to neighboring levels if the collisional time is shorter than the chopping period $1/f_1$ of the pump laser (Sect. 13.3). This causes additional double-resonance signals, which may impede the analysis (Fig. 10.17). On the other hand, they also can give much information on cross sections for collisional population and energy transfer (Chap. 13).

(c) The fluorescence from the level $|2\rangle$ into all accessible lower levels $|m\rangle$ is also modulated at f_1 and leads to a corresponding modulation of the population densities N_m.

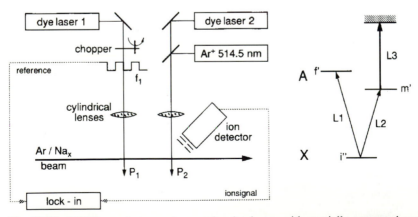

Fig. 10.18. OODR spectroscopy in a molecular beam with spatially separated pump and probe beams. The probe transition L2 can be monitored by ionizing the excited levels m' with L3

In order to avoid such secondary OODR signals, sub-Doppler excitation under collision-free conditions has to be realized. This can be achieved in a collimated molecular beam that is intersected by the two lasers L1 and L2 either at two different positions z_1 and z_2 (Fig. 10.18) or with two overlapping laser beams (Fig. 10.19). In the first arrangement the probe laser-induced fluorescence $I_{FI}(\lambda_2)$ can be imaged separately onto the detector, and chopping of the pump laser L1 with phase-sensitive detection of $I_{FI}(\lambda_2)$ yields the wanted OODR spectrum. In the second arrangement with superimposed pump and probe beams, the detected fluorescence $I_{FI} = I_{FI}(L1) + I_{FI}(L2)$ is the sum of pump and probe laser-induced fluorescence intensities, which are both modulated at the chopping frequency f_1. In order to eliminate the large background of $I_{FI}(L1)$ one can use a double-modulation technique, where L1 is modulated at f_1, L2 at f_2, and the fluorescence is monitored at the sum frequency $f_1 + f_2$. This selects the OODR signals from the background, as can be seen from the following consideration.

The fluorescence intensity $I_{FI}(\lambda_2)$ induced by the probe laser with intensity $I_2 = I_{20}(1 + \cos 2\pi f_2 t)$ on a transition starting from the lower level $|i\rangle$ of the pump transition is

$$I_{FI}(\lambda_2) \propto N_i I_2 = N_{i0}\{1 - a I_{10}[1 + \cos(2\pi f_1 t)]\} I_{20}[1 + \cos(2\pi f_2 t)] . \quad (10.9)$$

The nonlinear term

$$I_{10} I_{20} \cos(2\pi f_1 t) \cos(2\pi f_2 t)$$

$$= \frac{1}{2} I_{10} I_{20} [\cos 2\pi (f_1 + f_2) t + \cos 2\pi (f_1 - f_2) t] , \quad (10.10)$$

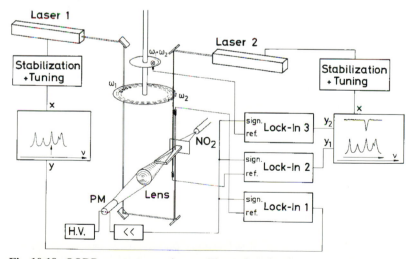

Fig. 10.19. OODR spectroscopy in a collimated molecular beam with overlapping pump and probe beams and detection at the sum frequency

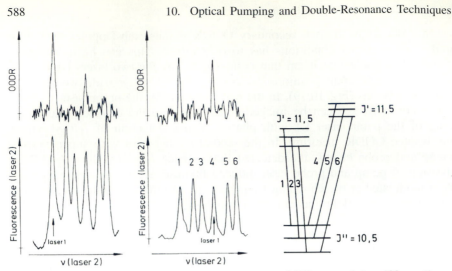

Fig. 10.20. Section of the linear excitation spectrum of NO_2 around $\lambda = 488$ nm (*lower spectrum*) and OODR signals obtained with the pump laser kept on transition 1 or 4, respectively [10.39]

which represents the OODR signal, is modulated at the frequencies $(f_1 + f_2)$ and $(f_1 - f_2)$ whereas the linear terms, representing the pump laser-induced fluorescence or the fluorescence induced by the probe laser on transitions from nonmodulated levels only contain the frequencies f_1 or f_2.

The application of OODR in molecular beams to the analysis of complex perturbed spectra is illustrated by Fig. 10.20, which shows a section of the visible NO_2 spectrum around $\lambda = 488$ nm. In spite of the small residual Doppler width of 15 MHz, not all lines are fully resolved and the analysis of the spectrum turns out to be very difficult, because the upper state is heavily perturbed. If in the OODR experiment the pump laser L1 is kept on line 1, one obtains the two OODR signals as shown in the upper left part of Fig. 10.20, which proves that lines 1 and 4 in the lower spectrum share a common lower level. The right part of Fig. 10.20 also shows that the lines 2 and 5 start from a common lower level. The whole lower spectrum consists of two rovibronic transitions $J'', K'' = 10, 5) \rightarrow (J', K') = (11, 5)$ (each with three hfs components), which end in two closely spaced rotational levels with equal quantum numbers (J', K') that belong to two different vibronic states coupled by a mutual interaction [10.39].

This V-type OODR has been used in many laboratories. Further examples can be found in [10.40–10.44].

10.4.2 Stepwise Excitation and Spectroscopy of Rydberg States

When the excited state $|2\rangle$ has been selectively populated by optical pumping with laser L1, transitions to still higher levels $|m\rangle$ can be induced by a second tunable laser L2 (Fig. 10.14b). This *two-step excitation* may be regarded as the special resonance case of the more general two-photon excitation with two different photons $\hbar\omega_2$ (Sect. 7.5). Because the upper levels $|m\rangle$ must have the same parity as the initial level $|1\rangle$, they cannot be reached by an allowed one-photon transition. The one-step excitation by a frequency-doubled laser with the photon energy $2\hbar\omega = \hbar(\omega_1 + \omega_2)$ therefore excites, in the same energy range, levels of opposite symmetry compared to those levels reached by the two-step excitation.

With two visible lasers, levels $|m\rangle$ with excitation energies up to 6 eV can be reached. Optical frequency doubling of both lasers allows even the population of levels up to 12 eV. This makes the Rydberg levels of most atoms and molecules accessible to detailed investigations. The population of Rydberg levels of species M can be monitored either by their fluorescence or by detecting the ions M^+ or the electrons e^- that are produced by photoionization, field ionization, collisional ionization, or autoionization of the Rydberg levels.

Rydberg states have some remarkable characteristics (Table 10.1). Their spectroscopic investigation allows one to study fundamental problems of quantum optics (Sect. 14.5), nonlinear dynamics, and chaotic behavior of quantum systems (see below). Therefore detailed studies of atomic and molecular Rydberg states have found increasing interest [10.45–10.57].

The term value T_n of a Rydberg level with principal quantum number n is given by the Rydberg formula

$$T_n = IP - \frac{R}{n^{*2}} , \quad \text{with} \quad n^* = n - \delta , \tag{10.11}$$

where IP is the ionization potential, R the Rydberg constant, and $\delta(n, \ell)$ the quantum defect that depends on n and on the angular momentum $\ell\hbar$ of the Rydberg electron. The quantum defect describes the deviation of the real po-

Table 10.1. Characteristic properties of Rydberg atoms

Property	n-dependence	Numerical values for		
		H $(n = 2)$	H $(n = 50)$	Na $(10d)$
Binding energy	$-R \cdot n^{-2}$	4 eV	5.4 meV	0.14 eV
$E(n+1) - E_n$	$\frac{R}{n^2} - \frac{R}{(n+1)^2} \propto \frac{1}{n^3}$	$\frac{5}{36} R \approx 2$ eV	0.2 meV $\hat{=} 2$ cm^{-1}	≈ 1.5 meV
Mean radius	$a_0 n^2$	$4a_0 \approx 0.2$ nm	132 nm	7 nm
Geometrical cross section	$\pi a_0^2 n^4$	$\approx 1.2 \times 10^{-16}$ cm^2	5×10^{-10} cm^2	$\approx 10^{-12}$ cm^2
Spontaneous lifetime	$\propto n^3$	5×10^{-9} s	1.5×10^{-4} s	$\approx 10^{-6}$ s
Critical field	$E_c = \pi\epsilon_0 R^2 e^{-3} n^{-4}$	5×10^9 V/m	5×10^3 V/m	3×10^6 V/m
Orbital period	$T_n \propto n^3$	10^{-15} s	2×10^{-11} s	2×10^{-13} s
Polarizability	$\alpha \propto n^7$	10^{-6} s^{-1}V^{-2}m^2	10^4 s^{-1}V^{-2}m^2	20 s^{-1}V^{-2}m^2

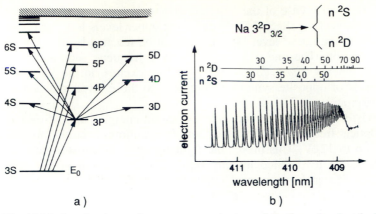

Fig. 10.21. Level scheme for two-step excitation of Rydberg levels of alkali atoms (**a**) and Rydberg series Na $3\,^2P_{3/2} \to n\,^2S$, n^2D measured by field ionization of the Rydberg states (**b**) [10.48]

tential (from the nucleus, including its shielding by the core electrons) from a pure Coulomb potential $V = Z_{\text{eff}} e^2/(4\pi\epsilon_0 r)$.

The investigation of atomic Rydberg states started with alkali atoms since they can be readily prepared in cells or atomic beams. Because of their relatively low ionization energy, their Rydberg states can be reached by stepwise excitation with two dye lasers in the visible spectral region. The first step generally uses the resonance transitions $nS \to nP$ to the first excited state, which has a large transition probability and therefore can be already saturated at laser intensities of $I(L1) \leq 0.1\,\text{W/cm}^2$! The second step requires higher intensities $I(L2) \approx 1-100\,\text{W/cm}^2$ (because the transition probability decreases with n^{-3} [10.50]), which, however, can be readily realized with cw dye lasers. For illustration, Fig. 10.21 exhibits the term diagram and measured Rydberg series $3p^2P \to nS$, and $3p^2P \to nD$ of Na atoms excited via the $3p^2P_{3/2}$ state.

Example 10.10

From (10.11) one can calculate that the term energy of Na*($n = 40$) is only about 0.005 eV below the ionization limit, and Fig. 10.22 shows that an external field with $E = 100\,\text{V/cm}$ is sufficient to ionize this state.

The external field E [V/m] decreases the ionization threshold of a Rydberg state down to the "appearance potential" AP of the ions:

$$AP = IP - \sqrt{e^3 E/\pi\epsilon_0} \,, \tag{10.12}$$

which can be readily derived from the maximum energy of the superposition of Coulomb potential and external potential $V_{\text{ext}} = Ex$ of the electric field in the x-direction.

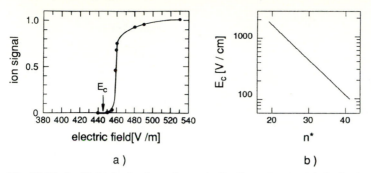

Fig. 10.22a,b. Field ionization of atomic Rydberg levels: (**a**) ionization rate of the Na $(31\,^2S)$ level; (**b**) threshold field E_c in dependence on the effective principal quantum number n^* for Na(n^*S) Rydberg levels [10.48]

> **Example 10.11**
>
> For $E = 1000\,\text{V/cm}$ the decrease $\Delta IP = IP - AP$ of the effective ionization potential is $\Delta IP = 20\,\text{meV} \triangleq 194\,\text{cm}^{-1}$. Rydberg levels with $n^* \geq 24$ are already field ionized.

Collisional deactivation of Rydberg levels under "field-free" conditions can be studied in the thermionic heat pipe (Sect. 6.4.5) with shielding grids or wires [10.52]. Since the excitation occurs in a field-free region, the effects of collisional broadening and shifts (Sect. 3.3) can be separately investigated up to very high quantum numbers [10.51–10.53]. In addition, the influence of the hyperfine energy of core electrons on the term values of Rydberg levels of the Sr atom could be studied in a heat pipe up to $n = 300$ [10.54].

When Rydberg levels are selectively excited under collision-free conditions in an atomic beam, it is found that after a time, which is short compared to the spontaneous lifetime, the population of neighboring Rydberg levels increases. The reason for this surprising result is the following: because of the large dipole moment of Rydberg atoms, the weak thermal radiation emitted by the walls of the apparatus is sufficient to induce transitions between the optically pumped level $|n, \ell\rangle$ and neighboring Rydberg levels $|n + \Delta n, \ell \pm 1\rangle$ [10.55].

The interaction of the Rydberg atoms with the thermal radiation field results in a small shift of the Rydberg levels (Lamb shift), which only amounts to $\Delta v/v \approx 2 \times 10^{-12}$ for rubidium. It has recently been measured with extremely well stabilized lasers [10.56]. In order to eliminate the influence of the thermal radiation field one has to enclose the interaction zone of the laser and atomic beam by walls cooled to a few degrees Kelvin.

On the other hand, the large dipole moment of Rydberg atoms offers the possibility to use them as sensitive detectors for microwave and submillimeter-wave radiation [10.57]. For the detection of radiation with frequency ω, a Rydberg level $|n\rangle$ is selectively excited by stepwise excitation with lasers in an external electric dc field. The field strength is adjusted in

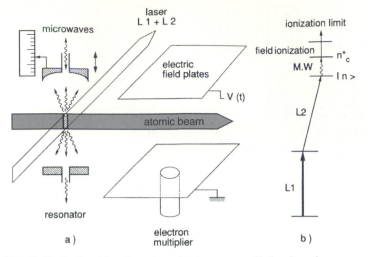

Fig. 10.23a,b. Sensitive detection of microwave radiation by microwave-photon ionization of Rydberg levels

such a way that the energy E_n of the Rydberg level $|n\rangle$ is just below the critical value E_c for field ionization, but $E_n + \hbar\omega$ is just above. Every absorbed microwave photon $\hbar\omega$ then produces an ion that can be detected with 100% efficiency (Fig. 10.23).

There is another interesting aspect of Rydberg levels: the Coulomb energy of a Rydberg electron in level $|n\rangle$ from the electric field of the core decreases with $1/n^{*2}$. For sufficiently large principle quantum numbers n^* the Zeeman energy in an external magnetic field $\boldsymbol{B}$ may become larger than the Coulomb energy.

The Lorentz force $\boldsymbol{F} = q(\boldsymbol{v} \times \boldsymbol{B})$ causes the electron to precess around the magnetic field direction $\boldsymbol{B}$, and even if the total energy of the electron lies above the field-free ionization limit, the electron cannot escape except into the direction of $\boldsymbol{B}$. This leads to relatively long lifetimes of such autoionizing states. The corresponding classical trajectories of the electron in such states are complicated and may even be chaotic. At present, investigations in several laboratories are attempting to determine how the chaotic behavior of the classical model is related to the term structure of the quantum states [10.58]. The question whether "quantum chaos" really exists is still matter of controversy [10.60–10.63].

Example 10.12

For $n = 100$ the Coulomb binding energy of the electron to the proton in a hydrogen Rydberg atom is $E_{\text{coul}} = \text{Ry}/n^2 = 1.3 \times 10^{-3}$ eV. The magnetic energy of the electron with an average orbital angular momentum of $50\hbar$ in a field of $B = 1\,\text{T}$ is $E_{\text{mag}} = \ell\mu_B = 2.8 \times 10^{-3}$ eV, where $\mu_B = 5.6 \times 10^{-5}$ eV/T is the Bohr magneton.

Until now detailed experiments on Rydberg atoms in crossed electric and magnetic fields have been performed on alkali atoms [10.59] and on Rydberg states of the H atom [10.64], which are excited either by direct two-photon transitions or by stepwise excitation via the $2\,^2p$ state. Since the ionization energy of H is 13.6 eV, one needs photon energies above 6.7 eV ($\lambda \le 190$ nm), which can be produced by frequency doubling of UV lasers in gases or metal vapors [10.65].

In the examples given so far only one single Rydberg electron was excited. Special laser excitation techniques allow the simultaneous excitation of two electrons into high-lying Rydberg states [10.66]. The total energy of this doubly excited system is far above the ionization energy. The correlation between the two electrons causes an exchange of energy and results in auto-ionization (Fig. 10.24). The population of the doubly-excited Rydberg state is achieved by two two-photon transitions: at first a one-electron Rydberg state is excited by a two-photon transition induced by a visible laser. A successive two-photon transition with UV photons excites a second electron from the core into a Rydberg state [10.67, 10.68]. Such doubly excited Rydberg atoms, which are named *planetary atoms* [10.69], offer the unique possibility to study the correlations between two electrons that are both in defined Rydberg orbits (n_1, ℓ_1, S_1) and (n_2, ℓ_2, S_2) by measuring the autoionization time for different sets of quantum numbers.

Molecular Rydberg series are by far more complex than those of atoms. The reason is that many more electronic states exist in molecules, where furthermore each of these states has a manifold of vibrational–rotational levels. Often vibronic levels of different electronic Rydberg states are coupled by several interactions (vibronic, spin-orbit, Coriolis interaction, etc.). This inter-

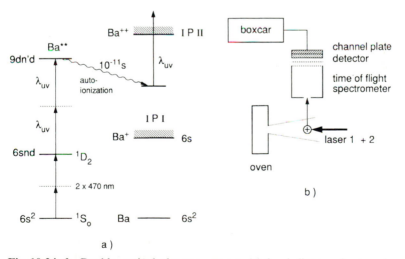

Fig. 10.24a,b. Double excited planetary atoms: (**a**) level diagram for two-step excitation of two electrons in the Ba atom with subsequent autoionization to Ba^{++}; (**b**) experimental arrangement [10.65]

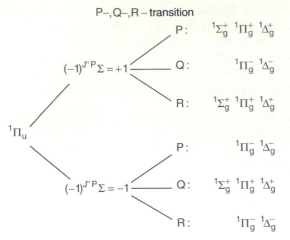

Fig. 10.25. Molecular Rydberg states of homonuclear alkali dimers, accessibly by one-photon transitions from the intermediate $B\,^1\Pi_u$ state

action shifts their energy, and such perturbed Rydberg series often show an irregular structure that is not easy to analyze [10.70].

Here, the stepwise excitation is, in particular, very helpful since only those Rydberg levels are excited that are connected by allowed electric dipole transitions to the known intermediate level, populated by the pump. This shall be illustrated by the example of a relatively simple molecule, the lithium dimer Li_2 [10.71, 10.72]. When the pump laser selectively populates a level (v'_k, J'_k) in the $B\,^1\Pi_u$ state, all levels $(v^*, J^* = J'_k \pm 1$ or $J^* = J'_k)$ in the Rydberg states $ns\,^1\Sigma$ with $\ell = 0$ and $nd\,^1$Delta, $nd\,^1\Pi$, or $nd\,^1\Sigma$ with $\ell = 2$ and $\lambda = 2, 1, 0$ are accessible by probe laser transitions with $\Delta J = +1$ (R-lines), $\Delta J = 0$ (Q-lines), or $\Delta J = -1$ (P-lines). This is shown in (Fig. 10.25), where all possible transitions from the two Λ components of the $B\,^1\Pi_u$ state with parity -1 and $+1$ into the different Rydberg states are compiled.

The excitation of Rydberg levels below the ionization potential IP can be monitored via the probe laser-induced fluorescence. There are, however, also many Rydberg levels above IP where higher vibrational or rotational levels are excited (Fig. 10.26). If the Rydberg electron can gain enough energy from the vibration or rotation of the molecular core it can leave the core. This *autoionization* demands a coupling between the kinetic energy of the nuclei and the energy of the Rydberg electron, which implies a breakdown of the adiabatic Born–Oppenheimer approximation [10.70]. This coupling is generally weak and the autoionization is therefore much slower than that in doubly excited atoms caused by electron–electron correlation. The autoionizing states are therefore "long-living" states and the corresponding transitions remarkably narrow. Since the spontaneous lifetimes increase proportionally to n^3, the autoionization may still be faster than the radiative decay of the levels. The ions produced by autoionization can therefore be used for sensitive detection of transitions to Rydberg levels above the ionization limit of molecules.

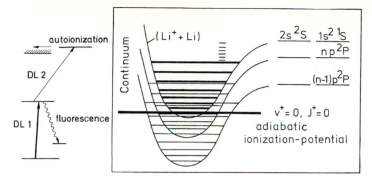

Fig. 10.26. Autoionization of molecular Rydberg levels (n^*, v^*, J^*) above the energy of the lowest level of the molecular ion M^+

When the wavelength λ (L2) of the second laser is tuned, one observes sharp intense resonances in the ion yield $N_{ion}(\lambda_2)$ superimposed on a continuous background of direct photoionization (Fig. 10.27). The analysis of such spectra demands the application of quantum defect theory [10.73–10.75].

A common experimental arrangement for the study of molecular Rydberg states is depicted in Fig. 10.28. The output beams of two pulsed narrow-band dye lasers, pumped by the same excimer laser, are superimposed and cross the molecular beam perpendicularly. The fluorescence emitted from the interme-

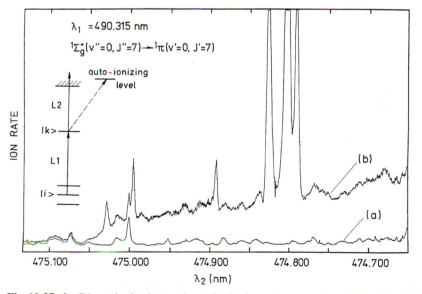

Fig. 10.27a,b. Direct ionization and autoionization resonances just above the ionization threshold observed through the ion yield $N_{ion}(\lambda_2)$ in Li_2 when the first laser excites the level $|k\rangle \triangleq (v' = 2, J' = 7)$ in the $B\,{}^1\Pi_u$ state: (**a**) with L1 off; (**b**) with L1 on

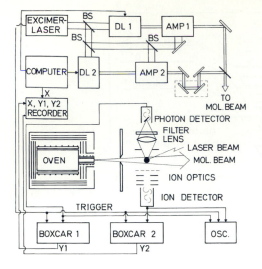

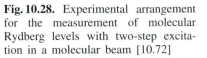

Fig. 10.28. Experimental arrangement for the measurement of molecular Rydberg levels with two-step excitation in a molecular beam [10.72]

diate level (v', J') or from the Rydberg levels (v^*, J^*) can be monitored by the photomultiplier PM1. The ions produced by autoionization (or for levels slightly below IP by field ionization) are extracted by an electric field and are accelerated onto an ion multiplier or channel plate. This allows the detection of single ions. In order to avoid electric Stark shifts of the Rydberg levels during their excitation, the extraction field is switched on only after the end of the laser pulse. Experimental details and more information on the structure and analysis of molecular Rydberg spectra can be found in [10.72–10.76].

A powerful technique for studying molecular ion states and ionization potentials of molecules is the zero-kinetic energy (ZEKE) electron spectroscopy where the photoelectrons with a very small kinetic energy, produced by laser excitation of ionic states just at the ionization limit, are selectively detected [10.78]. When the laser is tuned through the spectral range of interest, the excitation threshold for a new state is marked by the appearance of ZEKE electrons. Because of their very small kinetic energy, the electrons can be extracted by a small electric field and collected onto a detector with high efficiency. This ZEKE electron spectroscopy in combination with laser excitation and molecular beam techniques is a rapidly developing field of molecular Rydberg state spectroscopy [10.77–10.81].

From the optically pumped atomic or molecular Rydberg levels neighboring levels can be reached by microwave transitions, as was mentioned above. This "triple resonance" (two-step laser excitation plus microwave) is a very accurate method to measure quantum defects, fine-structure splitting, and Zeeman and Stark splitting in Rydberg states [10.82].

The Stark splitting and the field ionization of very high Rydberg levels provide sensitive indicators for measuring weak electric fields. These few examples demonstrate the variety of information obtained from Rydberg state spectroscopy. More examples can be found in Sect. 14.5 and in the literature on laser spectroscopy of Rydberg states [10.83–10.87].

10.4.3 Stimulated Emission Pumping

In the Λ-type OODR scheme (Fig. 10.29) the probe laser induces downward transitions from the upper level $k = |2\rangle$ of the pump transition to lower levels $f = |m\rangle$. This process, which is called *stimulated emission pumping* (SEP), may be regarded as a resonantly induced Raman-type transition. In case of monochromatic pump and probe lasers tuned to the frequencies ω_1 and ω_2, respectively, the resonance condition for a molecule moving with velocity $\boldsymbol{v}$ is

$$\omega_1 - \boldsymbol{k}_1 \cdot \boldsymbol{v} - (\omega_2 - \boldsymbol{k}_2 \cdot \boldsymbol{v}) = (E_m - E_1)/\hbar \pm \Gamma_{m1} \,, \tag{10.13}$$

where ω_i, $\boldsymbol{k}_i$ are frequency and wave vectors of laser i, and $\Gamma_{m1} = \gamma_1 + \gamma_m$ is the sum of the homogeneous widths of initial and final levels of the Raman transition. The probe laser is in resonance if its frequency is

$$\omega_2 = \omega_1 + (\boldsymbol{k}_2 - \boldsymbol{k}_1) \cdot \boldsymbol{v} - (E_m - E_1)/\hbar \pm \Gamma_{m1} \,. \tag{10.14}$$

For a collinear arrangement of the two laser beams $\boldsymbol{k}_1 \parallel \boldsymbol{k}_2$ the Doppler shift becomes very small for $|k_1| - |k_2| \ll |k_1|$, and the width of the OODR signal $S(\omega_2)$ is only determined by the sum of the level widths of the initial and final levels, while the width of the upper level $|2\rangle$ does not enter into the calculations. If the two levels $|1\rangle$ and $|m\rangle$ are, for example, vibrational–rotational levels in the electronic ground state of a molecule, their spontaneous lifetimes are very long (even infinite for homonuclear diatomics!) and the level widths are only determined by the transit time of the molecule through the laser beams. In such cases, the linewidth of the OODR signal may become smaller than the natural linewidth of the optical transitions $|1\rangle \to |2\rangle$ (Sect. 14.4).

A more thorough consideration starts similar to the discussion of saturation spectroscopy in Sect. 7.2 from the absorption coefficient of the probe wave

$$\alpha(\omega) = \int_{-\infty}^{+\infty} \sigma(v_z, \omega) \Delta N(v_z) \, dv_z \,, \tag{10.15}$$

with $\Delta N = (N_m - N_2)$. The population density $N_2(v_z)$ in the common upper level $|2\rangle$ is altered through optical pumping with the laser L1 from its thermal

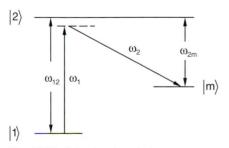

Fig. 10.29. Stimulated emission spectroscopy

equilibrium value $N_2^0(v_z)$ to

$$N_2^s(v_z) = N_2^0(v_z) + \frac{1}{2} \frac{[N_1(v_z) - N_2(v_z)]\gamma_2 S_0}{(\omega - \omega_{12} - k_1 v_z)^2 + (\Gamma_{12}/2)^2(1+S)} , \qquad (10.16)$$

where S is the saturation parameter of the pump transition, see (7.2). Inserting (10.16) into (10.15) and integrating over all velocity components v_z yields [10.88]

$$\alpha(\omega) = \alpha_0(\omega) \left(1 - \frac{1}{2} \frac{N_2^0 - N_1^0}{N_2^0 - N_m^0} \frac{k_2 S}{k_1 \sqrt{1+S}} \frac{\gamma_1 \Delta\Gamma}{[\Omega_2 \pm (k_2/k_1)\Omega_1]^2 + (\Delta\Gamma)^2} \right) , \qquad (10.17)$$

with $\Omega_1 = \omega_1 - \omega_2$, $\Omega_2 = \omega_{2m}$, and $\Delta\Gamma = \Gamma_{2m} + (k_2/k_1)\Gamma_{12}\sqrt{1+S}$.

This represents a Doppler profile $\alpha_0(\omega)$ with a narrow dip at the frequency

$$\omega_2 = (k_2/k_1)(\omega_1 - \omega_{12}) + \omega_{2m} .$$

If the second term in the bracket of (10.17) becomes larger than 1, $\alpha(\omega)$ becomes negative, that is, the probe wave is amplified instead of attenuated (Raman gain, see Sect. 5.7).

Chopping of the pump laser and detection of the probe laser Raman signal behind a lock-in amplifier yields a narrow OODR signal with a linewidth

$$\Delta\Gamma = \gamma_m + [(k_2/k_1)\gamma_1 + (1 \mp k_2/k_1)\gamma_2]\sqrt{1+S} , \qquad (10.18)$$

where the $-$ sign in the bracket of (10.18) stands for collinear, and the $+$ sign for anticollinear propagation of the two laser beams. This shows that for the collinear arrangement the level width γ_2 of the common upper level only enters with the fraction $(1 - k_2/k_1)$, which becomes very small for $k_2/k_1 \sim 1$ [10.89].

This Λ-type OODR can be used to probe high-lying vibrational–rotational levels close to the dissociaton limit of the electronic ground state. As an example, Fig. 10.30 shows the term diagram of the Cs_2 molecule where a high-lying vibrational level ($v' = 50$, $J' = 48$) in the $D\,^1\Sigma_u$ state is optically pumped at the inner turning point of the vibrating molecule and the probe laser induces downward transition with large Franck–Condon factors from the outer turning point into levels (v_3, J_3) closely below the dissociation energy of the $X^1\Sigma_g^+$ ground state. These levels show perturbations by hyperfine mixing with triplet levels induced by indirect nuclear spin–nuclear spin interactions, which are mediated by the electron spin. This OODR spectroscopic technique therefore can give very detailed information of atoms at large internuclear distances, which are gained increasing interest after the realization of Bose–Einstein condensation of alkali atoms, where such interactions cause spin-flips resulting in a loss of atoms from the condensate (see Sect. 14.1.9).

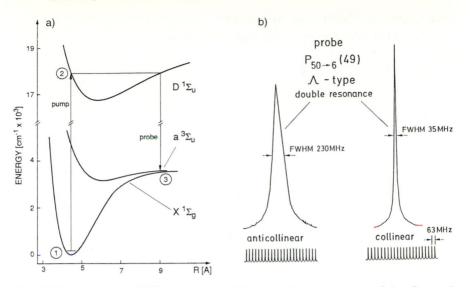

Fig. 10.30a,b. Λ-type OODR spectroscopy illustrated for the example of the Cs_2 molecule: (**a**) term diagram; (**b**) two OODR signals of the same transition measured with collinear and anticollinear propagation of pump and probe beams. For this example the homogeneous width of the upper (predissociating) level is $\gamma_2 \sim 100\,\mathrm{MHz}$ [10.95]

The linewidth of the OODR signal is much smaller than the level width of the predissociating upper level.

The downward transitions can be probed by the change in transmission of the probe wave by OODR–polarization spectroscopy or by ion-dip spectroscopy (Sect. 10.5.1).

Often the super-high spectral resolution of the induced resonant Raman transitions obtained with single-mode lasers is not neccessary if the levels $|m\rangle$ are separated by more than one Doppler width. Then pulsed lasers can be used for stimulated emission pumping [10.90]. Many experiments on high vibrational levels in the electronic ground state of polyatomic molecules have been performed so far by SEP with pulsed lasers. Compilations may be found in [10.91–10.94].

The stimulated emission pumping technique allows one to selectively populate high vibrational levels of molecules that are going to collide with other atoms or molecules. The investigation of the cross section for reactive collisions as a function of the excitation energy gives very valuable information on the reaction mechanism and the intermolecular potential (Chap. 13). Therefore, the SEP technique is a useful method for the selective population transfer into high-lying levels that are not accessible by direct absorption from the ground state.

An even more efficient scheme for population transfer is provided by the coherent stimulated rapid adiabatic passage (STIRAP) method, explained in Chap. 12.

10.5 Special Detection Schemes of Double-Resonance Spectroscopy

Because of their great advantages for the analysis of molecular spectra, a large variety of different detection schemes for double-resonance spectroscopy have been developed. In this section we present four examples.

10.5.1 OODR-Polarization Spectroscopy

Similar to polarization spectroscopy with a single tunable laser (Sect. 7.4), in OODR–polarization spectroscopy the sample cell is placed between two crossed polarizers and the transmitted probe laser intensity $I_T(\lambda_2)$ is measured as a function of the probe laser wavelength λ_2.

The optical pumping of the molecules is now performed with a separate laser beam, which is sent through the sample cell anticollinearly to the probe beam for V-type OODR, but collinearly for Λ-type OODR (Fig. 10.31). In order to keep the pump transition on the wanted selected transition, at first an ordinary Doppler-free polarization spectrum of the pump laser must be recorded. Therefore, the pump laser beam is split into a pump and a probe beam, and the spectrum is recorded while laser L2 is switched off. Now the pump laser is stabilized onto the wanted transition and the second (weak) probe laser L2 is simultaneously sent through the cell.

There are several experimental tricks to separate the two probe beams of L1 and L2: for Λ-type OODR these two probe beams travel into opposite directions and can therefore be detected by two different detectors. In case of V-type OODR the pump beam L1 and the probe beam L2 are anticollinear, which means that the two probe beams are collinear. If their wavelengths differ sufficiently they can be separated by a prism behind the analyzer. Although this is efficient for separating both probe beams, it has the disadvantage that the refraction angle of the prism changes with wavelength and the detector must be realigned when the probe wavelength is scanned over a larger spec-

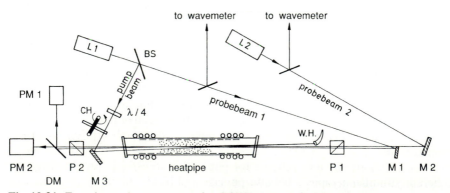

Fig. 10.31. Experimental arrangement for OODR Λ-type double resonance

tral range. If the wavelengths λ_1, λ_2 of the two probe beams differ sufficiently, the two probe beams can be separated by a dichroic mirror (Fig. 10.31).

The sum-frequency method is easier with respect to the optical arrangement but with more electronics involved. The pump beam L1 and the probe beam L2 are chopped at two different frequencies with the same chopper (Fig. 10.19). Both probe beams are detected by the same detector. The output signals are fed parallel into two lock-in detectors tuned to f_1 and $f_1 + f_2$, respectively. While the lock-in at f_1 records the polarization spectrum of L1 if λ_1 is tuned, the lock-in at $(f_1 + f_2)$ records the OODR spectrum induced by both lasers L1 and L2 at a fixed wavelength λ_1 while λ_2 is tuned.

For illustration, Fig. 10.32 displays a section of the polarization spectrum of Cs_2 with V-type and Λ-type OODR signals of Cs_2 transitions with a common upper or lower level, respectively [10.95, 10.96].

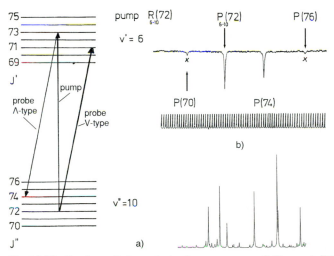

Fig. 10.32. Section of the polarization spectrum of Cs_2 and OODR signals. The *small lines* marked by *crosses* are collision-induced signals

10.5.2 Polarization Labeling

Often it is advantageous to gain an overlook over a wider range of the OODR spectra with lower resolution before Doppler-free scans over selected sections of this range are performed. Here, the polarization labeling technique, first introduced by Schawlow and his group [10.97, 10.98], turns out to be very useful. A polarized pump laser L1 orients molecules in a selected lower level $|i\rangle$ or an upper level $|k\rangle$ and therefore *labels* those levels. Instead of a single-mode probe laser, a linearly polarized spectral continuum is now sent through the sample between two crossed polarizers. Only those wavelengths λ_{im} or λ_{km} are affected in their polarization characteristics that correspond to molecular transitions starting from the labeled levels $|i\rangle$ or $|k\rangle$. These wavelengths are transmitted through the crossed analyzer, are

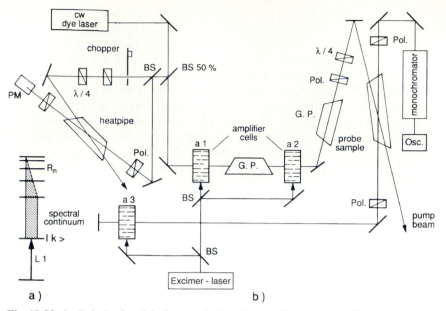

Fig. 10.33a,b. Polarization labeling method: (**a**) term diagram; and (**b**) experimental arrangement

separated by a spectrograph, and are simultaneously recorded by an optical multichannel analyzer or a CCD camera (Sect. 4.5). The spectral resolution is limited by the spectrograph and the spectral range by its dispersion and by the length of the diode array of the OMA.

A typical experimental arrangement used in our lab for studying molecular Rydberg states is exhibited in Fig. 10.33. The pump laser is a single-mode cw dye laser, which is amplified in two pulsed amplifier cells pumped by a nitrogen or excimer laser. The spectral continuum is provided by the broad fluorescence excited in a dye cell by part of the excimer laser beam. Since this spectral continuum is only formed within the small pump region in the dye cell, it can be transformed into a parallel beam by a lens and is sent antiparallel to the pump beam through a heat pipe containing lithium vapor. The OMA system behind the spectrograph is gated in order to make the detector sensitive only for the short time of the probe pulse.

Figure 10.34 illustrates a section of the polarization labeling spectrum of Rydberg transitions in Li_2 starting from the intermediate level $B\,{}^1\Pi_u$ ($v' = 1$, $J' = 27$), which is labeled by the pump laser [10.99]. In the upper spectrum the pump laser was circularly polarized, enhancing P and R transitions, while in the lower spectrum the pump laser was linearly polarized with an angle of $45°$ between the electric vectors of pump and probe. The Q-lines are now pronounced, while the P and R lines are weakened (Sect. 7.4). This demonstrates the advantage of polarization labelling for the analysis of the spectra.

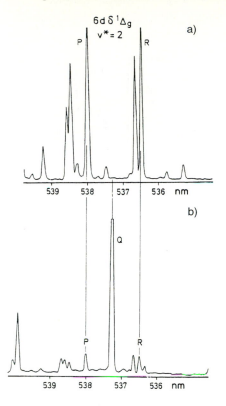

Fig. 10.34a,b. Section of *polarization-labeling* spectrum of Li_2 Rydberg transitions: (**a**) pump laser being circularly polarized; and (**b**) linear polarization. Note the difference in intensities for Q and P, R-transitions in both spectra

10.5.3 Microwave-Optical Double-Resonance Polarization Spectroscopy

A very sensitive and accurate double-resonance technique is microwave–optical double-resonance polarization spectroscopy (MOPS), developed by Ernst et al. [10.100]. This technique detects microwave transitions in a sample between crossed polarizers through the change in transmission of a polarized optical wave. The sensitivity of the method has been demonstrated by measurements of the hfs of rotational transitions in the electronic ground state of CaCl molecules that were produced by the reaction $2Ca + Cl_2 \rightarrow 2CaCl$ in an argon flow. In spite of the small concentrations of CaCl reaction products and the short absorption pathlength in the reaction zone, a good signal-to-noise ratio could be achieved at linewidths of $1-2\,MHz$ [10.101]. A more recent example is the application of this technique to Na_3 clusters in a cold molecular beam, where the hfs in the electronic groundstate of Na_3 has been measured [10.102].

10.5.4 Hole-Burning and Ion-Dip Double-Resonance Spectroscopy

Instead of keeping the pump laser at a fixed wavelength λ_1 and tuning the probe laser wavelength λ_2, one may also fix the probe transition $|i\rangle \rightarrow |k\rangle$

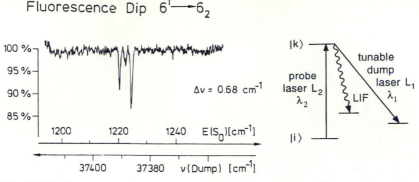

Fig. 10.35. Fluorescence-dip spectroscopy of C_6H_6 molecules. The upper rotational level $J' = 6$, $K' = 6$, $\ell'_6 = -1$ of the $6^1 v$ vibrational state has been selectively excited by L2 and the fluorescence is observed while the wavelength λ_1 of the dump laser L1 is tuned [10.103]

while the pump laser is scanned. The probe laser-induced fluorescence intensity I_{Fl} excited at λ_2 then exhibits a dip when the tuned pump radiation coincides with a transition that shares the common lower level $|i\rangle$ or upper level $|k\rangle$ with the probe laser (V-type or Λ-type OODR) (Fig. 10.35). With pulsed lasers, the time delay Δt between pump and probe laser pulses can be varied in order to study the relaxation of the hole that was burned into the population N_i by the pump or N_k by the dump laser [10.103].

If the absorption of the probe laser is monitored via one- or two-photon ionization of the upper probe transition (Fig. 10.36b), the ion signal shows a dip when the tunable pump beam burns a hole into the common lower level population N_i (*ion-dip spectroscopy*). This opens another way to detect stimulated emission pumping. Now three lasers are employed (Fig. 10.36b). The first laser L1 is kept on a transition $|i\rangle \rightarrow |k\rangle$. The molecules in the excited level $|k\rangle$ are ionized by a second laser L2. In favorable cases photons from L1 may also be used for ionization, thus making L2 unnecessary. A third tunable laser L3 induces downward transitions $|k\rangle \rightarrow |m\rangle$, which cause a de-

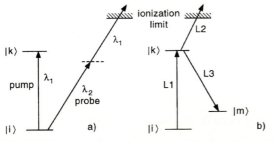

Fig. 10.36a,b. Ion-dip spectroscopy. The ion dips are monitored by the depletion of the lower level $|i\rangle$ (**a**) on the upper level $|k\rangle$ by stimulated emission pumping (**b**)

pletion of level $|k\rangle$. A dip in the ion signal $S(\lambda_3)$ is observed any time the wavelength λ_3 of the stimulated emission laser coincides with a transition $|k\rangle \rightarrow |m\rangle$ [10.103, 10.104].

10.5.5 Triple-Resonance Spectroscopy

For some spectroscopic problems it is necessary to use three lasers in order to populate molecular or atomic states that cannot be reached by two-step excitation. One example is the investigation of high-lying vibrational levels in excited electronic states, which give information about the interaction potential between excited atoms at large internuclear separations. This potential $V(R)$ may exhibit a barrier or hump, and the molecules in levels above the true dissociation energy $V(R = \infty)$ may tunnel through the potential barrier. Such a triple resonance is illustrated in Fig. 10.37a for the Na_2 molecule. A dye laser L1 excites to the selected level (v', J') in the $A\,^1\Sigma_u$ state. If a sufficiently large population $N(v', J')$ can be achieved, population inversion with respect to high vibrational levels (v'', J'') of the electronic ground state is reached and laser oscillation starts (Na_2 dimer laser), which populates the lower level (v'', J''). From this level a third laser induces transitions into levels (v^*, J^*) close to the dissociation limit of the $A\,^1\Sigma_u$ state. These levels could not have been populated from low vibrational levels (v'', J'') of the special state because the corresponding Franck–Condon factors are too small [10.105]. With this technique the very last bound level in the $X\,^1\Sigma_g^+$ ground state of Na_2 molecules could be measured [10.106]. These measurements yield the scattering length [10.107] for Na–Na collisions, an important quantity for the achievement of Bose–Einstein condensates (Sect. 14.1.9).

A second example, mentioned already, is microwave spectroscopy of Rydberg levels that have been excited by resonant two-step absorption of two dye lasers (Fig. 10.36b). The high accuracy of microwave spectroscopy allows

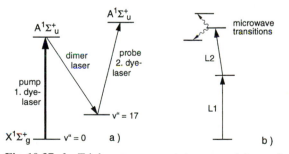

Fig. 10.37a,b. Triple-resonance spectroscopy: **(a)** population of high vibrational levels in the ground state by stimulated emission pumping and absorption of a third laser radiation, resulting in the population of levels just below and above the dissociation energy of the excited $A\,^1\Sigma_u$ state of Na_2 [10.106]; **(b)** stepwise excitation of Rydberg levels by OODR and microwave-induced transitions to neighboring levels

the precise determination of finer details, such as field-induced energy shifts of Rydberg levels, broadening of Rydberg transitions by blackbody radiation, or other effects that might not be resolvable with optical spectroscopy.

Problems

10.1 A linearly polarized laser beam passes in z-direction through a sample cell. Its wavelength is limited to a transition $|1\rangle \rightarrow |2\rangle$ with $J_1 = 0$, $J_2 = 1$. With the laser off, the upper level $|2\rangle$ is assumed to be empty. What is the relative population $N_2(M_2 = \pm 1)/N_2(M_2 = 0)$ of the M_2 levels and the corresponding degree of alignment, if the excitation rate is $10^7\,\text{s}^{-1}$, the effective lifetime is $\tau_2 = 10^{-8}\,\text{s}$, and the collision-induced mixing rate $(M_2 = \pm 1) \rightarrow M_2 = 0$ is $5 \times 10^6\,\text{s}^{-1}$?

10.2 Assume the absorption cross section $\sigma = 10^{-15}\,\text{cm}^2$ for a linearly polarized laser with $\boldsymbol{E} \parallel \boldsymbol{D}$. What is the angular distribution $N_2(\alpha)$ of excited atoms with the transition moment $\boldsymbol{D}$ forming the angle α against $\boldsymbol{E}$ on the transition $J_1 = 0 \rightarrow J_2 = 1$ for the laser intensity I?
 Calculate the saturation parameter $S(\alpha = 0)$ for $1 = 10^2\,\text{W/m}^2$.

10.3

(a) Molecules with a magnetic moment $\mu = 0.1\mu_B$ ($\mu_B = 9.27 \times 10^{-24}\,\text{J/T}$) pass with $v = 500\,\text{m/s}$ in the z-direction through an inhomogeneous magnetic field ($dB/dx = 1\,\text{T/m}$, $L_B = 0.1\,\text{m}$) of a Rabi molecular beam apparatus. Calculate the deflection angle of the molecules with mass $m = 50\,\text{AMU}$ ($1\,\text{AMU} = 1.66 \times 10^{-27}\,\text{kg}$).
(b) If the molecules enter the magnetic field in a nearly parallel beam with a divergence angle $\epsilon = 1°$ and a uniform density $N_0(x)$, they arrive at the detector 50-cm downstream of the end of the magnetic field with a transverse density distribution $N(x)$. What is the approximate profile of $N(x)$?
(c) Assume 60% of all molecules to be in level $|1\rangle$ with $\mu_1 = 0.1\mu B$ and 40% in level $|2\rangle$ with $\mu_2 = 0.11\mu_B$. What is the profile $N(x)$ with $N = N_1 + N_2$? How does $N(x)$ change, if an RF transition before the inhomogeneous magnet equalizes the population N_1 and N_2?

10.4 In an optical–microwave double-resonance experiment the decrease of the population N_i from excitation into level $|k\rangle$ by a cw laser is 20%. The lifetime of the upper level is $\tau_k = 10^{-8}\,\text{s}$, The collisional refilling rate of level $|i\rangle$ is $5 \times 10^7\,\text{s}^{-1}$. Compare the magnitude of the microwave signals $|k\rangle \rightarrow |m\rangle$ between rotational levels in the upper electronic state with that on the transitions $|i\rangle \rightarrow |n\rangle$ in the ground state, if without optical pumping a thermal ground-state population distribution is assumed and the cross section σ_{km} and σ_{in} are equal.

10.5 In a V-type OODR scheme molecules in a sample cell are excited by a chopped pump laser on the transition $|i\rangle \rightarrow |k\rangle$ with the absorption cross section $\sigma_{ik} = 10^{-17}\,\mathrm{cm}^2$. What pump intensity is required in order to achieve a modulation of 50% of the lower-level population N_i, if the relaxation rate into the level $|i\rangle$ is $10^7\,\mathrm{s}^{-1}$?

10.6 A potassium atom is excited into a Rydeberg level ($n = 50$, $\ell = 0$) with the quantum defect $\delta = 2.18$.

(a) Calculate the ionization energy of this level.
(b) What is the minimum electric field required for field ionization (the tunnel effect is neglected)?
(c) What frequency ω_{RF} is required for resonant RF transitions between levels $(n = 50, \ell = 0) \rightarrow (n = 50, \delta = 1.71)$?

10.7 In a Λ-type OODR scheme a transition is induced between the ground-state levels $|1\rangle$ and $|3\rangle$ of the Cs_2 molecule. The radiative lifetime of these levels is $\tau_1 = \tau_3 = \infty$. The upper level $|2\rangle$ has a lifetime of $\tau_2 = 1\,\mathrm{ns}$. Calculate the linewidth of the OODR signal

(a) for collinear, and
(b) for anticollinear propagation of pump and probe beams,

if the laser linewidth is $\delta\nu_L \leq 1\,\mathrm{MHz}$, the transit time of the molecules through the focused laser beam is $30\,\mathrm{ns}$, and the wavelengths are $\lambda_{pump} = 580\,\mathrm{nm}$, $\lambda_{probe} = 680\,\mathrm{nm}$.

11. Time-Resolved Laser Spectroscopy

The investigation of fast processes, such as radiative or collision-induced decays of excited levels, isomerization of excited molecules, or the relaxation of an optically pumped system toward thermal equilibrium, opens the way to study in detail the dynamic properties of excited atoms and molecules. A thorough knowledge of dynamical processes is of fundamental importance for many branches of physics, chemistry, or biology. Examples are predissociation rates of excited molecules, femtosecond chemistry, or the understanding of the visual process and its different steps from the photoexcitation of rhodopsin molecules in the retina cells to the arrival of electrical nerve pulses in the brain.

In order to study these processes experimentally, one needs a sufficiently good time resolution, which means that the resolvable minimum time interval Δt must still be shorter than the time scale T of the process under investigation. While the previous chapters emphasized the high *spectral* resolution, this chapter concentrates on experimental techniques that allow high *time* resolution.

The development of ultrashort laser pulses and of new detection techniques that allow a very high time resolution has brought about impressive progress in the study of fast processes. The achievable time resolution has been pushed recently into the femtosecond range (1 fs $= 10^{-15}$ s). Spectroscopists can now quantitatively follow up ultrafast processes, which could not be resolved ten years ago.

The *spectral* resolution $\Delta \nu$ of most time-resolved techniques is, in principle, confined by the Fourier limit $\Delta \nu = a/\Delta T$, where ΔT is the duration of the short light pulse and the factor $a \simeq 1$ depends on the profile $I(t)$ of the pulse. The spectral bandwidth $\Delta \nu$ of such *Fourier-limited pulses* is still much narrower than that of light pulses from incoherent light sources, such as flashlamps or sparks. Some time-resolved coherent methods based on regular trains of short pulses even circumvent the Fourier limit $\Delta \nu$ of a single pulse and simultaneously reach extremely high spectral and time resolutions (Sect. 12.3).

We will at first discuss techniques for the generation and detection of short laser pulses before their importance for different applications is demonstrated by some examples. Methods for measuring lifetimes of excited atoms or molecules and of fast relaxation phenomena are presented. These applications illustrate the relevance of pico- and femtosecond molecular physics and chemistry for our understanding of fundamental dynamical processes in molecules.

The special aspects of time-resolved coherent spectroscopy are covered in Sects. 12.2, 12.3. For a more extensive representation of the fascinating field of time-resolved spectroscopy some monographs [11.1–11.4], reviews [11.5–11.8], and conference proceedings [11.9–11.11] should be mentioned.

11.1 Generation of Short Laser Pulses

For incoherent, pulsed light sources (for example, flashlamps or spark discharges) the duration of the light pulse is essentially determined by that of the electric discharge. For a long time, microsecond pulses represented the shortest available pulses. Only recently could the nanosecond range be reached by using special discharge circuits with low inductance and with pulse-forming networks [11.12, 11.13].

For laser pulses, on the other hand, the time duration of the pulse is not necessarily limited by the duration of the pump pulse, but may be much shorter. Before we present different techniques to achieve ultrashort laser pulses, we will discuss the relations between the relevant parameters of a laser that determine the time profile of a laser pulse.

11.1.1 Time Profiles of Pulsed Lasers

In active laser media pumped by pulsed sources (for example, flashlamps, electron pulses, or pulsed lasers), the population inversion necessary for oscillation threshold can be maintained only over a time interval ΔT that depends on the duration and power of the pump pulse. A schematic time diagram of a pump pulse, the population inversion, and the laser output is shown in Fig. 11.1. As soon as threshold is reached, laser emission starts. If the pump power is still increasing, the gain becomes high and the laser power rises faster than the inversion, until the increasingly induced emission reduces the inversion to the threshold value.

The time profile of the laser pulse is not only determined by the amplification per round trip $G(t)$ (Sect. 5.2) but also by the relaxation times τ_i, τ_k of the upper and lower laser levels. If these times are short compared to the rise time of the pump pulse, quasi-stationary laser emission is reached, where the inversion $\Delta N(t)$ and the output power $P_L(t)$ have a smooth time profile, determined by the balance between pump power $P_p(t)$, which creates the inversion, and laser output power $P_L(t)$, which decreases it. Such a time behavior, which is depicted in Fig. 11.1a, can be found, for instance, in many pulsed gas lasers such as the excimer lasers (Sect. 5.7).

In some pulsed lasers (for example, the N_2 laser) the lower laser level has a longer effective lifetime than the upper level [11.14]. The increasing laser power $P_L(t)$ decreases the population inversion by stimulated emission. Since the lower level is not sufficiently quickly depopulated, it forms a bottleneck for maintaining threshold inversion. The laser pulse itself limits its duration and it ends before the pump pulse ceases (self-terminating laser, Fig. 11.1b).

If the relaxation times τ_i, τ_k are long compared to the rise time of the pump pulse, a large inversion ΔN may build up before the induced emis-

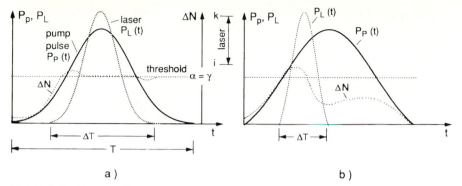

Fig. 11.1a,b. Time profiles of the pump power $P_p(t)$, the inversion density $\Delta N(t)$, and laser output power $P_L(t)$: (**a**) for sufficiently short lifetime τ_i of the lower laser level; and (**b**) for a self-terminating laser with $\tau_i^{\text{eff}} > \tau_k^{\text{eff}}$

sion is strong enough to deplete the upper level. The corresponding high gain leads to a large amplification of induced emission and the laser power P_L may become so high that it depletes the upper laser level faster than the pump can refill it. The inversion ΔN drops below threshold and the laser oscillation stops long before the pump pulse ends. When the pump has again built up a sufficiently large inversion, laser oscillations starts again. In this case, which is, for instance, realized in the flashlamp-pumped ruby laser, the laser output consists of a more or less irregular sequence of "spikes" (Fig. 11.2) with a duration of $\Delta T \simeq 1\,\mu s$ for each spike, which is much shorter than the pump–pulse duration $T \simeq 100\,\mu s$ to 1 ms [11.15, 11.16].

For time-resolved laser spectroscopy, pulsed dye lasers are of particular relevance due to their continuously tunable wavelength. They can be pumped by flashlamps ($T \simeq 1\,\mu s$ to 1 ms), by other pulsed lasers, for example, by copper-vapor lasers ($T \simeq 50\,ns$), excimer lasers ($T \simeq 15\,ns$), nitrogen lasers ($T = 2-10\,ns$), or frequency-doubled Nd:YAG lasers ($T = 5-15\,ns$). Because of the short relaxation times τ_i, $\tau_k (\simeq 10^{-11}\,s)$, no spiking occurs and the situation of Fig. 11.1a is realized (Sect. 5.7). The dye laser pulses have durations between 1 ns to $500\,\mu s$, depending on the pump pulses; typical peak powers range from 1 kW to 10 MW and pulse repetition rates from 1 Hz to 15 kHz [11.17].

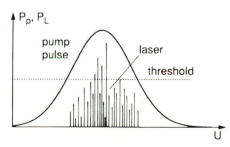

Fig. 11.2. Schematic representation of spikes in the emission of a flashlamp-pumped solid-state laser with long relaxation times τ_i, τ_k

11.1.2 Q-Switched Lasers

In order to obtain a single, powerful pulse of a flashlamp-pumped laser instead of the irregular sequence of many spikes, the technique of Q-switching was developed. Q-switching is based on the following principle:

Until a selected time t_0 after the start of the pump pulse at $t = 0$, the cavity losses of a laser are kept so high by a closed "optical switch" inside the laser resonator that the oscillation threshold cannot be reached. Therefore a large inversion ΔN is built up by the pump (Fig. 11.3). If the switch is opened at $t = t_0$ the losses are suddenly lowered (that is, the quality factor or Q-value of the cavity (Sect. 5.1) jumps from a low to a high value). Because of the large amplification $G \propto B_{ik}\rho\Delta N$ for induced emission, a quickly rising intense laser pulse develops, which depletes in a very short time the whole inversion that had been built up during the time interval t_0. This converts the energy stored in the active medium into a giant light pulse [11.16, 11.18, 11.19]. The time profile of the pulse depends on the rise time of Q-switching. Typical durations of these giant pulses are $1-20$ ns and peak powers up to 10^9 W are reached, which can be further increased by subsequent amplification stages.

Such an optical switch can be realized, for instance, if one of the resonator mirrors is mounted on a rapidly spinning motor shaft (Fig. 11.4). Only at that time t_0 where the surface normal of the mirror coincides with the resonator axis is the incident light reflected back into the resonator, giving a high Q-value of the laser cavity [11.20]. The optimum time t_0 can be selected by imaging the beam of a light-emitting diode (LED) after reflection at the spinning mirror onto the detector D, which provides the trigger signal for the flashlamp of the Q-switched laser. This technique, however, has some disad-

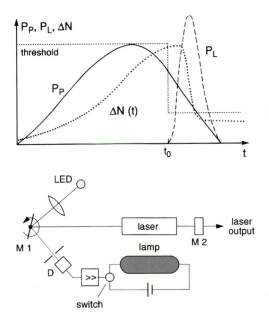

Fig. 11.3. Pump power $P_p(t)$, resonator losses $\gamma(t)$, inversion density $\Delta N(t)$, and laser output power $P_L(t)$ for a Q-switched laser

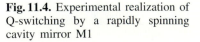

Fig. 11.4. Experimental realization of Q-switching by a rapidly spinning cavity mirror M1

vantages: the spinning mirror is not very stable and the switching time is not sufficiently short. Therefore other Q-switching methods have been developed that are based on electro-optical or acousto-optical modulators [11.21].

One example is given in Fig. 11.5, where a Pockels cell between two crossed polarizers acts as a Q-switch [11.22]. The Pockels cell consists of an optically anisotropic crystal that turns the plane of polarization of a transmitted, linearly polarized wave by an angle $\theta \propto |E|$ when an external electric field E is applied. Since the transmittance of the system is $T = T_0(1 - \cos^2 \theta)$, it can be changed by the voltage U applied to the electrodes of the Pockels cell. If for times $t < t_0$ the voltage is $U = 0$, the crossed polarizers have the transmittance $T = 0$ for $\theta = 0$. When at $t = t_0$ a fast voltage pulse $U(t)$ is applied to the Pockels cell, which causes a rotation of the polarization plane by $\theta = 90°$, the transmittance rises to its maximum value T_0. The linearly polarized lightwave can pass the switch, is reflected back and forth between the resonator mirrors M1 and M2, and is amplified until the inversion ΔN is depleted below threshold. The experimental arrangement shown in Fig. 11.5d works differently and needs only one polarizer P1, which is used as polarizing beam splitter. For $t < t_0$ a voltage U is kept at the Pockels cell, which causes circular polarization of the transmitted light. After reflection at M1 the light again passes the Pockels cell, reaches P1 as linearly polarized light with the plane of polarization turned by $\theta = 90°$, and is totally reflected by P1 and therefore lost for further amplification. For $t > t_0$ U is switched off, P1 now transmits, and the laser output is coupled out through M2.

The optimum choice of t_0 depends on the duration T of the pump pulse $P_p(t)$ and on the effective lifetime τ_k of the upper laser level. If $\tau_k \gg T \approx t_0$, only a small fraction of the energy stored in the upper laser

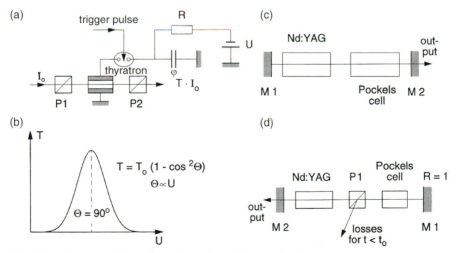

Fig. 11.5a–d. Q-switching with a Pockels cell inside the laser resonator: (**a**) Pockel's cell between two crossed polarizers, (**b**) transmission $T(\theta) \propto U$; (**c**),(**d**) possible experimental arrangements

level is lost by relaxation and the giant pulse can extract nearly the whole energy. For example, with the ruby laser ($\tau_k \approx 3\,$ms) the switching time t_0 can be chosen close to the end of the pump pulse ($t_0 = 0.1-1\,$ms), while for Nd:YAG-lasers ($\tau_k \approx 0.2\,$ms) the optimum switching time t_0 lies before the end of the pump pulse. Therefore only part of the pump energy can be converted into the giant pulse [11.16, 11.19, 11.23].

11.1.3 Cavity Dumping

The principle of Q-switching can also be applied to cw lasers. Here, however, an inverse technique, called *cavity dumping*, is used. The laser cavity consists of highly reflecting mirrors in order to keep the losses low and the Q-value high. The intracavity cw power is therefore high because nearly no power leaks out of the resonator. At $t = t_0$ an optical switch is activated that couples a large fraction of this stored power out of the resonator. This may again be performed with a Pockels cell (Fig. 11.5d), where now M1 and M2 are highly reflecting, and P1 has a large transmittance for $t < t_0$, but reflects the light out of the cavity for a short time Δt at $t = t_0$.

Often an acousto-optic switch is used, for example, for argon lasers and cw dye lasers [11.24]. Its basic principle is explained in Fig. 11.6. A short ultrasonic pulse with acoustic frequency f_s and pulse duration $T \gg 1/f_s$ is sent at $t = t_0$ through a fused quartz plate inside the laser resonator. The acoustic wave produces a time-dependent spatially periodic modulation of the refractive index $n(t, z)$, which acts as a Bragg grating with the grating constant $\Lambda = c_s/f_s$, equal to the acoustic wavelength Λ. When an optical wave $E_0 \cos(\omega t - \boldsymbol{k} \cdot \boldsymbol{r})$ with the wavelength $\lambda = 2\pi/k$ passes through the Bragg

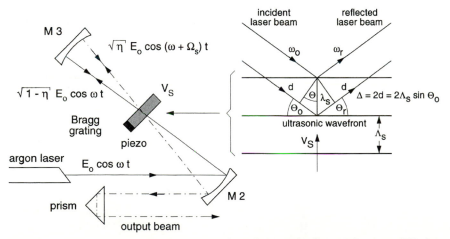

Fig. 11.6. Cavity dumping of a cw argon laser by a pulsed acoustic wave. The insert shows the Bragg reflection of an optical wave at a running ultrasonic wave with wavelength Λ_s

plate, the fraction η of the incident intensity I_0 is diffracted by an angle θ determined by the Bragg relation:

$$2\Lambda \sin\theta = \lambda/n . \tag{11.1}$$

The fraction η depends on the modulation amplitude of the refractive index and thus on the power of the ultrasonic wave.

When the optical wave is reflected at an acoustic wave front moving with the velocity v_s, its frequency ω suffers a Doppler shift, which is, according to (11.1)

$$\Delta\omega = 2\frac{nv_s}{c}\omega\sin\theta = 2n\frac{\Lambda\Omega}{\lambda\omega}\omega\sin\theta = \Omega , \tag{11.2}$$

and turns out to be equal to the acoustic frequency $\Omega = 2\pi f_s$. The amplitude of the deflected fraction is $E_1 = \sqrt{\eta}E_0\cos(\omega+\Omega)t$, that of the unaffected transmitted wave $E_2 = \sqrt{1-\eta}E_0\cos\omega t$. After reflection at the mirror M3 the fraction $\sqrt{1-\eta}E_1$ is transmitted and the fraction $\sqrt{\eta}E_2$ is deflected by the Bragg plate into the direction of the outcoupled beam. This time, however, the reflection occurs at receding acoustic wavefronts and the Doppler shift is $-\Omega$ instead of $+\Omega$. The total amplitude of the extracted wave is therefore

$$E_c = \sqrt{\eta}\sqrt{1-\eta}E_0[\cos(\omega+\Omega)t+\cos(\omega-\Omega)t] . \tag{11.3}$$

The average output power of the light pulse is then with $\langle\cos^2\omega t\rangle = 0.5$ and $\omega \gg \Omega$:

$$P_c(t) = c\epsilon_0\eta(t)[1-\eta(t)]E_0^2\cos^2\Omega t , \tag{11.4}$$

where the time-dependent efficiency $\eta(t)$ is determined by the time profile of the ultrasonic pulse (Fig. 11.7). During the ultrasonic pulse the fraction $2\eta(1-\eta)$ of the optical power $\frac{1}{2}\epsilon_0 E_0^2$, stored within the laser resonator, can be extracted in a short light pulse, which is still modulated at twice the acoustic frequency Ω. With $\eta = 0.3$ one obtains an extraction efficiency of $2\eta(1-\eta) = 0.42$.

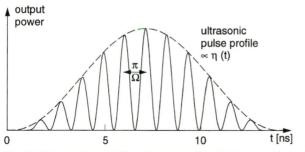

Fig. 11.7. Intensity profile of a cavity-dumped laser pulse, showing the intensity modulation at twice the ultrasonic frequency Ω

The repetition rate f of the extracted pulses can be varied within wide limits by choosing the appropriate repetition rate of the ultrasonic pulses. Above a critical frequency f_c, which varies for the different laser types, the peak power of the extracted pulses decreases if the time between two successive pulses is not sufficiently long to recover the inversion and to reach sufficient intracavity power.

The technique of cavity dumping is mainly applied to gas lasers and cw dye lasers. One achieves pulse durations $\Delta T = 10-100$ ns, pulse-repetition rates of $0-4$ MHz, and peak powers that may be $10-100$ times higher than for normal cw operation with optimized transmission of the output coupler. The average power depends on the repetition rate f. Typical values for $f = 10^4 - 4 \times 10^6$ Hz are $0.1-40\%$ of the cw output power. The disadvantage of the acoustic cavity dumper compared to the Pockels cell of Fig. 11.5 is the intensity modulation of the pulse at the frequency 2Ω.

Example 11.1

With an argon laser, which delivers 3-W cw power at $\lambda = 514.5$ nm, the cavity dumping yields a pulse duration $\Delta T = 10$ ns. At a repetition rate $f = 1$ MHz peak powers of 60 W are possible. The on/off ratio is then $f \Delta T = 10^{-2}$ and the average power $P \approx 0.6$ W is 20% of the cw power.

11.1.4 Mode Locking of Lasers

Without frequency-selective elements inside the laser resonator, the laser generally oscillates simultaneously on many resonator modes within the spectral gain profile of the active medium (Sect. 5.3). In this "multimode operation" no definite phase relations exist between the different oscillating modes, and the laser output equals the sum $\sum_k I_k$ of the intensities I_k of all oscillating modes, which are more or less randomly fluctuating in time (Sect. 6.2).

If coupling between the phases of these simultaneously oscillating modes can be established, a coherent superposition of the mode *amplitudes* may be reached, which leads to the generation of short output pulses in the picosecond range. This mode coupling or *mode locking* has been realized by optical modulators inside the laser resonator (*active* mode locking) or by saturable absorbers (*passive* mode locking) or by a combined action of both locking techniques [11.16, 11.25–11.29].

a) Active Mode Locking

When the intensity of a monochromatic lightwave

$$E = A_0 \cos(\omega_0 t - kx) ,$$

is modulated at the frequency $f = \Omega/2\pi$ (for example, by a Pockels cell or an acousto-optic modulator), the frequency spectrum of the optical wave con-

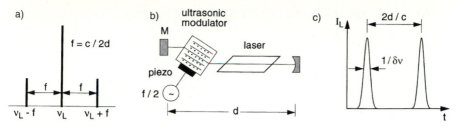

Fig. 11.8a–c. Active mode locking: (**a**) sideband generation; (**b**) experimental arrangement with a standing ultrasonic wave inside the laser resonator; (**c**) idealized output pulses

tains, besides the carrier at $\omega = \omega_0$, sidebands at $\omega = \omega_0 \pm \Omega$ (that is, $\nu = \nu_0 \pm f$) (Fig. 11.8).

If the modulator is placed inside the laser resonator with the mirror separation d and the mode frequencies $\nu_m = \nu_0 \pm m \cdot c/2d$ ($m = 0, 1, 2, \ldots$), the sidebands coincide with resonator mode frequencies if the modulation frequency f equals the mode separation $\Delta \nu = c/2d$. The sidebands can then reach the oscillation threshold and participate in the laser oscillation. Since they pass the intracavity modulator they are also modulated and new sidebands $\nu = \nu_0 \pm 2f$ are generated. This continues until all modes inside the gain profile participate in the laser oscillation. There is, however, an important difference from normal multimode operation: the modes do not oscillate independently, but are phase-coupled by the modulator. At a certain time t_0, the amplitudes of all modes have their maximum at the location of the modulator and this situation is repeated after each cavity round-trip time $T = 2d/c$ (Fig. 11.8c). We will discuss this in more detail:

The modulator has the time-dependent transmission

$$T = T_0[1 - \delta(1 - \cos \Omega t)] = T_0[1 - 2\delta \sin^2(\Omega/2)t], \qquad (11.5)$$

with the modulation frequency $f = \Omega/2\pi$ and the modulation amplitude $2\delta \leq 1$. Behind the modulator the field amplitude A_k of the kth mode becomes

$$A_k(t) = TA_{k0} \cos \omega_k t = T_0 A_{k0}[1 - 2\delta \sin^2(\Omega/2)t] \cos \omega_k t . \qquad (11.6)$$

This can be written as

$$A_k(t) = T_0 A_{k0} \left[(1 - \delta) \cos \omega_k t + \tfrac{1}{2}\delta[\cos(\omega_k + \Omega)t + \cos(\omega_k - \Omega)t] \right] . \quad (11.7)$$

For $\Omega = \pi c/d$, the sideband $\omega_k + \Omega$ corresponds to the next resonator mode and generates the amplitude

$$A_{k+1} = \tfrac{1}{2} A_0 T_0 \delta \cos \omega_{k+1} ,$$

which is further amplified by stimulated emission, as long as ω_{k+1} lies within the gain profile above threshold. Since the amplitudes of all three modes in (11.7) achieve their maxima at times $t = q2d/c$ ($q = 0, 1, 2, \ldots$), their phases are coupled by the modulation. A corresponding consideration applies to all other sidebands generated by modulation of the sidebands in (11.7).

Within the spectral width δv of the gain profile

$$N = \frac{\delta v}{\Delta v} = 2\delta v \frac{d}{c} ,$$

oscillating resonator modes with the mode separation $\Delta v = c/2d$ can be locked together. The superposition of these N phase-locked modes results in the total amplitude

$$A(t) = \sum_{k=-m}^{+m} A_k \cos(\omega_0 + k\Omega)t , \quad \text{with} \quad N = 2m+1 . \tag{11.8}$$

For equal mode amplitudes $A_k = A_0$, (11.8) gives the total time-dependent intensity

$$I(t) \propto A_0^2 \frac{\sin^2(\frac{1}{2}N\Omega t)}{\sin^2(\frac{1}{2}\Omega t)} \cos^2 \omega_0 t . \tag{11.9}$$

If the amplitude A_0 is time independent (cw laser), this represents a sequence of equidistant pulses with the separation

$$T = \frac{2d}{c} = \frac{1}{\Delta v} , \tag{11.10}$$

which equals the round-trip time through the laser resonator. The pulse width

$$\Delta T = \frac{2\pi}{(2m+1)\Omega} = \frac{2\pi}{N\Omega} = \frac{1}{\delta v} , \tag{11.11}$$

is determined by the number N of phase-locked modes and is inversely proportional to the spectral bandwidth δv of the gain profile above threshold (Fig. 11.9).

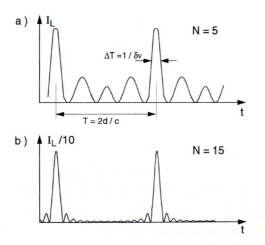

Fig. 11.9a,b. Schematic profile of the output of a mode-locked laser: (a) with 5 modes locked; (b) with 15 modes locked

The peak power of the pulses, which can be derived from the intensity maxima in (11.9) at times $t = 2\pi q/\Omega = q(2d/c)$ $(q = 0, 1, 2, \ldots)$, is proportional to N^2. The pulse energy is therefore proportional to $N^2 \Delta T \propto N$. In between the main pulses $(N-2)$ small maxima appear, which decrease in intensity as N increases.

Note: For equal amplitudes $A_k = A_0$ the time-dependent intensity $I(t)$ in (11.9) corresponds exactly to the spatial intensity distribution $I(x)$ of light diffracted by a grating with N grooves that are illuminated by a plane wave. One has to replace Ωt by the phase difference ϕ between neighboring interfering partial waves, see (4.28) in Sect. 4.1.3 and compare Figs. 4.21 and 11.9.

In real mode-locked lasers the amplitudes A_k are generally not equal. Their amplitude distribution A_k depends on the form of the spectral gain profile. This modifies (11.9) and gives slightly different time profiles of the mode-locked pulses, but does not change the principle considerations.

For *pulsed* mode-locked lasers the envelope of the pulse heights follows the time profile $\Delta N(t)$ of the inversion, which is determined by the pump power $P_p(t)$. Instead of a continuous sequence of equal pulses one obtains a finite train of pulses (Fig. 11.10).

For many applications a single laser pulse instead of a train of pulses is required. This can be realized with a synchronously triggered Pockels cell outside the laser resonator, which transmits only one selected pulse out of the pulse train. It is triggered by a mode-locked pulse just before the maximum of the train envelope, then it opens for a time $\Delta t < 2d/c$ and transmits only the next pulse following the trigger pulse [11.30]. Another method of single-pulse selection is cavity dumping of a mode-locked laser [11.31]. Here the trigger signal for the intracavity Pockels cell (Fig. 11.5) is synchronized by the mode-locked pulses and couples just one mode-locked pulse out of the cavity (Fig. 11.10b).

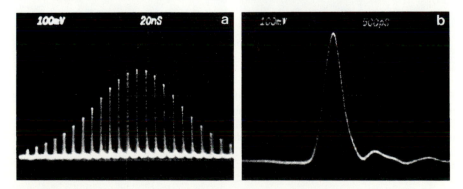

Fig. 11.10. (a) Pulse train of a mode-locked pulsed Nd:YAG laser; and **(b)** single pulse selected by a Pockel's cell. Note the different time scales in (a) and (b) [11.30]

Example 11.2

(a) The Doppler-broadened gain profile of the HeNe laser at $\lambda = 633$ nm has the spectral bandwidth $\delta v \approx 1.5$ GHz. Therefore, mode-locked pulses with durations down to $\Delta T \approx 500$ ps can be generated.

(b) Because of the higher temperature in the discharge of an argon-ion laser the bandwidth at $\lambda = 514.5$ nm is about $\delta v = 5-7$ GHz and one would expect pulses down to 150 ps. Experimentally, 200-ps pulses have been achieved. The apparently longer pulse width in Fig. 11.11 is limited by the time resolution of the detectors.

(c) The actively mode-locked Nd:glass laser [11.27, 11.28] delivers pulses at $\lambda = 1.06 \,\mu$m with durations down to 5 ps with high peak power ($\geq 10^{10}$ W), which can be frequency doubled or tripled in nonlinear optical crystals with high conversion efficiency. This yields powerful short light pulses in the green or ultraviolet region.

(d) Because of the large bandwidth δv of their spectral gain profile, dye lasers, Ti:sapphire, and color-center lasers (Sect. 5.7) are the best candidates for generating ultrashort light pulses. With $\delta v = 3 \times 10^{13}$ Hz (this corresponds to $\delta \lambda \sim 30$ nm at $\lambda = 600$ nm), pulse widths down to $\Delta T = 3 \times 10^{-14}$ s should be possible. This can, indeed, be realized with special techniques (Sect. 11.1.5). With active mode locking, however, one only reaches $\Delta T \geq 10-50$ ps [11.29]. This corresponds to the transit time of a light pulse through the modulator, which imposes a lower limit, unless new techniques are used.

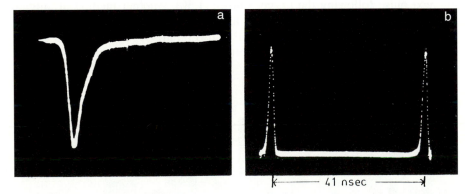

Fig. 11.11a,b. Measured pulses of a mode-locked argon laser at $\lambda = 488$ nm: (a) monitored with a fast photodiode and a sampling oscilloscope (500 ps/div). The small oscillations after the pulse are cable reflections. (b) The attenuated scattered laser light was detected by a photomultiplier (single-photon counting) and stored in a multichannel analyzer. The time resolution is limited by the pulse rise times of the photodiode and photomultiplier, respectively [11.32]

b) Passive Mode Locking

Passive mode locking is a technique that demands less experimental effort than active mode-locking; it can be applied to pulsed as well as to cw lasers. Pulse widths below 1 ps have been realized. Its basic principles can be understood as follows:

Instead of the active modulator, a saturable absorber is put inside the laser resonator, close to one of the end mirrors (Fig. 11.12). The absorbing transition $|k\rangle \leftarrow |i\rangle$ takes place between the levels $|i\rangle$ and $|k\rangle$ with short relaxation times τ_i, τ_k. In order to reach oscillation threshold in spite of the absorption losses the gain of the active medium must be correspondingly high. In the case of a pulsed pump source, the emission of the active laser medium at a time shortly before threshold is reached consists of fluorescence photons, which are amplified by induced emission. The peak power of the resulting photon avalanches (Sect. 5.2) fluctuates more or less randomly. Because of nonlinear saturation in the absorber (Sect. 7.1), the most intense photon avalanche suffers the lowest absorption losses and thus experiences the largest net gain. It therefore grows faster than other competing weaker avalanches, saturates the absorber more, and increases its net gain even more. After a few resonator round trips this photon pulse has become so powerful that it depletes the inversion of the active laser medium nearly completely and therefore suppresses all other avalanches.

This nonlinear interaction of photons with the absorbing and the amplifying media leads under favorable conditions to mode-locked laser operation

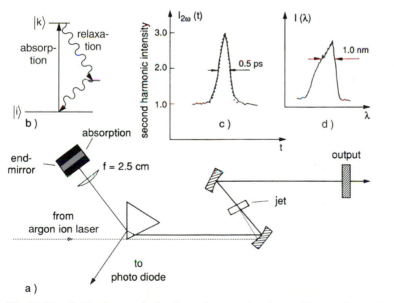

Fig. 11.12a–d. Passive mode locking of a cw dye laser: (**a**) experimental arrangement; (**b**) level scheme of absorber; (**c**) time profile; and (**d**) spectral profile of a mode-locked pulse [11.28]

starting from a statistically fluctuating, unstable threshold situation. After this short unstable transient state the laser emission consists of a stable, regular train of short pulses with the time separation $T = 2d/c$, as long as the pump power remains above threshold (which is now lower than at the beginning because the absorption is saturated).

This more qualitative representation illustrates that the time profile and the width ΔT of the pulses is determined by the relaxation times of absorber and amplifier. In order to suppress the weaker photon avalanches reliably, the relaxation times of the absorber must be short compared with the resonator round-trip time. Otherwise, weak pulses, which pass the absorber shortly after the stronger saturating pulse, would take advantage of the saturation and would experience smaller losses. The recovery time of the amplifying transition in the active laser medium, on the other hand, should be comparable to the round-trip time in order to give maximum amplification of the strongest pulse, but minimum gain for pulses in between. A more thorough analysis of the conditions for stable passive mode locking can be found in [11.16, 11.33, 11.34].

The Fourier analysis of the regular pulse train yields again the mode spectrum of all resonator modes participating in the laser oscillation. The coupling of the modes is achieved at the fixed times $t = t_0 + q2d/c$ when the saturating pulse passes the absorber. This explains the term *passive mode locking*. Different dyes can be employed as saturable absorbers. The optimum choice depends on the wavelength. Examples are methylene blue, diethyloxadicarbocyanine iodide DODCI, or polymethinpyrylin [11.35], which have relaxation times of $10^{-9} - 10^{-11}$ s.

Passive mode locking can also be realized in cw lasers. However, the smaller amplification restricts stable operation to a smaller range of values for the ratio of absorption to amplification, and the optimum conditions are more critical than in pulsed operation [11.36, 11.37]. With passively mode-locked cw dye lasers pulse widths down to 0.5 ps have been achieved [11.38].

More detailed representations of active and passive mode locking can be found in [11.16, 11.39, 11.40].

c) Synchronous Pumping with Mode-Locked Lasers

For synchronous pumping the mode-locked pump laser L1, which delivers short pulses with the time separation $T = 2d_1/c$, is employed to pump another laser L2 (for example, a cw dye laser or a color-center laser). This laser L2 then operates in a pulsed mode with the repetition frequency $f = 1/T$. An example, illustrated by Fig. 11.13, is a cw dye laser pumped by an acousto-optically mode-locked argon laser.

The optimum gain for the dye-laser pulses is achieved if they arrive in the active medium (dye jet) at the time of maximum inversion $\Delta N(t)$ (Fig. 11.14). If the optical cavity length d_2 of the dye laser is properly matched to the length d_1 of the pump laser, the round-trip times of the pulses in both lasers become equal and the arrival times of the two pulses in the amplifying dye jet are synchronized. Because of saturation effects, the dye-laser pulses be-

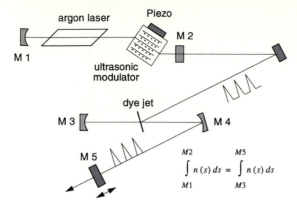

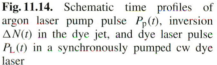

Fig. 11.13. Synchronously pumped cw dye laser

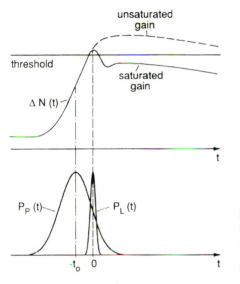

Fig. 11.14. Schematic time profiles of argon laser pump pulse $P_\mathrm{p}(t)$, inversion $\Delta N(t)$ in the dye jet, and dye laser pulse $P_\mathrm{L}(t)$ in a synchronously pumped cw dye laser

come much shorter than the pump pulses, and pulse widths below 1 ps have been achieved [11.41–11.43]. For the experimental realization of accurate synchronization one end mirror of the dye-laser cavity is placed on a micrometer carriage in order to adjust the length d_2. The achievable pulse width ΔT depends on the accuracy $\Delta d = d_1 - d_2$ of the optical cavity length matching. A mismatch of $\Delta d = 1\,\mu\mathrm{m}$ increases the pulsewidth from 0.5 to 1 ps [11.44].

For many applications the pulse repetition rate $f = c/2d$ (which is $f = 150\,\mathrm{MHz}$ for $d = 1$ m) is too high. In such cases the combination of synchronous pumping and cavity dumping (Sect. 11.1.2) is helpful, where only every kth pulse ($k \geq 10$) is extracted due to Bragg reflection by an ultrasonic pulsed wave in the cavity dumper. The ultrasonic pulse now has to be synchronized with the mode-locked optical pulses in order to assure that the

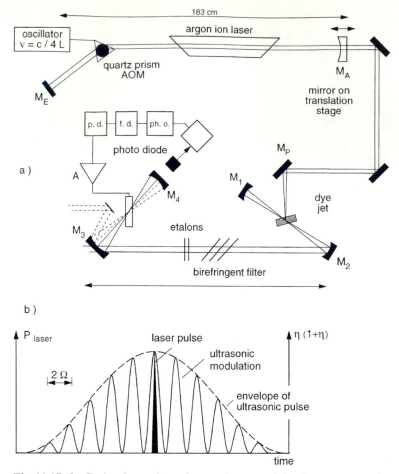

Fig. 11.15a,b. Cavity-dumped synchronously pumped dye laser system with synchronization electronics [Courtesy Spectra-Physics]: (**a**) experimental setup; (**b**) correct synchronization time of a mode-locked pulse with the time of maximum output coupling of the cavity dumping pulse

ultrasonic pulse is applied just at the time when the mode-locked pulse passes the cavity dumper (Fig. 11.15b).

The technical realization of this synchronized system is shown in Fig. 11.15a. The frequency $\nu_s = \Omega/2\pi$ of the ultrasonic wave is chosen as an integer multiple $\nu_s = q \cdot c/2d$ of the mode-locking frequency. A fast photodiode, which detects the mode-locked optical pulses, provides the trigger signal for the RF generator. This allows the adjustment of the phase of the ultrasonic wave in such a way that the arrival time of the mode-locked pulse in the cavity dumper coincides with its maximum extraction efficiency. During the ultrasonic pulse only one mode-locked pulse is extracted. The extraction repetition frequency $\nu_e = (c/2d)/k$ can be chosen between 1 Hz to 4 MHz by selecting the repetition rate of the ultrasonic pulses [11.45].

Table 11.1. Summary of different mode-locking techniques. With active mode locking an average power of 1 W can be achieved

Technique	Mode locker	Laser	Typical pulse duration	Typical pulse energy
Active mode locking	Acousto-optic modulator Pockels cell	Argon, cw HeNe, cw Nd:YAG, pulsed	300 ps 500 ps 100 ps	10 nJ 0.1 nJ 10 µJ
Passive mode locking	Saturable absorber	Dye, cw Nd:YAG	1 ps 1–10 ps	1 nJ 1 nJ
Synchronous pumping	Mode-locked pump laser and matching of resonator length	Dye, cw Color center	1 ps 1 ps	10 nJ 10 nJ
CPM	Passive mode locking and eventual synchronous pumping	Ring dye laser	< 100 fs	≈ 1 nJ
Kerr lens mode locking	optical Kerr effect	Ti:sapphire	< 10 fs	≈ 1–10 nJ

There are several versions of the experimental realizations of mode-locked or synchronously pumped lasers. Table 11.1 gives a short summary of typical operation parameters of the different techniques. More detailed representations of this subject can be found in [11.39–11.46].

11.1.5 Generation of Femtosecond Pulses

In the last sections it was shown that passive mode locking or synchronous pumping allows the generation of light pulses with a pulse width below 1 ps. Recently some new techniques have been developed that generate still shorter pulses. The shortest light pulses reported up to now are only 5-fs long [11.47]. At $\lambda = 600$ nm this corresponds to less than 3 oscillation periods of visible light! We will now discuss some of these new techniques.

a) The Colliding Pulse Mode-Locked Laser

A cw ring dye laser pumped by an argon laser can be passively mode locked by an absorber inside the ring resonator. The mode-locked dye laser pulses travel into both directions in the ring, clockwise and counterclockwise (Fig. 11.16). If the absorber, realized by a thin dye jet, is placed at a location where the path length A1–A2 between the amplifying jet and the absorbing jet is just one-quarter of the total ring length L, the net gain per round-trip is maximum when the counterpropagating pulses collide within the absorber. This can be seen as follows:

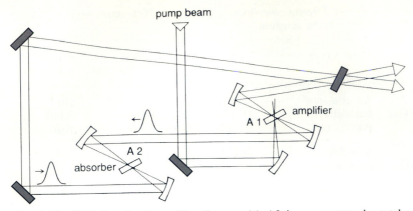

Fig. 11.16. CPM ring dye laser. The distance A1–A2 is one-quarter the total round-trip length L

For this situation the time separation $\Delta t = T/2$ between the passage of successive pulses through the amplifier achieves the maximum value of one-half of the round-trip time T. This means that the amplifying medium has a maximum time to recover its inversion after it was depleted by the previous pulse.

The total pulse intensity in the absorber where the two pulses collide is twice that of a single pulse. This means larger saturation and less absorption. Both effects lead to a maximum net gain if the two pulses collide in the absorber jet.

At a proper choice of the amplifying gain and the absorption losses, this situation will automatically be realized in the passively mode-locked ring dye laser. It leads to an energetically favorable stable operation, which is called *colliding-pulse mode* (CPM) *locking*, and the whole system is termed a *CPM laser*. This mode of operation results in particularly short pulses down to 50 fs. There are several reasons for this pulse shortening:

(i) The transit time of the light pulses through the thin absorber jet ($d < 100\,\mu m$) is only about 400 fs. During their superposition in the absorber the two colliding light pulses form, for a short time, a standing wave, which generates, because of saturation effects, a spatial modulation $N_i(z)$ of the absorber density N_i and a corresponding refractive index grating with a period of $\lambda/2$ (Fig. 11.17). This grating causes a partial reflection of the two incident light pulses. The reflected part of one pulse superimposes and interferes with the oppositely traveling pulse, resulting in a coupling of the two pulses. At $t = t_0$ the constructive interference is maximized (additive-pulse mode locking)

(ii) The absorption of both pulses has a minimum when the two pulse maxima just overlap. At this time the grating is most pronounced and the coupling is maximized. The pulses are therefore shortened for each successive round-trip until the shortening is compensated by other phenomena, which cause a broadening of the pulses. One of these effects is the dispersion due to the dielectric layers on the resonator mirrors, which causes a different round-trip time for the different wavelengths contained in the short pulse. The

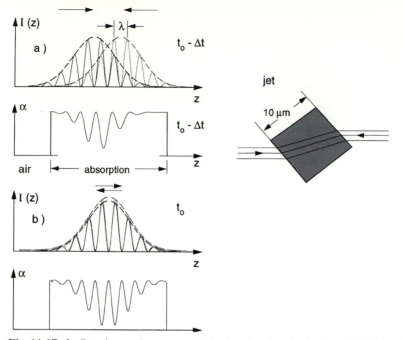

Fig. 11.17a,b. Density grating generated in the absorber jet by the colliding optical pulses at two different times: (**a**) partial overlap at time $t_0 - \Delta t$; (**b**) complete overlap at $t = t_0$

shorter the pulses the broader their spectral profile $I(\lambda)$ becomes and the more serious are dispersion effects.

The mirror dispersion can be compensated up to the first order by inserting into the ring resonator (Fig. 11.18) dispersive prisms, which introduce different optical path lengths $d_p n(\lambda)$ [11.48]. This dispersion compensation can be optimized by shifting the prisms perpendicularly to the pulse propagation, thus adjusting the optical path length $d_p n(\lambda)$ of the pulses through the prisms.

In principle, the lower limit ΔT_{min} of the pulse width is given by the Fourier limit $\Delta T_{min} = a/\delta\nu$, where $a \sim 1$ is a constant that depends on the time profile of the pulse (Sect. 11.2.2). The larger the spectral width $\delta\nu$ of the gain profile is, the smaller ΔT_{min} becomes. In reality, however, the dispersion effects, which increase with $\delta\nu$, become more and more important and pre-

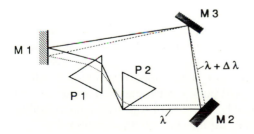

Fig. 11.18. Compensation of mirror dispersion by prisms within the cavity

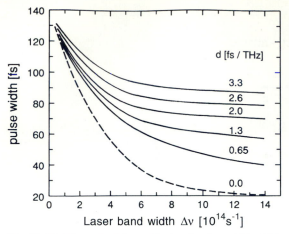

Fig. 11.19. Theoretical lower limit of pulse width ΔT as a function of the spectral bandwidth $\Delta \nu$ of a mode-locked laser for different values of the dispersion parameter D [fs/THz] [11.47]

vent reaching the principal lower limit of ΔT_{min}. In Fig. 11.19 the achievable limit ΔT_{min} is plotted against the spectral bandwidth for different dispersions [fs/cm], where the dashed curve gives the dispersion-free Fourier limit of the pulse width ΔT [11.48, 11.49].

Pulse widths below 100 fs can be reached with this CMP technique [11.50, 11.51]. If the CPM ring dye laser is synchronously pumped by a mode-locked argon laser stable operation over many hours can be realized [11.52]. Using a novel combination of saturable absorber dyes and a frequency-doubled mode-locked Nd:YAG laser as a pump, pulse widths down to 39 ps at $\lambda = 815$ nm have been reported [11.53].

b) Kerr Lens Mode Locking

For a long time dye solutions were the favorite gain medium for the generation of femtosecond pulses because of their broad spectral gain region. Meanwhile, different solid state gain materials have been found with very broad fluorescence bandwidths, which allow, in combination with new nonlinear phenomena, the realization of light pulses down to 5 fs.

For solid-state lasers typical lifetimes of the upper laser level range from 10^{-6} s to 10^{-3} s. This is much longer than the time between successive pulses in a mode-locked pulse train, which is about 10−20 ns. Therefore the saturation of the amplifying medium cannot recover within the time between two pulses and the amplifying medium therefore cannot contribute to the mode locking by dynamic saturation as in the case of CPM mode locking discussed before. One needs a fast saturable absorber, where the saturation can follow the short pulse profile of the mode-locked pulses. Such passively mode-locked solid-state lasers might not be completely stable with regard to pulse stability and pulse intensities and they generally do not deliver pulses below 1 ps.

The crucial breakthrough for the realization of ultrafast pulses below $100\,\mathrm{fs}$ was the discovery of a fast pulse-forming mechanism in 1991, *Kerr lens mode locking* (KLM), which can be understood as follows:

For large incident intensities I the refractive index n of a medium depends on the intensity. One can write

$$n(\omega, I) = n_0(\omega) + n_2(\omega)I .$$

This intensity-dependent change of the refractive index is caused by the non-linear polarization of the electron shell induced by the electric field of the optical wave and is therefore called the optical Kerr effect.

Because of the radial intensity variation of a Gaussian laser beam, the refractive index of the medium under the influence of a laser beam shows a radial gradient with the maximum value of n at the central axis. This acts as a focusing lens and leads to focusing of the incident laser beam, where the focal length depends on the intensity. Since the central part of the pulse time-profile has the largest intensiy, it is also focused more strongly than the outer parts, where the intensity is lower. A circular aperture at the right place inside the laser resonator transmits only this central part, that is, it cuts away the leading and trailing edges and the transmitted pulse is therefore shorter than the incident pulse (Fig. 11.20).

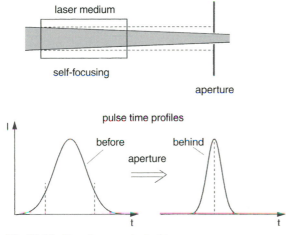

Fig. 11.20. Kerr lens mode locking

Example 11.3

For sapphire Al_2O_3 $n_2 = 3 \times 10^{-16}\,\mathrm{cm^2/W}$. At the intensity of $10^{14}\,\mathrm{W/cm^2}$ the refractive index changes by $\Delta n = 3 \times 10^{-2}\,n_0$. For a wave with $\lambda = 1000\,\mathrm{nm}$ this leads to a phase shift of $\Delta\Phi = (2\pi/\lambda)\Delta n = 300 \cdot 2\pi$ after a pathlength of $1\,\mathrm{cm}$, which results in a radius of curvature of the phase-front $R = 4\,\mathrm{cm}$ with the corresponding focal length of the Kerr lens.

Often the laser medium itself acts as Kerr medium and forms an additional lens inside the laser resonator. This is shown in Fig. 11.21, where the lenses with focal lengths f_1 and f_2 are in practice curved mirrors [11.54]. Without the Kerr lens the resonator is stable if the distance between the two lenses is $f_1 + f_2$. With the Kerr lens this distance has to be modified to $f_1 + f_2 + \delta$, where the quantity δ depends on the focal length of the Kerr lens, and therefore on the pulse intensity. If the distance between the two lenses is $f_1 + f_2 + \delta$ the resonator is only stable for values of δ within the limits

$$0 < \delta < \delta_1 , \quad \text{or} \quad \delta_2 < \delta < \delta_1 + \delta_2 ,$$

where

$$\delta_1 = \frac{f_2^2}{d_2 - f_2} , \quad \delta_2 = \frac{f_1^2}{d_1 - f_1} \tag{11.12}$$

Choosing the right value of δ makes the resonator stable only for the time interval around the pulse maximum.

In Fig. 11.22 the experimental setup for a femtosecond Ti:sapphire laser with Kerr lens mode locking is shown, where the amplifying laser medium acts simultaneously as a Kerr lens. The folded resonator is designed in such a way that only the most intense part of the pulse is sufficiently focused by the Kerr lens to always pass for every round-trip through the spatially confined

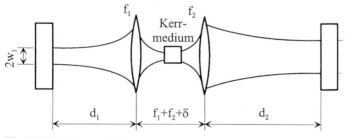

Fig. 11.21. Schematic illustration of Kerr lens mode locking inside the laser resonator [11.54]

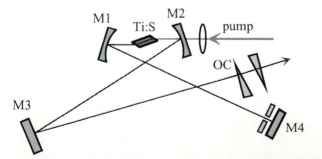

Fig. 11.22. Schematic diagram of the MDC Ti:sapphire oscillator used for soft-aperture and hard-aperture mode locking [11.54]

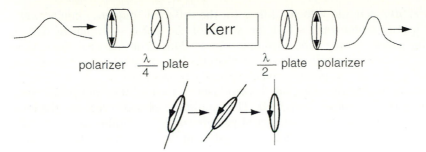

Fig. 11.23. Polarization-additive pulse mode-locking [11.55]

active region pumped by the argon-ion laser. Here the active gain medium defined by the pump focus acts as spatial "soft aperture." A mechanical aperture in front of mirror M4 in Fig. 11.22 realizes "hard aperture" Kerr lens mode locking. Output pulses with pulsewidths below 10 fs have been demonstrated with this design.

An alternative method for the realization of KLM uses the birefringent properties of the Kerr medium, which turns the plane of polarization of the light wave passing through the Kerr medium. This is illustrated in Fig. 11.23. The incident wave passes through a linear polarizer and is then elliptically polarized by a λ/4-plate. The Kerr medium causes a time dependent nonlinear polarization rotation. A λ/2-plate and a linear polarizer behind the Kerr medium can be arranged in such a way that the pulse transmission reaches its maximum at the peak of the incident pulse, thus shortening the pulse width [11.55]. This device acts similarly to a passive saturable absorber and is particularly useful for fiber lasers with ultrashort pulses.

11.1.6 Optical Pulse Compression

Since the principle lower limit $\Delta T_{\min} = 1/\delta\nu$ is given by the spectral bandwidth $\delta\nu$ of the gain medium, it is desirable to make $\delta\nu$ as large as possible. The idea of spectral broadening of optical pulses by *self-phase modulation* in optical fibers with subsequent pulse compression represented a breakthrough for achieving pulsewidths of only a few femtoseconds. The method is based on the following principle:

When an optical pulse with the spectral amplitude distribution $E(\omega)$ propagates through a medium with refractive index $n(\omega)$, its time profile will change because the group velocity

$$v_g = \frac{d\omega}{dk} = \frac{d}{dk}(v_{ph}k) = v_{ph} + k\frac{dv_{ph}}{dk} , \tag{11.13}$$

which gives the velocity of the pulse maximum, shows a dispersion

$$\frac{dv_g}{d\omega} = \frac{dv_g}{dk} \bigg/ \frac{d\omega}{dk} = \frac{1}{v_g}\frac{d^2\omega}{dk^2} . \tag{11.14}$$

For $d^2\omega/dk^2 \neq 0$, the velocity differs for the different frequency components of the pulse, which means that the shape of the pulse will change during its propagation through the medium (Fig. 11.24a). For negative dispersion ($dn/d\lambda < 0$), for example, the red wavelengths have a larger velocity than the blue ones, that is, the pulse becomes spatially broader.

If the optical pulse of a mode-locked laser is focused into an optical fiber, the intensity I becomes very high. The amplitude of the forced oscillations that the electrons in the fiber material perform under the influence of the optical field increases with the field amplitude and the refractive index becomes intensity dependent:

$$n(\omega, I) = n_0(\omega) + n_2 I(t) , \tag{11.15}$$

where $n_0(\omega)$ describes the linear dispersion (see Sect. 11.1.5). The phase $\phi = \omega t - kz$ of the optical wave $E = E_0 \cos(\omega t - kz)$

$$\phi = \omega t - \omega n \frac{z}{c} = \omega \left(t - n_0 \frac{z}{c} \right) - AI(t) , \quad \text{with} \quad A = n_2 \omega \frac{z}{c} , \tag{11.16}$$

now depends on the intensity

$$I(t) = c\epsilon_0 \int |E(\omega, t)|^2 \cos^2(\omega t - kz) \, d\omega . \tag{11.17}$$

Since the momentary optical frequency

$$\omega = \frac{d\phi}{dt} = \omega_0 - A \frac{dI}{dt} , \tag{11.18}$$

is a function of the time derivative dI/dt, (11.18) illustrates that the frequency decreases at the leading edge of the pulse ($dI/dt > 0$), while at the trailing

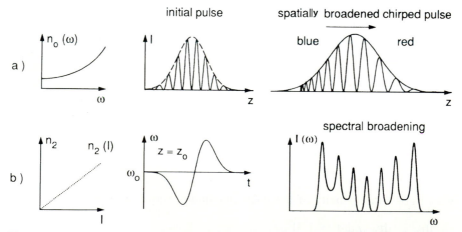

Fig. 11.24a,b. Spatial and spectral broadening of a pulse in a medium with normal linear (**a**) and nonlinear (**b**) refractive index

edge $(\mathrm{d}I/\mathrm{d}t < 0)$ ω increases (*self-phase modulation*). The leading edge is red shifted, the trailing edge blue shifted (*chirp*). The *spectral profile* of the pulse becomes broader (Fig. 11.24b).

The linear dispersion $n_0(\lambda)$ causes a *spatial* broadening, the intensity-dependent refractive index $n_2 I(t)$ a *spectral* broadening. The spatial broadening of the pulse (which corresponds to a broadening of its time profile) is proportional to the length of the fiber and depends on the spectral width $\Delta \omega$ of the pulse and on its intensity.

A quantitative description starts from the wave equation for the pulse envelope [11.56, 11.57]

$$\frac{\partial E}{\partial z} + \frac{1}{v_\mathrm{g}} \frac{\partial E}{\partial t} + \frac{\mathrm{i}}{2 v_\mathrm{g}^2} \frac{\partial v_\mathrm{g}}{\partial \omega} \frac{\partial^2 E}{\partial t^2} = 0 \,, \tag{11.19}$$

which can be derived from the general wave equation with the slowly varying envelope approximation $(\lambda \partial^2 E / \partial z^2 \ll \partial E / \partial z)$ [11.16].

For a pulse of initial width τ, which propagates with the group velocity v_g through a medium of length L, the solution of (11.19) yields the pulse width [11.58]

$$\tau' = \tau \sqrt{1 + (\tau_\mathrm{c}/\tau)^4} \,, \quad \text{with} \quad \tau_\mathrm{c} = 2^{5/4} \cdot \sqrt{(L/v_\mathrm{g})^2 (\partial v_\mathrm{g}/\partial \omega)} \,. \tag{11.20}$$

For $\tau = \tau_\mathrm{c}$, the initial pulse width τ increases by a factor $\sqrt{2}$. Pulses that are shorter than the critical pulse width τ_c become broader. After the length

$$L = \sqrt{\frac{3}{2} \frac{(\tau \cdot v_\mathrm{g}/2)^2}{\partial v_\mathrm{g}/\partial \omega}} \,, \tag{11.21}$$

the pulse width has doubled. The relative pulse broadening sharply increases with decreasing width τ of the incident pulse.

The compression of this spectrally and spatially broadened pulse can now be achieved by a grating pair with the separation D that has a larger pathlength for red wavelengths than for blue ones and therefore delays the leading red edge of the pulse compared to its blue trailing edge. This can be seen as follows [11.59]: the optical pathlength $S(\lambda)$ between two phase-fronts of a plane wave before and after the grating is, according to Fig. 11.25

$$S(\lambda) = S_1 + S_2 = \frac{D}{\cos \beta} (1 + \sin \gamma) \,, \quad \text{with} \quad \gamma = 90° - (\alpha + \beta). \tag{11.22}$$

This transforms with $\cos(\alpha + \beta) = \cos \alpha \cos \beta - \sin \alpha \sin \beta$ into

$$S(\lambda) = D \left[\frac{1}{\cos \beta} + \cos \alpha - \sin \alpha \tan \beta \right] \,.$$

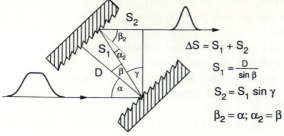

$$\Delta S = S_1 + S_2$$
$$S_1 = \frac{D}{\sin \beta}$$
$$S_2 = S_1 \sin \gamma$$
$$\beta_2 = \alpha; \ \alpha_2 = \beta$$

Fig. 11.25. Pulse compression by a grating pair

With the grating equation $d(\sin \alpha - \sin \beta) = \lambda$ of a grating with groove separation d and its dispersion $d\beta / d\lambda = 1/(d \cos \beta)$ for a given angle of incidence α (Sect. 4.1.3), we obtain the spatial dispersion

$$\frac{dS}{d\lambda} = \frac{dS}{d\beta} \frac{d\beta}{d\lambda} = \frac{-D\lambda}{d^2 [1 - (\lambda/d - \sin \alpha)^2]^{3/2}} . \tag{11.23}$$

This shows that the dispersion is proportional to the grating separation D and increases with λ. By choosing the correct value of D one can just compensate the chirp of the pulse generated in the optical fiber and obtain a compressed pulse.

A typical experimental arrangement is depicted in Fig. 11.26 [11.60]. The optical pulse from the mode-locked laser is spatially and spectrally broadened in the optical fiber and then compressed by the grating pair. The dispersion of the grating pair can be doubled if the pulse is reflected by the mirror M and passes the grating pair again. Pulse widths of 16 fs have been obtained with such a system [11.61].

With a combination of prisms and gratings (Fig. 11.27) not only the quadratic but also the cubic term in the phase dispersion

$$\phi(\omega) = \phi(\omega_0) + \left(\frac{\partial \phi}{\partial \omega}\right)_{\omega_0} (\omega - \omega_0) + \frac{1}{2}\left(\frac{\partial^2 \phi}{\partial \omega^2}\right)_{\omega_0} (\omega - \omega_0)^2$$
$$+ \frac{1}{6}\left(\frac{\partial^3 \phi}{\partial \omega^3}\right)_{\omega_0} (\omega - \omega_0)^3 ,$$

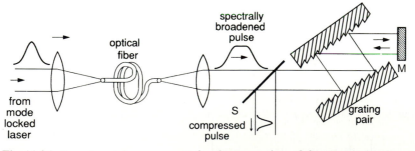

Fig. 11.26. Experimental arrangement for the generation of femtosecond pulses by self-phase-modulation with subsequent pulse compression by a grating pair [11.60]

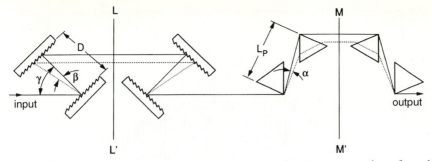

Fig. 11.27. Sequence of grating pairs and prism pairs for the compensation of quadratic and cubic phase dispersion. LL′ and MM′ are two phase-fronts. The solid line represents a reference path and the dashed line illustrates the paths for the wave of wavelength λ, which is diffracted by an angle β at the first grating and refracted by an angle α against the reference path in a prism [11.62]

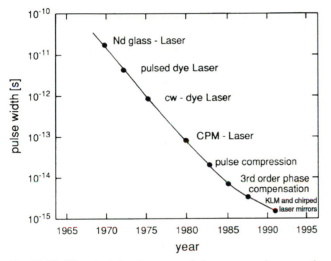

Fig. 11.28. Historical development of the progress in generating ultrashort pulses

can be compensated [11.62]. This allows one to reach pulsewidths of 6 fs. Figure 11.28, which illustrates the historical development on the way to the Fourier limit of ultrashort pulses, shows the progress achieved during the last 20 years.

Note: Extrapolation into the future might be erroneous [11.47].

11.1.7 Sub 10-fs Pulses with Chirped Laser Mirrors

The development of broadband saturable semiconductor absorber mirrors and of dispersion-engineered chirped multilayer dielectric mirrors has allowed the realization of self-starting ultrashort laser pulses, which routinely reach sub-10-fs pulsewidths and peak powers above the megawatt level.

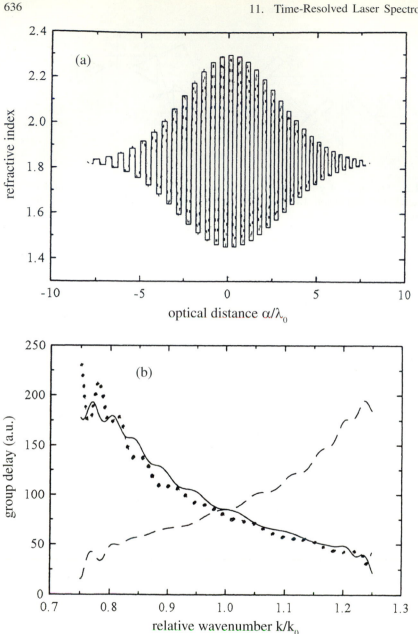

Fig. 11.29. (a) Refractive index profile of a discrete-valued chirped dielectric mirror; **(b)** group delay of chirped mirrors with positively (*dashed*) and negatively (*solid line*) chirped graded index profile. The *dotted curve* corresponds to the discrete index profile of (a) [11.64]

Femtosecond pulse generation relies on a net negative intracavity group-delay dispersion (GDD). Since solid-state gain media introduce a frequency-dependent positive dispersion, this must be overcompensated by media inside the laser cavity that have a correspondingly large negative dispersion. We saw in Sect. 11.1.5 that intracavity prisms can be used as such compensators. However, here the GDD shows a large wavelength dependence and for very short pulses (with a corresponding broad spectral range) this results in asymmetric pulse shapes with broad pedestals in the time domain. The invention of chirped dielectric low-loss laser mirrors has brought a substantial improvement [11.63].

These mirrors may be regarded as one-dimensional holograms that are generated when a chirped and an unchirped laser pulse from opposite directions are superimposed in a medium where they generate a refractive index pattern proportional to their total intensity [11.64]. When a chirped pulse is reflected by such a hologram, it becomes compressed, similar to the situation with phase-conjugated mirrors.

In practice, such mirrors are produced by evaporation techniques controlled by a corresponding computer program. In Fig. 11.29a the variation of the refractive index for the different dielectric layers is shown for a mirror with negative GDD, and in Fig. 11.29b the group delay is plotted as a function of wavelength for mirrors with negatively and positively chirped graded-index profiles. In combination with Kerr lens mode locking such chirped mirrors allow the generation of femosecond pulses down to 4 fs [11.54].

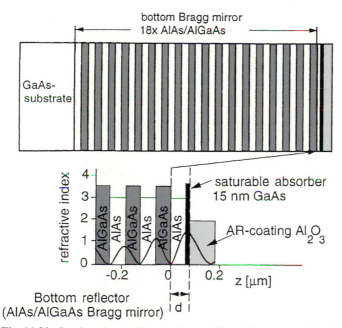

Fig. 11.30. Semiconductor Bragg mirror with a 15-nm saturable absorber layer of GaAs [11.65]

Another alternative for the generation of ultrafast pulses is the passive mode locking by fast semiconductor saturable absorbers in front of chirped mirrors in combination with Kerr lens mode locking [11.65]. The recovery time of the saturable absorber must be generally faster then the laser pulse width. This is provided by KLM, which may be regarded as artificial saturable absorber that is as fast as the Kerr nonlinearity following the laser intensity. Since the recovery time in a semiconductor material is given by the relaxation of the excited electrons into the initial state in the valence band, they represent fast saturable absorbers in the sub-picosecond range, but do not reach the 10-fs limit. Nevertheless, soliton-like pulses down to 13 fs have been achieved.

A combination of saturable semiconductor media in front of a chirped mirror and KLM techniques can be used for reliable operation of sub-10-fs pulses. In Fig. 11.30 such a multilayer saturable absorber is illustrated.

11.1.8 Fiber Lasers and Optical Solitons

In the Sect. 11.1.7 we discussed the self-phase modulation of an optical pulse in a fiber because of the intensity-dependent refractive index $n = n_0 + n_2 I(t)$. While the resultant spectral broadening of the pulse leads in a medium with normal negative dispersion $(dn_0/d\lambda < 0)$ to a spatial broadening of the pulse, an anomalous linear dispersion $(dn_0/d\lambda > 0)$ would result in a pulse compression. Such anomalous dispersion can be found in fused quartz fibers for $\lambda > 1.3\,\mu m$ [11.66, 11.67]. For a suitable choice of the pulse intensity the dispersion effects caused by $n_0(\lambda)$ and by $n_2 I(t)$ may cancel, which means that the pulse propagates through the medium without changing its time profile. Such a pulse is named a fundamental soliton [11.68, 11.69].

Introducing the refractive index $n = n_0 + n_2 I$ into the wave equation (11.19) yields stable solutions that are called *solitons* of order N. While the fundamental soliton $(N = 1)$ has a constant time profile $I(t)$, the higher-order solitons show an oscillatory change of their time profile $I(t)$: the pulsewidth decreases at first and then increases again. After a path length z_0, which depends on the refractive index of the fiber and on the pulse intensity, the soliton recovers its initial form $I(t)$ [11.70, 11.71].

Optical solitons in fused quartz fibers can be utilized to achieve stable femtosecond pulses in broadband infrared lasers, such as the color-center laser or the Ti:sapphire laser. Such a system is called a *soliton laser* [11.72–11.79]. Its experimental realization is shown in Fig. 11.31.

A KCl:Tl° color-center laser is synchronously pumped by a mode-locked Nd:YAG laser. The output pulses of the color-center laser at $\lambda = 1.5\,nm$ pass the beam splitter S. A fraction of the intensity is reflected by S and is focused into an optical fiber where the pulses propagate as solitons, because the dispersion of the fiber at $1.5\,\mu m$ is $dn/d\lambda > 0$. The pulses are compressed, are reflected by M5, pass the fiber again, and are coupled back into the laser resonator. If the length of the fiber is adjusted properly, the transit time along the path M^0–S–M_5–S–M_0 just equals the round-trip time $T = 2d/c$ through

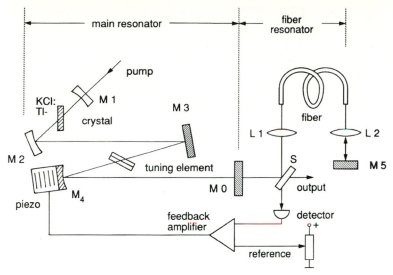

main resonator

fiber resonator

pump

KCl: Tl-

M 1

crystal

M 3

fiber

L 1

L 2

M 2

tuning element

S

M 5

piezo

M$_4$

M O

output

feedback amplifier

detector

reference

Fig. 11.31. Soliton laser [11.79]

the laser resonator M_0–M_1–M_0. In this case compressed pulses are always injected into the laser resonator at the proper times $t = t_0 + q2d/c$ ($q = 1, 2, \ldots$) to superimpose the pulses circulating inside the laser cavity. This injection of shortened pulses leads to a decrease of the laser pulse width until other broadening mechanisms, which increase with decreasing pulse width, compensate the pulse shortening.

In order to match the phase of the reflected fiber light pulse to that of the cavity internal pulse, the two resonator path lengths have to be equal within a small fraction of a wavelength. The output power, transmitted through S to the detector, critically depends on the proper matching of both cavity lengths and can therefore be used as feedback control for stabilizing the length of the laser resonator, which is controlled by the position of M4 mounted on a piezo cylinder. It turns out that the best stabilization can be achieved with solitons of order $N \geq 2$ [11.73].

With such a KCl:T1° color-center soliton laser, stable operation with pulse widths of 19 fs was demonstrated [11.78]. This corresponds at $\lambda = 1.5\,\mu m$ to only four oscillation periods of the infrared wave. More about soliton lasers can be found in [11.72–11.80].

The fabrication of rare-earth-doped optical fibers with a wide bandwidth gain have pushed the development of optical fiber amplifiers. This large bandwidth together with the low pump power requirements facilitated the realization of passively mode-locked femtosecond fiber lasers. The advantages of such fiber lasers are their compact setup with highly integrated optical components, their reliability, and their prealignment, which makes their daily operation more convenient [11.75].

The basic principle of a fiber ring laser is schematically shown in Fig. 11.32. The pump laser is coupled into the fiber ring laser through a fiber

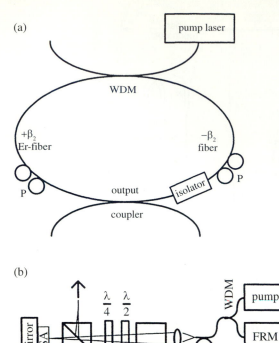

(a)

(b)

(c)

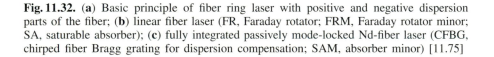

Fig. 11.32. (a) Basic principle of fiber ring laser with positive and negative dispersion parts of the fiber; **(b)** linear fiber laser (FR, Faraday rotator; FRM, Faraday rotator minor; SA, saturable absorber); **(c)** fully integrated passively mode-locked Nd-fiber laser (CFBG, chirped fiber Bragg grating for dispersion compensation; SAM, absorber minor) [11.75]

splice and the output power is extracted through a second splice. The fiber ring consists of negative $(-\beta_2)$ and positive $(+\beta_2)$ dispersion parts with an erbium amplifying medium. The isolator enforces unidirectional operation and fiber loops preserve the polarization state. Instead of ring fiber lasers, linear fiber lasers have also been realized, as shown in Fig. 11.32b. The saturable absorber in front of the end mirror allows passive mode locking and results in femtosecond operation. An example of a fully integrated fiber femtosecond laser is given in Fig. 11.32c, where the saturable absorber is butted directly to a fiber end, and a chirped fiber Bragg grating (CFBG) is used for dispersion compensation.

Output pulses with about 70-µJ pulse energy and pulse widths below 100 fs have been generated with such fiber lasers [11.76].

Soliton ring fiber lasers can be also realized with active mode locking by polarization modulation [11.77], or by additive-pulse mode locking (APM). In the latter technique the pulse is split into the two arms of an interferometer and the coherent superposition of the self-phase modulated pulses results in pulse shortening [11.55].

11.1.9 Shaping of Ultrashort Light Pulses

For many applications a specific time profile of short laser pulses is desired. One example is the coherent control of chemical reactions (see Sect. 15.2) Recently, some techniques have been developed that allow such pulse shaping and that work as follows:

The output pulses of a femtosecond laser are reflected by an optical diffraction grating. Because of the large spectral bandwidth of the pulse, the different wavelengths are diffracted into different directions (Fig. 11.33). The divergent beam is made parallel by a lens and is sent through a liquid crystal display (LCD), which is formed by a two-dimensional thin array of liquid crystal pixels with transparent electrodes. If a voltage is applied to a pixel, the liquid crystal changes its refractive index and with it the phase of the transmitted partial wave. Therefore the phase front of the transmitted pulse differs from that of the incident pulse, depending on the individual voltages applied to the different pixels. A second lens recombines the dispersed partial waves, and reflection by a second grating again overlaps the different wavelengths.

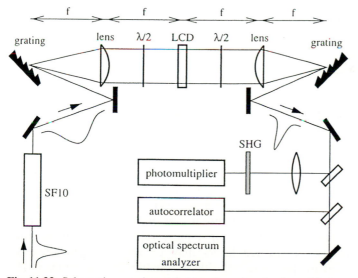

Fig. 11.33. Schematic experimental setup for pulse shaping of femtosecond pulses [11.83]

laser pulse E'(t)

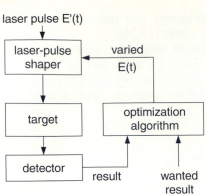

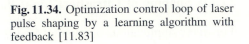

Fig. 11.34. Optimization control loop of laser pulse shaping by a learning algorithm with feedback [11.83]

This results in a light pulse with a time profile that depends on the phase differences between its spectral components, which in turn can be controlled by the LCD, driven by a special computer program (Fig. 11.34) [11.81, 11.82]. A self-learning algorithm can be incorporated into the closed loop, which compares the output pulse form with the wanted one and tries to vary the voltage at the different pixels in such a way that the wanted pulse form is approximated [11.83]. More details can be found in [11.84]

11.1.10 Generation of High-Power Ultrashort Pulses

The peak powers of ultrashort light pulses, which are generated by the techniques discussed in the previous section, are for many applications not high enough. Examples where higher powers are required are nonlinear optics and the generation of VUV ultrashort pulses, multiphoton ionization and excitation of multiply charged ions, the generation of high-temperature plasmas for optical pumping of X-ray lasers, or industrial applications in short-time material processing. Therefore, methods must be developed that increase the energy and peak power of ultrashort pulses. One solution of this problem is the amplification of the pulses in dye cells that are pumped by pulsed, powerful lasers, such as excimer, Nd:YAG, or Nd:glass lasers (Fig. 11.35). The dispersion of the amplifying cells leads to a broadening of the pulse, which can, however, be compensated by pulse compression with a grating pair. In order to suppress optical feedback by reflection from cell windows and amplification of spontaneous emission, saturable absorber cells are placed between the amplifying states, which are saturated by the wanted high-power pulses but suppress the weaker fluorescence [11.85–11.89].

A serious limitation is the low repetition rate of most pump lasers used for the amplifier chain. Although the input pulse rate of the pico- or femtosecond pulses from mode-locked lasers may be many megahertz, most solid-state lasers used for pumping only allow repetition rates below 1 kHz. Copper-vapor lasers can be operated up to 20 kHz. Recently, a multi-kilohertz

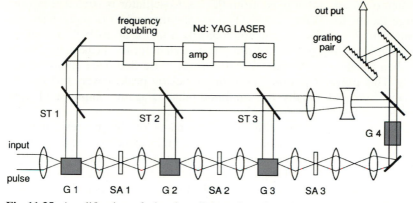

Fig. 11.35. Amplification of ultrashort light pulses through a chain of amplifier dye cells G1–G4, pumped by a frequency-doubled pulsed Nd:YAG laser. Saturable absorbers SA1–SA3 are placed between the amplifier cells in order to prevent feedback by reflection and to suppress amplified spontaneous emission

Ti:Al$_2$O$_3$ amplifier for high-power femtosecond pulses at $\lambda = 764$ nm has been reported [11.90].

Over the past ten years new concepts have been developed that have increased the peak power of short pulses by more than four orders of magnitude, reaching the terawatt (10^{12} W) or even the petawatt (10^{15} W) regime [11.89, 11.90–11.95]. One of these methods is based on chirped pulse amplification, which works as follows (Fig. 11.36):

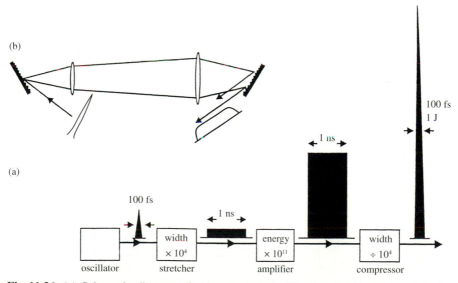

Fig. 11.36. (a) Schematic diagram of chirped pulse amplification (CPA); **(b)** possible design of pulse stretcher with a grating pair [11.95]

The femtosecond output pulse from the laser oscillator is stretched in time by a large factor, for example, 10^4. This means that a 100-fs pulse becomes 1-ns long. This long pulse is now amplified by a factor up to 10^{10}, which increases its energy, but, because of the pulse stretching, the peak power is far smaller than for the case where the initial short pulse had been amplified accordingly. This prevents destruction of the optics by peak powers that exceed the damage threshold. Finally, the amplified pulse is again compressed before it is sent to the target.

We will now discuss the different components of this process in more detail. The oscillator consists of one of the femtosecond devices discussed previously. The pulse stretcher uses a grating pair, where the two gratings, however, are not parallel as for pulse compression, but are tilted against each other (Fig. 11.36b). This increases the path difference between the blue and the red components in the pulse and stretches the pulse length. An aberration-free pulse stretcher with two curved mirrors and a grating is described in [11.96] and is depicted in Fig. 11.37.

The amplification is performed in a multipass amplifier system. Here the stretched laser pulse is sent many times through a gain medium, which is pumped by a nanosecond pump pulse from a Nd:YAG laser (Fig. 11.38). For amplification of Ti:sapphire pulses a highly doped Ti:sapphire crystal serves as the gain medium. After each transit of the stretched pulse through the gain medium, which depletes the inversion, the pump pulse regenerates it again. Often the system is designed in such a way that the different transits pass at slightly different locations through the amplifying medium. The number of

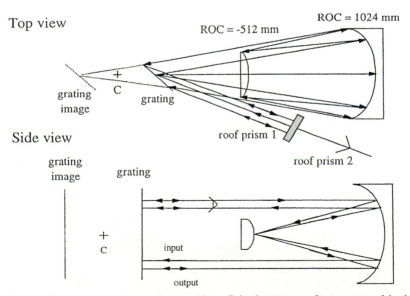

Fig. 11.37. Aberration-free pulse stretcher. C is the center of curvature of both mirrors, ROC is the radius of curvature [11.96]

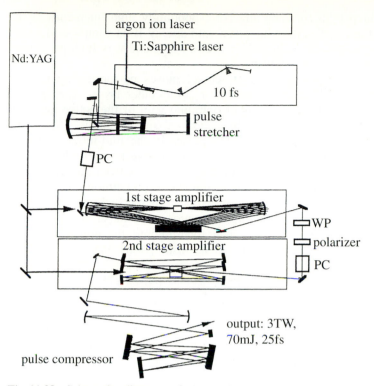

Fig. 11.38. Schematic diagram of the oscillator–amplifier system of a 3-TW 10-Hz Ti:sapphire CPA laser [11.89]

transits depends on the geometrical mirror arrangement and is limited by the duration of the pump pulse.

A second amplifier stage can be used for higher amplification. The different stages are separated by an optical diode, which prevents back reflections into the previous stage. A Pockels cell (PC) selects the amplified pulses with a repetition rate of a few kilohertz (limited by the pump laser) from the many more nonamplified pulses of the oscillator (at a repetition rate of about 80 MHz). The upper pulse energy that can be extracted from the amplifying medium is given by the saturation fluence, which depends on the emission cross section of the medium. For Ti:sapphire, for example, the highest achievable intensity is about $100\,\mathrm{TW/cm^2}$.

During the multipass transits the spatial mode quality of the pulse might be deteriorated, which means that the pulse cannot be tightly focused into the target. If the multipass design forms a true resonant cavity and the incoming pulses of the pump laser and the oscillator are both carefully mode-matched to the fundamental Gaussian mode of this cavity, the resonator will only support the $\mathrm{TEM_{00}}$ modes and the system acts as a spatial filter, because all other transverse modes will not be amplified. Such a regenerative amplifier preserves the spatial Gaussian pulse profile, which allows one to achieve diffraction-limited tight focusing and correspondingly high intensities in the focal plane. After amplification the pulse with energy W is compressed again, thus producing pulses with a duration $\tau = 20-100$ fs and very high peak powers $P = W/\tau$ up to several terawatts.

Most of the experiments on femtosecond pulses performed up to now have used dye lasers, Ti:sapphire lasers, or color-center lasers. The spectral ranges were restricted to the regions of optimum gain of the active medium. New spectral ranges can be covered by optical mixing techniques (Sect. 5.8). One example is the generation of 400-fs pulses in the mid-infrared around $\lambda = 5\,\mu m$ by mixing the output pulses from a CPM dye laser at 620 nm with pulses of 700 nm in a $LiIO_3$ crystal [11.93]. The pulses from the CPM laser are amplified in a six-pass dye amplifier pumped at a repetition rate of 8 kHz by a copper-vapor laser. Part of the amplified beam is focused into a traveling-wave dye cell [11.94], where an intense femtosecond pulse at $\lambda = 700$ nm is generated. Both laser beams are then focused into the nonlinear $LiIO_3$ mixing crystal, which delivers output pulses with 10-mJ energy and 400-fs pulsewidth.

For widely tunable, high-power femtosecond pulses either optical parametric oscillators [11.96] or "white-light" sources are used. By focusing high-power laser radiation into water, self-phase modulation leads to the generation of intense white-light pulses, which can be further amplified in laser-pumped amplifier chains [11.97]. The wanted spectral range can be selected either by choosing the proper amplifier medium, which only amplifies a restricted spectral interval, or by gratings or prisms. Such white-light femtosecond pulses are very useful for the investigation of ultrashort transient phenomena in molecules and solids (Sect. 11.4). For more information see [11.98, 11.99].

11.2 Measurement of Ultrashort Pulses

During recent years the development of fast photodetectors has made impressive progress. For example, PIN photodiodes (Sect. 4.5) are available with a rise time of 20 ps [11.100]. However, until now the only detector that reaches a time resolution slightly below 1 ps is the streak camera [11.101]. Femtosecond pulses can be measured with optical correlation techniques, even if the detector itself is much slower. Since such correlation methods represent the standard technique for measuring of ultrashort pulses, we will discuss them in more detail.

11.2.1 Streak Camera

The basic principle of a streak camera is schematically depicted in Fig. 11.39. The optical pulse with the time profile $I(t)$ is focused onto a photocathode, where it produces a pulse of photoelectrons $N_{PE}(t) \propto I(t)$. The photoelectrons are extracted into the z-direction by a plane grid at the high voltage U. They are further accelerated and imaged onto a luminescent screen at $z = z_s$. A pair

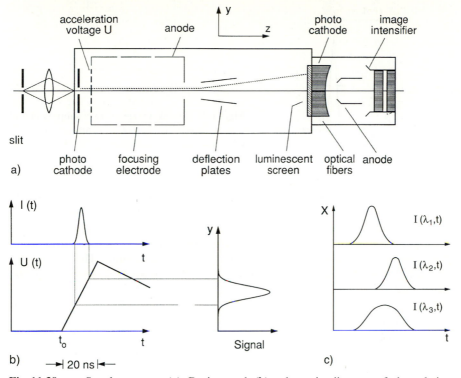

Fig. 11.39a–c. Streak camera. (**a**) Design and (**b**) schematic diagram of the relation between the time profile $I(t)$ and the spatial distribution $S(y)$ at the output plane; (**c**) spectrally resolved time profiles $I(\lambda, t)$

of deflection plates can deflect the electrons into the y-direction. If a linear voltage ramp $U_y(t) = U_0 \cdot (t - t_0)$ is applied to the deflection plates, the focal point of the electron pulse $(y_s(t), z_s)$ at the screen in the plane $z = z_s$ depends on the time t when the electrons enter the deflecting electric field. The spatial distribution $N_{PE}(y_s)$ therefore reflects the time profile $I(t)$ of the incident light pulse (Fig. 11.39b).

When the incident light is imaged onto a slit parallel to the x-direction the electron optics transfer the slit image to the luminescent screen S. This allows the visualization of an intensity–time profile $I(x, t)$, which might depend on the x-direction. For example, if the optical pulse is at first sent through a spectrograph with a dispersion $d\lambda/dx$, the intensity profile $I(x, t)$ reflects for different values of x the time profiles of the different spectral components of the pulse. The distribution $N_{PE}(x_s, y_s)$ on the screen S then yields the time profiles $I(\lambda, t)$ of the spectral components (Fig. 11.39c). The screen S is generally a luminescent screen (like the oscilloscope screen), which can be viewed by a video camera either directly or after amplification through an image intensifier. Often microchannel plates are used instead of the screen.

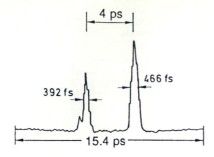

The start time t_0 for the voltage ramp $U_y = (t - t_0)U_0$ is triggered by the optical pulse. Since the electronic device that generates the ramp has a finite start and rise time, the optical pulse must be delayed before it reaches the cathode of the streak camera. This assures that the photoelectrons pass the deflecting field during the linear part of the voltage ramp. The optical delay can be realized by an extra optical path length such as a spectrograph (Sect. 11.4.1).

In commercial streak cameras the deflection speed can be selected between 1 cm/100 ps to 1 cm/10 ns. With a spatial resolution of 0.1 nm, a time resolution of 1 ps is achieved. A femtosecond streak camera has been developed [11.101] that has a time resolution selectable between 400 fs to 8 ps over a spectral range 200−850 nm. Figure 11.40 illustrates this impressive resolution by showing the streak camera screen picture of two femtosecond pulses which are separated by 4 ps. More details can be found in [11.100−11.103].

11.2.2 Optical Correlator for Measuring Ultrashort Pulses

For measurements of optical pulse widths below 1 ps the best choice is a correlation technique that is based on the following principle: the optical pulse with the intensity profile $I(t) = c\epsilon_0|E(t)|^2$ and the halfwidth ΔT is split into two pulses $I_1(t)$ and $I_2(t)$, which travel different path lengths s_1 and s_2 before they are again superimposed (Fig. 11.41). For a path difference $\Delta s = s_1 - s_2$ the pulses are separated by the time interval $\tau = \Delta s/c$ and their coherent superposition yields the total intensity

$$I(t, \tau) = c\epsilon_0[E_1(t) + E_2(t - \tau)]^2 .$$
(11.24)

A *linear* detector has the output signal $S_L(t) = aI(t)$. If the time constant T of the detector is large compared to the pulse length ΔT, the output signal is

$$S_L(\tau) = a\langle I(t, \tau)\rangle = \frac{a}{T} \int_{-T/2}^{+T/2} I(t, \tau)\, dt .$$
(11.25)

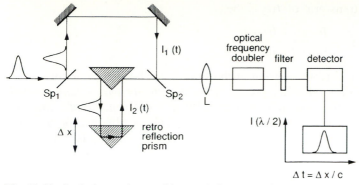

Fig. 11.41. Optical correlator with translation-retroreflecting prism and second-harmonic generation

For equal amplitudes of the two pulses $E_1(t) = E_2(t) = E_0(t)\cos\omega t$, the integrand becomes

$$I(t, \tau) = c\epsilon_0 [E_0(t)\cos\omega t + E_0(t - \tau)\cos\omega(t - \tau)]^2 , \tag{11.26}$$

and the integration yields

$$S_L(\tau) = 2c\epsilon_0 a \left\{ \langle E_0^2 \rangle + \frac{1}{T} \int\limits_{-\frac{1}{2}T}^{+\frac{1}{2}T} [E_0(t) E_0(t - \tau)\cos\omega t \cos\omega(t - \tau)]\, dt \right\} . \tag{11.27}$$

For strictly monochromatic cw light $(E_0(t) = \text{const})$ the integral equals $T E_0^2 \cos\omega\tau$, and the signal becomes for $T \to \infty$ the oscillatory function of τ

$$S_L(\tau) = 2ac\epsilon_0 E_0^2 (1 + \cos\omega\tau) , \tag{11.28}$$

with the period $\Delta\tau = \pi/\omega = 1/(2\nu) = \lambda/2c$ (two-beam interference, see Sect. 4.2).

Mode-locked pulses of duration ΔT and spectral width $\Delta\omega \approx 2\pi/\Delta T$ are composed of many modes with different frequencies ω. The oscillations of these modes have different periods $\Delta\tau(\omega) = \pi/\omega \ll T$, and after a time $t \geq \tau_c$ their phases differ by more than π and their amplitudes cancel. In other words, the coherence time is $\tau_c \leq \Delta T$ and interference can only be observed for delay times $\tau \leq \Delta T$.

The normalized signal (11.25) is

$$I(\tau) = \frac{\int |E(t) + E(t - \tau)|^2 \, dt}{2 \int |E(t)|^2 \, dt} = 1 + \frac{\int E^*(t) E(t - \tau) + E(t) E^*(t - \tau) \, dt}{2 \int |E(t)|^2 \, dt} , \tag{11.29}$$

which is independent of the pulse form (see Sect. 12.8.1).

The Fourier transform of $I(t)$ is then

$$F(I(\tau)) = F(1) + \frac{I(-\omega) + I(\omega)}{2 \int I(t)\,dt} , \qquad (11.30)$$

which gives the intensity spectrum of the laser pulse but does not contain the wanted information on the time profile. *A linear detector with a time constant $T \gg \tau$ would therefore give an output signal that is independent of τ and that yields no information on the time profile $I(t)$!* This is obvious, because the detector measures only the integral over $I_1(t) + I_2(t+\tau)$, that is, the sum of the energies of the two pulses, which is independent of the delay time τ as long as $T > \tau$. Therefore *linear detectors* with the time resolution T cannot be used for the measurement of time profiles of ultrashort pulses with $\Delta T < T$.

If, however, the two noncollinear pulses are focused into a nonlinear optical crystal which doubles the optical frequency, the intensity of the second harmonics $I(2\omega) \propto (I_1 + I_2)^2$ is proportional to the *square* of the incident intensity (Sect. 5.7) and the measured averaged signal $S(2\omega, \tau) \propto I(2\omega, \tau)$ becomes

$$\langle S_{\mathrm{NL}}(2\omega, \tau) \rangle = \frac{a}{T} \int\limits_{-T/2}^{+T/2} I(2\omega, t, \tau)\,dt ,$$

$$= a \left[\langle I_1^2 \rangle + \langle I_2^2 \rangle + 2\langle I_1(t) I_2(t+\tau) \rangle \right] . \qquad (11.31)$$

The first two terms are independent of τ and give a constant background. However, the third term depends on the delay time τ and contains information on the pulse profile $I(t)$. The detector signal $S(2\omega, \tau)$ measured versus the delay time τ therefore allows the determination of the time profile of $I(\omega, t)$.

Note the difference between linear detection (11.25) and nonlinear detection (11.31). With linear detection the *sum* $I_1(t) + I_2(t+\tau)$ is measured, which is independent of τ as long as $\tau < T$. The nonlinear detector measures the signal $S(2\omega, \tau)$ that contains the *product* $I_1(t) I_2(t+\tau)$, which does depend on τ as long as τ is smaller than the maximum width of the pulses.

All these devices, which are called *optical correlators*, measure the correlation between the field amplitude $E(t)$ or the intensity $I(t)$ at the time t and its values $E(t+\tau)$ or $I(t+\tau)$ at a later time. These correlations are mathematically expressed by normalized correlation functions of order k. The normalized first order correlation function

$$G^{(1)}(\tau) = \frac{\int_{-\infty}^{+\infty} E(t) \cdot E(t+\tau)\,dt}{\int_{-\infty}^{+\infty} E^2(t)\,dt} = \frac{\langle E(t) \cdot E(t+\tau) \rangle}{\langle E^2(t) \rangle} , \qquad (11.32)$$

describes the correlation between the field amplitudes at times t and $t+\tau$. From (11.32) we obtain $G^{(1)}(0) = 1$, and for pulses with a finite pulse duration ΔT (11.32) yields $G^{(1)}(\infty) = 0$.

The normalized second-order correlation function

$$G^{(2)}(\tau) = \frac{\int I(t) \cdot I(t+\tau)\,dt}{\int I^2(t) \cdot dt} = \frac{\langle I(t) \cdot I(t+\tau)\rangle}{\langle I^2(t)\rangle}, \tag{11.33}$$

describes the *intensity* correlation, where again $G^{(2)}(0) = 1$. The correlation signal (11.31) after the optical frequency doubler can be written in terms of $G^{(2)}(\tau)$ for $I_1 = I_2 = I/2$ as

$$S_{NL}(2\omega, \tau) = A[G^{(2)}(0) + 2G^{(2)}(\tau)] = A[1 + 2G^{(2)}(\tau)]. \tag{11.34}$$

There are two different techniques of measuring the time profiles of ultrashort laser pulses: the interferometric autocorrelation and the noncollinear intensity correlation.

a) Interferometric Autocorrelation

In interferometric autocorrelation the coherent superposition of the two collinear partial beams is realized. The basic principle is shown in Figs. 11.41 and 11.42. The incoming laser pulse is split by the beamsplitter Sp1 into two parts, which pass through two different pathlengths and are then collinearly superimposed at Sp2. When they are focused by the lens L into a nonlinear optical crystal, the output signal (11.31) is generated. Instead of the delay line arrangement in Fig. 11.41 a Michelson interferometer in Fig. 11.42 can also be used. The second harmonics are detected by a photomultiplier, while the fundamental wavelength is rejected by a filter.

The nonlinear crystal can be omitted, if the detector itself has a nonlinear response. This is, for instance, the case for a semiconductor detector with a band gap $\Delta E > h\nu$, where only two-photon absorption contributes to the signal.

A typical signal as a function of the delay time τ is shown in Fig. 11.43. The second-order autocorrelation (11.31) gives for $E_1 = E_2 = E_0(t)\exp(i\omega t)$ the signal

$$S(2\omega, \tau) = 2I_0 + 4I^{(1)}(\tau) + 4\,\mathrm{Re}\left[I^{(2)}(\tau) + I^{(2)*}(-\tau)\,e^{i\omega_0\tau}\right]$$

$$+ 2\,\mathrm{Re}\left[I^{(3)}(\tau)\,e^{2i\omega_0\tau}\right], \tag{11.35}$$

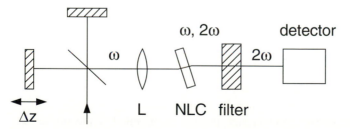

Fig. 11.42. Michelson interferometer for interferometric autocorrelation

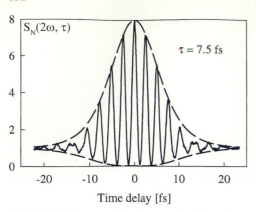

Fig. 11.43. Measured femtosecond pulse with upper and lower envelopes [11.99]

where the different terms are

$$I_0 = \int |E_0(t)|^4 \, dt \,, \quad I^{(1)}(\tau) = \int |E_0(t) E_0(t-\tau)|^2 \, dt \,,$$

$$I^{(2)}(\tau) = \int |E_0(t)|^2 E_0(t) E_0^*(t-\tau) \, dt \,,$$

$$I^{(3)}(\tau) = \int E_0^2(t)(E_0^2(t-\tau))^* \, dt \,.$$

Often the normalized interferometric autocorrelation function

$$I_N^{(2)}(\tau) = \frac{I_{2\omega}^{(2)}(\tau)}{I_{2\omega}^{(2)}(\infty)} \,, \tag{11.36}$$

is introduced, which yields the measured second-harmonic signal

$$S_N(2\omega, \tau) = 1 + \frac{1}{I^{(1)}(0)}\left[2I^{(1)}(\tau) + 2\,\mathrm{Re}\left\{ I^{(2)}(\tau) + I^{(2)}(-\tau)\,e^{i\omega_0\tau} \right\} \right.$$

$$\left. + \mathrm{Re}\left\{ I^{(3)}\, e^{i2\omega_0\tau} \right\} \right]. \tag{11.37}$$

Note: $S_N(2\omega, \tau)$ is symmetric, that is, $S_N(\tau) = S_N(-\tau)$. A possible asymmetry of the pulse **does not** show up in the signal S_N.

For $\tau = \infty$ the normalized signal becomes $S(2\omega, \infty) = 1$, while for $\tau = 0$ all terms in (11.37) are equal:

$$I^{(1)}(0) = I^{(2)}(0) = I^{(3)}(0) = I_0 = \int |E_0(t)|^4 \, dt \,.$$

The upper envelope of this interference pattern is obtained if the phase $\omega_0\tau$ is replaced by the constant phase 2π, the lower envelope for $\omega_0\tau = \pi$. The maximum signal is $S_N^{max}(2\omega, \tau) = 8$, while the background $S(2\omega, \infty) = 1$. For $\omega_0 \tau = \pi$, the minimum value is $S_N^{min}(2\omega, \tau) = 0$ (Fig. 11.43).

b) Noncollinear Intensity Correlation

Without background suppression (11.34) yields for completely overlapping pulses $S(2\omega, \tau = 0) = 3A$, and for completely separated pulses $S(2\omega, \tau \gg T) = A$.

The τ-independent background in (11.31) can be suppressed when the two beams are focused into the doubling crystal under different angles $\pm\beta/2$ against the z-direction (Fig. 11.44) where the signal $S(2\omega)$ is detected. If the phase-matching conditions for the doubling crystal (Sect. 5.7) are chosen in such a way that for two photons out of the same beam no phase matching occurs, but only for one photon out of each beam, then the two first terms in (11.31) do not contribute to the signal [11.104, 11.105]. In another method of background-free pulse measurement the polarization plane of one of the two beams in Fig. 11.44 is turned in such a way that a properly oriented doubling crystal (generally, a KDP crystal) fulfills the phase-matching condition only if the two photons each come from a different beam [11.106]. In this noncollinear scheme no interference occurs and the measured signal equals the envelope of the pulse profile in Fig. 11.43.For methods with background suppression no signal is obtained for $\tau \gg T$.

In the methods discussed above, one of the retroreflectors is mounted on a translational stage moved via micrometer screws by a step motor while the signal $S(2\omega, \tau)$ is recorded. Since τ must be larger than ΔT, the translational move should be at least $\Delta S = \frac{1}{2}c\tau \geq \frac{1}{2}c\Delta T$. For pulses of 10 ps this means $\Delta S \geq 1.5$ mm. With a rotating correlator (Fig. 11.45) the signal $S(2\omega, \tau)$ can be directly viewed on a scope, which is very useful when optimizing the pulse width. Two retroreflecting prisms are mounted on a rotating disc. During

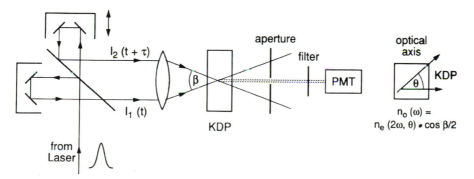

Fig. 11.44. Background-free measurement of the second-order correlation function $G^{(2)}(\tau)$ by choosing the phase-matching condition properly (SF, spectral filter neglecting scattered light of the fundamental wave at ω; KDP, potassium-dihydrogen-phosphate crystal for frequency doubling)

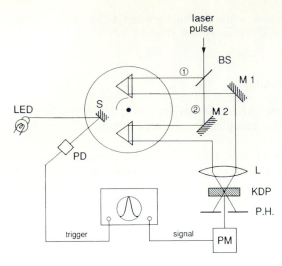

Fig. 11.45. Rotating autocorrelator, allowing the direct observation of the correlation signal $S(2\omega, \tau)$ on a scope that is triggered by the output signal of the photodiode PD

a certain fraction ΔT_{rot} of the rotation period T_{rot} the reflected beams reach the mirrors M1 and M2 and are focused into the KDP crystal. The viewing oscilloscope is triggered by a pulse obtained by reflecting the light of a LED onto the photodetector PD.

Instead of using optical frequency doubling other nonlinear effects can also be used, such as two-photon absorption in liquids or solids, which can be monitored by the emitted fluorescence. If the optical pulse is again split into two pulses traveling in the opposite $\pm z$-directions through the sample cell (Fig. 11.46), the spatial intensity profile $I_{\mathrm{FL}}(z) \propto I^2(\omega, \tau)$ can be imaged by magnifying optics onto a vidicon or an image intensifier. Since a pulse width

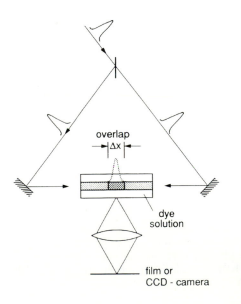

Fig. 11.46. Measurement of short pulses via two-photon induced fluorescence

Table 11.2. Ratios $\Delta\tau/\Delta T$ of the width $\Delta\tau$ of the autocorrelation profile and ΔT of the pulse $I(t)$, and products $\Delta\nu \cdot \Delta T$ of spectral widths and duration ΔT of pulses with different profiles $I(t)$

Pulse profile	Mathematical expression for $I(t)$	$\Delta\tau/\Delta T$	$\Delta\nu \cdot \Delta T$
Rectangular	$\begin{cases} I_0 \text{ for } 0 \le t \le \Delta T, \\ 0 \text{ elsewhere} \end{cases}$	1	0.886
Gaussian	$I_0 \exp[-t^2/(0.36\Delta T^2)]$	$\sqrt{2}$	0.441
Sech^2	$\mathrm{sech}^2(t/0.57\Delta T)$	1.55	0.315
Lorentzian	$[1 + (2t/\Delta T)^2]^{-1}$	2	0.221

$\Delta T = 1\,\mathrm{ps}$ corresponds to a path length of $0.3\,\mathrm{mm}$, this technique, which is based on the spatial resolution of the fluorescence intensity, is limited to pulse widths $\Delta T \ge 0.3\,\mathrm{ps}$. For shorter pulses the delay time τ between the pulses has to be varied and the total fluorescence

$$I_{\mathrm{FL}}(\tau) = \int I(z,\tau)\,\mathrm{d}z , \tag{11.38}$$

has to be measured as a function of τ [11.107, 11.108].

It is important to note that the profile $S(\tau)$ of the correlation signal depends on the time profile $I(t)$ of the light pulse. It gives the correct pulse width ΔT only if an assumption is made about the pulse profile. For illustration, Fig. 11.47a depicts the signal $S(2\omega, \tau)$ of Fourier-limited pulses with the Gaussian profile $I(t) = I_0 \exp(-t^2/0.36\Delta T^2)$ with and without background suppression. From the halfwidth $\Delta\tau$ of the signal the halfwidth ΔT of the pulses can only be derived if the pulse profile $I(t)$ is known. In Table 11.2 the ratio $\Delta\tau/\Delta T$ and the products $\Delta T \cdot \Delta\nu$ are compiled for different pulse profiles $I(t)$, while Fig. 11.47a–c illustrates the corresponding profiles and contrasts of $G^{(2)}(\tau)$. Even noise pulses and continuous random noise result in a maximum of the correlation function $G^{(2)}(\tau)$ at $\tau = 0$ (Fig. 11.47d), and the contrast becomes $G^{(2)}(0)/G^{(1)}(\infty) = 2$ [11.104, 11.109]. For the determination of the real pulse profile one has to measure the function $G^{(2)}(\tau)$ over a wider range of delay times τ. Generally, a model profile is assumed and the calculated functions $G^{(2)}(\tau)$ and even $G^{(3)}(\tau)$ are compared with the measured ones [11.110].

In Fig. 11.48 a chirped hyperbolic secant pulse

$$E(t) = [\mathrm{sech}(t/\Delta T)]^{(1-\mathrm{i}\beta)} ,$$

is shown for $\beta = 2$. Chirped pulses result in a more complex autocorrelation signal.

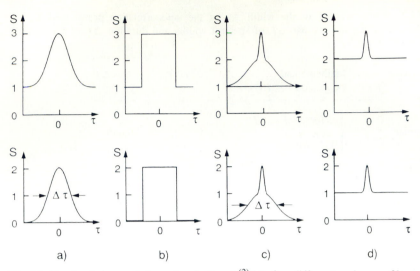

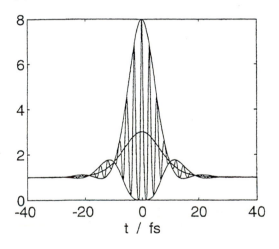

Fig. 11.47a–d. Autocorrelation signal $S \propto G^{(2)}(\tau)$ for different pulse profiles without background suppression (*upper part*) and with background suppression (*lower part*): (**a**) Fourier-limited Gaussian pulse; (**b**) rectangular pulse; (**c**) single noise pulse; and (**d**) continuous noise

Fig. 11.48. Interferometric autocorrelation of a pulse with $\Delta T = 10\,\text{fs}$ and a chirp of $\beta = 2$

The drawback of the techniques discussed so far is their lack of phase measurements. This can be overcome by the frequency-resolved optical gating technique (FROG) [11.111].

c) FROG Technique

We have seen that the second-order autocorrelations are symmetric and therefore do not provide any information on possible pulse asymmetries. Here the FROG technique is useful, since it allows measurements of the third-order

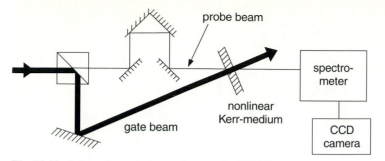

Fig. 11.49. Schematic experimental setup for FROG

autocorrelation. Its basic features are depicted in Fig. 11.49: as in the other autocorrelation techniques the incoming pulse is split by a polarizing beam-splitter into two partial beams with amplitudes E_1 and E_2. The probe pulse with amplitude $E_1(t)$ passes through a shutter (Kerr cell), which is opened by the delayed fraction $E_2(t - \tau)$ of the same pulse. The signal transmitted by the Kerr gate is then

$$\boldsymbol{E}_s(t, \tau) \propto \boldsymbol{E}(t) \cdot g(t - \tau) \,, \tag{11.39}$$

where g is the gate function $g(t - \tau) \propto I_2(t - \tau) \propto E_2^2(t - \tau)$.

If the transmitted pulse is sent through a spectrometer, where it is spectrally dispersed, a CCD camera will record the time dependence of the spectral components, which gives the two-dimensional function

$$I_t(\Omega, \tau) = \left| \int\limits_{-\infty}^{+\infty} \boldsymbol{E}(t) \cdot g(t - \tau) \, \mathrm{e}^{\mathrm{i}\Omega t} \right| \,. \tag{11.40}$$

In Fig. 11.50 this two-dimensional function is illustrated for a Gaussian pulse profile without frequency chirp and for a chirped pulse.

The integration of (11.39) over the delay time, which can be experimentally achieved by opening the gate for a time larger than all relevant delay times, yields the time profile of the pulse

$$\boldsymbol{E}(t) = \int\limits_{-\infty}^{+\infty} \boldsymbol{E}_s(t, \tau) \, \mathrm{d}\tau \,. \tag{11.41}$$

The two functions $E_s(t, \Omega)$ and $E_s(t, \tau)$ form a Fourier pair, related to each other by

$$\boldsymbol{E}_s(t, \Omega_\tau) = \frac{1}{2\pi} \int\limits_{-\infty}^{+\infty} \boldsymbol{E}_s(t, \tau) \, \mathrm{e}^{-\mathrm{i}\Omega_\tau \cdot \tau} \, \mathrm{d}\tau \,. \tag{11.42}$$

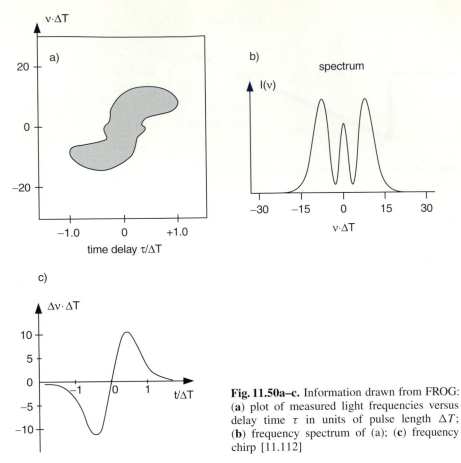

Fig. 11.50a–c. Information drawn from FROG: (a) plot of measured light frequencies versus delay time τ in units of pulse length ΔT; (b) frequency spectrum of (a); (c) frequency chirp [11.112]

The measured spectrogram $S_E(\Omega, \tau)$ can be expressed by

$$I_E(\Omega, \tau) = \left| \int\limits_{-\infty}^{+\infty} \int\limits_{-\infty}^{+\infty} E_s(t, \Omega_\tau) e^{-i\Omega t + i\Omega_\tau \cdot \tau} \, d\Omega_\tau \, dt \right|^2 . \tag{11.43}$$

The unknown signal $E(t, \Omega)$ can be extracted by a two-dimensional (the two dimensions are t and τ) phase retrieval. The reconstruction of the pulse $E(t, \Omega)$ yields the instantaneous frequency as a function of time and the pulse spectrum, shown in Fig. 11.50b [11.2].

11.3 Lifetime Measurement with Lasers

Measurements of lifetimes of excited atomic or molecular levels are of great interest for many problems in atomic, molecular, or astrophysics, as can be seen from the following three examples:

(i) From the measured lifetimes $\tau_k = 1/A_k$ of levels $|k\rangle$, which may decay by fluorescence into the lower levels $|m\rangle$, the absolute transition probability $A_k = \sum_m A_{km}$ can be determined (Sect. 2.7). From the measurements of relative intensities I_{km} of transitions $|k\rangle \rightarrow |m\rangle$ the absolute transition probabilities A_{km} can then be obtained. This yields the transition dipole matrix elements $\langle k|r|m\rangle$ (Sect. 2.7.4). The values of these matrix elements are sensitively dependent on the wave functions of the upper and lower states. Lifetime measurements therefore represent crucial tests for the quality of computed wave functions and can be used to optimize models of the electron distribution in complex atoms or molecules.

(ii) The intensity decrease $I(\omega, z) = I_0 e^{-\alpha(\omega)z}$ of light passing through absorbing samples depends on the product $\alpha(\omega)z = N_i \sigma_{ik}(\omega)z$ of the absorber density N_i and the absorption cross section σ_{ik}. Since σ_{ik} is proportional to the transition probability A_{ik} (2.22, 2.44), it can be determined from lifetime measurements, see item (i). Together with measurements of the absorption coefficient $\alpha(\omega)$ the density N_i of the absorbers can be determined. This problem is very important for testing of models of stellar atmospheres [11.113]. A well-known example is the measurement of absorption profiles of Fraunhofer lines in the solar spectrum. They yield density and temperature profiles and the abundance of the elements in the sun's atmosphere (photosphere and chromosphere). The knowledge of transition probabilities allows absolute values of these quantities to be determined.

(iii) Lifetime measurements are not only important to gain information on the dynamics of excited states but also for the determination of absolute cross sections for quenching collisions. The probability R_{kn} per second for the collision-induced transition $|k\rangle \rightarrow |n\rangle$ in an excited atom or molecule A

$$R_{kn} = \int_0^\infty N_B(v)\sigma_{kn}(v)v\,dv = N_B\langle \sigma_{kn}^{coll} \cdot v\rangle \approx N_B\langle \sigma_{kn}^{coll}\rangle \cdot \bar{v}\,, \tag{11.44}$$

depends on the density N_B of the collision partners B, the collision cross section σ_{kn}^{coll}, and the mean relative velocity $\bar{v}$. The total deactivation probability P_k of an excited level $|k\rangle$ is the sum of radiative probability $A_k = \sum_m A_{km} = 1/\tau^{rad}$ and the collisional deactivation probability R_k. Since the measured effective lifetime is $\tau_k^{eff} = 1/P_k$, we obtain the equation

$$\frac{1}{\tau_k^{eff}} = \frac{1}{\tau_k^{rad}} + R_k\,, \quad \text{with} \quad R_k = \sum_n R_{kn}\,. \tag{11.45}$$

In a gas cell at the temperature T the mean relative velocity between collision partners A and B with masses M_A, M_B is

$$\bar{v} = \sqrt{8kT/\pi\mu}\,, \quad \text{with} \quad \mu = \frac{M_A M_B}{M_A + M_B}\,. \tag{11.46}$$

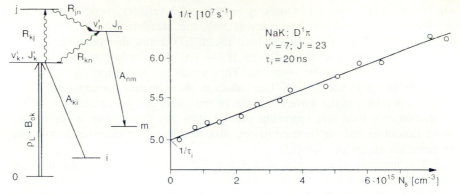

Fig. 11.51. Collisional depopulation of the excited level (v'_k, J'_k) of a molecule and example of a Stern–Vollmer plot for the NaK level $D^1\Pi_u$ $(v'=7, J'=13)$ depopulated by collisions with argon atoms at the density N_B

Using the thermodynamic equation of state $p = N \cdot kT$, we can replace the density N_B in (11.44) by the pressure p and obtain the *Stern–Vollmer equation*:

$$\frac{1}{\tau_k^{\text{eff}}} = \frac{1}{\tau_k^{\text{rad}}} + b\sigma_k p, \quad \text{with} \quad b = (\pi\mu/8kT)^{1/2}. \tag{11.47}$$

It represents a straight line when $1/\tau^{\text{eff}}$ is plotted versus p (Fig. 11.51). The slope $\tan\alpha = b\sigma_k$ yields the total quenching cross section σ_k and the intersect with the axis $p = 0$ gives the radiative lifetime $\tau_k^{\text{rad}} = \tau_k^{\text{eff}}$ $(p = 0)$.

In the following subsections we will discuss some experimental methods of lifetime measurements [11.114, 11.115]. Nowadays lasers are generally used for the selective population of excited levels. In this case, the induced emission, which contributes to the depletion of the excited level, has to be taken into account if the exciting laser is not switched off during the fluorescence detection. The rate equation for the time-dependent population density of the level $|k\rangle$, which gives the effective lifetime τ_k^{eff}, is then

$$\frac{dN_k}{dt} = +N_i B_{ik}\rho_L - N_k(A_k + R_k + B_{ki}\rho_L), \tag{11.48}$$

where ρ_L is the spectral energy density of the exciting laser, which is tuned to the transition $|i\rangle \to |k\rangle$. The solution $N_k(t) \propto I_{\text{Fl}}(t)$ of (11.48) depends on the time profile $I_L(t) = c\rho_L(t)$ of the excitation laser.

11.3.1 Phase-Shift Method

If the laser is tuned to the center frequency ω_{ik} of an absorbing transition $|i\rangle \to |k\rangle$, the detected fluorescence intensity I_{FL} monitored on the transition $|k\rangle \to |m\rangle$ is proportional to the laser intensity I_L as long as saturation can be neglected. In the *phase-shift method* the laser intensity is

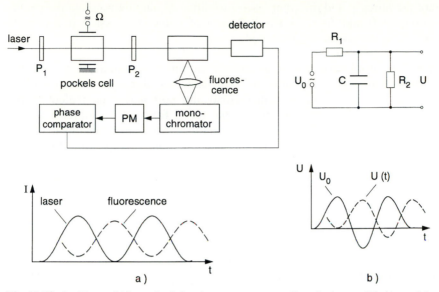

Fig. 11.52a,b. Phase-shift method for the measurement of excited-state lifetimes: (**a**) experimental arrangement; and (**b**) equivalent electric network

modulated at the frequency $f = \Omega/2\pi$ (Fig. 11.52a) according to

$$I_L(t) = \frac{1}{2}I_0(1 + a \sin \Omega t) \cos^2 \omega_{ik} t \ . \tag{11.49}$$

Inserting (11.49) with $I_L(t) = c\rho_L(t)$ into (11.48) yields the time-dependent population density $N_k(t)$ of the upper level, and therefore also the fluorescence power $P_{FL}(t) = N_k(t)A_{km}$ emitted on the transition $|k\rangle \to |m\rangle$. The result is

$$P_{FL}(t) = b\left[1 + \frac{a \sin(\Omega t + \Phi)}{[1 + (\Omega\tau_{eff})^2]^{1/2}}\right] \cos^2 \omega_{km} t \ , \tag{11.50}$$

where the constant $b \propto N_0\sigma_{0k}I_LV$ depends on the density N_0 of the absorbing molecules, the absorption cross section σ_{0k}, the laser intensity I_L, and the excitation volume V seen by the fluorescence detector. Since the detector averages over the optical oscillations ω_{km} we obtain $\langle\cos^2 \omega_{km}\rangle = 1/2$, and (11.50) gives similarly to (11.49) a sinewave-modulated function with a reduced amplitude and the phase shift ϕ against the exciting intensity $I_L(t)$. This plase shift depends on the modulation frequency Ω and the effective lifetime τ_{eff}. The evaluation yields

$$\tan \phi = \Omega\tau_{eff} \ . \tag{11.51}$$

According to (11.45–11.48) the effective lifetime is determined by the inverse sum of all deactivation processes of the excited level $|k\rangle$. In order to obtain the spontaneous lifetime $\tau_{spont} = 1/A_k$ one has to measure $\tau_{eff}(p, I_L)$ at different pressures p and different laser intensities I_L, and extrapolate the results toward $p \to 0$ and $I_L \to 0$. The influence of induced emission is a definite drawback of the phase-shift method.

Note: This problem of exciting atoms with sine wave-modulated light and determining the mean lifetime of their exponential decay from measurements of the phase shift ϕ is mathematically completely equivalent to the well-known problem of charging a capacitor C from an ac source with the voltage $U_0(t) = U_1 \sin \Omega t$ through the resistor R_1 with simultaneous discharging through a resistor R_2 (Fig. 11.52b). The equation corresponding to (11.48) is here

$$C\frac{dU}{dt} = \frac{U_0 - U}{R_1} - \frac{U}{R_2} , \tag{11.52}$$

which has the solution

$$U = U_2 \sin(\Omega t - \phi) , \quad \text{with} \quad \tan\phi = \Omega \frac{R_1 R_2 C}{R_1 + R_2} , \tag{11.53}$$

where

$$U_2 = U_0 \frac{R_2}{[(R_1 + R_2)^2 + (\Omega C R_1 R_2)^2]^{1/2}} .$$

A comparison with (11.50) shows that the mean lifetime τ corresponds to the time constant $\tau = RC$ with $R = R_1 R_2 / (R_1 + R_2)$ and the laser intensity to the charging current $I(t) = (U_0 - U)/R_1$.

Equation (11.51) anticipates a pure exponential decay. This is justified if a single upper level $|k\rangle$ is selectively populated. If several levels are simultaneously excited the fluorescence power $P_{FL}(t)$ represents a superposition of decay functions with different decay times τ_k. In such cases the phase shifts $\phi(\Omega)$ and the amplitudes $a/(1 + \Omega^2\tau^2)^{1/2}$ have to be measured for different modulation frequencies Ω. The mathematical analysis of the results allows one to separate the contributions of the simultaneously excited levels to the decay curve and to determine the different lifetimes of these levels [11.116]. A better solution is, however, if the fluorescence is dispersed by a monochromator and the detector selectively monitors a single transition from each of the different excited levels $|k_n\rangle$ separately.

11.3.2 Single-Pulse Excitation

The molecules are excited by a short laser pulse. The trailing edge of this pulse should be short compared with the decay time of the excited level, which is directly monitored after the end of the excitation pulse. Either the time-resolved LIF on transitions $|k\rangle \to |m\rangle$ to lower levels $|m\rangle$ is detected or the time-dependent absorption of a second laser, which is tuned to the transition $|k\rangle \to |j\rangle$, to higher levels $|j\rangle$.

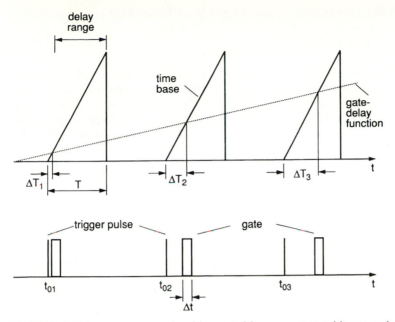

Fig. 11.53. Lifetime measurements with a gated boxcar system with successively increasing gate delay time

The time-dependent fluorescence can be viewed either with an oscilloscope or may be recorded by a transient recorder. Another method is based on a boxcar integrator, which transmits the signal through a gate that opens only during a selected time interval Δt (Sect. 4.5.5). After each successive excitation pulse the delay ΔT of the gate is increased by T/m. After m excitation cycles the whole time window T has been covered (Fig. 11.53). The direct observation of the decay curve on an oscilloscope has the advantage that nonexponential decays can be recognized immediately. For sufficiently intense fluorescence one needs only a single excitation pulse, although generally averaging over many excitation cycles will improve the signal-to noise ratio.

This technique of single-pulse excitation is useful for low repetition rates. Examples are the excitation with pulsed dye lasers pumped by Nd:YAG or excimer lasers [11.114, 11.117].

11.3.3 Delayed-Coincidence Technique

As in the previous method the delayed-coincidence technique also uses short laser pulses for the excitation of selected levels. However, here the pulse energy is kept so low that the detection probability P_D of a fluorescence photon per laser excitation pulse remains small ($P_D \leq 0.1$). If $P_D(t)\,dt$ is the probability of detecting a fluorescence photon in the time interval t to $t+dt$ after the excitation then the mean number $n_{Fl}(t)$ of fluorescence photons detected within this time interval for N excitation cycles $(N \gg 1)$ is

$$n_{\mathrm{Fl}}(t)\, dt = N P_{\mathrm{D}}(t)\, dt \ . \qquad (11.54)$$

The experimental realization is shown schematically in Fig. 11.54. Part of the laser pulse is send to a fast photodiode. The output pulse of this diode at $t = t_0$ starts a time-amplitude converter (TAC), which generates a fast-rising voltage ramp $U(t) = (t - t_0)U_0$. A photomultiplier with a large amplification factor generates for each detected fluorescence photon an output pulse that triggers a fast discriminator. The normalized output pulse of the discriminator stops the TAC at time t. The amplitude $U(t)$ of the TAC output pulse is proportional to the delay time $t - t_0$ between the excitation pulse and the fluorescence photon emission. These pulses are stored in a multichannel analyzer. The number of events per channel gives the number of fluorescence photons emitted at the corresponding delay time.

The repetition rate f of the excitation pulses is chosen as high as possible since the measuring time for a given signal-to-noise-ratio is proportional to $1/f$. An upper limit for f is determined by the fact that the time T between two successive laser pulses should be at least three times the lifetime τ_k of the measured level $|k\rangle$. This technique is therefore ideally suited for excitation with mode-locked or cavity-dumped lasers. There is, however, an electronic bottleneck: the input pulse rate of a TAC is limited by its dead time τ_{D} and should be smaller than $1/\tau_{\mathrm{D}}$. It is therefore advantageous to invert the functions of the start and stop pulses. The fluorescence pulses (which have a much smaller rate than the excitation pulses) now act as start pulses and the next

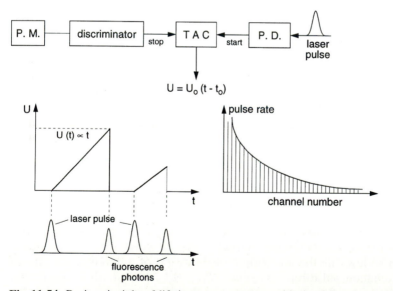

Fig. 11.54. Basic principle of lifetime measurements with the delayed-coincidence single-photon counting technique

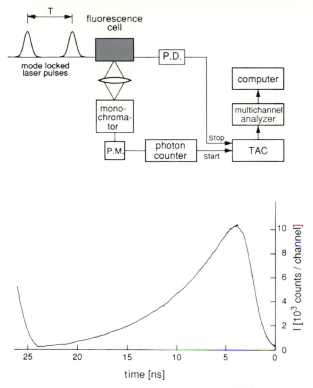

Fig. 11.55. Experimental arrangement for lifetime measurements with the delayed-coincidence single-photon counting technique and decay curve of the Na$_2$ ($B^1\Pi_u$ $v' = 6$, $J' = 27$) level [11.32]

laser pulse stops the TAC. This means that the time $(T - t)$ is measured instead of t. Since the time T between successive pulses of a mode-locked laser is very stable and can be accurately determined from the mode-locking frequency $f = 1/T$, the time interval between successive pulses can be used for time calibration of the detection system [11.32]. In Fig. 11.55 the whole detection system is shown together with a decay curve of an excited level of the Na$_2$ molecule, measured over a period of 10 min. More information about the delayed-coincidence method can be found in [11.118].

11.3.4 Lifetime Measurements in Fast Beams

The most accurate method for lifetime measurements in the range of $10^{-7} - 10^{-9}$ s is based on a modern version of an old technique that was used by N. Wien 70 years ago [11.119]. Here a time measurement is reduced to a pathlength and a velocity measurement:

The atomic or molecular ions produced in an ion source are accelerated by the voltage U and focused to form an ion beam. The different masses are

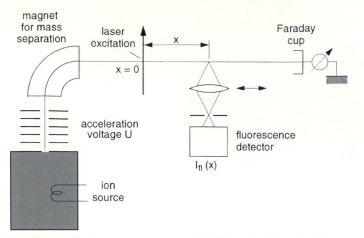

Fig. 11.56. Lifetime measurements of highly excited levels of ions or neutral atoms and molecules in a fast beam

separated by a magnet (Fig. 11.56) and the wanted ions are excited at the position $x = 0$ by a cw laser beam. The LIF is monitored as a function of the variable distance x between the excitation region and the position of a special photon detector mounted on a precision translational drive. Since the velocity $v = (2eU/m)^{1/2}$ is known from the measured acceleration voltage U, the time $t = x/v$ is determined from the measured positions x.

The excitation intensity can be increased if the excitation region is placed inside the resonator of a cw dye laser that is tuned to the selected transition. Before they reach the laser beam, the ions can be preexcited into highly excited long-living levels by gas collisions in a differentially pumped gas cell (Fig. 11.57a). This opens new transitions for the laser excitation and allows lifetime measurements of high-lying ionic states even with visible lasers [11.120].

The ions can be neutralized by charge-exchange collisions in differentially pumped alkali-vapor cells. Since charge exchange occurs with large collision cross sections at large impact parameters (grazing collisions), the momentum transfer is very small and the velocity of the neutrals is nearly the same as that of the ions. With this technique lifetimes of highly excited neutral atoms or molecules can be measured with high accuracy.

Collisional preexcitation has the drawback that several levels are simultaneously excited, which may feed by cascading fluorescence transitions the level $|k\rangle$ whose lifetime is to be measured. These cascades alter the time profile $I_{FL}(t)$ of the level $|i\rangle$ and falsify the real lifetime τ_i (Fig. 11.57b). This problem can be solved by a special measurement cycle: for each position x the fluorescence is measured alternately with and without laser excitation (Fig. 11.57c). The difference of both counting rates yields the LIF without cascade contributions. In order to eliminate fluctuations of the laser intensity or the ion beam intensity a second detector is installed at the fixed posi-

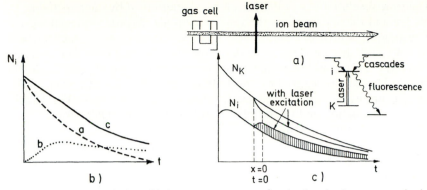

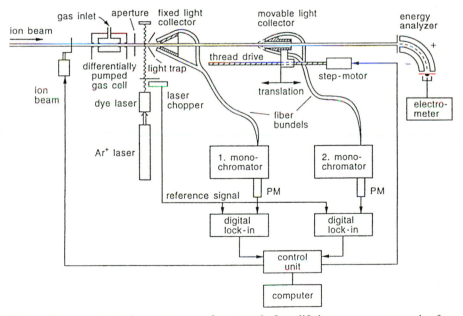

Fig. 11.57a–c. Cascade-free lifetime measurements despie the simultaneous excitation of many levels: (**a**) preexcitation by collisions in a gas cell with subsequent laser excitation; (**b**) decay of level $|i\rangle$ without cascading (*curve a*), its feeding by cascades (*curve b*), and resulting population $N_i(t)$ with cascading and decaying (*curve c*). (**c**) The fluorescence $I(x, \lambda)$ is measured alternately with and without selective laser excitation

tion x_0 (Fig. 11.58). The normalized ratios $S(x)/S(x_0)$, which are independent of these fluctuations, are then fed into a computer that fits them to a theoretical decay curve [11.121].

Fig. 11.58. Experimental arrangement for cascade-free lifetime measurements in fast beams of ions or neutrals with fluorescence collection by conically shaped optical-fiber bundles

The time resolution Δt of the detectors is determined by their spatial resolution Δx and the velocity v of the ions or neutrals. In order to reach a good time resolution, which is *independent* of the position x of the detector, one has to take care that the detector collects the fluorescence only from a small path interval Δx, but still sees the whole cross section of the slightly divergent ion beam. This can be realized by specially designed bundles of optical fibers, which are arranged in a conical circle around the beam axis (Fig. 11.58), while the outcoupling end of the fiber bundle has a rectangular form which is matched to the entrance slit of a spectrograph.

Lifetimes of atoms and ions have been measured very accurately with this technique. More experimental details and different versions of this laser beam method can be found in the extensive literature [11.120–11.124].

Example 11.4

Ne ions (23 atomic mass units, AMU) accelerated by $U = 150\,\text{kV}$ have a velocity $v = 10^6\,\text{m/s}$. In order to reach a time resolution of 1 ns the spatial resolution of the detector must be $\Delta x = 1\,\text{mm}$.

11.4 Pump-and-Probe Technique

For measurements of very fast relaxation processes that demand a time resolution below $10^{-10}\,\text{s}$ most detectors (except the streak camera) are not fast enough. Here the pump-and-probe technique is the best choice. It is based on the following principle shown in Fig. 11.59.

The molecules under investigation are excited by a fast laser pulse on the transition $|0\rangle \rightarrow |1\rangle$. A probe pulse with a variable time delay τ against the

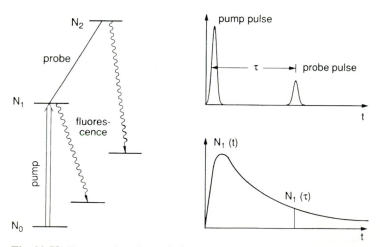

Fig. 11.59. Pump-and-probe technique

pump pulse probes the time evolution of the population density $N_1(t)$. The time resolution is only limited by the pulse width ΔT of the two pulses, but not by the time constants of the detectors!

In early experiments of this kind a fixed-frequency mode-locked Nd:glass or Nd:YAG laser was used. Both pulses came from the same laser and fortuitious coincidences of molecular transitions with the laser wavelength were utilized [11.125]. The time delay of the probe pulse is realized, as shown in Fig. 11.60, by beam splitting and a variable path-length difference. Since the pump and probe pulses coincide with the same transition $|i\rangle \rightarrow |k\rangle$, the absorption of the probe pulse measured as a function of the delay time τ, in fact, monitors the time evolution of the population difference $[N_k(t) - N_i(t)]$. A larger variety of molecular transitions becomes accessible if the Nd:YAG laser wavelength is Raman shifted (Sect. 5.9) into spectral regions of interest [11.126].

A broader application range is opened by a system of two independently tunable mode-locked dye lasers, which have to be pumped by the same pump laser in order to synchronize the pump and probe pulses [11.127]. For studies of vibrational levels in the electronic ground states of molecules the difference frequency generation of these two dye lasers can be used as a tunable infrared source for direct excitation of selected levels on infrared-active transitions. Raman-active vibrations can be excited by spontaneous or stimulated Raman transitions (Chap. 8).

In addition, short-pulse tunable optical parametric oscillators have been realized, where the pump wavelength and the signal or idler waves can be used for pump-and-probe experiments [11.128]. The wide tuning range allows more detailed investigations compared to the restricted use of fixed frequency lasers [11.129]. Another useful short-pulse source for these experiments is a three-wavelength Ti:sapphire laser, where two of the wavelengths can be indepently tuned [11.130].

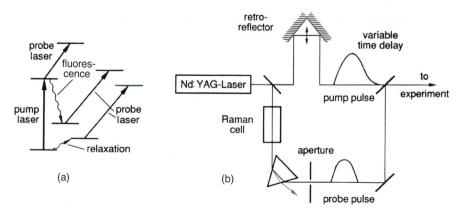

Fig. 11.60. Pump-and-probe technique for the measurements of ultrafast relaxation processes

Recently it has become possible to phase-lock two different femtosecond lasers. This opens many possibilities for spectroscopic applications. One example is the use of an infrared femtosecond pulse to excite nuclear vibrations of molecules and a second UV femtosecond pulse for electronic excitation. This allows one to study the influence of fast changes of the electron cloud on the nuclear oscillation period [11.131].

These developments widen the range of applications considerably. We will now give some examples of applications of the pump-and-probe technique.

11.4.1 Pump-and-Probe Spectroscopy of Collisional Relaxation in Liquids

Because of the high molecular density in liquids, the average time τ_c between two successive collisions of a selectively excited molecule A with other molecules A of the same kind or with different molecules B is very short ($10^{-12}-10^{-11}$ s). If A has been excited by absorption of a laser photon its excitation energy may be rapidly redistributed among other levels of A by collisions, or it may be transferred into internal energy of B or into translational energy of A and B (temperature rise of the sample). With the pump-and-probe technique this energy transfer can be studied by measurements of the time-dependent population densities $N_m(t)$ of the relevant levels of A or B. The collisions not only change the population densities but also the phases of the wave functions of the coherently excited levels (Sect. 2.9). These phase relaxation times are generally shorter than the population relaxation times.

Besides excitation and probing with infrared laser pulses, the CARS technique (Sect. 8.5) is a promising technique to study these relaxation processes. An example is the measurement of the dephasing process of the OD stretching vibration in heavy water D_2O by CARS [11.132]. The pump at $\omega = \omega_L$ is provided by an amplified 80-fs dye laser pulse from a CPM ring dye laser. The Stokes pulse at ω_s is generated by a synchronized tunable picosecond dye laser. The CARS signal at $\omega_{as} = 2\omega_L - \omega_s$ is detected as a function of the time delay between the pump and probe pulses.

Another example is the deactivation of high vibrational levels in the S_0 and S_1 singlet states of dye molecules in organic liquids pumped by a pulsed laser (Fig. 11.61). The laser populates many vibrational levels in the excited S_1 singlet state, which are accessible by optical pumping on transitions starting from thermally populated levels in the electronic ground state S_0. These excited levels $|v'\rangle$ rapidly relax by inelastic collisions into the lowest vibrational level $|v' = 0\rangle$ of S_1, which represents the upper level of the dye laser transition. This relaxation process can be followed up by measuring the time-dependent absorption of a weak probe laser pulse on transitions from these levels into higher excited singlet states.

Fluorescence and stimulated emission on transitions $(v' = 0 \rightarrow v'' > 0)$ lead to a fast rise of the population densities $N(v'')$ of high vibrational levels in the S_0 state. This would result in a self-termination of the laser oscillation if these levels were not depopulated quickly enough by collisions. The relaxation of $N(v'')$ toward the thermal equilibrium population $N_0(v'')$ can

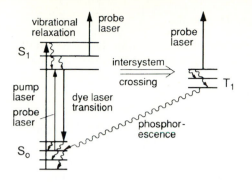

Fig. 11.61. Measurements of fast relaxation processes in excited and ground states

again be probed by a weak visible probe laser. Polarization spectroscopy (Sect. 7.4) with femtosecond pulses allows one to separately determine the decay times τ_{vib} of population redistribution and the dephasing times [11.133].

Of particular importance for dye laser physics is the intersystem crossing of dye molecules from the excited S_1 state into levels of the triplet state T_1. Because the population of these long-living triplet levels represents a severe loss for the dye laser radiation because of absorption on electronic transitions to higher triplet states, the time-dependent triplet concentration and possible quenching processes by triplet-quenching additives have been investigated in detail [11.134]. Furthermore, spin-exchange and transitions by collisions between excited S_1 dye molecules and triplet O_2 molecules or between T_1 dye molecules and excited O_2 ($^1\Delta$) molecules play a crucial role in photodynamical processes in cancer cells (Sect. 15.5).

11.4.2 Electronic Relaxation in Semiconductors

A very interesting problem is concerned with the physical limitations of the ultimate speed of electronic computers. Since any *bit* corresponds to a transition from a nonconducting to a conducting state of a semiconductor, or vice versa, the relaxation time of electrons in the conduction band and the recombination time certainly impose a lower limit for the minimum switching time. This electronic relaxation can be measured with the pump-and-probe technique. The electrons are excited by a femtosecond laser pulse from the upper edge of the valence band into levels with energies $E = \hbar\omega - \Delta E$ (ΔE: band gap) in the conduction band, from where they relax into the lower edge of the conduction band before they recombine with holes in the valence band. Since the optical reflectivity of the semiconductor sample depends on the energy distribution $N(E)$ of the free conduction electrons, the reflection of a weak probe laser pulse can be used to monitor the distribution $N(E)$ [11.135]. Because of their fast relaxaton semiconductors can be used as saturable absorbers for passive mode locking in femtosecond lasers [11.65]. In this case, a thin semiconductor sheet is placed in front of a resonator mirror (see Sect. 11.1.10. The characteristic time scales for interband and intraband electron relaxation are again measured with the pump-and probe technique [11.136].

11.4.3 Femtosecond Transition State Dynamics

The pump-and-probe technique has proved to be very well suited for studying short-lived transient states of molecular systems that had been excited by
a short laser pulse before they dissociate:

$$AB + h\nu \longrightarrow [AB]^* \longrightarrow A^* + B .$$

An illustrative example is the photodissociation of excited NaI molecules,
which has been studied in detail by Zewail et al. [11.137].

The adiabatic potential diagram of NaI (Fig. 11.62) is characterized by an
avoided crossing between the repulsive potential of the two interacting neutral atoms Na + I and the Coulomb potential of the ions $Na^+ + I^-$, which
is mainly responsible for the strong binding of NaI at small internuclear distances R. If NaI is excited into the repulsive state by a short laser pulse at the
wavelength λ_1, the excited molecules start to move toward larger values of R
with a velocity $v(R) = [(2/\mu)(E - E_{pot}(R)]^{1/2}$.

> **Example 11.5**
>
> For $E - E_{pot} = 1000\,cm^{-1}$ and $\mu = m_1 m_2/(m_1 + m_2) = 19.5\,AMU \rightarrow v \approx$
> $10^3\,m/s$. The time $\Delta T = \Delta R/v$ of passing through an interval of $\Delta R =$
> $0.1\,nm$ is then $\Delta T = 10^{-13}\,s = 100\,fs$.

When the excited system $[NaI]^*$ reaches the avoided crossing at $R = R_c$ it
may either stay on the potential $V_1(R)$ and oscillate back and forth between

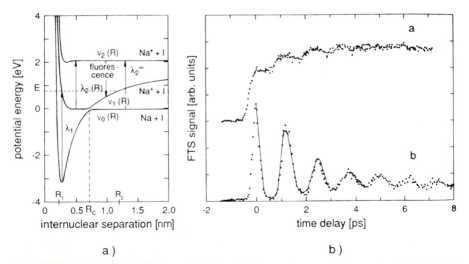

a)   b)

Fig. 11.62. (a) Potential diagram of NaI with the pump transition at λ_1 and the tunable probe pulse at $\lambda_2(R)$. **(b)** Fluorescence intensity $I_{Fl}(\Delta t)$ as a function of the delay
time Δt between pump and probe pulses: *(curve a)* with λ_2 tuned to the atomic Na* transition and *(curve b)* λ_2 tuned to $\lambda_2(R)$ with $R < R_c$ [11.137]

R_1 and R_2, or it may tunnel to the potential curve $V_0(R)$, where it separates into Na + I.

The time behavior of the system can be probed by a probe pulse with the wavelength λ_2 tuned to the transition from $V_1(R)$ into the excited state $V_2(R)$ that dissociates into Na* + I. At the fixed wavelength $\lambda_2 = 2\pi c/\omega_2$ the dissociating system absorbs the probe pulse only at that distance R where $V_1(R) - V_2(R) = \hbar\omega_2$. If λ_2 is tuned to the sodium resonance line $3s \rightarrow 3p$, absorption occurs for $R = \infty$.

Since the dissociation time is very short compared to the lifetime of the excited sodium atom Na*(3p), the dissociating (NaI)* emits nearly exclusively at the atomic resonance fluorescence. The atomic fluorescence intensity $I_{Fl}(Na^*, \Delta t)$, monitored in dependence on the delay time Δt between the pump-and-probe pulse, gives the probability for finding the excited system [NaI]* at a certain internuclear separation R, where $V_1(R) - V_2(R) = \hbar\omega_2$ (Fig. 11.62b).

The experimental results [11.138] shown in Fig. 11.62b reflect the oscillatory movement of Na*I(R) on the potential $V_1(R)$ between R_1 and R_2, which had been excited at the inner turning point R_1 by the pump pulse at $t = 0$. This corresponds to a periodic change between the covalent and ionic potential. The damping is due to the leakage into the lower-state potential around the avoided crossing at $R = R_c$. If λ_2 is tuned to the atomic resonance line, the accumulation of Na*(3p) atoms can be measured when the delay between the pump-and-probe pulse is increased.

11.4.4 Real-Time Observations of Molecular Vibrations

The time scale of molecular vibrations is on the order of $10^{-13} - 10^{-15}$ s. The vibrational frequency of the H_2 molecule, for example, is $\nu_{vib} = 1.3 \times 10^{14}$ s^{-1} $\rightarrow T_{vib} = 7.6 \times 10^{-15}$ s, that of the Na_2 molecule is $\nu_{vib} = 4.5 \times 10^{12}$ s^{-1} $\rightarrow T_{vib} = 2 \times 10^{-13}$ s, and even the heavy I_2 molecule still has $T_{vib} = 5 \times 10^{-13}$ s. With conventional techniques one always measures a time average over many vibrational periods.

With femtosecond pump-and-probe experiments "fast motion pictures" of a vibrating molecule may be obtained, and the time behavior of the wave packets of coherently excited and superimposed molecular vibrations can be mapped. This is illustrated by the following examples dealing with the dynamics of molecular multiphoton ionization and fragmentation of Na_2, and its dependence on the phase of the vibrational wave packet in the intermediate state [11.139]. There are two pathways for photoionization of cold Na_2 molecules in a supersonic beam (Fig. 11.63):

(i) One-photon absorption of a femtosecond pulse ($\lambda = 672$ nm, $\Delta T = 50$ fs, $I = 50$ GW/cm^2) leads to simultaneous coherent excitation of vibrational levels $v' = 10-14$ in the $A^1\Sigma_u$ state of Na_2 at the inner part of the potential $V_1(R)$. This generates a vibrational wave packet, which oscillates at a frequency of 3×10^{12} s^{-1} back and forth between the inner and outer turning point. Resonant enhanced two-photon ionization of the excited molecules

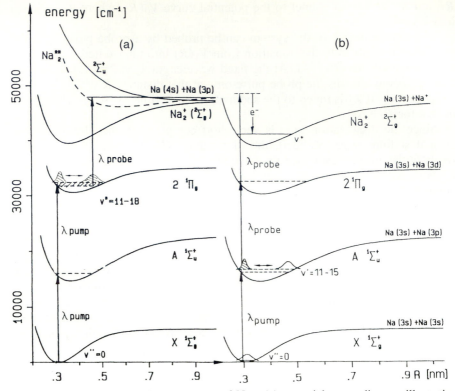

Fig. 11.63a,b. Femtosecond spectroscopy of Na$_2$: (**a**) potential curve diagram illustrating the preparation of a vibrational wave packet in the $2^1\Pi_g$ state of Na$_2$ due to coherent simultaneous two-photon excitation of vibrational levels ($v' = 11-18$). Further excitation by a third photon results in production of Na$_2^{**} \rightarrow$ Na$^+$ + Na* from the outer turning point of the wave packet. (**b**) One-photon excitation of a vibrational wave packet in the $A^1\Sigma_u$ state with subsequent two-photon ionization from the *inner* turning point [11.140]

by the probe pulse has a larger probability at the inner turning point than at the outer turning point, because of favorable Franck–Condon factors for transitions from the $A^1\Sigma_u$ state to the near-resonant intermediate state $2^1\Pi_g$, which enhances ionizing two-photon transitions at small values of the internuclear distance R (Fig. 11.63b). The ionization rate $N(\text{Na}_2^+, \Delta t)$ monitored as a function of the delay time Δt between the weak pump pulse and the stronger probe pulse, yields the upper oscillatory function of Fig. 11.64 with a period that matches the vibrational wave-packet period in the $A^1\Sigma_u$ state.

(ii) The second possible competing process is the two-photon excitation of wavepackets of the $v' = 11-18$ vibrational levels in the $2^1\Pi_g$ state of Na$_2$ by the pump pulse, with subsequent one-photon excitation into a doubly excited state of Na$_2^{**}$, which autoionizes according to

$$\text{Na}_2^{**} \longrightarrow \text{Na}_2^+ + e^- \longrightarrow \text{Na}^+ + \text{Na}^* + e^-,$$

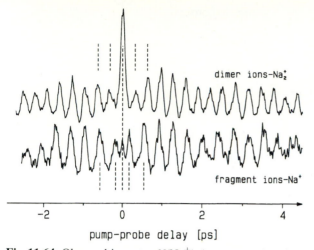

pump-probe delay [ps]

Fig. 11.64. Observed ion rates $N(Na_2^+)$ (*upper trace*) and $N(Na^+)$ (*lower trace*) as a function of the delay time Δt between the pump and probe pulses [11.140]

and results in the generation of Na^+ ions. The number $N(Na^+, \Delta t)$ of atomic ions Na^+, measured as a function of the delay time Δt between pump and probe pulses, shows again an oscillatory structure (Fig. 11.64, lower trace), but with a time shift of half a vibrational period against the upper trace. In this case, the ionization starts from the *outer* turning point of the $2^1\Pi_g$ levels and the oscillatory structure shows a $180°$ shift and slightly different oscillation period, which corresponds to the vibrational period in the $2^1\Pi_g$ state.

The photoelectrons and ions and their kinetic energies can be measured with two time-of-flight mass spectrometers arranged into opposite directions perpendicular to the molecular and the laser beams [11.141, 11.142].

11.4.5 Transient Grating Techniques

If two light pulses of different propagation directions overlap in an absorbing sample, they produce an interference pattern because of the intensity-dependent saturation of the population density (Fig. 11.65). When a probe pulse is sent through the overlap region in the sample, this interference pattern shows up as periodic change of the sample transmission and therefore acts as a grating that produces diffraction orders of the probe beam. The grating vector is $k_G = k_2 - k_1$, and the grating period depends on the angle Θ between the two pump beams. The grating amplitude can be inferred from the relative intensity of the different diffraction orders. This gives information on the saturation intensities. The grating will fade away if the delay times τ_1 and τ_2 are larger than the relaxation time of the sample molecules. Therefore this technique of transient gratings gives information on the dynamics of the sample [11.143].

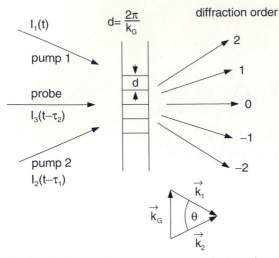

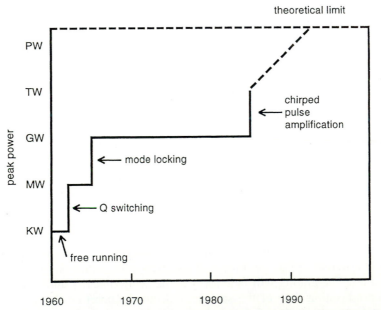

Fig. 11.65. Schematic diagram of a transient grating experiment

Most experiments have been performed in solid or liquid samples where the relaxation times are in the range of femto- to picoseconds [11.142].

There are numerous other examples where pico- and femtosecond spectroscopy have been applied to problems in atomic and molecular physics. Some of them are discussed in Sects. 12.5 and 15.2.

Fig. 11.66. The progress in achievable peak powers with the invention of different techniques for laser pulse generation [11.95]

In particular the high peak powers now available by the invention of new techniques (Fig. 11.66) allow a new class of experiments in nonlinear physics. Examples are the generation of high harmonics up to the 60th overtone, which generates XUV frequencies with $\lambda = 13$ nm from the fundamental wave with $\lambda = 800$ nm, or multiphoton ionization to produce highly charged ions from neutral atoms. The electric fields of such high power laser pulses exceed the inner-atomic field strength resulting in field ionization. The behavior of atoms and molecules in such strong electric ac fields have brought some surprises and many theoreticians are working on adequate models to describe such extreme situations.

Problems

11.1 A Pockels cell inside a laser resonator is used as a Q-switch. It has a maximum transmission of 95% for the applied voltage $U = 0$. What voltage U is required to prevent lasing before the gain $G_\alpha = \exp(\alpha L)$ of the active medium exceeds the value $\sigma_\alpha = 10$, when the "half-wave voltage" of the Pockels cell is 2 kV? What is the effective amplification factor G_{eff} immediately after the opening of the Pockels cell if the total cavity losses are $30°$ per round trip?

11.2 What is the actual time profile of mode-locked pulses from a cw argon laser if the gain profile is Gaussian with a halfwidth of 8 GHz (FWHM)?

11.3 An optical pulse with the Gaussian intensity profile $I(t)$, center wavelength $\lambda_0 = 600$ nm, and initial halfwidth $\tau = 500$ fs propagates through an optical fiber with refractive index $n = 1.5$.

(a) How large is its initial spatial extension?
(b) How long is the propagation length z_1 after which the spatial width of the pulse has increased by a factor of 2 from linear dispersion with $dn/d\lambda = 10^3$ per cm?
(c) How large is its spectral broadening at z_1 if its peak intensity is $I_p = 10^{13}$ W/m² and the nonlinear part of the refractive index is $n_2 = 10^{-20}$ m²/W?

11.4 Calculate the separation D of a grating pair that just compensates a spatial dispersion $dS/d\lambda = 10^5$ for a center wavelength of 600 nm and a groove spacing of 1 μm.

11.5 Calculate the pressure p of the argon buffer gas at $T = 500$ K that decreases the radiative lifetime $\tau = 16$ ns of an excited Na_2 level to 8 ns, if the total quenching cross section is $\sigma = 10^{-14}$ cm².

12. Coherent Spectroscopy

This chapter provides an introduction to different spectroscopic techniques that are based either on the coherent excitation of atoms and molecules or on the coherent superposition of light scattered by molecules and small particles. The coherent excitation establishes definite phase relations between the amplitudes of the atomic or molecular wave functions; this, in turn, determines the total amplitudes of the emitted, scattered, or absorbed radiation.

Either two or more molecular levels of a molecule are excited coherently by a spectrally broad, short laser pulse (level-crossing and quantum-beat spectroscopy) or a whole ensemble of many atoms or molecules is coherently excited simultaneously into identical levels (photon-echo spectroscopy). This coherent excitation alters the spatial distribution or the time dependence of the total, emitted, or absorbed radiation amplitude, when compared with incoherent excitation. Whereas methods of incoherent spectroscopy measure only the total intensity, which is proportional to the population density and therefore to the square $|\psi|^2$ of the wave function ψ, the coherent techniques, on the other hand, yield additional information on the amplitudes and phases of ψ.

Within the density-matrix formalism (Sect. 2.9) the coherent techniques measure the off-diagonal elements ρ_{ab} of the density matrix, called the *coherences*, while incoherent spectroscopy only yields information about the diagonal elements, representing the time-dependent population densities. The off-diagonal elements describe the atomic dipoles induced by the radiation field, which oscillate at the field frequency ω and which represent radiation sources with the field amplitude $A_k(\boldsymbol{r}, t)$. Under coherent excitation the dipoles oscillate with definite phase relations, and the phase-sensitive superposition of the radiation amplitudes A_k results in measurable interference phenomena (quantum beats, photon echoes, free induction decay, etc.).

After switching off the excitation sources, the phase relations between the different oscillating atomic dipoles are altered by different relaxation processes, which *perturb* the atomic dipoles. We may classify these processes into two categories:

- Population-changing processes: the decay of the population density in the excited level $|2\rangle$

$$N_2(t) = N_2(0)\, e^{-t/\tau_{\text{eff}}} ,$$

by spontaneous emission or by inelastic collisions decreases the intensity of the radiation with the time constant $T_1 = \tau_{\text{eff}}$, often called the *longitudinal relaxation time*.

- Phase-perturbing collisions (Sect. 3.3), or the different Doppler shifts of the frequencies of emitting atoms from their individual velocity v_k result in a change of the relative phases of the atomic dipoles, which also affects the total amplitudes of the superimposed, emitted partial waves. The time constant T_2 of this *phase decay* is called the *transverse relaxation time*. Generally, the phase decay is faster than the population decay and therefore $T_2 < T_1$. While phase-perturbing collisions result in a homogeneous line broadening $\gamma_2^{\text{hom}} = 1/T_2^{\text{hom}}$, the velocity distribution, which gives rise to Doppler broadening, adds the inhomogeneous contribution $\gamma_2^{\text{inhom}} = 1/T_2^{\text{inhom}}$ to the line profile, which has the total width

$$\Delta\omega = 1/T_1 + 1/T_2^{\text{hom}} + 1/T_2^{\text{inhom}} . \tag{12.1}$$

The techniques of coherent spectroscopy that are discussed below allow the elimination of the inhomogeneous contribution and therefore represent methods of "Doppler-free" spectroscopy, *although the coherent excitation may use spectrally broad radiation.* This is an advantage compared with the nonlinear Doppler-free techniques discussed in Chap. 7, where narrow-band single-mode lasers are required.

The combination of ultrafast light pulses with coherent spectroscopy allows, for the first time, the direct measurements of wave packets of coherently excited molecular vibrations and their decay (Sect. 12.3–12.6).

A rapidly expanding area of coherent spectroscopy is heterodyne and correlation spectroscopy, based on the interference between two coherent light waves. These two light waves may be generated by two stabilized lasers, or by one laser and the Doppler-shifted laser light scattered by a moving particle (atom, molecule, dust particle, microbes, living cells, etc). The observed frequency distribution of the *beat spectrum* allows a spectral resolution down into the millihertz range (Sect. 12.6).

Correlation fluorescence spectroscopy in combination with confocal microscopy has proved to be a very versatile tool for detection and temporal investigations of biomolecules in living cells and in the interaction of different molecules.

In the following section the most important techniques of coherent spectroscopy are discussed in order to illustrate more quantitatively the foregoing statements.

12.1 Level-Crossing Spectroscopy

Level-crossing spectroscopy is based on the measured change in the spatial intensity distribution or the polarization characteristics of fluorescence that is emitted from coherently excited levels when these levels cross under the influence of external magnetic or electric fields. Examples are fine or hyperfine levels with different Zeeman shifts, which may cross at a certain value B_c of the magnetic field B (Fig. 12.1). A special case of level crossing is the

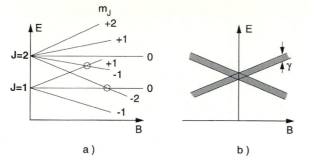

Fig. 12.1a,b. Schematic diagram of level crossing: (**a**) general case and (**b**) Hanle effect

zero-field level crossing (Fig. 12.1b) that occurs for a degenerate level with total angular momentum $J > 0$. For $B \neq 0$, the $(2J + 1)$ Zeeman components
split, which changes the polarization characteristics of the emitted fluorescence. This phenomenon was observed as early as 1923 by *Hanle* [12.1] and
is therefore called the Hanle effect.

12.1.1 Classical Model of the Hanle Effect

A typical experimental arrangement for level-crossing spectroscopy is depicted in Fig. 12.2. Atoms or molecules in a homogeneous magnetic field
$B = \{0, 0, B_z\}$ are excited by the polarized optical wave $E = E_y \cos(\omega t - kx)$
and tuned to the transition $|1\rangle \rightarrow |2\rangle$. The fluorescence $I_{Fl}(B)$ emitted from
the excited level $|2\rangle$ into the y-direction is observed behind a polarizer as
a function of the magnetic field B.

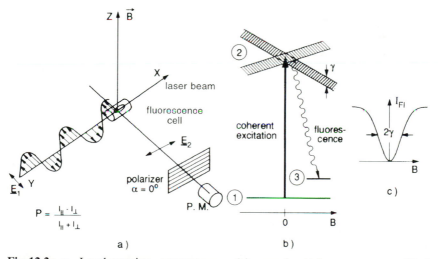

Fig. 12.2a–c. Level-crossing spectroscopy: (**a**) experimental arrangement; (**b**) level
scheme; and (**c**) Hanle signal

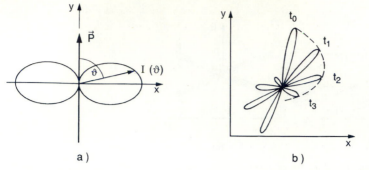

Fig. 12.3a,b. Spatial intensity distribution of the radiation emitted by a classical oscillating dipole: (**a**) stationary case for $B = 0$; and (**b**) time sequence of intensity distributions of a dipole with decay constant γ, precessing in a magnetic field $\boldsymbol{B} = \{0, 0, B_z\}$

In a classical, vivid description, the excited atom is represented by a damped oscillator that oscillates in the y-direction with a spatial emission charactistics $I(\vartheta) \propto \sin^2 \vartheta$, depending on the angle ϑ between observation direction and dipole axis (Fig. 12.3a). If the atom with an excited-state lifetime $\tau_2 = 1/\gamma_2$ is excited at $t = t_0$ by a short laser pulse, the amplitude of the emitted radiation is for $B = 0$:

$$\boldsymbol{E}(t) = \boldsymbol{E}_0 e^{-(i\omega + \gamma/2)(t - t_0)} .\tag{12.2}$$

With the angular momentum $\boldsymbol{J}$ in level $|2\rangle$, the atomic dipole with the magnetic dipole moment $\mu = g_J \mu_0$ (g_J: Landé factor; μ_0: Bohr's magneton) will process for $B \neq 0$ around the z-axis with the precession frequency

$$\Omega_p = g_J \mu_0 B/\hbar .\tag{12.3}$$

Together with the dipole axis the direction of maximum emission, which is perpendicular to the dipole axis, also precesses with Ω_p around the z-axis while the amplitude of the dipole oscillation decreases as $\exp[-(\gamma/2)t]$ (Fig. 12.3b). If the fluorescence is observed behind a polarizer with the transmission axis tilted by the angle α against the x-axis, the measured intensity $I_{Fl}(t)$ becomes

$$I_{Fl}(B, \alpha, t) = I_0 e^{-\gamma(t - t_0)} \sin^2[\Omega_p(t - t_0)] \cos^2 \alpha .\tag{12.4}$$

This intensity can either be monitored in a time-resolved fashion after pulsed excitation in the constant magnetic field B (quantum beats, Sect. 12.2), or the time-integrated fluorescence intensity is measured as a function of B (level crossing), where the excitation may be pulsed or cw.

In the case of cw excitation the fluorescence intensity $I(t)$ at any time t is the result of all atoms excited within the time interval between $t_0 = -\infty$ and $t_0 = t$. It can therefore be written as

$$I(B, \alpha) = C I_0 \cos^2 \alpha \int_{t_0 = -\infty}^{t} e^{-\gamma(t - t_0)} \sin^2[\Omega_p(t - t_0)] dt_0 .\tag{12.5}$$

The transformation $(t - t_0) \rightarrow t'$ of the variable t_0 shifts the integration limits to the interval from $t' = \infty$ to $t' = 0$, which shows that the integral is independent of t. It can be solved with the substitution $2\sin^2 x = 1 - \cos 2x$, yields the Hanle signal

$$I(B, \alpha) = C\frac{I_0 \cos^2 \alpha}{2\gamma}\left(1 - \frac{\gamma^2}{\gamma^2 + 4\Omega_p^2}\right) . \tag{12.6a}$$

Inserting (12.3) we obtain for $\alpha = 0$

$$I(B, \alpha = 0) = C\frac{I_0}{2\gamma}\left(1 - \frac{1}{1 + (2g\mu_0 B/\hbar\gamma)^2}\right) . \tag{12.7}$$

This intensity gives, as a function of the applied magnetic field, a Lorentzian-type signal (Fig. 12.2c) with the halfwidth (FWHM)

$$\Delta B_{1/2} = \frac{\hbar\gamma}{g\mu_0} = \frac{\hbar}{g\mu_0 \tau_{\text{eff}}} , \quad \text{with} \quad \tau_{\text{eff}} = 1/\gamma . \tag{12.8}$$

From the measured halfwidth $\Delta B_{1/2}$ the product $g_J \tau_{\text{eff}}$ of Landé factor g_J of the excited level $|2\rangle$ times its effective lifetime τ_{eff} can be derived. For *atomic* states the Landé factor g_J is generally known, and the measured value of $\Delta B_{1/2}$ determines the lifetime τ_{eff}. Measurements of $\tau_{\text{eff}}(p)$ as a function of pressure in the sample cell then yield by extrapolation $p \rightarrow 0$ the radiative lifetime τ_n (Sect. 11.3). The Hanle effect therefore offers an alternative method for the measurement of atomic lifetimes [12.2].

In case of excited *molecules* the angular momentum coupling scheme is often not known, in particular if hyperfine structure or perturbations between different electronic states affect the coupling of the various angular momenta. The total angular momentum $\boldsymbol{F} = \boldsymbol{N} + \boldsymbol{S} + \boldsymbol{L} + \boldsymbol{I}$ is composed of the molecular-rotation angular momentum $\boldsymbol{N}$, the total electron spin $\boldsymbol{S}$, the electronic orbital momentum $\boldsymbol{L}$, and the nuclear spin $\boldsymbol{I}$. The measured width $\Delta B_{1/2}$ can then be used to determine the Landé factor g and thus the coupling scheme, if the lifetime τ is already known from other independent measurements, for example, those discussed in Sect. 11.3 [12.3]. Note that the sign of the dispersion term in (12.6a) yields the sign of the Landé g-factor defined by (12.3) [12.4].

If a second detector is placed in the x-direction with a polarizer having the angle β against the y-axis, the measured Hanle signal is:

$$I_2(B, \beta) = C\frac{I_0 \cos^2 \beta}{2\gamma}\left(1 + \frac{\gamma^2}{\gamma^2 + 4\Omega_p^2}\right) . \tag{12.6b}$$

Detecting the difference $I_2(B, \beta = 0) - I_1(B, \alpha = 0)$ yields twice the signal heights and eliminates the background.

For illustration, a molecular Hanle signal is depicted in Fig. 12.4 that was obtained when Na_2 molecules were excited in a magnetic field B into the upper level $|2\rangle = B^1\Pi_u$ $(v' = 10, J' = 12)$ by an argon laser at

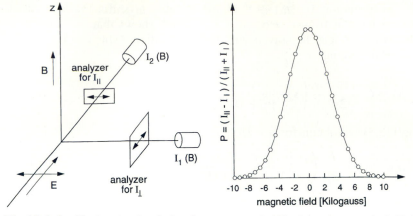

Fig. 12.4a,b. Hanle signal of the fluorescence $I_{Fl}(B)$ from laser-excited $Na_2^*(B^1\Pi_u,$ $v' = 10$, $J' = 12$) molecules: (**a**) schematic experimental setup for difference and ratio recording; and (**b**) measured polarization $P(B)$ [12.5]

$\lambda = 475$ nm [12.5]. For this Na_2 level with $S = 0$ Hund's coupling case a_α applies, and the Landé factor

$$g_F = \frac{F(F+1) + J(J+1) - I(I+1)}{2J(J+1)F(F+1)} \tag{12.9}$$

decreases strongly with increasing rotational quantum number J [12.6]. For larger values of J one therefore needs much larger magnetic fields than for atomic-level crossings.

12.1.2 Quantum-Mechanical Models

The quantum-mechanical treatment of level-crossing spectroscopy [12.4, 12.7] starts with the Breit formula

$$I_{Fl}(2 \rightarrow 3) = C \, |\langle 1| \, \boldsymbol{\mu}_{12} \cdot \boldsymbol{E}_1 \, |2\rangle|^2 \, |\langle 2| \, \boldsymbol{\mu}_{23} \cdot \boldsymbol{E}_2 \, |3\rangle|^2 \, , \tag{12.10}$$

for the fluorescence intensity excited on the transition $|1\rangle \rightarrow |2\rangle$ by a polarized light wave with the electric field vector $\boldsymbol{E}_1$ and observed on the transition $|2\rangle \rightarrow |3\rangle$ behind a polarizer with its axis parallel to $\boldsymbol{E}_2$ [12.8]. The spatial intensity distribution and the polarization characteristics of the fluorescence depend on the orientation of the molecular transition dipoles $\boldsymbol{\mu}_{12}$ and $\boldsymbol{\mu}_{23}$ relative to the electric vectors $\boldsymbol{E}_1$ and $\boldsymbol{E}_2$ of absorbed and emitted radiation.

The level $|2\rangle$ with the total angular momentum $\boldsymbol{J}$ has $(2J+1)$ Zeeman components with the magnetic quantum number M, which are degenerate at $B = 0$. The wave function of $|2\rangle$

$$\psi_2 = \sum_k c_k \psi_k \, e^{-i\omega_k t} \, , \tag{12.11}$$

is a linear combination of the wave functions $\psi_k \exp(-i\omega_k t)$ of all coherently excited Zeeman components. The products of matrix elements in (12.10) contain the interference terms $c_{M_i} c_{M_k} \psi_i \psi_k \exp[i(\omega_k - \omega_i)t]$.

For $B = 0$ all frequencies ω_k are equal. The interference terms become time independent and describe together with the other constant terms in (12.10) the spatial distribution of I_{Fl}. However, for $B \neq 0$ the phase factors $\exp[i(\omega_k - \omega_i)t]$ are time dependent. Even when all levels (J, M) for a given J have been excited coherently at time $t = 0$ and their wave functions $\psi_k(t = 0)$ all have the same phase $\psi(0) = 0$, the phases of ψ_k develop differently in time because of the different frequencies ω_k. The magnitudes and the signs of the interference terms change in time, and for $(\omega_i - \omega_k) \gg 1/\tau$ (which is equivalent to $g_J \mu_0 B/\hbar \gg \gamma$) the interference terms average to zero and the observed time-averaged intensity distribution becomes isotropic.

Note: Although the magnetic field alters the spatial distribution and the polarization characteristics of the fluorescence, it does not change the total fluorescence intensity!

Example 12.1

The quantum-mechanical description may be illustrated by a specific example, taken from [12.9], where optical pumping on the transition $(J, M) \rightarrow (J', M')$ is achieved by linearly polarized light with a polarization vector in the x–y-plane, perpendicular to the field direction along the z-axis (Fig. 12.5). The linearly polarized light may be regarded as a superposition of left- and right-hand circularly polarized components σ^+ and σ^-. The excited-state wave function for the excitation by light polarized in the x-direction is

$$|2\rangle_x = (-1/\sqrt{2})(a_{M+1} |M+1\rangle + a_{M-1} |M-1\rangle), \qquad (12.12)$$

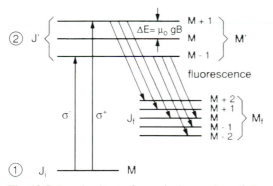

Fig. 12.5. Level scheme for optical pumping of Zeeman levels by light linearly polarized in the y-direction. Each of the two excited Zeeman levels can decay in three final-state sublevels. Superposition of the two different routes $(J'', M'') \rightarrow (J_f, M_f)$ generates interference effects [12.9]

and for excitation with y-polarization

$$|2\rangle_y = (-i\sqrt{2})(a_{M+1}|M+1\rangle - a_{M-1}|M-1\rangle) .\tag{12.13}$$

The coefficients a_M are proportional to the matrix element of the transition $(J, M) \to (J', M')$. The optical pumping process generates a coherent superposition of the eigenstates $(M' = M \pm 1)$ as long as the spectral width of the pump radiation is broader than the level splitting.

The time development of the excited-state wave function is described by the time-dependent Schrödinger equation

$$-\frac{\hbar}{i}\frac{\partial\psi_2}{\partial t} = \mathcal{H}\psi_2 ,\tag{12.14}$$

where the operator $\mathcal{H}$ has the eigenvalues $E_M = E_0 + \mu_0 g M B$. Including the spontaneous emission in a semiclassical way by the decay constant γ (Sect. 2.7.5), we may write the solution of (12.14)

$$\psi_2(t) = e^{-(\gamma/2)t} \exp(-i\mathcal{H}t/\hbar)\psi_2(0) ,\tag{12.15}$$

where the operator $\exp(-i\mathcal{H}t)$ is defined by its power-series expansion. From (12.12)–(12.15) we find for excitation with y-polarization

$$\psi_2(t) = \frac{-i}{\sqrt{2}} e^{-(\gamma/2)t} \exp(-i\mathcal{H}t/\hbar)[a_{M+1}|M+1\rangle - a_{M-1}|M-1\rangle] ,$$

$$= \frac{-i}{\sqrt{2}} e^{-(\gamma/2)t} e^{-i(E_0+\mu_0 g M B)t/\hbar}$$

$$\times \left[a_{M+1} e^{-i\mu_0 g B t/\hbar} |M+1\rangle - a_{M-1} e^{(i\mu_0 g B)t/\hbar} |M-1\rangle\right] ,$$

$$= e^{-(\gamma/2)t} e^{-i(E_0+\mu_0 g M B)t/\hbar}$$

$$\times \left[|\psi_2\rangle_x \sin(\mu_0 g B t/\hbar) + |\psi_2\rangle_y \cos(\mu_0 g B t/\hbar)\right] .\tag{12.16}$$

This shows that the excited wave function under the influence of the magnetic field B changes continuously from $|\psi_2\rangle_x$ to $|\psi_2\rangle_y$, and back. If the fluorescence is detected through a polarizer with the transmission axis parallel to the x axis, only light from the $|\psi_2\rangle_x$ component is detected. The intensity of this light is

$$I(E_x, t) = C |\psi_{2x}(t)|^2 = C e^{-\gamma t} \sin^2(\mu_0 g B t/\hbar) ,\tag{12.17}$$

which gives, after integration over all time intervals from $t = -\infty$ to the time t_0 of observation, the Lorentzian intensity profile (12.7) with $\gamma = 1/\tau$ observed in the y-direction behind a polarizer in the x-direction (Fig. 12.2)

$$I_y(E_x) = \frac{C\tau I_0}{2} \frac{(2\mu_0 g\tau B/\hbar)^2}{1+(2\mu_0 g\tau B/\hbar)^2} \, , \tag{12.18}$$

which turns out to be identical with the classical result (12.7).

12.1.3 Experimental Arrangements

Level-crossing spectroscopy was used in atomic physics even before the invention of lasers [12.1, 12.10–12.12]. These investigations were, however, restricted to atomic resonance transitions that could be excited with intense hollow-cathode or microwave atomic-resonance lamps. Only a very few molecules have been studied, where accidental coincidences between atomic resonance lines and molecular transitions were utilized [12.6].

Optical pumping with tunable lasers or even with one of the various lines of fixed-frequency lasers has largely increased the application possibilities of level-crossing spectroscopy to the investigation of molecules and complex atoms. Because of the higher laser intensity the population density in the excited state is much larger, and therefore the signal-to-noise ratio is higher. In combination with two-step excitation (Sect. 10.4), Landé factors and lifetimes of high-lying Rydberg levels can be studied with this technique. Furthermore, lasers have introduced new versions of this technique such as stimulated level-crossing spectroscopy [12.13].

Level-crossing spectroscopy with lasers has some definite experimental advantages. Compared with other Doppler-free techniques it demands a relatively simple experimental arrangement. Neither single-mode lasers and frequency-stabilization techniques nor collimated molecular beams are required. The experiments can be performed in simple vapor cells, and the experimental expenditure is modest. In many cases no monochromator is needed since sufficient selectivity in the excitation process can be achieved to avoid simultaneous excitation of different molecular levels with a resulting overlap of several level-crossing signals.

There are, of course, also some disadvantages. One major problem is the change of the absorption profile with the magnetic field. The laser bandwidth must be sufficiently large in order to assure that all Zeeman components can absorb the radiation independent of the field strength B. On the other hand, the laser bandwidth should not be too large, to avoid simultaneous excitation of different, closely-spaced transitions. This problem arises particularly in *molecular* level-crossing spectroscopy, where several molecular lines often overlap within their Doppler widths. In such cases a compromise has to be found for an intermediate laser bandwidth, and the fluorescence may have to be monitored through a monochromator in order to discriminate against other transitions. Because of the high magnetic fields required for Hanle signals from short-lived molecular levels with high rotational quantum numbers and therefore with small Landé factors, careful magnetic shielding of the photomultiplier is essential to avoid a variation of the muliplier gain factor with the magnetic field strength.

The level-crossing signal may only be a few percent of the field-independent background intensity. In order to improve the signal-to-noise ratio, either the field is modulated or the polarizer in front of the detector is made to rotate and the signal is recovered by lock-in detection.

Since the total fluorescence intensity is independent of the magnetic field, ratio recording can be used to eliminate possible field-dependent absorption effects. If the signals of the two detectors D_1 and D_2 in Fig. 12.4 are $S_1 = I_\parallel(B) \cdot f(B)$; $S_2 = I_\perp(B) \cdot f(B)$, the ratio

$$R = \frac{S_1 - S_2}{S_1 + S_2} = \frac{I_\parallel - I_\perp}{I_\parallel + I_\perp} ,$$

no longer depends on the field-dependent absorption $f(B)$ [12.5].

12.1.4 Examples

A large number of atoms and molecules have been investigated by level-crossing spectroscopy using laser excitation. A compilation of the measurements up to 1975 can be found in the review of Walther [12.14], up to 1990 in [12.15], and up to 1997 in [12.16].

The iodine molecule has been very thoroughly studied with electric and magnetic level-crossing spectroscopy. The hyperfine structure of the rotational levels affects the profile of the level-crossing curves [12.17]. A computer fit to the non-Lorentzian superposition of all Hanle curves from the different hfs levels allows simultaneous determination of the Landé factor g and the lifetime τ [12.18]. Because of different predissociation rates the effective lifetimes of different hfs levels differ considerably.

In larger molecules the phase-coherence time of excited levels may be shorter than the population lifetime because of perturbations between closely spaced levels of different electronic states, which cause a dephasing of the excited-level wave functions. One example is the NO_2 molecule, where the width of the Hanle signal turns out to be more than one order of magnitude larger than expected from independent measurements of population lifetime and Landé factors [12.19, 12.20]. This discrepancy is explained by a short intramolecular decay time (dephasing time), but a much larger radiative lifetime [12.21].

The energy levels $E_k(B)$, split by a magnetic field, can be made to recross again at $B \neq 0$ by an additional electric field. Such Zeeman–Stark recrossings allow the determination of magnetic and electric moments of selectively excited rotational levels. In the case of perturbed levels these moments are altered, and the g-values may vary strongly from level to level even if the level-energy separation is small [12.22].

An external electric field may also prevent level crossings caused by the magnetic field. Such anti-crossing effects in Rydberg states of Li atoms in the presence of parallel magnetic and electric fields have been studied in the region where the Stark and Zeeman components of adjacent principal quantum numbers n overlap. The experimental results give information on core effects and interelectronic coupling [12.23, 12.24].

By stepwise excitation with two or three lasers, highly excited states and Rydberg levels of atoms and molecules can be investigated. These techniques allow measurement of the natural linewidth, the fine structure, and hfs parameters of high-lying Rydberg states. Most experiments have been performed on alkali atoms. For example, Li atoms in an atomic beam were excited by a frequency-doubled pulsed dye laser and level-crossing signals of the excited Rydberg atoms were detected by field ionization [12.25].

An example of a more recent experiment is the study of hyperfine structure of highly excited levels of neutral copper atoms, which yielded the magnetic-dipole and electric-quadrupole interaction constants. The experimental results allowed a comparison with theoretical calculations based on multiconfiguration Hartree–Fock methods [12.26]. Measurements of lifetimes [12.27] and of state multipoles using level-crossing techniques can be found in [12.28].

12.1.5 Stimulated Level-Crossing Spectroscopy

So far we have considered level crossing monitored through spontaneous emission. A level-crossing resonance can also manifest itself as a change in absorption of an intense monochromatic wave tuned to the molecular transition when the absorbing levels cross under the influence of external fields. The physical origin of this stimulated level-crossing spectroscopy is based on saturation effects and may be illustrated by a simple example [12.13].

Consider a molecular transition between the two levels $|a\rangle$ and $|b\rangle$ with the angular momenta $J = 0$ and $J' = 1$ (Fig. 12.6). We denote the center frequencies of the $\Delta M = +1, 0, -1$ transitions by ω_+, ω_0, and ω_- and the corresponding matrix elements by μ_+, μ_0, and μ_-, respectively. Without an external field the M sublevels are degenerate and $\omega_+ = \omega_- = \omega_0$. The monochromatic wave $\boldsymbol{E} = \boldsymbol{E}_0 \cos(\omega t - kx)$, linearly polarized in the y-direction, induces transitions with $\Delta M = 0$ without an external field. The saturated absorption of the laser beam is then, according to (7.24),

$$\alpha_s(\omega) = \frac{\alpha_0(\omega_0)}{\sqrt{1 + S_0}} \, \mathrm{e}^{-[(\omega-\omega_0)/\Delta\omega_D]^2} , \tag{12.19}$$

where $\alpha_0 = (N_a - N_b)|\mu|^2 \omega / (\hbar \gamma^2)$ is the unsaturated absorption coefficient, and $S_0 = E_0^2 |\mu|^2 / (\hbar^2 \gamma_s)$ is the saturation parameter at the line center (Sect. 3.6). If an external electric or magnetic field is applied in the z-direction, the degenerate levels split and the laser beam, polarized in the y-direction, induces transitions $\Delta M = \pm 1$ because it can be composed of $\sigma^+ + \sigma^-$ contributions (see previous section). If the level splitting $(\omega_+ - \omega_-) \gg \gamma$, the

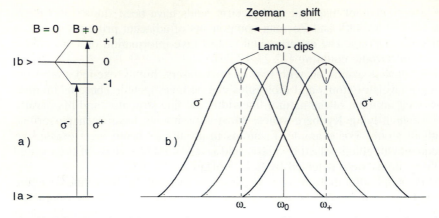

Fig. 12.6a,b. Stimulated level-crossing spectroscopy with a common lower level: (**a**) level scheme; and (**b**) saturation holes in the Doppler-broadened population distribution with and without magnetic field

absorption coefficient is now the sum of two contributions:

$$\alpha_s(\omega) = \frac{\alpha_0^+}{\left(1 + S_0^+\right)^{1/2}}\, e^{-[(\omega - \omega_+)/\Delta\omega_D]^2} + \frac{\alpha_0^-}{\left(1 + S_0^-\right)^{1/2}}\, e^{-[(\omega - \omega_-)/\Delta\omega_D]^2}\, .$$

(12.20)

For a $J = 1 \to 0$ transition $|\mu_+|^2 = |\mu_-|^2 = \frac{1}{2}|\mu_0|^2$. Neglecting the difference in the absorption coefficients $\alpha(\omega_+)$ and $\alpha(\omega_-)$ for $\omega_+ - \omega_- \ll \Delta\omega_D$, we may approximate (12.20) for $\hbar(\omega_+ - \omega_-) \gg \gamma$ by

$$\alpha_s(\omega) = \frac{\alpha_0^0}{\left(1 + \frac{1}{2} S_0\right)^{1/2}}\, e^{-[(\omega - \omega_0)/\Delta\omega_D]^2}\, ,$$

(12.21)

with $\quad S_0^+ \approx S_0^- = \frac{1}{2} S_0\, , \quad$ and $\quad \alpha_0^0 = \alpha_0^+ + \alpha_0^-\, ,$

which differs from (12.19) by the factor $\frac{1}{2}$ in the denominator. The difference of the absorption coefficient with and without a field (that is, for $\omega^+ = \omega^-$ and $\omega^+ - \omega^- > \gamma$) is

$$\Delta\alpha(\omega) = \alpha_0^0\, e^{-[(\omega - \omega_0)/\Delta\omega_D]^2} \left(\frac{1}{\sqrt{1 + \frac{1}{2} S^2}} - \frac{1}{\sqrt{1 + S^2}}\right)\, .$$

(12.22)

For $S \ll 1$, this becomes

$$\Delta\alpha(\omega) \approx \frac{1}{4} S^2 \alpha_0^0\, e^{-[(\omega - \omega_0)/\Delta\omega_D]^2}\, ,$$

(12.23)

where the saturation parameter $S(\omega)$ has a Lorentzian profile and width γ (see Sect. 3.6).

This demonstrates that the effect of level splitting on the absorption appears only in the saturated absorption and disappears for $S \to 0$. The absorption frequency is changed by altering the magnetic field. The absorption coefficient $\alpha_s(\omega_0, B)$, measured as a function of the magnetic field B while the laser is kept at ω_0, has a maximum at $B = 0$, and the transmitted laser intensity $I_t(B)$ shows a corresponding "Lamb dip." Although saturation effects may influence the line shape of the level-crossing signal, for small saturation it may still be essentially Lorentzian. The advantage of stimulated versus spontaneous level crossing is the larger signal-to-noise ratio and the fact that level crossings in the ground state can also be detected.

Most experiments on stimulated level crossing are based on intracavity absorption techniques because of their increased sensitivity (Sect. 6.2.2). Luntz and Brewer [12.29] demonstrated that even such small Zeeman splittings as in the molecular $^1\Sigma$ ground state can be precisely measured. They used a single-mode HeNe laser oscillating on the 3.39 μm line, which coincides with a vibration–rotation transition in the $^1\Sigma$ ground state of CH_4. Level crossings were detected as resonances in the laser output when the CH_4 transition was tuned by an external magnetic field. The rotational magnetic moment of the $^1\Sigma$ state of CH_4 was measured as $0.36 \pm 0.07 \mu_N$. In addition, Stark-tuned level-crossing resonances in the excited vibrational level of CH_4 have been detected with this method [12.30].

A number of stimulated level-crossing experiments have been performed on the active medium of gas lasers, where the gain of the laser transition is changed when sublevels of the upper or lower laser level cross each other. The whole gain tube is, for instance, placed in a longitudinal magnetic field, and the laser output is observed as a function of the magnetic field. Examples are the observation of stimulated hyperfine level crossings in a Xe laser [12.31], where accurate hyperfine splittings could be determined, or the measurement of Landé factors of atomic laser levels with high precision, as the determination of $g(^2P_4) = 1.3005 \pm 0.1\%$ in neon by Hermann et al. [12.32].

Two-photon induced level crossing [12.33], which relies on the OODR scheme of Raman-type transitions (Fig. 12.7), has been performed with the two neon transitions at $\lambda_1 = 632.8$ nm and $\lambda_2 = 3.39$ μm, which have the com-

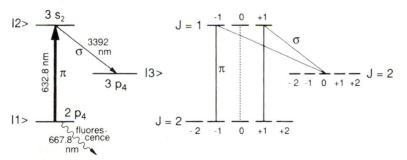

Fig. 12.7. Level scheme for two-photon stimulated Hanle effect. Only the sublevels $M = \pm 1$ of the lower state $|1\rangle$ are pumped by Raman-type transitions

mon upper $3s_2$ level. Here the Paschen notation of neon levels is used [12.34]. A HeNe laser in an external magnetic field B is simultaneously oscillating on both transitions. Each level splits into $(2J+1)$ Zeeman components, where J is the quantum number of the total angular momentum, which is the sum of the angular momenta of the core and the excited electron. The Hanle signal $S(B)$ is monitored via the fluorescence from the $3p_4$ level at $\lambda = 667.8\,\text{nm}$.

There is one important point to note. The width $\Delta B_{1/2}$ of the level-crossing signal reflects the average width $\gamma = \frac{1}{2}(\gamma_1 + \gamma_2)$ of the two crossing levels. If these levels have a smaller width than the other level of the optical transition, level-crossing spectroscopy allows a higher spectral resolution than, for example, saturation spectroscopy, where the limiting linewidth $\gamma = \gamma_a + \gamma_b$ is given by the sum of upper and lower level widths. Examples are all cw laser transitions, where the upper level always has a longer spontaneous lifetime than the lower level (otherwise inversion could not be maintained). Level-crossing spectroscopy of the upper level then yields a higher spectral resolution than the natural linewidth of the fluorescence between both levels. This is, in particular, true for level-crossing spectroscopy in electronic ground states, where the spontaneous lifetimes are infinite and other broadening effects, such as transit-time broadening or the finite linewidth of the laser, limit the resolution [12.35].

12.2 Quantum-Beat Spectroscopy

Quantum-beat spectroscopy represents not only a beautiful demonstration of the fundamental principles of quantum mechanics, but this Doppler-free technique has also gained increasing importance in atomic and molecular spectroscopy. Whereas commonly used spectroscopy in the frequency domain yields information on the stationary states $|k\rangle$ of atoms and molecules, which are eigenstates of the total Hamiltonian

$$\mathcal{H}\psi_k = E_k\psi_k , \tag{12.24a}$$

time-resolved spectroscopy with sufficiently short laser pulses characterizes nonstationary states

$$|\psi(t)\rangle = \sum c_k |\psi_k\rangle \, e^{-iE_k t/\hbar} , \tag{12.24b}$$

which can be described as a coherent time-dependent superposition of stationary eigenstates. Because of the different energies E_k this superposition is no longer independent of time. With time-resolved spectroscopy the time-dependence of $|\psi(t)\rangle$ can be measured in the form of a signal $S(t)$. The Fourier transformation of this time-dependent signal $S(t)$ yields the spectral information on the spectral components $c_k|\psi_k\rangle$ and their energies E_k. This is illustrated by the following subsections.

12.2.1 Basic Principles

If two closely spaced levels $|1\rangle$ and $|2\rangle$ are simultaneously excited from a common lower level $|i\rangle$ at time $t = 0$ by a short laser pulse with the pulse width $\Delta t < \hbar/(E_2 - E_1)$ (Fig. 12.8a), the wave function of the "coherent superposition state $|1\rangle + |2\rangle$" at $t = 0$

$$\psi(t = 0) = \sum c_k \psi_k(0) = c_1 \psi_1(0) + c_2 \psi_2(0) , \qquad (12.25a)$$

is represented by a linear combination of the wave functions ψ_k ($k = 1, 2$) of the "unperturbed" levels $|k\rangle$. The probability for the population of the level $|k\rangle$ at $t = 0$ is given by $|c_k|^2$. If the population N_k decays with the decay constant $\gamma_k = 1/\tau_k$ into a lower level $|m\rangle$, the time-dependent wave function of the coherent superposition state becomes

$$\psi(t) = \sum c_k \psi_k(0) \, \mathrm{e}^{-(\mathrm{i}\omega_{km} + \gamma_k/2)t} , \quad \text{with} \quad \omega_{km} = (E_k - E_m)/\hbar . \qquad (12.25b)$$

If the detector measures the total fluorescence emitted from both levels $|k\rangle$, the time-dependent signal $S(t)$ is

$$S(t) \propto I(t) = C \, |\langle\psi_m| \, \boldsymbol{\epsilon} \cdot \boldsymbol{\mu} \, |\psi(t)\rangle|^2 . \qquad (12.26)$$

C is a constant factor depending on the experimental arrangement, $\boldsymbol{\mu} = e \cdot \boldsymbol{r}$ is the dipole operator, and $\boldsymbol{\epsilon}$ gives the polarization direction of the emitted light. Inserting (12.25a) into (12.26) yields for equal decay constants $\gamma_1 = \gamma_2 = \gamma$ of both levels

$$I(t) = C \mathrm{e}^{-\gamma t}(A + B \cos \omega_{21} t) , \qquad (12.27a)$$

with

$$A = c_1^2 \, |\langle\psi_m| \, \boldsymbol{\epsilon} \cdot \boldsymbol{\mu} \, |\psi_1\rangle|^2 + c_2^2 \, |\langle\psi_m| \, \boldsymbol{\epsilon} \cdot \boldsymbol{\mu} \, |\psi_2\rangle|^2 ,$$
$$B = 2c_1 c_2 \, |\langle\psi_m| \, \boldsymbol{\epsilon} \cdot \boldsymbol{\mu} \, |\psi_1\rangle| \cdot |\langle\psi_m| \, \boldsymbol{\epsilon} \cdot \boldsymbol{\mu} \, |\psi_2\rangle| . \qquad (12.27b)$$

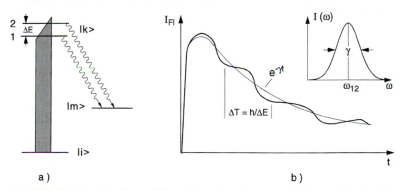

a)   b)

Fig. 12.8. (a) Level scheme illustrating the coherent excitation of levels $|1\rangle$ and $|2\rangle$ by a short pulse. (b) Quantum beats observed in the fluorescence decay of two coherently excited levels. *Insert:* Fourier spectrum $I(\omega)$ of (b) with $\omega_{12} = \Delta E/\hbar$

This represents an exponential decay $\exp(-\gamma t)$ superimposed by a modulation with the frequency $\omega_{21} = (E_2 - E_1)/\hbar$, which depends on the energy separation ΔE_{21} of the two coherently excited levels (Fig. 12.8b). This modulation is called *quantum beats*, because it is caused by the interference of the time-dependent wave functions of the two coherently excited levels.

The physical interpretation of the quantum beats is based on the following fact. When the molecule has reemitted a photon, there is no way to distinguish between the transitions $1 \rightarrow m$ or $2 \rightarrow m$ if the total fluorescence is monitored. As a general rule, in quantum mechanics the total probability amplitude of two indistinguishable processes is the sum of the two corresponding amplitudes and the observed signal is the square of this sum. This quantum-beat interference effect is analogous to Young's double-slit interference experiment. The quantum beats disappear if the fluorescence from only one of the upper levels is selectively detected.

The Fourier analysis of the time-dependent signal (12.27) yields a Doppler-free spectrum $I(\omega)$, from which the energy spacing ΔE as well as the width γ of the two levels $|k\rangle$ can be determined, even if ΔE is smaller than the Doppler width of the detected fluorescence (Fig. 12.8c). Quantum-beat spectroscopy therefore allows Doppler-free resolution [12.36].

12.2.2 Experimental Techniques

The experimental realization uses either short-pulse lasers, such as pulsed dye lasers (Sect. 5.7), or mode-locked lasers (Sect. 11.1). The time response of the detection system has to be fast enough to resolve the time intervals $\Delta t < \hbar/(E_2 - E_1)$. Fast transient digitizers or boxcar detection systems (Sect. 4.5) meet this requirement.

When atoms, ions, or molecules in a fast beam are excited and the fluorescence intensity is monitored as a function of the distance z downstream of the excitation point, the time resolution $\Delta t = \Delta z/v$ is determined by the particle velocity v and the resolvable spatial interval Δz from which the fluorescence is collected [12.37]. In this case, detection systems can be used that integrate over the intensity and measure the quantity

$$I(z)\Delta z = \left[\int_{t=0}^{\infty} I(t, z)\,dt \right] \Delta z \ .$$

The excitation can even be performed with cw lasers since the bandwidth necessary for coherent excitation of the two levels is assured by the short interaction time $\Delta t = d/v$ of a molecule with velocity v passing through a laser beam with the diameter d.

Example 12.2
With $d = 0.1$ cm and $v = 10^8$ cm/s $\rightarrow t \rightarrow 10^{-9}$ s. This allows coherent excitation of two levels with a separation up to 1000 MHz.

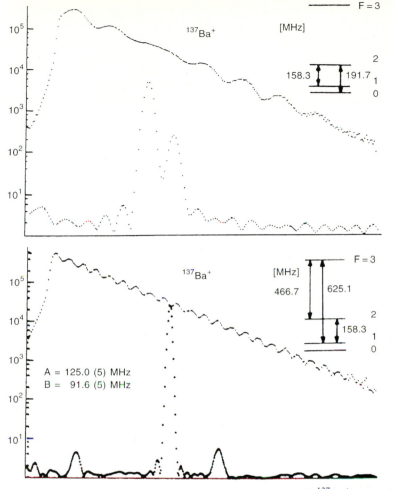

Fig. 12.9. Observed quantum beats in the fluorescence of ^{137}Ba$^+$ ions following an excitation of different sublevel groups at $\lambda = 455.4$ nm in a fast ion beam, and the corresponding Fourier transform spectra. The level schemes represent the hfs of the emitting upper level $6p^2P_{3/2}$ with the measured beat frequencies [12.37]

Figure 12.9 illustrates as an example quantum beats measured by Andrä et al. [12.37] in the fluorescence of ^{137}Ba$^+$ ions following the excitation of three hfs levels in the $6p^2P_{3/2}$ level. Either a tunable dye laser beam crossed perpendicularly with the ion beam or the beam of a fixed-frequency laser crossed under the tilting angle θ with the ion beam can be used for excitation. In the latter case, Doppler tuning of the ion transitions is achieved by tuning the ion velocity (Sect. 9.5) or the tilting angle θ, since the absorption frequency is $\omega = \omega_L - |\mathbf{k} \parallel \mathbf{v}| \cos \theta$. The lower spectra in Fig. 12.9 are the Fourier transforms of the quantum beats, which yield the hfs transitions depicted in the energy level diagrams.

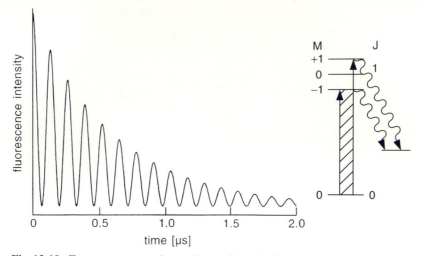

Fig. 12.10. Zeeman quantum beats observed in the fluorescence of Yb atoms in a magnetic field after pulsed excitation at $\lambda = 555.6\,\text{nm}$ [12.39]

By choosing the correct angle θ_i either the excitation $6s^2S_{1/2}(F'' = 1) \rightarrow 6p^2P_{3/2}(F' = 0, 1, 2)$ can be selected (upper part in Fig. 12.9) or the excitation $F'' = 2 \rightarrow F' = 1, 2, 3$ (lower part) can be selected.

Because of sub-Doppler resolution, quantum-beat spectroscopy has been used to measure fine or hyperfine structure and Lamb shifts of excited states of neutral atoms and ions [12.38].

If the Zeeman sublevels $|J, M = \pm 1\rangle$ of an atomic level with $J = 1$ are coherently excited by a pulsed laser, the fluorescence amplitudes are equal for the transitions $(J = 1, M = +1 \rightarrow J = 0)$ and $(J = 1, M = -1 \rightarrow J = 0)$. One therefore observes Zeeman quantum beats with a 100% modulation of the fluorescence decay (Fig. 12.10).

Quantum beats can be observed not only in emission but also in the transmitted intensity of a laser beam passing through a coherently prepared absorbing sample. This has first been demonstrated by Lange et al. [12.40, 12.41]. The method is based on time-resolved polarization spectroscopy (Sect. 7.4) and uses the pump-and-probe technique discussed in Sect. 11.4. A polarized pump pulse orientates atoms in a cell placed between two crossed polarizers (Fig. 12.11) and generates a coherent superposition of levels involved in the pump transition. This results in an oscillatory time dependence of the transition dipole moment with an oscillation period $\Delta T = 1/\Delta \nu$ that is determined by the splitting $\Delta \nu$ of the sublevels. When a probe pulse with variable delay Δt propagates through the sample, its transmittance $I_T(\Delta t)$ through the crossed analyzer shows this oscillation.

This time-resolved polarization spectroscopy has, quite similar to its cw counterpart, the advantage of a zero-background method, avoiding the problem of finding a small signal against a large background. In contrast to cw polarization spectroscopy, no narrow-band single-frequency lasers are required

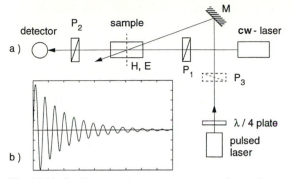

Fig. 12.11a,b. Quantum-beat spectroscopy of atomic or molecular ground states measured by time-resolved polarization spectroscopy: (**a**) experimental arrangement; and (**b**) Zeeman quantum beat signal of the Na $3^2S_{1/2}$ ground state recorded by a transient digitizer with a time resolution of 100 ns. (Single pump pulse, time scale 1 μs/div, magnetic field $B = 1.63 \times 10^{-4}$ T) [12.40]

and a broadband laser source can be utilized, which facilitates the experimental setup considerably. The *time* resolution of the pump-and-probe technique is *not* limited by that of the detector (Sect. 11.4). With a mode-locked cw dye laser a time resolution in the picosecond range can be achieved [12.42].

The *spectral* resolution of the Fourier-transformed spectrum is not limited by the bandwidth of the lasers or the Doppler width of the absorbing transition, but only by the homogeneous width of the levels involved. The amplitude of the quantum-beat signal is damped by collisions and by diffusion of the oriented atoms out of the interaction zone. If the diffusion time represents the limiting factor, the damping may be decreased by adding a noble gas, which slows down the diffusion. This decreases the spectral linewidth in the Fourier spectrum until collisional broadening becomes dominant. The decay time of the quantum-beat signal yields the phase relaxation time T_2 [12.43].

The basic difference of stimulated quantum beats in emission or absorption is illustrated by Fig. 12.12. The V-type scheme of Fig. 12.12a creates coherences in the excited state that can be observed by stimulated emission. The Λ-type scheme of Fig. 12.12b, on the other hand, describes coherence in the ground state, which can be monitored by absorption [12.41].

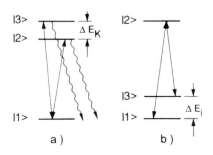

Fig. 12.12a,b. Preparation of coherences: (**a**) in the excited state; and (**b**) in the ground state of an atomic system

The time development of these coherences corresponds to a time-dependent susceptibility $\chi(t)$ of the sample, which affects the polarization characteristics of the probe pulse and appears as quantum beats of the transmitted probe pulse intensity.

An interesting technique for measuring hyperfine splittings of excited atomic levels by quantum-beat spectroscopy has been reported by Leuchs et al. [12.44]. The pump laser creates a coherent superposition of HFS sublevels in the excited state that are photoionized by a second laser pulse with variable delay. The angular distribution of photoelectrons, measured as a function of the delay time, exhibits a periodic variation because of quantum beats, reflecting the hfs splitting in the intermediate state.

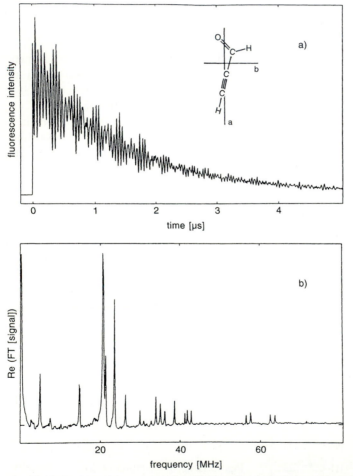

Fig. 12.13a,b. Complex quantum-beat decay of at least seven coherently excited levels of propynal (**a**) with the corresponding Fourier transform spectrum (**b**) [12.45]

12.2.3 Molecular Quantum-Beat Spectroscopy

Because quantum-beat spectroscopy offers Doppler-free spectral resolution, it has gained increasing importance in molecular physics for measurements of Zeeman and Stark splittings or of hyperfine structures and perturbations in excited molecules. The time-resolved measured signals yield not only information on the dynamics and the phase development in excited states but allow the determination of magnetic and electric dipole moments and of Landé g-factors.

One example is the measurement of hyperfine quantum beats in the polyatomic molecule propynal HC≡CCHO by Huber and coworkers [12.45]. In order to simplify the absorption spectrum and to reduce the overlap of absorbing transitions from different lower levels, the molecules were cooled by a supersonic expansion (Sect. 9.2). The Fourier analysis of the complex beat pattern (Fig. 12.13) showed that several upper levels had been excited coherently. Excitation with linear and circular polarization with and without an external magnetic field, allowed the analysis of this complex pattern, which is due to singlet–triplet mixing of the excited levels [12.45, 12.46].

Many other molecules, such as SO_2 [12.47], NO_2 [12.48], or CS_2 [12.49], have been investigated. A fine example of the capabilities of molecular quantum-beat spectroscopy is the determination of the magnitude and orientation of excited-state electric dipole moments in the vibrationless S_1 state of planar propynal [12.50].

More examples and experimental details as well as theoretical aspects of quantum-beat spectroscopy can be found in several reviews [12.36, 12.46, 12.51], papers [12.39–12.52, 12.54], and a book [12.53].

12.3 Excitation and Detection of Wave Packets in Atoms and Molecules

In the previous section we saw that the coherent excitation of several eigenstates by a short laser pulse leads to an excited nonstationary state

$$|\psi(t)\rangle = \sum c_k |\psi_k\rangle \exp(-\mathrm{i}E_k t/\hbar) \,,$$

which is described by a linear combination of stationary wave functions $|\psi_k\rangle$. Such a superposition is called a *wave packet*. Whereas quantum-beat spectroscopy gives information on the time development of this wave packet, it does not tell about the spatial localization of the system characterized by the wave packet $|\psi(x, t)\rangle$. This localization aspect is discussed in the present section.

We shall start with wave packets in atomic Rydberg states [12.55]: if a short laser pulse of duration τ brings free atoms from a common lower state into high Rydberg states (Sect. 10.4), all accessible levels within the energy interval $\Delta E = \hbar/\tau$ can be excited simultaneously. The total wave function describing the coherent superposition of excited Rydberg levels is the linear combination

$$\psi(x, t) = \sum_{n,\ell,m} a_{n,\ell,m} R_{n,\ell}(r) Y_{\ell,m}(\theta) \exp(-iE_n t/\hbar) ,\tag{12.28}$$

of stationary Rydberg state hydrogen-like functions with the principal quantum number n, angular momentum quantum number ℓ, and magnetic quantum number m. This superposition (12.28) represents a localized, nonstationary wave packet.

Depending on the kind of preparation, different types of Rydberg wave packets can be formed:

Radial wave packets are localized only with respect to the radial electronic coordinate r. They consist of a superposition, (12.28), with several different values of n but only a few values of ℓ or m.

Angular Rydberg wave packets, on the other hand, are formed by a superposition, (12.28), with many different values of ℓ and m but only a single fixed value of n.

When the Rydberg levels $|n\ell m\rangle$ are excited by a short laser pulse with the duration τ from the ground state $|i\rangle$, where the electron is localized within a few Bohr radii around the nucleus, the excitation process is fast compared to the oscillation period of a radial Rydberg wave packet, provided that $\tau \ll (E_n - E_{n-1})/h$. This fast excitation corresponds to a vertical transition in the potential diagram of Fig. 12.14a. A delayed probe pulse transfers the Rydberg electron into the ionization continuum (Fig. 12.14b). The probability of photoionization depends strongly on the radial coordinate r. Only in the near-core region (small r) can the Rydberg electron absorb a photon because

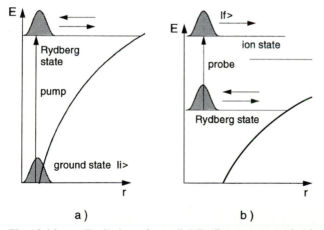

Fig. 12.14. (**a**) Excitation of a radial Rydberg wave packet by a short pump pulse from the ground state $|i\rangle$ into a Rydberg state at its inner turning point; and (**b**) its detection by a delayed photoionizing probe pulse [12.55]

the coupling to the nucleus can take the recoil and helps to satisfy energy and angular momentum conservation. For large values of r the electron moves like a free particle, which has a very small probability of absorbing a visible photon. Therefore the number of photoelectrons detected as a function of the delay time Δt between the pump and probe pulses will exhibit maxima each time the Rydberg electron is close to the nucleus, that is, if Δt is a multiple of the mean classical orbit time. Experiments performed on sodium Rydberg atoms confirm this oscillatory photoionization yield $N_{PE}(\Delta t)$ [12.56].

A second example, which is discussed in Sect. 11.4.4, concerns wave packets of molecular vibrations studied with femtosecond time resolution [11.140]. While stationary spectroscopy with narrow-band lasers excites the molecule into vibrational eigenstates corresponding to a time average over many vibrational periods, the excitation by short pulses produces nonstationary wave packets composed of all vibrational eigenfunctions within the energy range $\Delta E = h/\tau$. For high vibrational levels these wave packets represent the classical motion of the vibrating nuclei. As outlined in Sect. 11.4.4, the pump-and-probe technique with femtosecond resolution allows the real-time observation of the motion of vibrational wave packets. This is illustrated by Fig. 12.15, which shows schematically the level diagram of the I_2 molecule. A short pump pulse ($\Delta T \approx 70$ fs) at $\lambda = 620$ nm excites at least two vibrational levels v' in the $B^3\Pi_{ou}$ state simultaneously (coherently) from the $v'' = 0$

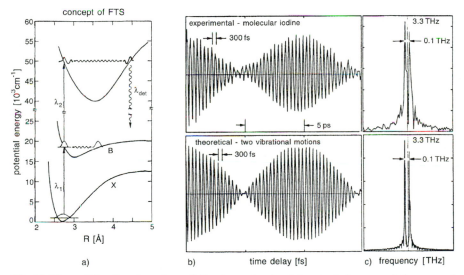

Fig. 12.15a–c. Vibrational motion of the I_2 molecule: (**a**) potential curve of the ground state, the excited $^3\Pi_{ou}$ state reached by the pump laser, and a higher excited state of I_2 populated by the probe laser. The probe laser-induced fluorescence is used to monitor the vibrational motion of the I_2 molecule in the $B^3\Pi_{ou}$ state depicted by the probability $\mathcal{P}(R)$ in a superpositional vibrational level v'. (**b**) Probe-laser induced fluorescence intensity as a function of the delay time between probe pulse and pump pulse, showing the oscillation of the wave packet. The short period gives the mean vibrational period in the two states, the long one represents the recurrence time. (**c**) Fourier spectrum of (b) [12.59]

vibrational level of the X-ground state. The probe pulse at $\lambda = 310\,\text{nm}$ excites the molecules further into a higher Rydberg state, where the excitation probability has a maximum at the inner turning point, since the Franck–Condon factor has here its maximum. The fluorescence from this state is monitored as a function of the delay time Δt between pump and probe pulse. The measured signal $I_{\text{Fl}}(\Delta t)$ shows fast oscillations with a frequency $\nu = (\nu_1 + \nu_2)/2$, which equals the mean vibrational frequency of the coherently excited vibrational levels in the $^3\Pi_{ou}$ state. Due to anharmonicity of the $^3\Pi_{ou}$ potential, the two vibrational frequencies ν_1 and ν_2 differ. The slowly varying envelope of the signal in Fig. 12.15 reflects the difference $\nu_1 - \nu_2$. After the recurrence time $\Delta T = 1/(\nu_1 - \nu_2)$, the two coherently excited vibrations are again "in phase." The Fourier transform of this beat signal gives the absolute vibrational energy E_{vib} and the separation ΔE_{vib} of the two coherently excited levels (Fig. 12.15c). This demonstrates that the "real time" vibrational motion of the molecule can be viewed with such a "stroboscopic" method. More examples can be found in [12.57–12.61].

12.4 Optical Pulse-Train Interference Spectroscopy

Let us consider atoms with optical transitions between a single level $|i\rangle$ and a state $|k\rangle$ that is split into two sublevels $|k_1\rangle$ and $|k_2\rangle$ (Fig. 12.8). If the atoms are irradiated by a short laser pulse of duration $\tau < \hbar/\Delta E = \hbar/(E_{k1} - E_{k2})$ and mean optical frequency $\omega = (E_i - E_k)/\hbar$, an induced dipole moment is produced, which oscillates at the frequency ω. The envelope of the damped oscillation shows a modulation with the beat frequency $\Delta\omega = \Delta E/\hbar$ (quantum beats, Sect. 12.2).

If the sample is now exposed not to a single pulse but to a regular train of pulses with the repetition frequency f such that $\Delta\omega = q \cdot 2\pi f$ $(q \in N)$, the laser pulses always arrive "in phase" with the oscillating dipole moments. With this synchronization the contributions of subsequent pulses in the regular pulse train add up coherently in phase and a macroscopic oscillating dipole moment is produced in the sample, where the damping between successive pulses is always compensated by the next pulse [12.62].

The regular pulse train of a mode-locked laser with the pulse-repetition frequency f corresponds in the frequency domain to a spectrum consisting of the carrier frequency $\nu_0 = \omega/2\pi$ and sidebands at $\nu_0 \pm q \cdot f$ $(q \in N)$ (Sect. 11.2). If a molecular sample with a level scheme depicted in Fig. 12.12b and a sublevel splitting $\Delta\nu = 2qf$ is irradiated by such a pulse train, where the frequeny ν_0 is chosen as $\nu_0 = (\nu_1 + \nu_2)/2$, the two frequencies $\nu_0 \pm qf$ are absorbed on the two molecular transitions ν_1, ν_2. This may be regarded as the superposition of two Raman processes ($|1\rangle \to |2\rangle \to |3\rangle$: Stokes process) and ($|3\rangle \to |2\rangle \to |1\rangle$: anti-Stokes process), where the population of the two sublevels $|1\rangle$ and $|3\rangle$ oscillates periodically with the frequency $\Delta\nu = \Delta E/h$. The level splitting ΔE can be obtained much more accurately from this oscillation frequency than from the difference of the two optical frequencies ν_1, ν_2.

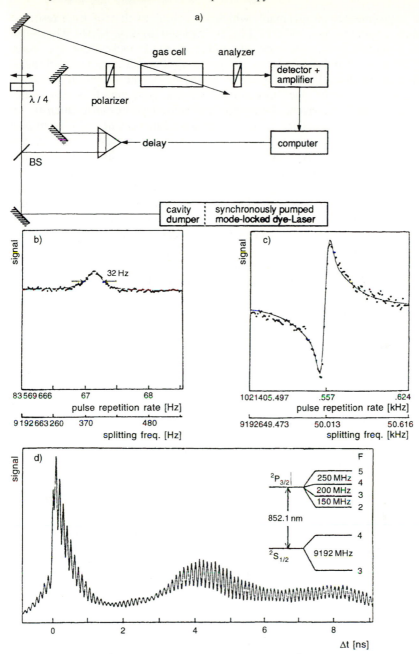

Fig. 12.16a–d. Measurements of the hfs splittings in the $7^2S_{1/2}$ ground state of Cs atoms with the pulse-train interference method. Excitation occurs on the D$_2$ line at $\lambda = 852.1$ nm with a repetition rate $f = q\Delta\nu$ with $q = 110$. (**a**) Experimental arrangement; (**b**) transmission of the probe pulse as a function of f; (**c**) fluorescene intensity $I_{Fl}(f)$ as a function of f with a modulated small external magnetic field; and (**d**) as a function of the delay time Δt at a fixed repetition frequency f [12.63]

The experimental arrangement, which is similar to that of time-resolved polarization spectroscopy (Fig. 12.11), is depicted in Fig. 12.16. The pulse train is provided by a synchronously pumped, mode-locked cw dye laser. A fraction of each pulse is split by the beam splitter BS and passes through an optical delay line. Pump and probe pulses propagate either in the same direction or in opposite ones through the sample cell, which is placed between two crossed polarizers. Either the transmitted probe-pulse intensity or the laser-induced fluorescence is monitored as a function of pulse-repetition frequency f and delay time Δt. For the resonance case $f = \Delta v/q$, the signal $S(f)$ becomes maximum (Fig. 12.16b). The halfwidth of $S(f)$ is determined by the homogeneous level widths of the levels $|1\rangle$ and $|3\rangle$ in Fig. 12.12b. If these are, for example, hfs levels in the electronic ground state, their spontaneous lifetime is very long (Problem 3.2b) and the halfwidth of $S(f)$ is mainly limited by collisional broadening and by the interaction time of the atoms with the laser. The latter can be increased by adding noble gases as buffer gas, which increases the time for diffusion of the pumped atoms out of the laser beam. As can be seen from Fig. 12.16d, linewidths down to a few hertz can be achieved. The center frequency of these signals $S(f)$ can be measured within uncertainties below 1 Hz, because the repetition frequency can be determined very precisely by digital counters [12.63]. The accuracy is mainly limited by the achievable signal-to-noise ratio and by asymmetries of the line profile $S(f)$.

This method allows the measurement of atomic level splittings by electronic counting of the pulse-repetition frequency! If the delay time Δt of the probe pulse is continuously varied at a fixed repetition frequency f, the oscillating atomic dipole moment becomes apparent through the time-dependent probe-pulse transmission $I_T(\Delta t)$, as shown in Fig. 12.16c for the Cs atom. The fast oscillation corresponds to the hfs splitting of the $7^2S_{1/2}$ ground state, the slowly damped oscillation to that of the excited state (quantum beats in transmission, Sect. 12.2.2). The Fourier transform of the quantum-beat signal yields the Doppler-free absorption spectrum of the D_2 line with the hfs splittings in the upper and lower states, and the decay time of the $^2P_{3/2}$ state.

12.5 Photon Echoes

Assume that N atoms have simultaneously been excited by a short laser pulse from a lower level $|1\rangle$ into an upper level $|2\rangle$. The total fluorescence intensity emitted on the transition $|2\rangle \rightarrow |1\rangle$ is given by (Sect. 2.7.4)

$$I_{Fl} = \sum_N \hbar\omega A_{21} = \frac{\omega^4}{3\pi\varepsilon_0 c^3} \frac{g_1}{g_2} \left| \sum_N \langle D_{21} \rangle \right|^2 , \tag{12.29}$$

where D_{21} is the dipole matrix element of the transition $|2\rangle \rightarrow |1\rangle$, and g_1, g_2 are the weight factors of levels $|1\rangle$, $|2\rangle$, respectively, see (2.50) and Table 2.2. The sum extends over all N atoms.

If the atoms are excited *incoherently*, no definite phase relations exist between the wave functions of the N excited atoms. The cross terms in the square of the sum (12.29) average to zero, and we obtain in the case of identical atoms

$$\left| \sum_N \langle D_{12} \rangle \right|^2 = \sum_N |\langle D_{12} \rangle|^2 = N |\langle D_{12} \rangle|^2 \rightarrow I_{Fl}^{incoh} = N\hbar\omega A_{21} . \quad (12.30)$$

The situation is drastically changed, however, under *coherent* excitation, where definite phase relations are established between the N excited atoms at the time $t = 0$ of excitation. If all N excited atomic states are *in phase* we obtain

$$\left| \sum_N \langle D_{12} \rangle \right|^2 = |N \langle D_{12} \rangle|^2 \rightarrow I_{Fl}^{coh} = N^2 |\langle D_{12} \rangle|^2 = N \cdot I_{Fl}^{incoh} . \quad (12.31)$$

This implies that the fluorescence intensity $I_{Fl}(t)$ at times $t \leq T_c$, where all excited atoms oscillate still *in phase*, is N times larger than in the incoherent case (Dicke super-radiance) [12.64].

This phenomenon of super-radiance is used in the photon-echo technique for high-resolution spectroscopy to measure population and phase decay times, expressed by the *longitudinal* and *transverse* relaxation times T_1 and T_2, see (12.1). This technique is analogous to the spin-echo method in nuclear magnetic resonance (NMR) [12.65]. Its basic principle may be understood in a simple model, transferred from NMR to the optical region [12.66].

Corresponding to the magnetization vector $\boldsymbol{M} = \{M_x, M_y, M_z\}$ in NMR spectroscopy, we introduce for our optical two-level system the *pseudopolarization vector*

$$\boldsymbol{P} = \{P_x, P_y, P_3\} , \quad \text{with} \quad P_3 = D_{12}\Delta N , \quad (12.32)$$

with the two components P_x, P_y of the atomic polarization, but a third component P_3 representing the product of the transition dipole moment D_{12} and the population density difference $\Delta N = N_1 - N_2$. Instead of the Bloch equation in NMR

$$\frac{d\boldsymbol{M}}{dt} = \boldsymbol{M} \times \boldsymbol{\Omega} , \quad (12.33a)$$

for the time variation of the magnetization $\boldsymbol{M}$ under the influence of a magnetic RF field with frequency $\boldsymbol{\Omega}$, one obtains from (2.80–2.85) the *optical Bloch equation*

$$\frac{d\boldsymbol{P}}{dt} = \boldsymbol{P} \times \boldsymbol{\Omega} - \{P_x/T_2; P_y/T_2; P_3/T_1\} , \quad (12.33b)$$

for the pseudopolarization vector under the influence of the optical field. The bracket contains the damping terms from phase relaxation and population relaxation that were neglected in (12.33a). The vector

$$\boldsymbol{\Omega} = \{(D_{12}/2\hbar), A_0, \Delta\omega\},\qquad\qquad\qquad\qquad (12.34)$$

is named the *optical nutation*. Its components represent the transition dipole D_{12}, the amplitude A_0 of the optical wave, and the frequency difference $\Delta\omega = \omega_{12} - \omega$ between the atomic resonance frequency $\omega_{12} = (E_2 - E_1/\hbar)$ and the optical field frequency ω. The time development of $\boldsymbol{P}$ describes the time-dependent polarization of the atomic system. This is illustrated by Fig. 12.17, which shows $\boldsymbol{P}$ in a $\{x, y, 3\}$ coordinate system, where the axis 3 represents the population difference $\Delta N = N_1 - N_2$. For times $t \leq -\tau$ all atoms are in their ground state where they are randomly oriented. This implies $P_x = P_y = 0$ and $\Delta N = N_1$. At $t = -\tau$ an optical pulse is applied that excites the atoms into level $|2\rangle$. For a proper choice of pulse intensity and pulse duration τ it is possible to achieve, at the end of the pump pulse ($t = 0$), equal populations $N_1 = N_2 \to \Delta N(t = 0) = 0$. This means the probabilities $|a_i|^2$ to find the system in level $|i\rangle$ change from $|a_1|^2 = 1$, $|a_2|^2 = 0$ before the pulse to $|a_1|^2 = |a_2|^2 = 1/2$ after the pulse. This implies $P_3(t = 0) = 0$, which means that the pseudopolarization vector now lies in the x–y-plane. Since such a pulse changes the phase of the probability amplitudes $a_i(t)$ by $\pi/2$, it is called a $\pi/2$-*pulse* (Sect. 2.7.6). At $t = 0$ all induced atomic dipoles oscillate in phase, resulting in the macroscopic polarization $\boldsymbol{P}$, which we have assumed to point into the y-direction (Fig. 12.17b).

Because of the finite linewidth $\Delta\omega$ of the transition $|1\rangle \to |2\rangle$ (for example, the Doppler width in a gaseous sample), the frequencies $\omega_{12} = (E_1 - E_2)/\hbar$ of the atomic transitions of our N dipoles are distributed within

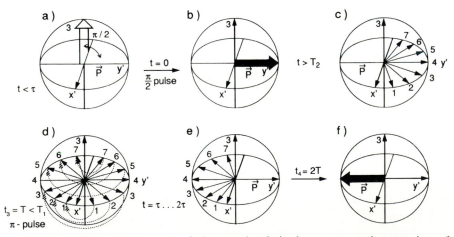

Fig. 12.17a–f. Time development of the pseudopolarization vector and generation of a photon echo observed at $t = 2\tau$ after applying a $\pi/2$-pulse at $t = 0$ and a π-pulse at $t = \tau$ (see text)

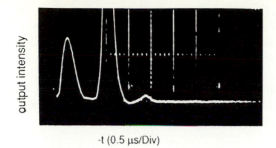

output intensity

-t (0.5 μs/Div)

Fig. 12.18. Oscilloscope trace of $\pi/2$- and π-pulses and a photon echo observed from SF_6 molecules that were excited by two CO_2 laser pulses [12.68]

the interval $\Delta\omega$. This causes the phases of the N oscillating dipoles to develop in time at different rates after the end of the $\pi/2$-pulse at $t = 0$. After a time $t > T_2$, which is large compared to the phase relaxation time T_2, the phases are again randomly distributed (Fig. 12.17c).

If a second laser pulse that has the proper intensity and duration to invert the phase of the induced polarization (π-pulse) is applied to the sample at a time $t_3 = T < T_1$, it causes a reversal of the phase development for each dipole (Fig. 12.17d–f). This means that after a time $t_4 = 2T$ all dipoles are again in phase (Fig. 12.17f). As discussed above, while these excited atoms are in phase they emit a superradiant signal at the time $t = 2T$ that is called *photon* echo (Fig. 12.18).

In the ideal case the magnitude of the photon echo is $N_2(2T)$ times larger than the incoherent fluorescence (which is emitted at all times $t > 0$), where $N_2(2T)$ is the number density of excited atoms at $t = 2T$.

There are, however, two relaxation processes that prevent the original state, as prepared just after the first $\pi/2$-pulse at $t = 0$, from being completely reestablished at the echo time $t = 2T$. Because of spontaneous or collision-induced decay, the population of the upper state decreases to

$$N_2(2T) = N_2(0)\,e^{-2T/T_1} \,. \tag{12.35}$$

This means that the echo amplitude decreases because of population decay with the longitudinal relaxation constant T_1.

A second, generally more rapid relaxation is caused by phase-perturbing collisions (Sect. 3.3), which change the phase development of the atoms and therefore prevent all atoms from being again in phase at $t = 2T$. Because such phase-perturbing collisions give rise to homogeneous line broadening (Sect. 3.5), the phase relaxation time due to these collisions is called T_2^{hom} in contrast to the inhomogeneous phase relaxation caused, for instance, by the different Doppler shifts of moving atoms in a gas (Doppler broadening).

The important point is that the *inhomogeneous* phase relaxation, which occurs between $t = 0$ and $t = 2T$ because of the random Doppler-shifted frequencies of atoms with different velocities, does *not* prevent a complete restoration of the initial phases by the π-pulse. If the velocity of a single atom does not change within the time $2T$, the different phase development of each atom between $t = 0$ and $t = T$ is exactly reversed by the π-pulse. This means that even in the presence of inhomogeneous line broadening, the homogeneous relaxation processes (that is, the homogeneous part of the broadening) can be measured with the photon-echo method. *This technique therefore allows Doppler-free spectroscopy.*

The production of the coherent state by the first pulse must, of course, be faster than these homogeneous relaxation processes. This implies that the laser pulse has to be sufficiently intense. From (2.93) we obtain the condition

$$D_{12}E_0 > \pi\hbar \left(1/T_1 + 1/T_2^{\text{hom}} \right) , \tag{12.36}$$

for the product of laser field amplitude E_0 and transition matrix element D_{12}. With typical relaxation times of 10^{-6} to 10^{-9} s the condition (12.36) requires power densities in the range kW/cm^2 to MW/cm^2, which can be readily achieved with pulsed or mode-locked lasers. With increasing delay time T between the first $\pi/2$-pulse and the second π-pulse the echo intensity I_e decreases exponentially as

$$I_e(2T) = I_e(0) \exp \left(-2T^{\text{hom}} \right) , \quad \text{with} \quad 1/T^{\text{hom}} = \left(1/T_1 + 1/T_2^{\text{hom}} \right) . \tag{12.37}$$

From the slope of a logarithmic plot of $I_e(2T)$ versus the delay time T the homogeneous relaxation times can be obtained.

The qualitative presentation of photon echoes, discussed above, may be put on a more quantitative base by using time-dependent perturbation theory. We outline briefly the basic considerations, which can be understood from the treatment in Sect. 2.9. For a more detailed discussion see [12.66].

A two-level system can be represented by the time-dependent wave function (2.58)

$$\psi(t) = \sum_{n=1}^{2} a_n(t) u_n \, e^{-E_n t/\hbar} . \tag{12.38}$$

Before the first light pulse is applied, the system is in the lower level $|1\rangle$, which means $|a_1| = 1$ and $|a_2| = 0$. The harmonic perturbation (2.46)

$$V = -D_{12}E_0 \cos \omega t , \quad \text{with} \quad \hbar\omega , = E_2 - E_1 \tag{12.39}$$

produces a linear superposition

$$\psi(t) = \cos \left(\frac{D_{12}E_0}{2\hbar} t \right) u_i \, e^{-iE_1 t/\hbar} + \sin \left(\frac{D_{12}E_0}{2\hbar} t \right) u_2 \, e^{-iE_2 t/\hbar} . \tag{12.40}$$

If this perturbation consists of a short intense light pulse of duration τ such that

$$(D_{12}E_0/\hbar)\tau = \pi/2 , \tag{12.41}$$

the total wave function becomes, with $\cos(\pi/4) = \sin(\pi/4) = 1/\sqrt{2}$,

$$\psi(t) = \frac{1}{\sqrt{2}} \left(u_1 e^{-iE_1 t/\hbar} + u_2 e^{-iE_2 t/\hbar} \right) . \tag{12.42}$$

After the time T, the phases have developed to $E_n T/\hbar$. If now a second π-pulse with $|(D_{12}E_0/\hbar)|\tau = \pi$ is applied, the wave functions of the upper and lower states are just interchanged, so that for time $t = T$

$$e^{-iE_1 T/\hbar} u_1 \rightarrow e^{-iE_1 T/\hbar} u_2 e^{-iE_2(t-T)/\hbar} ,$$

$$e^{-iE_2 T/\hbar} u_2 \rightarrow e^{-iE_2 T/\hbar} u_1 e^{-iE_1(t-T)/\hbar} . \tag{12.43}$$

The total wave function therefore becomes

$$\psi(t, T) = \frac{1}{\sqrt{2}} \left(u_2 e^{-i\omega_k(t-2T)/2} - u_1 e^{+i\omega_k(t-2T)/2} \right) , \tag{12.44}$$

and the dipole moment for each atom is

$$D_{12} = \langle \psi^* | er | \psi \rangle = - \langle u_2^* | er | u_1 \rangle e^{-i\omega_k(t-2T)} . \tag{12.45}$$

If the different atoms have slightly different absorption frequencies $\omega = (E_2 - E_1)/\hbar$ (for example, because of the different velocities in a gas), the phase factors for different atoms are different for $t \neq 2T$, and the macroscopic fluorescence intensity is the incoherent superposition (12.30) from all atomic contributions. However, for $t = 2T$ the phase factor is zero for all atoms, which implies that all atomic dipole moments are in phase, and super-radiance is observed.

Photon echoes were first observed in ruby crystals using two ruby laser pulses with variable delay [12.67]. The application of this technique to gases started with CO_2 laser pulses incident on a SF_6 sample. The collision-induced homogeneous relaxation time T_2^{hom} has been measured from the time decay of the echo amplitude. Figure 12.18 is an oscilloscope trace of the $\pi/2$- and π-pulses, which shows the echo obtained from a SF_6 cell at the pressure of 0.01 mb [12.68] as the small third pulse.

For sufficiently large transition-dipole matrix elements D_{12}, photon echoes can be also observed with cw lasers if the molecules are tuned for a short time interval into resonance with the laser frequency. There are two possible experimental realizations: the first uses an electro-optical pulsed modulator inside the laser cavity, which shifts the laser frequency $\omega \neq \omega_{12}$ for short time intervals τ into resonance ($\omega = \omega_{12}$) with the molecules. The second method shifts the absorption frequency ω_{12} of the molecules by a pulsed electric field (Stark shifting) into resonance with the fixed laser frequency ω_L. Figure 12.19 shows an example of an experimental arrangement [12.69] where a stable, tunable cw dye laser is used, and frequency switching is achieved with an ammonium dihydrogen phosphate (ADP) crystal driven by a sequence of low-voltage pulses. It produces a variation of the refractive index n and therefore a shift of the laser wavelength $\lambda = c/(2nd)$ in a resonator with mirror separation d.

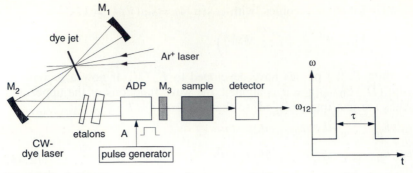

Fig. 12.19. Schematic of the laser-frequency switching apparatus for observing photon echoes and coherent optical transients. The intracavity ADP crystal is oriented in such a way that it changes the refractive index n and therefore the laser frequency without altering the polarization characteristics of the laser beam when a voltage is applied

Figure 12.20 illustrates the Stark switching technique, which can be applied to all those molecules that show a sufficiently large Stark shift [12.69]. In the case of Doppler-broadened absorption lines the laser of fixed frequency initially excited molecules of velocity v_z. A Stark pulse, which abruptly shifts the molecular absorption profile from the solid to the dashed curve, causes the velocity group v_z' to come into resonance with the laser frequencies Ω. We have assumed that the Stark shift of the molecular eigenfrequencies is larger than the homogeneous linewidth, but smaller than the Doppler width. With two Stark pulses the group v_z' emits an echo. This is shown in Fig. 12.20b, where the CH_3F molecules are switched twice into resonance with a cw CO_2 laser by 60-V/cm Stark pulses [12.70]. A more detailed discussion of photon echoes can be found in [12.66, 12.71–12.73].

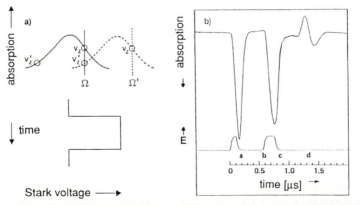

Fig. 12.20. (a) Stark switching technique for the case of a Doppler-broadened molecular transition. (b) Infrared photon echo for a $^{13}CH_3F$ vibration–rotation transition. The molecules are switched twice into resonance with a cw CO_2 laser by the two Stark pulses shown in the laser trace. The third pulse is the photon echo [12.69]

12.6 Optical Nutation and Free-Induction Decay

If the laser pulse applied to the sample molecules is sufficiently long and intense, a molecule (represented by a two-level system) will be driven back and forth between the two levels at the Rabi flopping frequency (2.96). The time-dependent probability amplitudes $a_1(t)$ and $a_2(t)$ are now periodic functions of time and we have the situation depicted in Fig. 2.21. Since the laser beam is alternately absorbed (induced absorption $E_1 \rightarrow E_2$) and amplified (induced emission $E_2 \rightarrow E_1$), the intensity of the transmitted beam displays an oscillation. Because of relaxation effects this oscillation is damped and the transmitted intensity reaches a steady state determined by the ratio of induced to relaxation transitions. According to (2.96) the flopping frequency depends on the laser intensity and on the detuning ($\omega_{12} - \omega$) of the molecular eigenfrequency ω_{12} from the laser frequency ω. This detuning can be performed either by tuning of the laser frequency ω (Fig. 12.19) or by Stark tuning of the molecular eigenfrequencies ω_{12} (Fig. 12.20).

We consider the case of a gaseous sample with Doppler-broadened transitions [12.73], where the molecular levels are shifted by a pulsed electric field. The Stark shift of the molecular eigenfrequencies is larger than the homogeneous linewidth but smaller than the Doppler width (Fig. 12.20a). During the steady-state absorption preceding the Stark pulse, the single-mode cw laser excites only a narrow velocity group of molecules around v_1 within the Doppler line shape. Sudden application of a Stark field shifts the transition frequency ω_{12} of this subgroup out of resonance with the laser field and another velocity subgroup around v_{z2} is shifted into resonance. At the end of the Stark pulse the first group of molecules is again switched into resonance. We can now realize two different situations:

(a) The two Stark pulses act for the optical excitation as $\pi/2$ and π-pulses. This generates a photon echo (Fig. 12.20b).
(b) The pulses are longer than the Rabi oscillation period. Then the molecules of each subgroup start their damped oscillation with the Rabi frequency $\Omega = D_{12}E_0/\hbar$ at the beginning of the excitation, giving the optical nutation patterns in Fig. 12.21. When the Stark pulse terminates at $t = \tau$, the initial velocity group is switched back into resonance and generates the second nutation pattern in Fig. 12.21a.

The amplitude $A(t)$ of this delayed nutation depends on the population of the first subgroup v_{z1} at the time τ where the Stark pulse ends. This population had partially been saturated before $t = 0$, but collisions during the time τ try to refill it. Berman et al. [12.74] have shown that

$$A(\tau) \propto N_1(\tau) - N_2(\tau) = \Delta N_0 + [\Delta N_0 - \Delta N(0)]e^{-\tau/T_1} , \tag{12.46}$$

where ΔN_0 is the unsaturated population difference in the absence of radiation, and $\Delta N(0)$, $\Delta N(\tau)$ are the saturated population difference at $t = 0$ or the partially refilled difference at $t = \tau$. The dependence of $A(\tau)$ on the length τ of the Stark pulse therefore allows measurement of the relaxation time T_1 for refilling the lower level and depopulating the upper one.

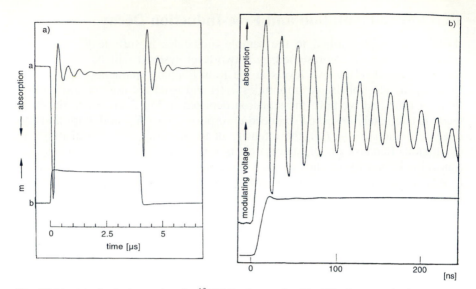

Fig. 12.21. (**a**) Optical nutation in $^{13}CH_3F$ observed with CO_2 laser excitation at $\lambda = 9.7\,\mu m$. The Rabi oscillations appear because the Stark pulse (*lower trace*) is longer than in Fig. 12.20. (**b**) Optical free-induction decay in I_2 vapor following resonant excitation with a cw dye laser at $\lambda = 589.6\,nm$. At the time $t = 0$ the laser is frequency-shifted with the arrangement depicted in Fig. 12.19 by $\Delta\omega = 54\,MHz$ out of resonance with the I_2 transition. The slowly varying envelope is caused by a superposition with the optical nutation of molecules in the velocity group $v_z = (\omega - \omega_0)/k$, which are now in resonance with the laser frequency ω. Note the difference in time scales of (a) and (b) [11.71]

After the end of the laser pulse the induced dipole moment of the co-herently prepared molecular system performs a damped oscillation at the frequency ω_{12}, where the damping is determined by the sum of all relaxation processes (spontaneous emission, collisions, etc.) that affect the phase of the oscillating dipole.

This *optical induction free decay* can be measured with a beat tech-nique: at time $t = 0$ the frequency ω of a cw laser is switched from $\omega = \omega_{12}$ to $\omega' \neq \omega_{12}$ out of resonance with the molecules. The superposition of the damped wave at ω_{12} emitted by the coherently prepared molecules with the wave at ω' gives a beat signal at the difference frequency $\Delta\omega = \omega_{12} - \omega'$, which is detected [12.70]. If $\Delta\omega$ is smaller than the Doppler width, the laser at ω' interacts with another velocity subgroup of molecules and produces optical nutation, which superimposes the free-induction decay and which is responsible for the slowly varying envelope in Fig. 12.21b.

These experiments give information on the polarizability of excited molecules (from the amplitude of the oscillation), on the transverse phase re-laxation times T_2 (which depend on the cross sections of phase-perturbing collisions, Sect. 3.3), and on the population decay time T_1. For more details see [12.72, 12.74].

12.7 Heterodyne Spectroscopy

Heterodyne spectroscopy uses two cw lasers with the frequencies ω_1 and ω_2 ($\Delta\omega = \omega_1 - \omega_2 \ll \omega_1, \omega_2$), which are stabilized onto two molecular transitions sharing a common level (Fig. 12.22). Measurements of the difference frequency of the two lasers then immediately yields the level splitting $\Delta E = E_1 - E_2 = \hbar\Delta\omega$ of the molecular levels $|1\rangle$ and $|2\rangle$.

For sufficiently low difference frequencies ($\Delta\nu = \Delta\omega/2\pi \leq 10^9$ Hz) fast photodiodes or photomultipliers can be used for detecting $\Delta\omega$. The two laser beams are superimposed onto the active area of the detector. The output signal S of the photodetector is proportional to the incident intensity, averaged over the time constant τ of the detector. For $\Delta\omega \ll 2\pi/\tau \ll \omega = (\omega_1 + \omega_2)/2$ we obtain the time-averaged output signal

$$\langle S \rangle \propto \left\langle (E_1 \cos\omega_1 t + E_2 \cos\omega_2 t)^2 \right\rangle = \frac{1}{2}\left(E_1^2 + E_2^2\right) + E_1 E_2 \cos(\omega_1 - \omega_2)t ,$$

(12.47)

which contains, besides the constant term $(E_1^2 + E_2^2)/2$, an ac term with the difference frequency $\Delta\omega$. Fast electronic counters can directly count frequencies up to $\Delta\nu \simeq 10^9$ Hz. For higher frequencies a mixing technique can be used. The superimposed two laser beams are focused onto a nonlinear crystal, which generates the difference frequency $\Delta\omega = \omega_1 - \omega_2$ (Sect. 5.8). The output of the crystal is then mixed with a microwave ω_{MW} in a Schottky diode or a point-contact MIM diode (Sects. 4.5.2c and 5.8.5). For a proper choice of ω_{MW} the difference frequency $\Delta\omega - \omega_{MW}$ falls into a frequency range $< 10^9$ Hz and thus can be counted directly. Often the output of the two lasers and the microwave can be mixed in the same device [12.75].

The accuracy of stabilizing the two lasers onto molecular transitions increases with decreasing linewidth. Therefore, the narrow Lamb dips of Doppler-broadened molecular transitions measured with saturation spectroscopy (Sect. 7.2) are well suited [12.76]. This was proved by Bridges and

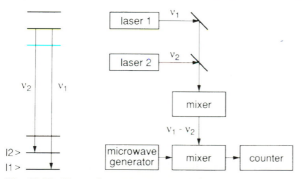

Fig. 12.22. Heterodyne spectroscopy with two lasers, stabilized onto two molecular transitions. The difference frequency, generated in a nonlinear crystal, is either measured or a second downconversion is used by mixing with a microwave

Chang [12.77] who stabilized two CO_2 lasers onto the Lamb dips of different rotational lines within the vibrational transitions $(00°1) \rightarrow (10°0)$ at $10.4\,\mu m$ and $(00°1) \rightarrow (02°0)$ at $9.4\,\mu m$. The superimposed beams of the two lasers were focused into a GaAs crystal, where the difference frequency was generated.

Ezekiel and coworkers [12.78] used the reduction of the Doppler width in a collimated molecular beam (Sect. 9.1) for accurate heterodyne spectroscopy. The beams of two argon lasers intersect the collimated beam of I_2 molecules perpendicularly. The laser-induced fluorescence is utilized to stabilize the laser onto the centers of two hfs components of a visible rotational transition. The difference frequency of the two lasers then yields the hfs splittings.

Instead of two different lasers, a single laser can be employed when its output is amplitude-modulated at the variable frequency f. Besides the carrier frequency ν_0 two tunable sidebands $\nu_0 \pm f$ appear. If the carrier frequency ν_0 is stabilized onto a selected molecular line, the sidebands can be tuned across other molecular transitions (sideband spectroscopy) [12.79, 12.80]. The experimental expenditure is smaller since only one laser has to be stabilized. The sideband tuning can be achieved with acousto-optical modulators at relatively small RF powers. The difference frequency $\Delta\omega$ is, however, restricted to achievable modulation frequencies ($\Delta\omega \leq 2\pi f_{max}$).

12.8 Correlation Spectroscopy

Correlation spectroscopy is based on the correlation between the measured frequency spectrum $S(\omega)$ of the photodetector output and the frequency spectrum $I(\omega)$ of the incident light intensity. This light may be the direct radiation of a laser or the light scattered by moving particles, such as molecules, dust particles, or microbes (homodyne spectroscopy). In many cases the direct laser light and the scattered light are superimposed on the photodetector, and the beat spectrum of the coherent superposition is detected (heterodyne spectroscopy) [12.81, 12.82].

12.8.1 Basic Considerations

We assume that the incident light wave has the amplitude $E(r, t)$, which does not vary over the area of the photocathode. The probability for the emission of one photoelectron within the time interval dt is

$$\mathcal{P}^{(1)}(t)\, dt = c \cdot \epsilon_0 \eta E^*(t) E(t)\, dt = \eta I(t)\, dt \,, \qquad (12.48)$$

where $\eta(\lambda)$ is the quantum efficiency of the photocathode, and $I(t) = c\epsilon_0 E^*(t) E(t)$ is the incident intensity. The measured photocurrent $i(t)$ at the detector output is then

$$i(t) = ea\mathcal{P}^{(1)}(t) \,, \qquad (12.49)$$

where a is the amplification factor of the detector, which is $a = 1$ for a vacuum photocell, but $a \simeq 10^6 - 10^8$ for a photomultiplier.

The joint probability per unit time that one photoelectron is emitted at time t *and also* another electron at time $t + \tau$ is described by the product

$$\mathcal{P}^{(2)}(t, t + \tau) = \mathcal{P}^{(1)}(t) \cdot \mathcal{P}^{(1)}(t + \tau) = \eta^2 I(t) \cdot I(t + \tau) . \qquad (12.50)$$

Most experimental arrangements generally detect the time-averaged photocurrent. For the description of the measured signals, therefore, the normalized correlation functions

$$G^{(1)}(\tau) = \frac{\langle E^*(t) \cdot E(t + \tau) \rangle}{\langle E^*(t) \cdot E(t) \rangle} = \lim_{T \to \infty} \frac{c\epsilon_0}{\langle I \rangle} \frac{1}{T} \int_0^T E^*(t) \cdot E(t + \tau) \, \mathrm{d}t , \quad (12.51)$$

$$G^{(2)}(\tau) = \frac{\langle E^*(t) \cdot E(t) \cdot E^*(t + \tau) \cdot E(t + \tau) \rangle}{[\langle E^*(t) \cdot E(t) \rangle]^2} = \frac{\langle I(t) \cdot I(t + \tau) \rangle}{\langle I \rangle^2} ,$$

$$= \lim_{T \to \infty} \frac{1}{\langle I \rangle^2} \frac{1}{T} \int_0^T I(t) \cdot I(t + \tau) \, \mathrm{d}t , \qquad (12.52)$$

of first and second order are introduced to describe the correlation between the field amplitudes and the intensities, respectively, at the times t and $t + \tau$. The first-order correlation function $G^{(1)}(t)$ is identical to the normalized mutual coherence function, defined by (2.119).

The Fourier transform of $G^{(1)}(\tau)$

$$F(\omega) = \lim_{T' \to \infty} \frac{1}{T'} \int_0^{T'} G^{(1)}(\tau) \, \mathrm{e}^{\mathrm{i}\omega\tau} \, \mathrm{d}\tau , \qquad (12.53)$$

can be calculated by inserting (12.51) into (12.53). This yields, as was first shown by Wiener [12.83]

$$F(\omega) = \frac{E^*(\omega) \cdot E(\omega)}{\langle E^* \cdot E \rangle} = \frac{I(\omega)}{\langle I \rangle} . \qquad (12.54)$$

The Fourier transform of the first-order correlation function $G^{(1)}(\tau)$ represents the normalized frequency spectrum of the incident light-wave intensity $I(\omega)$ (Wiener–Khintchine theorem) [12.82, 12.83].

Example 12.3

For completely uncorrelated light we have $G^{(1)}(\tau) = \delta(0 - \tau)$, where δ is the Kronnecker symbol, that is, $\delta(x) = 1$ for $x = 0$, and 0 elsewhere. For a strictly monochromatic light wave $E = E_0 \cos \omega t$, the first-order correlation function $G^{(1)}(\tau) = \cos(\omega t)$ becomes a periodic function of time τ oscillating with the period $\Delta\tau = 2\pi/\omega$ between the values $+1$ and -1 (Sect. 2.8.4).

The second-order correlation function is a constant $G^{(2)}(\tau) = 1$ for cases of both completely uncorrelated light and strictly monochromatic light.

Inserting $G^{(1)}(\tau) = \delta(0 - \tau)$ into (12.53) yields a constant intensity $I(\omega) = \langle I \rangle$ (white noise) for completely uncorrelated light, while $G^{(1)}(\tau) = \cos(\omega \tau)$ gives $F(\omega) = \cos^2 \omega t + \frac{1}{2}$, and therefore $I(\omega) = \langle I \rangle (\cos^2 \omega t + \frac{1}{2})$.

We must now discuss how the measured frequency spectrum $i(\omega)$ of the photocurrent is related to the wanted frequency spectrum $I(\omega)$ of the incident light wave.

Similar to (12.51) we define the correlation function

$$C(\tau) = \frac{\langle i(t) \cdot i(t + \tau) \rangle}{\langle i^2 \rangle} , \tag{12.55}$$

of the time-resolved photocurrent $i(t)$. The Fourier transform of (12.55) yields, analogously to (12.53), the power spectrum $P(\omega)$ of the measured photocurrent

$$P(\omega) = \int_0^\infty C(\tau) e^{i\omega\tau} \, d\tau . \tag{12.56}$$

The correlation function (12.55) of the photocurrent is determined by the two contributions

(a) The statistical process of emission of photoelectrons from the photocathode, resulting in a statistical fluctuation of the photocurrent, even when the incident light wave would consist of a completely regular and noise-free photon flux.

(b) The amplitude fluctuations of the incident light wave that are caused by the characteristics of the photon source or by the objects that scatter light onto the detector.

Let us consider these contributions for two different cases: for a constant light intensity I described by the probability function

$$\mathcal{P}(I) = \delta(I - \langle I \rangle) , \tag{12.57}$$

the mean photoelectron number within the time interval dt shall be $\langle n \rangle$. Then the probability $\mathcal{P}(n, dt)$ of detecting n photoelectrons within dt is given by the Poisson distribution

$$\mathcal{P}(n, dt) = \frac{1}{n!} \langle n \rangle^n e^{-\langle n \rangle} \, dt , \tag{12.58}$$

with the mean square fluctuation

$$\left\langle (\Delta n)^2 \right\rangle = \langle n \rangle , \tag{12.59}$$

of the rate $n = i_{ph}/e$ of the photoelectron emission [12.84, 12.85].

The light intensity scattered by statistically fluctuating particles can generally be described by a Gaussian intensity distribution [12.86]

$$\mathcal{P}(I) \, dI = \frac{1}{\langle I \rangle} e^{-I/\langle I \rangle} \, dI . \tag{12.60}$$

If a quasi-monochromatic light wave $I(\omega)$ with a statistically fluctuating intensity distribution (12.60) falls onto the photocathode, the probability of detecting n photoelectrons within the time interval dt is not described by (12.58) but by the Bose–Einstein distribution

$$\mathcal{P}(n)\,dt = \frac{dt}{1+\langle n\rangle(1+1/\langle n\rangle)^n}\,, \tag{12.61}$$

which leads to a mean-square deviation of the photoelectron rate of $\langle(\Delta n)^2\rangle = \langle n\rangle^2$ instead of (12.59).

Including both contributions a) and b), the mean-square deviation becomes [12.81]

$$\left\langle(\Delta n)^2\right\rangle = \langle n\rangle + \langle n\rangle^2\,, \tag{12.62}$$

and the correlation function

$$C(\tau) = e\,\langle n\rangle\,\delta(0-\tau) + \langle e\cdot n\rangle^2 = \langle i\rangle\,\delta(0-\tau) + i(t)\cdot i(t+\tau)\,,$$
$$= \langle i\rangle\,\delta(0-\tau) + \langle i\rangle^2 G^{(2)}(\tau)\,. \tag{12.63}$$

This shows that the autocorrelation function $C(\tau)$ of the photoelectron current is directly related to the second-order correlation function $G^{(2)}(\tau)$ of the light field.

As proved by Siegert [12.87] for optical fields with a Gaussian intensity distribution (12.60), the second-order correlation function $G^{(2)}(\tau)$ is related to $G^{(1)}(\tau)$ by the Siegert relation

$$G^{(2)}(\tau) = [G^{(2)}(0)]^2 + \left|G^{(1)}(\tau)\right|^2 = 1 + \left|G^{(1)}(\tau)\right|^2\,. \tag{12.64}$$

The procedure to obtain information on the spectral distribution of the light incident onto the detector is then as follows: from the time-resolved measured photocurrent $i(t)$ the correlation function $C(\tau)$ can be derived (12.63), which yields $G^{(2)}(\tau)$ and with (12.64) $G^{(1)}(\tau)$. The Fourier transform of $G^{(1)}(\tau)$ then gives the wanted intensity spectrum $I(\omega)$ according to (12.53) and (12.54).

Example 12.4

A time-dependent optical field with the amplitude

$$A(t) = A_0\,e^{-i\omega_0 t - (\gamma/2)t}\,, \tag{12.65}$$

has, according to (12.51), the first-order autocorrelation function

$$G^{(1)}(\tau) = e^{-i\omega_0\tau}\,e^{-(\gamma/2)\tau}\,. \tag{12.66}$$

The Fourier transformation (12.53), (12.54) yields the spectral distribution

$$I(\omega) = \frac{\langle I \rangle}{2\pi} \int_{-\infty}^{+\infty} e^{i(\omega-\omega_0)\tau - (\gamma/2)\tau} \, d\tau = \frac{\langle I \rangle \gamma/2\pi}{(\omega-\omega_0)^2 + (\gamma/2)^2} , \qquad (12.67)$$

of the Lorentzian line profile (Sect. 3.1). Inserting (12.66) into (12.63) yields the correlation function $C(\tau)$ of the photoelectron current which is, according to (12.56), related to the power spectrum $P_i(\omega)$ of the photocurrent

$$P_i(\omega \geq 0) = \frac{e}{\pi} \langle i \rangle + 2 \langle i \rangle^2 \delta(\omega) + 2 \langle i \rangle^2 \frac{\gamma/2\pi}{\omega^2 + (\gamma/2)^2} . \qquad (12.68)$$

Equation (12.68) reveals that the power spectrum has a peak at $\omega = 0$ [$\delta(\omega = 0) = 1$], giving the dc part $2\langle i \rangle^2$. The first term $(e/\pi)\langle i \rangle$ represents the shot-noise term, and the third term describes a Lorentzian frequency distribution peaked at $\omega = 0$ with the total power $(2/\pi\gamma)\langle i \rangle^2$. This represents the light-beating spectrum, which gives information on the intensity profile $I(\omega)$ of the incident light wave.

For more examples see [12.81, 12.82].

12.8.2 Correlation Spectroscopy of Light Scattered by Microparticles

The light scattered elastically by small particles contains information on the size, structure, and movement of the particles. The most simple case that can be treated theoretically is the model of homogeneous spherical particles.

We will distinguish three different situations:

(a) Randomly distributed static scatterers. In this case the correlation function is time independent, and one obtains for the normalized correlation function

$$G^{(1)} = \frac{\langle E(t) E(t+\tau) \rangle}{|E(t)|^2} = 1 .$$

The intensity spectrum of the elastically scattered radiation from N scatterers is

$$I(\omega) = N |E|^2 \delta(\omega - \omega_0) .$$

(b) Spherical scatterers with constant velocity v

$$G^{(1)}(\tau) = e^{i\boldsymbol{q} \cdot \boldsymbol{v}\tau} ,$$

where $\boldsymbol{q} = \boldsymbol{k}_0 - \boldsymbol{k}_s$ is the wave vector difference between incident and scattered light. The intensity of the scattered light is

$$I(\omega) = N |E|^2 \delta(\omega - \omega_0 + \boldsymbol{q} \cdot \boldsymbol{v}) .$$

(c) Spherical scatterers with translational diffusion

$$G^{(1)}(\tau) = e^{iD_T q^2 |\tau|} ,$$

$$I(\omega) = N |E|^2 \frac{D_\tau q^2 / \pi}{(\omega - \omega_0)^2 + (D_T q^2)^2} ,$$

where D_T is the diffusion coefficient (see also Sect. 5.6).

12.8.3 Homodyne Spectroscopy

The output of a laser does not represent a strictly monochromatic wave (even if the laser frequency is stabilized) because of frequency and phase fluctuations (Sect. 5.6). Its intensity profile $I(\omega)$ with the linewidth $\Delta \omega$ can be detected by homodyne spectroscopy. The different frequency contributions inside the line profile $I(\omega)$ interfere, giving rise to beat signals at many different frequencies $\omega_i - \omega_k < \Delta \omega$ [12.81]. If a photodetector is irradiated by the attenuated laser beam, the frequency distribution of the photocurrent (12.68) can be measured with an electronic spectrum analyzer. This yields, according to the discussion above, the spectral profile of the incident light. In the case of narrow spectral linewidths this correlation technique represents the most accurate measurement for line profiles [12.88].

This high accuracy can be further used to determine even slow motions of microparticles by measuring the spectral intensity profile of light scattered by these particles. When monochromatic light is scattered by moving particles that show thermal motion, the field amplitudes $E(\omega)$ show a Gaussian distribution. The correlation between $E(t)$, $E(t+\tau)$ of the scattered light can be described by the correlation function $G^{(1)}(\tau)$. The experimental arrangement for measuring the homodyne spectrum is shown in Fig. 12.23. The power spectrum $P(\omega)$ of the photocurrent (12.68),

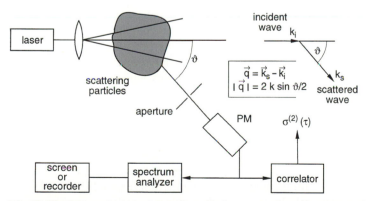

Fig. 12.23. Schematic experimental setup for measuring the autocorrelation function of scattered light (homodyne spectroscopy), with a correlator as an alternative to an electronic spectrum analyzer

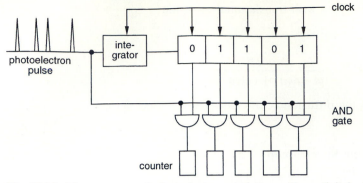

Fig. 12.24. Measurement of photoelectron statistics with a digital clipping correlator [12.88]

which is related to the spectral distribution $I(\omega)$, is measured either directly by an electronic spectrum analyzer, which mixes the frequencies with a local oscillator and determines the difference frequencies, or with a correlator, which determines the Fourier transform of the autocorrelation function $C(\tau) \propto \langle i(t) \rangle \langle i(t+\tau) \rangle$. According to (12.63), $C(\tau)$ is related to the intensity correlation function $G^{(2)}(\tau)$, which yields $G^{(1)}(\tau)$ (12.64), and $I(\omega)$.

The correlation function $C(\tau) = \langle i(t) \rangle \langle (t+\tau) \rangle / \langle i \rangle^2$ can be directly measured with a digital correlator, which measures the photoelectron statistics. A simple version of several possible realizations is depicted in Fig. 12.24. The time is divided into equal sections Δt by an internal clock. If the number $N \cdot \Delta t_i$ of photoelectrons measured within the ith time interval Δt_i exceeds a given number N_m, the correlator gives a normalized output pulse, counted as "one." For $N \Delta t_i < N_m$, the output gives "zero." The output pulses are transferred to a "shift register" and to "AND gates," which open for "one" and close for "zero", and are finally stored in counters (Malvern correlator [12.89, 12.90]).

An example of an application of homodyne spectroscopy is the measurement of the size distribution of small particles in the nanometer range that are dispersed in liquids or gases and fly through a laser beam. The intensity I_s of the scattered light depends in a nonlinear way on the size and the refractive index of the particles. For small homogeneous spheres with diameters d, which are small compared to the wavelength ($d \ll \lambda$), the relation $I_s \propto d^6$ holds. In Fig. 12.25 the measured intensity distribution of laser light scattered by a mixture of latex spheres with $d = 22.8$ nm and $d = 5.7$ nm (small squares) is compared with the size distribution obtained from electron microscopy, which can be used for calibration [12.91].

A further example is the light scattering by a liquid sample when the temperature is changed around the critical temperature and the sample undergoes a phase transition [12.92]. At such a phase transition long-range and short-range order generally change. This affects the correlation between the molecules.

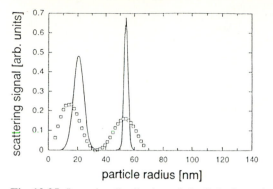

Fig. 12.25. Intensity distribution of the light intensity scattered by a mixture of two kinds of homogeneous small latex balls with diameters $d = 22.8$ nm and $d = 5.7$ nm (*squares*), compared with the size distribution of the balls measured by electron microscopy (*solid line*) [12.91]

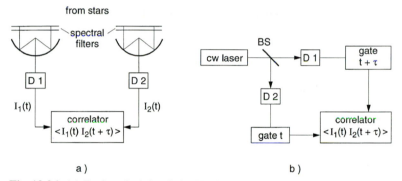

Fig. 12.26. (**a**) Basic principle of the Hanbury Brown–Twiss intensity-correlation interferometer in astronomy; (**b**) its application to measurements of spectral characteristics and statistics of laser radiation

A famous example for the application of intensity-correlation interferometry in astronomy is the Hanbury Brown–Twiss interferometer sketched in Fig. 12.26. In its original form it was intended to measure the degree of spatial coherence of starlight (Sect. 2.8) [12.93] from which diameters of stars could be determined. In its modern version it measures the degree of coherence and the photon statistics of laser radiation in the vicinity of the laser threshold [12.94].

12.8.4 Heterodyne Correlation Spectroscopy

For heterodyne correlation spectroscopy the scattered light that is to be analyzed is superimposed on the photocathode by part of the direct laser beam (Fig. 12.27). Assume the scattered light has the amplitude E_s at the detector and a frequency distribution around ω_s, whereas the direct laser radiation

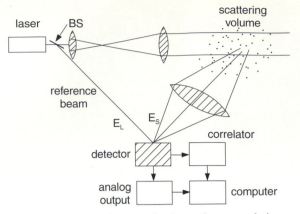

Fig. 12.27. Schematic setup for heterodyne correlation spectroscopy

$E_L = E_0 \exp(-i\omega_0 t)$ acts as monochromatic local oscillator with constant amplitude E_0. The total amplitude is then

$$E(t) = E_0 \exp(-i\omega_0 t) + E_s(t) . \tag{12.69}$$

The autocorrelation function of the photocurrent is, according to (12.55), (12.63)

$$C(\tau) = e \cdot \eta \delta(0-\tau) \langle E^*(t) \cdot E(t) \rangle + e^2 \eta^2 E^*(t) E(t) E^*(t+\tau) E(t+\tau) . \tag{12.70a}$$

Inserting (12.69) into (12.70a) we note that for $E_s \ll E_0$ the terms with E_s^2 can be neglected. Furthermore, the time average $\langle E_L E_s \rangle$ is zero. Therefore (12.70a) reduces to

$$C(\tau) = e \cdot i_L \delta(0-\tau) + i_L^2 + i_L \langle i_s \rangle \left(e^{i\omega_0 \tau} G^{(1)} + \text{c.c.} \right) . \tag{12.70b}$$

The power spectrum of the photocurrent is then obtained from (12.56) as

$$P_i(\omega) = \frac{e}{2\pi} i_L + i_L^2 \delta(0-\omega) + \frac{i_L}{2\pi} \langle i_s \rangle \int_{+\infty}^{-\infty} e^{i\omega\tau} \left[e^{i\omega_0 \tau} G_s^{(1)}(\tau) + \text{c.c.} \right] d\tau . \tag{12.71}$$

The first term represents the shot noise, the second is the dc term, and the third term gives the heterodyne beat spectrum with the difference $(\omega - \omega_0)$ and sum $(\omega + \omega_0)$ frequencies. The detector is not fast enough to detect the sum frequency. The output signal of the third term therefore contains only the difference frequency $(\omega - \omega_0)$. If this difference frequency lies in an inconvenient frequency range, the local oscillator can be shifted with an acousto-optic modulator to the frequency $\omega_L \pm \Delta\omega$ in order to bring the difference frequency $\omega_L \pm \Delta\omega - \omega_s$ into an easily accessible range [12.95].

Example 12.5

When we assume the correlation function

$$G_s^{(1)}(\tau) = e^{-(i\omega_s + \gamma)\tau} \, ,$$

for the scattered light, the power spectrum of (12.71) becomes

$$P_i(\omega) = \frac{ei_L}{\pi} + i_L^2 \delta(0 - \omega) + \frac{(\gamma/\pi)\langle i_s \rangle}{(\omega_s - \omega_L)^2 + \gamma^2} \, . \tag{12.72}$$

The heterodyne spectrum is a Lorentzian like the homodyne spectrum (12.67), but its maximum is shifted from ω_L to $\omega = (\omega_s - \omega_L)$.

12.8.5 Fluorescence Correlation Spectroscopy and Single Molecule Detection

The combination of confocal microscopy, laser excitation, and time-resolved fluorescence measurements allows the detection of single molecules or microparticles and the investigation of their movements in gases or liquids [12.96] (see Sect. 15.1.2). The exciting laser is focused by a microscope into the sample. If the focal diameter is smaller than the average distance between the molecules diluted in a liquid or a gas, a single molecule is excited (see Fig. 15.5). The laser-induced fluorescence or the light scattered by the microparticle is collected by the same microscope and imaged onto a small aperture in front of the detector. This aperture defines the volume from which light can be detected. The lifetime of the excited state is generally much smaller than the diffusion time of the molecule through the laser focus. In the case of microparticles the scattered light has no time delay against the excitation light. In this case the number of scattered photons depends on the transit time of the particle through the laser focus. Each particle moving through the excitation region gives a small burst pulse of scattered light. The random time sequence of these pulses yields information on the diffusion time, the transport coefficients, and the correlation between the movement of different particles. Measuring the fluorescence intensity signal $F(t)$ and $F(t+\tau)$ allows the determination of the first-order correlation function.

This chapter can only give a brief survey of some aspects of coherent spectroscopy. For more details, the reader is referred to the literature. Good representations of the fundamentals of coherent spectroscopy and its various applications can be found in several textbooks [12.82,12.97–12.99] and conference proceedings [12.100].

Problems

12.1 The Zeeman components of an excited atomic level with the quantum numbers $(J = 1, S = 0, L = 1, I = 0)$ and a radiative lifetime $\tau = 15$ ns cross at zero magnetic field. Calculate the Landé g-factor and the halfwidth $\Delta B_{1/2}$(HWHM) of the Hanle signal. Compare this with the halfwidth measured for a molecular level with $(J = 20, S = 0, I = 0, \tau = 15\,\text{ns})$.

12.2 What is the width $\Delta B_{1/2}$ of the Hanle signal from the crossing of the two Zeeman components of one of the hyperfine levels $F = 3$ or $F = 2$ in the ground state $5^2S_{1/2}$ of the rubidium atom $^{85}_{37}\text{Rb}$ with a nuclear spin $I = 5/2$, if the mean transit time of the atoms in a buffer gas through the excitation region is $T = 0.1\,\text{s}$? Explain why this method allows the realization of a sensitive magnetometer (see also [12.35]).

12.3 The two Zeeman components of an atomic level $(L = 1, S = 1/2, J = 1/2)$ are coherently excited by a short laser pulse. Calculate the quantum-beat period of the emitted fluorescence in a magnetic field of $10^{-2}\,\text{Tesla}$.

12.4 In a femtosecond pump-and-probe experiment the pump pulse excites the vibrational levels $v' = 10 - 12$ in the $A^1\Sigma_u$ state of Na_2 coherently. The vibrational spacings are $109\,\text{cm}^{-1}$ and $108\,\text{cm}^{-1}$. The probe laser pulse excites the molecules only from the inner turning point into a higher Rydberg state. If the fluorescence $I_{\text{Fl}}(\Delta t)$ from this state is observed as function of the delay time Δt between the pump and probe pulses, calculate the period ΔT_1 of the oscillating signal and the period ΔT_2 of its modulated envelope.

12.5 Two lasers are stabilized onto the Lamb dips of two molecular transitions. The width $\Delta \nu$ of the Lamb dip is $10\,\text{MHz}$ and the rms fluctuations of the two laser frequencies is $\delta \nu = 0.5\,\text{MHz}$. How accurately measured is the separation $\nu_1 - \nu_2$ of the two transitions, if the signal-to-noise ratio of the heterodyne signal of the two superimposed laser beams is 50?

12.6 Particles in a liquid show a random velocity distribution with $\sqrt{\langle v^2 \rangle} = 1\,\text{mm/s}$. What is the spectral line profile of the beat signal, if the direct beam of a HeNe laser at $\lambda = 630\,\text{nm}$ and the laser light scattered by the moving particles is superimposed on the photocathode of the detector?

13. Laser Spectroscopy of Collision Processes

The two main sources of information about atomic and molecular structure and interatomic interactions are provided by spectroscopic measurements and by the investigation of elastic, inelastic, or reactive collision processes. For a long time these two branches of experimental research developed along separate lines without a strong mutual interaction. The main contributions of classical spectroscopy to the study of collision processes have been the investigations of collision-induced spectral line broadening and line shifts (Sect. 3.3).

The situation has changed considerably since lasers were introduced to this field. In fact, laser spectroscopy has already become a powerful tool for studying various kinds of collision processes in more detail. The different spectroscopic techniques presented in this chapter illustrate the wide range of laser applications in collision physics. They provide a better knowledge of the interaction potentials and of the different channels for energy transfer in atomic and molecular collisions, and they give information that often cannot be adequately obtained from classical scattering experiments without lasers.

The high spectral resolution of various Doppler-free techniques discussed in Chaps. 7–9 has opened a new dimension in the measurement of collisional line broadening. While in Doppler-limited spectroscopy small line-broadening effects at low pressures are completely masked by the much larger Doppler width, Doppler-free spectroscopy is well suited to measure line-broadening effects and line shifts in the kilohertz range. This allows detection of *soft collisions* at large impact parameters of the collision partners that probe the interaction potential at large internuclear separations and that contribute only a small line broadening.

Some techniques of laser spectroscopy, such as the method of separated fields (*optical Ramsey fringes*, Sect. 14.4), coherent transient spectroscopy (Sect. 12.4), or polarization spectroscopy (Sect. 7.4) allow one to distinguish between phase-changing, velocity-changing, or orientation-changing collisions.

The high *time resolution* that is achievable with pulsed or mode-locked lasers (Chap. 11) opens the possibility for studying the dynamics of collision processes and relaxation phenomena. The interesting questions of how and how fast the excitation energy that is selectively pumped into a polyatomic molecule by absorption of laser photons, is redistributed among the various degrees of freedom by intermolecular or intramolecular energy transfer can be addressed by femtosecond laser spectroscopy.

One of the attractive goals of laser spectroscopy of reactive collision processes is the basic understanding of chemical reactions. The fundamental question in laser chemistry of how the excitation energy of the reactants influences the reaction probability and the internal state distribution of the reaction products can, at least, partly be answered by detailed laser-spectroscopic investigations. Section 13.4 treats some experimental techniques in this field.

The most detailed information on the collision process can be obtained from laser spectroscopy of *crossed-beam* experiments, where the initial quantum states of the collision partners before the collision are marked, and the scattering angle as well as the internal energy of the reactants is measured. In such an "ideal scattering experiment" all relevant parameters are known (Sect. 13.5).

The new and interesting field of *light-assisted collisions* (often called optical collisions), where absorption of laser photons by a collision pair results in an effective excitation of one of the collision partners, is briefly treated in the last section of this chapter. For further studies of the subject covered in this chapter, the reader is referred to books [13.1–13.3], reviews [13.4–13.10], and conference proceedings [13.11–13.14].

13.1 High-Resolution Laser Spectroscopy of Collisional Line Broadening and Line Shifts

In Sect. 3.3. we discussed how elastic and inelastic collisions contribute to the broadening and shifts of spectral lines. In a semiclassical model of a collision between partners A and B, the particle B travels along a definite path $r(t)$ in a coordinate system with its origin at the location of A. The path $r(t)$ is

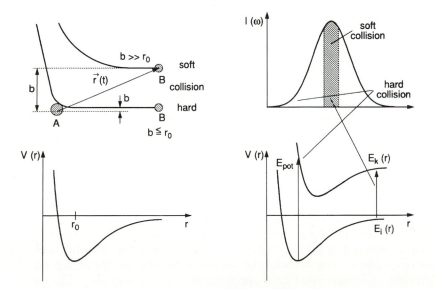

Fig. 13.1. Interaction potential $V(r)$ and semiclassical model for soft collisions with impact parameter $b \gg r_0$ and hard collisions $(b < r_0)$

completely determined by the initial conditions $r(0)$ and $(dr/dt)_0$ and by the interaction potential $V(r, E_A, E_B)$, which may depend on the internal energies E_A and E_B of the collision partners. In most models a spherically symmetric potential $V(r)$ is assumed, which may have a minimum at $r = r_0$ (Fig. 13.1). If the impact parameter b is large compared to r_0 the collision is classified as a *soft* collision, while for $b \leq r_0$ *hard* collisions occur.

For soft collisions B passes only through the long-range part of the potential and the scattering angle θ is small. The shift ΔE of the energy levels of A or B is accordingly small. If one of the collision partners absorbs or emits radiation during a soft collision, its frequency distribution will be only slightly changed by the interaction between A and B. Soft collisions therefore contribute to the kernel of a collision-broadened line, that is, the spectral range around the line center.

For *hard* collisions, on the other hand, the collision partners pass through the short-range part of their interaction potential, and the level shift ΔE during the collision is correspondingly larger. Hard collisions therefore contribute to the line wings (Fig. 3.1).

13.1.1 Sub-Doppler Spectroscopy of Collision Processes

In Doppler-limited spectroscopy the effect of collisions on the line kernel is generally completely masked by the much larger Doppler width. Any information on the collision can therefore be extracted only from the line wings of the Voigt profile (Sect. 3.2), by a deconvolution of the Doppler-broadened Gaussian profile and the Lorentzian profile of collisional broadening [13.15]. Since the collisional linewidth increases proportionally to the pressure, reliable measurements are only possible at higher pressures where collisional broadening becomes comparable to the Doppler width. At such high pressures, however, many-body collisions may no longer be negligible, since the probability that N atoms are found simultaneously within a volume $V \approx r^3$ increases with the Nth power of the density. This implies that not only two-body collisions between A and B but also many-body collisions $A + N \cdot B$ may contribute to line profile and line shift. In such cases the line profile no longer yields unambiguous information on the interaction potential $V(A, B)$ [13.16].

With techniques of sub-Doppler spectroscopy, even small collisional broadening effects can be investigated with high accuracy. One example is the measurement of pressure broadening and shifts of narrow Lamb dips (Sect. 7.2) of atomic and molecular transitions, which is possible with an accuracy of a few kilohertz if stable lasers are used. The most accurate measurements have been performed with stabilized HeNe lasers on the transitions at 633 nm [13.17] and 3.39 μm [13.18]. When the laser frequency ω is tuned across the absorption profiles of the absorbing sample inside the laser resonator, the output power of the laser $P_L(\omega)$ exhibits sharp Lamb peaks (inverse Lamb dip) at the line center of the absorbing transitions (Sect. 7.3). The line profiles of these peaks are determined by the pressure in the absorption cell, by saturation broadening, and by transit-time broad-

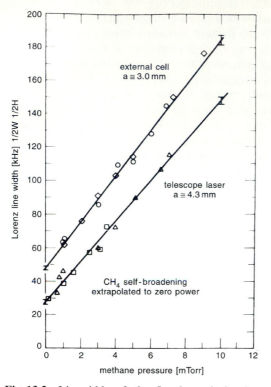

Fig. 13.2. Linewidth of the Lamb peak in the output power of a HeNe laser at $\lambda = 3.39\,\mu m$ with an intracavity CH_4 absorption cell and different beam waists of the expanded laser beam, causing a different transit-time broadening [13.18]

ening (Sect. 3.4). Center frequency ω_0, linewidth $\Delta\omega$, and line profile $P_L(\omega)$ are measured as a function of the pressure p (Fig. 13.2). The slope of the straight line $\Delta\omega(p)$ yields the line-broadening coefficient [13.19], while the measurement of $\omega_0(p)$ gives the collision-induced line shift.

A more detailed consideration of collisional broadening of Lamb dips or peaks must also take into account velocity-changing collisions. In Sect. 7.2 it was pointed out that only molecules with the velocity components $v_z = 0 \pm \gamma/k$ can contribute to the simultaneous absorption of the two counterpropagating waves. The velocity vectors $\boldsymbol{v}$ of these molecules are confined to a small cone within the angles $\beta \leq \pm\epsilon$ around the plane $v_z = 0$ (Fig. 13.3), where

$$\sin\beta = v_z/|\boldsymbol{v}| \Rightarrow \sin\varepsilon \leq \gamma/(k\cdot v)\,, \text{ with } \quad v = |\boldsymbol{v}| \,. \tag{13.1}$$

During a collision a molecule is deflected by the angle $\theta = \beta'' - \beta$ (Fig. 13.3). If $v\cdot\sin\theta < \gamma/k$, the molecule after the collision is still in resonance with the standing light wave inside the laser resonator. Such soft collisions with deflection angles $\theta < \epsilon$ therefore do not appreciably change the absorption

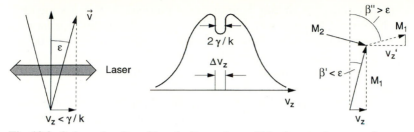

Fig. 13.3. Only molecules with velocity vectors within the angular range $\beta \leq \epsilon$ around the plane $z = 0$ contribute to the line profile of the Lamb dip. Velocity-changing collisions that increase β to values $\beta > \epsilon$ push the molecules out of resonance with the laser field

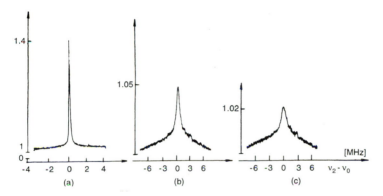

Fig. 13.4a–c. Line profiles of Lamb peaks of a HeNe laser at $\lambda = 3.39 \, \mu m$ with intra-cavity CH_4 absorption cell: (**a**) pure CH_4 at 1.4 mbar; (**b**) addition of 30 mbar He; and (**c**) 79 mbar He [13.20]

probability of a molecule. Because of their statistical phase jumps (Sect. 3.3) they do, however, contribute to the linewidth. The line profile of the Lamb dip broadened by soft collisions remains Lorentzian.

Collisions with $\theta > \epsilon$ may shift the absorption frequency of the molecule out of resonance with the laser field. After a hard collision the molecule, therefore, can only contribute to the absorption in the line wings.

The combined effect of both kinds of collisions gives a line profile with a kernel that can be described by a Lorentzian profile slightly broadened by soft collisions. The wings, however, form a broad background caused by velocity-changing collisions. The whole profile cannot be described by a single Lorentzian function. In Fig. 13.4 such a line profile is shown for the Lamb peak in the laser output $P_L(\omega)$ at $\lambda = 3.39 \, \mu m$ with a methane cell inside the laser resonator for different pressures of CH_4 and He [13.20].

13.1.2 Combination of Different Techniques

Often the collisional broadening of the Lamb dip and of the Doppler profile can be measured simultaneously. A comparison of both broadenings allows the separate determination of the different contributions to line broadening. For phase-changing collisions there is no difference between the broadenings of the two different line profiles. However, velocity-changing collisions do affect the Lamb-dip profile (see above), but barely affect the Doppler profile because they mainly cause a redistribution of the velocities but do not change the temperature.

Since the homogeneous width γ of the Lamb-dip profile increases with pressure p, the maximum allowed deflection angle ϵ in (13.1) also increases with p. A comparison of pressure-induced effects on the kernel and on the background profile of the Lamb dips and on the Doppler profile therefore yields more detailed information on the collision processes. Velocity-selective optical pumping allows the measurement of the shape of velocity-changing collisional line kernels over the full thermal range of velocity changes [13.21].

Collisions may also change the orientation of atoms and molecules (Sect. 10.1), which means that the orientational quantum number M of an optically pumped molecule is altered. This can be monitored by polarization spectroscopy (Sect. 7.4). The orientation of molecules within the velocity interval $\Delta v_z = (\omega_p \pm \gamma)/k$ induced by the polarized pump radiation with frequency ω_p determines the polarization characteristics of the transmitted probe laser wave at frequency ω and therefore the detected signal $S(\omega)$. Any collision that alters the orientation, the population density N_i of molecules in the absorbing level $|i\rangle$, or the velocity v of absorbing molecules affects the line profile $S(\omega)$ of the polarization signal. The velocity-changing collisions have the same effect on $S(\omega)$ as on the Lamb-dip profiles in saturation spectroscopy. The orientation-changing collisions decrease the magnitude of the signal $S(\omega)$, while inelastic collisions result in the appearance of new *satellite polarization signals* in neighboring molecular transitions (Sect. 13.2). Orientation-changing collisions may also be detected with OODR saturation spectroscopy (Sect. 10.4) if the Lamb-dip profile is measured in dependence on the pressure for different angles α between the polarization planes of the linearly polarized pump and probe waves.

A commonly used technique for the investigation of depolarizing collisions in *excited* states is based on the orientation of atoms or molecules by optical pumping with a polarized laser and the measurement of the degree of polarization $P = (I_\parallel - I_\perp)/(I_\parallel + I_\perp)$ of the fluorescence emitted from the optically pumped or the collisionally populated levels (Sect. 13.2).

Because of the large mean radius $\langle r_n \rangle \propto n^2$ of the Rydberg electron, Rydberg atoms or molecules have very large collision cross sections. Therefore optical transitions to Rydberg states show large collisional broadening, which can be studied with Doppler-free two-photon spectroscopy or with two-step excitation (Sect. 10.4). For illustration, Fig. 13.5 illustrates pressure broadening and shifts of a rotational transition to a Rydberg level of the Li_2 molecule measured with Doppler-free OODR polarization spectroscopy (Sect. 10.5) in a lithium/argon heat pipe [13.22], where the intermediate level $B(v', J')$ was pumped optically by a circularly polarized pump laser. For the chosen temperature and pressure conditions the argon is confined to the cooled outer parts of the heat pipe, and the center of the heat pipe contains pure lithium

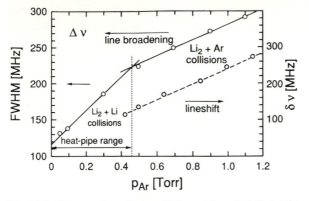

Fig. 13.5. Pressure broadening (*left scale*) and shift (*right scale*) of a Doppler-free rotational line in the Rydberg system $B\,^1\Sigma_u \rightarrow 6d\delta\,^1\Delta_g$ of the Li_2 molecule in a heat pipe for $Li_2^* + Li$ collisions at $p < 0.4\,mbar$ and for $Li_2^* + Ar$ collisions for $p > 0.4\,mbar$ [13.23]

vapor (98% Li atoms and 2% Li_2 molecules) with a total vapor pressure $p(L_i) = p(Ar)$ up to argon pressures of 0.7 mbar. The observed pressure broadening and shift in this range $p < 0.7\,mbar$ are therefore caused by $Li_2^* + Li$ collisions.

For $p(Ar) > 0.7\,mbar$ the argon begins to diffuse into the central part, if the temperature and thus the lithium vapor pressure remains constant while $p(Ar)$ increases. The slope of the curve $\Delta\omega(p)$ yields for $p > 0.7\,mbar$ the cross section for $Li_2^* + Ar$ collisions. For the example depicted in Fig. 13.5 the cross sections for line broadening are $\sigma(Li_2^* + Li) = 60\,nm^2$ and $\sigma(Li_2^* + Ar) = 41\,nm^2$, whereas the line shifts are $\partial\nu/\partial p = -26\,MHz/mbar$ for $Li_2^* + Ar$ collisions [13.23]. Similar measurements have been performed on Sr Rydberg atoms [13.24], where the pressure shift and broadenings of Rydberg levels $R(n)$ for the principle quantum numbers n in the range $8 \leq n \leq 35$ were observed.

13.2 Measurements of Inelastic Collision Cross Sections of Excited Atoms and Molecules

Inelastic collisions transfer the internal energy of an atom or molecule A either into internal energy of the collision partner B or into relative kinetic energy of both partners. In the case of atoms, either *electronic* internal energy can be transferred to the collision partner or the magnetic energy of spin-orbit interaction. In the first case this leads to collision-induced transitions between different electronic states, and in the second case to transitions between atomic fine-structure components [13.24]. For *molecules* many more possibilities for energy transfer by inelastic collisions exist, such as rotational–vibrational or electronic energy transfer, and collision-induced dissociation.

A large variety of different spectroscopic techniques has been developed for the detailed investigation of these various inelastic collisions. They are illustrated in the following sections.

13.2.1 Measurements of Absolute Quenching Cross Sections

In Sect. 11.3 we saw that the effective lifetime $\tau_k^{\text{eff}}(N_B)$ of an excited level $|k\rangle$ of an atom or molecule A depends on the density N_B of the collision partners B. From the slope of the Stern–Volmer plot

$$1/\tau_k^{\text{eff}} = 1/\tau_k^{\text{rad}} + \sigma_k^{\text{total}}\overline{v}N_B , \qquad (13.2)$$

the total deactivation cross section σ_k^{total} can be obtained, see (11.47). Since the collisional deactivation diminishes the fluorescence intensity emitted by $|k\rangle$, the inelastic collisions are called *quenching collisions* and σ_k^{total} is named the *quenching cross section*.

Several possible decay channels contribute to the depopulation of level $|k\rangle$, and the quenching cross section can be written as the sum

$$\sigma_k^{\text{total}} = \sum_m \sigma_{km} = \sigma_k^{\text{rot}} + \sigma_k^{\text{vib}} + \sigma_k^{\text{el}} , \qquad (13.3)$$

over all collision-induced transitions $|k\rangle \rightarrow |m\rangle$ into other levels $|m\rangle$, which might be rotational, vibrational, or electronic transitions (Fig. 13.6).

While measurements of the effective lifetime $\tau_k^{\text{eff}}(N_B)$ yield *absolute* values of the *total quenching cross section* σ_k^{total}, the different contributions in (13.3) have to be determined by other techniques, such as LIF spectroscopy. Even if only their *relative* magnitudes can be measured, this is sufficient to obtain, together with the absolute value of σ^{total}, their absolute magnitudes.

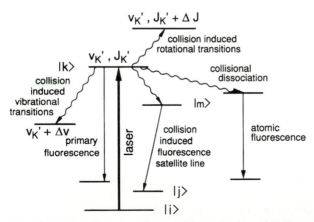

Fig. 13.6. Schematic term diagram for illustration of different possible inelastic collisional transitions of an optically pumped level $|k\rangle = (v', J')$ of a molecule M*

13.2.2 Collision-Induced Rovibronic Transitions in Excited States

When the level $|k\rangle = |v', J'_k\rangle$ of an excited molecule M* has been selectively populated by optical pumping, inelastic collisions M* + B, which occur during the lifetime τ_k, will transfer M*(k) into other levels $|m\rangle = |v'_k + \Delta v, J'_k + \Delta J\rangle$ of the same or of another electronic state:

$$M^*(v'_k, J'_k) + B \rightarrow M^*(v'_k + \Delta v, J'_k + \Delta J) + B^* + \Delta E_{\text{kin}}. \qquad (13.4)$$

The difference $\Delta E = E_k - E_m$ of the internal energies before and after the inelastic collision is transferred either into internal energy of B or into translation energy of the collisions partners.

Molecules in the collisionally populated levels $|m\rangle$ can decay by the emission of fluorescence or by further collisions. In the LIF spectrum new lines then appear besides the *parent lines*, which are emitted from the optically pumped level (Fig. 13.7). These new lines, called *collision-induced satellites* contain *the complete information on the collision process that has generated them*. Their wavelength λ allows the assignment of the upper level $|m\rangle = (v'_k + \Delta v, J'_k + \Delta J)$, their intensities yield the collisional cross sections σ_{km}, and their degree of polarization compared with that of the parent

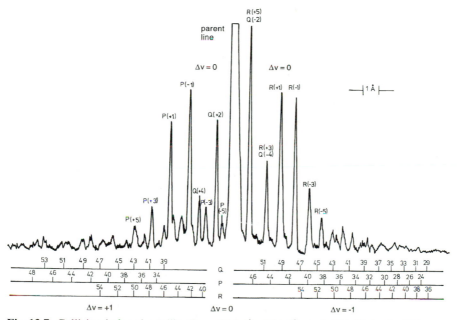

Fig. 13.7. Collision-induced satellite lines $Q(\Delta J')$, $R(\Delta J')$, and $P(\Delta J')$ in the LIF spectrum of Na$_2$ under excitation of the rotational level $B^1\Pi_u(v' = 6, J' = 43)$. The parent line is recorded with one-twentieth the sensitivity. The high satellite lines are related to $\Delta v = 0$ transitions and partly superimpose the weaker satellites from $\Delta v = \pm 1$ collision-induced transitions [13.29]

line gives the cross section for depolarizing, that is, orientation-changing collisions. This can be seen as follows:

Assume the upper level $|k\rangle$ is optically pumped on the transition $|i\rangle \rightarrow |k\rangle$ (Fig. 13.6) with a pump rate $N_i P_{ik}$ and collisions induce transitions $|k\rangle \rightarrow |m\rangle$. The rate equation for the population densities N_k, N_m can be written as

$$\frac{dN_k}{dt} = N_i P_{ik} - N_k \left(A_k + \sum_m R_{km} \right) + \sum_n N_n R_{nk} , \tag{13.5a}$$

$$\frac{dN_m}{dt} = N_k R_{km} - N_m \left(A_m + \sum_m R_{mn} \right) + \sum_n N_n R_{nm} , \tag{13.5b}$$

where the last two terms describe the collisional depopulation $|m\rangle \rightarrow |n\rangle$ and the repopulation $|n\rangle \rightarrow |k\rangle$ or $|n\rangle \rightarrow |m\rangle$ from other levels $|n\rangle$.

Under stationary conditions (optical pumping with a cw laser) $(dN_k/dt) = (dN_m/dt) = 0$. If the thermal population N_n is small (that is, $E_n \gg kT$) the last terms in (13.5) demand at least two successive collisional transitions $|k\rangle \rightarrow |m\rangle \rightarrow |n\rangle$ during the lifetime τ, which means a high collision rate. At sufficiently low pressures they can therefore be neglected.

From (13.5) the stationary population densities N_k, N_m are then obtained as

$$N_k = \frac{N_i P_{ik}}{A_k + \sum R_{km}} ,$$

$$N_m = \frac{N_k R_{km} + \sum N_n R_{nm}}{A_m + \sum R_{mn}} \simeq N_k \frac{R_{km}}{A_m} . \tag{13.6}$$

The ratio of the fluorescence intensities of satellite line $|m\rangle \rightarrow |j\rangle$ to parent line $|k\rangle \rightarrow |i\rangle$

$$\frac{I_{mj}}{I_{ki}} = \frac{N_m A_{mj} h \nu_{mj}}{N_k A_{ki} h \nu_{ki}} = R_{km} \frac{A_{mj} \nu_{mj}}{A_m A_{ki} \nu_{ki}} , \tag{13.7}$$

directly yields the probability R_{km} for collision-induced transitions $|k\rangle \rightarrow |m\rangle$, if the relative radiative transition probabilities A_{mj}/A_m and A_{ki}/A_k are known. Measurements of the spontaneous lifetime τ_k and τ_m (Sect. 11.3) allow the determination of $A_k = 1/\tau_k$ and $A_m = 1/\tau_m$. The absolute values of A_{mj} and A_{ki} can then be obtained by measuring the *relative* intensities of all fluorescence lines emitted by $|m\rangle$ and $|k\rangle$ under collision-free conditions.

The probability R_{km} is related to the collision cross section σ_{km} by

$$R_{km} = (N_B/\bar{v}) \int \sigma_{km}(v_{rel}) v_{rel} \, dv , \tag{13.8}$$

where N_B is the density of collision partners B, and v_{rel} the relative velocity between M and B.

When the experiments are performed in a cell at temperature T, the velocities follow a Maxwellian distribution, and (13.8) becomes

$$R_{km} = N_B \left(\frac{8kT}{\pi\mu}\right)^{-1/2} \langle\sigma_{km}\rangle , \qquad (13.9)$$

where $\mu = m_M m_B/(m_M + m_B)$ is the reduced mass, and $\langle\sigma_{km}\rangle$ means the average of $\sigma_{km}(v)$ over the velocity distribution. The cross sections σ_{km} obtained in this way represent integral cross sections, integrated over all scattering angles θ.

Such determinations of rotationally inelastic integral cross sections σ_{km} for collision-induced transitions in excited molecules obtained from measurements of satellite lines in the fluorescence spectrum have been reported for a large variety of different molecules, such as I_2 [13.25, 13.26], Li_2 [13.27, 13.28], Na_2 [13.29], or NaK [13.30]. For illustration, the cross section $\sigma(\Delta J)$ for the transition $J \rightarrow J + \Delta J$ in excited Na_2^* molecules induced by collisions $Na_2^* + He$ are plotted in Fig. 13.8. They rapidly decrease from a value $\sigma(\Delta J = \pm 1) \approx 0.3\,nm^2$ to $\sigma(\Delta J = \pm 8) \approx 0.02\,nm^2$. This decrease is essentially due to energy and momentum conservation, since the energy difference $\Delta E = E(J \pm \Delta J) - E(J)$ has to be transferred into the kinetic energy of the collision partners. The probability for this energy transfer is proportional to the Boltzmann factor $\exp[-\Delta E/(kT)]$ [13.31].

For interaction potentials $V(R)$ of the collision pair M + B with spherical symmetry, which only depend on the internuclear distance R, no internal angular momentum of M can be transferred. The absolute values of the cross sections $\sigma(\Delta J)$ are therefore a measure for the *nonspherical* part of the

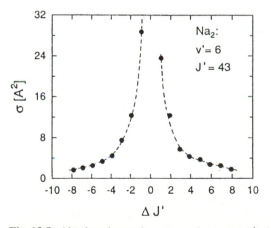

Fig. 13.8. Absolute integral cross sections $\sigma(\Delta J')$ for collision-induced rotational transitions when $Na_2^*(B\,^1\Pi_u, v' = 6, J' = 43)$ collides with the atoms at $T = 500\,K$ [13.29]

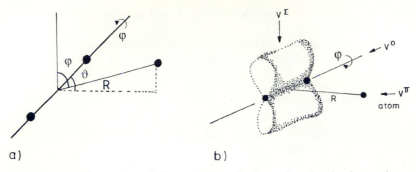

Fig. 13.9. (**a**) Illustration of the variables R, θ, and φ in the interaction potential $V(R, \theta, \varphi)$. (**b**) Schematic electron cloud distribution for excited homonuclear diatomic molecules in a Π state. Also shown are the two directions of φ for a Σ-type and a Π-type interaction potential

potential $V(M, B)$. This potential can be represented by the expansion

$$V(R, \theta, \phi) = V_0(R) + \sum_{\ell, m} a_{\ell} m Y_{\ell}^m (\phi) , \qquad (13.10)$$

where the Y_{ℓ}^m denote the spherical surface functions. A homonuclear diatomic molecule M in a Σ state has a cylindrically symmetric electron cloud. Its interaction potential must be independent of ϕ and furthermore has a symmetry plane perpendicular to the internuclear axis. For such symmetric states (13.10) reduces to

$$V(R, \theta) = a_0 V_0(R) + a_2 P_2(\cos \theta) + ... , \qquad (13.11)$$

where the coefficient a_2 of the Legendre polynomical $P_2(\cos \theta)$ can be determined from $\sigma(\Delta J)$, while a_0 is related to the value of the elastic cross section [13.32, 13.33].

In a Π state, however, the electron cloud possesses an electronic angular momentum and the charge distribution no longer has cylindrical symmetry (Fig. 13.9). The interaction potential $V(R, \theta, \phi)$ depends on all three variables. Since the two Λ components of a rotational level in a Π state ($\Lambda = 1$) correspond to different charge distributions, different magnitudes of the cross sections $\sigma(J, \pm \Delta J)$ for collision-induced rotational transitions are expected for the two Λ components of the same rotational level $|J\rangle$ [13.34, 13.35]. Such an asymmetry, which can be recognized in Fig. 13.7 (transitions with $\Delta J =$ even reach other Λ components than those with $\Delta J =$ odd), has indeed been observed [13.28, 13.34, 13.36].

In addition to LIF resonant two-photon ionization (Sect. 6.4) can also be used for the sensitive detection of collision-induced rotational transitions. This method represents an efficient alternative to LIF for those electronic states that do not emit detectable fluorescence because there are no allowed optical transitions into lower states. An illustrative example is the detailed investigation of inelastic collisions between excited N_2 molecules and different collision partners [13.37]. A vibration–rotation level (v', J') in the a $^1\Pi_g$

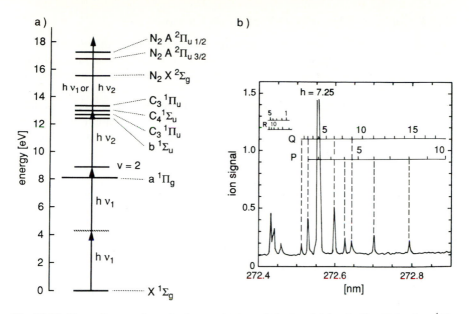

Fig. 13.10. Term diagram for selective excitation of the level ($v' = 2$, $J' = 7$) in the $a^1\Pi_g$ state of N_2 with the detection of collisional transitions by resonant two-photon ionization [13.37]

state of N_2 is selectively populated by two-photon absorption (Fig. 13.10). The collision-induced transitions to other levels ($v' + \Delta v$, $j' + \Delta J$) are monitored by resonant two-photon ionization (REMPI, Sect. 6.2) with a pulsed dye laser. The achievable good signal-to-noise ratio is demonstrated by the collisional satellite spectrum in Fig. 13.10b, where the optically pumped level was ($v' = 2$, $J' = 7$). This level is ionized by the $P(7)$ *parent line* in the spectrum, which has the signal height 7.2 on the scale of Fig. 13.10b.

If the parent level $|k\rangle = (v_k, J_k)$ of the optically pumped molecule has been oriented by polarized light, this orientation is partly transferred by collisions to the levels ($v_k + \Delta v$, $J_k + \Delta J$). This can be monitored by measuring the polarization ratio $R_p = (I_\parallel - I_\perp)/(I_\parallel + I_\perp)$ of the collisional satellite lines in the fluorescence [13.38].

The cross sections for collision-induced vibrational transitions are generally much smaller than those for rotational transitions. This is due to energy conservation (if $\Delta E_{vib} \gg kT$) and also to dynamical reasons (the collision has to be sufficiently nonadiabatic, that is, the collision time should be comparable or shorter than the vibrational period) [13.33]. The spectroscopic detection is completely analogous to that for rotational transitions. In the LIF spectrum new collision-induced bands ($v_k + \Delta_k$) → (v_m) appear (Fig. 13.7), with a rotational distribution that reflects the probabilities of collision-induced transitions (v_k, J_k) → ($v_k + \Delta_v$, $J_k + \Delta J$) [13.27, 13.29, 13.39, 13.40].

Vibrational transitions within the electronic ground state can be studied by time-resolved infrared spectroscopy (Sect. 13.3.1).

13.2.3 Collisional Transfer of Electronic Energy

Collisions may also transfer electronic energy. For instance, the electronic energy of an excited atom A^* or molecule M^* can be converted in a collision with a partner B into translational energy E_{kin} or, with higher probability, into internal energy of B [13.8].

For collisions at thermal energies the collision time $T_{coll} = d/\bar{v}$ is long compared to the time for an electronic transition. The interaction $V(A^*, B)$ or $V(M^*, B)$ can then be described by a potential. Assume that the two potential curves $V(A_i, B)$ and $V(A_k, B)$ cross at the energy $E(R_c)$ (Fig. 13.11). If the relative kinetic energy of the collision partners is sufficiently high to reach the crossing point, the collision pair may jump over to the other potential curve [13.41]. In Fig. 13.11, for instance, a collision $A_i + B$ can lead to electronic excitation $|i\rangle \rightarrow |k\rangle$ if $E_{kin} > E_2$, while for a collisional deexcitation $|k\rangle \rightarrow |i\rangle$ only the kinetic energy $E_{kin} > E_1$ is required.

The cross sections of electronic energy transfer $A^* + B \rightarrow A + B^* + \Delta E_{kin}$ are particularly large in cases of energy resonance, which means $\Delta E(A^* - A) \simeq \Delta E(B^* - B) \rightarrow \Delta E_{kin} \leq kT$. A well-known example is the collisional excitation of Ne atoms by metastable He^* atoms, which represents the main excitation mechanism in the HeNe laser.

The experimental proof for such electronic energy transfer ($E \rightarrow E$ transfer) is based on the selective excitation of A by a laser and the spectrally resolved detection of the fluorescence from B^* [13.42, 13.43].

In collisions between excited atoms and molecules either the atom A^* or the molecule M^* may be electronically excited. Although the two cases

$$M(v_i'', J_i'') + A^* \rightarrow M^*(v', J') + A , \tag{13.12a}$$

and

$$M^*(v_k', J_k') + A \rightarrow M(v'', J'') + A^* , \tag{13.12b}$$

represent inverse processes, their collision cross sections may differ considerably. This was demonstrated by the example $M = Na_2$, $A = Na$. Either the Na atoms were selectively excited into the 3P state and the fluorescence of Na_2^* was measured with spectral and time resolutions [13.44], or Na_2 was

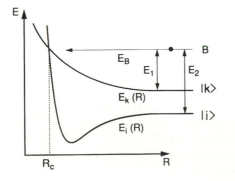

Fig. 13.11. Potential diagram for collision-induced transitions between the two electronic states $|i\rangle$ and $|k\rangle$ that can occur at the crossing point at $R = R_c$ of the two potential curves $E_i R = M_i + B$ and $E_k R = M_k + B$

excited into a definite level (v'_k, J'_k) in the $^1A\Sigma_u$ state and the energy transfer was monitored by time-resolved measurements of the shortened lifetime $\tau_k(v'_k, J'_k)$ and the population $N_{3p}(t)$ of the Na* $(3P)$ level [13.45]. There are two distinct processes that lead to the formation of Na*:

(a) direct energy transfer according to (13.12b), or
(b) collision-induced dissociation $Na_2^* + Na \rightarrow Na^* + Na + Na$.

This collision-induced dissociation of electronically excited molecules plays an important role in chemical reactions and has therefore been studied for many molecules [13.46–13.48]. The collision cross sections for this process may be sufficiently large to generate inversion between atomic levels. Laser action based on *dissociation pumping* has been demonstrated. Examples are the powerful iodine laser [13.49] and the Cs laser [13.50].

The transfer of electronic energy into rovibronic energy has much larger cross sections than the electronic to translational energy $(E \rightarrow T)$ transfer [13.51]. These processes are of crucial importance in photochemical reactions of larger molecules and detailed studies with time-resolved laser spectroscopy have brought much more insight into many biochemical processes [13.52–13.54].

One example of a state-selective experimental technique well suited for studies of electronic to vibrational $(E \rightarrow V)$ transfer processes is based on CARS (Sect. 9.4). This has been demonstrated by Hering et al. [13.55], who studied the reactions

$$Na^*(3P) + H_2(v = 0) \rightarrow Na(3S) + H_2(v = 1, 2, 3) , \tag{13.13}$$

where Na* was excited by a dye laser and the internal state distribution of $H_2(v'', J'')$ was measured with CARS.

An interesting phenomenon based on collisions between excited atoms and ground-state atoms is the macroscopic diffusion of optically pumped atoms realized in the *optical piston* [13.56]. It is caused by the difference in cross sections for velocity-changing collisions involving excited atoms A* or ground-state atoms A, respectively, and results in a spatial separation between optically pumped and unpumped atoms, which can be therefore used for isotope separation [13.57].

13.2.4 Energy Pooling in Collisions Between Excited Atoms

Optical pumping with lasers may bring an appreciable fraction of all atoms within the volume of a laser beam passing through a vapor cell into an excited electronic state. This allows the observation of collisions between two *excited* atoms, which lead to many possible excitation channels where the sum of the excitation energies is accumulated in one of the collision partners. Such *energy-pooling* processes have been demonstrated for Na* + Na*, where reactions

$$Na^*(3P) + Na^*(3P) \rightarrow Na^{**}(nL) + Na(3S) , \tag{13.14}$$

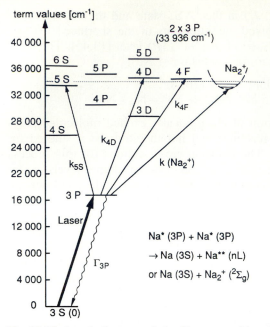

Fig. 13.12. Level diagram of the Na atom with schematic illustration of the different excitation channels in collisions Na*(3P) + Na*(3P) [13.60]

have been observed [13.58], leading to the excitation of high-lying levels $|n, L\rangle$ (Fig. 13.12).

Measurements of the fluorescence intensity $I_{Fl}(n, L)$ emitted by the levels Na** $(n, L = 4D$ or $5S)$ yields the collision rate

$$k_{n,L} = N^2(\text{Na}^*3p) \cdot \sigma_{n,L}\bar{v} \quad [\text{s/cm}^3],$$

of optically pumped Na atoms, which increases with the square N^2 of the density of excited Na* atoms. This density cannot be determined directly from the measured Na*(3P) fluorescence because radiation trapping falsifies the results [13.59]. The attenuation of the transmitted laser beam is a better measure because every absorbed photon produces one excited Na*(3P) atom.

Since the sum of the excitation energies of the two colliding Na*(3P) atoms is higher than the ionization limit of the Na$_2$ molecule, associative ionization

$$\text{Na}^*(3\text{P}) + \text{Na}^*(3\text{P}) \underset{k_{\text{Na}_2^+}}{\rightarrow} \text{Na}_2^+ + e^-, \tag{13.15}$$

can be observed. The measurement of Na$_2^+$ ions gives the reaction rate $k(\text{Na}_2^+)$ for this process [13.60].

A further experimental step that widens the application range is the excitation of two different species A$_1$ and A$_2$ with two dye lasers. An example is the simultaneous excitation of Na*(3P) and K*(4P) in a cell containing a mixture of sodium and potassium vapors. Energy pooling can lead to the excitation of high-lying Na** or K** states, which can be monitored by their fluorescence [13.61]. When both lasers are chopped at two different frequencies f_1 and f_2 (where $1/f$ is long compared to the collisional transfer time), lock-in detection of the fluorescence emitted from the highly excited levels at f_1, f_2 or $f_1 + f_2$ allows the distinction between the energy pooling processes Na* + Na*, K* + K*, and Na* + K* leading to the excitation of the levels $|n, L\rangle$ of Na** or K**. For further examples see [13.62, 13.63].

13.2.5 Spectroscopy of Spin-Flip Transitions

Collision-induced transitions between fine-structure components where the relative orientation of the electron spin with respect to the orbital angular momentum is changed have been studied in detail by laser-spectroscopic techniques [13.64]. One of the methods often used is *sensitized fluorescence*, where one of the fine-structure components is selectively excited and the fluorescence of the other component is observed as a function of pressure [13.65]. Either pulsed excitation and time-resolved detection is used [13.66] or the intensity ratio of the two fine-structure components is measured under cw excitation [13.67].

Of particular interest is the question into which of the fine-structure components of the atoms a diatomic molecule dissociates after excitation into predissociating states. The investigation of this question gives information on the recoupling of angular momenta in molecules at large intermolecular distances and about crossings of potential curves [13.68]. Early measurements of these processes in cells often gave wrong answers because collision-induced fine-structure transitions and also radiation trapping had changed the original fine-structure population generated by the dissociation process. Therefore, laser excitation in molecular beams has to be used to measure the initial population of a certain fine-structure component [13.69].

Collision-induced spin-flip transitions in molecules transfer the optically excited molecule from a singlet into a triplet state, which may drastically change its chemical reactivity [13.70]. Such processes are important in dye lasers and have therefore been intensively studied [13.71]. Because of the long lifetime of the lowest triplet state T_0, its population $N(t)$ can be determined by time-resolved measurements of the absorption of a dye laser tuned to the transition $T_0 \rightarrow T_1$.

A laser-spectroscopic technique for investigations of spin-flip transitions under electron impact excitation of Na atoms [13.72] is shown in Fig. 13.13. Optical pumping with σ^+ light on the $3S_{1/2} \rightarrow 3P_{1/2}$ transition in an external weak magnetic field orients the Na atoms into the $3S_{1/2}(m_s = +1/2)$ level. Starting from this level the Zeeman components $3P_{3/2}(M_J)$ are populated by electron impact. The population N_M is probed by a cw dye laser tuned to the transitions $3^2P_{3/2}(M_J) \rightarrow 5^2S_{1/2}(M_s = \pm 1/2)$, which can be monitored by the cascade fluorescence $5S_{1/2} \rightarrow 4P \rightarrow 3S_{1/2}$.

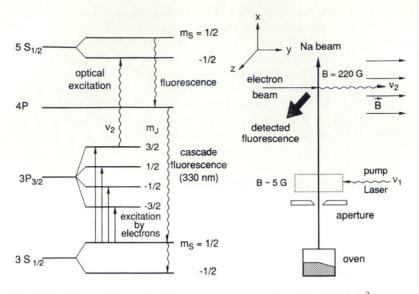

Fig. 13.13. The relative populations of Zeeman levels $|M_j\rangle$ in the $3\,^2P_{3/2}$ state of Na are probed by laser excitation of the $5\,^2S_{1/2}$ state, which is monitored via the cascade fluorescence into the $3\,S_{1/2}$ state [13.72]

Figure 13.14 summarizes all possible collision-induced transitions from a selectively excited molecular level $|v'_k, J'_k\rangle$ to other molecular or atomic levels.

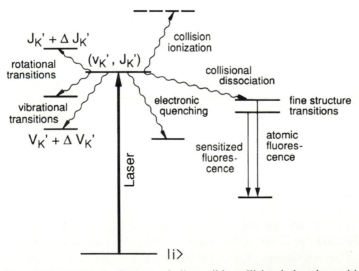

Fig. 13.14. Schematic diagram of all possible collision-induced transitions from a molecular level (v'_k, J'_k) in an excited electronic state selectively populated by optical pumping

13.3 Spectroscopic Techniques for Measuring Collision-Induced Transitions in the Electronic Ground State of Molecules

In the electronic ground states of molecules collision-induced transitions represent, for most experimental situations, the dominant mechanism for the redistribution of energy, because the radiative processes are generally too slow. In cases where a nonequilibrium distribution has been produced (for example, by chemical reactions or by optical pumping), these collisions try to restore thermal equilibrium. The relaxation time of the system is determined by the absolute values of collision cross sections.

For most infrared molecular lasers, such as the CO_2 and the CO laser, or for *chemical lasers*, such as the HF or HCl lasers, collisional energy transfer between vibrational–rotational levels of the lasing molecules plays a crucial role for the generation and maintenance of inversion and gain. These lasers are therefore called *energy-transfer lasers* [13.73]. Also in many visible molecular lasers, which oscillate on transitions between optically pumped levels (v', J') of the excited electronic state and high vibrational levels of the electronic ground state, collisional deactivation of the lower laser level is essential because the radiative decay is generally too slow. Examples are dye lasers [13.74] or dimer lasers, such as the Na_2 laser or the I_2 laser [13.75].

The internal energy $E_i = E_{vib} + E_{rot}$ of a molecule $M(v_i, J_i)$ in its electronic ground state may be transferred during a collision with another molecule AB

$$M(E_i) + AB(E_m) \rightarrow M(E_i - \Delta E_1) + AB^*(E_m + \Delta E_2) + \Delta E_{kin} , \quad (13.16)$$

into vibrational energy of AB^* ($V \rightarrow V$ transfer), rotational energy ($V \rightarrow R$ transfer), electronic energy ($V \rightarrow E$ transfer), or translational energy ($V \rightarrow T$ transfer). In collisions of M with an atom A only the last two processes are possible.

The experiments show that the cross sections are much larger for $V \rightarrow V$ or $V \rightarrow R$ transfer than for $V \rightarrow T$ transfer. This is particularly true when the vibrational energies of the two collision partners are near resonant.

A well-known example of such a near-resonant $V \rightarrow V$ transfer is the collisional excitation of CO_2 molecules by N_2 molecules

$$CO_2(0, 0, 0) + N_2(v = 1) \rightarrow CO_2(0, 0, 1) + N_2(v = 0) , \quad (13.17)$$

which represents the main excitation mechanism of the upper laser level in the CO_2 laser (Fig. 13.15) [13.76].

The experimental techniques for the investigation of inelastic collisions involving molecules in their electronic ground state generally differ from those discussed in Sect. 13.2. The reasons are the long spontaneous lifetimes of ground-state levels and the lower detection sensitivity for infrared radiation compared to those for the visible or UV spectrum. Although infrared fluorescence detection has been used, most of the methods are based on absorption measurements and double-resonance techniques.

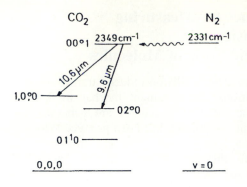

Fig. 13.15. Vibrational energy transfer from N_2 ($v = 1$) to the upper vibrational level ($v_1 = 0, v_2 = 0, l_{vib} = 0, v_3 = 1$) of the CO_2 laser transitions

13.3.1 Time-Resolved Infrared Fluorescence Detection

The energy transfer described by (13.16) can be monitored if M* is excited by a short infrared laser pulse and the fluorescence of AB* is detected by a fast cooled infrared detector (Sect. 4.5) with sufficient time resolution. Such measurements have been performed in many laboratories [13.6]. For illustration, an experiment carried out by Green and Hancock [13.77] is explained by Fig. 13.16a: a pulsed HF laser excites hydrogen fluoride molecules into the vibrational level $v = 1$. Collisions with other molecules AB (AB = CO, N_2) transfer the energy to excited vibrational levels of AB*. The infrared fluorescence emitted by AB* and HF* has to be separated by spectral filters. If two detectors are used, the decrease of the density $N(HF^*)$ of vibrationally excited HF molecules and the build-up and decay of $N(AB^*)$ can be monitored simultaneously.

For larger molecules M two different collisional relaxation processes have to be distinguished: collisions M* + AB may transfer the internal energy of M* to AB* (*inter*molecular energy transfer), or may redistribute the energy among the different vibrational modes of M* (*intra*molecular transfer)

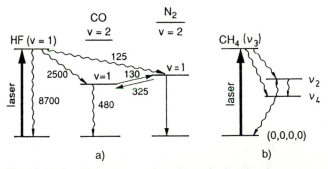

a)   b)

Fig. 13.16a,b. Fluorescence detection of vibrational energy transfer from an optically pumped molecular level: (**a**) intermolecular transfer HF* → CO, N_2. The mean numbers of collisions are given for the different collision-induced transitions. (**b**) *Intra*molecular transfer $CH_4^*(v_3) \rightarrow CH_4^*(v_2) + \Delta E_{kin}$ [13.77]

(Fig. 13.16b). One example is the molecule $M = C_2H_4O$, which can be excited on the C–H stretching vibration by a pulsed parametric oscillator at $3000\,cm^{-1}$ [13.78]. The fluorescence emitted from other vibrational levels which are populated by collision-induced transitions is detected through spectral filters. Further examples can be found in several reviews [13.79–13.81].

13.3.2 Time-Resolved Absorption and Double-Resonance Methods

While collision-induced transitions in excited electronic states can be monitored through the satellite lines in the *fluorescence* spectrum (Sect. 13.2.2), inelastic collisional transfer in electronic ground states of molecules can be studied by changes in the *absorption* spectrum. This technique is particularly advantageous if the radiative lifetimes of the investigated rotational–vibrational levels are so long that fluorescence detection fails because of intensity problems.

A successful technique for studying collision-induced transitions in electronic ground states is based on time-resolved double resonance [13.82]. The method is explained by Fig. 13.17. A pulsed laser L1 tuned to an infrared or

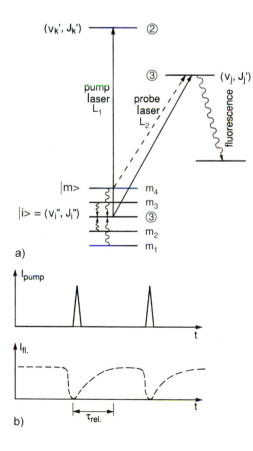

a)

b)

Fig. 13.17. (a) Level diagram for the determination of the refilling rate of a lower level (v_i'', J_i'') depleted by a pump laser pulse via measurements of the time-dependent probe laser absorption; **(b)** time dependence of pump pulses and level population $N_i(t)$ monitored via the probe laser-induced fluorescence

optical transition $|i\rangle \rightarrow |k\rangle$ depletes the lower level $|1\rangle = (v_i'', J_i'')$. The depleted level is refilled by collisional transitions from other levels. A second weak probe laser L2 is tuned to another transition $|1\rangle \rightarrow |3\rangle = (v_j', J_j')$ starting from the depleted lower level $|i\rangle$. If the pump and probe laser beams overlap within a path length Δz in the absorbing sample, the absorption of the probe laser radiation can be measured by monitoring the transmitted probe laser intensity

$$I(\Delta z, t) = I_0 e^{-\alpha \Delta z} \approx I_0[1 - N_i(t)\sigma_{ij}^{abs}\Delta z] . \tag{13.18}$$

The time-resolved measurement of $I(\Delta z, t)$ yields the time dependence of the population density $N_i(t)$. Using a cw probe laser the absorption can be measured by the transmitted intensity $I(\Delta z, t)$, or by the time-dependent fluorescence induced by the probe laser, which is proportional to $N_j(t) \propto N_i(t) \cdot P(L_2)$. If a pulsed probe laser is used, the time delay Δt between pump and probe is varied and the measured values of $\alpha(\Delta t)$ yield the refilling time of the level $|1\rangle$.

Without optical pumping the population densities are time independent at thermal equilibrium. From the rate equation

$$\frac{dN_i^0}{dt} = 0 = -N_i^0 \sum_m R_{im} + \sum_m N_m^0 R_{mi} , \tag{13.19}$$

we obtain the condition of *detailed balance*

$$N_i^0 \sum_m R_{im} = \sum_m N_m^0 R_{mi} , \tag{13.20}$$

where R_{im} are the relaxation probabilities for transitions $|i\rangle \rightarrow |m\rangle$. If the population N_i^0 is depleted to the saturated value $N_i^s < N_i^0$ by optical pumping at $t = 0$, the right side of (13.20) becomes larger than the left side and collision-induced transitions $|m\rangle \rightarrow |i\rangle$ from neighboring levels $|m\rangle$ refill $N_i(t)$. At a density N_B of collision partners the time dependence $N_i(t)$ after the end of the pump pulse is obtained from

$$\frac{dN_i}{dt} = \sum_m N_m R_{mi} - N_i(t) \sum_m R_{im} . \tag{13.21a}$$

If we assume that optical pumping of level $|i\rangle$ does not essentially affect the population densities N_m ($m \neq i$), we obtain from (13.20), (13.21a)

$$\frac{dN_i}{dt} = [N_i^0 - N_i(t)]K_i , \tag{13.21b}$$

with the relaxation constant

$$K_i = \sum_m R_{im} = \sqrt{\frac{8kT}{\pi\mu}} \langle \sigma_i^{total} \rangle . \tag{13.22}$$

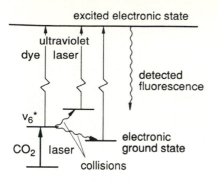

Fig. 13.18. Term diagram of IR–UV double resonance for measurements of collision-induced intramolecular vibrational transitions in the D_2CO molecule [13.83]

The relaxation constant K_i depends on the total cross section $\langle \sigma_i^{\text{total}} \rangle = \sum_m \langle \sigma_{im} \rangle$ averaged over the thermal velocity distribution and on the temperature T. Integration of (13.21) gives

$$N_i(t) = N_i^0 + [N_i^s(0) - N_i^0] e^{-K_i t} . \qquad (13.23)$$

This shows that after the end of the pump pulse at $t = 0$ the population density $N_i(t)$ returns exponentially from its saturated valued $N_i^s(0)$ back to its equilibrium value N_i^0 with the time constant

$$\tau = (K_i)^{-1} = \left(\bar{v}_{\text{rel}} \left\langle \sigma_i^{\text{total}} \right\rangle N_B \right)^{-1} , \qquad (13.24)$$

which depends on the averaged total refilling cross section $\langle \sigma_i^{\text{total}} \rangle$ and on the number density N_B of collision partners.

One example of this pump-and-probe technique is the investigation of collision-induced vibrational–rotational transitions in the different isotopes HDCO and D_2CO of formaldehyde by an infrared–UV double resonance [13.83]. A CO_2 laser pumpes the ν_6 vibration of the molecule (Fig. 13.18). The collisional transfer into other vibrational modes is monitored by the fluorescence intensity induced by a tunable UV dye laser with variable time delay.

The time resolution of the pump-and-probe technique is not limited by the rise time of the detectors. It can therefore be used in the pico- and femto-second range (Sect. 11.4) and is particularly advantageous for the investigation of ultrafast relaxation phenomena, such as collisional relaxation in liquids and solids [13.84, 13.85]. It is also useful for the detailed *real-time* study of the formation and dissociation of molecules where the collision partners are observed during the short time interval when forming or breaking a chemical bond [13.86].

13.3.3 Collision Spectroscopy with Continuous-Wave Lasers

State-selective measurements of the different contributions R_{im} in (13.3, 13.22) can be performed without time resolution if the absolute value of $\sigma_i^{\text{total}} = \sum \sigma_{im}$ is known from time resolved investigations, discussed in the previous

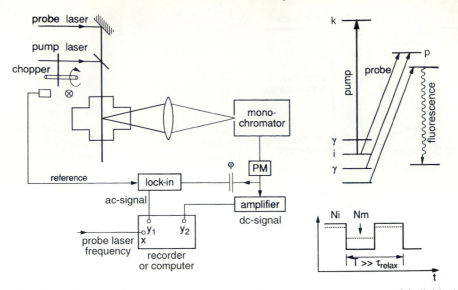

Fig. 13.19. Experimental arrangement and term diagram for measurements of individual collision-induced transitions in molecular electronic ground states with cw lasers

section. The pump laser may be a cw dye laser that is tuned to the wanted transition $|i\rangle \rightarrow |k\rangle$ (Fig. 13.19) and is chopped at a frequency $f \ll 1/\tau_i = K_i$. This always guarantees quasi-stationary population densities N_i^s during the on phase and N_i^0 during the off phase of the pump laser.

In this case the deviation $N_i^s - N_i^0$ from thermal equilibrium during the on phase does affect via collisional transfer the population densities N_m of neighboring levels $|m\rangle$ (13.21a). Under stationary conditions we obtain in analogy to (13.19, 13.21)

$$\frac{dN_m}{dt} = 0 = \sum_j (N_j R_{jm} - N_m R_{mj}) = (N_m - N_m^0)K_m \,, \tag{13.25}$$

where the right side of (13.25) represents the relaxation of N_m toward the equilibrium population N_m^0. Inserting the deviations $\Delta N = N - N^0$ into (13.25) yields with (13.19)

$$0 = \sum_j (\Delta N_j R_{jm} - \Delta N_m R_{mj}) \;\Rightarrow\; \Delta N_m = \frac{\sum \Delta N_j R_{jm}}{\sum R_{mj}} \,. \tag{13.26}$$

This equation relates the population change ΔN_m with that of all other levels $|j\rangle$, where the sum over j also includes the optically pumped level $|i\rangle$. For $j \neq i$ the quantities ΔN_j and R_{jm} are proportional to the density N_B of the collision partners. At sufficiently low pressures we can neglect all terms ΔN_j

for $j \neq i$ in (13.26) and obtain for the limiting case $p \to 0$ the relation

$$\frac{\Delta N_m}{\Delta N_i} = \frac{R_{im}}{\sum_j R_{mj}} = \frac{R_{im}}{K_m} . \tag{13.27}$$

The relaxation constant K_m can be obtained from a time-resolved measurement (Sect. 13.3.2). The signals from the chopped cw laser experiments give the following information: the ratio $\Delta N_m / \Delta N_i$ is directly related to the ac absorption signals of the probe laser tuned to the transition $|i\rangle \to |p\rangle$ and $|m\rangle \to |p\rangle$, respectively, when the lock-in detector is tuned to the chopping frequency f of the pump laser. The ratio

$$\frac{S_m^{ac}}{S_i^{ac}} = \frac{B_{mp}\Delta N_m}{B_{ip}\Delta N_i} , \tag{13.28}$$

of the ac signals is proportional to the ratio of the population changes and depends furthermore on the Einstein coefficients B_{mp} and B_{ip}.

If the time-averaged dc part of the probe laser signal

$$S_m^{dc} = C B_{mp} \frac{(N_m^s + N_m^0)}{2} , \tag{13.29}$$

for a square-wave modulation of the pump laser is also monitored, the relative changes $\Delta N_m / N_m$ of the population densities N_m can be obtained from the relations

$$\frac{\Delta N_m / N_m^0}{\Delta N_i / N_i^0} = \frac{S_m^{ac}}{S_m^{dc}} \frac{S_i^{dc}}{S_i^{ac}} . \tag{13.30}$$

With $N_i^0 / N_m^0 = (g_i / g_m) \exp(-\Delta E / kT)$, the absolute values of ΔN_m can be determined from (13.30).

13.3.4 Collisions Involving Molecules in High Vibrational States

Molecules in high vibrational levels $|v\rangle$ of the electronic ground state generally show much larger collision cross sections. If the kinetic energy E_{kin} of the collision partners exceeds the energy difference $(E_D - E_v)$ between the dissociation energy E_D and the vibrational energy E_v, dissociation can take place (Fig. 13.20a). Since the resulting fragments often have a larger reactivity than the molecules, the probability for the initiation of chemical reactions generally increases with increasing vibrational energy. The investigation of collisions involving vibrationally excited molecules is therefore of fundamental interest. Several spectroscopic methods have been developed for achieving a sufficient population of high vibrational levels: optical pumping with infrared lasers with one- or multiple-photon absorption is a possibility for infrared-active vibrational transitions [13.87]. For homonuclear diatomics or for infrared-inactive modes, this method is not applicable.

Here, stimulated emission pumping is a powerful technique to achieve a large population in selected vibrational levels [13.88, 13.89]. While the

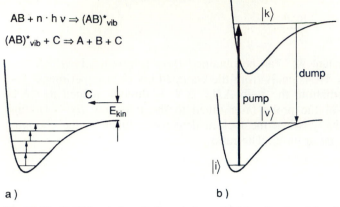

Fig. 13.20. Collision-induced dissociation of high vibrational levels populated either by multiple IR-photon absorption (**a**) or by stimulated emission pumping (**b**)

pump laser is kept on the transition $|i\rangle \rightarrow |k\rangle$, the probe laser is tuned to the downward transition $|k\rangle \rightarrow |v\rangle$ (Fig. 13.20b). Proper selection of the upper level $|k\rangle$ allows one to reach sufficiently large Franck–Condon factors for the transition $|k\rangle \rightarrow |v\rangle$. With pulsed lasers a considerable fraction of the initial population N_i in the level $|i\rangle$ can be transferred into the final level $|v\rangle$.

With a coherent stimulated Raman process (STIRAP), nearly the whole initial population N_i may be transferred into the final level $|v\rangle$ [13.90]. Here no exact resonance with the intermediate excited level $|k\rangle$ is wanted in order to avoid transfer losses by spontaneous emission from level $|k\rangle$. The population transfer can be explained by an adiabatic passage between *dressed states* (that is, states of the molecule plus the radiation field) [13.91, 13.92].

Investigations of collision processes with molecules in these selectively populated levels $|v\rangle$ yields the dependence of collision cross sections on the vibrational energy for collision-induced dissociation as well as for energy transfer into other bound levels of the molecule. Since the knowledge of this dependence is essential for a detailed understanding of the collision dynamics, a large number of theoretical and experimental papers on this subject have been published. For more information the reader is referred to some review articles and the literature given therein [13.81, 13.87, 13.93–13.95].

13.4 Spectroscopy of Reactive Collisions

A detailed understanding of reactive collisions is the basis for optimization of chemical reactions that is not dependent on "trial-and-error methods." Laser-spectroscopic techniques have opened a large variety of possible strategies to reach this goal. Two aspects of spectroscopic investigations for reactive collisions shall be emphasized:

(a) By selective excitation of a reactant the dependence of the reaction probability on the internal energy of the reactant can be determined much more accurately than by measuring the temperature dependence of the reaction. This is because, in the latter case, internal energy and translational energy both change with temperature.

(b) The spectroscopic assignment of the reaction products and measurements of their internal energy distributions allow the identification of the different reaction pathways and their relative probabilities under definite initial conditions for the reactants.

The experimental conditions for the spectroscopy of reactive collisions are quite similar to those for the study of inelastic collisions. They range from a determination of the velocity-averaged reaction rates under selective excitation of reactants in cell experiments to a detailed state-to-state spectroscopy of reactive collisions in crossed molecular beams (Sect. 13.5). Some examples shall illustrate the state of the art:

The first experiments on state-selective reactive collisions were performed for the experimentally accessible reactions

$$Ba + HF(v = 0, 1) \rightarrow BaF(v = 0-12) + H,$$
$$Ba + CO_2 \rightarrow BaO + CO,$$
$$Ba + O_2 \rightarrow BaO + O, \tag{13.31}$$

where the internal state distribution $N(v'', J'')$ of the reaction products was measured by LIF in dependence on the vibrational energy of the halogen molecule [13.96, 13.97].

The development of new infrared lasers and of sensitive infrared detectors has allowed investigations of reactions where the reaction-product molecules have known infrared spectra but do not absorb in the visible. An example is the endothermic reaction

$$Br + CH_3F(v_3) \rightarrow HBr + CH_2F, \tag{13.32}$$

where the CF stretching vibration v_3 of CH_3F is excited by a CO_2 laser [13.98]. The reaction probability increases strongly with increasing internal energy of CH_3F. If a sufficiently large concentration of excited $CH_3F(v_3)$ molecules is reached, energy-pooling collisions

$$CH_3F^*(v_3) + CH_3F^*(v_3) \rightarrow CH_3F^{**}(2v_3) + CH_3F, \tag{13.33}$$

result in an increase of E_{vib} in one of the collision partners. The term diagram relevant for (13.32) is shown in Fig. 13.21. If the sum of internal plus kinetic energy is larger than the reaction barrier, excited product molecules may appear that can be monitored by their infrared fluorescence.

A typical experimental arrangement for such investigations is depicted in Fig. 13.22. In a flow system, where the reactive collisions take place, the levels (v, J) of the reactant molecules are selectively excited by a pulsed infrared laser. The time-dependent population of excited levels in the reactants or the

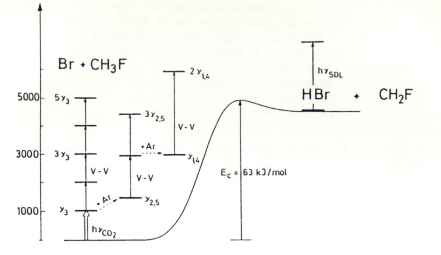

Fig. 13.21. Term diagram of the endothermic reaction $Br + CH_3F^*(\nu^3) \rightarrow HBr + CH_2F$ [13.98]

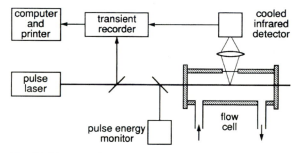

Fig. 13.22. Experimental arrangement for studies of infrared LIF in chemical reactions with spectral and time resolution

product molecules are monitored through their fluorescence, detected by fast, cooled infrared detectors (Sect. 4.5).

A longstanding controversy existed on the detailed reaction mechanism of the elementary exchange reaction of hydrogen

$$H_a + H_b H_c \rightarrow H_a H_b + H_c , \qquad (13.34a)$$

where accurate ab initio calculations of the H_3 potential surface have been performed. The experimental test of theoretical predictions is, however, very demanding. First, isotope substution has to be performed in order to distinguish between elastic and reactive scattering. Instead of (13.34a), the reaction

$$H + D_2 \rightarrow HD + D , \qquad (13.34b)$$

is investigated. Second, all electronic transitions fall into the VUV range. The excitation of H or D_2 and the detection of HD or D require VUV lasers. Third, hydrogen atoms have to be produced since they are not available in bottles.

The first experiments were carried out in 1983 [13.99, 13.100]. The H atoms were produced by photodissociation of HJ molecules in an effusive beam using the fourth harmonics of Nd:YAG lasers. Since the dissociated iodine atom is found in the two fine-structure levels $I(P_{1/2})$ and $I(P_{3/2})$, two groups of H atoms with translational energies $E_{kin} = 0.55\,eV$ or $1.3\,eV$ in the center-of-mass system $H + D_2$ are produced. If the slower H atoms collide with D_2 they can reach vibrational–rotational excitation energies in the product molecule up to $(v = 1, J = 3)$, while the faster group of H atoms can populate levels of HD up to $(v = 3, J = 8)$. The internal-state distribution of the HD molecules can be monitored either by CARS (Sect. 8.3) or by resonant multiphoton ionization [13.99]. Because of their fundamental importance, these measurements have been repeated by several groups with other spectroscopic techniques that have improved signal-to-noise ratios [13.101].

Ion–molecule reactions, which play an important role in interstellar clouds and in many chemical and biological processes, have increasing caught the attention of spectroscopists. As a recently investigated example, the charge-exchange reaction

$$N^+ + CO \rightarrow CO^+ + N \,,$$

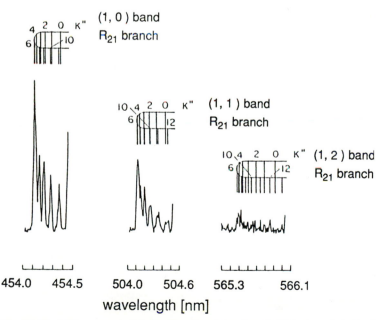

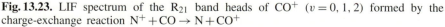

Fig. 13.23. LIF spectrum of the R_{21} band heads of CO^+ ($v = 0, 1, 2$) formed by the charge-exchange reaction $N^+ + CO \rightarrow N + CO^+$

is mentioned, where the rotational distribution $N(v, J)$ of CO^+ for different vibrational levels v has been measured by infrared fluorescence (Fig. 13.23) in dependence on the kinetic energy of the N^+ ions [13.102].

Photochemical reactions are often initiated by direct photodissociation or by collision-induced dissociation of a laser-excited molecule, where radicals are formed as intermediate products, which further react by collisions. The dynamics of photodissociation after excitation of the parent molecule by a UV laser has therefore been studied thoroughly [13.103]. While the first experiments were restricted to measurements of the internal-state distribution of the dissociation products, later more refined arrangements also allowed the determination of the angular distribution and of the orientation of the products for different polarizations of the photodissociating laser [13.104, 13.105]. The technique is illustrated by the example

$$ICN \underset{248\,\text{nm}}{\overset{h\nu}{\rightarrow}} CN + I$$

which was investigated with the apparatus depicted in Fig. 13.24. The ICN molecules are dissociated by the circularly polarized radiation of a KrF laser at $\lambda = 248$ nm. The orientation of the CN fragments is monitored through the fluorescence induced by a polarized dye laser. The dye laser is tuned through the $B^2\Sigma \leftarrow X^2\Sigma^+$ system of CN. This circular polarization is periodically switched between σ^+ and σ^- by a photoelastic polarization modulator (PEM). The corresponding change in the fluorescence intensity is a measure of the orientation of the CN fragments [13.106].

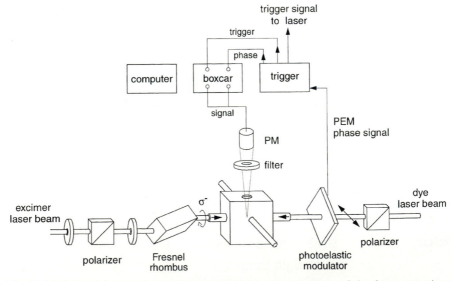

Fig. 13.24. Schematic experimental arrangement for measurements of the fragment orientation after photodissociation of the parent molecule [13.106]

A famous example is the photodissociation of H_2O leading to a preferentially populated Λ component of the OH fragment, which forms the basis of the astromonical OH maser observed in interstellar clouds [13.107]. There are many more examples of laser spectroscopy of chemical reactions, some of which can be found in [13.108–13.112].

13.5 Spectroscopic Determination of Differential Collision Cross Sections in Crossed Molecular Beams

The techniques discussed in Sects. 13.2–13.4 allowed the measurement of absolute rate constants of selected collision-induced transitions, which represent *integral* inelastic cross sections averaged over the thermal velocity distribution of the collision partners. Much more detailed information on the interaction potential can be extracted from measured *differential* cross sections, obtained in crossed molecular beam experiments [13.1, 13.113, 13.114].

Assume that atoms or molecules A in a collimated beam collide with atoms or molecules B within the interaction volume V formed by the overlap of the two beams (Fig. 13.25). The number of particles A scattered per second by the angle θ into the solid angle $d\Omega$ covered by the detector D is given by

$$\frac{dN_A(\theta)}{dt} d\Omega = n_A v_r n_B V \frac{d\sigma(\theta)}{d\Omega} d\Omega , \qquad (13.35)$$

where n_A is the density of the incident particles A, n_B the density of particles B, v_r is the relative velocity, and $(d\sigma(\theta)/d\Omega)$ the differential scattering cross section. The scattering angle θ is related to the impact parameter b and the interaction potential. It can be calculated for a given potential $V(r)$ [13.115]. Differential cross sections probe definite local parts of the potential, while integral cross sections only reflect the global effect of the potential on the deflected particles, averaged over all impact parameters. We will discuss which techniques can be used to measure differential cross sections for elastic, inelastic, and reactive collisions.

With classical techniques, the energy loss of the particle A during an inelastic collision is determined by measuring its velocity before and after the

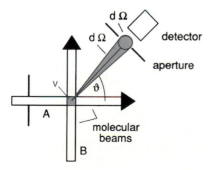

Fig. 13.25. Schematic diagram for the measurement of differential cross sections in crossed molecular beams

collision with velocity selectors or by time-of-flight measurements [13.33]. Since the velocity resolution $\Delta v/v$ and therefore the minimum detectable energy loss ΔE_{kin} is restricted by technical limitations, this method can be applied to only a limited number of real problems.

For polar molecules with an electric dipole moment, Rabi spectrometers with electrostatic quadrupole deflection fields offer the possibility to detect rotationally inelastic collisions since the focusing and deflecting properties of the quadrupole field depend on the quantum state $|J, M\rangle$ of a polar molecule. This technique is, however, restricted to small values of the rotational quantum number J [13.115].

Applications of laser-spectroscopic techniques can overcome many of the above-mentioned limitations. The energy resolution is higher by several orders of magnitude than that for time-of-flight measurements. In principle, molecules in arbitrary levels $|v, J\rangle$ can be investigated, as long as they have absorption transitions within the spectral range of available lasers. The essential progress of laser spectroscopy is, however, due to the fact that besides the scattering angle θ, the initial and final quantum states of the scattered particle A can also be determined. In this respect, laser spectroscopy of collisions in crossed beams represents the *ideal scattering experiment* in which all the relevant parameters are measured.

The technique is illustrated by Fig. 13.26. A collimated supersonic beam of argon, which contains Na atoms and Na_2 molecules, is crossed perpendicularly with a noble gas beam [13.116]. Molecules of Na_2 in the level $|v''_m J''_m\rangle$ that have been scattered by the angle θ are monitored by a *quantum-state-specific detector*. It consists of a cw dye laser focused into a spot behind the aperture A and tuned to the transition $|v''_m, J''_m\rangle \rightarrow |v'_j, J'_j\rangle$, and an optical system of a mirror and lens, which collects the LIF through an optical fiber bundle onto a photomultiplier. The entire detector can be turned around the scattering center.

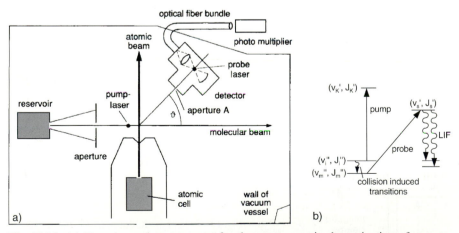

Fig. 13.26. (a) Experimental arrangement for the spectroscopic determination of state-to-state differential cross sections [13.117]. (b) Level scheme for pump-probe experiment

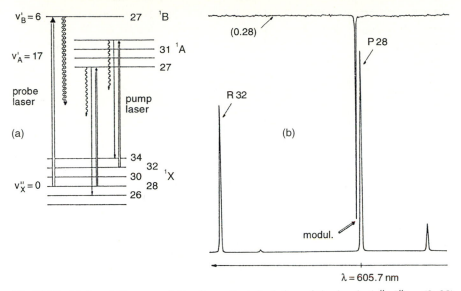

Fig. 13.27. (**a**) Level diagram of Na$_2$ for optical depletion of the levels $(v'', J'') = (0, 28)$ and optical probing of collisionally populated levels $(v'', J'' + \Delta J)$. (**b**) The measured spectrum in the right part gives the experimental proof that a level (v'', J'') can be completely depleted by optical pumping. The *lower* spectrum shows the pump laser-induced fluorescence I_{Fl} (L1) when the laser L1 is tuned, the *upper* trace the probe laser-induced fluorescence, where the probe laser was stabilized onto the transition $X(v'' = 0, J'' = 28) \rightarrow B(v' = 6, J' = 27)$. When the pump laser is tuned over the transition starting from $X(v'' = 0, J'' = 28)$ the probe laser-induced fluorescence drops nearly to zero. The *lower* spectrum has been shifted by 2 mm to the right [13.118]

The detected intensity of the LIF is a measure of the scattering rate $[\mathrm{d}N(v''_m, J''_m, \theta)/\mathrm{d}t]\mathrm{d}\Omega$ which has two contributions:

(a) The elastically scattered molecules $(v''_m, J''_m) \rightarrow (v''_m, J''_m, \theta)$, and
(b) the sum of all inelastically scattered molecules which have suffered collision-induced transitions $\sum_n [(v''_n, J''_n) \rightarrow (v''_m, J''_m)]$.

In order to select molecules with definite initial and final states, an OODR method is used (Sect. 10.4). Shortly before they reach the scattering volume, the molecules pass through the beam of a pump laser that induces transitions $(v''_i, J''_i) \rightarrow (v'_k, J'_k)$ and depletes the lower level (v''_i, J''_i) by optical pumping (Fig. 13.27). If the scattering rate is measured by the detector alternately with the pump laser on and off, the difference of the two signals just gives the contribution of those molecules with initial state (v''_i, J''_i), final state (v''_m, J''_m), and scattering angle θ [13.117].

Since the scattering angle θ is related to the impact parameter b, the experiment indicates which impact parameters contribute preferentially to collision-induced vibrational or rotational transitions, and how the transferred angular momentum ΔJ depends on impact parameters, initial state (v''_i, J''_i),

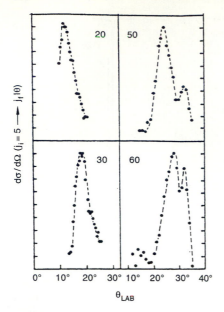

Fig. 13.28. Differential cross sections for collision-induced rotational transitions $[J_i = 5 \to J_f = J_i + \Delta J]$ in $Na_2 + Ne$ collisions for different values of J_f versus the scattering angle θ in the laboratory system [13.118]

and collision partner B [13.118]. In Fig. 13.28 some differential cross sections $\sigma(\theta)$ for inelastic $Na_2 + Ne$ collisions are plotted versus the scattering angle θ. The analysis of the measured data yields very accurate interaction potentials.

Of particular interest is the dependence of collision cross sections on the vibrational state, since the probability of inelastic and reactive collisions increases with increasing vibrational excitation [13.119]. Stimulated emission pumping before the scattering center allows a large population transfer into selected, high vibrational levels of the electronic ground state. For some molecules, such as Na_2 or I_2, an elegant realization of this technique is based on a molecular beam dimer laser [13.120], which uses the molecules in the primary beam as an active medium pumped by an argon laser or a dye laser (Fig. 13.29). Threshold pump powers below $1\,mW$ (!) could indeed be achieved.

Optical pumping also allows measurements of differential cross sections for collisions where one of the partners is electronically excited. The interaction potential $V(A^*, B)$ could be determined in this way for alkali–noble gas partners, which represent exciplexes since they are bound in an excited state but dissociate in the unstable ground state. One example is the investigation of $K + Ar$ collision partners [13.121]. In the crossing volume of a potassium beam and an argon beam, the potassium atoms are excited by a cw dye laser into the $4\,P_{3/2}$ state. The scattering rate for K atoms is measured versus the scattering angle θ with the pump laser on and off. The signal difference yields the contribution of the excited atoms to the elastic scattering if the fraction of excited atoms can be determined, which demands a consideration of the hyperfine structure. In order to reach the maximum concentration of excited K

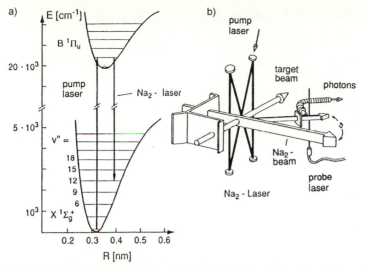

Fig. 13.29a,b. Selective pumping of a high vibrational level (v, J) by stimulated emission pumping in Na_2 with a Na_2 dimer laser: (**a**) level scheme; (**b**) experimental realization [13.120]

atoms, either a single-mode circularly polarized dye laser must be used, which is tuned to the transition $F'' = 2 \rightarrow F' = 3$ between selected hfs components, or a broad-band laser, which pumps all allowed hfs transitions simultaneously.

An example of measurements of *inelastic* collisions between excited atoms A^* and molecules M in crossed beams is the experiment by Hertel et al. [13.122], where laser-excited $Na^*(3P)$ atoms collide with molecules such as N_2 or CO and the energy transfer

$$A^*M(v'' = 0) \rightarrow A + M(v'' > 0) + \Delta E_{kin} , \qquad (13.36)$$

is studied. If the scattering angle θ and the velocity of the scattered Na atoms are measured, kinematic considerations (energy and momentum conservation) allow the determination of the fraction of the excitation energy that is converted into internal energy of M and that is transferred into kinetic energy ΔE_{kin}. The experimental results represent a crucial test of ab initio potential surfaces $V(r, \theta, \phi)$ for the interaction potential $Na^* - N_2$ ($v = 1, 2, 3, ...$) [13.123].

Scattering of electrons, fast atoms, or ions with laser-excited atoms A^* can result in elastic, inelastic, or superelastic collisions. In the latter case, the excitation energy of A^* is partly converted into kinetic energy of the scattered particles. Orientation of the excited atoms by optical pumping with polarized lasers allows investigations of the influence of the atomic orientation on the differential cross sections for $A^* + B$ collisions, which differs for collisions with electrons or ions from the case of neutral atoms [13.124]. An example of the measurement of differential cross sections of the reactive collisions $Na^* + HF$ is given in [13.125].

13.6 Photon-Assisted Collisional Energy Transfer

If two collision partners A and M absorb a photon $h\nu$ at a relative distance R_c, one of the partners may remain in an excited state after the collision:

$$A + M + h\nu \rightarrow A^* + M . \qquad (13.37)$$

This reaction can be regarded as the three step process:

$$A + M \rightarrow (AM) , \qquad (13.38)$$
$$(AM) \rightarrow h\nu \rightarrow (AM)^* , \qquad (13.39)$$
$$(AM)^* \rightarrow A^* + M . \qquad (13.40)$$

The two particles form a collision complex that is excited by absorption of the photon $h\nu$ at a relative distance R_c, where the energy difference $\Delta E = E(AM)^* - E(AM)$ between the two potential curves just equals the photon energy $h \cdot \nu$ (Fig. 13.30a) and the Franck–Condon factor has a maximum value [13.126].

If the potential is not monotonic, there are generally two trajectories with different impact parameters that lead to the same deflection angle θ in the center of mass system (Fig. 13.30b).

In a similar way collisions with excited atoms can be studied. The energy transfer by inelastic collisions between excited atoms A^* and ground-state atoms B

$$A^* + B \rightarrow B^* + A + \Delta E_{kin} , \qquad (13.41)$$

is governed by the conservation of energy and momentum. The energy difference $\Delta E_{int} = E(A^*) - E(B^*)$ is transferred into kinetic energy ΔE_{kin} of

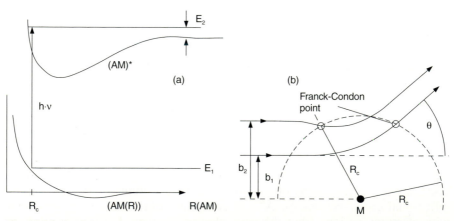

Fig. 13.30a,b. Schematic diagram of "optical collisions": (a) potential curves of ground and excited state of the collision complex; (b) classical trajectories in the center-of-mass system for two different impact parameters b_1, b_2

the collision partners. For $\Delta E_{\mathrm{kin}} \gg KT$, the cross section σ for the reaction (13.41) becomes very small, while for near-resonant collisions ($\Delta E_{\mathrm{kin}} \ll KT$), σ may exceed the gas kinetic cross section by several orders of magnitude.

If such a reaction (13.41) proceeds inside the intense radiation field of a laser, a photon may be absorbed or emitted during the collision, which may help to satisfy energy conservation for small ΔE_{kin} even if ΔE_{int} is large. Instead of (13.41), the reaction is now

$$A^* + B \pm \hbar\omega \rightarrow B^* + A + \Delta E_{\mathrm{kin}} . \tag{13.42}$$

For a suitable choice of the photon energy $\hbar\omega$, the cross section for a nonresonant reaction (13.41) can be increased by many orders of magnitude through the help of the photon, which makes the process near resonant. Such *photon-assisted* collisions will are in this section.

In the *molecular* model of Fig. 13.31a the potential curves $V(R)$ are considered for the collision pairs $A^* + B$ and $A + B^*$. At the critical distance R_c the energy difference $\Delta E = V(AB^*) - V(A^*B)$ may be equal to $\hbar\omega$. Resonant absorption of a photon by the collision pair $A^* + B$ at the distance R_c results in a transition into the upper potential curve $V(AB^*)$, which separates into $A + B^*$. The whole process (13.42) then leads to an energy transfer from A^* to B^*, where the initial and final kinetic energy of the collision partners depends on the slope dV/dR of the potential curves and the distance R_c of photon absorption.

If we start with $A + B^*$ the inverse process of stimulating photon emission will result in a transition from the upper into the lower potential, which means an energy transfer from B^* to A^* for the separated atoms.

For the experimental realization of such photon-assisted collisional energy transfer one needs two lasers: the pump laser L1 excites the atoms A into the

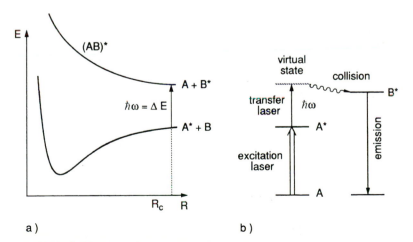

a)   b)

Fig. 13.31a,b. Photon-assisted collisional energy transfer: (**a**) molecular model; (**b**) dressed-atom model

excited state A* and the transfer laser L2 induces the transition between the two potential curves $V(A^*B) \rightarrow V(AB^*)$.

In the *atomic model* (Fig. 13.31b, often called the *dressed-atom model*) [13.127] the excited atom A* absorbes a photon $\hbar\omega$ of the transfer laser, which excites A* further into a *virtual state* $A^* + \hbar\omega$ that is near resonant with the excited state of B*. The collisional energy transfer $A^* + \hbar\omega \rightarrow B^* + \Delta E_{kin}$ with $\Delta E_{kin} \ll \hbar\omega$ then proceeds with a much higher probability than the off-resonance process without photon absorption [13.128].

The first experimental demonstration of photon-assisted collisional energy transfer was reported by Harris and coworkers [13.129], who studied the process

$$ Sr^*(5\,^1P) + Ca(4\,^1S) + \hbar\omega \rightarrow Sr(5\,^1S) + Ca(5^1D) \,, $$

$$ \rightarrow Sr(5\,^1S) + Ca(4p^2\,^1S) \,. $$

The corresponding term diagram is depicted in Fig. 13.32. The strontium atoms are excited by a pulsed pump laser at $\lambda = 460.7$ nm into the level $5s5p^1P_1^0$. During the collision with a Ca atom in its ground state the collision pair $Sr(5\,^1P_1^0)$–$Ca(4\,^1S)$ absorbs a photon $\hbar\omega$ of the transfer laser at $\lambda = 497.9$ nm. After the collision the excited $Ca^*(4p^2\,^1S)$ atoms are monitored through their fluorescence at $\lambda = 551.3$ nm.

It is remarkable that the transitions $4s^2\,^1S_1 \rightarrow 4p^2\,^1S$ and $4s^2\,^1S \rightarrow 4p^2\,^1D$ of the Ca atom do not represent allowed dipole transitions and are therefore forbidden for the isolated atom. Regarding the photon-absorption probability, the absorption by a collision pair with subsequent energy transfer to the atom is therefore called *collision-induced absorption* or collision-aided

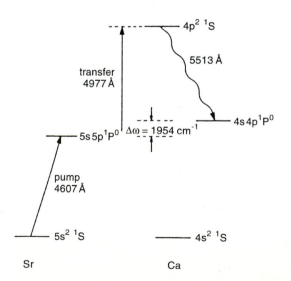

Fig. 13.32. Term diagram for photon-assisted collisional energy transfer from $Sr^*(^1P^0)$ to $Ca(4p^2\,^1S)$ [13.129]

radiative excitation [13.130], where a *dipole-allowed* transition of the molecular collision pair A^*B or AB^* takes place. The molecular transition has a dipole transition moment $\mu(R)$ that depends on the internuclear separation $R(A^*B)$ and that approaches zero for $R \rightarrow \infty$ [13.131]. It is caused by the induced dipole–dipole interaction from the polarizability of the collision partners.

Such collision-induced absorption of radiation is important not only for the initiation of chemical reactions (Sect. 15.2), but also plays an important role in the absorption of infrared radiation in planetary atmospheres and interstellar molecular clouds. By the formation of H_2–H_2 collision pairs, for instance, vibration–rotational transitions within the electronic ground state of H_2 become allowed although they are forbidden in the isolated homonuclear H_2 molecule [13.132, 13.133].

An interesting aspect of collision-aided radiative excitation is its potential for optical cooling of vapors. Since the change in kinetic energy of the collision partners per absorbed photon can be much larger than that transferred by photon recoil (Sect. 14.1), only a few collisions are necessary for cooling to low temperatures compared with a few thousand for recoil cooling [13.130].

Detailed investigations of the transfer cross section and its dependence on the wavelength λ_2 of the transfer laser L2 have been performed by Toschek and coworkers [13.134], who studied the reaction $Sr(5s5p) + Li(2s) + \hbar\omega \rightarrow Sr(5s^2) + Li(4d)$. The experimental setup is shown in Fig. 13.33. A mixture of strontium and lithium vapors is produced in a heat pipe [13.22]. The beams of two dye lasers pumped by the same N_2 laser are superimposed and focused into the center of the heat pipe. The first laser L1 at $\lambda_1 = 407$ nm excites the Sr atoms. The pulse of the second laser L2 is de-

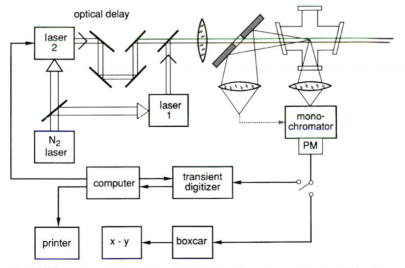

Fig. 13.33. Experimental setup for studies of photon-assisted collisional energy transfer [13.134]

layed by a variable time interval Δt and its wavelength λ_2 is tuned around $\lambda_1 = 700$ nm. The intensity $I_{Fl}(\lambda_2, \Delta t)$ of the fluorescence emitted by the excited Li* atoms is monitored in the transition $Li(d\,^2D \rightarrow 2p\,^2P)$ at $\lambda = 610$ nm as a function of the wavelength λ_2 and the delay time Δt. For the resonance case λ_2^{res} (that is, $\Delta E_{kin} = 0$ in (13.42)) energy transfer cross sections up to $\sigma_T > 2 \times 10^{-13}\,I \cdot cm^2$ (where I is measured in MW/cm^2) have been determined. For sufficiently high intensities I of the transfer laser, the photon-assisted collisional transfer cross sections σ_T exceed the gas-kinetic cross sections $\sigma \approx 10^{-15}$ cm^2 by 2−3 orders of magnitude.

Example 13.1

For saturation of the pump transition even moderate intensities ($I_p < 10^3$ kW/cm^2) are sufficient (Sect. 7.1). Assume an output power of 50 kW of the transfer laser L2 and a beam diameter of 1 mm in the focal plane, which yields $I_{transf} = 6.7$ MW/cm^2, energy transfer cross sections of $\sigma_T \approx 1.3 \times 10^{-12}$ cm^2, which is about 1000 times larger than the gas-kinetic cross section. The energy transfer $R_T = \sigma n \bar{v} h \nu_T$, where n is the density of collision partners, colliding with the excited atoms. With $n = 10^{16}$ cm^3, $\bar{v} = 10^5$ cm/s and $h\nu_T = 2$ eV, we obtain $R_T = 10^9$ eV/s.

Further information on this interesting field and its various potential application can be found in [13.135, 13.136].

13.7 Photoassociation Spectroscopy of Colliding Atoms

If two atoms A and B approach each other with a small relative velocity, they cannot form a bound molecule unless their kinetic energy is removed by a third collision partner. However, when the collision pair absorbs a photon on a transition into an excited bound state, this may lead to a stable molecule, if the optical excitation reaches the outer turning point of the vibrational level in the excited state. The molecule emits a fluorescence photon at the inner turning point, leading to a bound level in the electronic ground state (Fig. 13.34). Enhancing the last stabilizing step by stimulated emission might lead to stable cold molecules in low vibrational levels of the electronic ground state.

This photoassociation spectroscopy has made rapid progress recently after the realization of optically cooled atomic gases [13.137, 13.138] because it shows some very interesting aspects. First, it allows the formation of cold molecules in defined vibrational–rotational states. If the colliding atoms have a sufficiently small relative kinetic energy, the angular momentum of the collision complex is zero (S-wave scattering). This implies that the rotational quantum number of the resulting molecule in its electronic ground state is 0, 1, or 2, depending on the angular momentum transfer by the absorbed and emitted photon.

Second the photoassociation spectroscopy of cold atoms allows the detailed investigation of the long-range part of the difference potential between

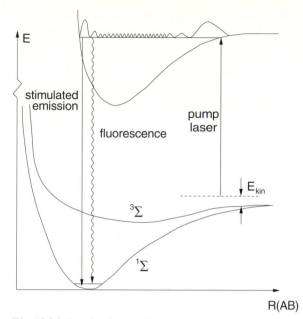

Fig. 13.34. Level scheme of photoassociation spectroscopy

the excited and ground state of the collision pair. Often the ground-state potential can be accurately calculated. Then the difference potential yields the potential of the excited state up to large internuclear distances. Here the potential energy is the sum of different contributions, such as induced and permanent dipole forces, and electron and nuclear spin contributions. Retardation effects may also play a role. The potential curve in excited states therefore has a more complicated R-dependence and may exhibit a hump or a shallow minimum at large internuclear separations.

Third, the spectroscopy of cold collisions may open the way to a new chemistry at ultralow temperatures, which allows more detailed insights into the mechanisms of molecular reactions and their dependence on the relative kinetic energy of the reactants [13.139].

A historical overview on the development of our understanding of the dynamics of molecular processes during the last 50 years can be found in [13.140].

Problems

13.1 Assume an inelastic cross section of $\sigma^{\text{inel}} = 10^{-17} \, \text{cm}^2$ for collisions between excited atoms A^* and partners B, and a velocity-changing cross section $\sigma_v(\Delta v) = \sigma_0 \exp(-\Delta v^2 / \langle v^2 \rangle)$ for velocity changes Δv of atom A. Discuss for $\sigma_0 = 10^{-16} \, \text{cm}^2$ the line profile of the Lamb dip in the transition $A \to A^*$ if the radiative lifetime is $\tau_{\text{rad}} = 10^{-8} \, \text{s}$, the pressure of partners B is 1 mbar, the temperature $T = 300\,\mathrm{K}$, the saturation parameter $S = 1$, and the reduced mass $\mu(\mathrm{AB}) = 40\,\mathrm{AMU}$.

13.2 The effective lifetime of an excited molecular level is $\tau_{\mathrm{eff}}(p = 5\,\mathrm{mbar}) = 8 \times 10^{-9}\,\mathrm{s}$ and $\tau_{\mathrm{eff}}(p = 1\,\mathrm{mbar}) = 12 \times 10^{-9}\,\mathrm{s}$ for molecules with the mass $M = 43\,\mathrm{AMU}$ in a gas cell with argon buffer gas at $T = 500\,\mathrm{K}$. Calculate the radiative lifetime, the collision-quenching cross section, and the homogeneous linewidth $\Delta\nu(p)$.

13.3 A single-mode laser beam with a Gaussian profile ($w = 3\,\mathrm{mm}$) and a power of $10\,\mathrm{mW}$, tuned to the center ω_0 of the sodium line ($3\,^2\mathrm{S}_{1/2} \to 3\,^2\mathrm{P}_{1/2}$) passes through a cell containing sodium vapor at $p = 10^{-3}\,\mathrm{mbar}$. The absorption cross section is $\sigma_{\mathrm{abs}} = 5 \times 10^{-11}\,\mathrm{cm}^2$, the natural linewidth is $\delta\nu_n = 10\,\mathrm{MHz}$, and the Doppler width $\delta\nu_{\mathrm{D}} = 1\,\mathrm{GHz}$. Calculate the saturation parameter S, the absorption coefficient $\alpha_s(\omega_0)$, and the mean density of ground-state and excited atoms that collide in an energy-pooling collision with $\sigma_{\mathrm{ep}} = 3 \times 10^{-14}\,\mathrm{cm}^2$. Which Rydberg levels are accessible by energy-pooling collisions $\mathrm{Na}^*(3P) + \mathrm{Na}^*(3P)$ if the relative kinetic energy is $E_{\mathrm{kin}} \leq 0.1\,\mathrm{eV}$?

13.4 A square-wave chopped cw laser depopulates the level (v_i'', J_i'') in the electronic ground state of the molecule M. What is the minimum chopping period T to guarantee quasi-stationary conditions for level $|i\rangle$ and its neighboring levels $|J_i \pm \Delta J\rangle$, if the total refilling cross section of level $|i\rangle$ is $\sigma = 10^{-14}\,\mathrm{cm}^2$ and the individual cross sections $\sigma_{ik} = 3 \times 10^{-16}$ per ΔJ for transitions between neighboring levels $|J\rangle$ and $|J \pm \Delta J\rangle$ of the same vibrational level at a pressure of $1\,\mathrm{mbar}$?

14. New Developments in Laser Spectroscopy

During the last few years several new ideas have been born and new spectroscopic techniques have been developed that not only improve the spectral resolution and increase the sensitivity for investigating single atoms but that also allow several interesting experiments for testing fundamental concepts of physics. In the historical development of science, experimental progress in the accuracy of measurements has often brought about a refinement of theoretical models or even the introduction of new concepts [14.1]. Examples include A. Einstein's theory of special relativity based on the interferometric experiments of Michelson and Morley [14.2], M. Planck's introduction of quantum physics for the correct explanation of the measured spectral distribution of blackbody radiation, the introduction of the concept of electron spin after the spectroscopic discovery of the fine structure in atomic spectra [14.3], or the test of quantum electrodynamics by precision measurements of the Lamb shift [14.4]. In this chapter some of these new and exciting developments are presented.

14.1 Optical Cooling and Trapping of Atoms

In order to improve the accuracy of spectroscopic measurements of atomic energy levels, all perturbing effects leading to broadening or shifts of these levels must be either eliminated or should be sufficiently well understood to introduce appropriate corrections. One of the largest perturbing effects is the thermal motion of atoms. In Chap. 9 we saw that in collimated atomic beams the velocity components v_x and v_y, which are perpendicular to the beam axes, can be drastically reduced by collimating apertures (geometrical cooling). The component v_z can be compressed by adiabatic cooling into a small interval $v_z = u \pm \Delta v_z$ around the flow velocity u. If this reduction of the velocity distribution is described by a *translational temperature* T_{trans}, values of $T_{\text{trans}} < 1\,\text{K}$ can be reached. However, the molecules still have a nearly uniform but large velocity u, and broadening effects such as transit-time broadening or shifts caused by the second-order Doppler effect (see below) are not eliminated.

In this section we discuss the new technique of *optical cooling*, which decreases the velocity of atoms to a small interval around $v = 0$. Optical cooling down to "temperatures" of a few microKelvin has been achieved; by combining optical and evaporative cooling even the nanoKelvin range was reached. This brought the discovery of quite new phenomena, such as Bose–Einstein condensation or atom–lasers, and atomic fountains. [14.5–14.7].

14.1.1 Photon Recoil

Let us consider an atom with rest mass M in the energy level E_i that moves with the velocity v. If this atom absorbs a photon of energy $\hbar\omega_{ik} \simeq E_k - E_i$ and momentum $\hbar k$, it is excited into the level E_k. Its momentum changes from $p_i = Mv_i$ before the absorption to

$$p_k = p_i + \hbar k \,, \tag{14.1}$$

after the absorption (the *recoil effect*, Fig. 14.1).

The relativistic energy conservation demands

$$\hbar\omega_{ik} = \sqrt{p_k^2 c^2 + (M_0 c^2 + E_k)^2} - \sqrt{p_i^2 c^2 + (M_0 c^2 + E_i)^2} \,. \tag{14.2}$$

When we extract $(M_0 c^2 + E_k)$ from the first root and $(M_0 c^2 + E_i)$ from the second, we obtain by a Taylor expansion the power series for the resonant absorption frequency

$$\omega_{ik} = \omega_0 + k \cdot v_i - \omega_0 \frac{v_i^2}{2c^2} + \frac{\hbar\omega_0^2}{2Mc^2} + \dots \,. \tag{14.3}$$

The first term represents the absorption frequency $\omega_0 = (E_k - E_i)$ of an atom at rest if the recoil of the absorbing atom is neglected. The second term describes the linear Doppler shift (first-order Doppler effect) caused by the motion of the atom at the time of absorption. The third term expresses the quadratic Doppler effect (second-order Doppler effect). Note that this term is independent of the direction of the velocity v. It is therefore *not* eliminated by the "Doppler-free" techniques described in Chaps. 7–10, which only overcome the *linear* Doppler effect.

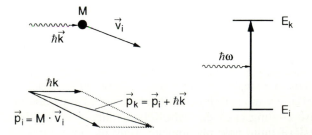

Fig. 14.1. Atomic recoil for absorption and emission of a photon

Example 14.1

A parallel beam of Ne ions accelerated by 10 keV moves with the velocity $v_z = 3 \times 10^5$ m/s. When the beam is crossed perpendicularly by a single-mode laser beam tuned to a transition with $\lambda = 500$ nm, even ions with $v_x = v_y = 0$ show a quadratic relativistic Doppler shift of $\Delta \nu / \nu = 5 \times 10^{-7}$, which yields an absolute shift of $\Delta \nu = 250$ MHz. This should be compared with the linear Doppler shift of 600 GHz, which appears when the laser beam is parallel to the ion beam (Example 9.6).

The last term in (14.3) represents the recoil energy of the atom from momentum conservation. The energy $\hbar \omega$ of the absorbed photon has to be larger than that for recoil-free absorption by the amount

$$\Delta E = \frac{\hbar^2 \omega_0^2}{2Mc^2} \quad \Rightarrow \quad \frac{\Delta E}{\hbar \omega} = \frac{1}{2} \frac{\hbar \omega}{Mc^2} . \tag{14.4}$$

When the excited atom in the state E_k with the momentum $\boldsymbol{p}_k = M\boldsymbol{v}_k$ emits a photon, its momentum changes to

$$\boldsymbol{p}_i = \boldsymbol{p}_k - \hbar \boldsymbol{k}$$

after the emission. The emission frequency becomes analogous to (14.3)

$$\omega_{ik} = \omega_0 + k v_k - \frac{\omega_0 v_k^2}{2c^2} - \frac{\hbar \omega_0^2}{2Mc^2} . \tag{14.5}$$

The difference between the resonant absorption and emission frequencies is

$$\Delta \omega = \omega_{ik}^{\text{abs}} - \omega_{ki}^{\text{em}} = \frac{\hbar \omega_0^2}{Mc^2} + \frac{\omega_0}{2c^2}(v_k^2 - v_i^2) \approx \frac{\hbar \omega_0^2}{Mc^2} , \tag{14.6}$$

since the second term can be written as $(\omega_0/Mc^2) \cdot (E_k^{\text{kin}} - E_i^{\text{kin}})$ and $\Delta E^{\text{kin}} \ll \hbar \omega$. It can be therefore neglected for atoms with thermal velocities.

The relative frequency shift between absorbed and emitted photons because of recoil

$$\frac{\Delta \omega}{\omega} = \frac{\hbar \omega_0}{Mc^2} , \tag{14.7}$$

equals the ratio of photon energy to rest-mass energy of the atom.

For γ-quanta in the MeV range this ratio may be sufficiently large that $\Delta \omega$ becomes larger than the linewidth of the absorbing transition. This means that a γ-photon emitted from a free nucleus cannot be absorbed by another identical nucleus at rest. The recoil can be greatly reduced if the atoms are embedded in a rigid crystal structure below the Debye temperature. This recoil-free absorption and emission of γ-quanta is called the *Mössbauer effect* [14.8].

In the *optical region* the recoil shift $\Delta\omega$ is extremely small and well below the natural linewidth of most optical transitions. Nevertheless, it has been measured for selected narrow transitions [14.9, 14.10].

14.1.2 Measurement of Recoil Shift

When the absorbing molecules with the resonant absorption frequency ω_0 are placed inside the laser resonator, the standing laser wave of frequency $\omega \neq \omega_0$ burns two Bennet holes into the population distribution $N_i(v_z)$ (Fig. 14.2b and Sect. 7.2), which, according to (14.3), appear at the velocity components

$$v_{zi} = \pm[\omega' - \hbar\omega_0^2/(2Mc^2)]/k, \quad \text{with} \quad \omega' = \omega - \omega_0(1 - v^2/2c^2). \quad (14.8)$$

The corresponding peaks in the population distribution $N_k(v_z)$ of molecules in the upper level $|k\rangle$ are shifted due to photon recoil (Fig. 14.2a). They show up, according to (14.5), at the velocity components

$$v_{zk} = \pm[\omega' + \hbar\omega_0^2/(2Mc^2)]/k. \quad (14.9)$$

For the example, illustrated by Fig. 14.2, we have chosen

$$\omega < \omega_0 \implies \omega' < 0.$$

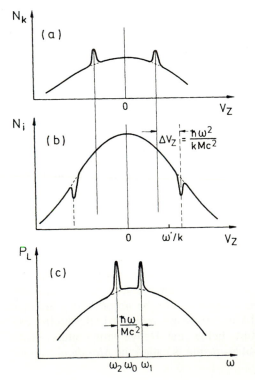

Fig. 14.2a–c. Generation of the recoil doublet of Lamb dips: (a) population peaks in the upper-state population for $\omega \neq \omega_0$; (b) Bennet holes in the lower-state population; (c) recoil doublet in the output power $P_L(\omega)$ of the laser

The two holes in the ground-state population coincide for $v_{zi} = 0$, which gives $\omega' = \hbar\omega_0^2/2Mc^2$. This happens, according to (14.8), for the laser frequency

$$\omega = \omega_1 = \omega_0[(1 - v^2/c^2) + \hbar\omega_0/(2Mc^2)] , \tag{14.10a}$$

while the two peaks in the upper level population coincide for another laser frequency

$$\omega = \omega_2 = \omega_0[(1 - v^2/c^2) - \hbar\omega_0/(2Mc^2)] . \tag{14.10b}$$

The absorption of the laser is proportional to the population difference $\Delta N = N_i - N_k$. This difference has two maxima at ω_1 and ω_2. The laser output therefore exhibits two Lamb peaks (*inverse Lamb dips*) (Fig. 14.2c) at the laser frequencies ω_1 and ω_2, which are separated by twice the recoil energy

$$\Delta\omega = \omega_1 - \omega_2 = \hbar\omega^2/Mc^2 . \tag{14.11}$$

Example 14.2

(a) For the transition at $\lambda = 3.39\,\mu m$ in the CH_4 molecule ($M = 16\,AMU$) the recoil splitting amounts to $\Delta\omega = 2\pi \cdot 2.16\,kHz$, which is still larger than the natural linewidth of this transition [14.10, 14.11].
(b) For the calcium intercombination line $^1S_0 \rightarrow {}^3P_1$ at $\lambda = 657\,nm$ we obtain with $M = 40\,AMU$ the splitting $\Delta\omega = 2\pi \cdot 23.1\,kHz$ [14.12].
(c) For a rotation–vibration transition in SF_6 at $\lambda = 10\,\mu m$ the frequency ω is one-third that of CH_4 but the mass 10 times larger than for CH_4. The recoil splitting therefore amounts only to about $0.02\,kHz$ and is not measurable [14.13]

Since such small splittings can only be observed if the width of the Lamb peaks is smaller than the recoil shift, all possible broadening effects, such as pressure broadening and transit-time broadening, must be carefully minimized. This can be achieved in experiments at low pressures and with expanded laser beam diameters. An experimental example is displayed in Fig. 14.3.

The transit-time broadening can greatly be reduced by the optical Ramsey method of separated fields. The best resolution of the recoil splittings has indeed been achieved with this technique (Sect. 14.4). The transit time can also be increased if only molecules with small transverse velocity components contribute to the Lamb dip. If the laser intensity is kept so small that saturation of the molecular transition is reached only for molecules that stay within the laser beam for a sufficiently long time, that is, molecules with small components v_x, v_y, the transit-time broadening is greatly reduced [14.11].

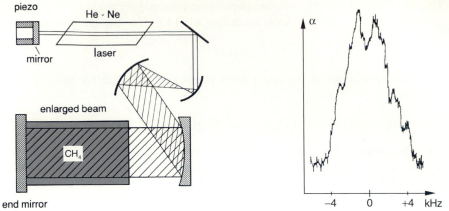

Fig. 14.3. Schematic experimental arrangement for measuring the recoil splitting and the measured signal of the recoil doublet of the hyperfine component $8 \rightarrow 7$ of the $(P(7), \nu_3)$ transition at $\lambda = 3.39 \, \mu$m in methane [14.11]

14.1.3 Optical Cooling by Photon Recoil

Although the recoil effect is very small when a single photon is absorbed, it can be used effectively for optical cooling of atoms by the cumulative effect of many absorbed photons. This can be seen as follows:

When atom A stays for the time T in a laser field that is in resonance with the transition $|i\rangle \rightarrow |k\rangle$, the atom may absorb and emit a photon $\hbar\omega$ many times, provided the optical pumping cycle is short compared to T and the atom behaves like a true two-level system. This means that the emission of fluorescence photons $\hbar\omega$ by the excited atom in level $|k\rangle$ brings the atom back only to the initial state $|i\rangle$, but never to other levels. With the saturation parameter $S = B_{ik}\rho(\omega_{ik})/A_{ik}$, the fraction of excited atoms becomes

$$\frac{N_k}{N} = \frac{S}{1+2S} .$$

The fluorescence rate is $N_k A_k = N_k/\tau_k$. Since N_k can never exceed the saturated value $N_k = (N_i + N_k)/2 = N/2$, the minimum recycling time for the saturation parameter $S \rightarrow \infty$ (Sect. 3.6) is $\Delta T = 2\tau_k$.

Example 14.3

When an atom passes with a thermal velocity $v = 500$ m/s through a laser beam with 2-mm diameter, the transit time is $T = 4 \, \mu$s. For a spontaneous lifetime $\tau_k = 10^{-8}$ s the atom can undergo $q \leq (T/2)/\tau_k = 400$ absorption–emission cycles during its transit time.

When a laser beam is sent through a sample of absorbing atoms, the LIF is generally isotropic, that is, the spontaneously emitted photons are randomly

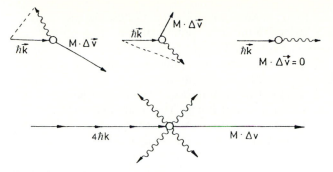

Fig. 14.4. Recoil momentum of an atom for a fixed direction of the absorbed photons but different directions of the emitted fluorescence photons

distributed over all spatial directions. Although each emitted photon transfers the momentum $\hbar k$ to the emitting atoms, the time-averaged momentum transfer tends to zero for sufficiently large values of $q = T/2\tau_k$.

The *absorbed* photons, however, all come from the same direction. Therefore, the momentum transfer for q absorptions adds up to a total recoil momentum $p = q\hbar k$ (Fig. 14.4). This changes the velocity v of an atom which flies against the beam propagation by the amount $\Delta v = \hbar k/M$ per absorption. For q absorption–emission cycles we get

$$\Delta v = q\frac{\hbar k}{M} = q\frac{\hbar \omega}{Mc} . \tag{14.12a}$$

Atoms in a collimated atomic beam can therefore be slowed down by a laser beam propagating anticollinearly to the atomic beam [14.14]. This can be expressed by the "cooling force"

$$F = M\frac{\Delta v}{\Delta T} = \frac{\hbar k}{\tau_k}\frac{S}{1+2S} . \tag{14.12b}$$

Note that the *recoil energy* transferred to the atom

$$\Delta E_{\text{recoil}} = q\frac{\hbar^2 \omega^2}{2Mc^2} , \tag{14.13}$$

is still very small.

Example 14.4

(a) For Na atoms with $M = 23$ AMU, which absorb photons $\hbar\omega \simeq 2\,\text{eV}$ on the transition $3\,^2S_{1/2} \rightarrow 3\,^2P_{3/2}$, (14.12) gives $\Delta v = 3\,\text{cm/s}$ per photon absorption. In order to reduce the initial thermal velocity of $v = 600\,\text{m/s}$ at $T = 500\,\text{K}$ to $20\,\text{m/s}$ (this would correspond to a temperature $T = 0.6\,\text{K}$ for a thermal distribution with $\bar{v} = 20\,\text{m/s}$),

this demands $q = 2 \times 10^4$ absorption–emission cycles. At the spontaneous lifetime $\tau_k = 16$ ns a minimum cooling time of $T = 2 \times 10^4 \cdot 2 \cdot$
16×10^{-9} s $\simeq 600$ µs is required. This gives a negative acceleration of
$a = -10^6$ m/s^2, which is 10^5 times the gravity acceleration on earth
$g = 9.81$ m/s^2! During the time T the atom has traveled the distance
$\Delta_z = v_0 T - \frac{1}{2} a T^2 \simeq 18$ cm. During this deceleration path length it always has to remain within the laser beam. The total energy transferred
by recoil to the atom is only -2×10^{-2} eV. It corresponds to the kinetic
energy $\frac{1}{2} M v^2$ of the atom and is very small compared to $\hbar \omega = 2$ eV.

(b) For Mg atoms with $M = 24$ AMU, which absorb on the singlet resonance transition at $\lambda = 285.2$ nm, the situation is more favorable
because of the higher photon energy $\hbar \omega \simeq 3.7$ eV and the shorter lifetime $\tau_k = 2$ ns of the upper level. One obtains: $\Delta v = -6$ cm/s per
absorbed photon; $q = 1.3 \times 10^4$. The minimum cooling time becomes
$T = 3 \times 10^{-5}$ s and the deceleration path length $\Delta z \simeq 1$ cm.

(c) In [14.15] a list of other atoms can be found that were regarded as
possible candidates for optical cooling. Some of them have since been
successfully tried.

The following remarks may be useful:

(a) Without additional tricks, this optical-cooling method is restricted to true
two-level systems. Therefore, *molecules cannot be cooled with this tech-
nique*, because after their excitation into level $|k\rangle$ they return by emission
of fluorescence into many lower vibrational–rotational levels and only
a small fraction goes back into the initial level $|i\rangle$. Thus, only one opti-

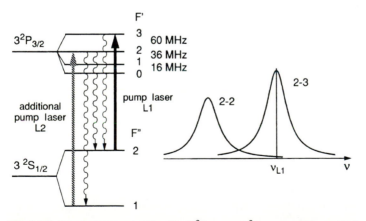

Fig. 14.5. Level diagram of the Na $3\,^2S_{1/2} \to 3\,^2P_{3/2}$ transition with hyperfine splittings.
Optical pumping on the hfs component $F'' = 2 \to F' = 3$ represents a true two-level system, provided any overlap between the hfs components can be avoided. The additional
pump laser L2 is necessary in order to pump atoms tranferred into level $F'' = 1$ by
spectral overlap between the components $2 \to 3$ and $2 \to 2$ back into level $F'' = 2$

cal pumping cycle is possible. There have been, however, other cooling mechanisms proposed and partly realized (see Sect. 14.1.5).

(b) The sodium transition $3S - 3P$, which represents the standard example for optical cooling, is in fact a multi level system because of its hyperfine structure (Fig. 14.5). However, after optical pumping with circularly polarized light on the hfs transition $3\,^2S_{1/2}(F'' = 2) \rightarrow 3\,^2P_{3/2}(F' = 3)$, the fluorescence can only reach the initial lower level $F'' = 2$. A true two-level system would be realized if any overlap of the pump transition with other hfs components could be avoided (see below).

(c) Increasing the intensity of the pump laser decreases the time ΔT for an absorption–emission cycle. However, for saturation parameters $S > 1$ this decrease is small and ΔT soon reaches its limit $2\tau_k$. On the other hand, the induced emission increases at the expense of spontaneous emission. Since the emitted induced photon has the same k-vector as the absorbed induced photon, the net momentum transfer is zero. The total deceleration rate therefore has a maximum at the optimum saturation parameter $S \simeq 1$.

14.1.4 Experimental Arrangements

For the experimental realization of optical cooling, which uses a collimated beam of atoms and a counterpropagating cw laser (dye laser or diode laser, Fig. 14.6) the following difficulties have to be overcome: during the deceleration time the Doppler-shifted absorption frequency $\omega(t) = \omega_0 + k \cdot v(t)$ changes with the decreasing velocity v, and the atoms would come out of resonance with the monochromatic laser. Three solutions have been successfully tried: either the laser frequency

$$\omega(t) = \omega_0 + k \cdot v(t) \pm \gamma , \tag{14.14}$$

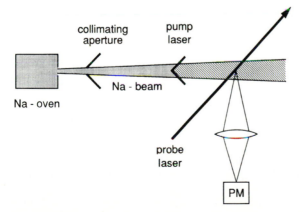

Fig. 14.6. Simplified experimental realization for the deceleration of atoms in a collimated beam by photon recoil

is synchronously tuned with the changing velocity $v(t)$ in order to stay within the linewidth γ of the atomic transition [14.16, 14.17], or the absorption frequency of the atom is appropriately altered along the deceleration path [14.18, 14.19]. A third alternative uses a broadband laser for cooling. We will discuss these methods briefly:

When the laser frequency ω for a counterpropagating beam $\mathbf{k}$ antiparallel to $\mathbf{v}$ is kept in resonance with the constant atomic frequency ω_0, it must have the time dependence

$$\omega(t) = \omega_0 \left(1 - \frac{v(t)}{c} \right) \quad \Rightarrow \quad \frac{d\omega}{dt} = -\frac{\omega_0}{c} \frac{dv}{dt} . \tag{14.15}$$

The velocity change per second for the optimum deceleration with a pump cycle period $T = 2\tau$ is, according to (14.12)

$$\frac{dv}{dt} = \frac{\hbar\omega_0}{2Mc\tau} . \tag{14.16}$$

Inserting this into (14.15) yields for the necessary time dependence of the laser frequency

$$\omega_L(t) = \omega(0)\, e^{\alpha t} \simeq \omega(0)(1 + \alpha t) , \quad \text{with} \quad \alpha = \frac{\hbar\omega_0}{2Mc^2\tau} \ll 1 . \tag{14.17}$$

This means that the pump laser should have a linear frequency chirp.

Example 14.5

For Na atoms with $v(0) = 1000\,\text{m/s}$, insertion of the relevant numbers into (14.17) yields $d\omega/dt = 2\pi \cdot 1.7\,\text{GHz/ms}$, which means that fast frequency tuning in a controlled way is required in order to fulfill (14.17).

An experimental realization of the controlled frequency chirp uses amplitude modulation of the laser with the modulation frequency Ω_1. The sideband at $\omega_L - \Omega_1$ is tuned to the atomic transition and can be matched to the time-dependent Doppler shift by changing $\Omega_1(t)$ in time. In order to compensate for optical pumping into other levels than $F'' = 2$ by overlap of the laser with other hfs components, the transition $F'' = 1 \rightarrow F' = 2$ is simultaneously pumped (Fig. 14.5). This can be achieved if the pump laser is additionally modulated at the second frequency Ω_2, where the sideband $\omega_L + \Omega_2$ matches the transition $F'' = 1 \rightarrow F' = 2$ [14.17, 14.20].

For the second method, where the laser frequency ω_L is kept constant, the atomic absorption frequency must be altered during the deceleration of the atoms. This can be realized by Zeeman tuning. In order to match the Zeeman shift to the changing Doppler shift $\Delta\omega(z)$, the longitudinal magnetic field must have the z-dependence

$$B = B_0 \sqrt{1 - 2az/v_0^2} , \tag{14.18}$$

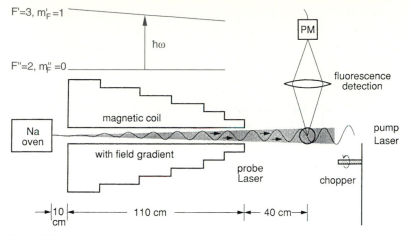

Fig. 14.7. Laser cooling of atoms in a collimated beam with a fixed laser frequency and Zeeman tuning of the atomic absorption frequency [14.19]

for atoms that enter the field at $z = 0$ with the velocity v_0 and experience the negative acceleration a [m/s^2] by photon recoil [14.19]. This field dependence $B(z)$ can be realized by a proper choice of the windings $N_W(z)$ per centimeters of the magnetic field coil (Fig. 14.7)

Most optical cooling experiments have been performed up to now on alkali atoms, such as Na or Rb, using a single-mode cw dye laser. The velocity decrease of the atoms is monitored with the tunable probe laser L2, which is sufficiently weak that it does not affect the velocity distribution. The probe-laser-induced fluorescence $I_{Fl}(\omega_2)$ is measured as a function of the Doppler shift. Experiments have shown that the atoms could be completely stopped and their velocity could even be reversed [14.16]. An example of the compression of the thermal velocity distribution into a narrow range around $v = 200$ m/s is illustrated in Fig. 14.8 for Na atoms.

A favorable alternative to dye lasers is a GaAs diode laser, which can cool rubidium or cesium atoms [14.21–14.23] and also metastable noble gas atoms, such as He* or Ar* [14.24]. The experimental expenditures are greatly reduced since the GaAs laser is much less expensive than an argon laser plus dye-laser combination. Furthermore, the frequency modulation is more readily realized with a diode laser than with a dye laser.

A very interesting alternative laser for optical cooling of atoms in a collimated beam is the *modeless* laser [14.25], which has a broad spectral emission (without mode structure, when averaged over a time of $T > 10$ ns, with an adjustable bandwidth and a tunable center frequency). Such a laser can cool all atoms regardless of their velocity if its spectral width $\Delta\omega_L$ is larger than the Doppler shift $\Delta\omega_D = v_0 k$ [14.26].

With the following experimental trick it is possible to compress the velocity distribution $N(v_z)$ of atoms in a beam into a small interval Δv_z around a wanted final velocity v_f. The beam from the modeless laser propagates an-

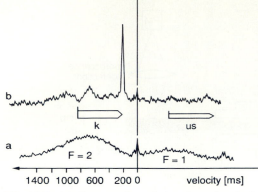

Fig. 14.8. (a) Compression of the thermal velocity distribution of Na atoms by optical cooling into a narrow velocity range around $v = 200$ m/s. **(b)** The sharp resonance at $v = 0$ is caused by the probe laser perpendicular to the atomic beam. The *arrow k* gives the tuning range of the cooling laser, the *arrow* µs gives that of the upper sideband, which pumps the transition $F'' = 1 \rightarrow F' = 2$ [14.16]

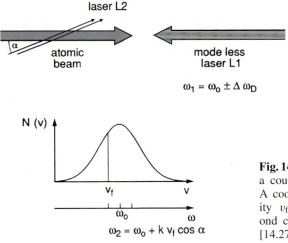

Fig. 14.9. Cooling of all atoms with a counterpropagating modeless laser. A cooling stop at a selectable velocity v_f can be realized with a second copropagating single-mode laser [14.27]

ticollinearly to the atomic beam and cools the atoms (Fig. 14.9). A second single-mode laser intersects the atomic beam under a small angle α against the beam axis. If it is tuned to the frequency

$$\omega_2 = \omega_0 + kv_f \cos\alpha \ ,$$

it accelerates the atoms as soon as they have reached the velocity v_f. This second laser therefore acts as a barrier for the lower velocity limit of cooled atoms [14.27].

The photon recoil can be used not only for the deceleration of collimated atomic beams but also for the deflection of the atoms, if the laser beam

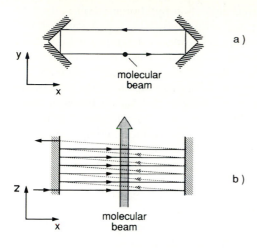

Fig. 14.10. Deflection of atoms in a collimated atomic beam using a multiple-path geometry. The molecular beam travels into the z-direction: (**a**) view into the z-direction; (**b**) view into the y-direction. On the *dashed* return paths the laser beam does not intersect the atomic beam

crosses the atomic beam perpendicularly [14.28–14.30]. In order to increase the transferred photon momentum, and with it the deflection angle, an experimental arrangement is chosen where the laser beam crosses the atomic beam many times in the same direction (Fig. 14.10). The deflection angle δ per absorbed photon, which is given by $\tan \delta = \hbar k / (m v_z)$, increases with decreasing atomic velocity v_z. Optically cooled atoms can therefore be deflected by large angles. Since the atomic absorption frequency differs for different isotopes, this deflection can be used for spatial isotope separation [14.31] if other methods cannot be applied (Sect. 15.1.6).

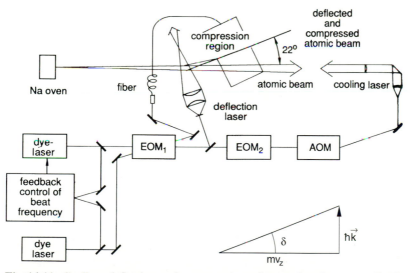

Fig. 14.11. Cooling, deflection and compression of atoms by photon recoil. The electro-optic modulators (EOM) and the acousto-optic modulator (AOM) serve for sideband generation and frequency tuning of the cooling laser sideband [14.30]

An interesting application of atomic deflection by photon recoil is the collimation and focusing of atomic beams with lasers [14.32]. Assume atoms with the velocity $v = \{v_x \ll v_z, 0, v_z\}$ pass through a laser resonator, where an intense standing optical wave in the $\pm x$-direction is present. If the laser frequency ω_L is kept slightly below the atomic resonance frequency ω_0 ($\gamma > \omega_0 - \omega_L > 0$), atoms with transverse velocity components v_x are always pushed back to the z-axis by photon recoil because that part of the laser wave with the k-vector antiparallel to v_x always has a larger absorption probability than the component with k parallel to v_x. The velocity component v_x is therefore reduced and the atomic beam is collimated.

If the atoms have been optically cooled before they pass the standing laser wave, they experience a large collimation in the maximum of the standing wave, but are not affected in the nodes. The laser wave acts like a transmission grating that "channels" the transmitted atoms [14.33].

A schematic diagram of an apparatus for optical cooling of atoms, deflection of the slowed-down atoms, and focusing is depicted in Fig. 14.11.

14.1.5 Threedimensional Cooling of Atoms; Optical Mollasses

Up to now we have only considered the cooling of atoms that all move into one direction. Therefore only one velocity component has been reduced by photon recoil. For cooling of atoms in a thermal gas where all three velocity components $\pm v_x$, $\pm v_y$, $\pm v_z$ have to be reduced, six laser beams propagating into the $\pm x$-, $\pm y$-, $\pm z$-directions are required [14.34]. All six beams are generated by splitting a single laser beam (Fig. 14.12). If the laser frequency is tuned to the red side of the atomic resonance ($\Delta\omega < 0$), a repulsive force is always acting on the atoms, because for atoms moving toward the laser wave

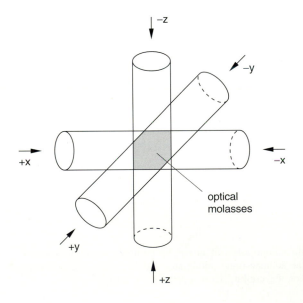

optical
molasses

Fig. 14.12. Optical molasses with six pairwise counterpropagating laser beams

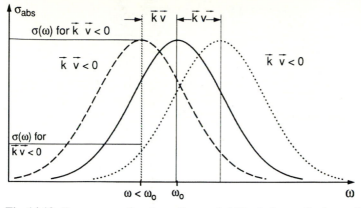

Fig. 14.13. For $\omega < \omega_0$, the absorption probability is larger for $\boldsymbol{k} \cdot \boldsymbol{v} < 0$ than for $\boldsymbol{k} \cdot \boldsymbol{v} > 0$

($\boldsymbol{k} \cdot \boldsymbol{v} < 0$) the Doppler-shifted absorption frequency is shifted toward the resonance frequency ω_0, whereas for the counterpropagating wave ($\boldsymbol{k} \cdot \boldsymbol{v} > 0$) it is shifted away from resonance (Fig. 14.13). For a quantitative description we denote the absorption rate of an atom with $\boldsymbol{k} \cdot \boldsymbol{v} > 0$ by $R^+(v)$, and for $\boldsymbol{k} \cdot \boldsymbol{v} < 0$ by $R^-(v)$. The net recoil force component F_i ($i = x, y, z$) is then

$$F_i = [R^+(v_i) - R^-(v_i)] \cdot \hbar k \,. \tag{14.19}$$

For a Lorentzian absorption profile with FHWM γ, the frequency dependence of the absorption rate is (Fig. 14.13)

$$R^{\pm}(v) = \frac{R_0}{1 + \left(\dfrac{\omega_L - \omega_0 \mp kv}{\gamma/2}\right)^2} \,. \tag{14.20}$$

Inserting (14.20) into (14.19) yields for $\boldsymbol{k} \cdot \boldsymbol{v} \ll \omega_L - \omega_0 = \delta$ the net force (Fig. 14.14)

$$F_i = -a \cdot v_i \,, \quad \text{with} \quad a = R_0 \frac{16 \delta \hbar k^2}{\gamma^2 \left[1 + (2\delta/\gamma)^2\right]^2} \,. \tag{14.21}$$

An atom moving within the overlap region of the six running laser waves therefore experiences a force $F_i(v_i) = -a v_i$ ($i = x, y, z$) that damps its velocity. From the relation $dv/dt = F/m \Rightarrow dv/v = -a/m \, dt$ we obtain the time-dependent velocity

$$v = v_0 e^{-(a/m)t} \,. \tag{14.22}$$

The velocity decreases exponentially with a damping time $t_D = m/a$.

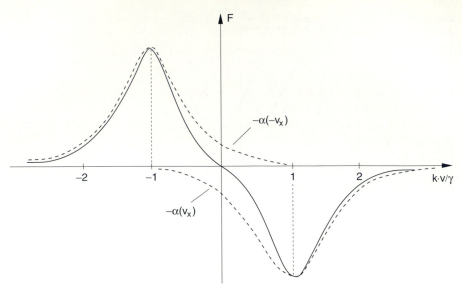

Fig. 14.14. Frictional force in an optical molasses (*solid curve*) for a red detuning $\delta = -\gamma$. The *dotted curve* shows the absorption profiles by a single atom moving with $v_x = \pm\gamma/k$ for a single laser beam propagating into the x-direction

Example 14.6

(a) For rubidium atoms ($m = 85$ AMU) the wavenumber is $k = 8 \times 10^6$ m^{-1}. At a detuning $\delta = \gamma$ and an absorption rate $R_0 = \gamma/2$ we obtain: $a = 4 \times 10^{-21}$ Ns/m. This gives a damping time $t = 35$ μs.

(b) For Na atoms with $\delta = 2\gamma$ one obtains $a = 1 \times 10^{-20}$ Ns/m and $t_D = 2.3$ μs. The atoms move in this light trap like particles in viscous molasses and therefore this atomic trapping arrangement is often called *optical molasses.*

Note: These optical trapping methods reduce the velocity components (v_x, v_y, v_z) to a small interval around $v = 0$. However, they do *not* compress the atoms into a spatial volume, except if the dispersion force for the field gradients $\nabla I \neq 0$ is sufficiently strong.

14.1.6 Cooling of Molecules

The optical cooling techniques discussed so far are restricted to true two-level systems because the cooling cycle of induced absorption and spontaneous emission has to be performed many times before the atoms come to rest. In molecules the fluorescence from the upper excited level generally ends in many rotational–vibrational levels in the electronic ground state that differ from the initial level. Therefore most of the molecules cannot be excited again with the same laser. They are lost for further cooling cycles.

Cold molecules are very interesting for several scientific and technical applications. One example is chemical reactions initiated by collisions between cold molecules where the collision time is very long and the reaction probability might become larger by several orders of magnitude. In addition interactions of cold molecules with surfaces where the sticking coefficient will be 100% open new insights into molecule–surface interactions and reactions between cold adsorbed molecules. Finally, the possibility to reach Bose–Einstein condensation of molecular gases opens new fascinating aspects of collective molecular quantum phenomena.

There have been several proposals how molecules might be cooled in spite of the above-mentioned difficulties [14.35–14.37]. An optical version of these proposals is based on a frequency comb laser, which oscillates on many frequencies, matching the relevant frequencies of the transitions from the upper to the lower levels with the highest transition probabilities [14.37]. In this case, the molecules can be repumped into the upper level from many lower levels, thus allowing at least several pumping cycles.

A very interesting optical cooling technique starts with the selective excitation of a collision pair of cold atoms into a bound level in an upper electronic state (Fig. 14.15). While this excitation occurs at the outer turning point of the upper-state potential, a second laser dumps the excited molecule down into a low vibrational level of the electronic ground state by stimulated emission pumping (photo-induced association). In favorable cases the level $v = 0$ can be reached. If the colliding atoms are sufficiently cold, the angular momentum of their relative movement is zero (S-wave scattering). Therefore the final

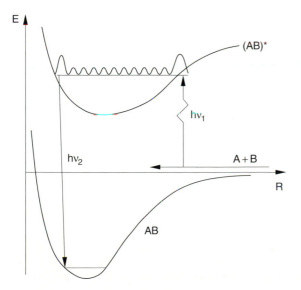

Fig. 14.15. Formation of cold molecules by photoassociation of a collision pair A+B

ground state has the rotational quantum number $J = 0$ if the two lasers transfer no angular momentum to the molecule (either two transitions with $\Delta J = 0$ or the absorbing and emission transition as R-transitions with $\Delta J = +1$ in absorption and -1 in emission) [14.38].

A promising nonoptical technique relies on cooling of molecules by collisions with cold atoms. If the gas mixture of atoms and molecules can be trapped in a sufficiently small volume long enough to achieve thermal equilibrium between atoms and molecules, the optically cooled atoms act as a heat sink for the molecules, which will approach the same temperature as the atoms (sympathetic cooling) [14.39].

An interesting proposal that could be realized uses a cold supersonic molecular beam with flow velocity u, which expands through a rotating nozzle (Fig. 14.16). We saw in Chap. 9 that in supersonic beams the velocity spread around the flow velocity u may become very small. The translational temperature in the frame moving with the velocity u can be as small as 0.1 K. If the nozzle moves with the speed $-v$, the molecules have the velocity $u - v$ in the laboratory frame. Tuning the angular velocity ω of the nozzle rotating on a circle with radius R makes it possible to reach any molecular velocity $v_m = u - \omega R$ between u and 0 in the laboratory frame. Since the beam must be collimated in order to reduce the other velocity components, there is only a small time interval per rotation period where the nozzle is in line with the collimating apertures. The cold molecules therefore appear as pulses behind the apertures.

An elegant technique has been developed in several laboratories, where cold helium clusters moving through a gas cell of atoms or molecules pick up these molecules, which then can diffuse into the interior of the helium cluster. The molecules then aquire the low temperature of the cluster. The binding energy is taken away by evaporation of He atoms from the cluster surface [14.40, 14.41].

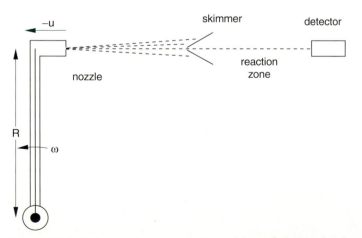

Fig. 14.16. Rotating nozzle for producing a beam of slow molecules

14.1.7 Optical Trapping of Atoms

The effectiveness of the optical molasses for cooling atoms anticipates that the atoms are trapped within the overlap region of the six laser beams for a sufficiently long time. This demands that the potential energy of the atoms shows a sufficiently deep minimum at the center of the trapping volume, that is, restoring forces must be present that will bring escaping atoms back to the center of the trapping volume.

We will briefly discuss the two most commonly used trapping arrangements. The first is based on induced electric dipole forces in inhomogeneous electric fields and the second on magnetic dipole forces in magnetic quadrupole fields. Letokhov proposed [14.42, 14.43] to use the potential minima of a three-dimensional standing optical field composed by the superposition of three perpendicular standing waves for spatial trapping of cooled atoms, whereas Ashkin and Gordon calculated [14.44] that the dispersion forces in focused Gaussian beams could be employed for trapping atoms.

a) Induced Dipole Forces in a Radiation Field

When an atom with the polarizability α is brought into an inhomogeneous electric field E, a dipole moment $p = \alpha E$ is induced and the force

$$F = -(p \cdot \mathrm{grad})E = -\alpha(E \cdot \nabla)E = -\alpha[(\nabla \tfrac{1}{2}E^2) - E \times (\nabla \times E)], \quad (14.23)$$

acts onto the induced dipole. The same relation holds for an atom in an optical field. However, when averaging over a cycle of the optical oscillation the last term in (14.23) vanishes, and we obtain for the mean dipole force [14.45]

$$\langle F_{\mathrm{D}} \rangle = -\frac{1}{2}\alpha \nabla(E^2) . \quad (14.24)$$

The polarizability $\alpha(\omega)$ depends on the frequency ω of the optical field. It is related to the refractive index $n(\omega)$ of a gas with the atomic density N (Sect. 3.1.3) by $\alpha \approx \epsilon_0(\epsilon - 1)/N$. With $(\epsilon - 1) = n^2 - 1 \approx 2(n - 1)$ we obtain

$$\alpha(\omega) = \frac{2\epsilon_0[n(\omega) - 1]}{N} . \quad (14.25)$$

Inserting (3.37b) for $n(\omega)$, the polarizability $\alpha(\omega)$ becomes

$$\alpha(\omega) = \frac{e^2}{2m_{\mathrm{e}}\omega_0} \frac{\Delta\omega}{\Delta\omega^2 + (\gamma_{\mathrm{s}}/2)^2} , \quad (14.26)$$

where m_{e} is the electron mass, $\Delta\omega = \omega - (\omega_0 + k \cdot v)$ is the detuning of the field frequency ω against the Doppler-shifted eigenfrequency $\omega_0 + kv$ of the atom, and $\gamma_{\mathrm{s}} = \delta\omega_{\mathrm{n}}\sqrt{1 + S}$ is the saturation-broadened linewidth characterized by the saturation parameter S (Sect. 3.6).

For $\Delta\omega \ll \gamma_{\mathrm{s}}$ the polarizability $\alpha(\omega)$ increases nearly linearly with the detuning $\Delta\omega$. From (14.24) and (14.26) it follows that in an intense laser beam

$(S \gg 1)$ with the intensity $I = \epsilon_0 c E^2$ the force $\boldsymbol{F}_\mathrm{D}$ on an induced atomic dipole is

$$\boldsymbol{F}_\mathrm{D} = -a\Delta\omega\nabla I \,, \quad \text{with} \quad a = \frac{e^2}{m\epsilon_0 c\gamma^2\omega_0 S} \,. \tag{14.27}$$

This reveals that in a homogeneous field (for example, an extended plane wave) $\nabla I = 0$ and the dipole force becomes zero. For a Gaussian beam with the beam waist w propagating in the z-direction, the intensity $I(r)$ in the x–y-plane is, according to (5.32)

$$I(r) = I_0\,\mathrm{e}^{-2r^2/w^2} \,, \quad \text{with} \quad r^2 = x^2 + y^2 \,.$$

The intensity gradient $\nabla I = -(4r/w^2)I(r)\hat{r}$ points into the radial direction and the dipole force $\boldsymbol{F}_\mathrm{D}$ is then directed toward the axis $r = 0$ for $\Delta\omega < 0$ and radially outwards for $\Delta\omega > 0$.

For $\Delta\omega < 0$ the z-axis of an intense Gaussian laser beam with $I(r = 0) = T_0$ represents a minimum of the potential energy

$$E_\mathrm{pot} = \int\limits_0^\infty F_\mathrm{D}\,\mathrm{d}r = +a\Delta\omega I_0 \,, \tag{14.28}$$

where atoms with sufficiently low radial kinetic energy may be trapped. In the focus of a Gaussian beam we have an intensity gradient in the r-direction as well as in the z-direction. If the two forces are sufficiently strong, atoms can be trapped in the focal region.

Besides this dipole force in the r- and z-directions the recoil force in the $+z$-direction acts onto the atom (Fig. 14.17). In a standing wave in $\pm z$-direction the radial force and the recoil force may be sufficiently strong to trap atoms in all directions. For more details see [14.46–14.48].

Example 14.7
In the focal plane of a Gaussian laser beam with $P_\mathrm{L} = 200\,\mathrm{mW}$ focused down to a beam waist of $w = 10\,\mu\mathrm{m}$ $(I_0 = 1.2 \times 10^9\,\mathrm{W/m^2})$, the radial intensity gradient is $\partial I/\partial r = 2r/w^2 I_0\,\mathrm{e}^{-2r^2/w^2}$, which gives for $r = w : (\partial I/\partial r)_{r=w} = 2I_0/w \cdot \mathrm{e}^2$. With the number above one obtains:

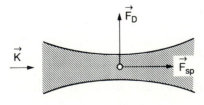

Fig. 14.17. Longitudinal and transverse forces on an atom in the focus of a Gaussian beam

$(\partial I/\partial r)_{r=w} = 2.4 \times 10^{13} \, \text{W/m}^3$ and the radial dipole force acting on a Na atom is for $\Delta\omega = -|\gamma| = -2\pi \cdot 10^7 \, \text{s}^{-1}$, $S = 0$, and $r = w$

$$F_D = +a\Delta\omega \frac{4r}{w^2} I_0 \hat{r}_0 = 1.5 \times 10^{-16} \, \text{N} \, ,$$

The axial intensity gradient is for a focusing lens with $f = 5 \, \text{cm}$ and a beam diameter $\partial I/\partial z = 4.5 \times 10^5 \, \text{W/m}^3$. This gives an axial dipole force of $F_D(z) = 3 \times 10^{-24} \, \text{N}$, while the recoil force is about

$$F_{\text{recoil}} = 3.4 \times 10^{-20} \, \text{N} \, .$$

In the axial direction the recoil force is many orders of magnitude larger than the axial dipole force. The potential minimum with respect to the radial dipole force is $E_{\text{pot}} \simeq -5 \times 10^{-7} \, \text{eV}$. In order to trap atoms in this minimum their radial kinetic energy must be smaller than $5 \times 10^{-7} \, \text{eV}$, which corresponds to the "temperature" $T \simeq 5 \times 10^{-3} \, \text{K}$.

Example 14.8

Assume a standing laser wave with $\lambda = 600 \, \text{nm}$, an average intensity $I = 10 \, \text{W/cm}^2$, and a detuning of $\Delta\omega = \gamma = 60 \, \text{MHz}$. The maximum force acting on an atom because of the intensity gradient $\nabla I = 6 \times 10^{11} \, \text{W/m}^3$ between maxima and nodes of the field becomes, according to (14.27), with a saturation parameter $S = 10$: $F_D = 10^{-17} \, \text{N}$. The trapping energy is then $1.5 \times 10^{-5} \, \text{eV} \hat{\approx} T = 0.15 \, \text{K}$.

Example 14.8 demonstrates that the negative potential energy in the potential minima (the nodes of the standing wave for $\Delta\omega > 0$) is very small. The atoms must be cooled to temperatures below 1 K before they can be trapped.

Another method of trapping cooled atoms is based on the net recoil force in a three-dimensional light trap, which can be realized in the overlap region of six laser beams propagating into the six directions $\pm x$, $\pm y$, $\pm z$.

b) Magneto-Optical Trap

A very elegant experimental realization for cooling and trapping of atoms is the magneto-optical trap (MOT), which is based on a combination of optical molasses and an inhomogeneous magnetic quadrupole field (Fig. 14.18). Its principle can be understood as follows:

In a magnetic field the atomic energy levels E_i experience Zeeman shifts

$$\Delta E_i = -\boldsymbol{\mu}_i \cdot \boldsymbol{B} = -\mu_{\text{B}} \cdot g_F \cdot m_F \cdot B \, , \tag{14.29}$$

which depend on the Lande g-factor g_F, Bohr's magneton μ_{B}, the quantum number m_F of the projection of the total angular momentum F onto the field direction, and on the magnetic field B.

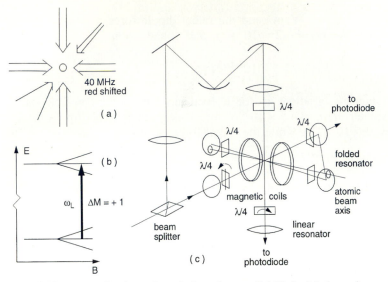

Fig. 14.18a–c. Realization of optical molasses [14.50a]: (**a**) laser beam arrangement; (**b**) Zeeman detuning of the $\Delta M = +1$ transition; and (**c**) schematic experimental setup

In the MOT the inhomogeneous field is produced by two equal electric currents flowing into opposite directions through two coils with radius R and distance $D = R$ (anti-Helmholtz arrangement). If we choose the z-direction as the symmetry axis through the center of the coils, the magnetic field around $z = 0$ in the middle of the arrangement can be described by the linear dependence

$$B = bz . \tag{14.30}$$

where the constant b depends on the size of the anti-Helmholtz coils. The Zeeman splittings of the transition from $F = 0$ to $F = 1$ are shown in Fig. 14.18c. Atoms in the center of this MOT are exposed to the six red-tuned laserbeams of the optical molasses. Let us at first only consider the two beams in the $\pm z$-direction, where the laserbeam in $+z$-direction is σ^+ polarized. Then the reflected beam in the $-z$ direction is σ^- polarized. For an atom at $z = 0$ where the magnetic field is zero, the absorption rates are equal for both laser beams, which means that the average momentum transferred to the atom is zero. For an atom at $z > 0$, however, the σ^--beam is preferentially absorbed because here the frequency difference $\omega_L - \omega_0$ is smaller than for the σ^+-beam. This means that the atom experiences a net momentum transfer into the $-z$-direction, back to the center.

In a similar way an atom at $z < 0$ shows a preferential absorption of σ^+-light and gets a net momentum in the $+z$-direction. This shows that the atoms in the MOT are compressed toward the trap center.

We will now discuss this spatially dependent restoring force more quantitatively.

From the above discussion the net force

$$F(z) = R_{\sigma+}(z)\hbar k_{\sigma+} + R_{\sigma-}(z)\hbar k_{\sigma-} , \tag{14.31}$$

is determined by the difference of the absorption rates $R_{\sigma+}$, $R_{\sigma-}$ (note that the wave vectors are antiparallel). For a Lorentzian absorption profile with halfwidth γ, the absorption rates become

$$R_{\sigma\pm} = \frac{R_0}{1 + \left[\dfrac{\omega_{\mathrm{L}} - \omega_0 \pm \mu b z/\hbar}{\gamma/2}\right]} . \tag{14.32}$$

Around $z = 0$ ($\mu b z \ll \hbar\delta$) this expression can be expanded as a power series of $\mu b z/\hbar\gamma$.

Taking only the linear term into account yields with $\delta = \omega_{\mathrm{L}} - \omega_0$:

$$F_z = -D \cdot z , \quad \text{with} \quad D = R_0\mu \cdot b \frac{16k \cdot \delta}{\gamma^2 \left(1 + (2\delta/\gamma)^2\right)^2} . \tag{14.33}$$

We therefore obtain a restoring force that increases linearly increasing with z. The potential around the center of the MOT can be then described (because of $F_z = -\partial V/\partial z$) as the harmonic potential

$$V(z) = \frac{1}{2}Dz^2 . \tag{14.34}$$

The atoms oscillate like harmonic oscillators around $z = 0$ and are spatially stabilized.

Note: There is a second force

$$F_\mu = -\boldsymbol{\mu} \cdot \operatorname{grad} \boldsymbol{B} ,$$

which acts on atoms with a magnetic moment in an inhomogeneous magnetic field.

Inserting the numbers for sodium atoms, it turns out that this force is negligibly small compared to the recoil force at laser powers in the milliwatt range. At very low temperatures, however, this force is essential to trap the atoms after the laser beams have been shut off (see Sect. 14.1.9)

In the discussion above we have neglected the velocity-dependent force in the optical molasses (Sect. 14.1.5).

The total force acting on an atom in the MOT

$$F_z = -Dz - av ,$$

results in a damped oscillation around the center with an oscillation frequency

$$\Omega_0 = \sqrt{D/m} , \tag{14.35}$$

and a damping constant

$$\beta = a/m \; .$$

So far we have only considered the movement in the z-direction. The anti-Helmholtz coils produce a magnetic quadrupole field with three components. From Maxwell's equation div $B = 0$ and the condition $\partial B_x/\partial x = \partial B_y/\partial y$, which follows from the rotational symmetry of the arrangement, we obtain the relations

$$\frac{\partial B_x}{\partial x} = \frac{\partial B_y}{\partial y} = -\frac{1}{2}\frac{\partial B_z}{\partial z} \; .$$

The restoring forces in the x- and y-directions are therefore half of the forces in z-directions. The trapped thermal cloud of atoms fills an ellipsoidal volume.

Example 14.9

With a magnetic field gradient of $0.04\,\mathrm{T/m}$ a sodium atom at $z = 0$ in a light trap with two counterpropagating σ^+-polarized laser beams L^+, L^- in the $\pm z$-direction with $I_+ = 0.8 I_{\mathrm{sat}}$, $\omega_+ = \omega_0 - \gamma/2$ and $I_- = 0.15 I_{\mathrm{sat}}$, $\omega_- = \omega_0 + \gamma/10$, the negative acceleration of a Na atom moving away from $z = 0$ reaches a value of $a = -3 \times 10^4 \,\mathrm{m/s^2}$ ($\hat{=} 3 \times 10^3 g!$), driving the atom back to $z = 0$.

Generally, the MOT is filled by slowing down atoms in an atomic beam (see Sect. 14.1.3). Spin-polarized cold atoms can also be produced by optical pumping in a normal vapor cell and trapped in a magneto-optic trap. This was demonstrated by Wieman and coworkers [14.50b], who captured and cooled 10^7 Cs atoms in a low-pressure vapor cell by six orthogonal intersecting laser beams. A weak magnetic field gradient regulates the light pressure in conjunction with the detuned laser frequency to produce a damped harmonic motion of the atoms around the potential minimum. This arrangement is far simpler than an atomic beam. Effective kinetic temperatures of $1\,\mu\mathrm{K}$ have been achieved for Cs atoms. For more details on MOT and their experimental realizations see [14.6, 14.7] and [14.51].

14.1.8 Optical Cooling Limits

The lowest achievable temperatures of the trapped atoms can be estimated as follows: because of the recoil effect during the absorption and emission of photons, each atom performs a statistical movement comparable to the Brownian motion. If the laser frequency ω_L is tuned to the resonance frequency ω_0 of the atomic transition, the net damping force becomes zero. Although the time average $\langle v \rangle$ of the atomic velocity approaches zero, the mean value of $\langle v^2 \rangle$ increases, analogous to the random-walk problem [14.52, 14.53]. The optical cooling for $\omega - \omega_0 < 0$ must compensate this "statistical heating" caused by the statistical photon scattering. If the velocity of the atoms has decreased to $v < \gamma/k$, the detuning $\omega - \omega_0$ of the laser frequency must be smaller than the homogeneous linewidth of the atomic transition γ in order to stay in resonance. This yields a lower limit of $\hbar\Delta\omega < k_B T_{min}$, or with $\Delta\omega = \gamma/2$

$$T_{min} = \hbar\gamma/2k_B \quad \text{(Doppler limit)}, \tag{14.36}$$

if the recoil energy $E_r = \hbar\omega^2(2Mc^2)^{-1}$ is smaller than the uncertainty $\hbar\gamma$ of the homogeneous linewidth.

Example 14.10

(a) For Mg^+ ions with $\tau = 2\,ns \rightarrow \gamma = 80\,MHz$ (14.36) yields $T_{min} = 2\,mK$.
(b) For Na atoms with $\gamma = 10\,MHz \rightarrow T_{min} = 240\,\mu K$.
(c) For Rb atoms with $\gamma = 5.6\,MHz \Rightarrow T_{min} = 140\,\mu K$.
(d) For Ca atoms cooled on the narrow intercombination line at $\lambda = 657\,nm$ with $\gamma = 20\,kHz$ the minimum temperature of $T_{min} = 240\,nK$ is calculated from (14.36).

Meanwhile, experiments have shown that, in fact, temperatures lower than this calculated Doppler limit can be reached [14.54–14.56]. How is this possible?

The experimental results can be explained by the following model of *polarization gradient cooling* [14.55–14.57]. If two orthogonally polarized light waves travel anticollinearly in $\pm z$-directions through the atoms of the optical molasses in a magnetic field, the total field amplitude acting on the atoms is

$$E(z, t) = E_1\hat{e}_x \cos(\omega t - kz) + E_2\hat{e}_y \cos(\omega t + kz). \tag{14.37}$$

This field shows a z-dependent elliptical polarization: for $z = 0$ it has linear polarization along the direction $\hat{e}_1 = (\hat{e}_x + \hat{e}_y)$ (assuming $E_1 = E_2$), for $z = \lambda/8$ it has elliptical σ^- polarization, for $z = \lambda/4$ again linear polarization along $\hat{e}_2 = (\hat{e}_x - \hat{e}_y)$, for $z = 3\lambda/8$ elliptical σ^+ polarization, etc. (Fig. 14.19a).

For an atom at rest with the level scheme of Fig. 14.19b, the energies and the populations of the two ground-state sublevels $g_{-1/2}$, $g_{+1/2}$ depend on the location z. For example, for $z = \lambda/8$ the atom rests in a σ^- light field and is therefore optically pumped into the $g_{-1/2}$ level, giving the stationary population probabilities $|(g_{-1/2})|^2 = 1$ and $|(g_{+1/2})|^2 = 0$, while for $z = (3/8)\lambda$ the atom is pumped by σ^+ light into the $g_{+1/2}$-level.

The electric field $E(z, t)$ of the standing light wave causes a shift and broadening of the atomic Zeeman levels (*ac Stark shift*) that depends on the saturation parameter, which in turn depends on the transition probability, the polarization of E, and the frequency detuning $\omega_L - \omega_0$. It differs for the different Zeeman transitions. Since the σ^- transition starting from $g_{-1/2}$ is three times as intense as that from $g_{+1/2}$ (Fig. 14.19b), the light shift Δ_- of $g_{-1/2}$ is three times the shift Δ_+ of $g_{+1/2}$. If the atom is moved to $z = (3/8)\lambda$, the situation is reversed, since now pumping occurs on a σ^+ transition.

The z-dependent energy shift of the ground-state sublevel therefore follows the curve of Fig. 14.19c. For those z values where a linearly polarized light

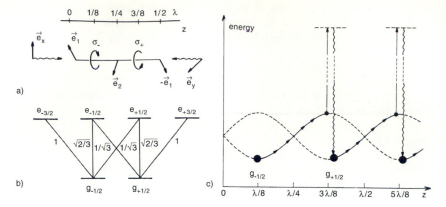

Fig. 14.19a–c. Schematic diagram of polarization gradient (Sisyphus) cooling: (**a**) two counterpropagating linearly polarized waves with orthogonal polarization create a standing wave with z-dependent polarization. (**b**) Atomic level scheme and Clebsch–Gordan coefficients for a $J_g = 1/2 \leftrightarrow J_e = 3/2$ transition. (**c**) Atomic Sisyphus effect in the lin $\perp$ lin configuration [14.55]

field is present, the transition probabilities and light shifts are equal for the two sublevels.

The cooling now proceeds as follows: the important point is that the optical pumping between the sublevels takes a certain time τ_p, depending on the absorption probability and the spontaneous lifetime of the upper level. Suppose the atom starts at $z = \lambda/8$ and moves to the right with such a velocity that it travels a distance of $\lambda/4$ during the time τ_p. Then it always climbs up the potential $E_{pot}^-(g_{-1/2})$. When the optical pumping takes place at $t = 3\lambda/8$ (which is the maximum transition probability for σ^+-light!), it is transferred to the minimum $E_{pot}^+(g_{+1/2})$ of the $g_{+1/2}$ potential and can again climb up the potential. During the optical pumping cycle it absorbs less photon energy than it emits, and the energy difference must be supplied by its kinetic energy. This means that its velocity decreases. The process is reminiscent of the Greek myth of Sisyphus, the king of Corinth, who was punished by the gods to roll a heavy rock uphill. Just before he reached the top, the rock slipped from his grasp and he had to begin again. Sisyphus was condemned to repeat this exhausting task for eternity. Therefore polarization gradient cooling is also called *Sisyphus cooling* [14.58].

Because the population density (indicated by the magnitude of the dots in Fig. 14.19c) is larger in the minimum than in the maxima of the potentials $E_{pot}(z)$, on the average the atom climbs uphill more than downhill. It transfers part of its kinetic energy to photon energy and is therefore cooled.

Depending on the polarization of the two counterpropagating laser beams the lin$\perp$lin configuration can be used with two orthogonal linear polarizations or the σ^+–σ^- configuration with a circularly polarized σ^+ wave, superimposed by the reflected σ^- wave. With Sisyphus cooling temperatures as low as $5-10\,\mu$K can be achieved.

14.1.9 Bose–Einstein Condensation

At sufficiently low temperatures where the de Broglie wavelength

$$\lambda_{DB} = \frac{h}{m \cdot v} , \tag{14.38}$$

becomes larger than the mean distance $d = n^{-1/3}$ between the atoms in the cold gas, a phase transition takes place for bosonic particles with integer total spin. More and more particles occupy the lowest possible energy state in the trap potential and are then indistinguishable, which means that all these atoms in the same energy state are described by the same wave function (note that for bosons the Pauli exclusion principle does not apply). Such a situation of a macroscopic state occupied by many indistinguishable particles is called a *Bose–Einstein condensate* (BEC, Nobel prize in physics 2001 for E. Cornell, W. Ketterle, and C. Wiemann).

More detailed calculations show that BEC is reached if

$$n \cdot \lambda_{DB}^3 > 2.612 . \tag{14.39}$$

With $v^2 = 3k_B T/m$ we obtain the de Broglie wavelength

$$\lambda_{DB} = \frac{h}{\sqrt{3m K_B T}} , \tag{14.40}$$

and the condition (14.40) for the critical density becomes

$$n > 13.57(m \cdot k_B T)^{3/2}/h^3 .$$

The minimum density for BEC depends on the temperature and decreases with $T^{3/2}$ [14.6, 14.59].

Example 14.12

For Na atoms at a temperature of $10\,\mu K$ the critical density would be $n = 6 \times 10^{14}\,/cm^3$, which at present not achievable. For experimentally realized densities of $10^{12}\,cm^{-3}$ the atoms have to be cooled below $100\,nK$ in order to reach BEC.

For rubidium atoms BEC was observed at $T = 170\,nK$ and a density of $3 \times 10^{12}\,cm^{-3}$.

14.1.10 Evaporative Cooling

The temperatures reached with the optical cooling methods discussed so far are not sufficiently low to obtain Bose–Einstein condensation. Here the very old and well-known technique of evaporation cooling leads to the desired goal. The principle of this method is as follows [14.60]:

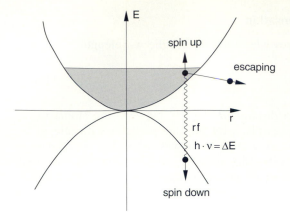

Fig. 14.20. Principle of evaporative cooling in a MOT

The optically cooled atoms are transferred from the MOT to a purely magnetic trap, where the restoring force

$$F = \mu \mathrm{grad}\, \boldsymbol{B} \,,$$

keeps the cold atoms in a cloud around the center of the magnetic trap. Now the particles with the highest energy are removed from the trap, thus perturbing the equilibrium velocity distribution. If the remaining atoms suffer sufficient mutual collisions, a new equilibrium state is reached with a lower temperature. Again, the upper part of the Maxwellian velocity distribution is thrown out of the trap. This is achieved by a radio-frequency field that induces spin flips and thus reverses the sign of the force, changing it from a restoring to a repellent force. This is shown in Fig. 14.20 in a potential diagram. Lowering the radio frequency continuously always results in a loss of the atoms with the highest kinetic energy, which reach the largest distance from the trap center and therefore have the largest Zeeman splittings.

The cooling process must be slow enough to maintain thermal equilibrium, but should be sufficiently fast in order not to lose particles from the trap by collision-induced spin flips.

The phase transition to BEC manifests itself by the sudden increase of the atomic density (Fig. 14.21), which can be monitored by measuring the absorption of an expanded probe laser beam.

The BEC can be kept only for a limited time. There are several loss mechanisms that result in a decay of the trapped particle density. These are spin-flip collisions, three-body recombination where molecules are formed, collisions with excited atoms where the excitation energy may be converted into kinetic energy, and collisions with rest gas atoms or molecules. The background pressure therefore has to be as low as possible (typical pressures are 10^{-10} to 10^{-11} mbar)

The scattering length a of the atoms in the BEC determines the elastic cross section $\sigma_{\mathrm{el}} = 8\pi a^2$. The value of a is positive for a repulsive mean po-

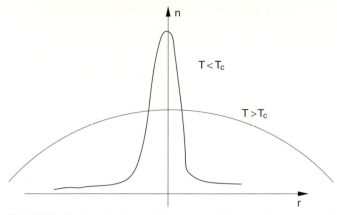

Fig. 14.21. Radial density profile for cooled atoms in a magnetic trap above and below the critical phase transition temperature T_c

tential of the condensate. For negative scattering lengths the condensate will finally collapse.

It is interesting to note that the value of the scattering length a for atoms A can be obtained from laser spectroscopy by measuring the energy and the vibrational wave function of the last bound vibrational level in the ground-state potential of the molecule A_2 [14.61].

14.1.11 Applications of Cooled Atoms and Molecules

Optical cooling and deflection of atoms open new areas of atomic and molecular physics. Collisions at very small relative velocities, where the deBroglie wavelength $\lambda_{DB} = h/(mv)$ is large, can now be studied. They give information about the long-range part of the interaction potential, where new phenomena arise, such as retardation effects and magnetic interactions from electron or nuclear spins. One example is the study of collisions between Na atoms in their $3\,^2S_{1/2}$ ground state. The interaction energy depends on the relative orientation of the two electron spins $S = \frac{1}{2}$. The atoms with parallel spins form a Na$_2$ molecule in a $^3\Sigma_u$ state, while atoms with antiparallel spins form a Na$_2(^1\Sigma_g^+)$ molecule. At large internuclear distances ($r > 1.5$ nm) the energy differences between the $^3\Sigma_u$ and $^1\Sigma_g$ potentials become comparable to the hyperfine splitting of the Na($3\,^2S_{1/2}$) atoms [14.62]. The interaction between the nuclear spins and the electron spins leads to a mixing of the $^3\Sigma_u$ and $^1\Sigma_u$ states, which corresponds in the atomic model of colliding atoms to *spin-flip* collisions (Fig. 14.22).

Collisions between cold atoms in a trap can be studied experimentally by measuring the loss rate of trapped atoms under various trap conditions (temperature, magnetic-field gradients, light intensity, etc.). It turns out that

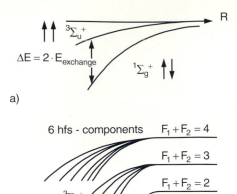

Fig. 14.22a,b. Interaction between two Na atoms at large internuclear distances R for different spin orientations: (**a**) without hyperfine structure and (**b**) including the nuclear spins $I = (3/2)\hbar$, which gives three dissociation limits

the density of excited atoms cannot be neglected compared with the density of ground-state atoms, and the interaction between excited- and ground-state atoms plays an essential role. For collisions at very low temperatures the absorption and emission of photons during the collisions is important, because the collision time $\tau_c = R_c/v$ becomes very long at low relative atomic velocities v. The two dominant energy-transfer processes are collision-induced fine-structure transitions in the excited state and radiative redistribution, where a photon is absorbed by an atom at the position r_1 in the trap potential $V(r)$ and another photon with a slightly different energy is reemitted after the atom has moved to another position r_2.

Another application is the deflection of atoms by photon recoil. For sufficiently good beam collimation, the deflection from single photons can be detected. The distribution of the transverse-velocity components contains information about the statistics of photon absorption [14.63]. Such experiments have successfully demonstrated the antibunching characteristics of photon absorption [14.64]. The photon statistic is directly manifest in the momentum distribution of the deflected atoms [14.65]. Optical collimation by radial recoil can considerably decrease the divergence of atomic beams and thus the beam intensity. This allows experiments in crossed beams that could not be performed before because of a lack of intensity.

A very interesting application of cold trapped atoms is their use for an optical frequency standard [14.66]. They offer two major advantages: reduction of the Doppler effect and prolonged interaction times on the order of 1 s or more. Optical frequency standards may be realized either by atoms in optical traps or by atomic fountains [14.67].

For the realization of an atomic fountain, cold atoms are released in the vertical direction out of an atomic trap. They are decelerated by the gravitational field and return back after having passed the culmination point with $v_z = 0$.

Example 14.12

Assume the atoms start with $v_{0z} = 5\,\text{m/s}$. Their upward flight time is then $t = v_{0z}/g = 0.5\,\text{s}$, their path length is $z = v_0 t - gt^2/2 = 1.25\,\text{m}$, and their total flight time is $1\,\text{s}$. Their transit time through a laser beam with the diameter $d = 1\,\text{cm}$ close to the culmination point is $T_{tr} = 90\,\text{ms}$, and the maximum transverse velocity is $v \leq 0.45\,\text{m/s}$. The transit-time broadening is then less than $10\,\text{Hz}$.

There are many possible applications of cold trapped molecules.

One example is the spectroscopy of highly forbidden transitions, which becomes possible because of the long interaction time. Another aspect is a closer look at the chemistry of cold trapped molecules, where the reaction rates and the molecular dynamics are dominated by tunneling and a manipulation of molecular trajectories seems possible. Experiments on testing time-reversal symmetry via a search for a possible electric dipole moment of the proton or the electron [14.68] are more sensitive when cold molecules are used [14.69, 14.70].

14.2 Spectroscopy of Single Ions

During recent years it has become possible to perform detailed spectroscopic investigations of single ions that are stored in electromagnetic (EM) traps and cooled by special laser arrangements. This allows tests of fundamental problems of quantum mechanics and electrodynamics and, furthermore, opens possibilities for precise frequency standards.

14.2.1 Trapping of Ions

Since ions show stronger interactions with EM fields than neutral atoms, which experience only a weak force because of their polarizability, they can be stored more effectively in EM traps. Therefore trapping of ions was achieved long before neutral particles were trapped [14.71, 14.72]. Two different techniques have been developed to store ions within a small volume: in the *radio frequency (RF) quadrupole trap* [14.72, 14.73] the ions are confined within a hyperbolic electric dc field superimposed by a RF field, while in the *Penning trap* [14.74] a dc magnetic field with a superimposed electric field of hyperbolic geometry is used to trap the ions.

The EM quadrupole trap (Paul trap) is formed by a ring electrode with a hyperbolic surface and a ring radius of r_0 as one pole, and two hyperbolic caps as the second pole (Fig. 14.23). The whole system has cylindrical sym-

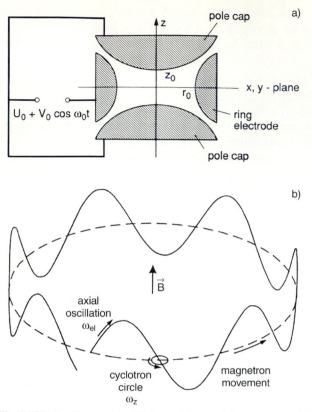

Fig. 14.23a,b. Quadrupole RF trap: (**a**) experimental setup; (**b**) ion movement

metry around the z-axis. The distance $2z_0$ between the two pole caps (which are at equal potential) is adjusted to be $2z_0 = r_0\sqrt{2}$.

With the voltage U between the caps and the ring electrode, the electric potential $\phi(r, z)$ is for points within the trap, [14.73]

$$\phi = \frac{U}{2r_0^2}(r^2 - 2z^2), \quad \text{with} \quad r^2 = x^2 + y^2 . \tag{14.41}$$

The applied voltage $U = U_0 + V_0 \cos(\omega_{\mathrm{RF}}t)$ is a superposition of the dc voltage U_0 and the RF voltage $V_0 \cos(\omega_{\mathrm{RF}}t)$. The equation of motion

$$m\ddot{\boldsymbol{r}} = -q \cdot \mathrm{grad}\,\phi , \tag{14.42}$$

for an ion with charge q inside the trap can be derived from (14.41). One obtains

$$\frac{\mathrm{d}^2}{\mathrm{d}t^2}\begin{pmatrix} x \\ y \\ z \end{pmatrix} + \frac{\omega_{\mathrm{RF}}^2}{4}(a + b\cos\omega_{\mathrm{RF}}t)\begin{pmatrix} x \\ y \\ -2z \end{pmatrix} = 0 . \tag{14.43}$$

The two parameters

$$a = \frac{4qU_0}{mr_0^2\omega_{RF}^2} \ , \quad b = \frac{4qV_0}{mr_0^2\omega_{RF}^2} \ , \tag{14.44}$$

are determined by the dc voltage U_0, the RF amplitude V_0, and the angular frequency ω_{RF} of the RF voltage, respectively. Because of the cylindrical symmetry, the same equation holds for the x- and y-components, while for the movement in z-direction $-2z$ appears in (14.43) because $r_0^2 = 2z_0^2$.

The equation of motion (14.43) is known as *Mathieu's differential equation*. It has stable solutions only for certain values of the parameters a and b [14.75]. Charged particles that enter the trap from outside cannot be trapped. Therefore, the ions have to be produced inside the trap. This is generally achieved by electron-impact ionization of neutral atoms.

The stable solutions of (14.43) can be described as a superposition of two components (Fig. 14.23b: A periodic "micromovement" of the ions with the RF frequency ω_{RF} around a "guiding center" that itself performs slower harmonic oscillations with the frequency Ω in the x–y-plane and with the frequency $\omega_z = 2\Omega$ in the z-direction (secular motion) [14.73]. The x- and z-components of the ion motion are

$$x(t) = x_0[1 + (b/4)\cos(\omega_{RF}t)]\cos\Omega t \ , \tag{14.45}$$

$$z(t) = z_0[1 + \sqrt{2}(\omega_z/\omega_{RF})\cos(\omega_{RF}t)]\cos(2\Omega t) \ . \tag{14.46}$$

The Fourier analysis of $x(t)$ and $z(t)$ yields the frequency spectrum of the ion movement, which contains the fundamental frequency ω_{RF} and its harmonics $n\omega_{RF}$, as well as the sidebands at $n\omega_{RF} \pm m\Omega$.

The trapped ions can be monitored either by laser-induced fluorescence [14.76, 14.77] or by the RF voltage that is induced in an outer RF circuit by the motion of the ions [14.73]. The LIF detection is very sensitive for a true two-level system, where the fluorescence photon rate R of a single ion with spontaneous lifetime τ_k of the upper level may reach the value $R = (2\tau_k)^{-1}[s^{-1}]$ (Sect. 14.1.4). Even for a three-level system, this can be achieved if a second laser is used that refills the ground-state level, depleted by optical pumping (Fig. 14.5). For $\tau_k = 10^{-8}$ s this implies that for sufficiently large laser intensities a single ion emits up to 5×10^7 fluorescence photons per second, which allows the detection of a single stored ion [14.78, 14.79].

14.2.2 Optical Sideband Cooling

Assume that an ion in the ion trap (absorption frequency ω_0 for $v = 0$) performing a harmonic motion in the x-direction with the velocity $v_x = v_0\cos\omega_v t$, is irradiated by a monochromatic wave propagating in the x-direction. In the frame of the oscillating ion, the laser frequency is modulated at the oscillation frequency due to the oscillating Doppler shift. If the linewidth γ of the absorbing transition is smaller than ω_v, the absorption

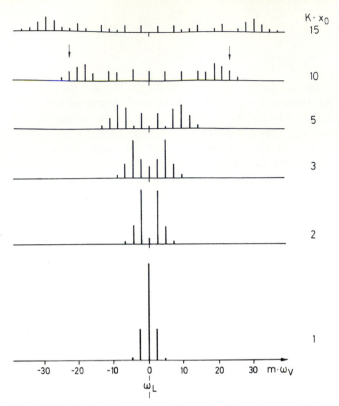

Fig. 14.24. Sideband spectrum of an oscillating ion as a function of the oscillating amplitude

spectrum of the oscillating ion consists of discrete lines at the frequencies $\omega_m = \omega_0 \pm m\omega_v$. The relative line intensities are given by the mth-order Bessel function $J_m(v_0\omega_0/c\omega_m)$ [14.80, 14.81], which depend on the velocity amplitude v_0 of the ion (Fig. 14.24).

If the laser is tuned to the frequency $\omega_L = \omega_0 - m\omega_v$ of a lower sideband, the atom can only absorb during that phase of the oscillation when the atom moves *toward* the laser wave ($\mathbf{k} \cdot \mathbf{v} < 0$). If its spontaneous lifetime is large compared to the oscillation period $T = 2\pi/\omega_v$, the fluorescence is uniformly averaged over all phases of the oscillation and its frequency distribution is symmetric to ω_0. On the average, the atom therefore loses more energy by emission than it gains by absorption. This energy difference is taken from its kinetic energy, resulting in an average decrease in its oscillation energy by $m\hbar\omega_v$ per absorption–emission cycle. As depicted in Fig. 14.24, the absorption spectrum narrows down to an interval around ω_0. This has the unwanted effect that the efficiency of sideband cooling at $\omega_L = \omega_0 - m\omega_v$ decreases with decreasing oscillation energy.

In a quantum-mechanical model, the ion confined to the trap can be described by a nearly harmonic oscillator with vibrational energy levels de-

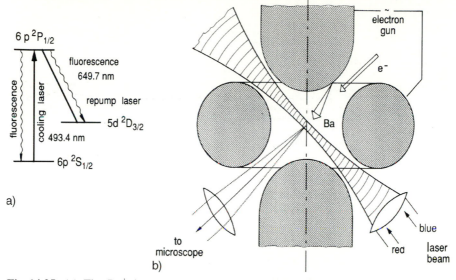

Fig. 14.25. (a) The Ba$^+$ ion as a three-level system. The second laser at $\lambda = 649.7$ nm pumps the 5d^{2}D$_{3/2}$ atoms back into the 6p^{2}P$_{1/2}$ level. **(b)** Experimental arrangement for trapping and cooling of Ba$^+$ ions in a Paul trap [14.77]

termined by the trap potential. Optical cooling corresponds to a compression of the population probability into the lowest levels.

Optical sideband cooling is quite analogous to the Doppler cooling by photon recoil discussed in Sect. 14.1. The only difference is that the confinement of the ion within the trap leads to discrete energy levels of the oscillating ion, whereas the translational energy of a free atom corresponds to a continuous absorption spectrum within the Doppler width.

Optical sideband cooling was demonstrated for Mg$^+$ ions in a Penning trap [14.80, 14.81] and for Ba$^+$ ions in an RF quadrupole trap (Paul trap, Nobel Prize in physics in 1989) [14.82]. The Mg$^+$ ions were cooled below 0.5 K on the $3s^2$S$_{1/2} - 3p^2$P$_{3/2}$ transition by a frequency-doubled dye laser at $\lambda = 560.2$ nm. The vibrational amplitude of the oscillating ions decreased to a few microns. With decreasing temperature the ions are therefore confined to a smaller and smaller volume around the trap center.

The cooling can be monitored with a weak probe laser that is tuned over the absorption profile. Its intensity must be sufficiently small to avoid heating of the ions for frequencies $\omega_{probe} > \omega_0$.

Recently, single Ba$^+$ ions could be trapped in a Paul trap, cooled down to the milliKelvin range, and observed with a microscope by their LIF. Since Ba$^+$ represents a three-level system (Fig. 14.25), two lasers have to be used that are tuned to the transitions $6p^2$S$_{1/2} \rightarrow 6p^2$P$_{1/2}$ (pump transition for cooling) and $5d^2$D$_{3/2} \rightarrow 6p^2$P$_{1/2}$ (repumping to avoid optical pumping into the $5d^2$D$_{3/2}$ level).

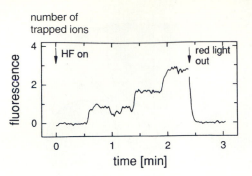

Fig. 14.26. Steps in the fluorescence intensity of a few trapped ions when the number of ions changes by 1. When the red repumping laser is switched off, the ions are transferred by optical pumping into the metastable $5D_{3/2}$ level and the pump laser-induced fluorescence vanishes [14.78]

Any change in the number N of trapped ions (from a change in electron impact ionization, or from collisions with residual gas molecules, which may throw an ion out of the trap) appears as a step in the LIF signal (Fig. 14.26).

Imaging of the spatial distribution of the LIF source by a microscope in combination with an image intensifier (Sect. 4.5) allows the measurement of the average spatial probability of a single ion as a function of its "temperature" [14.77].

14.2.3 Direct Observations of Quantum Jumps

In quantum mechanics the probability $\mathcal{P}_1(t)$ of finding an atomic system at the time t in the quantum state $|1\rangle$ is described by time-dependent wave functions. If one wants to find out whether a system is with certainty $[\mathcal{P}_1(t) = 1]$ in a well-defined quantum state $|1\rangle$ one has to perform a measurement that, however, changes this state of the system. Many controversial opinions have been published on whether it is possible to perform experiments with a single atom in such a way that its initial state and a possible transition to a well-defined final state can be unambiguously determined.

Laser-spectroscopic experiments with single ions confined in a trap have proved that such information can be obtained. The original idea was proposed by Dehmelt [14.83] and has since been realized by several groups [14.84–14.86]. It is based on the coupling of an intense allowed transition with a weak dipole-forbidden transition via a common level. For the example of the Ba$^+$ ion (Fig. 14.27) the metastable $5^2D_{5/2}$ level with a spontaneous lifetime of $\tau = (32\pm5)$ s can serve as a "shelf state." Assume that the Ba$^+$ ion is cooled by the pump laser at $\lambda = 493.4$ nm and the population leaking by fluorescence into the $5^2D_{3/2}$ level is pumped back into the $6^2P_{1/2}$ level by the second laser at $\lambda = 649.7$ nm. If the pump transition is saturated, the fluorescence rate is about 10^8 photons per second with the lifetime $\tau(6^2P_{1/2}) = 8$ ns.

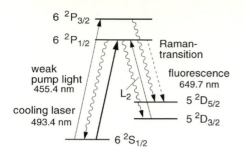

Fig. 14.27. More detailed level scheme of Ba^+ including the fine-structure splitting. The "dark level" $5\,^2D_{5/2}$ can be populated by off-resonance Raman transitions $6\,^2S_{1/2} \to 6\,^2P_{3/2} \to 5\,^2P_{5/2}$ induced by the cooling laser

If the metastable $5^2D_{5/2}$ level is populated (this can be reached, for example, by exciting the $6^2P_{3/2}$ level with a weak laser at $\lambda = 455\,nm$, which decays by fluorescence into the $5^2D_{5/2}$ level, or even without any further laser by a nonresonant Raman transition induced by the cooling laser), the ion is, on the average, for $\tau(5^2D_{5/2}) = 32\,s$ not in its ground state $6^2S_{1/2}$ and therefore cannot absorb the pump radiation at $\lambda = 493\,nm$. The fluorescence rate becomes zero but jumps to its value of 10^8 photons/s as soon as the $5^2D_{5/2}$ level returns by emission of a photon at $\lambda = 1.762\,\mu m$ back into the $6^2S_{1/2}$ level.

The allowed transition $6^2S_{1/2} - 6^2P_{1/2}$ serves as an amplifying detector for a single quantum jump on the dipole-forbidden transition $5^2D_{5/2} \to 6^2S_{1/2}$ [14.85]. In Fig. 14.28 the statistically occurring quantum jumps can be seen by the in-and-out phases of the fluorescence. The effective lifetime of the $5^2D_{5/2}$ level can be shortened by irradiating the Ba^+ ion with a third laser at $\lambda = 614.2\,nm$, which induces the transitions $5^2D_{5/2} \to 6^2P_{3/2}$ where the upper level decays into the $6^2S_{1/2}$ ground state.

Similar observations of quantum jumps have been made for Hg^+ ions confined in a Penning trap [14.86].

Of fundamental interest are measurements of the photon statistics in a three-level system, which can be performed by observing the statistics of quantum jumps. While the durations Δt_i of the "on-phases" or the "off-phases" show an exponential distribution, the probability $P(m)$ of m quantum jumps per second exhibits a Poisson distribution (Fig. 14.29). In a two-level system the situation is different. Here, a second fluorescence photon can be emitted after a first emission only, when the upper state has been reexcited

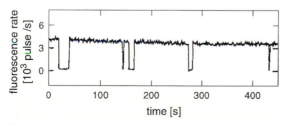

Fig. 14.28. Experimental demonstration of quantum jumps of a single ion [14.85]

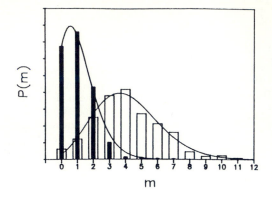

$P(m)$

m

Fig. 14.29. Distribution $P(m)$ of m quantum jumps per second measured over a period of 150 s (*black bars*) and 600 s (*open bars*). The curves are fits of Poisson distributions with two adapted parameters [14.85]

by absorption of a photon. The distribution $P(\Delta T)$ of the time intervals ΔT between successive emission of fluorescence photons shows a *sub-Poisson distribution* that tends to zero for $\Delta T \to 0$ (*photon antibunching*), because at least half of a Rabi period has to pass after the emission of a photon before a second photon can be emitted [14.87].

14.2.4 Formation of Wigner Crystals in Ion Traps

If several ions are trapped in an ion trap and are cooled by optical side-band cooling, a "phase transition" may occur at the temperature T_c where the ions arrange into a stable, spatially symmetric configuration like in a crystal [14.88–14.91]. The distances between the ions in this *Wigner crystal* are about $10^3 - 10^4$ times larger than those in an ordinary ion crystal such as NaCl.

This phase transition from the statistically distributed ions in a "gaseous" cloud to an ordered Wigner crystal can be monitored by the change in the fluorescence intensity. This is observed as a function of the detuning $\Delta \omega$ of the cooling laser. With decreasing red detuning the temperature of the ions is lowered until the phase transition occurs at the critical temperature T_c. The distribution $I_{Fl}(\Delta \omega)$ has a completely different shape for the ordered state of the Wigner crystal than for the disordered ion cloud (Fig. 14.30). For very small detunings the cooling rate becomes smaller then the heating rate and the crystal "melts" again.

When the detuning $\Delta \omega$ of the cooling laser or its intensity is changed, one observes a typical hysteresis (Fig. 14.31). At about 160-μW laser power and a fixed detuning of -120 MHz the fluorescence intensity increases by a factor of 4, indicating a phase transition to the ordered Wigner crystal. With further increases of the laser power, the system suddenly jumps back at $P_L \simeq 400\,\mu$W into the disordered state. Similar hysteresis curves can be found when the amplitude b of the trap's RF voltage is changed [14.91]. Using a microscope in combination with a sensitive image intensifier the location of the ions can be made visible on a screen (Fig. 14.32) and the transition from the disordered

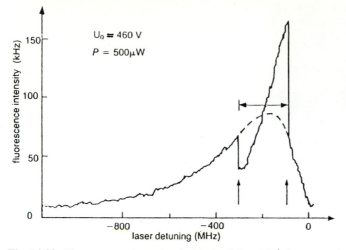

Fig. 14.30. Fluorescence intensity $I_{Fl}(\Delta\omega)$ of five Mg$^+$ ions as a function of laser detuning for a disordered ion cloud and for an ion crystal (*between the arrows*) [14.88]

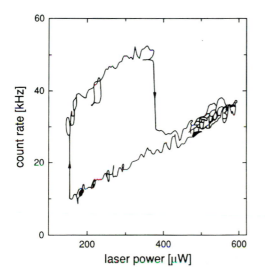

Fig. 14.31. Hysteresis curve for a phase transition. The laser-induced fluorescence rate is plotted as a function of laser power at a fixed detuning ($\Delta\nu = -120\,\text{MHz}$). At about 170-μW laser power the phase transition occurs from the disordered ion cloud to the ordered Wigner crystal, and at 400 μW the opposite phase transition is observed [14.90]

state of the ion cloud to the ordered Wigner crystal can be directly observed [14.88, 14.90].

Similar to the situation for a coupled pendulum, normal vibrations can be excited in a Wigner crystal. For example, the two-ion crystal has two normal vibrations where the two ions oscillate in the ion trap potential either in phase or with opposite phases. For the in-phase oscillations, the Coulomb repulsion between the ions does not influence the oscillation frequency because the distance between the ions does not change. The restoring force is solely due to the trap potential. One obtains two degenerate oscillations in

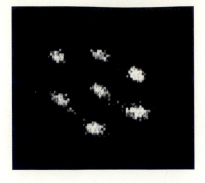

Fig. 14.32. Photograph of a Wigner crystal of seven trapped ions, taken with a microscope and an image intensifier. The distance between the ions is about 20 μm [14.89]

the x- and y-directions with the frequencies $\Omega_x = \Omega_y$, (14.45), and an oscillation in the z-direction with $\Omega_z \neq \Omega_x$. For the oscillation with opposite phases the vibrational frequency is additionally determined by the Coulomb repulsion. One obtains for the three components $\Omega_{2x} = \sqrt{3}\Omega_{1x} = \Omega_{2y}$ and $\Omega_{2z} = \sqrt{3}\Omega_{1z}$ [14.92].

These vibrational modes can be excited if an additional ac voltage with the proper frequency is applied to the trap electrodes. The excitation leads to heating of the Wigner crystal, which causes a decrease in the laser-induced fluorescence. By choosing the proper intensity and the detuning $\Delta\omega$ of the cooling laser, this heating can be kept stable. Such measurements allow the study of many-particle effects with samples of a selected small number of ions. They may give very useful information on solid-state problems [14.93].

14.2.5 Laser Spectroscopy in Storage Rings

In recent years, increasing efforts have been made to apply laser-spectroscopic techniques to high-energy ions in storage rings (Fig. 14.33) [14.94]. Laser cooling of these ions has become an effective alternative to electron cooling or stochastic cooling [14.95]. Beam cooling in storage rings not only results in a much better beam quality, but may also lead to a condensation of ion beams resulting in *one-dimensional Wigner crystals* [14.96a]. Possible candidates for such cooling experiments are metastable Li^+ ions, where the cooling transition $1s2s\,^3S_1 \rightarrow 1s2p\,^3P_1$ with $\lambda = 548.5$ nm lies conveniently within the tuning range of cw dye lasers. At 100-keV kinetic energy excited Li^+ ions in the $2p\,^3P_1$ level with a spontaneous lifetime of $\tau = 43$ ns travel about 8 cm. This allows 100 absorption–emission cycles per ion round-trip over the interaction length of 8 m between laser beam and ion beam.

In spite of large efforts, one-dimensional Wigner crystals in high-energy storage rings cannot yet be realized. However, phase transitions and ordered structures of laser-cooled Mg^+ ions stored in a small RF quadrupole storage ring with a diameter of 115 mm have been observed recently [14.96b]. Cooling of the ions results in a linear or a helical structure of ions aligned along the center line of the quadrupole field.

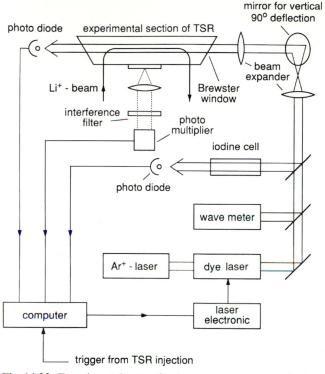

Fig. 14.33. Experimental setup for laser spectroscopy in the ion test storage ring TSR in Heidelberg [14.98]

There are, however, other interesting experiments that were successful. For instance, in the electron cooler ions and atoms move with nearly equal velocities and form an unusual kind of a plasma. Here laser spectroscopy may help to better understand the physics of highly ionized plasmas [14.97].

An interesting aspect is the spontaneous radiative recombination of electrons and ions, which can be enhanced by irradiation with resonant laser light. This has been demonstrated in the test storage ring in Heidelberg [14.98]. A proton beam of $E_k = 21$ MeV is superimposed by a cold electron beam. If a pulsed, tunable dye laser beam propagates antiparallel to the ion beam through a linear section of the storage ring, the laser wavelength of 450.5 nm is Doppler shifted into resonance with the downward transition $H^+ + e^- \rightarrow H(2s)$ from the ionization limit to the $2s$ state. The enhancement factor G, which represents the ratio of induced-to-spontaneous recombinations, depends on the laser wavelength and reaches values of $G \simeq 50$ for the resonance wavelength $\lambda = 450.5$ nm.

Further experiments of fundamental interest are precision measurements of transition frequencies of very fast moving ions, allowing tests of special relativity.

14.3 Optical Ramsey Fringes

In the previous sections we discussed how the interaction time of atoms or ions with laser fields can be greatly increased by cooling and trapping. Optical cooling cannot be applied to molecules, which do not represent two-level systems. Here another technique has been developed that increases the interaction time of atoms or molecules with EM fields by increasing the spatial interaction zone and thus decreasing the transit-time broadening of absorption lines (Sect. 3.4).

14.3.1 Basic Considerations

The problem of transit-time broadening was recognized many years ago in electric or magnetic resonance spectroscopy in molecular beams [14.99]. In these Rabi experiments [14.100], the natural linewidth of the radio frequency or microwave transitions is extremely small because the spontaneous transition probability is, according to (2.22), proportional to ω^3. The spectral widths of the microwave or RF lines are therefore determined mainly by the transit time $\Delta T = d/\overline{v}$ of molecules with the mean velocity $\overline{v}$ through the interaction zone in the C field (Fig. 10.9a) with length d.

A considerable reduction of the time-of-flight broadening could be achieved by the realization of Ramsey's ingenious idea of separated fields [14.101]. The molecules in the beam pass two phase-coherent fields that are spatially separated by the distance $L \gg d$, which is large compared with the extension d of each field (Fig. 14.34). The interaction of the molecules with the first field creates a dipole moment of each molecular oscillator with a phase that is dependent on the interaction time $\tau = d/\overline{v}$ and the detuning $\Omega = \omega_0 - \omega$ of the molecular transition (Sect. 2.8). After passing the first interaction zone, the molecular dipole precesses in the field-free region at its eigenfrequency ω_0. When it enters the second field, it therefore has accumulated the phase angle $\Delta\varphi = \omega_0 T = \omega_0 L/\overline{v}$. During the same time T the field phase has changed by ωT. The relative phase between the dipole and the field has therefore changed by $(\omega_0 - \omega)T$ during the flight time T through the field-free region.

The interaction between the dipole and the second field with the amplitude $E_2 = E_0 \cos\omega t$ depends on their relative phases. The observed signal is

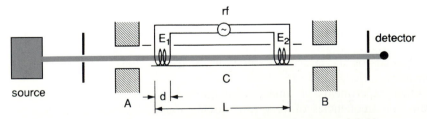

Fig. 14.34. Rabi molecular beam apparatus with Ramsey's separated fields

related to the power absorbed by the molecular dipoles in the second field and is therefore proportional to $E_2^2 \cos[(\omega - \omega_0)T]$. When we assume that all N molecules passing the field per second have the same velocity v, we obtain the signal

$$S(\omega) = aNE_2^2 \cos[(\omega_0 - \omega)L/v] \,, \tag{14.47}$$

where the constant a depends on the geometry of the beam and the field.

Measured as a function of the field frequency ω, this signal exhibits an oscillatory pattern called *Ramsey fringes* (Fig. 14.35). The full halfwidth of the central fringe, which is $\delta\omega = \pi(v/L)$, decreases with increasing separation L between the fields.

This interference phenomenon is quite similar to the well-known Young's interference experiment (Sect. 2.8.2), where two slits are illuminated by co-herent light and the superposition of light from both slits is observed as a function of the optical path difference Δs. The number of maxima observed in the two-slit interference pattern depends on the coherence length ℓ_c of the incident light and on the slit separation. The fringes can be seen if $\Delta s \leq \ell_c$.

A similar situation is observed for the Ramsey fringes. Since the velocities of the molecules in the molecular beam are not equal but follow a Maxwellian distribution, the phase differences $(\omega_0 - \omega)L/v$ show a corresponding distri-bution. The interference pattern is obtained by integrating the contributions to the signal from all molecules $N(v)$ with the velocity v

$$S = C \int N(v) E^2 \cos[(\omega_0 - \omega)L/v] dv \,. \tag{14.48}$$

Similar to Young's interference with partially coherent light, the velocity dis-tribution will smear out the interference pattern for the higher-order fringes, that is, large $(\omega_0 - \omega)$, but will essentially leave the central fringe narrow for small $(\omega_0 - \omega)$. With a halfwidth Δv of the velocity distribution $N(v)$, this restricts the maximum field separation to about $L \leq v^2/(\omega_0 \Delta v)$, since for larger L the higher interference orders for fast molecules overlap with the first order of slow molecules. Using supersonic beams with a narrow velocity distribution (Sect. 9.1), larger separations L are allowed. In general, however,

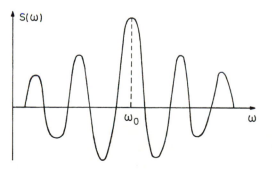

Fig. 14.35. Signal power absorbed by the molecules in the second field as a function of detuning $\Omega = \omega - \omega_0$ (Ramsey fringes) for a narrow velocity distribution $N(v)$

only the zeroth order of the Ramsey interference is utilized in high-resolution spectroscopy and the "velocity averaging" of the higher orders has the advantage that it avoids the overlap of different orders for two closely spaced molecular lines.

The extension of Ramsey's idea to the optical region seems quite obvious if the RF fields in Fig. 14.34 are replaced by two phase-coherent laser fields. However, the transfer from the RF region, where the wavelength λ is larger than the field extension d, to the optical range where $\lambda \ll d$, causes some difficulties [14.102]. Molecules with slightly inclined path directions traverse the optical fields at different phases (Fig. 14.36). Consider molecules starting from a point $x = 0$, $z = 0$ at the beginning of the first field. Only those molecules with flight directions within a narrow angular cone $\delta\theta \leq \lambda/2d$ around the z-axis experience phases at the end of the first field differing by less than π. These molecules, however, traverse the *second* field at a distance L downstream within the extension $\Delta x = L\delta\theta \leq L\lambda/2d$, where the phase φ of the optical field has a spatial variation up to $\Delta\varphi \leq L\pi/d$. If the method of separated fields is to increase the spectral resolution, L has to be large compared to d, which implies that $\Delta\varphi \gg \pi$. Although these molecules have experienced nearly the same phase in the first field, they *do not* generate observable Ramsey fringes when interacting with the second field because their interaction phases are all different, and the total signal is obtained by summing over all molecules present at the time t in the second field. This implies averaging over their different phases, which means that the Ramsey fringes are washed out. The same is true for molecules starting from different points $(x_1, z = 0)$ in the first interaction zone that arrive at the same point $(x_2, z = L)$ in the second zone (Fig. 14.36b). The phases of these molecules may differ by $\Delta\varphi \gg \pi$, and for all molecules starting at different x_1 they are randomly distributed and therefore no macroscopic polarization is observed at (x_2, L).

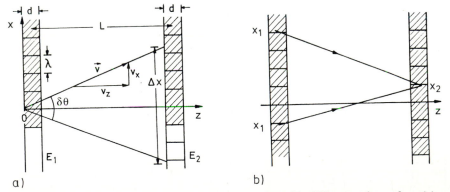

Fig. 14.36a,b. Molecules starting from the same point with different values of v_x (**a**) or from different points x_1, x_2 in the first zone (**b**) experience different phase differences in the second field

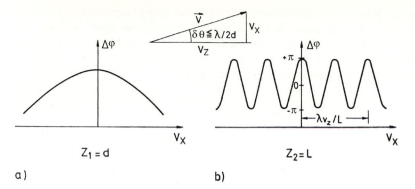

Fig. 14.37a,b. Relative phase $\Delta\varphi(v_x)$ between the oscillating dipole molecule and the EM field: (**a**) at the end of the first zone $z = d$; (**b**) in the second zone at $z = L \gg d$

Note: The requirement $\delta\theta < \lambda/2d$ for molecules that experience nearly the same phase in the first zone is equivalent to the condition that the residual Doppler width $\delta\omega_D$ of the absorption profile for a laser transverse to the beam axis does not exceed the transit-time broadening $\delta\omega_t = \pi v/d$. This can be seen immediately from the relations $v_x = v_z\delta\theta$ and $\delta\omega_D = \omega v_x/c = \omega\delta\theta v_z/c = \delta\theta v_z 2\pi/\lambda$. For

$$\delta\theta < \lambda/2d \quad \Rightarrow \quad \delta\omega_D < \pi v_z/d = \delta\omega_t . \tag{14.49}$$

The phase difference $\Delta\varphi(v_x)$ of a molecular dipole starting from the point $(x_1, 0)$ in the first zone can be plotted as a function of the transverse velocity v_x, as indicated in Fig. 14.37. Although $\Delta\varphi(v_x, z_1 = d)$ shows a flat distribution at the end of the first laser beam, it exhibits a modulation with the period $\Delta v_x = \lambda/2T = \lambda v_z/2L$ in the second zone. However, this modulation cannot be detected because it is washed out by integrating over all contributions from molecules with different starting points x_1 arriving at the point (x_2, L) with different transverse velocities v_x.

Fortunately, several methods have been developed that overcome these difficulties and that allow ultranarrow Ramsey resonances to be obtained. One of these methods is based on Doppler-free two-photon spectroscopy, while another technique uses saturation spectroscopy but introduces a third interaction zone at the distance $z = 2L$ downstream from the first zone to recover the Ramsey fringes [14.103–14.105]. We briefly discuss both methods.

14.3.2 Two-Photon Ramsey Resonance

In Sect. 7.4 we saw that the first-order Doppler effect can be exactly canceled for a two-photon transition if the two photons $\hbar\omega_1 = \hbar\omega_2$ have opposite wave vectors, i.e., $k_1 = -k_2$. A combination of Doppler-free two-photon absorption and the Ramsey method therefore avoids the phase dependence $\varphi(v_x)$ on the transverse velocity component. In the first interaction zone the molecular dipoles are excited with the transition amplitude a_1 and precess with their eigenfrequency $\omega_{12} = (E_2 - E_1)/\hbar$. If the two photons come from oppositely traveling waves with frequency ω, the detuning

$$\Omega = \omega + kv_x + \omega - kv_x - \omega_{12} = 2\omega - \omega_{12} , \tag{14.50}$$

of the molecular eigenfrequency ω_{12} from 2ω is independent of v_x. The phase factor $\cos(\Omega T)$, which appears after the transit time $T = L/v_z$ at the entrance of the molecular dipoles into the second field, can be composed as $\cos(\varphi_2^- + \varphi_2^+ - \varphi_1^- - \varphi_1^+)$, where each φ comes from one of the four fields (two oppositely traveling waves in each zone). If we denote the two-photon transition amplitudes in the first and second field zone by c_1 and c_2, respectively, we obtain the total transition probability as

$$\mathscr{P}_{12} = |c_1|^2 + |c_2|^2 + 2|c_1||c_2| \cos \Omega T . \tag{14.51}$$

The first two terms describe the conventional two-photon transitions in the first and second zones, while the third term represents the interference leading to the Ramsey resonance. Due to the *longitudinal* thermal velocity distribution $f(v_z)$, only the central maximum of the Ramsey resonance is observed with a theoretical halfwidth (for negligible natural width) of

$$\Delta\Omega = (2/3)\pi/T = 2\pi v/3L \tag{14.52}$$

$$\Rightarrow \quad \Delta\nu = \frac{1}{3T} , \quad \text{with} \quad T = L/v .$$

The higher interference orders are washed out.

> **Example 14.13**
> With a field separation of $L = 2.5\,\text{mm}$ and a mean velocity $\bar{v} = 270\,\text{m/s}$ at 400 K, we obtain a halfwidth (FWHM) of $\Delta\nu = 1/3T = 36\,\text{kHz}$ for the central Ramsey fringe if the other contributions to the linewidth are negligible.

The experimental arrangement is depicted in Fig. 14.38. Rubidium atoms are collimated in an atomic beam that traverses the two laser beams. The two counterpropagating waves in both fields are generated from a single laser beam by reflection from the resonator mirrors M1, M2, M3. The radiative lifetimes of the excited atomic levels must be longer than the transit time $T = L/v$. Otherwise, the phase information obtained in the first zone would be lost at the second zone. Therefore the method is applied to long-living states, such as Rydberg states or vibrational levels of the electronic ground state. The

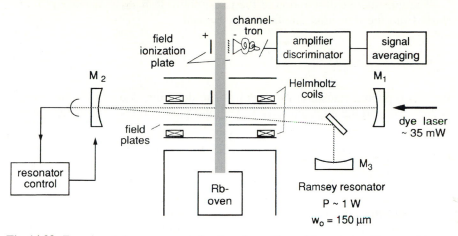

Fig. 14.38. Experimental arrangement for the observation of two-photon Ramsey fringes [14.107]

excited Rydberg atoms are detected by field ionization. The Helmholtz coils allow the investigation of Zeeman splittings of Rydberg transitions.

The achievable spectral resolution is demonstrated by Fig. 14.39, which shows a two-photon Ramsey resonance for a hyperfine component of the two-photon Rydberg transition $32\,^2S \leftarrow 5\,^2S$ of rubidium atoms ^{86}Rb [14.106, 14.107], measured with the arrangment of Fig. 14.38. The length of the Ramsey resonator must always be kept in resonance with the laser frequency. This is achieved by piezo-tuning elements and a feedback control system (Sect. 5.4.5). The halfwidth of the central Ramsey maximum was measured as $\Delta\nu = 37\,kHz$ for a field separation of 2.5 mm and comes close to the theoretical limit. With $L = 4.5\,mm$ an even narrower signal with $\Delta\nu = 18\,kHz$ could be achieved, whereas the two-photon resonance width in a single zone with a beam waist of $\omega_0 = 150\,\mu m$ was limited by transit-time broadening to about 600 kHz [14.107].

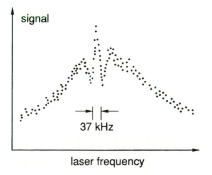

Fig. 14.39. Two-photon optical Ramsey resonance of the transition $32\,^2S \leftarrow 5\,^2S$, $F = 3$ in ^{86}Rb for a field separation of $L = 2.5\,mm$ [14.107]

The quantitative description of the two-photon Ramsey resonance [14.108] starts from the transition amplitude per second

$$c_{if}(t) = \frac{D_{if}I}{4\hbar^2 \Delta\omega}(e^{-i\Delta\omega t} - 1) \,, \tag{14.53}$$

for a two-photon transition $|i\rangle \to |f\rangle$ at the detuning $\Delta\omega = 2\omega - (\omega_{ik} + \omega_{kf}) = 2\omega - \omega_{if}$ and the laser intensity I. The two-photon transition dipole element is

$$D_{if} = \sum_k \frac{R_{ik}R_{kf}}{\omega - \omega_{ki}} \,, \tag{14.54}$$

where R_{ik}, R_{kf} are the one-photon matrix elements (2.67).

After a molecule has passed the first interaction zone with the transit time $\tau = d/v$, the amplitude for the transition $|i\rangle \to |f\rangle$ is

$$c_{if}(1) = \frac{D_{if}I_1\tau}{4\hbar^2 \Delta\omega}(e^{-i\Delta\omega\tau} - 1) \,. \tag{14.55}$$

After the transit time $T = L/v$ through the field-free region and a passage through the second interaction zone, the transition amplitude becomes

$$c_{if}(2) = c_{if}(1) + \frac{D_{if}I_2\tau}{4\hbar^2 \Delta\omega}(e^{-i\Delta\omega(T+\tau)} - e^{-i\Delta\omega T}) \,, \tag{14.56}$$

which yields the transition probability

$$\begin{aligned}
\mathscr{P}_{if}^{(2)}(2) &= |c_{if}(1)|^2 + |c_{if}(2)|^2 + 2c_{if}(1)c_{if}(2)\cos\Delta\omega T \,, \\
&= \frac{|D_{if}|^2\tau^2}{\hbar^2}[I_1^2 + I_2^2 + 2I_1 I_2\cos(\Delta\omega T)] \,, \tag{14.57}
\end{aligned}$$

which is identical to (14.51). The signal is proportional to the power $S(\Omega)$ absorbed in the second zone (Fig. 14.40).

When the upper level $|f\rangle$ has the spontaneous lifetime $\tau_f = 1/\gamma_f$, part of the excited molecules decay before they reach the second zone, and the transition amplitude becomes

$$c_{if}(2) = c_{if}(1)(e^{-\gamma_f T} + e^{-i\Delta\omega T}) \,, \tag{14.58}$$

which yields with $I_1 = I_2 = I$ the smaller signal

$$S(\Delta\omega) \propto \mathscr{P}^{(2)}(2) = \frac{|D_{if}|^2 I^2\tau^2}{\hbar^2}\left[1 + e^{-2\gamma_f T} + 2e^{-\gamma_f T}\cos\Delta\omega T\right] \,. \tag{14.59}$$

one interaction zone:

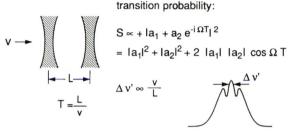

transit time linewidth:

$$\Delta v \propto \frac{v}{d}$$

two separated interaction zones:

transition probability:

$$S \propto + |a_1 + a_2 \, e^{-i\Omega T}|^2$$

$$= |a_1|^2 + |a_2|^2 + 2 \, |a_1| \, |a_2| \, \cos \Omega T$$

$$\Delta v' \propto \frac{v}{L}$$

Fig. 14.40. Illustration of two-photon resonance [14.107]

14.3.3 Nonlinear Ramsey Fringes Using Three Separated Fields

Another solution to restore the Ramsey fringes, which are generally washed out in the second field, is based on the introduction of a third field at the distance $2L$ downstream from the first field. The idea of this arrangement was first pointed out by Chebotayev and coworkers [14.103]. The basic idea may be understood as follows: In Sect. 7.2 it was discussed in detail that the nonlinear absorption of a molecule in a monochromatic standing wave field leads to the formation of a narrow Lamb dip at the center ω_0 of a Doppler-broadened absorption line (Fig. 7.6). The Lamb-dip formation may be regarded as a two-step process: the depletion of a small subgroup of molecules with velocity components $v_x = 0 \pm \Delta v_x$ by the pump wave (*hole burning*), and the successive probing of this depletion by a second wave. In the standing light wave of the second zone, the nonlinear saturation depends on the relative phase between the molecular dipoles and the field. This phase is determined by the starting point $(x_1, z = 0)$ in the first zone and by the transverse velocity component v_x. Figure 14.41a depicts the collision-free straight path of a molecule with transverse velocity component v_x, starting from a point $(x_1, z = 0)$ in the first zone, traversing the second field at $x_2 = x_1 + v_x T = x_1 + v_x L/v_z$, and arriving at the third field at $x_3 = x_1 + 2v_x T$. The relative phase between the molecule and the field at the entrance (L, x_2) into the second field is

$$\Delta\varphi = \varphi_1(x_1) + \Delta\omega \cdot T - \varphi_2(x_2) , \quad \text{with} \quad \Delta\omega = \omega_{12} - \omega .$$

The macroscopic polarization at (L, x_2), which equals the vector sum of all induced atomic dipoles, averages to zero because molecules with different ve-

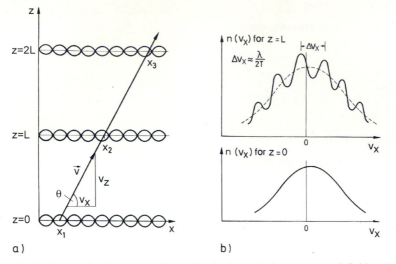

Fig. 14.41. (a) Straight path of a molecule through three separated fields at $z = 0$, $z = L$, and $z = 2L$. **(b)** Modulation of the population density $N(v_x)$ after the interaction with the second field [14.103]

locity components v_x arrive at x_2 from different points $(0, x_1)$. Note, however, that the population depletion ΔN_a in the second field depends on the relative phase $\Delta\varphi$ and therefore on v_x. If the phases $\varphi(x_1)$ and $\varphi(x_2)$ of the two fields are made equal for $x_1 = x_2$, the phase difference $\varphi(x_1) - \varphi(x_2) = \varphi(x_1 - x_2) = \varphi(v_x T)$ between the two fields at the intersection points depends only on v_x and not on x_1. *After the nonlinear interaction with the second field the number $n(v_x)$ of molecular dipoles shows a characteristic modulation* (Fig. 14.41b). This modulation cannot be detected in the second field because it appears in v_x but not in x. For the interaction of all molecules with the probe wave at $z = z_2$, which has a spatially varying phase $\varphi(x_2)$, this modulation is completely washed out. That is, however, not true in the third field. Since the intersection points x_1, x_2, and x_3 are related to each other by the transverse velocity v_x, the modulation $N(v_x)$ in the second beam results in a nonvanishing macroscopic polarization in the third beam, given by

$$P(\Delta\omega) = 2\,\mathrm{Re}\left\{ E_3 \int\limits_{x=0}^{x_0} \int\limits_{t=2T}^{2T+\tau} [P^0(z, t)\cos(kx + \varphi_3)\,\mathrm{e}^{\mathrm{i}\omega t}]\,\mathrm{d}x\,\mathrm{d}t \right\} . \qquad (14.60)$$

The detailed calculation, using third-order perturbation theory [14.103, 14.109], demonstrates that the power absorbed in the third field zone leads to the signal

$$S(\Delta\omega) = \frac{\hbar\omega}{2}|G_1 G_2^2 G_3|\tau^2 \cos^2(\Delta\omega T)\cos(2\varphi_2 - \varphi_1 - \varphi_3) , \qquad (14.61)$$

where $G_n = iD_{21}E_n/\hbar$ $(n = 1, 2, 3)$ and φ_1, φ_2, φ_3 are the spatial phases of the three fields

$$E_n(x, z, t) = 2E_n(z)\cos(k_n + \varphi_n)\cos\omega t \ . \tag{14.62}$$

Adjusting the phases φ_n properly, such that $2\varphi_2 = \varphi_1 + \varphi_3$, allows optimization of the signal in the third zone. A detailed calculation of nonlinear Ramsey resonances, based on the density matrix formalism, has been performed by Bordé [14.110].

The capability of this combination of saturation spectroscopy with optical Ramsey fringes has been demonstrated by Bergquist et al. [14.111] for the example of the neon transition $1s_5 \rightarrow 2p_2$ at $\lambda = 588.2$ nm (Fig. 14.42). A linewidth of 4.3 MHz for the Lamb dip has been achieved with the distance $L = 0.5$ cm between the interaction zones. This corresponds to the natural linewidth of the neon transition.

With four interaction zones the contrast of the Ramsey resonances can still be increased [14.112]. This was also demonstrated by Bordé and coworkers [14.113], who used four traveling waves instead of three standing waves, which crossed a supersonic molecular beam of SF_6 perpendicularly. The Ramsey signal for vibration–rotation transitions in SF_6 around $\lambda = 10\,\mu m$ was monitored with improved contrast by an optothermal detector (Sect. 6.3.3).

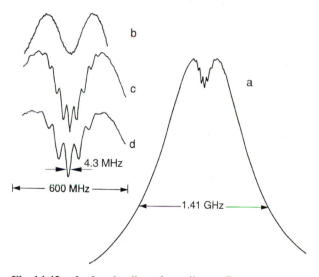

Fig. 14.42a–d. Lamb dip of nonlinear Ramsey resonance for the neon transition $1s_5 \rightarrow 2p_2$ at $\lambda = 588.2$ nm, measured in a fast beam of metastable neon atoms: (**a**) reduced Doppler profile in the collimated beam with the Lamb dip obtained with three laser beams; (**b**)–(**d**) expanded section of the Lamb dip for three different beam geometries: the atom interacts only with two standing waves (b), three equally spaced interaction zones (c), and four zones (d) [14.111]

For transitions with small natural linewidths the Ramsey resonances can be extremely narrow. For instance, the nonlinear Ramsey technique applied to the resolution of rotational–vibrational transitions in CH_4 in a methane cell at 2-mbar pressure yielded a resonance width of 35 kHz for the separation $L = 7$ mm of the laser beams. With the distance $L = 3.5$ cm the resonance width decreased to 2.5 kHz, and the Ramsey resonances of the well-resolved hyperfine components of the CH_4 transition at $\lambda = 3.39$ μm could be measured [14.109].

It is also possible to observe the Ramsey resonances at $z = 2L$ without the third laser beam. If two standing waves at $z = 0$ and $z = L$ resonantly interact with the molecules, we have a situation similar to that for photon echoes. The molecules that are coherently excited during the transit time τ through the first field suffer a phase jump at $t = T$ in the second zone, because of their nonlinear interaction with the second laser beam, which reverses the time development of the phases of the oscillating dipoles. At $t = 2T$ the dipoles are again in phase and emit coherent radiation with increased intensity for $\omega = \omega_i k$ (*photon echo*) [14.104, 14.109].

14.3.4 Observation of Recoil Doublets and Suppression of One Recoil Component

The increased spectral resolution obtained with the Ramsey technique because of the increased interaction time allows the direct observation of recoil doublets in atomic or molecular transitions (Sect. 14.1.1). An example is the Ramsey spectroscopy of the intercombination line $^1S_0 - {}^3P_1$ in calcium at $\lambda = 657$ nm [14.114], where a linewidth of 3 kHz was obtained for the central Ramsey maximum and the recoil doublet could be clearly resolved (Fig. 14.43).

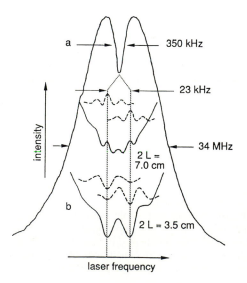

Fig. 14.43a,b. Ramsey resonances of the calcium intercombination line at $\lambda = 657$ nm, measured in a collimated Ca atomic beam: (**a**) Doppler profile with a reduced Doppler width and a central Lamb dip, if only one interaction zone is used; (**b**) expanded section of the Lamb dip with the two recoil components observed with three interaction zones with separations $L = 3.5$ cm and $L = 1.7$ cm. The *dashed curves* show results with partial suppression of one recoil component [14.114]

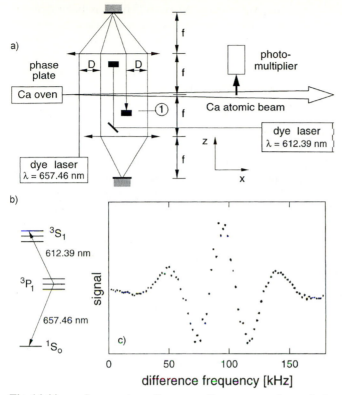

Fig. 14.44a–c. Suppression of one recoil component by optical pumping with a second laser: (**a**) experimental setup; (**b**) level scheme; and (**c**) Ramsey resonance of the remaining recoil component [14.115]

Although transit-time broadening is greatly reduced by the Ramsey technique, the quadratic Doppler effect is still present and may prevent the complete resolution of the recoil components. This may cause asymmetric line profiles where the central frequency cannot be determined with the desired accuracy. As was shown by Helmcke et al. [14.115, 14.116], one of the recoil components can be eliminated if the upper level $^3P_1$ of the Ca transition is depopulated by optical pumping with a second laser. In Fig. 14.44 the relevant level scheme, the experimental setup, and the measured central Ramsey maximum of the remaining recoil component is shown.

14.4 Atom Interferometry

Atomic particles moving with the momentum p can be characterized by their de Broglie wavelength $\lambda = h/p$. If beams of such particles can be split into coherent partial beams, which are recombined after traveling different path lengths, matter-wave interferometry becomes possible. This has been demonstrated extensively for electrons and neutrons, and recently also for neutral atoms [14.117, 14.118].

14.4.1 Mach–Zehnder Atom Interferometer

A beam of neutral atoms may be split into coherent partial beams by diffraction from two slits [14.119] or from microfabricated gratings [14.120], but also by photon recoil in laser beams. While the two former methods are analogous to the optical phenomena (Young's two-slit experiment) although with a much smaller wavelength λ, the latter technique has no counterpart in optical interferometers. Therefore, we will briefly discuss this method, which uses a four-zone Ramsey excitation as an atomic interferometer. It was proposed by Bordé [14.121], and experimentally realized by several groups. The explanation, which follows that of Helmcke and coworkers [14.122], is as follows: atoms in a collimated beam pass through the four interaction zones of the Ramsey arrangement, exhibited in Figs. 14.44 and 14.45. Atoms in state $|a\rangle$ that absorb a laser photon in the interaction zone 1 are excited into state $|b\rangle$ and suffer a recoil momentum $\hbar k$, which deflects them from their straight path. If these excited atoms undergo an induced emission in the second zone, they return to the state $|a\rangle$ and fly parallel to the atoms that have not absorbed a photon in both zones 1 and 2. In Fig. 14.45 the atoms in the different zones are characterized by their internal state, $|a\rangle$ or $|b\rangle$, and their transverse momentum $m\hbar k$. The figure illustrates that in the fourth zone there are two pairs of exit ports where the two components of each pair pass through the same lo-

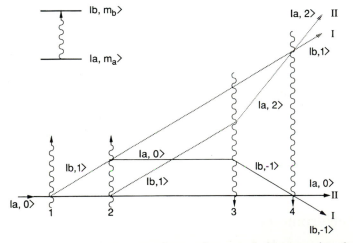

Fig. 14.45. Optical Ramsey scheme of an atomic beam passing through four traveling laser fields, interpreted as matter-wave interferometer. *Solid lines* represent the high-frequency recoil components, *dotted lines* the low-frequency components (only those traces leading to Ramsey resonances in the fourth zone are drawn) [14.122]

cation and can therefore interfere. The phase difference of the corresponding matter waves depends on the path difference and on the internal state of the atoms. The excited atoms in state $|b\rangle$ arriving at the fourth zone can be detected through their fluorescence. These atoms differ in the two interference ports by their transverse momentum $\pm\hbar k$, and their transition frequencies are therefore split by recoil splitting (Sect. 4.1.1). The signals in the exit ports in zone 4 depend critically on the difference $\Omega = \omega_L - \omega_0$ between the laser frequency ω_L and the atomic resonance frequency ω_0. The two trapezoidal areas in Fig. 14.45, which are each enclosed by the two interfering paths, may be regarded as two separate Mach–Zehnder interferometers (Sect. 4.2) for the two recoil components.

Matter-wave interferometry has found wide applications for testing basic laws of physics. One advantage of interferometry with massive particles, for instance, is the possibility of studying gravitational effects. Compared to the neutron interferometer, atomic interferometry can provide atomic fluxes that are many orders of magnitude higher than thermalized neutron fluxes from reactors. The sensitivity is therefore higher and the costs are much lower.

One example of an application is the measurement of the gravitational acceleration g on earth with an accuracy of $3 \times 10^{-8} g$ with a light-pulse atomic interferometer [14.123]. Laser-cooled wave packets of sodium atoms in an atomic fountain (Sect. 14.1.8) are irradiated by a sequence of three light pulses with properly chosen intensities. The first pulse is chosen as $\pi/2$-pulse, which creates a superposition of two atomic states $|1\rangle$ and $|2\rangle$ and results in a splitting of the atomic fountain beam at position 1 in Fig. 14.46 into two beams because of photon recoil. The second pulse is a π-pulse, which deflects the two partial beams into opposite directions; the third pulse finally is again a $\pi/2$-pulse, which recombines the two partial beams and causes the wave

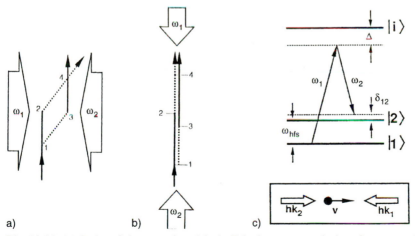

Fig. 14.46. (a) Paths of the atom in a Mach–Zehnder-type atomic interferometer; **(b)** momentum transfer by stimulated Raman transitions applied to rubidium atoms in an atomic fountain where the atoms move parallel to the laser beams [14.123]; **(c)** level scheme for Raman transitions

packets to interfere. This interference can, for example, be detected by the fluorescence of atoms in the upper state $|2\rangle$.

For momentum transfer stimulated Raman transitions between the two hyperfine levels $|1\rangle$ and $|2\rangle$ of the Na($3\,^2S_{1/2}$) state are used, which are induced by two laser pulses with light frequencies ω_1 and ω_2 ($\omega_1 - \omega_2 = \omega_{HFS}$) traveling into opposite directions (Fig. 14.46b). Each transition transfers the momentum $\Delta p \simeq 2\hbar k$. The gravitational field causes a deceleration of the upwards moving atoms in the fountain. This changes their velocity, which can be detected as a Doppler shift of the Raman transitions. Because the Raman resonance $\omega_1 - \omega_2 = \omega_{HFS}$ has an extremely narrow width (Sect. 14.6), even small Doppler shifts can be accurately measured.

14.4.2 Atom Laser

Atoms in a Bose–Einstein condensate are coherent because they are all described by the same wavefunction. If these atoms are released out of the trap (for instance, by a radio-frequency field (Fig. 14.47), or by switching off the magnetic field, which forms the trap potential) the atoms fall under the influence of gravity down as a parallel beam in the $-z$-direction, where all atoms are in the same coherent state. Since this is in analogy to the coherent beam of photons in a laser, this coherent beam of atoms is called an "atom laser." The first realization was a pulsed atom laser, where part of the BEC atoms were released by an RF pulse [14.124]. In addition, quasi-continuous atom lasers have also been realized [14.125]. Using a weak radio frequency field as output coupler with a small coupling strength, atoms could be continuously extracted from the BEC over a period of up to 100 ms. The duration was limited by the finite number of atoms in the BEC. There have been attempts to continuously load the BEC and continuously extract the atoms, which would give a true cw atom laser.

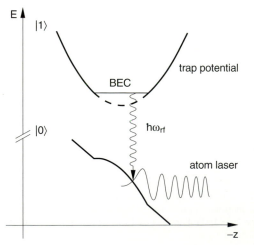

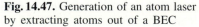

Fig. 14.47. Generation of an atom laser by extracting atoms out of a BEC

The divergence of this atom laser beam is very small and the brightness of such a very cold atomic beam source is larger by several orders of magnitude than that achievable with Zeeman slower (see Sect. 14.1.4). The brightness of the atom laser beam, defined as the integrated flux of atoms per source area divided by the velocity spread $\Delta v_x \Delta v_y \Delta v_z$ is about 2×10^{24} atoms $(s^2 m^{-5})$ [14.126].

Very well collimated atomic beams with such a high brightness can be used for the investigation of scattering, chemical reactions at very low relative velocities and for surface scattering experiments.

14.5 The One-Atom Maser

In Sect. 14.3 we discussed techniques to store and observe single ions in traps. We will now present some recently performed experiments that allow investigations of single atoms and their interaction with weak radiation fields in a microwave resonator [14.127]. The results of these experiments provide crucial tests of basic problems in quantum mechanics and quantum electrodynamics (often labeled "cavity QED"). Most of these experiments were performed with alkali atoms. The experimental setup is shown in Fig. 14.48.

Alkali atoms in a velocity-selected collimated beam are excited into Rydberg levels with large principal quantum numbers n either by stepwise excitation with two diode lasers, or by a single frequency-doubled dye laser. The spontaneous lifetime $\tau(n)$ increases with n^3 and is, for sufficiently large values of n, longer than the transit time of the atoms from the excitation point to the detection point. When the excited atoms pass through a microwave resonator that is tuned into resonance with a Rydberg transition frequency $v = (E_n - E_{n-1})/h$, a fluorescence photon emitted by a Rydberg atom during its passage through the resonator may excite a cavity mode [14.128].

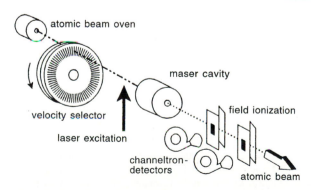

Fig. 14.48. Schematic experimental setup for the "one-atom maser" with resonance cavity and state-selective detection of the Rydberg atoms [14.127]

If the cavity resonator is cooled down to temperatures of a few Kelvin, the cavity walls become superconducting and its losses decrease drastically. Cavity Q-values of larger than 5×10^{10} can be achieved, which corresponds to a decay time of an excited mode of $T_R > 1\,s$ at a resonance frequency of $\nu = 21\,GHz$. This decay time is much longer than the transit time of the atom through the cavity. If the density of atoms in the beam is decreased, one can reach the condition that only a single atom is present in the cavity resonator during the transit time $T = d/c \simeq 100\,\mu s$. This allows investigations of the interaction of a single atom with the EM field in the cavity (Fig. 14.49). The resonance frequency of the cavity can be tuned continuously within certain limits by mechanical deformation of the cavity walls [14.129].

Because of its large transition dipole moment $D_{n,n-1} \propto n^2$ (!) the atom that emitted a fluorescence photon exciting the cavity mode may again absorb a photon from the EM field of the cavity and return to its initial state $|n\rangle$. This can be detected when the Rydberg atom passes behind the cavity through two static electric fields (Fig. 14.48). The field strength in the first field is adjusted to field-ionize Rydberg atoms in level $|n\rangle$ but not in $|n-1\rangle$, while the second, slightly stronger field also ionizes atoms in level $|n-1\rangle$. This allows one to decide in which of the two levels the Rydberg atoms have left the microwave resonator.

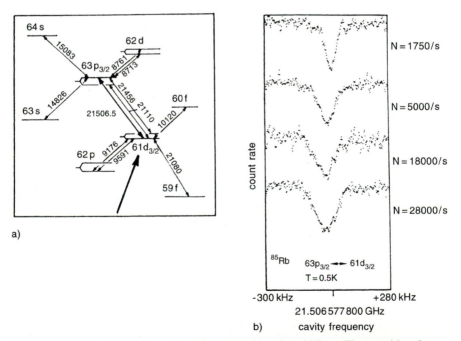

Fig. 14.49. (a) Level diagram of the maser transition in rubidium. The transition frequencies are given in MHz. **(b)** Measured ion signal, which is proportional to the number of Rydberg atoms in level $|n\rangle$, as a function of the cavity resonance frequency. Maser operation of the one-atom maser manifests itself in a decrease in the number of atoms in level $|n\rangle$ [14.130]

It turns out that the spontaneous lifetime τ_n of the Rydberg levels is shortened if the cavity is tuned into resonance with the frequency ω_0 of the atomic transition $|n\rangle \rightarrow |n-1\rangle$. It is prolonged if no cavity mode matches ω_0 [14.130]. This effect, which had been predicted by quantum electrodynamics, can intuitively be understood as follows: in the resonant case, that part of the thermal radiation field that is in the resonant cavity mode can contribute to stimulated emission in the transition $|n\rangle \rightarrow |n-1\rangle$, resulting in a shortening of the lifetime (Sect. 11.3). For the detuned cavity the fluorescence photon does not "fit" into the resonator. The boundary conditions impede the emission of the photon. If the rate dN/dt of atoms in the level $|n\rangle$ behind the resonator is measured as a function of the cavity resonance frequency ω, a minimum rate is measured for the resonance case $\omega = \omega_0$ (Fig. 14.49).

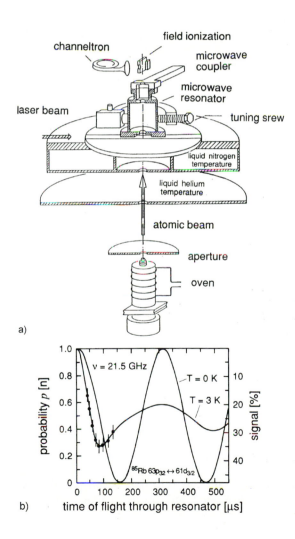

a)

b)

Fig. 14.50. (a) Schematic diagram of the cooled cavity; **(b)** calculated Rabi oscillation of an atom in the cavity field at $T = 0\,\mathrm{K}$ and at $T = 3\,\mathrm{K}$, damped by the statistical influence of the thermal radiation field. The experimental points are measured with a velocity selector that allowed transit times of $30-140\,\mu\mathrm{s}$ [14.132]

Without a thermal radiation field in the resonantly tuned cavity, the populations $N(n, T)$ and $N(n-1, T)$ of the Rydberg levels should be a periodic function of the transit time $T = d/v$, with a period T_R that corresponds to the Rabi oscillation period. The incoherent thermal radiation field causes induced emission and absorption with statistically distributed phases. This leads to a damping of the Rabi oscillation (Fig. 14.50). This effect can be proved experimentally if the atoms pass a velocity selector before they enter the resonator, which allows a continuous variation of the velocity and therefore of the transit time $T = d/v$.

The one-atom maser can be used to investigate the statistical properties of nonclassical light [14.131, 14.132]. If the cavity resonator is cooled down to $T \leq 0.5$ K, the number of thermal photons becomes very small and can be neglected. The number of photons induced by the atomic fluorescence can be measured via the fluctuations in the number of atoms leaving the cavity in the lower level $|n-1\rangle$. It turns out that the statistical distribution does not follow Poisson statistics, as in the output of a laser with many photons per mode, but shows a sub-Poisson distribution with photon number fluctuations 70% below the vacuum-state limit [14.133]. In cavities with low losses, pure photon number states of the radiation field (Fock states) can be observed [14.134], with photon lifetimes as high as 0.2 s!

Another interesting effect that is now accessible to measurements is the quantum electrodynamic energy shift of atomic levels when the atoms pass between two parallel metal plates (Casimir–Polder effect) [14.135].

14.6 Spectral Resolution Within the Natural Linewidth

Assume that all other line-broadening effects except the natural linewidth have been eliminated by one of the methods discussed in the previous chapters. The question that arises is whether the natural linewidth represents an insurmountable natural limit to spectral resolution. At first, it might seem that Heisenberg's uncertainty relation does not allow outwit the natural linewidth (Sect. 3.1). In order to demonstrate that this is not true, in this section we give some examples of techniques that do allow observation of structures *within* the natural linewidth. It is, however, not obvious that all of these methods may really increase the amount of information about the molecular structure, since the inevitable loss in intensity may outweigh the gain in resolution. We discuss under what conditions spectroscopy within the natural linewidth may be a tool that really helps to improve the quality of spectral information.

14.6.1 Time-Gated Coherent Spectroscopy

The first of these techniques is based on the selective detection of those excited atoms or molecules that have survived in the excited state for times $t \gg \tau$, which long compared to the natural lifetime τ.

If molecules are excited into an upper level with the spontaneous lifetime $\tau = 1/\gamma$ by a light pulse ending at $t = 0$, the time-resolved fluorescence amplitude is given by

$$A(t) = A(0)\,\mathrm{e}^{-(\gamma/2)t}\cos\omega_0 t\,. \tag{14.63}$$

If the observation time extends from $t = 0$ to $t = \infty$, a Fourier transformation of the measured intensity $I(t) \propto A^2(t)$ yields, for the line profile of the fluorescence emitted by atoms at rest, the Lorentzian profile (Sect. 3.1)

$$I(\omega) = \frac{I_0}{(\omega - \omega_0)^2 + (\gamma/2)^2}\,, \quad \text{with} \quad I_0 = \frac{\gamma}{2\pi}\int I(\omega)\,\mathrm{d}\omega\,. \tag{14.64}$$

If the detection probability for $I(t)$ is not constant, but follows the time dependence $f(t)$, the detected intensity $I_g(t)$ is determined by the *gate function* $f(t)$

$$I_g(t) = I(t)\,f(t)\,.$$

The Fourier transform of $I_g(t)$ now depends on the form of $f(t)$ and may no longer be a Lorentzian. Assume the detection is gated by a step function

$$f(t) = \begin{Bmatrix} 0 & \text{for } t < T \\ 1 & \text{for } t \geq T \end{Bmatrix}\,,$$

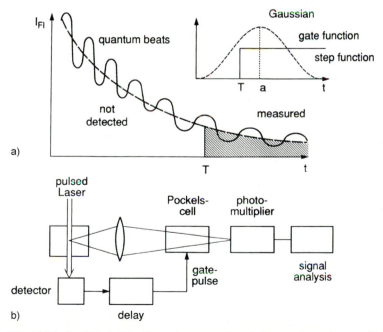

Fig. 14.51a,b. Gated detection of exponential fluorescence decay with a gate function $f(t)$: (**a**) schematic scheme; (**b**) experimental realization

which may simply be realized by a shutter in front of the detector that opens only for times $t \geq T$ (Fig. 14.51).

The Fourier transform of the amplitude $A(t)$ in (14.63) is now

$$A(\omega) = \int_T^\infty A(0)\, e^{-\gamma/2t} \cos(\omega_0 t)\, e^{-i\omega t}\, dt \;,$$

which yields in the approximation $\Omega = |\omega - \omega_0| \ll \omega_0$, with $\exp(-i\omega t) = \cos(\omega t) - i\sin(\omega t)$, the cosine and sine Fourier transforms:

$$A_c(\omega) = \frac{A_0}{2} \frac{e^{-(\gamma/2)T}}{\Omega^2 + (\gamma/2)^2} \left[\frac{\gamma}{2} \cos(\Omega T) - \Omega \sin(\Omega T) \right] \;,$$

$$A_s(\omega) = \frac{A_0}{2} \frac{e^{-(\gamma/2)T}}{\Omega^2 + (\gamma/2)^2} \left[\frac{\gamma}{2} \sin(\Omega T) + \Omega \cos(\Omega T) \right] \;. \tag{14.65}$$

If only the incoherent fluorescence intensity is observed without obtaining any information on the phases of the wave function of the excited state, we observe the intensity

$$I(\omega) \propto |A_c(\omega) - iA_s(\omega)|^2 = A_c^2 + A_s^2 = \frac{A_0^2}{2} \frac{e^{-\gamma T}}{\Omega^2 + (\gamma/2)^2} \;. \tag{14.66}$$

This is again a Lorentzian with a halfwidth (FWHM) of $\gamma = 1/\tau$, which is independent of the gate time T! Because of the delayed detection one loses the factor $\exp(-T/\tau)$ in intensity, since all fluorescence events occurring for times $t < T$ are missing. *This shows that with incoherent techniques no narrowing of the natural linewidth can be achieved, even if only fluorescence photons from selected long-living atoms with $t > T \gg \tau$ are selected* [14.136].

The situation changes, however, if instead of the intensity (14.66), the amplitude (14.63) or an intensity representing a coherent superposition of amplitudes can be measured, where the phase information and its development in time is preserved. Such measurements are possible with one of the coherent techniques discussed in Chap. 12.

One example is the quantum-beat technique. According to (12.27a) the fluorescence signal at the time t after the coherent excitation at $t = 0$ of two closely spaced levels separated by $\Delta\omega$, with equal decay times $\tau = 1/\gamma$, is

$$I(t) = I(0)\, e^{-\gamma t} (1 + a \cos \Delta\omega t) \;. \tag{14.67}$$

The term $\cos(\Delta\omega t)$ contains the wanted information on the phase difference

$$\Delta\varphi(t) = \Delta\omega t = (E_i - E_k)t/\hbar \;,$$

between the wave functions $\psi_n(t) = \psi_n(0)\, e^{-iE_n t/\hbar}$ $(n = i, k)$ of the two levels $|i\rangle$ and $|k\rangle$.

If the detector is gated to receive fluorescence only for $t > T$, the Fourier transform of (14.67) becomes

$$I(\omega) = \int_{T}^{\infty} I(0) \cdot e^{-\gamma t}(1 + a \cos \Delta\omega t)\, e^{-i\omega t}\, dt\ , \tag{14.68}$$

where $\omega \gg \Delta\omega$ is the mean light frequency of the fluorescence. The evaluation of the integral yields the cosine Fourier transform

$$I_c(\omega) = \frac{I_0}{2} \frac{e^{-\gamma T}}{(\Delta\omega - \omega)^2 + \gamma^2}\, [\gamma \cos(\Delta\omega - \omega)T - (\Delta\omega - \omega)\sin(\Delta\omega - \omega)T]\ . \tag{14.69}$$

For $T > 0$ the intensity $I_c(\omega)$ exhibits an oscillatory structure (Fig. 14.52) with a central maximum at $\omega \simeq \Delta\omega$ (because of the first two terms in (14.68) the center is not exactly at $\omega = \Delta\omega$) and the halfwidth

$$\Delta\omega_{12} = \frac{2\gamma}{\sqrt{1 + \gamma^2 T^2}}\ . \tag{14.70}$$

For $T = 0$ the width of the quantum beat signal is the sum of the level widths of the two coherently excited levels, contributing to the beat signal.

Example 14.14
For $T = 5\tau = 5/\gamma$ the *halfwidth* of the central peak has decreased from γ to 0.4γ. The peak intensity, however, has drastically decreased by the factor $\exp(-\gamma T) = \exp(-5) \simeq 10^{-2}$ to less than 1% of its value for $T = 0$.

The decrease of the peak intensity results in a severe decrease of the signal-to-noise ratio. This may, in turn, lead to a larger uncertainty in determining the line center.

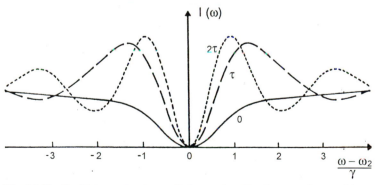

Fig. 14.52. Oscillatory structure of the cosine Fourier transform with narrowed central maximum for increasing values of the gate delay time $T = 0$, τ, and 2τ. The peak intensity has been normalized

The oscillatory structure can be avoided if the gate function $f(t)$ is not a step function but a Gaussian $f(t) = \exp[-(t-T)^2/b^2]$ with $b = (2T/\gamma)^{1/2}$ [14.137].

Another coherent technique that can be combined with gated detection is level-crossing spectroscopy (Sect. 12.1). If the upper atomic levels are excited by a pulsed laser and the fluorescence intensity $I_{\text{Fl}}(B, t \geq \tau)$ as a function of the magnetic field is observed with increasing delay times T for the opening of the detector gate, the central maximum of the Hanle signal becomes narrower with increasing T. This is illustrated by Fig. 14.53, which shows a comparison of measured and calculated line profiles for different gate-delay times T for Hanle measurements of the $\text{Ba}(6s6p\,^1P_1)$ level [14.138]. Similar measurements have been performed for the Na(3P) level [14.139].

It should be emphasized that line narrowing is only observed if the time development of the phase of the upper-state wave function can be measured. This is the case for all methods that utilize interference effects caused by the superposition of different spectral components of the fluorescence. Therefore, an interferometer with a spectral resolution better than the natural linewidth can be used, too. However, a narrowing of the observed fluorescence linewidth

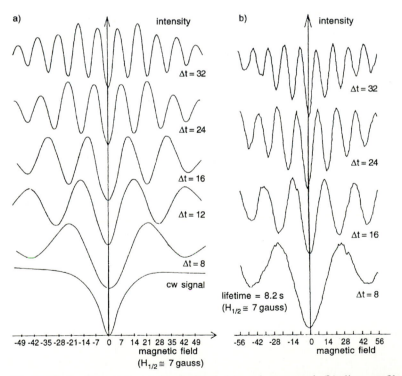

Fig. 14.53a,b. Comparison of calculated (**a**) and measured (**b**) line profiles of level-crossing signals under observation with different gate-delay times. The different curves have been renormalized to equal centered peak intensities [14.139]

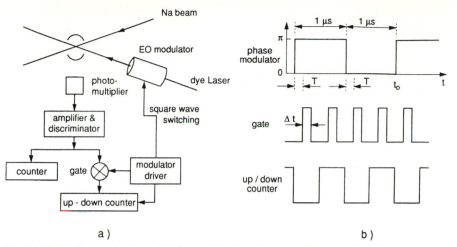

Fig. 14.54a,b. Laser phase modulation and gating time sequence for obtaining subnatural linewidths under cw excitation [14.140]

with increasing gate-delay time T is only observed if the gating device is placed between the interferometer and the detector; it is *not* observed if the gate is placed between the emitting source and the interferometer [14.136].

Instead of a gated switch in the fluorescence detector for pulsed excitation, one may also use excitation by a cw laser that is phase-modulated with a modulation amplitude of π (Fig. 14.54). The fluorescence generated under sub-Doppler excitation in a collimated atomic beam is observed during a short time interval Δt, which is shifted by the variable delay T against the time t_0 of the phase jump. If the fluorescence intensity $I_{Fl}(\omega, T)$ is monitored as a function of the laser frequency ω, a narrowing of the line profile is found with increasing T [14.140].

14.6.2 Coherence and Transit Narrowing

For transitions between two atomic levels $|a\rangle$ and $|b\rangle$ with the lifetimes $\tau_a \ll \tau_b$, a combination of level-crossing spectroscopy with saturation effects allows a spectral resolution that corresponds to the width $\gamma_b \ll \gamma_a$ of the longer-living level $|b\rangle$, although the natural linewidth of the transition should be $\gamma_{ab} = (\gamma_a + \gamma_b)$. The method was demonstrated by Bertucelli et al. [14.141] for the example of the $Ca(^3P_1 - {}^3S_1)$ transition. Calcium atoms in the metastable $^3P_1$ level in a collimated atomic beam pass through the expanded laser beam in a homogeneous magnetic field $\boldsymbol{B}$. The electric vector $\boldsymbol{E}$ of the laser beam is perpendicular to $\boldsymbol{B}$. Therefore transitions with $\Delta M = \pm 1$ are induced. For $B = 0$ all M-sublevels are degenerate and the saturation of the optical transition is determined by the square of the sum $\sum D_{M_a M_b}$ of the transition matrix elements $D_{M_a M_b}$ between the allowed Zeeman components $|a, M_a\rangle$ and $|b, M_b\rangle$. If the Zeeman splitting for $B \neq 0$ becomes larger

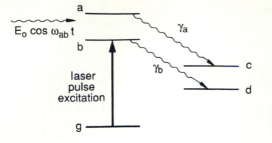

Fig. 14.55. Level scheme for transient line narrowing [14.142]

than the natural linewidth of the metastable levels in the ^3P state, the different Zeeman components saturate separately, and the saturation is proportional to $\sum |D_{M_a M_b}|^2$ instead of $|\sum D_{M_a M_b}|^2$ for the degenerate case. Measuring the intensity of the laser-induced fluorescence $I_{Fl}(B)$ as a function of the magnetic field B, one obtains a Hanle signal that depends nonlinearly on the laser intensity and that has a halfwidth $\gamma_B(I_L) = \gamma_B \times \sqrt{1+S}$, where $S = I_L/I_S$ is the saturation parameter (Sect. 7.2).

Another technique of subnatural linewidth spectroscopy is based on transient effects during the interaction of a two-level system with the monochromatic wave of a cw laser. Assume that the system is irradiated by the cw monochromatic wave $E = E_0 \cos \omega t$, which can be tuned into resonance with the energy separation $[\omega_{ab} = (E_a - E_b)/\hbar]$ of the two levels $|a\rangle$ and $|b\rangle$ with decay constants γ_a, γ_b (Fig. 14.55).

If the level $|b\rangle$ is populated at the time $t = 0$ by a short laser pulse one can calculate the probability $P_a(\Delta, t)$ to find the system at time t in the level $|a\rangle$ as a function of the detuning $\Delta = \omega - \omega_{ab}$ of the cw laser. Using time-dependent perturbation theory one gets [14.142, 14.143]:

$$\mathcal{P}(\Delta, t) = \left(\frac{D_{ab} E_0}{\hbar} \right)^2 \frac{e^{-\gamma_a t} + e^{-\gamma_b t} - 2\cos(t \cdot \Delta) e^{-\gamma_{ab} t}}{\Delta^2 + (\delta_{ab}/2)^2} , \qquad (14.71)$$

where the Lorentzian factor contains the *difference* $\delta_{ab} = (\gamma_a - \gamma_b)/2$ of the level widths instead of the sum $\gamma_{ab} = (\gamma_a + \gamma_b)$.

If the fluorescence emitted from $|a\rangle$ is observed only for times $t \geq T$, one has to integrate (14.71). This yields the signal

$$S(\Delta, T) \sim \gamma_a \int_T^\infty \mathcal{P}(\Delta, t)\, dt = \frac{\gamma_a (D_{ab} E_0/\hbar)^2}{\Delta^2 + (\delta_{ab}/2)^2}$$

$$\times \left\{ \frac{e^{-\gamma_a T}}{\gamma_a} + \frac{e^{-\gamma_b}}{\gamma_b} + \frac{2e^{-\gamma_{ab} T}}{\Delta^2 + (\gamma_{ab}/2)^2} [\Delta \sin(\Delta \cdot T) - \gamma_{ab} \cos(\Delta \cdot T)] \right\} ,$$

$$(14.72)$$

which represents an oscillatory structure with a narrowed central peak. For $T \to 0$ this again becomes a Lorentzian

$$S(\Delta, T = 0) = \frac{\gamma_{ab}}{\gamma_b} \frac{(D_{ab} E_0/\hbar)^2}{\Delta^2 + (\gamma_{ab}/2)^2} \cdot \tag{14.73}$$

14.6.3 Raman Spectroscopy with Subnatural Linewidth

A very interesting method for achieving subnatural linewidths of optical transitions is based on induced resonant Raman spectroscopy, which may be regarded as a special case of optical–optical double resonance (Sect. 10.4). The pump laser L1 is kept on the molecular transition $|1\rangle \to |2\rangle$ (Fig. 14.56), while the tunable probe laser L2 induces downward transitions. A double-resonance signal is obtained if the frequency ω_s of the probe laser matches the transition $|2\rangle \to |3\rangle$. This signal can be detected by monitoring the change in absorption or polarization of the transmitted probe laser.

When both laser waves with the wave vectors k_p and k_s are sent collinearly through the sample, energy conservation for the absorption of a photon $\hbar\omega_p$ and emission of a photon $\hbar\omega_s$ by a molecule moving with the velocity v demands

$$(\omega_p - k_p \cdot v) - (\omega_s - k_s \cdot v) = (\omega_{12} - \omega_{23}) \pm (\gamma_1 + \gamma_3), \tag{14.74}$$

where γ_i is the homogeneous width of level $|i\rangle$. The quadratic Doppler effect and the photon recoil have been neglected in (14.74). Their inclusion would, however, not change the argument.

Integration over the velocity distribution $N(v_z)$ of the absorbing molecules yields with (14.74) the width γ_s of the double-resonance signal [14.144]

$$\gamma_s = \gamma_3 + \gamma_1 (\omega_s/\omega_p) + \gamma_2 (1 \mp \omega_s/\omega_p), \tag{14.75}$$

where the minus sign holds for copropagating, the plus sign for counter-propagating laser beams. If $|1\rangle$ and $|3\rangle$ denote vibrational–rotational levels

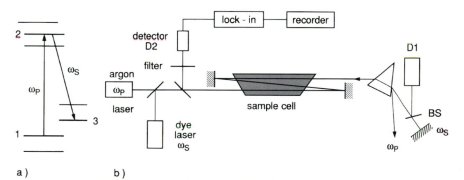

a)   b)

Fig. 14.56a,b. Stimulated resonant Raman transition spectroscopy with subnatural linewidth resolution: (a) level scheme; (b) experimental arrangement

in the electronic ground state of homonuclear diatomic molecules, their radiative lifetimes are very small and γ_1, γ_2 are mainly limited at higher pressures by collision broadening and at low pressures by transit-time broadening (Sect. 3.4). For $\omega_s \simeq \omega_p$ the contribution of the level width γ_2 to γ_s becomes very small for copropagating beams and the halfwidth γ_s of the signal may become much smaller than the natural linewidths of the transitions $|1\rangle \to |2\rangle$ or $|2\rangle \to |3\rangle$.

Of course, one cannot get accurate information on the level width γ_2 from (14.75) unless all other contributions to γ_s are sufficiently well known. The width γ_s is comparable to the width of the direct transition $|1\rangle \to |3\rangle$. However, if $|1\rangle \to |2\rangle$ and $|2\rangle \to |3\rangle$ are allowed electric dipole transitions, the direct transition $|1\rangle \to |3\rangle$ is dipole-forbidden.

Figure 14.56 exhibits the experimental arrangement used by Ezekiel et al. [14.145] to measure subnatural linewidths in I_2 vapor. An intensity-modulated pump beam and a continuous dye laser probe pass the iodine cell three times. The transmitted probe beam yields the forward scattering signal (detector D1) and the reflected beam the backward scattering (detector D2). A lock-in amplifier monitors the change in transmission from the pump beam, which yields the probe signal with width γ_s. Ezekiel et al. achieved a linewidth as narrow as 80 kHz on a transition with the natural linewidth $\gamma_n = 141$ kHz. The theoretical limit of $\gamma_s = 16.5$ kHz could not be reached because of dye laser frequency jitter. This extremely high resolution can be used to measure collision broad-

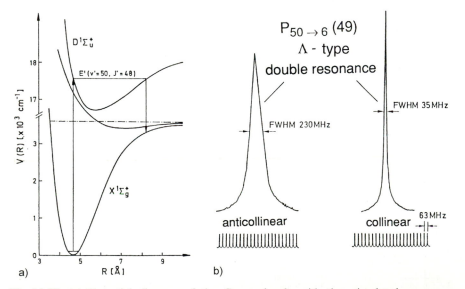

Fig. 14.57. (a) Potential diagram of the Cs_2 molecule with the stimulated resonance Raman transitions to high-lying vibration–rotation levels of the $X\,^1\Sigma_g^+$ ground state. (b) Comparison of linewidths of a Doppler-free transition $|1\rangle \to |2\rangle$ to a predissociating upper level $|2\rangle$ with an effective lifetime $\tau_{eff} \simeq 800$ ps and the much narrower linewidths of the Raman signal $S(\omega_s)$ [14.146]

ening at low pressures caused by long-distance collisions (Sect. 13.1) and to determine the collisional relaxation rates of the levels involved.

The small signal width of the OODR signal γ_s becomes essential if the levels $|3\rangle$ are high-lying rotation–vibration levels of heavy molecules just below the dissociation limit, where the level density may become very high. As an example, Fig. 14.57 shows such an OODR signal of a rotation–vibration transition $D\,^1\Sigma_u(v'=50, J'=48) \to X\,^2\Sigma_g(v''=125, J''=49)$ in the Cs_2 molecule. A comparison with a Doppler-free polarization signal on a transition $|1\rangle \to |2\rangle$ to the upper predissociating level $|2\rangle = D\,^1\Sigma_u(v'=50, J'=48)$, which has an effective lifetime of $\tau_{eff} \simeq 800$ ps because of predissociation, shows that the OODR signal is not influenced by γ_2 [14.146].

14.7 Absolute Optical Frequency Measurement and Optical Frequency Standards

In general, frequencies can be measured more accurately than wavelengths because of the deviation of light waves from the ideal plane wave caused by diffraction and local inhomogenities of the refractive index. This leads to local deviations of the phase fronts from ideal planes and therefore to uncertainties in measurements of the wavelength λ, which is defined as the distance between two phase fronts differing by 2π. While the wavelength $\lambda = c/v = c_0/(nv)$ of an EM wave depends on the refractive index n, the frequency v is independent of n.

In order to achieve the utmost accuracy for the determination of physical quantities that are related to the wavelength λ or frequency v of atomic transitions, it is desirable to measure the optical frequency v instead of the wavelength λ. From the relation $\lambda_0 = c_0/v$ the vacuum wavelength λ_0 can then be derived, since the speed of light in vacuum

$$c_0 \underset{\text{Def}}{=} 229\,792\,458 \text{ m/s} \,,$$

is now *defined* as a fixed value, taken from a weighted average of the most accurate measurements [14.147, 14.148]. In this section we will learn about some experimental techniques to measure frequencies of EM waves in the infrared and visible range.

14.7.1 Microwave–Optical Frequency Chains

With fast electronic counters, frequencies up to a few gigahertz can be measured directly and compared with a calibrated frequency standard, derived from the cesium clock, which is still the primary frequency standard [14.149]. For higher frequencies a heterodyne technique is used, where the unknown frequency v_x is mixed with an appropriate multiple mv_R of the reference fre-

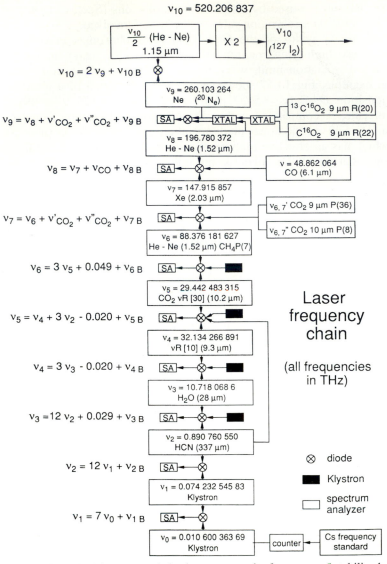

Fig. 14.58. Optical frequency chain that connects the frequency of stabilized optical lasers with the Cs frequency standard [14.150]

quency ν_R ($m = 1, 2, 3, ...$). The integer m is chosen such that the difference frequency $\Delta\nu = \nu_x - m\nu_R$ at the output of the mixer lies in a frequency range that can be directly counted.

This frequency mixing in suitable nonlinear mixing elements is the basis for building up a frequency chain from the Cs atomic beam frequency standard to the optical frequency of visible lasers. The optimum choice for the mixer depends on the spectral range covered by the mixed frequencies. When the output beams of two infrared lasers with known frequencies ν_1 and ν_2 are focused together with the superimposed radiation from the unknown frequency ν_x of another laser, the frequency spectrum of the detector output contains the frequencies

$$\nu = \pm\nu_x \pm m\nu_1 \pm n\nu_2 \quad (m, n : \text{integers}), \tag{14.76}$$

which can be measured with an electronic spectrum analyser if they are within its frequency range.

In the National Institute of Standards and Technology, NIST (the former National Bureau of Standards, NBS) the frequency chain, shown in Fig. 14.58, has been built [14.150], which starts at the frequency of the hyperfine transition in cesium at $\nu = 9.192\,631\,770\,0$ GHz, currently accepted now as the primary frequency standard. A klystron is frequency-offset locked (Sect. 7.3.4) to the Cs clock and stabilized with the accuracy of this clock at the frequency $\nu_0 = 10.600\,363\,690$ GHz. The seventh harmonic of ν_0 is mixed with the frequency ν_1 of another clystron in a nonlinear diode, and ν_1 is again frequency-offset locked to the frequency $\nu_1 = 7\nu_0 + \nu_{1B}$, where ν_{1B} is provided by a RF generator and can be directly counted. The 12th harmonic of ν_1 is then mixed with the radiation of a HCN laser at $\nu = 890$ GHz ($\lambda = 337$ μm). The 12th overtone of this radiation is mixed with the radiation of the H_2O laser at 10.7 THz ($\lambda = 28$ μm), etc. The frequency chain goes up to the frequencies of visible HeNe lasers stabilized to HFS components of a visible transition in the I_2 molecule (Chap. 5). MIM (metal–insulator–metal) mixing diodes (Fig. 5.118) consisting of a thin tungsten wire with a sharp-edged tip, which contacts a nickel surface coated by a thin oxyde layer [14.151], allow the mixing of frequencies above 1 THz [14.152]. Also, Schottky diodes have successfully been used for mixing optical frequencies with difference frequencies up to 900 GHz [14.153]. The physical processes relevant in such fast Schottky diodes have been understood only recently [14.154].

In addition, other frequency chains have been developed that start from stabilized CO_2 lasers locked to the cesium frequency standard in a similar way, but then use infrared color-center lasers to bridge the gap to the I_2-stabilized HeNe laser [14.155a].

A very interesting proposal for a frequency chain based on transition frequencies of the hydrogen atom has been made by Hänsch [14.155b]. The basic idea is illustrated in Fig. 14.59. Two lasers are stabilized onto two transitions of the hydrogen atoms at the frequencies f_1 and f_2. Their output beams are superimposed and focused into a nonlinear crystal, forming the sum frequency $f_1 + f_2$. The second harmonics of a laser oscillating at the frequency f_3 are phase-locked to the frequency $f_1 + f_2$. This stabilizes the frequency f_3 by a servo loop to the value $(f_2 + f_1)/2$. The original frequency difference $f_1 - f_2$ is therefore halved to $f_1 - f_3 = (f_1 - f_2)/2$. By connecting several such stages in cascade until the frequency difference becomes measurable by a counter, the difference $f_1 - f_2$ can be directly connected to the cesium clock. If a laser frequency f and its second harmonic are used as the starting frequencies, the beat signal at $f/2^n$ can be observed after n

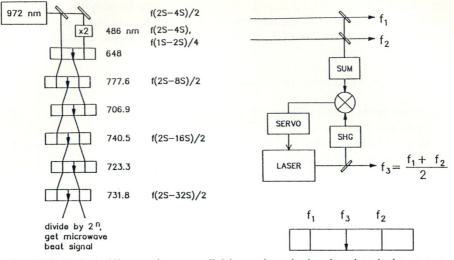

Fig. 14.59. Optical difference frequency-division and synthesizer based on hydrogen transition frequencies [14.155b]

stages. In order to link the $2s - 4s$ hydrogen transition at $\lambda = 486$ nm to the 9-GHz cesium frequency standard by this method, 16 stages would be required [14.155c].

14.7.2 Frequency Comb from Femtosecond Laser Pulses

Recently, a new technique has been developed [14.156] that allows the direct comparison of widely different reference frequencies and thus considerably simplifies the frequency chain from the cesium clock to optical frequencies by reducing it to a single step. Its basic principle can be understood as follows (Fig. 14.60):

The frequency spectrum of a mode-locked continuous laser emitting a regular train of short pulses consists of a comb of equally spaced frequency components (the modes of the laser resonator). The spectral width of this comb spectrum depends on the temporal width of the laser pulses (Fourier theorem). With femtosecond pulses the comb spectrum extends over more than 30 THz. The spectral width can be further increased by focusing the laser pulses into an optical fiber, where by self-phase modulation the spectrum is considerably broadened and extends over one decade (e.g., from 1064 nm to 532 nm). This corresponds to a frequency span of 300 THz [14.157]!

It turns out that the spectral modes of the comb are precisely equidistant, even in the far wings of the comb. This is also true when the spectrum is broadened by self-phase modulation. One example of frequency comparison over a broad spectral range is provided by the experimental setup in Fig. 14.60.

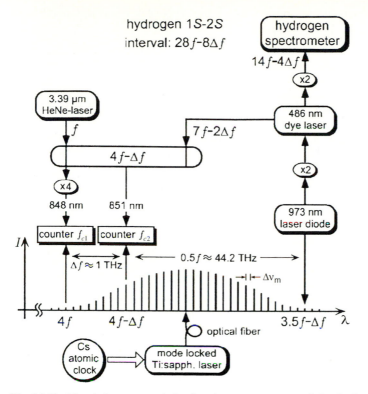

Fig. 14.60. Novel frequency chain for the measurement of the hydrogen $1S$–$2S$ interval. The frequency comb of a mode-locked laser is used to measure two large optical frequency differences [14.157]

The mode spacing $\Delta\nu_m$ is locked to the cesium clock frequency ν_{Cs} in such a way that $\nu_{Cs} = m \cdot \Delta\nu_m$. Therefore the frequency difference $N \cdot \Delta\nu_m$ between N comb modes is precisely known. The fourth harmonics of a stabilized HeNe laser at $\lambda = 3.39\,\mu m$ is now compared with the frequency of a mode of the frequency comb. The beat frequency f_{c1} between $4f_{HeNe}$ and the mode frequency f_1 is measured by a frequency counter. A dye laser at 486 nm is frequency doubled and excites the two-photon transition $1S$–$2S$ of the hydrogen atom. Its frequency is locked to the doubled frequency of a diode laser, which is in turn locked to a mode of the frequency comb. The dye laser frequency is not exactly 7 times the HeNe laser frequency f, but differs from $7f$ by $-2\Delta f$. A frequency divider chain (see Fig. 14.59) generates from f and $7f - 2\Delta f$ the frequency $4f - \Delta f$, which is just half of the sum $f + 7f - 2\Delta f$. This frequency, which corresponds to a wavelength of 851 nm, is compared with that of a mode of the frequency comb. The difference frequency f_{c2} is measured by a counter.

The final frequency of the hydrogen transition at $28f - 8\Delta f$ can now be determined by the two radiofrequencies f_{c1}, f_{c2} and the number of modes between the two selected modes onto which the HeNe laser and the diode laser are stabilized. This yields the relation

$$f_{1S-2S} = -8 f_{c1} - 64 f_{c2} + 2466.06384 \, \text{THz} \,,$$

where the last term gives the frequency difference between the two selected modes.

With this technique a relative uncertainty of the absolute frequency determination of less than 10^{-13} is possible. This means an accuracy of the $1S - S$ transition frequency at $2.4 \times 10^{15} \, \text{s}^{-1}$ of better than 300 Hz!

14.8 Squeezing

For very low light intensities the quantum structure of light becomes evident by statistical fluctuations of the number of detected photons, which lead to corresponding fluctuations of the measured photoelectron rate (Sect. 12.6). This *photon noise*, which is proportional to $\sqrt{N}$ at a measured rate of N photoelectrons per second, imposes a principal detection limit for experiments with low-level light detection [14.158]. Additionally, the frequency stabilization of lasers on the millihertz scale is limited by photon noise of the detector that activates the electronic feedback loop [14.159].

It is therefore desirable to further decrease the photon-noise limit. At first, this seems to be impossible because the limit is of principal nature. However, it has been shown that under certain conditions the photon-noise limit can be overcome without violating general physical laws. We will discuss this in some more detail, partly following the representation given in [14.160, 14.161].

Example 14.15

The shot-noise limit of an optical detector with the quantum efficiency $\eta < 1$ irradiated by N photons per second leads to a minimum relative fluctuation $\Delta S/S$ of the detector signal S, which is for a detection bandwidth Δf given by

$$\frac{\Delta S}{S} = \frac{\sqrt{N \eta \Delta f}}{N \eta} = \sqrt{\frac{\Delta f}{\eta N}} \,. \tag{14.77}$$

For a radiation power of 100 mW at $\lambda = 600 \, \mu\text{m} \rightarrow N = 3 \times 10^{17} \, \text{s}^{-1}$. With a bandwidth of $\Delta f = 100 \, \text{Hz}$ (time constant $\simeq 10 \, \text{ms}$) and $\eta = 0.2$, the minimum fluctuation is $\Delta S/S \geq 4 \times 10^{-8}$.

14.8.1 Amplitude and Phase Fluctuations of a Light Wave

The electric field of a single-mode laser wave can be represented by

$$E_L(t) = E_0(t) \cos[\omega_L t + k_L r + \phi(t)]$$
$$= E_1(t) \cos(\omega_L t + k_L r) + E_2(t) \sin(\omega_L t + k_L r) , \qquad (14.78)$$

with $\tan\phi = E_2/E_1$.

Even a well-stabilized laser, where all "technical noise" has been elimi-
nated (Sect. 5.6) still shows small fluctuations ΔE_0 of its amplitude and $\Delta\phi$ of
its phase, because of quantum fluctuations (Sect. 5.6). While technical fluctuations may be at least partly eliminated by difference detection (Fig. 5.44), this
is not possible with classical means for photon noise caused by uncorrelated
quantum fluctuations.

These fluctuations are illustrated in Fig. 14.61 in two different ways: the
time-dependent electric field $E(t)$ and its mean fluctuations of the amplitude E_0 and phase ϕ are shown in an $E(t)$ diagram and in a polar phase
diagram with the axes E_1 and E_2. In the latter, amplitude fluctuations cause an
uncertainty of the radius $r = |E_0|$, whereas phase fluctuations cause an uncer-
tainty of the phase angle ϕ (Fig. 14.61b). Because of Heisenberg's uncertainty
relation it is not possible that both uncertainties of amplitude and phase become simultaneously zero.

In order to gain a deeper insight into the nature of these quantum fluctuations, let us regard them from a different point of view: the EM field
of a well-stabilized single-mode laser can be described by a coherent state
(called a *Glauber state* [14.162])

$$\langle\alpha_k| = \exp(-|\alpha_k|^2/2) \sum_{N=0}^{\infty} \frac{\alpha_k^N}{\sqrt{N!}} |N_k\rangle , \qquad (14.79)$$

which is a linear combination of states with photon occupation numbers N_k.
The probability of finding N_k photons in this state is given by a Pois-

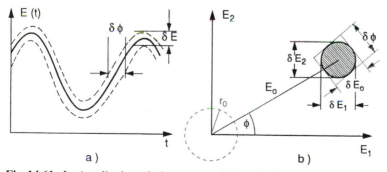

Fig. 14.61a,b. Amplitude and phase uncertainties of a laser wave shown in an amplitude–
time diagram (**a**) and in a polar phase diagram (**b**)

son distribution around the mean value $N = \langle N_k \rangle = \alpha_k^2$ with the width $\langle \Delta N_k \rangle = \sqrt{\langle N \rangle} = |\alpha_k|$.

Although the laser field is concentrated into a single mode, all other modes of the vacuum with different frequencies ω and wave vectors $\mathbf{k}$ still have an average occupation number of $N = 1/2$, corresponding to the zero-point energy $\hbar\omega/2$ of a harmonic oscillator. When the laser radiation falls onto a photodetector, all these other modes are also present (*vacuum zero-field fluctuations*) and beat signals are generated at the difference frequencies. The magnitudes of these beat signals are proportional to the product of the amplitudes of the two interfering waves. Since the amplitude of the laser mode is proportional to $\sqrt{N}$, and that of the vacuum modes to $\sqrt{1/2}$, the magnitude of the beat signals is proportional to $\sqrt{N/2}$. Adding up all beat signals in the frequency interval Δf of the detector bandwidth gives the shot noise treated in Example 14.14. In this model the shot noise is regarded as the beat signal between the occupied laser mode and all other vacuum modes. This view will be important for the understanding of interferometric devices used for squeezing experiments.

If the field amplitude E in (14.78) is normalized in such a way that

$$\langle E^2 \rangle = \langle E_1^2 \rangle + \langle E_2^2 \rangle = \frac{\hbar\omega}{2\epsilon_0 V} , \tag{14.80}$$

where V is the mode volume, and ϵ_0 the dielectric constant, the uncertainty relation can be written as [14.158, 14.162]

$$\Delta E_1 \cdot \Delta E_2 \geq 1 . \tag{14.81}$$

For coherent states (14.15) of the radiation field, and also for a thermal-equilibrium radiation field one obtains the symmetric relations

$$\Delta E_1 = \Delta E_2 = 1 , \tag{14.82}$$

and therefore the minimum possible value of the product $\Delta E_1 \cdot \Delta E_2$. In the phase diagram displayed in Fig. 14.61b the relation (14.82) yields a circle as the *uncertainty area*.

Coherent light shows phase-independent noise. This can be demonstrated by a two-beam interferometer, such as the Mach–Zehnder interferometer shown in Fig. 14.62. The monochromatic laser beam with the mean intensity I_0 is split by the beam splitter BS_1 into two partial beams b_1 and b_2 with amplitudes E_1 and E_2, which are superimposed again by BS_2. The beam b_2 passes a movable optical wedge, causing a variable phase shift ϕ between the two partial beams. The detectors PD1 and PD2 receive the intensities

$$\langle I \rangle = \frac{1}{2} c \cdot \epsilon_0 \left[\langle E_1^2 \rangle + \langle E_2^2 \rangle \pm 2 E_1 E_2 \cos\phi \right] ,$$

averaged over many cycles of the optical field with frequency ω. For equal amplitudes $E_1 = E_2$, the detected intensities are

$$\langle I_1 \rangle = \langle I_0 \rangle \cos^2 \phi/2 , \quad \langle I_2 \rangle = \langle I_0 \rangle \cdot \sin^2 \phi/2 \Rightarrow \langle I_1 \rangle + \langle I_2 \rangle = \langle I_0 \rangle .$$

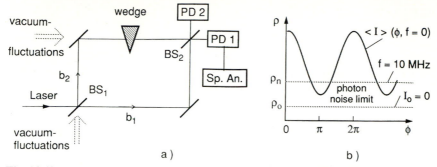

Fig. 14.62. (a) Mach–Zehnder interferometer with variable phase delay ϕ realized by an optical wedge. (b) Detected mean intensity $\langle I \rangle$ measured at $f = 0$, and phase-independent photon noise power density, measured at $f = 10\,\mathrm{MHz}$ with and without input intensity I_0

A variation of the phase ϕ by the wedge in one arm of the interferometer corresponds to a rotation of the vector $\boldsymbol{E}$ in the phase diagram of Fig. 14.61b. The two detectors in Fig. 14.62a measure the two projections of $\boldsymbol{E}$ onto the axes, E_1 and E_2. The arrangement of Fig. 14.62a therefore allows the separate determination of the fluctuations $\langle \delta E_1 \rangle$ and $\langle \delta E_2 \rangle$ by a proper choice of the phase ϕ.

If the frequency spectrum $I(f)$ of the detector signals is measured with a spectrum analyzer at sufficiently high frequencies f, where the technical noise is negligible, one obtains a noise power spectrum $\rho_n(f)$, which is essentially independent of the phase ϕ (Fig. 14.62b), but depends only on the number of photons entering the interferometer. It is proportional to $\sqrt{N}$. It is surprising that the noise power density $\rho_n(f)$ of each detector is independent of the phase ϕ. This can be understood as follows: the intensity fluctuations, because of the statistical emission of photons, are uncorrelated in the two partial beams. Although the mean intensities $\langle I_1 \rangle$ and $\langle I_2 \rangle$ depend on ϕ, their fluctuations do not! The detected noise power $\rho_n \propto \sqrt{N}$ shows the same noise level $\rho_n \propto \sqrt{I_0}$ for the minimum of $I(\phi)$ as for the maximum (Fig. 14.62b).

If the incident laser beam in Fig. 14.62a is blocked, the mean intensity $\langle I \rangle$ becomes zero. However, the measured noise power density $\rho_n(f)$ does *not* go to zero but approaches a lower limit ρ_0 that is attributed to the zero-point fluctuations of the vacuum field, which is also present in a dark room. The interferometer in Fig. 14.62a has two inputs: the coherent light field and a second field, which, for a dark input part, is the vacuum field. Because the fluctuations of these two inputs are uncorrelated, their noise powers add. Increasing the input intensity I_0 will increase the signal-to-noise-ratio

$$\frac{S}{\rho_n} \propto \frac{N}{\sqrt{N} + \rho_0},$$

where the fundamental limitation is set by the quantum noise ρ_0.

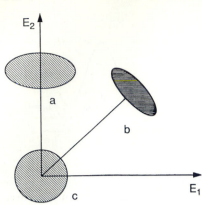

Fig. 14.63a–c. Uncertainty areas for different squeezing conditions: (**a**) $\langle E_1 \rangle = 0$, ΔE_2 is squeezed but $\Delta\phi$ is larger than in the nonsqueezed case; (**b**) general case of squeezing of ΔE at the expense of increasing $\Delta\phi$; (**c**) uncertainty area of zero-point fluctuations with $\langle E_1 \rangle = \langle E_2 \rangle = 0$

In the phase diagram of Fig. 14.63 the radius $r = \sqrt{\rho_0}$ of the circular uncertainty area around $E_1 = E_2 = 0$ corresponds to the vacuum fluctuation noise power density ρ_0.

The preparation of *squeezed* states tries to minimize the uncertainty of one of the two quantities ΔE or $\Delta\phi$ at the expense of the uncertainty of the other. Although the *uncertainty area* in Fig. 14.63, which is squeezed into an ellipse, does increase compared to the symmetric case (14.82), one may still gain in signal-to-noise ratio if the minimized quantity determines the noise level of the detected signal. This will be illustrated by some examples.

14.8.2 Experimental Realization of Squeezing

A typical layout of a squeezing experiment based on a Mach–Zehnder interferometer (Sect. 4.2.3) is shown in Fig. 14.64. The output of a well-stabilized laser is split into two beams, a pump beam b_1 and a reference beam b_2. The pump beam with the frequency ω_L generates by nonlinear interaction with a medium (e.g., four-wave mixing or parametric interaction) new waves at frequencies $\omega_L \pm f$. After superposition with the reference beam, which acts as a local oscillator, the resulting beat spectrum is detected by the photodetectors D1 and D2 as a function of the phase difference $\Delta\phi$, which can be controlled by a wedge in one of the interferometer arms. The difference between the two detector output signals is monitored as a function of the phase difference $\Delta\phi$. Contrary to the situation in Fig. 14.62, the spectral noise power density $\rho(f, \phi)$ $(= P_{NEP}$ per frequency interval $df = 1\,s^{-1})$ shows a periodic variation with ϕ. This is due to the nonlinear interaction of one of the beams with the nonlinear medium, which preserves phase relations. At certain values of ϕ the noise power density $\rho_n(f, \phi)$ drops below the photon noise limit

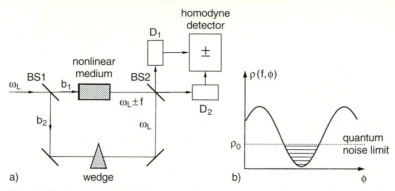

Fig. 14.64a,b. Schematic diagram of squeezing experiments with a nonlinear medium in a Mach–Zehnder interferometer: (**a**) experimental arrangement; (**b**) noise power density $\rho(\phi)$ and quantum-noise limit ρ_0, which is independent of the phase ϕ

$\rho_0 = (n \cdot h\nu/\eta\Delta f)^{1/2}$ that is measured without squeezing, when n photons fall onto the detector with bandwidth Δf and quantum efficienty η. This yields a signal-to-noise ratio

$$SNR_{\mathrm{sq}} = SNR_0 \frac{\rho_0}{\rho_{\mathrm{n}}} , \qquad (14.83)$$

which is larger than that without squeezing SNR_0.

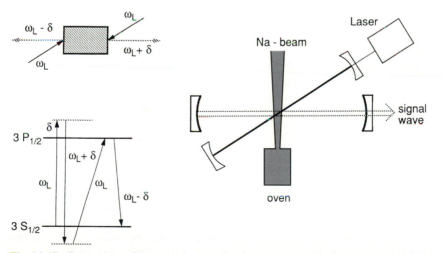

Fig. 14.65. Generation of squeezed states by four-wave mixing in an atomic Na beam. Both the pumped wave and signal and idler waves are resonantly enhanced by two optical resonators [14.163]

The degree of squeezing V_{sq} is defined as

$$V_{sq} = \frac{\rho_0 - \rho_{min}(f)}{\rho_0} . \qquad (14.84)$$

The first successful experimental realization of squeezing was reported by Slusher and coworkers [14.163], who used the four-wave mixing in a Na-atomic beam as nonlinear process (Fig. 14.65). The Na atoms are pumped by a dye laser at the frequency $\omega_L = \omega_0 + \delta$, which is slightly detuned from the resonance frequency ω_0. In order to increase the pump power, the Na beam is placed inside an optical resonator tuned to the pump frequency ω_L. Because of parametric processes (Sect. 5.8.8) of the two pump waves ($\omega_L, \pm k_L$) inside the resonator, new waves at $\omega_L \pm \delta$ are generated at this four-wave mixing process (signal and idler wave). Conservation of energy and momentum demands

$$2\omega_L \rightarrow \omega_L + \delta + \omega_L - \delta , \qquad (14.85a)$$

$$k_L + k_i = k_L + k_s . \qquad (14.85b)$$

A second resonator with a properly chosen length, such that the mode spacing is $\Delta \nu = \delta$, enhances the signal as well as the idler wave.

The essential point is that there are definite phase relations between pump, signal and idler waves, and this fact establishes a correlation between amplitude and phase of the signal wave. This correlation is illustrated in Fig. 14.66:

The four-wave mixing generates new sidebands. If the input contains the frequencies ω_L and $\omega_L + \delta$, the output has the additional sideband $\omega_L - \delta$. Thus, the amplitude of one sideband is increased at the cost of the other sideband until both sidebands have equal amplitudes. This maximizes the amplitude modulation of the output. Since the phases of the two sidebands are opposite, this transfer minimizes the phase modulation. As was shown in detail in [14.161], this correlation results in a phase-dependent noise that is, for certain phase ranges, below the quantum noise power ρ_0. In these experiments a squeezing degree of 0.1 was obtained, which means that $\rho_{min} = 0.9 \rho_0$.

The best squeezing results with a noise suppression of 60% (≈ -4 dB) below the quantum noise limit ρ_0 were obtained by Kimbel and coworkers [14.164] with an optical parametric oscillator, where the parametric interaction in a MgO:LiNbO$_3$ crystal was used for squeezing.

Another example, where a beam of squeezed light was realized with 3.2 mW and 52% noise reduction, is the second-harmonic generation with a cw Nd:YAG laser in a monolithic resonator [14.165]. The experimental setup is shown in Fig. 14.67a.

The output of the cw Nd:YAG laser is frequency doubled in a MgO:LiNbO$_3$ nonlinear crystal. The endfaces of the crystal form the resonator mirrors and by a special modulation technique the resonator is kept in resonance both for the fundamental and for the second harmonic wave. A balanced homodyne interferometric detection of the fundamental wave yields the intensity noise of the light beam (I_+ detected by D1) and the shot-noise level (I_- detected by D2) as a reference (Fig. 14.67b).

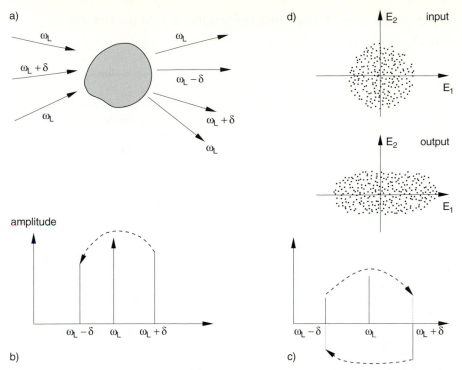

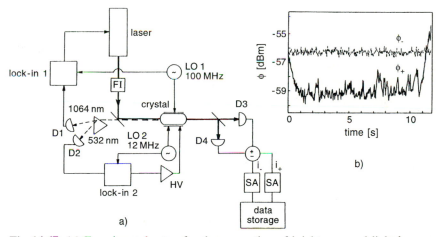

Fig. 14.66. (**a**) Schematic diagram of four-wave mixing; (**b**) amplitude transfer increasing amplitude modulation; (**c**) phase transfer decreasing phase modulation; (**d**) phase diagram

Fig. 14.67. (**a**) Experimental setup for the generation of bright squeezed light by second-harmonic generation in a monolithic resonator; (**b**) spectral power density of photocurrent fluctuations for squeezed light input to a balanced homodyne detector. The *upper curve* ϕ_- gives the shot-noise level [14.165]

14.8.3 Application of Squeezing to Gravitational Wave Detectors

Nowadays, it is believed that the most sensitive method for detecting gravitational waves is based on optical interferometry, where the change in length of an interferometer arm by gravitational waves can be monitored. Although until now no gravitational waves have been detected, there are great efforts in many laboratories to increase the sensitivity of laser gravitational wave detectors in order to discover those waves, even from very distant supernovae or rotating heavy binary neutron stars [14.166–14.168].

The basic part of such a detector (Fig. 14.68) is a Michelson interferometer with long arms (several kilometers) and an extremely well stabilized laser. If a gravitational wave causes a length difference ΔL between the two arms of the interferometer, a phase difference $\Delta\phi = (4\pi/\lambda)\Delta L$ appears between the two partial waves at the exit of the interferometer. The minimum detectable phase change $\Delta\phi$ is limited by the phase noise $\delta\phi_n$. If the laser delivers N photons $h\nu$ per second, the two detectors measure the intensities

$$I_1(\phi) = \frac{1}{2}Nh\nu\eta[1+\cos(\phi+\Delta\phi)] \,, \tag{14.86a}$$

$$I_2(\phi) = \frac{1}{2}Nh\nu\eta[1-\cos(\phi+\Delta\phi)] \,. \tag{14.86b}$$

This gives for $\phi = \pi/2$ a difference signal

$$S = \Delta I \approx N\hbar\nu\eta\Delta\phi \,. \tag{14.87}$$

The noise power of both detectors adds quadratically. The signal-to-noise ratio

$$\mathrm{SNR} = \frac{N\eta\Delta\phi}{\sqrt{N\Delta f}} \,, \tag{14.88}$$

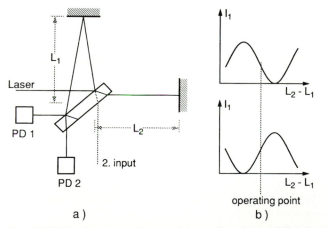

Fig. 14.68a,b. Gravitational wave detector based on a Michelson interferometer and a balanced homodyne detector

becomes larger than unity for $\Delta\phi > \sqrt{\Delta f/(\eta^2 \cdot N)}$. The minimum detectable phase change is therefore limited by the maximum available photon flux N.

In Sect. 14.8.1 it was shown that the noise limit without squeezing is caused by the zero-point field coupled into the second input part of the interferometer. If a beam of squeezed light is fed into this input part, the noise level can be decreased and the detectable length change ΔL can be further decreased [14.166].

For a more detailed representation of squeezing, the most recent experiments on the realization of squeezing, and for further applications of this interesting technique, the reader is referred to the literature [14.169–14.175].

15. Applications of Laser Spectroscopy

The relevance of laser spectroscopy for numerous applications in physics, chemistry, biology, and medicine, or to environmental studies and technical problems has rapidly gained enormous significance. This is manifested by an increasing number of books and reviews. This chapter can only discuss some examples that are selected in order to demonstrate how many fascinating applications already exist and how much research and development is still needed. For a detailed representation of more examples, the reader is referred to the references given in the sections that follow as well as to some monographs and reviews [15.1–15.4].

15.1 Applications in Chemistry

For many fields of chemistry, lasers have become indispensable tools [15.5–15.8]. They are employed in analytical chemistry for the ultrasensitive detection of small concentrations of pollutants, trace elements, or short-lived intermediate species in chemical reactions. Important analytical applications are represented by measurements of the internal-state distribution of reaction products with LIF (Sect. 6.3) and spectroscopic investigations of collision-induced energy-transfer processes (Sects. 13.3–13.6). These techniques allow a deeper insight into reaction paths of inelastic or reactive collisions, and their dependence on the interaction potential and the initial energy of the reactants.

Time-resolved spectroscopy with ultrashort laser pulses gives, for the first time, access to a direct view of the short time interval in which molecules are formed or fall apart during a collision. This femtosecond chemistry is discussed in Sect. 15.1.3.

The possibility of controlling chemical reactions by selective excitation of the reactands and by coherent control of the excited state wavefunction may open a new fascinating field of laser-induced chemistry.

15.1.1 Laser Spectroscopy in Analytical Chemistry

The first aspect of laser applications in analytical chemistry is the sensitive detection of small concentrations of impurity atoms or molecules. With the laser-spectroscopic techniques discussed in Chap. 6, detection limits down into the parts-per-billion (ppb) range can be achieved for molecules, which corresponds to a relative concentration of 10^{-9}. Atomic species and some favorable molecules can even be traced in concentrations within the parts-per-trillion

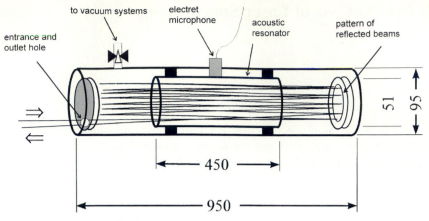

Fig. 15.1. Multipass cell of an optoacoustic spectrometer. All laser beams pass through regions in the acoustic resonator where the radial acoustic resonance has maximum amplitude. Measurementss are in millimeter

(ppt) ($\hat{=} 10^{-12}$) range. Recently "single molecule detection" in solids, solutions, and gases has become possible.

A very sensitive detection scheme is the photoacoustic method in combination with a multipass optical resonator, illustrated in Fig. 15.1. With this apparatus absorption coefficients down to $\alpha = 10^{-10}\,\mathrm{cm}^{-1}$ can be measured.

Example 15.1

With a diode laser spectrometer and a multipass absorption cell (Fig. 15.1) NO_2 concentrations down to the 50-ppt level in air were detected on a vibrational–rotational transition at $1900\,\mathrm{cm}^{-1}$, NO concentrations down to 300 ppt, while for SO_2 at $1335\,\mathrm{cm}^{-1}$ a sensitivity limit of 1 ppb was reached [15.9].

If the spectroscopic detection of atoms can be performed on transitions that represent a true two-level system (Sect. 14.1.5), atoms with the radiative lifetime τ may undergo up to $T/2\tau$ absorption–emission cycles during their transit time T through through the laser beam (photon burst). If the atoms are detected in carrier gases at higher pressures, the mean free path Λ becomes small ($\Lambda \ll d$) and T is only limited by the diffusion time. Although quenching collisions may decrease the fluorescence (Sect. 11.3), the ratio $T/2\tau$ becomes larger and the magnitude of the photon bursts may still increase inspite the decreasing quantum yield.

Example 15.2

For gases at low pressures where the mean free path Λ is larger than the diameter d of the laser beam, we obtain the typical value $T = d/\overline{v} = 10\,\mu s$

for $d = 5\,\mathrm{mm}$ and $\bar{v} = 5 \times 10^2\,\mathrm{m/s}$. For an upper-state lifetime of $\tau = 10\,\mathrm{ns}$ the atom emits 500 fluorescence photons (photon burst), allowing the detection of single atoms. With noble gas pressures of 1 mbar the mean-free path is $\approx 0.03\,\mathrm{mm}$ and the diffusion time through the laser beam may become 100 times longer. Although the lifetime is quenched to 5 ns, which means a fluorescence quantum yield of 0.5, this increases the photon burst to 5×10^4 photons.

Another very sensitive detection scheme is based on resonant two- or three-photon ionization of atoms and molecules in the gas phase (Sect. 6.3). With this technique even liquid or solid samples can be monitored if they can be vaporized in a furnace or on a hot wire. If, for instance, a heated wire or plate in a vacuum system is covered by the sample, the atoms or molecules are evaporated during the pulsed heating period and fly through the superimposed laser beams L1+L2 (+L3) in front of the heated surface (Fig. 15.2). The laser L1 is tuned to the resonance transition $|i\rangle \rightarrow |k\rangle$ of the wanted atom or molecule while L2 further excites the transition $|k\rangle \rightarrow |f\rangle$. Ions are formed if E_f is above the ionization potential IP. The ions are accelerated toward an ion multiplier. If L2 has sufficient intensity, all excited particles in the level $|k\rangle$ can be ionized and all atoms in the level $|i\rangle$ flying through the laser beam during the laser pulse can be detected (*single-atom detection*) [15.10–15.12]. If $E_f < \mathrm{IP}$ a third photon provided by L1 and L2 can ionize the species since there is no resonance condition for the ionizing step.

The selective detection sensitivity for spurious species with absorption spectra overlapping those of more abundant molecules or atoms can be enhanced further by a combination of mass spectrometer and resonant two-photon ionization (Fig. 15.3). This is, for instance, important for the detection of rare isotopic components in the presence of other more abundant isotopes [15.12, 15.13].

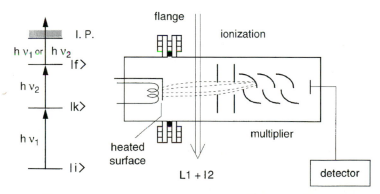

Fig. 15.2. Resonant multiphoton ionization (REMPI) as a sensitive detection technique for small quantities of atoms or molecules evaporated from a heated surface

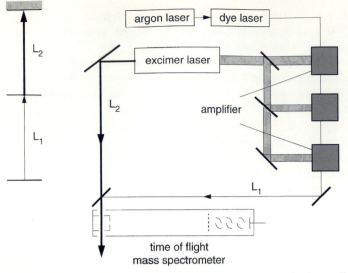

Fig. 15.3. Selective excitation of a wanted isotope by a pulsed, amplified single-mode dye laser with subsequent ionization by an excimer laser and mass-selective detection behind a time-of-flight spectrometer

15.1.2 Single-Molecule Detection

The combination of illuminating the sample with a strongly focused laser beam and observing the laser-induced fluorescence or the non-resonant scattered laser light with confocal microscopy allows the detection of single molecules and their diffusion through a liquid or gaseous medium. The experimental setup is illustrated in Fig. 15.4. The light scattered by molecules in a diluted solution with concentrations in the nanomole range, which diffuse into the laser focus, is imaged onto a small pinhole in front of the detector. The spatial resolution in the x–y-plane perpendicular to the incident laser beam can reach 500 nm. Using two-photon excitation the spatial resolution can be even improved to about 200 nm. The resolution in the z-direction (axial

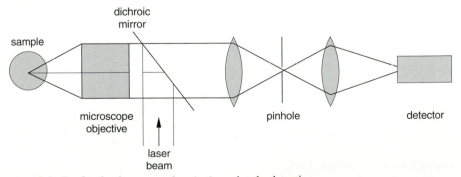

Fig. 15.4. Confocal microscopy for single-molecule detection

resolution) depends on the Rayleigh length of the focused laser beam. Typical values are $10-30\,\mu m$, where two-photon excitation also decreases this value below 1000 nm. This gives the extremely small detection volume of less than $10^{-15}\,\ell$. The fluorescence is detected by a high quantum efficiency avalanche photodiode [15.14]. Small concentrations with extinction coefficients of 10^{-4} are still detectable.

With time-resolved detection the temporal behavior of a single molecule can be followed. One example is the measurement of the relaxation time constant of intersystem crossing from the excited singlet state into the triplet state of a dye molecule during the diffusion time through the detection volume [15.15].

Molecular impurities in solids can now be detected down to the single-molecule level [15.16]. These molecules and their interaction with their surroundings can be probed by high-resolution laser spectroscopy. The relevance of such techniques in biology is obvious [15.17]. The development of single-molecule detection for the in vitro and in vivo quantification of biomolecular dynamics is essential for the understanding of biomolecular reactions and for the realization of evolutionary biotechnology [15.18].

More examples of lasers in analytical chemistry can be found in [15.19–15.21].

15.1.3 Laser-Induced Chemical Reactions

The basic principle of laser-induced chemical reactions is schematically illustrated in Fig. 15.5. The reaction is initiated by one- or multiphoton excitation of one or more of the reactants. The excitation can be performed before the reactants collide (Fig. 15.5a) or during the collision (Fig. 15.5b and Sect. 13.6).

For the selective enhancement of the wanted reaction channel by laser excitation of the reactants, the time span Δt between photon absorption and completion of the reaction is of fundamental importance. The excitation energy $n \cdot \hbar\omega$ ($n = 1, 2, \dots$) pumped by photon absorption into a selected excited molecular level may be redistributed into other levels by unwanted relaxation processes *before* the system ends in the wanted reaction channel. It can, for instance, be radiated by spontaneous emission, or it may be redistributed by intramolecular radiationless transitions due to vibrational or spin-orbit couplings onto many other nearly degenerate molecular levels. However, these levels may not lead to the wanted reaction channel. At higher pressures collision-induced intra- or intermolecular energy transfer may also play an important role in enhancing or suppressing a specific reaction channel.

We distinguish between three different time regimes:

(a) The excitation by the pump laser and the laser-induced reaction proceeds within a very short time Δt (femtosecond to picosecond range), which is shorter than the relaxation time for energy redistribution by fluorescence or inelastic nonreactive collisions. In this case, the above-mentioned loss mechanisms cannot play an important role, and the selectivity of driving the system into a wanted reaction channel by optical absorption is possible.

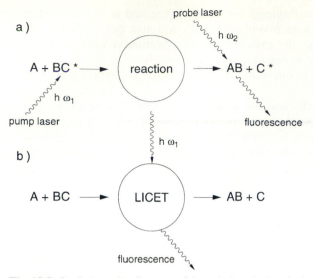

Fig. 15.5a,b. Schematic diagram of laser-induced chemical reactions with state-selective detection of the reaction products: (**a**) by excitation of the reactants; (**b**) by excitation of the collision pair (ABC)

(b) Within an intermediate time range (nanosecond to microsecond range), depending on the pressure in the reaction cell, the reaction may happen before the initial excitation energy is completely redistributed by collisions. If the excited reagent is a large molecule, the intramolecular energy transfer within the many accessible degrees of freedom is, however, fast enough to redistribute the excitation energy onto many excited levels. The excited molecule may still react with a larger probability than in the ground state, but the selectivity of reaction control is partly lost.

(c) For still longer time scales (microsecond to continuous wave excitation) the excitation energy is distributed by collisions statistically over all accessible levels and is finally converted into translational, vibrational, and rotational energy at thermal equilibrium. This raises the temperature of the sample, and the effect of laser excitation with respect to the wanted reaction channel does not differ much from that for thermal heating of the sample.

(d) Recently, the new technique of coherent control of chemical reactions has been developed, where definitely shaped femtosecond pulses are used (see Sect. 11.1) to prepare a coherent wavefunction in the excited state of a reactant molecule. The phase and amplitude of this wavefunction are chosen in a way that favors the decay of the molecule into the wanted reaction channel.

For the first time range ultrashort laser pulses, generated by mode-locked lasers (Chap. 11), are needed, whereas for the second class of experiments in the nanosecond to microsecond range pulsed lasers (Q-switched CO_2 lasers, excimer, or dye lasers) can be employed. Most experiments performed until now have used pulsed CO_2 lasers, chemical lasers, or excimer lasers. For femtosecond pulses Kerr-lens mode-locked Ti:sapphire lasers with subsequent amplifier stages are available (see Sect. 11.1).

Let us consider some specific examples: the first example is the laser-induced bimolecular reaction

$$HCl(v = 1, 2) + (O^3P) \rightarrow OH + Cl, \tag{15.1}$$

which proceeds after vibrational excitation of HCl with a HCl laser [15.22]. The internal-state distribution of the OH radicals is monitored by LIF, induced by a frequency-doubled dye laser, which is tuned to a specific rotational line of the $(v' = 0 \leftarrow v'' = 0)$ band in the $^2\Pi \leftarrow {}^2\Sigma$ system of OH at $\lambda = 308$ nm or the $(1 \leftarrow 1)$ band at 318 nm.

The second example is the spatially and temporally resolved observation of the explosion of an O_2/O_3 mixture in a cylindrical cell, initiated by a TEA CO_2 laser [15.23]. The progress of the reaction is monitored through the decrease of the O_3 concentration, which is detected by a time-resolved measurement of the UV absorption in the Hartley continuum of O_3. If the UV probe beam is split into several spatially separated beams with separate detectors (Fig. 15.6), the spatial progression of the explosion front in time can be monitored.

The high output power of pulsed CO_2 lasers allows excitation of high vibrational levels by multiphoton absorption, which eventually may lead to the dissociation of the excited molecule. In favorable cases the excited molecules or the dissociation fragments react selectively with other added components

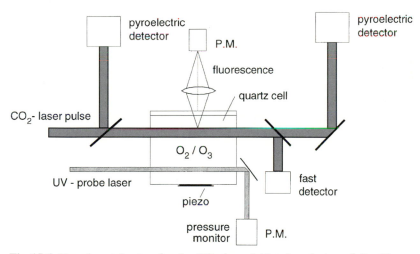

Fig. 15.6. Experimental setup for the CO_2 laser-initiated explosion of O_3. The expansion of the explosion front is monitored through the time-resolved O_3 absorption of several spatially separated UV probe laser beams. The piezo pressure detector allows a time-resolved measurement of the pressure [15.23]

[15.24]. Such selectively initiated chemical reactions induced by CO_2 lasers are particularly advantageous due to the large efficiency of these lasers, which makes the CO_2 photons cheap. As an example, let us consider the synthesis of SF_5NF_2 by multiphoton absorption of CO_2 photons in a mixture of S_2F_{10} and N_2F_4, which proceeds according to the following scheme:

$$S_2F_{10} + nh\nu \rightarrow 2SF_5 \,,$$

$$N_2F_4 + nh\nu \rightarrow 2NF_2 \,,$$

$$SF_5 + NF_2 \rightarrow SF_5NF_2 \,. \tag{15.2}$$

While the conventional synthesis without laser takes about $10-20$ at a temperature of 425 K and demands high pressures of S_2F_{10}, the laser-driven reaction proceeds much more quickly, even at the lower temperature of 350 K [15.25, 15.26].

Another example of CO_2 laser-initiated reactions is the gas-phase telomerization of methyl-iodide CF_3I with C_2F_4, which represents an exothermic radical chain reaction

$$(CF_3I)^* + nC_2F_4 \rightarrow CF_3(C_2F_4)_nI \,, \quad (n = 1, 2, 3) \,, \tag{15.3}$$

producing $CF_3(C_2F_4)_nI$ with low values of n. The CO_2 laser is in near resonance with the $\nu_2 + \nu_3$ band of CF_3I. The quantum yield for the reaction (15.3) increases with increasing pressure in the irradiated reaction cell [15.27].

In many cases, electronic excitation by UV lasers is necessary for laser-initiated reactions. An example is the photolysis of vinyl chloride

$$C_2H_3Cl + h\nu \rightarrow C_2H_3 + Cl \,, \tag{15.4a}$$

$$\rightarrow C_2H_2 + HCl \,, \tag{15.4b}$$

induced by a XeCl excimer laser (Sect. 5.7). In spite of the small absorption cross section ($\sigma = 10^{-24}$ cm^2 at $\lambda = 308$ nm) the yield ratio of the two reaction branches (15.4a, 15.4b) and its dependence on temperature could be accurately measured [15.28].

As another example, let us mention the chain reaction of organic bromides with olefins induced by a KrF excimer laser [15.29], which proceeds as follows:

$$
\begin{aligned}
&RBr + h\nu \rightarrow R + Br && \text{(initiation)} \,, \\
&R_n + C_2H_4 \rightarrow R_{n+2} && \text{(propagation)} \,, \\
&R_n + RBr \rightarrow R_nBr + R && \text{(chain transfer)} \,, \\
&Br + Br \rightarrow Br_2 && \text{(termination)} \,,
\end{aligned}
\tag{15.5}
$$

where R_n denotes any radical with n carbon atoms.

Many chemical reactions can be enhanced by catalytic effects at solid surfaces. The prospect of increasing these catalytic enhancements further by laser irradiation of the surface has initiated intense research activity [15.30, 15.31]. The laser may either excite atoms or molecules adsorbed at the surface, or it may excite desorbed molecules just above the surface. In both cases the desorption or adsorption process is altered because excited molecules have a different interaction potential between the molecule M* and the surface than the ground-state molecules. Furthermore, the laser may evaporate surface material, which can react with the molecules.

15.1.4 Coherent Control of Chemical Reactions

Coherent control is based on interference effects, which become essential when an excited state can be populated on two ore more excitation paths. The population rates then depend on the phase relations between the optical waves that are inducing these excitations. Coherent control has to be performed on a very short time scale in order to win over dephasing processes. Therefore femtosecond lasers are generally demanded [15.32].

Two different schemes have been proposed [15.33, 15.34]. In the first scheme (Fig. 15.7) two coupled levels $|a\rangle$ and $|b\rangle$ are simultaneously populated on two different excitation paths. Depending on the relative phase between the field amplitudes of the optical waves, there can be constructive interference, leading to maximum population of the mixed state $|a\rangle + |b\rangle$, or destructive interference, minimizing the population. An instructive example is the optical excitation of autoionizing states above the ionization limit, which interact with the ionization continuum (Fig. 15.7b). While the phase of the wavefunction of the discrete level changes rapidly when the optical frequency is tuned over the absorption profile, that of the continuum is only slightly affected. Therefore the phase difference of the two excitation paths depends on the optical frequency and changes in such a way that constructive interference occurs (maximum absorption) or destructive interference occurs, where the absorption is nearly zero. The resulting asymmetric absorption profile is

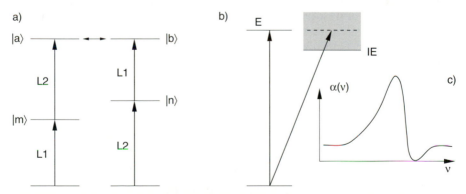

Fig. 15.7. (**a**) Coherent control of the population of the coupled levels $|a\rangle$ and $|b\rangle$ by interference between two different excitation paths; (**b**) excitation of autoionizing level; (**c**) Fano profile of absorption line

called the Fano profile. The produced ion rate can be therefore controlled by slight changes of the excitation frequency.

The second scheme of coherent control utilizes the time difference between two femtosecond laser pulses interacting with a molecule on two different transitions sharing a common level, which was illustrated in Sect. 11.4.4 by the example of the Na_2 molecule. Here the phase of the wave packet produced by the first pulse in the excited state develops in time, and the controlled time lag between the first and the second pulse selects a favorable phase for further excitation or deexcitation of the molecule by the second pulse.

In the first scheme it is not necessary to use two different lasers if a femtosecond pulse with a broad spectral range is used for excitation. The different spectral components in the pulse give rise to many different excitation paths. In order to achieve optimum population in the excited state, the relative phases of these different spectral components have to be optimized. This can be realized by the pulse-shaping techniques discussed in Sect. 11.1.9 where a feedback loop with a learning algorithm is used to maximize or minimize the wanted decay channel of the excited state [15.35, 15.36].

One example is the dissociation of laser-excited iron pentacarbonyl $Fe(CO)_5$, where the ratio $Fe(CO)_5/Fe$ can be varied between 0.06 to 4.8 by coherent control [15.37]. Another example is the photodissociation of $C_5H_5Fe(CO)_2Cl$ into selected fragments. Optimum shaped femtosecond pulses can alter the ratio $C_5H_5COCl/FeCl$ from 1 to 5 [15.38], or the selected bond dissociation of acetone $(CH_3)_2CO$ or acetophenone $C_6H_5COCH_3$ [15.39]. More examples and a detailed description of the technique can be found in [15.40] and [15.41].

15.1.5 Laser Femtosecond Chemistry

Chemical reactions are based on atomic or molecular collisions. These collisions, which may bring about chemical-bond formation or bond breaking, occur on a time scale of 10^{-11} to 10^{-13} s. In the past, the events that happened in the transition state between reagents and reaction products could not be time resolved. Only the stages "before" or "after" the reaction could be investigated [15.42].

The study of chemical dynamics concerned with the ultrashort time interval when a chemical bond is formed or broken may be called *real-time femtochemistry* [15.43]. It relies on ultrafast laser techniques with femtosecond time resolution [15.44].

Example 15.3

Assume a dissociating molecule with a fragment velocity of 10^3 m/s. Within 0.1 ps the fragment separation changes by

$$\Delta x = (10^3 \cdot 10^{-13} = 10^{-10})m = 0.1 \, nm \, .$$

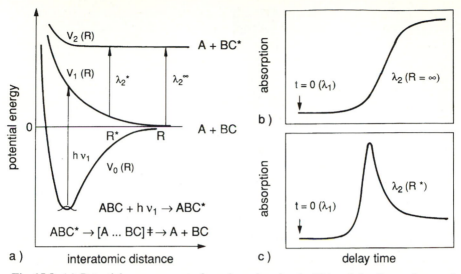

Fig. 15.8. (a) Potential energy curves for a bound molecule (V_0) and the first and second dissociative curves V_1, V_2; **(b)** the expected femtosecond transient signals $S(\lambda_2, t)$ versus the delay time t for $\lambda_2(R = \infty)$ and **(c)** for $\lambda_2(R^*)$ [15.43]

Let us regard the photodissociation process

$$ABC + h\nu \rightarrow [ABC]^* \rightarrow A + BC , \qquad\qquad (15.6)$$

which represents an unimolecular reaction. The real-time spectroscopy of bond breaking is explained by the potential diagram of Fig. 15.8.

The molecule ABC is excited by a pump photon $h\nu_1$ into the dissociating state with a potential curve $V_1(R)$. A second probe laser pulse with a tunable wavelength λ_2 is applied with the time delay Δt. If λ_2 is tuned to a value that matches the potential energy difference $h\nu = V_2(R) - V_1(R)$ at a selected distance R between A and the center of BC, the probe radiation absorption α (λ_2, Δt) shows a time dependence, as schematically depicted in Fig. 15.8b. When λ_2 is tuned to the transition $BC \rightarrow (BC)^* = V_2(R = \infty) - V_1(R = \infty)$ of the completely separated fragment BC, the curve in Fig. 15.19c is expected. These signals yield the velocity $v(R)$ of the dissociation products from which the energy difference $V_2(R) - V_1(R)$ can be derived.

The experimental arrangement for such femtosecond experiments is exhibited in Fig. 15.9. The output pulses from a femtosecond pulse laser (Sect. 11.1.5) are focused by the same lens into the molecular beam. The probe pulses are sent through a variable optical-delay line and the absorption $\alpha(\Delta t)$ of the probe pulse as a function of the delay time Δt is monitored via the laser-induced fluorescence. Cutoff filters suppress scattered laser light.

Another example is the real-time observation of ultrafast ionization and fragmentation of mercury clusters $(Hg)_n$ ($n \leq 110$) with femtosecond laser pulses. In pump–probe experiments short-time oscillatory modulations of the

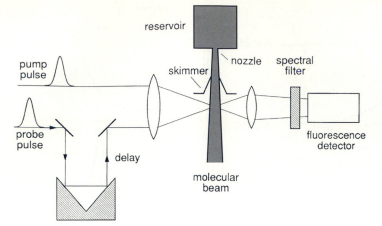

Fig. 15.9. Experimental arrangement for femtosecond spectrosopy in a molecular beam

transient Hg_n^+ and Hg_n^{++} signals indicate an intermediate-state dynamics common to all cluster sizes [15.45].

Photoinduced reactions in the liquid phase are influenced much more by collisions than those in the gas phase. In order to study such reactions on a time scale below the mean collision time, femtosecond spectroscopy is needed. One example is the exploration of transition-state dynamics and rotational dynamics of the fragments in the photolysis of mercuric iodide HgI_2 in ethanol solution [15.46, 15.47].

Another example is the detailed investigation of the femtosecond dynamics of ironcarbonyl $Fe(CO)_5$ [15.48], where the photodissociation after excitation with 267-nm pulses was studied by transient ionization. Five consecutive processes with time constants 21, 15, 30, 47, and 3300 fs were found. The first four short-time processes represent ionization from different excited configurations, which are reached by a pathway from the initially excited Franck–Condon region to other configurations through a chain of Jahn–Teller-induced conical intersections. The experiments also showed that intersystem crossing to the triplet ground states of $Fe(CO)_4$ and $Fe(CO)_3$ takes more than 500 ps. More detailed information on experiments on laser femtosecond chemistry can be found in [15.49].

15.1.6 Isotope Separation with Lasers

The classical methods of isotope separation on a large technical scale, such as the thermal diffusion or the gas-centrifuge techniques, are expensive because they require costly equipment or consume much energy [15.50]. Although the largest impetus for the development of efficient new methods for isotope separation was provided by the need for uranium ^{235}U separation, increasing demands exist for the use of isotopes in medicine, biology, geology, and hydrology. Independent of the future of nuclear reactors, it is therefore worthwhile to think about new, efficient techniques of isotope separation on a medium scale. Some of the novel techniques, which are based on a combination of isotope-selective excitation by lasers with subsequent photochemical reactions, represent low-cost processes that have already been proved feasible in laboratory experiments. Their extension to an industrial scale, however, demands still more efforts and improvements [15.51–15.56].

Most methods of laser isotope separation are based on the selective excitation of the desired atomic or molecular isotope in the gas phase. Some possible ways for separating the excited species are depicted schematically in Fig. 15.10, where A and B may be atoms or molecules, such as radicals. If the selectively excited isotope A_1 is irradiated by a second photon during the lifetime of the excited state, photoionization or photodissociation may take place if

$$E_0 + h\nu_1 + h\nu_2 > E(A^+) , \quad \text{or} \quad > E_{\text{Diss}} . \tag{15.7}$$

The ions can be separated from the neutrals by electric fields, which collect them into a Faraday cup. This technique has been used, for example, for the separation of ^{235}U atoms in the gas phase by resonant two-photon ionization with copper-vapor laser-pumped dye lasers at high repetition frequencies [15.57]. Since the line density in the visible absorption spectrum of ^{235}U is very high, the lasers are crossed perpendicularly with a cold collimated

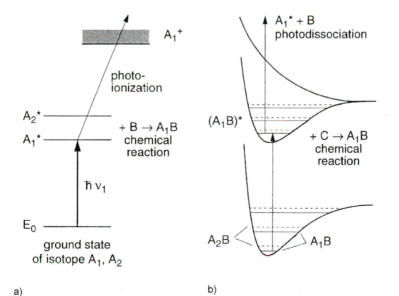

Fig. 15.10a,b. Different possible ways of isotope separation following selective excitation of the wanted isotope: (a) photoionization; and (b) dissociation or excitation of predissociating molecular states

beam of uranium atoms in order to reduce the line density and the absorption linewidth.

In the case of *molecular* isotopes, the absorption of the second photon may also lead to photodissociation. The fragments R are often more reactive than the parent molecules and may react with properly added scavenger reactants S to form new compounds RS, which can be separated by chemical means.

In favorable cases no second photon is necessary if reactants can be found that react with the excited isotopes M* with a much larger probability than with the ground-state molecules M. An example of such a chemical separation of laser-excited isotopes is the reaction

$$I^{37}Cl + h\nu \rightarrow (I^{37}Cl^*) ,$$

$$(I^{37}Cl)^* + C_6H_5Br \rightarrow {}^{37}ClC_6H_5Br + I ,$$

$$^{37}ClC_6H_5Br \rightarrow C_6H_5{}^{37}Cl + Br . \tag{15.8}$$

The isotope $I^{37}Cl$ can be selectively excited at $\lambda = 605$ nm by a cw dye laser. The excited molecules react in collisions with bromine benzene and form the unstable radical $^{37}ClC_6H_5Br$, which dissociates rapidly into $C_6H_5{}^{37}Cl + Br$. In a laboratory experiment, several milligrams of C_6H_5Cl were produced during a two hour exposure. Enrichment factors $K = {}^{37}Cl/{}^{35}Cl$ of $K = 6$ have been achieved [15.54].

Radioactive isotopes and isotopes with nuclear spin $I \neq 0$ play an important role in medical diagnostics. The radioactive technetium isotope Tc, for example, is now used instead of iodine ^{137}I for thyroid gland diagnostics and treatment because it has a shorter decay time and therefore the diagnosis can be performed with a smaller dose. The carbon isotope ^{13}C is employed, in addition to hydrogen ^1H, in NMR tomography for monitoring the brain, or to follow up the metabolism and its possible anomalies. The isotope ^{13}C can be separated by isotope-selective excitation of formaldehyde $^{13}CH_2O$ by an UV laser into predissociating states [15.58], or by multiphoton dissociation of Freon

$$CF_2HCl + n \cdot h\nu \rightarrow CF_2 + HCl , \tag{15.9}$$

with a CO_2 laser, which leads to an enrichment of $^{13}CF_2$ [15.59]. The efficiency of the process (15.9) has been improved by enhancing collisions between the fragments

$$^{13}CF_2 + {}^{13}CF_2 \rightarrow {}^{13}C_2F_4 ,$$

$$^{13}C_2F_4HCL \rightarrow {}^{13}CF_2HCl + {}^{13}CF_2 ,$$

which restores the parent molecule, now isotopically enriched (Fig. 15.11). This allows a continuous repetition of the enrichment cycle [15.60].

Even multiphoton dissociation of larger molecules such as SF_6 by CO_2 lasers may be isotope selective [15.52]. For the heavier molecule UF_6 the necessary selectivity of the first step, excited at $\lambda = 16\,\mu m$, can only be reached

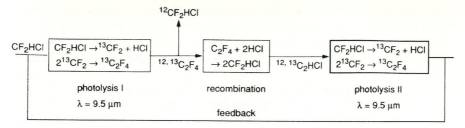

Fig. 15.11. Isotope enrichment of $^{13}CF_2$ by a cyclic process following isotope-enhanced multiphoton dissociation of Freon CF_2HCl [15.60]

in a collimated cold UF_6 beam. The vibrationally excited UF_6 isotope can then be ionized by a XeCl excimer laser at $\lambda = 308$ nm [15.55a]. The absolute amount of isotopes separated in this way is, however, very small [15.55b].

The last examples illustrate that the most effective method of isotope separation is a combination of isotope-selective excitation with selective chemical reactions. The laser plays the role of an isotope-selective initiator of the chemical reactions [15.61].

15.1.7 Summary of Laser Chemistry

The main advantages of laser applications in chemistry may be summarized as follows:

- Laser spectroscopy offers different techniques of increased sensitivity for the detection of tiny concentrations of impurities, pollutant gases, or rare isotopes down to the ultimate level of single-molecule detection.
- The combination of spectral and time resolutions allows detailed investigations of transition states and intermediate reaction transients. Femtosecond spectroscopy gives direct "real-time" information on the formation or breaking of chemical bonds during the collision process.
- Selective excitation of reaction products results (in favorable cases) in an enhancement of wanted reaction channels. It often turns out to be superior to temperature-enhanced nonselective reaction rates, particularly when coherent control techniques are used [15.61].
- The preparation of reagents in selected states and the study of the internal-state distribution of the reaction products by LIF or REMPI brings complete information on the "state-to-state" molecular dynamics [15.42].

More aspects of laser chemistry and many more examples of applications of laser spectroscopy in chemistry can be found in [15.1–15.6, 15.61–15.74].

15.2 Environmental Research with Lasers

A detailed understanding of our environment, such as the earth's atmosphere, the water resources, and the soil, is of fundamental importance for mankind.

Since in densely populated industrial areas air and water pollution has become a serious problem, the study of pollutants and their reactions with natural components of our environment is urgently needed [15.75]. Various techniques of laser spectroscopy have been successfully employed in atmospheric and environmental research: direct absorption measurements, laser-induced fluorescence techniques, photoacoustic detection, spontaneous Raman scattering and CARS (Chap. 8), resonant two-photon ionization, and many more of the sensitive detection techniques discussed in Chap. 6 can be applied to various environmental problems. This section illustrates the potential of laser spectroscopy in this field by some examples.

15.2.1 Absorption Measurements

The concentration N_i of atomic or molecular pollutants in the lower part of our atmosphere just above the ground can be determined by measurements of the direct absorption of a laser beam propagating through the atmosphere. The detector receives, at a distance L from the source, the fraction

$$P(L)/P_0 = e^{-a(\omega)L} , \qquad (15.10)$$

of the emitted laser power P_0. The attenuation coefficient

$$a(\omega) = \alpha(\omega) + S = N_i \sigma_i(\omega, p, T) + \sum_k N_k \sigma_k^{scat} , \qquad (15.11)$$

is represented by the sum of the absorption coefficient $\alpha(\omega) = N_i \sigma_i^{abs}$ (which equals the product of density N_i of absorbing molecules in level $|i\rangle$ and the absorption cross section σ_i^{abs}) and the scattering coefficient $S = \sum_k N_k \sigma_k^{scat}$ (which is due to light scattering by all particles present in the atmosphere). The dominant contribution to scattering is the Mie scattering [15.76] by small particles (dust, water droplets), and only a minor part is due to Rayleigh scattering by atoms and molecules.

The absorption coefficient $\alpha(\omega)$ assumes nonnegligible values only within the small spectral ranges $\Delta\omega$ of absorption lines (a few gigahertz around the center frequencies ω_0). On the other hand, the scattering cross section, which is proportional to ω^4 for Rayleigh scattering, does not vary appreciably over these small ranges $\Delta\omega$. Therefore, measurements of the laser beam attenuation at the two different frequencies ω_1 and ω_2 inside and just outside an absorption profile yields the ratio

$$\frac{P(\omega_1, L)}{P(\omega_2, L)} = e^{-[a(\omega_1) - a(\omega_2)]L} \simeq e^{-N_i[\sigma_i(\omega_1) - \sigma_i(\omega_2)]L} , \qquad (15.12)$$

of two transmitted laser powers, from which the wanted concentration N_i of the absorbers can be derived if the absorption cross sections σ_i are known. A possible experimental realization is illustrated in Fig. 15.12. The laser beam, enlarged by a telescope, reaches a cats-eye reflector at the distance $L/2$, which

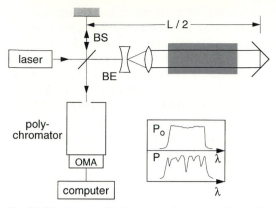

Fig. 15.12. Schematic diagram of an experimental setup for measuring the density of absorbing species integrated over the path length L. The polychromator with optical multichannel analyzer allows the simultaneous determination of several absorbing components

reflects the beam exactly back into itself. The reflected beam is sent to the detector by the beam splitter BS. For larger distances L, beam deviations from spatially inhomogeneous fluctuations of the refractive index of the air impose a severe problem. It may be partly solved by several measures: the switching cycle from ω_1 to ω_2 is performed in a statistical sequence, the switching frequency is chosen as high as possible, and the areas of the retroreflector and the detector are made so large that the detector still receives the full beam in spite of small beam deviations [15.77].

For such absorption measurements infrared lasers can be used which coincide with vibrational-rotational transitions of the investigated molecules (CO_2 laser, CO laser, HF or DF lasers, etc.). Particularly useful are tunable infrared lasers (diode lasers, color-center lasers, or optical parametric oscillators, Sect. 5.7), which may be tuned to selected transitions. The usefulness of diode lasers has been demonstrated by many examples [15.78]. For instance, with a recently-developed automated diode laser spectrometer, which is tuned by computer control through the spectral intervals of interest, up to five atmospheric trace gases can be monitored in unattended operation. Sensitivities down to 50 ppt for NO_2 and 300 ppt for NO have been reported [15.9].

The infrared lasers have the advantage that the contribution of scattering losses to the total beam attenuation is much smaller than in the visible range. For measurements of very low concentrations, on the other hand, visible dye lasers may be more advantageous because of the larger absorption cross sections for electronic transitions and the higher detector sensitivity.

Often a broadband laser (for example, a pulsed dye laser without etalons), or a multiline laser (for example, a CO_2 or CO laser without grating) may simultaneously cover several absorption lines of different molecules. In such cases the reflected beam is sent to a polychromator with a diode array or an optical multichannel analyzer (OMA, Sect. 4.5). If a fraction of the laser power $P_0(\omega)$ is imaged onto the upper part of the OMA detector and the transmitted power onto the lower part (Fig. 15.12b), electronic difference and ratio recording allows the simultaneous determination of the concentrations N_i of all absorbing species. A retroreflector arrangement is feasible for measurements at low altitudes above ground, where buildings or chimneys can support the construction. Examples are measurements of fluorine concentrations in an aluminum plant [15.79], or the detection of different constituents in the chimney emission of power plants, such as NO_x and SO_x components [15.80]. Often, ammonia is added to the exhaust of power stations in order to reduce the amount of NO_x emission. In such cases, the optimum concentration of NH_3 has to be controlled in situ. A recently developed detection system for these purposes has demonstrated its sensitivity and reliability [15.81].

In many cases samples of the air with its spurious pollutant molecules are taken and measured in an absorption cell. Here tunable diode lasers, which can be tuned over the vibrational bands of the molecules, have proved to be very useful. A review on recent work in this field can be found in [15.82].

For measurements over larger distances or in higher altitudes of the atmosphere, this absorption measurement with a retroreflector cannot be used. Here the LIDAR system, which is discussed in Sect. 15.2.2, has proven to be the best choice.

15.2.2 Atmospheric Measurements with LIDAR

The principle of LIDAR (light detection and ranging) is illustrated by Fig. 15.13. A short laser pulse $P_0(\lambda)$ is sent at time $t = 0$ through an expanding telescope into the atmosphere. A small fraction of $P_0(\lambda)$ is scattered back

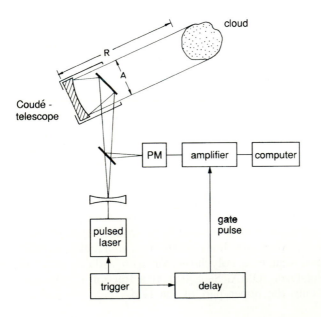

Fig. 15.13. Schematic diagram of a LIDAR system

into the telescope due to Mie scattering by droplets and dust particles, and Rayleigh scattering by the atmosopheric molecules. This backscattered light yields a photomultiplier signal $S(\lambda, t)$, which is measured with spectral and time resolution. The signal at time $t_1 = 2R/c$ depends on the scattering by particles at the distance R. If the detector is gated during the time interval $t_1 \pm \frac{1}{2}\Delta t$, the time-integrated measured signal

$$S(\lambda, t_1) = \int_{t_1 - \Delta t/2}^{t_1 + \Delta t/2} S(\lambda, t) \, \mathrm{d}t \, ,$$

is proportional to the light power scattered by particles within a distance $R \pm \frac{1}{2}\Delta R = \frac{1}{2}c(t_1 \pm \frac{1}{2}\Delta t)$. The magnitude of the received signal $S(\lambda, t)$ depends on the attenuation of the emitted power $P_0(\lambda)$ on its way back and forth, on the solid angle $\mathrm{d}\Omega = D^2/R^2$ covered by the telescope with the diameter D, and on the concentration N and the backscattering cross section σ^{scatt} of the scattering particles:

$$S(\lambda, t) = P_0(\lambda) \, \mathrm{e}^{-2a(\lambda)R} N \sigma^{\mathrm{scatt}}(\lambda) D^2/R^2 \, . \tag{15.13}$$

The quantity of interest is the factor $\exp[-2a(\lambda)R]$, where a is the sum of absorption and scattering coefficients, which gives with (15.11) the necessary information on the concentration of absorbing species. Similar to the method described in the previous section, the laser wavelength λ is tuned alternately to an absorption line at λ_1 and to a wavelength λ_2 where absorption by the wanted molecules is negligible. For sufficiently small values of $\Delta\lambda = \lambda_1 - \lambda_2$ the variation of the scattering cross section can be neglected. The ratio

$$Q(t) = \frac{S(\lambda_1, t)}{S(\lambda_2, t)} = \exp\left\{2 \int_0^R [\alpha(\lambda_2) - \alpha(\lambda_1)] \mathrm{d}R\right\} \, ,$$

$$\simeq \exp\left\{2 \int_0^R N_i(R)\sigma(\lambda_1) \mathrm{d}R\right\} \, , \tag{15.14}$$

yields the concentration $N_i(R)$ integrated over the total absorption path length. The dependence $N_i(R)$ on the distance R can be measured with a differential technique: the sequence of the quantities $S(\lambda_1, t)$, $S(\lambda_2, t)$ $S(\lambda_1, t+\Delta t)$, and $S(\lambda_2, t+\Delta t)$ is measured alternately. The ratio

$$\frac{Q(t+\Delta t)}{Q(t)} = \mathrm{e}^{-[\alpha(\lambda_2)-\alpha(\lambda_1)]\Delta R} \simeq 1 - [\alpha(\lambda_2) - \alpha(\lambda_1)]\Delta R \, ,$$

yields the absorption within the spatial interval $(R+\Delta R) - R = c\Delta t/2$. If the absorption coefficient under the atmospheric conditions (pressure, tempera- ture) within the detection volume $\Delta V = \Delta R \cdot A$ (A is the area of the laser beam at the distance R) is known, the concentration $N_i = \alpha_i/\sigma_i$ of the wanted species can be determined. The lower limit $\Delta R = c\Delta t/2$ is given by the time resolution Δt of the LIDAR system and by the attainable signal-to-noise ratio. This differential-absorption LIDAR *(DIAL) method* represents a very sensitive technique for atmospheric research.

In this way a complete *air-pollution map* of industrial and urban areas can be recorded and pollution sources can be localized. With pulsed dye lasers NO_2 concentrations in the parts-per-million (ppm) range at distances up to 5 km can be monitored [15.83]. Recent developments of improved LIDAR systems with frequency-doubled lasers have greatly increased the sensitivity as well as the spatial and spectral ranges that can be covered [15.84–15.86].

A further example of the application of the LIDAR technique is the measurement of the atmospheric *ozone* concentration, and its daily and annual variations as a function of altitude and latitude [15.87]. In order to reach reliable and stable laser operation, with reproducible wavelength switching even under unfavorable conditions (in an aircraft or on a ship), a XeCl excimer laser at $\lambda_1 = 308$ nm was used instead of a dye laser, and the second wavelength $\lambda_2 = 353$ nm was produced by Raman shifting in a hydrogen high-pressure cell. While radiation at λ_1 is strongly absorbed by O_3 the absorption of 353 nm is negligible [15.88, 15.89a]. However, one has to be sure that the absorption by other gaseous components is zero for both wavelengths, or at least equal. Otherwise, serious errors can arise [15.89b]. Besides the ozone layer at altitudes in the range 30−60 km, the ozone concentration just above ground is of vital interest. Here LIDAR measurements are able to provide a complete map of ozone concentrations in urban and rural areas and to find the different reactions that lead to ozone production and destruction as well as the sources of the reactants [15.90].

A third example of the usefulness of differential LIDAR is given by the spectroscopic determination of daily and annual variations of the temperature profile $T(h)$ of the atmosphere as a function of the height h above ground. The everywhere-present Na atoms can be used as tracer atoms since the Doppler width of the Na-D line, which is measured with a pulsed, narrow-band tunable dye laser, is a measure of the temperature [15.91, 15.92].

In the higher atmosphere, the aerosol concentration decreases rapidly with increasing altitudes. Mie scattering therefore becomes less effective and other techniques have to be used for the measurements of concentration profiles $N(h)$. Either the fluorescence induced by UV lasers or Raman scattering can provide the wanted signals. Fluorescence detection is sufficiently sensitive only if quenching collisions are not the dominant deactivation process for the excited levels. This means that either the radiative lifetime τ^{rad} of the excited levels must be sufficiently short or the pressure $p(h)$ should be low, that is, the altitude h high. If quenching collisions are not negligible, the effective lifetimes and the quenching cross sections have to be known in order to derive quantitative values of the concentration profiles $N_i(h)$ from measured fluorescence intensities. This difficulty may be overcome by Raman spectroscopy, which, however, has the disadvantage of smaller scattering cross sections [15.93–15.95].

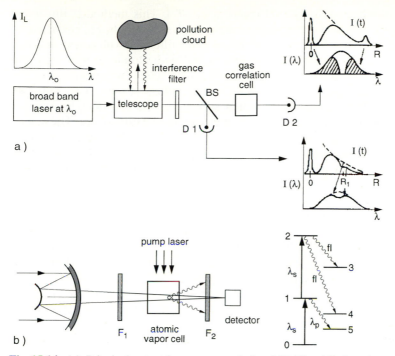

Fig. 15.14. (a) Principal setup for cross-correlation LIDAR with time-dependent signals and spectral distribution as detected by D1 and D2 [15.96]; (b) optically active atomic filter [15.97]

For daytime measurements the bright, continuous background of sunlight scattered in the atmosphere limits the attainable signal-to-noise ratio. With a narrow spectral filter, which is matched to the detected fluorescence wavelength, or in the case of LIDAR to the laser wavelength, the background can be substantially suppressed. An elegant technical trick of LIDAR measurements is based on a cross-correlation method (Fig. 15.14). The back-scattered light is split into two parts: one part is reflected by the beam splitter BS onto a detector D1, while the transmitted part is sent through an absorption cell of length ℓ, which contains the investigated molecular components under comparable conditions of total pressure and temperature as in the atmosphere. Their partial density N_i is, however, chosen so high that at the center of an absorption line $\alpha(\omega_0)\ell \gg 1$. The laser bandwidth is set to be slightly larger than the linewidth of the absorbing transition. The amplification of the two amplifiers following D1 and D2 is adjusted to obtain the difference signal $S_1 - S_2 = 0$ for a wanted concentration N_i. Any deviation of $N_i(R)$ from this balanced value is monitored behind the difference amplifier as the signal $\Delta S(N_i)$, which is essentially independent of fluctuations in laser intensity and frequency, since they affect both arms of the difference detection [15.96].

The cross-correlation technique represents a special case of a more general method, where the absorption lines of atomic vapors are used as narrow optical filters, matched to the special problem [15.97]. This method is illustrated in Fig. 15.14b. The backscattered laser light, with wavelength λ_L, collected by the telescope is sent through a narrow spectral filter F_1 with a transmission peak at λ_L and then through an absorption cell with atomic or molecular vapor that absorbs at λ_L. The atoms or molecules excited by absorption of the photons $\hbar\omega_L$ emit fluorescence with the wavelengths $\lambda_{Fl} > \lambda_L$. This radiation is detected behind a cutoff filter, which suppresses the wavelengths $\lambda < \lambda_{Fl}$. In this way the background radiation can essentially be eliminated. These *passive atomic filters* are restricted to those wavelengths λ_L that match atomic or molecular resonance transitions starting from the thermally populated ground state. The number of possible coincidences can be greatly enlarged if transitions between *excited* states of the filter atoms or molecules can be utilized. This can be achieved with a tunable pump laser that generates a sufficiently large population density of absorbing atoms or molecules in selectively excited states (*active atomic filters*, Fig. 15.14b). A further advantage of these active filters is the fact that the fluorescence induced by the signal wave may be shifted into the shorter wavelength range. This allows the detection of infrared radiation via visible or UV fluorescence.

A new interesting technique of air-pollution measurements is based on high-power terawatt femtosecond laser pulses that sent into the atmosphere. High intensities are reached in the focal area due to self-focusing, which leads to air breakdown and the production of a high-temperature plasma. This plasma represents a white-light source with a continuous spectrum extending

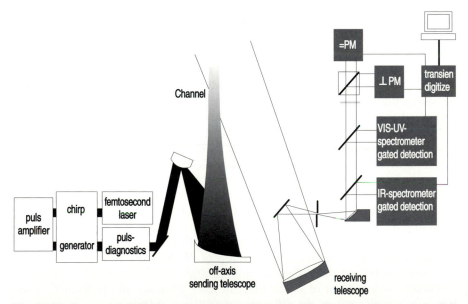

Fig. 15.15. Setup of fs-LIDAR: laser source and sending telescope are shown on the left side, receiving telescope and time resolved detection apparatus from the UV to the IR are depicted on the right side [15.98].

from the UV into the near infrared. In the plasma the laser beam is again de-focused. This restricts the plasma to a narrow range along the beam axis. The beam is then refocused again in air and generates a new plasma spot, which results in a series of bright white-light spots along the beam propagation axis like sausages on a straight cord. These white-light spots can be used as spec-troscopic light sources, where the continuum radiation propagates through the atmosphere to the detector (Fig. 15.15), which monitors the absorption spec-trum of the air constituents between sources and detector. Because of these many bright spots, the laser beam can be seen by the naked eye at distances up to some kilometers [15.98].

A detailed representation of the various laser-spectroscopic techniques for atmospheric research can be found in [15.94, 15.99]. Many examples were given in [15.99–15.103]. The basic physics of laser beam propagation through the atmosphere was discussed in [15.104, 15.105].

15.2.3 Spectroscopic Detection of Water Pollution

Unfortunately, the pollution of water by oil, gasoline, or other pollutants is increasing. Several spectroscopic techniques have been developed to measure the concentrations of specific pollutants. These techniques are not only helpful in tracing the polluting source but they can also be used to initiate measures against the pollution.

Often the absorption spectra of several pollutants overlap. It is therefore not possible to determine the specific concentrations of different pollutants from a single absorption measurement at a given wavelength λ. Either sev-eral well-selected excitation wavelengths λ_i have to be chosen (which is time consuming for in situ measurements) or time-resolved fluorescence excitation spectroscopy can be used. If the excited states of the different components have sufficiently different effective lifetimes, time-gated fluorescence detec-

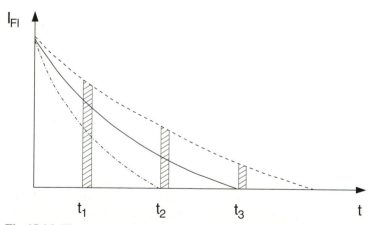

Fig. 15.16. Fluorescence decay curves of different oil species

tion at two or three time delays Δt_i after the excitation pulse allows a clear distinction between different components.

This was demonstrated by Schade [15.106], who measured laser-induced fluorescence spectra of diesel oil and gasoline, excited by a nitrogen laser at $\lambda = 337.1$ nm. Time-resolved spectroscopy exhibits two different lifetimes that are determined by measuring the intensity ratio of the LIF in two different time windows (Fig. 15.16). This time-resolved spectroscopy increases the detection sensitivity for field measurements. Mineral-oil pollutions of 0.5 mg/l in water or 0.5 mg/kg in polluted soil can be detected [15.106]. Using photoacoustic spectroscopy with a dye laser, concentrations down to $10^{-6} - 10^{-9}$ mol/l of transuranium elements could be detected in the ground water. Such techniques are important for safety control around military test grounds and plutonium recycling factories [15.107].

15.3 Applications to Technical Problems

Although the main application range of laser spectroscopy is related to basic research in various fields of physics, chemistry, biology, and medicine, there are quite a few interesting technical problems where laser spectroscopy offers elegant solutions. Examples include investigations and optimization of flames and combustion processes in fossil power stations, in car engines, or in steel plants; analytical spectroscopy of surfaces or liquid alloys for the production of high-purity solids; or measurements of flow velocities and turbulences in aerodynamics and hydrodynamic problems.

15.3.1 Spectroscopy of Combustion Processes

A detailed knowledge of the chemical reactions and gas-dynamical processes occurring during the combustion process is indispensible for the optimization of the thermodynamic efficiency and the minimization of pollutants. Spatially and time-resolved spectroscopy allow measurements of the concentrations of different reaction products during the combustion, which can lead to a detailed understanding of the different stages during its development and their dependence on temperature, pressure, and geometry of the combustion chamber.

The technical realization uses a one-, two-, or even three-dimensional grid of laser beams that pass through the combustion chamber. If the laser wavelength is tuned to absorption lines of atoms, molecules, or radicals, the spatial distribution of the LIF can be monitored with a video camera. With pulsed lasers and appropriately chosen gates in the electronic detection system, time-resolved spectroscopy is possible. The spatial distribution of the investigated reaction products during selected time intervals after the ignition can then be observed on a monitor. This gives direct information of the flame-front development, which can be followed on the screen in slow motion mode.

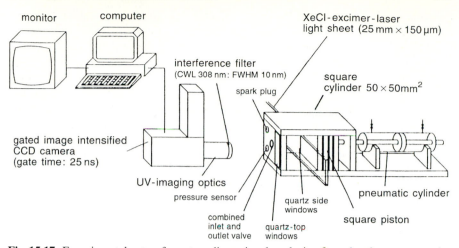

Fig. 15.17. Experimental setup for a two-dimensional analysis of combustion processes by measuring the spatial distribution of OH radicals by LIF [15.108]

In many combustion processes OH radicals are produced as an intermediate product. These radicals can be excited by a XeCl laser at 308 nm. The UV fluorescence of OH can be discriminated against the bright background of the flame by interference filters. A possible experimental setup is depicted in Fig. 15.17. The beam of the XeCl laser is imaged into the combustion chamber in such a way that it forms a cross section of $0.15 \times 25\,mm^2$. A CCD camera (Sect. 4.5) with UV optics and a gate time of 25 ns monitors the OH fluorescence with spatial and time resolution [15.108].

For quantitative measurements of molecular concentrations by LIF the ratio A_i/R_i of radiative and radiationless deactivation probabilities of the excited level $|i\rangle$ must be known. At high pressures the collisional quenching of $|i\rangle$ becomes important, which may change the ratio A_i/R_i considerably during the combustion process. However, if the laser excites predissociating levels with very short effective lifetimes (Fig. 15.18) the predissociation rate generally far exceeds the collisional quenching rate. Although the fluorescence rate is weakened, since most molecules predissociate before they emit a fluorescence photon, the fluorescence efficiency is not much influenced by collisions [15.109a]. With tunable excimer lasers such predissociating levels can be excited for most radicals that are relevant in combustion processes. The high intensity of excimer lasers has the additional advantage that the absorbing transition can be saturated. The LIF intensity is then independent of the absorption probability, but depends only on the concentration of the absorbing species [15.109b]. The experimental setup for measuring the concentrations of OH, NO, CO, DH, etc. radicals during the combustion in a car engine (Otto motor) is illustrated in Fig. 15.19, where the LIF is reflected by a mirror on the moving piston through an exit window onto the CCD camera [15.110].

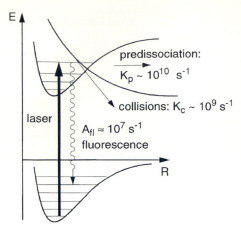

Fig. 15.18. LIF spectroscopy of predisso-ciating levels, where the predissociating rate is fast compared to the collisional quenching rates (A_{Fl}, fluorescence rate per molecule; k_{p}, predissociating rate; k_{c}, collisional quenching rate)

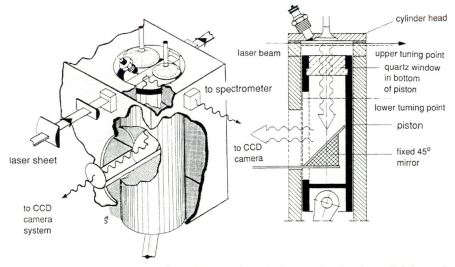

Fig. 15.19. Measurements of radical concentrations during combustion in a slightly mod-ified car engine with windows for laser entrance and fluorescence exit [15.110]

With picosecond or femtosecond lasers the excitation efficiency and the saturation of the absorbing level is nearly independent of collisions. At a pressure of 1 bar the mean time between inelastic collisions is about $10^{-9}-10^{-10}$ s, which is long compared to the laser pulse width.

The spatial temperature variation in flames and combustions can also be determined by CARS (Sect. 8.3), which yields the population distribution of rotational–vibrational levels over the combustion region [15.111, 15.112]. A computer transforms the spatially resolved CARS signal into a colorful temperature profile on the screen.

15.3.2 Applications of Laser Spectroscopy to Materials Science

For the production of materials for electronic circuits, such as chips, the demands regarding purity of materials, their composition, and the quality of the production processes become more and more stringent. With decreasing size of the chips and with increasing complexity of the electronic circuits, measurements of the absolute concentrations of impurities and dopants become important. The following two examples illustrate how laser spectroscopy can be successfully applied to the solution of problems in this field.

Irradiating the surface of a solid with a laser, material can be ablated in a controlled way by optimizing intensity and pulse duration of the laser (*laser ablation* [15.113]). Depending on the laser wavelength, the ablation is dominated by thermal evaporation (CO_2 laser) or photochemical processes (excimer laser). Laser-spectroscopic diagnostics can distinguish between the two processes. Excitation spectroscopy or resonant two-photon ionization of the sputtered atoms, molecules, or fragments allows their identification (Fig. 15.20). The velocity distribution of particles emitted from the surface can be obtained from the Doppler shifts and broadening of the absorption lines, and their internal energy distribution from the intensity ratios of different vibrational–rotational transitions [15.114]. With a pulsed ablation laser the measured time delay between ablation pulses and probe laser pulses allows the determination of the velocity distribution.

Resonant two-photon ionization in combination with a time-of-flight mass spectrometer gives the mass spectrum. In many cases, one observes a broad mass range of clusters. The question is whether these clusters were emitted from the solid or whether they were formed by collisions in the evaporated cloud just after emission. Measurements of the vibrational-energy distributions can answer this question. If the mean vibrational energy is much higher than the temperature of the solid, the molecules were formed in the gas phase, where an insufficient number of collisions cannot fully transfer the internal energy of molecules formed by recombination of sputtered atoms into kinetic energy [15.115].

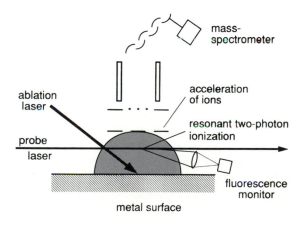

Fig. 15.20. Detection of atoms and molecules sputtered by ion bombardment or laser ablation from a surface and measurements of their energy distribution

Whereas laser ablation of graphite yields thermalized C_2 molecules with a rotational–vibrational energy distribution following a Boltzmann distribution at the temperature T of the solid, ablation of electrical isolators, such as AlO, produces AlO molecules with a large kinetic energy of 1 eV, but a "rotational temperature" of only 500 K [15.116].

For the production of thin amorphous silicon layers (for example, for solar photovoltaic energy converters) often the condensation of gaseous silane (SiH_4) or Si_2H_6, which is formed in a gas discharge, is utilized. During the formation of Si(H) layers, the radical SiH_2 plays an important role, which has absorption bands within the tuning range of a rhodamine 6G dye laser. With spectral- and time-resolved laser spectroscopy the efficiency of SiH_2 formation by UV laser photodissociation of stable silicon–hydrogen compounds can be investigated, as well as its reactions with H_2, SiH_4, or Si_2H_6. This gives information about the influence of SiH_2 concentration on the formation or elimination of dangling bonds in amorphous silicon [15.117].

Of particular interest for in situ determinations of the composition of alloys is the technique of laser microspectral analysis [15.118], where a microspot of the material surface is evaporated by a laser pulse and the fluorescence spectrum of the evaporated plume serves to monitor the composition.

Surface science is a rapidly developing field that has gained a lot from applications of laser spectroscopy [15.119, 15.120]. The sensitive technique of surface-enhanced Raman spectroscopy, which gives information of molecules adsorbed on surfaces, was discussed in Sect. 8.4.2.

15.3.3 Measurements of Flow Velocities in Gases and Liquids

For many technical problems of hydrodynamics or aerodynamics, the velocity profile $v(r, t)$ of a flowing medium in pipes or around solid bodies is of great importance. Doppler anemometry (Sect. 12.8) is a technique of heterodyne laser spectroscopy, where these velocity profiles can be determined from the measured Doppler shifts of the scattered light [15.121–15.123]. The beam of a HeNe or Ar^+ laser with wave vector k_L passes through a volume element dV of the flowing medium. The frequency ω' of light, scattered in the direction k_s by particles with velocity v (Fig. 15.21), is Doppler shifted to

$$\omega' = \omega_L - (k_L - k_s) \cdot v \,.$$

The scattered light is imaged onto a detector, where it is superimposed with part of the laser beam. The detector output contains the difference-frequency spectrum $\Delta\omega = \omega_L - \omega' = (k_L + k_s) \cdot v$, which is electronically monitored with a heterodyne technique. One example is an airborne CO_2 laser anemometer that was developed for measuring wind velocities in the stratosphere in order to improve long-term weather forecasts [15.124]. Further examples are measurements of the velocity profiles in the exhaust of turbine engines of planes, in pipelines for gases and liquids, or even in the arteries of the human body.

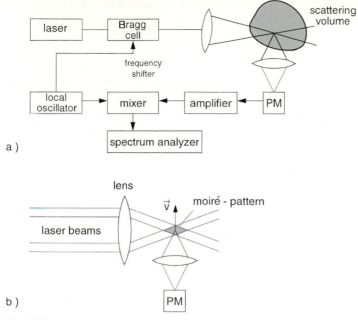

Fig. 15.21. Laser Doppler anemometry

15.4 Applications in Biology

Three aspects of laser spectroscopy are particularly relevant for biological applications. These are the high spectral resolution combined with high temporal resolution, and the high detection sensitivity. Laser light, focused into cells, also offers a high spatial resolution. The combination of LIF and Raman spectroscopy has proven very helpful in the determination of the structure of biological molecules, whereas time-resolved spectroscopy plays an essential role in the study of fast dynamical processes, such as the isomerization during photosynthesis or the formation of antenna molecules during the first steps of the visual process. Many of these spectroscopy techniques are based on the absorption of laser photons by a biological system, which brings the system into a nonequilibrium state. The time evolution of the relaxation processes that try to bring the system back to thermal equilibrium can be followed by laser spectroscopy [15.125].

   We will give several examples to illustrate the many possible applications of laser spectroscopy in the investigation of problems in molecular biology.

15.4.1 Energy Transfer in DNA Complexes

DNA molecules, with their double-helix structure, provide the basis of the genetic code. The four different bases (adenine, guanine, cytosine, and thymine), which are the building blocks of DNA, absorb light in the near UV at slightly different wavelengths, but overlapping absorption ranges. By inserting dye molecules between the bases, the absorption can be increased and shifted into the visible range. The absorption spectrum and fluorescence quantum yield of a dye molecule depend on the specific place within the DNA molecule where the dye molecule has been built in. After absorption of a photon, the excitation energy of the absorbing dye molecule may be transferred to the neighboring bases, which then emit their characteristic fluorescence spectrum (Fig. 15.22). On the other hand, excitation of the DNA by UV radiation may invert the direction of the energy transfer from the DNA bases to the dye molecule.

The efficiency of the energy transfer depends on the coupling between the dye molecule and its surroundings. Measurements of this energy transfer for different base sequences allow the investigation of the coupling strength and its variation with the base sequence [15.126]. For example, the base guanine within a DNA–dye complex can be selectively excited at $\lambda = 300$ nm without affecting the other bases. The energy transfer rate is determined from the ratio of quantum efficiencies under excitation with visible light (excitation of the acridine dye molecule) and UV light (direct excitation of guanine), respectively [15.127].

Since the use of dye molecules in cells plays an important role in the diagnosis and therapy of cancer (Sect. 15.5) a detailed knowledge of the relevant energy transfer processes and the photoactivity of different dyes is of vital interest.

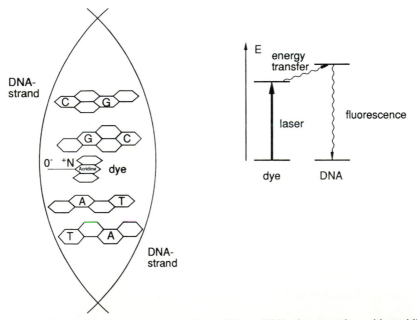

Fig. 15.22. Example of energy transfer within a DNA–dye complex with acridine dye molecules inserted between the bases adenine and guanine after laser excitation [15.127]

15.4.2 Time-Resolved Measurements of Biological Processes

The detailed knowledge of the different steps of biological processes on a molecular level is one of the ambitious goals of molecular biology. The importance of this field was underlined by the award of the Nobel Prize in chemistry in 1988 to J. Deisenhofer, R. Huber, and H. Michel for the elucidation of the primary steps in photosynthesis and the visual process [15.128]. This subsection illustrates the importance of time-resolved Raman spectroscopy in combination with pump-and-probe techniques (Sect. 11.4) for the investigation of fast biological processes.

Hemoglobine (Hb) is a protein that is used in the body of mammals for the transportation of O_2 and CO_2 through blood circulation. Although its structure has been uncovered by X-ray diffraction, not much is known about the structural change of Hb when it absorbs O_2 and becomes oxyhemoglobin HbO_2, or when it releases O_2 again. With laser Raman spectroscopy (Chap. 8) its vibrational structure can be studied, giving information on the force constants and the molecular dynamics. Based on high-resolution Raman spectroscopy with cw lasers, empirical rules have been obtained about the relationship between vibrational spectra and the geometrical structure of large molecules. The change in the Raman spectra of Hb before and after the attachment of O_2 therefore gives hints about the corresponding structural change. If HbO_2 is photodissociated by a short laser pulse, the Hb molecule is left in a nonstationary state. The relaxation of this nonequilibrium state back into the ground state of Hb can be followed by time-resolved Raman or LIF spectroscopy [15.129].

Excitation with polarized light creates a partial orientation of selectively excited molecules. The relaxation rate of these excited molecules and its dependence on the degree of orientation can be studied either by the time-resolved absorption of polarized light by the excited molecules, or by measuring the polarization of the fluorescence and its time dependence [15.130].

Of particular interest is the investigation of the primary processes of vision. The light-sensitive layer in the retina of our eye contains the protein *rhodopsin* with the photoactive molecule *retinal*. Rhodopsin is a membrane protein whose structure is not yet fully investigated. The polyene molecule retinal exists in several isomeric forms. Since the vibrational spectra of these isomers are clearly distinct, at present Raman spectroscopy provides the most precise information on the structure and dynamical changes of the different retinal configurations. In particular, it has allowed the assignment of the different retinal configurations present in the isomers rhodopsin and isorhodopsin before photoabsorption, and batho-rhodopsin after photoexcitation. With pico- and femtosecond Raman spectroscopy it was proved that the isomer batho-rhodopsin is formed within 1 ps after the photoabsorption. It transfers its excitation energy within 50 ns to transducin, which triggers an enzyme cascade that finally results, after several slower processes, in the signal transport by nerve conduction to the brain [15.131, 15.132].

Probably the most important biochemical process on earth is the photosynthesis within the chlorophyl cells in green plants. Recently, it was discovered that the primary processes within the reaction center of chlorophyl proceed on a time scale of 30 to 100 fs. The excitation energy is used for a proton transfer, which finally delivers the energy for the photosynthetic reaction [15.133].

These examples reveal that these extremely fast biochemical processes could not have been studied without time-resolved laser spectroscopy. It not only provides the necessary spectral resolution but also the sensitivity essential for those investigations. More examples and details on the spectroscopy of ultrafast biological processes can be found in [15.134–15.138].

15.4.3 Correlation Spectroscopy of Microbe Movements

The movement of microbes in a fluid can be observed with a microscope. They move for several seconds a straight path, but suddenly change direction. If they are killed by chemicals added to the fluid, their characteristic movement changes and can be described by a Brownian motion if there is no external perturbation. With correlation spectroscopy (Sect. 12.8), the average mean-square velocity $\langle v^2 \rangle$ and the distribution $f(v)$ of the velocities of living and dead microbes can be measured.

The sample is irradiated with a HeNe laser and the scattered light is superimposed with part of the direct laser light on the photocathode of a photomultiplier. The scattered light experiences a Doppler shift $\Delta v = v(v/c) \times (\cos \vartheta_1 - \cos \vartheta_2)$, where ϑ_1 and ϑ_2 are the angles of the velocity vector v against the incident laser beam and the direction of the scattered light, respectively. The frequency distribution of this heterodyne spectrum is a measure for the velocity distribution.

Measurements of the velocity distribution of E. coli bacteria in a temperature-stabilized liquid gave an average velocity of 15 µm/s (Fig. 15.23), where the maximum speed ranges up to 80 µm/s. Since the size of E. coli is only

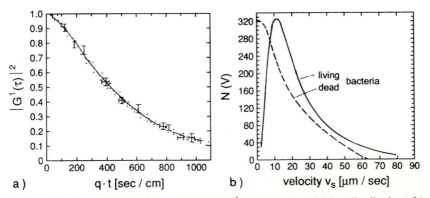

Fig. 15.23. Measured correlation function $G^1(\tau)$ (**a**) and velocity distribution (**b**) computed from $G^1(\tau)$ for living (*solid curve*) and dead (*dashed curve*) E. coli bacteria in solution. The *dashed curve* corresponds to the Poisson distribution of a random walk (Brownian motion) [15.139]

about 1 μm, this speed corresponds to 80 body lengths per second. In comparison, the swimming champion Ian Thorpe only reached 2 m/s, which corresponds to 1 body length per second. If $CuCl_2$ is added to the solution the bacteria die, and the velocity distribution changes to that of a Brownian motion, which shows another correlation spectrum $I(K, t) \propto \exp(-D\Delta K^2 t)$, where $\Delta K = K_0 - K_s$ is the difference between the wave vectors of incident and scattered light. From the correlation spectrum the diffusion coefficient $D = 5 \times 10^{-9}$ cm^2/s and a Stokes diameter of 1.0 μm can be derived [15.139].

Another technique uses a stationary Moiré fringe pattern produced by the superposition of two inclined beams of the same laser (Fig. 15.21b). The distance between the interference maxima is $\Delta = \lambda \sin(\frac{1}{2}\alpha)$, where α is the angle between the two wave vectors. If a particle moves with a velocity v across the maxima, the scattered light intensity $I_s(t)$ exhibits periodic maxima with a period $\Delta t = \Delta/(v \cos \beta)$, where β is the angle between v and $(k_1 + k_2)$.

15.4.4 Laser Microscope

A beam of a TEM$_{00}$ laser (Sect. 5.3) with a Gaussian intensity profile is focused to a diffraction-limited spot with the diameter $d \simeq 2\lambda f/D$ by an adapted lens system with the focal length f and the limiting aperture D. For example, with $f/D = 1$ at $\lambda = 500$ nm, a focal diameter of $d \simeq 1.0$ μm can be achieved with a corrected microscope lens system. This allows the spatial resolution of single cells and their selective excitation by the laser.

The LIF emitted by the excited cells can be collected by the same microscope and may be imaged either onto a video camera or directly observed visually. A commercial version of such a laser microscope is displayed in Fig. 15.24. For time-resolved measurements a nitrogen-laser-pumped dye laser can be used. The wavelength λ is tuned to the absorption maximum of the biomolecule under investigation. For absorption bands in the UV, the dye-laser output can be frequency doubled. Even if only a few fluorescence photons can be detected per laser pulse, the use of video-intensified detection and signal averaging over many pulses may still give a sufficiently good signal-to-noise ratio [15.140].

Many of the spectroscopic techniques discussed above can be applied in combination with the laser microscope, which gives the additional advantage of spatial resolution. One example is the spectrally and spatially resolved laser-induced fluorescence excited by the laser within a certain part of a living cell. The migration of the excitation energy through the cells to their membrane within a few seconds was observed. In addition, the migration of receptor cells through cell membranes can be studied with this technique [15.141]. One example is the measurement of the intracellular distribution of injected photosensitizing porphyrin and its aggregation [15.142]. Porphyrine fluorescence was localized in the plasma membrane, the cytoplasma, the nuclear membrane, and the nucleoli. A redistribution of the porphyrine molecules from the plasma membrane to the nuclear membrane and adjacent intracellular sites was observed with increasing incubation time.

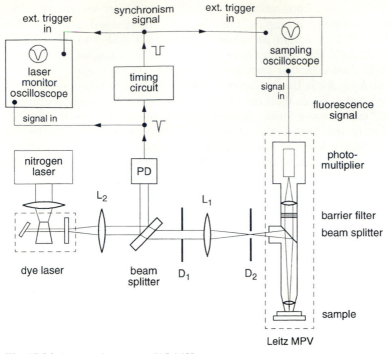

Fig. 15.24. Laser microscope [15.140]

Damage to the respiratory chain is correlated with a decrease in ATP production and a lack of certain enzymes and cytochromes. These defects can be detected by measuring the autofluorescence of flavine molecules in intact and respiratory-deficient yeast strains with advanced microscopic techniques [15.143a].

15.4.5 Time-Resolved Spectroscopy of Biological Processes

The combination of time-resolved laser spectroscopic techniques with a laser scanning microscope opens new possibilities for the investigation of dynamical processes with high spatial resolution. This is demonstrated in [15.143b] by time-resolved fluorescence measurements of carcinoma cells compared with normal cells.

The application of femtosecond lasers to the investigation of important biological processes, such as the photosynthesis or the visual process, has brought a very detailed understanding of the different steps between the photoexcitation and the final product of this processes [15.144]. For example, the primary reaction of sensory rhodopsin after the excitation of the S_1 state is the decay with a time constant of 4 ps into a redshifted photo product, while bacteriorhodopsin decays much faster with a time constant of 500 fs. The population decay can be probed with a second delayed pulse, which is absorbed by molecules in the excited state [15.145].

Another example is the femtosecond-transient absorption and fluorescence after two-photon excitation of carotenoids. The excited β-carotene decays with a time constant of 9 ± 0.2 ps. The energy transfer process from the excited S_1 state in light-harvesting proteins can be monitored by the observed chlorophyl fluorescence [15.146].

An important piece of information is the change in molecular structure during these fast processes. Here time-resolved Raman spectroscopy and X-ray diffraction with femtosecond laser-produced brilliant X-ray sources are powerful tools that are more and more applied to molecular biology.

15.5 Medical Applications of Laser Spectroscopy

Numerous books have been published on laser applications in medical research in hospital practice [15.147–15.150]. Most of these applications rely on the high laser-output power, which can be focused into a small volume. The strong dependence of the absorption coefficient of living tissue on the wavelength allows selection of the penetration depth of the laser beam by choosing the proper laser wavelength [15.149]. For example, skin carcinoma or portwine marks should be treated at wavelengths for a small penetration depth in order to protect the deeper layers of the epidermia from being damaged, while cutting of bones with lasers or treatment of subcutaneous cancer must be performed at wavelengths with greater penetration depth. The most spectacular outcomes of laser applications in medicine have been achieved in laser surgery, dermatology, ophthalmology, and dentistry.

There are, however, also very promising direct applications of laser *spectroscopy* for the solution of problems in medicine. They are based on new diagnostic techniques and are discussed in this section.

15.5.1 Applications of Raman Spectroscopy in Medicine

During surgery on a patient, the optimum concentration and composition of narcotic gases can be indicated by the composition of the respiratory gases, that is, with the concentration ratio of $N_2 : O_2 : CO_2$. This ratio can be measured in vivo with Raman spectroscopy [15.151]. The gas flows through a cell that is placed inside a multipass arrangement for an argon laser beam (Fig. 15.25). Several detectors with special spectral filters are arranged in a plane perpendicular to the beam axis. Each detector monitors a selected Raman line, which allows the simultaneous detection of all molecular components of the gas.

The sensitivity of the method is illustrated by Fig. 15.26, which depicts the time variation of the CO_2, O_2, and N_2 concentration in the exhaled air of a human patient. Note the variation of the concentrations with changing breathing periods. The technique can be used routinely in clinical practice for anesthetic control during operations and obviously also for alcohol tests of car drivers.

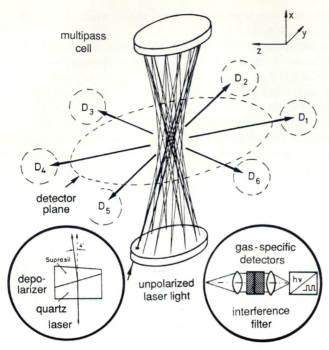

Fig. 15.25. Multipass cell and spectrally selective detector arrangement for sensitive Raman spectroscopy and diagnostics of molecular gases [15.151]

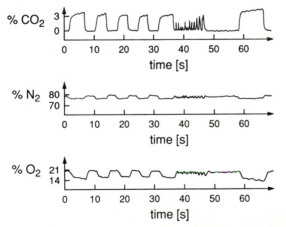

Fig. 15.26. CO_2, N_2, and O_2 concentrations of respiratory gases for varying breath periods, measured in vivo with the arrangement of Fig. 15.25 [15.151]

Instead of Raman spectroscopy, infrared absorption spectroscopy can be used in cases of infrared-active transition, for example, for CO, CO_2, NO, and CH_4. With cavity ring-down spectroscopy a high sensitivity can be reached and spurious molecular concentrations resulting from fermentation processes in the stomach can still be detected [15.152].

15.5.2 Heterodyne Measurements of Ear Drums

A large fraction of ear diseases of elderly people is due to changes in the frequency response of the ear drum. While until now investigations of such changes had to rely on the subjective response of the patient, novel laser-spectroscopic techniques allow objective studies of frequency-dependent vibrational amplitudes of the ear drum and their local variation for different locations on the drum with a laser Doppler vibrometer (Fig. 15.27). The experimental arrangement is illustrated in Fig. 15.28. The output of a diode laser is fed through an optical fiber to the ear drum. The light reflected by the drum is collected by a lens at the end of the fiber and is sent back through the fiber, where it is superimposed on a photodetector behind a beam splitter with part of the direct laser light. The ear is exposed to the sound waves of a loudspeaker with variable audio frequency f. The frequency ω of the light reflected by the vibrating ear drum is Doppler shifted. The amplitude $A(f)$ of the illuminated area of the vibrating drum can be derived from the frequency spectrum of the heterodyne signals (Sect. 12.7). In order to transfer the heterodyne spectrum in a region with less noise, the laser light is modulated at the frequency $\Omega \approx 40\,\mathrm{MHz}$ by an optoacoustic modulator producing sidebands at $\omega \pm \Omega$ [15.153].

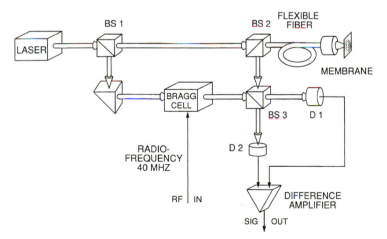

Fig. 15.27. Principle of laser Doppler vibrometer

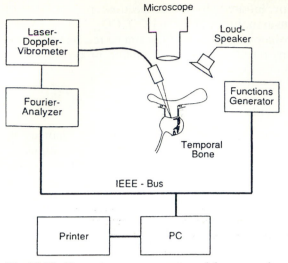

Fig. 15.28. Heterodyne measurements of frequency-dependent vibrations of the ear drum and their local variations [15.153]

15.5.3 Cancer Diagnostics and Therapy with the HPD Technique

Recently, a method for diagnostics and treatment of cancer has been developed that is based on photoexcitation of the fluorescing substance hemato-porphyrin derivative (HPD) [15.154, 15.155]. A solution of this substance is injected into the veins and is distributed in the whole body after a few hours. While HPD is released by normal cells after 2–4 days, it is kept by cancer cells for a longer time [15.156]. If a tissue containing HPD is irradiated by a UV laser, it emits a characteristic fluorescence spectrum, which can be used for a diagnostic of cancer cells. Figure 15.29 shows the emission spectrum of a tissue with and without HPD, and also the fluorescence of pure HPD in

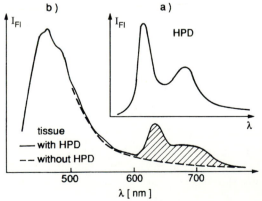

Fig. 15.29a,b. Nitrogen laser-excited fluorescence spectrum of HPD in solution (**a**) and of tissue (**b**) without HPD (*dashed curve*), and with HPD (*solid line*) two days after injection. The *hatched* area represents the additional absorption of HPD [15.157]

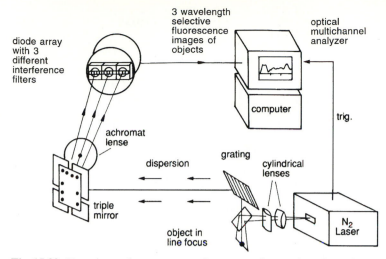

Fig. 15.30. Experimental arrangement for cancer diagnostics of rat tissue [15.157]

a liquid solution excited by a nitrogen laser at $\lambda = 337$ nm. The experimental arrangement for detecting cancerous tissue in rats is exhibited in Fig. 15.30 [15.157]. The fluorescence is spectrally resolved by a grating and spatially separated by three slightly folded mirrors, which image a cancer region and a region of normal cells onto different parts of the diode array of an optical multichannel analyzer (Sect. 4.5). A computer subtracts the fluorescence of the normal tissue from that of a cancerous tissue.

Absorption of photons in the range 620−640 nm brings HPD into an excited state, which reacts with normal oxygen in the $O_2(^3\Pi)$ state and transfer it into the $O_2(^1\Delta)$ state, which apparently reacts with the surrounding cells and destroys them. Although the exact mechanism of these processes is not yet completely understood, it seems that this HPD method allows a rather selective destruction of cancer cells without too much damage to the normal cells. The technique was developed in the USA, intensively applied in Japan [15.158], and has since been applied successfully to patients with esophageal cancer, cervical carcinoma, and other kinds of tumors that can be reached by optical fibers without invasive surgery [15.159].

15.5.4 Laser Lithotripsy

Thanks to the development of thin, flexible optical fibers with a high damage threshold for the radiation [15.160, 15.161], inner organs of the human body, such as the stomach, bladder, gallbladder or kidneys, can be selectively irradiated by laser radiation. A new technique for breaking kidneystones to pieces by irradiation with pulsed lasers (laser lithotripsy) has found increasing interest because it has several advantages compared to the ultrasonic shockwave lithotripsy [15.162–15.164].

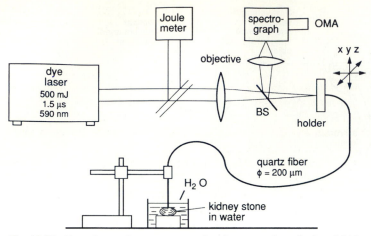

Fig. 15.31. Experimental arrangement for the spectral analysis of kidney stones for the determination of stone composition

An optical fused quartz fiber is inserted through the urinary tract until it nearly touches the stone that is to be broken. This can be monitored by X-ray diagnosis or by endoscopy through a fiber bundle that contains, besides the fiber for guiding the laser beam, other fibers for illumination, viewing and monitoring the laser-induced fluorescence (Fig. 15.31).

If the pulse of a flashlamp-pumped dye laser is transported through the fiber and focused onto the kidneystone, the rapid evaporation of the stone

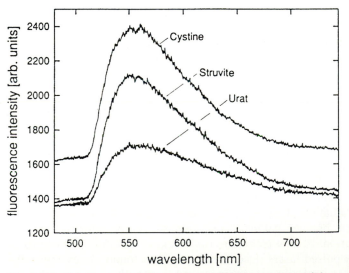

Fig. 15.32. Fluorescence of three different kidney stone materials excited with a dye laser at $\lambda = 497$ nm at low intensities to prevent plasma breakdown [15.165]

material results in a shock wave in the surrounding liquid, which leads to destruction of the stone after several laser shots [15.163]. The necessary laser power and the optimum wavelength depend on the chemical composition of the stone, which generally varies for different patients. It is therefore advantageous to know the stone composition before distruction in order to choose optimum laser conditions. This information can be obtained by collecting the fluorescence of the evaporated stone material at low laser powers through an optical fiber (Fig. 15.31). The fluorescence spectrum is monitored with an optical multichannel analyzer and a computer gives, within seconds, the wanted information about the stone composition [15.165].

First demonstrations of the capability of spectral analysis of kidney stones in vitro are illustrated in Fig. 15.32, where the fluorescence spectra of different kidney stones that had been irradiated in a water surrounding outside the body, and that were detected with the arrangement of Fig. 15.30 are shown [15.165]. Further information on laser lithotripsy and spectroscopic control of this technique can be found in [15.166, 15.167].

15.5.5 Laser-Induced Thermotherapy of Brain Cancer

Laser-induced interstitial thermotherapy represents a minimally invasive therapy, where the cancerous tissue is irradiated by laser light guided through an optical fiber. Planning the operation anticipates the knowledge of the absorption and scattering properties of cancerous tissue compared to healthy tissue. Several optical and computational techniques have been developed that can

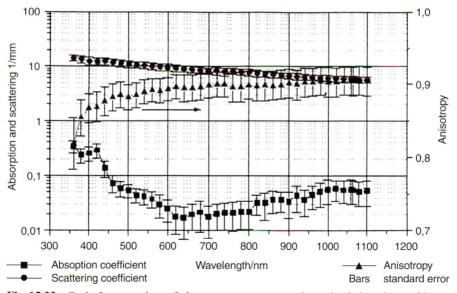

Fig. 15.33. Optical properties of human gray matter determined in vitro with an integrating-sphere setup and an inverse Monte Carlo technique [15.168]

be used to localize the cancerous tissue and to optimize the optical radiation dose. Here the wavelength dependence of the absorption and scattering coefficients is determined in foregoing experiments. Figure 15.33 shows for illustration the spectral dependence of both coefficients and also the anisotropy of the scattering in human brain tissue determimed in vitro [15.168].

15.5.6 Fetal Oxygen Monitoring

Monitoring the oxygen concentration of a baby during longer birth processes is of crucial importance for the lasting health of the child. Until now the available equipment was not well adapted for clinical applications. Here a laser-based technique, using light scattering measurements seems to be very successful. The laser light is transported to the skull of the baby through an optical fiber and the scattered light is collected by a second fiber, placed some centimeters distant from the first. The collected light is then monitored as a function of the wavelength [15.169]. Since the scattering cross section depends on the oxygen concentration, after calibration this method allows the determination of the wanted O_2 concentration.

15.6 Concluding Remarks

The preceeding selection of examples of laser spectroscopic applications is somewhat arbitrary and by no means complete. The progress in this field may be measured by the increasing variety of conferences and workshops on applications of laser spectroscopy in science and technology. A good survey can be found in many conference proceedings of the Society of Photo-optical Instrumentation and Engineering (SPIE) [15.170]. The reasons for this rapid expansion of applications are manifold:

- First, more types of reliable and "easy to handle" lasers in all spectral regions of interest are now commercially available.
- Second, spectroscopic equipment has greatly improved during recent years.
- Last, but not least, our understanding of many processes on a molecular level has become much more advanced. This allows a better analysis of spectral information and its transformation into reliable models of structures and processes.

References

Chapter 1

1.1   *Laser Spectroscopy I–XV*, Proc. Int. Confs. 1973–2001,
       I, Vale 1973, ed. by R.G. Brewer, A. Mooradian (Plenum, New York 1974);
       II, Megeve 1975, ed. by S. Haroche, J.C. Pebay-Peyroula, T.W. Hänsch,
       S.E. Harris, Lecture Notes Phys., Vol. 43 (Springer, Berlin, Heidelberg 1975);
       III, Jackson Lake Lodge 1977, ed. by J.L. Hall, J.L. Carlsten, Springer Ser. Opt.
       Sci., Vol. 7 (Springer, Berlin, Heidelberg 1977);
       IV, Rottach-Egern 1979, ed. by H. Walther, K.W. Rothe, Springer Ser. Opt. Sci.,
       Vol. 21 (Springer, Berlin, Heidelberg 1979);
       V, Jaspers 1981, ed. by A.R.W. McKellar, T. Oka, B.P. Stoichef, Springer Ser.
       Opt. Sci., Vol. 30 (Springer, Berlin, Heidelberg 1981);
       VI, Interlaken 1983, ed. by H.P. Weber, W. Lüthy, Springer Ser. Opt. Sci.,
       Vol. 40 (Springer, Berlin, Heidelberg 1983);
       VII, Maui 1985, ed. by T.W. Hänsch, Y.R. Shen, Springer Ser. Opt. Sci., Vol. 49
       (Springer, Berlin, Heidelberg 1985);
       VIII, Are 1987, ed. by W. Persson, S. Svanberg, Springer Ser. Opt. Sci., Vol. 55
       (Springer, Berlin, Heidelberg 1987);
       IX, Bretton Woods 1989, ed. by M.S. Feld, J.E. Thomas, A. Mooradian (Aca-
       demic, New York 1989);
       X, Font Romeau 1991, ed. by M. Ducloy, E. Giacobino, G. Camy (World Sci-
       entific, Singapore 1992);
       XI, Hot Springs, VA 1993, ed. by L. Bloomfield, T. Gallagher, D. Larson, AIP
       Conf. Proc. **290** (AIP, New York 1993);
       XII, Capri, Italy 1995, ed. by M. Inguscio, M. Allegrini, A. Sasso (World Sci-
       entific, Singapore 1995);
       XIII, Hangzhou, P.R. China 1997, ed. by Y.Z. Wang, Y.Z. Wang, Z.M. Zhang
       (World Scientific, Singapore 1997);
       XIV Innsbruck, Austria 1999, ed. by R. Blatt, J. Eschner, D. Leihfried,
       F. Schmidt-Kaler (World Scientific, Singapore 1999)
       XV, Snowbird, USA 2001, ed. by St. Chu (World Scientific, Singapore 2002)
1.2   *Advances in Laser Sciences I–IV*, Int. Conf. 1985–1989,
       I, Dallas 1985, ed. by W.C. Stwally, M. Lapp (Am. Inst. Phys., New York 1986)
       II, Seattle 1986, ed. by W.C. Stwalley, M. Lapp, G.A. Kennedy-Wallace (AIP,
       New York 1987);
       III, Atlantic City 1987, ed. by A.C. Tam, J.L. Gale, W.C. Stwalley (AIP, New
       York 1988);
       IV, Atlanta 1988, ed. by J.L. Gole et al. (AIP, New York 1989);
1.3   M. Feld, A. Javan, N. Kurnit (Eds.): *Fundamental and Applied Laser Physics*,
       Proc. Esfahan Symposium 1971 (Wiley, London 1973)
1.4   A. Mooradian, T. Jaeger, P. Stokseth (Eds.): *Tunable Lasers and Applications*,
       Springer Ser. Opt. Sci., Vol. 3 (Springer, Berlin, Heidelberg 1976)

1.5   R.A. Smith (Ed.): *Very High Resolution Spectroscopy* (Wiley Interscience, New York 1970)

1.6   *Int. Colloq. on Doppler-Free Spectroscopic Methods for Simple Molecular Systems, Aussois, May 1973* (CNRS, Paris 1974)

1.7   S. Martellucci, A.N. Chester (Eds.): *Analytical Laser Spectroscopy*, Proc. NATO ASI (Plenum, New York 1985)

1.8   Y. Prior, A. Ben-Reuven, M. Rosenbluth (Eds.): *Methods of Laser Spectroscopy* (Plenum, New York 1986)

1.9   A.C.P. Alves, J.M. Brown, J.M. Hollas (Eds.): *Frontiers of Laser Spectroscopy of Gases*, NATO ASI Series, Vol. 234 (Kluwer, Dordrecht 1988)
        T.W. Hänsch, M. Inguscio (Eds.): *Frontiers in Laserspectroscopy* (North Holland, Amsterdam 1994)

1.10   W. Demtröder, M. Inguscio (Eds.): *Applied Laser Spectroscopy*, NATO ASI Series, Vol. 241 (Plenum, New York 1991)

1.11   H. Walther (Ed.): *Laser Spectroscopy of Atoms and Molecules*, Topics Appl. Phys., Vol. 2 (Springer, Berlin, Heidelberg 1976)

1.12   K. Shimoda (Ed.): *High-Resolution Laser Spectroscopy*, Topics Appl. Phys., Vol. 13 (Springer, Berlin, Heidelberg 1976)

1.13   A. Corney: *Atomic and Laser Spectroscopy* (Clarendon, Oxford 1977)

1.14   V.S. Letokhov: *Laserspektroskopie* (Vieweg, Braunschweig 1977);
        V.S. Letokhov (Ed.): *Laser Spectroscopy of Highly Vibrationally Excited Molecules* (Hilger, Bristol 1989)

1.15   J.M. Weinberg, T. Hirschfeld (Eds.): *Unconventional Spectroscopy*, SPIE Proc. **82** (1976)

1.16   S. Jacobs, M. Sargent III, J. Scott, M.O. Scully (Eds.): *Laser Applications to Optics and Spectroscopy* (Addison-Wesley, Reading, MA 1975)

1.17   D.C. Hanna, M.A. Yuratich, D. Cotter: *Nonlinear Optics of Free Atoms and Molecules*, Springer Ser. Opt. Sci., Vol. 17 (Springer, Berlin, Heidelberg 1979)

1.18   M.S. Feld, V.S. Letokhov (Eds.): *Coherent Nonlinear Optics*, Topics Curr. Phys., Vol. 21 (Springer, Berlin, Heidelberg 1980)

1.19   S. Stenholm: *Foundations of Laser Spectroscopy* (Wiley, New York 1984)

1.20   J.I. Steinfeld: *Laser and Coherence Spectroscopy* (Plenum, New York 1978)

1.21   W.M. Yen, M.D. Levenson (Eds.): *Lasers, Spectroscopy and New Ideas*, Springer Ser. Opt. Sci., Vol. 54 (Springer, Berlin, Heidelberg 1987)

1.22   D.S. Kliger (Ed.): *Ultrasensitive Laser Spectroscopy* (Academic, New York 1983)

1.23   B.A. Garetz, J.R. Lombardi (Eds.): *Advances in Laser Spectroscopy, Vols. I and II* (Heyden, London 1982, 1983)

1.24   S. Svanberg: *Atomic and Molecular Spectroscopy*, 2nd edn., Springer Ser. Atoms Plasmas, Vol. 6 (Springer, Berlin, Heidelberg 1992)

1.25   D.L. Andrews, A.A. Demidov: *An Introduction to Laser Spectroscopy* (Plenum, New York 1989)
        D.L. Andrews (Ed.): *Applied Laser Spectroscopy* (VCH-Wiley, Weinheim 1992)

1.26   L.J. Radziemski, R.W. Solarz, J. Paissner: *Laser Spectroscopy and its Applications* (Dekker, New York 1987)

1.27   V.S. Letokhov (Ed.): *Lasers in Atomic, Molecular and Nuclear Physics* (World Scientific, Singapore 1989)

1.28   E.R. Menzel: *Laser Spectroscopy* (Dekker, New York 1994)

1.29   Z.-G. Wang, H.-R. Xia: *Molecular and Laser Spectroscopy*, Springer Ser. Chem. Phys., Vol. 50 (Springer, Berlin, Heidelberg 1991);
        R. Blatt, W. Neuhauser (Eds.): *High Resolution Laser Spectroscopy*. Appl. Phys. B **59** (1994)

1.30   J. Sneddon (Ed.): *Lasers in Analytical Atomic Spectroscopy* (Wiley, New York 1997)

1.31   R. Menzel: *Photonics: Linear and Nonlinear Interaction of Laser Light and Matter* (Springer, Heidelberg 2001)

1.32   J. Hecht: *Laser Pioneers* (Academic, Boston 1992)

1.33   Ch.H. Townes: *How the Laser Happened. Adventures of a Scientist* (Oxford Univ. Press, Oxford 1999)

1.34   J.C. Lindon, G.E. Trauter, J.L. Holmes: *Encyclopedia of Spectroscopy and Spectrometry, Vols. I–III* (Academic, London 2000)

Chapter 2

2.1   A. Corney: *Atomic and Laser Spectroscopy* (Clarendon, Oxford 1977)

2.2   A.P. Thorne, U. Litzén, S. Johansson: *Spectrophysics* (Springer, Heidelberg 1999)

2.3   I.I. Sobelman: *Atomic Spectra and Radiative Transitions*, 2nd edn., Springer Ser. Atoms Plasmas, Vol. 12 (Springer, Berlin, Heidelberg 1992)

2.4   H.G. Kuhn: *Atomic Spectra* (Longmans, London 1969)

2.5   M. Born, E. Wolf: *Principles of Optics*, 5th edn. (Pergamon, Oxford 1999)

2.6   R. Loudon: *The Quantum Theory of Light*, 3rd edn. (Clarendon, Oxford 2000)

2.7   W. Schleich: *Quantum Optics in Phase Space* (Wiley–VCH, Weinheim 2001)

2.8   S. Suter: *The Physics of Laser–Atom Interaction* (Cambridge Studies in Modern Optics, Cambridge 1997)

2.9   J.W. Robinson (Ed.): *Handbook of Spectroscopy, Vols. I–III* (CRC, Cleveland, Ohio 1974–81)
        Atomic Spectroscopy (Dekker, New York 1996)

2.10   L. May (Ed.): *Spectroscopic Tricks, Vols. I–III* (Plenum, New York 1965–73)

2.11   J.O. Hirschfelder, R. Wyatt, R.D. Coulson (Eds.): *Lasers, Molecules and Methods*, Adv. Chem. Phys., Vol. 78 (Wiley, New York 1986)

2.12   M. Cardona, G. Güntherodt (Eds.): *Light Scattering in Solids I–VI*, Topics Appl. Phys., Vols. 8, 50, 51, 54, 66, 68 (Springer, Berlin, Heidelberg 1983–91)

2.13   A. Stimson: *Photometry and Radiometry for Engineers* (Wiley-Interscience, New York 1974)

2.14   W.L. Wolfe: 'Radiometry'. In: *Appl. Optics and Optical Engineering, Vol. 8*, ed. by J.C. Wyant, R.R. Shannon (Academic, New York 1980)
        W.L. Wolfe: *Introduction to Radiometry* (SPIE, Bellingham, WA 1998)

2.15   D.S. Klinger, J.W. Lewis, C.E. Randull: *Polarized Light in Optics and Spectroscopy* (Academic, Boston 1997)

2.16   S. Huard, G. Vacca: *Polarization of Light* (Wiley, Chichester 1997)

2.17   E. Collet: *Polarized Light: Fundamentals and Applications* (Dekker, New York 1993)

2.18   D. Eisel, D. Zevgolis, W. Demtröder: Sub-Doppler laser spectroscopy of the NaK-molecule. J. Chem. Phys. **71**, 2005 (1979)

2.19   W.L. Wiese: 'Transition probabilities'. In: *Methods of Experimental Physics, Vol. 7a*, ed. by B. Bederson, W.L. Fite (Academic, New York 1968) p. 117;
        W.L. Wiese, M.W. Smith, B.M. Glennon: Atomic Transition Probabilities. Nat'l Standard Reference Data Series NBS4 and NSRDS-NBS22 (1966–1969), see Data Center on Atomic Transition Probabilities and Lineshapes, NIST Homepage (www.nist.org);
        P.L. Smith, W.L. Wiese: *Atomic and Molecular Data for Space Astronomy* (Springer, Berlin 1992)

2.20   C.J.H. Schutte: *The Wave Mechanics of Atoms, Molecules and Ions* (Arnold, London 1968);
        R.E. Christoffersen: *Basic Principles and Techniques of Molecular Quantum Mechanics* (Springer, Heidelberg 1989)

2.21   M.O. Scully, W.E. Lamb Jr., M. Sargent III: *Laser Physics* (Addison Wesley, Reading, MA 1974)

2.22   P. Meystre, M. Sargent III: *Elements of Quantum Optics*, 2nd edn. (Springer, Berlin, Heidelberg 1991)

2.23   G. Källen: *Quantum Electrodynamics* (Springer, Berlin, Heidelberg 1972)

2.24   C. Cohen-Tannoudji, B. Diu, F. Laloe: *Quantum Mechanics, Vols. I, II* (Wiley-International, New York 1977)
        C. Cohen-Tannoudji, J. Dupont-Roche, G. Grynberg: *Atom–Photon Interaction* (Wiley, New York 1992)

2.25   L. Mandel, E. Wolf: Coherence properties of optical fields. Rev. Mod. Phys. **37**, 231 (1965);
        L. Mandel, E. Wolf: *Optical Coherence and Quantum Optics* (Cambridge University Press, Cambridge 1995)

2.26   G.W. Stroke: *An Introduction to Coherent Optics and Holography* (Academic, New York 1969)

2.27   J.R. Klauder, E.C.G. Sudarshan: *Fundamentals of Quantum Optics* (Benjamin, New York 1968)

2.28   A.F. Harvey: *Coherent Light* (Wiley Interscience, London 1970)

2.29   H. Kleinpoppen: 'Coherence and correlation in atomic collisions'. In: *Adv. Atomic and Molecular Phys., Vol. 15*, ed. by D.R. Bates, B. Bederson (Academic, New York 1979) p. 423;
        H.J. Beyer, K. Blum, R. Hippler: *Coherence in Atomic Collision Physics* (Plenum, New York 1988)

2.30   R.G. Brewer: 'Coherent optical spectroscopy'. In: *Frontiers in Laser Spectroscopy, Vol. 1*, ed. by R. Balian, S. Haroche, S. Liberman (North-Holland, Amsterdam 1977) p. 342

2.31   B.W. Shore: *The Theory of Coherent Excitation* (Wiley, New York 1990)

Chapter 3

3.1   I.I. Sobelman, L.A. Vainstein, E.A. Yukov: *Excitation of Atoms and Broadening of Spectral Lines*, 2nd edn., Springer Ser. Atoms Plasmas, Vol. 15 (Springer, Berlin, Heidelberg 1995)

3.2   R.G. Breene: *Theories of Spectral Line Shapes* (Wiley, New York 1981)

3.3   K. Burnett: *Lineshapes Laser Spectroscopy* (Cambridge University Press, Cambridge 2000)

3.4   See, for instance, *Proc. Int. Conf. on Spectral Line Shapes*,
        Vol. 1, ed. by B. Wende (De Gruyter, Berlin 1981);
        Vol. 2, 5th Int. Conf., Boulder 1980, ed. by K. Burnett (De Gruyter, Berlin 1983);
        Vol. 3, 7th Int. Conf., Aussois 1984, ed. by F. Rostas (De Gruyter, Berlin 1985);
        Vol. 4, 8th Int. Conf., Williamsburg 1986, ed. by R.J. Exton (Deepak Publ., Hampton, VA 1987);
        Vol. 5, 9th Int. Conf., Torun, Poland 1988, ed. by J. Szudy (Ossolineum, Wroclaw 1989);
        Vol. 6, Austin 1990, ed. by L. Frommhold, J.W. Keto (AIP Conf. Proc. No. 216, 1990);
        Vol. 7, Carry Le Rovet 1992, ed. by R. Stamm, B. Talin (Nova Science, Paris 1994);
        Vol. 8, Toronto 1994, ed. by A.D. May, J.R. Drummond (AIP, New York 1995);
        Vol. 9, Florence 1996, ed. by M. Zoppi, L. Olivi (AIP, New York 1997);
        Vol. 10, State College, PA, USA, ed. by R.M. Herrmann (AIP, New York 1999);
        Vol. 11, Berlin 2000, ed. by J. Seidel (AIP, New York 2001)

3.5   C. Cohen-Tannoudji: *Quantum Mechanics* (Wiley, New York 1977)

3.6   S.N. Dobryakov, Y.S. Lebedev: Analysis of spectral lines whose profile is described by a composition of Gaussian and Lorentz profiles. Sov. Phys. Dokl. **13**, 9 (1969)

3.7   A. Unsöld: *Physik der Sternatmosphären* (Springer, Berlin, Heidelberg 1955)
        A. Unsöld, B. Baschek: *The New Cosmos*, 5th edn. (Springer, Berlin, Heidelberg 2001)

3.8   E. Lindholm: Pressure broadening of spectral lines. Ark. Mat. Astron. Fys. **32**A, 17 (1945)

3.9   A. Ben Reuven: The meaning of collisional broadening of spectral lines. The classical oscillation model. Adv. Atom. Mol. Phys. **5**, 201 (1969)

3.10   F. Schuler, W. Behmenburg: Perturbation of spectral lines by atomic interactions. Phys. Rep. C **12**, 274 (1974)

3.11   D. Ter Haar: *Elements of Statistical Mechanics* (Pergamon, New York 1977)

3.12   A. Gallagher: 'The spectra of colliding atoms'. In: *Atomic Physics, Vol. 4*, ed. by G. zu Putlitz, E.W. Weber, A. Winnaker (Plenum, New York 1975)

3.13   K. Niemax. G. Pichler: Determination of van der Waals constants from the red wings of self-broadened Cs principal series lines. J. Phys. B **8**, 2718 (1975)

3.14   N. Allard, J. Kielkopf: The effect of neutral nonresonant collisions on atomic spectral lines. Rev. Mod. Phys. **54**, 1103 (1982)

3.15   U. Fano, A.R.P. Rau: *Atomic Collisions and Spectra* (Academic, New York 1986)

3.16   K. Sando, Shi-I.: Pressure broadening and laser-induced spectral line shapes. Advanc. At. Mol. Phys. **25**, 133 (1988)

3.17   J.N. Murrel: *Introduction to the Theory of Atomic and Molecular Collisions* (Wiley, Chichester 1989)

3.18   R.J. Exton, W.L. Snow: Line shapes for satellites and inversion of the data to obtain interaction potentials. J. Quant. Spectrosc. Radiat. Transfer. **20**, 1 (1978)

3.19   H. Griem: *Principles of Plasma Spectroscopy* (Cambridge University Press, Cambridge 1997)

3.20   A. Sasso, G.M. Tino, M. Inguscio, N. Beverini, M. Francesconi: Investigations of collisional line shapes of neon transitions in noble gas mixtures. Nuov. Cimento D **10**, 941 (1988)

3.21   C.C. Davis, I.A. King: 'Gaseous ion lasers'. In: *Adv. Quantum Electronics, Vol. 3*, ed. by D.W. Godwin (Academic, New York 1975)

3.22   W.R. Bennett: *The Physics of Gas Lasers* (Gordon and Breach, New York 1977)

3.23   R. Moore: 'Atoms in dense plasmas'. In: *Atoms in Unusual Situations*, ed. by J.P. Briand, Nato ASI, Ser. B, Vol. 143 (Plenum, New York 1986)

3.24   H. Motz: *The Physics of Laser Fusion* (Academic, London 1979)

3.25   T.P. Hughes: *Plasmas and Laser Light* (Hilger, Bristol 1975)

3.26   A.S. Katzantsev, J.C. Hénoux: *Polarization Spectroscopy of Ionized Gases* (Kluwer Academ., Dordrecht 1995)

3.27   I.R. Senitzky: 'Semiclassical radiation theory within a quantum mechanical framework'. In: *Progress in Optics* **16** (North-Holland, Amsterdam 1978) p. 413

3.28   W.R. Hindmarsh, J.M. Farr: 'Collision broadening of spectral lines by neutral atoms'. In: *Progr. Quantum Electronics, Vol. 2, Part 4*, ed. by J.H. Sanders, S. Stenholm (Pergamon, Oxford 1973)

3.29   N. Anderson, K. Bartschat: *Polarization, Alignment and Orientation in Atomic Collisions* (Springer, Heidelberg 2001)

3.30   R.G. Breen: 'Line width'. In: *Handbuch der Physik, Vol. 27*, ed. by S. Flügge (Springer, Berlin 1964) p. 1

3.31   J. Hirschfelder, Ch.F. Curtiss, R.B. Bird: *Molecular Theory of Gases and Liquids* (Wiley, New York 1954)

3.32   S. Yi Chen, M. Takeo: Broadening and shift of spectral lines due to the presence of foreign gases. Rev. Mod. Phys. **29**, 20 (1957)

3.33   K.M. Sando, Shih-I. Chu: Pressure broadening and laser-induced spectral line shapes. Adv. At. Mol. Phys. **25**, 133 (1988)

3.34   R.H. Dicke: The effect of collisions upon the Doppler width of spectral lines. Phys. Rev. **89**, 472 (1953)

3.35   R.S. Eng, A.R. Calawa, T.C. Harman, P.L. Kelley: Collisional narrowing of infrared water vapor transitions. Appl. Phys. Lett. **21**, 303 (1972)

3.36   A.T. Ramsey, L.W. Anderson: Pressure Shifts in the ^{23}Na Hyperfine Frequency. J. Chem. Phys. **43**, 191 (1965)

3.37   K. Shimoda: 'Line broadening and narrowing effects'. In: *High-Resolution Spectroscopy*, Topics Appl. Phys., Vol. 13, ed. by K. Shimoda (Springer, Berlin, Heidelberg 1976) p. 11

3.38   J. Hall: 'The line shape problem in laser saturated molecular absorptions'. In: *Lecture Notes in Theor. Phys., Vol. 12A*, ed. by K. Mahanthappa, W. Brittin (Gordon and Breach, New York 1971)

3.39   V.S. Letokhov, V.P. Chebotayev: *Nonlinear Laser Spectroscopy*, Springer Ser. Opt. Sci., Vol. 4 (Springer, Berlin, Heidelberg 1977)

3.40   K.H. Drexhage: 'Structure and properties of laser dyes'. In: *Dye Lasers*, 3rd edn., Topics Appl. Phys., Vol. 1, ed. by F.P. Schäfer (Springer, Berlin, Heidelberg 1990)

3.41   D.S. McClure: 'Electronic spectra of molecules and ions in crystals'. In: *Solid State Phys., Vols. 8 and 9* (Academic, New York 1959)

3.42   W.M. Yen, P.M. Selzer (Eds.): *Laser Spectroscopy of Solids*, Springer Ser. Opt. Sci., Vol. 14 (Springer, Berlin, Heidelberg 1981)

3.43   A.A. Kaminskii: *Laser Crystals*, 2nd edn., Springer Ser. Opt. Sci., Vol. 14 (Springer, Berlin, Heidelberg 1991)

3.44   C.H. Wei, K. Holliday, A.J. Meixner, M. Croci, U.P. Wild: Spectral hole-burning study of BaFClBrSm$^{(2+)}$. J. Lumin. **50**, 89 (1991)

3.45   W.E. Moerner: *Persistent Spectral Hole-Burning: Science and Applications*, Topics Curr. Phys., Vol. 44 (Springer, Berlin, Heidelberg 1988)

Chapter 4

4.1   R. Kingslake, B.J. Thompson (Eds.): *Applied Optics and Optical Engineering, Vols. 1–10* (Academic, New York 1969–1985);
        M. Bass, E. van Skryland, D. Williams, W. Wolfe (Eds.): *Handbook of Optics, Vols. I and II* (McGraw-Hill, New York 1995)

4.2   E. Wolf (Ed.): *Progress in Optics, Vols. 1–42* (North-Holland, Amsterdam 1961–2001)

4.3   M. Born, E. Wolf: *Principles of Optics*, 4th edn. (Pergamon, Oxford 1970)

4.4   A.P. Thorne, U. Litzen, S. Johansson: *Spectrophysics*, 2nd edn. (Springer, Berlin 1999);
        G.L. Clark (Ed.): *The Encyclopedia of Spectroscopy* (Reinhold, New York 1960)

4.5   (a) L. Levi: *Applied Optics* (Wiley, London 1980);
        (b) D.F. Gray (Ed.): *Am. Inst. Phys. Handbook* (McGraw-Hill, New York 1980)

4.6   R.D. Guenther: *Modern Optics* (Wiley, New York 1990)

4.7   F. Graham-Smith, T.A. King: *Optics and Photonics* (Wiley, London 2000)

4.8   H. Lipson: *Optical Physics*, 3rd edn. (Cambridge University Press, Cambridge 1995)

4.9   K.I. Tarasov: *The Spectroscope* (Hilger, London 1974)

4.10   S.P. Davis: *Diffraction Grating Spectrographs* (Holt, Rinehard & Winston, New York 1970)

4.11   A.B. Schafer, L.R. Megil, L. Dropleman: Optimization of the Czerny-Turner spectrometer. J. Opt. Soc. Am. **54**, 879 (1964)

4.12 *Handbook of Diffraction Gratings, Ruled and Holographic* (Jobin Yvon Optical Systems, Metuchen, NJ 1970)
Bausch and Lomb Diffraction Grating Handbook (Bausch & Lomb, Rochester, NY 1970)

4.13 G.W. Stroke: 'Diffraction gratings'. In: *Handbuch der Physik, Vol. 29*, ed. by S. Flügge (Springer, Berlin, Heidelberg 1967)

4.14 M.C. Hutley: *Diffraction Gratings* (Academic, London 1982);
E. Popov, E.G. Loewen: *Diffraction Gratings and Applications* (Dekker, New York 1997)

4.15 See, for example, E. Hecht: *Optics*, 4th edn. (Addison-Wesley, London 2002)

4.16 G. Schmahl, D. Rudolph: 'Holographic diffraction gratings'. In: *Progress in Optics* **14**, 195 (North-Holland, Amsterdam 1977)

4.17 E. Loewen: 'Diffraction gratings: ruled and holographic'. In: *Applied Optics and Optical Engineering, Vol. 9* (Academic, New York 1980)

4.18 M.D. Perry, et al.: High-efficiency multilayer dielectric diffraction gratings. Opt. Lett. **20**, 940 (1995)

4.19 Basic treatments of interferometers may be found in general textbooks on optics. A more detailed discussion has, for instance, been given in S. Tolansky: *An Introduction to Interferometry* (Longman, London 1973);
W.H. Steel: *Interferometry* (Cambridge Univ. Press, Cambridge 1967);
J. Dyson: *Interferometry* (Machinery Publ., Brighton 1970);
M. Francon: *Optical Interferometry* (Academic, New York 1966)

4.20 H. Polster, J. Pastor, R.M. Scott, R. Crane, P.H. Langenbeck, R. Pilston, G. Steingerg: New developments in interferometry. Appl. Opt. **8**, 521 (1969)

4.21 K.M. Baird, G.R. Hanes: 'Interferometers'. In: [4.1], Vol. 4, pp. 309–362

4.22 P. Hariharan: *Optical Interferometry* (Academic, New York 1986)
W.S. Gornall: The world of Fabry–Perots. Laser Appl. **2**, 47 (1983)

4.23 M. Francon, J. Mallick: *Polarisation Interferometers* (Wiley, London 1971)

4.24 H. Welling, B. Wellingehausen: High resolution Michelson interferometer for spectral investigations of lasers. Appl. Opt. **11**, 1986 (1972)

4.25 P.R. Saulson: *Fundamentals of Interferometric Gravitational Wave Detectors* (World Scientific, Singapore 1994)

4.26 R.W.P. Drever, J.L. Hall, F.V. Kowalski, J. Hough, G.M. Ford, A.J. Munley, H. Ward: Laser phase and frequency stabilization using an optical resonator. Appl. Phys. B **31**, 97 (1983)
A. Wicht, K. Danzmann, M. Fleischhauer, M. Scully, G. Müller, R.-H. Rinkleff: White-light cavities, atomic phase coherence and gravitational wave detectors. Opt. Commun. **134**, 431 (1997)

4.27 R.J. Bell: *Introductory Fourier Transform Spectroscopy* (Academic, New York 1972)

4.28 P. Griffiths, J.A. de Haseth: *Fourier-Transform Infrared Spectroscopy* (Wiley, New York 1986)

4.29 V. Grigull, H. Rottenkolber: Two beam interferometer using a laser. J. Opt. Soc. Am. **57**, 149 (1967);
W. Schumann, M. Dubas: *Holographic Interferometry*, Springer Ser. Opt. Sci., Vol. 16 (Springer, Berlin, Heidelberg 1979);
W. Schumann, J.-P. Zürcher, D. Cuche: *Holography and Deformation Analysis*, Springer Ser. Opt. Sci., Vol. 46 (Springer, Berlin, Heidelberg 1986);

4.30 W. Marlow: Hakenmethode. Appl. Opt. **6**, 1715 (1967)

4.31 I. Meroz (Ed.): *Optical Transition Probabilities. A Representative Collection of Russian Articles* (Israel Program for Scientific Translations, Jerusalem 1962)

4.32 J.P. Marioge, B. Bonino: Fabry–Perot interferometer surfacing. Opt. Laser Technol. **4**, 228 (1972)

4.33 M. Hercher: Tilted etalons in laser resonators. Appl. Opt. **8**, 1103 (1969)

4.34   W.R. Leeb: Losses introduced by tilting intracavity etalons. Appl. Phys. **6**, 267 (1975)

4.35   W. Demtröder, M. Stock: Molecular constants and potential curves of Na$_2$ from laser-induced fluorescence. J. Mol. Spectrosc. **55**, 476 (1975)

4.36   P. Connes: L'etalon de Fabry–Perot spherique. Phys. Radium **19**, 262 (1958); P. Connes: *Quantum Electronics and Coherent Light*, ed. by P.H. Miles (Academic, New York 1964) p. 198

4.37   D.A. Jackson: The spherical Fabry–Perot interferometer as an instrument of high resolving power for use with external or with internal atomic beams. Proc. Roy. Soc. (London) A **263**, 289 (1961)

4.38   J.R. Johnson: A high resolution scanning confocal interferometer. Appl. Opt. **7**, 1061 (1968)

4.39   M. Hercher: The spherical mirror Fabry–Perot interferometer. Appl. Opt. **7**, 951 (1968)

4.40   R.L. Fork, D.R. Herriot, H. Kogelnik: A scanning spherical mirror interferometer for spectral analysis of laser radiation. Appl. Opt. **3**, 1471 (1964)

4.41   F. Schmidt-Kaler, D. Leibfried, M. Weitz, T.W. Hänsch: Precision measurements of the isotope shift of the 1s–2s transition of atomic hydrogen and deuterium. Phys. Rev. Lett. **70**, 2261 (1993)

4.42   J.R. Johnson: A method for producing precisely confocal resonators for scanning interferometers. Appl. Opt. **6**, 1930 (1967)

4.43   P. Hariharan: *Optical Interferometry* (Academic, New York 1985); G.W. Hopkins (Ed.): *Interferometry*. SPIE Proc. **192** (1979); R.J. Pryputniewicz (Ed.): *Industrial Interferometry*. SPIE Proc. **746** (1987); R.J. Pryputniewicz (Ed.): *Laser Interferometry*. SPIE Proc. **1553** (1991); J.D. Briers: Interferometric testing of optical systems and components. Opt. Laser Techn. (February 1972) p. 28

4.44   J.M. Vaughan: *The Fabry–Perot Interferometer* (Hilger, Bristol 1989); Z. Jaroscewicz, M. Pluta (Eds.): *Interferometry 89: 100 Years after Michelson: State of the Art and Applications*. SPIE Proc. **1121** (1989)

4.45   J. McDonald: *Metal Dielectric Multilayer* (Hilger, London 1971)

4.46   A. Thelen: *Design of Optical Interference Coatings* (McGraw-Hill, New York 1988) Z. Knittl: *Optics of Thin Films* (Wiley, New York 1976)

4.47   V.R. Costich: 'Multilayer dielectric coatings'. In: *Handbook of Lasers*, ed. by R.J. Pressley (CRC, Cleveland, Ohio 1972)

4.48   H.A. MacLeod (Ed.): Optical interference coatings. Appl. Opt. **28**, 2697–2974 (1989); R.E. Hummel, K.H. Guenther (Eds.): *Optical Properties, Vol. 1: Thin Films for Optical Coatings* (CRC, Cleveland, Ohio 1995)

4.49   A. Musset, A. Thelen: 'Multilayer antireflection coatings'. In: *Progress in Optics* **3**, 203 (North-Holland, Amsterdam 1970)

4.50   J.T. Cox, G. Hass: In: *Physics of Thin Films, Vol. 2*, ed. by G. Hass (Academic, New York 1964)

4.51   E. Delano, R.J. Pegis: 'Methods of synthesis for dielectric multilayer filters'. In: *Progress in Optics, Vol. 7*, 69 (North-Holland, Amsterdam 1969)

4.52   H.A. Macleod: *Thin Film Optical Filter*, 3rd edn. (Inst. of Physics Publ., London 2001)

4.53   J. Evans: The birefringent filter. J. Opt. Soc. Am. **39**, 229 (1949)

4.54   H. Walther, J.L. Hall: Tunable dye laser with narrow spectral output. Appl. Phys. Lett. **17**, 239 (1970)

4.55   M. Okada, S. Iliri: Electronic tuning of dye lasers by an electro-optic birefringent Fabry–Perot etalon. Opt. Commun. **14**, 4 (1975)

4.56   B.H. Billings: The electro-optic effect in uniaxial crystals of the type XH_2PO_4.
        J. Opt. Soc. Am. **39**, 797 (1949)

4.57   R.L. Fork, D.R. Herriot, H. Kogelnik: A scanning spherical mirror interferometer
        for spectral analysis of laser radiation. Appl. Opt. **3**, 1471 (1964)

4.58   V.G. Cooper, B.K. Gupta, A.D. May: Digitally pressure scanned Fabry–Perot
        interferometer for studying weak spectral lines. Appl. Opt. **11**, 2265 (1972)

4.59   J.M. Telle, C.L. Tang: Direct absorption spectroscopy, using a rapidly tunable
        cw-dye laser. Opt. Commun. **11**, 251 (1974)

4.60   P. Cerez, S.J. Bennet: New developments in iodine-stabilized HeNe lasers. IEEE
        Trans. IM-**27**, 396 (1978)

4.61   K.M. Evenson, J.S. Wells, F.R. Petersen, B.L. Danielson, G.W. Day, R.L. Barger,
        J.L. Hall: Speed of light from direct frequency and wavelength measurements of
        the methane-stabilized laser. Phys. Rev. Lett. **29**, 1346 (1972)

4.62   K.M. Evenson, D.A. Jennings, F.R. Petersen, J.S. Wells: 'Laser frequency
        measurements: a review, limitations and extension to 197 THz'. In: *Laser Spec-
        troscopy III*, ed. by J.L. Hall, J.L. Carlsten, Springer Ser. Opt. Sci., Vol. 7
        (Springer, Berlin, Heidelberg 1977)

4.63   K.M. Evenson, J.S. Wells, F.R. Petersen, B.L. Davidson, G.W. Day, R.L. Barger,
        J.L. Hall: The speed of light. Phys. Rev. Lett. **29**, 1346 (1972)

4.64   A. DeMarchi (Ed.): *Frequency Standards and Metrology* (Springer, Berlin, Hei-
        delberg 1989)

4.65   P.R. Bevington: *Data Reduction and Error Analysis for the Physical Sciences*
        (McGraw-Hill, New York 1969)

4.66   J.R. Taylor: *An Introduction to Error Analysis* (Univ. Science Books, Mill Valley
        1982)

4.67   J.L. Hall, S.A. Lee: Interferometric real time display of CW dye laser wave-
        length with sub-Doppler accuracy. Appl. Phys. Lett. **29**, 367 (1976)

4.68   J.J. Snyder: 'Fizeau wavelength meter'. In: *Laser Spectroscopy III*, ed. by
        J.L. Hall, J.L. Carlsten, Springer Ser. Opt. Sci., Vol. 7 (Springer, Berlin, Hei-
        delberg 1977) p. 419

4.69   R.L. Byer, J. Paul, M.D. Duncan: 'A wavelength meter'. In: *Laser Spectroscopy
        III*, ed. by J.L. Hall, J.L. Carlsten, Springer Ser. Opt. Sci., Vol. 7 (Springer,
        Berlin, Heidelberg 1977) p. 414

4.70   A. Fischer, H. Kullmer, W. Demtröder: Computer-controlled Fabry–Perot-
        wavemeter. Opt. Commun. **39**, 277 (1981)

4.71   N. Konishi, T. Suzuki, Y. Taira, H. Kato, T. Kasuya: High precision wavelength
        meter with Fabry–Perot optics. Appl. Phys. **25**, 311 (1981)

4.72   F.V. Kowalski, R.E. Teets, W. Demtröder, A.L. Schawlow: An improved
        wavemeter for CW lasers. J. Opt. Soc. Am. **68**, 1611 (1978)

4.73   R. Best: *Theorie und Anwendung des Phase-Locked Loops* (AT Fachverlag,
        Stuttgart 1976)

4.74   F.M. Gardner: *Phase Lock Techniques* (Wiley, New York 1966);
        Phase-Locked Loop Data Book (Motorola Semiconductor Prod., Inc. 1973)

4.75   B. Edlen: Dispersion of standard air. J. Opt. Soc. Am. **43**, 339 (1953)

4.76   J.C. Owens: Optical refractive index of air: Dependence on pressure, temperature
        and composition. Appl. Opt. **6**, 51 (1967)

4.77   R. Castell, W. Demtröder, A. Fischer, R. Kullmer, K. Wickert: The accuracy of
        laser wavelength meters. Appl. Phys. B **38**, 1 (1985)

4.78   J. Cachenaut, C. Man, P. Cerez, A. Brillet, F. Stoeckel, A. Jourdan, F. Hartmann:
        Description and accuracy tests of an improved lambdameter. Rev. Phys. Appl.
        14, 685 (1979)

4.79   J. Viqué, B. Girard: A systematic error of Michelson's type lambdameter. Rev.
        Phys. Appl. **21**, 463 (1986)

4.80   J.J. Snyder: 'An ultrahigh resolution frequency meter'. *Proc. 35th Ann. Freq.*
          Control USAERADCOM May 1981. Appl. Opt. **19**, 1223 (1980)
4.81   P. Juncar, J. Pinard: Instrument to measure wavenumbers of CW and pulsed laser
          lines: The sigma meter. Rev. Sci. Instrum. **53**, 939 (1982);
          P. Jacquinot, P. Juncar, J. Pinard: 'Motionless Michelson for high precision laser
          frequency measurements'. In: *Laser Spectroscopy III*, ed. by J.L. Hall, J.L. Carl-
          sten, Springer Ser. Opt. Sci., Vol. 7 (Springer, Berlin, Heidelberg 1977) p. 417
4.82   J.J. Snyder: Fizeau wavemeter. SPIE Proc. **288**, 258 (1981)
4.83   M.B. Morris, T.J. McIllrath, J. Snyder: Fizeau wavemeter for pulsed laser wave-
          length measurement. Appl. Opt. **23**, 3862 (1984)
4.84   J.L. Gardner: Compact Fizeau wavemeter. Appl. Opt. **24**, 3570 (1985)
4.85   J.L. Gardner: Wavefront curvature in a Fizeau wavemeter. Opt. Lett. **8**, 91
          (1983)
4.86   J.J. Keyes (Ed.): *Optical and Infrared Detectors*, 2nd edn., Topics Appl. Phys.,
          Vol. 19 (Springer, Berlin, Heidelberg 1980)
4.87   P.N. Dennis: *Photodetectors* (Plenum, New York 1986)
4.88   M. Bleicher: *Halbleiter-Optoelektronik* (Hüthig, Heidelberg 1976)
4.89   E.L. Dereniak, G.D. Boreman: *Infrared Detectors and Systems* (Wiley, New
          York 1996)
4.90   G.H. Rieke: *Detection of Light: From the Ultraviolet to the Submillimeter* (Cam-
          bridge University Press, Cambridge 1994)
4.91   J. Wilson, J.F.B. Hawkes: *Optoelectronics* (Prentice Hall, London 1983)
4.92   R. Paul: *Optoelektronische Halbleiterbauelemente* (Teubner, Stuttgart 1985)
4.93   T.S. Moss, G.J. Burell, B. Ellis: *Semiconductor Opto-Electronics* (Butterworth,
          London 1973)
4.94   R.W. Boyd: *Radiometery and the Detection of Optical Radiation* (Wiley, New
          York 1983)
4.95   E.L. Dereniak, D.G. Crowe: *Optical Radiation Detectors* (Wiley, New York
          1984)
4.96   F. Stöckmann: Photodetectors, their performance and limitations. Appl. Phys. **7**,
          1 (1975)
4.97   F. Grum, R.L. Becher: *Optical Radiation Measurements, Vols. 1 and 2* (Aca-
          demic, New York 1979 and 1980)
4.98   R.H. Kingston: *Detection of Optical and Infrared Radiation*, Springer Ser. Opt.
          Sci., Vol. 10 (Springer, Berlin, Heidelberg 1978)
4.99   E.H. Putley: 'Thermal detectors'. In: [4.86], p. 71
4.100   T.E. Gough, R.E. Miller, G. Scoles: Infrared laser spectroscopy of molecular
          beams. Appl. Phys. Lett. **30**, 338 (1977)
          M. Zen: Cryogenic bolometers, in *Atomic and Molecular Beam Methods*, ed. by
          G. Scoles (Oxford Univ. Press, New York 1988) Vol. 1
4.101   D. Bassi, A. Boschetti, M. Scotoni, M. Zen: Molecular beam diagnostics by
          means of fast superconducting bolometer. Appl. Phys. B **26**, 99 (1981)
4.102   J. Clarke, P.L. Richards, N.H. Yeh: Composite superconducting transition edge
          bolometer. Appl. Phys. Lett. B **30**, 664 (1977)
4.103   M.J.E. Golay: A Pneumatic Infra-Red Detector. Rev. Scient. Instrum. **18**, 357
          (1947)
4.104   B. Tiffany: Introduction and review of pyroelectric detectors. SPIE Proc. **62**, 153
          (1975)
4.105   C.B. Boundy, R.L. Byer: Subnanosecond pyroelectric detector. Appl. Phys. Lett.
          21, 10 (1972)
4.106   L.E. Ravich: Pyroelectric detectors and imaging. Laser Focus **22**, 104 (1986)
4.107   H. Melchior: 'Demodulation and photodetection techniqes'. In: *Laser Handbook,*
          Vol. 1, ed. by F.T. Arrecchi, E.O. Schulz-Dubois (North-Holland, Amsterdam
          1972) p. 725

4.108   H. Melchior: Sensitive high speed photodetectors for the demodulation of visible and near infrared light. J. Lumin. **7**, 390 (1973)

4.109   D. Long: 'Photovoltaic and photoconductive infrared detectors'. In: [4.86], p. 101

4.110   E. Sakuma, K.M. Evenson: Characteristics of tungsten nickel point contact diodes used as a laser harmonic generation mixers. IEEE J. QE-**10**, 599 (1974)

4.111   K.M. Evenson, M. Ingussio, D.A. Jennings: Point contact diode at laser frequencies. J. Appl. Phys. **57**, 956 (1985);
H.D. Riccius, K.D. Siemsen: Point-contact diodes. Appl. Phys. A **35**, 67 (1984);
H. Rösser: Heterodyne spectroscopy for submillimeter and far-infrared wavelengths. Infrared Phys. **32**, 385 (1991)

4.112   H.-U. Daniel, B. Maurer, M. Steiner: A broad band Schottky point contact mixer for visible light and microwave harmonics. Appl. Phys. B **30**, 189 (1983);
T.W. Crowe: GaAs Schottky barrier mixer diodes for the frequency range from $1-10\,\text{THz}$. Int. J. IR and Millimeter Waves **10**, 765 (1989);
H.P. Röser, R.V. Titz, G.W. Schwab, M.F. Kimmitt: Current-frequency characteristics of submicron GaAs Schottky barrier diodes with femtofarad capacitances. J. Appl. Phys. **72**, 3194 (1992)

4.113   F. Capasso: Band-gap engineering via graded-gap structure: Applications to novel photodetectors. J. Vac. Sci. Techn. B**12**, 457 (1983)

4.114   F. Capasso (Ed.): *Physics of Quantum Electron Devices*, Springer Ser. Electron. Photon., Vol. 28 (Springer, Berlin, Heidelberg 1990)

4.115   F. Capasso: Multilayer avalanche photodiodes and solid state photomultipliers. Laser Focus **20**, 84 (July 1984)

4.116   G.A. Walter, E.L. Dereniak: Photodetectors for focal plane arrays. Laser Focus **22**, 108 (March 1986)

4.117   A. Tebo: IR detector technology. Arrays. Laser Focus **20**, 68 (July 1984);
E.L. Dereniak, R.T. Sampson (Eds.): *Infrared Detectors, Focal Plane Arrays and Imaging Sensors*, SPIE Proc. **1107** (1989);
E.L. Dereniak (Ed.): Infrared Detectors and Arrays. SPIE Proc. **930** (1988)

4.118   D.F. Barbe (Ed.): *Charge-Coupled Devices*, Topics Appl. Phys., Vol. 38 (Springer, Berlin, Heidelberg 1980)

4.119   see special issue on CCDs of Berkeley Lab **23**, 3 (Fall 2000) and G.C. Holst: *CCD Arrays, Cameras and Display* (Sofitware, ISBN 09640000024, 2000)
K.P. Proll, J.M. Nivet, C. Voland: Enhancement of the dynamic range of the detected intensity in an optical measurement system by a three channel technique. Appl. Opt. **41**, 130 (2002)

4.120   H. Zimmermann: *Integrated Silicon Optoelectronics* (Springer, Berlin, Heidelberg 2000)

4.121   R.B. Bilborn, J.V. Sweedler, P.M. Epperson, M.B. Denton: Charge transfer device detectors for optical spectroscopy. Appl. Spectrosc. **41**, 1114 (1987)

4.122   I. Nin, Y. Talmi: CCD detectors record multiple spectra simultaneously. Laser Focus **27**, 111 (August 1991)

4.123   H.R. Zwicker: Photoemissive detectors. In *Optical and Infrared Detectors*, 2nd edn., ed. by J. Keyes, Topics Appl. Phys., Vol. 19 (Springer, Berlin, Heidelberg 1980)

4.124   C. Gosh: Photoemissive materials. SPIE Proc. **346**, 62 (1982)

4.125   R.L. Bell: *Negative Electron Affinity Devices* (Clarendon, Oxford 1973)

4.126   L.E. Wood, T.K. Gray, M.C. Thompson: Technique for the measurement of photomultiplier transit time variation. Appl. Opt. **8**, 2143 (1969)

4.127   J.D. Rees, M.P. Givens: Variation of time of flight of electrons through a photomultiplier. J. Opt. Soc. Am. **56**, 93 (1966)

4.128   (a) B. Sipp. J.A. Miehe, R. Lopes Delgado: Wavelength dependence of the time resolution of high speed photomultipliers used in single-photon timing experiments. Opt. Commun. **16**, 202 (1976)

(b) G. Beck: Operation of a 1P28 photomultipier with subnanosecond response time. Rev. Sci. Instrum. **47**, 539 (1976)

(c) B.C. Mongan (Ed.): *Adv. Electronics and Electron Physics, Vol. 74* (Academic, London 1988)

4.129 A. van der Ziel: *Noise in Measurements* (Wiley, New York 1976)

4.130 A.T. Young: Undesirable effects of cooling photomultipliers. Rev. Sci. Instrum. **38**, 1336 (1967)

4.131 J. Sharpe, C. Eng: Dark Current in Photomultiplier Tubes (EMI Ltd. information document, ref. R-P021470)

4.132 Phototubes and Photocells. In: *An Introduction to the Photomultiplier* (RCA Manual, EMI Ltd. information sheet, 1966)

4.133 E.L. Dereniak, D.G. Crowe: *Optical Radiation Detectors* (Wiley, New York 1984)

4.134 R.W. Boyd: *Radiometry and the Detection of Optical Radiation* (Wiley, New York 1983)

4.135 G. Pietri: Towards picosecond resolution. Contribution of microchannel electron multipiers to vacuum tube design. IEEE Trans. NS-**22**, 2084 (1975);
J.L. Wiza: Microchannel plate detectors (Galileo information sheet, Sturbridge, MA, 1978)

4.136 L.M. Bieberman, S. Nudelman (Eds.): *Photoelectronic Imaging Devices* (Plenum, New York 1971);
I.P. Csonba (Ed.): *Image Intensification*, SPIE Proc. **1072** (1989)

4.137 *Proc. Topical Meeting on Quantum-Limited Imaging and Image Processing* (Opt. Soc. Am., Washington, DC 1986)

4.138 T.P. McLean, P. Schagen (Eds.): *Electronic Imaging* (Academic, London 1979)

4.139 H.K. Pollehn: 'Image intensifiers'. In: [4.1], Vol. 6 (1980) p. 393

4.140 S. Jeffers, W. Weller: 'Image intensifier optical multichannel analyser for astronomical spectroscopy'. In: *Adv. Electronics and Electron Phys. B* **40** (Academic, New York 1976) p. 887

4.141 L. Perko, J. Haas, D. Osten: Cooled and intensified array detectors for optical spectroscopy. SPIE Proc. **116**, 64 (1977)

4.142 J.L. Hall: 'Arrays and charge coupled devices'. In: [4.1], Vol. 8 (1980) p. 349

4.143 J.L. Weber: Gated optical multichannel analyzer for time resolved spectroscopy. SPIE Proc. **82**, 60 (1976)

4.144 R.G. Tull: A comparison of photon-counting and current measuring techniques in spectrometry of faint sources. Appl. Opt. **7**, 2023 (1968)

4.145 J.F. James: On the use of a photomultiplier as a photon counter. Monthly Notes R. Astron. Soc. **137**, 15 (1967)

4.146 D.V. O'Connor, D. Phillips: *Time-Correlated Photon-Counting* (Academic, London 1984);
G.F. Knoll: *Radiation Detectors and Measurement* (Wiley, New York 1979)

4.147 P.W. Kruse: 'The photon detection process'. In: *Optical and Infrared Detectors*, 2nd edn., ed. by J.J. Keyes, Topics Appl. Phys., Vol. 19 (Springer, Berlin, Heidelberg 1980)

4.148 Signal Averagers. (Information sheet, issued by Princeton Appl. Res., Princeton, NJ, 1978)

4.149 Information sheet on transient recorders, Biomation, Palo Alto, CA

4.150 C. Morgan: Digital signal processing. Laser Focus **13**, 52 (Nov. 1977)

4.151 Handshake: Information sheet on waveform digitizing instruments (Tektronic, Beaverton, OR 1979)

4.152 Hamamatsu photonics information sheet (February 1989)

4.153 H. Mark: *Principles and Practice of Spectroscopic Calibration* (Wiley, New York 1991)

4.154   A.C.S. van Heel (Ed.): *Advanced Optical Techniques* (North-Holland, Amsterdam 1967)

4.155   W. Göpel, J. Hesse, J.N. Zemel (Eds.): *Sensors, A Comprehensive Survey* (Wiley-VCH, Weinheim 1992)

4.156   D. Dragoman, M. Dragoman: *Advanced Optical Devices* (Springer, Heidelberg 1999)

4.157   F. Grum, R.L. Becherer (Eds.): *Optical Radiation Measurements, Vols. I, II* (Academic, New York 1979, 1980)

4.158   C.H. Lee: *Picosecond Optoelectronics Devices* (Academic, New York 1984)

4.159   *The Photonics and Application Handbook* (Laurin, Pittsfield, MA 1990)

Chapter 5

5.1   A.E. Siegman: *An Introduction to Lasers and Masers* (McGraw-Hill, New York 1971);
A.E. Siegman: *Lasers* (Oxford Univ. Press, Oxford 1986)

5.2   I. Hecht: *The Laser Guidebook*, 2nd edn. (McGraw-Hill, New York 1992)

5.3   *Lasers, Vols. 1–4*, ed. by A. Levine (Dekker, New York 1966–76)

5.4   A. Yariv: *Quantum Electrons* (Wiley, New York 1975);
A. Yariv: *Optical Electronics*, 3rd edn. (Holt, Rinehart, Winston, New York 1985)

5.5   O. Svelto: *Principles of Lasers*, 4th edn. (Plenum, New York 1998)

5.6   *Laser Handbook, Vols. I–V* (North-Holland, Amsterdam 1972–1985)
M.J. Weber: *Handbook of Lasers* (CRC, New York 2001)
M.J. Weber: *Handbook of Laser Wavelengths* (CRC, New York 1999)

5.7   A. Maitland, M.H. Dunn: *Laser Physics* (North-Holland, Amsterdam 1969);
P.W. Milona, J.H. Eberly: *Lasers* (Wiley, New York 1988)

5.8   F.K. Kneubühl, M.W. Sigrist: *Laser*, 5th edn. (Teubner, Stuttgart 1999)

5.9   I.T. Verdeyen: *Laser Electronics*, 2nd edn. (Prentice Hall, Englewood Cliffs, NJ 1989)

5.10   C.C. Davis: *Lasers and Electro-Optic* (Cambridge University Press, Cambridge 1996)

5.11   M.O. Scully, W.E. Lamb Jr., M. Sargent III: *Laser Physics* (Addison Wesley, Reading, MA 1974);
P. Meystre, M. Sargent III: *Elements of Quantum Optics*, 2nd edn. (Springer, Berlin, Heidelberg 1991)

5.12   H. Haken: *Laser Theory* (Springer, Berlin, Heidelberg 1984)

5.13   D. Eastham: *Atomic Physics of Lasers* (Taylor & Francis, London 1986)

5.14   R. Loudon: *The Quantum Theory of Light* (Clarendon, Oxford 1973)

5.15   A.F. Harvey: *Coherent Light* (Wiley, London 1970)

5.16   E. Hecht: *Optics*, 3rd edn. (Addison Wesley, Reading, MA 1997)

5.17   M. Born, E. Wolf: *Principles of Optics*, 7th edn. (Cambridge University Press, Cambridge 1999)

5.18   G. Koppelmann: Multiple beam interference and natural modes in open resonators. *Progress in Optics* 7 (North-Holland, Amsterdam 1969) pp. 1–66

5.19   A.G. Fox, T. Li: Resonant modes in a maser interferometer. Bell System Techn. J. **40**, 453 (1961)

5.20   G.D. Boyd, J.P. Gordon: Confocal multimode resonator for millimeter through optical wavelength masers. Bell Syst. Techn. J. **40**, 489 (1961)

5.21   G.D. Boyd, H. Kogelnik: Generalized confocal resonator theory. Bell Syst. Techn. J. **41**, 1347 (1962)

5.22   A.G. Fox, T. Li: Modes in maser interferometers with curved and tilted mirrors. Proc. IEEE **51**, 80 (1963)

5.23   H.K.V. Lotsch: The Fabry–Perot resonator. Optik **28**, 65, 328, 555 (1968);
        H.K.V. Lotsch: Optik **29**, 130, 622 (1969)
        H.K.V. Lotsch: The confocal resonator system. Optik **30**, 1, 181, 217, 563
        (1969/70)

5.24   H. Kogelnik, T. Li: Laser beams and resonators. Proc. IEEE **54**, 1312 (1966)

5.25   N. Hodgson, H. Weber: *Optical Resonators* (Springer, Berlin, Heidelberg, New
        York 1997)

5.26   A.E. Siegman: Unstable optical resonators. Appl. Opt. **13**, 353 (1974)

5.27   W.H. Steier: 'Unstable resonators'. In: *Laser Handbook III*, ed. by M.L. Stitch
        (North-Holland, Amsterdam 1979)

5.28   R.L. Byer, R.L. Herbst: The unstable-resonator YAG laser. Laser Focus **14**, 48
        (July 1978)

5.29   Y.A. Anan'ev: *Laser Resonators and the Beam Divergence Problem* (Hilger,
        Bristol 1992)

5.30   N. Hodgson, H. Weber: Unstable resonators with excited converging wave. IEEE
        J. QE-**26**, 731 (1990);
        N. Hodgson, H. Weber: High-power solid state lasers with unstable resonators.
        Opt. Quant. Electron. **22**, 39 (1990)

5.31   W. Magnus, F. Oberhettinger, R.P. Soni: *Formulas and Theories for the Special
        Functions of Mathematical Physics* (Springer, Berlin, Heidelberg 1966)

5.32   T.F. Johnston: Design and performance of broadband optical diode to enforce
        one direction travelling wave operation of a ring laser. IEEE J. QE-**16**, 483
        (1980);
        T.F. Johnson: Focus on Science 3, No. 1, (1980) (Coherent Radiation, Palo Alto,
        Calif.);
        G. Marowsky: A tunable flash lamp pumped dye ring laser of extremely narrow
        bandwidth. IEEE J. QE-**9**, 245 (1973)

5.33   I.V. Hertel, A. Stamatovic: Spatial hole burning and oligo-mode distance control
        in CW dye lasers. IEEE J. QE-**11**, 210 (1975)

5.34   D. Kühlke, W. Diehl: Mode selection in cw-laser with homogeneously broad-
        ened gain. Opt. Quant. Electron. **9**, 305 (1977)

5.35   W.R. Bennet Jr.: *The Physics of Gas Lasers* (Gordon and Breach, New York
        1977)

5.36   R. Beck, W. Englisch, K. Gürs: *Table of Laser Lines in Gases and Vapors*, 2nd
        edn., Springer Ser. Opt. Sci., Vol. 2 (Springer, Berlin, Heidelberg 1978);
        M.J. Weber: *Handbook of Laser Wavelengths* (CRC, New York 1999)

5.37   K. Bergmann, W. Demtröder: A new cascade laser transition in a He-Ne mix-
        ture. Phys. Lett. **29** A, 94 (1969)

5.38   B.J. Orr: A constant deviation laser tuning device. J. Phys. E **6**, 426 (1973)

5.39   L. Allen, D.G.C. Jones: The helium-neon laser. Adv. Phys. **14**, 479 (1965)

5.40   C.E. Moore: Atomic Energy Levels, Nat. Stand. Ref. Ser. **35**, NBS Circular 467
        (U.S. Dept. Commerce, Washington, DC 1971)

5.41   P.W. Smith: On the optimum geometry of a 6328 Å laser oscillator. IEEE J. QE-
        2, 77 (1966)

5.42   W.B. Bridges, A.N. Chester, A.S. Halsted, J.V. Parker: Ion laser plasmas. IEEE
        Proc. **59**, 724 (1971)

5.43   A. Ferrario, A. Sirone, A. Sona: Interaction mechanisms of laser transitions in
        argon and krypton ion-lasers. Appl. Phys. Lett. **14**, 174 (1969)

5.44   C.C. Davis, T.A. King: 'Gaseous ion lasers'. In: *Adv. Quantum Electronics,
        Vol. 3*, ed. by D.W. Goodwin (Academic, London 1975)

5.45   G. Herzberg: *Molecular Spectra and Molecular Structure, Vol. II* (Van Nostrand
        Reinhold, New York 1945)

5.46   D.C. Tyle: 'Carbon dioxyde lasers". In: *Adv. Quantum Electronics, Vol. 1*, ed. by
        D.W. Goodwin (Academic, London 1970)

5.47   W.J. Witteman: *The CO_2 Laser*, Springer Ser. Opt. Sci., Vol. 53 (Springer, Berlin, Heidelberg 1987)

5.48   H.W. Mocker: Rotational level competition in CO_2-lasers. IEEE J. QE-**4**, 769 (1968)

5.49   J. Haisma: Construction and properties of short stable gas lasers. Phillips Res. Rpt., Suppl. No. 1 (1967) and Phys. Lett. **2**, 340 (1962)

5.50   M. Hercher: Tunable single mode operation of gas lasers using intra-cavity tilted etalons. Appl. Opt. **8**, 1103 (1969)

5.51   P.W. Smith: Stabilized single frequency output from a long laser cavity. IEEE J. QE-**1**, 343 (1965)

5.52   P. Zory: Single frequency operation of argon ion lasers. IEEE J. QE-**3**, 390 (1967)

5.53   V.P. Belayev, V.A. Burmakin, A.N. Evtyunin, F.A. Korolyov, V.V. Lebedeva, A.I. Odintzov: High power single-frequency argon ion laser. IEEE J. QE-**5**, 589 (1969)

5.54   P.W. Smith: Mode selection in lasers. Proc. IEEE **60**, 422 (1972)

5.55   W.W. Rigrod, A.M. Johnson: Resonant prism mode selector for gas lasers. IEEE J. QE-**3**, 644 (1967)

5.56   R.E. Grove, E.Y. Wu, L.A. Hackel, D.G. Youmans, S. Ezekiel: Jet stream CW-dye laser for high resolution spectroscopy. Appl. Phys. Lett. **23**,. 442 (1973)

5.57   H.W. Schröder, H. Dux, H. Welling: Single mode operation of CW dye lasers. Appl. Phys. **1**, 347 (1973)

5.58   T.W. Hänsch: Repetitively pulsed tunable dye laser for high resolution spectroscopy. Appl. Opt. **11**, 895 (1972)

5.59   J.P. Goldsborough: 'Design of gas lasers'. In: *Laser Handbook I*, ed. by F.T. Arrecchi, E.O. Schulz-Dubois (North-Holland, Amsterdam 1972) p. 597

5.60   B. Peuse: 'New developments in CW dye lasers'. In: *Physics of New Laser Sources*, ed. by N.B. Abraham, F.T. Annecchi, A. Mooradian, A. Suna (Plenum, New York 1985)

5.61   M. Pinard, M. Leduc, G. Trenec, C.G. Aminoff, F. Laloc: Efficient single mode operation of a standing wave dye laser. Appl. Phys. **19**, 399 (1978)

5.62   E. Samal, W. Becker: *Grundriß der praktischen Regelungstechnik* (Oldenbourg, München 1996)

5.63   P. Horrowitz, W. Hill: *The Art of Electronics*, 2nd edn. (Cambridge Univ. Press, Cambridge 1989)

5.64   Schott Information Sheet (Jenaer Glaswerk Schott & Gen., Mainz 1972)

5.65   R.W. Cahn, P. Haasen, E.J. Kramer (Eds.): *Materials Science and Technology, Vol. 11* (Wiley-VCH, Weinheim 1994)

5.66   J.J. Gagnepain: *Piezoelectricity* (Gordon & Breach, New York 1982)

5.67   W. Jitschin, G. Meisel: Fast frequency control of a CW dye jet laser. Appl. Phys. **19**, 181 (1979)

5.68   D.P. Blair: Frequency offset of a stabilized laser due to modulation distortion. Appl. Phys. Lett. **25**, 71 (1974)

5.69   J. Hough, D. Hills, M.D. Rayman, L.-S. Ma, L. Holbing, J.L. Hall: Dye-laser frequency stabilization using optical resonators. Appl. Phys. B **33**, 179 (1984)

5.70   F. Paech, R. Schmiedl, W. Demtröder: Collision-free lifetimes of excited NO_2 under very high resolution. J. Chem. Phys. **63**, 4369 (1975)

5.71   S. Seel, R. Storz, G. Ruosa, J. Mlynek, S. Schiller: Cryogenic optical resonators: a new tool for laser frequency stabilization at the 1 kHz lLevel. Phys. Rev. Lett. **78** 4741 (1997)

5.72   G. Camy, B. Decomps, J.-L. Gardissat, C.J. Bordé: Frequency stabilization of argon lasers at 582.49 THz using saturated absorption in $^{127}I_2$. Metrologia **13**, 145 (1977)

5.73   F. Spieweck: Frequency stabilization of Ar$^+$ lasers at 582 THz using expanded beams in external $^{127}I_2$ cells. IEEE Trans. IM-**29**, 361 (1980)

5.74   S.N. Bagayev, V.P. Chebotajev: Frequency stability and reproducibility at the 3.39 nm He-Ne laser stabilized on the methane line. Appl. Phys. **7**, 71 (1975)

5.75   J.L. Hall: 'Stabilized lasers and the speed of light'. In: *Atomic Masses and Fundamental Constants, Vol. 5*, ed. by J.H. Sanders, A.H. Wapstra (Plenum, New York 1976) p. 322

5.76   M. Niering et al.: Measurement of the hydrogen 1S–2S transition frequency by phase coherent comparison with a microwave cesium fountain clock. Phys. Rev. Lett. **84**, 5496 (2000)

5.77   T.W. Hänsch: 'High resolution spectroscopy of hydrogen'. In: *The Hydrogen Atom*, ed. by G.F. Bassani, M. Inguscio, T.W. Hänsch (Springer, Berlin, Heidelberg 1989)

5.78   P.H. Lee, M.L. Skolnik: Saturated neon absorption inside a 6328 Å laser. Appl. Phys. Lett. **10**, 303 (1967)

5.79   H. Greenstein: Theory of a gas laser with internal absorption cell. J. Appl. Phys. **43**, 1732 (1972)

5.80   T.N. Niebauer, J.E. Faller, H.M. Godwin, J.L. Hall, R.L. Barger: Frequency stability measurements on polarization-stabilized HeNe lasers. Appl. Opt **27**, 1285 (1988)

5.81   C. Salomon, D. Hils, J.L. Hall: Laser stabilization at the millihertz level. J. Opt. Soc. Am. B **5**, 1576 (1988)

5.82   D.G. Youmans, L.A. Hackel, S. Ezekiel: High-resolution spectroscopy of I_2 using laser-molecular-beam techniques. J. Appl. Phys. **44**, 2319 (1973)

5.83   D.W. Allan: 'In search of the best clock: An update'. In: *Frequency Standards and Metrology* (Springer, Berlin, Heidelberg 1989)

5.84   E.A. Gerber, A. Ballato (Eds.): *Precision Frequency Control* (Academic, New York 1985)

5.85   D. Hils, J.L. Hall: 'Ultrastable cavity-stabilized laser with subhertz linewidth'. In: [5.86], p. 162

5.86   *Frequency Standards and Metrology*, ed. by A. De Marchi (Springer, Berlin, Heidelberg 1989);
        M. Zhu, J.L. Hall: Short and long term stability of optical oscillators. J. Opt. Soc. Am. B **10**, 802 (1993)

5.87   K.M. Baird, G.R. Hanes: Stabilisation of wavelengths from gas lasers. Rep. Prog. Phys. **37**, 927 (1974)

5.88   T. Ikegami, S. Sudo, Y. Sakai: *Frequency Stabilization of Semiconductor Laser Diodes.* (Artech House, Boston 1995)

5.89   J. Hough, D. Hils, M.D. Rayman, L.S. Ma, L. Hollberg, J.L. Hall: Dye-laser frequency stabilization using optical resonators. Appl. Phys. B **33**, 179 (1984)

5.90   J.C. Bergquist (Ed.): Proc. 5th Symposium on Frequency Standards and Metrology (World Scientific, Singapore 1996)

5.91   M. Ohtsu (Ed.): Frequency Control of Semiconductor Lasers. (Wiley, New York 1991)

5.92   L.F. Mollenauer, J.C. White, C.R. Pollock (Eds.): *Tunable Lasers*, 2nd. edn., Topics Appl. Phys., Vol. 59 (Springer, Berlin, Heidelberg 1992)

5.93   H.G. Danielmeyer: Stabilized efficient single-frequency Nd:YAG laser. IEEE J. QE-**6** 101 (1970)

5.94   J.P. Goldsborough: Scanning single frequency CW dye laser techniques for high-resolution spectroscopy. Opt. Eng. **13**, 523 (1974)

5.95   S. Gerstenkorn, P. Luc: *Atlas du spectre d'absorption de la molecule d'iode* (Edition du CNRS, Paris 1978) with corrections in Rev. Phys. Appl. **14**, 791 (1979)

5.96   A Giachetti, R.W. Stanley, R. Zalibas: Proposed secondary standard wavelengths
        in the spectrum of thorium. J. Opt. Soc. Am. **60**, 474 (1970)

5.97   B.A. Palmer, R.A. Keller, F.V. Kovalski, J.L. Hall: Accurate wave-number mea-
        surements of uranium spectral lines. J. Opt. Soc. Am. **71**, 948 (1981)

5.98   (a) W. Jitschin, G. Meisel: 'Precise frequency tuning of a single mode dye laser'.
        In: *Laser'77, Opto-Electronics*, ed. by W. Waidelich (IPC Science and Technol-
        ogy, Guildford, Surrey 1977)
        (b) J.L. Hall: 'Saturated absorption spectroscopy'. In: *Atomic Physics, Vol. 3*, ed.
        by S. Smith, G.K. Walters (Plenum, New York 1973)

5.99   H.M. Nussenzweig: *Introduction to Quantum Optics* (Gordon & Breach, New
        York 1973)

5.100   W. Brunner, W. Radloff, K. Junge: *Quantenelektronik* (VEB Deutscher Verlag
        der Wissenschaften, Berlin 1975) p. 212

5.101   A.L. Schawlow, C.H. Townes: Infrared and optical masers. Phys. Rev. **112**, 1940
        (1958)

5.102   C.J. Bordé, J.L. Hall: 'Ultrahigh resolution saturated absorption spectroscopy'.
        In: *Laser Spectroscopy*, ed. by R.G. Brewer, H. Mooradian (Plenum, New York
        1974) pp. 125–142

5.103   J.L. Hall, M. Zhu, P. Buch: Prospects focussing laser-prepared atomic fountains
        for optical frequency standards applications. J. Opt. Soc. Am. B **6**, 2194 (1989)

5.104   R.W.P. Drever, J.L. Hall, F.V. Kowalski, J. Hough, G.M. Ford, A.J. Munley,
        H.W. Ward: Laser phase and frequency stabilization using an optical resonator.
        Appl. Phys. B **31**, 97 (1983)

5.105   S.N. Bagayev, A.E. Baklarov, V.P. Chebotayev, A.S. Dychkov, P.V. Pokasov:
        'Super high resolution laser spectroscopy with cold particles'. In: *Laser Spec-
        troscopy VIII*, ed. by W. Persson, J. Svanberg, Springer Ser. Opt. Sci., Vol. 55
        (Springer, Berlin, Heidelberg 1987);
        V.P. Chebotayev: 'High resolution laser spectroscopy'. In: *Frontier of Laser
        Spectroscopy*, ed. by T.W. Hänsch, M. Inguscio (North-Holland, Amsterdam
        1994)

5.106   K. Ueda, N. Uehara: Laser diode pumped solid state lasers for gravitational
        wave antenna. Proc. SPIE **1837**, 337 (1992)

5.107   G. Ruoso, R. Storz, S. Seel, S. Schiller, J. Mlynek: Nd:YAG laser frequency
        stabilization to a supercavity at the 0.1 Hz instability level. Opt. Comm. **133**,
        259 (1997)

5.108   M.J. Colles, C.R. Pidgeon: Tunable lasers. Rep. Prog. Phys. **38**, 329 (1975)

5.109   P.F. Moulton: 'Tunable paramagnetic-ion lasers'. In: *Laser Handbook, Vol. 5*, ed.
        by M. Bass, M.L. Stitch (North-Holland, Amsterdam 1985) p. 203

5.110   R.S. McDowell: 'High resolution infrared spectroscopy with tunable lasers'.
        In: *Advances in Infrared and Raman Spectroscopy, Vol. 5*, ed. by R.J.H. Clark,
        R.E. Hester (Heyden, London 1978)

5.111   K.W. Nill: Spectroscopy with tunable diode lasers. Laser Focus **13**, 32 (February
        1977)

5.112   E.D. Hinkley, K.W. Nill, F.A. Blum: 'Infrared spectroscopy with tunable lasers'.
        In: *Laser Spectroscopy of Atoms and Molecules*, ed. by H. Walther, Topics Appl.
        Phys. Vol. 2 (Springer, Berlin, Heidelberg 1976)

5.113   G.P. Agraval (Ed.): *Semiconductor Lasers* (AIP, Woodbury 1995)

5.114   G.P. Agrawal, N.K. Dutta: *Long Wavelength Semiconductor Lasers* (Van Nos-
        trand, Reinhold, New York 1986)

5.115   A. Mooradian: 'High resolution tunable infrared lasers'. In: *Very High Resolu-
        tion Spectroscopy*, ed. by R.A. Smith (Academic, London 1976)

5.116   I. Melgailis, A. Mooradian: 'Tunable semiconductor diode lasers and applica-
        tions'. In: *Laser Applications in Optics and Spectroscopy* ed. by S. Jacobs,
        M. Sargent, M. Scully, J. Scott (Addison Wesley, Reading, MA 1975) p. 1

5.117   H.C. Casey, M.B. Panish: *Heterostructure Lasers* (Academic, New York 1978)

5.118   C. Vourmard: External-cavity controlled 32 MHz narrow band CW GaAs-diode laser. Opt. Lett. **1**, 61 (1977)

5.119   W. Fleming, A. Mooradian: Spectral characteristics of external cavity controlled semiconductor lasers. IEEE J. QE-**17**, 44 (1971)

5.120   W. Fuhrmann, W. Demtröder: A widely tunable single-mode GaAs-diode laser with external cavity. Appl. Phys. B **49**, 29 (1988)

5.121   H. Tabuchi, H. Ishikawa: External grating tunable MQW laser with wide tuning range at 240 nm. Electron. Lett. **26**, 742 (1990);
M. de Labachelerie, G. Passedat: Mode-hop suppression of Littrow grating-tuned lasers. Appl. Opt. **32**, 269 (1993)

5.122   H. Wenz, R. Großkloß, W. Demtröder: Kontinuierlich durchstimmbare Halbleiterlaser. Laser & Optoelektronik **28**, 58 (Febr. 1996)
C.J. Hawthorne, K.P. Weber, R.E. Schulten: Littrow configuration tunable external cavity diode laser with fixed direction output beam. Rev. Sci. Instrum. **72**, 4477 (2001)

5.123   Sh. Nakamura, Sh.F. Chichibu: *Introduction to Nitride Semiconductor Blue Lasers* (Taylor and Francis, London 2000)

5.124   N.W. Carlson: *Monolythic Diode Laser Arrays* (Springer, Berlin, Heidelberg, New York 1994)
R. Diehl (Ed.): *High Power Diode Lasers* (Springer, Berlin, Heidelberg, New York 2000)

5.125   R.C. Powell: *Physics of Solid-State Laser Materials* (Springer, Berlin, Heidelberg, New York 1998)

5.126   A.A. Kaminskii: *Laser Crystals*, 2nd edn., Springer Ser. Opt. Sci., Vol. 14 (Springer, Berlin, Heidelberg 1990); A.A. Kaminsky: *Crystalline Lasers* (CRC, New York 1996)

5.127   M. Inguscio, R. Wallenstein (Eds.): *Solid State Lasers; New Developments and Applications* (Plenum, New York 1993)

5.128   U. Dürr: Vibronische Festkörperlaser: Der Übergangsmetallionen-Laser. Laser Optoelectr. **15**, 31 (1983)

5.129   A. Miller, D.M. Finlayson (Eds.): *Laser Sources and Applications* (Institute of Physics, Bristol 1996)

5.130   F. Gan: *Laser Materials* (World Scientific, Singapore 1995)

5.131   S.T. Lai: Highly efficient emerald laser. J. Opt. Soc. Am. B **4**, 1286 (1987)

5.132   G.T. Forrest: Diode-pumped solid state lasers have become a mainstream technology. Laser Focus **23**, 62 (1987)

5.133   W.P. Risk, W. Lenth: Room temperature continuous wave 946 nm Nd:YAG laser pumped by laser diode arrays and intracavity frequency doubling to 473 nm. Opt. Lett. **12**, 993 (1987)

5.134   P. Hammerling, A.B. Budgor, A. Pinto (Eds.): *Tunable Solid State Lasers I*, Springer Ser. Opt. Sci., Vol. 47 (Springer, Berlin, Heidelberg 1984)

5.135   A.B. Budgor, L. Esterowitz, L.G. DeShazer (Eds.): *Tunable Solid State Lasers II*, Springer Ser. Opt. Sci., Vol. 52 (Springer, Berlin, Heidelberg 1986)

5.136   W. Koechner: *Solid-State Laser Enginering*, 4th edn., Springer Ser. Opt. Sci., Vol. 1 (Springer, Berlin, Heidelberg 1996)

5.137   N.P. Barnes: 'Transition Metal Solid State Lasers'. In: Tunable Laser Handbook, ed. by F.J. Duarte (Academic, San Diego 1995)

5.138   W.B. Fowler (Ed.): *Physics of Color Centers* (Academic, New York 1968)

5.139   F. Lüty: 'F_A-Centers in Alkali Halide Crystals'. In: *Physics of Color Centers*, ed. by W.B. Fowler (Academic, New York 1968)

5.140   L.F. Mollenauer, D.H. Olsen: Broadly tunable lasers using color centers. J. Appl. Phys. **46**, 3109 (1975)

5.141   L.F. Mollenauer: 'Color center lasers'. In: *Laser Handbook IV*, ed. by M. Bass, M.L. Stitch (North-Holland, Amsterdam 1985) p. 143

5.142   V. Ter-Mikirtychev: Diode pumped tunable room-temparature LiF:F_2^- color center laser. Appl. Opt. **37**, 6442 (1998)

5.143   H.W. Kogelnik, E.P. Ippen, A. Dienes, C.V. Shank: Astigmatically compensated cavities for CW dye lasers. IEEE J. QE-**8**, 373 (1972)

5.144   R. Beigang, G. Litfin, H. Welling: Frequency behavior and linewidth of CW single mode color center lasers. Opt. Commun. **22**, 269 (1977)

5.145   G. Phillips, P. Hinske, W. Demtröder, K. Möllmann, R. Beigang: NaCl-color center laser with birefringent tuning. Appl. Phys. B **47**, 127 (1988)

5.146   G. Litfin: Color center lasers. J. Phys. E **11**, 984 (1978)

5.147   L.F. Mollenauer, D.M. Bloom, A.M. Del Gaudio: Broadly tunable CW lasers using F_2^+-centers for the 1.26−1.48 µm and 0.82−1.07 µm bands. Opt. Lett. **3**, 48 (1978)

5.148   L.F. Mollenauer: Room-temperature stable F_2^+-like center yields CW laser tunable over the 0.99−1.22 µm range. Opt. Lett. **5**, 188 (1980)

5.149   W. Gellermann, K.P. Koch, F. Lüty: Recent progess in color center lasers. Laser Focus **18**, 71 (1982)

5.150   M. Stuke (Ed.): *25 Years Dye Laser*, Topics Appl. Phys., Vol. 70 (Springer, Berlin, Heidelberg 1992)

5.151   F.P. Schäfer (Ed.): *Dye Lasers*, 3rd edn., Topics Appl. Phys., Vol. 1 (Springer, Berlin, Heidelberg 1990)

5.152   F.J. Duarte, L.W. Hillman: *Dye Laser Principles* (Academic, Boston 1996);
        F.J. Duarte: *Tunable Lasers Handbook* (Academic, New York 1996)

5.153   G. Marowsky, R. Cordray, F.K. Tittel, W.L. Wilson, J.W. Keto: Energy transfer processes in electron beam excited mixtures of laser dye vapors with rare gases. J. Chem. Phys. **67**, 4845 (1977)

5.154   B. Steyer, F.P. Schäfer: Stimulated and spontaneous emission from laser dyes in the vapor phase. Appl. Phys. **7**, 113 (1975)

5.155   F.J. Duarte (Ed.): *High-Power Dye Lasers*, Springer Ser. Opt. Sci., Vol. 65 (Springer, Berlin, Heidelberg 1991)

5.156   W. Schmidt: Farbstofflaser. Laser **2**, 47 (1970)

5.157   G.H. Atkinson, M.W. Schuyler: A simple pulsed laser system, tunable in the ultraviolet. Appl. Phys. Lett. **27**, 285 (1975)

5.158   A. Hirth, H. Fagot: High average power from long pulse dye laser. Opt. Commun. **21**, 318 (1977)

5.159   J. Jethwa, F.P. Schäfer, J. Jasny: A reliable high average power dye laser. IEEE J. QE-**14**, 119 (1978)

5.160   J. Kuhl, G. Marowsky, P. Kunstmann, W. Schmidt: A simple and reliable dye laser system for spectroscopic investigations. Z. Naturforsch. **27a**, 601 (1972)

5.161   H. Walther, J.L. Hall: Tunable dye laser with narrow spectral output. Appl. Phys. Lett. **6**, 239 (1970)

5.162   P.J. Bradley, W.G.I. Caugbey, J.I. Vukusic: High efficiency interferometric tuning of flashlamp-pumped dye lasers. Opt. Commun. **4**, 150 (1971)

5.163   M. Okada, K. Takizawa, S. Ieieri: Tilted birefringent Fabry–Perot etalon for tuning of dye lasers. Appl. Opt. **15** 472 (1976)

5.164   G.M. Gale: A single-mode flashlamp-pumped dye laser. Opt. Commun. **7**, 86 (1973)

5.165   J.J. Turner, E.I. Moses, C.L. Tang: Spectral narrowing and electro-optical tuning of a pulsed dye-laser by injection-locking to a CW dye laser. Appl. Phys. Lett. **27**. 441 (1975)

5.166   M. Okada, S. Ieiri: Electronic tuning of dye-lasers by an electro-optical birefringent Fabry–Perot etalon. Opt. Commun. **14**, 4 (1975)

5.167   J. Kopainsky: Laser scattering with a rapidly tuned dye laser. Appl. Phys. **8**, 229 (1975)

5.168   F.P. Schäfer, W. Schmidt, J. Volze: Appl. Phys. Lett. **9** 306 (1966)

5.169   P.P. Sorokin, J.R. Lankard: IBM J. Res. Develop. **10**, 162 (1966);
        P.P. Sorokin: Organic lasers. Sci. Am. **220**, 30 (February 1969)

5.170   F.B. Dunning, R.F. Stebbings: The efficient generation of tunable near UV radiation using a N_2-pumped dye laser. Opt. Commun. **11**, 112 (1974)

5.171   T.W. Hänsch: Repetitively pulsed tunable dye laser for high resolution spectroscopy. Appl. Opt. **11**, 895 (1972)

5.172   R. Wallenstein: Pulsed narrow band dye lasers. Opt. Acta **23**, 887 (1976)

5.173   I. Soshan, N.N. Danon, V.P. Oppenheim: Narrowband operation of a pulsed dye laser without intracavity beam expansion. J. Appl. Phys. **48**, 4495 (1977)

5.174   S. Saikan: Nitrogen-laser-pumped single-mode dye laser. Appl. Phys. **17**, 41 (1978)

5.175   K.L. Hohla: Excimer-pumped dye lasers – the new generation. Laser Focus **18**, 67 (1982)

5.176   O. Uchino, T. Mizumami, M. Maeda, Y. Miyazoe: Efficient dye lasers pumped by a XeCl excimer laser. Appl. Phys. **19**, 35 (1979)

5.177   R. Wallenstein: 'Pulsed dye lasers'. In: *Laser Handbook Vol.3*, ed. by M.L. Stitch (North-Holland, Amsterdam 1979) pp. 289–360

5.178   W. Hartig: A high power dye laser pumped by the second harmonics of a Nd:YAG laser. Opt. Commun. **27**, 447 (1978)

5.179   F.J. Duarte, J.A. Piper: Narrow linewidths, high pulse repetition frequency copper-laser-pumped dye-laser oscillators. Appl. Opt. **23**, 1991 (1984)

5.180   M.G. Littman: Single-mode operation of grazing-incidence pulsed dye laser. Opt. Lett. **3**, 138 (1978)

5.181   K. Liu, M.G. Littmann: Novel geometry for single-mode scanning of tunable lasers. Opt. Lett. **6**, 117 (1981)

5.182   M. Littmann, J. Montgomery: Grazing incidence designs improve pulsed dye lasers: Laser Focus **24**, 70 (February 1988)

5.183   F.J. Duarte, R.W. Conrad: Diffraction-limited single-longitudinal-mode multiple-prism flashlamp-pumped dye laser oscillator. Apl. Opt. **26**, 2567 (1987);
        F.J. Duarte, R.W. Conrad: Multiple-prism Littrow and grazing incidence pulsed CO_2 lasers. Appl.Opt. **24**, 1244 (1985)

5.184   F.J. Duarte, J.A. Piper: Prism preexpanded gracing incidence grating cavity for pulsed dye lasers. Appl. Opt. **20**. 2113 (1981)

5.185   F.J. Duarte: Multipass dispersion theory of prismatic pulsed dye lasers. Opt. Acta **31**, 331 (1984)

5.186   Lambda Physik information sheet, Göttingen 2000
        and http://www.lambdaphysik.com

5.187   S. Leutwyler, E. Schumacher, L. Wöste: Extending the solvent palette for CW jet-stream dye lasers. Opt. Commun. **19**, 197 (1976)

5.188   P. Anliker, H.R. Lüthi, W. Seelig, J. Steinger, H.P. Weber: 33 watt CW dye laser. IEEE J. QE-**13**, 548 (1977)

5.189   P. Hinske: Untersuchung der Prädissoziation von Cs_2-Molekülen. Diploma Thesis, F.B. Physik, University of Kaiserslautern (1988)

5.190   A. Bloom: Modes of a laser resonator, containing tilted birefringent plates. J. Opt. Soc. Am. **64**, 447 (1974)

5.191   H.W. Schröder, H. Dux, H. Welling: Single-mode operation of CW dye lasers. Appl. Phys. **7**, 21 (1975)

5.192   H. Gerhardt, A. Timmermann: High resolution dye-laser spectrometer for measurements of isotope and isomer shifts and hyperfine structure. Opt. Commun. **21**, 343 (1977)

5.193   H.W. Schröder, L. Stein, D. Fröhlich, F. Fugger, H. Welling: A high power single-mode CW dye ring laser. Appl. Phys. **14**, 377 (1978)

5.194   D. Kühlke, W. Diehl: Mode selection in CW laser with homogeneously broadened gain. Opt. Quantum Electron. **9**, 305 (1977)

5.195   J.D. Birks: Excimers. Rep. Prog. Phys. **38**, 903 (1977)

5.196   C.K. Rhodes (Ed.): *Excimer Lasers*, 2nd edn., Topics Appl. Phys., Vol. 30 (Springer, Berlin, Heidelberg 1984)

5.197   H. Scheingraber, C.R. Vidal: Discrete and continuous Franck–Condon factors of the $Mg_2 A\,^1\Sigma_\mu^+ - X\,^1\Sigma_g^+$ system. J. Chem. Phys. **66**, 3694 (1977)

5.198   D.J. Bradley: 'Coherent radiation generation at short wavelengths'. In: *High-Power Lasers and Applications*, ed. by K.L. Kompa, H. Walther, Springer Ser. Opt. Sci., Vol. 9 (Springer, Berlin, Heidelberg 1978) pp. 9–18

5.199   R.C. Elton: *X-Ray Lasers* (Academic, New York 1990)

5.200   H.H. Fleischmann: High current electron beams. Phys. Today **28**, 34 (May 1975)

5.201   C.P. Wang: Performance of XeF/KrF lasers pumped by fast discharges. Appl. Phys. Lett. **29**, 103 (1976)

5.202   H. Pummer, U. Sowada, P. Oesterlin, U. Rebban, D. Basting: Kommerzielle Excimerlaser. Laser u. Optoelektronik **17**, 141 (1985)

5.203   S.S. Merz: Switch developments could enhance pulsed laser performance. Laser Focus **24**, 70 (May 1988)

5.204   P. Klopotek, V. Brinkmann, D. Basting, W. Mückenheim: A new excimer laser producing long pulses at 308 nm. Lambda-Physik Mitteilung (Göttingen 1987)

5.205   D. Basting: Excimer lasers – new perspectives in the UV. Laser and Elektro-Optic **11**, 141 (1979)

5.206   K. Miyazak: T. Fukatsu, I. Yamashita, T. Hasama, K. Yamade, T. Sato: Output and picosecond amplifcation characteristics of an efficient and high-power discharge excimer laser. Appl. Phys. B **52**, 1 (1991)

5.207   J.M.J. Madey: Stimulated emission of Bremsstrahlung in a periodic magnetic field. J. Appl. Phys. **42**, 1906 (1971)

5.208   E.L. Saldin, E. Schneidmiller, M. Yurkow: *The Physics of Free Electron Lasers* (Springer, Berlin, Heidelberg, New York 2000)

5.209   T.C. Marshall: *Free-Electron Lasers* (MacMillan, New York 1985)

5.210   .P. Freund, T.M. Antonsen: *Principles of Free-electron Lasers* (Chapmand Hall, London 1992)

5.211   *Report of the Committee of the National Academy of Science on Free Electron Lasers and Other Advanced Sources of Light* (National Academy Press, Washington 1994);

5.212   A. Yariv, P. Yeh: *Optical waves in crystals* (Wiley, New York 1984)
        C.R. Vidal: Coherent VUV sources for high resolution spectroscopy. Appl. Opt. **19**, 3897 (1980)

5.213   J.R. Reintjes: 'Coherent ultraviolet and VUV-sources'. In: *Laser Handbook V*, ed. by M. Bass, M.L. Stitch (North-Holland, Amsterdam 1985)

5.214   N. Bloembergen: *Nonlinear Optics*, 4th edn. (World Scientific, Singapore 1996)

5.215   A. Newell, J.V. Moloney: *Nonlinear Optics* (Addison-Wesley, Reding, MA 1992)

5.216   D.L. Mills: *Nonlinear Optics*, 2nd edn. (Springer, Berlin, Heidelberg 1998)

5.217   G.C. Baldwin: *An Introduction to Nonlinear Optics* (Plenum, New York 1969)

5.218   F. Zernike, J.E. Midwinter: *Applied Nonlinear Optics* (Academic, New York 1973)

5.219   A.C. Newell, J.V. Moloney: *Nonlinear Optics* (Addison Wesley, Redwood City 1992);
        Y.P. Svirko, N.I. Zheluder: *Polarization of Light in Nonlinear Optics* (Wiley, Chichester 1998)

5.220   P.G. Harper, B.S. Wherrett (Eds.) *Nonlinear Optics* (Academic, London 1977)

5.221   Laser Analytical Systems GmbH, Berlin, now available at Spectra Physics, Palo Alto, USA

5.222   P. Günther (Ed.): *Nonlinear Optical Effects and Materials* (Springer, Berlin, Heidelberg, New York 2000)

5.223   D.A. Kleinman, A. Ashkin, G.D. Boyd: Second harmonic generation of light by focussed laser beams. Phys. Rev. **145**, 338 (1966)

5.224   P. Lokai, B. Burghardt, S.D. Basting: W. Mückenheim: Typ-I Frequenzverdopplung und Frequenzmischung in β-BaB_2O_4. Laser Optoelektr. **19**, 296 (1987)

5.225   R.S. Adhav, S.R. Adhav, J.M. Pelaprat: BBO's nonlinear optical phase-matching properties. Laser Focus **23**, 88 (1987)

5.226   G.G. Gurzadian et al.: *Handbook of Nonlinear Optical Crystals* (Springer, Berlin, Heidelberg, New York 1999)

5.227   H. Schmidt, R. Wallenstein: Beta-Bariumbort: Ein neues optisch-nichtlineares Material. Laser Optoelektr. **19**, 302 (1987)

5.228   Ch. Chuangtian, W. Bochang, J. Aidong, Y. Giuming: A new-type ultraviolet SHG crystal: β-BaB_2O_4. Scientia Sinica B **28**, 235 (1985)

5.229   I. Shogi, H. Nakamura, K. Obdaira, T. Kondo, R. Ito: Absolute measurement of second order nonlinear-optical cefficient of beta-BaB_2O_4 for visible and ultraviolet second harmonic wavelengths. J. Opt. Soc. Am. B **16**, 620 (1999)

5.230   H. Kouta, Y. Kuwano: Attaining 186 nm light generation in cooled beta BaB_2O_4 crystal. Opt Lett. **24**, 1230 (1999)

5.231   J.C. Baumert, J. Hoffnagle, P. Günter: High efficiency intracavity frequency doubling of a styril-9 dye laser with $KNbO_3$ crystals. Appl. Opt. **24**, 1299 (1985)

5.232   T.J. Johnston Jr.: 'Tunable dye lasers'. In: *The Encyclopedia of Physical Science and Technology*, Vol. 14 (Academic, San Diego 1987)

5.233   W.A. Majewski: A tunable single frequency UV source for high resolution spectroscopy. Opt. Commun. **45**, 201 (1983)

5.234   S.J. Bastow, M.H. Dunn: The generation of tunable UV radiation from 238–249 nm by intracavity frequency doubling of a coumarin 102 dye laser. Opt. Commun. **35**, 259 (1980)

5.235   D. Fröhlich, L. Stein, H.W. Schröder, H. Welling: Efficient frequency doubling of CW dye laser radiation. Appl. Phys. **11**, 97 (1976)

5.236   A. Renn, A. Hese, H. Busener: Externer Ringresonator zu Erzeugung kontinuierliche UV-Strahlung. Laser Optoelektr. **3**, 11 (1982)

5.237   F.V. Moers, T. Hebert, A. Hese: Theory and experiment of CW dye laser injection locking and its application to second harmonic generation. Appl. Phys. B **40**, 67 (1986)

5.238   N. Wang, V. Gaubatz: Optical frequency doubling of a single-mode dye laser in an external resonator. Appl. Phys. B **40**, 43 (1986)

5.239   J.T. Lin, C. Chen: Chosing a nonlinear crystal. Laser and Optronics **6**, 59 (November 1987)

5.240   Castech-Phoenix Inc. Fujian, China, information sheets

5.241   L. Wöste: 'UV-generation in CW dye lasers'. In: *Advances in Laser Spectroscopy* ed. by F.T. Arrechi, F. Strumia, H. Walther (Plenum, New York 1983)

5.242   H.J. Müschenborn, W. Theiss, W. Demtröder: A tunable UV-light source for laser spectroscopy using second harmonic generation in β-BaB_2O_4. Appl. Phys. B **50**, 365 (1990)

5.243   H. Dewey: Second harmonic generation in $KB_4OH \cdot 4H20$ from 217 to 315 nm. IEEE J. QE-**12**, 303 (1976)

5.244   M. Feger, G.M. Magel, D.H. Hundt, R.L. Byer: Quasi-phase-matched second harmonic generation: tuning and tolerances. IEEE J. Quant. Electr. **28**, 2631 (1992)

5.245   M. Pierrou, F. Laurell, H. Karlsson, T. Kellner, C. Czeranowsky, G. Huber: Generation of 740 mW of blue light by intracavity frequency doubling with a first order quasi-phase-matched KTiOPO$_4$ crystal. Opt. Lett. **24**, 205 (1999)

5.246   F.B. Dunnings: Tunable ultraviolet generation by sum-frequency mixing. Laser Focus **14**, 72 (1978)

5.247   S. Blit, E.G. Weaver, F.B. Dunnings, F.K. Tittel: Generation of tunable continuous wave ultraviolet radiation from 257 to 329 nm. Opt. Lett. **1**, 58 (1977)

5.248   R.F. Belt, G. Gashunov, Y.S. Liu: KTP as an harmonic generator for Nd:YAG lasers. Laser Focus **21**, 110 (1985)

5.249   J. Halbout, S. Blit, W. Donaldson, Ch.L. Tang: Efficient phase-matched second harmonic generation and sum-frequency mixing in urea. IEEE J. QE-**15**, 1176 (1979)

5.250   G. Nath, S. Haussühl: Large nonlinear optical coefficient and phase-matched second harmonic generation in LiIO$_3$. Appl. Phys. Lett. **14**, 154 (1969)

5.251   F.B. Dunnings, F.K. Tittle, R.F. Stebbings: The generation of tunable coherent radiation in the wavelength range 230 to 300 nm using lithium formate monohydride. Opt. Commun. **7**, 181 (1973)

5.252   F.B. Dunnings: Tunable ultraviolet generation by sum-frequency mixing. Laser Focus **14**, 72 (May 1978)

5.253   G.A. Massey, J.C. Johnson: Wavelength tunable optical mixing experiments between 208 and 259 nm. IEEE J. QE-**12**, 721 (1976)

5.254   A. Borsutzky, R. Brünger, Ch. Huang, R. Wallenstein: Harmonic and sum-frequency generation of pulsed laser radiation in BBO, LBO and KD$^+$P. Appl. Phys. B **52**, 55 (1991)

5.255   K. Matsubara, U. Tanaka, H. Imago, M. Watanabe: All-solid state light source for generation of continious-wave coherent radiation near 202 nm. J. Opt. Soc. Am. B **16**, 1668 (1999)

5.256   C.R. Vidal: Third harmonic generation of mode-locked Nd:glass laser pulses in phase-matched Rb-Xe-mixtures. Phys. Rev. A **14**, 2240 (1976)

5.257   R. Hilbig, R. Wallenstein: Narrow band tunable VUV-radiation generated by nonresonant sum- and difference-frequency mixing in xenon and krypton. Appl. Opt. **21**, 913 (1982)

5.258   C.R. Vidal: 'Four-wave frequency mixing in gases'. In: *Tunable Lasers*, 2nd edn., ed. by I.F. Mollenauer, J.C. White, C.R. Pollock, Topics Appl. Phys., Vol. 59 (Springer, Berlin, Heidelberg 1992)

5.259   G. Hilber, A. Lago, R. Wallenstein: Broadly tunable VUV/XUV-adiation generated by resonant third-order frequency conversion in Kr. J. Opt. Soc. Am. B **4**, 1753 (1987);
A. Lago, G. Hilber, R. Wallenstein: Optical frequency conversion in gaseous media. Phys. Rev. A **36**, 3827 (1987)

5.260   S.E. Harris, J.F. Young, A.H. Kung, D.M. Bloom, G.C. Bjorklund: 'Generation of ultraviolet and VUV-radiation'. In: *Laser Spectroscopy I*, ed. by R.G. Brewer, A. Mooradian (Plenum, New York 1974)

5.261   A.H. Kung, J.F. Young, G.C. Bjorklund, S.E. Harris: Generation of Vacuum Ultraviolet Radiation in Phase-matched Cd Vapor. Phys. Rev. Lett. **29**, 985 (1972)

5.262   W. Jamroz, B.P. Stoicheff: 'Generation of tunable coherent vacuum-ultraviolet radiation'. In: *Progress in Optics* **20**, 324 (North-Holland, Amsterdam 1983)

5.263   A. Timmermann, R. Wallenstein: Generation of tunable single-frequency CW coherent vacuum-ultraviolet radiation. Opt. Lett. **8**, 517 (1983)

5.264   T.P. Softley, W.E. Ernst, L.M. Tashiro, R.Z. Zare: A general purpose XUV laser spectrometer: Some applications to N$_2$, O$_2$ and CO$_2$. Chem. Phys. **116**, 299 (1987)

5.265   J. Bokon, P.H. Bucksbaum, R.R. Freeman: Generation of 35.5 nm coherent radiation. Opt. Lett. **8**, 217 (1983)

5.266   P.F. Levelt, W. Ubachs: XUV-laser spectroscopy in the $Cu\,^1\Sigma_u^+$, $v = 0$ and
        $c_3\,^1\Pi_u$, $v = 0$ Rydberg states of N_2. Chem. Phys. **163**, 263 (1992)

5.267   H. Palm, F. Merkt: Generation of tunable coherent XUV radiation beyond 19 eV
        by resonant four wave mixing in argon. Appl. Phys. Lett. **73**, 157 (1998)

5.268   U. Hollenstein, H. Palm, F. Merkt: A broadly tunable XUV laser source with a
        $0.008\,cm^{-1}$ bandwidth. Rev. Sci. Instrum. **71**, 4023 (2000)

5.269   D.L. Matthews, R.R. Freeman (Eds.): The generation of coherent XUV and soft
        X-Ray radiation. J. Opt. Soc. Am. B **4**, 533 (1987)

5.270   C. Yamanaka (Ed.): *Short-Wavelength Lasers*, Springer Proc. Phys., Vol. 30
        (Springer, Berlin, Heidelberg 1988)

5.271   D.L. Matthews, M.D. Rosen: Soft X-ray lasers. Sci. Am. **259**, 60 (1988)

5.272   T.J. McIlrath (Ed.): *Laser Techniques for Extreme UV Spectroscopy, AIP Conf.
        Proc.* **90** (1986)

5.273   B. Wellegehausen, M. Hube, F. Jim: Investigation on laser plasma soft X-ray
        sources generated with low-energy laser systems. Appl. Phys. B **49**, 173 (1989)

5.274   E. Fill (Guest Ed.): X-ray lasers. Appl. Phys. B **50**, 145–226 (1990);
        E. Fill: Appl. Phys. B **58**, 1–56 (1994)

5.275   A.L. Robinson: Soft X-ray laser at Lawrence Livermore Lab. Science **226**, 821
        (1984)

5.276   E.E. Fill (Guest Ed.): X-ray lasers. Appl. Phys. B **50** (1990);
        E.E. Fill (Ed.): *X-Ray Lasers* (Institute of Physics, Bristol 1992)

5.277   A.S. Pine: 'IR-spectroscopy via difference-frequency generation'. In: *Laser
        Spectroscopy III*, ed. by J.L. Hall, J.L. Carlsten, Springer Ser. Opt. Sci., Vol. 21
        (Springer, Berlin, Heidelberg 1977) p. 376

5.278   A.S. Pine: High-resolution methane ν_3-band spectra using a stabilized tunable
        difference-frequency laser system. J. Opt. Soc. Am. **66**, 97 (1976);
        A.S. Pine: J. Opt. Soc. Am. **64**, 1683 (1974)

5.279   U. Simon, S. Waltman, I. Loa, L. Holberg, T.K. Tittel: External cavity differ-
        ence frequency source near 3.2 μm based on mixing a tunable diode laser with
        a diode-pumped Nd:YAG-laser in $AgGaS_2$. J. Opt. Soc. Am. B **12**, 322 (1995)

5.280   A.H. Hilscher, C.E. Miller, D.C. Bayard, U. Simon, K.P. Smolka, R.F. Curl,
        F.K. Tittel: Optimization of a midinfrared high-resolution difference frequency
        laser spectrometer. J. Opt. Soc. Am. B **9**, 1962 (1992)

5.281   M. Seiter, D. Keller, M.W. Sigrist: Broadly tunable difference-frequency spec-
        trometry for trace gas detection with noncollinear critical phase-matching in
        $LiNbO_3$. Appl. Phys. B **67**, 351 (1998)

5.282   W. Chen, J. Buric, D. Boucher: A widely tunable cw laser difference frequency
        source for high resolution infrared spectroscopy. Laser Physics **10**, 521 (2000)

5.283   R.Y. Shen (Ed.): *Nonlinear Infrared Generation*, Topics Appl. Phys., Vol. 16
        (Springer, Berlin, Heidelberg 1977)

5.284   K.M. Evenson, D.A. Jennings, K.R. Leopold, L.R. Zink: 'Tunable far infrared
        spectroscopy'. In: *Laser Spectroscopy VII*, ed. by T.W. Hänsch, Y.R. Shen,
        Springer Ser. Opt. Sci., Vol. 49 (Springer, Berlin, Heidelberg 1985) p. 366

5.285   M. Inguscio: Coherent atomic and molecular spectroscopy in the far-infrared.
        Physica Scripta **37**, 699 (1988)

5.286   L.R. Zink, M. Prevedelti, K.M. Evenson, M. Inguscio: 'High resolution far-
        infrared spectroscopy'. In: *Applied Laser Spectroscopy X*, ed. by W. Demtröder,
        M. Inguscio (Plenum, New York 1990)

5.287   M. Inguscio, P.R. Zink, K.M. Evenson, D.A. Jennings: Sub-Doppler tunable far-
        infrared spectroscopy. Opt. Lett. **12**, 867 (1987)

5.288   S.E. Harris: Tunable optical parametric oscillators. Proc. IEEE **57**, 2096 (1969)

5.289   R.L. Byer: 'Parametric oscillators and nonlinear materials'. In: *Nonlinear Optics*,
        ed. by P.G. Harper, B.S. Wherret (Academic, London 1977)

5.290   R.L. Byer, R.L. Herbet, R.N. Fleming: 'Broadly tunable IR-source'. In: *Laser Spectroscopy II*, ed. by S. Haroche, J.C. Pebay-Peyroula, T.W. Hänsch, S.E. Harris, Lect. Notes Phys., Vol. 43 (Springer, Berlin, Heidelberg 1974) p. 207

5.291   C.L. Tang, L.K. Cheng: *Fundamentals of Optical Parametric Processes and Oscillators* (Harwood, Amsterdam 1995)

5.292   A. Yariv: 'Parametric processes'. In: *Progress in Quantum Electronics Vol. 1, Part 1*, ed. by J.H. Sanders, S. Stenholm (Pergamon, Oxford 1969)

5.293   R. Fischer: Vergleich der Eigenschaften doppelt-resonanter optischer parametrischer Oszillatoren. Exp. Technik der Physik **21**, 21 (1973)

5.294   C.L. Tang, W.R. Rosenberg, T. Ukachi, R.J. Lane, L.K. Cheng: Advanced materials aid parametric oscillator development. Laser Focus **26**, 107 (1990)

5.295   A. Fix, T. Schröder, R. Wallenstein: New sources of powerful tunable laser radiation in the ultraviolet, visible and near infrared. Laser und Optoelektronik **23**, 106 (1991)

5.296   R.L. Byer, M.K. Oshamn, J.F. Young, S.E. Harris: Visible CW parametric oscillator. Appl. Phys. Lett. **13**, 109 (1968)

5.297   J. Pinnard, J.F. Young: Interferometric stabilization of an optical parametric oscillator. Opt. Commun. **4**, 425 (1972)

5.298   J.G. Haub, M.J. Johnson, B.J. Orr, R. Wallenstein: Continuously tunable, injection-seeded β-barium borate optical parametric oscillator. Appl. Phys. Lett. **58**, 1718 (1991)

5.299   S. Schiller, J. Mlynek (Eds.): Special issue on cw optical parametric oscillators. Appl. Phys. B **25** (June 1998) and J. Opt. Soc. Am. B **12**, (11) (1995)

5.300   M.E. Klein, M. Scheidt, K.J. Boller, R. Wallenstein: Dye laser pumped, continous-wave KTP optical parametric oscillators. Appl. Phys. B **25**, 727 (1998)

5.301   R. Al-Tahtamouni, K. Bencheik, R. Sturz, K. Schneider, M. Lang, J. Mlynek, S. Schiller: Long-term stable operation and absolute frequency stability of a doubly resonant parametric oscillator. Appl. Phys. B **25**, 733 (1998)

5.302   Information leaflet Coherent Laser Group, Palo Alto, Calif.; and Linos Photonics GmbH, Göttingen, Germany

5.303   V. Wilke, W. Schmidt: Tunable coherent radiation source covering a spectral range from 185 to 880 nm. Appl. Phys. **18**, 177 (1979)

5.304   W. Hartig, W. Schmidt: A broadly tunable IR waveguide Raman laser pumped by a dye laser. Appl. Phys. **18**, 235 (1979)

5.305   C. Lin, R.H. Stolen, W.G. French, T.G. Malone: A CW tunable near-infrared (1.085−1.175 μm) Raman oscillator. Opt. Lett. **1**, 96 (1977)

5.306   D.J. Brink, D. Proch: Efficient tunable ultraviolet source based on stimulated Raman scattering of an excimer-pumped dye laser. Opt. Lett. **7**, 494 (1982)

5.307   A.Z. Grasiuk, I.G. Zubarev: High power tunable IR Raman lasers. Appl. Phys. **17**, 211 (1978)

5.308   K. Suto: *Semiconductor Raman Lasers* (Boston Artech House, Boston 1994)

5.309   B. Wellegehausen, K. Ludewigt, H. Welling: Anti-Stokes Raman lasers. SPIE Proc. **492**, 10 (1985)

5.310   A. Weber (Ed.): *Raman Spectroscopy of Gases and Liquids*, Topics Curr. Phys., Vol. 11 (Springer, Berlin, Heidelberg 1979)

Chapter 6

6.1   R.J. Bell: *Introductory Fourier Transform Spectroscopy* (Academic, New York 1972);
P. Griffiths, J.A. de Haset: *Fourier Transform Infrared Spectroscopy* (Wiley, New York 1986);

J. Kauppinen, J. Partanen: *Fourier Transforms in Spectroscopy* (Wiley, New York 2001)

6.2   D.G. Cameron, D.J. Moffat: A generalized approach to derivative spectroscopy. Appl. Spectrosc. **41**, 539 (1987);
G. Talsky: *Derivative Spectrophotometers* (VCH, Weinheim 1994)

6.3   G.C. Bjorklund: Frequency-modulation spectroscopy: A new method for measuring weak absorptions and dispersions. Opt. Lett. **5**, 15 (1980)

6.4   M. Gehrtz, G.C. Bjorklund, E. Whittaker: Quantum-limited laser frequency-modulation spectroscopy. J. Opt. Soc. Am. B **2**, 1510 (1985)

6.5   G.R. Janik, C.B. Carlisle, T.F. Gallagher: Two-tone frequency-modulation spectroscopy. J. Opt. Soc. Am. B **3**, 1070 (1986)

6.6   F.S. Pavone, M. Inguscio: Frequency- and wavelength-modulation spectroscopy: Comparison of experimental methods, using an AlGaAs diode laser. Appl. Phys. B **56**, 118 (1993)

6.7   R. Grosskloss, P. Kersten, W. Demtröder: Sensitive amplitude and phase-modulated absorption spectroscopy with a continuously tunable diode laser. Appl. Phys. B **58**, 137 (1994)

6.8   P.C.D. Hobbs: Ultrasensitive laser measurements without tears. Appl. Opt. **36**, 903 (1997)

6.9   P. Wehrle: A review of recent advances in semiconductor laser gas monitors. Spectrochim. Acta, Part A **54**, 197 (1998)

6.10   J.A. Silver: Frequency modulation spectroscopy for trace species detection. Appl. Opt. **31**, 707 (1992)

6.11   W. Brunner, H. Paul: On the theory of intracavity absorption. Opt. Commun. **12**, 252 (1974)

6.12   K. Tohama: A simple model for intracavity absorption. Opt. Commun. **15**, 17 (1975)

6.13   A. Campargue, F. Stoeckel, M. Chenevier: High sensitivity intracavity laser spectroscopy: applications to the study of overtone transitions in the visible range. Spectrochimica Acta Rev. **13**, 69 (1990)

6.14   A.A. Kaschanov, A. Charvat, F. Stoeckel: Intracavity laser spectroscopy with vibronic solid state lasers. J. Opt. Soc. Am. B **11**, 2412 (1994)

6.15   V.M. Baev, T. Latz, P.E. Toschek: Laser intracavity absorption spectroscopy. Appl. Phys. B **69**, 172 (1999);
V.M. Baev: Intracavity spectroscopy with diode lasers. Appl. Phys. B **55**, 463 (1992)

6.16   V.R. Mironenko, V.I. Yudson: Quantum noise in intracavity laser spectroscopy. Opt. Commun. **34**, 397 (1980);
V.R. Mironenko, V.I. Yudson: Sov. Phys. JETP **52**, 594 (1980)

6.17   P.E. Toschek, V.M. Baev: '"One is not enough": Intracavity laser spectroscopy with a multimode laser'. In: *Laser Spectroscopy and New Ideas*, ed. by W.M. Yen, M.D. Levenson, Springer Ser. Opt. Sci., Vol. 54 (Springer, Berlin, Heidelberg 1987)

6.18   E.M. Belenov, M.V. Danileiko, V.R. Kozuborskii, A.P. Nedavnii, M.T. Shpak: Ultrahigh resolution spectroscopy based on wave competition in a ring laser. Sov. Phys. JETP **44**, 40 (1976)

6.19   E.A. Sviridenko, M.P. Frolov: Possible investigations of absorption line profiles by intracavity laser spectroscopy. Sov. J. Quant. Electron. **7**, 576 (1977)

6.20   T.W. Hänsch, A.L. Schawlow, P. Toschek: Ultrasensitive response of a CW dye laser to selective extinction. IEEE J. Quantum Electron. **8**, 802 (1972)

6.21   R.N. Zare: Laser separation of isotopes. Sci. Am. **236**, 86 (February 1977)

6.22   R.G. Bray, W. Henke, S.K. Liu, R.V. Reddy, M.J. Berry: Measurement of highly forbidden optical transitions by intracavity dye laser spectroscopy. Chem. Phys. Lett. **47**, 213 (1977)

6.23   H. Atmanspacher, B. Baldus, C.C. Harb, T.G. Spence, B. Wilke, J. Xie, J.S. Harris, R.N. Zane: Cavity-locked ring-down spectroscopy. J. Appl. Phys. **83**, 3991 (1998)

6.24   W. Schrepp, H. Figger, H. Walther: Intracavity spectroscopy with a color-center laser. Lasers and Applications **77** (July 1984)

6.25   V.M. Baev, K.J. Boller, A. Weiler, P.E. Toschek: Detection of spectrally narrow light emission by laser intracavity spectroscopy. Opt. Commun. **62**, 380 (1987)

6.26   V.M. Baev, A. Weiler, P.E. Toschek: Ultrasensitive intracavity spectroscopy with multimode lasers. J. Phys. (Paris) **48**, C7, 701 (1987)

6.27   T.D. Harris: 'Laser intracavity-enhanced spectroscopy'. In: *Ultrasensitive Laser Spectroscopy*, ed. by D.S. Kliger (Academic, New York 1983)

6.28   E.H. Piepmeier (Ed.): *Analytical Applications of Lasers* (Wiley, New York 1986)

6.29   H. Atmanspacher, H. Scheingraber, C.R. Vidal: Dynamics of laser intracavity absorption. Phys. Rev. A **32**, 254 (1985);
        H. Atmanspacher, H. Scheingraber, C.R. Vidal: Mode-correlation times and dynamical instabilities in a multimode CW dye laser. Phys. Rev. A **33**, 1052 (1986)

6.30   H. Atmanspacher, H. Scheingraber, V.M. Baev: Stimulated Brillouin scattering and dynamical instabilities in a multimode laser. Phys. Rev. A **35**, 142 (1987)

6.31   P. Zalicki, R.N. Zare: Cavity ringdown spectroscopy for quantitative absorption measurements. J. Chem. Phys. **102**, 2708 (1995)

6.32   D. Romanini, K.K. Lehmann: Ring-down cavity absorption spectroscopy of the very weak HCN overtone bands with six, seven and eight stretching quanta. J. Chem. Phys. **99**, 6287 (1993)

6.33   M.D. Levenson, B.A. Paldus, T.G. Spence, C.C. Harb, J.S. Harris, R.N. Zare: Optical heterodyne detection in cavity ring-down spectroscopy. Chem. Phys. Lett. **290**, 335 (1998)

6.34   B.A. Baldus, R.N. Zare, et al.: Cavity-locked ringdown spectroscopy. J. Appl. Phys. **83**, 3991 (1998)

6.35   J.J. Scherer, J.B. Paul, C.P. Collier, A. O'Keefe, R.J. Saykally: Cavity-ringdown laser absorption spectroscopy and time-of-flight mass spectroscopy of jet-cooled gold silicides. J. Chem. Phys. **103**, 9187 (1995)

6.36   K.H. Becker, D. Haaks, T. Tartarczyk: Measurements of C_2-radicals in flames with a tunable dye lasers. Z. Naturforsch. **29a**, 829 (1974)

6.37   A.O'Keefe: Integrated cavity output analysis of ultraweak absorption. Chem. Phys. Lett. **293**, 331 (1998)

6.38   J.J. Scherer, J.B. Paul, C.P. Collier, A.O'Keefe, R.J. Saykally: Cavity ringdown laser absorption spectroscopy history, development and application to pulsed molecular beams. Chem. Rev. **97**, 25 (1997)

6.39   G. Berden, R. Peeters, G. Meijer: Cavity ringdown spectroscopy: experimental schemes and applications. Int. Rev. Phys. Chemistry **19**, 565 (2000)

6.40   W.M. Fairbanks, T.W. Hänsch, A.L. Schawlow: Absolute measurement of very low sodium-vapor densities using laser resonance fluorescence. J. Opt. Soc. Am. **65**, 199 (1975)

6.41   H.G. Krämer, V. Beutel, K. Weyers, W. Demtröder: Sub-Doppler laser spectroscopy of silver dimers Ag_2 in a supersonic beam. Chem. Phys. Lett. **193**, 331 (1992)

6.42   P.J. Dagdigian, H.W. Cruse, R.N. Zare: Laser fluorescence study of AlO, formed in the reaction $Al + O_2$: Product state distribution, dissociation energy and radiative lifetime. J. Chem. Phys. **62**, 1824 (1975)

6.43   W.E. Moerner, L. Kador: Finding a single molecule in a haystack. Anal. Chem. **61**, 1217A (1989);
        W.E. Moerner: Examining nanoenvironments in solids on the scale of a single, isolated impurity molecule. Science **265**, 46 (1994)

6.44   K. Kneipp, S.R. Emory, S. Nie: Single-molecule Raman-spectroscopy: Fact or fiction?. Chimica **53**, 35 (1999)

6.45   T. Plakbotnik, E.A. Donley, U.P. Wild: Single molecule spectroscopy. Ann. Rev. Phys. Chem. **48**, 181 (1997)

6.46   References to the historical development can be found in H.J. Bauer: Son et lumiere or the optoacoustic effect in multilevel systems. J. Chem. Phys. **57**, 3130 (1972)

6.47   Yoh-Han Pao (Ed.): *Optoacoustic Spectroscopy and Detection* (Academic, New York 1977)

6.48   A. Rosencwaig: *Photoacoustic and Photoacoustic Spectroscopy* (Wiley, New York 1980)

6.49   V.P. Zharov, V.S. Letokhov: *Laser Optoacoustic Spectroscopy*, Springer Ser. Opt. Sci., Vol. 37 (Springer, Berlin, Heidelberg 1986)

6.50   M.W. Sigrist (Ed.): *Air Monitoring by Spectroscopic Techniques*. (Wiley, New York 1994);
        J. Xiu, R. Stroud: *Acousto-Optic Devices: Principles, Design and Applications*. (Wiley, New York 1992)

6.51   P. Hess, J. Pelzl (Eds.): *Photoacoustic and Photothermal Phenomena*, Springer Ser. Opt. Sci., Vol. 58 (Springer, Berlin, Heidelberg 1988)

6.52   P. Hess (Ed.): *Photoacoustic, Photothermal and Photochemical Processes in Gases*, Topics Curr. Phys., Vol. 46 (Springer, Berlin, Heidelberg 1989)

6.53   J.C. Murphy, J.W. Maclachlan Spicer, L.C. Aamodt, B.S.H. Royce (Eds.): *Photoacoustic and Photothermal Phenomena II*, Springer Ser. Opt. Sci., Vol. 62 (Springer, Berlin, Heidelberg 1990)

6.54   L.B. Kreutzer: Laser optoacoustic spectroscopy: A new technique of gas analysis. Anal. Chem. **46**, 239A (1974)

6.55   W. Schnell, G. Fischer: Spectraphone measurements of isotopes of water vapor and nitricoxyde and of phosgene at selected wavelengths in the CO- and CO_2-laser region. Opt. Lett. **2**, 67 (1978)

6.56   C. Hornberger, W. Demtröder: Photoacoustic overtone spectroscopy of acetylene in the visible and near infrared. Chem. Phys. Lett. **190**, 171 (1994)

6.57   C.K.N. Patel: Use of vibrational energy transfer for excited-state opto-acoustic spectroscopy of molecules. Phys. Rev. Lett. **40**, 535 (1978)

6.58   G. Stella, J. Gelfand, W.H. Smith: Photoacoustic detection spectroscopy with dye laser excitation. The 6190 Å CH_4 and the 6450 NH_3-bands. Chem. Phys. Lett. **39**, 146 (1976)

6.59   A.M. Angus, E.E. Marinero, M.J. Colles: Opto-acoustic spectroscopy with a visible CW dye laser. Opt. Commun. **14**, 223 (1975)

6.60   E.E. Marinero, M. Stuke: Quartz optoacoustic apparatus for highly corrosive gases. Rev. Sci. Instrum. **50**, 31 (1979)

6.61   A.C. Tam: 'Photoacoustic spectroscopy and other applications'. In: *Ultrasensitive Laser Spectroscopy*, ed. by D.S. Kliger (Academic, New York 1983) pp. 1–108

6.62   V.Z. Gusev, A.A. Karabutov: *Laser Optoacoustics*. (Springer, Berlin, Heidelberg, New York 1997)

6.63   A.C. Tam, C.K.N. Patel: High-resolution optoacoustic spectroscopy of rare-earth oxide powders. Appl. Phys. Lett. **35**, 843 (1979)

6.64   T.E. Gough, G. Scoles: Optothermal infrared spectroscopy. In: *Laser Spectroscopy V*, ed. by A.R.W. McKeller, T. Oka, B.P. Stoicheff, Springer Ser. Opt. Sci., Vol. 30 (Springer, Berlin, Heidelberg 1981) p. 337
        T.E. Gough, R.E. Miller, G. Scoles: Sub-Doppler resolution infrared molecular beam spectroscopy. Faraday Disc. **71**, 6 (1981)

6.65   M. Zen: 'Cyrogenic bolometers'. In: *Atomic and Molecular Beams Methods, Vol. 1* (Oxford Univ. Press, London 1988) Vol. 1

6.66   R.E. Miller: Infrared laser spectroscopy. In: *Atomic and Molecular Beam Methods*, ed. by G. Scoles, (Oxford Univ. Press, London 1992) pp. 192 ff.;
        D. Bassi: Detection principles. In: *Atomic and Molecular Beam Methods*, ed. by G. Scoles (Oxford Univ. Press, London 1992) pp. 153 ff.

6.67   T.B. Platz, W. Demtröder: Sub-Doppler optothermal overtone spectroscopy of ethylene. Chem. Phys. Lett. **294**, 397 (1998)

6.68   K.K. Lehmann, G. Scoles: Intramolecular dynamics from Eigenstate-resolved infrared spectra. Ann. Rev. Phys. Chem. **45**, 241 (1994)

6.69   H. Coufal: Photothermal spectroscopy and its analytical application. Fresenius Z. Anal. Chem. **337**, 835 (1990)

6.70   F. Träger: Surface analysis by laser-induced thermal waves. Laser u. Optoelektronik **18**, 216 Sept. (1986);
        H. Coufal, F. Träger, T.J. Chuang, A.C. Tam: High sensitivity photothermal surface spectroscopy with polarization modulation. Surf. Sci. **145**, L504 (1984)

6.71   P.E. Siska: Molecular-beam studies of Penning ionization. Rev. Mod. Phys. **65**, 337 (1993)

6.72   Y.Y. Kuzyakov, N.B. Zorov: Atomic ionization spectrometry. CRC Critical Rev. Anal. Chem. **20**, 221 (1988)

6.73   G.S. Hurst, M.G. Payne, S.P. Kramer, J.P. Young: Resonance ionization spectroscopy and single atom detection. Rev. Mod. Phys. **51**, 767 (1979)

6.74   G.S. Hurst, M.P. Payne, S.P. Kramer, C.H. Cheng: Counting the atoms. Physics Today **33**, 24 (September 1980)

6.75   M. Keil, H.G. Krämer, A. Kudell, M.A. Baig, J. Zhu, W. Demtröder, W. Meyer: Rovibrational structures of the pseudo-rotating lithium trimer Li$_3$. J. Chem. Phys. **113**, 7414 (2000)

6.76   L. Wöste: Zweiphotonen-Ionisation. Laser u. Optoelektronik **15**, 9 (February 1983)

6.77   G. Delacretaz, J.D. Garniere, R. Monot, L. Wöste: Photoionization and fragmentation of alkali metal clusters in supersonic molecular beams. Appl. Phys. B **29**, 55 (1982)

6.78   H.J. Foth, J.M. Gress, C. Hertzler, W. Demtröder: Sub-Doppler laser spectroscopy of Na$_3$. Z. Physik D **18**, 257 (1991)

6.79   V.S. Letokhov: *Laser Photoionization Spectroscopy* (Academic, Orlando 1987)

6.80   G. Hurst, M.G. Payne: *Principles and Applications of Resonance Ionization Spectroscopy*, ed. by D.S. Kliger (Academic, New York 1983)

6.81   D.H. Parker: 'Laser ionization spectroscopy and mass spectrometry'. In: *Ultrasensitive Laser Spectroscopy*, ed. by D.S. Kliger (Academic, New York 1983)

6.82   V. Beutel, G.L. Bhale, M. Kuhn, W. Demtröder: The ionization potential of Ag$_2$. Chem. Phys. Lett. **185**, 313 (1991)

6.83   H.J. Neusser, U. Boesl, R. Weinkauf, E.W. Schlag: High-resolution laser mass spectrometer. Int. J. Mass Spectrom. **60**, 147 (1984)

6.84   J.E. Parks, N. Omeneto (Eds.): *Resonance Ionization Spectroscopy*. Inst. Phys. Conf. Ser. **114** (1990)
        D.M. Lübman (Ed.): *Lasers and Mass Spectrometry* (Oxford Univ. Press, London 1990)

6.85   P. Peuser, G. Herrmann, H. Rimke, P. Sattelberger, N. Trautmann, W. Ruster, F. Ames, J. Bonn, H.J. Kluge, V. Krönert, E.W. Otten: Trace detection of plutonium by three-step photoionization with a laser system pumped by a copper vapor laser. Appl. Phys. B **38**, 249 (1985)

6.86   D. Popescu, M.L. Pascu, C.B. Collins, B.W. Johnson, I. Popescu: Use of space charge amplification techniques in the absorption spectroscopy of Cs and Cs$_2$. Phys. Rev. A **8**, 1666 (1973)

6.87   K. Niemax: Spectroscopy using thermionic diode detectors. Appl. Phys. B **38**, 1 (1985)

6.88   R. Beigang, W. Makat, A. Timmermann: A thermionic ring diode for high res-
        olution spectroscopy. Opt. Commun. **49**, 253 (1984)
6.89   R. Beigang, A. Timmermann: The thermionic annular diode: a sensitive detector
        for highly excited atoms and molecules. Laser u. Optoelektronik **4**, 252 (1984)
6.90   D.S. King, P.K. Schenck: Optogalvanic spectroscopy. Laser Focus **14**, 50 (March
        1978)
6.91   J.E.M. Goldsmith, J.E. Lawler: Optogalvanic spectroscopy. Contemp. Phys. **22**,
        235 (1981)
6.92   B. Barbieri, N. Beverini, A. Sasso: Optogalvanic spectroscopy. Rev. Mod. Phys.
        62, 603 (1990)
6.93   K. Narayanan, G. Ullas, S.B. Rai: A two step optical double resonance study of
        a Fe–Ne hollow cathode discharge using optogalvanic detection. Opt. Commun.
        184, 102 (1991)
6.94   C.R. Webster, C.T. Rettner: Laser optogalvanic spectroscopy of molecules. Laser
        Focus **19**, 41 (February 1983)
        D. Feldmann: Optogalvanic spectroscopy of some molecules in discharges: NH_2,
        NO_2, A_2 and N_2. Opt. Commun. **29**, 67 (1979)
6.95   K. Kawakita, K. Fukada, K. Adachi, S. Maeda, C. Hirose: Doppler-free opto-
        galvanic spectrum of $He_2(b\,^3\Pi_g - f\,^3\Delta_u)$ transitions. J. Chem. Phys. **82**, 653
        (1985)
6.96   K. Myazaki, H. Scheingraber, C.R. Vidal: 'Optogalvanic double-resonance spec-
        troscopy of atomic and molecular discharge'. In: *Laser Spectroscopy VI*, ed. by
        H.P. Weber, W. Lüthy, Springer Ser. Opt. Sci., Vol. 40 (Springer, Berlin, Heidel-
        berg 1983) p. 93
6.97   J.C. Travis: 'Analytical optogalvanic spectroscopy in flames'. In: *Analytical
        Laser Spectroscopy*, ed. by S. Martellucci, A.N. Chester (Plenum, New York
        1985) p. 213
6.98   D. King, P. Schenck, K. Smyth, J. Travis: Direct calibration of laser wavelength
        and bandwidth using the optogalvanic effect in hollow cathode lamps. Appl. Opt.
        16, 2617 (1977)
6.99   V. Kaufman, B. Edlen: Reference wavelength from atomic spectra in the range
        15 Å to 25 000 Å. J. Phys. Chem. Ref. Data **3**, 825 (1974)
6.100   A. Giacchetti, R.W. Stanley, R. Zalubas: Proposed secondary standard wave-
        lengths in the spectrum of thorium. J. Opt. Soc. Am. **60**, 474 (1969)
6.101   J.E. Lawler, A.I. Ferguson, J.E.M. Goldsmith, D.J. Jackson, A.L. Schawlow:
        'Doppler-free optogalvanic spectroscopy'. In: *Laser Spectroscopy IV*, ed. by
        H. Walther, K.W. Rothe, Springer Ser. Opt. Sci., Vol. 21 (Springer, Berlin, Hei-
        delberg 1979) p. 188
6.102   W. Bridges: Characteristics of an optogalvanic effect in cesium and other gas
        discharge plasmas. J. Opt. Soc. Am. **68**, 352 (1978)
6.103   R.S. Stewart, J.E. Lawler (Eds.): *Optogalvanic Spectroscopy* (Hilger, London 1991)
6.104   R.J. Saykally, R.C. Woods: High resolution spectroscopy of molecular ions. Ann.
        Rev. Phys. Chem. **32**, 403 (1981)
6.105   C.S. Gudeman, R.J. Saykally: Velocity modulation infrared laser spectroscopy
        of molecular ions. Am. Rev. Phys. Chem. **35**, 387 (1984)
6.106   C.E. Blom, K. Müller, R.R. Filgueira: Gas discharge modulation using fast elec-
        tronic switches. Chem. Phys. Lett. **140**, 489 (1987)
6.107   M. Gruebele, M. Polak, R. Saykally: Velocity modulation laser spectroscopy of
        negative ions: The infrared spectrum of SH^-. J. Chem. Phys. **86**, 1698 (1987)
6.108   J.W. Farley: Theory of the resonance lineshape in velocity-modulation spec-
        troscopy J. Chem. Phys. **95**, 5590 (1991)
6.109   G. Lan, H.D. Tholl, J.W. Farley: Double-modulation spectroscopy of molecular
        ions: Eliminating the background in velocity-modulation spectroscopy. Rev. Sci.
        Instrum. **62**, 944 (1991)

6.110 M.B. Radunsky, R.J. Saykally: Electronic absorption spectroscopy of molecular ions in plasmas by dye laser velocity modulation spectroscopy. J. Chem. Phys. **87**, 898 (1987)

6.111 K.J. Button (Ed.): *Infrared and Submillimeter Waves* (Academic, New York 1979)

6.112 K.M. Evenson, R.J. Saykally, D.A. Jennings, R.E. Curl, J.M. Brown: 'Far infrared laser magnetic resonance'. In: *Chemical and Biochemical Applications of Lasers*, ed. by C.B. Moore (Academic, New York 1980) Chapt. V

6.113 P.B. Davies, K.M. Evenson: 'Laser magnetic resonance (LMR) spectroscopy of gaseous free radicals'. In: *Laser Spectroscopy II*, ed. by S. Haroche, J.C. Pebay-Peyroula, T.W. Hänsch, S.E. Harris, Lect. Notes Phys., Vol. 43 (Springer, Berlin, Heidelberg 1975)

6.114 W. Urban, W. Herrmann: Zeeman modulation spectroscopy with spin-flip Raman laser. Appl. Phys. **17**, 325 (1978)

6.115 K.M. Evenson, C.J. Howard: 'Laser Magnetic Resonance Spectroscopy'. In: *Laser Spectroscopy*. R.G. Brewer, ed. by A. Mooradian (Plenum, New York 1974)

6.116 A. Hinz, J. Pfeiffer, W. Bohle, W. Urban: Mid-infrared laser magnetic resonance using the Faraday and Voigt effects for sensitive detection. Mol. Phys. **45**, 1131 (1982)

6.117 Y. Ueda, K. Shimoda: 'Infrared laser Stark spectroscopy'. In: *Laser Spectroscopy II*, ed. by S. Haroche, J.C. Pebay-Peyroula, T.W. Hänsch, Lecture Notes Phys., Vol. 43 (Springer, Berlin, Heidelberg 1975) p. 186

6.118 K. Uehara, T. Shimiza, K. Shimoda: High resolution Stark spectroscopy of molecules by infrared and far infrared masers. IEEE J. Quantum Electron. **4**, 728 (1968)

6.119 K. Uchara, K. Takagi, T. Kasuya: Stark Modulation Spectrometer, Using a Wideband Zeeman-Tuned He-Xe Laser. Appl. Phys. **24** (1981)

6.120 L.R. Zink, D.A. Jennings, K.M. Evenson, A. Sasso, M. Inguscio: New techniques in laser Stark spectroscopy. J. Opt. Soc. Am. B **4**, 1173 (1987)

6.121 K.M. Evenson, R.J. Saykally, D.A. Jennings, R.F. Curl, J.M. Brown: 'Far infrared laser magnetic resonance'. In: *Chemical and Biochemical Applications of Lasers*, Vol. V, ed. by C.B. Moore (Academic, New York 1980)

6.122 M. Inguscio: Coherent atomic and molecular spectroscopy in the far infrared. Phys. Scripta **37**, 699 (1989)

6.123 W.H. Weber, K. Tanaka, T. Kanaka (Eds.): Stark and Zeeman techniques in laser spectroscopy. J. Opt. Soc. Am. B **4**, 1141 (1987)

6.124 J.L. Kinsey: Laser-induced fluorescence. Ann. Rev. Phys. Chem. **28**, 349 (1977)

6.125 A. Delon, R. Jost: Laser-induced dispersed fluorescence spectroscopy of 107 vibronic levels of NO_2 ranging from 12 000 to 17 600 cm^{-1}. J. Chem. Phys. **114**, 331 (2001)

6.126 M.A. Clyne, I.S. McDermid: Laser-induced fluorescence: electronically excited states of small molecules. Adv. Chem. Phys. **50**, 1 (1982)

6.127 J.R. Lakowicz: *Topics in Fluorescence Spectroscopy* (Plenum, New York 1991); J.N. Miller: *Fluorescence Spectroscopy* (Ellis Harwood, Singapore 1991); O.S. Wolflich (Ed.): *Fluorescence Spectroscopy* (Springer, Berlin, Heidelberg 1992)

6.128 C. Schütte: *The Theory of Molecular Spectroscopy* (North-Holland, Amsterdam 1976)

6.129 G. Herzberg: *Molecular Spectra and Molecular Structure, Vol. I* (Van Nostrand, New York 1950)

6.130 G. Höning, M. Cjajkowski, M. Stock, W. Demtröder: High resolution laser spectroscopy of Cs_2. J. Chem. Phys. **71**, 2138 (1979)

6.131   C. Amiot, W. Demtröder, C.R. Vidal: High resolution Fourier-spectroscopy and
        laser spectroscopy of Cs_2. J. Chem. Phys. **88**, 5265 (1988)
6.132   C. Amiot: Laser-induced fluorescence of Rb_2. J. Chem. Phys. **93**, 8591 (1990)
6.133   R. Bacis, S. Chunassy, R.W. Fields, J.B. Koffend, J. Verges: High resolution and
        sub-Doppler Fourier transform spectroscopy. J. Chem. Phys. **72**, 34 (1980)
6.134   R. Rydberg: Graphische Darstellung einiger bandenspektroskopischer Ergeb-
        nisse. Z. Physik **73**, 376 (1932)
6.135   O. Klein: Zur Berechnung von Potentialkurven zweiatomiger Moleküle mit Hilfe
        von Spekraltermen. Z. Physik **76**, 226 (1938)
6.136   A.L.G. Rees: The calculation of potential-energy curves from band spectroscopic
        data. Proc. Phys. Soc. London, Sect. A **59**, 998 (1947)
6.137   R.N. Zare, A.L. Schmeltekopf, W.J. Harrop, D.L. Albritton: J. Mol. Spectrosc.
        46, 37 (1973)
6.138   G. Ennen, C. Ottinger: Laser fluorescence measurements of the $^7LiD(X\,^1\Sigma^+)$-
        potential up to high vibrational quantum numbers. Chem. Phys. Lett. **36**, 16
        (1975)
6.139   M. Raab, H. Weickenmeier, W. Demtröder: The dissociation energy of the ce-
        sium dimer. Chem. Phys. Lett. **88**, 377 (1982)
6.140   C.E. Fellows: The NaLi $1\,^1\Sigma + (X)$ electronic ground state dissociation limit. J.
        Chem. Phys. **94**, 5855 (1991)
6.141   A.G. Gaydon: *Dissociation Energies and Spectra of Diatomic Molecules* (Chap-
        man and Hall, London 1968)
6.142   H. Atmanspacher, H. Scheingraber, C.R. Vidal: Laser-induced fluorescence of
        the MgCa molecule. J. Chem. Phys. **82**, 3491 (1985)
6.143   R.J. LeRoy: *Molecular Spectroscopy, Specialist Periodical Reports, Vol. 1*
        (Chem. Soc., Burlington Hall, London 1973) p. 113
6.144   W. Demtröder, W. Stetzenbach, M. Stock, J. Witt: Lifetimes and Franck–Condon
        factors for the $B\,^1\Pi_u \rightarrow X\,^1\Sigma_g^+$-system of Na_2. J. Mol. Spectrosc. **61**, 382
        (1976)
6.145   E.J. Breford, F. Engelke: Laser-induced fluorescence in supersonic nozzle beams:
        applications to the NaK $D\,^1\Pi \rightarrow X\,^1\Sigma$ and $D\,^1\Pi \rightarrow X\,^3\Sigma$ systems. Chem. Phys.
        Lett. **53**, 282 (1978);
        E.J. Breford, F. Engelke: J. Chem. Phys. **71**, 1949 (1979)
6.146   J. Tellinghuisen, G. Pichler, W.L. Snow, M.E. Hillard, R.J. Exton: Analaysis of
        the diffuse bands near 6100 Å in the fluorescence spectrum of Cs_2. Chem. Phys.
        50, 313 (1980)
6.147   H. Scheingraber, C.R. Vidal: Discrete and continuous Franck–Condon factors of
        the Mg_2 $A\,^1\Sigma_u - X\,^1I_s$ system and their J dependence. J. Chem. Phys. **66**, 3694
        (1977)
6.148   C.A. Brau, J.J. Ewing: 'Spectroscopy, kinetics and performance of rare-gas
        halide lasers'. In: *Electronic Transition Lasers*, ed. by J.I. Steinfeld (MIT Press,
        Cambridge, Mass. 1976)
6.149   D. Eisel, D. Zevgolis, W. Demtröder: Sub-Doppler laser spectroscopy of the
        NaK-molecule. J. Chem. Phys. **71**, 2005 (1979)
6.150   E.V. Condon: Nuclear motions associated with electronic transitions in diatomic
        molecules. Phys. Rev. **32**, 858 (1928)
6.151   J. Tellinghuisen: The McLennan bands of I_2: A highly structured continuum.
        Chem. Phys. Lett. **29**, 359 (1974)
6.152   H.J. Vedder, M. Schwarz, H.J. Foth, W. Demtröder: Analysis of the perturbed
        NO_2 $^2B_2 \rightarrow {}^2A_1$ system in the $591.4-592.9$ nm region based on sub-Doppler
        laser spectroscopy. J. Mol. Spectrosc. **97**, 92 (1983)
6.153   A. Delon, R. Jost: Laser-induced dispersed fluorescence spectra of jet-cooled
        NO_2. J. Chem. Phys. **95**, 5686 (1991)

6.154   Th. Zimmermann, H.J. Köppel, L.S. Cederbaum, G. Persch, W. Demtröder: Confirmation of random-matrix fluctuations in molecular spectra. Phys. Rev. Lett. **61**, 3 (1988)

6.155   K.K. Lehmann, St.L. Coy: The optical spectrum of NO_2: Is it or isn't it chaotic? Ber. Bunsenges. Phys. Chem. **92**, 306 (1988)

6.156   J.M. Gomez-Llorentl, H. Taylor: Spectra in the chaotic region: A classical analysis for the sodium trimer. J. Chem. Phys. **91**, 953 (1989)

6.157   K.L. Kompa: *Chemical Lasers*, Topics Curr. Chem., Vol. 37 (Springer, Berlin, Heidelberg 1975)

6.158   R. Schnabel, M. Kock: Time-Resolved nonlinear LIF-techniques for a combined lifetime and branching fraction measurements. Phys. Rev. A **63**, 125 (2001)

6.159   P.J. Dagdigian, H.W. Cruse, A. Schultz, R.N. Zare: Product state analysis of BaO from the reactions $Ba + CO_2$ and $Ba + O_2$. J. Chem. Phys. **61**, 4450 (1974)

6.160   J.G. Pruett, R.N. Zare: State-to-state reaction rates: $Ba + HF(v = 0) \rightarrow BaF(v = 0 - 12) + H''$. J. Chem. Phys. **64**, 1774 (1976)

6.161   H.W. Cruse, P.J. Dagdigian, R.N. Zare: Crossed beam reactions of barium with hydrogen halides. Faraday Discuss. Chem. Soc. **55**, 277 (1973)

6.162   Y. Nozaki, et al.: Identification of Si and SiH. J. Appl. Phys. **88**, 5437 (2000)

6.163   V. Hefter, K. Bergmann: 'Spectroscopic detection methods'. In: *Atomic and Molecular Beam Methods, Vol. I*, ed. by G. Scoles (Oxford Univ. Press, New York 1988) p. 193

6.164   J.E.M. Goldsmith: 'Recent advances in flame diagnostics using fluorescence and ionisation techniques'. In: *Laser Spectroscopy VIII*, ed. by S. Svanberg, W. Persson, Springer Ser. Opt. Sci., Vol. 55 (Springer, Berlin, Heidelberg 1987) p. 337

6.165   J. Wolfrum (Ed.): Laser diagnostics in combustion. Appl. Phys. B **50**, 439 (1990)

6.166   T.P. Hughes: *Plasma and Laser Light* (Hilger, Bristol 1975)

6.167   M. Bellini, P. DeNatale, G. DiLonardo, L. Fusina, M. Inguscio, M. Prevedelli: Tunable far infrared spectroscopy of $^{16}O_3$ ozone. J. Mol. Spectrosc. **152**, 256 (1992)

Chapter 7

7.1   W.R. Bennet, Jr.: Hole-burning effects in a He-Ne-optical maser. Phys. Rev. **126**, 580 (1962)

7.2   V.S. Letokhov, V.P. Chebotayev: *Nonlinear Laser Spectroscopy*, Springer Ser. Opt. Sci., Vol. 4 (Springer, Berlin, Heidelberg 1977)

7.3   S. Mukamel: *Principles of nonlinear optical spectroscopy* (Oxford Univ. Press, Oxford 1999)

7.4   M.D. Levenson: *Introduction to Nonlinear Spectroscopy* (Academic, New York 1982)

7.5   W.E. Lamb: Theory of an optical maser. Phys. Rev. A **134**, 1429 (1964)

7.6   H. Gerhardt, E. Matthias, F. Schneider, A. Timmermann: Isotope shifts and hyperfine structure of the $6s - 7p$-transitions in the cesium isotopes 133, 135 and 137. Z. Phys. A **288**, 327 (1978)

7.7   See, for instance: S.L. Chin: *Fundamentals of Laser Optoelectronics* (World Scientific, Singapore 1989) pp. 281 ff.

7.8   M.S. Sorem, A.L. Schawlow: Saturation spectroscopy in molecular iodine by intermodulated fluorescence. Opt. Commun. **5**, 148 (1972)

7.9   M.D. Levenson, A.L. Shawlow: Hyperfine interactions in molecular iodine. Phys. Rev. A **6**, 10 (1972)

7.10   H.J. Foth: Sättigungsspektroskopie an Molekülen. Diplom thesis, University of Kaiserslautern, Germany (1976)

7.11   R.S. Lowe, H. Gerhardt, W. Dillenschneider, R.F. Curl, Jr., F.K. Tittel: Intermodulated fluorescence spectroscopy of BO_2 using a stabilized dye laser. J. Chem. Phys. **70**, 42 (1979)

7.12   A.S. Cheung, R.C. Hansen, A.J. Nerer: Laser spectroscopy of VO: analysis of the rotational and hyperfine structure. J. Mol. Spectrosc. **91**, 165 (1982)

7.13   L.A. Bloomfield, B. Couillard, Ph. Dabkiewicz, H. Gerhardt, T.W. Hänsch: Hyperfine structure of the $2^3S - 5^3P$ transition in ^3He by high resolution UV laser spectroscopy. Opt. Commun. **42**, 247 (1982)

7.14   Ch. Hertzler, H.J. Foth: Sub-Doppler polarization spectra of He, N_2 and Ar^+ recorded in discharges. Chem. Phys. Lett. **166**, 551 (1990)

7.15   H.J. Foth, F. Spieweck. Hyperfine structure of the R(98), (58-1)-line of I_2 at $\lambda = 514.5$ nm. Chem. Phys. Lett. **65**, 347 (1979)

7.16   W.G. Schweitzer, E.G. Kessler, R.D. Deslattes, H.P. Layer, J.R. Whetstone: Description, performance and wavelength of iodine stabilised lasers. Appl. Opt. **12**, 2927 (1973)

7.17   R.L. Barger, J.B. West, T.C. English: Frequency stabilization of a CW dye laser. Appl. Phys. Lett. **27**, 31 (1975)

7.18   C. Salomon, D. Hills, J.L. Hall: Laser stabilization at the millihertz level. J. Opt. Soc. B **5**, 1576 (1988)

7.19   V. Bernard, et al.: CO_2-Laser stabilization to 0.1 Hz using external electro-optic modulation. IEEE J. Quantum Electron. **33**, 1288 (1997)

7.20   J.C. Hall, J.A. Magyar: 'High resolution saturation absorption studies of methane and some methyl-halides'. In: *High-Resolution Laser Spectroscopy*, ed. by K. Shimoda, Topics Appl. Phys., Vol. 13 (Springer, Berlin, Heidelberg 1976) p. 137

7.21   J.L. Hall: 'Sub-Doppler spectroscopy, methane hyperfine spectroscopy and the ultimate resolution limit'. In: *Colloq. Int. due CNRS, No. 217* (Edit. due CNRS, 15 quai Anatole France, Paris 1974) p. 105

7.22   B. Bobin, C.J. Bordé, J. Bordé, C. Bréant: Vibration-rotation molecular constants for the ground and ($\nu_3 = 1$) states of SF_6 from saturated absorption spectroscopy. J. Mol. Spectrosc. **121**, 91 (1987)

7.23   M. de Labachelerie, K. Nakagawa, M. Ohtsu: Ultranarrow $^{13}C_2H_2$ saturated absorption lines at 1.5 µm. Opt. Lett. **19**, 840 (1994)

7.24   C. Wieman, T.W. Hänsch: Doppler-free laser polarization spectroscopy. Phys. Rev. Lett. **36**, 1170 (1976)

7.25   R.E. Teets, F.V. Kowalski, W.T. Hill, N. Carlson, T.W. Hänsch: 'Laser polarization spectroscopy'. In: *Advances in Laser Spectroscopy*, SPIE Proc. **113**, 80 (1977)

7.26   M.E. Rose: *Elementary Theory of Angular Momentum* (Wiley, New York 1957)

7.27   R.N. Zare: *Angular Momentum: Understanding Spatial Aspects in Chemistry and Physics* (Wiley, New York 1988)

7.28   V. Stert, R. Fischer: Doppler-free polarization spectroscopy using linear polarized light. Appl. Phys. **17**, 151 (1978)

7.29   H. Gerhardt, T. Huhle, J. Neukammer, P.J. West: High resolution polarization spectroscopy of the 557 nm transition of KrI. Opt. Commun. **26**, 58 (1978)

7.30   M. Raab, G. Höning, R. Castell, W. Demtröder: Doppler-free polarization spectroscopy of the Cs_2 molecule at $\lambda = 6270$ Å. Chem. Phys. Lett. **66**, 307 (1979)

7.31   M. Raab, G. Höning, W. Demtröder, C.R. Vidal: High resolution laser spectroscopy of Cs_2. J. Chem. Phys. **76**, 4370 (1982)

7.32   W. Ernst: Doppler-free polarization spectroscopy of diatomic molecules in flame reactions. Opt. Commun. **44**, 159 (1983)

7.33   M. Francesconi, L. Gianfrani, M. Inguscio, P. Minutolo, A. Sasso: A new approach to impedance atomic spectroscopy. Appl. Phys. B **51**, 87 (1990)

7.34   L. Gianfrani, A. Sasso, G.M. Tino, F. Marin: Polarization spectroscopy of atomic oxygen by dye and semiconductor diode lasers. Il Nuovo Cimento **D10**, 941 (1988)

7.35   M. Göppert-Mayer: Über Elementarakte mit zwei Quantensprüngen. Ann. Physik **9**, 273 (1931)

7.36   W. Kaiser, C.G. Garret: Two-photon excitation in LLCA F_2: E_u^{2+}. Phys. Rev. Lett. **7**, 229 (1961)

7.37   J.J. Hopfield, J.M. Worlock, K. Park: Two-quantum absorption spectrum of KI. Phys. Rev. Lett. **11**, 414 (1963)

7.38   P. Bräunlich: 'Multiphoton spectroscopy'. In: *Progress in Atomic Spectroscopy*, ed. by W. Hanle, H. Kleinpoppen (Plenum, New York 1978)

7.39   J.M. Worlock: 'Two-photon spectroscopy'. In: *Laser Handbook*, ed. by F.T. Arrecchi, E.O. Schulz-Dubois (North-Holland, Amsterdam 1972)

7.40   B. Dick, G. Hohlneicher: Two-photon spectroscopy of dipole-forbidden transitions. Theor. Chim. Acta **53**, 221 (1979);
        B. Dick, G. Hohlneicher: J. Chem. Phys. **70**, 5427 (1979)

7.41   J.B. Halpern, H. Zacharias, R. Wallenstein: Rotational line strengths in two- and three-photon transitions in diatomic molecules. J. Mol. Spectrosc. **79**, 1 (1980)

7.42   K.D. Bonin, T.J. McIlrath: Two-photon electric dipole selection rules. J. Opt. Soc. Am. B **1**, 52 (1984)

7.43   G. Grynberg, B. Cagnac: Doppler-free multiphoton spectroscopy. Rep. Progr. Phys. **40**, 791 (1977)

7.44   F. Biraben, B. Cagnac, G. Grynberg: Experimental evidence of two- photon transition without Doppler broadening. Phys. Rev. Lett. **32**, 643 (1974)

7.45   G. Grynberg, B. Cagnbac, F. Biraben: 'Multiphoton resonant processes in atoms'. In: *Coherent Nonlinear Optics*, ed. by M.S. Feld, V.S. Letokhov, Topics Curr. Phys., Vol. 21 (Springer, Berlin, Heidelberg 1980)

7.46   T.W. Hänsch, K. Harvey, G. Meisel, A.L. Shawlow: Two-photon spectroscopy of Na 3s-4d without Doppler-broadening using CW dye laser. Opt. Commun. **11**, 50 (1974)

7.47   M.D. Levenson, N. Bloembergen: Observation of two-photon absorption without Doppler-broadening on the $3s-5s$ transition in sodium vapor. Phys. Rev. Lett. **32**, 645 (1974)

7.48   A. Timmermann: High resolution two-photon spectroscopy of the $6p^2 {}^3P_0 - 7p^3 P_0$ transition in stable lead isotopes. Z. Physik A **286**, 93 (1980)

7.49   S.A. Lee, J. Helmcke, J.L. Hall, P. Stoicheff: Doppler-free two-photon transitions to Rydberg levels. Opt. Lett. **3**, 141 (1978)

7.50   R. Beigang, K. Lücke, A. Timmermann: Singlet–Triplet mixing in $4s$ and Rydberg states of Ca. Phys. Rev. A **27**, 587 (1983)

7.51   S.V. Filseth, R. Wallenstein, H. Zacharias: Two-photon excitation of CO ($A^1 \Pi$) and N_2 ($a^1 \Pi_g$). Opt. Commun. **23**, 231 (1977)

7.52   E. Riedle, H.J. Neusser, E.W. Schlag: Electronic spectra of polyatomic molecules with resolved individual rotational transitions: benzene. J. Chem. Phys. **75**, 4231 (1981)

7.53   H. Sieber, E. Riedle, J.H. Neusser: Intensity distribution in rotational line spectra I: Experimental results for Doppler-free $S_1 \leftarrow S_0$ transitions in benzene. J. Chem. Phys. **89**, 4620 (1988);
        E. Riedle: Doppler-freie Zweiphotonen-Spektroskopie an Benzol. Habilitation thesis, Inst. Physikalische Chemie, TU München, Germany (1990)

7.54   E. Riedle, H.J. Neusser: Homogeneous linewidths of single rotational lines in the "channel three" region of C_6H_6. J. Chem. Phys. **80**, 4686 (1984)

7.55   U. Schubert, E. Riedle, J.H. Neusser: Time evolution of individual rotational states after pulsed Doppler-free two-photon excitation. J. Chem. Phys. **84**, 5326 and **84**, 6182 (1986)

7.56   W. Bischel, P.J. Kelley, Ch.K. Rhodes: High-resolution Doppler-free two-photon spectroscopic studies of molecules. Phys. Rev. A **13**, 1817 and **13**, 1829 (1976)

7.57   R. Guccione-Gush, H.P. Gush, R. Schieder, K. Yamada, C. Winnewisser: Doppler-free two-photon absorption of NH_3 using a CO_2 and a diode laser. Phys. Rev. A **23**, 2740 (1981)

7.58   G.F. Bassani, M. Inguscio, T.W. Hänsch (Eds.): *The Hydrogen Atom* (Springer, Berlin, Heidelberg 1989)

7.59   M. Weitz, F. Schmidt-Kaler, T.W. Hänsch: Precise optical Lamb-shift measurements in atomic hydrogen. Phys. Rev. Lett. **68**, 1120 (1992);
S.A. Lee, R. Wallenstein, T.W. Hänsch: Hydrogen 1S-2S-isotope shift and 1S Lamb shift measured by laser spectroscopy. Phys. Rev. Lett. **35**, 1262 (1975)

7.60   J.R.M. Barr, J.M. Girkin, J.M. Tolchard, A.I. Ferguson: Interferometric measurement of the $1S_{1/2} - 2S_{1/2}$ transition frequency in atomic hydrogen. Phys. Rev. Lett. **56**, 580 (1986)

7.61   M. Niering, et al.: Measurement of the hydrogen $1S - 2S$ transition frequency by phase coherent comparison with a microwave cesium fountain clock. Phys. Rev. Lett. **84**, 5496 (2000)

7.62   F. Biraben, J.C. Garreau, L. Julien: Determination of the Rydberg constant by Doppler-free two-photon spectroscopy of hydrogen Rydberg states. Europhys. Lett. **2**, 925 (1986)

7.63   F.H.M. Faisal, R. Wallenstein, H. Zacharias: Three-photon excitation of xenon and carbon monoxide. Phys. Rev. Lett. **39**, 1138 (1977)

7.64   B. Cagnac: 'Multiphoton high resolution spectroscopy'. In: *Atomic Physics 5*, ed. by R. Marrus, M. Prior, H. Shugart (Plenum, New York 1977) p. 147

7.65   V.I. Lengyel, M.I. Haylak: Role of autoionizing states in multiphoton ionization of complex atoms. Adv. At. Mol. Phys. **27**, 245 (1990)

7.66   E.M. Alonso, A.L. Peuriot, V.B. Slezak: CO_2-laser-induced multiphoton absorption of CF_2Cl_2. Appl. Phys. B **40**, 39 (1986)

7.67   V.S. Lethokov: Multiphoton and multistep vibrational laser spectroscopy of molecules. Commen. At. Mol. Phys. **8**, 39 (1978)

7.68   W. Fuss, J. Hartmann: IR absorption of SF_6 excited up to the dissociation limit. J. Chem. Phys. **70**, 5468 (1979)

7.69   F.V. Kowalski, W.T. Hill, A.L. Schawlow: Saturated-interference spectroscopy. Opt. Lett. **2**, 112 (1978)

7.70   R. Schieder: Interferometric nonlinear spectroscopy. Opt. Commun. **26**, 113 (1978)

7.71   S. Tolanski: *An Introduction to Interferometry* (Longman, London 1973)

7.72   C. Delsart, J.C. Keller: 'Doppler-free laser induced dichroism and birefringence'. In: *Laser Spectroscopy of Atoms and Molecules*, ed. by H. Walther, Topics Appl. Phys., Vol. 2, (Springer, Berlin, Heidelberg 1976) p. 154

7.73   M.D. Levenson, G.L. Eesley: Polarization selective optical heterodyne detection for dramatically improved sensitivity in laser spectroscopy. Appl. Phys. **19**, 1 (1979)

7.74   M. Raab, A. Weber: Amplitude-modulated heterodyne polarization spectroscopy. J. Opt. Soc. Am. B **2**, 1476 (1985)

7.75   K. Danzmann, K. Grützmacher, B. Wende: Doppler-free two-photon polarization spectroscopy measurement of the Stark-broadened profile of the hydrogen H_α line in a dense plasma. Phys. Rev. Lett. **57**, 2151 (1986)

7.76   T.W. Hänsch, A.L. Schawlow, C.W. Series: The spectrum of atomic hydrogen. Sci. Am. **240**, 72 (1979)

7.77   R.S. Berry: How good is Niels Bohrs atomic model? Contemp. Phys. **30**, 1 (1989)

7.78   F. Schmidt-Kalen, D. Leibfried, M. Weitz, T.W. Hänsch: Precision measurement of the isotope shift of the 1S–2S transition of atomic hydrogen and deuterium. Phys. Rev. Lett. **70**, 2261 (1993)

7.79   V.S. Butylkin, A.E. Kaplan, Y.G. Khronopulo: *Resonant Nonlinear Interaction of Light with Matter* (Springer, Berlin, Heidelberg 1987)

7.80   J.J.H. Clark, R.E. Hester (Eds.): *Advances in Nonlinear Spectroscopy* (Wiley, New York 1988)

7.81   S.S. Kano: *Introduction to Nonlinear Laser Spectroscopy* (Academic, New York 1988)

7.82   T.W. Hänsch: 'Nonlinear high-resolution spectroscopy of atoms and molecules'. In: *Nonlinear Spectroscopy, Proc. Int. School of Physics "Enrico Fermi" Course LXIV* (North-Holland, Amsterdam 1977) p. 17

7.83   D.C. Hanna, M.Y. Yunatich, D. Cotter: *Nonlinear Optics of Free Atoms and Molecules*, Springer Ser. Opt. Sci., Vol. 17 (Springer, Berlin, Heidelberg 1979)

7.84   St. Stenholm: *Foundations of Laser Spectroscopy* (Wiley, New York 1984)

7.85   R. Altkorn, R.Z. Zare: Effects of saturation on laser-induced fluorescence measurements. Ann. Rev. Phys. Chem. **35**, 265 (1984)

7.86   B. Cagnac: 'Laser Doppler-free techniques in spectroscopy'. In: *Frontiers of Laser Spectroscopy of Gases*, ed. by A.C.P. Alves, J.M. Brown, J.H. Hollas, Nato ASO Series C, Vol. 234, (Kluwer, Dondrost 1988)

7.87   S.H. Lin (Ed.): *Advances in Multiphoton Processes and Spectroscopy* (World Scientific, Singapore 1985-1992)

Chapter 8

8.1   A. Anderson: *The Raman Effect, Vols. 1, 2* (Dekker, New York 1971, 1973)

8.2   D.A. Long: *Raman Spectroscopy* (McGraw-Hill, New York 1977)

8.3   B. Schrader: *Infrared and Raman Spectroscopy* (Wiley VCH, Weinheim 1993); M.J. Pelletier (Ed.): *Analytical Applications of Raman Spectroscopy* (Blackwell Science, Oxford 1999)

8.4   J.R. Ferraro, K. Nakamato: *Introductory Raman Spectroscopy* (Academic, New York 1994)

8.5   I.R. Lewis, H.G.M. Edwards (Eds.): *Handbook of Raman Spectroscopy* (Dekker, New York 2001)

8.6   M.C. Tobin: *Laser Raman Spectroscopy* (Wiley Interscience, New York 1971)

8.7   A. Weber (Ed.): *Raman Spectroscopy of Gases and Liquids*, Topics Curr. Phys., Vol. 11 (Springer, Berlin, Heidelberg 1979)

8.8   G. Placzek: 'Rayleigh-Streuung und Raman Effekt'. In: *Handbuch der Radiologie, Vol. VI*, ed. by E. Marx (Akademische Verlagsgesellschaft, Leipzig 1934)

8.9   L.D. Barron: 'Laser Raman spectroscopy, in *Frontiers of Laser Spectroscopy of Gases*, ed. by A.C.P. Alves, J.M. Brown, J.M. Hollas, NATO ASI Series, Vol. 234 (Kluwer, Dordrecht 1988)

8.10   N.B. Colthup, L.H. Daly, S.E. Wiberley: *Introduction to Infrared and Raman Spectroscopy*, 3rd edn. (Academic, New York 1990)

8.11   R.J.H. Clark, R.E. Hester (Eds.): *Advances in Infrared and Raman Spectroscopy, Vols. 1–17* (Heyden, London 1975–1990)

8.12   J. Popp, W. Kiefer: 'Fundamentals of Raman spectroscopy'. In: *Encyclopedia of Analytical Chemistry* (Wiley, New York 2001)

8.13   P.P. Pashinin (Ed.): *Laser-Induced Raman Spectroscopy in Crystals and Gases* (Nova Science, Commack 1988)

8.14   J.R. Durig, J.F. Sullivan (Eds.): *XII Int. Conf. on Raman Spectroscopy* (Wiley, Chichester 1990)

8.15   H. Kuzmany: *Festkörperspektroskopie* (Springer, Berlin, Heidelberg 1989)

8.16   M. Cardona (Ed.): *Light Scattering in Solids*, 2nd edn., Topics Appl. Phys. Vol. 8 (Springer, Berlin, Heidelberg 1983);
M. Cardona, G. Güntherodt (Eds.): *Light Scattering in Solids II-VI*, Topics Appl. Phys., Vols. 50, 51, 54, 66, 68 (Springer, Berlin, Heidelberg 1982, 1984, 1989, 1991)

8.17   K.W. Szymanski: *Raman Spectroscopy I & II* (Plenum, New York 1970)

8.18   G. Herzberg: *Molecular Spectra and Molecular Structure, Vol. II, Infrared and Raman Spectra of Polyatomic Molecules* (van Nostrand Reinhold, New York 1945)

8.19   H.W. Schrötter, H.W. Klöckner: 'Raman scattering cross sections in gases and liquids'. In: *Raman Spectroscopy of Gases and Liquids*, ed. by A. Weber, Topics Curr. Phys., Vol. II (Springer, Berlin, Heidelberg 1979) pp. 123 ff.

8.20   D.L. Rousseau: 'The resonance Raman effect'. In: *Raman Spectroscopy of Gases and Liquids*, ed. by A. Weber, Topics Curr. Phys., Vol. II (Springer, Berlin, Heidelberg 1979) pp. 203 ff.

8.21   S.A. Acher: UV resonance Raman studies of molecular structures and dynamics. Ann. Rev. Phys. Chem. **39**, 537 (1988)

8.22   R.J.H. Clark, T.J. Dinev: 'Electronic Raman spectroscopy'. In: *Advances in Infrared and Raman Spectroscopy, Vol. 9*, ed. by R.J.H. Clark, R.E. Hester (Heyden, London 1982) p. 282

8.23   J.A. Koningstein: *Introduction to the theory of the Raman effect* (Reidel, Dordrecht 1972)

8.24   D.J. Gardner: *Practical Raman Spectroscopy* (Springer, Berlin, Heidelberg, New York 1989)

8.25   A. Weber: 'High-resolution rotational Raman spectra of gases'. In: *Advances in Infrared and Raman Spectroscopy, Vol. 9*, ed. by R.J.H. Clark, R.E. Hester (Heyden, London 1982) Chapt. 3

8.26   E.B. Brown: *Modern Optics* (Krieger, New York 1974) p. 251

8.27   J.R. Downey, G.J. Janz: 'Digital methods in Raman spectroscopy'. In: *XII Int. Conf. on Raman Spectroscopy*, ed. by J.R. Durig, J.F. Sullivan (Wiley, chichester 1990) pp. 1–34

8.28   W. Knippers, K. van Helvoort, S. Stolte: Vibrational overtones of the homonuclear diatomics N_2, O_2, D_2. Chem. Phys. Lett. **121**, 279 (1985)

8.29   K. van Helvoort, R. Fantoni, W.L. Meerts, J. Reuss: Internal rotation in CH_3CD_3: Raman spectroscopy of torsional overtones. Chem. Phys. Lett. **128**, 494 (1986);
K. van Helvoort, R. Fantoni, W.L. Meerts, J. Reuss: Chem. Phys. **110**, 1 (1986);

8.30   W. Kiefer: 'Recent techniques in Raman spectroscopy'. In: *Adv. Infrared and Raman Spectroscopy, Vol. 3*, ed. by R.J.H. Clark, R.E. Hester (Heyden, London 1977)

8.31   G.W. Walrafen, J. Stone: Intensification of spontaneous Raman spectra by use of liquid core optical fibers. Appl. Spectrosc. **26**, 585 (1972)

8.32   H.W. Schrötter, J. Bofilias: On the assignment of the second-order lines in the Raman spectrum of benzene. J. Mol. Struct. **3**, 242 (1969)

8.33   D.A. Long: 'The polarisability and hyperpolarisability tensors'. In: *Nonlinear Raman Spectroscopy and its Chemical Applications*, ed. by W. Kiefer, D.A. Long (Reidel, Dordrecht 1982)

8.34   L. Beardmore, H.G.M. Edwards, D.A. Long, T.K. Tan: 'Raman spectroscopic measurements of temperature in a natural gas laser flame'. In: *Lasers in Chemistry*, ed. by M.A. West (Elsevier, Amsterdam 1977)

8.35   A. Leipert: Laser Raman-Spectroskopie in der Wärme- und Strömungstechnik. Physik in unserer Zeit **12**, 107 (1981)

8.36   K. van Helvoort, W. Knippers, R. Fantoni, S. Stolte: The Raman spectrum of ethane from 600 to 6500 cm^{-1} Stokes shifts. Chem. Phys. **111**, 445 (1987)

8.37   J. Lascombe, P.V. Huong, (Eds.): *Raman Spectroscopy: Linear and Nonlinear* (Wiley, New York 1982)

8.38   E.J. Woodbury, W.K. Ny: IRE Proc. **50**, 2367 (1962)

8.39   G. Eckardt: Selection of Raman laser materials. IEEE J. Quantum Electron. **2**, 1 (1966)

8.40   A. Yariv: *Quantum Electronics*, 3rd edn. (Wiley, New York 1989)

8.41   W. Kaiser, M. Maier: 'Stimulated Rayleigh, Brillouin and Raman spectroscopy'. In: *Laser Handbook*, ed. by F.T. Arrecchi, E.O. Schulz-Dubois (North-Holland, Amsterdam 1972) pp. 1077 ff.

8.42   E. Esherik, A. Owyoung: 'High resolution stimulated Raman spectroscopy'. In: *Adv. Infrared and Raman Spectroscopy Vol. 9* (Heyden, London 1982)

8.43   H.W. Schrötter, H. Frunder, H. Berger, J.P. Boquillon, B. Lavorel, G. Millet: 'High Resolution CARS and Inverse Raman spectroscopy'. In: *Adv. Nonlinear Spectroscopy* **3**, 97 (Wiley, New York 1987)

8.44   R.S. McDowell, C.W. Patterson, A. Owyoung: Quasi-CW inverse Raman spectroscopy of the ω_1 fundamental of $^{13}CH_4$. J. Chem. Phys. **72**, 1071 (1980)

8.45   E.K. Gustafson, J.C. McDaniel, R.L. Byer: CARS measurement of velocity in a supersonic jet. IEEE. J. Quantum Electron. **17**, 2258 (1981)

8.46   A. Owyoung: 'High resolution CARS of gases'. In: *Laser Spectroscopy IV*, ed. by H. Walther, K.W. Roth, Springer, Ser. Opt. Sci., Vol. 21 (Springer, Berlin, Heidelberg, 1979) p. 175

8.47   N. Bloembergen: *Nonlinear Optics*, 3rd ptg. (Benjamin, New York 1977); D.L. Mills: *Nonlinear Optics* (Springer, Berlin, Heidelberg 1991)

8.48   C.S. Wang: 'The stimulated Raman process'. In: *Quantum Electronics: A Treatise, Vol. 1*, ed. by H. Rabin, C.L. Tang (Academic, New York 1975) Chapt. 7

8.49   M. Mayer: Applications of stimulated Raman scattering. Appl. Phys. **11**, 209 (1976)

8.50   G. Marowski, V.V. Smirnov (Eds.): *Coherent Raman Spectroscopy*, Springer Proc. Phys., Vol. 63 (Springer, Berlin, Heidelberg 1992)

8.51   W. Kiefer: 'Nonlinear Raman Spectroscopy: Applications'. In: *Encyclopedia of Spectroscopy and Spectrometry* (Academic, New York 2000) p. 1609

8.52   J.W. Nibler, G.V. Knighten: 'Coherent anti-Stokes Raman spectroscopy'. In: *Raman Spectroscopy of Gases and Liquids*, ed. by A. Weber, Topics Curr. Phys., Vol. II (Springer, Berlin, Heidelberg 1979) Chapt. 7

8.53   J.W. Nibler: 'Coherent Raman spectroscopy: Techniques and recent applications'. In: *Applied Laser Spectroscopy*, ed. by W. Demtröder, M. Inguscio, NATO ASI, Vol. 241 (Plenum, London 1990) p. 313

8.54   S.A.J. Druet, J.P.E. Taran: CARS spectroscopy. Progr. Quantum Electron. **7**, 1 (1981)

8.55   I.P.E. Taran: 'CARS spectroscopy and applications'. In: *Appl. Laser Spectroscopy*, ed. by W. Demtröder, M. Inguscio (Plenum, London 1990) pp. 313–328

8.56   W. Kiefer, D.A. Long (Eds.): *Nonlinear Raman Spectroscopy and its Chemical Applications* (Reidel, Dordrecht 1982)

8.57   F. Moya, S.A.J. Druet, J.P.E. Taran: 'Rotation-vibration spectroscopy of gases by CARS'. In: *Laser Spectroscopy II*, ed. by S. Haroche, J.C. Pebay-Peyroula, T.W. Hänsch, S.E. Harris, Springer Notes Phys., Vol. 34 (Springer, Berlin, Heidelberg 1975) p. 66

8.58   S.A. Akhmanov, A.F. Bunkin, S.G. Ivanov, N.I. Koroteev, A.I. Kourigin, I.L. Shumay: 'Development of CARS for measurement of molecular parameters'. In: *Tunable Lasers and Applications*, ed. by A. Mooradian, T. Jaeger, P. Stokseth, Springer Ser. Opt. Sci., Vol. 3 (Springer, Berlin, Heidelberg 1976)

8.59   J.P.E. Taran: 'Coherent anti-Stokes spectroscopy'. In: *Tunable Lasers and Applications*, ed. by A. Mooradian, T. Jaeger, P. Stokseth, Springer Ser. Opt. Sci., Vol. 3 (Springer, Berlin, Heidelberg 1976) p. 315

8.60   Q.H.F. Vremen, A.J. Breiner: Spectral properties of a pulsed dye laser with monochromatic injection. Opt. Commun. **4**, 416 (1972)

8.61   T.J. Vickers: Quantitative resonance Raman spectroscopy. Appl. Spectrosc. Rev. **26**, 341 (1991)

8.62   B. Attal, Debarré, K. Müller-Dethlets, J.P.E. Taran: Resonant coherent anti-Stokes Raman spectroscopy of C_2. Appl. Phys. B **28**, 221 (1982)

8.63   A.C. Eckbreth: BOX CARS: Crossed-beam phase matched CARS generation. Appl. Phys. Lett. **32**, 421 (1978)

8.64   Y. Prior: Three-dimensional phase matching in four-wave mixing. Appl. Opt. **19**, 1741 (1980)

8.65   S.J. Cyvin, J.E. Rauch, J.C. Decius: Theory of hyper-Raman effects. J. Chem. Phys. **43**, 4083 (1965)

8.66   P.D. Maker: 'Nonlinear light scattering in methane'. In: *Physics of Quantum Electronics*, ed. by P.L. Kelley, B. Lax, P.E. Tannenwaldt (McGraw-Hill, New York 1960) p. 60

8.67   K. Altmann, G. Strey: Enhancement of the scattering intensity for the hyper-Raman effect. Z. Naturforsch. **32a**, 307 (1977)

8.68   S. Nie, L.A. Lipscomb, N.T. Yu: Surface-enhanced hyper-Raman spectroscopy. Appl. Spectrosc. Rev. **26**, 203 (1991)

8.69   J. Reif, H. Walther: Generation of Tunable 16 µm radiation by stimulated hyper-Raman effect in strontium vapour. Appl. Phys. **15**, 361 (1978)

8.70   M.D. Levenson, J.J. Song: Raman-induced Kerr effect with elliptical polarization. J. Opt. Soc. Am. **66**, 641 (1976)

8.71   S.A. Akhmanov, A.F. Bunkin, S.G. Ivanov, N.I. Koroteev: Polarization active Raman spectroscopy and coherent Raman ellipsometry. Sov. Phys. JETP **47**, 667 (1978)

8.72   J.W. Nibler, J.J. Young: Nonlinear Raman spectroscopy of gases. Ann. Rev. Phys. Chem. **38**, 349 (1987)

8.73   Z.Q. Tian, B. Ren (Eds.): *Progress in Surface Raman Spectroscopy* (Xiaman Univ. Press, Xiaman, China 2000)

8.74   B. Eckert, H.D. Albert, H.J. Jodl: Raman studies of sulphur at high pressures and low temperatures. J. Phys. Chem. **100**, 8212 (1996)

8.75   P. Dhamelincourt: 'Laser molecular microprobe'. In: *Lasers in Chemistry*, ed. by M.A. West (Elsevier, Amsterdam 1977) p. 48

8.76   G. Mariotto, F. Ziglio, F.L. Freire, Jr.: Light-emitting porous silicon: a structural investigation by high spatial resolution Raman spectroscopy J. Non-Crystalline Solids **192**, 253 (1995)

8.77   L. Quin, Z.X. Shen, S.H. Tang, M.H. Kuck: The modification of a spex spectrometer into a micro-Raman spectrometer Asian J. Spectrosc. **1**, 121 (1997)

8.78   W. Kiefer: Femtosecond coherent Raman spectroscopy. J. Raman Spectrosc. **31**, 3 (2000)

8.79   M. Danfus, G. Roberts: Femtosecond transition state spectroscopy and chemical reaction dynamics. Commen. At. Mol. Phys. **26**, 131 (1991)

8.80   L. Beardmore, H.G.M. Edwards, D.A. Long, T.K. Tan: 'Raman spectroscopic measurements of temperature in a natural gas/air-flame'. In: *Lasers in Chemistry*, ed. by M.A. West (Elsevier, Amsterdam 1977) p. 79

8.81   M.A. Lapp, C.M. Penney: 'Raman measurements on flames'. In: *Advances in Infrared and Raman Specroscopy, Vol. 3*, ed. by R.S.H. Clark, R.E. Hester (Heyden, London 1977) p. 204

8.82   J.P. Taran: 'CARS: Techniques and applications'. In: *Tunable Lasers and Applications*, ed. by A. Mooradian, P. Jaeger, T. Stokseth, Springer Ser. Opt. Sci., Vol. 3 (Springer, Berlin, Heidelberg 1976) p. 378

8.83   T. Dreier, B. Lange, J. Wolfrum, M. Zahn: Determination of temperature and concentration of molecular nitrogen, oxygen and methane with CARS. Appl. Phys. B **45**, 183 (1988)

8.84   H.D. Barth, C. Jackschath, T. Persch, F. Huisken: CARS spectroscopy of molecules and clusters in supersonic jets. Appl. Phys. B **45**, 205 (1988)

8.85   F. Adar, J.E. Griffith (Eds.): Raman and luminescent spectroscopy in technology. SPIE Proc. **1336** (1990)

8.86   A.C. Eckbreth: 'Laser diagnostics for combustion temperature and species'. In: *Energy and Engineering Science*, ed. by A.K. Gupta, D.G. Lilley (Abacus Press, Cambridge 1988)

Chapter 9

9.1   R. Abjean, M. Leriche: On the shapes of absorption lines in a divergent atomic beam. Opt. Commun. **15**, 121 (1975)

9.2   R.W. Stanley: Gaseous atomic beam light source. J. Opt. Soc. Am. **56**, 350 (1966)

9.3   J.B. Atkinson, J. Becker, W. Demtröder: Hyperfine structure of the 625 nm band in the $a^3\Pi_u \leftarrow X^1\Sigma_g$ transition for Na_2. Chem. Phys. Lett. **87**, 128 (1982); J.B. Atkinson, J. Becker, W. Demtröder: Chem. Phys. Lett. **87**, 92 (1982)

9.4   R. Kullmer, W. Demtröder: Sub-Doppler laser spectroscopy of SO_2 in a supersonic beam. J. Chem. Phys. **81**, 2919 (1984)

9.5   W. Demtröder, F. Paech, R. Schmiedle: Hyperfine-structure in the visible spectrum of NO_2. Chem. Phys. Lett. **26**, 381 (1974)

9.6   R. Schmiedel, I.R. Bonilla, F. Paech, W. Demtröder: Laser spectroscopy of NO_2 under very high resolution. J. Mol. Spectrosc. **8**, 236 (1977)

9.7   U. Diemer: Dissertation, Universität Kaiserslautern, Germany (1990); U. Diemer, H.M. Greß, W. Demtröder: The $2^3\Pi_g \leftarrow X^3\Sigma_u$-triplet system of Cs_2. Chem. Phys. Lett. **178**, 330 (1991); H. Bovensmann, H. Knöchel, E. Tiemann: Hyperfine structural investigations of the excited AO^+ state of TlI. Mol. Phys. **73**, 813 (1991)

9.8   C. Duke, H. Fischer, H.J. Kluge, H. Kremling, T. Kühl, E.W. Otten: Determination of the isotope shift of ^{190}Hg by on line laser spectroscopy. Phys. Lett. A **60**, 303 (1977)

9.9   G. Nowicki, K. Bekk, J. Göring, A. Hansen, H. Rebel, G. Schatz: Nuclear charge radii and nuclear moments of neutrons deficient Ba-isotopes from high resolution laser spectroscopy. Phys. Rev. C **18**, 2369 (1978)

9.10   L.A. Hackel, K.H. Casleton, S.G. Kukolich, S. Ezekiel: Observation of magnetic octople and scalar spin-spin interaction in I_2 using laser spectroscopy. Phys. Rev. Lett. **35**, 568 (1975); L.A. Hackel, K.H. Casleton, S.G. Kukolich, S. Ezekiel: J. Opt. Soc. Am. **64**, 1387 (1974)

9.11   P. Jacquinot: 'Atomic beam spectroscopy'. In: *High-Resolution Laser Spectroscopy*, ed. by K. Shimoda, Topics Appl. Phys., Vol. 13 (Springer, Berlin, Heidelberg 1976) p. 51

9.12   W. Lange, J. Luther, A. Steudel: Dye lasers in atomic spectroscopy. In: *Adv. Atomic and Molecular Phys., Vol. 10* (Academic, New York 1974)

9.13   G. Scoles (Ed.): *Atomic and Molecular Beam Methods, Vols. I and II* (Oxford Univ. Press, New York 1988, 1992)

C. Whitehead: Molecular beam spectroscopy. Europ. Spectrosc. News **57**, 10 (1984)

9.14   J.P. Bekooij: High resolution molecular beam spectroscopy at microwave and optical frequencies. Dissertation, University of Nijmwegen, The Netherlands (1983)

9.15   W. Demtröder: 'Visible and ultraviolet spectroscopy'. In: *Atomic and Molecular Beam Methods II*, ed. by G. Scoles (Oxford Univ. Press, New York 1992);
W. Demtröder, H.J. Foth: Molekülspektroskopie in kalten Düsenstrahlen. Phys. Blätter **43**, 7 (1987)

9.16   S.A. Abmad, et al. (Eds.): *Atomic, Molecular and Cluster Physics* (Narosa Publ. House, New Delhi 1997)

9.17   R. Campargue (Ed.): *Atomic and Molecular Beams – The State of the Art 2000* (Springer, Berlin, Heidelberg, New York 2001)

9.18   P.W. Wegner (Ed.): *Molecular Beams and Low Density Gas Dynamics* (Dekker, New York 1974)

9.19   K. Bergmann, W. Demtröder, P. Hering: Laser diagnostics in molecular beams. Appl. Phys. **8**, 65 (1975)

9.20   H.-J. Foth: Hochauflösende Methoden der Laserspektroskopie zur Interpretation des NO_2-Moleküls. Dissertation, F.B. Physik, Universität Kaiserslautern, Germany (1981)

9.21   K. Bergmann, U. Hefter, P. Hering: Molecular beam diagnostics with internal state selection. Chem. Phys. **32**, 329 (1978);
K. Bergmann, U. Hefter, P. Hering: J. Chem. Phys. **65**, 488 (1976)

9.22   G. Herzberg: *Molecular Spectra and Molecular Structure* (van Nostrand, New York 1950)

9.23   N. Ochi, H. Watanabe, S. Tsuchiya: Rotationally resolved laser-induced fluorescence and Zeeman quantum beat spectroscopy of the V^1B state of jet-cooled CS_2. Chem. Phys. **113**, 271 (1987)

9.24   D.H. Levy, L. Wharton, R.E. Smalley: 'Laser spectroscopy in supersonic jets'. In: *Chemical and Biochenical Applications of Lasers, Vol. II*, ed. by C.B. Moore (Academic, New York 1977)

9.25   H.J. Foth, H.J. Vedder, W. Demtröder: Sub-Doppler laser spectroscopy of NO_2 in the $\lambda = 592-5$ nm region. J. Mol. Spectrosc. **88**, 109 (1981)

9.26   D.H. Levy: The spectroscopy of supercooled gases. Sci. Am. **251**, 86 (1984)

9.27   E. Pebay-Peyroula, R. Jost: S_1-S_0 laser excitation spectra of glyoxal in a supersonic jet. J. Mol. Spectr. **121**, 167 (1987)
B. Soep, R. Campargue: 'Laser spectroscopy of biacetyl in a supersonic jet and beam'. In: *Rarefied Gas Dynamics, Vol. II*, ed. by R. Campargue (Commissariat A L'Energie Atomique, Paris 1979)

9.28   M. Ito: 'Electronic spectra in a supersonic jet'. In: *Vibrational Spectra and Structure, Vol. 15*, ed. by J.R. Durig (Elsevier, Amsterdam 1986);
M. Ito, T. Ebata, N. Mikami: Laser spectroscopy of large polyatomic molecules in supersonic jets. Ann. Rev. Phys. Chem. **39**, 123 (1988)

9.29   W.R. Gentry: 'Low-energy pulsed beam sources'. In: *Atomic and Molecular Beam Methods I*, ed. by G. Scoles (Oxford Univ. Press, New York 1988) p. 54

9.30   S.B. Ryali, J.B. Fenn: Clustering in free jets. Ber. Bunsenges. Phys. Chem. **88**, 245 (1984)

9.31   P. Jena, B.K. Rao, S.N. Khanna (Eds.): *Physics and Chemistry of Small Clusters* (Plenum, New York 1987)

9.32   P.J. Sarre: Large gas phase clusters. Faraday Transactions **13**, 2343 (1990)

9.33   G. Benedek, T.P. Martin, G. Paccioni (Eds.): *Elemental and Molecular Clusters*, Springer Ser. Mater. Sci., Vol. 6 (Springer, Berlin, Heidelberg 1987);
U. Kreibig, M. Vollmer: *Optical Properties of Metal Clusters*, Springer Ser. Mater. Sci., Vol. 25 (Springer, Berlin, Heidelberg 1995)

9.34   M. Kappes, S. Leutwyler: 'Molecular beams of clusters'. In: *Atomic and Molec-
        ular Beam Methods, Vol. 1*, ed. by G. Scoles (Oxford Univ. Press, New York
        1988) p. 380

9.35   H.J. Foth, J.M. Greß, C. Hertzler, W. Demtröder: Sub-Doppler spectroscopy of
        Na_3. Z. Physik D **18**, 257 (1991)

9.36   M.M. Kappes, M. Schär, U. Röthlisberger, C. Yeretzian, E. Schumacher: Sodium
        cluster ionization potentials revisited. Chem. Phys. Lett. **143**, 251 (1988)

9.37   C. Brechnignac, P. Cahuzac, J.P. Roux, D. Davolini, F. Spiegelmann: Adiabatic
        decomposition of mass-selected alkali clusters. J. Chem. Phys. **87**, 3694 (1987)

9.38   J.M. Gomes Llorente, H.S. Tylor: Spectra in the chaotic region: A classical
        analysis for the sodium trimer. J. Chem. Phys. **91**, 953 (1989)

9.39   M.M. Kappes: Experimental studies of gas-phase main-group metal clusters.
        Chem. Rev. **88**, 369 (1988)

9.40   M. Broyer, G. Delecretaz, P. Labastie, R.L. Whetten, J.P. Wolf, L. Wöste: Spec-
        troscopy of Na_3. Z. Physik D **3**, 131 (1986)

9.41   C. Brechignac, P. Cahuzac, F. Carlier, M. de Frutos, J. Leygnier: Alkali-metal
        clusters as prototype of metal clusters. J. Chem. Soc. Faraday Trans. **86**, 2525
        (1990)

9.42   J. Blanc, V. Boncic-Koutecky, M. Broyer, J. Chevaleyre, P. Dugourd, J. Koutecki,
        C. Scheuch, J.P. Wolf, L. Wöste: Evolution of the electronic structure of lithium
        clusters between four and eight atoms. J. Chem. Phys. **96**, 1793 (1992)

9.43   W.D. Knight, W.A. deHeer, W.A. Saunders, K. Clemenger, M.Y. Chou,
        M.L. Cohen: Alkali metal clusters and the jellium model. Chem. Phys. Lett. **134**,
        1 (1987)

9.44   V. Bonacic-Koutecky, P. Fantucci, J. Koutecky: Systemic ab-initio configuration–
        interaction studies of alkali-metal clusters. Phys. Rev. B **37**, 4369 (1988)

9.45   A. Kiermeier, B. Ernstberger, H.J. Neusser, E.W. Schlag: Benzene clusters in
        a supersonic beam. Z. Physik D **10**, 311 (1988)

9.46   K.H. Fung, W.E. Henke, T.R. Hays, H.L. Selzle, E.W. Schlag: Ionization poten-
        tial of the benzene–argon complex in a jet. J. Phys. Chem. **85**, 3560 (1981)

9.47   C.M. Lovejoy, M.D. Schuder, D.J. Nesbitt: Direct IR laser absorption spec-
        troscopy of jet-cooled CO_2HF complexes. J. Chem. Phys. **86**, 5337 (1987)

9.48   E.L. Knuth: Dimer-formation rate coefficients from measurements of terminal
        dimer concentrations in free-jet expansions. J. Chem. Phys. **66**, 3515 (1977)

9.49   J.M. Philippos, J.M. Hellweger, H. van den Bergh: Infrared vibrational predis-
        sociation of van der Waals clusters. J. Phys. Chem. **88**, 3936 (1984)

9.50   J.B. Hopkins, P.R. Langridge-Smith, M.D. Morse, R.E. Smalley: Supersonic
        metal cluster beams of refractory metals: Spectral investigations of ultracold
        Mo_2. J. Chem. Phys. **78**, 1627 (1983);
        J.M. Hutson: Intermolecular forces and the spectroscopy of van der Waals
        molecules. Ann. Rev. Phys. Chem. **41**, 123 (1990)

9.51   H.W. Kroto, J.R. Heath, S.C. O'Brian, R.F. Curl, R.E. Smalley: C_{60}: Buckmin-
        sterfullerene. Nature **318**, 162 (1985)

9.52   S. Grebenev, M. Hartmann, M. Havenith, B. Sartakov, J.P. Toennies, A.F. Vilesov:
        The rotational spectrum of single OCS molecules in liquid ^{4}He droplets. J.
        Chem. Phys. **112**, 4485 (2000)

9.53   S. Grebenev, et al.: Spectroscopy of molecules in helium droplets. Physica B
        280, 65 (2000)

9.54   S. Grebenev, et al.: Spectroscopy of OCS-hydrogen clusters in He-droplets. Proc.
        Nobel Symposium 117 (World Scientific, Singapore 2001) pp. 123 ff.

9.55   S. Grebenev, et al.: The structure of OCS-H_2 van der Waals complexes embed-
        ded in ^{4}He/^{3}He-droplets. J. Chem. Phys. **114**, 617 (2001)

9.56   F. Stienkemeyer, W.E. Ernst, J. Higgins, G. Scoles: On the use of liquid He-cluster beams for the preparation and spectroscopy of alkali dimers and often weakly bound complexes. J. Chem. Phys. **102**, 615 (1995)

9.57   J. Higgins, et al.: Photo-induced chemical dynamics of high spin alkali trimers. Science **273**, 629 (1996)

9.58   R. Michalak, D. Zimmermann: Laser-spectroscopic investigation of higher excited electronic states of the KAr molecules. J. Mol. Spectrosc. **193**, 260 (1999)

9.59   F. Bylicki, G. Persch, E. Mehdizadeh, W. Demtröder: Saturation spectroscopy and OODR of NO_2 in a collimated molecular beam. Chem. Phys. **135**, 255 (1989)

9.60   T. Kröckertskothen, H. Knöckel, E. Tiemann: Molecular beam spectroscopy on FeO. Chem. Phys. **103**, 335 (1986)

9.61   G. Meijer, B. Janswen, J.J. ter Meulen, A. Dynamus: High resolution Lamb-dip spectroscopy on OD and SiCl in a molecular beam. Chem. Phys. Let. **136**, 519 (1987)

9.62   G. Meijer: Structure and dynamics of small molecules studied by UV laser spectroscopy. Dissertation, Katholicke Universiteit te Nijmegen, Holland (1988)

9.63   H.D. Barth, C. Jackschatz, T. Pertsch, F. Huisken: CARS spectroscopy of molecules and clusters in supersonic jets. Appl. Phys. B **45**, 205 (1988)

9.64   E.K. Gustavson, R.L. Byer: 'High resolution CW CARS spectroscopy in a supersonic expansion'. In: *Laser Spectroscopy VI*, ed. by H.P. Weber, W. Lüthy, Springer Ser. Opt. Sci., Vol. 40 (Springer, Berlin, Heidelberg 1983) p. 326

9.65   J.W. Nibler, J. Yang: Nonlinear Raman spectroscopy of gases. Ann. Rev. Phys. Chem. **38**, 349 (1987)

9.66   J.W. Nibler: 'Coherent Raman spectroscopy: techniques and recent applications'. In: *Applied Laser Spectroscopy*, ed. by W. Demtröder, M. Inguscio, NATO ASI Series B, Vol. 241 (Plenum, New York 1991) p. 313

9.67   J.W. Nibler, G.A. Puhanz: *Adv. Nonlinear Spectroscopy* **15**, 1 (Wiley, New York 1988)

9.68   S.L. Kaufman: High resolution laser spectroscopy in fast beams. Opt. Commun. **17**, 309 (1976)

9.69   W.H. Wing, G.A. Ruff, W.E. Lamb, J.J. Spezeski: Observation of the infrared spectrum of the hydrogen molecular ion HD^+. Phys. Rev. Lett. **36**, 1488 (1976)

9.70   M. Kristensen, N. Bjerre: Fine structure of the lowest triplet states in He_2. J. Chem. Phys. **93**, 983 (1990)

9.71   D.C. Lorents, S. Keiding, N. Bjerre: Barrier tunneling in the He_2 $c^3\Sigma_g^+$ state. J. Chem. Phys. **90**, 3096 (1989)

9.72   H.J. Kluge: 'Nuclear ground state properties from laser and mass spectroscopy'. In: *Applied Laser Spectroscopy*, ed. by W. Demtröder, M. Inguscio, NATO ASI Series B, Vol. 241 (Plenum, New York 1991)

9.73   E.W. Otten: 'Nuclei far from stability'. In: *Treatise on Heavy Ion Science, Vol. 8* (Plenum, New York 1989) p. 515

9.74   R. Jacquinot, R. Klapisch: Hyperfine spectroscopy of radioactive atoms. Rept. Progr. Phys. **42**, 773 (1979)

9.75   J. Eberz, et al.: Collinear laser spectroscopy of $^{108g108m}In$ using an ion source with bunched beam release. Z. Physik A **328**, 119 (1986)

9.76   B.A. Huber, T.M. Miller, P.C. Cosby, H.D. Zeman, R.L. Leon, J.T. Moseley, J.R. Peterson: Laser-ion coaxial beam spectroscopy. Rev. Sci. Instrum. **48**, 1306 (1977)

9.77   M. Dufay, M.L. Gaillard: 'High-resolution studies in fast ion beams', In: *Laser Spectroscopy III*, ed. by J.L. Hall, J.L. Carlsten, Springer Ser. Opt. Sci., Vol. 7 (Springer, Berlin, Heidelberg 1977) p. 231

9.78   S. Abed, M. Broyer, M. Carré, M.L. Gaillard, M. Larzilliere: High resolution spectroscopy of N_2O^+ in the near ultraviolet, using FIBLAS (Fast-Ion-Beam Laser Spectroscopy). Chem. Phys. **74**, 97 (1983)

9.79   D. Zajfman, Z. Vager, R. Naaman, et al.: The structure of C_2H^+ and $C_2H_2^+$ as measured by Coulomb explosion. J. Chem. Phys. **94**, 6379 (1991)

9.80   L. Andric, H. Bissantz, E. Solarte, F. Linder: Photofragment spectroscopy of molecular ions: design and performance of a new apparatus using coaxial beams. Z. Phys. D **8**, 371 (1988)

9.81   J. Lermé, S. Abed, R.A. Hold, M. Larzilliere, M. Carré: Measurement of the fragment kinetic energy distribution in laser photopredissociation of N_2O^+. Chem. Phys. Lett. **96**, 403 (1983)

9.82   H. Stein, M. Erben, K.L. Kompa: Infrared photodissociation of sulfur dioxide ions in a fast ion beam. J. Chem. Phys. **78**, 3774 (1983)

9.83   N.J. Bjerre, S.R. Keiding: Long-range ion-atom interactions studied by field dissociation spectroscopy of molecular ions. Phys. Rev. Lett. **56**, 1458 (1986)

9.84   D. Neumark: 'High resolution photodetachment studies of molecular negative ions'. In: *Ion and Cluster Ion Spectroscopy and Structure*, ed. by J.P. Maier (Elsevier, Amsterdam 1989) pp. 155 ff.

9.85   R.D. Mead, V. Hefter, P.A. Schulz, W.C. Lineberger: Ultrahigh resolution spectroscopy of C_2^-. J. Chem. Phys. **82**, 1723 (1985)

9.86   O. Poulsen: 'Resonant fast-beam interactions: saturated absorption and two-photon absorption'. In: *Atomic Physics 8*, ed. by I. Lindgren, S. Svanberg, A. Rosén (Plenum, New York 1983) p. 485

9.87   D. Klapstein, S. Leutwyler, J.P. Maier, C. Cossart-Magos, D. Cossart, S. Leach: The $B^2A_2'' \rightarrow \tilde{X}^2F''$ transition of $1,3,5\text{-}C_6F_3H_3^+$ and $1,3,5\text{-}C_6F_3D_3^+$ in discharge and supersonic free jet emission sources. Mol. Phys. **51**, 413 (1984)

9.88   S.C. Foster, R.A. Kennedy, T.A. Miller: 'Laser spectroscopy of chemical intermediates in supersonic free jet expansions'. In: *Frontiers of Laser Spectroscopy*, ed. by A.C.P. Alves, J.M. Brown, J.M. Hollas, NATO ASI Series C, Vol. 234 (Kluwer, Dordrecht 1988)

9.89   P. Erman, O. Gustafssosn, P. Lindblom: A simple supersonic jet discharge source for sub-Doppler spectroscopy. Phys. Scripta **38**, 789 (1988)

9.90   D. Pflüger, W.E. Sinclair, A. Linnartz, J.P. Maier: Rotationally resolved electronic absorption spectra of triacethylen cation in a supersonic jet

9.91   M.A. Johnson, R.N. Zare, J. Rostas, L. Leach: Resolution of the $\tilde{A}$ photoionization branching ratio paradox for the $^{12}CO_2$ state. J. Chem. Phys. **80**, 2407 (1984)

9.92   A. Kiermeyer, H. Kühlewind, H.J. Neusser, E.W. Schlag: Production and unimolecular decay of rotationally selected polyatomic molecular ions. J. Chem. Phys. **88**, 6182 (1988)

9.93   U. Boesl: Multiphoton excitation and mass-selective ion detection for neutral and ion spectroscopy. J. Phys. Chem. **95**, 2949 (1991)

9.94   K. Walter, R. Weinkauf, U. Boesl, E.W. Schlag: Molecular ion spectroscopy: mass-selected resonant two-photon dissociation spectra of CH_3I^+ and CD_3I^+. J. Chem. Phys. **89**, 1914 (1988)

9.95   J.P. Maier: 'Mass spectrometry and spectroscopy of ions and radicals'. In: *Encyclopedia of Spectroscopy and Spectrometry*, ed. by J.C. Lindon, G.E. Trauter, J.L. Holm (Academic, New York 1999) p. 2181

9.96   E.J. Bieske, M.W. Rainbird, A.E.W. Knight: Suppression of fragment contribution to mass-selected resonance enhanced multiphoton ionization spectra of van der Waals clusters. J. Chem. Phys. **90**, 2086 (1989);
E.J. Bieske, M.W. Rainbird, A.E.W. Knight: ibid. **94**, 7019 (1991)

9.97   U. Boesl, J. Grotemeyer, K. Walter, E.W. Schlag: Resonance ionization and time-of-flight mass spectroscopy. Anal. Instrum. **16**, 151 (1987)

9.98   C.W.S. Conover, Y.J. Twu, Y.A. Yang, L.A. Blomfield: A time-of-flight mass spectrometer for large molecular clusters produced in supersonic expansions. Rev. Scient. Instrum. **60**, 1065 (1989)

9.99   J.A. Syage, J.E. Wessel: 'Molecular multiphoton ionization and ion fragmentation spectroscopy'. In: *Appl. Spectrosc. Rev.* **24**, 1 (Dekker, New York 1988)

Chapter 10

10.1   R.A. Bernheim: *Optical Pumping, an Introduction* (Benjamin, New York 1965)

10.2   B. Budick: 'Optical pumping methods in atomic spectroscopy'. In: *Adv. At. Mol. Phys.* **3**, 73 (Academic, New York 1967)

10.3   R.N. Zare: 'Optical pumping of molecules'. In: *Int'l Colloquium on Doppler-Free Spectroscopic Methods for Simple Molecular Systems* (CNRS, Paris 1974) p. 29

10.4   M. Broyer, G. Gouedard, J.C. Lehmann, J. Vigue: 'Optical pumping of molecules'. In: *Adv. At. Mol. Phys.* **12**, 164 (Academic, New York 1976)

10.5   G. zu Putlitz: 'Determination of nuclear moments with optical double resonance'. *Springer Tracts Mod. Phys.* **37**, 105 (Springer, Berlin, Heidelberg 1965)

10.6   C. Cohen-Tannoudji: 'Optical pumping with lasers.' In: *Atomic Physics IV*, ed. by G. zu Putlitz, E.W. Weber, A. Winnacker (Plenum, New York 1975) p. 589

10.7   R.N. Zare: *Angular Momentum* (Wiley, New York 1988)

10.8   R.E. Drullinger, R.N. Zare: Optical pumping of molecules. J. Chem. Phys. **51**, 5532 (1969)

10.9   K. Bergmann: 'State selection via optical methods'. In: *Atomic and Molecular Beam Methods*, ed. by G. Scoles (Oxford Univ. Press, Oxford 1988) p. 293

10.10   H.G. Weber, P. Brucat, W. Demtröder, R.N. Zare: Measurement of NO_2 $^2B_2$ state g-values by optical radio frequency double-resonance. J. Mol. Spectrosc. **75**, 58 (1979)

10.11   W. Happer: Optical pumping. Rev. Mod. Phys. **44**, 168 (1972)

10.12   B. Decomps, M. Dumont, M. Ducloy: 'Linear and nonlinear phenomena in laser optical pumping'. In: *Laser Spectroscopy of Atoms and Molecules*, ed. by H. Walther, Topics Appl. Phys., Vol. 2 (Springer, Berlin, Heidelberg 1976) p. 284

10.13   G.W. Series: Thirty years of optical pumping. Contemp. Phys. **22**, 487 (1981)

10.14   I.I. Rabi: Zur Methode der Ablenkung von Molekularstrahlen. Z. Physik **54**, 190 (1929)

10.15   H. Kopfermann: *Kernmomente* (Akad. Verlagsanstalt, Frankfurt 1956)

10.16   N.F. Ramsay: *Molecular Beams*, 2nd edn. (Clarendon, Oxford 1989)

10.17   J.C. Zorn, T.C. English: 'Molecular beam electric resonance spectroscopy'. In: *Adv. At. Mol. Phys.* **9**, 243 (Academic, New York 1973)

10.18   D.D. Nelson, G.T. Fraser, K.I. Peterson, K. Zhao, W. Klemperer: The microwave spectrum of $K = O$ states of $Ar - NH_3$. J. Chem. Phys. **85**, 5512 (1986)

10.19   A.E. DeMarchi (Ed.): *Frequency Standards and Metrology* (Springer, Berlin, Heidelberg 1989) pp. 46 ff.

10.20   W.J. Childs: Use of atomic beam laser RF double resonance for interpretation of complex spectra. J. Opt. Soc. Am. B **9**, 191 (1992)

10.21   S.D. Rosner, R.A. Holt, T.D. Gaily: Measurement of the zero-field hyperfine structure of a single vibration-rotation level of Na_2 by a laser-fluorescence molecular-beam resonance. Phys. Rev. Lett. **35**, 785 (1975)

10.22   A.G. Adam: Laser-fluorescence molecular-beam-resonance studies of Na_2 lineshape due to HFS. PhD. thesis, Univ. of Western Ontario, London, Ontario (1981);
        A.G. Adam, S.D. Rosner, T.D. Gaily, R.A. Holt: Coherence effects in laser-fluorescence molecular beam magnetic resonance. Phys. Rev. A **26**, 315 (1982)

10.23   W. Ertmer, B. Hofer: Zerofield hyperfine structure measurements of the metastable states $3d^2 4s^4 F_{3/2} 9/2$ of *SC using laser-fluorescence-atomic beam magnetic resonance technique. Z. Physik A **276**, 9 (1976)

10.24   J. Pembczynski, W. Ertmer, V. Johann, S. Penselin, P. Stinner: Measurement of the hyperfine structure of metastable atomic states of ^{55}Mm, using the ABMR-LIRF-method. Z. Physik A **291**, 207 (1979);
J. Pembczynski, W. Ertmer, V. Johann, S. Penselin, P. Stinner: Z. Physik A **294**, 313 (1980)

10.25   N. Dimarca, V. Giordano, G. Theobald, P. Cérez: Comparison of pumping a cesium beam tube with D_1 and D_2 lines. J. Appl. Phys. **69**, 1159 (1991)

10.26   G.W. Chantry (Ed.): *Modern Aspects of Microwave Spectroscopy* (Academic, London 1979)

10.27   K. Shimoda: 'Double resonance spectroscopy by means of a laser'. In: *Laser Spectroscopy of Atoms and Molecules*, ed. by H. Walther, Topics Appl. Phys., Vol. 2 (Springer, Berlin, Heidelberg 1976) p. 197

10.28   K. Shimoda: 'Infrared-microwave double resonance'. In: *Laser Spectroscopy III*, ed. by J.L. Hall, H.L. Carlsten, Springer Ser. Opt. Sci., Vol. 7 (Springer, Berlin, Heidelberg 1975) p. 279

10.29   H. Jones: Laser microwave-double-resonance and two-photon spectroscopy. Commen. At. Mol. Phys. **8**, 51 (1978)

10.30   F. Tang, A. Olafson, J.O. Henningsen: A study of the methanol laser with a 500 MHz tunable CO_2 laser. Appl. Phys. B **47**, 47 (1988)

10.31   R. Neumann, F. Träger, G. zu Putlitz: 'Laser microwave spectroscopy'. In: *Progress in Atomic Spectroscopy*, ed. by H.J. Byer, H. Kleinpoppen (Plenum, New York 1987)

10.32   J.C. Petersen, T. Amano, D.A. Ramsay: Microwave-optical double resonance of DND in the $A\,^1 A''(000)$ state. J. Chem. Phys. **81**, 5449 (1984)

10.33   R.W. Field, A.D. English, T. Tanaka, D.O. Harris, P.A. Jennings: Microwave-optical double resonance with a CW dye laser, BaO $X\,^1\Sigma$ and $A\,^1\Sigma$. J. Chem. Phys. **59**, 2191 (1973)

10.34   R.A. Gottscho, J. Brooke-Koffend, R.W. Field, J.R. Lombardi: OODR spectroscopy of BaO. J. Chem. Phys. **68**, 4110 (1978); R.A. Gottscho, J. Brooke-Koffend, R.W. Field, J.R. Lombardi: J. Mol. Spectrosc. **82**, 283 (1980)

10.35   J.M. Cook, G.W. Hills, R.F. Curl: Microwave-optical double resonance spectrum of NH_2. J. Chem. Phys. **67**, 1450 (1977)

10.36   W.E. Ernst, S. Kindt: A molecular beam laser-microwave double resonance spectrometer for precise measurements of high temperature molecules. Appl. Phys. B **31**, 79 (1983)

10.37   W.J. Childs: The hyperfine structure of alkaline-earth monohalide radicals: New methods and new results 1980–82. Comments At. Mol. Phys. **13**, 37 (1983)

10.38   W.E. Ernst, S. Kindt, T. Törring: Precise Stark-effect measurements in the $^2\sigma$-ground state of CaCl. Phys. Rev. Phys. Lett. **51**, 979 (1983); W.E. Ernst, S. Kindt, T. Törring: Phys. Rev. A **29**, 1158 (1984)

10.39   W. Demtröder, D. Eisel, H.J. Foth, G. Höning, M. Raab, H.J. Vedder, D. Zevgolis: Sub-Doppler laser spectroscopy of small molecules. J. Mol. Structure **59**, 291 (1980)

10.40   F. Bylicki, G. Persch, E. Mehdizadeh, W. Demtröder: Saturation spectroscopy and OODR of NO_2 in a collimated molecular beam. Chem. Phys. **135**, 255 (1989)

10.41   M.A. Johnson, C.R. Webster, R.N. Zare: Rotational analysis of congested spectra: Application of population labelling to the BaI C-X system. J. Chem. Phys. **75**, 5575 (1981)

10.42   M.A. Kaminsky, R.T. Hawkins, F.V. Kowalski, A.L. Schawlow: Identifiction of absorption lines by modulated lower-level population: Spectrum of Na_2. Phys. Rev. Lett. **36**, 671 (1976)

10.43   A.L. Schawlow: Simplifying spectra by laser labelling. Phys. Scripta **25**, 333 (1982)

10.44   D.P. O'Brien, S. Swain: Theory of bandwidth induced asymmetry in optical double resonances. J. Phys. B **16**, 2499 (1983)

10.45   S.A. Edelstein, T.F. Gallagher: 'Rydberg atoms'. In: *Adv. At. Mol. Phys.* **14**, 365 (Academic, New York 1978)

10.46   I.I. Sobelman: *Atomic Spectra and Radiative Transitions*, 2nd edn., Springer Ser. Atoms and Plasmas, Vol. 12 (Springer, Berlin, Heidelberg 1992)

10.47   R.F. Stebbings, F.B. Dunnings (Eds.): *Rydberg States of Atoms and Molecules* (Cambridge Univ. Press, Cambridge 1983)

10.48   H. Figger: Experimente an Rydberg-Atomen und Molekülen. Phys. in unserer Zeit **15**, 2 (1984)

10.49   J.A.C. Gallas, H. Walther, E. Werner: Simple formula for the ionization rate of Rydberg states in static electric fields. Phys. Rev. Lett. **49**, 867 (1982)

10.50   C.E. Theodosiou: Lifetimes of alkali-metal-atom Rydberg states. Phys. Rev. A **30**, 2881 (1984)

10.51   J. Neukammer, H. Rinneberg, K. Vietzke, A. König, H. Hyronymus, M. Kohl, H.J. Grabka: Spectroscopy of Rydberg atoms at $n = 500$. Phys. Rev. Lett. **59**, 2847 (1987)

10.52   K.H. Weber, K. Niemax: Impact broadening of very high Rb Rydberg levels by Xe. Z. Physik A **312**, 339 (1983)

10.53   K. Heber, P.J. West, E. Matthias: Pressure shift and broadening of SnI Rydberg states in noble gases. Phys. Rev. A **37**, 1438 (1988)

10.54   R. Beigang, W. Makat, A. Timmermann, P.J. West: Hyperfine-induced n-mixing in high Rydberg states of ^{87}Sr. Phys. Rev. Lett. **51**, 771 (1983)

10.55   T.F. Gallagher, W.E. Cooke: Interaction of blackbody radiation with atoms. Phys. Rev. Lett. **42**, 835 (1979)

10.56   L. Holberg, J.L. Hall: Measurements of the shift of Rydberg energy levels induced by blackbody radiation. Phys. Rev. Lett. **53**, 230 (1984)

10.57   H. Figger, G. Leuchs, R. Strauchinger, H. Walther: A photon detector for sub-millimeter wavelengths using Rydberg atoms. Opt. Commun. **33**, 37 (1980)

10.58   D. Wintgen, H. Friedrich: Classical and quantum mechanical transition between regularity and irregularity. Phys. Rev. A **35** 1464 (1987)

10.59   G. Raithel, M. Fauth, H. Walther: Quasi-Landau resonances in the spectra of rubidium Rydberg atoms in crossed electric and magnetic fields. Phys. Rev. A **44**, 1898 (1991)

10.60   G. Wunner: Gibt es Chaos in der Quantenmechanik? Phys. Blätter **45**, 139 (Mai 1989);
        M. Gutzwiller: *Chaos in Classical and Quantum Mechanics* (Springer, Berlin, Heidelberg 1990)

10.61   A. Holle, J. Main, G. Wiebusch, H. Rottke, K.H. Welge: 'Laser spectroscopy of the diamagnetic hydrogen atom in the chaotic region'. In: *Atomic Spectra and Collisions in External Fields*, ed. by K.T. Taylor, M.H. Nayfeh, C.W. Clark (Plenum, New York 1988)

10.62   P. Meystre, M. Sargent III: *Elements of Quantum Optics*, 2nd edn. (Springer, Berlin, Heidelberg 1991)

10.63   H. Held, J. Schlichter, H. Walther: Quantum chaos in Rydberg atoms. Lecture Notes in Physics **503**, 1 (1998)

10.64   A. Holle, G. Wiebusch, J. Main, K.H. Welge, G. Zeller, G. Wunner, T. Ertl, H. Ruder: Hydrogenic Rydberg atoms in strong magnetic fields. Z. Physik D **5**, 271 (1987)

10.65   H. Rottke, K.H. Welge: Photoionization of the hydrogen atom near the ioniza-
         tion limit in strong electric field. Phys. Rev. A **33**, 301 (1986)

10.66   C. Fahre, S. Haroche: 'Spectroscopy of one- and two-electron Rydberg atoms'.
         In: *Rydberg States of Atoms and Molecules*, ed. by R.F. Stebbings, F.B. Dun-
         nings (Cambridge Univ. Press, Cambridge 1983)

10.67   J. Boulmer, P. Camus, P. Pillet: *Autoionizing Double Rydberg States in Barium*,
         ed. by H.B. Gilbody, W.R. Newell, F.H. Read, A.C. Smith (Elsevier, Amsterdam
         1988)

10.68   J. Boulmer, P. Camus, P. Pillet: Double Rydberg spectroscopy of the barium
         atom. J. Opt. Soc. Am. B **4**, 805 (1987)

10.69   I.C. Percival: Planetary atoms. Proc. Roy. Soc. London A **353**, 289 (1977)

10.70   R.S. Freund: 'High Rydberg molecules'. In: *Rydberg States of Atoms and
         Molecules*, ed. by R.F. Stebbing, F.B. Dunning (Cambridge Univ. Press, Cam-
         bridge 1983);
         G. Herzberg: Rydberg molecules. Ann. Rev. Phys. Chem. **38**, 27 (1987)

10.71   R.A. Bernheim, L.P. Gold, T. Tipton: Rydberg states of $^7Li_2$ by pulsed optical-
         optical double resonance spectroscopy. J. Chem. Phys. **78**, 3635 (1983);
         D. Eisel, W. Demtröder, W. Müller, P. Botschwina: Autoionization spectra of Li_2
         and the $X^2\Sigma_g^+$ ground state of Li_2^+. Chem. Phys. **80**, 329 (1983)

10.72   M. Schwarz, R. Duchowicz, W. Demtröder, C. Jungen: Autoionizing Rydberg
         states of Li_2: analysis of electronic-rotational interactions. J. Chem. Phys. **89**,
         5460 (1988)

10.73   C.H. Greene, C. Jungen: 'Molecular applications of quantum defect theory'. In:
         Adv. At. Mol. Phys. **21**, 51 (Academic, New York 1985)

10.74   F. Merkt: Molecules in high Rydberg states. Ann. Rev. Phys. Chemistry **48**, 675
         (1997);
         F. Merkt: Chimica **54**, 89 (2000)

10.75   A. Osterwalder, F. Merkt: High resolution spectroscopy of high Rydberg states.
         Chimica **54**, 89 (2000)

10.76   S. Fredin, D. Gauyacq, M. Horani, C. Jungen, G. Lefevre, F. Masnou-Seeuws: *S*
         and *d* Rydberg series of NO probed by double resonance multiphoton ionization.
         Mol. Phys. **60**, 825 (1987)

10.77   U. Aigner, L.Y. Baranov, H.L. Selzle, E.W. Schlag: Lifetime enhancement of
         ZEKE-states in molecular clusters and cluster fragmentation. J. Electron. Spec-
         trosc. Rel. Phenom. **112**, 175 (2000)

10.78   M. Sander, L.A. Chewter, K. Müller-Dethlefs, E.W. Schlag: High-resolution
         zero-kinetic-energy photoelectron spectroscopy of NO. Phys. Rev. A **36**, 4543
         (1987)

10.79   K. Müller-Dethlefs, E.W. Schlag: High-resolution ZEKE photoelectron spec-
         troscopy of molecular systems. Ann. Rev. Phys. Chem. **42**, 109 (1991);
         E.R. Grant, M.G. White: ZEKE threshold photoelectron spectroscopy. Nature
         354, 249 (1991)

10.80   C.E.H. Descent, K. Müller-Dethlefs: Hydrogen-bonding and van der Waals Com-
         plexes Studies by ZEKE and REMP Spectroscopy. Chem. Rev. **100**, 3999 (2000)

10.81   R. Signorelli, U. Hollenstein, F. Merkt: PFI–ZEKE photo electron spectroscopy
         study of the first electronic states of Kr_2^+. J. Chem. Phys. **114**, 9840 (2001)

10.82   P. Goy, M. Bordas, M. Broyer, P. Labastie, B. Tribellet: Microwave transitions
         between molecular Rydberg states. Chem. Phys. Lett. **120**, 1 (1985)

10.83   P. Filipovicz, P. Meystere, G. Rempe, H. Walther: Rydberg atoms, a testing
         ground for quantum electrodynamics. Opt. Acta **32**, 1105 (1985)

10.84   C.J. Latimer: Recent experiments involving highly excited atoms. Contemp.
         Phys. **20**, 631 (1979)

10.85   J.C. Gallas, G. Leuchs, H. Walther, H. Figger: 'Rydberg atoms: High resolution
         spectroscopy'. In: *Adv. At. Mol. Phys.* **20**, 414 (Academic, New York 1985)

10.86   G. Alber, P. Zoller: Laser-induced excitation of electronic Rydberg wave packets. Contemp. Phys. **32**, 185 (1991)

10.87   K. Harth, M. Raab, H. Hotop: Odd Rydberg spectrum of ^{20}Ne: High resolution laser spectroscopy and MQDT analysis. Z. Physik D **7**, 219 (1987)

10.88   V.S. Letokhov, V.P. Chebotayev: *Nonlinear Laser Spectroscopy*, Springer Ser. Opt. Sci., Vol. 4 (Springer, Berlin, Heidelberg 1977) Chap. 5

10.89   T. Hänsch, P. Toschek: Theory of a three-level gas laser amplifier. Z. Physik **236**, 213 (1970)

10.90   C. Kitrell, E. Abramson, J.L. Kimsey, S.A. McDonald, D.E. Reisner, R.W. Field, D.H. Katayama: Selective vibrational excitation by stimulated emission pumping. J. Chem. Phys. **75**, 2056 (1981)

10.91   Hai-Lung Da (Guest Ed.): Molecular spectroscopy and dynamics by stimulated-emission pumpings. J. Opt. Soc. Am. B **7**, 1802 (1990)

10.92   G. Zhong He, A. Kuhn, S. Schiemann, K. Bergmann: Population transfer by stimulated Raman scattering with delayed pulses and by the stimulated-emission pumping method: A comperative study. J. Opt. Soc. Am. B **7**, 1960 (1990)

10.93   K. Yamanouchi, H. Yamada, S. Tsuciya: Vibrational levels structure of highly excited SO$_2$ in the electronic ground state as studied by stimulated emission pumping spectroscopy. J. Chem. Phys. **88**, 4664 (1988)

10.94   U. Brinkmann: Higher sensitivity and extended frequency range via stimulated emission pumping SEP. Lamda Physik Highlights (June 1990) p. 1

10.95   H. Weickenmeier, V. Diemer, M. Wahl, M. Raab, W. Demtröder, W. Müller: Accurate ground state potential of Cs$_2$ up to the dissociation limit. J. Chem. Phys. **82**, 5354 (1985)

10.96   H. Weickemeier, U. Diemer, W. Demtröder, M. Broyer: Hyperfine interaction between the singlet and triplet ground states of Cs$_2$. Chem. Phys. Lett. **124**, 470 (1986)

10.97   R. Teets, R. Feinberg, T.W. Hänsch, A.L. Schawlow: Simplification of spectra by polarization labelling. Phys. Rev. Lett. **37**, 683 (1976)

10.98   N.W. Carlson, A.J. Taylor, K.M. Jones, A.L. Schawlow: Two step polarization-labelling spectroscopy of excited states of Na$_2$. Phys. Rev. A **24**, 822 (1981)

10.99   B. Hemmerling, R. Bombach, W. Demtröder, N. Spies: Polarization labelling spectroscopy of molecular Li$_2$ Rydberg states. Z. Physik D **5**, 165 (1987)

10.100   W.E. Ernst: Microwave optical polarization spectroscopy of the X^2S state of SrF. Appl. Phys. B **30**, 2378 (1983)

10.101   W.E. Ernst, T. Törring: Hyperfine Structure in the X^2S state of CaCl, measured with microwave optical polarization spectroscopy. Phys. Rev. A **27**, 875 (1983)

10.102   W.E. Ernst, O. Golonska: Microwave transitions in the Na$_3$ cluster. Phys. Rev. Lett., submitted (2002)

10.103   Th. Weber, E. Riedle, H.J. Neusser: Rotationally resolved fluorescence dip and ion-dip spectra of single rovibronic states of benzene. J. Opt. Soc. Am. B **7**, 1875 (1990)

10.104   M. Takayanagi, I. Hanazaki: Fluorescence dip and stimulated emission-pumping laser-induced-fluorescence spectra of van der Waals molecules. J. Opt. Soc. Am. B **7**, 1878 (1990)

10.105   H.S. Schweda, G.K. Chawla, R.W. Field: Highly excited, normally inaccessible vibrational levels by sub-Doppler modulated gain spectroscopy. Opt. Commun. **42**, 165 (1982)

10.106   M. Elbs, H. Knöckel, T. Laue, C. Samuelis, E. Tiemann: Observation of the last bound levels near the Na$_2$ ground state asymptote. Phys. Rev. A **59**, 3665 (1999)

10.107   A. Crubellier, O. Dulieu, F. Masnou-Seeuws, M. Elbs, H. Knöckel, E. Tiemann: Simple determination of Na$_2$ scattering lengths using observed bound levels of the ground state asymptote. Eur. Phys. J. D **6**, 211 (1999)

Chapter 11

11.1   J. Herrmann, B. Wilhelmi: *Lasers for Ultrashort Light Impulses* (North Holland, Amsterdam 1987)

11.2   J.C. Diels, W. Rudolph: *Ultrashort Laser Pulse Phenomena* (Academic Press, San Diego 1996);
C. Rulliere (Ed.): *Femtosecond Laser Pulses* (Springer, Berlin, Heidelberg, New York 1998)

11.3   S.A. Akhmanov, V.A. Vysloukhy, A.S. Chirikin: *Optics of Femtosecond Laser Pulses* (AIP, New York 1992)

11.4   V. Brückner, K.H. Felle, V.W. Grummt: *Application of Time-Resolved Optical Spectroscopy* (Elsevier, Amsterdam 1990)

11.5   J.G. Fujimoto (Ed.): Special issue on ultrafast phenomena. IEEE J. QE-**25**, 2415 (1989)

11.6   G.R. Fleming: Sub-picosecond spectroscopy. Ann. Rev. Phys. Chem. **37**, 81 (1986)

11.7   W.H. Lowdermilk: 'Technology of bandwidth-limited ultrashort pulse generation'. In: *Laser Handbook*, ed. by M.L. Stitch (North Holland, Amsterdam 1979) Vol. 3, Chapt. B1, pp. 361–420

11.8   L.P. Christov: 'Generation and propagation of ultrashort optical pulses'. In: *Progress in Optics* **24**, 201 (North Holland, Amsterdam 1991)

11.9   W. Kaiser (Ed.): *Ultrashort Laser Pulses*, 2nd edn., Topics Appl. Phys., Vol. 60 (Springer, Berlin, Heidelberg 1993)
S.L. Shapiro (Ed.): *Ultrashort Light Pulses*. Topics Appl. Phys., Vol. 18 (Springer, Berlin, Heidelberg 1977)

11.10   *Picosecond/Ultrashort Phenomena I-IX*, Proc. Int'l Confs. 1978–1994:
Picosecond Phenomena I, ed. by K.V. Shank, E.P. Ippen, S.L. Shapiro, Springer Ser. Chem. Phys., Vol. 4 (Springer, Berlin, Heidelberg 1978);
Picosecond Phenomena II, ed. by R.M. Hochstrasser, W. Kaiser, C.V. Shank, Springer Ser. Chem. Phys., Vol. 14 (Springer, Berlin, Heidelberg 1980);
Picosecond Phenomena III, ed. by K.B. Eisenthal, R.M. Hochstrasser, W. Kaiser, A. Laubereau, Springer Ser. Chem. Phys., Vol. 38 (Springer, Berlin, Heidelberg 1982);
Ultrashort Phenomena IV, ed. by D.H. Auston, K.B. Eisenthal, Springer Ser. Chem. Phys., Vol. 38 (Springer, Berlin, Heidelberg 1984);
Ultrashort Phenomena V, ed. by G.R. Fleming, A.E. Siegman, Springer Ser. Chem. Phys., Vol. 46 (Springer, Berlin, Heidelberg 1986);
Ultrashort Phenomena VI, ed. by T. Yajima, K. Yoshihara, C.B. Harris, S. Shionoya, Springer Ser. Chem. Phys., Vol. 48 (Springer, Berlin, Heidelberg 1988);
Ultrashort Phenomena VII, ed. by E. Ippen, C.B. Harris, A. Zewail, Springer Ser. Chem. Phys., Vol. 53 (Springer, Berlin, Heidelberg 1990);
Ultrafast Phenomena VIII, ed. by J.-L. Martin, A. Migus, G.A. Mourou, A.H. Zewail, Springer Ser. Chem. Phys., Vol. 55 (Springer, Berlin, Heidelberg 1993);
Ultrafast Phenomena IX, ed. by P.F. Barbara, W.H. Knox, G.A. Mourou, A.H. Zewail, Springer Ser. Chem. Phys., Vol. 60 (Springer, Berlin, Heidelberg 1994);
Ultrafast Phenomena X, ed. by P.F. Barbard, J.G. Fujimoto Springer Ser. Chem. Phys. (Springer, Berlin, Heidelberg 1996);
Ultrafast Phenomena XI, ed. by T. Elsaesser, J.G. Fujimoto, D.A. Wiersma, W. Zinth Springer Ser. Chem. Phys. (Springer, Berlin, Heidelberg 1998);
Ultrafast Phenomena XII, ed. by T. Elsaesser, S. Mukamel, M.M. Murnane Springer Ser. Chem. Phys. (Springer, Berlin, Heidelberg 2000)

11.11   T.R. Gosnel, A.J. Taylor (Eds.): *Ultrafast Laser Technology*. SPIE Proc. **44** (1991)

11.12   E. Niemann, M. Klenert: A fast high-intensity-pulse light source for flash-photolysis. Appl. Opt. **7**, 295 (1968)

11.13   L.S. Marshak: *Pulsed Light Sources* (Consultants Bureau, New York 1984)

11.14   P. Richter, J.D. Kimel, G.C. Moulton: Pulsed nitrogen laser: dynamical UV behaviour. Appl. Opt. **15**, 756 (1976)

11.15   D. Röss: *Lasers, Light Amplifiers and Oscillators* (Academic, London 1969)

11.16   A.E. Siegman: *Lasers* (University Science Books, Mill Valey, CA 1986)

11.17   F.P. Schäfer (Ed.): *Dye Lasers*, 3rd edn., Topics Appl. Phys., Vol. 1 (Springer, Berlin, Heidelberg 1990);
F.J. Duarte (Ed.): *High Power Dye Lasers*, Springer Ser. Opt. Sci., Vol. 65 (Springer Berlin, Heidelberg 1991)

11.18   F.J. McClung, R.W. Hellwarth: Characteristics of giant optical pulsation from ruby. IEEE Proc. **51**, 46 (1963)

11.19   R.B. Kay, G.S. Waldman: Complete solutions to the rate equations describing Q-spoiled and PTM laser operation. J. Appl. Phys. **36**, 1319 (1965)

11.20   O. Kafri, S. Speiser, S. Kimel: Doppler effect mechanism for laser Q-switching with a rotating mirror. IEEE J. QE-**7**, 122 (1971)

11.21   G.H.C. New: The generation of ultrashort light pulses. Rpt. Progr. Phys. **46**, 877 (1983)

11.22   E. Hartfield, B.J. Thompson: 'Optical modulators'. In: *Handbook of Optics*, ed. by W. Driscal, W. Vaughan (McGraw Hill, New York 1974)

11.23   W.E. Schmidt: Pulse stretching in a Q-switched Nd:YAG laser. IEEE J. QE-**16**, 790 (1980)

11.24   Spectra Physics: Instruction Manual on Model 344S Cavity Dumper

11.25   A. Yariv: *Quantum Electronics* (Wiley, New York 1975)

11.26   P.W. Smith, M.A. Duguay, E.P. Ippen: 'Mode-locking of lasers'. In: *Progr. Quantum Electron.*, Vol. 3 (Pergamon, Oxford 1974)

11.27   M.S. Demokan: *Mode-Locking in Solid State and Semiconductor-Lasers* (Wiley, New York 1982)

11.28   W. Koechner: *Solid-State Laser Engineering*, 4th edn, Springer Ser. Opti. Sci, Vol. 1 (Springer, Berlin, Heidelberg 1996)

11.29   C.V. Shank, E.P. Ippen: 'Mode-locking of dye lasers'. In: *Dye Lasers*, 3rd edn., ed. by F.P. Schäfer (Springer, Berlin, Heidelberg 1990) Chap. 4

11.30   W. Rudolf: Die zeitliche Entwicklung von Mode-Locking-Pulsen aus dem Rauschen. Dissertation, Fachbereich Physik, Universität Kaiserslautern (1980)

11.31   P. Heinz, M. Fickenscher, A. Lauberau: Electro-optic gain control and cavity dumping of a Nd:glass laser with active passive mode-locking. Opt. Commun. **62**, 343 (1987)

11.32   W. Demtröder, W. Stetzenbach, M. Stock, J. Witt: Lifetimes and Franck–Condon-factors for the $B\hat{O}X$ system of Na_2. J. Mol. Spectrosc. **61**, 382 (1976)

11.33   H.A. Haus: *Waves and Fields in Optoelectronics* (Prentice Hall, New York 1982)

11.34   R. Wilbrandt, H. Weber: Fluctuations in mode-locking threshold due to statistics of spontaneous emission. IEEE J. QE-**11**, 186 (1975)

11.35   B. Kopnarsky, W. Kaiser, K.H. Drexhage: New ultrafast saturable absorbers for Nd:lasers. Opt. Commun. **32**, 451 (1980)

11.36   E.P. Ippen, C.V. Shank, A. Dienes: Passive mode-locking of the cw dye laser. Appl. Phys. Lett. **21**, 348 (1972)

11.37   G.R. Flemming, G.S. Beddard: CW mode-locked dye lasers for ultrashort spectroscopic studies. Opt. Laser Technol. **10**, 257 (1978)

11.38   D.J. Bradley: 'Methods of generations'. In: *Ultrashort Light Pulses*, ed. by S.L. Shapiro, Topics Appl. Phys., Vol. 18 (Springer, Berlin, Heidelberg 1977) Chap. 2

11.39   P.W. Smith: Mode-locking of lasers. Proc. IEEE **58**, 1342 (1970)
11.40   L. Allen, D.G.C. Jones: 'Mode-locking of gas lasers'. In: *Progress in Optics* **9**, 179 (North-Holland, Amsterdam 1971)
11.41   C.K. Chan: Synchronously pumped dye lasers. Laser Techn. Bulletin **8**, Spectra Physics (June 1978)
11.42   J. Kühl, H. Klingenberg, D. von der Linde: Picosecond and subpicosecond pulse generation in synchroneously pumped mode-locked CW dye lasers. Appl. Phys. **18**, 279 (1979)
11.43   G.W. Fehrenbach, K.J. Gruntz, R.G. Ulbrich: Subpicosecond light pulses from synchronously pumped mode-locked dye lasers with composite gain and absorber medium. Appl. Phys. Lett. **33**, 159 (1978)
11.44   D. Kühlke, V. Herpers, D. von der Linde: Characteristics of a hybridly mode-locked CW dye lasers. Appl. Phys. B **38**, 159 (1978)
11.45   R.H. Johnson: Characteristics of acousto-optic cavity dumping in a mode-locked laser. IEEE J. QE-**9**, 255 (1973)
11.46   B. Couillaud, V. Fossati-Bellani: Mode locked lasers and ultrashort pulses I and II. Laser and Applications **4**, 79 (January 1985) and 91 (February 1985)
11.47   W.H. Knox, R.S. Knox, J.F. Hoose, R.N. Zare: Observation of the O-fs pulse. Opt. & Photon. News **1**, 44 (April 1990)
11.48   R.L. Fork, O.E. Martinez, J.P. Gordon: Negative dispersion using pairs of prisms. Opt. Lett. **9**, 150 (1984);
         D. Kühlke: Calculation of the colliding pulse mode locking in CW dye ring lasers. IEEE J. QE-**19**, 526 (1983)
11.49   S. DeSilvestri, P. Laporta, V. Magni: Generation and applications of femtosecond laser-pulses. Europhys. News **17**, 105 (Sept. 1986)
11.50   R.L. Fork, B.T. Greene, V.C. Shank: Generation of optical pulses shorter than 0.1 ps by colliding pulse mode locking. Appl. Phys. Lett. **38**, 671 (1981)
11.51   K. Naganuma, K. Mogi: 50 fs pulse generation directly from a colliding-pulse mode-locked Ti:sapphire laser using an antiresonant ring mirror. Opt. Lett. **16**, 738 (1991)
11.52   M.C. Nuss, R. Leonhardt, W. Zinth: Stable operation of a synchronously pumped colliding pulse mode-locking ring dye laser. Opt. Lett. **10**, 16 (1985)
11.53   P.K. Benicewicz, J.P. Roberts, A.J. Taylor: Generation of 39 fs pulses and 815 nm with a synchronously pumped mode-locked dye laser. Opt. Lett. **16**, 925 (1991)
11.54   L. Xu, G. Tempea, A. Poppe, M. Lenzner, C. Spielmann, F. Krausz, A. Stingl, K. Ferencz: High-power sub-10-fs Ti:Sapphire oscillators. Appl. Phys. B **65**, 151 (1997)
         A. Poppe, A. Führbach, C. Spielmann, F. Krausz: 'Electronics on the time scale of the light oscillation period'. In: *OSA Trends in Optics and Photonics*, Vol. 28 (Opt. Soc. Am., Washington 1999)
11.55   L.E. Nelson, D.J. Jones, K. Tamura, H.A. Haus, E.P. Ippen: Ultrashort-pulse fiber ring lasers. Appl. Phys. B **65**, 277 (1997)
11.56   G.P. Agrawal: *Nonlinear Fiber Optics* (Academic, London 1989)
11.57   S.A. Akhmanov, A.P. Sukhonukov, A.S. Chirkin: Nonstationary nonlinear optical effects and ultrashort light pulse formation. IEEE J. QE-**4**, 578 (1968);
         W.J. Tomlinson, R.H. Stollen, C.V. Shank: Compression of optical pulses chirped by self-phase modulation in fibers. J. Opt. Soc. Am. B **1**, 139 (1984)
11.58   D. Marcuse: Pulse duration in single-mode fibers. Appl. Opt. **19**, 1653 (1980)
11.59   E.B. Treacy: Optical pulse compression with diffraction gratings. IEEE J. QE-**5**, 454 (1969)
11.60   C.V. Shank, R.L. Fork, R. Yen, R.H. Stolen, W.J. Tomlinson: Compression of femtosecond optical pulses. Appl. Phys. Lett. **40**, 761 (1982)

11.61   J.G. Fujiimoto, A.M. Weiners, E.P. Ippen: Generation and measurement of op-
        tical pulses as short as 16 fs. Appl. Phys. Lett. **44**, 832 (1984)

11.62   R.L. Fork, C.H. BritoCruz, P.C. Becker, C.V. Shank: Compression of optical
        pulses to six femtoseconds by using cubic phase compensation. Opt. Lett. **12**,
        483 (1987)

11.63   R. Szipöcs, K. Ferencz, C. Spielmann, F. Krausz: Chirped mutilayer coatings for
        broadband dispersion control in femtosecond lasers. Opt. Lett. **19**, 201 (1994)

11.64   R. Szipöcz, A. Köbázi-Kis: Theory and designs of chirped dielectric laser mir-
        rors. Appl. Phys. B **65**, 115 (1997)

11.65   I.D. Jung, F.X. Kärtner, N. Matuschek, D.H. Sutter, F. Morier-Genoud, Z. Shi,
        V. Scheuer, M. Milsch, T. Tschudi, U. Keller: Semiconductor saturable absorber
        mirrors supporting sub 10-fs pulses. Appl. Phys. B **65**, 137 (1997)

11.66   J.E. Midwinter: *Optical Fibers for Transmission* (Wiley, New York 1979)

11.67   E.G. Neumann: *Single-Mode Fibers*, Springer Ser. Opt. Sci., Vol. 57 (Springer,
        Berlin, Heidelberg 1988)

11.68   V.E. Zakharov, A.B. Shabat: Exact theory of two-dimensional self-focussing and
        one-dimensional self-modulation of waves in nonlinear media. Sov. Phys. JETP
        37, 823 (1973)

11.69   A. Hasegawa: *Optical Solitons in Fibers*, 2nd edn. (Springer, Berlin, Heidelberg
        1990)

11.70   J.R. Taylor: *Optical Solitons – Theory and Experiment* (Cambridge Univ. Press,
        Cambridge 1992)

11.71   G.P. Agrawal: *Nonlinear Fiber Optics* (Academic Press, San Diego 1989)

11.72   L.F. Mollenauer, R.H. Stolen: The soliton laser. Opt. Lett. **9**, 13 (1984)

11.73   F.M. Mitschke, L.F. Mollenauer: Stabilizing the soliton laser. IEEE J. QE-**22**,
        2242 (1986)

11.74   E. Dusuvire: *Erbium-doped Fiber Amplifiers* (Wiley, New York 1994)

11.75   M.E. Fermann, A. Galvanauskas, G. Sucha, D. Harter: Fiber-lasers for ultrafast
        optics. Appl. Phys. B **65**, 259 (1997);
        U. Keller: Ultrafast all solid-state laser technology. Appl. Phys. B **58**, 349 (1994)

11.76   K. Tamura, H.A. Haus, E.P. Ippen: Self-starting additive pulse mode-locked er-
        bium fiber ring laser. Electron. Lett. **28**, 2226 (1992)

11.77   E.P. Ippen, D.J. Jones, L.E. Nelson, H.A. Haus: 'Ultrafast fiber lasers'. In: T. El-
        saesser et al. (Ed.): *Ultrafast Phenomena XI*, (Springer, Berlin, Heidelberg 1998)
        p. 30

11.78   F.M. Mitschke, L.F. Mollenauer: Ultrashort pulses from the soliton laser. Opt.
        Lett. **12**, 407 (1987)

11.79   F.M. Mitschke: Solitonen in Glasfasern. Laser und Optoelektronik **4**, 393 (1987)

11.80   B. Wilhelmi, W. Rudolph (Eds.): *Light Pulse Compression* (Harwood Academic,
        Chur 1989)

11.81   T. Brixner, M. Strehle, G. Gerber: Feedback-controlled optimization of amplified
        femtosecond laser pulses. Appl. Phys. B **68**, 281 (1999)

11.82   T. Hornung, R. Meier, M. Motzkus: Optimal control of molecular states in
        a learning loop with a parametrization in frequency and time domain. Chem.
        Phys. Lett. **326**, 445 (2000)

11.83   T. Baumert, T. Brixner, V. Seyfried, M. Strehle, G. Gerber: Femtosecond pulse
        shaping by an evolutionary algorithm with feedback. Appl. Phys. B **65**, 779
        (1997)

11.84   A. Pierce, M.A. Dahleh, H. Rubitz: Optimal control of quantum-mechanical sys-
        tems. Phys. Rev. A **37**, 4950 (1988)

11.85   R.W. Schoenlein, J.Y. Gigot, M.T. Portella, C.V. Shank: Generation of blue-
        green 10 fs pulses using an excimer pumped dye amplifier. Appl. Phys. Lett.
        58, 801 (1991)

11.86   C.V. Shank, E.P. Ippen: Subpicosecond kilowatt pulses from a mode-locked CW dye laser. Sov. Phys. JETP **34**, 62 (1972)

11.87   R.L. Fork, C.V. Shank, R.T. Yen: Amplification of 70-fs optical pulses to gigawatt powers. Appl. Phys. Lett. **41**, 233 (1982)

11.88   S.R. Rotman, C. Roxlo, D. Bebelaar, T.K. Yee, M.M. Salour: Generation, stabilization and amplification of subpicosecond pulses. Appl. Phys. B **28**, 319 (1982)

11.89   A. Rundquist, et al.: Ultrafast laser and amplifier sources. Appl. Phys. B **65**, 161 (1997)

11.90   E. Salin, J. Squier, G. Mourov, G. Vaillancourt: Multi-kilohertz Ti:Al$_2$O$_3$ amplifier for high power femtosecond pulses: Opt. Lett. **16**, 1964 (1991)

11.91   G. Sucha, D.S. Chenla: Kilohertz-rate continuum generation by amplification of femtosecond pulses near 1.5 μm. Opt. Lett. **16**, 1177 (1991)

11.92   A. Sullivan, H. Hamster, H.C. Kapteyn, S. Gordon, W. White, H. Nathel, R.J. Blair, R.W. Falcow: Multiterawatt, 100 fs laser. Opt. Lett. **16**, 1406 (1991)

11.93   T. Elsässer, M.C. Nuss: Femtosecond pulses in the mid-infrared generated by downconversion of a travelling-wave dye laser. Opt. Lett. **16**, 411 (1991)

11.94   J. Heling, J. Kuhl: Generation of femtosecond pulses by travelling-wave amplified spontaneous emission. Opt. Lett. **14**, 278 (1991)

11.95   G.A. Mourou, C.P.J. Barty, M.D. Pery: Ultrahigh-intensity lasers: physics of the extreme on a tabletop. Physics Today, Jan. 1998, p. 22

11.96   T. Wilhelm, J. Piel, E. Riedle: Sub-20 fs pulses, tunable across the visible from a blue-pumped single pass noncollinear parameter oscillator. Opt. Lett. **22**, 1494 (1997)

11.97   S. Svanberg, et al.: 'Applications of terrawatt lasers'. In: *Laser Spectroscopy XI*, ed. by L. Bloomfield, T. Gallagher, D. Lanson (AIP, New York 1993)

11.98   R.R. Alfano (Ed.): *The Supercontinuum Laser Source* (Springer, New York 1989); J.D. Kmetec, J.I. MacKlin, J.F. Young: 0.5 TW, 125 fs Ti:sapphire laser. Opt. Lett. **16**, 1001 (1991)

11.99   T. Brabec, F. Krausz: Intense few cycle laser fields: frontiers of nonlinear optics. Rev. Mod. Phys. **77**, 545 (2000)

11.100   C.H. Lee: *Picosecond Optoelectronic Devices* (Academic, New York 1984)

11.101   Hamamatsu: FESCA (Femtosecond Streak camera 2908, information sheet, August 1988) and actual information under http://usa.hamamatsu.com/sys-streak/guide.htm

11.102   F.J. Leonberger, C.H. Lee, F. Capasso. H. Morkoc (Eds.): *Picosecond Electronics and Optoelectronics II*, Springer Ser. Electron. Photon., Vol. 28 (Springer, Berlin, Heidelberg 1987)

11.103   D.J. Bradley: 'Methods of generation'. In: *Ultrashort Light Pulses*, ed. by S.L. Shapiro, Topics Appl. Phys., Vol. 18 (Springer, Berlin, Heidelberg 1977) Chap. 2

11.104   D.J. Bowley: Measuring ultrafast pulses. Laser and Optoelectronics **6**, 81 (1987)

11.105   H.E. Rowe, T. Li: Theory of two-photon measurement of laser output. IEEE J. QE-**6**, 49 (1970)

11.106   H.P. Weber: Method for pulsewidth measurement of ultrashort light pulses, using nonlinear optics. J. Appl. Phys. **38**, 2231 (1967)

11.107   J.A. Giordmaine, P.M. Rentze, S.L. Shapiro, K.W. Wecht: Two-photon Excitation of fluorescence by picosecond light pulses. Appl. Phys. Lett. **11**, 216 (1967); see also [11.26]

11.108   W.H. Glenn: Theory of the two-photon absorption-fluorescence method of pulsewidth measurement. IEEE J. QE-**6**, 510 (1970)

11.109   E.P. Ippen, C.V. Shank: 'Techniques for measurement'. In: *Ultrashort Light Pulses*, ed. by S.L. Shapiro, Topics Appl. Phys., Vol. 18 (Springer, Berlin, Heidelberg 1977) Chap. 3

11.110 D.H. Auston: Higher order intensity correlation of optical pulses. IEEE J. QE-**7**, 465 (1971)

11.111 R. Trebino, D.J. Kane: Using phase retrieval to measure the intensity and phase of ultrashort pulses: frequency resolved optical gating. J. Opt. Soc. Am. A **11**, 2429 (1993);

11.112 D.J. Kane, R. Trebino: Single-shot measurement of the intensity and phase of an arbitrary ultrashort pulse using frequency-resolved optical gating. Opt. Lett. **18**, 823 (1993)

11.113 A. Unsöld, B. Baschek: *The New Cosmos*, 5th edn. (Springer, Berlin, Heidelberg 1991)

11.114 R.E. Imhoff, F.H. Read: Measurements of lifetimes of atoms, molecules. Rep. Progr. Phys. **40**, 1 (1977)

11.115 M.C.E. Huber, R.J. Sandeman: The measurement of oscillator strengths. Rpt. Progr. Phys. **49**, 397 (1986)

11.116 J.R. Lakowvicz, B.P. Malivatt: Construction and performance of a variable-frequency phase-modulation fluorometer. Biophys. Chemistry **19**, 13 (1984) and Biophys. J. **46**, 397 (1986)

11.117 J. Carlson: Accurate time resolved laser spectroscopy on sodium and bismuth atoms. Z. Physik D **9**, 147 (1988)

11.118 D.V. O'Connor, D. Phillips: *Time Correlated Single Photon Counting* (Academic, New York 1984)

11.119 W. Wien: Über Messungen der Leuchtdauer der Atome und der Dämpfung der Spektrallinien. Ann. Physik **60**, 597 (1919)

11.120 P. Hartmetz, H. Schmoranzer: Lifetime and absolute transition probabilities of the $^2P_{10}$ ($^3S_1$) level of NeI by beam-gas-dye laser spectroscopy. Z. Physik A **317**, 1 (1984)

11.121 D. Schulze-Hagenest, H. Harde, W. Brandt, W. Demtröder: Fast beam-spectroscopy by combined gas-cell laser excitation for cascade free measurements of highly excited states. Z. Physik A **282**, 149 (1977)

11.122 L. Ward, O. Vogel, A. Arnesen, R. Hallin, A. Wännström: Accurate experimental lifetimes of excited levels in NaII, SdII. Phys. Scripta **31**, 149 (1985)

11.123 H. Schmoranzer, P. Hartmetz, D. Marger, J. Dudda: Lifetime measurement of the $B\,^2\Sigma_u^+$ ($v = 0$) state of $^{14}N_2^+$ by the beam-dye-laser method. J. Phys. B **22**, 1761 (1989)

11.124 A. Arnesen, A. Wännström, R. Hallin, C. Nordling, O. Vogel: Lifetime in KII with the beam-laser method. J. Opt. Soc. Am. B **5**, 2204 (1988)

11.125 A. Lauberau, W. Kaiser: 'Picosecond investigations of dynamic processes in polyatomic molecules and liquids'. In: *Chemical and Biochemical Applications of Lasers II*, ed. by C.B. Moore (Academic, New York 1977)

11.126 W. Zinth, M.C. Nuss, W. Kaiser: 'A picosecond Raman technique with resolution four times better than obtained by spontaneous Raman spectroscopy'. In: *Picosecond Phenomena III*, ed. by K.B. Eisenthal, R.M. Hochstrasser, W. Kaiser, A. Lauberau, Springer Ser. Chem. Phys., Vol. 38 (Springer, Berlin, Heidelberg 1982) p. 279

11.127 A. Seilmeier, W. Kaiser: 'Ultrashort intramolecular and intermolecular vibrational energy transfer of polyatomic molecules in liquids'. In: *Picosecond Phenomena III*, ed. by K.B. Eisenthal, R.M. Hochstrasser, W. Kaiser, A. Lauberau, Springer Ser. Chem. Phys., Vol. 38 (Springer, Berlin, Heidelberg 1982) p. 279

11.128 M. Nisoli, et al.: Highly efficient parametric conversion of femtosecond Ti:Sapphire laser pulses at 1 kHz. Opt. Lett. **19**, 1973 (1994)

11.129 E. Riedle, M. Beutter, S. Lochbrunner, J. Piel, S. Schenkl, S. Spörlein, W. Zinth: Generation of 10 to 50 fs pulses, tunable through all of the visible and the NIR. Appl. Phys. B **71**, 457 (2000)

11.130 W. Shuicai, H. Junfang, X. Dong, Z. Changjun, H. Xun: A three-wavelength Ti:sapphire femtosecond laser for use with the multi-excited photosystem II. Appl. Phys. B **72**, 819 (2001)

11.131 Long-Sheng Ma, et al.: Synchronization and phase-locking of two independent femtosecond lasers. Conference Proceedings of ICOLS 2001 (Snowbird, Utah 2001) p. 2–26

11.132 W. Zinth, W. Holzapfel, R. Leonhardt: Femtosecond dephasing processes of molecular vibrations, in [Ref. 11.10, VI, p. 401 (1988)]

11.133 G. Angel, R. Gagel, A. Lauberau: Vibrational dynamics in the S_1 and S_0 states of dye molecules studied separately by femtosecond polarization spectroscopy, in [Ref. 11.10, VI, p. 467 (1988)]

11.134 F.J. Duarte (Ed.): *High-Power Dye Lasers*, Springer Ser. Opt. Sci., Vol. 65 (Springer, Berlin, Heidelberg 1991)

11.135 W. Kütt, K. Seibert, H. Kurz: High density femtosecond excitation of hot carrier distributions in InP and InGaAs, in [Ref. 11.10, VI, p. 233 (1988)]

11.136 W.Z. Lin, R.W. Schoenlein, M.J. LaGasse, B. Zysset, E.P. Ippen, J.G. Fujimoto: Ultrafast scattering and energy relaxation of optically excited carriers in GaAs and AlGaAs, in [Ref. 11.10, VI, p. 210 (1988)]

11.137 L.R. Khundkar, A.H. Zewail: Ultrafast molecular reaction dynamics in real-time. Ann. Rev. Phys. Chem. **41**, 15 (1990);
A.H. Zewail (Ed.): *Femtochemistry: Ultrafast Dynamics of the Chemical Bond, I and II* (World Scientific, Singapore 1994)

11.138 A.H. Zewail: Femtosecond transition-state dynamics. Faraday Discuss. Chem. Soc. **91**, 207 (1991)

11.139 T. Baumert, M. Grosser, R. Thalveiser, G. Gerber: Femtosecond time- resolved molecular multiphoton ionisation: The Na$_2$ system. Phys. Rev. Lett. **67**, 3753 (1991)

11.140 T. Baumert, B. Bühler, M. Grosser, R. Thalweiser, V. Weiss, E. Wiedemann, R. Gerber: Femtosecond time-resolved wave packet motion in molecular multiphoton ionization and fragmentation. J. Phys. Chem. **95**, 8103 (1991)

11.141 E. Schreiber: *Femtosecond Real Time Spectroscopy of Small Molecules and Clusters* (Springer, Berlin, Heidelberg, New York 1998)

11.142 O. Svelto, S. DeSilvestry, G. Denardo (Eds.): *Ultrafast Processes in Spectroscopy* (Plenum, New York 1997)

11.143 H.J. Eichler, P. Günther, D.W. Pohl: *Laser-Induced Dynamic Gratings*. Springer Series in Optical Sciences Vol. 50 (Springer 1986)

Chapter 12

12.1 W. Hanle: Über magnetische Beeinflussung der Polarisation der Resonanzfluoreszenz. Z. Physik **30**, 93 (1924)

12.2 M. Norton, A. Gallagher: Measurements of lowest-S-state lifetimes of gallium, indium and thallium. Phys. Rev. A **3**, 915 (1971)

12.3 F. Bylicki, H.G. Weber, H. Zscheeg, M. Arnold: On NO$_2$ excited state lifetime and g-factors in the 593 nm band. J. Chem. Phys. **80**, 1791 (1984)

12.4 H.H. Stroke, G. Fulop, S. Klepner: Level crossing signal line shapes and ordering of energy levels. Phys. Rev. Lett. **21**, 61 (1968)

12.5 M. McClintock, W. Demtröder, R.N. Zare: Level crossing studies of Na$_2$, using laser-induced fluorescence. J. Chem. Phys. **51**, 5509 (1969)

12.6 R.N. Zare: Molecular level crossing spectroscopy. J. Chem. Phys. **45**, 4510 (1966)

12.7 P. Franken: Interference effects in the resonance fluorescence of "crossed" excited states. Phys. Rev. **121**, 508 (1961)

12.8   G. Breit: Quantum theory of dispersion. Rev. Mod. Phys. **5**, 91 (1933)

12.9   R.N. Zare: Interference effects in molecular fluorescence. Accounts Chem. Res. **4**, 361 (November 1971)

12.10   G.W. Series: 'Coherence effects in the interaction of radiation with atoms'. In: *Physics of the One- and Two-Electron Atoms*, ed. by F. Bopp, H. Kleinpoppen (North-Holland, Amsterdam 1969) p. 268

12.11   W. Happer: Optical pumping. Rev. Mod. Phys. **44**, 168 (1972)

12.12   J.N. Dodd, R.D. Kaul, D.M. Warrington: The modulation of resonance fluorescence excited by pulsed light. Proc. Phys. Soc. **84**, 176 (1964)

12.13   M.S. Feld. A. Sanchez, A. Javan: 'Theory of stimulated level crossing'. In: *Int'l Colloq. on Doppler-Free Spectroscopic Methods for Single Molecular Systems*, ed. by J.C. Lehmann, J.C. Pebay-Peyroula (Ed. du Centre National Res. Scient., Paris 1974) p. 87

12.14   H. Walther (Ed.): *Laser Spectroscopy of Atoms and Molecules*, Topics Appl. Phys., Vol. 2 (Springer, Berlin, Heidelberg 1976)

12.15   G. Moruzzi, F. Strumia (Eds.): *The Hanle Effect and Level Crossing Spectroscopy* (Plenum, New York 1992)

12.16   P. Hannaford: Oriented atoms in weak magnetic fields. Physica Scripta T **70**, 117 (1997)

12.17   J.C. Lehmann: 'Probing small molecules with lasers'. In: *Frontiers of Laser Spectroscopy, Vol. 1*, ed. by R. Balian, S. Haroche, S. Liberman (North-Holland, Amsterdam 1977)

12.18   M. Broyer, J.C. Lehmann, J. Vigue: G-factors and lifetimes in the *B*-state of molecular iodine. J. Phys. **36**, 235 (1975)

12.19   H. Figger, D.L. Monts, R.N. Zare: Anomalous magnetic depolarization of fluorescence from the NO_2 $^2B_2$-state. J. Mol. Spectrosc. **68**, 388 (1977)

12.20   J.R. Bonilla, W. Demtröder: Level crossing spectroscopy of NO_2 using Doppler-reduced laser excitation in molecular beams. Chem. Phys. Lett. **53**, 223 (1978)

12.21   H.G. Weber, F. Bylicki: NO_2 lifetimes by Hanle effect measurements. Chem. Phys. **116**, 133 (1987)

12.22   F. Bylicki, H.G. Weber, G. Persch, W. Demtröder: On g factors and hyperfine structure in electronically excited states of NO_2. J. Chem. Phys. **88**, 3532 (1988)

12.23   H.J. Beyer, H. Kleinpoppen: 'Anticrossing spectroscopy'. In: *Progr. Atomic Spectroscopy*, ed. by W. Hanle, H. Kleinpoppen (Plenum, New York 1978) p. 607

12.24   P. Cacciani, S. Liberman, E. Luc-Koenig, J. Pinard, C. Thomas: Anticrossing effects in Rydberg states of lithium in the presence of parallel magnetic and electric fields. Phys. Rev. A **40**, 3026 (1989)

12.25   G. Raithel, M. Fauth, H. Walther: Quasi-Landau resonances in the spectra of rubidium Rydberg atoms in crossed electric and magnetic fields. Phys. Rev. A **44**, 1898 (1991)

12.26   J. Bengtsson, J. Larsson, S. Svanberg, C.G. Wahlström: Hyperfine-structure study of the $3d^{10}p^2P_{3/2}$ level of neutral copper using pulsed level crossing spectroscopy at short laser wavelengths. Phys. Rev. A **41**, 233 (1990)

12.27   G. Hermann, G. Lasnitschka, J. Richter, A. Scharmann: Determination of lifetimes and hyperfine splittings of Tl states $nP_{3/2}$ by level crossing spectroscopy with two-photon excitation. Z. Physik D **10**, 27 (1988)

12.28   G. von Oppen: Measurements of state multipoles using level crossing techniques. Commen. At. Mol. Phys. **15**, 87 (1984)

12.29   A.C. Luntz, R.G. Brewer: Zeeman-tuned level crossing in $^1\Sigma$ CH_4. J. Chem. Phys. **53**, 3380 (1970)

12.30   A.C. Luntz, R.G. Brewer, K.L. Foster, J.D. Swalen: Level crossing in CH_4 observed by nonlinear absorption. Phys. Rev. Lett. **23**, 951 (1969)

12.31   J.S. Levine, P. Boncyk, A. Javan: Observation of hyperfine level crossing in stimulated emission. Phys. Rev. Lett. **22**, 267 (1969)

12.32   G. Hermann, A. Scharmann: Untersuchungen zur Zeeman-Spektroskopie mit Hilfe nichtlinearer Resonanzen eines Multimoden Lasers. Z. Physik **254**, 46 (1972)

12.33   W. Jastrzebski, M. Kolwas: Two-photon Hanle effect. J. Phys. B **17**, L855 (1984)

12.34   L. Allen, D.G. Jones: The helium-neon laser. Adv. Phys. **14**, 479 (1965)

12.35   C. Cohen-Tannoudji: Level-crossing resonances in atomic ground states. Commen. At. Mol. Phys. **1**, 150 (1970)

12.36   S. Haroche: 'Quantum beats and time resolved spectroscopy'. In: *High Resolution Laser Spectroscopy*, ed. by K. Shimoda, Topics Appl. Phys., Vol. 13 (Springer, Berlin, Heidelberg 1976) p. 253

12.37   H.J. Andrä: Quantum beats and laser excitation in fast beam spectroscopy, in *Atomic Physics 4*, ed. by G. zu Putlitz, E.W. Weber, A. Winnacker (Plenum, New York 1975) p. 635

12.38   H.J. Andrä: Fine structure, hyperfine structure and Lamb-shift measurements by the beam foil technique. Phys. Scripta **9**, 257 (1974)

12.39   R.M. Lowe, P. Hannaford: Observation of quantum beats in sputtered metal vapours. 19th EGAS Conference, Dublin (1987)

12.40   W. Lange, J. Mlynek: Quantum beats in transmission by time resolved polarization spectroscopy. Phys. Rev. Lett. **40**, 1373 (1978)

12.41   J. Mlynek, W. Lange: A simple method of observing coherent ground-state transients. Opt. Commun. **30**, 337 (1979)

12.42   H. Harde, H. Burggraf, J. Mlynek, W. Lange: Quantum beats in forward scattering: subnanosecond studies with a mode-locked dye laser. Opt. Lett. **6**, 290 (1981)

12.43   J. Mlynek, K.H. Drake, W. Lange: 'Observation of transient and stationary Zeeman coherence by polarization spectroscopy'. In: *Laser Spectroscopy IV*, ed. by A.R.W. McKellar, T. Oka, B.P. Stoicheff, Springer Ser. Opt. Sci., Vol. 30 (Springer, Berlin, Heidelberg 1981) p. 616

12.44   G. Leuchs, S.J. Smith, E. Khawaja, H. Walther: Quantum beats observed in photoionization. Opt. Commun. **31**, 313 (1979)

12.45   M. Dubs, J. Mühlbach, H. Bitto, P. Schmidt, J.R. Huber: Hyperfine quantum beats and Zeeman spectroscopy in the polyatomic molecule propynol CHOC-CHO. J. Chem. Phys. **83**, 3755 (1985)

12.46   H. Bitto, J.R. Huber: Molecular quantum beat spectroscopy. Opt. Commun. **80**, 184 (1990)

12.47   W. Scharfin, M. Ivanco, St. Wallace: Quantum beat phenomena in the fluorescence decay of the $C(^1B_2)$ State of SO_2. J. Chem. Phys. **76**, 2095 (1982)

12.48   P.J. Brucat, R.N. Zare: NO_2 $A\,^2B_2$ state properties from Zeeman quantum beats. J. Chem. Phys. **78**, 100 (1983); J. Chem. Phys. **81**, 2562 (1984)

12.49   N. Ochi, H. Watanabe, S. Tsuchiya, S. Koda: Rotationally resolved laser-induced fluorescence and Zeeman quantum beat spectroscopy of the $V\,^1B_2$ state of jet cooled CS_2. Chem. Phys. **113**, 271 (1987)

12.50   P. Schmidt, H. Bitto, J.R. Huber: Excited state dipole moments in a polyatomic molecule determined by Stark quantum beat spectroscopy. J. Chem. Phys. **88**, 696 (1988)

12.51   J.N. Dodd, G.W. Series: 'Time-resolved fluorescence spectroscopy'. In: *Progress in Atomic Spectroscopy*, ed. by W. Hanle, H. Kleinpoppen (Plenum, New York 1978)

12.52   J. Mlynek: Neue optische Methoden der hochauflösenden Kohärenzspektroskopie an Atomen. Phys. Blätter **43**, 196 (1987)

12.53   A. Corney: *Atomic and Laser Spectroscopy* (Oxford Univ. Press, London 1977)

12.54   B.J. Dalton: Cascade Zeeman quantum beats produced by stepwise excitation using broad-line laser pulses. J. Phys. B **20**, 251, 267 (1987)

12.55   G. Alber, P. Zoller: Laser-induced excitation of electronic Rydberg wave packets. Contemp. Phys. **32**, 185 (1991)

12.56   A. Wolde, I.D. Noordam, H.G. Müller, A. Lagendijk, H.B. van Linden: Observation of radially localized atomic electron wave packets. Phys. Rev. Lett. **61**, 2099 (1988)

12.57   T. Baumert, V. Engel, C. Röttgermann, W.T. Strunz, G. Gerber: Femtosecond pump-probe study of the spreading and recurrance of a vibrational wave packet in Na_2. Chem. Phys. Lett. **191**, 639 (1992)

12.58   M. Gruebele, A.H. Zewail: Ultrashort reaction dynamics. Phys. Today **43**, 24 (May 1990)

12.59   M. Gruebele, G. Roberts, M. Dautus, R. M Bowman, A.H. Zewail: Femtosecond temporal spectroscopy and direct inversion to the potentials: application to iodine. Chem. Phys. Lett. **166**, 459 (1990)

12.60   F.C. deSchryver, S.E. Fyter, G. Schweitzer: *Femtochemistry* (Wiley & Sons, New York 2001)

12.61   E. Schreiber: *Femtosecond Real-Time Spectroscopy of Small Molecules and Clusters*, Springer Tracts in Modern Physics Vol. 143 (Springer, Berlin, Heidelberg, New York 1999)

12.62   J. Mlyneck, W. Lange, H. Harde, H. Burggraf: High resolution coherence spectroscopy using pulse trains. Phys. Rev. A **24**, 1099 (1989)

12.63   H. Lehmitz, W. Kattav, H. Harde: 'Modulated pumping in Cs with picosecond pulse trains'. In: *Methods of Laser Spectroscopy*, ed. by Y. Prior, A. Ben-Reuven, M. Rosenbluth (Plenum, New York 1986) p. 97

12.64   R.H. Dicke: Coherence in spontaneous radiation processes. Phys. Rev. **93**, 99 (1954)

12.65   E.L. Hahn: Spin echoes. Phys. Rev. **80**, 580 (1950);
         C.P. Slichter: *Principles of Magnetic Resonance*, 3rd edn., Springer Ser. Solid-State Sci., Vol. 1 (Springer, Berlin, Heidelberg 1990)

12.66   I.D. Abella: 'Echoes at optical frequencies'. In: *Progress in Optics* **7**, 140 (North-Holland, Amsterdam 1969)

12.67   S.R. Hartmann: 'Photon echoes'. In: *Lasers and Light, Readings from Scientific American* (Freeman, San Francisco 1969) p. 303

12.68   C.K.N. Patel, R.E. Slusher: Photon echoes in gases. Phys. Rev. Lett. **20**, 1087 (1968)

12.69   R.G. Brewer: Coherent optical transients. Phys. Today **30**, 50 (May 1977)

12.70   R.G. Brewer: A.Z. Genack: Optical coherent transients by laser frequency switching. Phys. Rev. Lett. **36**, 959 (1976)

12.71   R.G. Brewer: 'Coherent optical spectroscopy'. In: *Frontiers in Laser Spectroscopy*, ed. by R. Balian, S. Haroche, S. Lieberman (North-Holland, Amsterdam 1977)

12.72   L.S. Vasilenko, N.Y. Rubtsova: Coherent spectroscopy of gaseous media: ways of increasing spectral resolution. Bull. Acad. Sci. USSR Phys. Ser. **53**, No. 12, 54 (1989)

12.73   R.G. Brewer, R.L. Shoemaker: Photon echo and optical nutation in molecules. Phys. Rev. Lett. **27**, 631 (1971)

12.74   P.R. Berman, J.M. Levy, R.G. Brewer: Coherent optical transient study of molecular collisions. Phys. Rev. A **11**, 1668 (1975)

12.75   C. Freed, D.C. Spears, R.G. O'Donnell: 'Precision heterodyne calibration'. In: *Laser Spectroscopy*, ed. by R.G. Brewer, A. Mooradian (Plenum, New York 1974) p. 17

12.76   F.R. Petersen, D.G. McDonald, F.D. Cupp, B.L. Danielson: Rotational constants of $^{12}C^{16}O_2$ from beats between Lamb-dip stabilized laser lines. Phys. Rev. Lett.

31, 573 (1973); also in *Laser Spectroscopy*, ed. by R.G. Brewer, A. Mooradian (Plenum, New York 1974) p. 555

12.77 T.J. Bridge, T.K. Chang: Accurate rotational constants of CO_2 from measurements of CW beats in bulk GaAs between CO_2 vibrational-rotational laser lines. Phys. Rev. Lett. **22**, 811 (1969)

12.78 L.A. Hackel, K.H. Casleton, S.G. Kukolich, S. Ezekiel: Observation of magnetic octupole and scalar spin-spin interaction in I_2 using laser spectroscopy. Phys. Rev. Lett. **35**, 568 (1975)

12.79 W.A. Kreiner, G. Magerl, E. Bonek, W. Schupita, L. Weber: Spectroscopy with a tunable sideband laser. Physica Scripta **25**, 360 (1982)

12.80 J.L. Hall, L. Hollberg, T. Baer, H.G. Robinson: Optical heterodyne saturation spectroscopy. Appl. Phys. Lett. **39**, 680 (1981)

12.81 H.Z. Cummins, H.L. Swinney: Light beating spectroscopy. *Progress in Optics* **8**, 134 (North-Holland, Amsterdam 1970)

12.82 E.O. DuBois (Ed.): *Photon Correlation Techniques*, Springer Ser. Opt. Sci., Vol. 38 (Springer, Berlin, Heidelberg 1983)

12.83 N. Wiener: Generalized harmonic analysis. Acta Math. **55**, 117 (1930)

12.84 C.L. Mehta: 'Theory of photoelectron counting'. In: *Progress in Optics* **7**, 373 (North Holland, Amsterdam 1970)

12.85 B. Saleh: *Photoelectron Statistics*, Springer Ser. Opt. Sci., Vol. 6 (Springer, Berlin, Heidelberg 1978)

12.86 L. Mandel: 'Fluctuation of light beams'. In: *Progress in Optics* **2**, 181 (North-Holland, Amsterdam 1963)

12.87 A.J. Siegert: MIT Rad. Lab. Rpt. No. 465 (1943)

12.88 E.O. Schulz-DuBois: High-resolution intensity interferometry by photon correlation, in [Ref. 12.82, p. 6]

12.89 P.P.L. Regtien (Ed.): *Modern Electronic Measuring Systems* (Delft Univ. Press, Delft 1978)
P. Horrowitz, W. Hill: *The Art of Electronics* (Cambridge Univ. Press, Cambridge 1980)

12.90 H.Z. Cummins, E.R. Pike (Eds.): *Photon Correlation and Light Spectroscopy* (Plenum, New York 1974)

12.91 E. Stelzer, H. Ruf, E. Grell: Analysis and resolution of polydispersive systems, in [Ref. 12.82, p. 329]

12.92 N.C. Ford, G.B. Bennedek: Observation of the spectrum of light scattered from a pure fluid near its critical point. Phys. Rev. Lett. **15**, 649 (1965)

12.93 R. Hanbury Brown: *The Intensity Interferometer* (Taylor and Francis, London 1974)

12.94 F.T. Arecchi, A. Berné, P. Bulamacchi: High-order fluctuations in a single mode laser field. Phys. Rev. Lett. **88**, 32 (1966)

12.95 M. Adam, A. Hamelin, P. Bergé: Mise au point et étude d'une technique de spectrographie par battements de photons hétérodyne. Opt. Acta **16**, 337 (1969)

12.96 R. Riegler: *Fluorescence Correlation Spectroscopy* (Springer, Berlin, Heidelberg, New York 2001)

12.97 A.F. Harvey: *Coherent Light* (Wiley, London 1970)

12.98 J.I. Steinfeld (Ed.): *Laser and Coherence Spectroscopy* (Plenum, New York 1978)

12.99 B.W. Shore: *The Theory of Coherent Atomic Excitation, Vols. 1, 2* (Wiley, New York 1990)

12.100 L. Mandel, E. Wolf: *Coherence and Quantum Optics I-V*, Proc. Rochester Conferences (Plenum, New York 1961, 1967, 1973, 1978, 1984)

Chapter 13

13.1   R.B. Bernstein: *Chemical Dynamics via Molecular Beam and Laser Techniques* (Clarendon, Oxford 1982)

13.2   J.T. Yardle: *Molecular Energy Transfer* (Academic, New York 1980)

13.3   W.H. Miller (Ed.): *Dynamics of Molecular Collisions* (Plenum, New York 1976)

13.4   P.R. Berman: Studies of collisions by laser spectroscopy. Adv. At. Mol. Phys. **13**, 57 (1977)

13.5   G.W. Flynn, E. Weitz: Vibrational energy flow in the ground electronic states of polyatomic molecules. Adv. Chem. Phys. **47**, 185 (1981)

13.6   V.E. Bondeby: Relaxation and vibrational energy redistribution processes in polyatomic molecules. Ann. Rev. Phys. Chem. **35**, 591 (1984)

13.7   S.A. Rice: Collision-induced intramolecular energy transfer in electronically excited polyatomic molecules. Adv. Chem. Phys. **47**, 237 (1981)

13.8   P.J. Dagdigian: State-resolved collision-induced electronic transitions. Ann. Rev. Phys. Chem. **48**, 95 (1997)

13.9   J.J. Valentini: State-to-State chemical reaction dynamics in polyatomic systems. Ann. Rev. Phys. Chem. **52**, 15 (2001)

13.10   C.A. Taatjes, J.F. Herschberger: Recent progress in infrared absorption techniques for elementary gas phase reaction kinetics. Ann. Rev. Phys. Chem. **52**, 41 (2001)

13.11   See for instance: *Proc. Int'l Conf. on the Physics of Electronic and Atomic Collisions, ICPEAC I – XVII* (North-Holland, Amsterdam)

13.12   *Proc. Int'l Conf. on Spectral Line Shapes: Vols. I–XI*, Library of Congress Catalog

13.13   J. Hinze (Ed.): *Energy Storage and Redistribution in Molecules* (Plenum, New York 1983)

13.14   F. Aumeyer, H. Winter (Eds.): *Photonic, Electronic and Atomic Collisions* (World Scientific, Singapore 1998)

13.15   J. Ward, J. Cooper: Correlation effects in the theory of combined Doppler and pressure broadening. J. Quant. Spectr. Rad. Transf. **14**, 555 (1974)

13.16   J.O. Hirschfelder, Ch.F. Curtiss, R.B. Bird: *Molecular Theory of Gases and Liquids* (Wiley, New York 1954)

13.17   Th.W. Hänsch, P. Toschek: On pressure broadening in a He-Ne laser. IEEE J. QE-**6**, 61 (1969)

13.18   J.L. Hall: 'Saturated absorption spectroscopy'. In: *Atomic Physics, Vol. 3*, ed. by S.J. Smith, G.K. Walthers (Plenum, New York 1973) pp. 615 ff.

13.19   J.L. Hall: *The Line Shape Problem in Laser-Saturated Molecular Absorption* (Gordon & Breach, New York 1969)

13.20   S.N. Bagayev: 'Spectroscopic studies into elastic scattering of excited particles'. In: *Laser Spectroscopy IV*, ed. by H. Walther, K.W. Rothe, Springer Ser. Opt. Sci., Vol. 21 (Springer, Berlin, Heidelberg 1979) p. 222

13.21   K.E. Gibble, A. Gallagher: Measurements of velocity-changing collision kernels. Phys. Rev. A **43**, 1366 (1991)

13.22   C.R. Vidal, F.B. Haller: Heat pipe oven applications: production of metal vapor-gas mixtures. Rev. Sci. Instrum. **42**, 1779 (1971)

13.23   R. Bombach, B. Hemmerling, W. Demtröder: Measurement of broadening rates, shifts and effective lifetiems of Li_2 Rydberg levels by optical double-resonance spectroscopy. Chem. Phys. **121**, 439 (1988)

13.24   K.D. Heber, P.J. West, E. Matthias: Collisions between Sr-Rydberg atoms and intermediate and high principal quantum number and noble gases. J. Phys. D **21**, 563 (1988)

13.25   R.B. Kurzel, J.I. Steinfeld, D.A. Hazenbuhler, G.E. LeRoi: Energy transfer pro-
        cesses in monochromatically excited iodine molecules. J. Chem. Phys. **55**, 4822
        (1971)

13.26   J.I. Steinfeld: 'Energy transfer processes'. In: *Chemical Kinetics Phys. Chem.
        Ser. One, Vol. 9*, ed. by J.C. Polany (Butterworth, London 1972)

13.27   G. Ennen, C. Ottinger: Rotation-vibration-translation energy transfer in laser ex-
        cited Li$_2$ ($B\,^1\Pi_u$). Chem. Phys. **3**, 404 (1974)

13.28   Ch. Ottinger, M. Schröder: Collision-induced rotational transitions of dye laser
        excited Li$_2$ molecules. J. Phys. B **13**, 4163 (1980)

13.29   K. Bergmann, W. Demtröder: Inelastic cross sections of excited molecules. J.
        Phys. B **5**, 1386, 2098 (1972)

13.30   D. Zevgolis: Untersuchung inelastischer Stoßprozesse in Alkalidämpfen mit
        Hilfe spektral- und zeitaufgelöster Laserspektroskopie. Dissertation, Faculty of
        Physics, Kaiserslautern (1980)

13.31   T.A. Brunner, R.D. Driver, N. Smith, D.E. Pritchard: Rotational energy transfer
        in Na$_2$-Xe collisions. J. Chem. Phys. **70**, 4155 (1979);
        T.A. Brunner, et al.: Simple scaling law for rotational energy transfer in Na$_2^*$-Xe
        collisions. Phys. Rev. Lett. **41**, 856 (1978)

13.32   R. Schinke: *Theory of Rotational Transitions in Molecules, Int'l Conf. Phys. El.
        At. Collisions XIII* (North-Holland, Amsterdam 1984) p. 429

13.33   M. Faubel: Vibrational and rotational excitation in molecular collisions. Adv. At.
        Mol. Phys. **19**, 345 (1983)

13.34   K. Bergmann, H. Klar, W. Schlecht: Asymmetries in collision-induced rotational
        transitions. Chem. Phys. Lett. **12**, 522 (1974)

13.35   H. Klar: Theory of collision-induced rotational energy transfer in the ¼-state of
        diatomic molecule. J. Phys. B **6**, 2139 (1973)

13.36   A.J. McCaffery, M.J. Proctor, B.J. Whitaker: Rotational energy transfer: polar-
        ization and scaling. Annu. Rev. Phys. Chem. **37**, 223 (1986)

13.37   G. Sha, P. Proch, K.L. Kompa: Rotational transitions of N$_2$ ($a\,^1\Pi_g$) induced by
        collisions with Ar/He studied by laser REMPI spectroscopy. J. Chem. Phys. **87**,
        5251 (1987)

13.38   C.R. Vidal: Collisional depolarization and rotational energy transfer of the Li$_2$
        ($B\,^1\Pi_u$)-Li(^{2}S$_{1/2}$) system from laser-induced fluorescence. Chem. Phys. **35**, 215
        (1978)

13.39   W.B. Gao, Y.Q. Shen, H. Häger, W. Krieger: Vibrational relaxation of ethylene
        oxide and ethylene oxide-rare-gas mixtures. Chem. Phys. **84**, 369 (1984)

13.40   M.H. Alexander, A. Benning: Theoretical studies of collision-induced energy
        transfer in electronically excited states. Ber. Bunsengesell. Phys. Chem. **14**, 1253
        (1990)

13.41   E. Nikitin, L. Zulicke: *Theorie Chemischer Elementarprozesse* (Vieweg, Braun-
        schweig 1985)

13.42   E.K. Kraulinya, E.K. Kopeikana, M.L. Janson: Excitation energy transfer in
        atom-molecule interactions of sodium and potassium vapors. Chem. Phys. Lett.
        39, 565 (1976); Opt. Spectrosc. **41**, 217 (1976)

13.43   W. Kamke, B. Kamke, I. Hertel, A. Gallagher: Fluorescence of the Na* + N$_2$
        collision complex. J. Chem. Phys. **80**, 4879 (1984)

13.44   L.K. Lam, T. Fujiimoto, A.C. Gallagher, M. Hessel: Collisional excitation tranfer
        between Na und Na$_2$. J. Chem. Phys. **68**, 3553 (1978)

13.45   H. Hulsman, P. Willems: Transfer of electronic excitation in sodium vapour.
        Chem. Phys. **119**, 377 (1988)

13.46   G. Ennen, Ch. Ottinger: Collision-induced dissociation of laser excited Li$_2$
        $B\,^1\Pi_\mu$. J. Chem. Phys. **40**, 127 (1979); ibid. **41**, 415 (1979)

13.47   J.E. Smedley, H.K. Haugen, St.R. Leone: Collision-induced dissociation of laser-
        excited Br$_2$ [$B\,^3\Pi(O_u^+)$; v', J']. J. Chem. Phys. **86**, 6801 (1987)

13.48   E.W. Rothe, U. Krause, R. Dünen: Photodissociation of Na_2 and Rb_2: analysis of atomic fine structure of ^2P products. J. Chem. Phys. **72**, 5145 (1980)

13.49   G. Brederlow, R. Brodmann, M. Nippus, R. Petsch, S. Witkowski, R. Volk, K.J. Witte: Performance of the Asterix IV high power iodine laser. IEEE J. QE-**16**, 122 (1980)

13.50   V. Diemer, W. Demtröder: Infrared atomic Cs laser based on optical pumping of Cs_2 molecules. Chem. Phys. Lett. **176**, 135 (1991)

13.51   St. Lemont, G.W. Flynn: Vibrational state analysis of electronic-to-vibrational energy transfer processes. Annu. Rev. Phys. Chem. **28**, 261 (1977)

13.52   A. Tramer, A. Nitzan: Collisional effects in electronic relaxation. Adv. Chem. Phys. **47** (2), 337 (1981)

13.53   S.A. Rice: Collision-induced intramolecular energy transfer in electronically excited polyatomic molecules. Adv. Chem. Phys. **47** (2), 237 (1981)

13.54   K.B. Eisenthal: 'Ultrafast chemical reactions in the liquid state'. In: *Ultrashort Laser Pulses*, 2nd edn., ed. by W. Kaiser, Topics Appl. Phys., Vol. 60 (Springer, Berlin, Heidelberg, New York 1993)

13.55   P. Hering, S.L. Cunba, K.L. Kompa: Coherent anti-Stokes Raman spectroscopy study of the energy paritioning in the $Na(3P)$-H_2 collision pair with red wing excitation. J. Phys. Chem. **91**, 5459 (1987)

13.56   H.G.C. Werij, J.F.M. Haverkort, J.P. Woerdman: Study of the optical piston. Phys. Rev. A **33**, 3270 (1986)

13.57   A.D. Streater, J. Mooibroek, J.P. Woerdman: Light-induced drift in rubidium: spectral dependence and isotope separation. Opt. Commun. **64**, 1 (1987)

13.58   M. Allegrini, P. Bicchi, L. Moi: Cross-section measurements for the energy transfer collisions $Na(3P) + Na(3P) \to Na(5S, 4D) + Na(3S)$. Phys. Rev. A **28**, 1338 (1983)

13.59   J. Huenneckens, A. Gallagher: Radiation diffusion and saturation in optically thick Na vapor. Phys. Rev. A **28**, 238 (1983)

13.60   J. Huenneckens, A. Gallagher: Associative ionization in collisions between two $Na(3P)$ atoms. Phys. Rev. A **28**, 1276 (1983)

13.61   S.A. Abdullah, M. Allegrini, S. Gozzini, L. Moi: Three-body collisions between laser excited Na and K-atoms in the presence of buffer gas. Nuov. Cimento D **9**, 1467 (1987)

13.62   H.G.C. Werij, M. Harris, J. Cooper, A. Gallagher: Collisional energy transfer between excited Sr atoms. Phys. Rev. A **43**, 2237 (1991)

13.63   S.G. Leslie, J.T. Verdeyen, W.S. Millar: Excitation of highly excited states by collisions between two excited cesium atoms. J. Appl. Phys. **48**, 4444 (1977)

13.64   A. Ermers, T. Woschnik, W. Behmenburg: Depolarization and fine structure effects in halfcollisions of sodium-noble gas systems. Z. Physik D **5**, 113 (1987)

13.65   P.W. Arcuni, M.L. Troyen, A. Gallagher: Differential cross section for Na fine structure transfer induced by Na and K collisions. Phys. Rev. A **41**, 2398 (1990)

13.66   G.C. Schatz, L.J. Kowalenko, S.R. Leone: A coupled channel quantum scattering study of alignment effects in $Na(^2P_{3/2}) + He \to Na(^2P_{1/2}) + He$ collisions. J. Chem. Phys. **91**, 6961 (1989)

13.67   T.R. Mallory, W. Kedzierski, J.B. Atkinson, L. Krause: $9\,^2D$ fine structure mixing in rubidium by collisions with ground-state Rb and noble-gas atoms. Phys. Rev. A **38**, 5917 (1988)

13.68   A. Sasso, W. Demtröder, T. Colbert, C. Wang, W. Ehrlacher, J. Huennekens: Radiative lifetimes, collisional mixing and quenching of the cesium 5 Dy levels. Phys. Rev. A **45**, 1670 (1992)

13.69   D.L. Feldman, R.N. Zare: Evidence for predissociation of Rb_2^* ($C\,^1\Pi_u$) into Rb^* ($^2P_{3/2}$) and Rb ($^2S_{1/2}$). Chem. Phys. **15**, 415 (1976);

E.J. Breford, F. Engelke: Laser induced fluorescence in supersonic nozzle beams: predissociation in the Rb$_2$ $C\,^1\Pi_u$ and $D\,^1\Pi_u$ states. Chem. Phys. Lett. **75**, 132 (1980)

13.70   K.F. Freed: Collision-induced intersystem crossing. Adv. Chem. Phys. **47**, 211 (1981)

13.71   J.P. Webb, W.C. McColgin, O.G. Peterson, D. Stockman, J.H. Eberly: Intersystem crossing rate and triplet state lifetime for a lasing dye. J. Chem. Phys. **53**, 4227 (1970)

13.72   X.L. Han, G.W. Schinn, A. Gallagher: Spin-exchange cross sections for electron excitation of Na 3S–3P determined by a novel spectroscopic technique. Phys. Rev. A **38**, 535 (1988)

13.73   T.A. Cool: 'Transfer chemical laser'. In: *Handbook of Chemical Lasers*, ed. by R.W.F. Gross, J.F. Bott (Wiley, New York 1976)

13.74   S.A. Ahmed, J.S. Gergely, D. Infaute: Energy transfer organic dye mixture lasers. J. Chem. Phys. **61**, 1584 (1974)

13.75   B. Wellegehausen: Optically pumped CW dimer lasers. IEEE J. QE-**15**, 1108 (1979)

13.76   W.J. Witteman: *The* CO$_2$ *Laser*, Springer Ser. Opt. Sci., Vol. 53 (Springer, Berlin, Heidelberg, New York 1987)

13.77   W.H. Green, J.K. Hancock: Laser excited vibrational energy exchange studies of HF, CO and NO. IEEE J. QE-**9**, 50 (1973)

13.78   W.B. Gao, Y.Q. Shen, J. Häger, W. Krieger: Vibrational relaxation of ethylene oxide and ethylene oxide-rare gas mixtures. Chem. Phys. **84**, 369 (1984)

13.79   S.R. Leone: State-resolved molecular reaction dynamics. Annu. Rev. Phys. Chem. **35**, 109 (1984)

13.80   J.O. Hirschfelder, R.E. Wyatt, R.D. Coalson (Eds.): *Lasers, Molecules and Methods* (Wiley, New York 1989)

13.81   E. Hirota, K. Kawaguchi: High resolution infrared studies of molecular dynamics. Annu. Rev. Phys. Chem. **36**, 53 (1985)

13.82   R. Feinberg, R.E. Teets, J. Rubbmark, A.L. Schawlow: Ground-state relaxation measurements by laser-induced depopulation. J. Chem. Phys. **66**, 4330 (1977)

13.83   J.G. Haub, B.J. Orr: Coriolis-assisted vibrational energy transfer in D$_2$CO/D$_2$CO and HDCO/HDCO-collisions. J. Chem. Phys. **86**, 3380 (1987)

13.84   A. Lauberau, W. Kaiser: Vibrational dynamics of liquids and solids investigated by picosecond light pulses. Rev. Mod. Phys. **50**, 607 (1978)

13.85   T. Elsaesser, W. Kaiser: Vibrational and vibronic relaxation of large polyatomic molecules in liquids. Annu. Rev. Phys. Chem. **43**, 83 (1991)

13.86   L.R. Khunkar, A.H. Zewail: Ultrafast molecular reaction dynamics in real-time. Annu. Rev. Phys. Chem. **41**, 15 (1990)

13.87   R.T. Bailey, F.R. Cruickshank: Spectroscopic studies of vibrational energy transfer. Adv. Infr. Raman Spectrosc. **8**, 52 (1981)

13.88   Ch.E. Hamilton, J.L. Kinsey, R.W. Field: Stimulated emission pumping: new methods in spectroscopy and molecular dynamics. Annu. Rev. Phys. Chem. **37**, (1986)

13.89   A. Geers, J. Kappert, F. Temps, J.W. Wiebrecht: Preparation of single rotation-vibration states of CH$_3$O(C($\,^2$E) above the H-CH$_2$O dissociation threshold by stimulated emission pumping. Ber. Bunsenges. Phys. Chem. **94**, 1219 (1990)

13.90   M. Becker, U. Gaubatz, K. Bergmann, P.L. Jones: Efficient and selective population of high vibrational levels by stimulated near resonance Raman scattering. J. Chem. Phys. **87**, 5064 (1987)

13.91   N.V. Vitanov, T. Halfmann, B.W. Shore, K. Bergmann: Laser-induced population transfer by adiabatic passage techniques. Ann. Rev. Phys. Chem. **52**, 763 (2001)

13.92   G.W. Coulston, K. Bergmann: Population transfer by stimulated Raman scattering with delayed pulses. J. Chem. Phys. **96**, 3467 (1992)

13.93   W.R. Gentry: 'State-to-state energy transfer in collisions of neutral molecules'.
        In: *ICPEAC XIV* (1985) (North-Holland, Amsterdam 1986) p. 13

13.94   See the special issue on "Molecular Spectroscopy and Dynamics by Stimulated
        Emission Pumping", H.L. Dai (Guest Ed.) J. Opt. Soc. Am. B **7**, 1802 (1990)

13.95   P.J. Dagdigian: 'Inelastic scattering: optical methods'. In: *Atomic and Molecular
        Methods*, ed. by G. Scoles (Oxford Univ. Press, Oxford 1988) p. 569

13.96   J.B. Pruett, R.N. Zare: State-to-state reaction rates: $Ba + HF(v = 0.1) \rightarrow BaF(v =
        0 - 12) + H$. J. Chem. Phys. **64**, 1774 (1976)

13.97   P.J. Dagdigian, H.W. Cruse, A. Schultz, R. N Zare: Product state analysis of
        BaO from the reactions $Ba + CO_2$ and $Ba + O_2$. J. Chem. Phys. 61, 4450 (1974)

13.98   K. Kleinermanns, J. Wolfrum: Laser stimulation and observation of elementary
        chemical reactions in the gas phase. Laser Chem. **2**, 339 (1983)

13.99   K.D. Rinnen, D.A.V. Kliner, R.N. Zare: The $H + D_2$ reaction: prompt HD dis-
        tribution at high collision energies. J. Chem. Phys. **91**, 7514 (1989)

13.100 D.P. Gerrity, J.J. Valentini: Experimental determination of product quantum state
        distributions in the $H + D_2 \rightarrow HD + D$ reaction. J. Chem. Phys. **79**, 5202 (1983)

13.101 H. Buchenau, J.P. Toennies, J. Arnold, J. Wolfrum: $H + H_2$: the current status.
        Ber. Bunsenges. Phys. Chem. **94**, 1231 (1990)

13.102 V.M. Bierbaum, S.R. Leone: 'Optical studies of product state distribution in
        thermal energy ion-molecule reactions'. In: *Structure, Reactivity and Thermo-
        chemistry of Ions*, ed. by P. Ausloos, S.G. Lias (Reidel, New York 1987) p. 23

13.103 M.N.R. Ashfold, J.E. Baggott (Eds.): *Molecular Photodissociation Dynamics,
        Adv. in Gas-Phase Photochemistry* (Roy. Soc. Chemistry, London 1987)

13.104 E. Hasselbrink, J.R. Waldeck, R.N. Zare: Orientation of the CN $X^2\Sigma^+$ frag-
        ment, following photolysis by circularly polarized light. Chem. Phys. **126**, 191
        (1988)

13.105 F.J. Comes: Molecular reaction dynamics: Sub-Doppler and polarization spec-
        troscopy. Ber. Bunsenges. Phys. Chem. **94**, 1268 (1990)

13.106 J.F. Black, J.R. Waldeck, R.N. Zare: Evidence for three interacting potential en-
        ergy surfaces in the photodissociation of ICN at 249 nm. J. Chem. Phys. **92**,
        3519 (1990)

13.107 P. Andresen, G.S. Ondrey, B. Titze, E.W. Rothe: Nuclear and electronic dynam-
        ics in the photodissociation of water. J. Chem. Phys. **80**, 2548 (1984)

13.108 St.R. Leone: Infrared fluorescence: A versatile probe of state selected chemical
        dynamics. Acc. Chem. Res. **16**, 88 (1983)

13.109 See, for instance, many contributions in: Laser Chemistry, an International Jour-
        nal (Harwood, Chur)

13.110 G.E. Hall. P.L. Houston: Vector correlations in photodissociation dynamics.
        Annu. Rev. Phys. Chem. **40**, 375 (1989)

13.111 See special issue on "Dynamics of Molecular Photofragmentation", R.N. Dixon,
        G.G. Balint-Kurti, M.S. Child, R. Donovun, J.P. Simmons (Guest Eds.) Faraday
        Discuss. Chem. Soc. **82** (1986)

13.112 A. Gonzales Urena: Influence of translational energy upon reactive scattering
        cross sections of neutral-neutral collisions. Adv. Chem. Phys. **66**, 213 (1987)

13.113 M.A.D. Fluendy, K.P. Lawley: *Chemical Applications of Molecular Beam Scat-
        tering* (Chapman and Hall, London 1973)

13.114 K. Liu: Crossed-beam studies of neutral reactions: state specific differential cross
        sections. Ann. Rev. Phys. Chem. **52**, 139 (2001)

13.115 V. Borkenhagen, M. Halthau, J.P. Toennies: Molecular beam measurements of
        inelastic cross sections for transition between defiend rotational staes of CsF.
        J. Chem. Phys. **71**, 1722 (1979)

13.116 K. Bergmann, R. Engelhardt, U. Hefter, J. Witt: State-resolved differential cross
        sections for rotational transition in $Na_2 + Ne$ collisions. Phys. Rev. Lett. **40**, 1446
        (1978)

13.117   K. Bergmann: 'State selection via optical methods'. In: *Atomic and Molecular Beam Methods, Vol. 12*, ed. by G. Coles (Oxford Univ. Press, Oxford 1989)

13.118   K. Bergmann, U. Hefter, J. Witt: State-to-state differential cross sections for rotational transitions in $Na_2 + He$-collisions. J. Chem. Phys. **71**, 2726 (1979); K. Bergmann, U. Hefter, J. Witt: J. Chem. Phys. **72**, 4777 (1980)

13.119   H.G. Rubahn, K. Bergmann: The effect of laser-induced vibrational band stretching in atom-molecule collisions. Annu. Rev. Phys. Chem. **41**, 735 (1990)

13.120   V. Gaubatz, H. Bissantz, V. Hefter, I. Colomb de Daunant, K. Bergmann: Optically pumped supersonic beam lasers. J. Opt. Soc. Am. B **6**, 1386 (1989)

13.121   R. Düren, H. Tischer: Experimental determination of the $K(4\,^2P_{3/3}$-Ar) potential. Chem. Phys. Lett. **79**, 481 (1981)

13.122   I.V. Hertel, H. Hofmann, K.A. Rost: Electronic to vibrational-rotational energy transfer in collisions of $Na(3\,^2P)$ with simple molecules. Chem. Phys. Lett. **47**, 163 (1977)

13.123   P. Botschwina, W. Meyer, I.V. Hertel, W. Reiland: Collisions of excited Na atoms with H_2-molecules: ab initio potential energy surfaces. J. Chem. Phys. **75**, 5438 (1981)

13.124   E.E.B. Campbell, H. Schmidt, I.V. Hertel: Symmetry and angular momentum in collisions with laser excited polarized atoms. Adv. Chem. Phys. **72**, 37 (1988)

13.125   R. Düren, V. Lackschewitz, S. Milosevic, H. Panknin, N. Schirawski: Differential cross sections for reactive and nonreactive scattering of electronically excited Na from HF molecules. Chem. Phys. Lett. **143**, 45 (1988)

13.126   J. Grosser, O. Hoffmann, S. Klose, F. Rebentrost: Optical excitation of collision pairs in crossed beams: determination of the NaKr $B\,^2\Sigma$-potential. Europhys. Lett. **39**, 147 (1997)

13.127   C. Cohen-Tannoudji, S. Reynaud: Dressed-atom description of absorption spectra of a multilevel atom in an intense laser beam. J. Phys. B **10**, 345 (1977)

13.128   S. Keynaud, C. Cohen-Tannoudji: 'Collisional effects in resonance fluorescence'. In: *Laser Spectroscopy V*, ed. by A.R.W. McKellar, T. Oka, B.P. Stoicheff, Springer Ser. Opt. Sci., Vol. 30 (Springer, Berlin, Heidelberg 1981) p. 166

13.129   S.E. Harris, R.W. Falcone, W.R. Green, P.B. Lidow, J.C. White, J.F. Young: 'Laser-induced collisions'. In: *Tunable Lasers and Applications*, ed. by A. Mooradian, T. Jaeger, P. Stockseth, Springer Ser. Opt. Sci., Vol. 3 (Springer, Berlin, Heidelberg 1976) p. 193

13.130   S.E. Harris, J.F. Young, W.R. Green, R.W. Falcone, J. Lukasik, J.C. White, J.R. Willison, M.D. Wright, G.A. Zdasiuk: 'Laser-induced collisional and radiative energy transfer'. In: *Laser Spectroscopy IV*, ed. by H. Walther, K.W. Rothe, Springer Ser. Opt. Sci., Vol. 21 (Springer, Berlin, Heidelberg 1979) p. 349

13.131   E. Giacobino, P.R. Berman: Cooling of vapors using collisionally aided radiative excitation. NBD special publication No. 653 (US Dept. Commerce, Washington, DC 1983)

13.132   A. Gallagher, T. Holstein: Collision-induced absorption in atomic electronic transitions. Phys. Rev. A **16**, 2413 (1977)

13.133   A. Birnbaum, L. Frommhold, G.C. Tabisz: 'Collision-induced spectroscopy: absorption and light scattering'. In: *Spectral Line Shapes 5*, ed. by J. Szudy (Ossolineum, Wroclaw 1989) p. 623; A. Bonysow, L. Frommhold: Collision-induced light scattering. Adv. Chem. Phys. **35**, 439 (1989)

13.134   F. Dorsch, S. Geltman, P.E. Toschek: Laser-induced collisional and energy in thermal collisions of lithium and stontium. Phys. Rev. A **37**, 2441 (1988)

13.135   K. Burnett: 'Spectroscopy of collision complexes'. In: *Electronic and Atomic Collisions*, ed. by J. Eichler, I.V. Hertel, N. Stoltenfoth (Elsevier, Amsterdam 1984) p. 649

13.136   N.K. Rahman, C. Guidotti (Eds.): *Photon-Associated Collisions and Related Topics* (Harwood, Chur 1982)
13.137   J. Weiner: Advances in ultracold collisions. Adv. At. Mol. Opt. Phys. **35**, 332 (1995)
13.138   W.C. Stwalley, He Wang: Photoassociation of ultracold atoms: a new spectroscopic technique. J. Mol. Spectrosc. **194**, 228 (1999)
13.139   T. Walker, P. Feng: Measurements of collisions between laser-cooled atoms. Adv. At. Mol. Opt. Phys. **34**, 125 (1994)
13.140   G. Boato, G.G. Volpi: Experiments on the dynamics of molecular processes: a chronicle of fifty years. Ann. Rev. Phys. Chem. **50**, 23 (1999)

Chapter 14

14.1   J.L. Hall: 'Some remarks on the interaction between precision physical measurements and fundamental physical theories'. In: *Quantum Optics, Experimental Gravity and Measurement Theory*, ed. by P. Meystre, M.V. Scully (Plenum, New York 1983)
14.2   A.I. Miller: *Albert Einstein's Special Theory of Relativity* (Addison-Wesley, Reading, MA 1981);
        J.L. Heilbron: *Max Planck* (Hirzel, Stuttgart 1988)
14.3   H.G. Kuhn: *Atomic Spectra*, 2nd edn. (Longman, London 1971);
        I.I. Sobelman: *Atomic Spectra and Radiative Transitions*, 2nd edn., Springer Ser. Atoms Plasmas, Vol. 12 (Springer, Berlin, Heidelberg, New York 1992)
14.4   W.E. Lamb Jr., R.C. Retherford: Fine-structure of the hydrogen atom by a microwave method. Phys. Rev. **72**, 241 (1947); Phys. Rev. **79**, 549 (1959)
14.5   C. Salomon, J. Dalibard, W.D. Phillips, A. Clairon, S. Guellati: Laser cooling of cesium atoms below 3 μK. Europhys. Lett. **12**, 683 (1990)
14.6   H.J. Metcalf, P. van der Straaten: *Laser Cooling and Trapping* (Springer, Berlin, Heidelberg, New York 1999)
14.7   K. Sengstock, W. Ertmer: Laser manipulation of atoms. Adv. At. Mol. Opt. Phys. **35**, 1 (1995)
14.8   H. Frauenfelder: *The Mössbauer Effect* (Benjamin, New York 1963);
        U. Gonser (Ed.): *Mössbauer Spectroscopy*, Topics Appl. Phys., Vol. 5 (Springer, Berlin, Heidelberg 1975)
14.9   J.L. Hall: 'Sub-Doppler spectroscopy: methane hyperfine spectroscopy and the ultimate resolution limit'. In: *Laser Spectroscopy II*, ed. by S. Haroche, J.C. Pebay-Peyroula, T.W. Hänsch, S.E. Harris, Lecture Notes Phys., Vol. 43 (Springer, Berlin, Heidelberg 1975) p. 105
14.10   C.H. Bordé: 'Progress in understanding sub-Doppler-line shapes'. In: *Laser Spectroscopy III*, ed. by J.L. Hall, J.L. Carlsten, Springer Ser. Opt. Sci., Vol. 7 (Springer, Berlin, Heidelberg 1977) p. 121
14.11   S.N. Bagayev, A.E. Baklanov, V.P. Chebotayev, A.S. Dychkov, P.V. Pokuson: 'Superhigh resolution laser spectroscopy with cold particles'. In: *Laser Spectroscopy VIII*, ed. by W. Pearson, S. Svanberg, Springer Ser. Opt. Sci., Vol. 55 (Springer, Berlin, Heidelberg, New York 1987) p. 95
14.12   J.C. Berquist, R.L. Barger, D.L. Glaze: 'High resolution spectroscopy of calcium atoms'. In: *Laser Spectroscopy IV*, ed. by H. Walther, K.W. Rothe, Springer Ser. Opt. Sci., Vol. 21 (Springer, Berlin, Heidelberg 1979) p. 120
14.13   B. Bobin, C. Bordé, C. Breaut: Vibration-rotation molecular constants for the ground state of SF_6 from saturated absorption spectroscopy. J. Mol. Spectrosc. **121**, 91 (1987)
14.14   T.W. Hänsch, A.L. Schawlow: Cooling of gases by laser radiation. Opt. Commun. **13**, 68 (1975)

14.15   W. Ertmer, R. Blatt, J.L. Hall: Some candidate atoms and ions for frequency standards research using laser radiative cooling techniques. Progr. Quantum Electron. **8**, 249 (1984)

14.16   W. Ertmer, R. Blatt, J.L. Hall, M. Zhu: Laser manipulation of atomic beam velocities: demonstration of stopped atoms and velocity reversal. Phys. Rev. Lett. **54** 996 (1985)

14.17   R. Blatt, W. Ertmer, J.L. Hall: Cooling of an atomic beam with frequency-sweep techniques. Progr. Quantum Electron. **8**, 237 (1984)

14.18   W.O. Phillips, J.V. Prodan, H.J. Metcalf: 'Neutral atomic beam cooling, experiments at NBS'. In: *NBS Special Publication No. 653* (US Dept. of Commerce, June 1983); Phys. Lett. **49**, 1149 (1982)

14.19   H. Metcalf: 'Laser cooling and magnetic trapping of neutral atoms'. In: *Methods of Laser Spectroscopy*, ed. by Y. Prior, A. Ben-Reuven, M. Rosenbluth (Plenum, New York 1986) p. 33

14.20   J.V. Prodan, W.O. Phillips: 'Chirping the light-fantastic?' In: *Laser Cooled and Trapped Atoms, NBS Special Publication No. 653* (US Dept. Commerce, June 1983)

14.21   D. Sesko, C.G. Fam, C.E. Wieman: Production of a cold atomic vapor using diode-laser cooling. J. Opt. Soc. Am. B **5**, 1225 (1988)

14.22   R.N. Watts, C.E. Wieman: Manipulating atomic velocities using diode lasers. Opt. Lett. **11**, 291 (1986)

14.23   B. Sheeby, S.Q. Shang, R. Watts, S. Hatamian, H. Metcalf: Diode laser deceleration and collimation of a rubidium beam. J. Opt. Soc. Am. B **6**, 2165 (1989)

14.24   H. Metcalf: Magneto-optical trapping and its application to helium metastables. J. Opt. Soc. Am. B **6**, 2206 (1989)

14.25   I.C.M. Littler, St. Balle, K. Bergmann: The CW modeless laser: spectral control, performance data and build-up dynamics. Opt. Commun. **88**, 514 (1992)

14.26   J. Hoffnagle: Proposal for continuous white-light cooling of an atomic beam. Opt. Lett. **13**, 307 (1991)

14.27   I.C.M. Littler, H.M. Keller, U. Gaubatz, K. Bergmann: Velocity control and cooling of an atomic beam using a modeless laser. Z. Physik D **18**, 307 (1991)

14.28   R. Schieder, H. Walther, L. Wöste: Atomic beam deflection by the light of a tunable dye laser. Opt. Commun. **5**, 337 (1972)

14.29   I. Nebenzahl, A. Szöke: Deflection of atomic beams by resonance radiation using stimulated emission. Appl. Phys. Lett. **25**, 327 (1974)

14.30   J. Nellesen, J.M. Müller, K. Sengstock, W. Ertmer: Large-angle beam deflection of a laser cooled sodium beam. J. Opt. Soc. Am. B **6**, 2149 (1989)

14.31   S. Villani (Ed.): *Uranium Enrichment*, Topics Appl. Phys., Vol. 35 (Springer, Berlin, Heidelberg 1979)

14.32   C.E. Tanner, B.P. Masterson, C.E. Wieman: Atomic beam collimation using a laser diode with a self-locking power buildup-cavity. Opt. Lett. **13**, 357 (1988)

14.33   J. Dalibard, C. Salomon, A. Aspect, H. Metcalf, A. Heidmann, C. Cohen-Tannoudji: 'Atomic motion in a standing wave'. In: *Laser Spectroscopy VIII*, ed. by S. Svanberg, W. Persson, Springer Ser. Opt. Sci., Vol. 55 (Springer, Berlin, Heidelberg, New York 1987) p. 81

14.34   St. Chu, J.E. Bjorkholm, A. Ashkin, L. Holberg, A. Cable: 'Cooling and trapping of atoms with laser light'. In: *Methods of Laser Spectroscopy*, ed. by Y. Prior, A. Ben-Reuven, M. Rosenbluth (Plenum, New York 1986) p. 41

14.35   T. Baba, I. Waki: Cooling and mass analysis of molecules using laser-cooled atoms. Jpn. J. Appl. Phys. **35**, 1134 (1996)

14.36   J.T. Bahns, P.L. Gould, W.C. Stwalley: Formation of Cold ($T < 1$ K) Molecules. Adv. At. Mol. Opt. Phys. **42**, 171 (2000)

14.37   W.C. Stwalley: 'Making Molecules at Microkelvin'. In: R. Campargue (Ed.): *Atomic and Molecular Beams* (Springer, Berlin, Heidelberg, New York 2001) p. 105

14.38   P. Pillet, F. Masnou-Seeuws, A. Crubelier: 'Molecular photoassociation and ultracold molecules'. In: R. Campargue (Ed.): *Atomic and Molecular Beams* (Springer, Berlin, Heidelberg, New York 2001) p. 113

14.39   J.M. Doyle, B. Friedrich, J. Kim, D. Patterson: Buffer-gas loading of atoms and molecules into a magnetic trap. Phys. Rev. A **52**, R2515 (1995)

14.40   E. Lusovoj, J.P. Toennies, S. Grebenev, et al.: 'Spectroscopy of molecules and unique clusters in superfluid He-droplets'. In: R. Campargue (Ed.): *Atomic and Molecular Beams* (Springer, Berlin, Heidelberg, New York 2001) p. 775

14.41   S. Grebenev, M. Hartmann, M. Havenith, B. Sartakov, J.P. Toennies, A.F. Vilesov: The rotational spectrum of single OCS molecules in liquid ^{4}He droplets. J. Chem. Phys. **112**, 4485 (2000)

14.42   V.S. Letokhov, V.G. Minogin, B.D. Pavlik: Cooling and capture of atoms and molecules by a resonant light field. Sov. Phys. JETP **45**, 698 (1977); Opt. Commun. **19**, 72 (1976)

14.43   V.S. Letokhov, B.D. Pavlik: Spectral line narrowing in a gas by atoms trapped in a standing light wave. Appl. Phys. **9**, 229 (1976)

14.44   A. Ashkin, J.P. Gordon: Cooling and trapping of atoms by resonance radiation pressure. Opt. Lett. **4**, 161 (1979)

14.45   J.P. Gordon: Radiation forces and momenta in dielectric media. Phys. Rev. A **8**, 14 (1973)

14.46   M.H. Mittelman: *Introduction to the Theory of Laser-Atom Interaction* (Plenum, New York 1982)

14.47   J.E. Bjorkholm, R.R. Freeman, A. Ashkin, D.B. Pearson: 'Transverse resonance radiation pressure on atomic beams and the influence of fluctuations'. In: *Laser Spectroscopy IV*, ed. by H. Walther, K.W. Rothe, Springer Ser. Opt. Sci., Vol. 21 (Springer, Berlin, Heidelberg 1979) p. 49

14.48   R. Grimm, M. Weidemüller, Y.B. Ovchinnikov: Optical Dipole Traps for Neutral Atoms. Adv. At. Mol. Opt. Phys. **42**, 95 (2000)

14.49   D.E. Pritchard, E.L. Raab, V. Bagnato, C.E. Wieman, R.N. Watts: Light traps using spontaneous forces. Phys. Rev. Lett. **57**, 310 (1986);
        H. Metcalf: Magneto-optical trapping and its application to helium metastables. J. Opt. Soc. Am. B **6**, 2206 (1989)

14.50   (a) J. Nellessen, J. Werner, W. Ertmer: Magneto-optical compression of a monoenergetic sodium atomic beam. Opt. Commun. **78**, 300 (1990);
        (b) C. Monroe, W. Swann, H. Robinson, C. Wieman: Very cold trapped atoms in a vapor cell. Phys. Rev. Lett. **65**, 1571 (1990)

14.51   A.M. Steane, M. Chowdhury, C.J. Foot: Radiation force in the magneto-optical trap. J. Opt. Soc. Am. B **9**, 2142 (1992)

14.52   See, for instance, *Feynman Lectures on Physics I* (Addison-Wesley, Reading, MA 1965)

14.53   S. Stenholm: The semiclassical theory of laser cooling. Rev. Mod. Phys. **58**, 699 (1986)

14.54   A. Aspect, E. Arimondo, R. Kaiser, N. Vansteenkiste, C. Cohen-Tannoudji: Laser cooling below the one-photon recoil energy by velocity-selective coherent population trapping. J. Opt. Soc. Am. B **6**, 2112 (1989)

14.55   J. Dalibard, C. Cohen-Tannoudji: Laser cooling below the Doppler limit by polarization gradients: simple theoretical model. J. Opt. Soc. Am. B **6**, 2023 (1989);
        C. Cohen-Tannoudji: 'New laser cooling mechanisms'. In: *Laser Manipulation of Atoms and Ions*, ed. by A. Arimondo, W.D. Phillips, F. Strumia (North-Holland, Amsterdam 1992) p. 99

14.56   D.S. Weiss, E. Riis, Y. Shery, P. Jeffrey Ungar, St. Chu: Optical molasses and multilevel atoms: experiment. J. Opt. Soc. Am. B **6**, 2072 (1989)

14.57   P.J. Ungar, D.S. Weiss, E. Riis, St. Chu: Optical molasses and multilevel atoms: theory. J. Opt. Soc. Am. B **6**, 2058 (1989)

14.58   C. Cohen-Tannoudji, W.D. Phillips: New mechanisms for laser cooling. Physics Today **43**, 33 (October 1990)

14.59   S. Martelucci (Ed.): *Bose–Einstein Condensates and Atom Laser* (Kluwer Academic, New York 2000);
A. Griffin, D.W. Snoke, S. Stringari (Eds.): *Bose–Einstein Condensation* (Cambridge Univ. Press, Cambridge 1995)

14.60   W. Ketterle, N.J. van Druten: Evaporative cooling of trapped atoms. Adv. At. Mol. Opt. Phys. **37**, 181 (1996)

14.61   A. Crubellier, O. Dulieu, F. Masnou-Seeuws, H. Knöckel, E. Tiemann: Simple determination of scattering length using observed bound levels at the ground state asymptote. Europhys. J. D **6**, 211 (1999)

14.62   H. Weickenmeier, U. Diemer, W. Demtröder, M. Broyer: Hyperfine-interaction between the singlet and triplet ground states and $Cs_2<$. Chem. Phys. Lett. **124**, 470 (1986)

14.63   K. Rubin, M.S. Lubell: 'A proposed study of photon statistics in fluorescence through high resolution measurements of the transverse deflection of an atomic beam'. In: *Laser Cooled and Trapped Atoms, NBS Special Publ. No. 653* (June 1983) p. 119

14.64   Y.Z. Wang, W.G. Huang, Y.D. Cheng, L. Liu: 'Test of photon statistics by atomic beam deflection'. In: *Laser Spectroscopy VII*, ed. by T.W. Hänsch, Y.R. Shen, Springer Ser. Opt. Sci., Vol. 49 (Springer, Berlin, Heidelberg, New York 1985) p. 238

14.65   V.M. Akulin, F.L. Kien, W.P. Schleich: Deflection of atoms by a quantum field. Phys. Rev. A **44**, R1462 (1991)

14.66   W. Ertmer, S. Penselin: Cooled atomic beams for frequency standards. Metrologia **22**, 195 (1986);
C. Salomon: 'Laser cooling of atoms and ion trapping for frequency standards'. In: *Metrology at the Frontiers of Physics and Technology*, ed. by L. Crovini, T.J. Quinn (North-Holland, Amsterdam 1992) p. 405

14.67   J.L. Hall, M. Zhu, P. Buch: Prospects for using laser prepared atomic fountains for optical frequency standards applications. J. Opt. Soc. Am. B **6**, 2194 (1989)

14.68   E.D. Commins: Electric dipole moments of leptons. Adv. At. Mol. Opt. Phys. **40**, 1 (1999)

14.69   F.M.H. Crompfoets, H.L. Bethlem, R.T. Jongma, G. Meyer: A prototype storage ring for neutral molecules. Nature **411**, 174 (2001)

14.70   B. Friedrich: Slowing of supersonically cooled atoms and molecules by time-varying nonresonant dipole forces. Phys. Rev. A **61**, 025403 (2000)

14.71   W. Paul, M. Raether: Das elektrische Massenfilter. Z. Physik **140**, 262 (1955);
W. Paul: Elektromagnetische Käfige für geladene und neutrale Teilchen. Phys. Blätter **46**, 227 (1990)

14.72   E. Fischer: Die dreidimensionale Stabilisierung von Ladungsträgern in einem Vierpolfeld. Z. Physik **156**, 1 (1959)

14.73   G.H. Dehmelt: Radiofrequency spectroscopy of stored ions. Adv. At. Mol. Phys. **3**, 53 (1967); Adv. At. Mol. Phys. **5**, 109 (1969)

14.74   R.E. Drullinger, D.J. Wineland: Laser cooling of ions bound to a penning trap. Phys. Rev. Lett. **40**, 1639 (1978)

14.75   See, for instance, E.T. Whittacker, S.N. Watson: *A Course of Modern Analysis* (Cambridge Univ. Press, Cambridge 1963);
J. Meixner, F.W. Schaefke: *Mathieusche Funktionen und Sphäroidfunktionen* (Springer, Berlin, Göttingen, Heidelberg 1954)

14.76   P.E. Toschek, W. Neuhauser: 'Spectroscopy on localized and cooled ions'. In: *Atomic Physics Vol. 7*, ed. by D. Kleppner, F.M. Pipkin (Plenum, New York 1981)

14.77   W. Neuhauser, M. Hohenstatt, P.E. Toschek, H.G. Dehmelt: Visual observation and optical cooling of electrodynamically contained ions. Appl. Phys. **17**, 123 (1978)

14.78   P.E. Toschek, W. Neuhauser: Einzelne Ionen für die Doppler-freie Spektroskopie. Phys. Blätter **36**, 1798 (1980)

14.79   T. Sauter, H. Gilhaus, W. Neuhauser, R. Blatt, P.E. Toschek: Kinetics of a single trapped ion. Europhys. Lett. **7**, 317 (1988)

14.80   R.E. Drullinger, D.J. Wineland: 'Laser cooling of ions bound to a Penning trap'. In: *Laser Spectroscopy IV*, ed. by H. Walther, K.W. Rother, Springer Ser. Opt. Sci., Vol. 21 (Springer, Berlin, Heidelberg 1979) p. 66; Phys. Rev. Lett. **40**, 1639 (1978)

14.81   D.J. Wineland, W.M. Itano: Laser cooling of atoms. Phys. Rev. A **20**, 1521 (1979)

14.82   W. Neuhauser, M. Hohenstatt, P.E. Toschek, H. Dehmelt: Optical sideband cooling of visible atom cloud confined in a parabolic well. Phys. Rev. Lett. **41**, 233 (1978)

14.83   H.G. Dehmelt: Proposed $10^{14} \Delta\nu < \nu$ laser fluorescence spectroscopy on a T1$^+$ mono-ion oscillator. Bull. Am. Phys. **20**, 60 (1975)

14.84   P.E. Toschek: Absorption by the numbers: recent experiments with single trapped and cooled ions. Phys. Scripta T **23**, 170 (1988)

14.85   T. Sauter, R. Blatt, W. Neuhauser, P.E. Toschek: Quantum jumps in a single ion. Phys. Scripta **22**, 128 (1988); Opt. Commun. **60**, 287 (1986)

14.86   W.M. Itano, J.C. Bergquist, R.G. Hulet, D.J. Wineland: 'The observation of quantum jumps in Hg$^+$'. In: *Laser Spectroscopy VIII*, ed. by S. Svanberg, W. Persson, Springer Ser. Opt. Sci., Vol. 55 (Springer, Berlin, Heidelberg, New York 1987) p. 117

14.87   F. Diedrich, H. Walther: Nonclassical radiation of a single stored ion. Phys. Rev. Lett. **58**, 203 (1987)

14.88   R. Blümel, J.M. Chen, E. Peik, W. Quint, W. Schleich, Y.R. Chen, H. Walther: Phase transitions of stored laser-cooled ions. Nature **334**, 309 (1988)

14.89   F. Diedrich, E. Peik, J.M. Chen, W. Quint, H. Walther: Ionenkristalle und Phasenübergänge in einer Ionenfalle. Phys. Blätter **44**, 12 (1988)

14.90   F. Diedrich, E. Peik, J.M. Chen, W. Quint, H. Walther: Observation of a phase transition of stored laser-cooled ions. Phys. Rev. Lett. **59**, 2931 (1987)

14.91   R. Blümel, C. Kappler, W. Quint, H. Walther: Chaos and order of laser-cooled ions in a Paul trap. Phys. Rev. A **40**, 808 (1989)

14.92   J. Javamainen: Laser cooling of trapped ion-clusters. J. Opt. Soc. Am. B **5**, 73 (1988)

14.93   D.J. Wineland, J.C. Bergquist, W.M. Itano, J.J. Bollinger, C.H. Manney: Atomic-ion Coulomb clusters in an ion trap. Phys. Rev. Lett. **59**, 2935 (1987);
        W. Quint: Chaos und Ordnung von lasergekühlten Ionen in einer Paulfalle. Dissertation, MPQ-Berichte 150, MPQ für Quantenoptik, Garching (1990)

14.94   Th.V. Kühl: Storage ring laser spectroscopy. Adv. At. Mol. Opt. Phys. **40**, 113 (1999)

14.95   H. Poth: Applications of electron cooling in atomic, nuclear and high energy physics. Nature **345**, 399 (1990)

14.96   (a) J.P. Schiffer: Layered structure in condensed cold one-component plasma confined in external fields. Phys. Rev. Lett. **61**, 1843 (1988)
        (b) I. Waki, S. Kassner, G. Birkl, H. Walther: Observation of ordered structures of laser-cooled ions in a quadrupole storage ring. Phys. Rev. Lett. **68**, 2007 (1992)

14.97   J.S. Hangst, M. Kristensen, J.S. Nielsen, O. Poulsen, J.P. Schiffer, P. Shi: Laser cooling of a stored ion beam to 1 mK. Phys. Rev. Lett. **67**, 1238 (1991)

14.98   U. Schramm, et al.: Observation of laser-induced recombination in merged electron and proton beams. Phys. Rev. Lett. **67**, 22 (1991)

14.99   T.C. English, J.C. Zorn: 'Molecular beam spectroscopy'. In: *Methods of Experimental Physics, Vol. 3*, ed. by D. Williams (Academic, New York 1974)

14.100   I.I. Rabi: Zur Methode der Ablenkung von Molekularstrahlen. Z. Physik **54**, 190 (1929)

14.101   N.F. Ramsey: *Molecular Beams*, 2nd edn. (Clarendon, Oxford 1989)

14.102   J.C. Bergquist, S.A. Lee, J.L. Hall: 'Ramsey fringes in saturation spectroscopy'. In: *Laser Spectroscopy III*, ed. by J.L. Hall, J.L. Carlsten (Springer, Berlin, Heidelberg 1977)

14.103   Y.V. Baklanov, B.Y. Dubetsky, V.P. Chebotayev: Nonlinear Ramsey resonance in the optical region. Appl. Phys. **9**, 171 (1976)

14.104   V.P. Chebotayev: The method of separated optical fields for two level atoms. Appl. Phys. **15**, 219 (1978)

14.105   C. Bordé: Sur les franges de Ramsey en spectroscopie sans élargissement Doppler. C.R. Acad. Sc. (Paris) Serie B **282**, 101 (1977)

14.106   S.A. Lee, J. Helmcke, J.L. Hall, P. Stoicheff: Doppler-free two-photon transitions to Rydberg levels. Opt. Lett. **3**, 141 (1978)

14.107   S.A. Lee, J. Helmcke, J.L. Hall: 'High-resolution two-photon spectroscopy of Rb Rydberg levels'. In: *Laser Spectroscopy IV*, ed. by H. Walther, K.W. Rothe, Springer Ser. Opt. Sci., Vol. 21 (Springer, Berlin, Heidelberg 1979) p. 130

14.108   Y.V. Baklanov, V.P. Chebotayev, B.Y. Dubetsky: The resonance of two-photon absorption in separated optical fields. Appl. Phys. **11**, 201 (1976)

14.109   S.N. Bagayev, V.P. Chebotayev, A.S. Dychkov: Continuous coherent radiation in methane at $\lambda = 3.39 \,\mu m$ in spatially separated fields. Appl. Phys. **15**, 209 (1978)

14.110   C.J. Bordé: 'Density matrix equations and diagrams for high resolution nonlinear laser spectroscopy: application to Ramsey fringes in the optical domain'. In: *Advances in Laser Spectroscopy*, ed. by F.T. Arrecchi, F. Strumia, H. Walther (Plenum, New York 1983) p. 1

14.111   J.C. Bergquist, S.A. Lee, J.L. Hall: Saturated absorption with spatially separated laser fields. Phys. Rev. Lett. **38**, 159 (1977)

14.112   J. Helmcke, D. Zevgolis, B.U. Yen: Observation of high contrast ultra narrow optical Ramsey fringes in saturated absorption utilizing four interaction zones of travelling waves. Appl. Phys. B **28**, 83 (1982)

14.113   C.J. Bordé, C. Salomon, S.A. Avrillier, A. Van Lerberghe, C. Breant, D. Bassi, G. Scoles: Optical Ramsey fringes with travelling waves. Phys. Rev. A **30**, 1836 (1984)

14.114   J.C. Bergquist, R.L. Barger, P.J. Glaze: 'High resolution spectroscopy of calcium atoms'. In: *Laser Spectroscopy IV*, ed. by H. Walther, K.W. Rothe, Springer Ser. Opt. Sci., Vol. 21 (Springer, Berlin, Heidelberg 1979) p. 120

14.115   J. Helmcke, J. Ishikawa, F. Riehle: 'High contrast high resolution single component Ramsey fringes in Ca'. In: *Frequency Standards and Metrology*, ed. by A. De Marchi (Springer, Berlin, Heidelberg, New York 1989) p. 270

14.116   F. Riehle, J. Ishikawa, J. Helmcke: Suppression of recoil component in nonlinear Doppler-free spectroscopy. Phys. Rev. Lett. **61**, 2092 (1988)

14.117   See, for instance, J. Mlynek, V. Balykin, P. Meystere (Guest Eds.): Atom interferometry. Appl. Phys. B **54**, 319–368 (1992);
C.S. Adams, M. Siegel, J. Mlynek: Atom optics. Phys. Rpt. **240**, 144 (1994)

14.118   P. Bermann (Ed.): *Atom Interferometry* (Academic, San Diego 1997)

14.119   O. Carnal, J. Mlynek: Young's double slit experiment with atoms: a simple atom interferometer. Phys. Rev. Lett. **66**, 2689 (1991)

14.120   D.W. Keith, C.R. Ekstrom, Q.A. Turchette, D.E. Pritchard: An interferometer for atoms. Phys. Rev. Lett. **66**, 2693 (1991)

14.121   C.J. Bordé: Atomic interferometry with internal state labelling. Phys. Lett. A **140**, 10 (1989)

14.122   F. Riehle, A. Witte, T. Kisters, J. Helmcke: Interferometry with Ca atoms. Appl. Phys. B **54**, 333 (1992)

14.123   M. Kasevich, S. Chu: Measurement of the gravitational acceleration of an atom with a light-pulse atom interferometer. Appl. Phys. B **54**, 321 (1992)

14.124   M.R. Andrews, C.G. Townsend, H.J. Miesner, D.S. Durfee, D.M. Kurn, W. Ketterle: Observation of interference between two Bose condensates. Science **275**, 637 (1997)

14.125   I. Block, T.W. Hänsch, T. Esslinger: Atom laser with a cw output coupler. Phys. Rev. Lett. **82**, 3008 (1999)

14.126   S. Martellucci, A.N. Chester, A. Aspect, M. Inguscio (Eds.): *Bose–Einstein Condensates and Atom Lasers* (Kluwer/Plenum, New York 2000)

14.127   F. Diedrich, J. Krause, G. Rempe. M.O. Scully, H. Walther: Laser experiments on single atoms and the test of basic physics. Physica B **151**, 247 (1988); IEEE J. QE-**24**, 1314 (1988)

14.128   S. Haroche, J.M. Raimond: Radiative properties of Rydberg states in resonant cavities. Adv. At. Mol. Phys. **20**, 347 (1985)

14.129   G. Rempe, H. Walther: 'The one-atom maser and cavity quantum electrodynamics'. In: *Methods of Laser Spectroscopy*, ed. by Y. Prior, A. Ben-Reuven, M. Rosenbluth (Plenum, New York 1986)

14.130   H. Walther: Single-atom oscillators. Europhys. News **19**, 105 (1988)

14.131   G. Rempe, M.O. Scully, H. Walther: 'The one-atom maser and the generation of nonclassical light'. In: *Proc. ICAP 12*, Ann Arbor (1990)

14.132   P. Meystre, G. Rempe, H. Walther: Very low temperature behaviour of a micromaser. Opt. Lett. **13**, 1078 (1988)

14.133   G. Rempe, H. Walther: Sub-Poissonian atomic statistics in a micromaser. Phys. Rev. A **42**, 1650 (1990)

14.134   B.T. Varcoe, S. Brattke, M. Weidinger, H. Walther: Preparing pure photon number states of the radiatium field. Nature **403**, 743 (2000)

14.135   M. Marrocco, M. Weidinger, R.T. Sang, H. Walther: Quantum electrodynamic shifts of Rydberg energy levels between two parallel plates. Phys. Rev. Lett. **81**, 5784 (1998)

14.136   H. Metcalf, W. Phillips: Time resolved subnatural width spectroscopy. Opt. Lett. **5**, 540 (1980)

14.137   J.N. Dodd, G.W. Series: 'Time-resolved fluorescence spectroscopy'. In: *Progr. Atomic Spectroscopy A*, ed. by W. Hanle, H. Kleinpoppen (Plenum, New York 1978)

14.138   S. Schenk, R.C. Hilburn, H. Metcalf: Time resolved fluorescence from Ba and Ca, excited by a pulsed tunable dye laser. Phys. Rev. Lett. **31**, 189 (1973)

14.139   H. Figger, H. Walther: Optical resolution beyond the natural linewidth: a level crossing experiment on the $3\,^2P_{3/2}$ level of sodium using a tunable dye laser. Z. Physik **267**, 1 (1974)

14.140   F. Shimizu, K. Umezu, H. Takuma: Observation of subnatural linewidth in Na D_2-lines. Phys. Rev. Lett. **47**, 825 (1981)

14.141   G. Bertuccelli, N. Beverini, M. Galli, M. Inguscio, F. Strumia: Subnatural coherence effects in saturation spectroscopy using a single travelling wave. Opt. Lett. **10**, 270 (1985)

14.142   P. Meystre, M.O. Scully, H. Walther: Transient line narrowing: a laser spectroscopic technique yielding resolution beyond the natural linewidth. Opt. Commun. **33**, 153 (1980)

14.143 A. Guzman, P. Meystre, M.O. Scully: 'Subnatural spectroscopy'. In: *Adv. Laser Spectroscopy*, ed. by F.T. Arecchi, F. Strumia, H. Walther (Plenum, New York 1983) p. 465

14.144 V.S. Letokhov, V.P. Chebotayev: *Nonlinear Laser Spectroscopy*, Springer Ser. Opt. Sci., Vol. 4 (Springer, Berlin, Heidelberg 1977)

14.145 R.P. Hackel, S. Ezekiel: Observation of subnatural linewidths by two-step resonant scattering in I_2-vapor. Phys. Rev. Lett. **42**, 1736 (1979); and in [Ref. 1.11b, p. 88]

14.146 H. Weickenmeier, U. Diemer, W. Demtröder, M. Broyer: Hyperfine interaction between the singlet and triplet ground states of Cs_2. Chem. Phys. Lett. **124**, 470 (1986)

14.147 E.R. Cohen, B.N. Taylor: The 1986 CODATA recommended values of the fundamental physical constants. J. Phys. Chem. Ref. Data **17**, 1795 (1988)

14.148 F. Bayer-Helms: Neudefinition der Basiseinheit Meter im Jahr 1983. Phys. Blätter **39**, 307 (1983);
Documents concerning the new definition of the metre. Metrologia **19**, 163 (1984)

14.149 K.M. Baird: Frequency measurements of optical radiation. Phys. Today **36**, 1 (January 1983)

14.150 K.M. Evenson, D.A. Jennings, F.R. Peterson, J. S Wells: 'Laser frequency measurements: A. Review, limitations, extension to 197 THz (1.5 μm)'. In: *Laser Spectroscopy III*, ed. by J.L. Hall, J.L. Carlsten, Springer Ser. Opt. Sci., Vol. 7 (Springer, Berlin, Heidelberg 1977);
D.A. Jennings, F.R. Peterson, K.M. Evenson: 'Direct frequency measurement of the 260 THz (1.15 μm) ^{20}Ne laser: And beyond'. In: *Laser Spectroscopy IV*, ed. by H. Walther, K.W. Rothe, Springer Ser. Opt. Sci., Vol. 21 (Springer, Berlin, Heidelberg 1979) p. 39

14.151 K.M. Evenson, M. Inguscio, D.A. Jennings: Point contact diode at laser frequencies. J. Appl. Phys. **57**, 956 (1985)

14.152 L.R. Zink, M. Prevedelli, K.M. Evenson, M. Inguscio: 'High resolution far infrared spectroscopy'. In: *Applied Laser Spectroscopy*, ed. by M. Inguscio, W. Demtröder (Plenum, New York 1991) p. 141

14.153 H.V. Daniel, B. Maurer, M. Steiner: A broadband Schottky point contact mixer for visible laser light and microwave harmonics. J. Appl. Phys. B **30**, 189 (1983)

14.154 H.P. Roeser, R.V. Titz, G.W. Schwaab, M.F. Kimmit: Current-frequency characteristics of submicron GaAs Schottky barrier diodes with femtofarad capacitances. J. Appl. Phys. **72**, 3194 (1992)

14.155 (a) B.G. Whitford: 'Phase-locked frequency chains to 130 THz at NRC'. In: *Frequency Standards and Metrology*, ed. by A. De Marchi (Springer, Berlin, Heidelberg, New York 1989);
(b) T.W. Hänsch: 'High resolution spectroscopy of hydrogen'. In: *The Hydrogen Atom*, ed. by G.F. Bussani, M. Inguscio, T.W. Hänsch (Springer, Berlin, Heidelberg, New York 1989);
(c) S.G. Karshenboim, F.S. Pavone, G.F. Bussani, M. Inguscio, T.W. Hänsch (Eds.): *The Hydrogen Atom* (Springer, Berlin, Heidelberg, New York 2001)

14.156 J. Reichert, M. Niering, R. Holzwarth, M. Weitz, T. Udem, T.W. Hänsch: Phase coherent vacuum ultraviolet to radiofrequency comparison with a mode-locked laser. Phys. Rev. Lett. **84**, 3232 (2000)

14.157 S.A. Diddams, T.W. Hänsch, et al.: Direct link between microwave and optical frequencies with a 300 THz femtosecond pulse. Phys. Rev. Lett. **84**, 5102 (2000)

14.158 R. Loudon: *The Quantum Theory of Light* (Clarendon, Oxford 1973)

14.159 H. Gerhardt, H. Welling, A. Güttner: Measurements of laser linewidth due to quantum phase and quantum amplitude noise above and below threshold. Z. Physik **253**, 113 (1972);

M. Zhu, J.L. Hall: Stabilization of optical phase/frequency of a laser system. J. Opt. Soc. Am. B **10**, 802 (1993)

14.160   H.A. Bachor, P.J. Manson: Practical implications of quantum noise. J. Mod. Opt. **37**, 1727 (1990);
H.A. Bachor, P.T. Fisk: Quantum noise - a limit in photodetection. Appl. Phys. B **49**, 291 (1989)

14.161   H.A. Bachor: *A Guide to Experiments in Quantum Optics* (Wiley VCH, Weinheim 1998)

14.162   R.J. Glauber: 'Optical coherence and photon statistics'. In: *Quantum Optics and Electronics*, ed. by C. DeWitt, A. Blandia, C. Cohen-Tannoudji (Gordon & Breach, New York 1965) p. 65;
J.D. Cresser: Theory of the spectrum of the quantized light field. Phys. Rpt. **94**, 48 (1983);
H. Paul: Squeezed states – nichtklassische Zustände des Strahlungsfeldes. Laser und Optoelektronik **19**, 45 (März 1987)

14.163   R.E. Slusher, L.W. Holberg, B. Yorke, J.C. Mertz, J.F. Valley: Observation of squeezed states generated by four wave mixing in an optical cavity. Phys. Rev. Lett. **55**, 2409 (1985)

14.164   M. Xiao, L.A. Wi, H.J. Kimble: Precision measurements beyond the shot noise limit. Phys. Rev. Lett. **59**, 278 (1987)

14.165   H.J. Kimble, D.F. Walls (Guest ceds.): Feature issue on squeezed states of the electromagnetic field. J. Opt. Soc. Am. B **4**, 1449 (1987);
P. Kurz, R. Paschotta, K. Fiedler, J. Mlynek: Bright squeezed light by second harmonic generation and monolytic resonator. Europhys. Lett. **24**, 449 (1993)

14.166   T.M. Niebaum, A. Rüdiger, R. Schilling, L. Schnupp, W. Winkler, K. Danzmann: Pulsar search using data compression with the Garching gravitational wave detector. Phys. Rev. D **47**, 3106 (1993)

14.167   P.G. Blair (Ed.): *The Detection of Gravitational Waves* (Cambridge Univ. Press, Cambridge 1991)

14.168   P.S. Saulson: *Fundamentals of Interferometric Gravitational Wave Detectors* (World Scientific, Singapore 1994)

14.169   K. Zaheen, M.S. Zubairy: Squeezed states of the radiation field. Adv. At. Mol. Phys. **28**, 143 (1991)

14.170   H.J. Kimble: Squeezed states of light. Adv. Chem. Phys. **38**, 859 (1989)

14.171   P. Tombesi, E.R. Pikes (Eds.): *Squeezed and Nonclassical Light* (Plenum, New York 1989)

14.172   E. Giacobino, C. Fabry (Guest Eds.): Quantum noise reduction in optical systems. Appl. Phys. B **55**, 187–297 (1992)

14.173   D.F. Walls, G.J. Milburn: *Quantum Optics*, study edn. (Springer, Berlin, Heidelberg, New York 1995)

14.174   H.A. Haus: *Electromagnetic Noise and Quantum Optical Measurements* (Springer, Berlin, Heidelberg, New York 2000)

14.175   H.J. Carmichael, R.J. Glauber, M.O. Scully (Eds.): *Directions in Quantum Optics* (Springer, Berlin, Heidelberg, New York 2001)

Chapter 15

15.1   A. Mooradian, T. Jaeger, P. Stokseth (Eds.): *Tunable Lasers and Applications*, Springer Ser. Opt. Sci., Vol. 3 (Springer, Berlin, Heidelberg 1976)

15.2   C.T. Lin, A. Mooradian (Eds.): *Lasers and Applications*, Springer Ser. Opt. Sci., Vol. 26 (Springer, Berlin, Heidelberg 1981)

15.3   J.F. Ready, R.K. Erf (Eds.): *Lasers and Applications, Vols. 1–5* (Academic, New York 1974–1984)

15.4   S. Svanberg: *Atomic and Molecular Spectroscopy*, 2nd edn., Springer Ser. Atoms
         Plasmas, Vol. 6 (Springer, Berlin, Heidelberg, New York 1991)

15.5   C.B. Moore: *Chemical and Biochemical Applications of Lasers, Vols. 1–5* (Aca-
         demic, New York 1974-1984)

15.6   D.K. Evans: *Laser Applications in Physical Chemistry* (Dekker, New York
         1989);
         D.L. Andrews: *Lasers in Chemistry* (Springer, Berlin, Heidelberg, New York
         1986);
         A.H. Zewail (Ed.): *Advances in Laser Chemistry*, Springer Ser. Chem. Phys.,
         Vol. 3 (Springer, Berlin, Heidelberg, New York 1978)

15.7   G.R. van Hecke, K.K. Karukstis: *A Guide to Lasers in Chemistry* (Jones &
         Bartlett Publ., Boston 1997)

15.8   R.T. Rizzo, A.B. Myers: *Laser Techniques in Chemistry* (Wiley, New York 1995)

15.9   G. Schmidtke, W. Kohn, U. Klocke, M. Knothe, W.J. Riedel, H. Wolf: Diode
         laser spectrometer for monitoring up to five atmospheric trace gases in unat-
         tended operation. Appl. Opt. **28**, 3665 (1989)

15.10   G.S. Hurst, M.P. Payne, S.P. Kramer, C.H. Cheng: Counting the atoms. Phys.
         Today **33**, 24 (Sept. 1980)

15.11   V.S. Letokhov: *Laser Photoionization Spectroscopy* (Academic, Orlando, FL
         1987)

15.12   P. Peuser, G. Herrmann, H. Rimke, P. Sattelberger, N. Trautmann: Trace detec-
         tion of plutonium by three-step photoionization with a laser system pumped by
         a copper vapor laser. Appl. Phys. B **38**, 249 (1985)

15.13   T. Whitaker: Isotopically selective laser measurements. Lasers Appl. **5**, 67 (Aug.
         1986)

15.14   H. Kano, H.T.M. van der Voort, M. Schrader, G.M.P. van Kampen, S.W. Hell:
         Avalanche photodiode detection with object scanning and image restoration
         provides 2−4 fold resolution increase in two-photon fluorescence microscopy.
         Bioimaging **4**, 187 (1996)

15.15   J. Widengren, Ü. Mets, R. Rigler: Fluorescence correlation spectroscopy of
         triplet states in solution. J. Chem. Phys. **99**, 13368 (1995)

15.16   W.E. Moerner, R.M. Dickson, D.J. Norris: Single-molecule spectroscopy and
         quantum optics in solids. Adv. At. Mol. Opt. Phys. **38**, pp. 193 ff. (1997)

15.17   G. Jung, J. Wiehler, B. Steipe, C. Bräuchle, A. Zumbusch: Single-molecule
         microscopy of the green fluorescent protein using two-color excitation. Chem.
         Phys. Chem. **2**, 392 (2001)

15.18   P. Schwille, U. Haupts, S. Maiti, W.W. Web: Molecular dynamics in living cells
         observed by fluorescence correlation spectroscopy. Biophys. J. **77**, 2251 (1999)

15.19   E.H. Piepmeier (Ed.): *Analytical Applicability of Lasers* (Wiley, New York 1986)

15.20   J. Sneddon, T.L. Thiem, Y. Lee (Eds.): *Lasers in Analytical Atomic Spectroscopy*
         (Wiley VCH, Weinheim 1997)

15.21   K. Niemax: *Analytical Aspects of Atomic Laser Spectrochemistry* (Harwood
         Acad. Publ., Philadelphia 1989)

15.22   A. Baronarski, J.W. Butler, J.W. Hudgens, M.C. Lin, J.R. McDonald, M.E. Um-
         stead: 'Chemical Applications of Lasers'. In: A.H. Zewail (Ed.): *Advances in
         Laser Chemistry*, Springer Ser. Chem. Phys., Vol. 3 (Springer, Berlin, Heidelberg
         New York 1986) p. 62

15.23   B. Raffel, J. Wolfrum: Spatial and time resolved observation of CO_2-laser in-
         duced explosions of O_2-O_3-mixtures in a cylindrical cell. Z. Phys. Chem. (NF)
         161, 43 (1989)

15.24   R.L. Woodin, A. Kaldor: Enhancement of chemical reactions by infrared lasers.
         Adv. Chem. Phys. **47**, 3 (1981)
         M. Quack: Infrared laser chemistry and the dynamics of molecular multiphoton
         excitation. Infrared Phys. **29**, 441 (1989)

15.25   C.D. Cantrell (Ed.): *Multiple-Photon Excitation and Dissociation of Polyatomic Molecules*, Springer Topics. Curr. Phys., Vol. 35 (Springer, Berlin, Heidelberg, New York 1986);
V.N. Bagratashvili, V.S. Letokhov, A.D. Makarov, E.A. Ryabov: *Multiple Photon Infrared Laser Photophysics and Photochemistry* (Harwood, Chur 1985)

15.26   J.H. Clark, K.M. Leary, T.R. Loree, L.B. Harding: 'Laser synthesis chemistry and laser photogeneration of catalysis'. In: D.K. Evans: *Laser Applications in Physical Chemistry* (Dekker, New York 1989) p. 74

15.27   Gong Mengxiong, W. Fuss, K.L. Kompa: CO_2 laser induced chain reaction of $C_2F_4 + CF_3I$. J. Phys. Chem. **94**, 6332 (1990)

15.28   M. Schneider, J. Wolfrum: Mechanisms of by-product formation in the dehydrochlorination of dichlorethane. Ber. Bunsenges. Phys. Chem. **90**, 1058 (1986)

15.29   Zhang Linyang, W. Fuss, K.L. Kompa: KrF laser induced telomerization of bromides with olefins. Ber. Bunsenges. Phys. Chem. **94**, 867 (1990)

15.30   K.L. Kompa: Laser photochemistry at surfaces. Angew. Chem. **27**, 1314 (1988)

15.31   M.S. Djidjoev, R.V. Khokhlov, A.V. Kieselev, V.I. Lygin, V.A. Namiot, A.I. Osipov, V.I. Panchenko, YB.I. Provottorov: 'Laser chemistry at surfaces'. In: D.K. Evans: *Laser Applications in Physical Chemistry* (Dekker, New York 1989) p. 7

15.32   de Vivie-Riedle, H. Rabitz, K.L. Kompa (Eds.): Laser Control of Quantum Dynamics. Special Issue of Chemical Physics **267** (2001)

15.33   P. Brumer, M. Shapiro: Control of unimolecular reactions using coherent light. Chem. Phys. Lett. **126**, 541 (1986);
M. Shapiro, P. Brumer: Coherent control of atomic, molecular and electronic processes. Adv. At. Mol. Opt. Phys. **42**, 287 (2000)

15.34   D.J. Tannor, R. Kosloff, S.A. Rice: Coherent pulse sequence induced control of selectivity reactions. J. Chem. Phys. **85**, 5805 (1986)

15.35   M. Shapiro: Association, dissociation and the acceleration and suppression of reactions by laser pulses. Adv. Chem. Phys. **114**, 123–192 (1999);
A. Assion, T. Baumert, M. Bergt, T. Brixner, B. Kiefer, V. Seyfried, M. Strehle, G. Gerber: Control of chemical reactions by feedback-optimized phase-shaped femtosecond laser pulses. Science **282**, 919 (1998)

15.36   D. Zeidler, S. Frey, K.L. Kompa, M. Motzkus: Evolutionary algorithms and their applications to optimal control studies. Phys. Rev. A **64**, O23420 (2001)

15.37   M. Bergt, T. Brixner, B. Kiefer, M. Strehle, G. Gerber: Controlling the femtochemistry of $Fe(CO)_5$. J. Phys. Chem. **103**, 10381 (1999)

15.38   T. Brixner, B. Kiefer, G. Gerber: Problem complexity in femtosecond quantum control. Chem. Phys. **267**, 241 (2001)

15.39   R.J. Levis, G.M. Menkir, H. Rabitz: Selective bond dissociation and rearrangement with optimally tailored, strong field laser pulses. Science **292**, 709 (2001)

15.40   A. Rice, M. Zhao: *Optical Control of Molecular Dynamics* (Wiley, New York 2000)

15.41   P. Gaspard, I. Burghardt (Eds.): Chemical reactions and their control on the femtosecond time scale. Adv. Chem. Phys. **101**, (1997)

15.42   R.N. Zare, R.B. Bernstein: State to state reaction dynamics. Phys. Today **3**, 43 (Nov. 1980)

15.43   A.H. Zewail: Laser femtochemistry. Science **242**, 1645 (1988);
A.H. Zewail: The birth of molecules. Sci. Am. **262**, 76 (Dec. 1990);
J. Manz, L. Wöste (Eds.): *Femtosecond Chemistry, Vols. I and II* (VCH, Weinheim 1995)

15.44   A.H. Zewail: *Femtochemistry* (World Scientific, Singapore 1994)

15.45   B. Bescós, B. Lang, J: Weiner, V. Weiss, E. Wiedemann, G. Gerber: Real-time observation of ultrafast ionization and fragmentation of mercury clusters. Eur. Phys. J. D **9**, 399 (1999)

15.46   H. Bürsing, P. Vöhringer: Transition state probing and fragment rotational dynamics of HgI$_2$. Phys. Chem. Chem. Phys. **2**, 73 (2000)

15.47   St. Hess, H. Bürsing, P. Vöhringer: Dynamics of fragment recoil in the femtosecond photodissociation of triiodide ions. J. Chem. Phys. **111**, 5461 (1999)

15.48   S.A. Trushin, W. Fuss, K.L. Kompa, W.E. Schmid: Femtosecond dynamics of Fe(CO)$_5$ photodissociation at 267 nm studied by transient ionization. J. Phys. Chem. A **104**, 1997 (2000)

15.49   A.H. Zewail: Femtosecond transition-state dynamics. Faraday Discuss. Chem. Soc. **91**, 1 (1991)

15.50   S. Villani: *Isotope Separation* (Am. Nucl. Soc., Hinsdale, Ill. 1976);
S. Villani: *Uranium Enrichment*, Topics Appl. Phys., Vol. 35 (Springer, Berlin, Heidelberg 1979);
W. Ehrfeld: *Elements of Flow and Diffusion Processes in Separation Nozzles*, Springer Tracts Mod. Phys., Vol. 97 (Springer, Berlin, Heidelberg 1983)

15.51   C.D. Cantrell, S.M. Freund, J.L. Lyman: 'Laser induced chemical reactions and isotope separation'. In: *Laser Handbook, Vol. 3*, ed. by M.L. Stitch (North-Holland, Amsterdam 1979);
R.N. Zare: Laser separation of isotopes. Sci. Am. **236**, 86 (Feb. 1977);
F.S. Becker, K.L. Kompa: Laser isotope separation. Europhys. News **12**, 2 (July 1981);
R.D. Alpine, D.K. Evans: Laser isotope separation by the selective multiphoton decomposition process. Adv. Chem. Phys. **60**, 31 (1985)

15.52   J.P. Aldridge, J.H. Birley, C.D. Cantrell, D.C. Cartwright: 'Experimental and studies of laser isotope separation'. In: *Laser Photochemistry, Tunable Lasers*, ed. by S.E. Jacobs, S.M. Sargent, M.O. Scully, C.T. Walker (Addison-Wesley, Reading, MA 1976)

15.53   J.I. Davies, J.Z. Holtz, M.L. Spaeth: Status and prospects for lasers in isotope separation. Laser Focus **18**, 49 (Sept. 1982)

15.54   M. Stuke: Isotopentrennung mit Laserlicht. Spektrum Wissenschaft. **4**, 76 (1982)

15.55   (a) F.S. Becker, K.L. Kompa: The practical and physical aspects of uranium isotope separation with lasers. Nuc. Technol. **58**, 329 (1982);
(b) C.P. Robinson, R.J. Jensen: 'Laser methods of uranium isotope separation'. In: S. Villani: *Uranium Enrichment*, Topics Appl. Phys., Vol. 35 (Springer, Berlin, Heidelberg 1979) p. 269

15.56   L. Mannik, S.K. Brown: Laser enrichment of carbon 14. Appl. Phys. B **37**, 79 (1985)

15.57   A. von Allmen: *Laser-Beam Interaction with Materials*, 2nd edn., Springer Ser. Mater. Sci., Vol. 2 (Springer, Berlin, Heidelberg, New York 1995);
P.N. Bajaj, K.G. Manohar, B.M. Suri, K. Dasgupta, R. Talukdar, P.K. Chakraborti, P.R.K. Rao: Two colour multiphoton ionization spectroscopy of uranium from a metastable state. Appl. Phys. **B47**, 55 (1988)

15.58   A. Outhouse, P. Lawrence, M. Gauthier, P.A. Hacker: Laboratory scale-up of two stage laser chemistry separation of ^{13}C from CF$_2$HCL. Appl. Phys. B **36**, 63 (1985);
I. Deac, V. Cosma, D. Silipas, L. Muresan, V. Tosa: Parametric study of the IRMPD of CF$_2$HCl molecules with the 9P22 CO$_2$ laser time. Appl. Phys. B **51**, 211 (1990)

15.59   C. D'Ambrosio, W. Fuss, K.L. Kompa, W.E. Schmid, S. Trushin: ^{13}C separation by a continuous discharge CO$_2$ laser Q-switched at 10 kHz. Infrared Phys. **29**, 479 (1989); Appl. Phys. B **47**, 19 (1988)

15.60   K. Kleinermanns, J. Wolfrum: Laser in der Chemie – Wo stehen wir heute? Angew. Chemie **99**, 38 (1987);
J. Wolfrum: Laser spectroscopy for studying chemical processes. Appl. Phys. B **46**, 221 (1988)

15.61   D.J. Neshitt, St.R. Leone: Laser-initiated chemical chain reactions. J. Chem. Phys. **72**, 1722 (1980)

15.62   D. Bäuerle: *Chemical Processing with Lasers*, Springer Ser. Mater. Sci., Vol. 1 (Springer, Berlin, Heidelberg, New York 1986)

15.63   V.S. Letokhov (Ed.): *Laser Analytical Spectrochemistry* (Hilger, Bristol 1985)

15.64   K. Peters: Picosecond organic photochemistry. Annu. Rev. Phys. Chem. **38**, 253 (1987)

15.65   M. Gruehele, A.H. Zewail: Ultrafast reaction dynamics. Phys. Today, **13**, 24 (May 1990)

15.66   J. Wolfrum: Laser spectroscopy for studying chemical processes. Appl. Phys. B **46**, 221 (1988);
        J. Wolfrum: Laser stimulation and observation of simple gas phase radical reactions. Laser Chem. **9**, 171 (1988)

15.67   E. Hirota: From high resolution spectroscopy to chemical reactions. Ann. Rev. Phys. Chem. **42**, 1 (1991)

15.68   J.I. Steinfeld: *Laser-Induced Chemical Processes* (Plenum, New York 1981)

15.69   L.J. Kovalenko, S.L. Leone: Innovative laser techniques in chemical kinetics. J. Chem. Educ. **65**, 681 (1988)

15.70   A. Ben-Shaul, Y. Haas, K.L. Kompa, R.D. Levine: *Lasers and Chemical Change*, Springer Ser. Chem. Phys., Vol. 10 (Springer, Berlin, Heidelberg 1981)

15.71   K.L. Kompa, S.D. Smith (Eds.): *Laser-Induced Processes in Molecules*, Springer Ser. Chem. Phys., Vol. 6 (Springer, Berlin, Heidelberg 1979)

15.72   K.L. Kompa, J. Warner: *Laser Applications in Chemistry* (Plenum, New York 1984)

15.73   A.H. Zewail: Femtosecond transition state dynamics. Faraday Discuss. Chem. Soc. **91**, 1 (1991)

15.74   L.R. Khundar, A.H. Zewail: Ultrafast reaction dynamics in real times. Annu. Rev. Phys. Chem. **41**, 15 (1990)

15.75   J. Steinfeld: *Air Pollution* (Wiley, New York 1986)

15.76   C.F. Bohren, D.R. Huffman: *Absorption and Scattering of Light by Small Particles* (Wiley, New York 1983)

15.77   R. Zellner, J. Hägele: A double-beam UV-laser differential absorption method for monitoring tropospheric trace gases. Opt. Laser Technol. **17**, 79 (April 1985)

15.78   E.D. Hinkley (Ed.): *Laser Monitoring of the Atmosphere*, Topics Appl. Phys., Vol. 14 (Springer, Berlin, Heidelberg 1976);
        B. Stumpf, D. Göring, R. Haseloff, K. Herrmann: Detection of carbon monoxide, carbon dioxide with pulsed tunable $Pb_{1-x}Se_x$ diode lasers. Collect. Czech. Chem. Commun. **54**, 284 (1989)

15.79   A. Tönnissen, J. Wanner, K.W. Rothe, H. Walther: Application of a CW chemical laser for remote pollution monitoring and process control. Appl. Phys. **18**, 297 (1979)

15.80   W. Meinburg, H. Neckel, J. Wolfrum: Lasermeßtechnik und mathematische Simulation von Sekundärmaßnahmen zur NO_x-Minderung in Kraftwerken. Appl. Phys. B **51**, 94 (1990);
        A. Arnold, H. Becker, W. Ketterle, J. Wolfrum: Combustion diagnostics by two dimensional laser-induced fluorescence using tunable excimer lasers. SPIE Proc. **1602**, 70 (1991)

15.81   W. Meienburg, H. Neckel, J. Wolfrum: In situ measurement of ammonia with a $^{13}CO_2$-waveguide laser system. Appl. Phys. B **51**, 94 (1990)

15.82   P. Wehrle: A review of recent advances in semiconductor laser based gas monitors. Spectrochim. Acta, Part A **54**, 197 (1998)

15.83   K.W. Rothe, U. Brinkmann, H. Walther: Remote measurement of NO_2-emission from a chemical factory by the differential absorption technique. Appl. Phys. **4**, 181 (1974)

15.84   H.J. Kölsch, P. Rairoux, J.P. Wolf, L. Wöste: Simultaneous NO and NO$_2$ DIAL measurements using BBO crystals. Appl. Opt. **28**, 2052 (1989)

15.85   J.P. Wolf, H.J. Kölsch, P. Rairoux, L. Wöste: 'Remote detection of atmospheric pollutants using differential absorption LIDAR techniques'. In: *Applied Laser Spectroscopy*, ed. by W. Demtröder, M. Inguscio (Plenum, New York 1991) p. 435

15.86   A.L. Egeback, K.A. Fredrikson, H.M. Hertz: DIAL techniques for the control of sulfur dioxide emissions. Appl. Opt. **23**, 722 (1984)

15.87   J. Werner, K.W. Rothe, H. Walther: Monitoring of the stratospheric ozone layer by laser radar. Appl. Phys. B **32**, 113 (1983)

15.88   W. Steinbrecht, K.W. Rothe, H. Walther: Lidar setup for daytime and night-time probing of stratospheric ozone and measurements in polar and equitorial regimes. Appl. Opt. **28**, 3616 (1988)

15.89   (a) J. Shibuta, T. Fukuda, T. Narikiyo, M. Maeda: Evaluation of the solarblind effect in ultraviolet ozone lidar with Raman lasers. Appl. Opt. **26**, 2604 (1984); (b) C. Weitkamp, O. Thomsen, P. Bisling: Signal and reference wavelengths for the elimination of SO$_2$ cross sensitivity in remote measurements of tropospheric ozone with lidar. Laser Optoelectr. **24**, 246 (April 1992)

15.90   A. Asmann, R. Neuber, P. Rairoux (Eds.): *Advances in Atmospheric Remote Sensing with LIDAR* (Springer, Berlin, Heidelberg, New York 1997)

15.91   U. v. Zahn, P. von der Gathen, G. Hansen: Forced release of sodium from upper atmospheric dust particles. Geophys. Res. Lett. **14**, 76 (1987)

15.92   F.J. Lehmann, S.A. Lee, C.Y. She: Laboratory measurements of atmospheric temperature and backscatter ratio using a high-spectral-resolution lidar technique. Opt. Lett. **11**, 563 (1986)

15.93   M.M. Sokolski (Ed.): *Laser Applications in Meterology and Earth- and Atmospheric Remote Sensing*. SPIE Proc. **1062** (1989)

15.94   R.M. Measure: *Laser Remote Sensing: Fundamentals and Applications* (Wiley, Toronto 1984)

15.95   J. Looney, K. Petri, A. Salik: Measurements of high resolution atmospheric water vapor profiles by use of a solarblind Raman lidar. Appl. Opt. **24**, 104 (1985)

15.96   H. Edner, S. Svanberg, L. Uneus, W. Wendt: Gas-correlation LIDAR. Opt. Lett. **9**, 493 (1984)

15.97   J.A. Gelbwachs: Atomic resonance filters. IEEE J. QE-**24**, 1266 (1988)

15.98   P. Rairoux, H. Schillinger, S. Niedermeier, M. Rodriguez, F. Ronneberger, R. Sauerbrey, B. Stein, D. Waite, C. Wedekind, H. Wille, L. Wöste: Remote sensing of the atmosphere, using ultrashort laser pulses. Appl. Phys. B **71**, 573 (2000)

15.99   S. Svanberg: 'Fundamentals of atmospheric spectroscopy'. In: *Surveillance of Environmental Pollution and Resources by El. Mag. Waves*, ed. by I. Lund (Reidel, Dordrecht 1978)
        Ph.N. Slater: *Remote Sensing* (Addison-Wesley, London 1980)

15.100 R.M. Measures: *Laser Remote Chemical Analysis* (Wiley, New York 1988)

15.101 D.K. Killinger, A. Mooradian (Eds.): *Optical and Laser Remote Sensing*, Springer Ser. Opt. Sci., Vol. 39 (Springer, Berlin, Heidelberg 1983)

15.102 R.N. Dubinsky: Lidar moves towards the 21st century. Laser Optron. **7**, 93 (April 1988);
        S. Svanberg: 'Environmental monitoring using optical techniques'. In: *Applied Laser Spectroscopy*, ed. by W. Demtröder, M. Inguscio (Plenum, New York 1991) p. 417

15.103 H. Walther: Laser investigations in the atmosphere. *Festkörperprobleme* **20**, 327 (Vieweg, Braunschweig 1980)

15.104 E.J. McCartney: *Optics of the Atmosphere* (Wiley, New York 1976)

15.105   J.W. Strohbehn (Ed.): *Laser Beam Propagation in the Atmosphere*, Topics Appl.
         Phys., Vol. 25 (Springer, Berlin, Heidelberg 1978)

15.106   W. Schade: Experimentelle Untersuchungen zur zeitaufgelösten Fluoreszenzspek-
         troskopie mit kurzen Laserpulsen. Habilitation-Thesis, Math.-Naturw. Fakultät,
         Univ. Kiel, Germany (1992)

15.107   J. Ilkin, R. Stumpe, R. Klenze: Laser-induced photoacoustic spectroscopy for
         the speciation of transuranium elements in natural aquatic systems. Topics Curr.
         Chem. **157**, 129 (Springer, Berlin, Heidelberg, New York 1990)

15.108   R. Suntz, H. Becker, P. Monkhouse, J. Wolfrum: Two-dimensional visualization
         of the flame front in an internal combustion engine by laser-induced fluorescence
         of OH radicals. Appl. Phys. B **47**, 287 (1988)

15.109   (a) A.M. Wodtke, L. Hüwel, H. Schlüter, H. Voges, G. Meijer, P. Andresen:
         High sensitivity detection of NO in a flame using a tunable Ar-F-laser. Opt. Lett.
         13, 910 (1988);
         (b) M. Schäfer, W. Ketterle, J. Wolfrum: Saturated 2D-LIF of OH and 2D deter-
         mination of effective collisional lifetimes in atmospheric pressure flames. Appl.
         Phys. B **52**, 341 (1991)

15.110   P. Andresen, G. Meijer, H. Schlüter, H. Voges, A. Koch, W. Hentschel,
         W. Oppermann: Zweidimensionale Konzentrationsmessungen im Brennraum des
         Transparentmotors mit Hilfe von Laser-Fluoreszenzverfahren. Bericht 11/1989,
         MPI für Strömungsforschung Göttingen (1989);
         Combustion optimization pushed forward by excimer LIF-methods. Lambda-
         Physic Highlights No. 14 (December 1988)

15.111   M. Alden, K. Fredrikson, S. Wallin: Application of a two-colour dye laser in
         CARS experiments for fast determination of temperatures. Appl. Opt. **23**, 2053
         (1984)

15.112   J.P. Taran: 'CARS spectroscopy and applications'. In: *Applied Laser Spec-
         troscopy*, ed. by W. Demtröder, M. Inguscio (Plenum, New York 1991) p. 365;
         A. D'Allescio, A. Cavaliere: 'Laser spectroscopy applied to combustion'. In: *Ap-
         plied Laser Spectroscopy*, ed. by W. Demtröder, M. Inguscio (Plenum, New York
         1991) p. 393

15.113   R.W. Dreyfus: 'Useful macroscopic phenomena due to laser ablation'. In: *Des-
         orption Induced by Electronic Transitions DIET IV*, Springer Ser. Surf. Sci.,
         Vol. 19 (Springer, Berlin, Heidelberg, New York 1990) p. 348;
         J.C. Miller, R.F. Haglund (Eds.): *Laser Ablation: Mechanisms and Applications*,
         Lecture Notes Phys., Vol. 389 (Springer, Berlin, Heidelberg 1991);
         J.C. Miller (Ed.): *Laser Ablation*, Springer Ser. Mater. Sci., Vol. 28 (Springer,
         Berlin, Heidelberg, New York 1994)

15.114   R. DeJonge: Internal energy of sputtered molecules. Comm. At. Mol. Phys. **22**,
         1 (1988)

15.115   H.L. Bay: Laser induced fluorescence as a technique for investigations of sput-
         tering phenomena. Nucl. Instrum. Meth. B **18**, 430 (1987)

15.116   R.W. Dreyfus, J.M. Jasinski, R.E. Walkup, G. Selwyn: Laser spectroscopy in
         electronic materials processing research. Laser Focus **22**, 62 (Dec. 1986);
         R.W. Dreyfus, R.W. Walkup, R. Kelly: Laser-induced fluorescence studies of
         excimer laser ablation of Al_2O_3. J. Appl. Phys. **49**, 1478 (1986)

15.117   J.M. Jasinski, E.A. Whittaker, G.C. Bjorklund, R.W. Dreyfus, R.D. Estes,
         R.E. Walkup: Detection of SiH_2 in silane and disilane glow discharge by fre-
         quency modulated absorption spectroscopy. Appl. Phys. Lett. **44**, 1155 (1984)

15.118   H. Moenke, L. Moenke-Blankenburg: *Einführung in die Laser Mikrospektral-
         analyse* (Geest und Portig, Leipzig 1968)

15.119   D. Bäuerle: *Laser Processing and Chemistry*, 3rd edn. (Springer, Berlin, Heidel-
         berg, New York 2000)

15.120 Hai-Lung, H. Wilson (Eds.): *Laser Spectroscopy and Photochemistry on Metal Surfaces* (World Scientific, Singapore 1995)

15.121 F. Durst, A. Melling, J.H. Whitelaw: *Principles and Practice of Laser-Doppler Anemometry*, 2nd edn (Academic, New York 1981)

15.122 T.S. Durrani, C.A. Greated: *Laser Systems in Flow Measurement* (Plenum, New York 1977)

15.123 L.E. Drain: *The Laser Doppler Technique* (Wiley, New York 1980)

15.124 F. Durst, G. Richter: 'Laser Doppler measurements of wind velocities using visible radiation'. In: *Photon Correlation Techniques in Fluid Mechanics*, ed. by E.O. Schulz-Dubois, Springer Ser. Opt. Sci., Vol. 38 (Springer, Berlin, Heidelberg 1983) p. 136

15.125 R.M. Hochstrasser, C.K. Johnson: Lasers in biology. Laser Focus **21**, 100 (May 1985)

15.126 A. Anders: Dye-laser spectroscopy of bio-molecules. Laser Focus **13**, 38 (Feb. 1977);
A. Anders: Selective laser excitation of bases in nucleic acids. Appl. Phys. **20**, 257 (1979)

15.127 A. Anders: Models of DNA-dye-complexes: energy transfer and molecular structure. Appl. Phys. **18**, 373 (1979);
M.E. Michel-Beyerle (Ed.): *Antennas and Reaction Centers of Photosynthetic Bacteria*, Springer Ser. Chem. Phys., Vol. 42 (Springer, Berlin, Heidelberg 1983)

15.128 R.R. Birge, B.M. Pierce: 'The nature of the primary photochemical events in bacteriorhodopsin and rhodopsin'. In: *Photochemistry and Photobiology*, ed. by A.H. Zewail (Harwood, Chur 1983) p. 841

15.129 P. Cornelius, R.M. Hochstrasser: 'Picosecond processes involving CO, O_2 and NO derivatives of hemoproteins'. In: *Picosecond Phenomena III*, ed. by K.B. Eisenthal, R.M. Hochstrasser, W. Kaiser, A. Laberau, Springer Ser. Chem. Phys., Vol. 23 (Springer, Berlin, Heidelberg 1982)

15.130 D.P. Millar, R.J. Robbins, A.H. Zewail: Torsion and bending of nucleic acids, studied by subnanosecond time resolved depolarization of intercalated dyes. J. Chem. Phys. **76**, 2080 (1982)

15.131 L. Stryer: The molecules of visual excitation. Sci. Am. **157**, 32 (July 1987)

15.132 R.A. Mathies, S.W. Lin, J.B. Ames, W.T. Pollard: From femtoseconds to biology: mechanisms of bacterion rhodopsin's light driven proton pump. Annu. Rev. Biophysics Biophys. Chem. **20**, 1000 (1991)

15.133 D.C. Youvan, B.L. Marrs: Molecular mechanisms of photosynthesis. Sci. Am. **256**, 42 (June 1987)

15.134 A.H. Zewail (Ed.): *Photochemistry and Photobiology* (Harwood, London 1983)
V.S. Letokhov: *Laser Picosecond Spectroscopy and Photochemistry of Biomolecules* (Hilger, London 1987);
R.R. Alfano (Ed.): *Biological Events Probed by Ultrafast Laser Spectroscopy* (Academic, New York 1982)

15.135 W. Kaiser (Ed.): *Ultrashort Laser Pulses*, 2nd edn., Topics Appl. Phys., Vol. 60 (Springer, Berlin, Heidelberg, New York 1993)

15.136 E. Klose, B. Wilhelmi (Eds.): *Ultrafast Phenomena in Spectroscopy*, Springer Proc. Phys., Vol. 49 (Springer, Berlin, Heidelberg, New York 1990)

15.137 J.R. Lakowicz (Ed.): *Time-Resolved Laser Spectroscopy in Biochemistry*. SPIE Proc. **909** (1988)

15.138 R.R. Birge, L.A. Nufie (Eds.): *Biomolecular Spectroscopy*. SPIE Proc. **1432** (1991)

15.139 R. Nossal, S.H. Chen: Light scattering from mobile bacteria. J. Physique Suppl. **33**, C1-171 (1972)

15.140 A. Andreoni, A. Longoni, C.A. Sacchi, O. Svelto: 'Laser-induced fluorescence of biological molecules'. In: A. Mooradian, T. Jaeger, P. Stokseth (Eds.): *Tun-

able Lasers and Applications, Springer Ser. Opt. Sci., Vol. 3 (Springer, Berlin, Heidelberg 1976) p. 303

15.141   G.N. McGregor, H.G. Kaputza, K.A. Jacobsen: Laser-based fluorescence microscopy of living cells. Laser Focus **20**, 85 (Nov. 1984)

15.142   H. Schneckenburger, A. Rück, B. Baros, R. Steiner: Intracellular distribution of photosensitizing porphyrins measured by video-enhanced fluorescence microscopy. J. Photochem. Photobiol. B **2**, 355 (1988)

15.143   (a) H. Schneckenburger, A. Rück, O. Haferkamp: Energy transfer microscopy for probing mitochondrial deficiencies. Analyt. Chimica Acta **227**, 227 (1988); (b) P. Fischer: Time-resolved methods in laser scanning microscopy. Laser Opt. Elektr. **24**, 36 (Febr. 1992)

15.144   H. Scheer: 'Chemistry and spectroscopy of chlorophylls'. In: *CRC Handbook of Organic Photochemistry and Photobiology*, ed. by W.M. Horspool, P.S. Song (CRC, New York 1995) p. 1402; P. Mathis: 'Photosynthetic reaction centers'. ibid. p. 1412

15.145   I. Lutz, W. Zinth, et al.: 'Primary reactions of sensory rhodopsins'. In: *Ultrafast Phenomena XII*, ed. by T. Elsäser, et al., Springer Series in Chem. Phys., Vol. 66 (Springer, Berlin, Heidelberg, New York 2000) p. 677+680; W. Zinth, et al.: 'Femtosecond spectroscopy and model calculations for an understanding of the primary reactions in bacterio-rhodopson'. ibid. p. 680

15.146   P.J. Walla, P.A. Linden, G.R. Fleming: 'Fs-transient absorption and fluorescence upconversion after two-photon excitation of carotenoids in solution and in LHC II'. ibid. p. 671

15.147   L. Goldstein (Ed.): *Laser Non-Surgical Medicine. New Challenges for an Old Application* (Lancaster, Basel 1991)

15.148   G. Biamino, G. Müller (Eds.): *Advances in Laser Medicine I* (Ecomed. Verlagsgesell., Berlin 1988)

15.149   S.L. Jacques (Ed.): *Proc. Laser Tissue Interaction II*. SPIE Proc. **1425** (1991); A. Anders, I. Lamprecht, H. Schacter, H. Zacharias: The use of dye lasers for spectroscopic investigations and photodynamics therapy of human skin. Arch. Dermat. Res. **255**, 211 (1976)

15.150   H.P. Berlien, G. Müller (Eds.): *Angewandte Lasermedizin* (Ecomed, Landsberg 1989)

15.151   H. Albrecht, G. Müller, M. Schaldach: Entwicklung eines Raman- spektroskopisches Gasanalysesystems. Biomed. Tech. **22**, 361 (1977); *Proc. VII Int'l Summer School on Quantum Optics*, Wiezyca, Poland (1979)

15.152   M. Mürtz, T. Kayser, D. Kleine, S. Stry, P. Hering, W. Urban: Recent developments on cavity ringdown spectroscopy with tunable cw lasers in the mid-infrared. Proc. SPIE **3758**, 7 (1999)

15.153   H.J. Foth, N. Stasche, K. Hörmann: Measuring the motion of the human tympanic membrane by laser Doppler vibrometry. SPIE Proc. **2083**, 250 (1994)

15.154   T.J. Dougherty, J.E. Kaulmann, A. Goldfarbe, K.R. Weishaupt, D. Boyle, A. Mittleman: Photoradiation therapy for the treatment of malignant tumors. Cancer Res. **38**, 2628 (1978); D. Kessel: Components of hematoporphyrin derivates and their tumor-localizing capacity. Cancer Res. **42**, 1703 (1982)

15.155   G. Jori: 'Photodynamic therapy: basic and preclinical aspects'. In: *CRC Handbook of Organic Photochemistry and Photobiology*, ed. by W.M. Horspool, P.S. Song (CRC, New York 1995) p. 1379; T.J. Dougherty: 'Clinical applications of photodynamic therapy'. ibid. p. 1384

15.156   P.J. Bugelski, C.W. Porter, T.J. Dougherty: Autoradiographic distribution of HPD in normal and tumor tissue in the mouse. Cancer Res. **41**, 4606 (1981)

15.157   A.S. Svanberg: Laser spectroscopy applied to energy, environmental and medical research. Phys. Scr. **23**, 281 (1988)

15.158   Y. Hayata, H. Kato, Ch. Konaka, J. Ono, N. Takizawa: Hematoporphyrin deriva-
         tive and laser photoradiation in the treatment of lung cancer. Chest **81**, 269
         (1982)

15.159   A. Katzir: *Optical Fibers in Medicine IV.* SPIE Proc. **1067** (1989); ibid. **906**
         (1988)

15.160   L. Prause, P. Hering: Lichtleiter für gepulste Laser: Transmissionsverhalten,
         Dämpfung und Zerstörungsschwellen. Laser Optoelektron. **19**, 25 (January
         1987); ibid. **20**, 48 (May 1988)

15.161   A. Katzir: Optical fibers in medicine. Sci. Am. **260**, 86 (May 1989)

15.162   H. Schmidt-Kloiber, E. Reichel: 'Laser lithotripsy'. In: H.P. Berlien, G. Müller
         (Eds.): *Angewandte Lasermedizin* (Ecomed, Landsberg 1989) VI, Sect. 2.12.1

15.163   R. Steiner (Ed.): *Laser Lithotripsy* (Springer, Berlin, Heidelberg, New York
         1988);
         R. Pratesi, C.A. Sacchi (Eds.): *Lasers in Photomedicine and Photobiology*,
         Springer Ser. Opt. Sci., Vol. 31 (Springer, Berlin, Heidelberg 1982);
         L. Goldmann (Ed.): *The Biomedical Laser* (Springer, Berlin, Heidelberg York
         1981)

15.164   W. Simon, P. Hering: Laser-induzierte Stoßwellenlithotripsie an Nieren- und Gal-
         lensteinen. Laser Optoelektron. **19**, 33 (January 1987)

15.165   D. Beaucamp, R. Engelhardt, P. Hering, W. Meyer: 'Stone identification dur-
         ing laser-induced shockwave lithotripsy'. In: *Proc. 9th Congress Laser 89*, ed.by
         W. Waidelich (Springer, Berlin, Heidelberg, New York 1990)

15.166   R. Engelhardt, W. Meyer, S. Thomas, P. Oehlert: Laser-induzierte Schockwellen-
         Lithotripsie mit Mikrosekunden Laserpulsen. Laser Optoelektr. **20**, 36 (April
         1988)

15.167   S.P. Dretler: Techniques of laser lithotripsy. J. Endourology **2**, 123 (1988);
         B.C. Ihler: Laser lithotripsy: system and fragmentation processes closely exam-
         ined. Laser Optoelektron. **24**, 76 (April 1992)

15.168   S. Willmann, A. Terenji, I.V. Yaroslavsky, T. Kahn, P. Hering: Determination of
         the optical properties of a human brain tumor using a new microspectrophoto-
         metric technique. Proc. SPIE **3598**, 233 (1999)

15.169   A.N. Yaroslavsky, I.V. Yaroslavsky, T. Goldbach, H.J. Schwarzmaier: Influence
         of the scattering phase function approximation on the optical properties of blood.
         J. Biomedical Optics **4**, 47 (1999)

15.170   Check-Yin Ng (Ed.): *Optical Methods for Time- and State Resolved Chemistry*,
         SPIE Proc. **1638** (1992)
         B.L. Feary (Ed.): *Optical Methods for Ultrasensitive Dilution and Analysis.*
         SPIE Proc. **1435** (1991);
         J.L. McElroy, R.J. McNeal: *Remote Sensing of the Atmosphere.* SPIE Proc. **1491**
         (1991);
         S.A. Akhmanov, M. Poroshina (Eds.): *Laser Applications in Life Sciences.* SPIE
         Proc. **1403** (1991);
         L.O. Jvassand (Ed.): *Future Trends in Biomedical Applications of Lasers.* SPIE
         Proc. **1535** (1991);
         J.R. Lakowicz (Ed.): *Time-Resolved Laser Spectroscopy in Biochemistry.* SPIE
         Proc. **1204** (1991);
         R.R. Birge, L.A. Nafie: *Biomolecular Spectroscopy.* SPIE Proc. **1432** (1991)

Subject Index

Printing (Computer to Film): Saladruck Berlin
Binding: Stürtz AG, Würzburg